6th International Conference on Advances in Mechanical Engineering (ICAME 2019)

AA007717

IOP Conference Series: Materials Science and Engineering
Volume 834

Kota Kinabalu, Malaysia
14 - 16 August 2019

Editors:

Muhammad Hussain Ismail
Alias Mohd Saman
Rizal Effendy Mohd Nasir
Jamaluddin Mahmud
Mohd Azmi Yunus
Mimi Azlina Abu Bakar
Amir Radzi Ab Ghani
Ramlan Zailani
Kausalyah A.P. Venkatason
Mohd Nor Azmi Ab Patar
Abdul Halim Abdullah
Nor Hayati Saad
Siti Mariam Abdul Rahman
Baljit Singh A.L. Bhathal Singh
Ahmad Khushairy Makhtar
Rosnadiah Bahsan

ISBN: 978-1-7138-4727-4
ISSN: 1757-8981

Printed from e-media with permission by:

Curran Associates, Inc.
57 Morehouse Lane
Red Hook, NY 12571

Some format issues inherent in the e-media version may also appear in this print version.

This work is licensed under a Creative Commons Attribution 3.0 International Licence.
Licence details: http://creativecommons.org/licenses/by/3.0/.

No changes have been made to the content of these proceedings. There may be changes to pagination and minor adjustments for aesthetics.

Printed with permission by Curran Associates, Inc. (2022)

For permission requests, please contact the Institute of Physics at the address below.

Institute of Physics
Dirac House, Temple Back
Bristol BS1 6BE UK

Phone: 44 1 17 929 7481
Fax: 44 1 17 920 0979

techtracking@iop.org

Additional copies of this publication are available from:

Curran Associates, Inc.
57 Morehouse Lane
Red Hook, NY 12571 USA
Phone: 845-758-0400
Fax: 845-758-2633
Email: curran@proceedings.com
Web: www.proceedings.com

TABLE OF CONTENTS

IOP Conf. Series: Materials Science and Engineering **834** (2020) 011001 doi:10.1088/1757-899X/834/1/011001

Preface

Chairman, ICAME2019

First and foremost, I would like to extend our appreciation to the participants of the 6th International Conference on Advances in Mechanical Engineering, ICAME 2019.

Society of Mechanical Engineering Liveliness (SOMEL), Faculty of Mechanical Engineering, UiTM Shah Alam is proud and very pleased to continuously organize this international conference. Back in 2005, the faculty had organized a national conference, NAME 2005 followed by international conferences of ICAME 2009, ICAME 2010, ICAME 2013, ICAME 2015 and ICAME 2017. The experiences and responses of these conferences have helped the planning and execution of ICAME 2019. To some of the participants, you may have come before, and we thank you for your continuous support.

The main objective of ICAME 2019 was to bring together researchers, academician, engineers and practitioners from Malaysia and abroad to present their latest research findings and share new innovative ideas, knowledge, and technologies related to advances in mechanical engineering. ICAME 2019 promoted multi-disciplinary collaboration and partnership as the way forward in this 21st century. The diversity of the topics in this conference includes Aeronautics, Automotive and Aerodynamics, Computational Modelling and Analysis, Design, Manufacturing and Industrial Engineering, Dynamics and Control, Energy and Tribology, Material and Processes, Robotics and Mechatronics, Sensor and Image Processing, Structure and Solid Mechanics, and Thermofluids.

This year ICAME 2019 includes Malaysia Automotive Robotics, IoT Institute (MARii) as our key collaborator. Malaysia Automotive Robotics, IoT Institute (MARii), is an agency under the Ministry of International Trade and Industry (MITI) serves as the focal point, coordination centre and think tank for the nation's automotive industry. MARii functions to enhance technology, human capital, supply chain, market outreach and aftersales capabilities of all automotive stakeholders and ecosystems. MARii's main objective is to enhance the competitiveness of the automotive industry and overall mobility including Intelligent Transportation Systems (ITS) and Related Services through the adoption of robotics & IoT.

Another key collaborator for ICAME 2019 is Metal Forming Research Center (MFRC), Korea. MFRC has unceasingly developed the metal forming simulator of AFDEX, standing for Advisor for metal forming process design experts, based on rigid elasto-thermoviscoplastic finite element method with intelligent remeshing technology embedded. MFRC also supports and consults their customers in various ways of both online and offline education, sharing application experiences and useful findings from annual conference on metal forming CAE technologies which covers special short term or long-term lectures and publication of books and papers as well.

Finally, I take this opportunity to express my sincere gratitude to the International and National Advisory Committees, and the organizing committees for such an excellent job in organizing this major event in support of the advancement in mechanical engineering field. I am also grateful to all the reviewers for the great efforts in maintaining the standard of the conference papers. My thanks go to all the technical sponsors and those involved in making this conference a success.

Professor Dr. Salmiah Kasolang @ Kasalung
Chairman, ICAME2019
Society of Mechanical Engineering Liveliness SOMEL
Faculty of Mechanical Engineering
Universiti Teknologi MARA

Content from this work may be used under the terms of the Creative Commons Attribution 3.0 licence. Any further distribution of this work must maintain attribution to the author(s) and the title of the work, journal citation and DOI.
Published under licence by IOP Publishing Ltd

IOP Conf. Series: Materials Science and Engineering **834** (2020) 011001 doi:10.1088/1757-899X/834/1/011001

Vice Chairman, ICAME2019

It gives me great pleasure to welcome all participants to the 6th International Conference on Advances in Mechanical Engineering (ICAME 2019). First of all, I would like to thank all the ICAME 2019 committee members for their tireless effort and time spent to make this conference a reality.

Society of Mechanical Engineering Liveliness (SOMEL), Faculty of Mechanical Engineering, UiTM Shah Alam is proud and very pleased to continuously organize this international conference. We are honoured to have **Malaysia Automotive Robotics and IoT Institute (MARii)** as our **Platinum Sponsor**, **Metal Forming Research Corporation (MFRC)** from Korea as our **Gold Sponsor** and **Festech Sdn Bhd.** We are also proud to be recognized by the **Board of Engineers (BEM)** with **16 Continuing Professional Development (CPD) hours.** We would like to express our appreciation to **Tourism Malaysia, Sabah Tourism Board** and **Smart Manufacturing Research Institute (SMRI)** for promoting and supporting ICAME2019. The involvement of all the sponsors and all parties in the 6th ICAME2019 is a tremendous value- added to the conference especially in sharing and discussing the latest Industry4Ward status and advancement of technology. Greatest gratitude for their willingness to be a part of ICAME2019.

ICAME2019 is delighted to welcome the world class keynote speakers: CEO of MARii, Dato' Ts. Madani Sahari, Professor ManSoo Joun from Gyeongsang National University South Korea, Professor Akihiko Hanafusa from Shibaura Institute of Technology Japan, Dr. Ing. Marcel Graf from Technische Universitat Chemnitz Germany and Assoc. Prof. Dr. Nor Hayati Saad from Faculty of Mechanical Engineering, Universiti Teknologi MARA Malaysia.

This year, ICAME received 164 papers submission. Each paper, regardless of the 10 tracks, received at least 2 reviews. Congratulations to participants of accepted papers to be published in IOP Scopus indexed proceeding as well as Scopus and high impact journals: Journal of Mechanical Engineering and Evergreen Journal. The success of ICAME2019 depends completely on the effort, talent, and energy of researchers from all across the globe: Malaysia, Indonesia, India, Germany, South Korea, China, Japan, Netherlands, Austria, Czech Republic, Thailand, Nigeria, Russia and Iraq who have written and submitted papers on a variety of topics in the field of mechanical engineering.

The Faculty of Mechanical Engineering UiTM and MARii, Shibaura Institute of Technology Japan, MFRC Korea and Virtual Production Engineering of Technische Universitat Chemnitz Germany will ink the Memorandum of Understanding (MoU), Memorandum of Agreement (MoA) and Letter of Intent (LOI) in ICAME2019 to collaborate towards the capacity building of the future industry leaders in setting a bold path for a sustainable, healthy, secure, connected, and creative humanity.

In addition to the 6th ICAME, three programs are introduced- the 1st Workshop on Engineering Education: brought by Society of Mechanical Engineering Liveliness (SOMEL) and Workshop on Metal Forming CAE Technology: brought by MFRC South Korea as well as excursion to MARii's first centre outside of the Klang Valley opened in Kota Kinabalu- its first ever under the MARii Satellite network targeted to allow for more extensive reach for technology adoption within the automotive, robotics and IoT (Internet of Things) sectors as well as overall mobility.

Ir. Dr. Noor Azlina Mohd. Salleh
Vice Chairman, ICAME2019
Society of Mechanical Engineering Liveliness SOMEL
Faculty of Mechanical Engineering
Universiti Teknologi MARA

IOP Conf. Series: Materials Science and Engineering **834** (2020) 011001 doi:10.1088/1757-899X/834/1/011001

Co-Chairman, ICAME2019

It gives me great pleasure to welcome all participants to the 6th International Conference on Advances in Mechanical Engineering (ICAME 2019). First of all, I would like to thank all the ICAME 2019 committee members for their tireless effort and time spent to make this conference a reality. Not to forget, special thank also to our sponsors for their supports.

Since 2005, ICAME has been recognised as a platform for academicians, researchers and industry players as well as postgraduate students from all over the world to share and exchange their knowledge and expertise in the field of Mechanical Engineering. Our concept of "knowledge sharing with leisure" has not only captured the hearts of our loyal participants but has also attracted many newcomers. This is evident by the increasing number of participants to the conference since 2013. The venues are carefully chosen to meet the concept such as Malacca, the historical city in Malaysia in 2013; Bali, Indonesia in 2015; Krabi, Thailand in 2017; and this year in Sabah, Malaysia. Sabah is one of the famous tourist attraction state in Malaysia, known for its beaches, rainforest, coral reefs, abundant wildlife and diving destinations.

From 2013 to 2017, ICAME has successfully drawn close to 700 participants and more than 600 accepted papers have been published in national and international indexed journals. Innovations are continuously introduced to the conference as added value and to attract more participants. The conference theme "*Engineering for Humanity*" supports the true purpose of engineering and technology which is to further improve and enhance the quality of life for mankind. This year, ICAME managed to gain two key partners; the Malaysia Automotive Robotics and IoT Institute (MARii) and the Metal Forming Research Centre (MFRC) Korea. With these collaborations, ICAME 2019 is able to offer special topics parallel sessions and also an excursion to MARii's satellite office, both included in the conference package.

Finally, I wish all participants a good time conferencing and exploring Sabah. Thank you for your continuous support and hope to see you at ICAME2021.

Best wishes,

Dr Mohd. Azman Yahaya
Co-Chairman of ICAME2019
Society of Mechanical Engineering Liveliness SOMEL
Faculty of Mechanical Engineering
Universiti Teknologi MARA

IOP Conf. Series: Materials Science and Engineering **834** (2020) 011001 doi:10.1088/1757-899X/834/1/011001

Chief of Editors, ICAME2019

This conference proceeding contains a compilation of 120 accepted papers from the 6th International Conference on Advances in Mechanical Engineering (ICAME 2019) which was held at the Pacific Sutera Harbour, Kota Kinabalu, Sabah from 14-16 August. This year, ICAME was organized by the Society of Mechanical Engineering Liveliness (SoMEL), Faculty of Mechanical Engineering, UiTM Shah Alam. It is a continuation from a series of conference organized by the Faculty of Mechanical Engineering, which was initially at a national level, namely NAME 2005 followed by 5 series at the international levels; ICAME 2009 (Shah Alam), ICAME 2010 (Kuala Lumpur), ICAME 2013 (Melaka), ICAME 2015 (Bali, Indonesia) and ICAME 2017 (Krabi, Thailand).

With the conference theme "Engineering for Humanity", it was aimed to promote a platform for multi-disciplinary research and share new innovative ideas, knowledge, and technologies related to advances in mechanical engineering which is to further improve and enhance the quality of life for mankind. The diversity of the topics in this conference includes Aeronautics, Automotive and Aerodynamics, Computational Modelling and Analysis, Design, Manufacturing and Industrial Engineering, Dynamics and Control, Energy and Tribology, Materials and Processes, Robotics and Mechatronics, Sensor and Image Processing, Structure and Solid Mechanics, and Thermofluids.

As the Chief Editor, I would like to take this great opportunity to thanks to all participants of ICAME for promptly revising the technical papers according to the guideline provided. It is a great pleasure to all of us when the response for ICAME 2019 was overwhelming. All submitted papers were then peer- reviewed, revised according to reviewers' comment and accepted for publication in this proceeding. Thanks to all reviewers for your tireless effort in reviewing the submitted papers in maintaining the standard of the technical papers. My thanks go to all the sponsors and those involved in making this conference a success.

Thank you

Kota Kinabalu, August 2019
Assoc. Prof. Dr. Muhammad Hussain Ismail
Chief of Editors, ICAME2019
Faculty of Mechanical Engineering
Universiti Teknologi MARA

IOP Conf. Series: Materials Science and Engineering **834** (2020) 011002 doi:10.1088/1757-899X/834/1/011002

MEDIA COVERAGE

N7-LANGSUNG: SABAH [15 OGOS 2019]

N7-LANGSUNG: SABAH [15 OGOS 2019]

N7-LANGSUNG: SABAH [15 OGOS 2019]

N7-LANGSUNG: SABAH [15 OGOS 2019]

ICAME2019 Coverage in RTM TV Channel (https://youtu.be/8CpPMYKxXlY?t=231)

ICAME2019 6th International Conference on Advances in Mechanical Engineering

ICAME2019 6th International Conference on Advances in Mechanical Engineering

ICAME2019 6th International Conference on Advances in Mechanical Engineering

ICAME2019 6th International Conference on Advances in Mechanical Engineering

ICAME2019 Coverage in MARii Youtube Channel (https://youtu.be/RizqKDuG5Vw)

Content from this work may be used under the terms of the Creative Commons Attribution 3.0 licence. Any further distribution of this work must maintain attribution to the author(s) and the title of the work, journal citation and DOI.
Published under licence by IOP Publishing Ltd

IOP Conf. Series: Materials Science and Engineering **834** (2020) 011002 doi:10.1088/1757-899X/834/1/011002

ICAME2019 Coverage in Malaysia Newspapers

IOP Conf. Series: Materials Science and Engineering **834** (2020) 011002 doi:10.1088/1757-899X/834/1/011002

MEDIA PRESS RELEASE STATEMENT

FOR IMMEDIATE RELEASE
14th August 2019

ICAME 2019 TO EMPHASISE ON TECHNOLOGIES TO FULFILL MODERN HUMANITY NEEDS

Kota Kinabalu, 14 AUGUST 2019 – the 6th International Conference on Advances in Mechanical Engineering (ICAME 2019), was officiated today and will be held from 14th -16th August 2019 in Pacific Sutera Hotel, Kota Kinabalu, Sabah.

The biennial international conference is a platform for academia, industry leaders and policy makers to share and discuss latest researches, ideas, and survey reports in the theoretical and practical areas regarding all aspects of mechanical engineering and related areas.

The theme for the 2019 edition of ICAME will be "Engineering for Humanity", emphasizing the role of engineering in providing technologies to fulfill the needs of modern humanity. The conference was officiated by Dato' Ts. Madani Sahari, CEO of Malaysia Automotive, Robotics and IoT Institute (MARii).

ICAME is organised by Society of Mechanical Engineering Liveliness (SoMEL), a society under the Faculty of Mechanical Engineering, Universiti Teknologi MARA (UiTM), Shah Alam, in collaboration with the Malaysia Automotive, Robotics and IoT Institute (MARii).

"The pace of current technological advancement has led to the interdependence of various disciplines and specialisations within the engineering field. It is important that the exchange of ideas and knowledge be promoted across the nation, bringing forth novel technology solutions that address human needs and enrich the lives of society.", said Dato' Madani in his keynote speech.

"ICAME 2019 is one of many continuing collaborations between MARii and UiTM to enhance competitiveness of the Malaysian industry. I also congratulate UiTM for taking this conference of international stature to different locations in Malaysia and spreading knowledge across the country", he added.

This is MARii's second involvement in ICAME 2019, and was announced during the opening ceremony of MARii Satellite Sabah in February 2019. The announcement was made by Minister of International and Industry (MITI), YB Datuk Darell Leiking during the MARii Satellite launch.

Industry experts from Malaysia, Germany, Japan and Korea will be speaking on various topics such as human capital development in the era of Industry 4.0, advanced manufacturing technology and simulation and design applications within the automotive industry.

Industry experts and researches who will be speaking in ICAME 2019 include:

- Dato' Ts. Madani Sahari, CEO MARii
- Professor ManSoo Joun, Gyeongsang National University, South Korea
- Professor Akihiko Hanafusa, Shibaura Institute of Technology Japan
- Dr. Ing. Marcel Graf, Technische Universitat Chemnitz Germany
- Assoc. Prof. Dr. Nor Hayati Saad, Faculty of Mechanical Engineering, UiTM

Participants from Malaysia, Indonesia, India, Germany, South Korea, China, Japan, Netherlands, Austria, Czech Republic, Thailand, Nigeria, Russia and Iraq submitted a total of 164 research papers. A total of 140 papers are accepted (an acceptance rate of 85.5%) based on the reviews from the program committee and external reviewers. The selected papers will be published in IOP scopus indexed proceeding as well as Scopus and high impact journals: Journal of Mechanical Engineering and Evergreen Journal.

The following activities will also be conducted in conjunction with ICAME 2019:

- Workshop on Engineering Education in Promoting 21st Learning Skills by SoMEL
- Workshop on Metal Forming CAE Technology by Metal Forming Research Corporation (MFRC) South Korea
- An excursion to MARii's first centre outside of the Klang Valley – MARii Satellite Sabah

Vice Chairman of ICAME2019, Ir. Dr. Noor Azlina Mohd. Salleh said, "The research areas, workshops and activities at ICAME 2019 are strategized to reach out to those within in academic community, industry, government and public, in driving the needed technologies for humanity. SoMEL and ICAME2019 believes that the spirit of giving in liveliness is the basis for effective and impactful action".

In conjunction with ICAME 2019, three Memorandum of Understandings (MoUs) and Letters of Intent (LOI) will be exchanged:

- between MARii and MFRC Korea on capacity building in Metal Forming capabilities
- between UiTM and MFRC Korea on capacity building in Metal Forming capabilities
- between UiTM and Shibaura Institute of Technology Japan, on the Global Course Program

IOP Conf. Series: Materials Science and Engineering **834** (2020) 011002 doi:10.1088/1757-899X/834/1/011002

- between UiTM and Virtual Production Engineering of Technische Universitat Chemnitz Germany in areas of Virtual Manufacturing, Metal Additive Manufacturing and Product Lifecycle Management.

-END-

About the Malaysia Automotive Robotics and IoT Institute (MARii)

Malaysia Automotive, Robotics and IoT Institute (MARii) – formally known as the Malaysia Automotive Institute (MAI) – is an agency under the Ministry of International Trade and Industry (MITI). MARii serves as the focal point, coordinating centre and think tank towards enhancing the competitiveness of the automotive industry and overall mobility, including intelligent transportation system and related services through adoption of robotics & IoT.

About Society of Mechanical Engineering Liveliness (SoMEL), Faculty of Mechanical Engineering, Universiti Teknologi MARA Malaysia

ICAME is a biennial international conference organized by Society of Mechanical Engineering Liveliness (SoMEL), Faculty of Mechanical Engineering, Universiti Teknologi MARA Malaysia for disseminating current research on key areas of mechanical engineering and other related areas. For more information, visit www.icame.my.

Media Contact

Ir. Dr. Noor Azlina Mohd Salleh / M: 019-5592390 / E:noorazlinamohdsalleh@uitm.edu.my
Sarah Mamat / M: 011-26900611 / E: sarah@marii.my

IOP Conf. Series: Materials Science and Engineering **834** (2020) 011003 doi:10.1088/1757-899X/834/1/011003

ORGANIZED BY

SoMEL
SOCIETY OF MECHANICAL ENGINEERING LIVELINESS

PLATINUM SPONSOR

MARii
INSTITUT AUTOMOTIF ROBOTIK DAN IoT MALAYSIA
MALAYSIA AUTOMOTIVE ROBOTICS AND IoT INSTITUTE

GOLD SPONSOR

MFRC
Metal Forming Research Corporation
www.afdex.com | mfrc@afdex.com

SUPPORTED BY

LEMBAGA JURUTERA MALAYSIA
Smart Manufacturing Research Institute

FESTECH

TOURISM MALAYSIA

SABAH

Content from this work may be used under the terms of the Creative Commons Attribution 3.0 licence. Any further distribution of this work must maintain attribution to the author(s) and the title of the work, journal citation and DOI.
Published under licence by IOP Publishing Ltd

IOP Conf. Series: Materials Science and Engineering **834** (2020) 011003 doi:10.1088/1757-899X/834/1/011003

ORGANISING COMMITTEE

CHAIRMAN
Professor Salmiah Kasolang @ Kasalung

CO-CHAIRMAN
Dr. Mohd Azman Yahaya

VICE CHAIRMAN
Ir. Dr. Noor Azlina Mohd Salleh

INTERNATIONAL COMMITTEE
Professor Man Soo Joun
Professor Akihiko Hanafusa
Dr.-Ing. Marcel Graf

INDUSTRIAL COMMITTEE
Dato' Ts. Mohamad Madani Sahari

SECRETARIAT
Ir. Dr. Irnie Azlin @ Nur Aqilah Zakaria - Head
Nor Merlisa Ali - Secretary

TREASURER
Dr. Nor Hafiez Mohamad Nor

PROGRAM, TECHNICAL, LOGISTICS & AWARDS
Dr. Ahmad Hussein Abdul Hamid – Head
Dr. Mohd Hafiz Mohd Noh
Mohd Harysham Zainal Abidin
Mohamad Asrul Mahamad Razali
Mohd Helmi Mohd Mustamil

SPONSORSHIP, PROMOTION & WEBSITE
Dr. Nik Roselina Nik Roseley – Head
Assoc. Prof. Dr. Nor Hayati Saad
Dr. Muhad Rozi Mat Nawi
Dr. Bibi Intan Suraya Murat
Dr. Siti Khadijah Alias

PUBLICATION
Assoc. Prof. Dr. Muhamad Hussain Ismail – Head
Dr. Siti Mariam Abdul Rahman
Ir. Dr. Baljit Singh A/L Bhathal Singh
Dr. Ahmad Khushairy Makhtar
Rosnadiah Bahsan

IOP Conf. Series: Materials Science and Engineering **834** (2020) 011004 doi:10.1088/1757-899X/834/1/011004

BOARD OF REVIEWERS

EDITOR IN CHIEF

Assoc. Prof. Dr. Muhammad Hussain Ismail – Universiti Teknologi MARA, Malaysia

EDITORIAL BOARD

1. Dr. Alias Mohd Saman - Universiti Teknologi MARA, Malaysia
   (Advanced Manufacturing Technology and Simulation)
2. Dr. Rizal Effendy Mohd Nasir - Universiti Teknologi MARA, Malaysia
   (Aerospace Science and Technology)
3. Professor Ir. Dr. Jamaluddin Mahmud - Universiti Teknologi MARA, Malaysia
   (Advanced Materials)
4. Ir. Dr. Mohd Azmi Yunus - Universiti Teknologi MARA, Malaysia
   (Structural Dynamics and Vibration)
5. Assoc. Prof. Dr. Mimi Azlina Abu Bakar - Universiti Teknologi MARA, Malaysia
   (Adhesion, Friction, Wear and Lubrication in Surface Interaction)
6. Ir. Dr. Amir Radzi Ab Ghani - Universiti Teknologi MARA, Malaysia
   (Automotive Engineering)
7. Assoc. Prof. Dr. Ramlan Zailani - Universiti Teknologi MARA, Malaysia
   (Thermofluids and Energy)
8. Ir. Dr. Kausalyah A/ P Venkatason - Universiti Teknologi MARA, Malaysia
   (Computer Aided Design, Engineering and Technology)
9. Dr. Mohd Nor Azmi Ab Patar - Universiti Teknologi MARA, Malaysia
   (Mechatronics, Control and Intelligent System)
10. Ts. Dr. Abdul Halim Abdullah - Universiti Teknologi MARA, Malaysia
    (Biomechanical, Clinical and Sports Engineering)
11. Assoc. Prof. Ts. Dr. Nor Hayati Saad - Universiti Teknologi MARA, Malaysia
    (Engineering Education)

EDITORIAL EXECUTIVE

1. Dr. Siti Mariam Abdul Rahman - Universiti Teknologi MARA, Malaysia
2. Ir. Ts. Dr. Baljit Singh A/L Bhathal Singh - Universiti Teknologi MARA, Malaysia
3. Dr. Ahmad Khushairy Makhtar - Universiti Teknologi MARA, Malaysia
4. Rosnadiah Bahsan - Universiti Teknologi MARA, Malaysia

INTERNATIONAL AND INDUSTRIAL ADVISORY PANEL

1. Dato' Ts. Mohamad Madani Sahari, Malaysia Automotive, Robotics & IoT Institute
2. Professor Man Soo Joun, Gyeongsang National University
3. Professor Akihiko Hanafusa, Shibaura Institute of Technology
4. Prof. Dr. Ali.Erdemir, Argonne National Laboratory
5. Dr.-Ing. Marcel Graf, Technische Universitat Chemnitz Germany

Content from this work may be used under the terms of the Creative Commons Attribution 3.0 licence. Any further distribution of this work must maintain attribution to the author(s) and the title of the work, journal citation and DOI.
Published under licence by IOP Publishing Ltd

IOP Conf. Series: Materials Science and Engineering **834** (2020) 011004 doi:10.1088/1757-899X/834/1/011004

LIST OF REVIEWERS

Prof. Ir. Dr. Jaharah A Ghani, Universiti Kebangsaan Malaysia
Prof. Ir. Dr. Jamaluddin Mahmud, Universiti Teknologi MARA
Prof. Ir. Dr. Wahyu Kuntjoro, Universiti Teknologi MARA
Prof. Dr. Mariyam Jameelah Ghazali, Universiti Kebangsaan Malaysia
Prof. Dr. Salmiah Kasolang, Universiti Teknologi MARA
Assoc. Prof. Ir. Ts. Dr. Al Emran Bin Ismail, Universiti Tun Hussein Onn Malaysia
Assoc. Prof. Ir. Dr. Bulan Abdullah, Universiti Teknologi MARA
Assoc. Prof. Ir. Dr. Hamdan Ya, Universiti Teknologi PETRONAS
Assoc. Prof. Ir. Dr. Mohd Rizal Alkahari, Universiti Teknikal Malaysia Melaka
Assoc. Prof. Ir. Dr. Salina Budin, Universiti Teknologi MARA
Assoc. Prof. Ir. Dr. Shasthri Sivaguru, Heriot-Watt University Malaysia
Assoc. Prof. Ts. Dr. Nor Hayati Saad, Universiti Teknologi MARA
Assoc. Prof. Dr. Aman Mohd Ihsan Mamat, Universiti Teknologi MARA
Assoc. Prof. Dr. Fairuz Izzuddin Romli, Universiti Putra Malaysia
Assoc. Prof. Dr. Ing. Shahrul Azam Abdullah @ Ab. Aziz, Universiti Teknologi MARA
Assoc. Prof. Dr. Koay Mei Hyie, Universiti Teknologi MARA
Assoc. Prof. Dr. Ramlan Zailani, Universiti Teknologi MARA
Assoc. Prof. Dr. Tee Boon Tuan, Universiti Teknikal Malaysia Melaka
Assoc. Prof. Dr. Wan Mohd Nazri Wan Abdul Rahman, Universiti Teknologi MARA
Assoc. Prof. Dr. Yap Tze Chuen, Heriot-Watt University Malaysia
Assoc. Prof. Ismail Nasiruddin Ahmad, Universiti Teknologi MARA
Engr. Dr. Fadly Jashi Darsivan Ridhuan Siradj, International Islamic University Malaysia
Ir. Dr. Amir Radzi Ab Ghani, Universiti Teknologi MARA
Ir. Dr. Amirul Abd Rashid, Universiti Teknologi MARA
Ir. Dr. Irnie Azlin Zakaria, Universiti Teknologi MARA
Ir. Dr. Kausalyah a/p Venkatason, Universiti Teknologi MARA
Ir. Dr. Noor Azlina Mohd Salleh, Universiti Teknologi MARA
Ir. Dr. Noriah Yusoff, Universiti Teknologi MARA
Ir. Mohd Firdaus Mohamad Idris, Rapid Rail Sdn Bhd
Ir. Ts. Dr. Baljit Singh a/l Bathal Singh, Universiti Teknologi MARA
Ir. Ts. Dr. Mohamad Ali Ahmad, Universiti Teknologi MARA
Ts. Dr. Abdul Halim Abdullah, Universiti Teknologi MARA
Ts. Dr. Azli Abd Razak, Universiti Teknologi MARA
Ts. Dr. Lee Chang Chuan, Universiti Malaysia Perlis
Ts. Dr. Muhamad Nasir Murad, Universiti Malaysia Perlis
Ts. Dr. Muhammad Hazli Mazlan, Universiti Tun Hussein Onn Malaysia
Ts. Dr. Nor Fazli Adull Manan, Universiti Teknologi MARA
Ts. Dr. Zulkifli Mohamed, Universiti Teknologi MARA
Dr Azizul Hakim Samsudin, Alam Dingin Air Conditioning Engineering Sdn. Bhd.
Dr. Abang Mohammad Nizam Abang Kamaruddin, Universiti Malaysia Sarawak
Dr. Abdul Rahman Hemdi, Universiti Teknologi MARA
Dr. Ahmad Humaizi Hilmi, Universiti Malaysia Perlis
Dr. Ahmad Hussein Abdul Hamid, Universiti Teknologi MARA
Dr. Ahmad Jazlan Haja Mohideen, International Islamic University Malaysia
Dr. Ahmad Khushairy Bin Makhtar, Universiti Teknologi MARA

Dr. Ahmad Shahir Jamaludin, Universiti Malaysia Pahang
Dr. Alhassan Tijani Salami, Universiti Teknologi MARA
Dr. Alias Mohd Saman, Universiti Teknologi MARA
Dr. Anizah Kalam, Universiti Teknologi MARA
Dr. Anwar P.P. Abdul Majeed, Universiti Malaysia Pahang
Dr. Asnida Abdul Wahab, Universiti Teknologi Malaysia
Dr. Azianti Ismail, Universiti Teknologi MARA
Dr. Bibi Intan Suraya Murat, Universiti Teknologi MARA
Dr. Fauziah Binti Jerai @ Junaidi, Universiti Teknologi MARA
Dr. Hafizan Hashim, Universiti Teknologi MARA
Dr. Helmi Rashid, Universiti Teknologi MARA
Dr. Juri Saedon, Universiti Teknologi MARA
Dr. Mohd Fadzli Bin Abdollah, Universiti Teknikal Malaysia Melaka
Dr. Mohd Fauzi Ismail, Universiti Teknologi MARA
Dr. Mohd Nor Azmi Ab Patar, Universiti Teknologi MARA
Dr. Mohd Rozaiman Aziz, Universiti Teknologi MARA
Dr. Mohd Sanusi Abdul Aziz, Universiti Teknikal Malaysia Melaka
Dr. Mohd Shahriman Adenan, Universiti Teknologi MARA
Dr. Mohd. Sultan Ibrahim Shaik Dawood, International Islamic University Malaysia
Dr. Muhamad Sukri Hadi, Universiti Teknologi MARA
Dr. Muhammad Fairuz Remeli, Universiti Teknologi MARA
Dr. Muhammad Hanif Ramlee, Universiti Teknologi Malaysia
Dr. Muhammad Hilmi Jalil, Universiti Kebangsaan Malaysia
Dr. Muhammad Izzat Nor Bin Ma'Arof, INTI International University
Dr. Napisah Sapiai, Universiti Teknologi MARA
Dr. Nik Roselina Nik Roseley, Universiti Teknologi MARA
Dr. Nik Rozlin Nik Mohd Masdek, Universiti Teknologi MARA
Dr. Noor Ayuni Che Zakaria, Universiti Teknologi MARA
Dr. Nor Afifah Yahaya, Universiti Teknologi MARA
Dr. Nor Hafiez Mohamad Nor, Universiti Teknologi MARA
Dr. Norazean Shaari, Universiti Selangor
Dr. Norhashimah Mohd Shaffiar, International Islamic University Malaysia
Dr. Norliana Mohd Abbas, Universiti Teknologi MARA
Dr. Nurin Wahidah Mohd Zulkifli, University of Malaya
Dr. Nursalbiah Nasir, Universiti Teknologi MARA
Dr. Ong Jiunn Chit, Universiti Teknologi MARA
Dr. Praveena Nair Sivasankaran, KDU University College
Dr. Rizal Effendy Mohd Nasir, Universiti Teknologi MARA
Dr. Siti Humairah Kamarul Bahrain, Universiti Teknologi MARA
Dr. Siti Khadijah Alias, Universiti Teknologi MARA
Dr. Siti Mariam Abdul Rahman, Universiti Teknologi MARA
Dr. Siti Nur Liyana Mamauod, PoCResT, Universiti Teknologi MARA
Dr. Siti Rabizah Makhsin, Universiti Teknologi MARA
Dr. Syahrullail Samion, Universiti Teknologi Malaysia
Dr. Wan Emri Wan Abdul Rahman, Universiti Teknologi MARA
Dr. Wan Mazlina Wan Mohamed, Universiti Teknologi MARA
Dr. Wan Muhammad Syahmi Wan Fauzi, METAL, Universiti Teknologi MARA

Dr. Yana Mazwin Binti Mohmad Hassim, Universiti Tun Hussein Onn Malaysia
Dr. Zainudin A. Rasid, Universiti Teknologi Malaysia
Adila Nalisa Mohd Roslan, Universiti Teknologi MARA
Ahmad Najmie Rusli, Universiti Teknologi MARA
Ana Syahidah Mohd Rodzi, Universiti Teknologi MARA
Che Faridah Mat Taib, Universiti Teknologi MARA
Haszeme Abu Kasim, Universiti Teknologi MARA
Hazriel Faizal Pahroraji, Universiti Teknologi MARA
Khairul Imran Sainan, Universiti Teknologi MARA
Mohamad Mali, Multimedia University
Mohd Arif Mat Norman, Universiti Selangor
Nur Aini Sabrin Manssor, Universiti Teknologi MARA
Nurrul Amilin Zainal Abidin, Universiti Teknologi MARA
Nurul Nadiah Azmi, Universiti Teknologi MARA
Shahrul Hisyam Marwan, Universiti Teknologi MARA

IOP Conf. Series: Materials Science and Engineering **834** (2020) 011005 doi:10.1088/1757-899X/834/1/011005

PHOTOGRAPHS

Participant Registration Counter

Content from this work may be used under the terms of the Creative Commons Attribution 3.0 licence. Any further distribution of this work must maintain attribution to the author(s) and the title of the work, journal citation and DOI.
Published under licence by IOP Publishing Ltd

IOP Conf. Series: Materials Science and Engineering **834** (2020) 011005 doi:10.1088/1757-899X/834/1/011005

Opening Ceremony

ICAME 2019 Participants

IOP Conf. Series: Materials Science and Engineering **834** (2020) 011005 doi:10.1088/1757-899X/834/1/011005

Keynote Speakers

IOP Conf. Series: Materials Science and Engineering **834** (2020) 011005 doi:10.1088/1757-899X/834/1/011005

MoU Exchange Ceremony

IOP Conf. Series: Materials Science and Engineering **834** (2020) 011005 doi:10.1088/1757-899X/834/1/011005

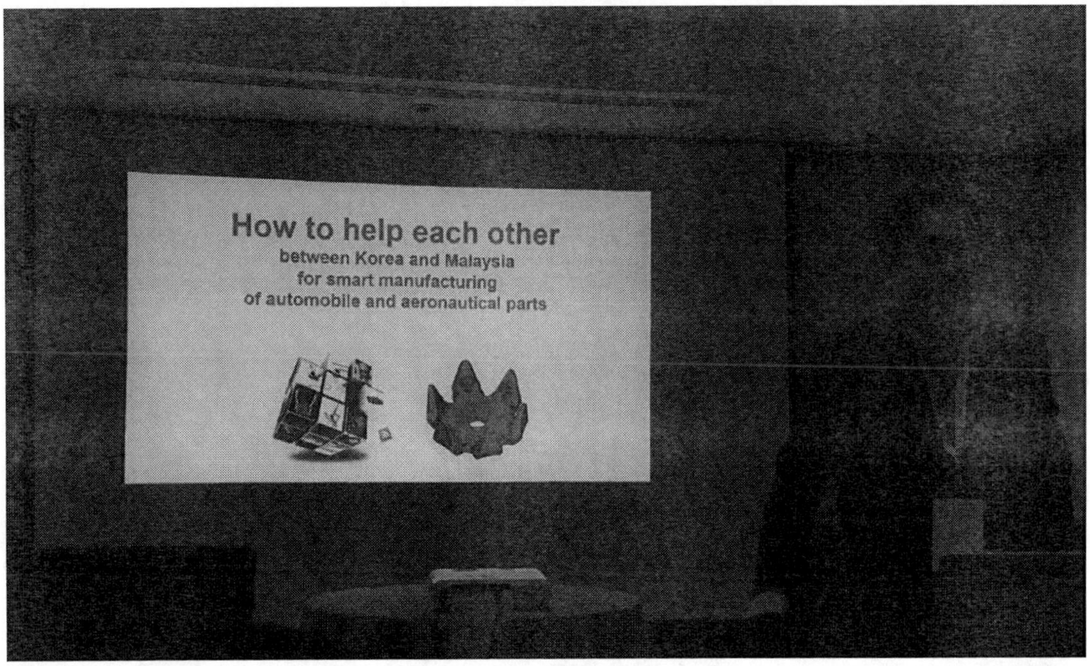

Metal Forming Workshop

Engineering Education Workshop

IOP Conf. Series: Materials Science and Engineering **834** (2020) 011005 doi:10.1088/1757-899X/834/1/011005

ICAME 2019 Visit to MARii Satellite Office in Sabah

IOP Conf. Series: Materials Science and Engineering **834** (2020) 011005 doi:10.1088/1757-899X/834/1/011005

Best Paper and Best Presenter Awards

IOP Conf. Series: Materials Science and Engineering **834** (2020) 011005 doi:10.1088/1757-899X/834/1/011005

MARii and ICAME 2019 Media Coverage

IOP Conf. Series: Materials Science and Engineering **834** (2020) 011005 doi:10.1088/1757-899X/834/1/011005

Tourism Malaysia

ICAME 2019 Committee Members

Peer review statement

All papers published in this volume of *IOP Conference Series: Materials Science and Engineering* have been peer reviewed through processes administered by the proceedings Editors. Reviews were conducted by expert referees to the professional and scientific standards expected of a proceedings journal published by IOP Publishing.

Content from this work may be used under the terms of the Creative Commons Attribution 3.0 licence. Any further distribution of this work must maintain attribution to the author(s) and the title of the work, journal citation and DOI.
Published under licence by IOP Publishing Ltd

IOP Conf. Series: Materials Science and Engineering **834** (2020) 012001 doi:10.1088/1757-899X/834/1/012001

Design of the Portable Motorcycle Cover

Mohammad Azzeim Mat Jusoh*, Muhamad Sufian Che Ashari and Helmi Rashid

Fakulti Kejuruteraan Mekanikal (FKM), Universiti Teknologi MARA (UiTM), 40450, Shah Alam, Selangor, MALAYSIA

*m_azzeim@yahoo.com

Abstract. As fuel cost increases, majority of road users in Malaysia turns to motorcycle as an economical method of transportation. Unfortunately, the weather condition in this country is unpredictable especially with sudden rainy season. Due to this reason, the idea of a portable motorcycle cover has been developed. The objective of this project is to design a motorcycle cover, which can protect the user from rain and other weather condition. Currently, we are focusing the usage in a non-mobile (static) mode. The design shall focus on workability and ease of installation, followed by the affordability factor. This design was generated with the aid of CATIA software as the design tools. FEA simulation was performed in order to validate the performance. Result shows that the safety factor of the most critical section (rear compartment) is around 7.15, which is within safety range. Performance check on the prototype shows reliable result. In near future, more improvement shall be made towards the design to enhance the function and performance factor.

1. Introduction

Based from statistics obtained from the official website of Road Transport Department of Malaysia (JPJ) in 2014, the total number of motorcycle registration have increased to 1.16 million, coincides with the increase of population in this country [1]. This means that more than 46% of Malaysian road users are motorcyclists. One of the main reasons is the increase of fuel cost, which largely effects the low and averaged income users.

In general, majority of user prefer to use their motorcycle as the daily transport as it can save their budget, since any type of four wheeled vehicles shall consumes more fuel. In addition, the heavy traffic condition in most cities, especially during festive seasons is one of the culprits. Unfortunately, most common motorcycle does not provide any type of shelter or roofing to their user. This becomes a problem, due to the location of this country within the equatorial region, which contributes to the unpredictable heavy rainy season [2].

Statistically, motorcyclists are more exposed to critical accidents compared to other method of transportation [3]. The choice to continue the journey even during bad weather condition further increases this possibility. Result from a case study performed in Malaysia between 2010 and 2012, shows that the time of day and weather conditions contributes to fatal crash cases [4]. In rainy day for example, we can observe the behavior of most motorcyclists to drive in a high-risk driving pattern. They are exposed to risks such as by riding fast on slippery road or rushing to find shelter on unnecessary area. From here, we can conclude that the main factor that contributes to motorcycle crashes is the behavior of the motorcyclists themselves [5]. For these reason, we are aiming to support the motorcyclist

Content from this work may be used under the terms of the Creative Commons Attribution 3.0 licence. Any further distribution of this work must maintain attribution to the author(s) and the title of the work, journal citation and DOI.
Published under licence by IOP Publishing Ltd

community, by providing a simple yet affordable solution. A portable motorcycle cover is seen as the most suitable option to solve this problem.

2. Literature Review

In this segment, we shall discuss on the main factors, which must be considered in designing the portable motorcycle cover. This includes market study of current motorcycle cover, the study on Malaysia's climate condition, basic design specification and more.

2.1. Review on Current Product

Based from market study performed, our observation result shows that the number of product with similar concept is very minimum. In addition, these products also have a very limited number of users. Each product has their own advantages and disadvantages. Below are the summary of three related products, in terms of its features, specification plus other contents; with visuals shown in Table 1.

i) The fabric outdoor motorcycle cover

The weight of this motorcycle cover is around 0.45 kg, which is very light. The general idea of the product is to cover the motorcycle during storage. It is suitable for 125-155 cc motorcycles with rear box; which means that the maximum dimension covered is around 1.05 x 1.25 x 2.45-meter length (WxHxL). It also has a storage bag, with anti-theft lock hole and buckle strap design. This can prevent the motorcycle cover from being blow away by the wind [6].

Pros: Portable, lightweight & easy to use
Cons: Only able to cover the motorcycle (during storage mode) and does not protect the motorcyclist.

ii) The biker raincoat type

This product features a full body and rear mirror covered design. The weight of this motorcycle cover is around 1.0 kg with thickness of 0.24 mm. The maximum dimension that can be covered is 1.45 m for open length, 1.0 m for rear width, 1.2 m for front width and 0.3 m for header part height [7].

Pros: Portable, lightweight & easy to use
Cons: The motorcycle cover is easy to blow away by the wind. It also has a low aesthetic value (the design is too simple). The safety measures (anti-theft) is also not provided.

iii) The mobility scooter umbrella type

Finally, we have a product with most similar concept. The size of this motorcycle cover is around 1.4 x 0.5 m (Length and width). The overall weight is around 1.05 kg.

Pros: Portable and can be used in motion (but within limited speed limit)
Cons: Parachute effect may occur if over speed. Also, it needs more time to assemble and dismantle.

A concept score matrix was summarised based from these products. The score was made based from the engineering specifications and customer requirement. This includes certain criteria such as size, weight, reliability, assemble and disassemble factor, safety and cost. The score matrix was defined from 1~5 incremental range, with the highest number representing excellent value vice versa. As a result, the fabric outdoor motorcycle cover and mobility scooter umbrella gain the highest score. However, as the fabric outdoor motorcycle cover concept is more towards storage purposes, we shall focus more on the mobility scooter umbrella as the main benchmark instead.

IOP Conf. Series: Materials Science and Engineering **834** (2020) 012001 doi:10.1088/1757-899X/834/1/012001

Table 1. Concept Flow Matrix.

	Fabric Outdoor Motorcycle Cover	Biker Raincoat	Mobility Scooter Umbrella
	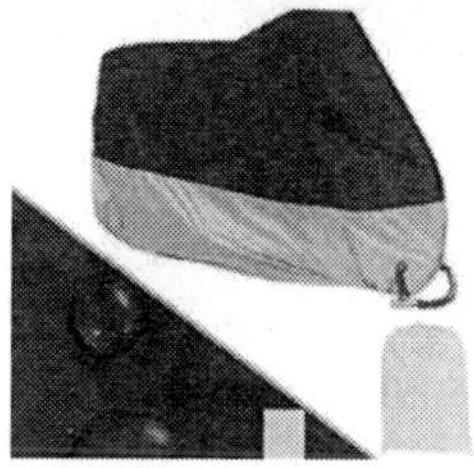	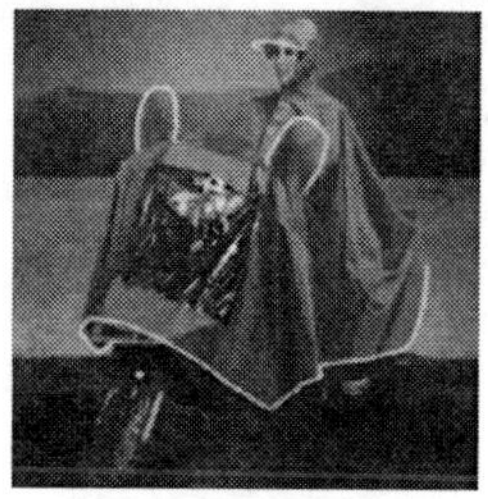	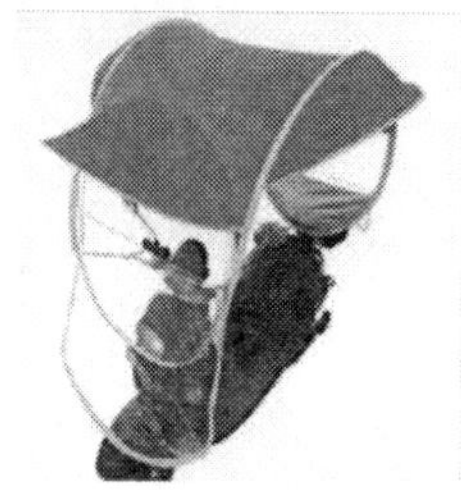
Size	3	3	3
Weight	4	4	4
Reliability	1	1	3
Ease of use	2	3	2
Safety	2	1	1
Cost	3	3	4
Availability	4	3	2
TOTAL SCORE	**19**	**18**	**19**

2.2. Main Issues of Current Product

The application of currently available motorcycle covers is quite complex. A cover which can protect both motorcycle and motorcyclist usually cannot be portable. It needs to be permanent installed or requires too much time to set up. Normally, a motorcyclist will ride as far as possible before they coming to a halt, either due to bad weather such as heavy rain or due to fatigue. Therefore, a simple assembly operation is required.

Another problem is space. In order to produce a good convertible or foldable motorcycle cover, space is needed to act as storage and mounting. However, most motorcycles have limited space. Therefore, the selection of space is one the main priority as it will effect the space consumption. The comparison of safety, reliability and ease to use factors need to be considered during the selection. In addition, the coverage area of the cover when opened must be optimized.

2.3. Malaysian Climate

Uniform temperature, high humidity and copious rainfall are the main characteristic of Malaysian climate. In general, the weather is changing from time to time in a single year, where the wind over the country is generally light and variable with some uniform periodic changes in the wind flow patterns. This wind flow patterns can be classified as the northeast monsoon, southwest monsoon and two shorter periods' inter-monsoon seasons. The northeast monsoon commonly has winds of 10 to 30 knots, southeast monsoon has prevailing wind flow that below 15 knots and inter-monsoon winds are generally light and variable [2]. Since 1 knot is equal to 0.514 m/s, so the highest prevailing wind flow is around 15.433 m/s. Since the maximum wind velocity is around 16 m/s, so the approximation of wind load is around 154 Pa or equivalent [8].

3. Methodology

3.1. The Engineering Design Process

The development of this product can be divided into three major categories, which are the conceptual design stage, the design stage and the manufacturing stage. The flow process of this motorcycle cover shall be using Deming Cycle method, which involves 'plan-do-check-action' (Refer Fig. 1).

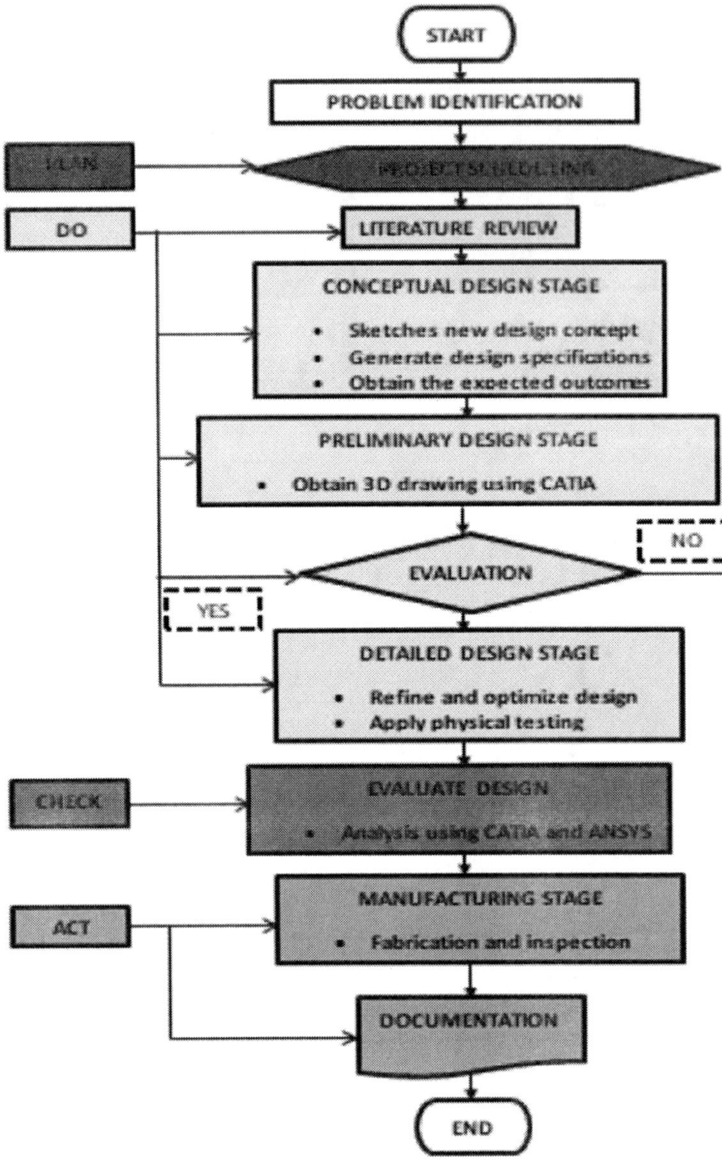

Figure 1. The design process flow

At the beginning, identifying the problems is done by performing study on the requirement of motorcycle cover and by comparing to products that is available in the market. The process is done after thorough discussions with designers and colleagues, product benchmarking process, by reviewing of related journals, patents and other documents. At the Plan stage, a Gantt chart is made to schedule this project, after having deep understanding of the problem statements. Since the period to complete this project has been decided, optimization of time is important.

Next stage is the Do stage. It consists of literature review, conceptual design, preliminary design and detail design parts. During literature review, the data or information is gathered to initiate possible solutions from earlier problems. Accumulation of information related to this theme such as road safety regulations, design safety factors, benchmark study of existing motorcycle cover, general summary of motorcycle type and its dimension, the weather condition in Malaysia and others. The source of this information is either from reference books, general observation, websites, journals and articles.

The conceptual design is the process where suitable concept is identified for this motorcycle cover design such as its shape and materials used. Aluminum has been chosen as the main material in order to reduce overall weight. This also includes consideration of overall size of the product. The final conceptual design is selected using Pugh chart. In addition, the 3D design of the product shall be made via CATIA during the preliminary design stage.

The next step is to analyze the design. This is where the evaluation process begins, where confirmation of material used, specific dimension and shape of the motorcycle cover obtained during the preliminary and detail design phase. The engineering detail drawing shall be made for each component, sub-assembly and assembly. In addition, the final drawing of this product shall reflect on the main information about it such as its dimensions, fabrication process, material type and more. One of the expected result is to reduce the assembly and disassembly time of the motorcycle cover.

Check stage is a phase after gaining the feedback from designer or potential user, either from meetings or discussion. It is important as the fabrication of the prototype for a motorcycle cover can be done if the final adjustment is made.

Finally, the Act stage is the final phase where fabrication of a tangible prototype for a motorcycle cover will take place. Machines shall be used during fabrication, which includes milling, cutter and column drill. Next, inspection shall be made on the product including the evaluation on dimensions, material specifications, manufacturing details and others; before proceed to the documentation stage where the overall report shall be made. Performance check is done during testing and inspection.

As for this product, several design revision is expected in order to come out with the best reliable output. Reason given, the final product shall be used in a harsh and unpredictable open-environment. The design must passes through severe checking before it can be used. In this study, we shall focus on the first design cycle, which we define under the initial preliminary phase.

3.2. FBD (Free Body Diagram) analysis of the motorcycle cover

The objective of this calculation is to estimate the load and stresses, which is experienced by the motorcycle cover and indirectly the motorcycle itself. Next, suitable material shall be selected based on their properties.

At first, considering that the height and the width of the front windshield section is around 0.6 m and 0.4 m, resulting in affected area of A = 0.24 [m^2]. As for the incoming wind speed, the average wind speed in Malaysia is considered around 20 [m/s] with allowable operating speed of vehicle around 40 [km/h]. The combination of both factor shall resulting in the maximum velocity of V_{max} = 31.11 [m/s].

$$\Delta P = ½ \rho v^2 \tag{1}$$

$$F = PA \tag{2}$$

As for the convertible cover, the front windshield section shall receive direct load from upcoming wind. The dynamic pressure can be viewed from (1) which resulting in ΔP = 604.9 Pa (using ρ=1.25 [kg/m^3]). Referring to (2), considering both affected area (A) and dynamic pressure, shall resulting in concentrated force of F=145.18 N.

From Fig. 2, we can see the free body diagram of the motorcycle cover from side view, where line AB represents the front windshield section, line BC represents the top cover and finally line CD represents the mainframe section. Through the calculation of equilibrium, we are able to define the loading at each critical position by considering the calculated force F as a distributed force applied at windshield area AB. As a result, the resultant force acting on point A, moving towards the vertical direction (upwards) is around F_{Ay} = 111.34 N, whereas the horizontal load also at point A (moving to the left) is around F_{Ax} = 43.66 N. Finally, the resultant load at point D is around F_{Dy} = 15.34 N, which considered for vertical load only. This is because the rear position is fixed. From the calculation, we can conclude that the most critical area is the front lower portion. However, in the FEA simulation, we shall be focusing on the reliability of the rear compartment area. Reason given, our plan is to use a standard polycarbonate based front windshield as support, which is more reliable and firm. Whereas the rear compartment needs to be more robust, as it support the load during storage and also during assembled mode.

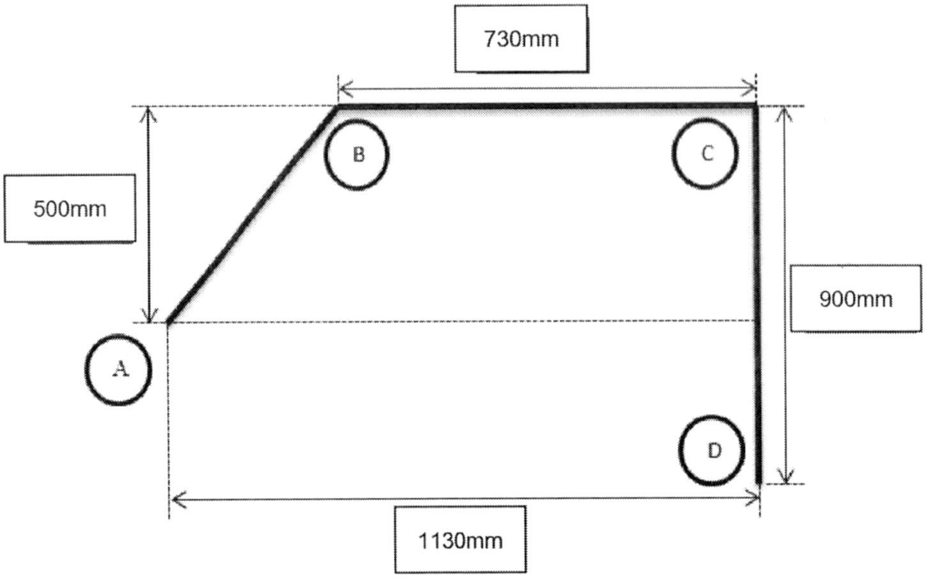

Figure 2. FBD of the motorcycle cover (from side view).

4. Result and Discussion

4.1. FEA result

As informed previously, the rear compartment or the monorack has been selected for FEA analysis, since majority of the load of the product is concentrated on the rear side. From Fig. 3, the maximum deflection of the product is 0.172 mm, which is occurring at the rear side of the monorack, whereas the von Mises stress is around 11.6 MPa. Considering the yield strength of aluminum is 83 MPa, which brings to the safety factor of 7.15. Although the safety factor is within the safety range, we must take caution that this simulation was performed in static condition. A more thorough study needs to be performed towards the prototype before detailed design improvement can be made. We are expected to proceed with the dynamic analysis in near future.

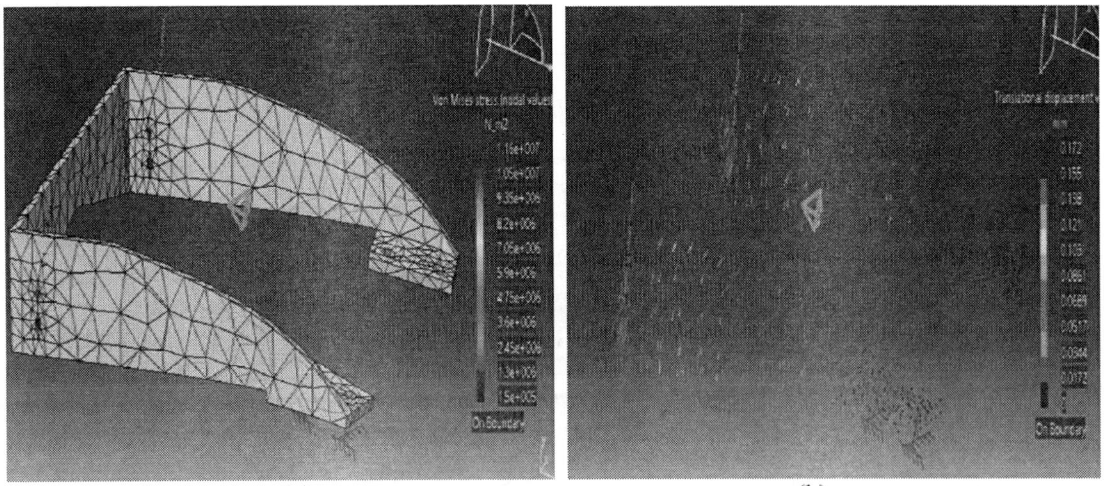

(a) (b)

Figure 3. FEA analysis result of the rack (a) von Mises stress and (b) displacement.

4.2. Validation of prototype
Referring to Fig. 4, we can simulate the prototype during folded mode and fully-opened mode. Basically, the estimated time to set up (open) the product to use is less than 4 minutes. To simplify the fabrication process and to reduce the part cost, the front windshield segment was made using thin plastic sheet. As we are focusing current usage in static mode, this condition able to serve the purpose. In future, we shall replace the design of the front windshield using permanent polycarbonate based.

(a) (b)

Figure 4. Final condition of the product in (a) folded-mode and (b) fully-open mode.

The main frame was fabricated using extendable stainless steel mechanism. The main purpose is to monitor the mechanism during opening and closing state. When fully closed, majority of the parts including mainframe and top cover shall be situated on the monorack section behind passenger seat. Currently, the enclosed box design is still in progress. In future, the stainless steel mechanism shall be replaced by aluminum material in order to reduce the weight. Overall, the design of all parts especially rear structure section shall be polished and improved.

4.3. Workability and physical test
Several inspections and testing have been performed towards the prototype. The first one is the physical inspection on the product, including the evaluation of its dimensions, material specifications and manufacturing details. Basically, the structure and position of the cover frame needs to be symmetrical. Next, the prototype also undergoes a simple weather testing (in non-mobile or static mode). As the portable motorcycle cover is fully installed on the motorcycle and being set up (open), the spray water is applied at 3 positions (front, side and rear) as a simple rain simulation (Refer Fig. 5(a)). The main purpose of this test is to monitor the area coverage during rain. As a result, the product was able to cover up the minimum number of area. However, more consideration needs to be made to increase the coverage in case of side wind. Also, we are able to test the durability of the cover, as there was no water leakage coming through the fabric during test.

In addition, we also proceed with another additional test, which was to conduct an on-road test during maximum velocity of 40 km/h (Refer Fig. 5(b)). This was to experience the actual feel of the user, at the same time to monitor the stability of the motorcycle with cover attached during use. The test was conducted on a closed road section without any bystanders nearby. This is to ensure the safety of the user as well as the surroundings. The speed was increased gradually before reaching the maximum speed. As a result, the product maintained attached even during maximum velocity.

However, displacement still occurs especially at the front position due to upcoming wind. As informed previously, we shall replace the design of the front windshield using permanent polycarbonate based, as this will increase the stability during motion. Also, there was no parachute-effect experienced during motion. Considering the preliminary design target, which was to confirm the reliability during static mode, we can consider this on-road test result as success.

(a) (b)

Figure 5. Physical testing during (a) water splash/rain test and (b) on road test.

5. Conclusion

As the fabrication of the motorcycle cover mock-up design has been completed, the goal of this project has been successfully met. The basic calculation and simulation has been performed to monitor the performance of the critical region. An actual on-road test was performed to validate the reliability and performance of the product. In future, a wind tunnel test shall be conducted, as the velocity from the wind can be fully controlled and monitored, producing a better output.

We can also conclude that motorcyclists should be given more option to protect themselves from incoming weather via a more robust product such as the portable motorcycle cover. In term of safety, we are hoping that this motorcycle cover can ease the traveling journey, preventing from fatigue and possible accidents.

In future, further improvement towards the design is positively viewed. Consideration of a mobile mode is required as the current design is limited to static or non-moving condition. To ensure that the product can be used widely and legally, the study on the stability factor of the motorcycle should be improved. A more detail simulation on the design could be added in order to improve its reliability and performance. Finally, reduction of the overall part cost shall be made after considering the final design concept.

Acknowledgments

The authors would like to thank the Faculty of Mechanical Engineering of Universiti Teknologi MARA (UiTM) for supporting this research in terms of facilities and resources.

References

[1] Laporan tahunan. (2014). Malaysia. Retrieved from www.jpj.gov.my/documents/10157/cc6788cf-fe2d-48ed-80ac-a121675665b6

[2] Malaysian Meteorological Department. (2013). Malaysian Meteorological Department. Climate General Information Malaysia _Retrieved October 1, 2017, from http://www.met.gov.my/en/web/metmalaysia. 1-21

[3] MIROS. (2016). Official Website of Malaysian Institute of Road Safety Research (MIROS). Malaysia. Retrieved from https://www.miros.gov.my.1-2

[4] Abdul Manan, M. M., Várhelyi, A., Çelik, A. K., & Hashim, H. H. (2017). Road characteristics and environment factors associated with motorcycle fatal crashes in Malaysia. IATSS Research, (risk factors contributing to traffic crashes in 9,176 fatal cases involving motorcycle in Malaysia between 2010 and 2012), 1–4.

[5] Ibrahim, M. K. A.-H., Nor, S. M. M., Mohamad, N. A., & Yusoff, M. F. M. (2012). A Case Study on Risk-taking Behaviours Among Motorcyclists A Case Study on Risk-taking Behaviours Among Motorcyclists in Klang Valley, Malaysia (MRR 07/201). Klang Valley. 1-26

[6] "BA020 Motorcycle Cover 190T Waterproof", https://www.lelong.com.my/ba020-motorcyclecover-190t-waterproof-dustproof-uv-protection-cover-happyestore-I5784092C2007-01Sale-I.htm, Accessed on 1st Nov 201

[7] "Biker Raincoat Full Body Cover Moto", https://www.lelong.com.my/biker-raincoat-full-bodycover-motorcycle-slicker-baju-hujan-motor-kavenneoh-I5786220-2007-01-Sale-I.htm, Accessed on 1st Nov 2017

[8] "Universal Motor Scooter Umbrella Mobility Sunshade & Rain Cover Waterproof Sale", https://www.banggood.com/Universal-Motor-Scooter-Umbrella-Mobility-Sun-Shade-Rain Cover-Waterproof-p-1194234.html?cur_warehouse=CNLaporan tahunan. (2014). Malaysia. Retrieved from www.jpj.gov.my/documents/10157/cc6788cf-fe2d-48ed-80ac-a121675665b6

The Effect of Zn on the Properties of Mg-10Al-6Ca-1Sn Alloy

Rosmamuhamadani, R.*, Mahat, M.M., Sabrina, M.Y

Faculty of Applied Sciences, Universiti Teknologi MARA, 40450, Malaysia

*rosma614@uitm.edu.my

Abstract. In this study, Mg-10Al-6Ca-1Sn and Mg-17Al-4Zn-3Ca-1Sn composite alloys were prepared. They were prepared by melting them in argon atmosphere on a water-cooled copper hearth. The sample were then analyse to determine the microstructure and mechanical properties using scanning electron microscope (SEM) and hardness testing, respectively. Moreover, a few of tests had been done to observe and determine microstructure and mechanical properties of the alloys. Mg-17Al-4Zn-3Ca-1Sn alloy, the microstructure of Mg-17Al-4Zn-3Ca-1Sn alloy distinctly portrays fine grains, resulting abundant of grain boundaries, which consist of eutectic phase, surround of primary magnesium grains. In contrast, the Mg-10Al-6Ca-1Sn alloy exhibits large differences in grain size. As a control, the sample without the addition of Zn element portrays large particles and creating very small amount of grain boundaries in the captured area of 30 μm. In perspective of mechanical properties, Mg-17Al-4Zn-3Ca-1Sn composite alloy showed higher Vickers hardness value with average of 81.8 H_v compared to Mg-10Al-6Ca-1Sn composite alloy, which was 70.5 H_V. In conclusion, the addition of Zn has effectively improved the hardness of the Mg alloy. This is supported by XRD analysis, which confirms the phase of new alloys that composed of Zinc element.

1. Introduction

Magnesium (Mg) based alloys are becoming some major industrial materials for today's advancement technology and latest manufacturing application. These alloys are favourable, as it possesses superior properties. With outstanding combination characteristics of high elastic modulus, good damping behaviour, very light in weight and low density, Mg alloys stand as one of the most crucial lightweight materials in structural application [1]. Besides, Mg alloys are often used in motor vehicles and aircraft application including crankcase, tack coves and gearbox housing [2]. Mg alloys are essential to be developed and improved in order to achieve great performance of alloy with excellent workability.

For instance, it is very lightweight compared to aluminium (Al) and steels. Engineers and scientist have showed so much interest of Mg alloys for decades due to these properties. Study of Mg alloys lead to reasonable prices and properties enhancement of Mg and its alloys will lead to great use of magnesium. This is because compared to using alternative materials, Mg alloys results in a 22% to 70% weight reduction. To add their competitive edge, Mg alloy started to be used in racing cars. Fuel efficiency, increase sustainability and performance are the most important issues. Mg possess low melting temperature (T_m = 650 °C) and hexagonal crystal structure (HCP). Inadequate slip system causes it to soften and weak when being exposed to high surrounding temperature. Alloying helps the materials having good room temperature strength. New composition of Mg alloys with addition of element have been developed. Most frequently used alloying elements are zinc (Zn) and calcium (Ca). Zn will increase the strength of the material at high temperature exposure because of precipitation hardening. Addition of Ca in Mg and Al will act as grain refiner [3]. By alloying, Mg properties may be improving not only

Content from this work may be used under the terms of the Creative Commons Attribution 3.0 licence. Any further distribution of this work must maintain attribution to the author(s) and the title of the work, journal citation and DOI.
Published under licence by IOP Publishing Ltd

at high temperature but also in ambient state. In this study, microstructural and hardness properties on Mg-10Al-6Ca-1Sn and Mg-17Al-4Zn-3Ca-1Sn are carried out. The characterization will be carried out by X-ray Diffraction (XRD) technique, Vickers Hardness Testing and Scanning Electron Microscopes (SEM).

Mg alloys are light in weight and own superior damping capacity. However, [4]. stated that the low strength and ductility of Mg alloys to Al alloys have limited their application and the higher demands in advance application requires better mechanical properties. Besides, [5] highlighted that Mg-Al suffers poor mechanical properties due to instability of ß ($Mg17Al_{12}$) phase at elevated temperatures thus limit the application of this alloy. Even though those elements added in are to improve mechanical properties of the alloy, not all properties are developed and some bad side effects can also be seen from the combination made. Al has large affinity towards micro porosity while Ca has low protection towards oxidation process especially in its molten condition. [6]. stated that Ca is one of the important elements in Mg alloy as it could improve the ignition-proof of pure Mg. However, the study of Ca addition into Mg alloy of Mg-10Al-6Ca-1Sn and Mg-17Al-4Zn-3Ca-1Sn have not been studied. Moreover, [7] stated that Mg_2Sn alloy superior heat resistance properties can fix the weakness of Mg that tends to soften easily at elevated temperature. Therefore, it is interesting to study the effect of Zn and Ca addition into Mg alloy.

[8] stated that increasing strength of Mg alloys at high or room temperature are based on Mg and Al with additional combination of Zn, Ca, Si and radical earth (RE). The ternary element Zn is added to the binary alloy to develop the response process of precipitation hardening [9]. This is because Zn is almost three times stronger than Al as it possesses better solid solution strengthening properties especially after being treated and under quenched state. [10] highlighted that fracture toughness of pure Mg was low which reported to be only 17.8 MPa $m^{1/2}$. However, in Mg-1.6 at.% Zn solid solution strengthening, the value rise by 5.9 MPa $m^{1/2}$ to 23.7 MPa $m^{1/2}$.

Sn is beneficial in alloying when it is combined with Mg and small amount of Al by substituting into few metallic phases [11]. In this study, 1 wt.% of Sn is added to improve properties of Mg alloys. Previously, [12] highlighted that (1-10) % of Sn can improve the creep resistance in Mg alloy. They added that Sn content helps in refining secondary dendritic arm spacing of the primary α-Mg phase. The precipitation of Mg_2Sn results in increment of compressive and tensile strength by which affected to finer grain refinement of Mg alloys [12]; [13]. Addition of Sn into Mg alloys will increase the strength of the materials and improve the ductility. By the addition of Sn, yield strength and hardness will increase due to intermetallic phases. Grain refinement and solid solution strengthening of Mg-Sn give strengthening effect in single-phase alloy where this will result in better mechanical properties to Mg alloys.

By the addition of Sn, it is expected to give superior properties to the alloy. However, Sn element is quite expensive which can cause increasing of the production cost. Hence, over addition of Sn in the alloy can bring limitation in cost aspect to the production too. New findings of Mg alloys are very crucial in order to develop the materials to have close tolerance towards requirement for commercial purposes. In this study, Mg-10Al-6Ca-1Sn and Mg-17Al-4Zn-3Ca-1Sn samples were carried out so that Mg alloy mechanical properties can be improved. It is expected to obtain better microstructure, good damping behaviour, increment of hardness and elastic modulus in Mg alloys by the addition of few elements like Sn, Al, Ca and Zn. Good mechanical properties and high strength will result in high durability and long-life span of the products. Improvement of Mg alloy can bring enhancement in the latest technology production while at the same time widened the field of Mg studies

2. Materials and Methods

2.1 Sample Preparation

Mg-10Al-6Ca-1Sn and Mg-17Al-4Zn-3Ca-1Sn composites were prepared using 99.5% Mg, 70 % Mg-30 % Ca master alloy ingot, Sn ingot, Zn chips and Al shots. The samples were measured and weight accordingly, then melted using an arc-meting furnace in argon atmosphere on a water-cooled copper hearth. Then sample were placed inside the furnace. The chamber were vacuumed for at least five minutes and argon gas were flushed into the chamber. This process were done twice to ensure the chamber atmosphere is filled with argon gas. Airflow need to be in control to avoid sample was blown away due to high speed of airflow and the tightly closed door of the furnace to avoid air from diffusing into the chamber and causing the sample to oxidize. Once purging completed, titanium (Ti) getter and the samples were then be melted in order to eliminate impurity gases such as oxygen and nitrogen. Once completed, melting process will continue by melting of the samples. Samples were be mounted using cold mounting epoxy and hardener according the ratio of 9:1. The mixture of epoxy and hardener were transferred into a polystyrene cup and stirred slowly. The samples were left for 4-5 hours in order to let it harden. The mounting which the specimen is embedded in the solution/solvent have purposes to ease in handling and protect the specimen physically.

After the mounting processes, the specimen were grounded using silicon carbide abrasive paper. Figure 1 shows Metkon Gripo 2V Grinder-Polisher that used in this process. The purposes of this step is to eliminate the damage from cutting, remove contaminants and also repair the surface of the specimen. The grinding processes starts from 120, 240, 600, 800, 1000 and 1200 grit consequently. Proper grinding can help minimizing time spending for polishing process. For polishing process, polishing cloths and diamond slurry were used to gain a better surface finish that possess high reflectivity before any microscopic testing can be done. Diamond slurry were used as it offers highest level of accuracy that can either be water-based or oil-based slurries. In this experiment, water-based slurry, which is harmless to the environment with super cleaning properties, were used to polish the metallic specimen.

Figure 1. Grinder Polisher (Metkon Gripo 2V)

After grinding and polishing, the samples were be etched by strong acid to remove the surface of the exposed area in order to achieve a better microscopic view. Etching is done in order to remove oxide layer that might form at the surface area or any impurities adhered onto it. The etchants help to cut the unnecessary parts and reveal the protected parts. Etching is done by immersing sample into the strong acid and is left for a moment in range 3-60 seconds. The etchants than can be used for Mg alloys is glycol solution.

2.2 Mechanical Characterization

For mechanical behaviour of the specimen finding, hardness testing were conducted by vickers hardness tester, Shimadzu HMZ Series. By using load of 500g and dwell of 15 seconds, a square based pyramid shape were used to lift the load. An indenter having an angle of 136° must sufficiently harder than the specimen tested so that it cannot easily deformed by the force applied.

Once the testing over, the width or depth of the indentation were measured to determine the hardness of the sample tested. The vickers hardness is the quotient of the tested load applied (500g) for the area of the indentation in millimetre, mm unit by considering the upside down square base pyramid. vickers hardness is identified according to equation (1) and (2);

$$HV = \frac{2F sin\frac{136^\circ}{2}}{d^2} \tag{1}$$

$$HV = 1.854 \frac{F}{d^2} \qquad (2)$$

$where; F = load$ (kgf), and D = average value of two diagonals, d_1 and d_2 (mm)

2.3 Microstructural and Phase Analysis

Microstructural examination of samples was conducted by using SEM, INCA, SUPRA 40V 40VP) Zeiss. It was used in order to observe the microstructure of that composites. The standard grinding and polishing method were used to prepare the sample for SEM observation. The sample were grind using silicon carbide paper from 120 grit to 1200 grit and then followed by polishing process that use diamond slurry from 6 µm to 1 µm. The formation of new phases in this work was monitored by Shimadzu X-ray diffraction XRD-600 with Cu-Kα radiation operated at 30 kV and 20 mA and 2θ between 20 to 100 °C at the scanning rate of 2°/ min.

3. Results and discussion

3.1 Hardness Testing

The vickers hardness test was performed to measure the micro hardness of the two different alloys at least three different regions were indented for each sample under a load of 0.4 kg for 15 seconds time dwelling. Mg-17Al-4Zn-3Ca-1Sn composite alloy shows a higher Vickers hardness value with average of 81.8 H_v. Meanwhile, Mg-10Al-6Ca-1Sn composite alloy exhibits 70.5 H_v. It is hypothesised that Alloy with the addition of element Zn is predicted to have higher vickers hardness than the one without Zinc. Zn is approximately three times affective on an atomic per cent basis than aluminium in the yield strength for alloys. It is shown that Mg-4Zn-6Ca-1Sn alloy has higher resistance to the localization of plastic deformation compared to Mg-10Al-6Ca-1Sn alloy. Figure 2 shows the Vickers hardness value both of Mg-10Al-6Ca-1Sn and Mg-17Al-4Zn-3Ca-1Sn composite alloys. [14] found that studied by adding Zn as the alloying element into Mg-based composite, the microhardness value has increased about 12%, suggesting that zinc shows a greater effect on the hardness value. In another work by [15], the Mg alloy portrays higher hardness value when the Zn contents increased. This is due to the intermetallic components and formation of precipitate in the alloy.

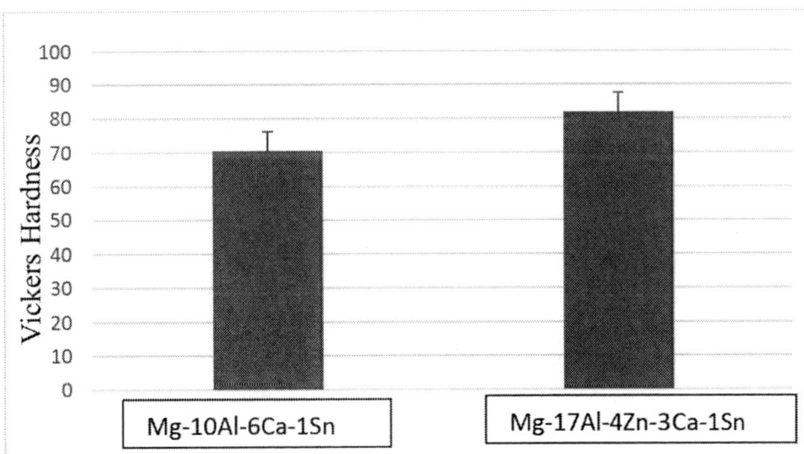

Figure 2. Vickers hardness value both of Mg-10Al-6Ca-1Sn and Mg-17Al-4Zn-3Ca-1Sn alloys

3.2 Microstructure

Microstructural analysis was perform at the magnification of 2000x for both samples in size of 30µm. Based on the morphological in Figure 2, there are two different types of eutectic morphology in both two samples. Mg-17Al-4Zn-3Ca-1Sn, was found to have one with brighter white contrast, which involves lamellar phase richer in terms of Mg-Ca-Zn. On the other hand, the lamellae of the second

IOP Conf. Series: Materials Science and Engineering **834** (2020) 012002 doi:10.1088/1757-899X/834/1/012002

eutectic with the grey contrast was determined as a phase richer in other elements like Mg-Al-Ca-Sn. Meanwhile, in Mg-10Al-6Ca-1Sn, the Mg-Ca exists in lamellar phase with brighter white contrast and the remaining elements in the grey contrast. [16] stated that the eutectic is a common pattern when it comes to Mg – Al alloy and Zn addition is related to the changes of the phases in the alloy.

For Mg-17Al-4Zn-3Ca-1Sn alloy, the microstructure distinctly portrays fine grains and therefore resulting abundant of grain boundaries, which consist of eutectic phase, surround of primary magnesium grains. In contrast, alloy sample of Mg-10Al-6Ca-1Sn alloy shows large differences in grain size. With the same magnification, the alloy without addition of Zn element tends to have large particles and creating very small amount of grain boundaries in the area of 30μm captured. This is due to the group of alloys that present based mainly on the Mg-Al system, which tends to have large and variable grain size. The addition of Zn to Mg metal alloys has increased the amount and continuity of the compound at the grain boundaries [11]. Zn acts as grain refiner that attributed to inoculation of the particles in the alloy. The presence of Zn atoms has contributed to the solid solution strenghtening, as the presence of grain boundary phases reduces grain boundary sliding [3]. Therefore, an easiest way to improve the strength of a material is to make grain small as possible, increasing the amount of grain boundary. Smaller grains have greater ratios of surface area to volume, which result in greater ratio of grain boundary to dislocation.

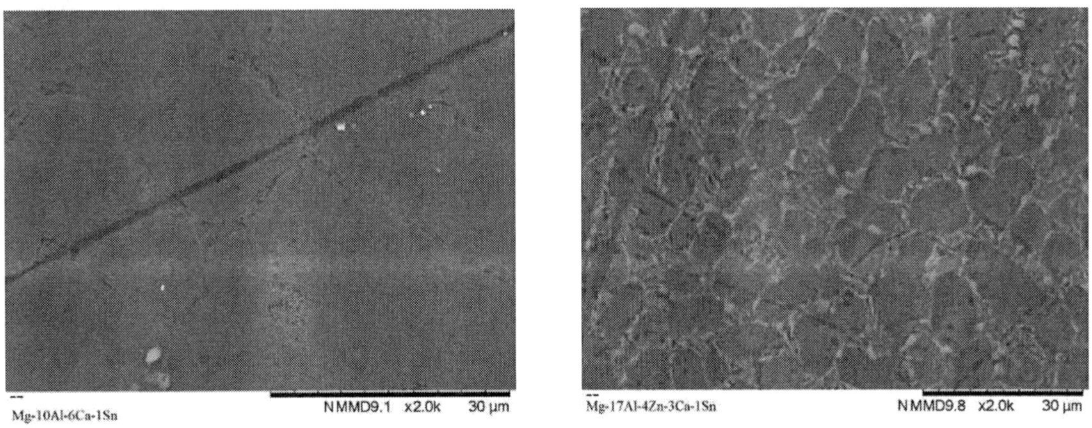

(a) (b)

Figure 3. Microstructure of (a) Mg-10Al-6Ca-1Sn and (b) Mg-17Al-4Zn-3Ca-1Sn observed by SEM with 2000x magnification

3.3 X-ray diffraction

Figures 4(a) and (b) show the XRD pattern of Mg alloys without and with the presence of Zn metals, respectively. Coherent scattering by diffraction of x-ray can detect crystal structure of materials shows the two XRD pattern are polycrystalline with sharp peaks in each graph. We found that patterns correspond to the peaks of Mg alloys with and without Zn element as Mg-17Al-4Zn-3Ca-1Sn has more peaks than Mg-10Al-6Ca-1Sn alloy. As Mg composition is the highest in both alloys, it also shows the highest peak in XRD pattern. Furthermore, the lower intensity peaks of Zn and Sn elements is lower compared to Mg, Al and Ca. This is due to small percentage of the elements content added into the alloy.

The Mg peak was detected in all samples. From XRD pattern the results shown that the addition of 4 wt% of Zn does not lead to the production of intermetallic phase. [16] stated that some intermetallic phase will form if the Zn composition reach in ranging of 29 wt% to 53 wt%. Intermetallic phases such as Mg_2Zn_{11} and MgO were detected and existed in Mg-17Al-4Zn-3Ca-1Sn composite alloy.

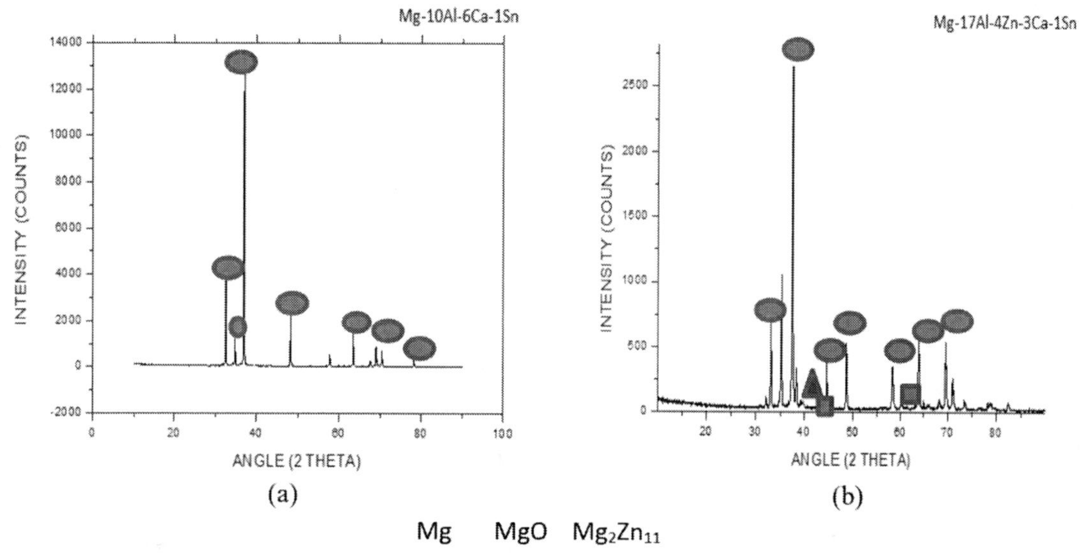

Figure 4. XRD pattern of (a) Mg 10Al-6Ca-1Sn and (b) Mg-17Al-4Zn-3Ca-1Sn composite alloys

4. Conclusions

We conclude that;

i. Mg-17Al-4Zn-3Ca-1Sn composite alloy exhibits higher hardness value compared to Mg-10Al-6Ca-1Sn. This is due to the addition of Zn element that has induced the formation of finer grains within the alloys.

ii. Intermetallic phases such as Mg_2Zn_{11} and MgO were found in the Mg-17Al-4Zn-3Ca-1Sn composite alloy, which confirmed by XRD.

Acknowledgments

The authors acknowledge the financial support (LESTARI 148/2017) from Research Management Institue (IRMI), Universiti Teknologi MARA, Shah Alam, Selangor, Malaysia.

References

[1] S. You, Y. Huang, K.U, and N. Hort, "Recent research and developments on wrought magnesium alloys," *Journal of Magnesium and Alloys, vol* 5(3), pp. 239-253, 2017.

[2] W. Darham, M.K. Talari, M. K., A.L. Anis, and N.K. Babu, "Impression creep of Mg-1.5Ca and Mg-3.0Ca alloys," *Journal of Engineering and Applied Sciences* vol. 11(12), pp. 2573-2576, 2016.

[3] G. Wang, G. Huang, X. Chen, Q. Deng, A. Tang, B. Jiang, and F. Pan, "Effects of Zn addition on the mechanical properties and texture of extruded Mg-Zn-Ca-Ce magnesium alloy sheets," *Materials Science and Engineering* A 705, pp. 46-54, 2017.

[4] S.M. Liang, Y.Q. Ma, R.S. Chen, and E.H. Han, "Effect of composition on the microstructure and mechanical properties of Mg-Al-Zn alloys," *The Minerals, Metals & Materials Society,* pp. 331-336, 2008.

[5] J. Jun, "Damping behaviour of Mg-Zn-Al casting alloy," *Materials Sciences and Engineering* A665, pp 86-89, 2016.

[6] D. Wan, J. Wang, L. Lin, Z. Feng, and G. Yang, "Damping properties of Mg–Ca binary alloys," *Physica B: Condensed Matter*, 403, vol (13-16), pp. 2438-2442, 2008.

[7] H, Pan, G. Qin, M. Xu, H. Fu, Y. Ren, F. Pan, and B. Song, "Enhancing mechanical properties of Mg–Sn alloys by combining addition of Ca and Zn," *Materials & Design* 83, pp. 736-744, 2015.

[8] H. Son, J. Lee, I. Oh, D. Kim, K. Yoshimi, and K. Maruyama, K. "Microstructure and mechanical properties of Mg-Al-Ca-Nd alloys fabricated by gravity casting and extrusion process," *Materials Transactions* 49, vol. 5, pp. 1025-1031, 2008.

[9] J. Nie, and B. Muddle, "Precipitation hardening of Mg-Ca -Zn alloys," *Scripta Materialia* 37, vol. 10, pp. 1475-1481, 1997.

[10] H. Somekawa and T. Mukai, "High strength and fracture toughness balance on the extruded Mg Ca–Zn alloy," *Materials Science and Engineering*: A, 459, vol. (1-2), pp. 366-370, 2007.

[11] T. Padfield, "Metallography and microstructures of magnesium and its alloys," *Metallography and Microstructures,* pp. 810-805, 2004.

[12] P. Poddar, A. Kamaraj, A. Murugesan, S. Bagui, and K. Sahoo, "Microstructural features of Mg 8%Sn alloy and its correlation with mechanical properties," *Journal of Magnesium and Alloys* 5, vol. 3, pp.348-354, 2017.

[13] W.L. Cheng, S.S. Park, B.S. You, and B.H. Koo, "Microstructure and mechanical properties of binary Mg–Sn alloys subjected to indirect extrusion," *Material Science and Engineering A* 527, pp. 4650-4653, 2010.

[14] Z.N. Hussain, N, M. Isa, & B.K. Dhindaw, "Effect of Alloying Elements on Properties of Biodegradable Magnesium Composite for Implant," *Application Powder Metallurgy & Mining, 6, vol.* 3, pp.171-183, 2017.

[15] K. Ponappa, S. Aravindan, and P.V Rao, "Influenece of Y_2O_3 on Mechanical Properties of Magnesium and Magnesium Alloy (AZ91D)," *Journal of Composite Materials 47* vol. 10, pp. 1231-1239, 2012.

[16] L. Li. S. Ji, Q. Zhu, Y. Wang, X. Dong, W. Yang, S. Midson, Y. Kang, "Effect of Zn Concentration on the Microstructure and Mechanical Properties of Al-Mg-Si-Zn Alloys Processed by Gravity Die," Casting, *Metallurgical and Materials Transactions* A 49, vol.8, pp. 3247–3256, 2018.

[17] D. Annur, P. Franciska, A. Erryani, M.I Amal, L.S. Sitorus, and I. Kartika, "The Synthesis and Characterization of Mg-Zn-Ca Alloy by Powder Metallurgy Process," *AIP Conference Proceedings, 1725,* 2016.

Study on Rheological and Dimensional Properties of Polypropylene on End-Of-Life Vehicles

Mohd Hafiz Izzuden Bin Abdul Hamid*, Noor Azlina Binti Mohd Salleh, Nor Hafiez Mohamad Nor, Muhammad Hussain Bin Ismail, Muhammad Firdaus Bin Yahya, Mat Isa Bin Saari, Siti Maisarah Mohd Pisal

Malaysia Automotive, Robotics and IoT Institute
Faculty of Mechanical Engineering, Universiti Teknologi MARA, Shah Alam

*izzuden94@gmail.com, noorazlinamohdsalleh@gmail.com

Abstract. As the global population increased, the production of vehicles also has increased. It is estimated that the population of vehicles in the world has exceeded 1.32 billion units at the end of 2016. The increased vehicle production brings a lot of negative impact to the environment such as noise pollution and the increase of end-of-life vehicles (ELVs). This study aims on the potential of recycled polypropylene that has been used in ELVs. The recycled materials are obtained from vehicles dismantling center at DRB-HICOM Environmental Services Sdn Bhd. The materials were then cleaned and crushed into the pellet before undergoing rheological test at the heat treatment laboratory, Faculty of Mechanical Engineering in Uitm Shah Alam. The purpose of rheological test is to study the melt flow behavior of recycled polypropylene in order to obtain the optimum temperature to be applied in plastic injection molding process. The graph of shear viscosity vs. shear rate represents the behavior of both materials at 190^{0}C, 200^{0}C and 210^{0}C. Then materials are produced as a new product. The process of plastic injection molding is carried out at Sheikh Brothers Industries Sdn Bhd in Pelabuhan Klang.

1. Introduction

Vehicle production has seen drastically increased for the past 20 years. This scenario is heavily influenced by the increase in world population and transport needs in human life. The goal of mass production is low-cost manufacturing [1]. In Malaysia, as the population is increasing (equivalent to 0.42% of total world population), the demand for the vehicles is also increased [4]. The sudden increase in vehicles production contributes to the increase of end-of-life vehicles (ELVs). Motor vehicles have been around for little more than one hundred years, however, due to the mass production of cars, it is estimated that around 1.6 - 2 million end-of-life vehicles (ELVs) are arising in the UK each year [3]. Not only aggravate the scenery, but also ELVs complicate the process of conserving the environment.

The recycled polypropylene is collected and segregated from ELVs and undergo rheological test. The purpose of the test is to examine the liquid recycling polypropylene behaviour at optimum temperatures so that the materials will yield the best results in the plastic injection molding process. Then, the process is analysed. Wavelike flow marks are a kind of surface defect that can arise during the filling stage of the injection molding process [8]. Cavity pressure control during filling, packing, and cooling phases is imperative for maintaining product quality in injection molding process [7]. The recycled materials then produced as new product and all dimensions of the product is recorded and compared with the actual design. Overall, this study provides information on the proper way to recycle ELVs and the potential of materials used in ELVs.

Content from this work may be used under the terms of the Creative Commons Attribution 3.0 licence. Any further distribution of this work must maintain attribution to the author(s) and the title of the work, journal citation and DOI.
Published under licence by IOP Publishing Ltd

IOP Conf. Series: Materials Science and Engineering **834** (2020) 012003 doi:10.1088/1757-899X/834/1/012003

2. Methodology

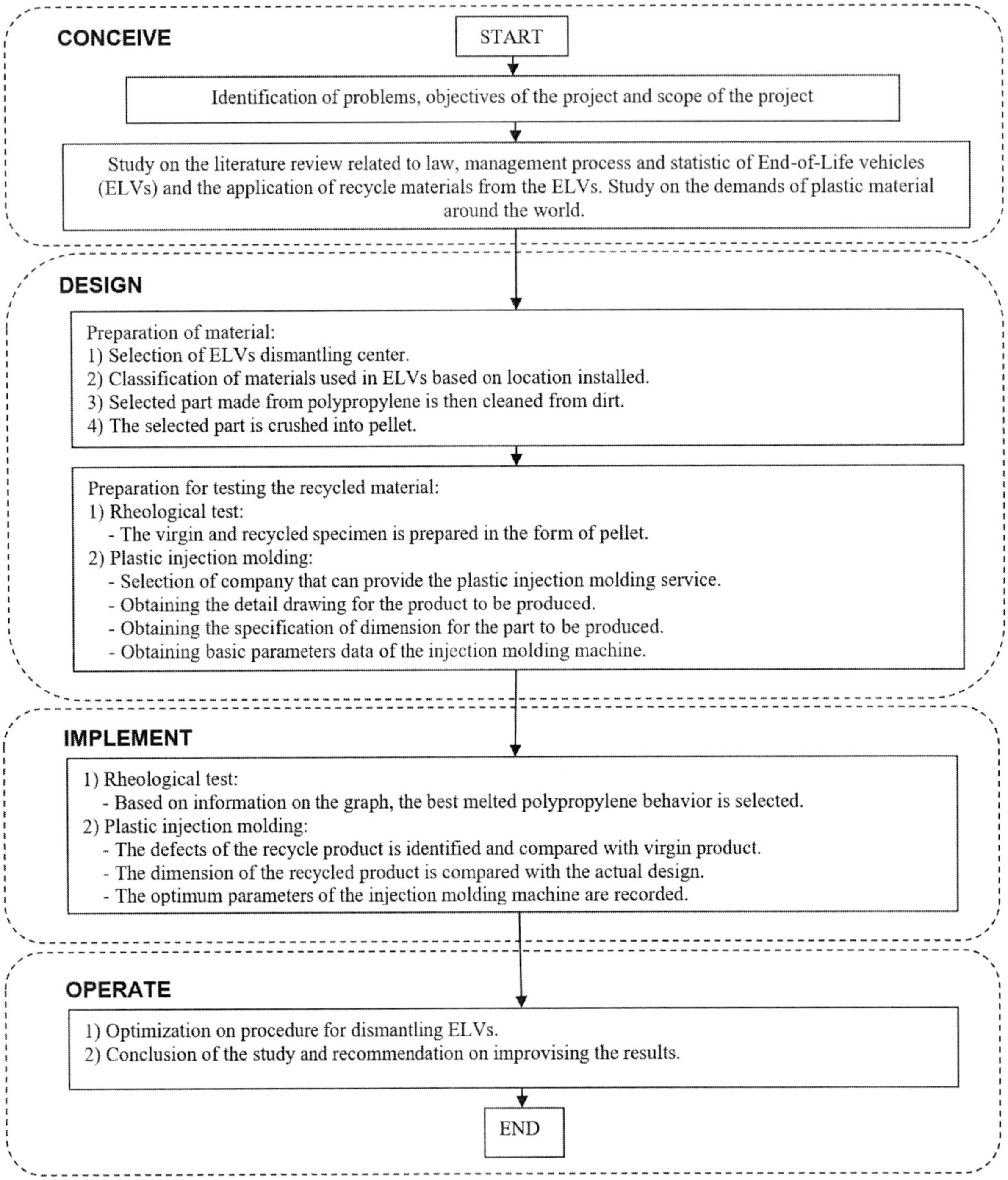

Figure 1. Research methodologies (CDIO)

3. Result and Discussion

All data for the rheological test is calculated automatically by the software. For the rheological test, software that has been used is Flowmaster. For the injection molding process, the results are obtained manually.

3.1 Rheological Test

The results for this test is based on temperature applied and types of materials. The result for Shear Viscosity (Pa.s) vs. Shear Rate (/s) is then plotted.

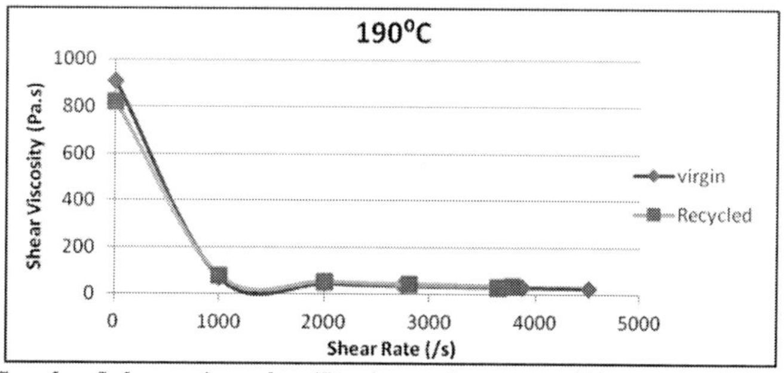

Figure 2. Graph of shear viscosity (Pa.s) vs. shear rate (/s) at temperature of 190°C

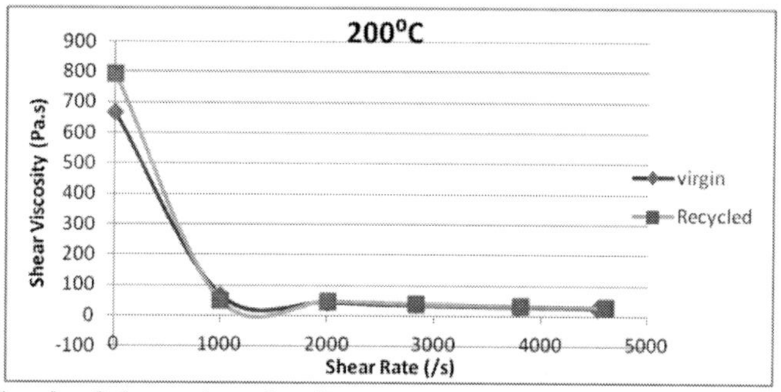

Figure 3. Graph of shear viscosity (Pa.s) vs. shear rate (/s) at temperature of 200°C

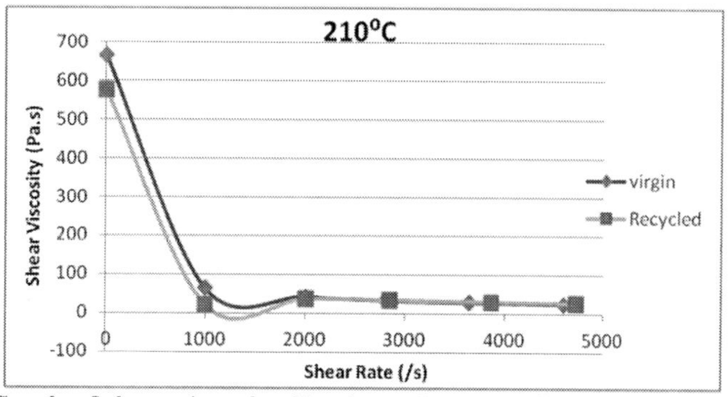

Figure 4. Graph of shear viscosity (Pa.s) vs. shear rate (/s) at temperature of 210°C

3.2 Discussion

Both results are showing the degradation of recycled polypropylene. For viscosity test, the results presented as graph of shear viscosity vs. shear rate. For plastic injection molding, the results are taken from the different of dimension compared to actual design. The defects of product were also studied.

For rheological test, the temperatures applied are 190°C, 200°C and 210°C. Theoretically, viscosity of a fluid depends on the temperature. For liquid, as the temperature is high, the viscosity is low. Fluid flow can be divided into two, Newtonian fluid and non-Newtonian fluid. Newtonian fluid is the fluid that satisfies the Newton's Law of Viscosity. For polypropylene, fluids are shear thinning when the viscosity is decreased and the shear rate is increased. This behaviour is known as pseudo-plastic.

For the first graph (Figure 3.1.2), the test was carried out using temperature of 190°C. Shear viscosity of recycled polypropylene is lower than shear viscosity of virgin polypropylene for the first 1000 shear rate. Then, the graph of recycled polypropylene remains almost the same as the graph of virgin polypropylene. For the second graph (Figure 3.1.3), the test was carried out using temperature of 200°C. Shear viscosity of recycled polypropylene is higher than shear viscosity of virgin polypropylene for the first 1000 shear rate. Then, at range 1000 to 2000 of shear rate, the shear viscosity of recycled polypropylene is lower than shear viscosity of virgin polypropylene. The graph of recycled polypropylene remains almost the same as the graph of virgin polypropylene. The last graph (Figure 3.1.4) obtained from test at 210°C. Shear viscosity of recycled polypropylene is lower than shear viscosity of virgin polypropylene for the first 2000 shear rate. Then, the graph of recycled polypropylene remains almost the same as the graph of virgin polypropylene.

The injection molding process is carried out by using recycled polypropylene at three different nozzle temperatures, 190°C, 200°C and 230°C. Major flow marks appear when applying nozzle temperature of 230°C. This is due to the viscosity of the melted polypropylene is too low. Thus, the most suitable temperature to be applied in the injection molding process is 190°C.

Then, the best product is selected and the dimension is recorded. Based on the results on the details drawing, from 45 points of dimension that have been checked, there is one dimension that is not satisfying the standard dimension. The dimension is out of ranged allowed because the part is warped due to several reasons such as pressure in the cavity, fill rate and cooling rate. Warping is caused by too much stress acting on the product. The reason of the stress acting on the product is uneven pressure in the mold and temperature of mold is too low. To improve this defect, the parameter of the machine needs to be changed.

4. Conclusion and Recommendation

The study has successfully satisfy the objectives which are to identify materials used in ELVs, to compare rheological behaviour of recycled polypropylene from ELVs with virgin polypropylene and to produce product by using recycled polypropylene from ELVs.

Rheological test indicates that at temperature of 190°C, the behaviour of melted recycled polypropylene is nearly the same as compared to virgin polypropylene. As the temperature is increased, the shear viscosity of the recycled polypropylene dropped as compared to virgin polypropylene due to several factors such as location of polypropylene parts that have been installed and life cycle of the polypropylene. To increase the properties of the recycled materials, some improvement can be made. For example, additive such as talc will increase the strength of the material. The product that has been produced by using three different temperatures have shown that the recycled polypropylene is best operates at 190°C. This proves that the result in rheological test is applicable in real manufacturing process. In order to make the recycling process of ELVs more effective, proper standard operating procedures (SOPs) need to be followed. For example, at the dismantling center, the parts need to be sorted according to type of materials. For plastic, the parts need to be isolated by type of plastic. With this procedure, process of recycling recycled material will be more effective.

Acknowledgement

The author would like to thank the Faculty of Mechanical Engineering, Universiti Teknologi MARA Shah Alam for giving this opportunity to further study on this topic. Also, a big appreciation to the Malaysia Automotive Institute, DRB-HICOM Environmental Services and Sheikh Brothers Industries Sdn Bhd for supporting the author in order to finish this study

IOP Conf. Series: Materials Science and Engineering **834** (2020) 012003 doi:10.1088/1757-899X/834/1/012003

References

[1] Yoram Koren, (2010). *Mass Production and Lean Manufacturing;*

[2] Peter Wad, (2001) *Transforming Industrial Relations: the Case of the Malaysian Auto Industry.*

[3] End of Life Vehicles, *What is an end of life vehicles?* Retrieved by: 27/10/2017 https://www.ciwm.co.uk/ciwm/knowledge/end-of-life-vehicles.aspx

[4] Worldometers, *Malaysia Population.* Retrieved by: 30/10/2017: http://www.worldometers.info/world-population/malaysia-population/

[5] Future for Growth- *The Malaysian Plastic Industry.* Retrieved by: 10/11/2017: http://www.mpma.org.my/Documents/Roadshow_2015_PT2%20Future%20for%20Growth.pdf

[6] Solomon, M. J., Almusallam, A. S., Seefeldt, K. F., Somwangthanaroj, A., & Varadan, P. (2001). *Rheology of polypropylene/clay hybrid materials. Macromolecules, 34*(6), 1864-1872.

[7] B. Pramujati, R. Dubay, (2006) *Cavity pressure control during cooling in plastic injection molding*

[8] Sung Yong Kang, Seung Mo Kim, Woo Il Lee, (2007) *Finite Element Analysis for wavelike flow marks in injection molding.*

IOP Conf. Series: Materials Science and Engineering **834** (2020) 012004 doi:10.1088/1757-899X/834/1/012004

Competitive Molecular Interaction between Zinc Dialkydithiophosphate (ZDDP) and Methyl Oleate along Boundary Lubrication Regime

S. H. Hamdan[1], M. B. Lee[2] and W. W. F. Chong*[2,3]

[1]Bio-Engineering Technology Department, University Kuala Lumpur Malaysian Institute of Chemical & Bioengineering Technology (UniKL MICET), Alor Gajah, Melaka, Malaysia
[2]School of Mechanical Engineering, Faculty of Engineering, Universiti Teknologi Malaysia (UTM), Johor Bahru, Johor, Malaysia
[3]UTM Centre for Low Carbon Transport in Cooperation with Imperial College London, Universiti Teknologi Malaysia (UTM), Johor Bahru, Johor, Malaysia

*william@utm.my

Abstract. Fatty acid methyl ester (FAME) possesses good lubricity properties, prompting numerous research work to explore their potential as lubricant additives. However, fully formulated lubricants consist of additive packages, which includes essential anti-wear agents, such as zinc dialkyldithiophosphate (ZDDP). Adding FAME into a lubricant could lead to adverse tribological effects if the competitive molecular interaction between different additives are not well understood. Therefore, the present study determines the tribological impact of molecular interaction between ZDDP and methyl oleate along boundary lubrication regime. Lateral force microscopy, coupled with fluid imaging approach, is adopted, allowing for the analysis to focus on asperity level interaction. A silicon nitride tip is used to slide against a mirror-polished steel substrate while being fully submerged in the tested fluids. The tested fluids are mixtures of ZDDP and methyl oleate, blended at different volumetric percentages. It is observed that the lowest friction force is measured for ZDDP-methyl oleate mixture containing 70-vol% of methyl oleate. Interestingly, the friction force measured for such mixture is found to be lower than that of neat methyl oleate. This finding indicates that an optimum blending ratio between FAME and ZDDP is essential in achieving better boundary lubrication performance for tribological conjunctions.

1. Introduction

Fatty acid methyl ester (FAME) is commonly used as an alternative biodegradable fuel in compression ignition engines. Such application has resulted in significant emissions reduction from compression ignition engines [1, 2]. More importantly, FAME is also known to have good lubricity properties [3], which has been evidenced by the improved fuel lubricity after being added to low-sulphur petrol-diesel. Typically, tribological properties of FAME have been investigated using tribometers, such as the high frequency reciprocating rig (HFRR)[4], four-ball tester [5], pin-on-disk [6] and ball-on-disk [7] type

Content from this work may be used under the terms of the Creative Commons Attribution 3.0 licence. Any further distribution of this work must maintain attribution to the author(s) and the title of the work, journal citation and DOI.
Published under licence by IOP Publishing Ltd

tribometers. Coefficient of friction and wear scar properties are commonly reported for FAME derived from different vegetable oils. Alternatively, based on the classical Stribeck curve analysis, for various types of tested vegetable oil and animal fat derived FAMEs, Hamdan et al. found that FAME has good friction modifier property, which delays the onset of mixed and boundary lubrication regime [6].

To date, most of the studies of FAME lubricity focuses on its tribological properties with little emphasis on the use of FAME as an additive. When being considered as lubricant additive, it is essential to be able to understand the molecular interaction of FAME with other types of additives, such as anti-wear agents. This is because the lack of such fundamental knowledge could lead to less desired tribological performances due to the competitive interactions between various types of additives. As an example, in a recent study on FAME dilution of engine lubricants, Hamdan et al. showed that the existence of FAME in the engine lubricants does in fact enhance the friction modifier effect of the lubricant [7]. However, care must also be taken because beyond a certain threshold amount of FAME dilution, adverse tribological performance are observed, mainly due to reduced load carrying ability of the lubrication film, which could result in more significant wear.

The present study intends to determine the tribological properties of anti-wear agent in the presence of FAME along boundary lubrication regime. The anti-wear agent selected in this study is zinc dialkyldithiophosphate while the selected FAME is in the form of methyl oleate. The selection of methyl oleate is because this type of FAME has been shown to consistently produce better tribological properties in terms of wear scar diameter and coefficient of friction as compared to the other types of FAME [3]. The molecular structures for ZDDP and methyl oleate are given in Figure 1. It is shown that ZDDP has a much more complex structure as compared to the fairly linear structure of methyl oleate.

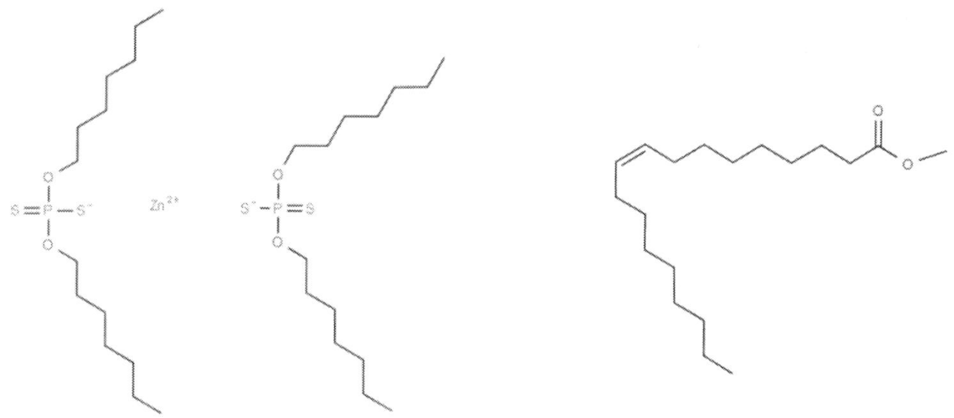

(a) (b)

Figure 1. Molecular structure for (a) zinc dialkyldithiophosphate, $C_{28}H_{60}O_4P_2S_4Zn$ (ZDDP) and (b) methyl oleate, $C_{19}H_{36}$ (MO) (generated from http://www.chemspider.com)

2. Experimental Approach

In the present study, commercially available ZDDP and methyl oleate are obtained. Typical friction measurements using tribometers often produces results that couples macro-, micro- and nanotribological effects. To isolate only the boundary lubrication characteristics of the tested fluids, it is imperative that measurements are done at relevant length scale, which is at asperity level. Therefore, lateral force microscopy coupled with fluid imaging approach is adopted, allowing for the boundary lubrication properties of the ZDDP and methyl oleate mixtures to be investigated at asperity level. The asperity level friction measurement approach in this study follows the procedure proposed by Chong and Ng [8]. A silicon AFM probe with tip radius of 10 nm is used to slide against a mirror-polished stainless-steel substrate while being fully submerged in the tested fluids. The tested fluids are blends of ZDDP with methyl oleate mixtures at different volumetric percentages. During the test, applied normal load on the silicon AFM probe is varied between 1 nN and 10 nN at a constant sliding velocity of 10 μm.s^{-1}. The measured frictional parameters are then calibrated and converted into physical values following the method given in reference [9]. The setup of the friction measurement using lateral force microscopy approach is provided in Figure 2.

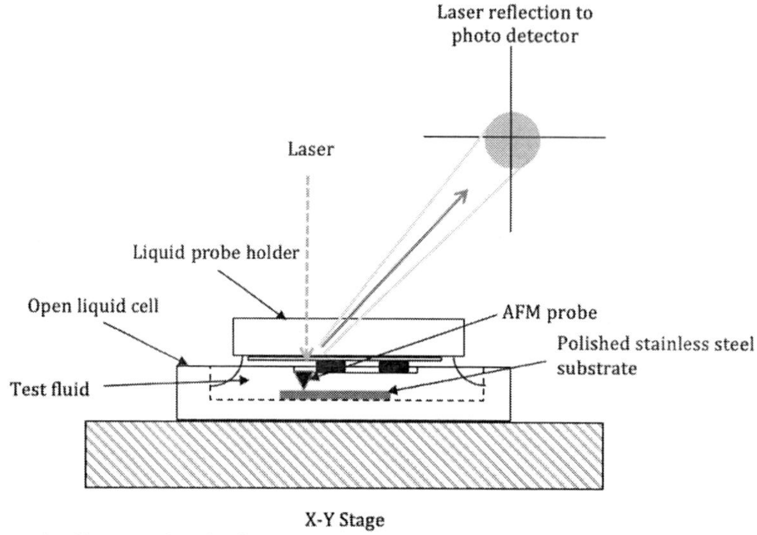

Figure 2. Schematic diagram for the lateral force microscopy setup based on fluid imaging approach

3. Results and Discussions

As mentioned above, the present study intends to determine the extent of competitive molecular interactions between ZDDP and methyl oleate along boundary lubrication regime. Before the friction test is conducted using lateral force microscopy, dynamic viscosity values for both ZDDP and methyl oleate is measured using a rheometer at different temperatures. The change of dynamic viscosities for ZDDP and methyl oleate with temperature is given in Figure 3. It is shown that ZDDP has a significantly higher viscosity when compared with methyl oleate. For Newtonian fluids, shear stress is proportional to the viscosity of the fluid. Therefore, hydrodynamically, ZDDP is expected to generate higher friction than methyl oleate. On top of this, adding methyl oleate into ZDDP is expected to result in a mixture with lower viscosity value, possibly leading to lower frictional losses.

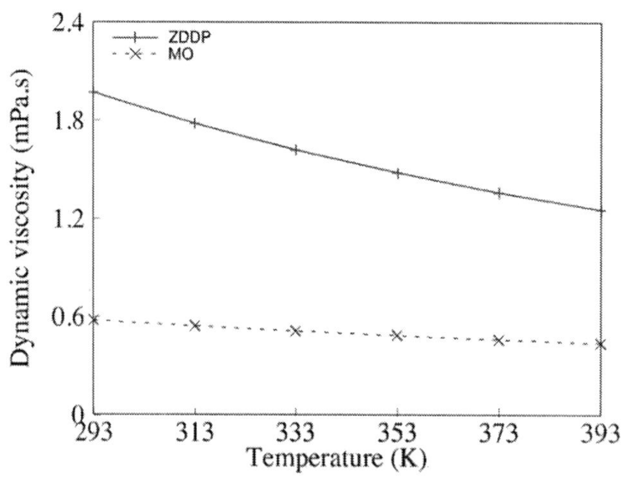

Figure 3. Measured dynamic viscosity for zinc dialkyldithiophosphate (ZDDP) and methyl oleate (MO) at different temperatures

However, along boundary lubrication regime, lubricants are most often known to operate under Non-Newtonian behaviour. The proportional relationship between shear stress and viscosity is no longer valid along this lubrication regime. Therefore, measuring friction force at asperity level is essential in obtaining the tribological behaviour of lubricants along boundary lubrication regime. Figure 4 shows the friction force measured for different ZDDP-methyl oleate mixtures at applied normal loads between 1 nN and 10 nN at 1 $\mu m.s^{-1}$. It is observed that all the measured friction forces increase linearly with larger applied normal loads. Neat ZDDP is shown to generate the highest friction force. When the volumetric percentage of methyl oleate is increased from 30-vol% to 70-vol%, friction force is found to reduce significantly. More interestingly, the friction force even drops lower than that of neat methyl oleate for ZDDP-methyl oleate mixtures with 50-vol% and 70-vol% methyl oleate. The trend observed indicates that the addition of methyl oleate molecules at an appropriate amount not only complements the tribological behaviour of the contact lubricated by ZDDP, it could also result in a much-improved frictional performance as observed in Figure 4.

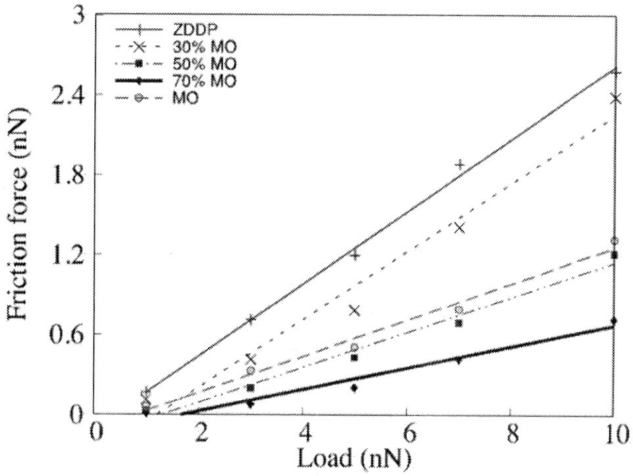

Figure 4. Measured friction force for zinc dialkyldithiophosphate (ZDDP) added with different vol% of methyl oleate (MO)

Figure 5 illustrates the coefficient of friction (CoF) for ZDDP-methyl oleate mixtures obtained from the friction force-load curve measured in Figure 4. The CoF also indicates the change in friction force with applied normal load. At asperity level, neat ZDDP lubricated contact generates the largest CoF value at 0.272. However, the addition of methyl oleate to ZDDP is shown to lead to a drop in CoF. Such reduction in CoF reaches a minimum value of 0.081 with the addition of 70-vol% methyl oleate to the ZDDP-methyl oleate mixture. Increase in the amount of methyl oleate beyond this volumetric percentage is shown to no further reduce the CoF value of the mixture. The minimum CoF value is also depicted to be lower than that of neat methyl oleate. Again, such tribological improvement indicates that both molecules of ZDDP and methyl oleate is capable of complementing each other's boundary lubrication properties so long as the blending ratio is properly optimised. Similar trend is also being reported for ZDDP containing SAE grade engine lubricants when diluted with methyl esters [8].

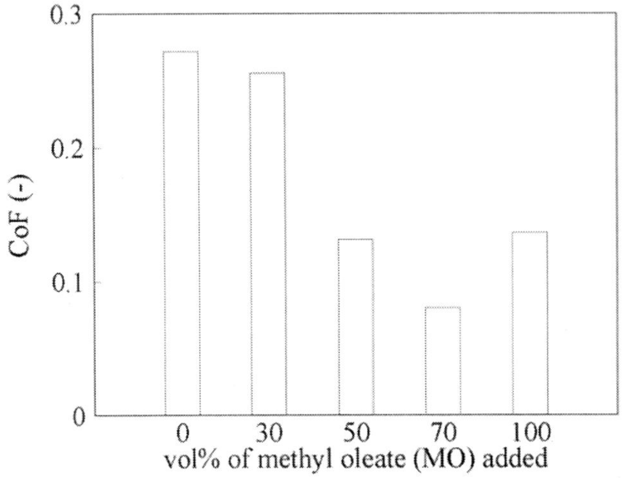

Figure 5. Coefficient of friction (CoF) variation for zinc dialkyldithiophosphate (ZDDP) added with different vol% of methyl oleate (MO)

4. Conclusion

The present study emphasises on the molecular interaction between ZDDP and methyl oleate along boundary lubrication regime at asperity level. Frictional properties are measured using lateral force microscopy coupled with fluid imaging for mixtures of ZDDP and methyl oleate at different volumetric percentages. It is observed that friction forces reduce when the amount of methyl oleate increases in the ZDDP-methyl oleate mixture. More interestingly, the addition of methyl oleate at 50-vol% and 70-vol% resulted in friction forces lower than that of the neat methyl oleate. When comparing the coefficient of friction (CoF) values, the lowest value is achieved for the ZDDP-methyl oleate mixture with 70-vol% methyl oleate, which is 0.081. This value is lower than the CoF of neat methyl oleate at 0.137. The results indicate that a much improved tribological performance along boundary lubrication regime could be obtained as long as an optimum blending ratio between ZDDP and methyl oleate is determined. This finding allows for a platform to further investigate the feasibility of using FAME as lubricant additives.

Acknowledgements

The authors would like to acknowledge the support provided by the Ministry of Education, Malaysia through the Fundamental Research Grant Scheme (FRGS) Phase 2018/1, awarded to Universiti Teknologi Malaysia (UTM) under project vot no. 5F055.

References

[1] Graboski MS, McCormick RL. Combustion of fat and vegetable oil derived fuels in diesel engines. Progress in energy and combustion science. 1998;24(2):125-64.

[2] Murillo S, Miguez JL, Porteiro J, Granada E, Moran JC. Performance and exhaust emissions in the use of biodiesel in outboard diesel engines. Fuel. 2007;86(12-13):1765-71.

[3] Knothe G, Steidley KR. Lubricity of components of biodiesel and petrodiesel. The origin of biodiesel lubricity. Energy & fuels. 2005;19(3):1192-200.

[4] Maru MM, Trommer RM, Cavalcanti KF, Figueiredo ES, Silva RF, Achete CA. The Stribeck curve as a suitable characterization method of the lubricity of biodiesel and diesel blends. Energy. 2014;69:673-81.

[5] Dandan MA, Samion S. Tribological Analysis of Palm Kernel Methyl Ester Containing Polymeric Viscosity Improver. Green Materials. 2019:1-33.

[6] Hamdan SH, Chong WWF, Ng J-H, Ghazali M, Wood RJ. Influence of fatty acid methyl ester composition on tribological properties of vegetable oils and duck fat derived biodiesel. Tribology International. 2017;113:76-82.

[7] Maru MM, Trommer RM, Almeida FA, Silva RF, Achete CA. Assessment of the lubricant behaviour of biodiesel fuels using Stribeck curves. Fuel processing technology.

2013;116:1304.

[8] Hamdan SH, Chong WWF, Ng J-H, Chong CT, Rajoo S. A study of the tribological impact of biodiesel dilution on engine lubricant properties. Process Safety and Environmental Protection. 2017;112:288-97.

[9] Chong WWF, Ng J-H. An atomic-scale approach for biodiesel boundary lubricity characterisation. International Biodeterioration & Biodegradation. 2016;113:34-43.

[10] Chong WWF, Rahnejat H. Nanoscale friction as a function of activation energies. Surface Topography: Metrology and Properties. 2015;3(4):044002.

T6 Solutionizing Heat Treatment Parameter of A356 Alloy by Investment Casting

N. F. Bazilah*[1], M. R. M. Kamal[1], N. A. Maidin[1], Z. Marjom[1], M. A. M. Ali[2], U. H. Ahmad[1]

[1]Faculty of Mechanical and Manufacturing Engineering Technology, Universiti Teknikal Malaysia Melaka
[2]Faculty of Manufacturing Engineering, Universiti Teknikal Malaysia Melaka

*nurfarah@utem.edu.my

Abstract. The heat treatment process is important to conduct on A356 Aluminum Alloy that broadly used in an automotive application. Analysis of the effect grain refinement, eutectic silicon modification of T6 solutionizing heat treatment parameter on strength development of A356 Aluminum Alloy were performed to become an important prospect of controlling the behaviour of materials as well as duration and temperature. In this study, there are three parameters condition were conducted which is duration Time (Hour) and Temperature (˚C) which is 465˚C in 8 hours, 500˚C in 7 hours and 535˚C in 6 hours, respectively. The study generated three responses that are tensile, hardness test and microstructure analysis of the different temperature and duration time, respectively following by water-quenching. The result showed that the modification of duration time and high temperature of solutionizing heat treatment have superior mechanical properties while the microstructure analysis showed the clear formation of refinement α-Al dendrite and eutectic silicon particle.

1. Introduction

Al-Si matrix alloys have been used widely as wear resistant materials in a varies engineering field and lighter than Al which is consists good in corrosion resistance and castability characteristics. The morphological of eutectic Si particles reaction to heat treatment remains bounded with good strength properties due to binary eutectic Al-Si [1]. To achieve the optimum physical and mechanical properties, the heating process called as heat treatment followed by cooling is the process of modifying the structure of crystalline and morphology of materials [2]. The study of the responses and influences of the Al-Si aluminium alloy makes it interesting from technical and academic viewpoints. The factors that control the optimum of the material's behaviour are respect to the modification of parameter and duration of heat treatment.

Nowadays, the material application for the transportation field is critical research where researchers studied that lightweight aluminium alloys are important to improve energy efficiency. The researchers also discuss the effects of thermal conductivity effects on alloying elements and analyze the precipitation in the heat treatment process decreasing thermal conductivity [3]. Besides that, the particular characteristics of aluminum-silicon cast alloy is low with density and low cost, high- capability with temperature and great stability, the alloys are applied in automotive applications widely. Nevertheless, these alloys are suffering from the low tensile force and fatigue resistance, low ductility, low wear resistance and poor machinability. Many researchers approach to adopt additional of the heat treatment to refine the sound structure of Si particles due to the presence of porosity and coarse grain [4].

Content from this work may be used under the terms of the Creative Commons Attribution 3.0 licence. Any further distribution of this work must maintain attribution to the author(s) and the title of the work, journal citation and DOI.
Published under licence by IOP Publishing Ltd

A few studies have been conducted to investigate the effect of T6 heat treatment on mechanical properties such as tensile properties and the trend of fracture behaviour, also the microstructure of the modifies A356 alloy. The finding that the size of eutectic silicon particles was reduces compared to unmodified alloy due to the aspect ratio of eutectic silicon particles, mean diameter, and roundness [5]. [2] review the other research study on the influence on mechanical properties of Al- FA-Sic Hybrid MMCS when applied with heat treatment; they were concluded that it could be improving the ultimate tensile strength by adding Sic and fly ash. It improving much better by a heat treatment process that is the tensile strength increase from 120 MPa to 210 MPa for composition Al6061. It is well study that the eutectic modification the initial elemental composition impressed the morphology structure of microstructure and the size of the eutectic phase. [6] testified that during the T6 heat treatment experiment, the A356 aluminium alloy facing the fragmentation, necking and spheroidization on coarse plate-like eutectic silicon and long particles in that alloy. However, the prospect ratio of silicon particles and the average area fraction were decreased simultaneously. When the T6 heat treatment was applied, it is improving the tensile strength of the A356 aluminium alloy, hence it can eliminate the segregation of solute elements from dendritic solidification. The interfacial shear stress also improved by about 76.7 MPa to 102.7 MPa [6]. The study of improving the abrasive wear by researchers also presents the behaviour of the heat treated Al-Si alloy over the as-cast alloy was attributed to decreasing the fracture of inclination or disposition of the alloys because of the presentation of spheroidal Si particles after heat treatment. Also, the Si particles provide the protection to weaker and soften the α-Al phase after the heat treatment, it is increasing the hardness and be a greater strength [1]. Thus, this research will emphasize and focus on the influence and improvement modifying T6 solutionizing heat treatment parameter on A356 aluminium alloy in terms of morphology and mechanical properties by investment casting. The temperature and duration of heat treatment be an aspect of concern that important control for the behaviour of materials.

2. Experimental procedure

2.1. Samples Preparation

The experimental study has been extended out to look into the influence of modification of T6 solutionizing heat treatment on microstructure and mechanical properties on an Al-Si alloy. Aluminum-Silicon alloy (A356) is taken as matrix alloys with the chemical composition are presented in Table 1.

Table 1. Chemical compositions of the material A356 alloy (wt.%) [6].

	Chemical compositions (wt.%)								
Alloys	Si	Cu	Mg	Mn	Zn	Fe	Ti	Other	Al
A356	7.5	0.2	0.4	0.1	0.1	0.2	0.2	0.15	Rest

A pattern of wax is made exactly the shape of the specimen to be cast and shape of mould are used for investment casting process with cross-sectional area 20mm width x 12.5mm thickness x 200mm length as shown in Figure 1.

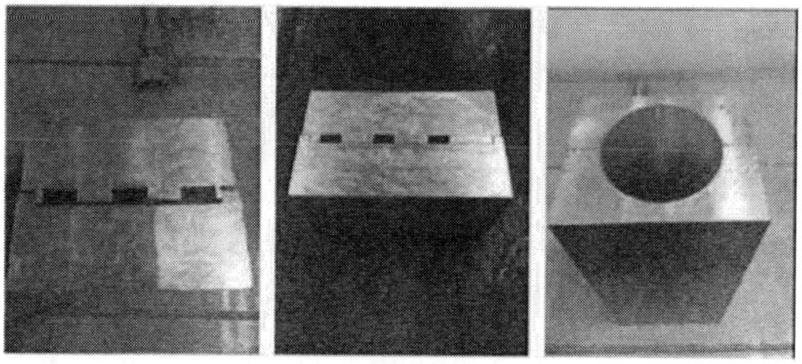

Figure 1. Mould for the preparing of the wax pattern.

The fully wax pattern is dipped in a refractory slurry which is water or grained silica and binders as a composition mixer of them with the viscosity of composition were measured by Zhan's cup and take time around 30 to 40 seconds. The surface of the pattern is dipped by a layer of ceramic. The process continuously repeats until the pattern is dipped for eight layers to make sure the thickness of the ceramic shell is increased and each layer needs to be layered with stucco. Besides, each layer is allowed to dry in room temperature for 6 to 12 hours as shown in Figure 2. This is an important step to be considered to avoid the ceramic shell from any crack condition. The hardened ceramic shell is turned upside down in the furnace and needs to be heated to a temperature around 150°C as the temperature rise about 1 hour and then uniformly for 1 hour and 30 minutes for the dewaxing purpose.

The ceramic mould then needs to be sintered around 800°C for 3 hours for the temperature to rise and 2 hours as it to uniform the temperature for the sintering process. The mould becomes strengthen and any leftover wax or any contaminants inside the ceramic mould be eliminated as shown in Figure 3.

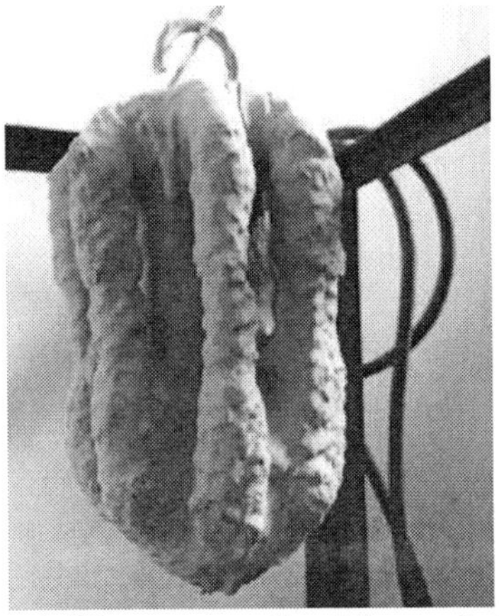

Figure 2. The pattern after 8 layers between dipped in slurry and stucco

Figure 3. The process of dewaxing and sintering to remove leftover wax

A356 Aluminum alloy was heated and melted in a crucible by using an electric furnace for 1 to 2 hours. The shell mould was preheated to 550°C and the pouring temperature was set at 750°C. The shell mould is cooling down and the shell is removed and knocking off from solidifying material by using the suitable hammer as shown in Figure 4.

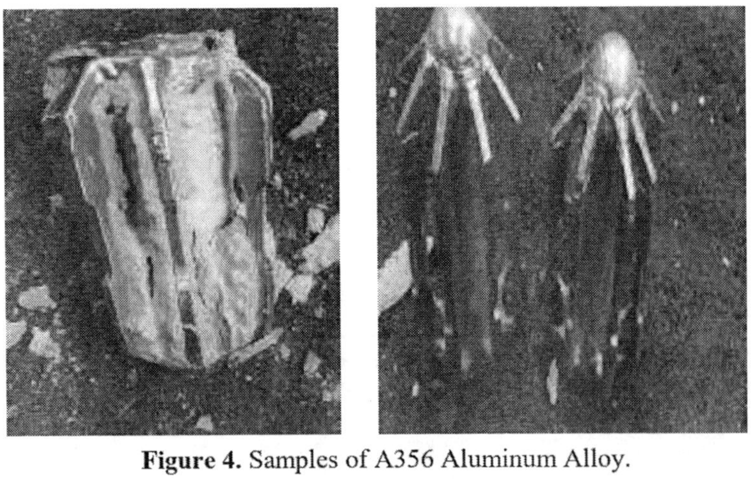

Figure 4. Samples of A356 Aluminum Alloy.

IOP Conf. Series: Materials Science and Engineering **834** (2020) 012005 doi:10.1088/1757-899X/834/1/012005

2.2. Solutionizing heat treatment
The matrix alloy was subjected to T6 heat treatment with varied modifying parameter involving solutionizing following water-quenching. The modifying solutionizing parameter consists of 465 ˚C at 8 hours, 500 ˚C at 7 hours and 535 ˚C at 6 hours and quenched in water at ambient temperature, respectively.

2.3. Microstructural characterization
The samples of A356 alloy as-cast and solutionizing heat-treated conditions of dimension 25 mm thickness and 40 mm diameter cut for microstructural analysis. These samples polished following standard metallographic procedure and they etched with Keller's reagent. The observation of microstructural was done by with optical microscopy (Model: Axioscope 2 MAT).

2.4. Tensile properties
The ultimate tensile strength and tensile strain of samples alloy as-cast and heat treated condition were carried out according to ASTM B557M as shown in Figure 5. Those samples were measured using Universal Testing Machine at room temperature under tensile loading with a load of 1000 N and speed of 2 mm/min.

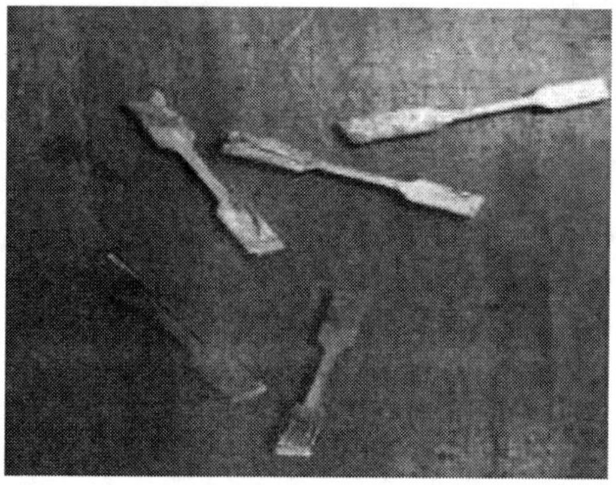

Figure 5. Samples of A356 alloy for the tensile test.Hardness properties

Hardness measurements readings (nine) were carried out according to ASTM E-18 for each as-cast and heat treated samples in Hv scale using Matsuzawa Micro Vickers Hardness Tester at a constant load of 100 N for 10 seconds and the average value was taken. The metallographic polishing procedure of the samples must be done and before the Vickers hardness measurement.

3. Results and discussion

3.1. Microstructural characterization
Pictures of material microstructures before and after T6 solutionizing heat treatment presented in Figure 6. The optical microstructures of A356 alloy indicate phases in the microstructure are α-Al primary, eutectic silicon and Mg_2Si for as-cast and heat-treated samples. Solidification process creates nucleation and this nucleation resulting in the combination of Al and Si microstructure called as a eutectic phase. By contrast, the transformation of the eutectic phase for the unmodified A356 alloy contains a coarse grain of α-Al dendrites primary and long needle-like eutectic Si particles, which causes poor ductility and brittleness due to higher contains of Magnesium that is increase the size of grain during solidification process.

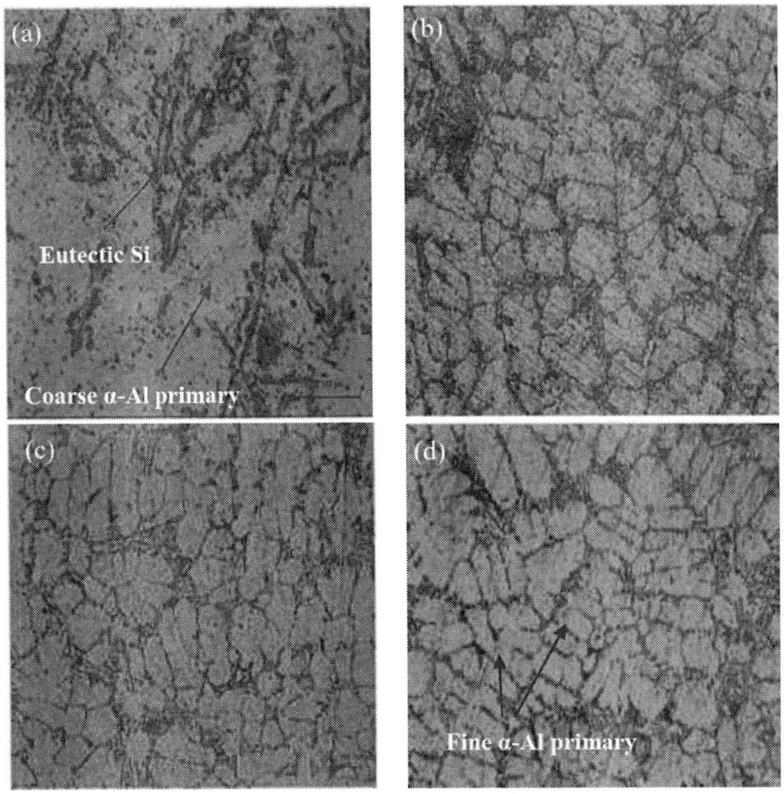

Figure 6. Optical microscopy of (a) as-cast A356 alloy (b) at 465°C in 8 hours (c) at 500°C in 7 hours and (d) at 535°C in 6 hours.

The heat treatment for all temperatures and hours resulted in a breakdown the coarse of Al dendrites and Si particles. Figure 6(d) show the high temperature with short hours create near spherical Si particle in the Al matrix but also uniform distribution of fine Si particles. The high temperature for a short period followed by rapid cooling (quenching) cause Magnesium diffuse more homogeneously across the interface. Modifying parameters of T6 heat-treated samples of A356 alloy indicates an alteration in morphological from long needle-like in the as-cast condition to spheroidal shape in the eutectic Si particles when the duration and temperature of heat treatment are varied.

3.2. Hardness properties

Figure 7 shows the Vickers hardness of the A356 alloy as a function of the solutionizing parameter. The result shows the values of hardness in as-cast condition and modified heat treated condition (465°C in 8 hours, 500°C in 7 hours and at 535°C in 6 hours) were 60.58 Hv, 52.69 Hv, 83.31 Hv and 91.79 Hv, respectively. When the T6 solutionizing heat treatment is applied, the elimination process happened and it results the segregation of solute elements from dendritic solidification [6]. During the high rapid diffuse process at high temperature with a short period and rapid cooling, the distance between particles decrease until achieved at critical point of value, there are increasing the emphasize of stress needed to move dislocations and the material becomes hardened. It can show the highest hardness and tensile are achieved at high temperature with a short period of hour. Modifying the temperature and duration of T6 solutionizing heat treatment increasing the interfacial bonding strength of the eutectic phase after T6 heat-treated.

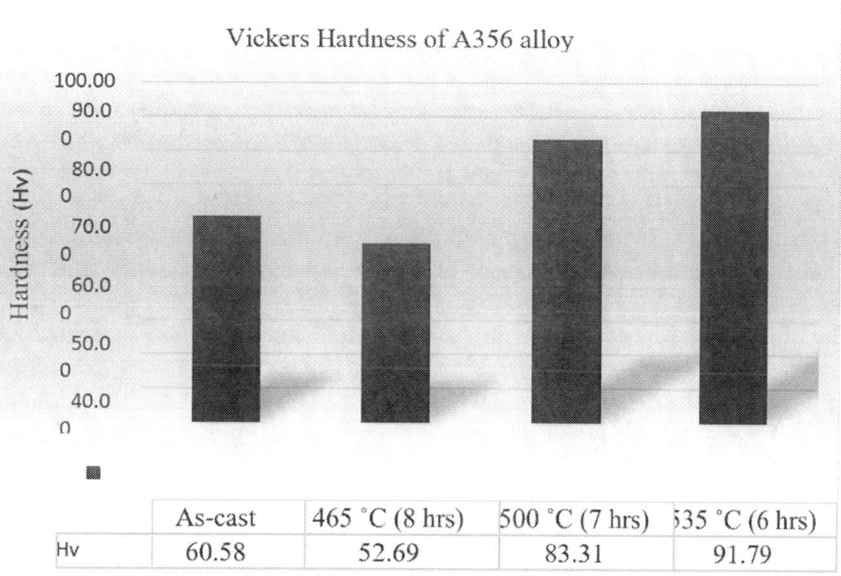

Figure 7. Vickers hardness of the sample.

3.3. Tensile Properties

The ultimate tensile strength (UTS) and tensile strain of samples alloy as-cast and heat treated condition were carried out as shown in Figure 8. After T6 solutionizing heat-treated, the ultimate tensile strength (UTS) of samples alloy in modifying heat-treated condition (465°C in 8 hours, 500°C in 7 hours and at 535°C in 6 hours) were 117.27 MPa, 169.23 MPa and 195.23 MPa, respectively while for as-cast is 171.12 MPa. The highest temperature with the short heat-treated period show the significant improvement of tensile properties due to the refinement of microstructure, homogeneously uniform distribution of second phase particles (Mg) and elimination of porosities. The tensile properties of A356 alloy clear can be improved by modifying the interfacial bonding strength of Al-Si eutectic phase by the parameter of T6 heat treatment since the applying service loads on materials, the tensile strength is emphasize important in duty of resisting deformation.

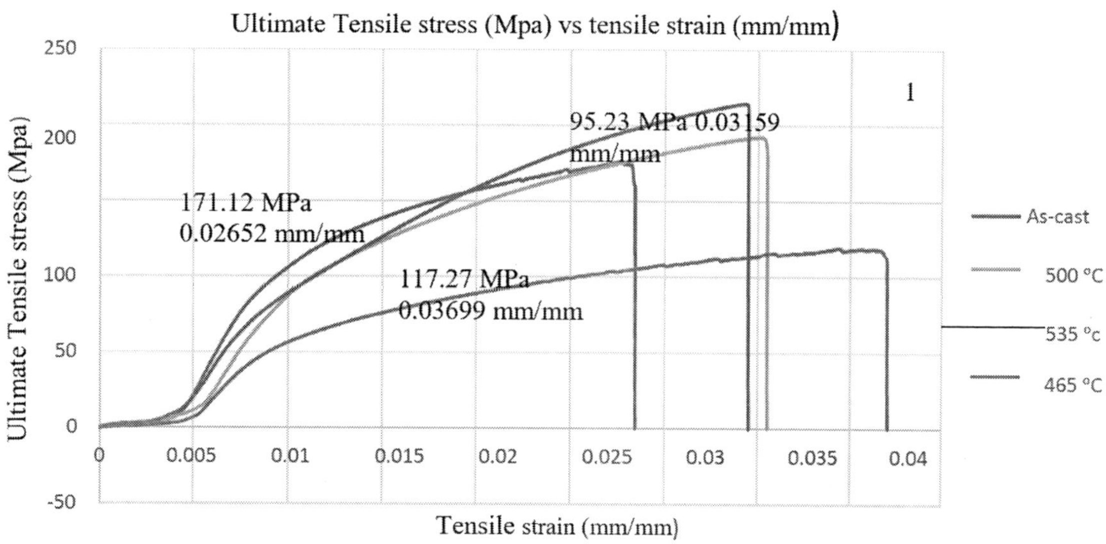

Figure 8. Ultimate tensile stress (UTS) vs Tensile strain (mm/mm).

4. Conclusion

This experimental study shows the influence of microstructure features on strength and hardness development of modifying in T6 solutionizing heat treatment A356 alloy was performed. The following conclusions are drawn:

- The as-cast A356 alloy present with needle-like eutectic silicon and α-Al. However, modifying T6 solutionizing heat treatment for 535°C (6 hrs) change the morphology of the silicon particles become obvious spheroidal and homogeneously distributed in the α-Al matrix grain boundary compared to 65°C (8 hrs) and 500°C (7 hrs).
- The hardness of A356 alloy with modifying T6 solutionizing heat treatment for 535°C (6 hrs) condition is improved and reaches 91.79 Hv compared to the as-cast condition, 65°C (8 hrs) and 500°C (7 hrs).
- The ultimate tensile strength with tensile strain for modifies T6 solutionizing heat treatment for 535°C (6 hrs) reaches 195.23 MPa with 0.03159 mm/mm strain higher compared to the as-cast condition, modifying T6 heat treatment for 65°C (8 hrs) and 500°C (7 hrs).
- T6 heat treatment with modifying solutionizing temperature and time with high temperature and short period hours found to be optimum for A356 alloy in modifying the morphology of eutectic silicon and refining the size of primary α-Al.

Acknowledgement

The authors would like to thank Universiti Teknikal Melaka Malaysia (UTeM) for funded support through grant PJP/2018/FTK(2C)/S01586 by Ministry of Higher Education Malaysia.

References

[1] Gupta A K, Prasad B K, Pajnoo R K and Das S 2012 Effects of T6 heat treatment on mechanical, abrasive and erosive-corrosive wear properties of eutectic Al-Si alloy *Transactions of Nonferrous Metals Society of China* **22** 1041-50

[2] Tiwari S K, Soni S, Rana R S and Singh A 2017 Effect of heat treatment on mechanical properties of aluminum alloy-fly ash metal matrix composite *Materials Today: Proceedings* **4** 3458-65

[3] Chen J K, Hung H Y, Wang C F and Tang N K 2017 Effects of casting and heat treatment processes on the thermal conductivity of an Al-Si-Cu-Fe-Zn alloy *International Journal of Heat and Mass Transfer* **105** 189-195

[4] Guru P R, Khan M D F, Panighari S K and Janaki Ram G D 2015 Enhancing strength, ductility and machinability of a Al-Si cast alloy by friction stir processing *Journal of Manufacturing Processes* **18** 67-74

[5] Nullusamy S 2016 A review on the effects of casting quality, microstructure and mechanical properties of cast Al-Si-0.3Mg Alloy *International Journal of Performability Engineering* **12** 143-154

[6] Liu G, Wang Q, Liu T, Ye B, Jiang H and Ding W 2017 Effect of T6 heat treatment on microstructure and mechanical property of 6101/A356 bimetal fabricated by squeeze casting *Material Science & Engineering A* **696** 208-215

IOP Conf. Series: Materials Science and Engineering **834** (2020) 012006 doi:10.1088/1757-899X/834/1/012006

Environmentally Friendly Lubricating Greases from Coconut Oils and Palm Oil

D. Gasni*, I. Fajar, H. Adinda

Mechanical Engineering Department, Faculty of Engineering, Universitas Andalas Kampus Limau Manis, 25162 Padang, Indonesia

*d.gasni@ft.unand.ac.id

Abstract. The purpose of this study is to compare the lubricating grease from coconut oils and palm oil as base oils. The consistency, dropping point, and tribological properties of lubricating greases were investigated and compared between coconut oils and palm oil. The consistency and dropping point were measured to investigate physical properties. Whereas, the pin on disc apparatus was used to investigate wear and coefficient of friction. The results indicate that the coconut oil greases show superior tribological properties when compare to the palm oil grease.

1. Introduction

In grease technology, lubricant base fluids was derived from the refining of crude oil downstream petroleum raw materials. Due to the environment concerns, the worldwide trend of lubricating grease is eco-friendly. The eco-friendly lubricating grease is renewable, biodegradable, and non toxic [1]. Biodegradability of greases is related to the ability of their base oil to biodegradability. The sources of the base oils could come from vegetable oils or animal fats. In this paper investigated the characteristics of greases with different base oils. The coconut oils and palm oil was used as base oils. The characteristics of grease were investigated consisted of mechanical stability, lubricity and dropping point.

2. Methodology

Coconut oils and palm oil were used as based oil and these oils are a major component to produce grease. There are two kinds of coconut oils: refined coconut oil (RCO) and virgin coconut oil (VCO). Calcium soap was used as a thickener as based oils. Analysis of physical properties of greases consists of penetration by using ASTM D 217-10, NLGI Number by using ASTM D 217-10, and dropping point by using ASTM D 566-09.

Tribological properties were used a pin on disc apparatus. The pins were 7.9 mm diameter and the disks were 170 mm diameter and 5 mm thick. The flat faces of disk were ground to a surface finish of about 0.972 Ra. The surface hardness of the pin was about 610 BHN and that of the disks about 135 BHN. The disk was prepared from AISI 1015 and stainless steel 440C used to prepare the pin. The disk was rotated at speed of 500 rpm and load of 300 N.

Content from this work may be used under the terms of the Creative Commons Attribution 3.0 licence. Any further distribution of this work must maintain attribution to the author(s) and the title of the work, journal citation and DOI.
Published under licence by IOP Publishing Ltd

IOP Conf. Series: Materials Science and Engineering **834** (2020) 012006 doi:10.1088/1757-899X/834/1/012006

3. Results and Discussion

3.1 Physical properties

The penetration value and dropping point of coconut oils and palm oil lubricating greases are shown in Table 1. The penetration tests evaluate consistency of lubricating greases. From the table, it shows that the penetration value of coconut oils is lower than that of palm oil and the dropping point of coconut oils lubricating grease is higher than that of palm oil. This is due to effect of fatty acid chain length in palm and coconut oils. Longer fatty acid chain makes stronger interlocking fibers, resulting in harder grease matrix [2].

Table 1 Physical properties of base oils

Base oil	**Penetration (x 0.1mm)**	**NLGI** *Number*	*Dropping Point* (°C)
Palm oil	292	2	110
RCO	213	4	120
VCO	177	4	120

3.2 Tribological properties

3.2.1 Wear

The wear test was carried out at speed of 500 rpm and load of 300 N load under room temperature by using pin on disc apparatus for a test duration of 50 minutes. Figure 1 shows that wear rate of the disc with different lubricating greases. Wear rate of palm oil is higher than that of coconut oil at the first 10 minutes and after 50 minutes wear rate of palm oil is almost coincide with the RCO oil. Total wear mass of lubricating greases for a test duration of 50 minutes is shown in Figure 2. The wear mass of the VCO is the lowest.

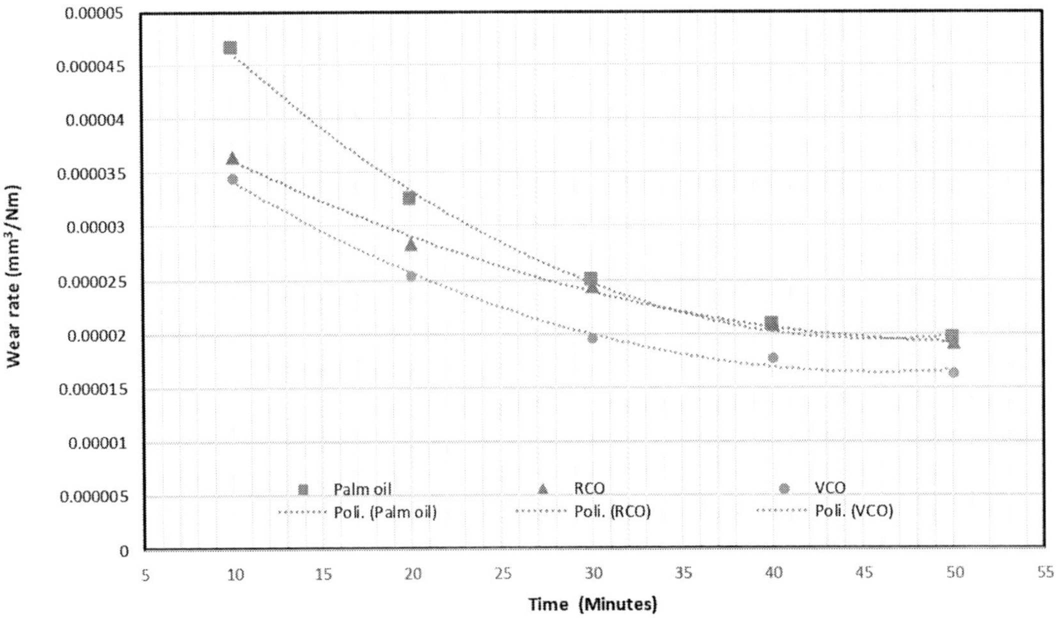

Figure 1. Comparison of wear rate of lubricating grease from coconut oil and palm oil

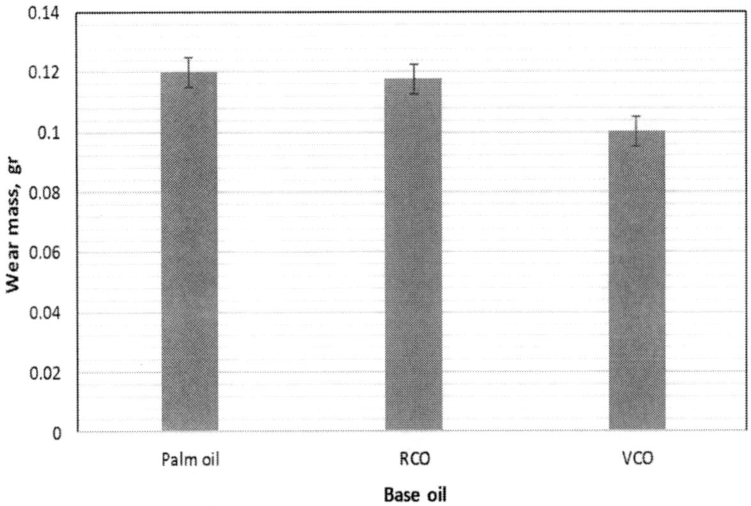

Figure 2. Total wear mass of lubricating greases from coconut oils and palm oil

3.2.2 Scar width and scar diameter

Scar width and scar diameter of disc and pin are shown in Figure 3 and 4 for different lubricating greases. The VCO has the good lubricating greases with 3.889 and 2.282 μm scar width and scar diameter, respectively.

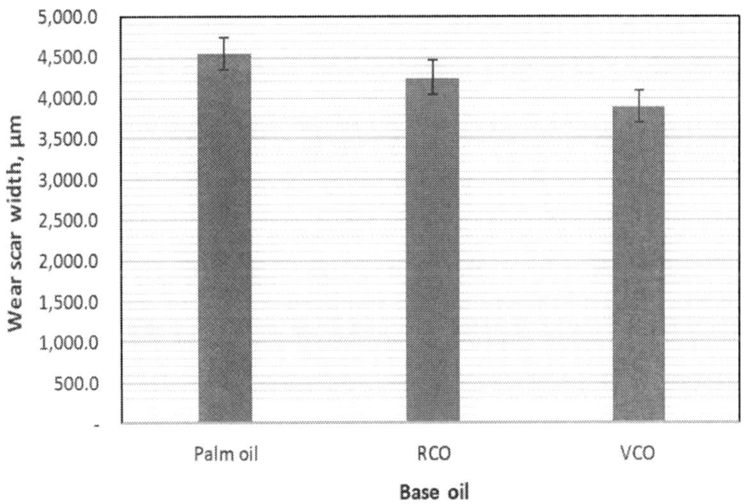

Figure 3. Wear scar width of disc with different greases

3.2.3 Coefficient of friction.

Coefficient of friction (COF) was measured at the steady state condition. From Figure 5, it shows that CoF of VCO is the lowest among the base oils.

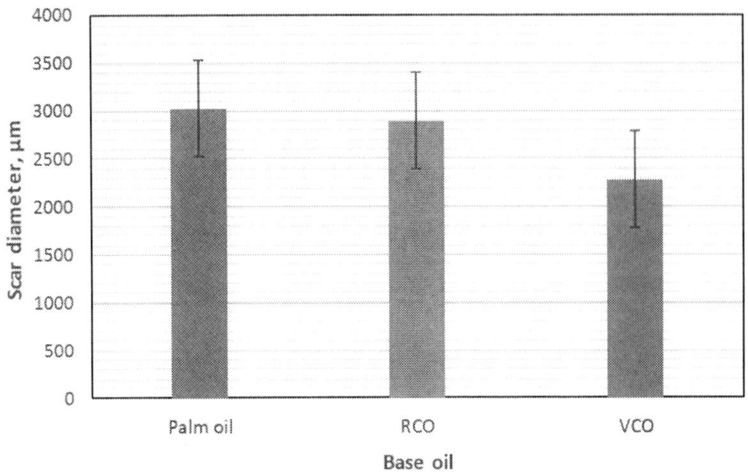

Figure 4. Scar diameter of pin with different lubricating greases

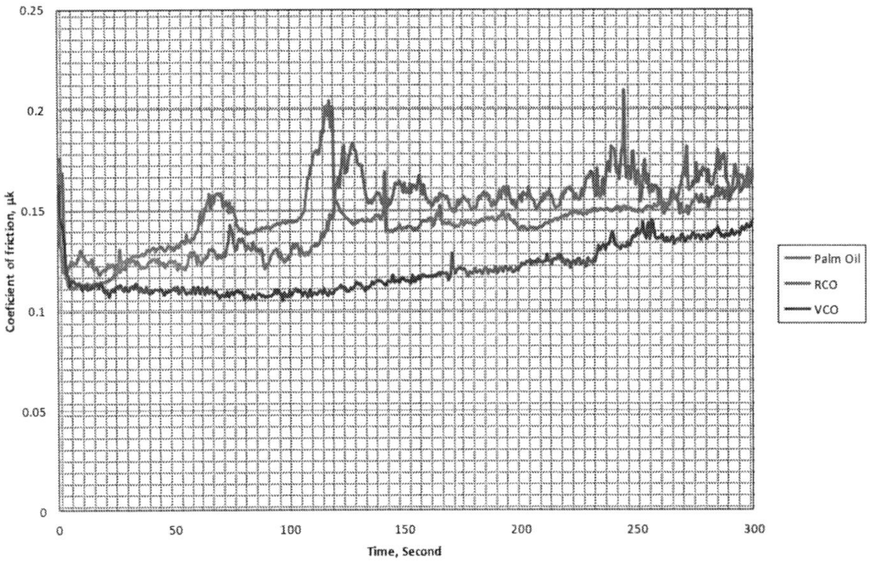

Figure 5. Coefficient of friction of lubricating greases

4. Conclusion

Eco-friendly lubricating greases from coconut oils and palm oil have been developed. The performance parameters of wear and coefficient of friction of coconut oil lubricating grease is better than that of palm oil.

Acknowledgement
This work was funded Research Scheme of Mechanical Engineering Department, Universitas Andalas 2019.

References

[1] Ponnekanti Nagendramma and Prashant Kumar, Eco-friendly multipurpose lubricating grease from residual oils, Lubricants 3, 826-636, 2015.

[2] Atanu Adhvaryu, Sevim Z. Herhan and Joseph M. Perez, Preparation of Soybean Oil-Based Greases: Effect of Composition and Structure on Physical Properties, J. Agric. Food Chem., 52 (21), 6456-6459, 2004.

IOP Conf. Series: Materials Science and Engineering **834** (2020) 012007 doi:10.1088/1757-899X/834/1/012007

Investigation of Oil Palm Empty Fruit Bunch (OPEFB) Embedded with Artocarpus Odorattisimus Mechanical Behaviour as an Alternative Replacement for Raw Material in Wood Industry

M. F. Jundam*, M. K. Mohd Shah, M. H. Abdul Karim and L. Madato

Faculty of Engineering, Universiti Malaysia Sabah, 88700, Kota Kinabalu, Sabah, Malaysia

*fazil.adtec.bintulu@gmail.com

Abstract. This paper presented the investigation of Oil Palm Empty Fruit Bunch (OPEFB) reinforced with Artocarpus Odorattisimus (Mahang/Tarap) board in which fabricated manually using hand lay method and hot pressed in order to determine a suitable alternative fiber board as a replacement for the usage of woods as raw materials in various types of industry. The idea of conducting this research came due to the worldwide community attention on major deforestation which may lead to natural disaster throughout the world. The effects of adding Artocarpus Odorattisimus (Mahang/Tarap) to the oil palm empty fruit bunch (OPEFB) on certain dry weight ratios are being studied and the main objective of this research is to determine certain mechanical properties of the board especially tensile strength to be compared to the pure oil palm empty fruit bunch (OPEFB) board (4.712 Nmm-2)[22]. The experimental process is carried out in accordance to test standard of ASTM D3039/3039m-17. Morphological structure study by using Scanning Electron Microscope (SEM) also conducted on the tested samples to further understand the board properties.Practically, aluminium swarf (chips) collected from an automotive production line is less likely to have any contaminants once the lubricants are removed. In theory, metals do not degrade in value and can be used infinitely.

1. Introduction

Forest has been one of precious source of woods since decades ago in which woods has become main raw materials various construction and much more. As the world became more advance in terms of technology, the demands of woods as raw material has arisen to critical level where it gives a big impact on major deforestation. This matter has come into worldwide community attention in which it gives a lot of negatives impact to the world stability. Based on United Nation Foods and Agriculture Organisation (FAO), the global forest reserve worldwide in 1990 is 4128 million ha while in 2015 worldwide forest reserve has decreased to 3999 million ha in which shows a change in percentage from 31.6% to 30.6% throughout stated time period [1]. As for that, an alternative resources are needed to replace woods as main raw materials needed by the global industries while on the same time continuously supporting the technological advancement momentum. Oil Palm Empty Fruit Bunch (OPEFB) board is one of alternative sources that can be introduced as a replacement for woods and throughout time there are lot of experiments carried out using Oil Palm Empty Fruit Bunch (OPEFB) in order to acquire its properties [2-11]. Malaysia alone produced about 30 million tons annually of palm oil biomass in which included fronds, trunks and also empty fruit bunches. On the other hand, the

Content from this work may be used under the terms of the Creative Commons Attribution 3.0 licence. Any further distribution of this work must maintain attribution to the author(s) and the title of the work, journal citation and DOI.
Published under licence by IOP Publishing Ltd

utilization of the oil palm processing scrap such as empty fruit bunch will gives benefits to the nation [12].

In Malaysia, production of alternatives board as an alternative replacement for woods has been done such as the fabrication of Medium Density Board (MDF) built from Oil Palm Empty Fruit Bunch (OPEFB) incorporated with rubber woods, rice husk, waste paper and much more, but up until now there is no hybridization from Oil Palm Empty Fruit Bunch (OPEFB) incorporated with Artocarpus Odorattisimus been done [13]. [22] in her research titled " Developing and prototyping new high density board" found out the pure Oil Palm Empty Fruit Bunch (OPEFB) board has a tensile strength of (4.712 Nmm-2) while comparing it with Oil Palm Empty Fruit Bunch (OPEFB) that embedded with waste paper. While tarap/mahang or known as Artocarpus Odorattisimus in which belong to Moraceae plant family where it can be found either growing in wild or been planted in various region in Borneo include Sabah and Sarawak [14]. The fruit has become more popular nowadays as it said to be rich with anti-oxidantsand possess other benefits when consumed [15]. A low density, non-abrasiveness and bio-degradability are among the advantages gained from using oil palm fibres. Hybrid composites which very often combined with natural fibers as Oil Palm Empty Fruit Bunch (OPEFB) usually demonstrates a good mechanical properties [16].

2. Materials and Methods

2.1. Materials

Oil Palm empty fruit bunch (OPEFB) was supplied by Makmal Konkrit, Fakulti Kejuruteraan, Universiti Malaysia Sabah. Artocarpus Odorattisimus (Mahang/Tarap) gathered from Kg. Tagasan Tani, Semporna, Sabah. Urea Formaldehyde that been used as a binder in this research supplied by Sepanggar Chemicals Sdn. Bhd (Sabah, Malaysia).

Figure 1. Palm Oil Empty Fruit Bunch (OPEFB)

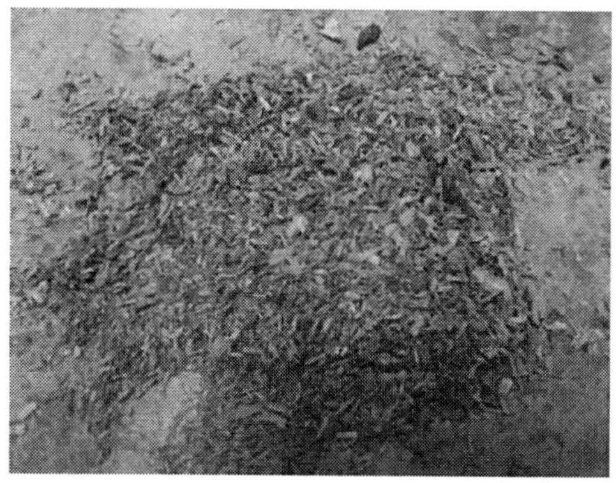

Figure 2. Tarap/Mahang Tree Chip

Table 1. Physical and mechanical properties of oil palm fibre

Properties	OPEFB Fibre
Density (g/cm3)	0.7 - 1.55
Tensile strength (MPa)	0.1 - 0.4
Young's Modulus (GPa)	1 – 9
Elongation at break (%)	8 – 18
Cellulose content (%)	49.6
Lignin content (%)	21.2

Sources: Bismarck A. Mishra S, Lample T. Plant fibers as reinforcement for green composites.

Table 2. Properties of Urea Formaldehyde

Viscosity / 30°C	164 cps
Specific Gravity / 30°C	1.1935
pH value	7.6
Free Formaldehyde	1.20%
Pot Life	75 Min
Solid Content	51.4%

Sources: Sepanggar Chemicals Sdn. Bhd, Kota Kinabalu, Sabah, Malaysia.

2.2. *Methodology*

2.2.1. *Specimen Fabrication*

Empty fruit bunch as a solid waste from oil palm processing procedure was taken from oil palm mill in a single strands shape and then shredded in order to acquire size ranged from 10mm to 15mm. The fiber then, was treated by using a mixture of detergent powder in a boiled oil to remove impurities for about approximately 10 minutes. After that, the fibers being rinsed using potable water to remove the detergent and then kept in air tight container in order to avoid additional increase of humidity to the fibers. A branches of Artocarpus Odorattisimus (Mahang/Tarap) tree in age of approximately 10 to 15 years was taken and the chopped by using a chainsaw. The chip as a results from chain sawing process of the Mahang branches in sizes of 10mm to 15mm are collected and then are kept under air tight container to avoid additional moisture to the chip.

A cast iron mould in size of 150mm x 150mm x 150mm selected to fabricate the specimen using hot press method. The empty fruit bunch and Artocarpus oddorattisimus are mixed together in 4 dry weight ratios that is 100:0, 93.3:6.7, 86.6:13.4, 66.7:33.3 (empty fruit bunch : arto) where in total the combination of each ratios will results 100 gram each. The mixture then added with 50 gram Urea Formaldehyde (UF) and blended together so that the Urea Formaldehyde (UF) will able to reach every part of the mixture. The mixture then placed in the hot cast iron mould in which the mould was heated for 24 hours in an oven until it reach 200 ° C and then pressed using KENCO compressive strength machine under 300KN load to produce a fibre board in size of 150mm x 150mm x 5mm. The board the reshaped into size of 150mm x 25mm x 5mm in accordance to ASTM D3039/3039m-17. For each type of mixture, there will be 3 specimens prepared.

Figure 3. Cast Iron mould

Figure 4. Weighing equipment

IOP Conf. Series: Materials Science and Engineering **834** (2020) 012007 doi:10.1088/1757-899X/834/1/012007

Figure 5. Oven

Figure 6. Fabricated specimen

Figure 7. Press Machine

2.2.2. Specimen Testing

The fabricated specimen of 150mm x 150mm will be reshaped into a size of 150mm x 25mm using universal cutting machine in accordance with ASTM requirements. The reshaped specimen then are tested to gain its mechanical properties using universal testing machine (Tensile). The machine that been used in this experiment is GOTECH/AI-7000M.

Figure 8. Cutting Machine

Figure 9. Reshaped specimen

Figure 10. Reshaped specimen

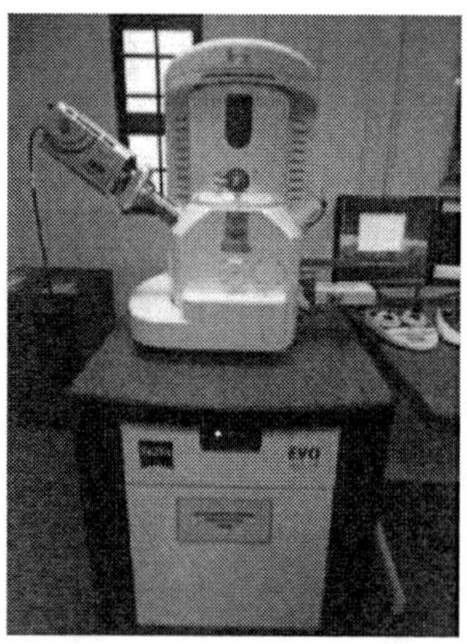

Figure 11. ZEISS EVO MA 10 Scanning Electron Microscope (SEM)

After the specimen tested for its tensile properties, the cracked specimen then are brought to Scanning Electron Microscope (SEM) machine in order to investigate the morphological structure of the specimen especially on the cracked area. In this research, ZEISS EVO MA 10 Scanning Electron Microscope (SEM) are used to conduct the image sampling.

3. Result & Analysis

Table 3. Results of Tensile strength Test

No	Specimen Ratio	Max Load (N)	Tensile Strength (N/mm^2)	Elongation (mm)
1	Specimen A (100g OPEFB: 0g ARTO)	579.35	3.84	2.89
2	Specimen B (93.3g OPEFB: 6.7g ARTO)	731.00	5.98	1.80
3	Specimen C (86.6g OPEFB: 13.7g ARTO)	774.74	5.17	2.21
4	Specimen D (66.7g OPEFB: 33.3g ARTO)	954.52	7.61	2.26

*OPEFB: Oil Palm Empty Fruit Bunch, ARTO: Artocarpus Oddorattisimus

3.1. Tensile Test

Based from the plotted graph, the highest tensile strength of the mixture of oil palm empty fruit bunch (OPEFB) and artocarpus odorattisimus (ARTO) gained at specimen D in which the mixture ratios of 66.7% OPEFB to 33.3% ARTO, this followed by specimen B (93.3% OPEFB to 6.7% ARTO), specimen C (86.6% OPEFB to 13.4% ARTO) and lastly is specimen D in which made from pure OPEFB (no addition of ARTO). The results shows that the addition of artocarpus odorattisimus (ARTO) gives significance improvement in terms of tensile strength to the fabricated board. The pure OPEFB board only possess 3.84 Nmm^{-2} of tensile strength value while addition of ARTO into the board shows that an increased value of tensile strength on all types of ratios which is Specimen B tensile strength value is 5.98 Nmm^{-2}, Specimen C tensile value is 5.17 Nmm^{-2} and Specimen D tensile value is 7.61 Nmm^{-2}. This matter of situation also been discussed by H Ku et al. in which said that addition of fiber to composites matrices will improve the tensile strength due to the stiffness value and higher strength of the fibers [17-19].

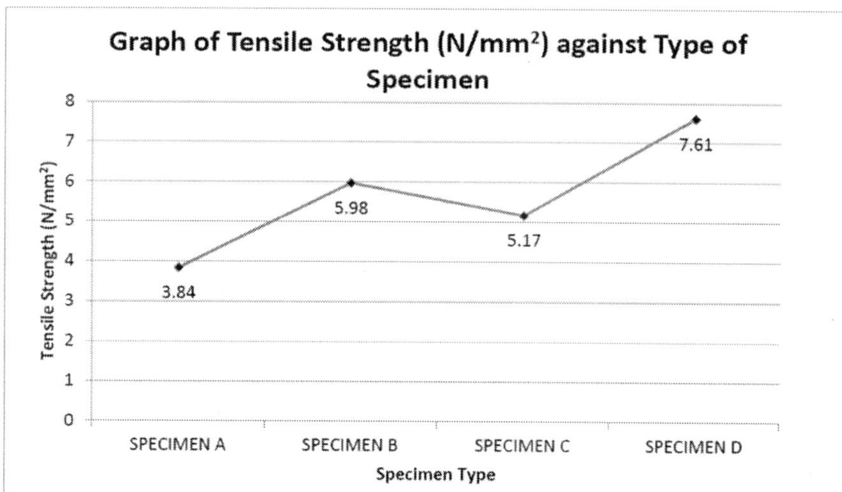

Figure 12. Graph of Tensile strength (Nmm-2) against Type of Specimen

By comparing the tensile strength value between the ratios of OPEFB and ARTO (excluding pure OPEFB), it is determined that the more addition of ARTO into the mixture will gives an increase to the tensile strength value of the fabricated board. The results shows that the highest tensile strength value obtained from the specimen in which contain 33.3% of ARTO (Specimen D) that is 7.61 Nmm-2 while the other specimen (specimen B (6.7% ARTO) and specimen C (13.4% ARTO)) although shows an increase amount of tensile strength compared to pure OPEFB board, still possess lower tensile strength value if compared to specimen D. This matter shows that the addition of artocarpus odorattisimus into oil palm empty fruit bunch board improve inter locking bonding between materials thus improve its tensile strength [17].

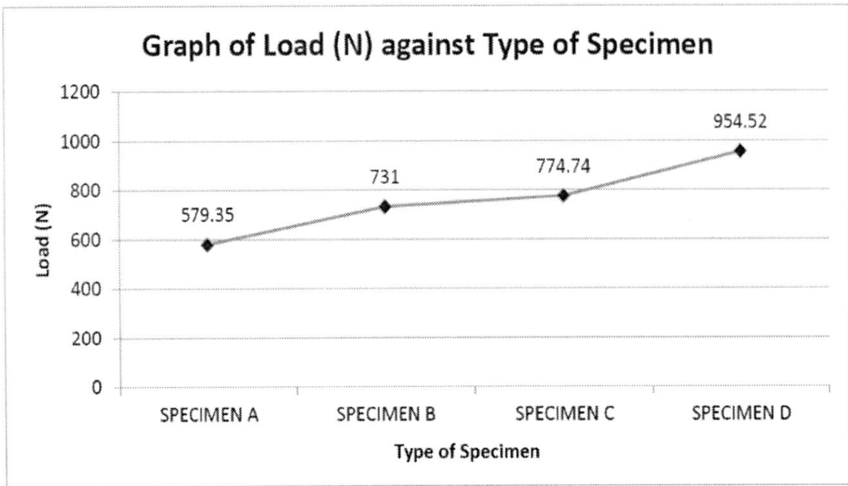

Figure 13. Graph of Load (N) against Type of Specimen

FIgure 13 shows the maximum load needed to be applied until the tested specimen reach crack propagation. Based on the chart, it shows the highest maximum load need to be applied is at specimen D (66.7% OPEFB and 33.3% ARTO) that is 954.52 N and followed by specimen C (86.6% OPEFB and 13.4% ARTO) 774.74N, specimen B (93.3% OPEFB and 6.7% ARTO) 731.00N and lastly specimen A (pure OPEFB) 579.35N. From the gathered data, it can be said that the addition of ARTO into the specimen will increase the inter bonding relation within the materials as the maximum load applied keep on increasing as the portion of ARTO increased in the specimen. Irina M.M.W et al [20] stated that hybrid composites has the ability to increase stiffness, reduced weight an also balancing up modulus properties and thus explaining the increasing of applied forces needed to break the specimen.

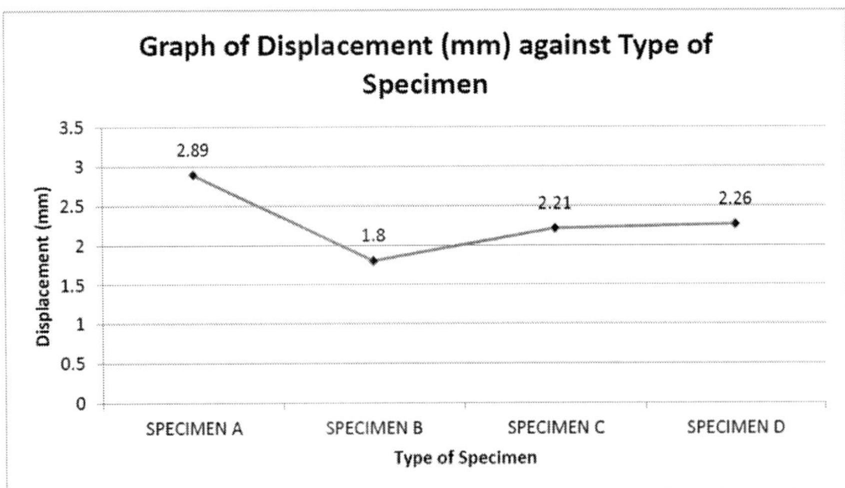

Figure 14. Graph of Displacement (mm) versus type of Specimen

Figure 14 represents the elongation (displacement) for each type of specimen before break when pulled with certain of forces during tensile testing procedure. Based on the data obtained, specimen A in which made from 100% pure OPEFB gives the longest elongation that is 2.89 mm before the specimen started to break, this followed by specimen D (66.7% OPEFB and 33.3% ARTO) have an elongation of 2.26 mm before breaking, specimen C (86.6% OPEFB and 13.4% ARTO) 2.21mm and lastly specimen B (93.3% OPEFB and 6.7% ARTO) which gives the lowest elongation among all the tested specimen that is 1.8mm. According to the data, it is clear that the addition of artocarpus odorattisimus to the empty fruit bunch board will decrease the ductility of the board while on the same time increase the brittleness of the fabricated board. But as the ratio of artocarpus odorattisimus portion increased in the specimen mixture, the ductility of the fabricated board can be seen increased even the value of the ductility still below the ductility value of the pure OPEFB board. G. Jordi et al stated that the addition of any filler to composites will lead to decrease of elongation at yield and break [21].

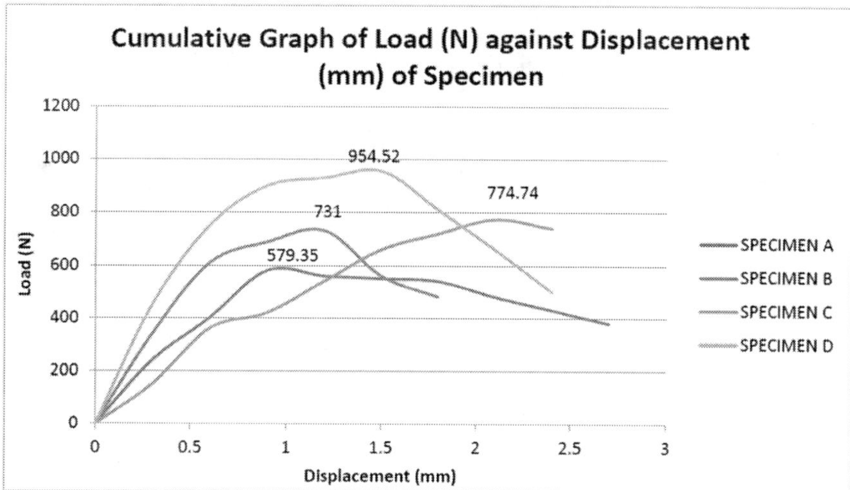

Figure 15. Cumulative graph of Load (N) against Displacement (mm) of Specimen

Figure 15 shows the graph of Load (N) against Displacement (mm) of all 4 specimen that been combined in one graph. Based from Graph 4, it can be seen that Specimen D (66.7% OPEFB: 33.3% ARTO) possess the highest maximum load during the tensile test, followed by Specimen C (86.6% OPEFB and 13.4% ARTO), Specimen B (93.3% OPEFB and 6.7% ARTO) and then Specimen A (100% OPEFB).The graph also show that Specimen A (pure OPEFB) reach its maximum load faster than other specimen. This was followed by Specimen B, Specimen D and Specimen C. But in contrast of that, the displacement of specimen before break indicates that pure OPEFB has the longest displacement, followed by Specimen D, Specimen C and lastly Specimen B. Form the graph, it can be said that addition of ARTO to OPEFB board will increase the maximum load of the specimen and in terms of displacement even though the addition of ARTO to OPEFB will decrease the displacement before break but as the portion of ARTO increase in OPEFB, the displacement also can be seen improving. Scanning Electron Microscope (SEM) used to further understand the mechanical behavior of the tested specimen. All 4 types of specimen with different ratios of Oil Palm Empty Fruit Bunch (OPEFB) and artocarpus oddorattisimus (ARTO) which is (100:0, 93.3:6.7, 86.6:13.4, and 66.7:33.3) are scanned using SEM to determine their morphological structure.

Figure 16 show the results from Scanning Electron Microscope (SEM) of Specimen A (100% OPEFB) under 25X magnification and also 80X magnification. Based on the picture we can see that there are several dimple and voids consists within the materials. Small crack growth appeared as on picture 12(a) on the surface and edge of the materials. Delamination also can be seen as a results from tension during the testing session. Picture 12(c) shows that the existence of voids within the fiber. It can be seen that, the occurrence of voids within the fabricated specimen (pure OPEFB) are much likely to occur even when the matrix and the fiber are trying to hold up against another. The matrix will transfer the stresses to the fiber at the fiber-matrix interface by shear stresses that possibly causes the delamination within the materials and resulting the strong stiffness and tensile strength but has a low fracture toughness within the composites [23].

Figure 16. Scanning Electron Microscope (SEM) of Specimen A - pure (100% OPEFB)

Picture 17 show the morphology structure of Specimen B (93.3% OPEFB and 6.7% ARTO) under 80X magnification. Based on the picture, it can be seen that the occurrence of fiber bridging within the specimen as in picture 13(a). The addition of artocarpus oddorattisimus into OPEFB resulting the bridging scenario within the materials. This can be seen to strengthen the tensile strength of the fabricated materials as bridging fiber tends to slow the crack propagation. [24] Porosity also can be seen on the fabricated specimen as the addition of fibers to the specimen. Addition of artocarpus oddorattisimus into the specimen has reduce the gap between the matrix and fibers. Delamination still can be seen as a result from tension during testing process as in picture 13(c). Several fiber pull out (picture 13(d)) still can be seen on the specimen but on a lesser amount if compared to specimen A.

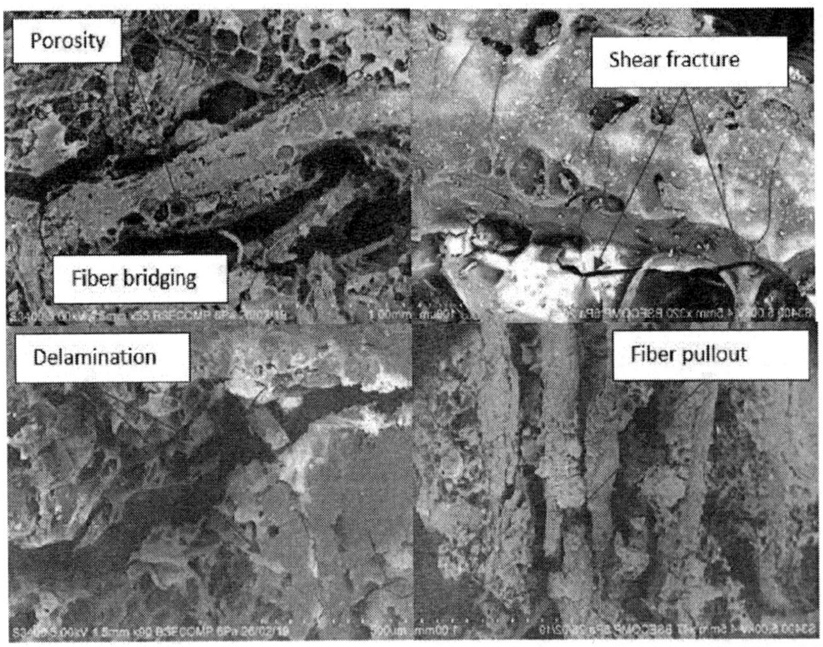

Figure 17. Scanning Electron Microscope (SEM) of Specimen B - (93.3% OPEFB; 6.7% ARTO)

Figure 18 shows the morphological structure of specimen C in which fabricated using 86.6% OPEFB and 13.4% ARTO under 80X magnification. According to the picture 14, much lesser voids can be seen on the microstructure. Picture 14(b) shows that the addition of higher portion of fiber (artocarpus oddorattisimus) has increased the fibers bridging within the materials structure. Formation of fine dimple as in picture 14(d) occur near the crack propagation. The existence of lesser voids within the specimen microstructure shows that the specimen C are much tougher to crack if compared to specimen A and specimen B.

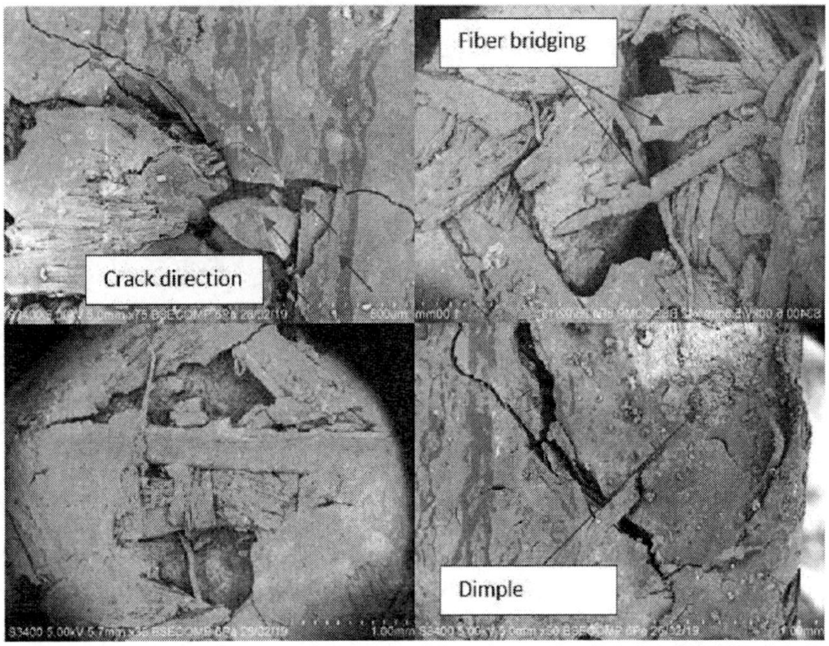

Figure 18. Scanning Electron Microscope (SEM) of Specimen C - (86.6% OPEFB; 13.4% ARTO)

Figure 19 show the morphological structure of specimen D in which fabricated from 66.7% OPEFB and 33.3% ARTO. Based on the picture, it can be seen that the formation of micro buckling and ply splitting as in picture 15(a). This situation occur due to the intensity amount of fiber (*artocarpus oddorattisimus*) that added to the structures. But still much lesser voids can be seen and the amount of fiber bridging within the structures also increased. The fiber bridging that occurred near the crack initiation will slows the crack progress thus the materials has the ability to with stand much higher stress.

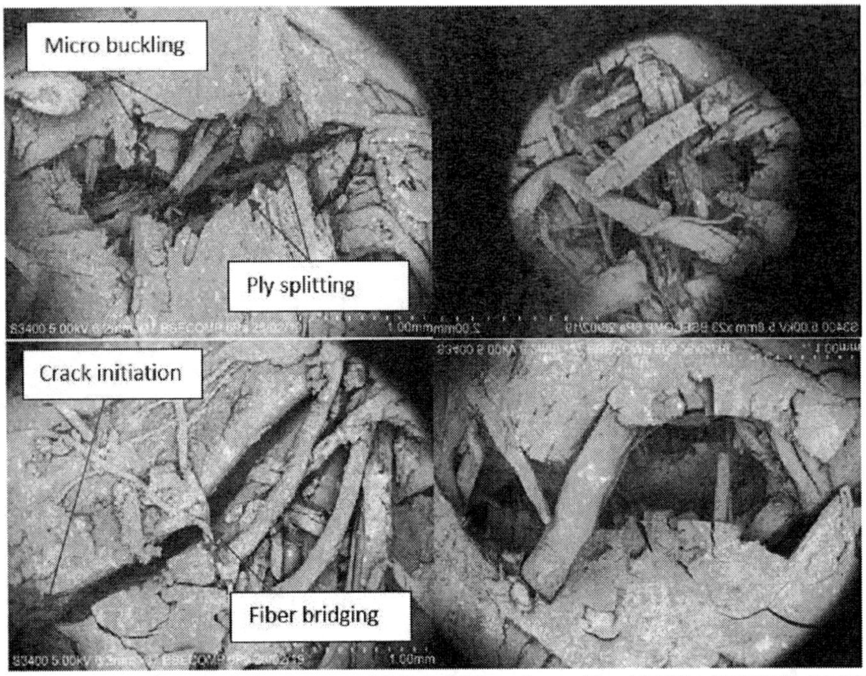

Figure 19. Scanning Electron Microscope (SEM) of Specimen D - (66.7% OPEFB; 33.3% ARTO)

4. Conclusion

Oil pam empty fruit bunch board has become one of suitable candidate in order to replace the usage of conventional woods in nowadays industry. There are various experiment conducted such as hybridization of OPEFB with paper, rice husk, bamboo and much more in order to determine the best board properties to be introduced to the industry especially on its mechanical properties. The addition of artocarpus odorattisimus (Mahang/Tarap) into OPEFB board has shown an improvement in terms of its mechanical properties especially on tensile strength value and this value keep on increasing as the portion of artocarpus odorattisimus added to the mixture. While on the same time the ductility of the fabricated board with addition of ARTO decreased if compared to pure OPEFB board, but as the portion of ARTO increased in the mixture, the ductility of the materials slowly increase but still below the ductility of pure OPEFB.

References

[1] "Food and Agriculture Organisation of United Nations (FAO)", Forestry paper 179, National Socio-economic surveys in forestry, 2018.

[2] H. D. Rozman, K. R. Ahmadhilmi, and A. Abubakar, "Polyurethane (PU) - Oil palm empty fruit bunch (EFB) composites: The effect of EFBG reinforcement in mat form and isocyanate treatment on the mechanical properties," Polym. Test., vol. 23, no. 5, pp. 559–565, 2004.

[3] H. Nasution et al., "Impact Strength and Thermal Degradation of Waste Polypropylene (wPP)/Oil Palm Empty Fruit Bunch (OPEFB) Composites: Effect of Maleic Anhydride -g-polypropylene (MAPP) Addition," Procedia Chem., vol. 16, pp. 432–437, 2015.

[4] R. Roslan et al., "Physico-mechanical properties of resol phenolic adhesives derived from liquefaction of oil palm empty fruit bunch fibres," Ind. Crops Prod., vol. 62, pp. 119–124, 2014.

[5] M. Jawaid, H. P. S. Abdul Khalil, and A. Abu Bakar, "Mechanical performance of oil palm empty fruit bunches/jute fibres reinforced epoxy hybrid composites," Mater. Sci. Eng. A, vol. 527, no. 29–30, pp. 7944–7949, 2010.

[6] H. D. Rozman et al., "Polypropylene ± oil palm empty fruit bunch ± glass ® bre hybrid composites: a preliminary study on the flexural and tensile properties," Eur. Polym. J., vol. 37, no. 6, pp. 1283–1291, 2001.

[7] F. Sulaiman et al., "An outlook of Malaysian energy, oil palm industry and its utilization of wastes as useful resources," Biomass and Bioenergy, vol. 35, no. 9, pp. 3775–3786, 2011.
[8] S. Yusoff, "Renewable energy from palm oil - Innovation on effective utilization of waste," J. Clean. Prod., vol. 14, no. 1, pp. 87–93, 2006.
[9] L. Tan et al., "Fractionation of oil palm empty fruit bunch by bisulfite pretreatment for the production of bioethanol and high value products," Bioresour. Technol., vol. 200, pp. 572 578, 2016.
[10] N. Zulkiple, M. Y. Maskat, and O. Hassan, "Pretreatment of Oil Palm Empty Fruit Fiber (OPEFB) with Aquaeous Ammonia for High Production of Sugar," Procedia Chem., vol. 18, no. Mcls 2015, pp. 155–161, 2016.
[11] M. H. Ahmad et al., "Effects of Oil Palm Empty Fruit Bunch Filler on the Electrical Tree Propagation in Silicone Rubber," APCBEE Procedia, vol. 3, no. May, pp. 147–153, 2012.
[12] Malaysia Palm Oil Berhad (MPOB), "Statistical report: Overview of Malaysian Oil Palm Industry", 2017.
[13] H.P.S. Abdul Khalil et al., "Development and material properties of new hybrid medium density fibreboard from empty fruit bunch and rubberwood", University Sains Malaysia, Pulau Pinang, 2010.
[14] M. Noorfarahzilah et al., "Proximate Composition, Mineral Content and Functional Properties of Sabah Tarap (Artocarpus Odoratissimus) Flour", Universiti Malaysia Sabah, 2016.
[15] B. M. Fadzelly et al.,"Phytochemicals and antioxidant activity of different parts of bambangan (Mangifera pajang) and tarap (Artocarpus odoratissimus)", Food Chemistry, Vol. 113, pp. 479–483, 2009.
[16] P. Wambua, J. Ivens and I. Verpoest, "Natural fibers: Can they replace glass in fiber reinforced plastics?", Compos Sci. Technol., vol. 63, pp. 1259-64, 2003.
[17] H Ku et al.,"A review on the tensile properties of natural fibre reinforced polymer composites" Centre of Excellence in Engineered Fibre Composites and Faculty of Engineering, University of Southern Queensland, pp. 3 – 22, 2011.
[18] R. Malkapuram, V. Kumar and S. N. Yuvraj, "Recent Development in Natural Fibre Reinforced Polypropylene Composites", Journal of Reinforced Plastics and Composites, vol. 28, pp. 1169-1189, 2008.
[19] J. Holbery and D. Houston, "Natural-Fiber-Reinforced Polymer Composites in Automotive Applications", JOM, vol. 58, no. 11, pp. 80-86, 2006.
[20] M.M.W. Irina et al., "Evaluation of Mechanical Properties of Hybrid Fiber Reinforced Polymer Composites and Their Architecture", Universiti Malaysia Perlis (UniMaP), 2015.
[21] G. Jordi et al., "Natural filler-reinforced composites: comparison of reinforcing potential among technical fibers, stem fragments and industrial by-products', Mines ParisTech, PLS – Research University, 2018.
[22] M.N. Nuratiqah, "Developing and Prototyping New High Density Board", School of Engineering, Universiti Malaysia Sabah (UMS), 2014
[23] E. E. Gdoutos, K. Pilakoutas and C. Rodopoulos, "Failure Analysis of Industrial Composite Materials", McGraw Hill, 2000.
[24] T. Reza, " Nano Composites in Dielectrics", Faculty of Chemical & Materials engineering, Shahrood University of Technology, Iran, 2018.

Prediction of Crack Propagation Rate and Stress Intensity Factor of Fatigue and Welded Specimen with a Two-Dimensional Finite Element Method

Y. O. Busari[1,3], A. Ariri[1,2], Y. H. P. Manurung*[1], D. Sebayang[2], M. Leitner[4], W. S. B. W. Zaini [1], M. A. B. M. Kamilzukairi[1], E. Celik[1,5]

[1]Advanced Manufacturing Technology Excellence Center (AMTEx), Faculty of Mechanical Engineering Universiti Teknologi MARA (UiTM), Shah Alam, Malaysia.
[2]Faculty of Mechanical Engineering, Mercubuana University, Jakarta Indonesia.
[3]Materials & Metallurgical Engineering Department, University of Ilorin, Ilorin, Nigeria.
[4]Montanuniversität Leoben, Chair of Mechanical Engineering, 8700 Leoben, Austria
[5]Faculty of Mechanical and Bioprocess Engineering, Hochschule Hannover, 30459, Hannover, Germany.

*yupiter.manurung@uitm.edu.my

Abstract. This paper presents the fundamental investigation on crack propagation rate (CPR) and Stress Intensity Factor (SIF) for a typical fatigue and welded specimens which are Compact Tension (CT) and Single Edge Notch Tension (SENT) as well as Butt and longitudinal T-joint. The material data of austenitic stainless steel SS316L was used to observe crack propagation rate with different initial crack length and different tensile load was used for the fracture mechanics investigation. The geometry of the specimens was modelled by using open source software CASCA while Franc 2D was used for post processing based on Paris Erdogan Law with different crack increment steps. The analysis of crack propagation using fracture mechanics technique requires an accurate calculation of the stress intensity factor SIF and comparison of the critical strength of the material (K_{IC}) was used to determine the critical crack length of the specimens. it can be concluded that open source finite element method software can be used for predicting of fatigue life on simplified geometry.

1. Introduction

The weldment of structural steel under load can fracture not only due to imperfection, but the process can be strongly influenced by the presence of certain components such as voids and other defects of different size which possibly evolve into crack are likely to be present in a welded joint, under increasing load the cracks begin to propagate depending on the tip of the crack.

Corrosion resistance stainless steel is used to improve the strength of a components because of its lightweight in design. Components subjected to variety of loads can be found in many applications. Over the time, a small crack will appear due to the tensile stresses that can lead the propagation of crack up to the point where the crack cannot be controlled, this is classified as failure or sudden fracture [1].

The weld material is more brittle than the parent material making it more susceptible to failure from repeated loadings, because welding process induce voids, high tensile residual stress, local stress concentration at the weld geometry, that encourage the initiation and growth of cracks in the welded

Content from this work may be used under the terms of the Creative Commons Attribution 3.0 licence. Any further distribution of this work must maintain attribution to the author(s) and the title of the work, journal citation and DOI.
Published under licence by IOP Publishing Ltd

structures. However, predicting the fatigue life of welded structures can be a challenge, despite the amount of research that has been carried out in this field even when the weld stress calculation is obtained and the S-N fatigue curve is plotted in its logarithms.

Access to fatigue life information early in the design phase results in shorter time to market, improve the product reliability, and customer confidence. Recently, many experimental approaches were used to investigate the crack propagation in welded structure but most of these experimental procedures are destructive, expensive and require a lot of time to reach a significant solution. However, researchers developed various analytical and numerical approach to reach results quickly [2-6].

Since the manufacturing cost is increasing day by day, the real systems become impractical. It is involving prototyping and testing cost. Trial and error method to get the best parameters is becoming irrelevant, because it demands extreme cost and has reliability issues. Thus, the investigation on fatigue crack growth and critical crack length by using a two-dimensional finite element method (FEM) such as freeware Franc 2D with CASCA. The issue is now what sample specimen best represent the real crack propagation. The existing fracture mechanic sample specimen of CT and SENB was simulated using CASCA and FRANC2D as finite element analysis (FEA) method compared with the SENT sample specimen. However, the study on crack growth and evaluation of the critical position for the specimen to fracture based on critical stress intensity factor and the remaining life of a certain structural component, rigorous numerical analyses were performed to evaluate SIFs [4]. The fatigue crack will begin on the point of the notch and extend through the sample, the fatigue crack is a good representation of real life inconsistencies which are introduced to a material due to processing techniques or welding [5].

Since no crack propagation takes place at the SIF range near to a lower threshold values ΔKth [7]. The linear portion of the Paris law provides a good approximation to the extant experimental data, the law relates the crack growth and SIF range and the K_{max} tends to the critical SIF, K_{IC} where the crack propagation becomes more rapid and the crack becomes more unstable. Therefore, the crack propagation software Franc 2D can give a good representation to establish the best specimen for crack propagation.

The purpose of the crack propagation measurements is to understand the crack behaviour of such materials and set load cycles limit at a particular crack depth a. The measurements of displacement to cycles points are normally evaluated in the form of a crack growth rate (da/dN) versus Stress Intensity Factor (ΔK) curve. It is not always the case that the crack propagation can be described by the range of stress-intensity factor ΔK. If it cannot be so described, other laws can be applied, e.g. crack growth rate as a function of Kmax.

However, the crack growth behaviour depends on a number of parameters (such as temperature and environment, residual stress and load spectrum). The framework within which the test is to be carried out needs to be precisely defined in order to avoid undesired effects on the results. The crack-growth measurements are generally used for investigating the influence of fatigue-crack growth on the predicted life of a component, or for evaluating the crack-growth resistance of a material or heat-treat condition, defining the requirements of NDT testing and macroscopic quantitative determination of various factors (e.g. load, microstructure, manufacture, etc.).

Few different methods to get the stress intensity factor calculation that has been implemented in FRANC2D that can produce accurate results [8]. The methods are the displacement correlation technique [9][10] and the modified crack closure integral [11] and the equivalent domain version of the elastic J-integral [12]. FRANC2D is finite element based two-dimensional crack propagation simulator and is specifically employed for predicting the crack propagation assuming material having isotropic nature are following the manual guide from Cornell University [13]. Application of FRANC2D proved useful and time saving for predicting the edge crack propagation behaviour of stainless steel 316L. The availability and applicability of such innovative simulation tools encourage research and development of applied fracture mechanics to address crack growth problems in a variety of engineering alloys [14]. Single Edge Notch Tension (SENT) specimen has been tested by [15].

A.M. Al-Mukhtar et al in 2009 investigated the calculation of stress intensity factor (SIF) of cruciform and butt joint from some geometrical parameters since he agreed, that stress concentrations occur at the weld toe and at the weld root, which make these regions the points from which fatigue cracks may initiate. To calculate the fatigue life of welded structures and to analyse the progress of these cracks using fracture mechanics technique requires an accurate calculation of the stress intensity factor SIF.

[16] said that the requirements of endurable light-weight structures are challenging tasks for engineers. his research was on comparison of crack gauge and optical measurement for a specimen made of S355 and S960 at constant amplitude loading using SENT sample specimen. [17] modelled the fatigue process in welded joints for example butt welded joint. A fracture mechanics model has been proposed to describe the entire fatigue process. The model is calibrated to fit the crack growth measurements carried out on fillet welded joints. The objective is to establish a unified approach, which is consistent with rules and regulation both based on the S-N approach and applied fracture mechanics. High-cycle fatigue tests were conducted to investigate the effects of temperature, stress ratio (R), specimen orientation, welding and specimen size on the fatigue behaviour of type SS316L stainless steel. The high-cycle fatigue test results indicated that the fatigue limits significantly decreased, when the stress ratio (R) decreased [18].

The stress intensity factors during the crack propagation phase were calculated by using the software FRANC2D, which is shown to be highly accurate, with the direction of crack propagation being predicted by using the maximum normal stress criterion. The SIF results from FRANC2D were compared with those from the International Institute of Welding-IIW, and literature. A good correlation was obtained and the work results have bench marked, which made it possible to use FRANC2D to simulate different weld geometries.

Al-Mukhtar et al used Franc 2D FE Based to show that no difference between the quarter and complete model. The different in mesh size has no or negligible effect on the SIF calculations by Franc 2D [19]. This study aims at applying different 2D FEM fatigue sample model to evaluate the practicability of these concept in order to assess the critical displacement crack growth in each case for butt weld and longitudinal T- joint in a stainless steel 316.

2. FEM modelling & simulation for fatigue crack propagation

Stress intensity factor of a material is used to predict the stress state near the tip of a crack caused by a loads or residual stresses. It describes the stress on crack tip and it is related with the crack growth. Stress intensity factor can also locate the critical displacement of crack distance on specimen. It is the most fundamental and useful parameters in all of fracture mechanics. Hence, lot of information was gathered from the previous researchers to justify the value of critical displacement crack growth using CASCA and Franc2D.

The parameters involved are thickness of the work piece, the loads, the initial crack condition, yield strength, and critical stress intensity factor.

2.1 Geometry Modelling

The geometry was two-dimensionally design and modelled using CASCA then imported into FRANC2D software. The model of specimens for Compact Tension (CT), Single Edge Notch Tension (SENT), Longitudinal T-joint and Butt-Welded Joint were designed with dimension based on the standard of ASTM and ISO [2,3]. The thickness is 10mm. Figure 1 to 4 show the dimension used for Compact Tension (CT), Single Edge Notch Tension (SENT) and Butt Welded Joint. For the Butt-welded Joint specimen, the weld bead is 10mm while the weld height is 2mm. In this simulation, X-groove weld bead was used. Stress Intensity Factor is being considered.

The longitudinal T-joint weld had a web plate dimension of length, breadth and thickness of (147x55x10) mm and the base plate of (20.5x54x10) mm respectively. The reason is the variety of parameter selection can be determined according to several standards. In this stage CASCA was used as pre-processor software to create specimen geometrical layout according to its dimension and a mesh pattern was generated as shown below.

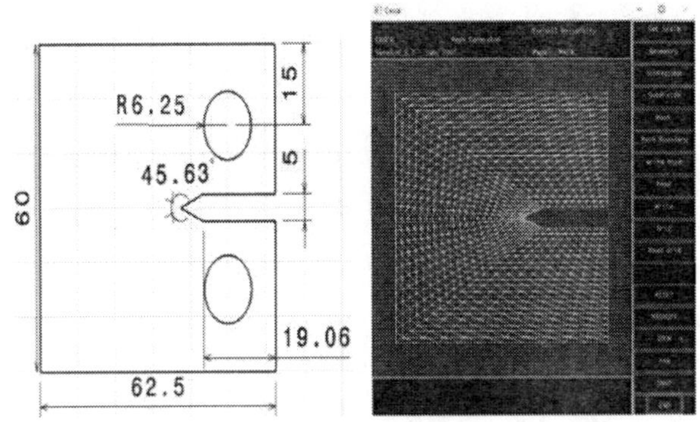

Figure 1. Dimension of Compact Tension (CT) Specimen ASTM E647-00[2] and developed CASCA model.

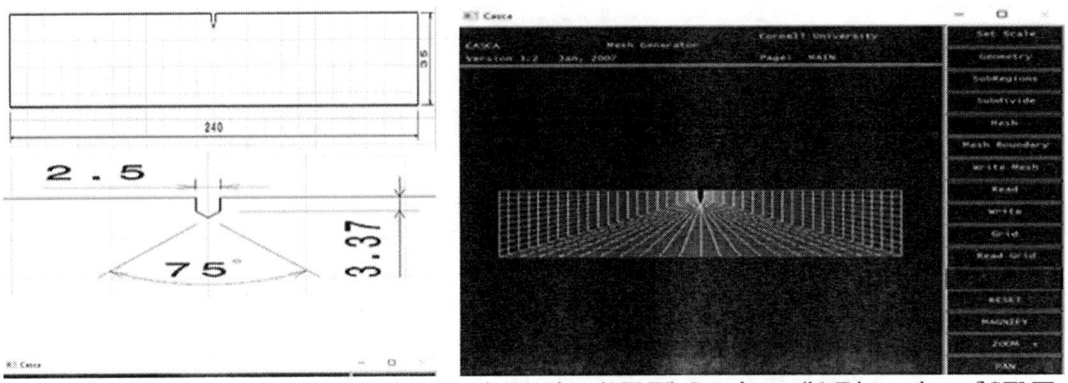

Figure 2. a) Dimension of Single Edge Notch Tension (SENT) Specimen (b) Dimension of SENT Notching.

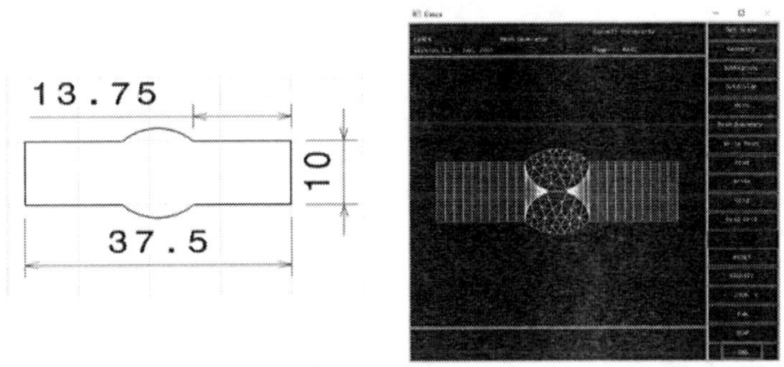

Figure 3. a) Dimension of Butt Welded Joint and CASCA Model

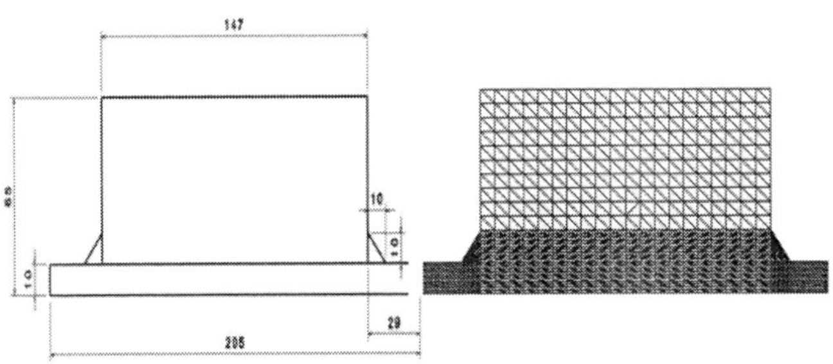

Figure 4. Welded T-joint specimen geometry and Longitudinal T-Joint mesh geometry

2.2 Material Modelling

The materials that represent the specimens are stainless steel SS316L. The properties are listed in the table 1 below:

Table 1. Physical property of SS316L in Franc 2D.

Young's Modulus (GPa)	Poisson Ratio	K_{IC} (MPa. √m)	Density (g/m^3)
205	0.275	112	8.07

2.3 Simulation Procedure for Fatigue Crack Growth

The material properties of stainless steel S316L, specimen fixity point, load, pre-crack and crack increment was embedded in the FRANC2D software to start the propagation of crack.

Pre-crack or initial crack value was inserted in FRANC2D before the propagation of crack. The crack will start propagate at the tip of pre-crack as a starting point. As shown in the figure 6 below:

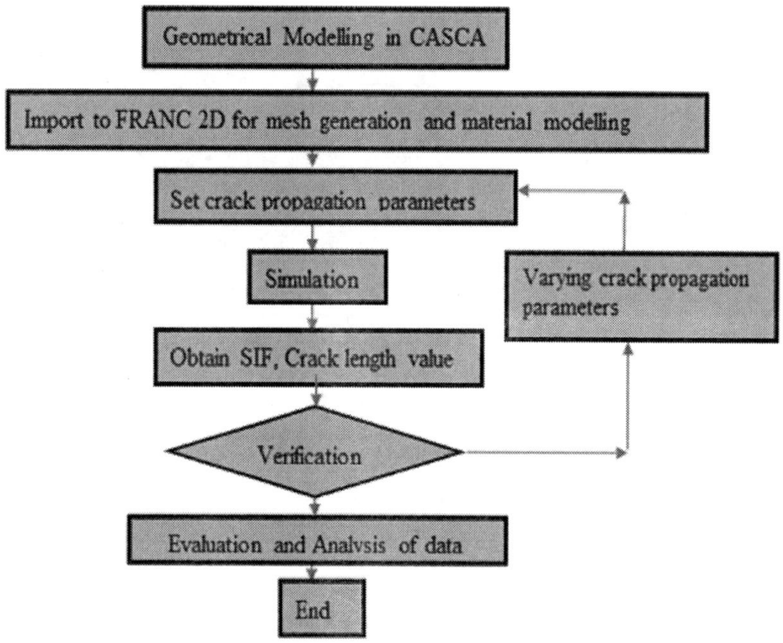

Figure 5. Schematic Flowchart of the Simulation

IOP Conf. Series: Materials Science and Engineering **834** (2020) 012008 doi:10.1088/1757-899X/834/1/012008

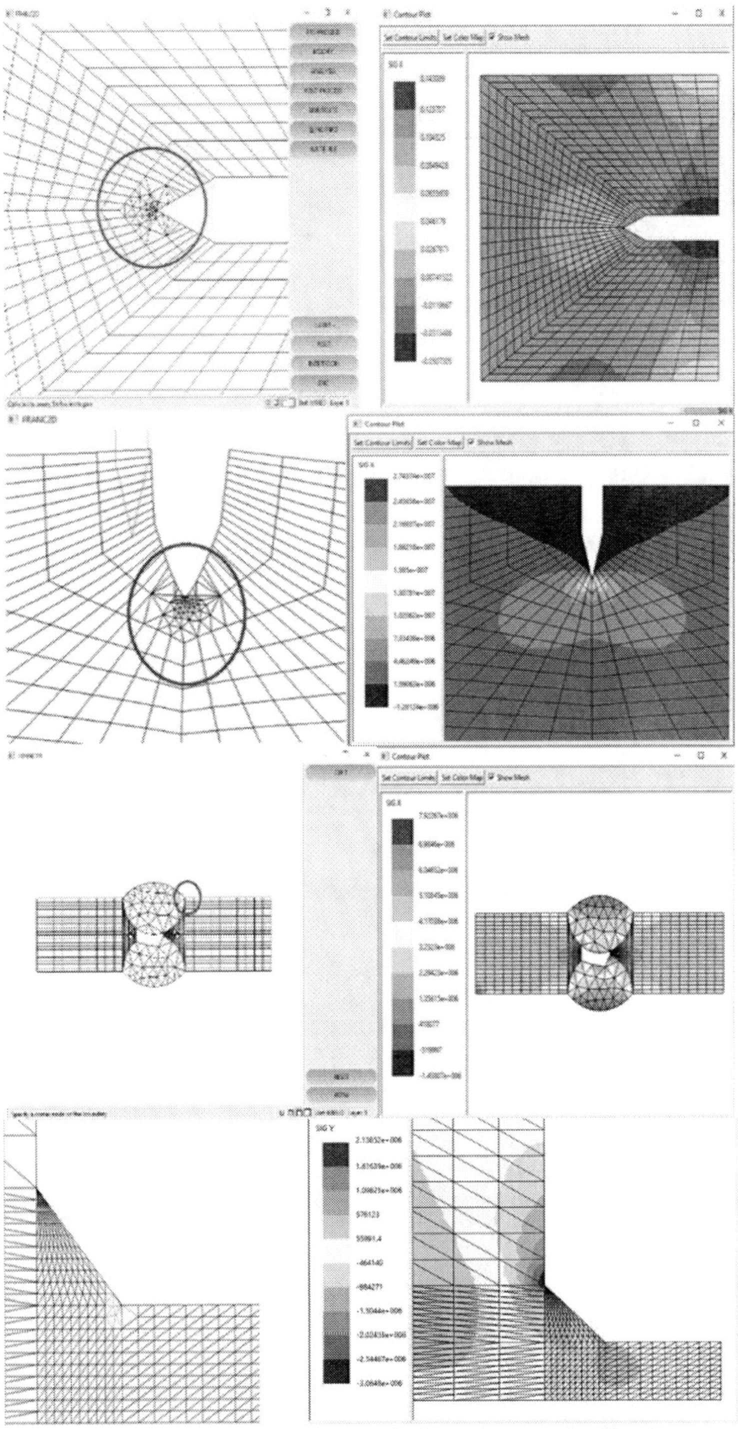

Figure 6. Pre-crack at the notch tip and stress distribution.

2.3.1 Crack Propagation

After the value of pre-crack was defined such as 0.1mm,0.2mm and 0.5mm in FRANC2D and the load with 3MPa, 5MPa and 10MPa started to propagate the crack according to the number of steps. If the number of crack propagation step exceeds the specimen boundary, the crack will stop to propagate automatically [8]. The result in terms of stress intensity factor and crack length was calculated in the software. The same processes sequence was conducted for all sample specimens as shown in figure 7 and 8 below

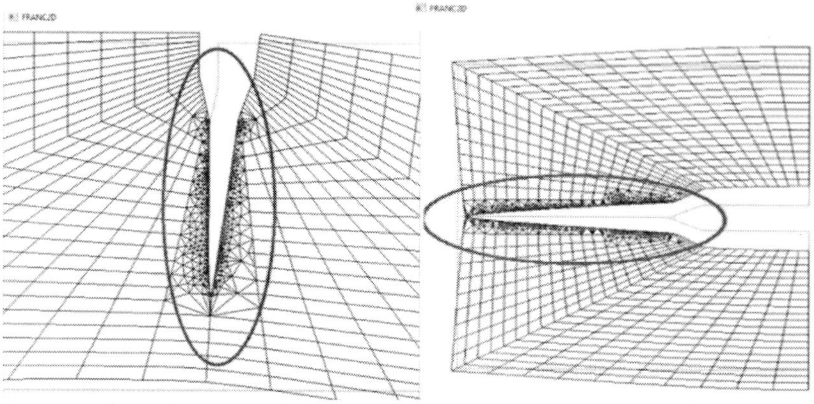

Figure 7. Propagation of crack in SENT and CT specimen to determine the critical crack length.

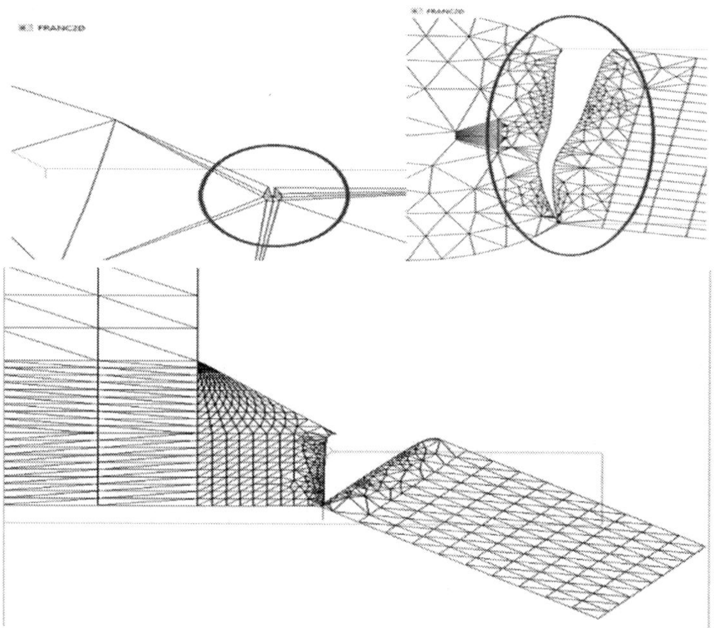

Figure 8. Propagation of crack in Butt weld joint and longitudinal T-joint to determine the critical crack length.

3. Results and Discussion

The simulation in FRANC2D created an initial crack, crack propagation with a certain load. The SIF results is then used to determine the critical displacement of crack growth. The simulations for Compact Tension (CT), Single Edge Notch Tension (SENT), Butt Welded Joint and longitudinal T-joint weld joint specimens was conducted on CASCA and FRANC2D and the effect of loads, initial crack and crack growth on the critical displacement crack length. Tensile loads of 3MPa, 5MPa and 10MPa with an initial crack of 0.1mm, 0.2mm and 0.5mm along with crack propagation of 0.5mm, 0.6mm and 0.7mm with 10mm thickness was used. Blue, red and green indicator indicates the 3MPa, 5MPa and 10MPa respectively.

3.1. Comparison of SENT specimen with CT specimen

The simulation for Single Edge Notch Tension (SENT) specimen was conducted on FRANC2D and the effect of loads, initial crack and crack growth on the critical displacement crack length was observed. Loads of 3MPa, 5MPa and 10MPa with an initial crack of 0.1mm, 0.2mm and 0.5mm along with crack propagation of 0.5mm, 0.6mm and 0.7mm with 10mm thickness compared to the same conditions of the compact tension CT specimen. The result of SENT specimen show more conservative critical crack length compared to that of the CT specimen because with 0.1mm initial crack,0.5mm pre-crack and 3MPa load for the same SS316L steel model in Franc2D, the SENT specimen has 14.7mm critical crack length while CT specimen is 17.9mm. other critical crack length were examined as shown in the figure 9 and 10 below.

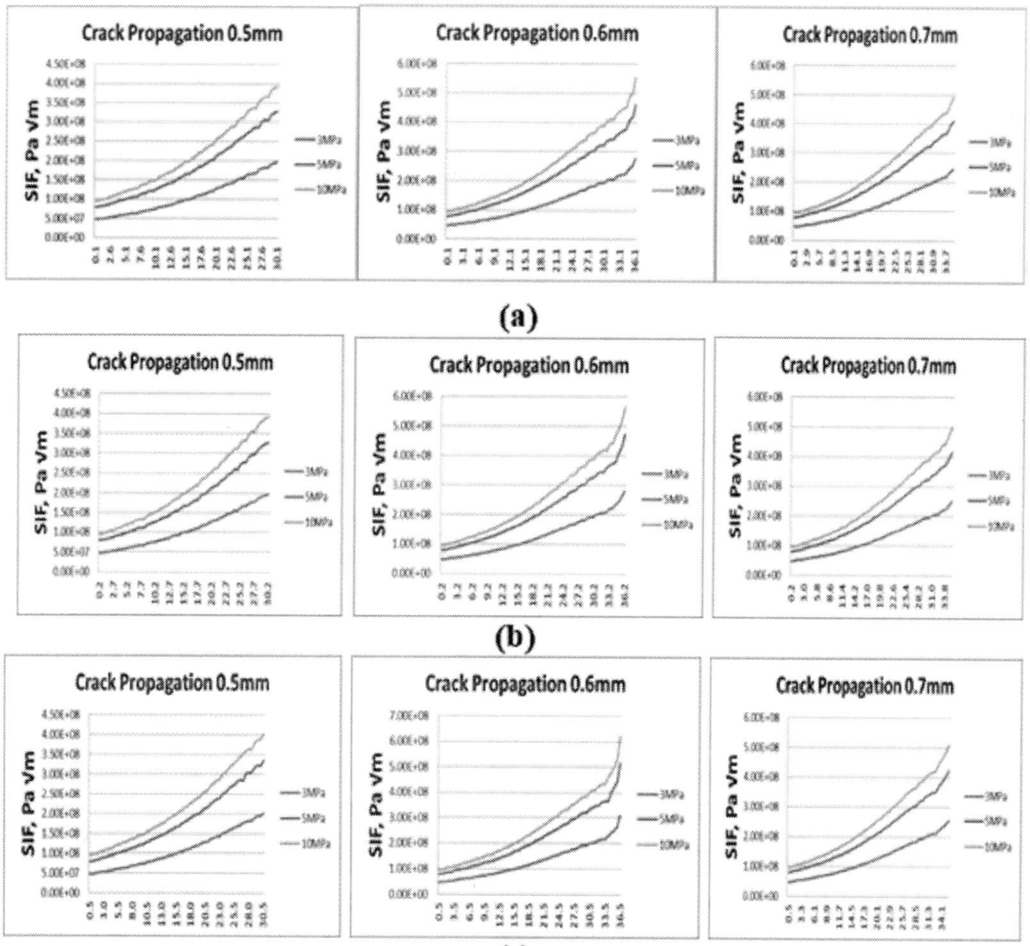

Figure 9. a) 0.1 mm initial crack (b) 0.2 mm initial crack (c) 0.5 mm initial crack for CT Specimen

IOP Conf. Series: Materials Science and Engineering **834** (2020) 012008 doi:10.1088/1757-899X/834/1/012008

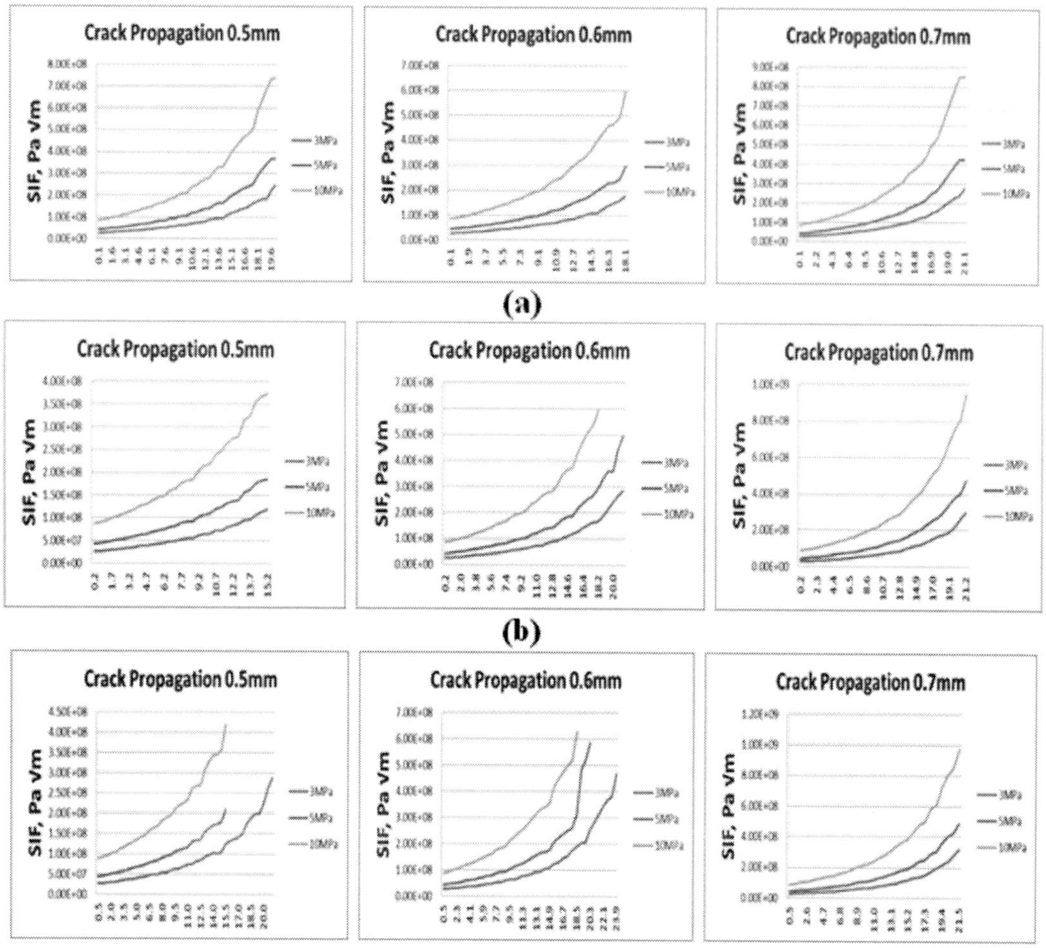

Figure 10. a) 0.1 mm initial crack (b) 0.2 mm initial crack (c) 0.5 mm initial crack for SENT Specimen

However, the critical crack length was obtained from the extrapolated value between the upper and lower limit value of the SIF range (K_I) derived from the simulation, given that the yield strength (K_{IC}) of the material is 112 MPa√m in the case of S316L and its equivalent crack length was obtained and tabularised in table 2 below;

The value of the SENT specimen is more conservative because of the level reliability in fatigue and fracture analysis of component because of the return of investment or life that may be involved.

Table 2. Value of Critical Displacement Crack Growth of Compact Tension and SENT

Pre-Crack (mm)	Crack Propagation (mm)	Load (MPa)	Critical Length(mm) (CT specimen)	Critical Length(mm) (SENT specimen)
0.1	0.5	3	17.9	14.7
		5	7.8	10.4
		10	3.9	3.0
	0.6	3	17.8	15.2
		5	7.7	10.1
		10	4.0	3.0

		3	17.8	14.9
	0.7	5	7.7	10.2
		10	4.0	3.0
		3	18.0	14.6
	0.5	5	8.3	10.3
		10	4.1	3.1
		3	17.8	14.6
0.2	0.6	5	7.7	10.2
		10	4.0	3.1
		3	17.9	14.7
	0.7	5	7.7	10.2
		10	3.9	3.1
		3	17.8	15.2
	0.5	5	7.8	10.4
		10	4.0	3.0
		3	17.7	15.0
0.5	0.6	5	7.7	10.0
		10	4.0	3.1
		3	17.8	14.7
	0.7	5	7.7	10.4
		10	3.9	3.1

3.2. *SENT specimen with Butt weld and longitudinal T-joint weld specimen*

The approach of linear elastic fracture mechanics as a fatigue assessment tool such as Franc2D, gives more in-depth method resulting in the evaluation of both the crack growth direction and the crack propagation. The advantages of this method can be combined with the implementation of the initial crack which represents the weld toe in this case to evaluated the critical crack length of weld such as butt and longitudinal T-Joint welds.

Since cracks occur because of the stress concentration in the weld toe region, the initial crack and plays an important role in the fatigue life of these joints [4, 9]. Therefore, the weld model has an initial crack size at the weld toe and the effects of varying size of pre-crack was used to determine critical crack length with different loads and varying crack increment as shown in figure below 11 for butt weld joint and figure 12 for Longitudinal T-joint weld below

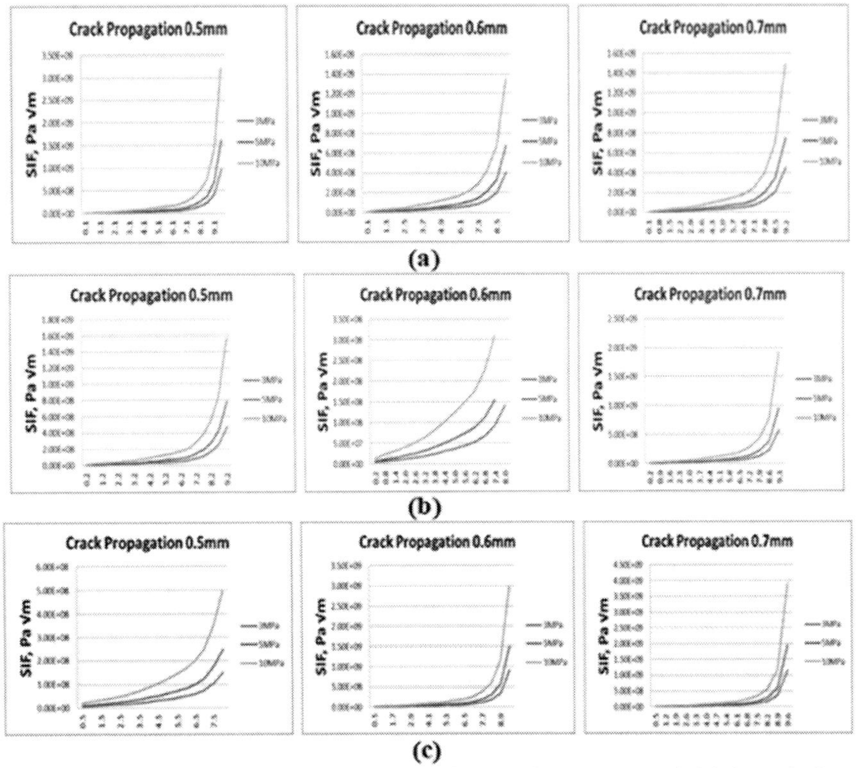

Figure 11. a) 0.1 mm initial crack (b) 0.2 mm initial crack (c) 0.5 mm initial crack for Butt weld Specimen

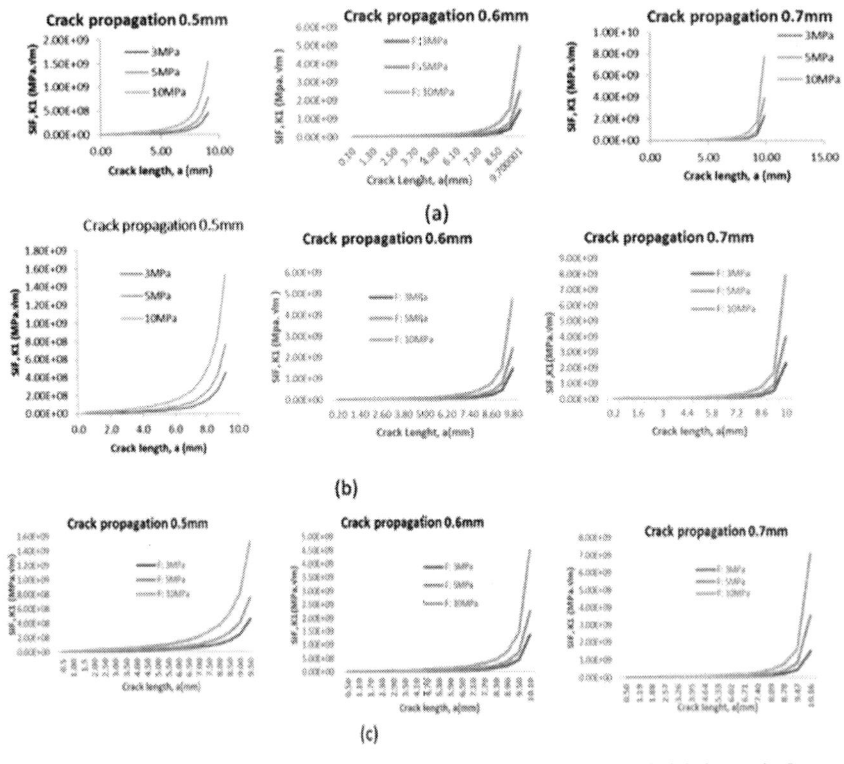

Figure 12. a) 0.1 mm initial crack (b) 0.2 mm initial crack (c) 0.5 mm initial crack for Longitudinal T-joint weld Specimen

Moreover, the weld metal is more brittle than the base metal therefore the critical crack length obtained in the butt and T- joint welds with varying parameters and conditions shows a more reduced length as shown in the table 3 below

Table 3. Value of Critical Displacement Crack Growth of Butt Welded Joint and Longitudinal T-Joint weld specimen

Pre-Crack (mm)	Crack Propagation (mm)	Load (MPa)	Critical Length(mm) (Butt Weld Joint)	Critical Length(mm) (Longitudinal T-Joint Weld)
0.1	0.5	3	7.696	7.56
		5	6.797	6.71
		10	4.679	5.20
	0.6	3	7.671	7.53
		5	6.781	6.71
		10	4.709	5.20
	0.7	3	7.678	7.54
		5	6.838	6.68
		10	4.677	5.21
0.2	0.5	3	7.650	7.66
		5	6.815	6.80
		10	4.681	5.32
	0.6	3	7.640	7.64
		5	6.776	6.82
		10	4.686	5.31
	0.7	3	7.632	7.66
		5	6.745	6.79
		10	4.696	5.31
0.5	0.5	3	7.591	8.00
		5	6.750	7.10
		10	4.711	5.62
	0.6	3	7.612	7.98
		5	6.760	7.13
		10	4.674	5.61
	0.7	3	7.595	8.01
		5	6.703	7.01
		10	4.693	5.56

4. Conclusion

Accurate predictions of critical crack length and fatigue life require accurate SIF estimates because of the nature of the Paris crack growth law. However, the Single Edge Notched tension (SENT) specimen model shows a more cautiously moderate representation of fracture mechanic crack growth with respect to loading thus it is more reliable compare to the Compact Tension (CT) specimen in S316L steel.

Since Fatigue limit increased with decreasing the grain size and the lower crack growth rate also yielded an increase critical crack length for all the specimen samples this further confirms Finite element analysis solution for calculation of SIF for the initial crack steps by A. Al-Mukhtar et al., [9] which agrees with Frank and fisher, BS.7910 and IIW recommendations. However, it can be concluded that open source finite element method software can be used for predicting of fatigue life on simplified geometry.

Acknowledgments
The authors would like to express their gratitude to staff member of Research Interest Group: Advanced Manufacturing Technology (RIG: AMT) and Advanced Manufacturing Technology Excellence Centre (AMTEx) at Faculty of Mechanical Engineering, Universiti Teknologi MARA (UiTM) for encouraging this research. This research and conference participation are also financially supported by ASEA-UNINET grant with the project number ASEA 2019/Montan/1, ERASMUS+ (Montan University in Leoben) and Technogerma Engineering & Consulting Sdn. Bhd.

References

[1] Boyer H E, 1986, Atlas of Fatigue Curves. Ohio: ASM International
[2] Shahani, A. R., & Fasakhodi, M. A. (2009). Finite element analysis of dynamic crack propagation using remeshing technique. Materials & design, 30(4), 1032-1041.
[3] Ooi, E. T., & Yang, Z. J. (2011). Modelling dynamic crack propagation using the scaled boundary finite element method. International journal for numerical methods in engineering, 88(4), 329 349.
[4] Ye, J., He, Y., Chen, X., Zhai, Z., Wang, Y., & He, Z. (2010). Pipe crack identification based on finite element method of second-generation wavelets. Mechanical Systems and Signal Processing, 24(2), 379-393.
[5] Kumar, S., Singh, I. V., Mishra, B. K., & Singh, A. (2016). New enrichments in XFEM to model dynamic crack response of 2-D elastic solids. International Journal of Impact Engineering, 87, 198-211.
[6] Ghaffari, M. A., & Hosseini-Toudeshky, H. (2013). Fatigue crack propagation analysis of repaired pipes with composite patch under cyclic pressure. Journal of Pressure Vessel Technology, 135(3), 031402.
[7] Paris P.C., Erdogan F., 1963, A critical analysis of crack propagation laws. In: Journal Basic Engineering 85, p. 528-534
[8] Forman R.G., Mettu S.R., 1992, Behaviour of surface and corner cracks subjected to tensile and bending loads in Ti-6Al-4V alloy. In: Fracture Mechanics: 22nd Symposium, Vol. 1 (Eds H.A. Ernst, A. Saxena, D.L. McDowell), ASTM STP 1131, American Society for Testing and Materials, Philadelphia, 519-546.
[9] Al-Mukhtar A.M., (2010), The Safety Analysis Concept of Welded Components Under Cyclic Loads Using Components Under Cyclic Loads Using, 219
[10] Krueger R. Virtual crack closure technique: History, approach, and applications. Applied Mechanics Reviews 2004; 57(1-6): p. 109-143.
[11] Hellen T.K., 1975 On the method of virtual crack extension. International Journal for Numerical Methods in Engineering, 9(1): p. 187-207.
[12] Rybicki E. F. and Kanninen M.F., 1977 "A finite element calculation of stress intensity factors by a modified crack closure integral," Eng. Fract. Mech., 9, 931-938
[13] www.cfg.cornell.edu/software/software.htm , Cornell University Fracture Group, 2015
[14] Wawrzynek P.A, Ingraffea A.R. (1993). FRANC2D A Two-Dimensional Crack Propagation Simulator, User's Guide, Version 3.1, Cornell University, Ithaca.
[15] Shen G. and Tyson W. 2009, Crack Size Evaluation Using Unloading Compliance in Single Specimen Single-Edge-Notched Tension Fracture Toughness Testing, Journal of Testing and Evaluation 34.
[16] Simunek D., Leitner M., and Grün F., 2018, "In-situ crack propagation measurement of high strength steels including overload effects," Procedia Eng., 213, pp. 335–345.
[17] Philippe D., Diego S., Naman R. and Tom L., 2004 "A fracture mechanics approach for the crack growth in welded joints with reference to BS 7910," ECF15, Stock

[18] Crawford J.M. et al., 2012, "From the Cover information in biochemical networks" 109(20) 2012.
[19] Al-Mukhtar A.M.,Henkel S., Biermann H., Hübner P., 2009. A finite element calculation of stress intensity factors of cruciform and butt-welded joints for some geometrical parameters, Jordan Journal of Mechanical and Industrial Engineering, 3 (4), pp. 236 - 245

IOP Conf. Series: Materials Science and Engineering **834** (2020) 012009 doi:10.1088/1757-899X/834/1/012009

Grain Growth Prediction of Bead-on-Plate with Filler Wire SS316L using FEM

Muhd Faiz Mat[1], Yupiter H. P. Manurung*[1], Norasiah Muhammad[2], Anton Ditler[3], Mohd Syakir Abd Ghani[1], Martin Leitner[4]

[1]Faculty of Mechanical Engineering, Universiti Teknologi MARA (UiTM), Malaysia.
[2]Mechanical Engineering Department, Politeknik Sultan Salahuddin Abdul Aziz Shah, Malaysia.
[3]Hochschule Osnabrück, Osnabrück, Germany.
[4]Chair of Mechanical Engineering, Montanuniversität Leoben, Austria.

*yupiter.manurung@uitm.edu.my

Abstract. This fundamental investigation presents the study on the grain growth prediction for bead-on-plate of stainless steel SS316L. At first, the numerical model is developed by using basic equation with ordinary differential equation for calculating the grain growth with the presence of growing precipitates. The modified kinetic constant (M_o^*) is defined based on experimental investigation using quench and deformation dilatometer for peak temperature of 1200°C and holding time of 30 seconds. The FEM simulation is conducted based on temperature dependant materials properties using MSC Marc/Mentat. As the outcome various temperature ranges for calculating the grain growth are presented. These results are to be used for further investigation using experimental bead-on-plate.

1. Introduction

The microstructure and mechanical properties of the heat-affected zone (HAZ) of welded metal is significantly affected by the continuous welding heat input. This will affect the grain structure that affects the strength, toughness, ductility and corrosion resistance of the alloys [1]. The high operating temperature leads to change of grain size. Therefore, the attempt to predict the grain size is crucial to evaluate the effect of grain growth to degree of sensitization (DOS) as the inverse co-relationship of grain size and DOS was conclude in [2]. The modelling of the microstructural changes within the welded parts are the upmost important for engineers. It should be noted that certain software, based on thermal and metallurgical modelling have been developed to predict the grain size at HAZ's in multipass welding and their effects on mechanical properties. The objective is to predict the component properties and if applicable used the numerical simulation for other potential application. Thus, it is important that the temperature-dependant material behaviour is known.

Due to its good weldability, corrosion resistance and adequate high temperature mechanical properties, austenitic stainless steel (ASS) is used in different industries, such as petrochemical, shipbuilding, and nuclear industries. These steels are non-magnetic, stable at room temperature, and cannot be hardened by heat treatment [3–5]. In recent years, significant contributions have been made to predict the grain behaviour of surface treated component rather than welded components using simulation software. As a preliminary investigation towards grain behaviour of welded components using simulation software for welding process, material information is very important as grain behaviour is highly dependent on the material chemical composition. Different technique and models have been employed to simulate arc welding process with various physical phenomena was observed and predicted.

Content from this work may be used under the terms of the Creative Commons Attribution 3.0 licence. Any further distribution of this work must maintain attribution to the author(s) and the title of the work, journal citation and DOI.
Published under licence by IOP Publishing Ltd

H. Jamshidi have investigate the grain growth kinetics of SS304 to predict the grain size [6]. Investigation of grain size after Gas Metal Arc Welding on dissimilar material SS304 and SS316 (filler) has proven that alloy segregation and prolong cooling time also generate coarse austenite grain structures next to the sensitized region at the fusion zone, and it became weaker than the surrounding structure [7]. Thus, the prediction of the grain size will focus at the HAZ zone as it is the area adjacent to the weld that was heated high enough to affect its microstructure but not enough to melt it. By undergoing microstructural changes, the HAZ has different mechanical and physical properties than the weld and the adjacent base metal [8].

In this study, an attempt has been made to estimate the grain size in the HAZ of filler wire SS316L using bead-on-plate process. In industries like aeronautical, aerospace, chemical, petrochemical, and marine one of the most important materials are the nickel-based alloys. High corrosion resistance, high strength at both ambient and elevated temperatures, and good ductility and toughness at low temperatures are among the main mechanical properties of the nickel-based alloys [9]. Grain size of SS316L has been experimentally determined to obtain the preliminary modified kinetic constant ($\boldsymbol{M_0}^*$) of the said material. Then, a subroutine for MSC Marc/Mentat grain size calculation have been developed based on the average grain size in the presence of growing precipitates.

2. Numerical Computational and FEM Simulation

2.1 Numerical Computation for Grain Growth prediction.

The grain size behaviour of type 304 stainless steel in low cycle fatigue at elevated temperature was investigated by Yada based on the smooth specimens with various grain sizes [10]. As default the grain size calculation within MSC Marc/Mentat are based on Yada's model is more suitable for forming process. However, the capability to use a different analytical modelling of grain growth is available by developing a subroutine integrated within MSC Marc/Mentat software. As such, there is a need to implement a new mathematical model to simulate this phenomenon. The mathematical model for grain growth has been selected from [11]. This preliminary investigation focuses on grain growth in the presence of growing precipitates. Since the HAZ is heated to temperature approaching the solidus temperature of the alloy, many precipitates that are present in the base metal may dissolve. Carbides and nitrides are the most likely precipitate form in the HAZ of austenitic stainless steel [12].

General equation for grain growth.

$$\frac{\partial \underline{g}}{\partial t} = M_0{}^* exp\left[-\frac{Q_{app}}{RT(t)}\right]\left[\frac{1}{\underline{g}} - \frac{1}{\underline{g_{lim}}}\right]^{\left(\frac{1}{n}-1\right)} \tag{1}$$

Where;

$M_0{}^*$= Modified kinetic constant ($\mu m^2 s^{-1}$)

Q_{app}= Activation energy ($kJ\ mol^{-1}$)

R= Gas constant (8314 J $K^{-1} mol^{-1}$)

T= Absolute temperature (K)

$\underline{g}$= Average/initial grain size (μm)

The $\underline{g_{lim}}$ value depends on the different cases.

The grain growth in the presence of growing precipitates,

Where;

$$\underline{g_{lim}} = \left[\left(\underline{g_{lim}}^0\right) + \left(\frac{k}{f_0}\right)^3 I_2\right]^{\frac{1}{3}}$$

$$\underline{g_{lim}}^0 = k\frac{r_0}{f_0}$$

$$I_2 = C_5 \int_0^t \frac{1}{T(t)} exp\left[-\frac{Q_s}{RT(t)}\right] dt \tag{2}$$

r_0 = Initial radius of precipitate (μm)
f_0 = Initial volume fraction of precipitate
I_2 = Kinetic strength of the thermal cycle particle coarsening (μm^3)
C_5 = Kinetic constant
Q_s = Activation energy for coarsening process ($kJ\ mol^{-1}$)

The implementation of the mathematical model is established by writing a user subroutine to be used by the finite element analysis (FEA) software. The newly developed grain growth model based on the models of Zener, Hellman and Hillbert and Gladman estimate the limiting austenite grain size D̄lim. in the transformed parts of the weld HAZ when the oxygen and sulphur contents of the as-deposited weld metal are 0.04 and 0.01 wt%, respectively [11]. The grain growth model was suitable to be used as the material selected was the same category as type 316 austenitic stainless steel. The data used was compiled from miscellaneous source and experiment shown in Table 1.

Table 1. Data used for type 316 austenitic stainless steel

n	Q_{app} (kJ mol^{-1})	ΔH^* (kJ mol^{-1})	Q_d (kJ mol^{-1})	$\underline{D}^{\circ}_{lim.}$ (μm)	$\underline{D}^{\circ}$ (μm)	M_o^* (μm^2 s^{-1})	Ω^* (μm^2 s^{-1})
0.5^b	224^b	60^b	240^b	18^a	18^a	2000 x 10$^{9\ a}$	2.1 x 10$^{14\ b}$

^{a} Value obtained from experiment for preliminary Modified kinetic constant (M_o)
^{b} Value obtained from miscellaneous sources

To find the parameters of grain size for 316 stainless steels, an experiment has been conducted. The chemical composition is given in Table 2. Subsequent experiment is needed to identify and verify the data used based on the raw material used for the investigation. Firstly, the specimens as shown in Figure 2 had been through heat treatment process using DIL 805A/D quenching and deformations dilatometers. This regulated experiment controls the exposed temperatures for the heat treatment process were 1200 °C (peak temperature) with a holding time of 30 s. Afterwards an epoxy mount has been used for the specimens. After 12 hours, the specimens were grinded by forcipol grinding machine. The specimens were polished for better surface texture. Then a metallographic observation was conducted, the specimens were etched with V2A solution for 10 minutes and consequently the grain size of the 316 stainless steel specimens were defined.

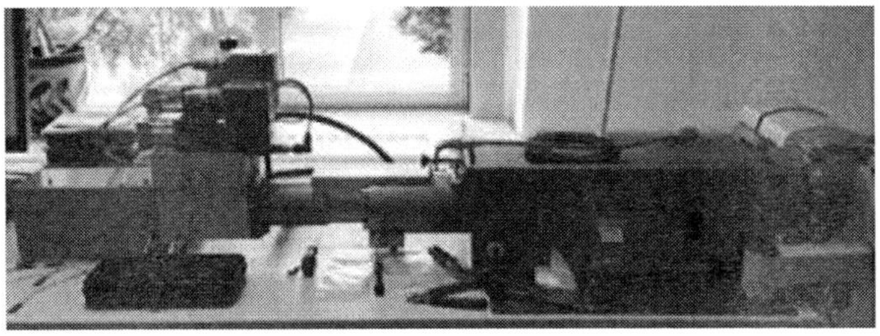

Figure 1. DIL 805A/D Dilatometer

Figure 2. SS316L specimens

2.2 FEM Simulation for Grain Growth prediction on bead-on-plate using MSC Marc/Mentat

The thermo-mechanical simulation of the welding process is applied using MSC MSC Marc/Mentat. The simulation is divided into 3 stages; preprocessing, solving and post-processing. The general flow of the simulation is displayed in Figure 3 below, where the first seven stages are preprocessing followed by solving and post processing. The chosen type of meshing is quad-mesh due to the geometric shape of the work piece to be relatively simplified. Since not much strain and change of shape is expected, using simplified mesh is done to save computing time and ease the modelling effort.

IOP Conf. Series: Materials Science and Engineering **834** (2020) 012009 doi:10.1088/1757-899X/834/1/012009

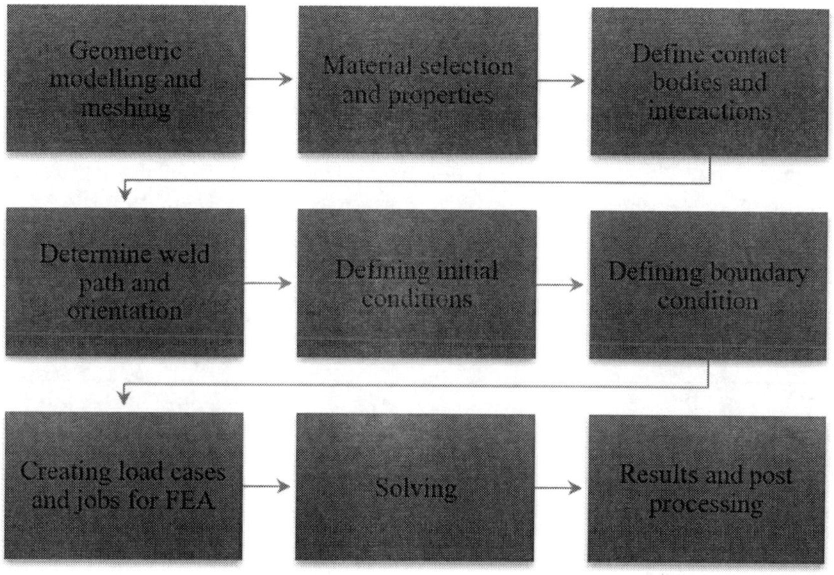

Figure 3. General Flowchart of simulation procedure using MSC Marc/Mentat [10]

2.2.1 Geometrical Modelling

A schematic illustration of Finite Element (FE) models of bead on plate is displayed in Figure 4. To build the model as accurate as possible the entire FE model of bead on plate components consists of a Table (rigid solid), the substrate material (60 mm x 140 mm x 4 mm) and weld beads was developed. A rectangular-shape model of 90 mm of length along with 6 mm in width was modelled as weld filler. Simulating the experiment, the welding trajectory is located at the upper layer of the base plate. The location and forces of the clamping conditions is important to be as identical to the experiment also has been included in the simulation. This is important as clamping conditions such as clamping location, clamping forces and clamping connection has been previously studied will affect the distortion value of WAAM process [14–17]. Finer elements also have been considered along the welding HAZ for research purposes.

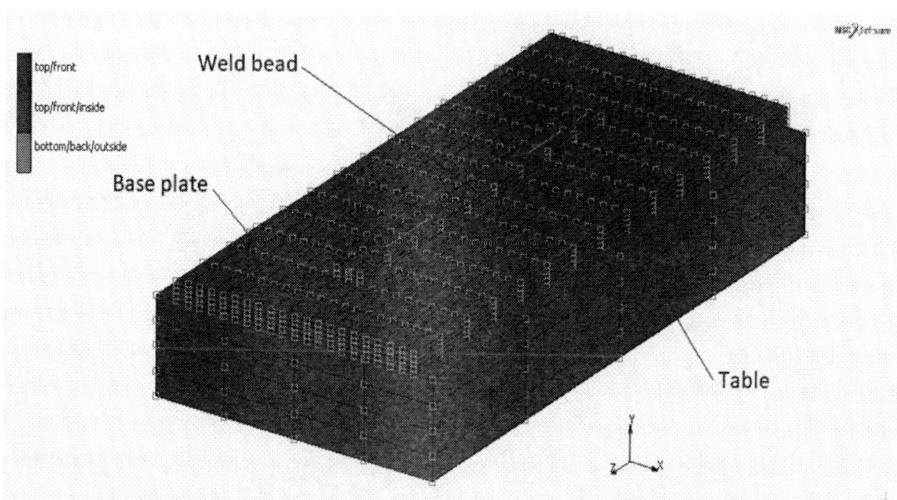

Figure 4. FEM Geometrical modelling

2.2.2 Material characterisation and modelling

SS316L Stainless Steel has been used as the filler material, and the base plate material to predict the grain size. This type of stainless steel has been widely used in many applications, combining good strength with good welding properties. It is nonmagnetic in the annealed condition but can become slightly magnetic because of cold working or welding [18]. Table 2 shows the chemical composition based on international standard and the experimental results obtained using EDX with scanning electron microscope (SEM). In this simulation both thermo-physical and thermo-mechanical material properties should be imported into FEM simulation using existing database from the software library, but as SS316L material wasn't in the default database, it has been added manually by referring to Simufact additive software. The thermo-physical and thermo-mechanical properties are as presented in Figure 5, 6, 7 and 8.

Table 2. SS316L chemical composition

	Catalog	SS316L Filler Wire					
SiK	0.9	0.4	0.7	0.9	0.7	0.8	0.8
	Average			**0.7**			
MoL	2.6	0.7	1.3	1.4	1.3	1.4	1.1
	Average			**1.2**			
CrK	18.4	18.4	17.9	17.6	17.4	17.7	17.4
	Average			**17.7**			
MnK	1.8	2.6	2.5	2.6	2.3	2.1	2.5
	Average			**2.40**			
FeK	Balance	65.5	66.7	67	67.6	66.5	67.5
	Average			**66.8**			
NiK	12.2	12.8	11.6	12	12.2	10.6	12.3
	Average			**11.91**			

The mechanical and physical properties are as presented in Table 3.

Table 3. Mechanical and physical properties of SS316L

Properties	Values
Yield Strength [MPa]	25,000
Ultimate Tensile Strength [Mpa]	70,000
Poisson's ratio [v]	0.275
Melting Range [°C]	1390 - 1440
Modulus of Elasticity [GPa]	200

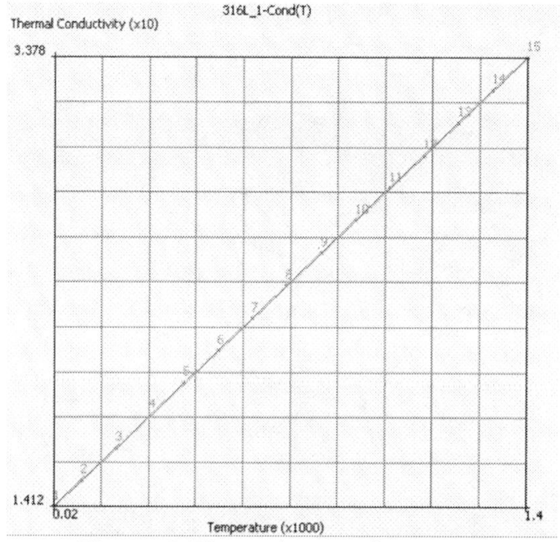

Figure 5. Thermal expansion SS316L

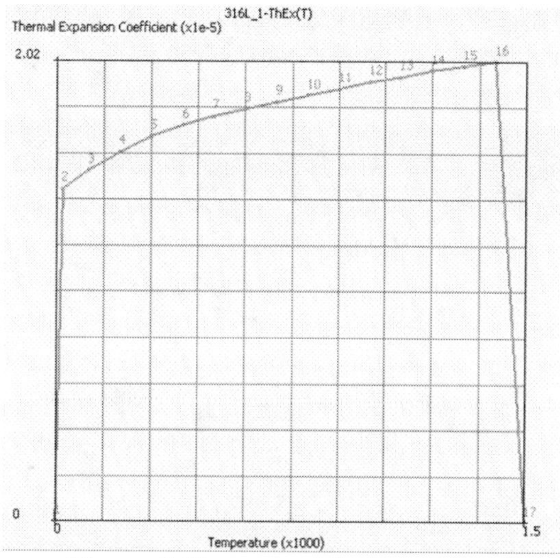

Figure 6. Thermal conductivity SS316L

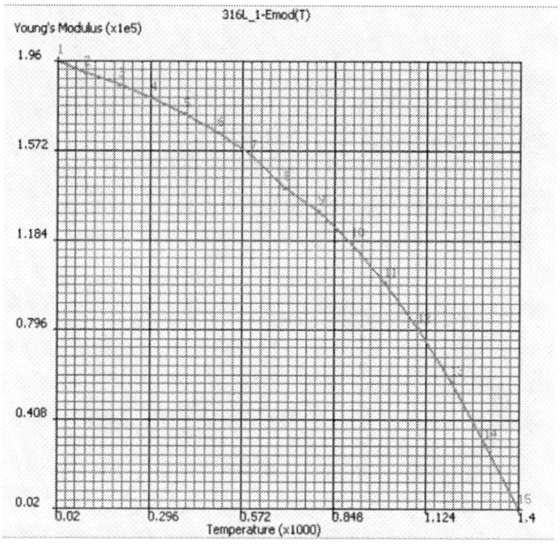

Figure 7. Young's modulus SS316L

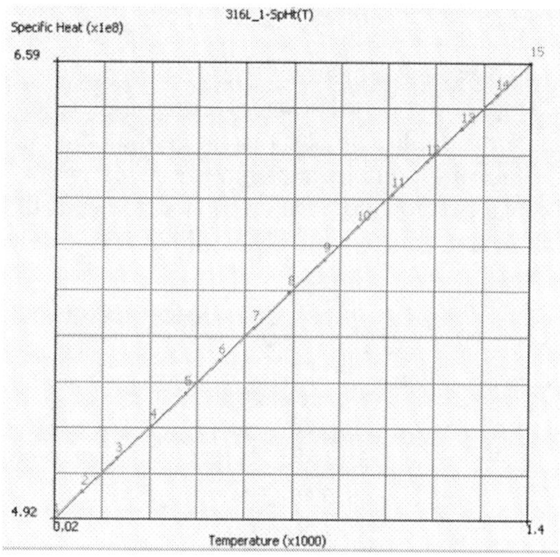

Figure 8. Specific heat SS316L

2.2.3 Welding parameter and heat source model

Table 4 displays the parameters which are implemented on the commercial software MSC Marc/Mentat. Therefore, the Current (I) and the Voltage (V) are considered under the power equation. The assigned travel speed was based on the experimental welding speed to have a more accurate model.

Table 4. Welding parameters

Welding parameters	Value
Current, I (A)	172
Voltage, V (V)	20
Travel speed, v (mm/s)	5.0

In a welding simulation using FEM, an important factor to be considered is the heat source model. The Goldak double ellipsoid model is mostly used as the heat source model in a typical simulation of an arc welding process as in Figure 6. This is due to its accuracy and reliability in representing the shape and distribution of the heat flux proposed by Goldak et al [19]. The heat input model is by no mean

IOP Conf. Series: Materials Science and Engineering **834** (2020) 012009 doi:10.1088/1757-899X/834/1/012009

constrained to the double ellipsoid. A thorough description of this heat source model can be found in [20]. The schematic of the heat source model is shown in Figure 9. The heat source parameters were obtained from previous studies [21].

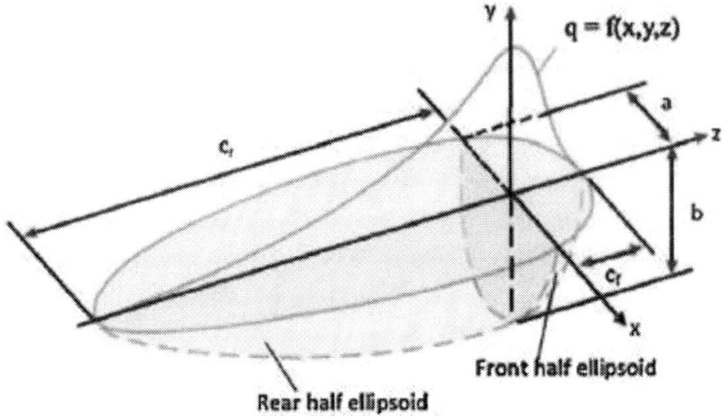

Figure 9. Goldak's double ellipsoid heat source model [16].

3. FEM Simulation on bead-on-plate

3.1 Temperature distribution

Temperature distribution of bead-on-plate process is presented in Figure 10. The temperature by previous studies suggest that the cooling time from peak temperature until 900 °C of the welding process is the most influential temperature for grain size transformation [22,23]. However, the grain size transformation has been calculated successfully by implementing a numerical model for predicting the grain size transformation for various temperature specifically.

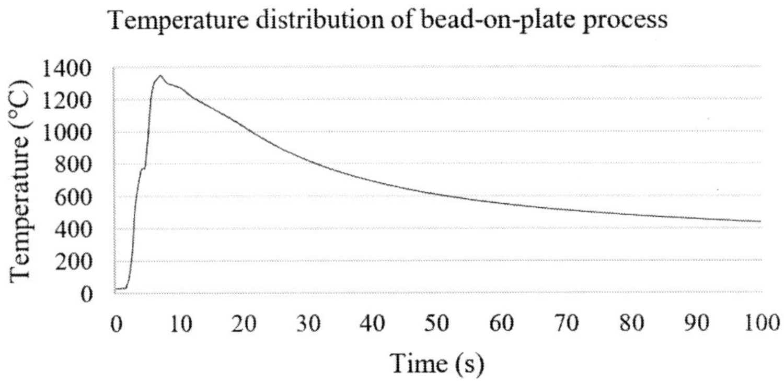

Figure 10. Temperature distribution

3.2 Grain Size

The algorithm of numerical model was developed as a preliminary study for a continuous arc welding application, where the thermal history consists of complex heating and cooling periods. Here, the temperature is transient. The numerical model can also be used for the average grain size calculations in the presence of stable alloying elements/precipitates or growing precipitates or dissolving precipitates by modifying the rate equation. The maximum average grain size depends on the peak temperature of the thermal cycle. This allow expansion and progress for more complex cases of grain size transformation. Modelling and experimental verification will focus on grain growth with growing precipitation with value of M_o^{*}= 2000 x 10^9 m^2 s^{-1}.

4. Results and Discussion

Grain size prediction using the numerical model has been conducted successfully for various temperature range is shown in Table 5.

Table 5. Grain size prediction

Initial temperature (°C)	Final temperature (°C)	Final grain size (μm)
1200	Cooling temperature	50.42
1200	500	50.21
1200	900	50.07
1300	Cooling temperature	50.37
1300	500	50.01
1300	900	49.89

5. Conclusions

The quality of any weld joint is highly influenced by the microstructure of the weldment. The layer by layer process of additive manufacturing will expose the heat affected zone (HAZ) to a complex succession of reheating cycle during the process. The progressive high temperature environment will deteriorate the quality of weldment due to metallurgical changes such as grain growth in the heat affected zone and loss of material by vaporization. Numerical simulation to predict grain size has been conducted successfully. Grain size obtained from numerical model shows good result compared to experimental results. However, by using finer mesh and greater number of increment it can be greatly improving accuracy of the simulation, but it will consume more computational time. Hence, meshing should be considered wisely. Simulation of welding process is very important not only to reduce defect of welded, but also to achieve good and expected mechanical properties. Further investigation of grain size effect towards mechanical properties of actual bead-on-plate process would give beneficial for the future studies.

Acknowledgement

The authors would like to express their gratitude to staff member of Research Interest Group: Advanced Manufacturing Technology (RIG:AMT) and Advanced Manufacturing Technology Excellence Centre (AMTEx) at Faculty of Mechanical Engineering, Universiti Teknologi MARA (UiTM) for encouraging this research. This research and conference participation are also financially supported by ASEA-UNINET grant with the project number ASEA 2019/Montan/1, ERASMUS+ (Montan University in Leoben) and Technogerma Engineering & Consulting Sdn. Bhd.

References

[1] Lewandowski J J and Seifi M 2016 Metal Additive Manufacturing: A Review of Mechanical Properties *Annu. Rev. Mater. Res.*
[2] Taiwade R V., Shukla R, Vashishtha H, Ingle A V. and Dayal R K 2014 Effect of Grain Size on Degree of Sensitization of Chrome-Manganese Stainless Steel *ISIJ Int.*
[3] Chen X, Li J, Cheng X, He B, Wang H and Huang Z 2017 Microstructure and mechanical properties of the austenitic stainless steel 316L fabricated by gas metal arc additive manufacturing *Mater. Sci. Eng. A*
[4] Mohammad Soltani H and Tayebi M 2018 Comparative study of AISI 304L to AISI 316L stainless steels joints by TIG and Nd:YAG laser welding *J. Alloys Compd.* **767** 112–21
[5] Feng Y, Luo Z, Liu Z, Li Y, Luo Y and Huang Y 2015 Keyhole gas tungsten arc welding of AISI 316L stainless steel *Mater. Des.* **85** 24–31
[6] Jamshidi Aval H, Serajzadeh S and Kokabi A H 2009 Prediction of grain growth behavior in haz during gas tungsten arc welding of 304 stainless steel *J. Mater. Eng. Perform.*
[7] Saha S, Mukherjee M and Pal T K 2015 Microstructure, Texture, and Mechanical Property Analysis of Gas Metal Arc Welded AISI 304 Austenitic Stainless Steel *J. Mater. Eng. Perform.*

IOP Conf. Series: Materials Science and Engineering **834** (2020) 012009 doi:10.1088/1757-899X/834/1/012009

[8] Kou S and Le Y 1988 Welding parameters and the grain structure of weld metal — A thermodynamic consideration *Metall. Trans. A* **19** 1075–82

[9] DuPont J N, Lippold J C and Kiser S D 2009 *Welding Metallurgy and Weldability of Nickel-Base Alloys*

[10] Okazaki M, Yada T and Endoh T 1989 Surface small crack growth behavior in low-cycle fatigue at elevated temperature and application limit of macroscopic crack growth law *Nucl. Eng. Des.*

[11] Grong Ø 1997 *Metallurgical Modelling of Welding* (The Institute of Materials)

[12] Lippold J C and Kotecki D J 2005 *Welding Metallurgy and Weldability of Stainless Steel* (Hoboken, NJ: John Wiley)

[13] Yahya O, Manurung Y H P, Sulaiman M S and Keval P 2018 Virtual Manufacturing for Prediction of Martensite Formation and Hardness Value induced by Laser Welding Process using Subroutine Algorithm in MSC Marc / Mentat **15** 107–25

[14] Montevecchi F, Venturini G, Scippa A and Campatelli G 2016 Finite Element Modelling of Wire-arc-additive-manufacturing Process *Procedia CIRP*

[15] Patel L S, Patel T C, Professor A and Student M E 2014 Optimization Of Welding Parameters For TIG Welding Of 304 L Stainless Steel Using Taguchi Approach *Int. J. Eng. Dev. Res.*

[16] Lundbäck A 2010 *Modelling and Simulation of Welding and Metal Deposition*

[17] Ganesh K C, Vasudevan M, Balasubramanian K R, Chandrasekhar N and Vasantharaja P 2014 Thermo-mechanical analysis of TIG welding of AISI 316LN stainless steel *Mater. Manuf. Process.*

[18] Ziętala M, Durejko T, Polański M, Kunce I, Płociński T, Zieliński W, Łazińska M, Stępniowski W, Czujko T, Kurzydłowski K J and Bojar Z 2016 The microstructure, mechanical properties and corrosion resistance of 316 L stainless steel fabricated using laser engineered net shaping *Mater. Sci. Eng. A*

[19] Goldak J, Chakravarti A and Bibby M 1984 A new finite element model for welding heat sources *Metall. Trans. B*

[20] Lindgren L E 2007 *Computational Welding Mechanics: Thermomechanical and Microstructural Simulations*

[21] Hou Z B and Komanduri R 2000 General solutions for stationary/moving plane heat source problems in manufacturing and tribology *Int. J. Heat Mass Transf.* **43** 1679–98

[22] Choi J and Mazumder J 2002 Numerical and experimental analysis for solidification and residual stress in the GMAW process for AISI 304 stainless steel *J. Mater. Sci.*

[23] Wang Y, Ding M, Zheng Y, Liu S, Wang W and Zhang Z 2016 Finite-Element Thermal Analysis and Grain Growth Behavior of HAZ on Argon Tungsten-Arc Welding of 443 Stainless Steel *Metals (Basel).*

IOP Conf. Series: Materials Science and Engineering **834** (2020) 012010 doi:10.1088/1757-899X/834/1/012010

Friction and Wear Study of Fe-Cu-C-CaF_2 Self-lubricating Composite at High Speed and High Temperature

Sanjay Mohan*[1], Ankush Anand[1], R. Arvind Singh[2], S. Jayalakshmi[2], Xizhang Chen[2], S. Konovalov[3]

[1]School of Mechanical Engineering, Shri Mata Vaishno Devi University, Katra, India
[2]School of Mechanical and Electrical Engineering, Wenzhou University, Wenzhou, Zhejiang 325035, China
[3]Institute of Rocket and Space Technology, Samara National Research University, Samara 443086, Russia

*sanjay.mohan@smvdu.ac.in

Abstract. A novel Fe-Cu-C based self-lubricating composite is developed, wherein Calcium fluoride (CaF_2) has been used as a solid lubricant. CaF_2 is added in varying weight percentages of 3, 6, 9 and 12% to the base matrix comprising of Iron, Copper and Graphite (Fe-2Cu-0.8C). The composites were fabricated through Powder Metallurgy using uni-axial compaction and sintering. The developed composites were tested for friction and wear characteristics using a pin-on-disc configuration, conducted at a speed and load of 10 m/s and 20 N respectively. All tests were conducted at high temperature of 500°C for a constant sliding distance of 4000 m. Results show low coefficient of friction for the composites with 3-9 wt% CaF_2 making them self-lubricating. Due to testing at high temperature, weight gain was observed in all the composites because of oxidation. The increase in weight gain was observed to be dependent on the CaF_2 content. Adhesion, ploughing and delamination were identified to be the prominent wear mechanisms of the developed self-lubricating composites as revealed by SEM analysis.

1. Introduction

The continuous use of sliding/moving parts relative to each other in various industries, house hold appliances, automotive sectors etc., has led to developments in the area of self-lubricating materials. Self-lubricating materials comprises of either one or more number of solid lubricants as reinforcement incorporated in the base matrix consisting of a metal, polymer ceramic, etc. Solid lubricants such as h-BN, graphite, molybdenum disulfide (MoS_2), etc. are being widely used in these matrices [1-5]. The lubricating behavior of solid lubricants also depends upon working environment, e.g., graphite is more effective in moist conditions whereas dry conditions are more suitable for MoS_2. Researchers have also tried Boron Nitride (BN), Calcium Fluoride (CaF_2), Antimony (Sb), Tin, (Sn) and Silver (Ag), Tungsten disulphide (WS_2) as solid lubricants [6-10] with different base matrices. Also, fluorides of calcium, barium etc. have shown high thermal and oxidation resistance when compared to other solid lubricants [11-12]. It has been observed that there is an increase in the lubrication capability of fluorides at higher temperatures. This is due to the transition from brittle to ductile phases of fluorides causing low shear strength, thus resulting in lubrication [13]. Iron, being economical amongst other metals is used as base matrix in numerous applications. Copper, nickel, graphite etc. are alloyed in iron to obtain enhanced mechanical properties. The addition of antimony as a reinforcement in iron matrix has shown increase in hardness and compressive strength of the composites [14]. The tribological behavior of various steels

Content from this work may be used under the terms of the Creative Commons Attribution 3.0 licence. Any further distribution of this work must maintain attribution to the author(s) and the title of the work, journal citation and DOI.
Published under licence by IOP Publishing Ltd

was enhanced by using solid lubricants as additives such as h-BN and MoS_2 [15,16]. Fluorides such as CaF_2 and BaF_2 when tried with sintered Fe based composites have resulted in low friction and wear [17-19]. The wear loss of high speed steels is found to decrease on adding TiC, whereas the addition of CaF_2 and MnS to steels led to increased lubrication properties [20]. Fe has been associated with many base matrices, and Fe-Cu-C is the one which is being widely used in different applications due to its properties such as high hardness, strength, etc. However, Fe-Cu-C alloys are not very effective tribological candidates and hence solid lubricant additives are incorporated with the view of enhancing their tribological characteristics. Various metals and lubricants have been tried with Fe-Cu-C in order to improve both, tribological and mechanical properties [15,21-22]. Along with different reinforcement materials, the operating parameters also play an important role, e.g., working temperature, speeds, load etc, in influencing the wear and friction properties [23-26].

In the present work, Fe-Cu-C composites are investigated for their tribological behavior at high speed and high temperature. The Fe-Cu-C base matrix has been reinforced with CaF_2 solid lubricant in different weight percentages. Tribological tests were carried out at speed and load of 10 m/s and 20 N respectively, at the high temperature of 500°C. The wear mechanism and the surface morphology were investigated using optical microscope and scanning electron microscope (SEM).

2. Materials and Methods

Powder Metallurgy technique comprising of uni-axial compaction and sintering has been used to prepare samples from the compositions as shown in Table 1. In order to obtain a homogeneous blend of powders, the mixing was done in two steps: (i) the powders of Fe, Cu, C and CaF_2 were initially blended in a mortar pestle for a duration of 15 minutes and (ii) followed by mixing in a double cone mixer for 30 minutes at a speed of 100 rpm. The lubrication between the compact and the die walls, and also within the powder particles has been ensured by mixing zinc stearate to the powders of various compositions. The average particle size of Fe, Cu, C and CaF_2 powders were around 75 µm, 50 µm, 50 µm and 10 µm respectively. The powder after mixing were compacted in a cylindrical die at a pressure of 600 MPa. The green compacts were then obtained in the form of cylindrical pins and were sintered at 1120 °C in an inert atmosphere consisting of 90 % nitrogen and 10 % hydrogen for 30 minutes.

Table 1. Compositions used in the study

Material	**Compositions (wt%)**			
Elements	Fe	Cu	C	CaF_2
Sample 1	97.2	2	0.8	0
Sample 2	94.2	2	0.8	3
Sample 3	91.2	2	0.8	6
Sample 4	88.2	2	0.8	9
Sample 5	85.2	2	0.8	12

The density of the prepared composite samples was determined using Archimedes principle and hardness was measured using Vickers hardness tester. Hardness testing was carried out at a load of 50 N, and for each test sample, an average of 5 measurements was taken. The tribological testing was carried out using a Pin on Disc (POD) tribometer. En 31 steel disc having hardness of 60±2HRC was used as the counterface material for wear and friction testing. Cylindrical samples of 8 mm diameter were used for friction and wear tests. Prior to friction and wear tests, each sample was polished with 800 and 1000 grade emery papers followed by cleaning of the samples with acetone.The weight of each cleaned sample was recorded. Tribological testing was carried out at a constant load of 20 N and speed of 10 m/s at high temperature of 500 °C. During friction and wear testing, sliding distance was maintained at 4000 m. Each test was repeated for three times, and average value is reported. The wear gain or loss tabulated was determined from the change in the weight of the sample prior to and after the tests under similar test conditions. Frictional force was obtained directly from the machine. Optical and scanning electron microscopy were used to study the surface morphology and wear mechanisms. The images of the worn samples were analysed using a JEOL JSM 5600 LV scanning electron microscope.

3. Results and Discussions

3.1. Optical Microscopy

The optical microscopy carried out on the samples revealed grain refinement in the CaF_2 added samples. The difference in the grain refinement is shown in Fig.1, where grain refinement in sample with 3 wt% CaF_2 (Fig.1 (b)) is clearly observed as compared to the base matrix (Fig.1 (a)).

***Fig. 1.** Optical images of a) Sample 1 (Base matrix) b) Sample 2 (with 3 wt.% CaF_2)*

3.2. Density, Hardness and Porosity

Fig. 2 shows the density of the composites determined using Archimedes'principle. There is a slight decrease in the density of the sample with 3 wt% CaF_2 as compared to the sample with 0 wt% CaF_2 i.e., base matrix. The samples with weight percentage of 6, 9 and 12 CaF_2 has shown further decrease in density. This decrease is density is mainly attributed to the addition of low density CaF_2 to the base matrix.

Fig.3 shows the variation of hardness with increasing CaF_2 addition. A decreasing trend is observed in the hardness values of the samples, except for sample 2 as shown in Fig.3. Sample 2 that has 3% CaF_2 shows higher hardness due to grain refinement as depicted in Fig.1 because of the addition of CaF_2 reinforcement [18, 27]. The decreasing trend in the hardness of rest of the samples is attributed to the increasing porosity as shown in Fig.4. The addition of fluoride resulting in increased porosity has been reported earlier [28].

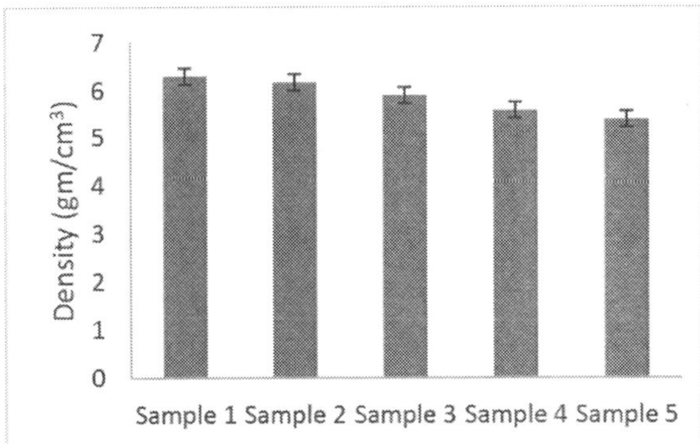

Fig.2. Density of Fe-Cu-C matrix and Fe-Cu-C-CaF2 composites

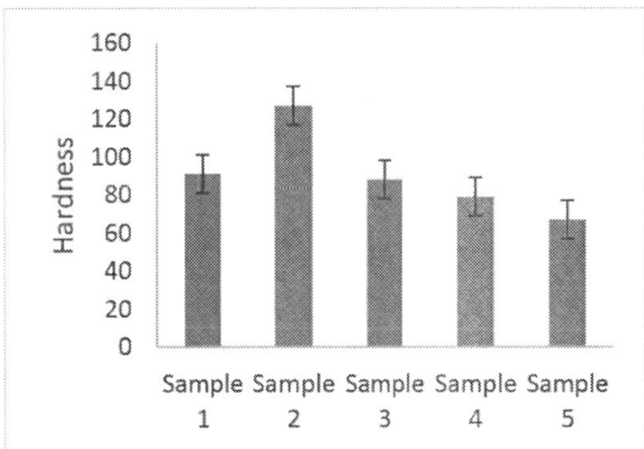

Fig.3. Hardness of Fe-Cu-C matrix and Fe-Cu-C-CaF2 composites

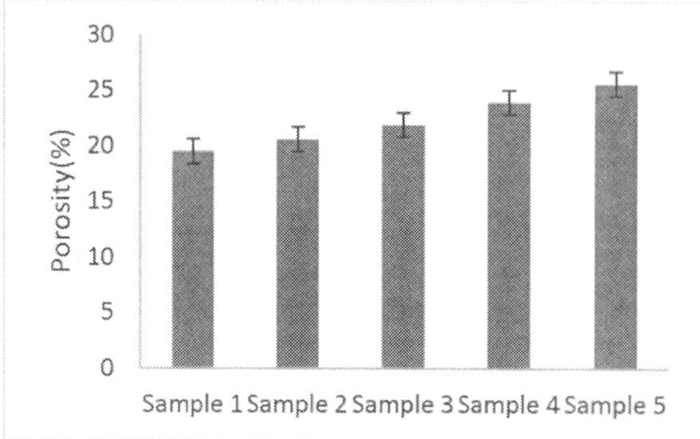

Fig.4. Porosity of Fe-Cu-C matrix and Fe-Cu-C-CaF2 composites

3.3. Tribological testing at High Speed and High Load

Fig.5 shows the coefficient of friction of the test samples tested at 550 °C. From the figure, it could be seen that at the tested high speed (10 m/s) and high load (20 N) conditions waviness is observed in the friction curves as shown in Fig. 5. The waviness or fluctuation in the curves is observed to increase initially, and this is attributed to the frequent formation and removal of oxides. However, stability in the curves is observed after sliding through a distance of 1500 m. Fig. 6 shows the variation in the average values of frictional coefficient of all the samples. Initially, a decreasing trend is observed from base matrix to sample with 6 wt % CaF_2, followed by an increase up to sample with 12 wt % CaF_2. The decrease in COF from base matrix to sample with 6 wt % CaF_2 is due to the dominant effect of solid lubricant over hard iron oxides. Upto 6 wt % CaF_2, dominance of solid lubricant was observed, and thus decrease in friction values is seen. The hard iron oxides are abrasive in nature and they tend to increase the friction, which is the case in the samples with > 6 wt.% CaF_2. SEM images show less ploughing for 3% CaF_2 (shown in Fig.8(a)) as compared to Fig.8(b) and Fig 8(c) for samples with higher CaF_2 contents. Increased ploughing is ascribed to the increased formation of hard iron oxides. The increase in the presence of oxides is due to the increased porosity, thus indicating the dominant effect of hard oxides over the solid lubricant from 6 wt % to 12 wt % CaF_2 added samples. The average value of COF lies between 0.331-0.352.

Also due to high temperature, weight gain resulted during wear as shown in Fig.7. It is observed that except for the base matrix, all CaF_2 added samples exhibited increased weight gain, which is primarily due to the compacted and sintered glaze layers formed on the worn surfaces. The upward slope indicates weight loss which is due to the oxide formation, i.e., oxide formation in sample with 3wt% CaF_2 is less when compared with rest of the samples. The increase in weight gain from 3 wt % to 12 wt % CaF_2 added samples is mainly due to the increased oxidation resulting from increased porosity. Despite weight gain, the samples with weight percentage of 9 and 12 CaF_2 shows more wear than oxidation, which may be attributed to their initial low hardness values.

IOP Conf. Series: Materials Science and Engineering **834** (2020) 012010 doi:10.1088/1757-899X/834/1/012010

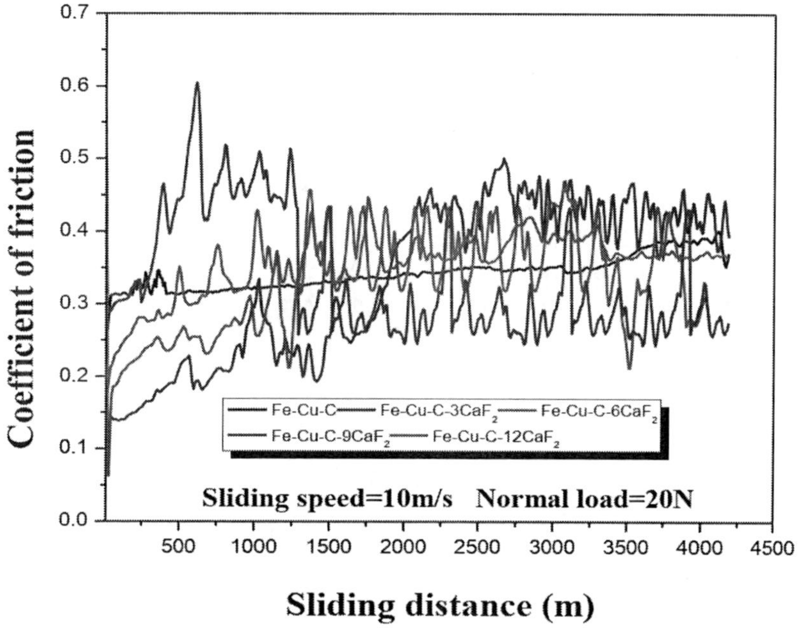

Fig. 5: High Temperature COF of all compositions at high speed-high load

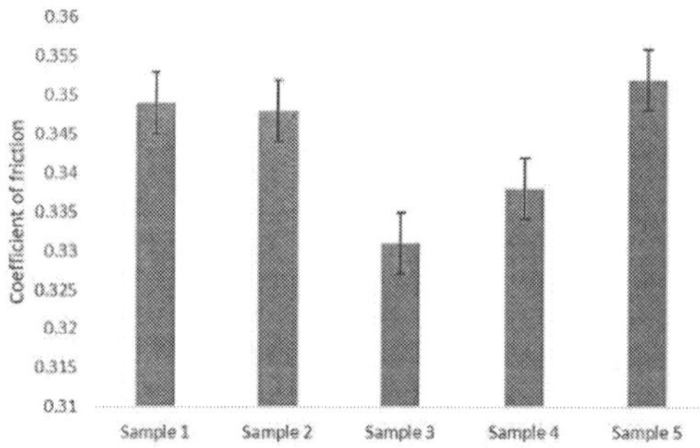

Fig.6.:COF at high speed-high load

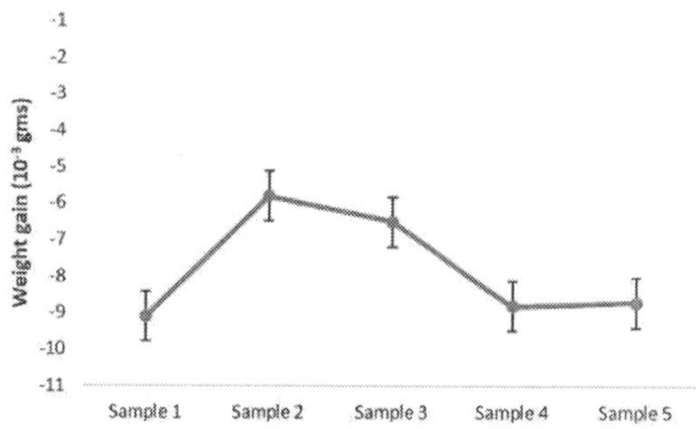

Fig.7: Weight gain at high speed-high load

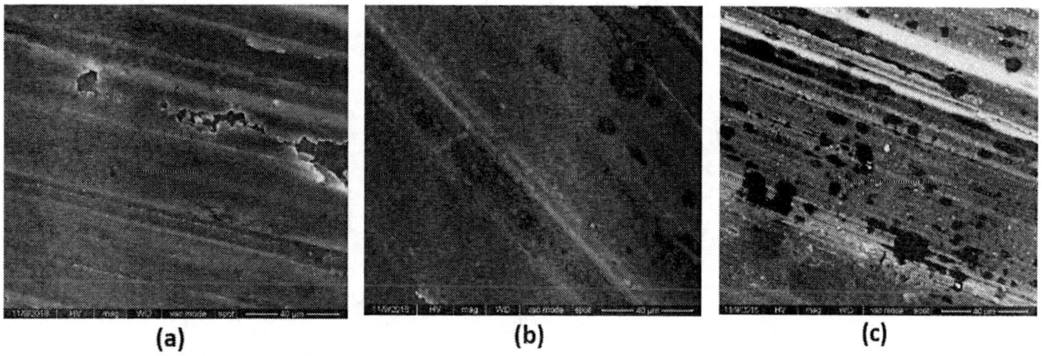

Fig. 8. SEM Images of a) Sample 2 b) Sample 3 c) Sample 4

4. Conclusions

In this paper, Fe-Cu-C based self-lubricating composites have been developed wherein CaF_2 as a solid lubricant was reinforced to the base matrix using single-stage compaction and sintering process. The developed composites were investigated for density, porosity, hardness, friction and wear. The following are the conclusions drawn:

- The density of fabricated composites decreases with the increase in CaF_2 content and the porosity increased with the addition of CaF_2.
- The hardness of the composites also decreased with the increase in CaF_2 addition except for the composite with 3 wt% CaF_2. This is due to the grain refinement due to addition of small amount of CaF_2.
- The friction coefficient of the developed composites with CaF_2 up to 6 wt% is less as compared to rest of the composites which is due to the role of solid lubricant and hard iron oxides.
- Weight gain resulted in the composites due to oxidation. However, weight gain of CaF_2 added composites is more than that of the base matrix.
- Based upon the above stated facts, it is evident that the developed composites can be used in antifriction applications at high temperature and high speed.

References

[1] Kato H, Takama M, Iwai Y, Washida K, Sasaki Y. Wear and mechanical properties of sintered copper–tin composites containing graphite or molybdenum disulfide. Wear. 2003 Aug 1;255(1-6):573-8.

[2] Dhanasekaran S, Gnanamoorthy R. Dry sliding friction and wear characteristics of Fe–C–Cu alloy containing molybdenum di sulphide. Materials & design. 2007 Jan 1;28(4):1135-41.

[3] C.H.Ding, C.H. Liu, Z.M.Yang, Y.P. Wang, Z.B. Sun, and L. Yu, Effect of size refinement and distribution of lubricants on friction coefficient of high temperature self-lubricating composites, *Compos. Sci. Technol.*, 70(2010), 1000-1005.

[4] Haq, M. I. U., & Anand, A. (2018). Dry sliding friction and wear behaviour of Hybrid AA7075/Si3N4/Gr self lubricating composites. *Materials Research Express*.5(6), 066544

[5] Anand, A., Haq, M. I. U., Raina, A., Vohra, K., Kumar, R., & Sharma, S. M. (2017). Natural Systems and Tribology-Analogies and Lessons. *Materials Today: Proceedings*, *4*(4), 5228 5232.

[6] Sharma, S. M., and A. Anand. "Solid Lubrication in Iron Based Materials--A Review." Tribology in Industry 38, no. 3 (2016).

[7] Singh N, Mir IU, Raina A, Anand A, Kumar V, Sharma SM. Synthesis and tribological investigation of Al-SiC based nano hybrid composite. Alexandria Engineering Journal. 2017 May 27.

[8] Sliney HE. Solid lubricant materials for high temperatures—a review. Tribology International. 1982 Oct 1;15(5):303-15.

[9] Sliney HE. The use of silver in self-lubricating coatings for extreme temperatures. ASLE transactions. 1986 Jan 1;29(3):370-6.

[10] Sharma SM, Anand A. Friction and wear behaviour of Fe-Cu-C based self lubricating material with CaF2 as solid lubricant. Industrial Lubrication and Tribology. 2017 Sep 4;69(5):715-22.

[11] Fedorchenko IM, Zozulya VD, Shevchuk YF. Lubricating properties of calcium fluoride. Soviet Powder Metallurgy and Metal Ceramics. 1970 Oct 1;9(10):818-20.

[12] Y.F.Shevchuk, Sintered antifriction materials for use at elevated temperatures, *Powder. Metall. Met. Ceram.*, 40(2001), 40-43

[13] Wang HM, Yu YL, Li SQ. Microstructure and tribological properties of laser clad CaF2/Al2O3 self-lubrication wear-resistant ceramic matrix composite coatings. Scripta Materialia. 2002 Jul 5;47(1):57-61.

[14] Verghese R, Gopinath K. The influence of antimony additions on sintered iron-copper bearing materials. InKey Engineering Materials 1988 (Vol. 29, pp. 457-464). Trans Tech Publications.

[15] Dhanasekaran S, Gnanamoorthy R. Dry sliding friction and wear characteristics of Fe–C–Cu alloy containing molybdenum di sulphide. Materials & design. 2007 Jan 1;28(4):1135-41.

[16] Mahathanabodee, S., Palathai, T., Raadnui, S., Tongsri, R., & Sombatsompop, N. (2013). Comparative studies on wear behaviour of sintered 316L stainless steels loaded with h-BN

and MoS2. In *Advanced Materials Research* (Vol. 747, pp. 307-310). Trans Tech Publications.

[17] Shevchuk, Y. F., Zozulya, V. D., & Khrienko, A. F. (1968). Friction of materials with additions of substances acting as solid lubricants. *Powder Metallurgy and Metal Ceramics*, *7*(12), 985 988.

[18] Han, J., Jia, J., Lu, J., & Wang, J. (2009). High temperature tribological characteristics of Fe–Mo based self-lubricating composites. *Tribology letters*, *34*(3), 193-200.

[19] Kostornov, A. G., Fushchich, O. I., Gorban, V. F., Chevychelova, T. M., & Kostenko, O. D. (2012). Effect of dry friction parameters on the tribosynthesis of secondary structures on composite antifriction iron-based material. *Powder Metallurgy and Metal Ceramics*, *51*(7-8), 412-419.

[20] Zuomin, L., & Childs, T. H. C. (2004). The study of wear characteristics of sintered high speed steels containing CaF2, MnS and TiC additives at elevated temperature. *Wear*, *257*(3-4), 435 440.

[21] S. Dhanasekaran and R. Gnanamoorthy: 'Microstructure, strength and tribological behavior of Fe–C–Cu–Ni sintered steels prepared with MoS2 addition, J. Mater. Sci., 42 (2007), 4659 4666.

[22] Merie, V. V., Cândea, V. C., Bîrleanu, C., Păşcuţă, P., & Popa, C. O. (2014). The influence of titanium dioxide on the tribological characteristics of a Fe-based friction composite material. *Journal of Composite Materials*, *48*(2), 235-243.

[23] Thakare, M. R., Mason, J. F., Owen, A. K., Gillespie, D. R. H., Wilkinson, A. J., & Franceschini, G. (2016). Effect of sliding speed and counterface properties on the tribo-oxidation of brush seal material under dry sliding conditions. *Tribology International*, *96*, 373-381.

[24] Kim, D. W., & Kim, K. W. (2014). Effects of sliding velocity and ambient temperature on the friction and wear of a boundary-lubricated, multi-layered DLC coating. *Wear*, *315*(1-2), 95 102.

[25] Earles, S. W. E., & Kadhim, M. J. (1965). Friction and wear of unlubricated steel surfaces at speeds up to 655 ft/s. *Proceedings of the Institution of Mechanical Engineers*, *180*(1), 531 548.

[26] Gunes, I. (2015). Effect of sliding speed on the frictional behavior and wear performance of borided and plasma-nitrided w9mo3cr4v high-speed steel. *Materiali in tehnologije*, *49*(1), 111-116.

[27] Sharma, S. M., & Anand, A. (2017). Friction and wear behaviour of Fe-Cu-C based self-lubricating material with CaF2 as solid lubricant. *Industrial Lubrication and Tribology*, *69*(5), 715-722.

[28] Du, L., Huang, C., Zhang, W., Zhang, J., & Liu, W. (2010). Effect of NiCr Clad BaF 2· CaF 2 Addition on Wear Performance of Plasma Sprayed Chromium Carbide-Nichrome Coating. *Journal of thermal spray technology*, *19*(3), 551-557.

Distortion Analysis of SLM Product of SS316L using Inherent Strain Method

Thoufeili Taufek[1], Yupiter HP Manurung*[1], Stephan Lüder[2], Marcel Graf[2] and Farrahshaida Mohd. Salleh[1]

[1]Faculty of Mechanical Engineering, Universiti Teknologi Mara, Selangor, Malaysia
[2]Professorship of Virtual Production Engineering (VIF), TU Chemnitz, Chemnitz, Germany

*yupiter.manurung@uitm.edu.my

Abstract. As one of additive manufacturing (AM) technologies, selective laser melting (SLM) which uses higher energy input enabling fully molten powder bed materials is nowadays increasingly applied to build full dense components without post processing. In the present work, specimens made of stainless steel powder SS 316L were to be processed using SLM Solutions 280 HL, characterized and compared to the simulation from Finite Element Analysis (FEA) Software. The powders have been melted to produce cantilever specimens with different build orientation by using fixed major process parameters such as laser power, hatching distance and layer thickness as well as scan speed. The experiment starts with the FEA simulation which focused on Inherent Strain Method (ISM) and followed by experimental process. While the distortion of the specimens after the process to be the main concern and to be discussed based on the build orientation of the specimens. The measurement distortion of the specimens was conducted by using KEYENCE 3D scanner. It is found out that the build orientation play a big role on distortion of the specimens made from SLM. Meanwhile, result of distortion from the experiment and FEA simulation are close to each other. It is expected that this basic FEA study will provide basic information and estimation towards the distortion and computational time by simulating product prior to commencement of real product manufacturing. Thus, by using ISM will reduce the computational time and reduce the waste material by simulating the process before the experiment have been held.

1. Introduction

Additive manufacturing (AM) is a processes that joint materials by adding layer-by-layer which follows the geometry from computer aided design (CAD) drawing in order to form final product. The process are varies which depends on the heat source such as electron beam, laser or arc and also how the material is deposited [1]. AM also allows production of complex parts directly from the design without using expensive tooling or forms such as punches, dies or casting molds thus it reduces the needs for many conventional processing steps. For these reasons, AM is now widely recognized as a new standard for the design and production of high performance components for aerospace, medical, energy and automotive applications. Aerospace examples include complex fuel injector nozzles that previously required assembly of multiples parts and lightweight engineered structures that result in significant cost savings [2]. In this paper, the distortion of the produced part from SLM has been investigated. SLM is a process uses a layer-wise building principle where the powder is deposited in a bed and molten locally

Content from this work may be used under the terms of the Creative Commons Attribution 3.0 licence. Any further distribution of this work must maintain attribution to the author(s) and the title of the work, journal citation and DOI.
Published under licence by IOP Publishing Ltd

in a layer according to CAD drawing [3]. The material is locally and rapidly heated above its melting temperature and then allowed to solidify and cool in order to form a dense geometry [4]. This process involves large thermal gradients[5]. This complex thermal process can lead to large thermal strain and consequently large residual stress or distortion in the manufactured metal parts [6]. According to [7], AISI 316L and PH 17-4 stainless steels, Maraging steel Grade 300, Inconel 625 and 718, Al10SiMg, CoCr and Ti6Al4V alloys are among the most used and studied materials for SLM. In this study, commercial stainless steel 316L (SS 316L) powder have been used to produce parts from SLM process. According to [8], it is almost impossible to minimize the distortion of the final product from SLM without simulation tools. In order to predict the residual distortion and stress in the AM process, finite element method (FEM) have been used to model the AM process [6]. Based on the numerical approaches, finite element method (FEM) simulation can be implemented for optimization SLM parameters such as laser scanning strategies, build-up directions and design support structures [9]. In this paper, inherent strain method (ISM) have been proposed in order to investigate the distortion of the product which prior to SLM process. In ISM, only pure mechanical computation has been considered which the phase transformation and microstructure predictions have been neglected in order to gain solution speed [10].

The work presented in this paper is focused on the development and application of an alternative in order to predict SLM distortion based on ISM approach. In the simulation, ISM is a mechanical approach where no transient thermal analysis is involved. A commercial stainless steel 316L (SS 316L) powder have been fed into SLM Solutions 280 HL in order to produce cantilever specimen. SIMUFACT Additive has been chosen as FEM software for modelling the whole SLM process. It has been previously developed for calculating of welding distortion of large parts and it requires to adjust the volume of the shrinking elements and their initial temperature.

2. Simulation Procedure using Inherent Strain Method (ISM)

2.1. Theory of Inherent Strain Method (ISM)

In the micro-welding process, the material along the weld path will first be heated, melted, and then solidified in a short time span, which would result in large temperature gradient and complex deformation path. After welding is completed, the welded part cools to the ambient temperature. Figure 1 shows the basic deformation of nodes and element of the specimen during SLM process.

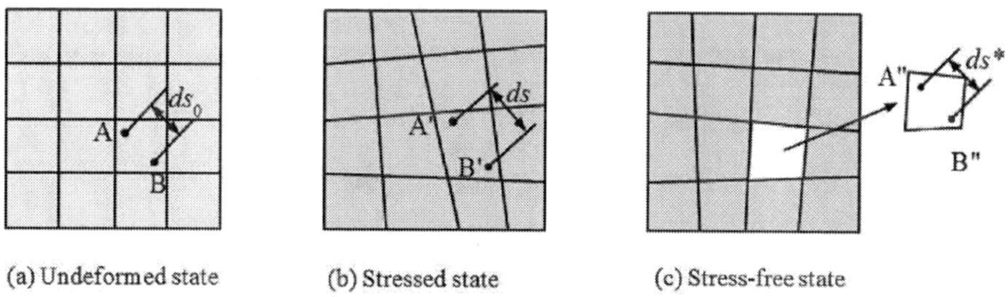

Figure 1. Original definition of inherent strain for welding mechanics [9].

By definition, the inherent strain in the element is defined as the residual strain in the stress-relieved state with reference to the under formed state in Figure 1 (a) before the welding take place:

$$\varepsilon^* = (ds^* - ds_0) / ds_0 \quad (1)$$

After the welding process, the metal part is cooled down to the ambient temperature and thermal strain in the part vanishes. Since the reference temperature employed here is always the ambient temperature of the entire system, thermal strain is not concerned and only mechanical strain is involved.

$$\varepsilon^* = \underbrace{\frac{(ds-ds_0)}{ds_0}}_{\varepsilon} - \underbrace{\frac{(ds-ds^*)}{ds}}_{\varepsilon_e} \qquad (2)$$

The first term on the left-hand side is the total mechanical strain ε after the welding is finished, while the second term on the right-hand is the mechanical elastic strain ε_e, which is directly proportional to the stress released. Practical application of the original inherent strain theory to welding problems makes a key assumption that the elastic strain insignificant compared to the plastic strain. Hence it can be assumed that all the elastic strains vanish and inherent strain ε^* becomes equal to the plastic strain ε_p generated from the welding process:

$$\varepsilon^* = \varepsilon_p \qquad (3)$$

2.2. Inherent Strain Method (ISM) using pure mechanical simulation

In simulation, the inherent strains are to be considered during building process by thermal strains, plastic strains and phase transformation. These strains can be calibrated from experiments by measuring the distortion of printed cantilever beam at certain cutting height and running simulations with the same parameters from the experimental in order to match printed cantilever as a reference. Some process parameters such as layer parameters and inherent strain estimation are to be counted, therefore, material properties which were taken at room temperature is pure mechanical computation. During simulation of mechanical calibration method, no temperature effects are shown or calculated. Unlike Detailed Transient Analysis (DTA) which considered the thermal effects such as in thermo-mechanical simulation which increasing the computational time and limited to small scale specimen. Inherent Shrinkage Method (ISM) takes the strains that are usually applied during cooling of molten metal down to room temperature. Since the inherent strains value are from thermal shrinkage which consist of compressive stress as mentioned from [11], the value with respect to x, y and z axis are to be negative.

The strain increment, ε, in a thermo-mechanical equation consist of an elastic strain, ε^e, plastic strain, ε^p, thermal strain, $\varepsilon^{thermal}$, and phase transformation, ε^{phase}. Thus, total residual strain, ε_{tot}, can be calculate by using Equation 1.

$$\varepsilon_{tot} = \varepsilon^e + \varepsilon^p + \varepsilon^{thermal} + \varepsilon^{phase} \qquad (4)$$

Unlike in mechanical simulation which during numerical calculation, thermal effects have been neglected. The value of inherent strain, ε^* is defined as the difference between the total residual strain and elastic strain as seen in Equation 2.

$$\varepsilon^* = \varepsilon_{tot} - \varepsilon^e \qquad (5)$$

By using Equation 5, the inherent strain can be used to calculate the residual stresses without the information of the thermal effect such as thermal strain and phase transformation. Where [K] is the elastic stiffness matrix, [u] is the nodal displacement vector, f^* is the nodal force vector induced by the inherent strains. When the displacement vector [u] has been solved using Equation 3, the total strain, ε, and consequently the residual stress, σ, can be calculated.

$$[K][u] = [f^*] \qquad (6)$$

$$f^* = \int [B]\,[D]\,[\varepsilon^*]\, dV \qquad (7)$$

Where [B] denotes the nodal deformation matrix and [D] denotes the material elastic matrix. The inherent strain, shown in Figure 2, can be seen as the input for the FEM computation of the stresses and can be determined in either of the two ways, through a thermal-elasto-plastic simulation or experimental measurements.

$$\varepsilon = [B]\ [u] \tag{8}$$

$$\sigma = [D^e]\ (\ [\varepsilon_{tot}] - [\varepsilon^*]\ ) \tag{9}$$

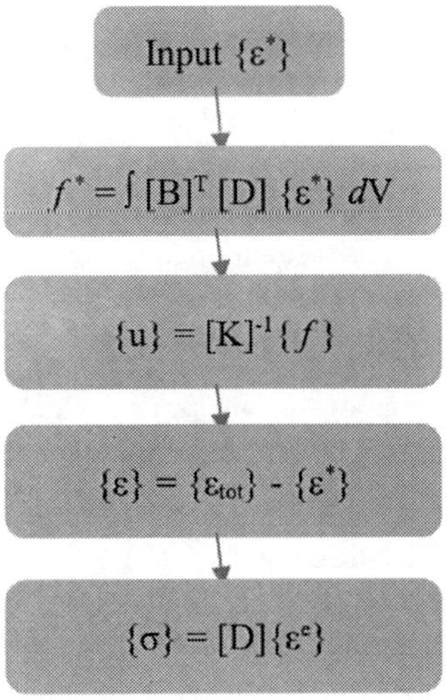

Figure 2. Flow chart of FEM computation of stress with the inherent strain as input.

3. Experimental Setup

Cantilever shape and dog bone specimens were fabricated by SLM Solution 280 HL. Both specimens and the base plate were made from SS 316L. Dog bone specimens were manufactured in a SLM Solutions 280 HL machine meanwhile base plate has been fabricated by using Computer Numerical Control (CNC) milling machine in order to get specified dimension to fit into SLM 280 HL machine. Figure 3 shows the fabricated specimens by SLM process, consist of 8 dog bone samples with 2 different build direction: horizontal and vertical, followed by 3 cantilever specimens with three different build orientation: 0°, 45°, and 90°. All printed samples were fabricated by using process parameters showed in Table 1.

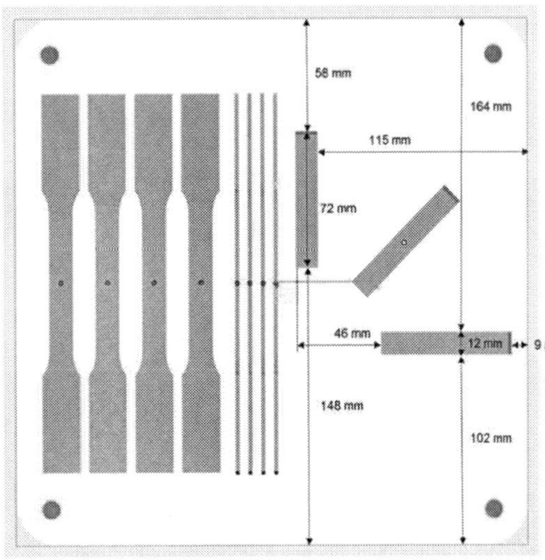

Figure 3. Fabricated specimens by using SLM

IOP Conf. Series: Materials Science and Engineering **834** (2020) 012011 doi:10.1088/1757-899X/834/1/012011

Figure 4. Cantilever specimens after cutting process

Table 1: SLM process parameters

No.	Description	Parameter
1	Laser power	275W
2	Hatch distance	0.12mm
3	Layer thickness	50μm
4	Scan speed	700mm/s
5	Scan strategy (Hatch style)	Bi-diretional
6	Oxygen content	< 0.1%

After manufacturing, cantilever specimens were partially cut 1mm from the baseplate by using water jet cutting machine and electrical discharge machine (EDM). The final vertical distortions (Z direction) of each samples were measured and the value of the distortion have been collected as ΔZmax.

Figure 5 shows the image of fabricated dog bone specimens. Both dog bone specimens have two different build direction which plays an important role which affected the final geometry of printed SLM sample. In this paper, sample with horizontal build orientation have been considered since the value of Z distortion can be clearly seen on a sample with horizontal build orientation. All fabricated specimens by using SLM process have been removed from the base plate by cutting the support structure using hand cutter tools. Figure 6 shows the image of the base plate after removing all the specimens.

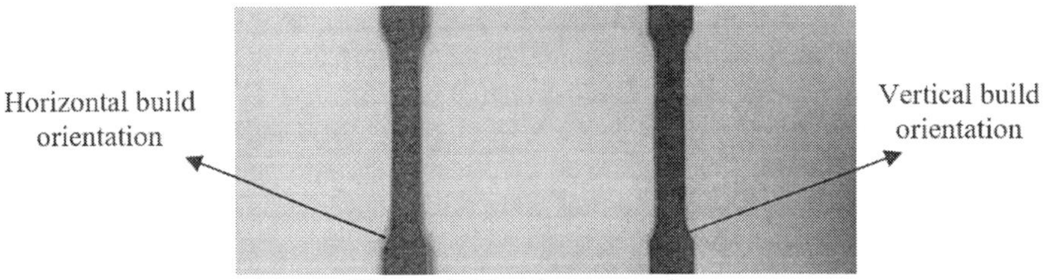

Figure 5. Dog bone specimens

Figure 6. Base plate after removing samples

4. Simulation Setup & Procedure

In this paper, FEM approach have been used in order to calculate the pure mechanical simulation. Purely mechanical analysis approach of Simufact Additive requires the input of inherent strain values prior to the simulation. These inherent strain values depends on material properties and process parameters. Figure 7 shows the information of basic inherent strain method (ISM). From Figure 7, thermal strains and phase transformation have been neglected thus it will reduce the computational time since in mechanical simulation, only value of inherent strain are needed in order to get the value of stress, strains, and distortion as the outputs.

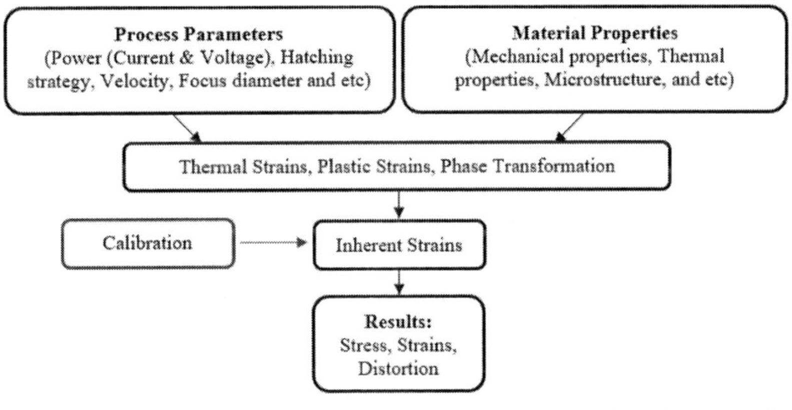

Figure 7. Basic information for Inherent Strain Method (ISM) by using FEA simulation

All values for inherent strain can be obtained from the experimental work which cantilever specimens have been used for calibration method. Figure 8 shows the image of flow chart methodology for the whole process for ISM simulation and experimental.

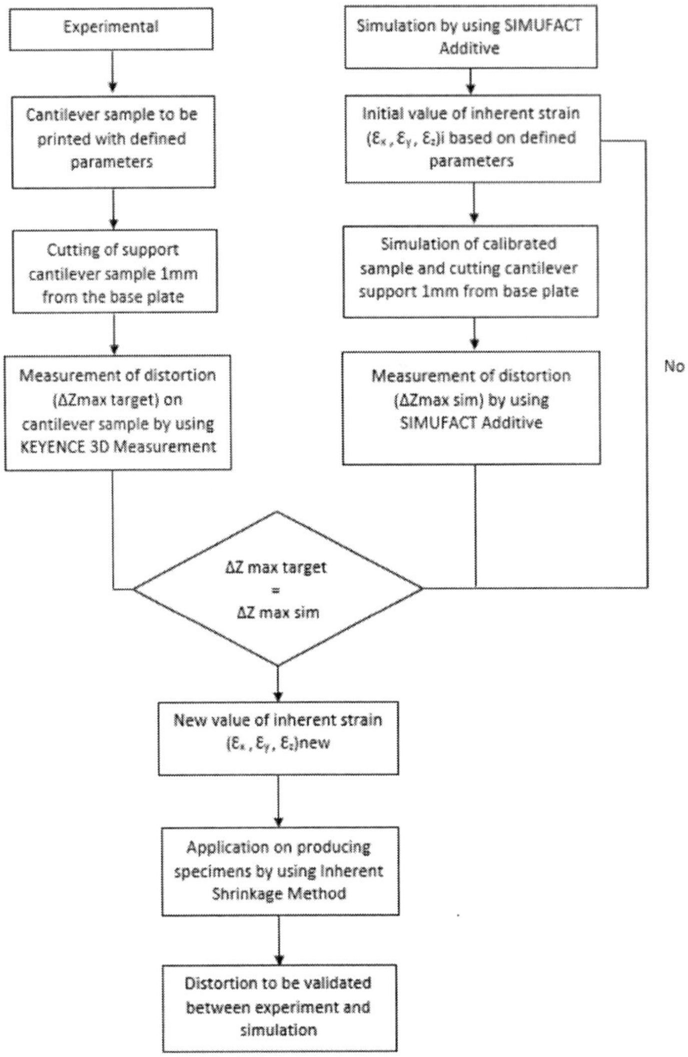

Figure 8. Inherent strain method (ISM) flow chart.

4.1 Calibration Method

For this study, 3 cantilever beams with three different build direction (0°, 45°, 90°) have been printed in order to investigate the value of the distortion for each cantilever beam specimens, values of the distortion then to be used as *ΔZ*max. Reference height have been set as 9mm since the cutting height between the cantilever beam specimen to the base plate is 1mm. Figure 9 shows the result of the distortion of each specimens have been investigated by using KEYENCE 3D scanner and Table 2 is the value of *ΔZ*max.

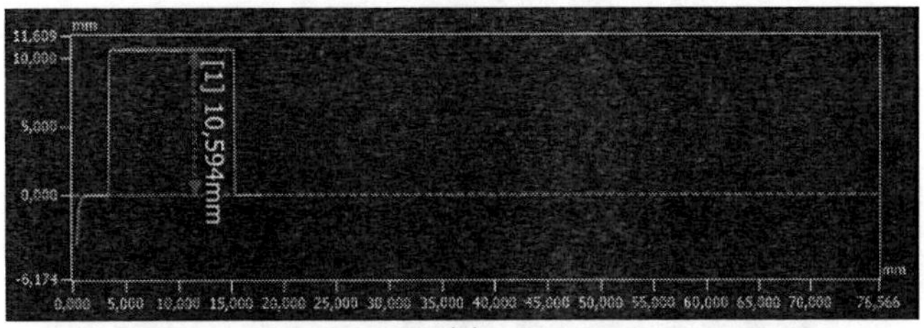

(a)

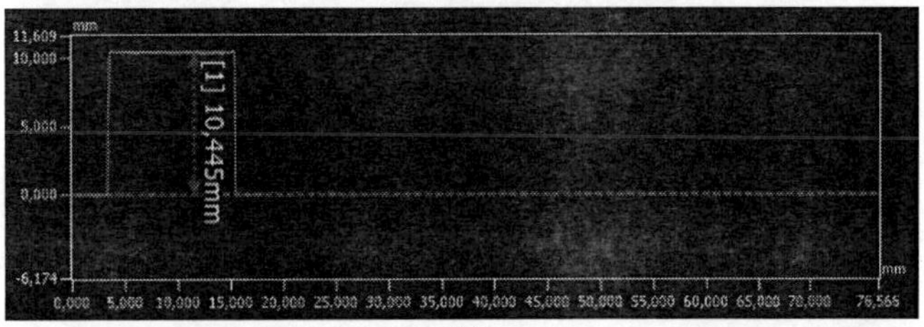

(b)

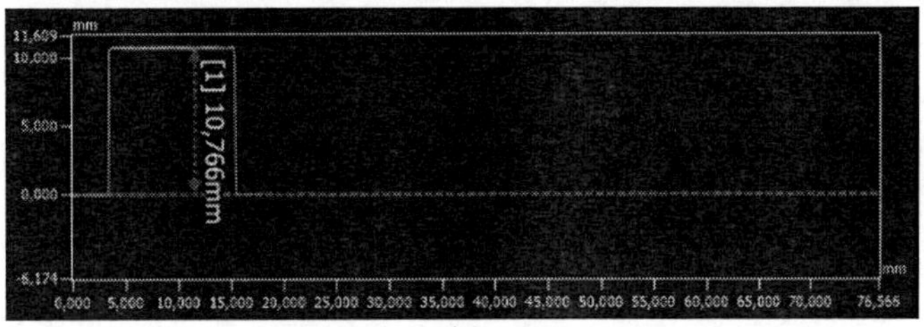

(c)

Figure 9. Result of distortion for each cantilever specimens with different build orientation: (a) 0°, (b) 45°, (c) 90°.

Table 2. Value of *ΔZ*max.

Build Orientation (°)	0	45	90
***ΔZ*max (mm)**	1.594	1.445	1.766

Figure 10 shows the image of the printed cantilever inside the red circle indicator before the sample have been removed from the base plate. In the calibration method, the parameters used in the simulation are shown in Table 3. The cutting parameter have been set 1mm from the base plate which is similar to the experimental. Figure 14 shows the result of the calibration simulation and the value of new inherent strain which respect to each cantilever beam specimens are to be considered as (ε_x, ε_y, ε_z)new. The value of new inherent strain then to be collected and the value have been used in the mechanical simulation in order to predict big scale part simulation.

Figure 10. Printed SS316L cantilever beams.

Table 3. Calibration method simulation parameters

Parameter	Value
Layer thickness, mm	0.05
Scan width, mm	10
Beam width, mm	0.08
Laser power, W	275
Velocity, mm/s	700
Efficiency, %	25
Material	SS 316L

4.2 Mechanical Calibration Method

Cantilever specimens with three different build orientation have been simulate. In order to get the new value of inherent strain, apple to apple comparison between experimental and simulation have to be done such as position, material, and cutting height of the cantilever specimen. Figure 11 shows the simulated cantilever specimen and the base plate. In this study, material of SS316L have been assigned to the cantilever specimen and also the base plate. All the general properties, chemical composition, thermal properties and mechanical properties and flow curve of the material SS316L powder have been determined by Simufact Material 2019. All the information about SS316L powder have been used for SIMUFACT Additive to solve the numerical equation in finite element analysis.

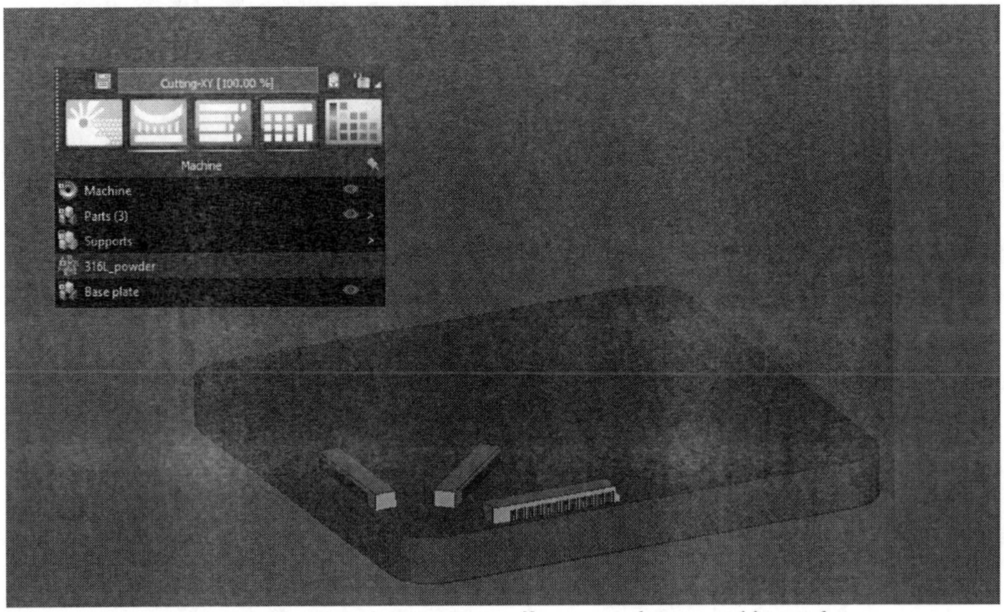

Figure 11. Simulated SS316L cantilever specimens and base plate.

4.3 Build Properties

In SIMUFACT Additive, build properties consist of two sections to be defined. Figure 12 shows the calibration parameters for the whole process such as measuring point coordinate and ΔZ_{max} for every cantilever specimens. Next, the build parameters such as machine parameters including beam width, velocity, power and efficiency and the material properties will be taken into account in order to estimate the inherent strain value which prior to the calibration process. Figure 13 shows the image of build parameters and the red rectangular shape shows the value of inherent strain after the machine process parameters and material properties have been calculated by the software. In this study, the type of inherent strain distribution has been defined as orthotropic distribution since the value of inherent strain which respect to Z-axis were to be counted.

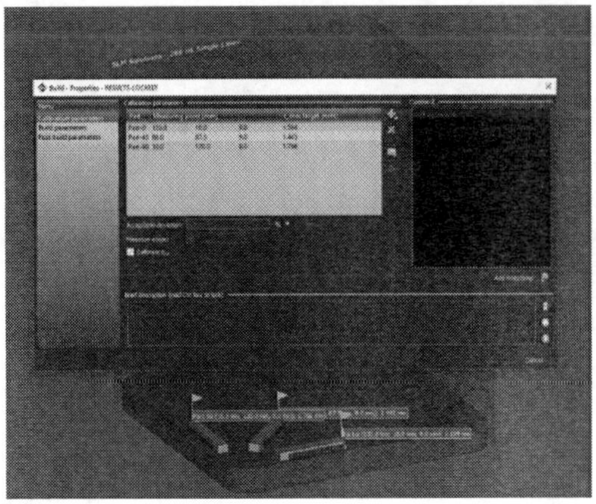

Figure 12. Calibration parameters during mechanical calibration process.

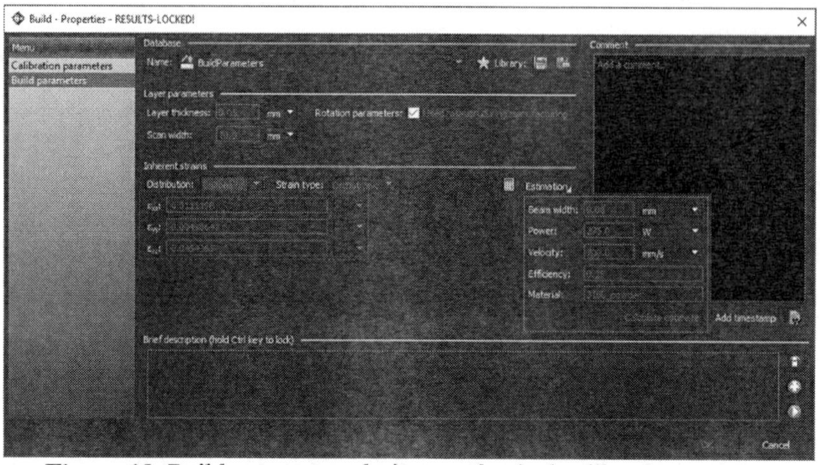

Figure 13. Build parameters during mechanical calibration process.

4.4 *Calibration result*

In this section of calibration results, status of every current step are shown in graph ΔZ_{max} over the number of calibration steps. In this study, the overall step are set to 50 steps in order to complete the calibration process. From the graph ΔZmax over the number of calibration steps illustrated in Figure 14, once the process parameters and material properties have been calibrated, SIMUFACT Additive have been able to estimate the value of new inherent strain by solving the Equation. 10.

$$S = \sum_{i=1}^{n} \quad r_i^2 \tag{10}$$

Where r_i is the difference between $z_{i,sim}$ and $z_{i,\mathrm{exp}}$. i is the number of running index, n represents the number of sample points, $Z_{i,sim}$ is the simulated distortion and $Z_{i,exp}$ is the measured distortion from the experiments. In this research, the new value of inherent strain is ε_{xx}= -0.0140362, ε_{yy}= -0.0139978, and ε_{zz} = -0.0455615. Next, the new value of inherent strains will be used for mechanical simulation in order to simulate SLM product with same material which is SS316L. .shows the image of printed cantilever specimens in SIMUFACT Additive after the samples have been cut 1mm from the base plate.

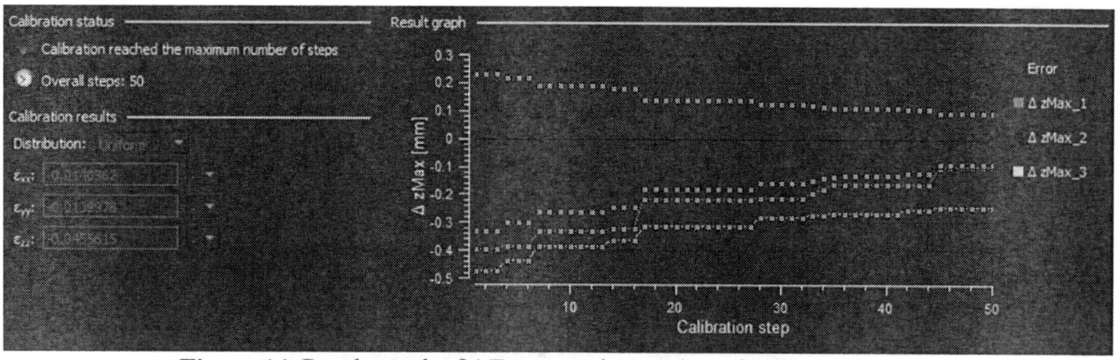

Figure 14. Result graph of ΔZ_{max} over the number of calibration steps.

Figure 15. Cantilever specimens after simulation of calibration by SIMUFACT Additive.

5. Result and discussion

Mechanical simulation in SIMUFACT Additive used macroscopic scale so called layer model. All SLM parts are extremely large compared to the actual size of every micro-weld and each weld seam. Thermal history and phase transformation are to be neglected in mechanical simulation in order to reduce the computational time. The value of inherent strains have been determined during mechanical calibration and the values of inherent strain then will be used as input for mechanical simulation. Noted that, the inherent strains from mechanical calibration can only be used on components with same material and built with the same parameters. Figure 16 shows the build parameters assigned in SIMUFACT Additive during mechanical simulation. In this study, two dog bone samples with different build orientation have been simulate in order to investigate the value of distortion. The value of new inherent strain, ε_{xx} = -0.0140362, ε_{yy} = -0.0139978, and ε_{zz} = -0.0455615 have been assigned in build parameters as input for SIMUFACT Additive to complete the mechanical simulation. The parameters used for both dog bone samples in this mechanical simulation are set similar with parameters from Table 1.

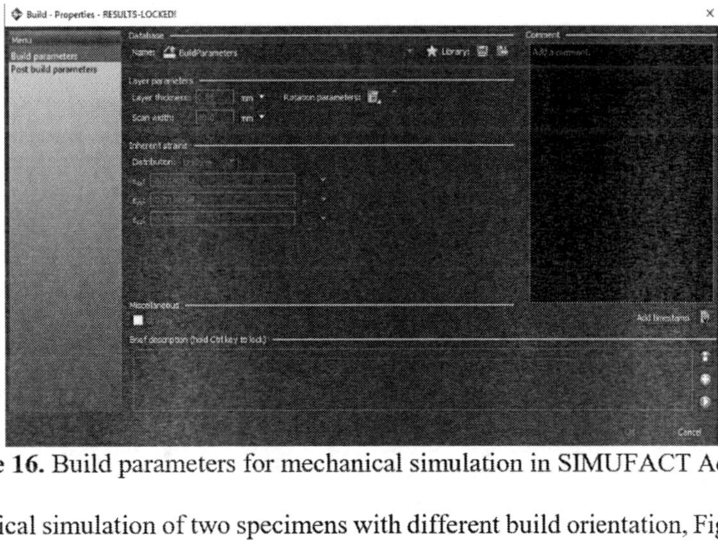

Figure 16. Build parameters for mechanical simulation in SIMUFACT Additive.

From the mechanical simulation of two specimens with different build orientation, Figure 17 shows that specimen with horizontal build orientation produce more value of distortion compared to sample with vertical build orientation. Moving heat source for horizontal build specimen have more time exposure compared to vertical build specimen during metal powder melted layer by layer during build up process. This is due to horizontal build specimen have bigger area of surface in the process chamber compared to vertical build specimen. Result from simulation made by SIMUFACT Additive shows that the maximum value of distortion from the horizontal build specimen is 3.40mm meanwhile vertical build specimen have only 1.81mm. From the value of distortion from both simulated specimens, it is clearly shown that build orientation of the parts during manufacturing process will affect the final geometry of the SLM product. The value of new inherent strain from mechanical calibration simulation by SIMUFACT Additive can be used to other SLM parts with bigger and complex geometry. It is clearly shown that result of new value of inherent strain from calibration process can be used for mechanical simulation process since the simulation of both samples with two different build orientation was completed without any error.

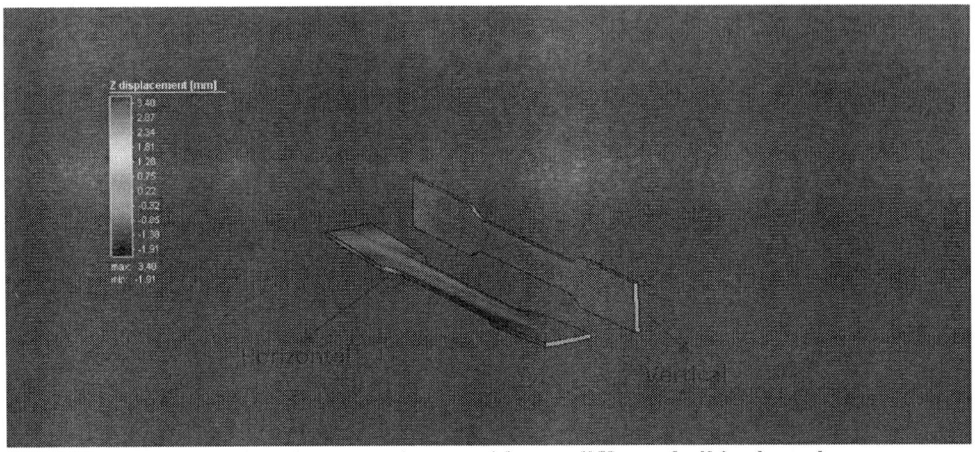

Figure 17. Dog bone specimens with two different build orientation.

Acknowledgement

The authors would like to express their gratitude to staff member of Welding Laboratory, Advanced Manufacturing Laboratory, Advanced Manufacturing Technology Excellence Centre (AMTEx), Research Interest Group: Advanced Manufacturing Technology (RIG:AMT) and Smart Manufacturing Research Institute (SMRI) at Faculty of Mechanical Engineering, Universiti Teknologi MARA (UiTM) as well as Professorship of Virtual Production Engineering and Chair of Welding Engineering at Chemnitz University of Technology in Germany for conducting this research. This research and conference participation are also financially supported by DAAD Germany (2019) and Technogerma Engineering & Consulting Sdn. Bhd.

References

[1] J. J. Lewandowski and M. Seifi, "Metal Additive Manufacturing: A Review of Mechanical Properties," *Annu. Rev. Mater. Res.*, vol. 46, no. 1, pp. 151–186, 2016.

[2] T. DebRoy *et al.*, "Additive manufacturing of metallic components – Process, structure and properties," *Prog. Mater. Sci.*, vol. 92, pp. 112–224, 2018.

[3] M. L. Pace *et al.*, "3D additive manufactured 316L components microstructural features and changes induced by working life cycles," *Appl. Surf. Sci.*, vol. 418, pp. 437–445, 2017.

[4] M. S. Sebastian, I. Setien, A. M. Mancisidor, and A. Echeverria, "SLM (Near)-Net-Shape Part Design Optimization Based on Numerical Prediction of Process Induced Distortions," pp. 117–126, 2018.

[5] N. Keller and V. Ploshikhin, "NEW METHOD FOR FAST PREDICTIONS OF RESIDUAL STRESS AND DISTORTION OF AM PARTS."

[6] X. Liang, Q. Chen, L. Cheng, Q. Yang, and A. To, "A Modified Inherent Strain Method for Fast Prediction of Residual Deformation in Additive Manufacturing of Metal Parts," *Solid Free. Fabr. Symp.*, no. September, pp. 2539–2545, 2017.

[7] R. Casati, J. Lemke, and M. Vedani, "Microstructure and Fracture Behavior of 316L Austenitic Stainless Steel Produced by Selective Laser Melting," *J. Mater. Sci. Technol.*, vol. 32, no. 8,

pp. 738–744, 2016.
[8] C. Seidel and M. F. Zaeh, "Multi-scale Modelling Approach for Contributing to Reduced Distortion in Parts Made by Laser-based Powder Bed Fusion," *Procedia CIRP*, vol. 67, pp. 197–202, 2018.
[9] X. Liang, L. Cheng, Q. Chen, Q. Yang, and A. C. To, "A modified method for estimating inherent strains from detailed process simulation for fast residual distortion prediction of single-walled structures fabricated by directed energy deposition," *Addit. Manuf.*, vol. 23, pp.471–486, 2018.
[10] A. Mendizabal, J. B. González-Díaz, M. San Sebastián, and A. Echeverría, "Improved Accuracy of the Inherent Shrinkage Method for Fast and More Reliable Welding Distortion Calculations," *J. Mater. Eng. Perform.*, vol. 25, no. 7, pp. 2670–2678, 2016.
[11] "Welding distortion of a thin-plate panel structure," *Weld. J. (Miami, Fla)*, vol. 78, no. 5, pp. 156–165, 1999.

IOP Conf. Series: Materials Science and Engineering **834** (2020) 012012 doi:10.1088/1757-899X/834/1/012012

Investigation of Tensile Properties of the Eco-Board of Hybrid Composite that Consist Of Oil Palm Empty Fruit Bunch (OPEFB) Fiber Added with Rice Husk

M. H. Abdul Karim*, M. K. Mohd Shah, M. F. Jundam and S. Abdullah

Faculty of Engineering, Universiti Malaysia Sabah, 88700, Kota Kinabalu, Sabah, Malaysia

*afezmcilroy@gmail.com

Abstract. This project focuses on the oil palm empty fruit bunch (OPEFB) fibre that be added with rice husk (RH) as an alternative raw material to replace wood in a several types of industries. ASTM D3039 standard for tensile test was used to design and compose a composite of EFB fiber reinforced rice husk. In this study, the specimen were be created by the compression method based on the numerous ratio that be chosen which is specimen A, pure EFB, specimen B (93.3g EFB: 6.7g RH), Specimen C (86.6g EFB: 13.4g RH) and specimen D (66.7g EFB: 33.3g RH). The objective of this research is to identify the tensile properties of the Eco-board of hybrid composite and to determine the best properties to be introduced to the industry. The failure pattern and fractography failure mode were investigated using Scanning Electron Microscope (SEM). The tensile test was performed to characterize the hybrid composite properties using a GOTECH/AI-7000M Electronic Mechanical Testing Machine. Result showed that the higher the rice husk added to the EFB fiber, the higher its tensile strength up to certain ratio. As the conclusion, the rice husk help the EFB fiber to increase strength however at the optimum value ratio where the best properties of hybrid composite is specimen C with 9.685 MPa and suitable for many application.

1. Introduction

At present, there are various new composite materials are investigated to make products that able to facilitate human daily activities. Generally, the composites are made of two main types of fiber that are synthetic and natural fiber. Synthetic fibers generally come from synthetic materials such as petrochemicals, glass and carbon fiber. Natural fibers include those produced by plants, animals, and from agricultural by product [1]. These agricultural resources can be dated back in the primitive age such as wood, is used to make shelter, cook food, construct tools and make weapons. Natural fibers have been considerable demand in recent years due to its eco-friendly and renewable nature [2]. In addition, the advantages of using such resources is that, they are widely distributed in all zones of the world, low cost, strong, aesthetic and biodegradability compared to synthetic fiber [3]. Agricultural by product such as palm oil trunk, bamboo fiber, coconut fiber, and pineapple leaf fiber, has been studied as reinforcement for composite.

As we approach the 21st century, there is a larger consciousness of the need for materials in getting higher world population and increasing affluence [2]. At the same period, we have an awareness that our landfills are satisfying. Our resources are being used up, our planet is existence polluted, and that non-renewable resources will not last forever. Wood will continue to be a major source that has higher density than annual plants so there will be more bulk when using agricultural crop fiber. However, in this country where there is little or no wood possessions left or where area are in place to limit the use

Content from this work may be used under the terms of the Creative Commons Attribution 3.0 licence. Any further distribution of this work must maintain attribution to the author(s) and the title of the work, journal citation and DOI.
Published under licence by IOP Publishing Ltd

of wood, alternate sources of fiber are needed if there is to be a natural fiber industry. Empty fruit bunch (EFB) is one of the natural fibers selected to make a board and to replace a wood. EFB is obtained after the elimination of oil seeds from the fruit bunch for oil extraction. When the palm oil seed are harvested, a large amount of EFB remains because each oil palm plant cannot be used for the next harvest [4], and expansion of exploration of oil palm plantations has resulted enormous amounts of vegetable waste, creating problems in replanting process and exceptional environmental concerns. It is reported that, during recent past years, Malaysia alone produced about 30 million tons annually of oil palm biomass, including trunks, fronds, and empty fruit bunches [5]. Hence, without any addition cost EFB fibers as a waste product can be used for industrial purposes and will benefit the environment significantly.

The rice husk (RH) is also the natural fibers that be added with oil palm empty fruit bunch (OPEFB) as an alternative raw materials to replace wood in a several types of industries. Rice husk is an economical by-product of rice processing and is separated from rice grain during the rice milling process. Rice husk incineration is discouraged due to producing hazes, ash and toxic gases, which lead to serious air pollution [6]. The land area of the paddy planting in Malaysia is approximately 680,000 ha and 840,000 tons rice husk were produced every year [7]. Annually about 3.66 million tons of rice husk waste is left in the field. This values is predicted to increase to 7 million tons per year towards year 2020 due to evolving technology development in agriculture industries [8]. As a way to reduce it, the by product can be widely used in composite manufacturing such as cardboard, roofing tiles and for making some compound like silica silicon compounds [9]. A different ratio between oil palm empty fruit bunch and rice husk were includes in this project to make an Eco-board as a specimen and a urea formaldehyde uses as a resin.

The tensile properties on Eco-board of hybrid composite will be observed using tensile tests equipment. The results of tensile tests are used in choosing materials for engineering applications [10]. Tensile properties commonly are included in material specifications to ensure quality. Tensile properties frequently are measured during development of new materials and processes, so that different materials and processes can be compared. In this study, the experimental of tensile test are be conducted with different ratio of EFB fiber and RH. Therefore, the best properties of tensile strength will be found to introduce to industry and the failure mode was investigated using scanning electron microscope (SEM).

2. Material and Methods

2.1. *Materials*

Oil palm empty fruit bunch and rice husk were supplied by Makmal Konkrit, Fakulti Kejuruteraan, Universiti Malaysia Sabah. The empty fruit bunches as solid waste have been discharged at palm oil after the elimination of oil seeds for oil extraction. The fibers are partially loose and partly paralleled lapping to each other in thickening formation through soft composition. It is not truly rounded. The EFB been cut to have requires 3 - 8 mm long fibers and then immediately treated with boiling water mixed with detergent powder to removes impurities and change the dull blunt brown colour to light-golden colour, which is similar to natural wood colour. However, the range of fiber length that be compressed is between 0.3 mm and 0.5 mm as shown in Figure 1. Rice husk is a rice processing obtained from the factory after it was separated from rice grain during the rice milling process. Figure 2 to shows the rice husk that have been dried before used as reinforcement material with EFB. The specimen of Eco-Board was formed by using cast iron mold with molding size is 150 mm x 150 mm x 150 mm as shown in Figure 3. Figure 4 was the oven that be used to heat the mold at temperature 120°C - 200°C.

IOP Conf. Series: Materials Science and Engineering **834** (2020) 012012 doi:10.1088/1757-899X/834/1/012012

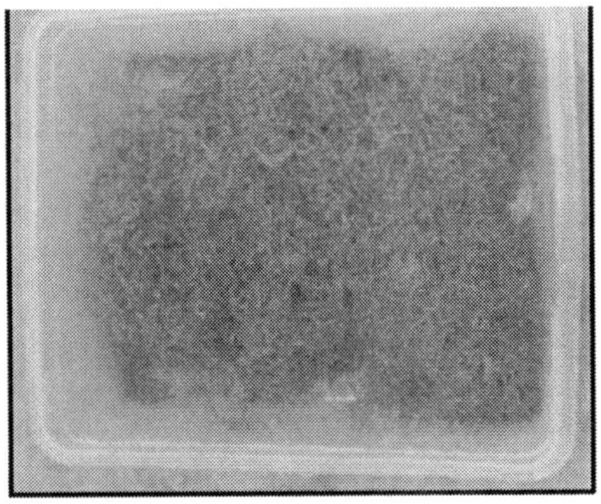

Figure 1. Empty fruit bunch (EFB) fibre

Figure 2. Rice husk (RH)

Figure 3. Cast iron mold

Figure 4. Oven

Table 1 shows the physical and mechanical properties of oil palm fibre that get from the Bismarck et al., Plant fibers as reinforcement for green composites. Table 2 is the properties of urea formaldehyde that get from Sepanggar Chemicals Sdn. Bhd. Kota Kinabalu, Sabah.

Table 1. Physical and mechanical properties of oil palm fibre

Properties	OPEFB Fibre
Density (g/cm3)	0.7 - 1.55
Tensile strength (MPa)	0.1 - 0.4
Young's Modulus (GPa)	1 – 9
Elongation at break (%)	8 – 18
Cellulose content (%)	49.6
Lignin content (%)	21.2

Sources: Bismarck A. Mishra S, Lample T. Plant fibers as reinforcement for green composites.

Table 2. Properties of Urea Formaldehyde

Viscosity / 30°C	164 cps
Specific Gravity / 30°C	1.1935
pH value	7.6
Free Formaldehyde	1.20%
Pot Life	75 Min
Solid Content	51.4%

Sources: Sepanggar Chemicals Sdn. Bhd, Kota Kinabalu, Sabah, Malaysia.

2.2. *Methodology*

2.2.1. *Specimen Preparation*
The specimens of Eco-Board were manufactured in the Makmal Konkrit, Fakulti Kejuruteraan, Universiti Malaysia Sabah through compression mold. Matrix material which is urea formaldehyde (UF) reinforced with empty fruit bunches (EFB) and rice husk (RH). These materials are mixed well in ratio of EFB to RH 10:0, 93.3:6.7, 86.6:13.4, 66.7:33.3 then mixed with 50 g of UF. This ratio determine by different weight of EFB and RH as shown in Table 3.

Table 3. The Ratio between Empty Fruit Bunches and the Rice Husk

Specimen	Empty Fruit Bunches (EFB), g	Rice Husk (RH), g
A	100	0
B	93.3	6.7
C	86.6	13.4
D	66.7	33.3

In the first process of manufacturing the Eco-board, EFB and RH were weight based on the ratio above and mixed well with the UF with manually hand to make sure it were in proper mixing. Then, the composite materials were put into a mould that has been heated in the oven before uniformly as shown in the Figure 5. After that, a steel plate is placed on the composite materials as a cover for a compression process to obtain flat surfaces. Compressive strength machine was used to compress the composite materials as a compression method for load around 300 kN. The mould was placed at the centre point position in the compressive strength machine and rod was put in the mold as a load to compress the composites as shown in Figure 6.

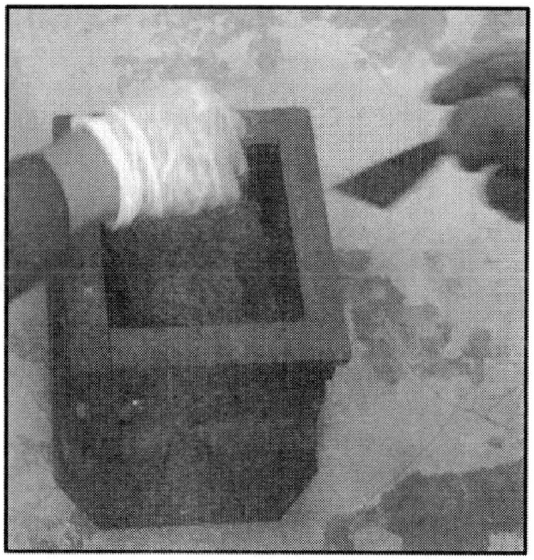

Figure 5. EFB fiber for specimen A that put into a mould

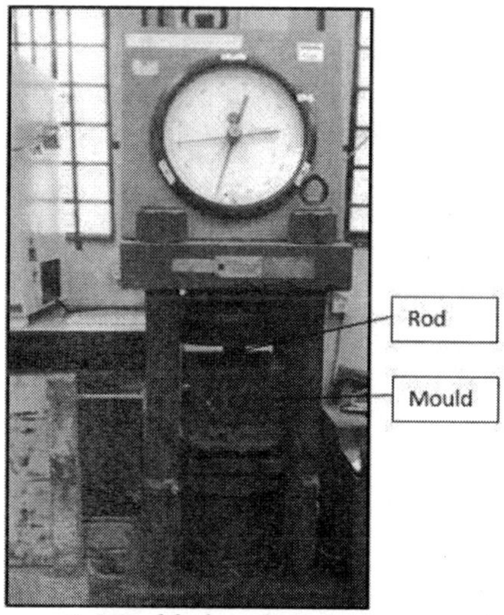

Figure 6. Mould placed on the compressive Strength machine

The composites was taken out from the mould and left at room temperature for a while. Then, the 4 board of dimensions of 150 mm X 150 mm X 5 mm with different ratios were manufactured as in Figure 7.

IOP Conf. Series: Materials Science and Engineering **834** (2020) 012012 doi:10.1088/1757-899X/834/1/012012

Figure 7. Four specimens of Eco-Board

2.2.2. Specimen Testing

The Figure 8 shows the Universal Cutting machine that been used to get the reshaped specimen with 150 mm X 25 mm X 5 mm dimension with 50 mm gauge length for tensile testing in accordance with ASTM requirements [11]. All the specimens are tested using GOTECH/AI-7000M Electronic Mechanical Testing machine with speed rate 1 mm/min. Specimens are placed in the grips of the machine at specified grip separation and pulled until failure. The machines will automatically produces a stress versus strain diagram, thus the mechanical behaviour of the composites can be interpreted from the diagram. The specimen will elongate as the tensile test starts. The load value (F) is recorded up to the point the specimen breaks. The instrument software which provided with the machine equipment will calculate the tensile properties such as the tensile, yield strength and elongation.

Figure 8 a). Universal Cutting

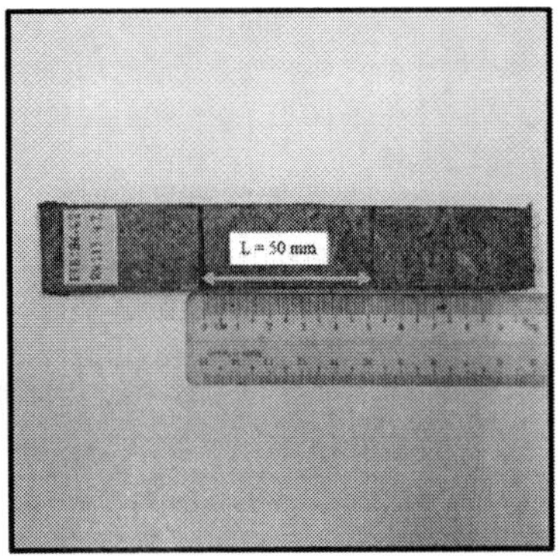

Figure 8 b). Reshaped specimen with 50mm gauge length

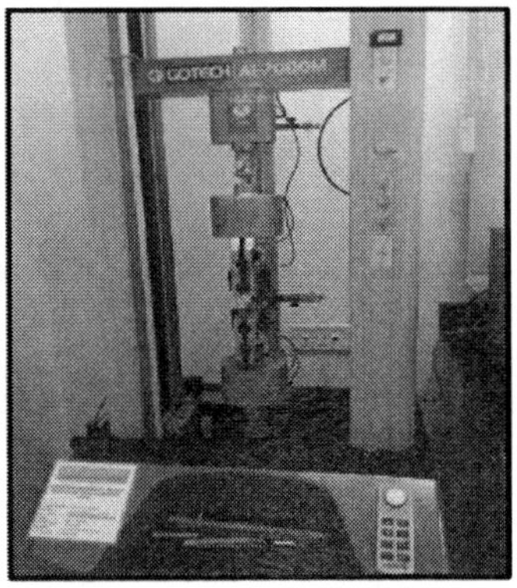

Figure 8 c). GOTECH/AI-7000M Electronic Mechanical Testing Machine

2.2.3. Scanning Electron Microscope (SEM)

Scanning electron microscope as shown in Figure 9 was be used in this project to study the surface of the four different composites samples after the tensile testing. SEM is a type of electron microscope that images of the sample surface is scanned with a high-energy beam of electrons in a raster scan pattern. The electrons interact with the atoms that make up the sample producing signals that contain information about the sample's surface topography, composition and other properties such as electrical conductivity.

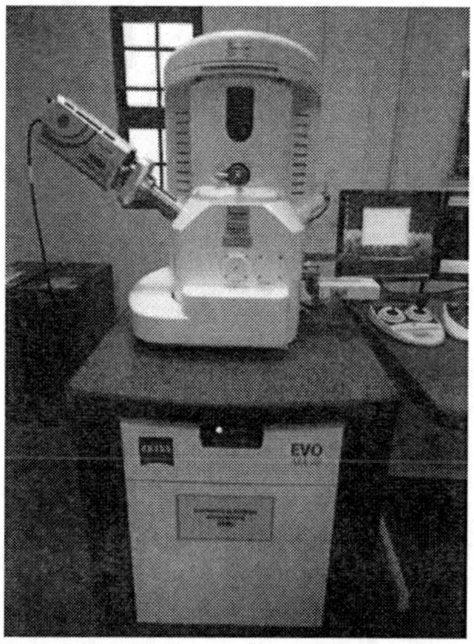

Figure 9. ZEISS EVO MA 10 Scanning Electron Microscope (SEM)

3. Result & Analysis

3.1. Tensile Properties

The maximum load, tensile strength and elongation of specimen as shown in Table 4 was interpreted from the stress versus strain diagram of automatically machine after the tensile test be done.

Table 4. Results of the Tensile Test

No	Specimen	Max Load (N)	Tensile Strength (N/mm^2)	Elongation (mm)
1	Specimen A (100g OPEFB: 0g ARTO)	581.79	4.65	2.94
2	Specimen B (93.3g OPEFB: 6.7g ARTO)	686.74	5.49	2.41
3	Specimen C (86.6g OPEFB: 13.7g ARTO)	1210.62	9.68	1.86
4	Specimen D (66.7g OPEFB: 33.3g ARTO)	713.14	5.71	1.45

Figure 10 represents the line graph load that applied to the specimen with different ratio of the empty fruit bunch (EFB) fiber and rice husk (RH) against the displacement. Based on the graph, it shows the load and displacement slightly increase and the testing machine performed under control of loading rate until it reach the maximum load level, a sudden failure of the specimens occurs. The machines gives small increments of load to the specimen and resultant deformation is measures, as result, when the incremental load goes over the maximum level, the specimen fracture suddenly. From the gathered data, it shows that the highest maximum load applied is at specimen C (86.6% EFB: 13.4% RH) which is 1210.62 N where the specimen can support and followed by specimen D (66.7% EFB: 33.3% RH) 713.14 N, specimen B (93.3% EFB: 6.7% RH) 686.74 N and the lowest is specimen A with pure EFB which is 481.79 N. It can be said that the addition of rice husk into the specimen will increase the inter-bonding relation within the materials as the maximum load applied keep on increasing. This can be supported by Irina M.M.W et al. [12] stated that hybrid composites has the ability to increase stiffness, reduced weight an also balancing up modulus properties and thus explaining the increasing of applied forces needed to break the composites specimen.

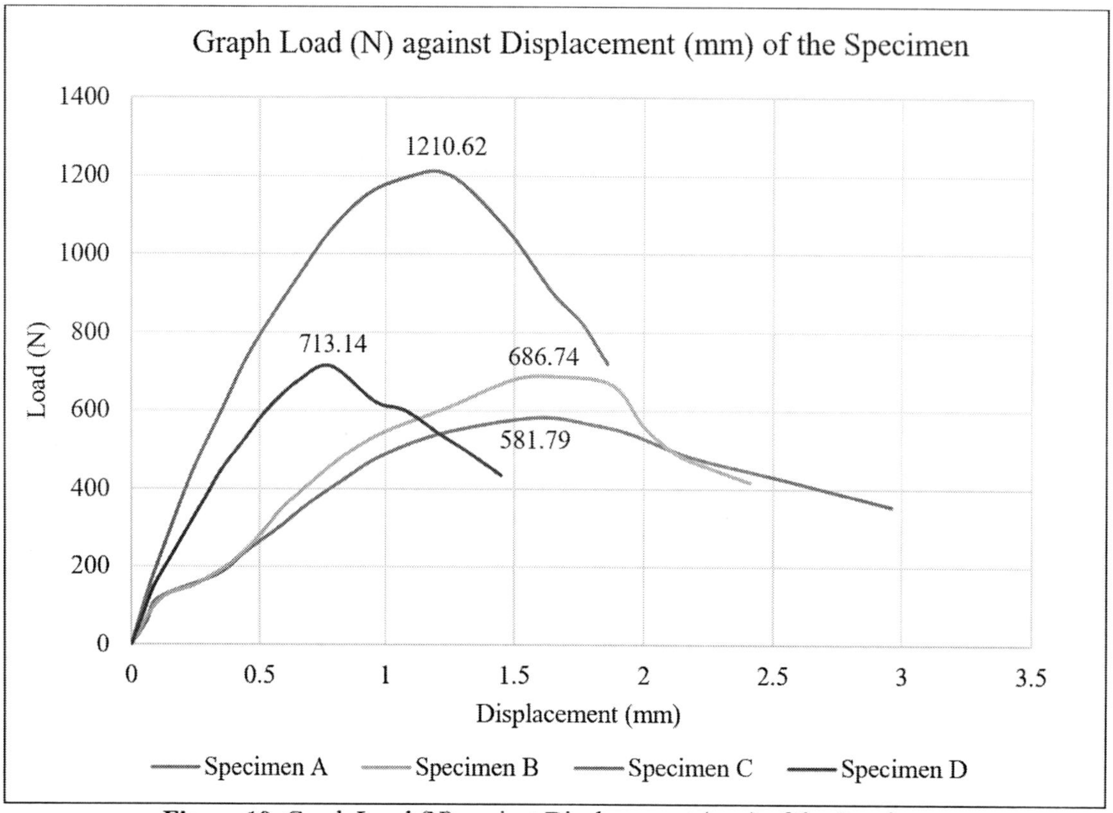

Figure 10. Graph Load (N) against Displacement (mm) of the Specimen.

Figure 11 shows the graph of tensile strength against type of specimen. The tensile strength of the specimen get from the calculation of maximum load from Figure 10 over the area, where an area of the each specimen is 125 mm2. A specimen gives the lowest maximum load so that it have the lowest tensile strength which is 4.65 MPa. As the specimen load move up from the yield strength, the stress on the specimen increases until it reach a maximum applied stress, where it can support the specimen. Then, the tensile strength follows by the specimen B (93.3% EFB: 6.7% RH) which is 5.49 MPa, specimen D, 5.71 MPa (66.7% EFB: 33.3% RH) and the highest maximum tensile strength is specimen C with 86.6% of EFB fiber added with 13.4% RH equal to 9.68 MPa.

From the gathered data, the pure empty fruit bunch (specimen A) gives the lowest tensile strength because it only have one mechanical properties of fiber acts on tensile testing. Different with the other specimen which the hybrid composite have the increasing stiffness, and balancing up the modulus properties of the specimen. Thus, the tensile strength is increase when it was hybrid with other composite rather it stand in pure EFB. This is supported by H Ku et al. [13] that said addition of fibre to composites matrices will improve the tensile strength due to the stiffness value and higher the strength of the fibres. In addition, with the addition rice husk into empty fruit bunch fibre improve inter locking bonding between materials thus improve its tensile strength. Khoathane et al. [14] found that the tensile strength and Young's modulus of composites reinforced with bleached hemp fibers increased in credibly with increasing fiber loading.

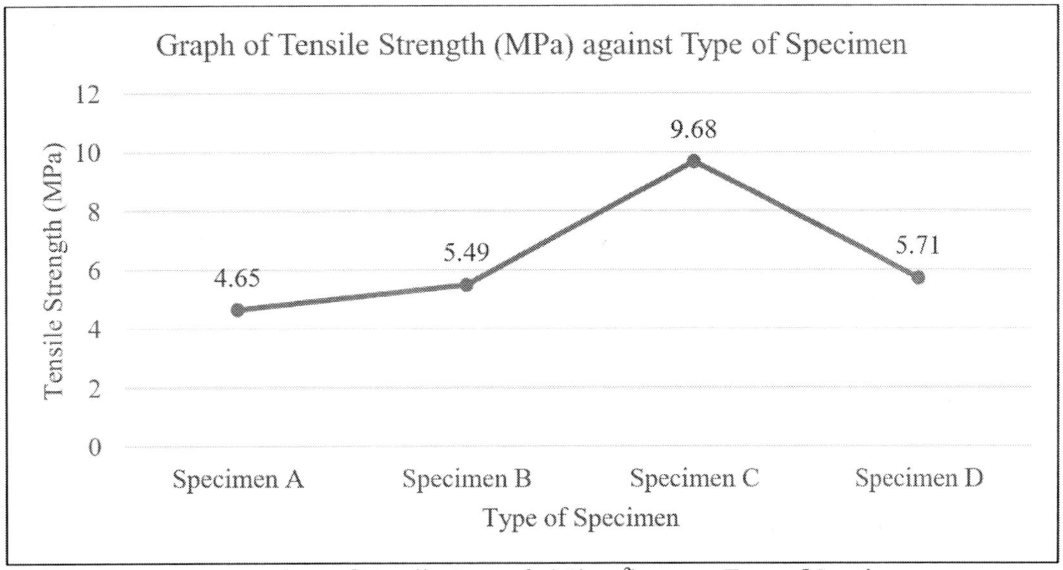

Figure 11. Chart of Tensile Strength (N/mm^2) versus Type of Specimen.

Figure 12 represents the chart of elongation of the Eco-Board until it breaks versus the type of specimen ratio. This data interpreted from the Figure 10, graph load versus the displacement of the specimen. The ductile specimen experiences significant deformation before it eventually breaks. The brittle of specimen typically have higher yield strengths than ductile materials. Based on the chart, it shows that a specimen A with 100% pure EFB gives the longest elongation which is 2.96 mm before the specimen start to break. Then, it followed by specimen B (93.3% EFB: 6.7% RH) with 2.41 mm, specimen C (86.6% EFB: 13.4% RH) 1.86 mm and the lowest elongation until it breaks is specimen D (66.7g EFB: 33.3g RH) with 1.45 mm. Based on the data obtained, it is clearly shows that the increasing of rice husk to the empty fruit bunch board will decrease the ductility of the Eco-Board on the same time increase the brittleness of the specimen. G. Jordi et al. [15] found that the addition of any filler to composites will lead to decrease of elongation at yield and break. Therefore, a specimen D is a most brittle and the specimen A has most ductility material.

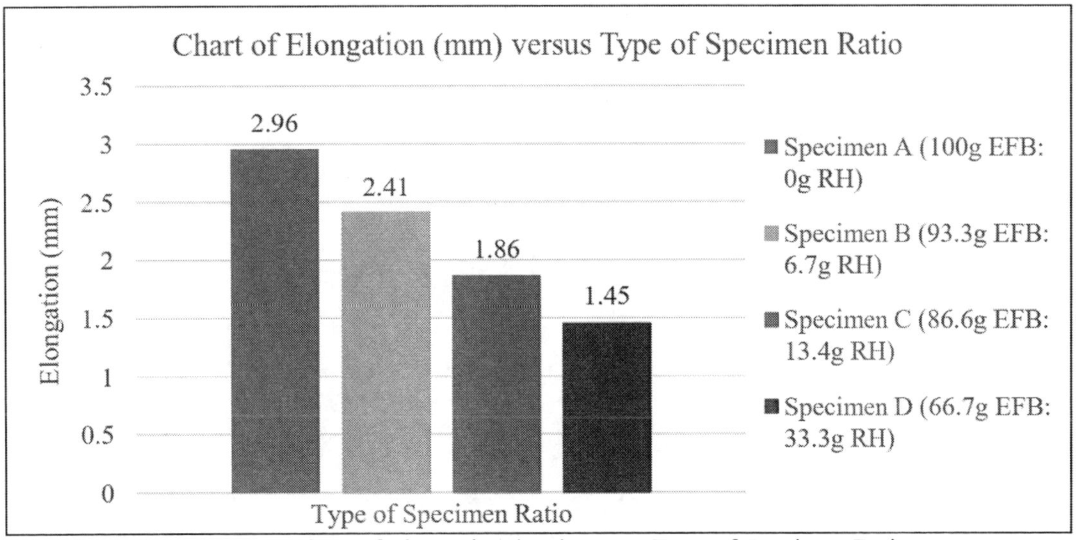

Figure 12. Chart of Elongation (mm) versus Type of Specimen Ratio

3.2. *Scanning Electron Microscope (SEM)*

The crack propagation and the morphology of surface structure of oil palm empty fruit bunches with rice husk in four different ratios (100:0, 93.3:6.7, 86.6:13.4, and 66.7:33.3) were investigated by SEM. Figure 13 shows the micrograph of surface failure under tensile test for specimen A (100g EFB: 0g RH). The Figure 13 (a) is under 80X in magnification and the other figures are under 25X in magnification. Based on these figure, the small crack growth appeared on the surface. For this pure 100% EFB shows that there are much void that occurred on the surface structure. Without any combination of RH, the void for specimen is much easy to be occurred. From this figure, the 100% pure EFB explained that the matrix and the fibre are more likely to having gap/void between them even the fibre tries to hold up the uniform matrix by carrying loads. Hence, this figure also shows that the fibre is not good in orientation. The stresses transfer from matrix to fibre in a composite that take place by shear at the fibre-matrix interface. Thus, strong interfaces will give high strength and stiffness but low in fracture toughness composites [16].

It is difficult to assess the fibre breakage and roughness of the surface with macroscopy to confirm the identification of the fracture failure. Therefore, microscopic is necessary to continue the analysis. At crack origin, a set of fibre bundles was found to be broken, shows on Figure 13(c). This acted as a stress concentrator for initiation and propagation of crack under tensile test. On Figure 13(d), shows that the fibre pulled out or no matrix between the fibre. The failure occurred along the fibre-matrix interface leads the fibre to pull out on further straining [17]. It show that, the strong the interfacial bond strength causing strong stress concentration and a tendency for crack to pass through large bundles of fibre without much deviation [18]. By analyzing figure above, it was observed that the random discontinuous of fibres can decrease the interfacial bond strength between fibres and matrix [19]. Hence, it is seen that the composite were not bonded strongly with each other and has high possibility to separate as shown in figure above.

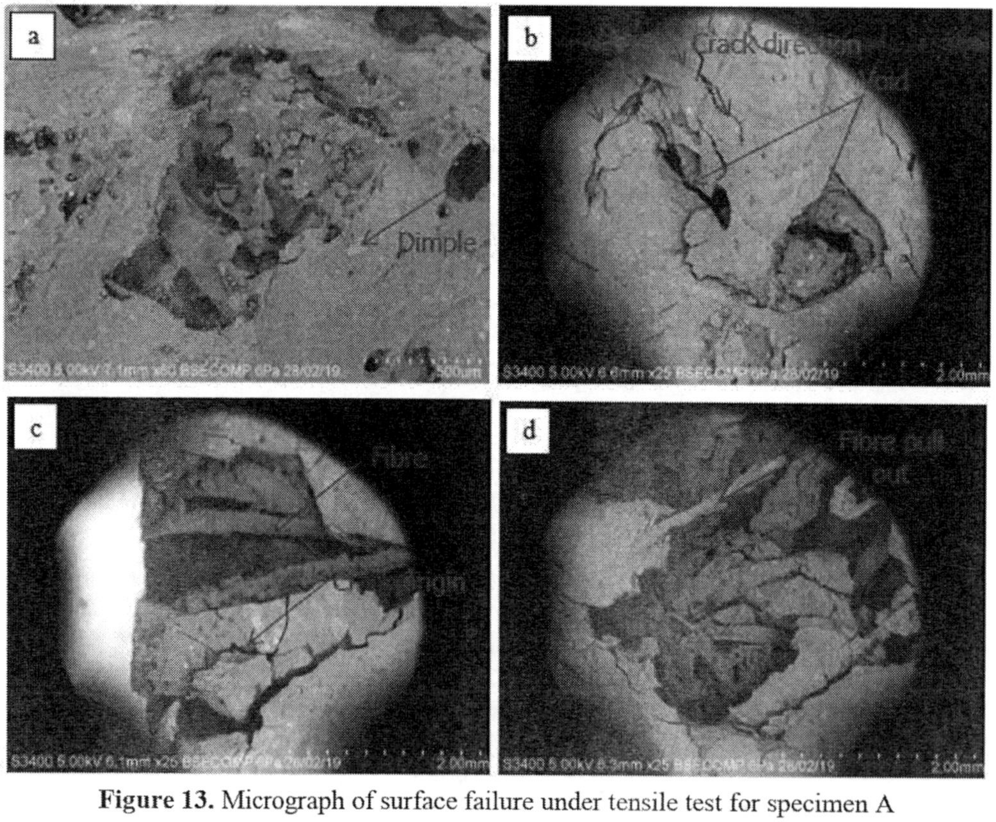

Figure 13. Micrograph of surface failure under tensile test for specimen A

Figure 14, shows the morphology structure on surface for specimen B (93.3g EFB: 6.7g RH) under 55X in magnification for Figure 14(a) and (b). While the Figure 14(c) and (d) under 75X and 40X in magnification respectively. Based on this figure, it shows that the void is less occurred on the surface structure for this ratio. This is because of the mixed with rice husk. The combination of oil palm empty fruit bunches with rice husk reduced the gap between the fibre and the matrix itself.Based on Figure 14(a), it shows that the matrix and the fibre are in good combination with no fibre pulled out or broke. However, the fine cleavage still occurred on the surface structure. Figure 14(c) the fibre cannot sustain the increased load, so the fibre will fail, leading to a single fracture surface. This figure is in unidirectional fibre-reinforced composite. As can be seen, the failure leads to fibre pulled out and thus further energy absorption [17]. Figure 14(d), represents that the fibre fails in fibre splitting. It is uncommon for fibre to develop kink band or splitting on the compression of side bent fibres. These cases, the energy absorbing mechanisms that contribute to the failure process are fibre pull out and fibre-matrix debonding [17].

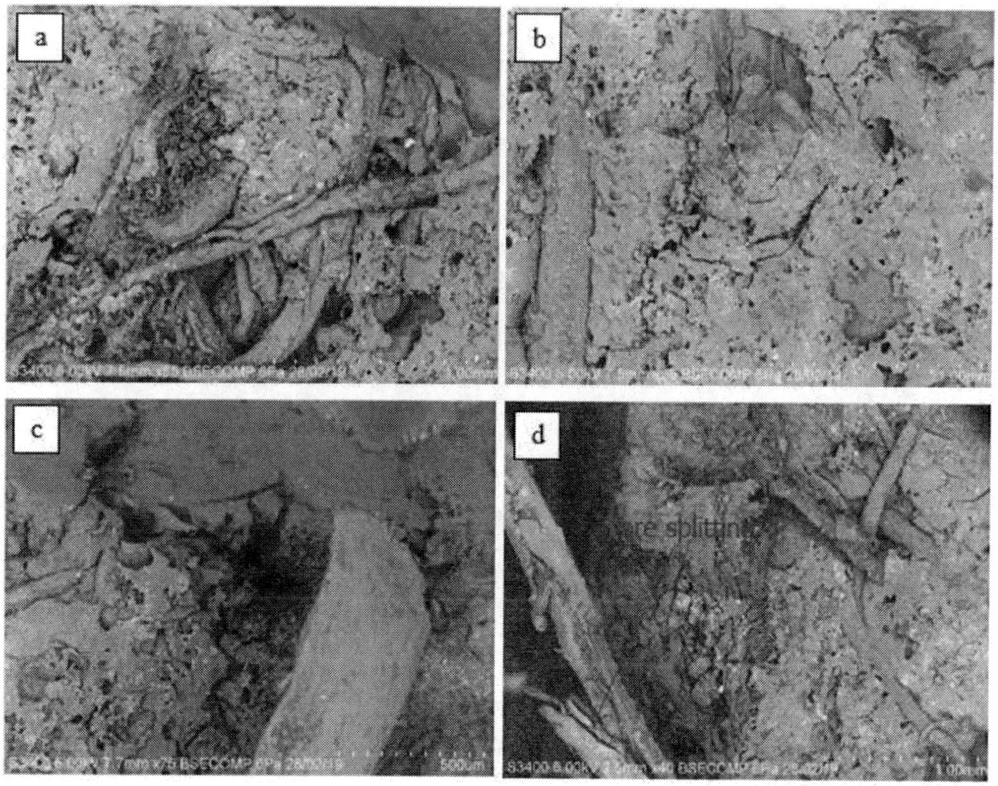

Figure 14. Micrograph of surface failure under tensile test for specimen B

Figure 15, above shows the morphology structure of the surface for specimen C (86.6g EFB: 13.4g RH). Figure 15(a) shows the SEM micrograph is under 47X in magnification, while Figure 15(b) under 65X, (c) under 30X and (d) is in under 32X in magnification. Based on this figure, it shows that there is no void occurred surface structure. However, it can be clearly see that fine dimples are occurred on the Figure 15(b). The region of initial crack growth was surrounded with fine dimples Figure 15(a). It shows that the small arrows indicate cracking on the surface. This combination ratio OPEFB with RH shows that, the RH reduced the gap between the fibre and the matrix. Chevron is usually formed from river line or also known as textured microflow. The riverline are found on this figure in the matrix adjacent to a fibre (in mode I fracture) interlaminar tensile fracture or in resin-rich regions [20].

Based on Figure 15(c), it shows that fracture occurs because of defect fibre pull out. It indicate that the random discontinuous improved the interfacial bond strength between fibre and matrix decreasing the value of the flexural strength of the composite. Some load transfer between fibres and matrix is possible by interfacial forces due to matrix shrinkage into the fibres. After the fracture, the composite typically shows a matrix-crack plane. Analyzing Figure 15(c) it was observed the random discontinuous of the composite fibre, which can decrease of the interfacial bond strength between fibre and matrix [19]. Figure 15(d) shows, the fracture occurred because of the fibre bridging [20]. When the bundle of fibres is tied between two adjacent surfaces, it is a phenomenon of fibres bridging. It is prevalent in unidirectional ply surfaces and is one of the main features in mode I interlaminar tensile fracture. It is depends highly on the matrix and fibre interface quality and on the amount of resin in this region [20]. Matrix-wing phenomenon can also happen due to fibre debonding. Pieces of matrix adhere to the fibre after debonding.

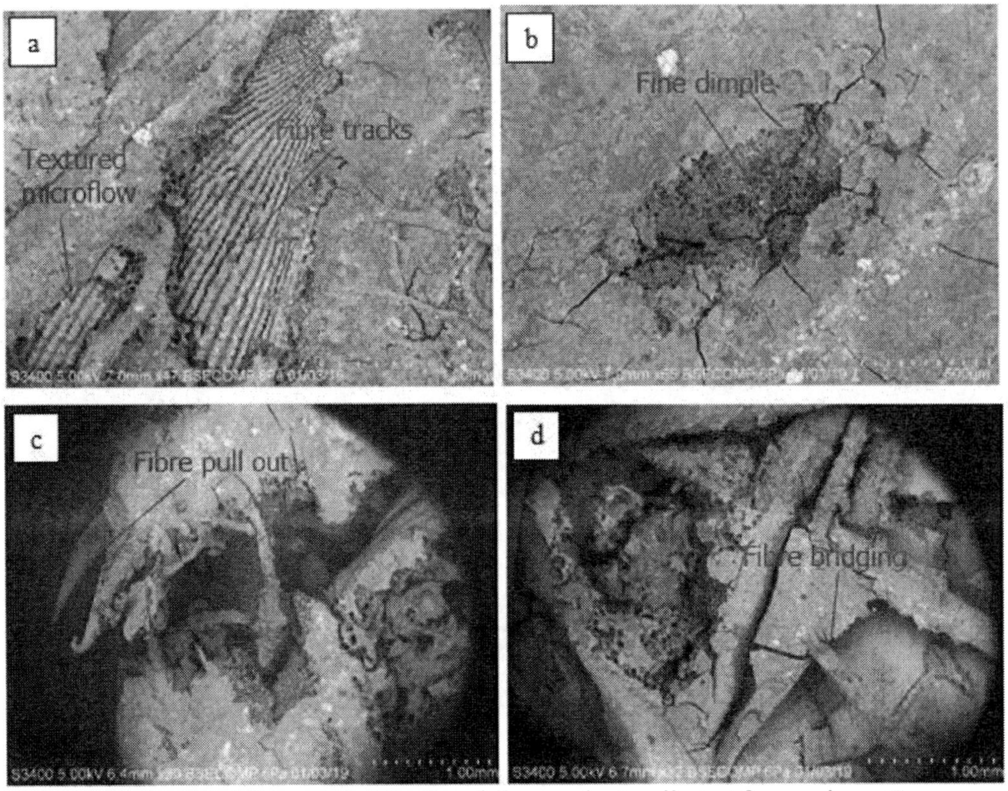

Figure 15. Micrograph of surface failure under tensile test for specimen C

Figure 16(a), (b) was analyzed their fracture surface under SEM, with 85X in magnification while Figure 16(c) and (d) are under 40X and 27X in magnification respectively. As seen in figure (a), the fracture occurred because of the fine dimple. Appearance dimple on the surface structure will causing weak boundary. Figure 16(d) shows that it is well known that the fracture surface morphologies render the interfacial characteristic of the composites. Extensive fibres pull out usually indicate that a relatively week fibre or matrix interfacial bonding, while little fibres pull out and short fibre pull out length illustrate a strong fibre or matrix interfacial bonding. Some load transfer between fibres and matrix is possible by interfacial forces due to matrix shrinkage into the fibre. The friction produces a non-uniform stress along the debonded fibres [19]. This will cause the fibre easily to break.

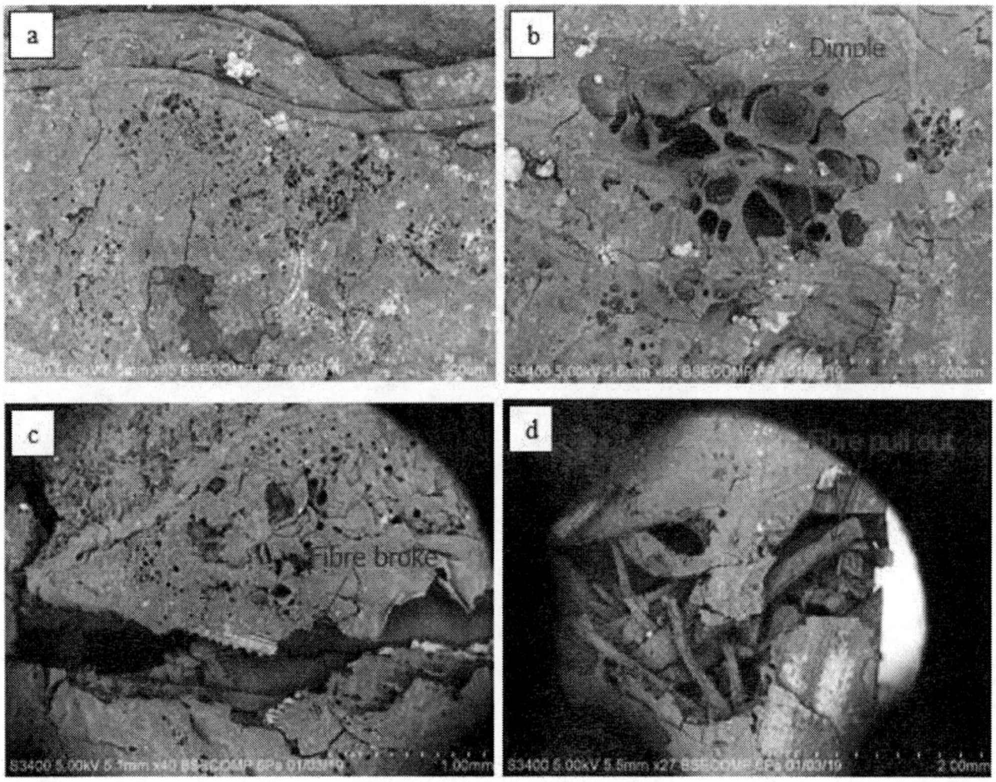

Figure 16. Micrograph of surface failure under tensile test for specimen D

4. Conclusion

It is determined that addition of the rice husk to Eco-Board increase the tensile properties of the board and at the same time decreasing the ductility of the specimen as a elongation of the Eco-Board keep on decreasing proportionate to the increasing the amount of rice husk added to the specimen. The ultimate forces also can be seen that increase on the hybridize specimen if compared to the pure specimen up until certain ratio where addition of more rice husk will decrease the ultimate forces. From the bonding and failure mode between fibre and matrix that studied through SEM, it can be conclude that the poor interfacial bonding will lead the fibre and matrices to break. Overall, the hybridization of Eco-Board with rice husk definitely increase some mechanical properties of material. So that it can be used to replace the usage of conventional woods in industry and the possible application that suitable for this composite material is fibre board, decoration and as furniture.

References

[1] Jawaid, M. (2010). Mechanical Performance of Oil Palm Empty Fruit Bunches/Jute Fibers Reinforced Epoxy Hybrid Composites. Materials Science and Enginering. 572(29-30)

[2] Rowell, R.M., Han, J.S. & Rowell, J.S. (2000). Characterization and Factors Effecting Fiber Properties. Natural Polymers and Agro Fibers Composite.

[3] Sanjay M. R., Arphita G. R., Yogesha B. (2016). Studies on Mechanical Properties of Jute/E Glass Fiber Reinforced Epoxy Hybrid Composites. 15-25.

[4] Ramesh M., Logesh R., Manikandan M. et al. (2017). Mechanical and Water Intake Properties of Banana-Carbon Hybrid Fiber Reinforced Polymer Composites. Material Research, 20(2), 365 376.

[5] BioCentrum-DTU. (2008). Ethanol Potential for Empty Fruit Bunches Pre-treated by Wet Explosion. A report prepared under the Malaysian – Danish Environmental Cooperation Programme Renewable Energy and Energy Efficiency Component, Denmark's Technical University.

[6] Arjmandi R., Hassan A., Zakaria Z. (2017). Rice Husk and Kenaf Fiber Reinforced Polypropylene Biocomposites. Lignocellulosic Fibre and Biomass-Based Composite Materials: Processing, Properties and Applications, 77-94.

[7] Isa, K., Daud, S., Hamidin N. et al., (2011). Thermogravimetric Analysis and The Optimisation of Bio-oil Yield from Fixed-bed Pyrolysis of Rice Husk using Response Surface Methodology. Industrial Crops and Products, 33 (2): 481-487.

[8] Shafie S.M. (2015). Paddy Residue Based Power Generation in Malaysia: Environmental Assessment Using LCA Approach. 10(15), 6643-6648.

[9] Temitope A., Onaopemipo A., Abayomi O. et al. (2015). Recycling of Rice Husk into a Locally Made Water-Resistant Particle Board. Industrial Engineering & Management, 04(03).

[10] Davis J.R. (2004). Tensile Testing, Second Edition.

[11] American Society for Testing and Materiasl (ASTM) International Standard D3039 (2018).

[12] Irina M.M.W , A.I. Azmi, Tan C.L. , Lee C.C. and A.N.M Khalil, "Evaluation of Mechanical Properties of Hybrid Fiber Reinforced Polymer Composites and Their Architecture", Universiti Malaysia Perlis (UniMaP), 2015.

[13] H Ku , H Wang, N Pattarachaiyakoop and M Trada, "A review on the tensile properties of natural fibre reinforced polymer composites" Centre of Excellence in Engineered Fibre Composites and Faculty of Engineering, University of Southern Queensland, pp. 3 – 22, 2011

[14] Khoathane, M C, Vorster, O C and Sadiku E.R, Hemp Fiber-Reinforced 1-Pentene/Polypropylene Copolymer: The Effect of Fiber Loading on the Mechanical and Thermal Characteristics of the Composites, Journal of Reinforced Plastics andComposites, 2008, Vol. 27, pp.1533-1544.

[15] G. Jordi, Loan T. T. Vo, D.G. Erika and Patrick Navard, "Natural filler-reinforced composites: comparison of reinforcing potential among technical fibers, stem fragments and industrial by products', Mines ParisTech, PLS – Research University, 2018.

[16] E. E. Gdoutos, K. Pilakoutas, Chris rodopoulos, Failure Analysis of Industrial Composite Materials, McGraw Hill (2000).

[17] Charles Brooks, Ashok Choudhury, Charlie R. Brooks, Failure Analysis of Engineering Materials, McGraw Hill (2002).

[18] D. Hull and T. W. Clyne, "An Introduction to Composite Materials", Cambridge University Press, Cambridge, 1996.

[19] Muftil Badri, Sugiman, Aguswandi, Jayadi, SEM Observation on Fracture Surface of Coconut Fibers Reinforced Polyester Composite, Journal of Ocean, Mechaanical and Aerospace, Science and Engineering, Vol.40, 2017.

[20] Baas, S. J. W., (1994). AG14, Fractography of composites, final report, s.l.: GARTEUR

A Hybrid Haptic Feedback Stimulation Device to Recover the Missing Sensation of the Upper Limb Amputees

Mohammed Najeh Nemah[1,2], Omar Hammad Aldulaymi[1], Cheng Yee Low*[1], Noor Ayuni Che Zakaria[3] , Shahrol Mohamaddan [4]

[1]Faculty of Mechanical and Manufacturing Engineering, University Tun Hussein Onn Malaysia, 86400 Parit Raja, Batu Pahat, Johor, Malaysia.
[2]Engineering Technical College-Najaf, Al-Furat Al-Awsat Technical University, 32001, Najaf, Iraq.
[3]Faculty of Mechanical Engineering, Universiti Teknologi MARA, 40450 Shah Alam, Selangor, Malaysia.
[4]Department of Mechanical and Manufacturing Engineering, Faculty of Engineering, University Malaysia Sarawak, Malaysia.

*cylow@uthm.edu.my

Abstract. Providing sensory feedback for prostheses users increases the manipulation ability, as well as introduces a feeling of the embodiment. A novel hybrid haptic feedback stimulation system of the amputees of upper limb mutilation was designed, developed, and evaluated in this study. The haptic wearable device was built by mean of single servomotor as a pressure feedback display and a pair of vibration motors as a vibration feedback display. Accordingly, the pressure display is responsible for conveying the missing sensing of the contact pressure and its force level. While the vibration display has utilized to provide an indication about the continued contact pressure. Able-body participants are engaged to evaluate the performance of the proposed feedback stimulation system. The results show that the participants are able to distinguish the touch, the start of touch, the end of touch, and the range of force with 100 % accuracy. While 96 % and 88 % of the able-body participants are capable to identify the grasp and the slipping objects, respectively. Furthermore, the ability of the proposed haptic wearable device to convey the tactile sensory information to the user's brain without confusing or pre-training was improved.

Keywords: Feeling recover; Contact pressure detection; sensory system; Tactile glove; Upper limb prostheses.

1. Introduction

In the US alone there are more than 1.6 million people living with the loss of a limb with upper-limb loss accounting for 68.6% of trauma related amputations and 58.5% of congenital birth defects [1]. The upper limb prostheses have been used to help the amputees who are unfortunately missing there healthy hands. The prosthetic hands were designed to assist the amputees to touch and grasp objects, however, the lack of feeling through the use of prosthetic hand still the main design's issue, in order to make the amputees in continues interacting with their surroundings.

Content from this work may be used under the terms of the Creative Commons Attribution 3.0 licence. Any further distribution of this work must maintain attribution to the author(s) and the title of the work, journal citation and DOI.
Published under licence by IOP Publishing Ltd

IOP Conf. Series: Materials Science and Engineering **834** (2020) 012013 doi:10.1088/1757-899X/834/1/012013

Several previous works developed the technologies of the tactile displays that used to stimulate the residual skin and nerves system of the amputees [2]. These technologies have ranged from non-invasive approaches using vibrators [3] or pressure [4] to more complex approaches providing spatially-mapped tactile displays of pressure, vibration, shear force, and temperature [5] in subjects who have undergone targeted re-innervation surgery [6]. Furthermore, the previous researchers are going to gathering two or more types of the haptic feedback system named the hybrid feedback stimulation system [7-11]. In general, a common requirement from amputees is to be able to operate prostheses without constant visual attention [12, 13], and proprioceptive feedback has been shown to improve targeting accuracy under non-sighted conditions [14].

The main purpose of this study is to present the first steps towards design a hybrid haptic feedback stimulation system (HHFSS), which has the ability to help the patients of the upper limb amputation to recognize multi-information about their surroundings without brain confusing or a pre-training requiring. The haptic device was designed to deliver the sensation of the contact pressure to the amputees' brain by a multi-site pressure and vibration displays applied on specific spots on their residual limb. The pressure display used a single servomotor where the contact pressures on the fingertips of the prostheses were delivered as pressure on the skin. Meanwhile, the vibration display translated the pressures on the fingertips into vibration sensation by mean of two vibration motors. To the best of our knowledge, this is the first study to combine the haptic pressure and vibration feedback displays for upper limb prostheses, in the way that allows the stimulators to operate in different directions and various excitations signals but performing the same purpose, i.e. delivering of the tactile sensory information to the users in a quick manner.

2. Design conception of haptic feedback stimulation system

In general, a HHFSS of the prosthetic hand can be sorted into three principal parts: collect the data of the amputees' environment by using different kinds of sensors in the haptic system, process the sensitive data, and control the feedback stimulators to stimulate the amputee's understanding using a computer system, as shown in Figure 1. Indeed, the HHFSS excites the amputees' nervous system by stimulating the skin of their remaining parts such as the forearm, upper limb, or any additional parts of the body.

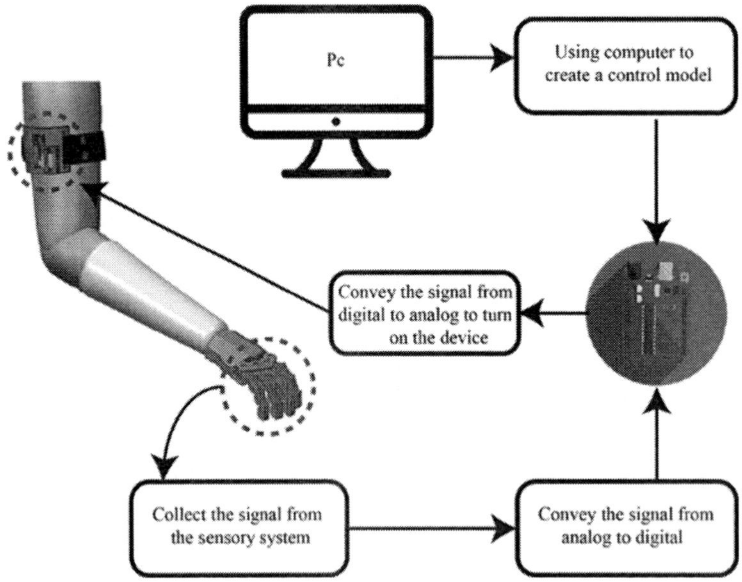

Figure 1. The design concept of HHFSS.

Accordingly, six Quantum Tunnelling Composites (QTC) force sensors were utilized to detect the contact pressure, five sensors were attached to fingertips and one force sensor was fixed on the hand's palm. The device was designed to excite the amputees' nervous system with a hybrid sensation of pressure and vibration stimulations at the same time by a mean of two circular vibration motors and one mini servomotor.

The obtained signals from the haptic glove system were conveyed to the control process system by utilizing an Arduino Mega microcontroller. Subsequently, those six force signals received from the QTC force sensors were compared to produce the highest instant output signal. The highest instant signal of entire sensory signals was passed through the control system as the input actuators signal of the haptic system. Depending on the input actuators signal, the control processor system manages the actuators of the HHFSS by exciting both the pressure feedback display and the vibration feedback display. The PID controller was used to control the excitation actuators at the high response and acceptable accuracy [15].

The purpose of the hybrid haptic feedback stimulation system is to excite the residual upper limb of amputees and convey the feeling of the touch, the start of touch, end of touch, grasp, slippage, and force range to the patient's brain. Therefore, the hybrid haptic feedback system has been designed to increase the ability of the amputees' brain to understand the huge quantity of data produced by the tactile glove system.

Regarding the haptic wearable device, three actuators have been fixed on 70 cm diameter and 8mm depth curve case, as displayed in Figure 2. The curve case was designed using Solidwork 2018 and fabricated by a 3D printer using ABS material [16]. The feedback stimulator device includes one servomotor [17] for the pressure excitation and two Linear Resonant Actuator (LRA) vibration motors [18] for the vibration excitation. An auxiliary of piston of 15 mm diameter and a piston's arm of 33 mm length were used with the servomotor to produce the pressing force.

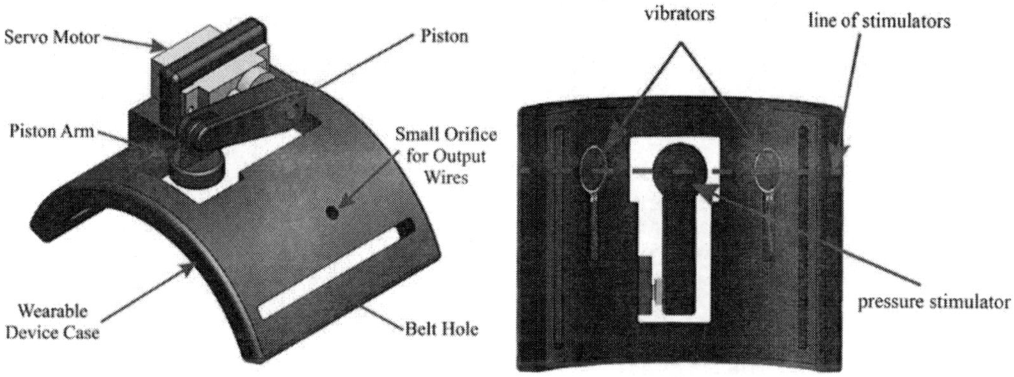

Figure 2. The design conception of the haptic wearable device.

3. Fabrication of the tactile pressure glove

A plastic glove with QTC pressure sensors was created as the tactile glove system of the HHFSS. The tactile glove's design, fabrication, and evaluation examinations were considered in our previous work [19]. In general, the haptic glove was fabricated from a plastic material. While each finger covered with hard support material under the pressure sensors, in order to limit the elasticity of volunteer's skin from changing the recording of the pressure sensors. Three significant advantages based on this kind of fabrication of the tactile system are: (i) the volunteers can wear the haptic glove system easily and push their fingers' links evenly because the glove is manufactured from the flexible material, (ii) the hard material which covered of fingertips and the ball hand form the basis of the pressure sensors satisfying to work with great efficiency, and (iii) the glove is suitable for different hand sizes because of its expandability performance.

The solid shield was designed using the Solidwork 2018 software and printed by a 3D printer. The QTC force sensors with 10 mm diameter and 0.1 N to 20N acting force limit from Peratech [20]; were attached on critical points of the hand. However, five sensors have been fixed on each fingertip, and another sensor was placed on the hand's palm, as described in Figure 3. Such arrangement is used to ensure that all critical areas of the hand are successfully covered, that's lead to enhance the possibility to detect the contact pressure.

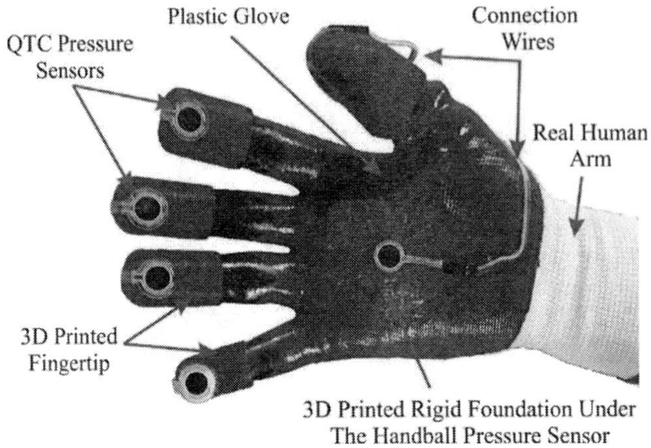

Figure 3. The fabrication of the tactile glove.

4. The experiment setup and functionality test

The pressure level experiment has been accomplished to evaluate the functionality of the HHFSS. The evaluation of pressure level experiment has been performed by covering the finger of the haptic glove to a column cantilever. After that, a hanging mass was attached vertically to the force sensor, which was fixed on the finger. The hanging mass was increased from 0 to 2 kg (i.e. from 0 to 20N) by adding 200g for each increasing step. The main goal of this experiment is to check the responses of the servomotor and the two vibration motors due to the rising in the force sensor signal.

For the stimuli identification experiment, twenty-five healthful participants have been volunteered in this project, mean age (25-34 years). The volunteers wore the haptic glove and HHFSS on their right arm, as shown in Figure 4. The eyes cover was utilized to avoid any visual vision during the test, in order to impact the brain for completely depending on the haptic stimulation.

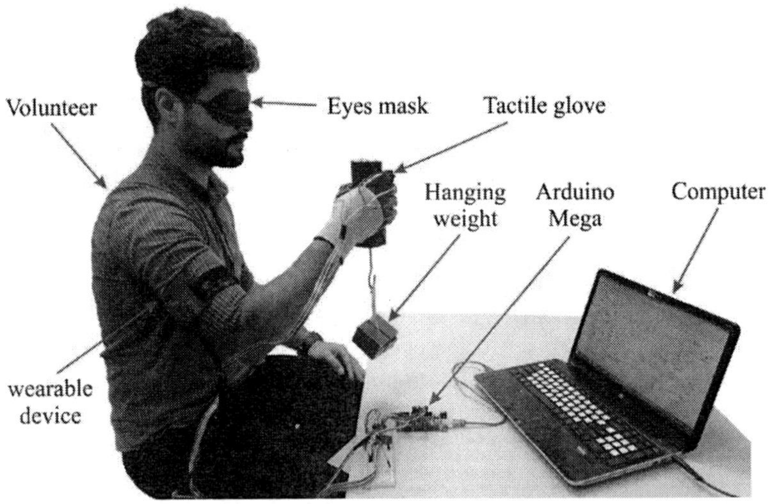

Figure 4. The experimental setup.

The main goal of the experiments is to examine the volunteers' stimuli identification rate (SIR), which is the rate that the participants could rightly recognize the haptic information through simply using the hybrid haptic feedback stimulation system. Three different tests were performed to rightly check the effectiveness of the proposed design and its ability to convey multi tactile information to the user's brain at the same operation time.

At the first test, irregular forces were applied to the force sensors of the tactile glove. The volunteers were asked to press their fingers on the table and touch several objects. The volunteers have to answer whether they can distinguish the touch, start of touch, and end of touch. This is to evaluate the efficiency of the wearable device in transferring information to the volunteer's brain by mean of the haptic wearable device.

At the second test, evaluating the ability of the participants to identify the grasping objects was verified. The examiner asked the participants to grasp objects in different sizes and make a handshake with him. This is to test the functionality of a HHFSS to identify the contact pressure during grasping objects.

At the third test: the slipping object experiment was performed by using a columnar pipe equipped with a steel hook from its bottom side, as described in Figure 5. The columnar pipe was carried up by the volunteer. Then, variable masses were attached to the hook of the cylindrical pipe. The hanging load was increased gradually from 0 to 5.5 kg, 0.5 kg for each time. The main purpose of this test is to verify the functionality of the HHFSS to recognize the slipping objects. The slip detection test was conducted by asking the volunteers to grasp and handle the columnar pipe, while the suspended weight was raised slowly till the pipe slid down from hand. On the other hand, the second purpose of this test is to measure the ability of the volunteer to distinguish the sense of the increase in the contact pressure.

The experimenter evaluated the experimental observations as follows: yes, if the participants were able to identify the kind and the level of the excitation through the test, no, if the participants were unclear or inefficient to separate the haptic information.

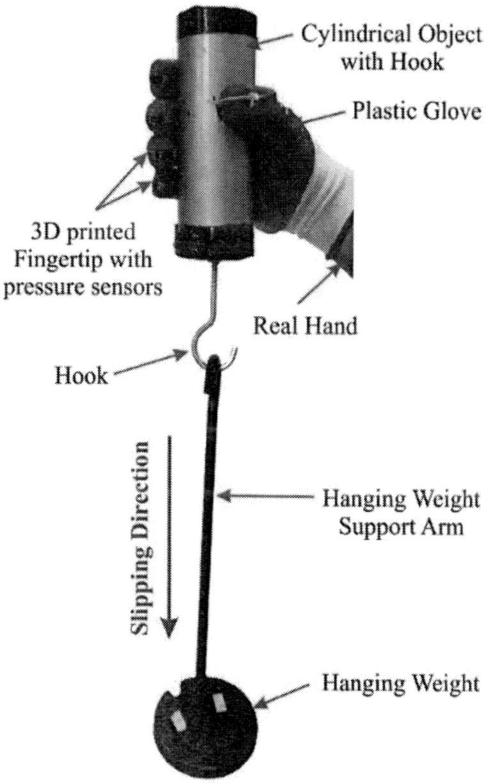

Figure 5. The slipping object experiment.

5. Results

For the pressure feedback display of a hybrid haptic feedback stimulation system, Figure 6 presents the linear association between the servomotor rotational position and the highest haptic glove signal during the pressure level experiment. The rotational action of the servomotor was varied from 0-35 degree when the force signal of the tactile sensory system raised from 0-5 V, i.e. the load was increased from

0-20N. Based on the servomotor's maximum rotational movement, which is 35 degree, and the length of servomotor's piston arm, the perpendicular displacement of the servomotor's piston was calculated, which is equal to15.6 mm. This indicated that the pressure feedback display of a HHFSS is able to deform the skin of the user's upper arm within 15.6 mm perpendicular displacement. This amount of displacement has the ability to stimulate the patient's nervous system and notify the patient's brain about the changing weight. This also means that this is the highest value to overcome the dead layer of the patient's skin.

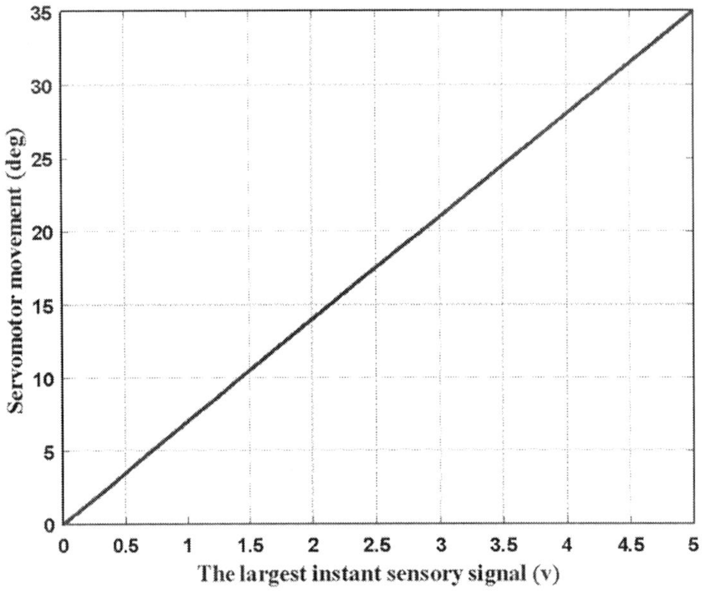

Figure 6. The increasing of the servomotor angle against contact pressure racing.

For the vibration feedback display, a repeatable pulses signal with 3V maximum amplitude, 0.2 sec period time, and a pulse width equal to 50% of the period time was designed to drive the two vibration motors, as shown in Figure 7. The special pulses vibration motor signal was designed based on several stimulation trials, in order to obtain the best feedback stimulation from the vibration feedback display without any interference with the noise frequency of the servomotor.

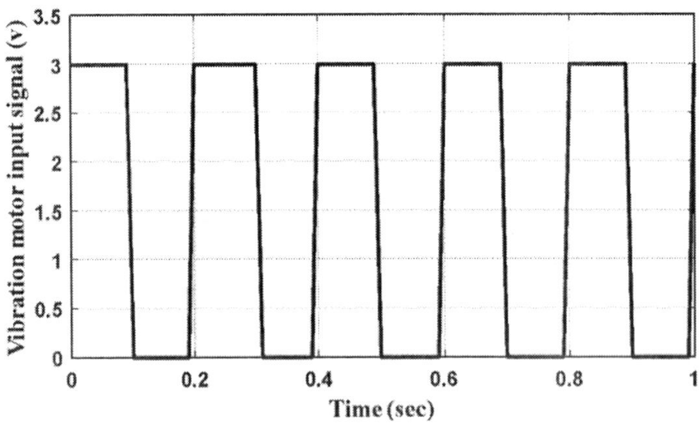

Figure 7. The excitation signal of the vibration motors during 1 sec.

The excitation signal of the vibration motors was designed to work with constant frequency when the largest instant pressure signal varying from 0-5V. In other words, the excitation frequency of the vibration motors works undependably on applied contact pressure. The main reason behind this type of the excitation is to inform the user that the touch or grasp action is started and the still continues. The excitation signal was programmed to start working when the sensory signal exceeds 0.5V, in order to avoid activating the vibration feedback display when the phantom signals generated by the pressure sensors. The behaviour of the excitation pulsing signal and the vibration motors' excitation signals among the large instant pressure signal of the tactile sensory system were described in Figure 8.

The stimuli identification rate (SIR) was measured for all experimental tests, which evaluated the performance of a hybrid haptic feedback stimulation system for conveying the tactile data to the volunteers' brain, as illustrated in Figure 9. The results showed that all volunteers were able to realize the touch, start of touch, end of touch, and the force range with 100% accuracy. Furthermore, the lowest recognition accuracies of 88% and 96% were attained for the test cases of grasping objects and the slippage, respectively. The low identification percentage of the slippage test was related to the design of the sensory glove itself. A stable sliding sensor is not available in the sensory glove in order to notice the slippage. It simply depends on the organization among the pressure sensors and the replies of the participant's nervous system.

The contributions of this work can be proved by performing a comparison study between the SIR results of the current study and previous studies, as summarized in table 1. The selected previous studies were selected as similar as possible to this work, in which spot pressure sensors were utilized in the respective works. On the other hand, previous studies used different types of feedback stimulation system, such as pressure, vibration, electro, squeeze, and skin stretch to compare the functionality of the designed wearable device with all types of the existing feedback stimulation systems. In addition, table1 also summarizes the information about the installation position of the previous haptic feedback displays and type of subjects. The SIR results indicated that the current work present the best results in all the experiment examinations.

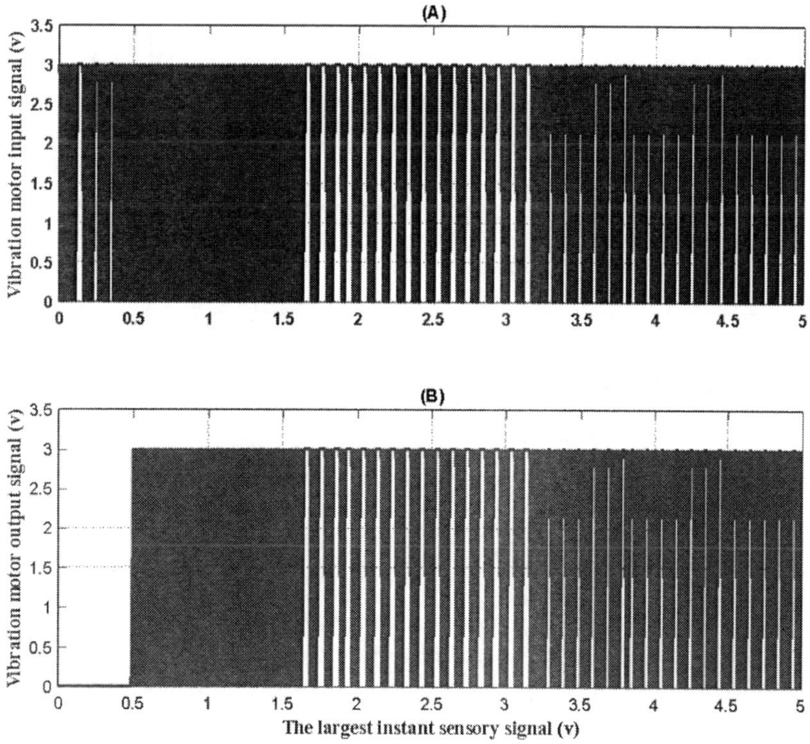

Figure 8. The excitation signal of the vibration motors: A) the vibration motor input signal, and B) the vibration motor output signal.

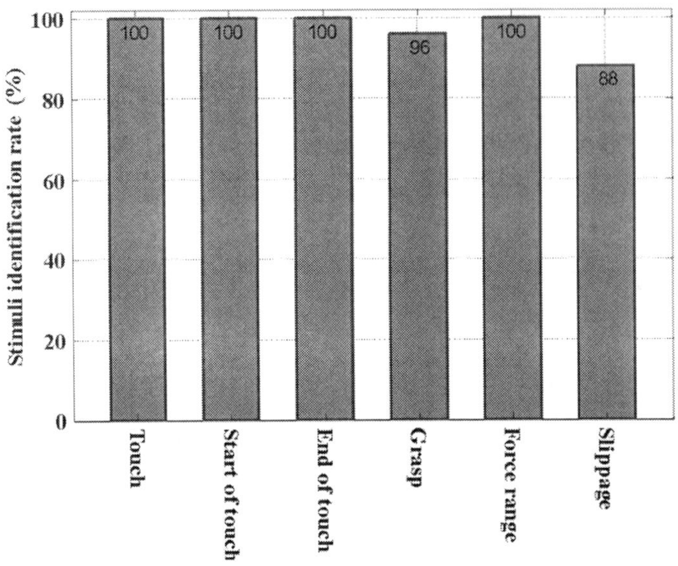

Figure 9. The volunteers' stimuli identification rate during the entire experiments.

Table 1. Comparison between the stimuli identification rate of the current study with the previous works.

Refe renc es	Type of sensory system	Type of Feedback display	Setup of Feedbac k display	Stimuli identification rate (SIR)			
				Touc h	**Gras p**	**Force range**	**Slipp age**
Curren t Study	Pressure	Hybrid pressure & vibration	Upper limb	100 %	96 %	100 %	88 %
[21]	Pressure	Vibration	Forearm	99.3 %		94.7 %	
[22]	Pressure	Vibration	Upper limb	94 %		75 %	
[23]	Pressure	Vibration	Upper limb	93.9 %	85 %		
[24]	Pressure	Pressure and Vibration	Upper limb	93%		90 %	
[25]	Pressure	Electro	Forearm	90 %			
[26]	Pressure	Squeeze	Forearm	70.3 %			62.2 %

6. Conclusion

This paper presents a new hybrid haptic feedback stimulation system of the upper limb prostheses to assist the amputees of upper limb mutilation to recover the missing feeling through their prostheses. The hybrid haptic device was created to excite the user's skin using two types of stimulators: pressure and vibration stimulators. The sensory glove created from a plastic element. Five QTC pressure sensors were installed on each fingertip and another sensor was placed on the palm. The structure of a hybrid haptic feedback stimulation system device consists of a servomotor and a pair of vibrators, which operate concurrently to excite the patients' nervous system.

The evaluation experiment tests indicated that the hybrid haptic feedback stimulation system works with a high level of precision. In overall, the SIR of the evaluation experiments confirmed the functionality and effectiveness of the hybrid haptic feedback stimulation system to convey multi tactile information to the volunteer's brain without pre-training. The SIR rate of 100% were achieved for most of the experimental tests, i.e. start of touch, end of touch, touch and the force range. However, only 88% and 96% of the volunteers passed the slippage test and were able to recognize the grasping targets, respectively. In addition, the SIR of current work was compared to the results of the previous studies, in order to highlight the work contribution. The comparison results showed that the results of the current work are better than of the previous works, in spite of the current experiment was performed without using any kind of a pre-training to improve the perception precision.

References

[1] Ziegler-Graham K, MacKenzie E J, Ephraim P L, Travison T G and Brookmeyer R 2008 Estimating the prevalence of limb loss in the United States: 2005 to 2050 Archives of physical medicine and rehabilitation **89** 422-9
[2] Nemah M N, Low C Y, Aldulaymi O H, Ong P and Qasim A A 2019 A Review of Non-Invasive Haptic Feedback stimulation Techniques for Upper Extremity Prostheses International Journal of Integrated Engineering **11**
[3] Pylatiuk C, Kargov A and Schulz S 2006 Design and evaluation of a low-cost force feedback system for myoelectric prosthetic hands JPO: Journal of Prosthetics and Orthotics **18** 57-61
[4] Antfolk C, Björkman A, Frank S-O, Sebelius F, Lundborg G and Rosen B 2012 Sensory feedback from a prosthetic hand based on air-mediated pressure from the hand to the forearm skin Journal of rehabilitation medicine **44** 702-7
[5] Kim K, Colgate J E and Peshkin M A 2007 A pilot study of a thermal display using a miniature tactor for upper extremity prosthesis. In: Frontiers in the Convergence of Bioscience and Information Technologies, 2007. FBIT 2007: IEEE) pp 531-6
[6] Marasco P D, Schultz A E and Kuiken T A 2009 Sensory capacity of reinnervated skin after redirection of amputated upper limb nerves to the chest Brain **132** 1441-8
[7] Antfolk C, D'Alonzo M, Controzzi M, Lundborg G, Rosén B, Sebelius F and Cipriani C 2013 Artificial redirection of sensation from prosthetic fingers to the phantom hand map on transradial amputees: vibrotactile versus mechanotactile sensory feedback IEEE transactions on neural systems and rehabilitation engineering **21** 112-20
[8] Huang H, Li T, Bruschini C, Enz C, Justiz J, Antfolk C and Koch V M 2017 Multi-modal Sensory Feedback System for Upper Limb Amputees. In: 2017 New Generation of CAS (NGCAS), (Genova, Genoa: IEEE) pp 193-6
[9] Romero E and Elias D 2016 Design of a non invasive haptic feedback device for transradial myoelectric upper limb prosthesis. In: ANDESCON, 2016 IEEE, (Arequipa, Peru: IEEE) pp 1-4
[10] Ajoudani A, Godfrey S B, Bianchi M, Catalano M G, Grioli G, Tsagarakis N and Bicchi A 2014 Exploring teleimpedance and tactile feedback for intuitive control of the pisa/iit softhand IEEE transactions on haptics **7** 203-15
[11] Jimenez M C and Fishel J A 2014 Evaluation of force, vibration and thermal tactile feedback in prosthetic limbs. In: Haptics Symposium (HAPTICS), 2014 IEEE, (Houston, TX, USA: IEEE) pp 437-41
[12] Pylatiuk C, Schulz S and Döderlein L 2007 Results of an Internet survey of myoelectric prosthetic hand users Prosthetics and orthotics international **31** 362-70
[13] Biddiss E, Beaton D and Chau T 2007 Consumer design priorities for upper limb prosthetics Disability and Rehabilitation: Assistive Technology **2** 346-57

[14] Blank A, Okamura A M and Kuchenbecker K J 2010 Identifying the role of proprioception in upper-limb prosthesis control: Studies on targeted motion ACM Transactions on Applied Perception (TAP) **7** 15
[15] Hussein M T and Nemah M N 2015 Control of a two-link (rigid-flexible) manipulator. In: 2015 3rd RSI International Conference on Robotics and Mechatronics (ICROM): IEEE) pp 720-4
[16] Savioz G and Perriard Y 2012 Self-sensing of linear short-stroke actuators for multi-finger haptic interfaces using induced high frequency oscillations. In: Advanced Intelligent Mechatronics (AIM), 2012 IEEE/ASME International Conference on: IEEE) pp 764-9
[17] Bimbo J, Pacchierotti C, Aggravi M, Tsagarakis N and Prattichizzo D 2017 Teleoperation in cluttered environments using wearable haptic feedback. In: Intelligent Robots and Systems (IROS), 2017 IEEE/RSJ International Conference on: IEEE) pp 3401-8
[18] Huang H, Li T, Antfolk C, Enz C, Justiz J and Koch V M 2016 Experiment and investigation of two types of vibrotactile devices. In: Biomedical Robotics and Biomechatronics (BioRob), 2016 6th IEEE International Conference on, (Singapore, Singapore: Ieee) pp 1266-71
[19] Nemah M N, Low C Y, Ong P and Zakaria N A C 2018 Development and evaluation of a spot sensor glove for the tactile prosthetic hand International Journal of Engineering and Technology (UAE) **7** 63-9
[20] Peratech November 2015 QTC SP200 Series Datasheet SP200-05 Series, Single Point Sensors.
[21] Culjat M O, Son J, Fan R E, Wottawa C, Bisley J W, Grundfest W S and Dutson E P 2010 Remote tactile sensing glove-based system. In: Engineering in Medicine and Biology Society (EMBC), 2010 Annual International Conference of the IEEE, (Buenos Aires, Argentina: IEEE) pp 15504
[22] Mohammed Najeh Nemah C Y L, Joerg Hoffmann, Winfried Kraft, Martin Benjak, Gerrit Lange, Pauline Ong, and Ching Theng Koh 2018 A hybrid haptic feedback stimulation system for upper limb prostheses using vibration stimulators. (International Journal of Advanced Robotic Systems p (In process)
[23] Fontana J M, O'Brien R, Laciar E, Maglione L S and Molisani L 2018 Vibrotactile Stimulation in the Upper-Arm for Restoring Individual Finger Sensations in Hand Prosthesis Journal of Medical and Biological Engineering **38** 782–9
[24] Huang H, Li T, Bruschini C, Enz C, Justiz J, Antfolk C and Koch V M 2017 Multi-modal Sensory Feedback System for Upper Limb Amputees. In: CAS (NGCAS), 2017 New Generation of: IEEE) pp 193-6
[25] Štrbac M, Belić M, Isaković M, Kojić V, Bijelić G, Popović I, Radotić M, Došen S, Marković M and Farina D 2016 Integrated and flexible multichannel interface for electrotactile stimulation Journal of neural engineering **13** 046014
[26] Damian D D, Ludersdorfer M, Kim Y, Arieta A H, Pfeifer R and Okamura A M 2012 Wearable haptic device for cutaneous force and slip speed display. In: Robotics and Automation (ICRA), 2012 IEEE International Conference on, (Saint Paul, MN, USA: IEEE) pp 1038-43

The Significant Study of using Vegetable Oil as a Cutting Lubricant on Conventional Lathe Machine

U. H. Ahmad*, A. H. M. Azman, N. A. Maidin, N. F. B. Abdul Wakhi Anuar and M. H. Abd Rahman

Faculty of Mechanical and Manufacturing Engineering Technology, Universiti Teknikal Malaysia Melaka, Hang Tuah Jaya, 76100 Durian Tunggal, Melaka, Malaysia

*umihayati@utem.edu.my

Abstract. The occurrence of tool wear in machining is a natural phenomenon that cause failure. The deformation during cutting at the interface between the tool face and workpiece tends to generate high cutting temperature. The application of flood coolant to reduce the friction at the tool-work piece may create several environmental problems. The introduction of Minimum Quantity Lubrication (MQL) as an alternative technique which is the process of pulverizing a very small amount of oil (< 30ml/h) can be regarded as replacement of dry machining while it may also be considered as an alternative to flood cooling. The effect of vegetable oil lubricant and cutting speed on tool wear and surface integrity were the research scope. Three machine speed; 120, 141, and 174 m/min were used in the experimental setup on mild steel and carbon steel on work material using cemented carbide tools. The result of surface roughness and tool wear using canola oil mixing with 2 weight percent (wt%) of Zinc-dialkyl-dithio-phosphate (ZDDP) were compared against synthetic oil coolant. Comparison results show that canola oil mix can perform better compare to the synthetic oil coolant in term of surface finish. From tool wear perspective, canola oil mix show 26.5% smaller area compare to the synthetic oil coolant.

1. Introduction

The interface during machining process between tool bit and the work piece tends to create high cutting temperature. At this point the cutting liquid/coolant or cutting oil take part in the machining procedure. The present of cutting fluid or coolant is to decrease the cutting temperature and friction, additionally to facilitate the removal of the chip introduce during the process and also improving surface finish. Flood distribution is a normal method to cool down the work material. This method typically consumes large volumes of cutting fluid at low velocity and with huge volume of cutting fluid used will rise an environment issue towards global [1]. Improper handling of cutting fluids may damage soil and water resources which will cause a huge loss to the environment, besides that for companies, the expenses related to cutting fluids represent a large amount of the total machining costs [2]. Thus MQL was introduced to overcome this issue. In this method, the lubricant used in the form of spray and aerosols is subjected to high thermal and mechanical loads. It is highly important to ensure non-toxic and harmless lubricants are used in MQL [3].

The cutting liquid or coolant that widely used in machining industry had cause some environmental concern such as harming the soil and water contamination when using coolant. Every year, for European Union alone using roughly 320,000 tons MWF (Metal Working Fluid), no less than 66% should be disposed and it is expanding a years by years due to high request of the machining process [4]. Research study on dry machining and (MQL) minimum quantity lubrication found that utilization of cutting

Content from this work may be used under the terms of the Creative Commons Attribution 3.0 licence. Any further distribution of this work must maintain attribution to the author(s) and the title of the work, journal citation and DOI.
Published under licence by IOP Publishing Ltd

IOP Conf. Series: Materials Science and Engineering **834** (2020) 012014 doi:10.1088/1757-899X/834/1/012014

lubricant gravely impacts nature and human wellbeing both during usage and transferred [5]. MQL is introduced by manufacture to overcome the cutting lubricant used in the industry, which is a low amount of lubricant strategy contrasted with the coursed oil method technique with emulsions. Usually the type of cutting lubricant used in MQL is biodegradable oil that not harmful to the environment. This implies utilizing clean oils that are fatty alcohol or ester based [6]. This shows that the synthetic cutting lubricant is not guarantee that is safe to environment or human.

Cutting fluids have been utilized as a part of machining procedures to diminish the temperature during machining by splashing the coolant into the machining zone straight forwardly on the cutting tool and the part, however the amount used of the lubricant due to the toxic content is worrisome [7]. Thus, achieving sustainable manufacturing are priorities and the possible solution in this problem are dry machining. But due to huge capital investment on dry machining method will not be cost effective solution. Cutting fluid are unpredictable in their composition which may lead to other issues. Microbial poisons are likewise produced by microscopic organisms and growths exhibit, especially in water-solvent cutting liquids [8]. The toxins and bacteria can cause harmful to the user in long term period. Regular contact with the lubricant would increase the probability of getting affected to the bacteria and lead to other discomfort. However, in this study canola oil that been used in cooking will be used as cutting lubricant in a machining process using conventional lathe machine. The vegetable oil has become frequent option to be used as cutting lubricant in most of the industry. Vegetable oil innovation has been appeared to convey extensive headways in profitability and longer tool life. Utilizing vegetable oil-based coolants amid machining can accomplish surface completions and dimensional resistances with agreeable edges [9]. The purpose of this study is to investigate whether the alternative environmentally friendly lubricant that replace cutting fluid will affect surface roughness and tool wear.

Tool wear has a great impact to tooling expense and part quality due to contact among tool and work piece or between the tool and chip. Flank wear which is one of tool wear affected the surface roughness of the machined material by using MQL with the present strategy has reduce flank wear and thus able to enhance instrument life [7], [10]. For work material, mild steel and carbon steel will be used to compare the level of surface roughness and tool wear. Mild steel is a standout among the most adaptable, versatile and imperative materials. It is widely used due to from its great structure and mechanical properties, and flexibility [11], [12].

2. Methodology

This study consists of 4 stages that involved to investigate whether the vegetable oil will give any positive result to this study. The study starts with selection of vegetable oil. The canola oil was selected based on several literature review that had been done related to this study.

2.1. Experimental Setup

In this experiment cemented carbide had been used. Cemented carbide is quite a common tool bit used that been used in the industry. This experiment only focuses on single type of tool against the vegetable oil coolant and the synthetic coolant to identify the tool wear. Four faces tool bit of cemented carbide had been used in this experiment for both vegetable oil and synthetic coolant; two sides for the vegetable oil coolant and other two side for the synthetic coolant. The usage of these tool bit side is divided as according to Figure 1 and Table 1.

In this stage, the conventional lathe machine parameter will be set into the parameter that assigned. The original mild steel diameter is 50 mm. During the facing process the mild steel and carbon steel work material had been machined to 30mm diameter. Turning process were machined on mild steel and carbon steel become from 100 mm to35 mm length. The process had been repeated three times on mild steel and carbon steel for each parameter in Table 2.

Figure 1. Cemented carbide tool bit

Table 1. Tool bit side

Side	Material	Cutting Lubricant
1	Mild Steel	Synthetic Oil Coolant
2	Mild Steel	Canola Oil + 2 wt % ZDDP
3	Carbon Steel	Synthetic Oil Coolant
4	Carbon Steel	Canola Oil + 2 wt % ZDDP

Table 2. Machine Parameter

Cutting Speed (m/min)	Feed Rate (mm/rev)	Depth of Cut (mm)
120	0.10	1.0
141	0.10	1.0
174	0.10	1.0

After the experiment complete, two tests had been done on the work material and cemented carbide tool; surface roughness and tool wear test.

3. Result and Discussion

This section presents obtained results and following by discussion.

3.1. Surface Roughness Test

Surface roughness test has been accomplished by using Mitutoyo Surftest SJ-401 to determine the smoothness of the surface of the part after being machined using same tool bit with different type of lubricant. Below is the result and the comparison of mild steel that are being machined with synthetic and canola oil.

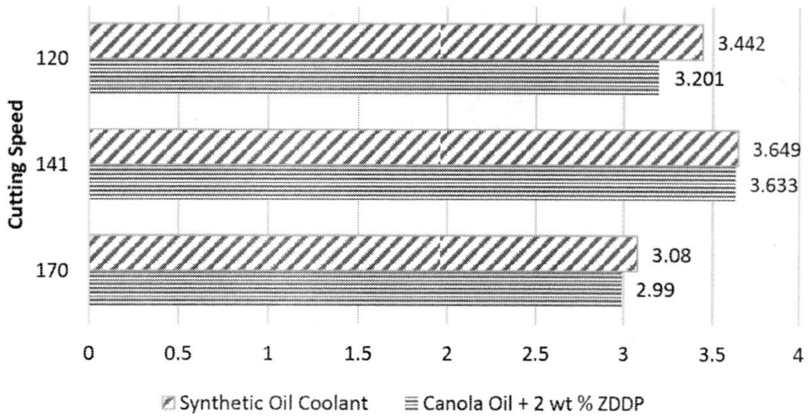

Figure 2. Result of surface roughness on different Cutting Speed using mild steel

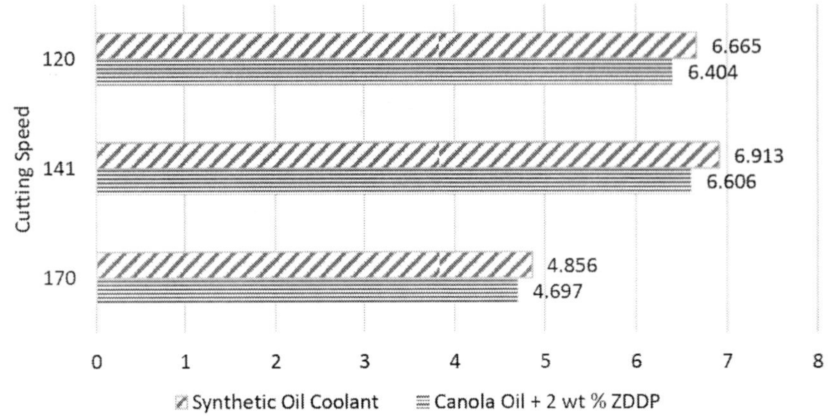

Figure 3. Result of surface roughness on different Cutting Speed using carbon steel

Figure 2 and Figure 3 shows that canola oil + 2 wt % ZDDP can perform better than the synthetic oil coolant. This show that, canola oil can perform better compare to the synthetic oil coolant in term of surface finish. Even though the different is small between these two coolants, but overall, canola oil can perform as good as synthetic oil coolant for machining process. Based on Figure 3, the greatest surface finish is achieved at 174 m/min with value of 4.697 ɥm on carbon steel work material. While the value of surface roughness on mild steel work material, obtained at 174 m/min with value of 2.990 ɥm. Theoretically, every machining process with high spindle speed, can get a good surface finish compare to the slow one. But in term of coolant that use in machining process, canola oil + 2 wt % ZDDP can get a smoother surface finish compare to the synthetic oil coolant.

3.2. *Tool Wear*

Tool wear is regular occurrence in a cutting tool after going through a machining process. Tool wear conditions also formed due to the distribution of stress. Tool wear usually appear due to the friction in between a cutting tool area and the workpiece material itself. Tool wear can be measure using a tool wear tester, and for this study optical microscope Nikon Measuring Microscope MM-800 with highly specification was used to obtain the wear of cemented carbide cutting tool.

Material: Mild Steel
Type of oil: Synthetic oil coolant
Area of wear (mm x mm): 0.6702 mm2
Length: 5.276 mm

Material: Mild Steel
Type of oil: Canola Oil + 2 wt % ZDDP
Area of wear (mm x mm) = 0.4050 mm2
Length = 3.9875 mm

Figure 4. Tool wear area

IOP Conf. Series: Materials Science and Engineering **834** (2020) 012014 doi:10.1088/1757-899X/834/1/012014

Material: Carbon Steel
Type of oil: Synthetic oil coolant
Area of wear (mm x mm): 1.8762 mm2
Length: 13.6112 mm

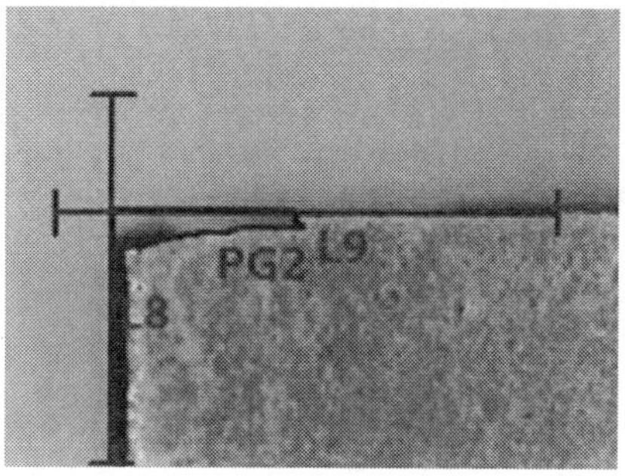

Material: Carbon Steel
Type of oil: Canola Oil + 2 wt % ZDDP
Area of wear (mm x mm) = 1.1834 mm2
Length = 9.2743 mm

Figure 4. Tool wear area (cont.)

The result in Figure 4 shows the area of wear for canola oil + 2 wt % ZDDP have lower area of wear compare to the synthetic oil coolant. This indicate that the usage of coolant also plays a major role to ensure long lasting tool life. Canola oil + 2 wt % ZDDP have about 26.5 % smaller area of wear compare to the synthetic oil coolant. This proves that additives such as ZDDP able to prolong the tool life with the ability to reduce tool wear. As for carbon steel, Figure 5 also shows that canola oil + 2 wt % ZDDP have lower area of wear compared to the synthetic coolant oil. Canola oil + 2 wt % ZDDP is 69.3 % lower than synthetic oil coolant wear.

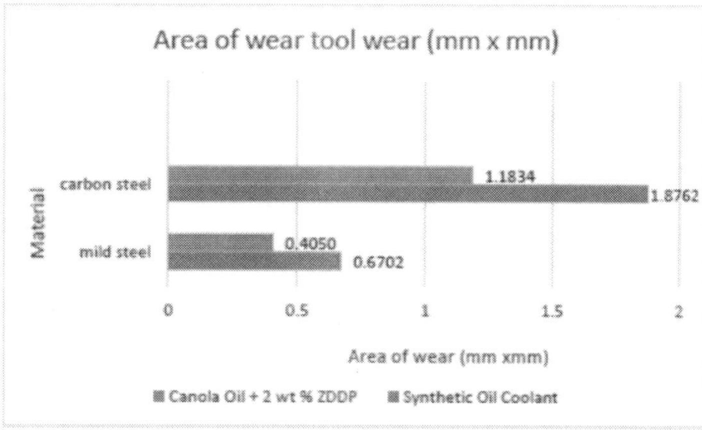

Figure 5. Comparison of area tool wear

Figure 5 indicate that the lowest area of wear can be obtained when using mild steel as the material and canola oil + 2 wt % ZDDP as the cutting lubricant to perform the machining process. This is due to mechanical properties of mild steel that is less stiff compared to carbon steel. That is why the tool wear on carbon steel is larger than mild steel. Overall, canola oil + 2 wt % ZDDP can perform better in tool life. Cutting lubricant used in the machining process plays an important in tool life period beside the set feed rate. In this experiment, feed rate and depth of cut are constant variable which is 0.1 mm/rev for feed rate and 1 mm for depth of cut but cutting conditions and machining parameter will also influence the output. The cutting contact level stress also could be the reason of the reduced area of wear.

4. Conclusion

This paper has presented the minimum quantity lubrication method as an alternative technique of dry machining which using very small amount of oil using vegetable oil lubricant. From the results, it can be concluded that canola oil mix can perform better compare to synthetic oil coolant in term of surface finish, but from tool wear perspective, canola oil mix show 26.5% smaller area compare to synthetic oil coolant.

Acknowledgement

Authors gratefully acknowledge the support of Universiti Teknikal Malaysia Melaka through Short Term Grant PJP/2018/FTK (14D)/S01640.

References

[1] S. Ebbrell et al., "Effects of Cutting Fluid Application Methods on the Grinding Process", International Journal of Machine Tools and Manufacture, vol. 40, no. 22, pp. 209–223, 2000.
[2] M. M. A. Khan and N. R. Dhar, "Performance Evaluation of Minimum Quantity Lubrication by Vegetable Oil in Terms of Cutting Force, Cutting Zone Temperature, Tool Wear, Job Dimension and Surface Finish in Turning AISI-1060 Steel", Journal of Zhejiang University SCIENCE A, vol. 7, no. 11, pp. 1790–1799, 2006.
[3] S. Delfini, A. D. Nicolantonio, and J. Roser, "Minimum Quantity Lubrication", pp. 55–60, 2011.
[4] S. A. Lawal, I. A. Choudhury and Y. Nukman, "Application of Vegetable Oil-Based Metalworking Fluids in Machining Ferrous Metals - A Review", International Journal of Machine Tools and Manufacture, vol. 52, no. 1, pp. 1–12, 2012.
[5] A. K. Sharma, A. K. Tiwari and A. R. Dixit, "Effects of Minimum Quantity Lubrication (MQL) in Machining Processes Using Conventional and Nanofluid Based Cutting Fluids: A Review". Journal of Cleaner Production, vol. 127, pp. 1–18, 2015.
[6] V. N. Gaitonde, R. S. Karnik and J. P. Davim, "Minimum Quantity Lubrication in Machining. Sustainable Manufacturing", pp. 79–110, November, 2013.
[7] S. A. Lawal, I. A. Choudhury and Y. Nukman, "A Critical Assessment of Lubrication Techniques in Machining Processes: A Case for Minimum Quantity Lubrication Using Vegetable Oil Based Lubricant", Journal of Cleaner Production, vol. 41, pp. 210–221, 2013.
[8] S. Safian et al., "Feasibility Study of Using Vegetable Oil As A Cutting Lubricant Through The Use Of Minimum Quantity Lubrication During Machining. Universiti Teknologi Malaysia, pp. 1– 37, 2009.
[9] B. U. Krahenbuhl, and N. D., "Vegetable Oil-Based Coolants Improve Cutting Performance", pp. 1–2.
[10] N. R. Dhar, M. T. Ahmed and S. Islam, "An Experimental Investigation on Effect of Minimum Quantity Lubrication in Machining AISI 1040 Steel", International Journal of Machine Tools and Manufacture, vol. 47, no. 5, pp. 748–753, 2007.
[11] M. Prabakaran et al., "Aster Koraiensis as Nontoxic Corrosion Inhibitor for Mild Steel in Sulfuric Acid", The Korean Society of Industrial and Engineering Chemistry, Journal of Industrial and Engineering Chemistry, 2017.
[12] M. Prakash et al., "Effect of Machining Configuration on the Corrosion of Mild Steel", Journal of Materials Processing Technology, vol. 219, pp. 70–83, 2015.

Aerodynamics Analysis on the Effect of Canard Aspect Ratio on Blended Wing Body Aircraft using CFD Simulation

Z. M. Ali*[1], W. Kuntjoro[2], W. Wisnoe[2], I. S. Ishak[3], M. Z. Katon[4] and N. Ahmad[2]

[1]Fakulti Kejuruteraan Mekanikal, Universiti Teknologi MARA Pasir Gudang, Johor.
[2]Fakulti Kejuruteraan Mekanikal, Universiti Teknologi MARA, Shah Alam, Selangor
[3]Univesiti Teknologi Malaysia, Univerisiti Teknologi Malaysia, Skudai, Johor
[4]Fakulti Kejuruteraan Mekanikal, Universiti Teknologi MARA Bukit Besi, Terengganu

*zurriatimohdali@uitm.edu.my

Abstract. Blended Wing Body Aircraft (BWB) is an unconventional aircraft that combined the conventional and flying wing aircraft's design. The unique design of the BWB offers a high fuel efficiency, low noise, and large payload volume for the size of aircraft. However, the BWB is unstable due to the absence of the its vertical stabilizer. A possible solution to improve this problem is using a horizontal control surface, the canard. For this purpose, the numerical simulation using NUMECA Computational Fluid Dynamic (CFD) software were conducted to obtain the aerodynamics forces subjected on this aircraft. The lift coefficient (C_L), drag coefficient (C_D) and moment coefficient (C_M) are analysed at Mach number 0.1. The study is specifically to discuss the effect of canard setting angle ™ of BWB from ™ = -10° to 10° in that particular of aspect ratio (AR 2, 4, 6 and 8) towards aerodynamics parameters, C_L, C_D, and C_M.

1. Introduction

The idea of BWB comes from the flying wing (FW) concepts. FW is an aircraft with the absence of fuselage, tail and only a wing. Due to its design, it has low drag, limited payload and is unstable. FW enables an entire payload or its portion to be accommodated in the wing. From the hybrid FW studied by [1] the maximum lift-to drag achieved by the airplane was approximately 24.5 at Mach = 0.85. It confirmed preliminary assumption on the possibility of achieving higher technical and economical characteristics of FW configuration airplane as compared with conventional airplanes. However, FW was a not a blended wing body designs because the fuselage contributed no lift and had little cargo and passenger room. The interest in tailless aircraft has emerged since then. The requirement to have a large transport with economical design and low efficiency has driven the engineer to come out with a new aircraft concept. The blended wing body, was an aircraft that blends the fuselage, wing and engines into a lifting surface, has been introduced by [2]. It features wide curved fuselage and a thick delta wing. The BWB (X-48A), the prototype has been constructed by Boeing Company and Air Force team for wind tunnel testing. Then, the Boeing and Cranfield Aerospace, a United Kingdom firm has introduced two X-49Bs BWB. The X-49Bs has a 6.4 m wingspan, weighs 181 kg and flies at up to 240 km/h at 3 km. Three engines mounted above the wing and use a carbon-fiber airframe and carbon-composite skin. The movable surface on the trailing edge of wing are responsible for all aircraft control [3]. In initial design of 800-passenger BWB (Boeing BWB-450), the reduction of 15% of takeoff weight and 27% fuel burn per seat mile shows the BWB has a better performance over the conventional aircraft.

Content from this work may be used under the terms of the Creative Commons Attribution 3.0 licence. Any further distribution of this work must maintain attribution to the author(s) and the title of the work, journal citation and DOI.
Published under licence by IOP Publishing Ltd

Among all the advantages of the BWB, there is still shortcoming. The BWB's pitch control has a short lever arm to the center of gravity compared to a conventional aircraft. This affects flight path control during rotation and landing since flare pitch changes are accompanied by unwanted initial plunging. [4] studied the advantages using a belly-flaps to enhance the lift and pitching moment coefficient during landing, go-around and take off. A belly flaps is located near the mid-section on the lower surface of wing. The wind tunnel model result using a generic BWB-450 model has been used for the study. Due to the study, the lift and moment coefficient increases with the employment of the belly flaps. The increase of lift improves the landing field length, takeoff field length and pitch lagging in go-around.

Determining the performance of BWB with canard using numerical methods has proven to be extremely difficult due to the viscous flow region that needs to be modeled. The flow over a BWB has greatly been complicated by the rounded trailing edge and the introduction of the canard. Many numerical studies were conducted which examined the possibility of using Navier-Stokes equations to predict the characteristics of aircraft, BWB or an aircraft with canard. [5] has been developed an unsteady three-dimensional Navier-Stokes analysis procedure has applied to Circulation Control Wing (CCW) configurations. The solver can be used in both a 2-D and a 3-D mode and can thus model airfoils as well as finite wings. [6] used a steady state, CFD code FLUENT to calculate a BWB model using the standard one equation turbulence model, Spalart-Allmaras. [7] investigates the design of conventional and unconventional wings with winglet. One of the methods that have been used in the design process is by using a CFD to analyze the flow mechanisms and optimize winglet design.

The Universiti Teknologi MARA (UiTM) BWB's research and development has started since 2005. The first design known as BWB-Baseline I. It has sharp edges, broad body, small wing area and elevator as a pitching moment control. Due to ineffective control of elevator, the BWB was then modified, while maintaining the wing span, body length and reference planform area. The BWB-Baseline-II was a completely-revised and redesigned version of BWB-Baseline- I [8]. Its feature has a simpler planform, broader-chord wing and slimmer body than previous. The small control canard is incorporated on Baseline-II since the center body elevator located at the aft of the body of BWB-Baseline I is ineffective as the pitching moment produced, causing change of trim impossible [9]. After that, the studies of BWB with canard were aggressively done through experimental and computer simulation [10]. This research, therefore, is considered a continuation of earlier efforts. However, the BWB being used for this study has a rectangular shape of canard and the aspect ratios having constant area were varied.

2. Methodology

2.1 Geometry Set-up and Grid Generation

The half body of BWB's geometry, is constructed in CATIA V5 software and has been scaled down to 1:6 from its original size, as shown in Figure 1. The scaling procedure is performed from the central body to the wing tip at the interval 349.43mm (~350mm), and nose (leading edge) to tail (trailing edge) about 348mm respectively. The advantages of scaling down the model is, it can save the time when calculating in the CFD simulation.

The boundary condition has been constructed with the dimension of 1 meters from nose to internal wall and 3 meters from external wall. The top and the bottom wall are 1 meter far from the BWB body, as shown in Figure 2. The unstructured grid generator, HEXPRESSTM software is used to generate the flowfield and surface grid of the BWB model. The HEXPRESS™ is unstructured hexahedral mesh generator software designed to automatically generate meshes in complex 2D and 3D geometries. A major obstacle to the treatment of arbitrarily complex configurations is the difficulty and cost of mesh generation. This can be mitigated by the use of unstructured meshes. The advantage of unstructured mesh is it provides flexibility to handle complex geometries. In this grid generator, a three dimensional (3D) unstructured hexahedral meshes are utilized for computing a flow around the aircraft configuration. The finer mesh was created near the aircraft model, to ensure the accuracy of the result can be achieved. Figure 3 shows the baseline surface grids generated for the wing-body without the canard configuration. The meshing size is coarser as it is away from model. The figure also shows a typical mesh employed

IOP Conf. Series: Materials Science and Engineering **834** (2020) 012015 doi:10.1088/1757-899X/834/1/012015

around the BWB for the computation. The number of cells of the computed flow field for the BWB was 500000.

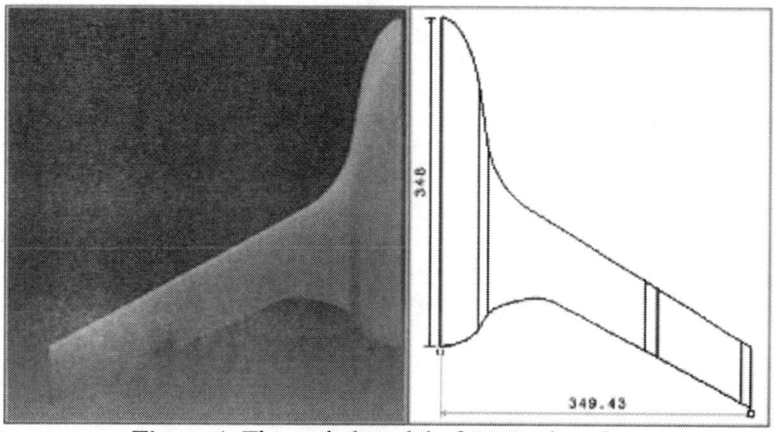

Figure 1. The scaled model of BWB aircraft

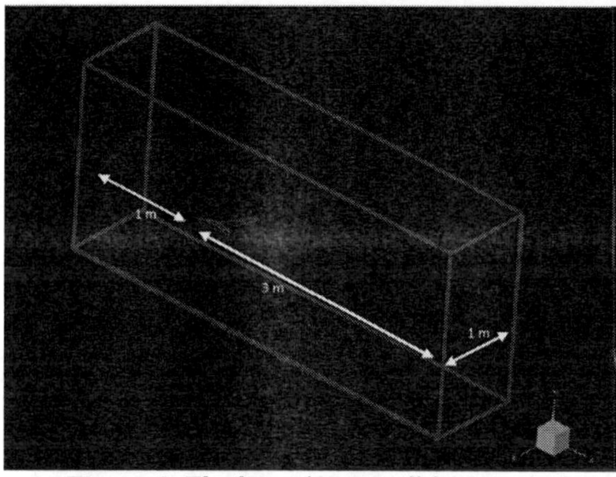

Figure 2. The boundary condition set-up

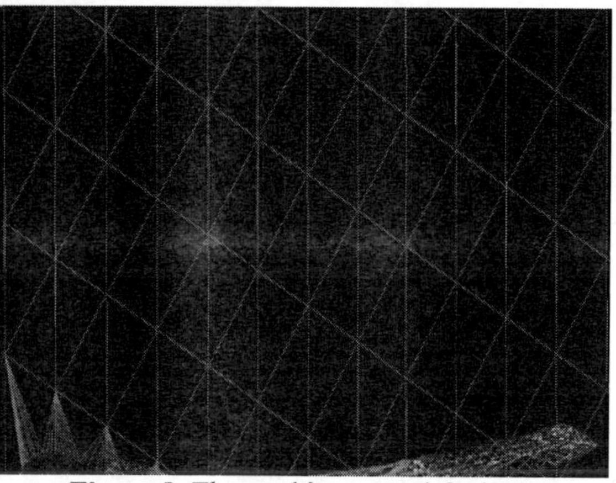

Figure 3. The meshing around the body

2.2 Computation

The CFD result was obtained using FINE™/Hexa flow solver. Table 1 shows the BWB-Baseline aircraft main characteristics and Table 2 is the pre-processing properties for the simulation setup. For the present study, the segregated solver is selected to solve the equation. The implicit formulation takes less computer time [9]. The effects of turbulence are accounted for by the Spalart-Allmaras model. To solve the governing equation, comprising the model, boundary conditions are needed at each part of the domain boundary [10]. No slip boundary condition is imposed at the solid boundary.

Table 1. Model Geometry Characteristics (based on scaled model)

References area	*0.03995m²*
Mean Aerodynamics Chord	*0.114m²*
Body Length	*0.348m*
Half Wing Span	*0.348*
Canard Area	*0.005m²*

Table 2: Pre-processing properties

Fluid	air

IOP Conf. Series: Materials Science and Engineering **834** (2020) 012015 doi:10.1088/1757-899X/834/1/012015

Reference Pressure	101325 Pa
Reference Temperature	297 K
Reference Velocity	35 m/s
Turbulence Model	Spalart Allmaras

2.3 Post-Processing – Parameter Validation

Figure 4 (a) and (b) show the comparison between lift and drag coefficient with respect to angles of attack with the experimental result obtained by [11]. It is observed that the lift and drag coefficient have the same trend with the experimental. The lift is increases as the angles of attack, but the CFD is not being able to predict the stall that occurs around 8-degree angles of attack. The differences between the CFD and experimental result is small at angles of attack ranged between -10 to 10 degree. After that, the CFD result predict higher lift coefficient compare to the experimental result. The same trend is also observed in drag coefficient graph. Since the study is only fixed at one angle of attack (10 degree), the result obtained from the CFD is assumed to be validated.

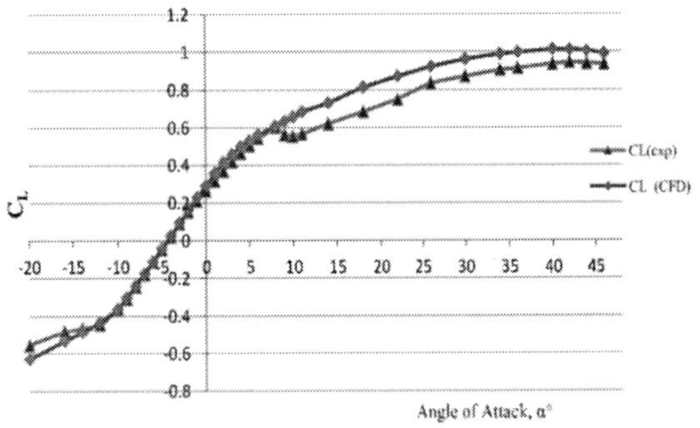

(a)

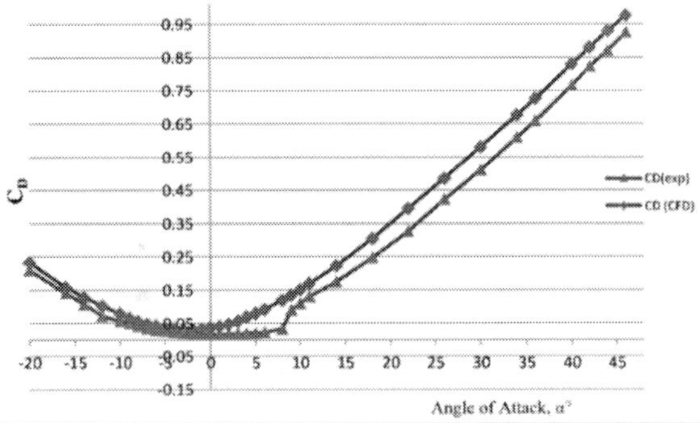

(b)

Figure 4. (a) Lift Coefficient vs Angles of Attack. (b). Drag Coefficient vs Angles of Attack

2.4 Grid Independent Study

The Grid Independence Study (GIS) is performed to observe the effect of number of cells towards the computation results. The GIS is achieved when the computation results has no effect by the change of parameters given (number of cells) . In this study, the size of the meshing has been change from coarse to finer on BWB aircraft model at ⟨= 0°. Six different sizes of unstructured grids were used for flow solver FINE™/Hexa computation. The number of cells, nb represents the increasing in grid size from the coarsest to the finest. The convergence of lifts and drag coefficient with regard to number of cells is illustrated in Figure 5(a) and (b). From the result, it can be seen that the computed lift and drag force converges with increasing of grid density. It was clearly indicated at 1841955 numbers of cells; the calculation is converged.

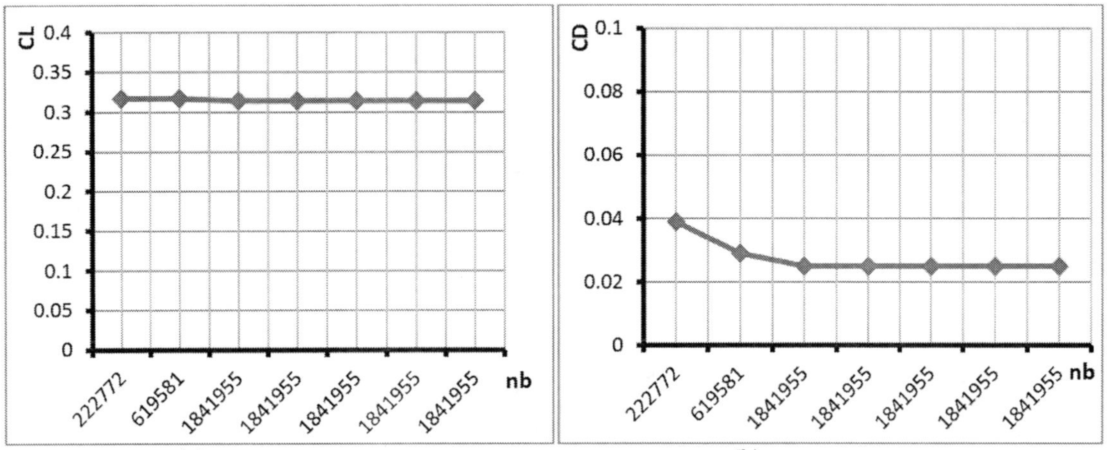

(a) (b)

Figure 5. (a) lift coefficient vs angles of attack, (b) drag coefficient vs angles of attack

3. Result and Discussion

3.1 Lift Coefficient

The lift coefficient versus canard deflection angle for each canard has been presented in Figure 6. The lift coefficient trend for the BWB aircraft with all four canard ratios almost the same particularly at a low angle of attack, ⟨ = -4° to 0°. It can be observed that at angle of attack ⟨= 0°, the lift coefficient of BWB with canard AR 2, 4, 6 and 8 is continuously increases as the canard deflection is increasing. The maximum lift coefficient ($C_{L,max}$) for the BWB with canard (except AR2) was 0.4 when canard was deflected to ™ = 10°. The canard with aspect ratio 2 starts to stall starting at ™ = 5°. As the ⟨ increased, the lift coefficient increases. The lift performance of BWB with canard AR 2, 6 and 8 starts to decrease at ™ = 5° when the BWB aircraft is notch up to ⟨= 4° and 10°. This result eliminates the ability of the canard at AR 6 and 8 to increase the lift of the aircraft. The BWB with canard AR4 however, shows the increasing of lift when the deflection is increases for each increasing angles of attacks. Therefore, this configuration has a good potential mechanism in increasing a lift force.

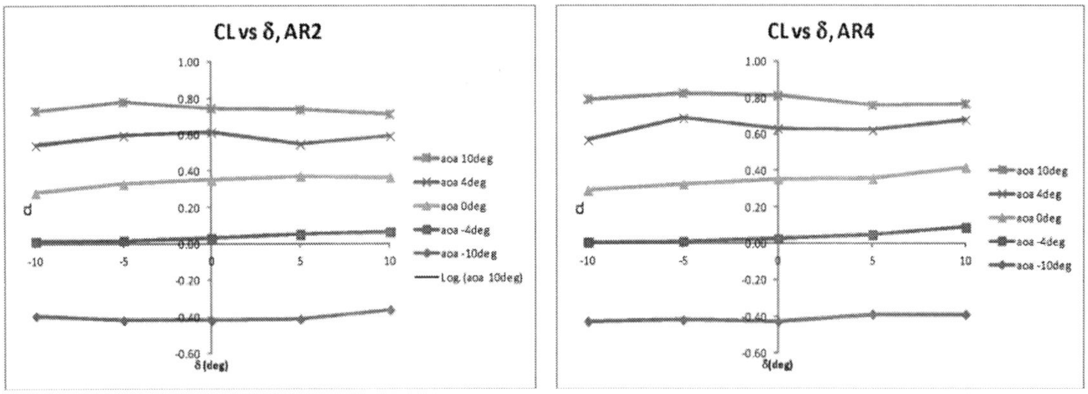

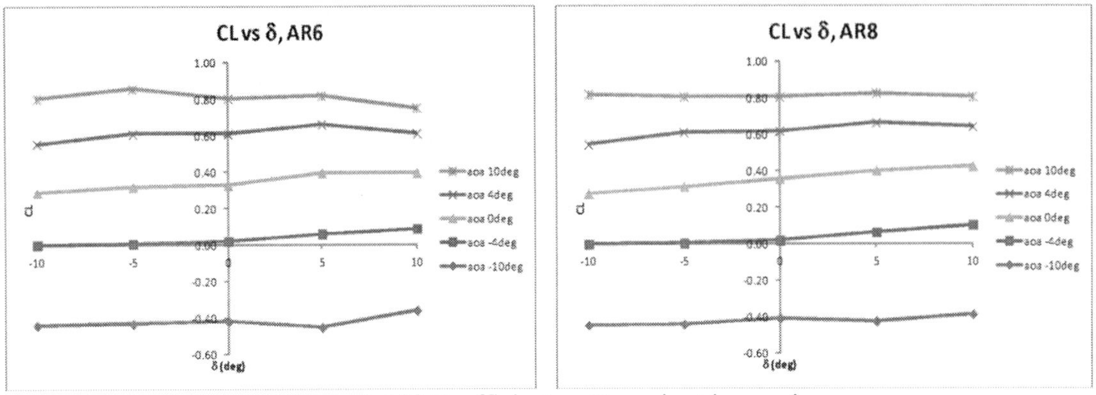

Figure 6. Lift Coefficient vs Canard setting angles

3.2 Drag Coefficient

The drag coefficient, C_D versus canard setting angles, ™ plots is presented in Figs. 7. Overall graphs show the same trend, where the drag is increases as the angles of attacks were increasing, as well as canard deflection angle. At angle of attack 〈 = 0°, the BWB with canard has an average drag (C_D = 0.04) when it ™ = 10°. However, at the same canard deflection angle, the drag was greatly increase to 0.08 (for BWB with canard AR2, 6 and 8) when the aircraft is increase to 〈 = 4°. The BWB with canard AR 4 has a lower drag that is 0.06. At angle of attack 〈 = 8°, the drag was drastically increase. This is due to the fact that, the flow on the upper surface of the canards was deteriorated.

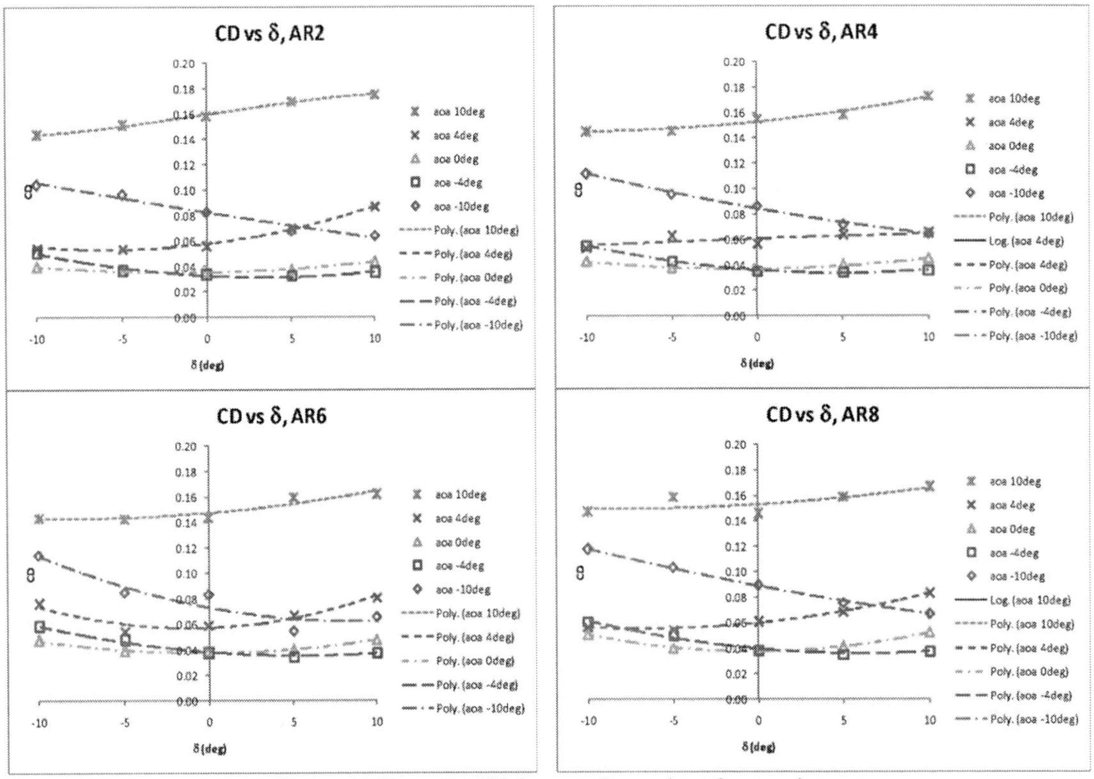

Figure 7. Drag Coefficient vs Canard setting angles

3.3 Flow Visualization

The physical flow around the BWB aircraft, particularly on the canard can be clearly seen through the flow visualization, as shown in Figure 8. From observation, it can be seen that the flow on the upper surface of canard has an increasing velocity as the canard setting angles are increased. The flow from canard is passing over the upper surface of wing and lower pressure area which indicates the high velocity, is seen near the leading edge of centre wing, as shown Fig. 8 (b). The flow is smooth and streamline around the BWB when canard is deflected to -0°, as seen in Fig. 8 (c). At d = c°, the lower pressure area starts to develop near the leading edge of canard, as shown in Fig. 8 (d). The flow at the upper surface of wing is streamline and it can be observed that the lower pressure area near the leading edge of main wing is getting bigger. The lower pressure area is getting bigger and spread to the whole wing and canard, except at the trailing edge, as the canard is increase to 3°. However, the flow is streamline and the flow at the tip of canard start to curl up to produce small vortex.

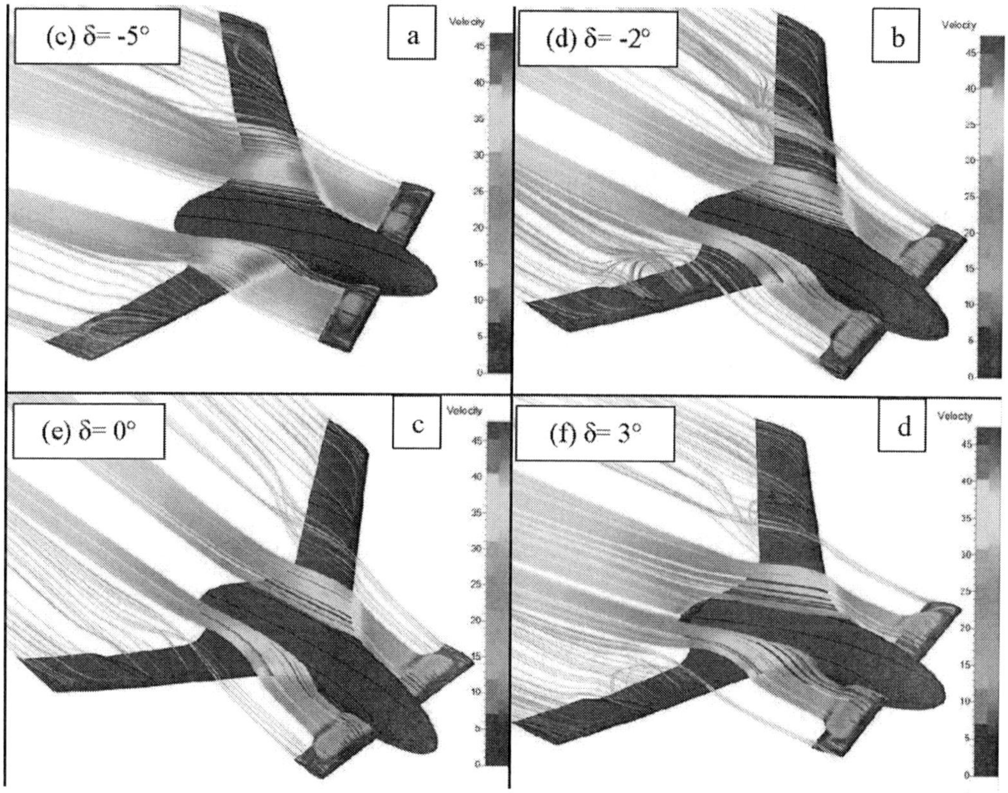

Figure 8. Flow Visualization

4. Conclusion

Aerodynamic analysis of UiTM's BWB-UAV aircraft with different canard aspect ratio was completed through CFD. The lift, drag and pitching moment coefficients were obtained and discussed. Grid independence study was performed showing almost independent results for the range of number of cells used. Parameter validation indicates the CFD results agreed with the wind tunnel test results. The lift, drag and moment coefficient has shown a similar trend among different canard aspect ratio. The results may be summarized as follows:

- The BWB with canard increases the lift coefficient particularly at higher angles of attack.
- The increasing of canard setting angle caused an increment of the lift curve depending on the aspect ratio of the canard itself.

- Canard caused a higher drag on the BWB configuration.
- The BWB without canard has a maximum lift-to-drag ratio. The existing of canard decreases the maximum L/D of the BWB. The BWB with canard AR 2 and 4 has maximum L/D compared to others when canard setting angle is not deflected. However, the L/D continuously reduces as the canard setting angle is increased depending to the canard aspect ratio

Acknowledgement
This work is part of the research program of BWB aircraft in Fakulti Kejuruteraan Mekanikal UiTM, which is financed by the Kementerian Pendidikan Malaysia (FRGS-173/2019).

References

[1] L. Bolsunovsky, N. P. Buzoverya, B. I. G. V. E. Denisov, A. I. Dunaevsky, L. M. Shkadov, O. V. Sonin, *et al.*, "Flying Wing problems & decision " *Aircraft Design,* vol. 4, pp. 193-219, 2001.

[2] R. H. Liebeck, "Design of the Blended Wing Body Subsonic Transport," *Journal Of Aircraft,* vol. 41, pp. 10-25, 2004.

[3] S. J. Mraz, " A makeover for transport aircraft?," in *Special Section: Aerospace,* Machine Designed, 2007, pp. 2-10

[4] Y. D. Staelens and R. F. Blackwelder, "Computer Simulation of Landing, Takeoff and Go-around of a Blended-Wing-Body Airplane with Belly-Flaps," presented at the 46th AIAA Aerospace Sciences Meeting and Exhibit, Reno, Nevada, 2008.

[5] Y. Liu, "Numerical Simulations of the Aerodynamic Characteristics of Circulation Control Wing Sections," Doctor of Philosophy, School of Aerospace Engineering, Georgia Institute of Technology, 2003.

[6] A M I Mamat, R. E. M. Nasir, W. Kuntjoro, W. Wisnoe, Z. Ngah, and R. Ramly, "Aerodynamics of Blended Wing Body (BWB) Unmanned Aerial Vehicle (UAV) Using Computational Fluid Dynamics (CFD)," *Journal of Mechanical Engineering* ,vol. 5, 2011.

[7] D. R. K. Nangia, D. M. E. Palmer, and M. R. H. Doe, "Aerodynamic Design Studies of Conventional &Unconventional Wings with Winglets," presented at the 25th Applied Aerodynamics Conference, San Francisco, CA, USA, 2006.

[8] R. E. M. Nasir, K. W, W. Wisnoe, Z. M. Ali, N. F. Reduan, F. Mohamad, *et al.*, "Preliminary Design of "Baseline-II" Blended Wing-Body (BWB) Unmanned Aerial Vehicle (UAV): Achieving Higher Aerodynamic Efficiency Through Planform Redesign and Low-Fidelity Inverse Twist Method," in *3rd Engineering Conference on Advancement in Mechanical and Manufacturing For Sustainable Environment (EnCon 2010)*, Sarawak, Malaysia, 2010.

[9] R. E. M. Nasir, W. Kuntjoro, and W. Wisnoe, "Longitudinal Static Stability of a Blended Wing Body Unmanned Aircraft with Canard as Longitudinal Control Surface," *Journal of Mechanical Engineering,* vol. 9, pp. 99-121, 2012.

[10] F D Nasri, W Wisnoe , R E M Nasir, E S S Askari," Effect of Wing Locations to the Aerodynamic of UiTM's Blended Wing Body-Unmanned Aerial Vehicle (BWB-UAV) Prototype, "International Journal of Engineering & Technology, 7 (4.25) (2018) 119-125

[11] Z M Ali, W Kuntjoro, W Wisnoe, "Wind Tunnel Testing for Blended Wing Body Aircraft with Canard," Proc. of The Fourth Intl. Conf. On Advances in Mechanical, Aeronautical and Production Techniques - MAPT 2015

IOP Conf. Series: Materials Science and Engineering **834** (2020) 012016 doi:10.1088/1757-899X/834/1/012016

Recent Advances in Metal Forming Simulation Technology for Automobile Parts by AFDEX

Mansoo Joun

ReCAFT, School of Mechanical and Aerospace Engineering, Gyeongsang National University, 501 Jinjudaero, Jinju, Republic of Korea, 52828

msjoun@gnu.ac.kr

Abstract. In this presentation, hot issues of metal forming simulation are presented, including intelligent remeshing, flow stress modelling and material identification, friction, material model, complete analysis, quality quantification and optimal process design, multi-body forming simulation technology, sheet and plate forging by bulk finite element, etc. Some selected success stories for recent two decades are also introduced with a lot of industrial applications, especially to automobile industries. Finally, some futuristic research works are introduced for weight reduction and maximized life-span of auto-parts to save the global environments and some suggestion of international cooperation for the common pursuit is stated in conclusion.

1. Introduction

Metal forming technology has been everlastingly developed since human beings started to utilize metals about 5000 years ago. It had been occupied by the governing societies even after the Industrial Revolution because it had governed the strength of weapons and life quality. Note that it has still been dominated by the industrialized countries.

Since metal forming industry should be supported by so many infra-structural industries and companies as well as experience-based technologies and techniques, it had been very difficult for the developing or under-developed countries to be competitive in the related business sector. As consequence, limited number of industrialized countries have become independent of metal forming technology. Many countries involved in the global top ten of automobile or electronic industry suffer from the lack of metal forming technology and their automobile or electric industries are much dependent on developed countries.

Metal forming technology is composed of material and/or steel fabrication technology, press and equipment technology, process design technology, die material and making technology, surface and heat treatment technology, lubricant technology, manufacturing and automation technology, etc. From standpoint of engineers in metal forming companies, the process design can be an active source of creativity. Of course, one of obstacles pressing the metal forming companies in the developing countries is the experience-based technology, which can be efficiently come over by the metal forming simulation technology [1]. The metal forming simulation technology has become important in the related industry because it is innovating the costly experience-based approach of process design. In the past, this technology had been used just for being advised while currently it has occupied the major spot of the process design procedure.

Content from this work may be used under the terms of the Creative Commons Attribution 3.0 licence. Any further distribution of this work must maintain attribution to the author(s) and the title of the work, journal citation and DOI.
Published under licence by IOP Publishing Ltd

Metal forming simulation technology is still growing together with metal forming technology to replace the experience-based approach to process design in metal forming. In this paper, some recent advances in hot issues of metal forming simulation technology are discussed with the state-of-the-art technologies, including solution accuracy, remeshing, modeling technology of material and friction, complete analysis and special applications to plate and sheet metal forming processes and special incremental metal forming process.

2. Hot issues of metal forming simulation technology

2.1. Solution accuracy

The major factors affecting deformation behaviors of material include elastic strain, plastic strain, strain rate, temperature, recrystallization, damage, anisotropy, etc. It is not easy to reflect all the factors in the metal forming simulation. The less influential factors should be thus disregarded from standpoint of engineering solution for the sense of practical application. Plastic strain is a major factor of metal forming at the room temperature while strain rate is a major factor and temperature is the secondary factor at the elevated temperature. In most simulations of bulk metal forming processes, engineers and researchers are interested in solving the material filling phenomena into die cavity affecting the under-filling defect and forming load. In such simulations, the appropriate major factors can give process design engineers and researchers some acceptable sorts of predictions because they are much affected by the volume consistency, that is, even rigid-plastic finite element analysis for cold forging and even rigid-viscoplastic finite element analysis may not discourage them if they are interested in global material flow itself. Typical application examples are shown in Fig. 1. The left figure in Fig. 1(a) is the first application of forging simulation made by industry in Korea [1] and Korea could become independent of forging process design and the related engineering from the developed countries in the mid nineties. The procedure of forging technology development with forging simulation technology in Korea can be seen from the homepage of AFDEX.

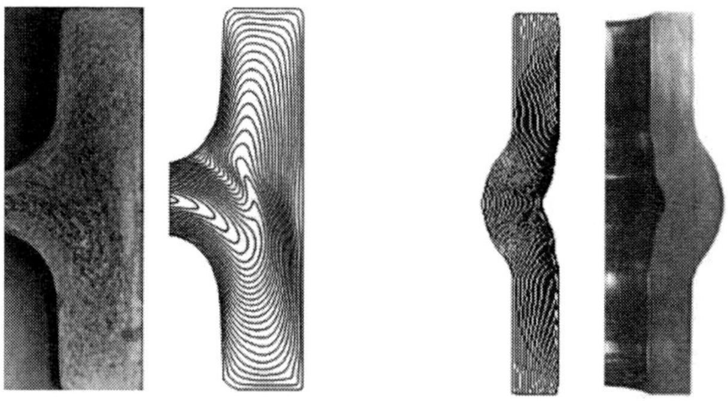

(a) 2D (Left: hot; Right: cold)

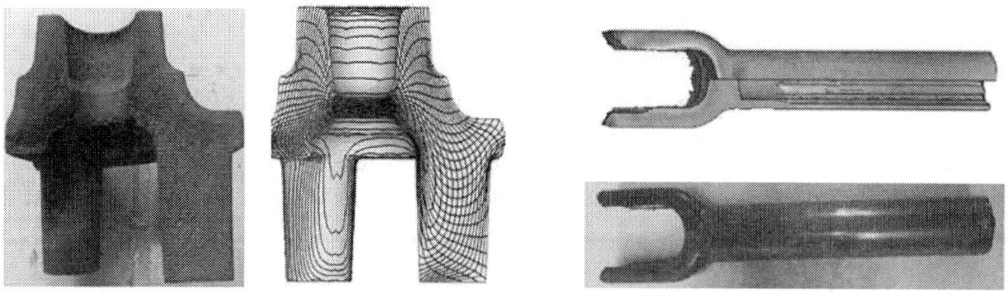

(b) 3D (Left: hot, Right: cold)

Figure 1. Accuracy of forging simulations

2.2. Remeshing

Remeshing is one of key factors affecting solution accuracy on bulk metal forming simulation [2, 3]. Remeshing is inevitable when the mesh system is too severely distorted or it does not describe the material within acceptable tolerance. It should be noted that the remeshing during metal forming simulation hurts considerably solution accuracy because it causes numerical smoothing of state variables as well as artificial change of the surface of material. Smaller elements near die corners in metal forming simulation can be easily and frequently damaged, causing frequent remeshing. Therefore, in metal forming, there is an appropriate number of finite elements which reduces number of remeshings at a little expense of numerical error. In addition, number of finite elements affects the computational time, implying that mesh density control is also of great importance in metal forming simulation. Fig. 2 shows some typical examples of intelligent finite element mesh systems. Fig. 3 shows the predictions of a coining process with emphasis on mesh density, which is an extreme case in terms of finite element mesh system [4].

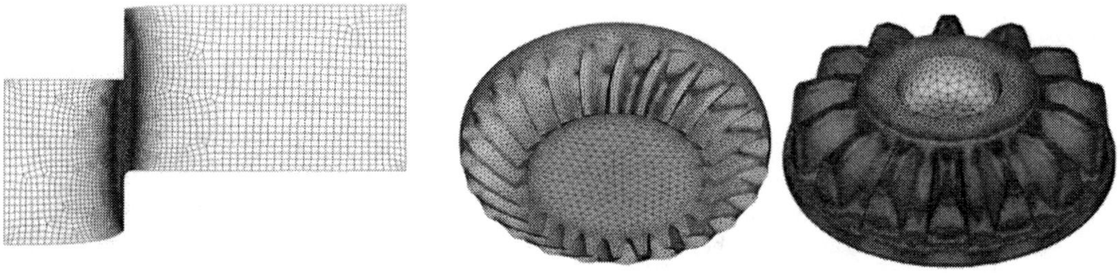

(a) 2D (b) 3D

Figure 2. Intelligent finite element mesh systems for metal forming simulation

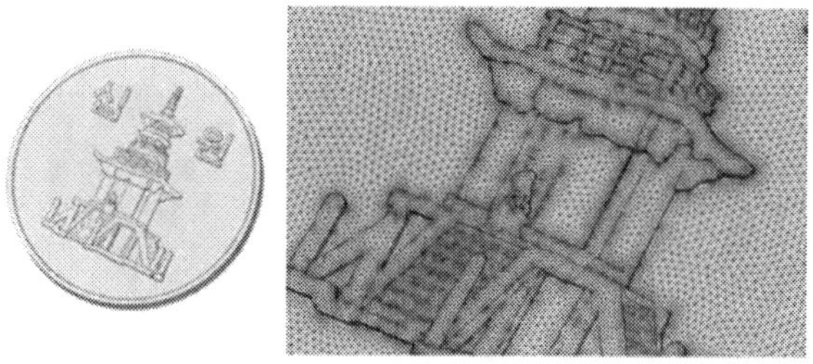

Figure 3. Coining process

2.3. Material identification

Remeshing is one of key factors affecting solution accuracy on bulk metal forming simulation. Particularly, plastic behavior of the material at the elevated temperature is very complicated. Its features involve strain softening due to recrystallization. When the coefficient and exponent in the power-law of flow stress with respect to effective strain rate are formulated as functions of strain and temperature, its unique features can be described as can be seen in Fig. 4 in which Voce model [5, 6], its variants including Ebrahimi et al. model [7], CFF and PLF models proposed by Razali et al. [8] and Hensel-Spittel model [9] are compared with experiments. However, these models except Voce model are weak in expressing the flow stress at extremely low strain rate which occurs in deal metal zone, etc. and cannot be neglected in many cases, for example, rolling with front and back tensions. A systematic approach to identify the flow stress of commercial metals at room temperature and to understand the plastic deformation behaviors of materials can be referred to the related works done by the author's group [10, 11].

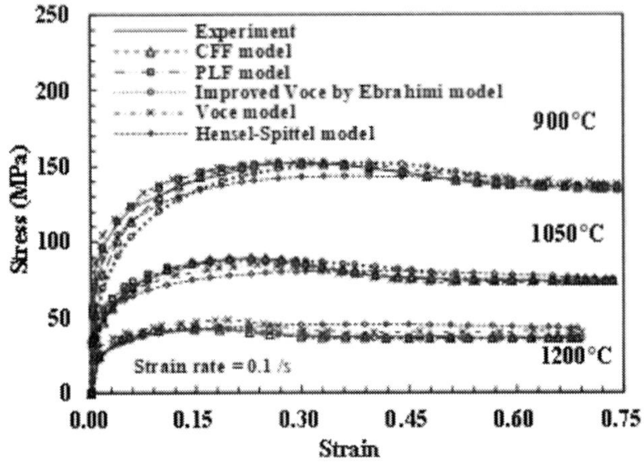

Figure 3. Fitted flow stresses and experimental flow stress curve

2.4. Friction model

Traditionally, laws of Coulomb friction and constant shear friction have been employed for metal forming simulation [12]. It is noted that Coulomb friction and its variants are better for both cold and hot metal forming processes. For well-lubricated cases in cold metal forming, both friction models can be helpful for revealing the grain flow or metal flow lines. To the contrary, for example, when a metal forming process of a pancake shape of material at the elevated temperature is predicted, two laws show much difference from each other and the law of Coulomb friction is better in terms of metal flow lines.

Recently, a new concept of critical surface strain was presented by the author's group, which is used to explain drastic change of frictional phenomena. Fig. 5(a) shows an experiment. Fig. 5(b) is one of unacceptable solutions. It should be noted that all the predictions were wrong when they were obtained using law of Coulomb friction with constant friction coefficients or law of constant shear friction with constant friction factors.

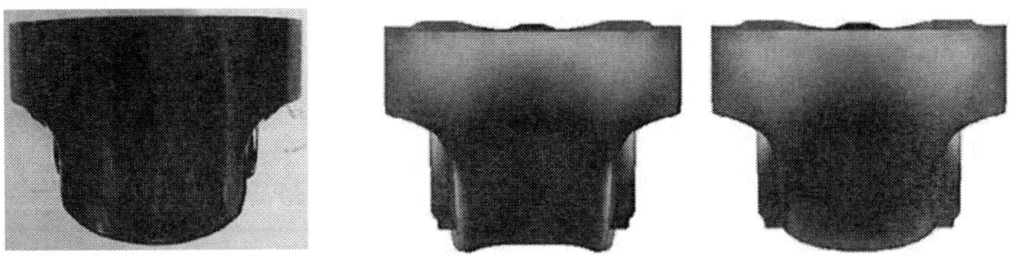

(a) Experiments (b) Constant friction (c) Varying friction

Figure 5. Comparison of experiment and predictions

2.5. Complete simulation

In most cases of practical metal forming simulation, dies or tools have been assumed rigid in metal forming simulation. With much better information about material and friction models, the solution accuracy has been much enhanced and metal forming simulation without any assumption or with minimized assumption on material and die deformation, called complete simulation has been becoming more and more important. However, the numerical complexity has hindered the advance in the complete simulation technology. Recently fully complete simulation of forging processes became realized and Fig. 6 is a typical example of complete simulation of a cold forging process considering pre-stressed dies owing to shrink fit. It is noted that the die deflection is coupled with deformation of material.

(a) Final shape of material (b) Pre-stressed die parts (c) Die stress at final stroke

Figure 6. Typical example of complete simulation considering pre-stressed die parts

2.6. Quantification of grain flow or metal flow lines

Grain flow has been predicted by a finite element method for practical use by forging simulation technologies [1,13] for nearly 25 years. Traditionally, in the finite element method, the desired mesh defined on a planar section is traced to show the grain flow. It should be noted that traditional numerical approaches do not give a solution to quantify the quality of grain flow. The quantification of grain flow will become of great importance in the near-future because it will become essential for realizing process design optimization to meet strict quality specifications in simulation analyses. Recently the author of this work presented a unique method [14] of predicting grain flow or metal flow lines, which are contour lines of grain flow functions and new concepts of grain flow density and grain flow density gradient are presented to quantify the grain flow and its quality. A typical application example is given in Fig. 7, which shows the change of grain flow lines and the overlapping index. Fig. 7(a) and Fig. 7(b) are grain flow lines and overlapping index in the cross-section A-A defined in Fig. 7(c), respectively. Fig. 8 shows an example of optimal design considering normality of grain flow lines to the surface where taper rollers contact in service.

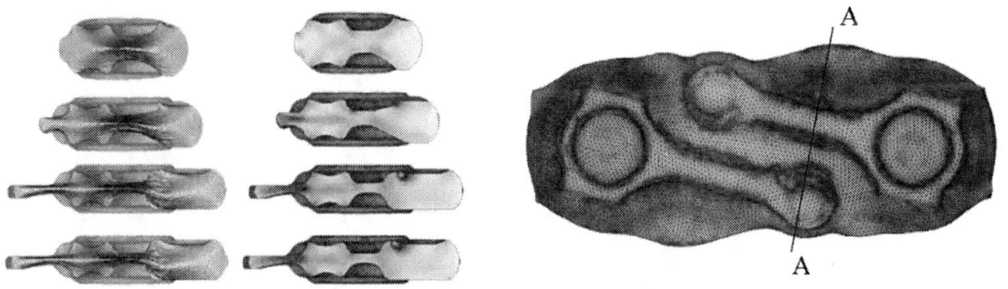

(a) Grain flow (b) Overlapping index (c) Overlapping index with 3D mesh system

Figure 7. Typical example of quantifying grain flow

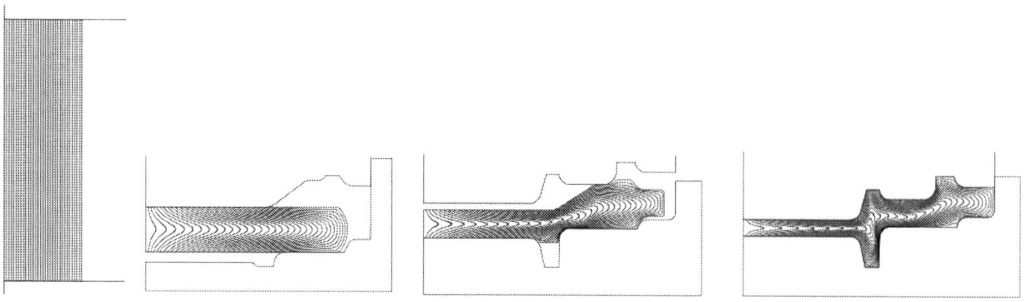

(a) Optimized process design

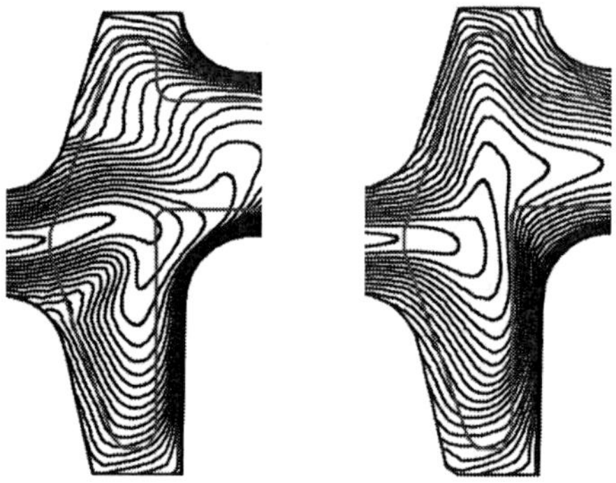

(b) Comparison of initial design (left) and optimized design (right)

Figure 8. Optimal process design in terms of normality of grain flow to the surface

2.7. Multi-body metal forming process simulation

Metal forming of multi-body processes, especially for assembling mechanical parts into assemblies, is one of hottest issues. Because residual interfacial stresses are important in these cases, elastoplastic finite element method is essential. These sorts of simulation will be becoming more and more important to guarantee the quality of assembly and optimize structural design of the assembly by the numerical way.

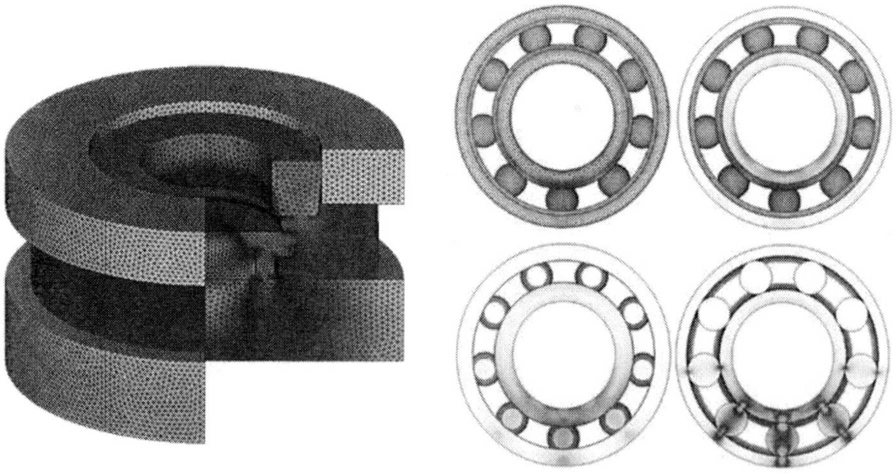

(a) Process of assembling three parts (b) Stresses due to increase in lateral force

Figure 9. Simulation of multi-body metal forming and mechanical processes

Fig. 9 shows two typical examples of multi-body metal forming process. Fig. 9(a) is an assembly process of three metal parts in which the residual interfacial stresses are of great importance and Fig. 9(b) shows the change of contact stresses owing to lateral force in ball bearing assembly.

2.8. Thick plate blanking simulation

Shearing, piercing, trimming and blanking processes are essential in metal forming. However, their simulation techniques are not so easy because of fracture phenomena and complicated remeshing during the process. Traditional theory of ductile fracture can be employed to separate the material. To the contrary, remeshing is not so simple in these cases. An example of typical blanking simulation is given in Fig. 10.

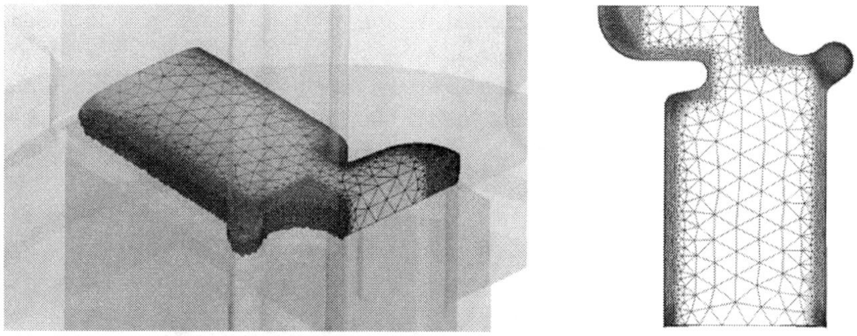

(a) Bird's view (b) Top view

Figure 10. Blanking simulation

2.9. Sheet metal forming simulation by solid element

Traditionally sheet metal forming processes have been solved by finite element methods with shell or plate finite-elements, which give quite reliable solutions especially when the processes rely on the stretching of thin sheet materials. However, these conventional approaches have shown decisive defects when thickness change of sheet or plate due to forming is one of major issues in the process design. In these cases, solid elements can give much better solutions and they can be inevitable in many cases. Fig. 11 shows a typical application of solid element approach to sheet metal forming of an oil pan.

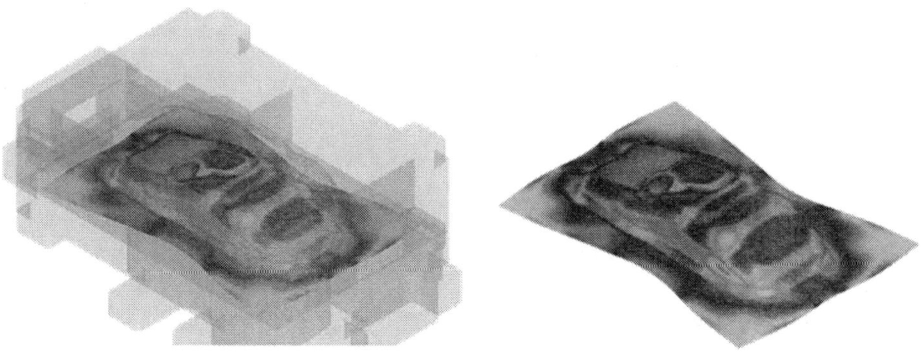

Figure 11. Sheet metal forming simulation by solid finite-element

Fig. 12 shows comparison of experiments and predictions of a sheet metal forming process. It is noteworthy that the material is a sheet of titanium-clad aluminum. The experimental dimensions of a and b were 126.9 and 67.9 mm, respectively while their predictions were 126.7 and 68.3 mm, respectively.

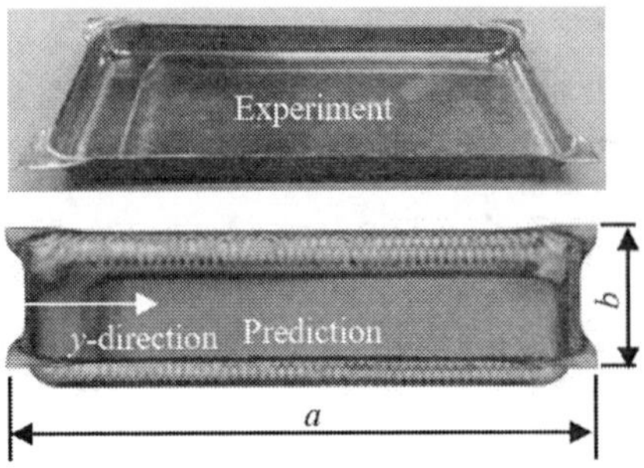

Figure 12. Comparison of predictions of a clad-metal sheet forming process with experiments

2.10. Incremental metal forming process simulation

Incremental metal forming processes are attractive because they can form very complicated shapes of product with reduced stages in a quite creative way. From standpoint of their simulation, they are very characterized by long computational time and special functions which should be user-friendly.

A typical example is shown in Fig. 13, which forms the stepped bar and the end spline. The stepped bar was formed by around 1000 blows in the lateral forging or swaging while the spline at an end is formed by the punch with oscillatory motion which is aimed to reduce the forming load.

(a) Experiment

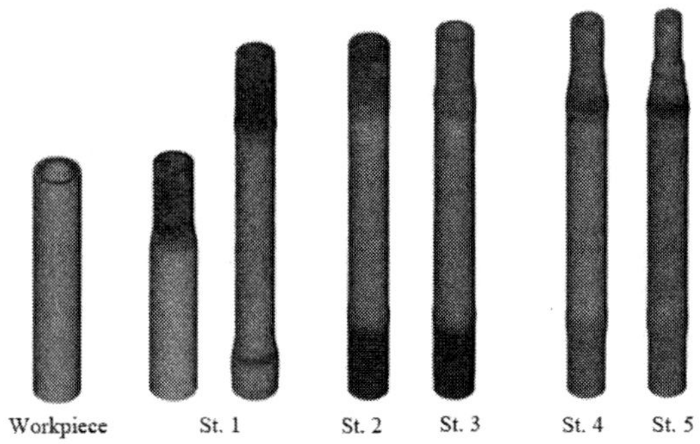

(b) Swaging process

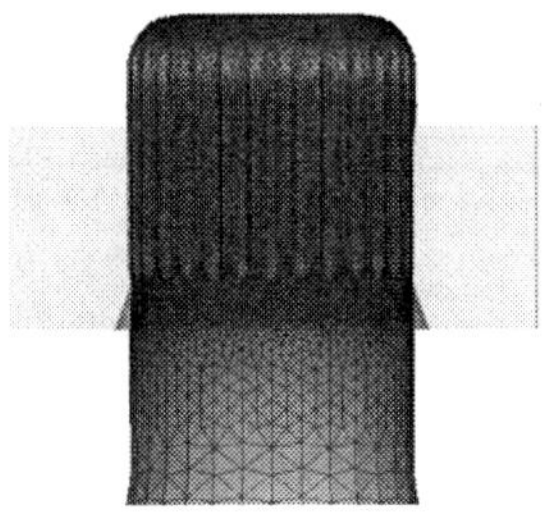

(c) Spline forming

Figure 13. Special incremental forming process of a stepped bar with spline at an end

3. Summary

In this study, some recent hot issues and special topics in metal forming simulation were presented with typical examples, which show some advances and/or characteristics related with accuracy, remeshing, material and friction modeling, multi-body process simulation, complete simulation considering effects of die elastic deformation, solid-element approach to sheet metal forming simulation, process design optimization and extreme case of incremental forming process. The hottest interests from standpoint of developers of metal forming simulators for engineers and researchers include process optimal design and sheet metal forming process simulation using the solid-elements.

The importance of metal forming technologies were emphasized in terms of automobile and electronic industries together with the difficulties in developing the nationwide metal forming industry. The characteristics of metal forming technologies were explained, revealing that nationwide cooperation is essential to develop the metal forming industry because it should be supported by so many infra-structural industries and companies. The importance of the role of academic sides to develop the metal forming industry in the developing countries was also emphasized.

In Korea, AFDEX was started to be used by the leading Korean forging companies in the middle of nineties when most commercial software could be applied mainly for academic purposes by universities and research institutes and they have been able to innovate the imported or dead-copies forging processes by themselves. It should be noted that the forging process design engineers could not dare to change the traditional forging processes in those days. Nowadays most forging companies are using forging simulation technology and web services are also being given to small-sized forging companies by Korea Institute of Industrial Technology established by Korean government.

The examples shown in this study showed how Korean metal forming societies have been everlastingly evolving and how academic side can prepare or lead the potential development of metal forming industry in the developing countries. International cooperation for sharing experiences in this field is needed for saving global environments because metal forming is the most influential green manufacturing technology.

Acknowledgement
This work is carried out by the support of the BK 21 PLUS project and Industrial Core Technology Development Project (2003950) from KEIT.

References

[1] Joun MS (Ed.), 1996, Workshop on metal forming CAE technology, Jinju Korea.

[2] Joun MS and Lee MC, 1997, Quadrilateral finite-element generation and mesh quality control for metal forming simulation, *Int. J. Num. Meth. Eng.* **40(21)** 4059-4075.

[3] Lee MC, Lee JK and Joun MS, 2007, Adaptive tetrahedral element generation and refinement to improve the quality of bulk metal forming simulation, *Finite Elem. Anal. Des.* **43(10)** 788-802.

[4] Jeong SW, Eom JG, Lee MC, Lee JE, Joun MS, Finite element analysis of a coining process with extreme number of finite elements, *Proc. KSTP*, Joun MS, Daegu, Korea.

[5] Meyer LW, Weise A, Hahn F, 1997, Comparison of constitutive flow curve relations in cold and hot forming, *J. Phys.* **7** 3-20.

[6] Razali MK, Irani M, Joun MS, 2018, Consideration of Voce's flow stress model and its variants, *Proc. KSTP*, Joun MS, Seoul, Korea 138–139.

[7] Ebrahimi R, Zahiri SH, Najafizadeh A, 2006, Mathematical modelling of the stress–strain curves of Ti-IF steel at high temperature, *J. Mater. Process. Technol.* **171** 301–305.

[8] Razali MK, Irani M, Joun MS, 2019, General modelling of flow stress curves of alloys at elevated temperatures using bi-linearly interpolated or closed-form functions for material parameters, *J. Mater. Res. Technol.* **8(3)** 2710-2720.

[9] Opěla P, Schindler I, Kawulok P, Kawulok R, Rusz S, Petrek T, 2015, Hot flow stress models of the steel C45, *Metallurgical* **54** 469-472.

[10] Joun MS, Eom JG, Lee MC, 2008, A new method for acquiring true stress–strain curves over a large range of strains using a tensile test and finite element method, *Mech. Mater.* **40(7)** 586-593.

[11] Eom JG, Son YH, Jeong SW, Ahn ST, Jang SM, Yoon DJ, Joun MS, 2014, Effect of strain hardening capability on plastic deformation behaviors of material during metal forming, *Mater. Design.* **54** 1010-1018.

[12] Joun MS, Moon HG. Choi IS, Lee MC, Jun BY, 2009, Effects of friction laws on metal forming processes, *Tribol. Int.* **42(2)** 311-319.

[13] Joun MS, Lee SW, Jung JH, 1998, Finite element analysis of a multi-stage axisymmetric forging process having a spring-attached die for controlling metal flow lines, *Int. J. Mach. Tools Manu.* **38(7)** 843-854.

[14] Joun MS, Lee MC, Eom JG, 2016, Simulation device for object to be plastically deformed, *Japanese patent*.

IOP Conf. Series: Materials Science and Engineering **834** (2020) 012017 doi:10.1088/1757-899X/834/1/012017

Cycle Time Reduction for Coating Process in Manual Assembly Manufacturing Towards Economic Sustainability

Shamini Janasekaran[1], Vijaya Prakash Vijayasree[2] and Marcus K H Chong[1]

[1]Centre for Intelligent Manufacturing System, Department of Mechanical Engineering, Faculty of Engineering and Built Environment, SEGi University, Petaling Jaya, Selangor, Malaysia
[2]Faculty of Environmental studies, Universiti Putra Malaysia, Serdang, Selangor, Malaysia

shaminijanasekaran@segi.edu.my

Abstract. Lean manufacturing has been one of the techniques to eliminate waste in production to increase productivity. Improper usage of cycle time can extend lead-time of delivery for the production and incur more cost than needed. Determining sufficient cycle time is important to ensure the productivity of the each assembly line is within the delivery time span. Improvement of productivity contributes in maintaining sustainable environment that attracts customer and government supports. In this paper, a case study in a manufacturing assembly line was conducted to identify the operator's motion waste as non-value added activity. As a result, the cycle time was reduced from 60 seconds to 50 seconds in a multiple engaging production activities besides improving the ergonomic posture of the operators. A double four stages rack was designed to reduce the cycle time by 16.7%, which leads to increment of 1000 units of products per month. The rack was built-in together within the assembly process and reduced movement of operators.

1. Introduction

In current era, customers are looking for higher productivity output at shorter lead-time without compensating the quality. Small and medium enterprises are eager to supply their products at short notice of delivery to compete within each other. Only by doing that, the SMEs can sustain for longer period in the business-oriented society. SMEs are slammed with challenges by the rapid changing market requirements where even seconds' delay can contribute to the losses of business [1]. These SMEs industries are playing huge increasing role in Malaysia economy where they contribute 37.1% in 2017 compared to 36.6% in 2016. Their contribution is remarkable with RM435.1 billion in 2017 to Malaysia's economy with a real gross domestic product growth of 7.2% compared with 5.2% in 2016 [2]. Therefore, continuous improvement also known as kaizen is always been in demand through lean manufacturing. Kaizen activities include elimination of seven waste and 5S activities. These seven wastes include motion, waiting, overproduction, inventory, defects, over processing and transportation, which can be, eliminate by sorting Value Added (VA) activity and Non Value Added (NVA) activity. The NVA waste which is the vital obstacle for VA activity will definitely give customer non-satisfaction with the respective industry [3]. NVA categorized into seven wastes according lean manufacturing practices. They are wastes from overproduction, waiting time, transportation, inventory, over-processing, motion and defects. Despite the extensive knowledge and available lean resources, many SMEs are struggling to stay "lean". The main intention in lean manufacturing are to reduce waste in human effort and inventory, which able products to reach market on time, while producing quality products in the most efficient and economical manner besides being responsive towards customer query [4]. Modification of current production process with implementation of lean will be

Content from this work may be used under the terms of the Creative Commons Attribution 3.0 licence. Any further distribution of this work must maintain attribution to the author(s) and the title of the work, journal citation and DOI.
Published under licence by IOP Publishing Ltd

leading path towards adaption of Industrial Revolution 4.0 (IR 4.0) within shorter lead time [5]. However, besides focusing on IR 4.0, the environment awareness plays high importance to all responsible manufacturers. For global scale, this considered as competitive when their production is aligned and adopted by sustainable techniques such as minimizing number of manufacturing steps. This can be done by employing alternative methods with minimizing energy consumption including human energy and ergonomically stable [6, 7]. The adaption of new techniques is important despite of all the effectiveness in the manufacturing due being instigated by the stakeholders such as customers, policy makers, regulatory authorities and society [8].

In this case study, motion of workers to turn at improper position has impact on the cycle time which can be categorized as NVA. Simplification of movement surely reduces the motion waste. From the previous research done by others [9-11], waste in production has been highlighted but specific studies on motion related to ergonomics and sustainability has yet to be conducted and presented clearly. Therefore, this research would propose one of the optimum designs of coating rack that improves the ergonomic motion of operators in aid to improve the productivity and reduction of cycle time for a particular process which leads to reduction of lead time overall.

2. Experimental Procedure

Conformal coating process is one of the essential processes in completion of loaded printed circuit board (PCB) in most of the manufacturing practice. Coating process eliminates the possibilities of the electronics components' lead pin from touching unnecessary printed copper circuit lines on the loaded boards. This process cycle time is counted from the time the PCB taken from conveyor until it is laid again on conveyor after the coating and drying process complete. For this particular model, the longest process was coating and needs four operators to work simultaneously to get cycle time for 60 seconds. Total time given was 240 seconds (this study was done separately using tact time studies adding each steps from taking the PCB to start to work on until it is placed back on the conveyor) even though the total value added activity only needs 190 seconds. This is because 50 seconds needed for the operators' movement such as bending to take the uncoated PCB from the rack, which is placed beside the conveyor. Coating liquid is applied in bidirectional at the bottom of each PCB as shown in Figure 1.

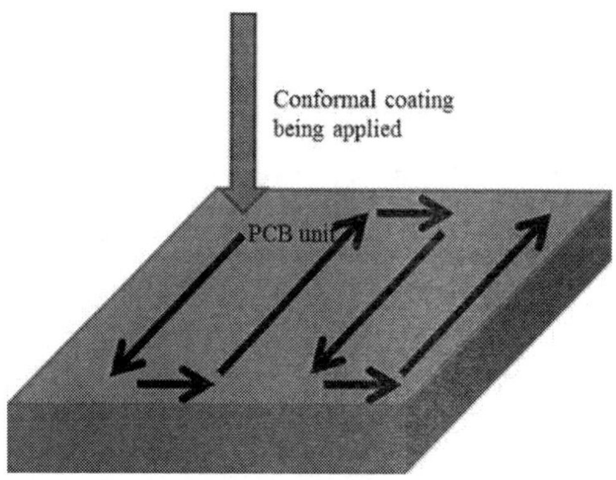

Figure 1. Bi-directional coating orientation on bottom of PCB that takes 240 sec to complete

Figure 2 shows the possible bending of 90^0 to 145^0 for every PCB. Eight PCB were kept on the waiting rack beside the conveyor at the left position of the operator. First, the operator needs to take the PCB, coat the bottom layer of PCB with adhesive liquid, and place it back on the rack for it to dry before sending it on the conveyor for next process. The repetitive movement distress the working ergonomic posture and tend to increase fatigue in the operator which can lead to poor performance and errors [12]. The most common error detected was placing not fully dried PCB on the conveyor, which will be categorized as defects if the coating is sticky for the next process. Previous studies had shown ergonomic issues can produce poor quality assembled products and defects to customers [13, 14].

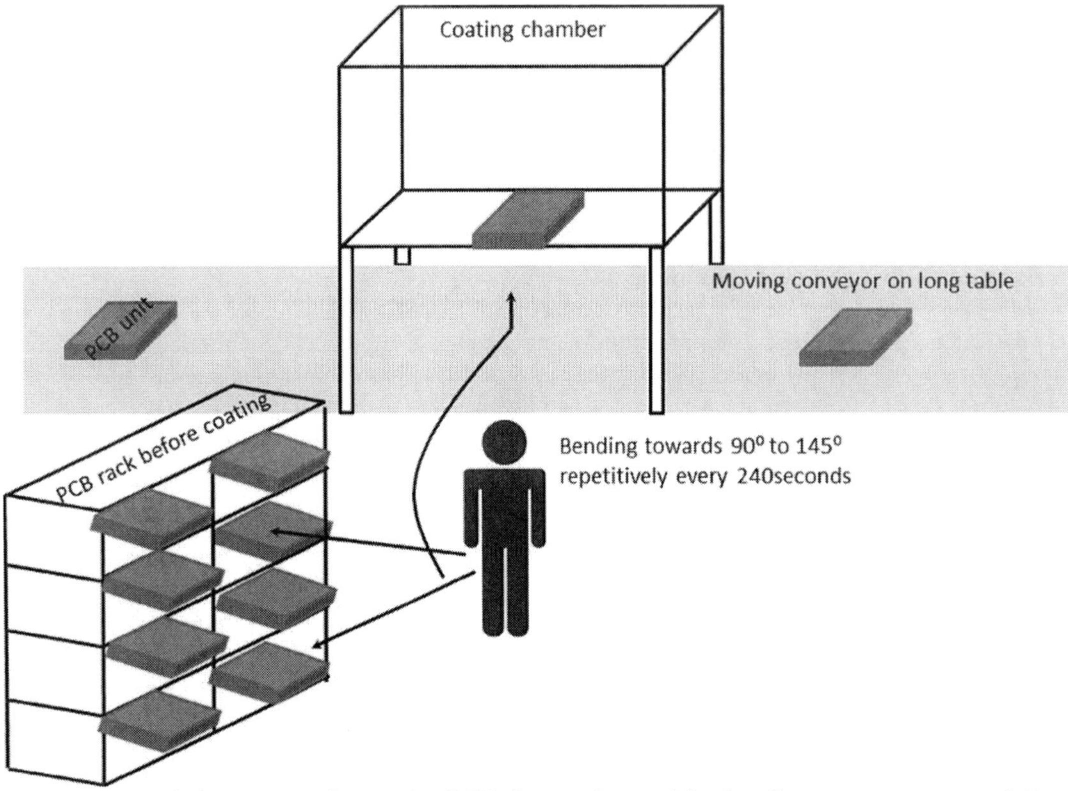

Figure 2. Existing process for coating PCB that needs repetitive bending movement every 240 seconds

3. Results and Discussion

3.1 Implementation of new coating rack

In this specific study, fixing the stages of PCB rack beside the coating chamber throughout the operation had reduced the motion involved. The ergonomically studied motion had reduced the cycle time to 50 seconds which overall increases the productivity. Figure 3 shows the best rack design obtained to reduce the operators' movement for every PCB. In this proposal, operators need to turn slight (approximately 20^0) left and right each time to grab a PCB to do the coating process. The operator takes uncoated PCB from left and proceeds for coating within 190 seconds and place the PCB at the right side of the coating rack for drying. Once it has dried, the operator placed it back to the conveyor for next process. The total process had reduced to the cycle time for the whole lot by 10 seconds from the time observation study. Initial rack position consumed operator 50 seconds to take and return the PCB meanwhile with new rack design, 40 seconds were reduced, which is 80% time reduction.

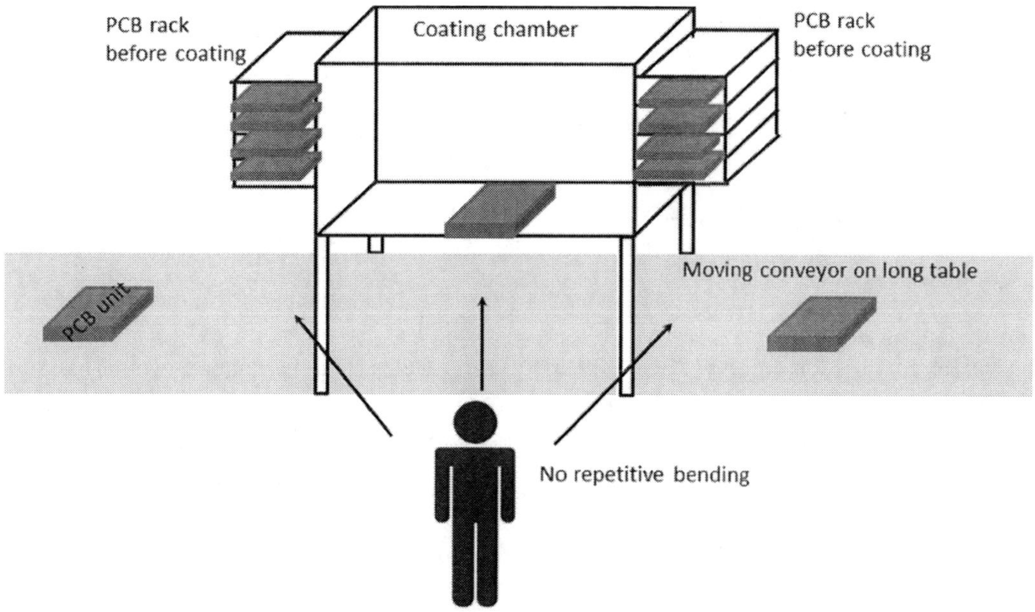

Figure 3. Implementation of double four stages rack for coating process

3.2 Improvement in productivity

The implementation of new coating rack has reduced the cycle time for the whole lot that can be calculated as shown in Equation 1-4. A lot of 5000 units of PCB were considered for one week production. The existing required cycle time (tact time) is 60 seconds which reduced to 50 seconds after implementation of 4 stages coating rack.

$$Time\ taken\ to\ produce\ y\ number\ of \frac{PCB}{week} = y * tact\ time \tag{1}$$

From Eq 1, time taken for 5000 PCB was previously 300000 seconds and reduced to 250000 seconds per week. Considering the increment of PCB using same 300000 seconds per week will produce 6000 units (additional of 1000 units) using Equation 2.

$$New\ PCB\ units = \frac{Time\ taken\ to\ produce\ y\ PCB\ previously}{new\ tact\ time} \tag{2}$$

The coating rack improvement has increased the productivity of the coating process by 1000 units, which can be converted into cost savings within the same period as shown in Equation 3.

$$Profit = cost\ per\ unit * additional\ units \tag{3}$$

$$Cycle\ time\ reduction = \frac{300000 - 250000}{300000} x100\% = 16.7\% \tag{4}$$

The detailed results have been tabulated in Table 1.

Table 1. Detailed results obtained before and after implementation of double four stages rack for coating process

Details	Implementation of coating rack	
	Before	After
Lead time for 4 PCBs coating (seconds)	240	200
PCB coated weekly (pieces)	5000	6000
Cycle time for whole production line for each output (seconds)	60	50

4. Conclusion

From this case study, following can be concluded:

a) Productivity had increased by 1000 units of product per month after elimination of non-value added activity such as motion.
b) The cycle time of the coating process workstation was reduced by 16.7% after implementation of double 4 stages rack.
c) Motion of operators during production can be manipulated to reduce the ergonomics discomfort such as repetitive bending and turning with 80% of NVA has been reduced.
d) Minimizing operators' movement has proved manufacturers' commitment towards economy sustainability.

Acknowledgments
This research was supported by SEGi University, Kota Damansara and authors would like to thank anonymous reviewers for their comments.

References

[1] Kumar, M. and R. Vaishya, *Real-Time Monitoring System to Lean Manufacturing.* Procedia Manufacturing, 2018. **20**: p. 135-140.
[2] Jaafar, S.S., *SME contribution to Malaysia's economy rose to 37% in 2017*, in *The Edge Market.* 2018, Edge Media Group.
[3] Arya, A.K. and S. Choudhary, *Assessing the application of Kaizen principles in Indian small-scale industry.* International Journal of Lean Six Sigma, 2015. **6**(4): p. 369-396.
[4] Rahman, N.A.A., S.M. Sharif, and M.M. Esa, *Lean manufacturing case study with Kanban system implementation.* Procedia Economics and Finance, 2013. **7**: p. 174-180.
[5] Balasingham, K., *Industry 4.0: securing the future for german manufacturing companies.* 2016, University of Twente.
[6] Gupta, K., et al., *Recent developments in sustainable manufacturing of gears: a review.* Journal of Cleaner Production, 2016. **112**: p. 3320-3330.
[7] Longoni, A. and R. Cagliano, *Cross-functional executive involvement and worker involvement in lean manufacturing and sustainability alignment.* International Journal of Operations & Production Management, 2015. **35**(9): p. 1332-1358.
[8] Qureshi, M.I., et al., *Sustainability: A new manufacturing paradigm.* Jurnal Teknologi, 2015. **77**(22).
[9] Poppendieck, M., *Principles of lean thinking.* IT Management Select, 2011. **18**(1): p. 1-7.
[10] Elnamrouty, K. and M.S. Abushaaban, *Seven wastes elimination targeted by lean manufacturing case study "gaza strip manufacturing firms".* Seven wastes elimination targeted by lean manufacturing case study "gaza strip manufacturing firms", 2013. **1**(2).
[11] Modi, D.B. and H. Thakkar, *Lean thinking: reduction of waste, lead time, cost through lean manufacturing tools and technique.* International Journal of Emerging Technology and Advanced Engineering, 2014. **4**(3): p. 339-334.
[12] Bohgard, M., et al., *Work and technology on human terms.* 2009: Prevent.
[13] Falck, A.C., R. Örtengren, and D. Högberg, *The impact of poor assembly ergonomics on product quality: A cost–benefit analysis in car manufacturing.* Human Factors and Ergonomics in Manufacturing & Service Industries, 2010. **20**(1): p. 24-41.
[14] Baraldi, E.C., *Ergonomic planned supply in an automotive assembly line.* Human Factors and Ergonomics in Manufacturing & Service Industries, 2011. **21**(1): p. 104-119.

Modification of a Square Cylinder Wake through a Secondary Upstream Cylinder

Hamid, A.H.A., Jali, M.S.M, Azmi, A.M., Noh, M.H.M.

Faculty of Mechanical Engineering Universiti Teknologi MARA, 40450, Malaysia

a.hussein.uitm@gmail.com

Abstract. One of the inherent characteristics of a bluff body is the large drag coefficient even at a relatively low velocity. The condition is made worse when the flow separates from the body, and a time-dependent flow occurs downstream of the body; i.e. a vortex shedding phenomenon. It has been shown that suppression of the vortex shedding can significantly reduce drag coefficient of the bluff body. The present study investigates the intensity of the resulting laminar wake of a single and two square cylinders in tandem arrangement, where the intensity was measured through the lift force signals. The findings indicated that the presence of the secondary cylinder significantly altered the magnitude of lift force of the cylinder system, particularly at high Reynolds numbers under the laminar flow regime. Furthermore, the magnitude of lift force was lower when the secondary cylinder was separated from the main cylinder as compared to the case where both cylinders were attached.

1. Introduction

The first Studies on vortex shedding of bluff bodies and analysis of flow separation have been a topic of interest due to their fundamental significance and practical importance in some engineering applications such as aerodynamics and hydrodynamics. Examples of such applications are vibration of pipelines lying on the sea bottom under the influence of currents and waves, bridges, interaction of currents and waves with offshore structures, flow tubular or tube banks heat exchangers, skyscrapers, chimneystacks, suspension bridges and chimneys near tall buildings, structures in the atmospheric boundary layers and submarine pipe see e.g., [1], [2], [3], [4] and [5].

These bluff bodies might experience unsteady loads due to the flow separation and formation of the wake region and vortices. By controlling the wake structures and flow separations, the unsteady forces experienced by the bluff bodies is significantly reduced. The unsteady forces due to the vortex shedding induce various effects such as mixing enhancement, vibration and noise production, structural system resonance and scouring. Depending on the target application, the effects can be beneficial or harmful, and others could be even detrimental. In the latter case, the strategy of vortex shedding suppression should be implemented clearly.

Various control methods have been suggested to control the wake structure and vortex shedding, such as using current injection [6], utilizing an external small element [7], using rotary oscillations [8], generating a secondary flow (suction, blowing, bleed, and synthetic jets) [9], modifying surface roughness [10], using thermal effects [11] and adding add on devices [12]–[14]. These approaches fall into two major categories, which are passive and active control methods. The active control method requires some external energy to alter the fluid flow characteristics, while passive method does not require any external energy which is simpler to implement than that of the active method.

[15] has performed a numerical study of vortex shedding behind a bluff body for flow at low Reynolds number by placing a relatively smaller control cylinder placed in the wake region of the main

Content from this work may be used under the terms of the Creative Commons Attribution 3.0 licence. Any further distribution of this work must maintain attribution to the author(s) and the title of the work, journal citation and DOI.
Published under licence by IOP Publishing Ltd

bluff body. Computed results showed that the shear layer stabilized with the presence of the control cylinder and the local favorable pressure gradient was produced by the control cylinder in the wake region of the flow. [16] also employ a similar approach to suppress the vortex shedding of a bluff bady by using a rectangular or square cylinders placed upstream of a square bluff body. The results showed that the drag force decreased with the height of the upstream bluff body for a specific gap height (distance between the cylinder and wall).

In the context of passive flow control approach, the effective zone (location element the wake where vortex shedding can be suppressed, and the drag and lift fluctuation can massively decrease) is still not well configured or understood due to the sensitivity of the passive method to the bluff body arrangements. In addition. there is still lack of square bluff bodies wake control studies in the literatures. The present work seeks to examine the wake of and forces acting on a square cylinder subjected to an upstream flow condition due to the presence of a relatively small square control cylinder.

2. Methodology

In this investigation an internal flow of viscous fluid over square cylinders in tandem arrangement in a rectangular duct was considered (as depicted in Figure 1). A benchmark case (Case 0) of a single square cylinder with side length d in the domain was also considered in this study. For the case of two square cylinders in tandem, a small cylinder of size $0.5d$ was placed upstream of a large cylinder of size d. For the first case, both cylinders were attached (Case 1), as shown in Figure 1(a), while for the second case (Case 2), both cylinders were separated with a centre-to-centre gap of $1.25d$, , as shown in Figure 1(b). A fully-developed flow was imposed at the inlet boundary and a Neumann boundary condition of zero pressure gradient was imposed at the outlet boundary. All other boundaries were set to no-slip condition.

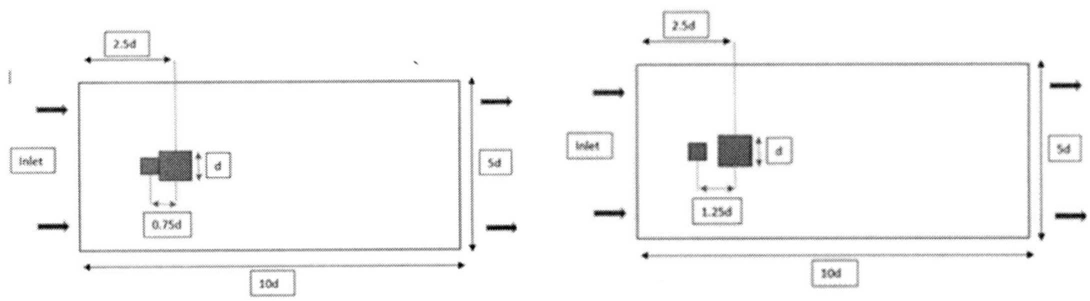

(a) (b)

Figure 1. Schematic diagram of the system under investigation. The control cylinder is either (a) attached (Case 1) to or (b) separated from the main cylinder (Case 2).

The meshes were constructed in such a way that elements are adequately small near boundaries in order to properly resolve high velocity gradients at the boundaries, as depicted in Figure 2. The flow was set as laminar and the governing equation was solved using a SIMPLE scheme for pressure-velocity coupling.

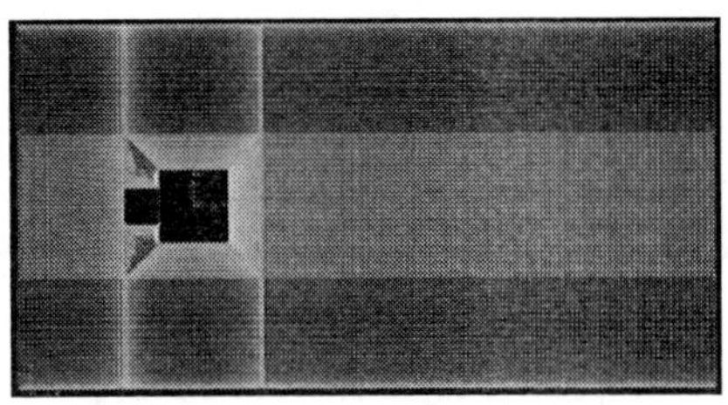

Figure 2. A typical mesh with two cylinders in tandem configuration, extends 2.5d upstream and 7.5d downstream.

3. Results and Discussion

Flows over cylinders with different configurations were simulated for Reynolds numbers ranging between 50 and 200, with the increment of 50. At $Re = 50$, flows in all configurations were stable, i.e. no time-dependent wakes behind the cylinder. This can be clearly seen in the plot of lift coefficient (a typical example for Cases 0 (indicated by a blue solid line) and 1 (indicated by grey solid and dashed lines) are shown in Figure 3(a) and Figure 3(b)), where there is no fluctuation in the signal of lift force. When Re is increased to 100, the single cylinder case shows a time-periodic wake in the downstream flow. This wake flow caused a periodic fluctuation of positive and negative lift forces experienced by the cylinder, as indicated by the sinusoidal fluctuation of the lift force in Figure 3(c). However, for Case 1, the lift force signals for both cylinders show very little periodicity with much smaller magnitude than that of Case 0. This observation indicates that the presence of a small cylinder attached to the upstream side of the main large cylinder tends to delay the onset of full periodicity of vortex shedding. For Case 2 however, there is no sign of fluctuating lift force even until $t = 60$s, where t is flow simulation time (as shown in Figure 3(d)).

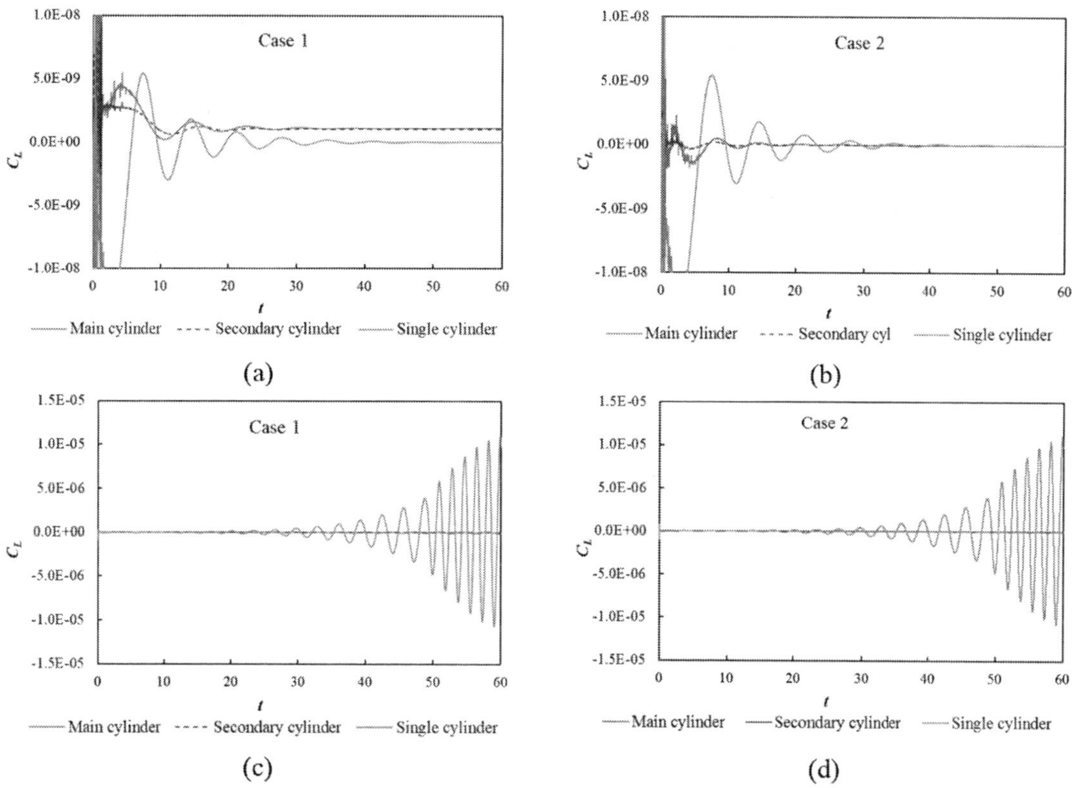

Figure 3. Plots of lift coefficient time history for (a-b) $Re = 50$ and (c-d) $Re = 100$.

When Re is increased to 150, it is observed that the flows reach a fully-periodic state as early as $t \approx$ 30s, as depicted in Figure 4(a). For Case 1, the main cylinder experiences a fluctuating lift force with an amplitude similar to that of the single cylinder case. The amplitude of lift force for the secondary cylinder is, however, much lower than that of the main cylinder. This observation is attributed to the fact that the presence of the large cylinder downstream of the secondary small cylinder tends to inhibit the wake flow of the secondary cylinder, thus reducing the amplitude of lift force. The influence of the secondary cylinder is even more pronounced for Case 2, where the vortex shedding of both cylinders is almost attenuated at $Re = 150$ (the amplitude of the lift force is two to three orders of magnitudes smaller than that of Case 1 – as shown in Figure 4(b)).

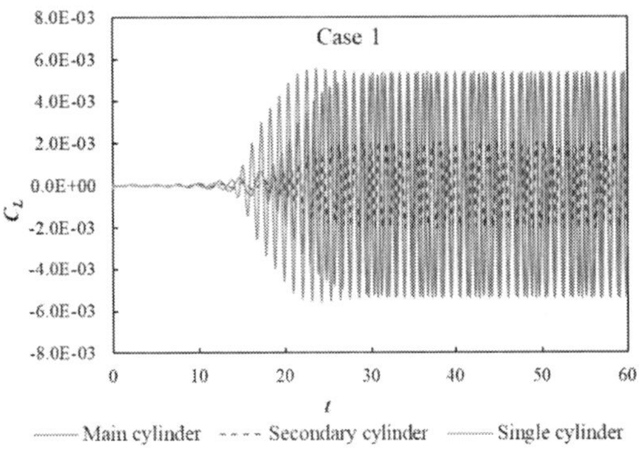

(a)

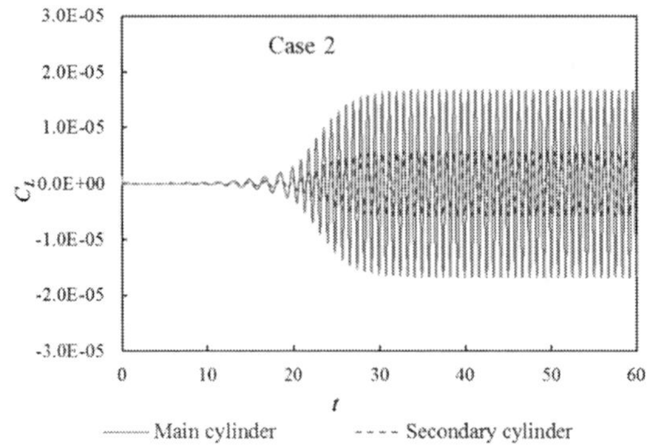

(b)

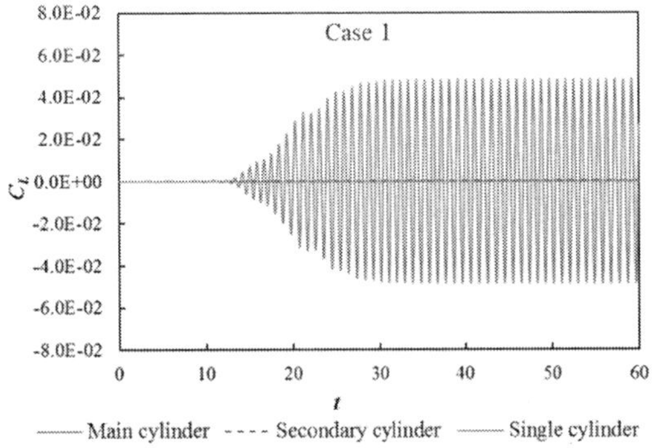

(c)

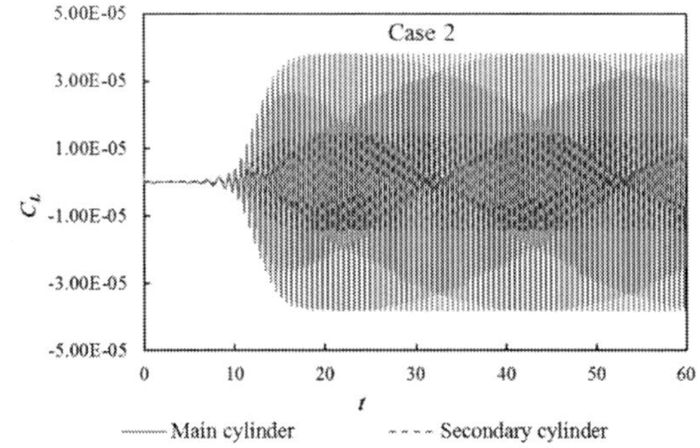

(d)

Figure 4. Plots of lift coefficient time history for (a-b) $Re = 150$ and (c-d) $Re = 200$.

When flow reaches Re of 200, the amplitude of lift force exerted on the cylinder of Case 0 increases considerably, as depicted in Figure 4(c). For Case 1, the amplitude of lift force for both cylinders decreases by two orders of magnitudes compared to that of the $Re = 150$ case. In contrast, the amplitude of lift force exerted on both cylinders for Case 2 doubled when Reynolds number is increased from 150 to 200, as shown in Figure 4(d). However, it is important to note that the rms amplitude of lift force exerted on both cylinders for Case 2 is just 0.65% lower than that of Case 1 at this Reynolds number. This is due to the fact that the vortices shed at almost the same downstream distance for both cases and that the intensity of shed vortices is almost identical (as indicated in Figure 5).

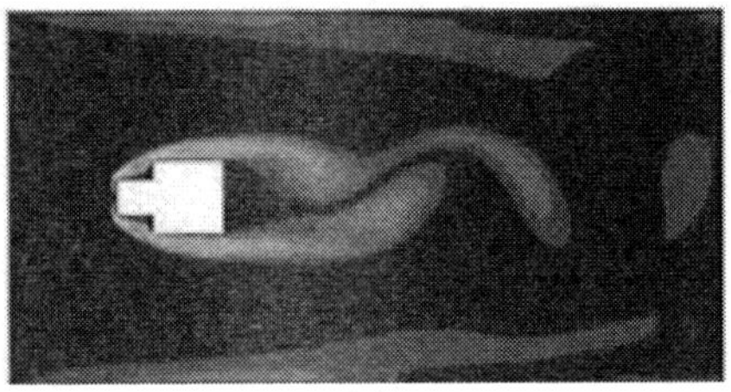

(a)

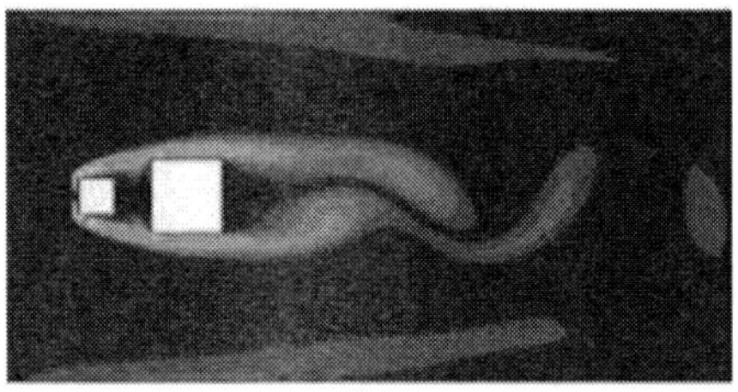

(b)

Figure 5. Contours of vorticity magnitude at Re = 200 for (a) Case 1 and (b) Case 2. The contour ranges from 0 to $125 s^{-1}$.

4. Conclusion

Flow over a single and two square cylinders in tandem arrangement has been studied. Low Reynolds number producing laminar wakes were considered. It was shown that the presence of an upstream secondary cylinder modified the amplitude of lift force exerted on the main cylinder, except for when the secondary cylinder is attached to the main cylinder at $Re = 150$ and for Reynolds number in which the wake of a single cylinder is steady. Furthermore, the presence of this secondary cylinder tends to delay the onset of full periodicity of vortex shedding at certain Reynolds number. It was also shown that the placement of a separated secondary cylinder upstream of the main cylinder was better at inhibiting time-dependent wakes as compared to the attached counterparts within the investigated Reynolds number range.

References

[1] Z. P. Zang, F. P. Gao, and J. S. Cui, "Physical modeling and swirling strength analysis of vortex shedding from near-bed piggyback pipelines," *Appl. Ocean Res.*, vol. 40, pp. 50–59, 2013.

[2] Z. P. Zang and F. P. Gao, "Steady current induced vibration of near-bed piggyback pipelines: Configuration effects on VIV suppression," *Appl. Ocean Res.*, vol. 46, pp. 62–69, 2014.

[3] M. Bovand, S. Rashidi, and J. A. Esfahani, "Enhancement of heat transfer by nanofluids and orientations of the equilateral triangular obstacle," *Energy Convers. Manag.*, vol. 97, pp. 212 223, 2015.

[4] S. Rashidi, M. Bovand, I. Pop, and M. S. Valipour, "Numerical Simulation of Forced Convective Heat Transfer Past a Square Diamond-Shaped Porous Cylinder," *Transp. Porous Media*, vol. 102, no. 2, pp. 207–225, 2014.

[5] S. Rashidi, A. Nouri-Borujerdi, M. S. Valipour, R. Ellahi, and I. Pop, "Stress-jump and Continuity Interface Conditions for a Cylinder Embedded in a Porous Medium," *Transp. Porous Media*, vol. 107, no. 1, pp. 171–186, 2015.

[6] A. H. A. Hamid, W. K. Hussam, and G. J. Sheard, "Combining an obstacle and electrically driven vortices to enhance heat transfer in a quasi-two-dimensional MHD duct flow," *J. Fluid Mech.*, vol. 792, pp. 364–396, 2016.

[7] C. H. Kuo and C. C. Chen, "Passive control of wake flow by two small control cylinders at Reynolds number 80," *J. Fluids Struct.*, vol. 25, no. 6, pp. 1021–1028, 2009.

[8] W. K. Hussam, M. C. Thompson, and G. J. Sheard, "Enhancing heat transfer in a high Hartmann number magnetohydrodynamic channel flow via torsional oscillation of a cylindrical obstacle," *Phys. Fluids*, vol. 24, no. 11, 2012.

[9] L. Hao Feng, J. J. Wang, and C. Pan, "Effect of novel synthetic jet on wake vortex shedding modes of a circular cylinder," *J. Fluids Struct.*, vol. 26, no. 6, pp. 900–917, 2010.

[10] E. Achenbach and E. Heinecke, "On vortex shedding from smooth and rough cylinders in the range of Reynolds numbers 6×103 to 5×10^{6}," *J. Fluid Mech.*, vol. 109, no. 12, pp. 239–251, 1981.

[11] J. C. Lecordier, L. W. B. Browne, S. Le Masson, F. Dumouchel, and P. Paranthoën, "Control of vortex shedding by thermal effect at low Reynolds numbers," *Exp. Therm. Fluid Sci.*, vol. 21, no. 4, pp. 227–237, 2000.

[12] A. M. Azmi, H. Jamil, and T. Zhou, "Effect of Hole-uniformity of Shrouds on Vortex Shedding Suppression behind a Circular Cylinder," *IOP Conf. Ser. Mater. Sci. Eng.*, vol. 417, no. 1, 2018.

[13] A. M. Azmi, T. Zhou, Y. Zhou, H. Wang, and L. Cheng, "Coherent structures in a screen cylinder wake," *Phys. Rev. Fluids*, vol. 3, no. 7, p. 074702, 2018.

[14] A. M. Azmi and T. Zhou, "Effect of screen geometry on the vortex formation behind a circular cylinder," *J. Adv. Res. Fluid Mech. Therm. Sci.*, vol. 44, no. 1, pp. 47–54, 2018.

[15] S. Mittal and A. Raghuvanshi, "Control of vortex shedding behind circular cylinder for flows at low Reynolds numbers," *Int. J. Numer. Methods Fluids*, vol. 35, no. 4, pp. 421–447, 2001.

[16] D. K. Maiti and R. Bhatt, "Vortex shedding suppression and aerodynamic characteristics of square cylinder due to offsetting of rectangular cylinders towards a plane," *Ocean Eng.*, vol. 82, pp. 91–104, 2014.

IOP Conf. Series: Materials Science and Engineering **834** (2020) 012019 doi:10.1088/1757-899X/834/1/012019

Improvement of Insulation Material for Cool Box Application

M. Khairulmaini[1], Z. Michael[1], M. A. M. Shah[1], M. S. Zakaria[2], B. Abdullah*[2] and A. A. Rashid[2]

[1]Faculty of Mechanical Engineering, Universiti Teknologi MARA, Cawangan Johor Kampus Pasir Gudang, Johor Darul Takzim, Malaysia
[2]Faculty of Mechanical Engineering, Universiti Teknologi MARA Shah Alam, 40450 Selangor Darul Ehsan, Malaysia

*bulan_abd@uitm.edu.my

Abstract. Current conventional cool boxes which use expanded polystyrene (EPS) as its insulating material cannot seem to maintain storage at low temperature for long periods. This study aims to determine the effect for different types of insulating materials which can be used in cool boxes to improve its thermal insulation. Using theoretical and simulation approach, three out of six potential insulating materials were chosen in terms of their heat transfer rate, and temperature at the outer wall. The three best insulating materials, polyurethane (PU), expanded polystyrene (EPS) and poly-glass fibre (PGF) were then tested at three different room temperatures experimentally using the thermal insulation test to determine the best insulating material for a cool box. Tensile and density were also conducted to identify the mechanical behaviour and properties of these insulating materials. From the results obtained from the experiment, polyurethane (PU) recorded the longest time taken for ice to change into water at the three different temperatures when compared to other insulating materials. The results showed that when thermal conductivity is low, the heat transferred through the walls of the cool box was also low thus resulting in better thermal insulation of the cool box. This study gives a better understanding in term of thermal insulation and the results obtained from this study can help with the production of better cool boxes with improved thermal insulation.

1. Introduction

A cool box, portable ice chest, icebox or cooler box is an insulated box most commonly used to keep perishable foods or drinks at a relatively cold temperature. The use of cooler boxes ranges from small applications such a family picnics, to large applications such as cold chain logistics. Conventional cool box is made from expandable polystyrene (EPS), can only maintain a cool temperature of its storage for 24 hours with a surrounding temperature of 32°C. To maintain a cool temperature within the cool box, heat transfer to and from the cool box and its surrounding must be relatively low. This can be achieved by providing better thermal insulation for the cool box.

Thermal insulation is the reduction of heat transfer (the transfer of thermal energy between objects of differing temperature) between objects in thermal contact. Heat transfer is all about the transfer of heat from one point to another where heat is transferred from high temperature to low temperature. For example, if a system is at a higher temperature compared to its surroundings, heat will transfer from the system to the surroundings [1]. Thermal insulation provides a layer of protection in which thermal conduction is reduced or thermal radiation is deflected rather than absorbed by the lower-temperature body [2]. The capability of a material to provide insulation is measured by its thermal conductivity (K) value. Materials with high insulating capability, resistance (R) will have a lower thermal conductivity

Content from this work may be used under the terms of the Creative Commons Attribution 3.0 licence. Any further distribution of this work must maintain attribution to the author(s) and the title of the work, journal citation and DOI.
Published under licence by IOP Publishing Ltd

value [3]. In thermal engineering application, besides thermal conductivity, other important properties of material with the purpose of insulation are density (ρ) and specific heat capacity (c) [2].

Thermal insulation of an object is very much influenced by the insulation material. For this study, six insulation materials have been selected, expanded polystyrene (EPS), polyurethane (PU), glass wool, asbestos, poly glass fibre, and coconut fibre. With regards to these six insulating materials, several studies have shown that they can be used to reduce heat transfer, hence improving the thermal insulation for the object [4, 5, and 6].

Expanded polystyrene (EPS) is the most commonly used insulating material used for cool boxes today. Known for its lightweight, rigidity, and easy for formability, EPS in the form of moulded sheets is also used for building insulation [7] and depending on the density of the EPS board, has a thermal conductivity value ranging from 0.032 to 0.038 W/m.K [8]. EPS performance as an insulating material can be further improved by pairing it with other materials such as aluminium [9].

Polyurethane (PU) as an insulating material has become one of the most important materials that can be used to minimize the energy needed to keep refrigerators and freezers cold [10]. Secondly, PU panels have also been widely used in marine-based and commercial or industrial based projects to build walk-in cold room on board of ships, restaurants, supermarkets and as construction material for factory building and also for applications such as telecommunication cabin shelters, food, fruits, vegetables & meat storage room, marine containers [11]. Similarly, with EPS, PU can also be used as an insulating material for building insulation [12]. When used as insulation, in the case of a fire breakout, PU exhibits two important characteristics which are it produces non-hazardous gasses, and PU foam is completely biodegradable [13].

Fibreglass or glass wool is an insulation material made of glass fibres which are prepared with binders into the same texture similar to wool. Thermal insulation is produced by air between the glasses. Normally produced in the form of rolls or slabs, it can also be produced as a material that can be sprayed or applied in place, on the surface to be insulated [14]. Its application ranges from duct and roof insulation to facings used in metal buildings [15].

Asbestos is a fibrous mineral that has been known since ancient times in resistance to heat. Known for its resistance to heat, wear, alkalis and acids, and its flexibility makes it a suitable material for use as insulation in industrial textiles and many other fields [16]. As reported by Barbalace (2004), asbestos is a very good insulator, strong, resists acid, chemically inactive and it also can be woven into cloth. The thermal conductivity for asbestos is 0.08 W/m.K, and it is readily available at low cost [17].

Poly glass fibre as an insulation material was tested in by following ASTM C518: Standard Test Method for steady-state thermal transmission properties using the heat flow meter. Results showed that when compared with thermal conductivity of typical insulation materials (0.035 – 0.16 W/m.K) or rock wool insulation (0.035 – 0.038 W/m.K) at mean temperatures of approximately 20°C, Poly glass fibre provides low thermal conductivity, meaning they are highly effective in terms of reducing heat transfer [18].

Coconut fibre is also widely as an insulating material. The interest of using natural fibres as reinforcements in polymer composites replacing conventional synthetic fibres have risen due to increasing environmental consciousness and awareness of the need for sustainable development [19]. Other advantages of natural fibres include low cost, low density, unlimited and sustainable availability, and low abrasive wear of processing machinery.

2. Materials and Methods

2.1. *Preliminary material selection*

Three of the best out of a possible six insulating material which can be used as an insulator for a cool box is determined by using both calculation and simulation method. The best insulating material is judged by its heat transfer rate, $\dot{Q}$ and temperature at the outer wall, T_w of the cool box. The six insulating materials are polyurethane (PU), expanded polystyrene (EPS), glass wool, asbestos, poly-glass fibre (PGF), and coconut fibre. The schematic diagram for the cool box for calculation purposes is shown in Figure 1.

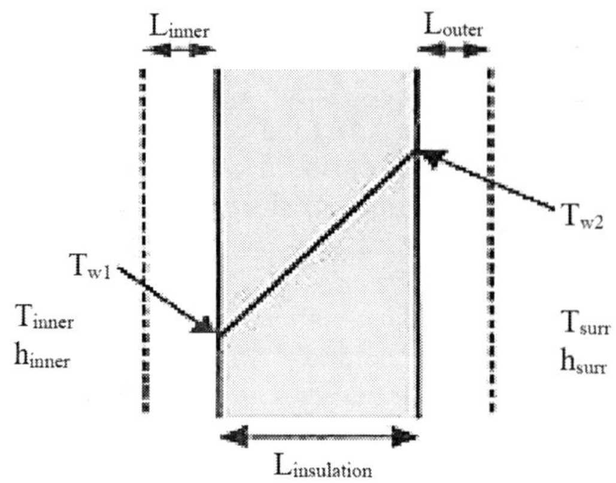

Figure 1. The schematic diagram for conduction and convection analysis, T = temperature, h = convective heat transfer coefficient, L = thickness

For calculation, the conduction and convection formula is used to determine the heat transfer rate, and temperature at the outer wall of the cool box. The formula is as follow:

Heat transfer rate, $\dot{Q}$ [20]:

$$\dot{Q} = \frac{T_{surr} - T_{inner}}{\Delta R} \tag{1}$$

Where;

T_{surr} is the surrounding air temperature (°C)
T_{inner} is the inner cool box temperature (°C)
ΔR is the overall thermal resistance (K/W)

For T_{surr}, three different surrounding air temperatures were used to take into consideration different climate conditions around the world. The temperatures were 20°C, 30°C, and 40°C respectively. For T_{inner}, a constant temperature of 15°C was used throughout the calculation. The overall thermal resistance (ΔR) is determined using the following formula:

Overall thermal resistance, ΔR [20];

$$\Delta R = \frac{1}{h_{inner}A} + \frac{L_1}{Ak_1} + \ldots + \frac{L_n}{Ak_n} + \frac{1}{h_{surr}A} \tag{2}$$

Where;

A is the surface area of the cool box (m^2)
h_{inner} is inner convective heat transfer coefficient ($W/m^2.K$)
h_{surr} is surrounding convective heat transfer coefficient ($W/m^2.K$)
L is the thickness of the insulating and cool box material (m)
k is the thermal conductivity of the insulating and cool box material (W/m.K)

The values for thermal conductivity, K for the insulating material used were taken for various studies that have been conducted. These values are shown in Table 1.

IOP Conf. Series: Materials Science and Engineering **834** (2020) 012019 doi:10.1088/1757-899X/834/1/012019

Table 1. Thermal conductivity, *k* value for various insulating materials.

Insulating material	Thermal conductivity, *k* ($Wm^{-1}K^{-1}$)
Polyurethane (PU)	0.023 [21]
Expanded polystyrene (EPS)	0.033 [8]
Glass wool	0.040 [14]
Asbestos	0.080 [17]
Poly-glass fiber (PGF)	0.035 [18]

Thickness of the insulating material and cool box, the other parameters are considered constant during the calculation procedure. The values are as shown in Table 2.

Table 2. Constant parameters for thermal resistance, ΔR calculation

Parameter	Value
Area, A	$0.048m^2$
Inner convective heat transfer coefficient, h_{inner}	$10\ Wm^{-2}K^{-1}$
Outer convective heat transfer coefficient, h_{outer}	$30\ Wm^{-2}K^{-1}$

The following formula is used to determine the temperature at the outer wall, T_{w2} of the cool box,

$$\dot{Q} = Ah_{surr}(T_{surr} - T_{w,2}) \tag{3}$$

The results obtained using the conduction and convection formula were then compared using the simulation method. Simulation is achieved by using the ANSYS software. The software specification is shown in Table 3.

Table 3. Simulation software specifications.

Item	Remark
Software name	ANSYS Mechanical
Version	14.0
Analysis type	Steady-state
Thermal modeling	Conduction and Convection

ANSYS software allows the user to simulate the heat transfer rate and temperature at the wall of the cool box. Based on the results from both of these methods, three of the best insulating material is selected.

2.2. *Sample, cool box preparation, and testing method*

Samples for the best three insulating material based on the calculation and simulation results were prepared to test its performance under actual conditions. Some raw materials used for sample preparation were obtained from a local company, while some were obtained from the local hardware store. Figure 2 shows the example of raw materials used as insulation for this study.

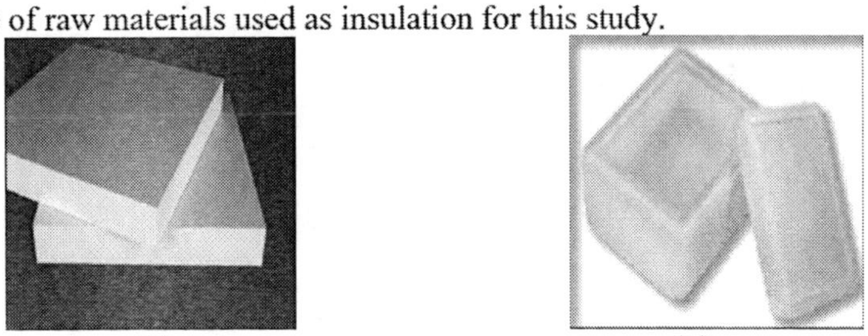

(a) (b)

Figure 2. Examples of raw material a) Polyurethane (PU), b) Expanded polystyrene (EPS)

To determine the tensile strength, sample materials were cut to a rectangular shape and the dimension follows the standard test method (ASTM). The dimension of sample materials was 250mm (L) x 20mm (w) x 12mm (h). To grip the specimens on the tensile test machine, the end of the specimen was supported with aluminium or sandpaper to avoid specimen fracture at the grip. Example for tensile test specimen is as shown in Figure 3.

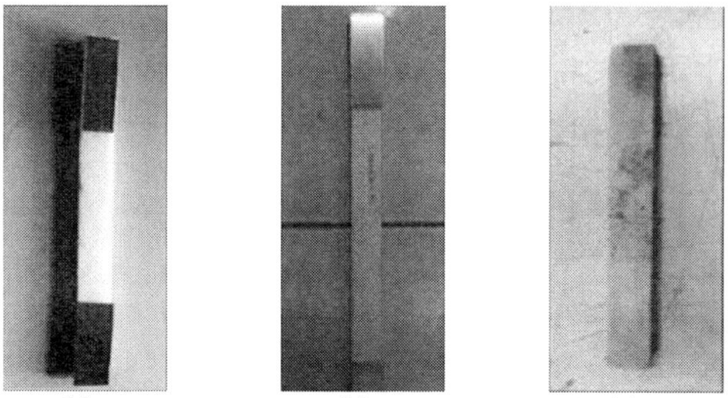

(a) (b) (c)

Figure 3. Tensile test specimen a) Polyurethane (PU), b) Expanded polysty-rene (EPS), c) Poly-glass fiber (PGF)

Tensile testing is a way to determine how a material reacts when force is applied to it in tension. It is one of the simplest and most widely used mechanical tests. This testing was conducted by using the tensile testing machine INSTRON 3382 Floor Model Universal Testing Systems as shown in Figure 4. This machine ideal is for tension or compression applications for tests up to 100kN (22,500lbf) and provides simplicity, reliability, and affordability for heavy-duty quality control and production testing. The sample of each material must be prepared at least five samples and take the average to get an accurate result. The speed must follow the standard to ensure the accuracy of the results.

Figure 4. INSTRON 3382 (100kN) floor model universal testing systems

For the density test, sample materials were cut into cubes with a dimension of 30mm (L) x 30mm (w) x 30mm (h). Material samples cut to the required dimension are shown in Figure 5.

(a) (b)

Figure 5. Density test specimen a) Polyurethane (PU), b) Expanded polystyrene (EPS)

The density test was conducted to determine the density of the insulating materials of this project. The weight of the insulating material was obtained using an electronic balance machine as shown in Figure 6. Using the following equation, the density and porosity percentage of the insulating materials were obtained.

For density [23], ρ_r:

$$\rho_r = \frac{W_{insulating\ material,air}}{W_{insulating\ material,air} - W_{insulating\ material,water}} \times \rho_w \tag{4}$$

For porosity percentage [21], ε:

$$\epsilon = \left(1 - \frac{\rho_b}{\rho_r}\right) \times 100\% \tag{5}$$

The density test was conducted twice for the three best insulating materials and the average value was calculated in order to get an accurate value of the density of the insulating material.

Figure 6. Electronic balance machine

For the thermal insulation test, an inner and outer box was made using an aluminium sheet with a thickness of 0.5mm as shown in Figure 7. The dimension for inner box was 208mm (L) x 208mm (w) x 208mm (h) and for outer box was 220mm (L) x 220mm (w) x 220mm (h). This cubic design was chosen for ease in changing the insulating materials for the testing. The inner box will be placed inside the outer box, and the insulating material will be placed in between the boxes.

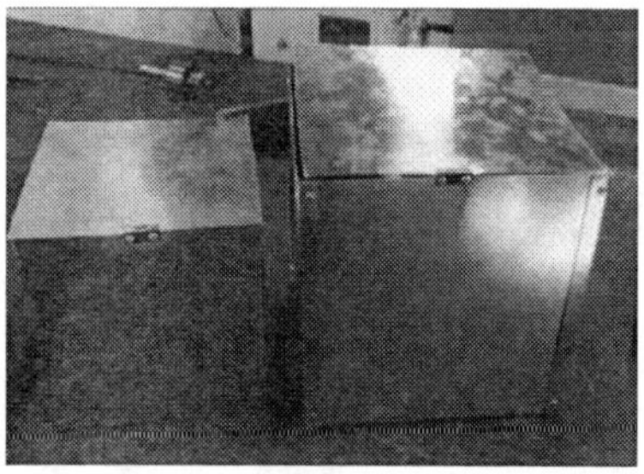

Figure 7. Inner (left) and outer (right) cool box for thermal insulation test

Thermal insulation test was conducted at three different temperatures, 20°C, 30°C, and 40°C. For 20°C, the test was conducted in the cold room to obtain a steady temperature reading. For 30°C, the test was conducted inside a standard laboratory room, and for 40°C, testing was conducted by using Memmert Drying Oven as shown in Figure 8.

Figure 8. Memmert Drying Oven for thermal insulation test

Using a digital thermometer as shown in Figure 9 via the use of thermocouples, the temperature on the inside and outside of the cool box was recorded. Temperature reading was taken every two hours to record the temperature drop with respect to time for the ice to change into water. Temperature change from cold water to normal water temperature was also recorded.

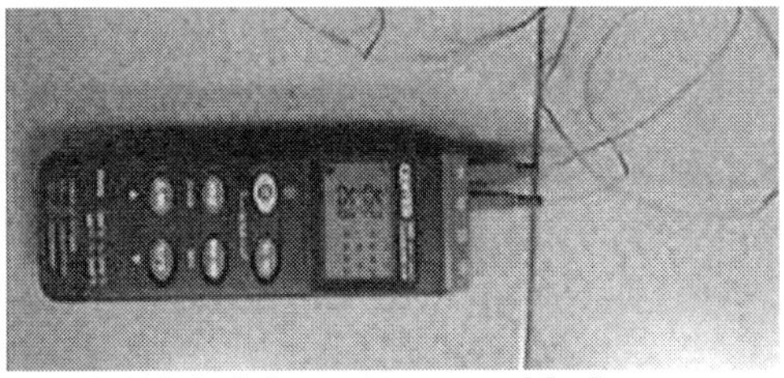

Figure 9. Digital thermometer with thermocouples

3. Results and discussion

Using the conduction and convection heat transfer formula, the heat transfer and the temperature at the outer wall of the cool box at different temperature conditions were determined. Table 4 and Table 5 shows the results obtained using the calculation method.

Table 4. Heat transfer across the cool box at different temperature conditions (theoretical).

Insulating material	Thermal conductivity, k ($Wm^{-1}K^{-1}$)	20°C $\dot{Q}(W)$	30°C $\dot{Q}(W)$	40°C $\dot{Q}(W)$
Polyurethane (PU)	0.023	0.37	1.10	1.83
Expanded polystyrene (EPS)	0.033	0.48	1.45	2.42
Poly-glass Fiber (PGF)	0.035	0.50	1.51	2.53
Glass wool	0.040	0.55	1.66	2.76
Coconut Fiber	0.048	0.62	1.88	3.14
Asbestos	0.080	0.94	2.53	4.20

Table 5. Outer wall temperature of the cool box at different temperature conditions (theoretical).

Insulating material	Thermal conductivity, k ($Wm^{-1}K^{-1}$)	20°C T_{outer} (°C)	30°C T_{outer} (°C)	40°C T_{outer} (°C)
Polyurethane (PU)	0.023	19.7	29.2	39.4
Expanded polystyrene (EPS)	0.033	19.6	29.0	39.2
Poly-glass Fiber (PGF)	0.035	19.7	29.0	39.1
Glass wool	0.040	19.6	28.8	39.1
Coconut Fiber	0.048	19.8	29.3	38.9
Asbestos	0.080	19.9	29.2	38.6

The results obtained from the calculation method were then verified by using the simulation method. Table 6 and Table 7 shows the results obtained from the simulation method.

Table 6. Heat transfer across the cool box at different temperature conditions (simulation).

Insulating material	Thermal conductivity, k ($Wm^{-1}K^{-1}$)	20°C $\dot{Q}(W)$	30°C $\dot{Q}(W)$	40°C $\dot{Q}(W)$
Polyurethane (PU)	0.023	0.3664	1.0991	1.8318
Expanded polystyrene (EPS)	0.033	0.4829	1.4488	2.4146
Poly-glass Fiber (PGF)	0.035	0.5040	1.5120	2.5200
Glass wool	0.040	0.5538	1.6615	2.7692
Coconut Fiber	0.048	0.6261	1.8782	3.1304
Asbestos	0.080	0.9416	2.5411	4.2352

Table 7. Outer wall temperature of the cool box at different temperature conditions (simulation).

Insulating material	Thermal conductivity, k ($Wm^{-1}K^{-1}$)	20°C T_{outer} (°C)	30°C T_{outer} (°C)	40°C T_{outer} (°C)
Polyurethane (PU)	0.023	19.873	29.618	39.364
Expanded polystyrene (EPS)	0.033	19.832	29.497	39.162
Poly-glass Fiber (PGF)	0.035	19.825	29.475	39.125
Glass wool	0.040	19.808	29.423	39.038
Coconut Fiber	0.048	19.783	29.348	38.913
Asbestos	0.080	19.952	29.118	38.529

Overall comparison between theoretical and simulation results for heat transfer rate across the cool box wall and temperature at the outer wall of the cool box is shown in Figures 10 through 15.

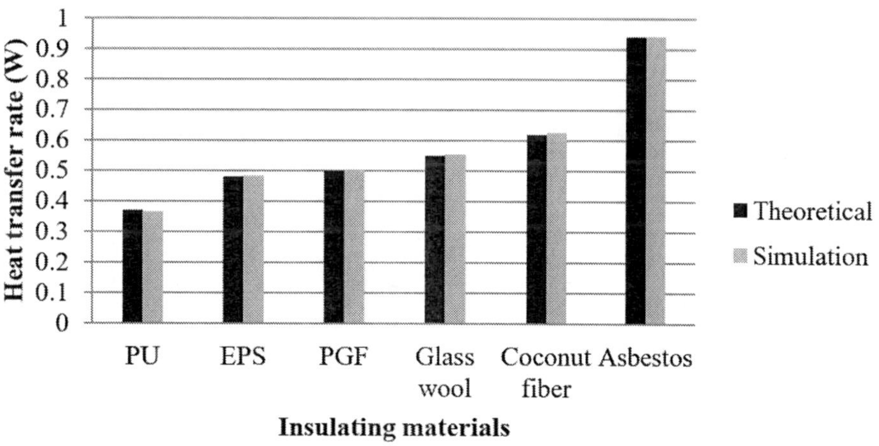

Figure 10. Comparison for heat transfer rate at 20°C

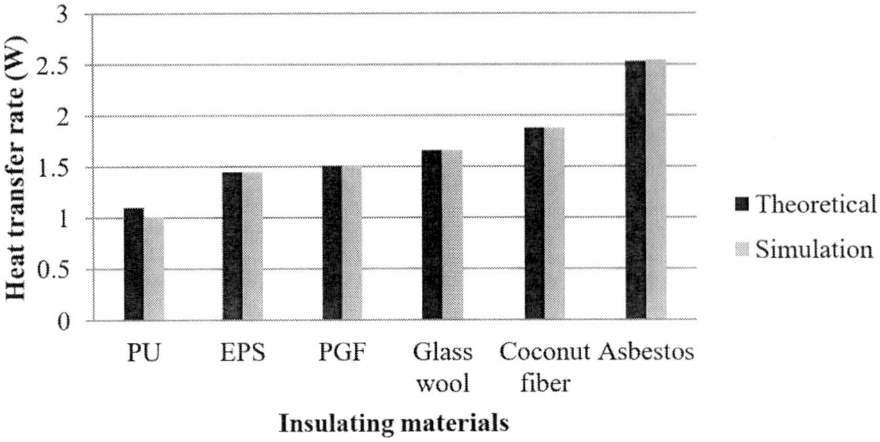

Figure 11. Comparison for heat transfer rate at 30°C

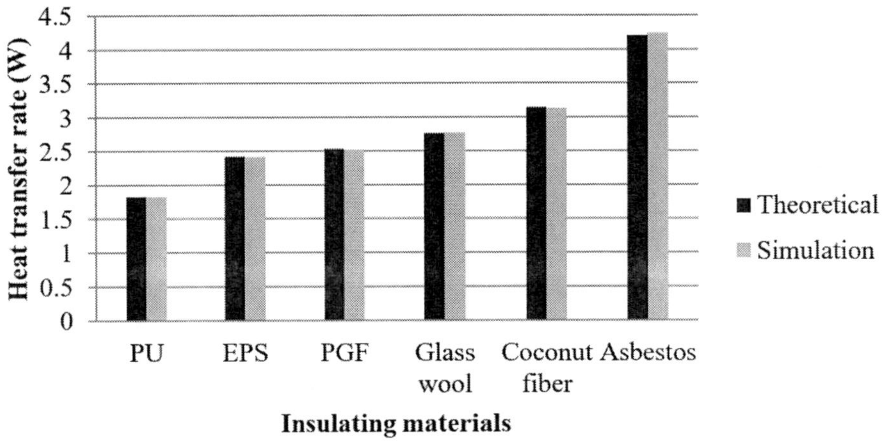

Figure 12. Comparison for heat transfer rate at 40°C

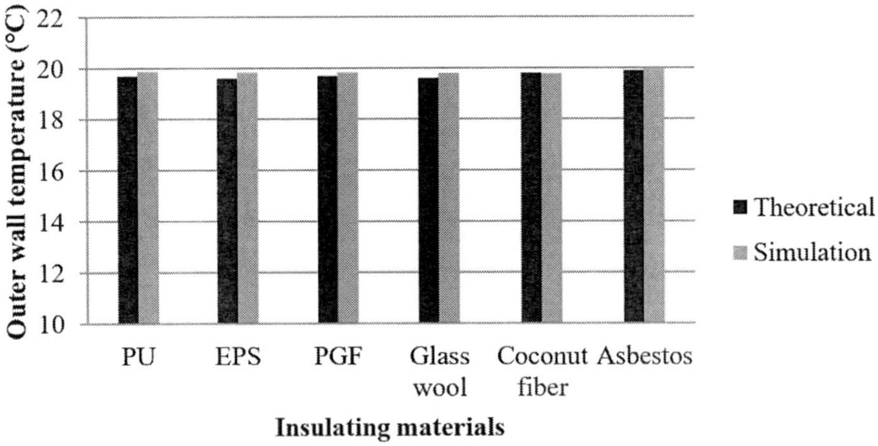

Figure 13. Comparison for temperature for outer wall at 20°C

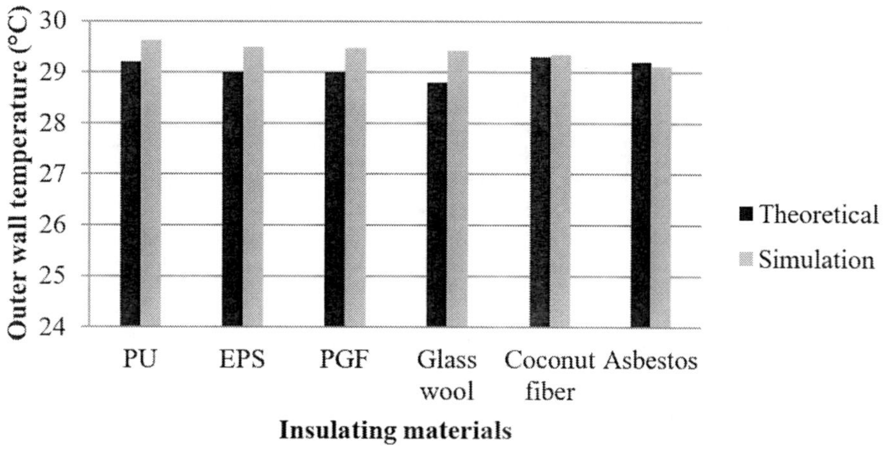

Figure 14. Comparison for temperature for outer wall at 30°C

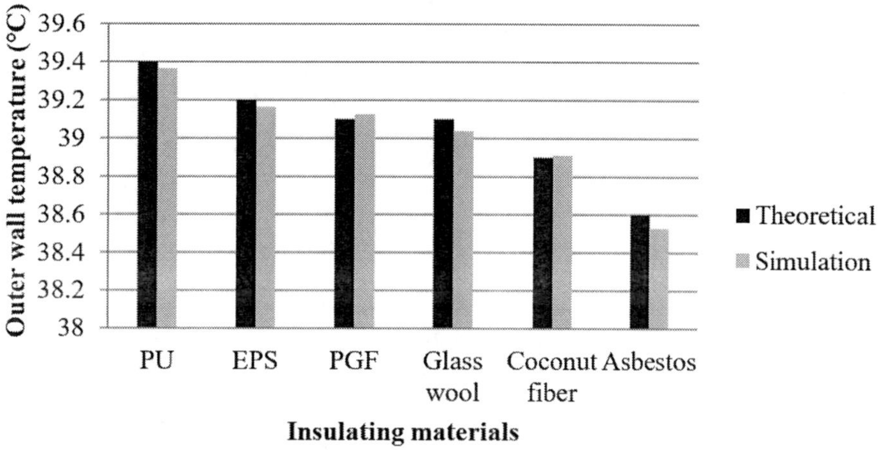

Figure 15. Comparison for temperature for outer wall at 40°C

Although there are minor discrepancies between the results obtained from both theoretical and simulation method. It can be concluded that the three best insulating materials across all three temperature conditions in terms of heat transfer were polyurethane (PU), expanded polystyrene (EPS), and poly-glass fibre (PGF). The worst insulating material was Asbestos. The results obtained concurs with [2, 24] who found that the lower thermal conductivity of the material, the lower the rate of heat transfer across the material. Heat transfer across materials of higher thermal conductivity occurs at a faster rate than across materials of low thermal conductivity. As a result, insulating material that has a lower thermal conductivity will have a lower heat transfer rate. The three best insulating material was being further tested experimentally using the thermal insulation test.

During the thermal insulation test, the three best insulating material was placed between the inner and outer layer of the cool box which was prepared previously. By placing ice inside the inner cool box, the time for the ice to change from solid phase to liquid phase (cold) was recorded. Also recorded was the time for the liquid (cold) too reached room temperature (normal). The normal temperature for 20°C condition was also set at 20°C, while the normal temperature for 30°C and, 40°C was set at 27°C. Similarly, with the previous methods, three different temperature conditions were used for the thermal insulation test. The results from the thermal insulation test are shown in Table 8 and Figure 16.

Table 8. Thermal insulation test result.

Insulating material	20°C		30°C		40°C	
	P1 (min)	P2 (min)	P1 (min)	P2 (min)	P1 (min)	P2 (min)
PU	630	445	400	265	240	110
EPS	535	420	360	240	215	93
PGF	525	413	345	233	205	90

*P1 – Time taken for solid (ice) to change into liquid (cold)
*P2 – Time taken for liquid (cold) to reach normal room temperature

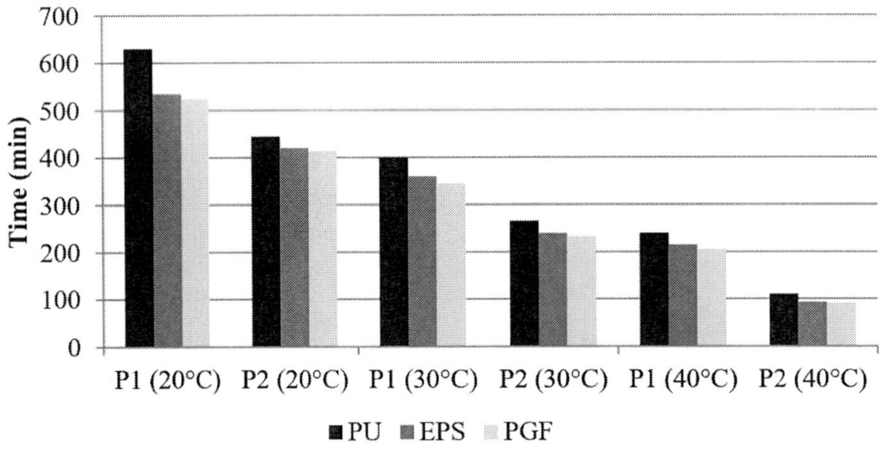

Figure 16. Thermal insulation test result

From the results obtained, Polyurethane (PU) recorded the longest time for both scenarios P1 and P2. The time recorded for PU to change from solid (ice) into liquid (cold) for the first scenario, P1, was 630 minutes (at 20°C), 400 minutes (at 30°C), and 240 (at 40°C) respectively. For the second scenario, P2, the time recorded was 445 minutes (at 20°C), 265 minutes (at 30°C), and 110 (at 40°C) respectively. All though there was not much difference in terms of time recorded between expanded polystyrene (EPS) and, poly-glass fibre, PGF was seen to be the worst performing insulating material. The reason behind the lengthy time recorded by the cool box insulated with PU is because of its low thermal conductivity when compared to cool box insulated with either EPS or, PGF. As calculated and, simulated previously PU heat transfer across the cool box wall insulated with PU is low, which resulted in longer periods for ice to change phases from solid to liquid (cold), and for liquid (cold) to reach normal room temperature conditions. This makes PU a suitable candidate to replace the conventional insulating material, which is EPS used in today's cool boxes.

A tensile test was conducted on the three insulating materials to determine their ultimate tensile strength. The tensile test involved five specimens for each insulating material to obtain a more accurate result. The mean (average) ultimate tensile strength for each insulating material was calculated. The result for ultimate tensile strength is shown in Figure 17.

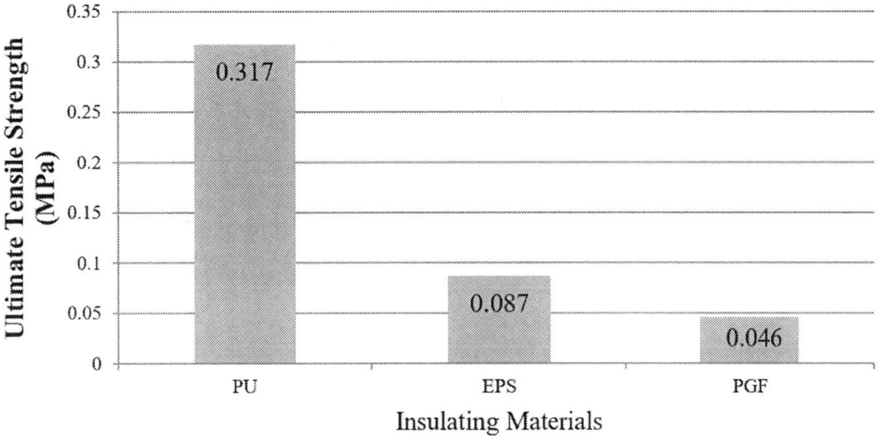

Figure 17. Ultimate tensile strength for insulating materials

Polyurethane (PU) foam has the highest ultimate tensile strength which was 0.317MPa when compared with the expandable polystyrene (EPS), and poly-glass fibre (PGF). The ultimate tensile strength of EPS and PGF were 0.087MPa and 0.046MPa. The PU foam has higher ultimate tensile strength have the highest ability to withstand a pulling (tensile) force and PGF is lowest. The ability to resist breaking under tensile stress is one of the most important and widely measured properties of materials used in structural applications [25].

For the density test, only two insulating material was tested which were, polyurethane (PU), and expanded polystyrene (EPS) this is because, for poly glass fibre (PGF), it tends to dissolve in water [18]. The results from the density and porosity percentage conducted for this study are shown in Table 9.

Table 9. Result for density test and porosity percentage

Insulating material	Mass, m (g)	Volume, V (cm^3)	Density, ρ_r (g/cm^3)	Bulk density, ρ_b (g/cm^3)	Porosity percentage (%)
PU	27.0	27.0	1.0000	0.72	28.00
EPS	24.5	27.0	0.9074	0.64	29.44

From the results obtained, even though the volumes for both insulating materials were identical, they do differ in mass. As a result, the density obtained for both PU and EPS were different, 1.000g/cm^3 and, 0.9074g/cm^3, respectively. Having a higher density resulted in PU having a lower porosity percentage (28.00%) when compared to EPS (29.44%). For an application involving insulation, a material having a low porosity value is recommended as the increase in porosity percentage will affect the heat transfer rate, by reducing the resistance of heat across the insulating material.

4. Conclusion

In this study, improvement of insulating material used in cool boxes was conducted both theoretically and experimentally. Six insulating materials were selected initially which were polyurethane (PU), expanded polystyrene (EPS), poly-glass fibre, asbestos, coconut fibre, and glass wool. Out of the six tested insulating materials, three of the best insulation materials which can be used for cool box application predicted from the calculation and simulation method were PU, EPS, and PGF. These three insulating materials were then tested further experimentally using the thermal insulation test. Experiment results showed that PU was the best performing insulating material in terms of heat transfer rate and recorded the longest time for ice (solid) to change into liquid (cool). The time recorded for liquid (cool) to reach normal room temperature was also longer for PU. Mechanical properties such as ultimate tensile strength, density, and porosity percentage for the respective insulating materials were also determined and results showed that when comparing PU and EPS, PU has higher density and lower porosity percentage. Based on these findings, it is apparent that PU has the potential to replace EPS which is the current insulating material used in a conventional cool box today due to its lower heat transfer rate, and resulted in the longer time taken for material (ice) to change from solid to liquid and down to normal room temperature.

Acknowledgement

We would like to thank our colleagues from the Faculty of Mechanical Engineering, UiTM Shah Alam who provided many thoughtful insights and shared their expertise which had greatly helped in completing this study. We would also like to thank the technical staff from various laboratories at the Faculty of Mechanical Engineering, UiTM Shah Alam as without their assistance this study would not have been possible.

References

[1] S.A. Zavattoni and A. G., "A novel CSP receiver based on airlight energy technology - optimization of the thermal insulation system by means of CFD analysis", Energy Procedia, vol. 49, pp. 589-598, 2014.
[2] S. Gustafsson, "Transient plane source techniques for thermal conductivity and thermal diffusivity measurements of solid materials", The Review of Scientific Instruments, vol. 62 no. 3, pp. 797–804, 1991.
[3] R. Bolot and Q. J., "Effect of thermal treatment on the effective thermal conductivity of YPSZ coatings", Surface and Coatings Technology, vol. 205, no. 4, pp. 1034-1038, 2010.
[4] Al-Homoud and D. M., "Performance characteristics and practical applications of common building thermal insulation materials", Building and Environment, vol. 40, no. 3, pp. 353- 366, 2015.
[5] Ozel and M., "Thermal performance and optimum insulation thickness of building walls with different structure materials", Applied Thermal Engineering, vol. 31, no. 17-18, pp. 3854-3863, 2011.
[6] Manohar et al., "Biodegradable fibrous thermal insulation". Journal of the Brazilian Society of Mechanical Sciences and Engineering, 28(1), 45-47, 2006.
[7] L. L. Jiao, "A thermal degradation study of insulation Extruded Polystyrene", Procedia Engineering, vol. 71, pp. 622-628, 2014.
[8] S. A. González, "Lightweight material made with gypsum and extruded polystyrene waste with enhanced thermal behaviour", Construction and Building Materials, pp. 57–63, 2015.
[9] O. Almanza and R. P. M., "Applicability of the transient plane source method to measure the thermal conductivity of low-density polyethylene foams", Journal of Polymer Science Part B: Polymer Physics, vol. 42, no. 7, pp. 1226–1234, 2004.
[10] D. Randall and S. Lee, "The polyurethanes book" Wiley, 2002.
[11] "Velox Series Cold Room", (n.d.). [Online]. Available: http://www.saicond.com.my/product_velox.html?gclid=Cj0KCQiA05zhBRCMARIsACKDWjfvvNNIOqrxWAjXHoDZvxvhgPIq1RUMqcsIiGfPvZMXQzoBuPhMMEaAsepEALw_wcB. [Accessed: 13-12-2014]
[12] G. Cardoso, "Characterization of rigid polyurethane foam (PU) extracted from castor oil (Ricinus communis) for building insulation", VDM Verlag Dr.Muller, Saarbrucken, Germany, vol. 84, 2011.
[13] C. Zhang and H. W., "High bio-content polyurethane composites with urethane modified lignin as filler", Polymer vol. 69, pp. 52–57, 2014.
[14] Miotto and J. L., "Structural efficiency of full-scale timber–concrete composite beams strengthened with fiberglass reinforced polymer", Composite Structures, pp. 145–154, 2015.
[15] Z. Salleh and M. B., "Open Hole Tensile Properties of Kenaf Composite and Kenaf/Fibreglass Hybrid Composite Laminates", Procedia Engineering, vol. 68, pp. 399–404, 2013.
[16] M. Dallas and T. J. Cowan, "Analysis of workplace compliance measurements of asbestos by the U.S. Occupational Safety and Health Administration", Regulatory Toxicology and Pharmacology, vol. 72, pp. 615–629, 2014.

[17] Barbalace and R. C., "Asbestos, its Chemical and Physical Properties", Environmental Chemistry.com. [Online]. Available: http://environmentalchemistry.com/yogi/environmental/asbestosproperties20. [Accessed: 13-02-2015]
[18] "Formaldehyde Free Glass Mineral Wool Insulation Products & Solutions for Commercial, Mechanical, Residential, Industrial and Technical Applications", PGF Insulation® | (n.d.), [Online]. Available:http://www.ecowool.com.my/. [Accessed: 13-02-2015]
[19] D. Verma and P. G., "10 – The use of coir/coconut fibers as reinforcements in composites", Biofiber Reinforcements in Composite Materials, pp. 285–319, 2015.
[20] Çengel and Y. A. "Heat transfer, a practical approach". Boston (Mass.): McGraw-Hill, 2006.
[21] S. A. Al-Ajlan, "Measurements of thermal properties of insulation materials by using transient plane source technique", Applied Thermal Engineering, vol. 26, no. 17–18, pp. 2184-2191, 2006.
[22] Ali and M., "Experimental investigations on coconut-fibre rope tensile strength and pullout from coconut fibre reinforced concrete", Construction and Building Materials, vol. 41, pp. 681-690, 2013.
[23] Dullien, and F. A. "Porous media: fluid transport and pore structure", Academic press, 2012.
[24] J. Guangqi and C. Z., "Conical Calorimeter Fireproof Test for Outer-wall External Thermal Insulation System Building", Science, vol. 24, pp. 4-8, 2008.
[25] D. Chen and C. Z. M., "Analysis of several test methods about heat insulation capabilities of ceramic thermal barrier coatings", Physics Procedia, vol. 50, pp. 248-252, 2013.

IOP Conf. Series: Materials Science and Engineering **834** (2020) 012020 doi:10.1088/1757-899X/834/1/012020

FEM-based Virtual Manufacturing and Testing Procedure for Resistance Spot Welding

M. N. Yusuf[1], **W. E. W. Rahman**[1] **and Y. H. P. Manurung***[1]

[1]Faculty of Mechanical Engineering, UiTM Shah Alam, Selangor, Malaysia

*yupiter.manurung@salam.edu.my

Abstract. This research focused on the development of numerical approximation procedure applied for resistance spot welding process which includes testing procedure to evaluate the tensile strength of a spot welded joint. The spot weld is modelled according to standardized dimension for tensile test with main material properties of Cu-sw as electrode and low carbon steel S235 as sheets with 1mm thickness. It is including electric conductivity, resistivity and heat transfer coefficient in solid body as well as contact interface. The FEM simulation is conducted using the process parameter of current 15,000A, force 5,000 N and different stages of time following the welding process and tensile test which is simulated after released both electrodes and material reaches the initial temperature with contact clamp velocity of 5mm/min. It is expected that the resistance spot welding process can be successfully simulated and producing result within acceptable range of errors.

1. Introduction

Resistance spot welding (RSW) known as one of the most common techniques applied for joining metal sheets. This metal joining techniques is broadly used in manufacturing industry such as automotive which vehicle body and parts are assembled. This welding technique was preferred by the manufacturers for metal sheet related products due to the advantage of time and cost efficient [1]. Among of the important parameters required in resistance spot welding process are heat, pressure and time [2]. The fusion of resistance spot welded joint is represented by weld nugget which is formed as a result of heat generated due to resistance to the high current flow at low voltage [2]. The quality of spot welded joint is highly influenced by the welding current, current flow time, welding tips contact surface, cross-sectional area and sheet thickness.

In the present study, Finite Element Method (FEM) is used to model and simulate the resistance spot welding process to investigate the thermal, electrical and mechanical behaviour of a spot welded joint. Finite Element Method (FEM) was known as one of efficient methods to analyse different phenomenon of engineering problem as it has the capability to produce a much more detailed set of results than experimental investigations.

In spot welding, suitable process parameter is crucial to be used on a specific set of material properties to achieve a desired and acceptable result. For example, size of weld nugget and electrode indentation are depending on the electrode force, weld current and weld time. These parameters are also crucial to achieve high strength of spot welded joint [3]. It was one of the reasons that many researchers were focused their studies on parameter optimization in resistance spot welding.

Content from this work may be used under the terms of the Creative Commons Attribution 3.0 licence. Any further distribution of this work must maintain attribution to the author(s) and the title of the work, journal citation and DOI.
Published under licence by IOP Publishing Ltd

Resistance spot welding physics is governed according to Ohm's law, where I represent the current, E is voltage drop across the electrodes and R is the resistance through the material in ohms. R represents the total resistances in the form of contact and material resistance. Therefore, for a given value of R, the magnitude of I is determined by E. The transformer controlled the current to convert from alternating current to a low voltage which then delivered to a weld of a given resistance [4]. The resistance spot welding process is defined using the following mathematic equation:

$$E = IR \quad (1)$$

The total heat (H) generated in watt-seconds at the spot welded joint, can be expressed using the formula as follow:

$$H = I^2RT \text{ or } H = IET$$
$$T \text{ is time in seconds.} \quad (2)$$

Figure 1 below shows type of resistance which exist during the spot welding process:

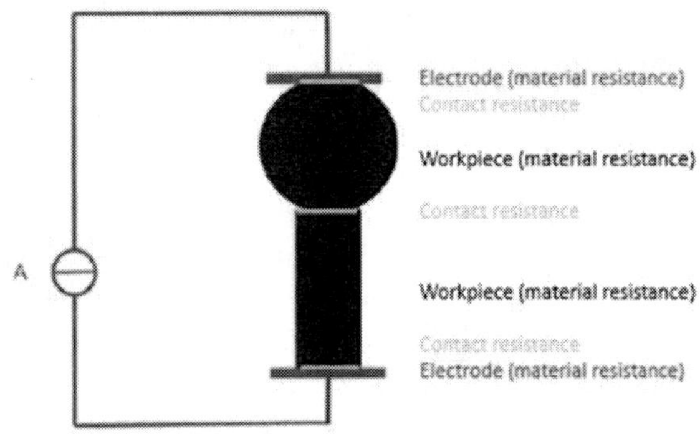

Figure 1. Example model of resistance spot welding process

2. Virtual Manufacturing Simulation of Resistance Spot Welding process using MSC. Marc.

MSC. Marc is an advanced non-linear finite element software which capable to analyze and simulate product behavior under static, dynamic and multi-physics loading condition [5]. MSC. Marc versatility in modeling nonlinear material behaviors and transient environmental conditions makes it ideal to solve different kind of complex design and engineering problems. Founded since 1971, the software was widely used in several industries such as automotive, aerospace, machinery, electronics, biomedical, oil and gas etc.

For resistance spot welding, the simulation in MSC. Marc involved 3 stages; pre-processing, solving and post-processing. The process is illustrated in figure 2 below.

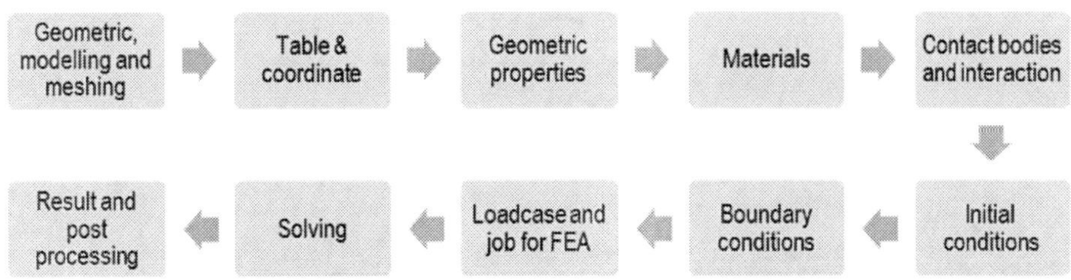

Figure 2. Resistance spot welding procedure in MSC. Marc

The component involved for spot welding modelling in MSC. Marc are two metal sheets overlap each other. The length of the specimen is 100mm along the x-axis while the width is 50mm along the z-axis with thickness of 1mm each. Symmetrical electrode (truncated shape) is drawn with the tip surface of the electrode facing the outer surface of the sheets. Dimension of the electrode used in simulation is

15mm length along the y-axis with 16mm body diameter and 4mm tip diameter. Figure 3 shows the modelling of resistance spot welding in MSC. Marc:

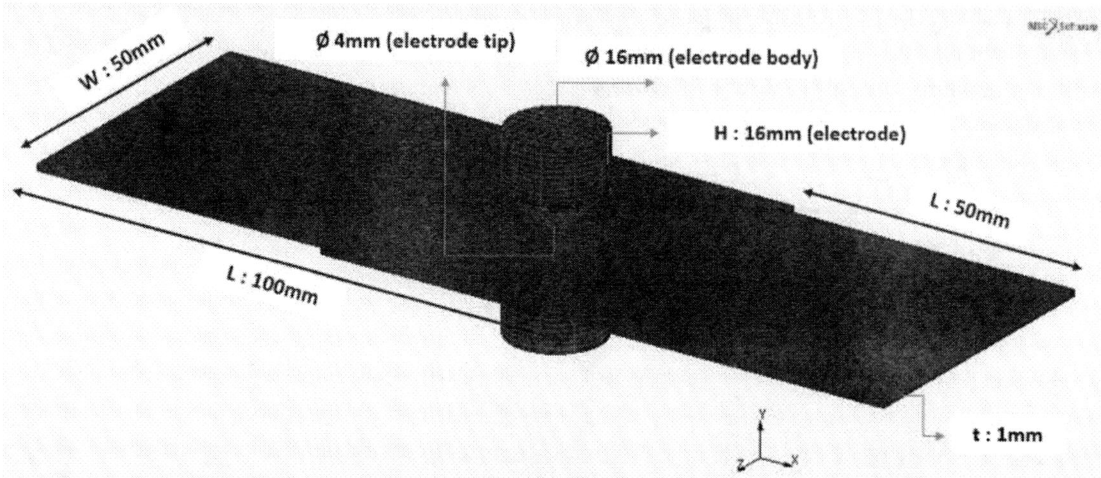

Figure 3. Details modelling of resistance spot welding in MSC. Marc

Tetrahedral-mesh is applied to the electrodes to accommodate the truncated shape while hexahedron-mesh is applied to the specimen. For resistance spot welding process, analysis class required in the simulation is current, thermal & structural.

To accommodate the process parameter of resistance spot welding in MARC, parameter table & coordinate system were created due to most of the properties used are temperature dependent and not constant along the processes. Independent variables which is required to be defined in the simulation are current, electrode force, thermal contact heat transfer coefficient, electricity contact conductivity, young modulus, thermal expansion coefficient, thermal conductivity, specific heat, electric resistivity, electric conductivity.

In this simulation material used for the specimen is S235-JMP-MPM-sw low carbon steel while Cu-Sw copper is used for both of the electrodes. Table 1 shows the S235 chemical composition while table 2 indicates the main parameter setting used in simulating the resistance spot welding process.

Table 1. S235 chemical composition (%)

C	Cu	Fe	Mn	N	P	S
0.15	0.4	98.293	1.1	0.01	0.02	0.027

Table 2. RSW simulation parameters assigned in MSC. Marc

Parameters	Value
Force (kN)	5
Current (kA)	15
Hold time (cycle)	2
Weld time (cycle)	15
Squeeze time (cycle)	2

The plasticity data used for the metal sheet material is based on S235-JMP-MPM-sw flow curve which was created and converted into .mat file. Figure 4 shows the flow curve for S235 material.

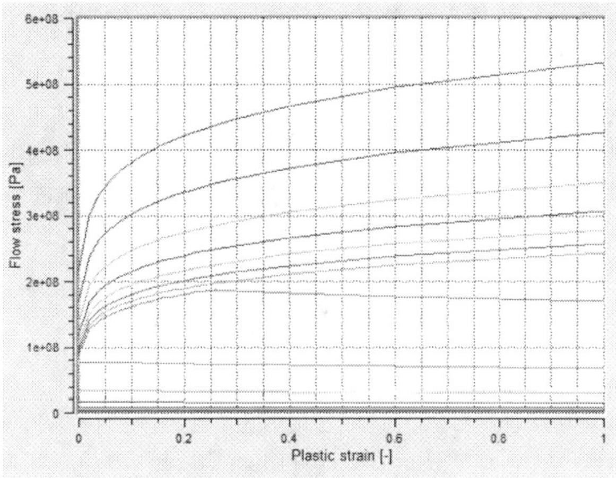

Figure 4. S235 Flow curve

Contact body and interaction are another key parameter to be defined when simulating resistance spot welding process in MSC. Marc especially with RSW process involved analysis of electrical, thermal & structural. The heat generated to form a fusion was because of resistance to the high value of current flow in the form of contact and material resistance. Other than resistance which exist in the material, contact resistance/conductance for both electrical & thermal are vital in RSW simulation. The relationship was defined through contact table option.

Before the simulation of RSW process can be run, the relevant boundary conditions are to be defined in MSC. Marc. The boundary input forms all physical instruction to the welding process. Among of the boundary condition required to be set for RSW are boundary on positioning of the specimen and movement of the electrode, point load for the electrode pressure at 5kN, value of the weld current at 15kA, fix temperature after cooling, tensile test displacement at constant speed of 5mm/min.

Prior to running the simulation, load cases will be created, and the job parameter will be defined. 3 load cases were created each for welding, cooling and tensile test. For S235 material, user subroutine UACT GLUE will activate the glue function at the node of specimen that reach welding temperature of approximately 1,500-1600 ^{0}C. This will define the fusion area for the spot welding at the end of the simulation process.

3. Results and Discussion

Figure 5 shows the contact condition between each of the components, top electrode, top sheet, bottom sheet and bottom electrode. The contact status at the end of the simulation indicates that the contact is exist between each of the components which is crucial for contact resistance to generate the heat at the spot-welded point.

Electric potential result as shown in figure 6 indicates the current flow through the top electrode, specimen and bottom electrode. The point current was applied to the top surface of the top electrode while zero voltage is set at the bottom surface of the bottom electrode.

Figure 7 shows the formation of weld nugget at temperature >1500^{0}C during the resistance spot welding process. The peak temperature is also used to define the fusion area with the application of UACT GLUE user subroutine. All nodes at the contact surface between the two sheets will be subjected to glue deactivation condition except those nodes that reach the pre-defined temperature in the user subroutine.

Figure 8 shows the sheets which were subjected to tensile shear force at the end of the simulation. High stress region is represented with yellow and orange region when the contact clamp velocity of 5mm/min is applied.

IOP Conf. Series: Materials Science and Engineering **834** (2020) 012020 doi:10.1088/1757-899X/834/1/012020

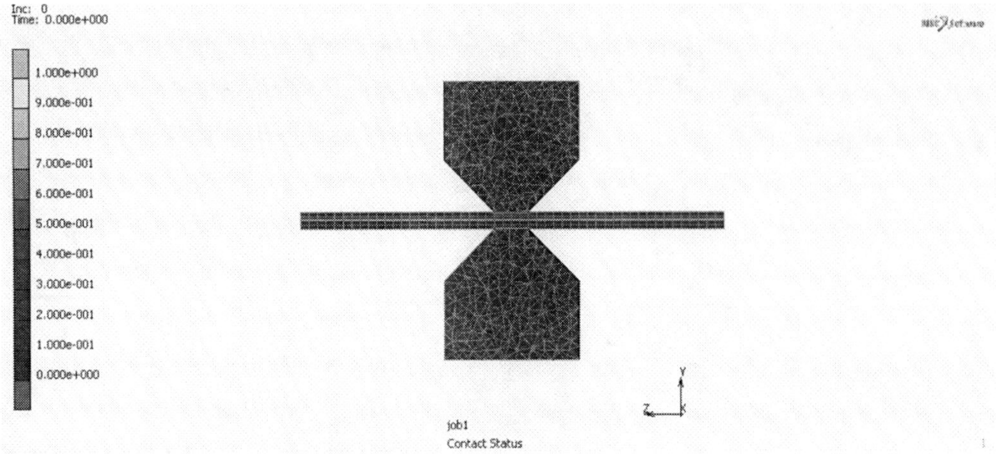

Figure 5. Contact Status

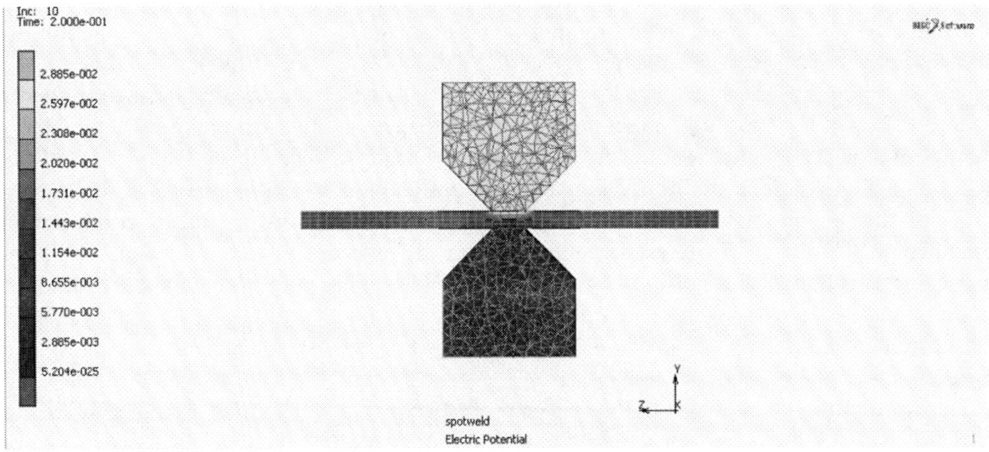

Figure 6. Electric Potential

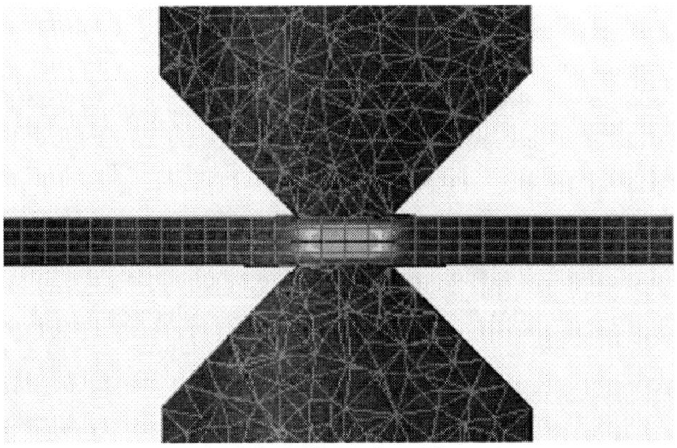

Figure 7. Formation of weld nugget

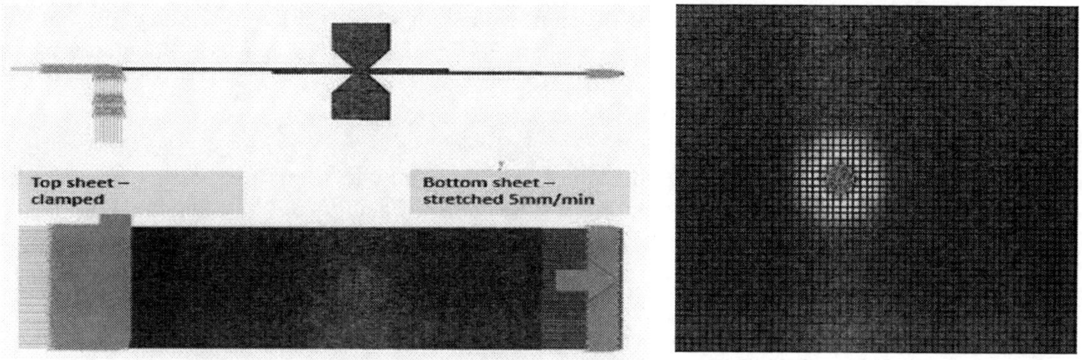

Figure 8. Tensile modelling in MSC Marc

For resistance spot welding process simulation in MSC. Marc, process parameter was recognized as key to obtain successful result for welding simulation and tensile test. After the simulation, it was found that welding current *(I)* was known to have highest impact on the simulation result while each of the process welding time had also play significant factor to the result.

Other influential factor that was recognized after the simulation is the electrical and thermal contact conductivity or resistance. In most cases the electrical contact resistance is 30 time larger than the electrical material resistance. Electrical material resistance and electrical contact resistance are strongly depending on temperature. Additionally, the electrical contact resistance is also depending on the contact pressure. During the welding process the electrical contact resistance decreases rapidly whereas the electrical material resistance increases. Figure 9 shows the time curves of electrical material resistance, electrical contact resistance and the complete resistance of the system during the spot welding process.

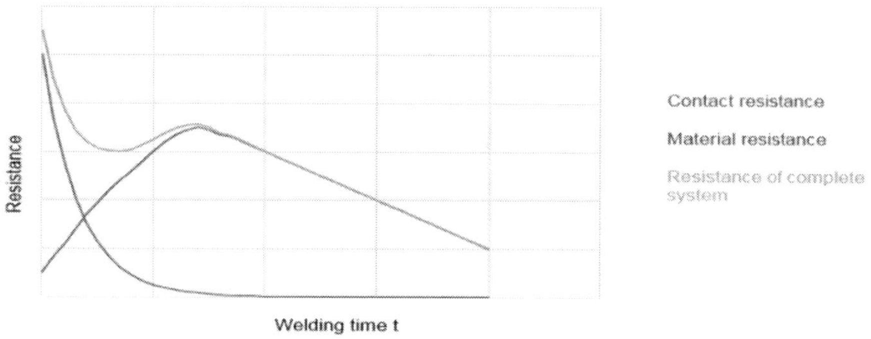

Figure 9. Time curves for electrical resistances during a resistance welding process

4. Conclusion

Through Finite Element Method (FEM), resistance spot welding process with tensile test had been modelled. Result produced from the simulation highlights several critical findings as follows:

1. The resistance spot welding process was successfully simulated using MSC. Marc mentat software with suitable materials and process parameters were defined in clear and structured procedure to achieve the expected and desire result.
2. Other than material properties and boundary condition setting, key parameters such as weld current, electrode pressure, and welding time is found to be significance and even small changes in the value, resulting to simulation error and in some cases consume longer simulation time.

The result data obtain through the FEM simulation is useful to establish deeper understanding on the resistance spot welding process and behaviour. Information collected in the simulation is also useful to minimize the gap between simulation and experimental result. The procedures implemented in the simulation will be further developed to explore more knowledge in resistance spot welding behaviour such as effect of parameter change to fatigue strength and use of suitable damage model to investigate and characterize the strength of the spot welded joint under different load conditions. This research will contribute towards further development of research in Virtual Manufacturing by considering the material behaviour and modelling.

Acknowledgement

The collaborating authors would like to express highly gratitude to the associates in Advanced Manufacturing Technology Excellence Centre (AMTEx), Faculty of Mechanical Engineering, Universiti Teknologi MARA (UiTM) in Malaysia for their contribution in this research.

References

[1] Joel Andersson, Jonatan Deleskog, Fatigue Life and Stiffness of the Spider Spot Weld Model, Master's thesis, ISSN 1652-8557, 2014:22

[2] Ario Sunar Baskoro, M. Rizky Trianda, Jos Istiyanto, Sugeng Supriyadi, Danardono A. Sumarsono, Gandjar Kiswanto, Effects of Welding Time and Welding Current to Weld Nugget and Shear Load on Electrical Resistance Spot Welding of Cold Rolled Sheet for Body Construction, 2014 IEEE International Conference.

[3] Manoj Raut and Vishal Achwal, Optimization Of Spot Welding Process Parameters for Maximum Tensile Strength, International Journal of Mechanical Engineering and Robotic Research, Vol. 3, No. 4, October 2014

[4] Miller Technical Handbook for Resistance Spot Welding, 2012-003 335D, Miller Electric Mfg. Co.

[5] Qilei Sun, Zerui Liu, Long Yang, Application of Finite Element Analysis Software MSC.Marc in Material Tests, Advanced Materials Research, Vols. 945-949, pp 95-98, 2014

IOP Conf. Series: Materials Science and Engineering **834** (2020) 012021 doi:10.1088/1757-899X/834/1/012021

Steady-state heat flow through hollow clay bricks

H Ghailane[1], M A Ahamat[1*], M Md Padzi[1] and S Beddu[2]

[1] Universiti Kuala Lumpur, 1016, Jalan Sultan Ismail, 50250 Kuala Lumpur, Malaysia.
[2] Universiti Tenaga Nasional, Jalan Ikram-Uniten, 43000 Kajang, Selangor, Malaysia.

*asmidzam@unikl.edu.my

Abstract. Clay bricks are commonly used for building wall. One of the functions of a wall is to create a habitable internal environment, such as enabling internal temperature to be controlled at a human comfort level (~24°C) which is lower than the outside temperature for hot and humid climate countries. Thus, it is important for clay bricks to minimize heat flow into the building to reduce the energy consumption by air-conditioning system. The aim of this paper is to investigate the reduction of heat flow through hollow clay bricks relative to solid bricks. Finite element method was used to simulate steady-state heat transfer through solid and hollow clay bricks. Several arrangements of cavities filled with air, wool and expanded foam were simulated. It was found that the effective thermal conductivity reached almost a constant when the void volume and reduction in cross-sectional area exceeds 10%. Generally, heat transfer through hollow clay brick is expected to be significantly lower than the solid clay brick. It can be concluded that building wall made from hollow clay bricks could reduce the energy consumption by its air conditioning system. Future work is on the transient heat flow analysis and evaluation of mechanical properties of hollow clay bricks.

1. Introduction

Buildings consume almost 40% of the world total energy consumption and emits more than 36% of total carbon dioxide [1]. One of the factors that leads to a high energy consumption in buildings is the cooling effect requirement for building in hot countries. For buildings in cold climate countries, energy for heating is significant. It can be concluded that heat flow through building's wall contribute to the huge energy consumption in buildings [2]. Thus, this paper presents one method that could reduce heat flow through building's wall by using hollow bricks filled with air, wool and expanded foam.

Since the heating and cooling systems in buildings consume huge amount of energy, a solution that can reduce energy consumption while maintaining a comfortable internal environment in tropical and cold climate countries is required. It is reported that 60 to 80% of heat loss occurs through building envelopes [3], a higher thermal resistance is required by building envelops. It is also reported that almost 30% of the energy loss in building during winter is through wall [4]. Nowadays, the common method to improve wall insulation is to incorporate insulating layers in building's envelope. A proper selection method of properties of building's wall contribute to the reduction in energy consumption in buildings [5]. Well-insulated building walls can reduce cooling energy up to 23% [6].

Lately, clay hollow bricks have been widely used for the construction of single and double walls. Hollow bricks have different widths according to their cavity distribution. Such bricks are characterized

Content from this work may be used under the terms of the Creative Commons Attribution 3.0 licence. Any further distribution of this work must maintain attribution to the author(s) and the title of the work, journal citation and DOI.
Published under licence by IOP Publishing Ltd

by significant differences in thermal insulation properties depending on the model, the number and the distribution of holes. Several studies have been conducted to determine the correlation between the bricks' holes and its properties. Walls made of hollow bricks are characterized by their thermal resistance which decreases when the size of the hole increases and its row decreases [7] [8] [9]. In another study, it has been concluded that as long as thermal bridges are avoided and hole rows are increased, a better thermal performance is guaranteed [10] [11]. A brick with a smaller diameter hole also has lower thermal conductivity [12].

Beside the choice of the appropriate building material, building insulation is an essential technology for energy saving. Incorporating insulating materials in hollow bricks cavities minimizes the thermal transfer from the outer to the inner space to reduce heat loss or gain [2]. Many researches advised to fill in the cavities of hollow bricks with different types of insulating materials. Polystyrene filled cavities decreased the heat transfer by 36% [13]. The study that used compressed straw filling cavities can lower heat transfer within a brick [14]. EPS filling showed that thermal improvement was noticed with the increase of EPS filling ratio [15].

To understand the properties of materials, several numerical methods and types of simulations have been elaborated. Advanced software was used to simulate different materials and display the distribution and behavior of their properties through a sequence of computer-generated experiments. Microstructure-Based Simulation has been adopted by many researchers. Kikuchi et al predicted the properties of Mo fiber-Cu composites through a finite element method of microstructure modelling [16]. Zbynek and Marek used CFD analyses to investigate the heat transfer of hollow bricks in vertical direction [17]. Mahmouda et al developed a CFD model for a thermal analysis of hollow autoclaved aerated concrete blocks used for hot climate buildings [18].

Thus, this paper presents work on further investigation of effect of void volume and reduction in cross-sectional area for heat transfer within hollow bricks. Effective thermal conductivity of bricks with various number of holes, different hole diameters and filling materials are presented.

2. Methodology

This section presents the preparation of models of bricks followed by method of steady-state heat flow simulation based on finite element model.

2.1. *Preparation of models of bricks*

The size of the brick models is conformed to the Malaysian Standard MS 1064-Part 08 [19], where the dimensions are 190 mm x 90 mm x 90 mm. Typical arrangement of holes in the brick is as in Figure 1. A total of 11 geometries of brick models were constructed as in Table 1. The holes were filled with air, wool or expanded form (Table 2).

Figure 1. One of the brick models (10 holes).

Table 1. Number and diameter of holes in bricks.

Brick geometry	Number of holes	Diameter of hole (mm)
1	3	40
2	3	35
3	3	30
4	6	30
5	6	25
6	6	20
7	8	30
8	8	25
9	8	20
10	10	25
11	10	20

Table 2. Thermal conductivity of materials used in the simulation.

Material	Thermal Conductivity (W/m.K)
Brick	1.1
Air	0.0246
Wool	0.04
Expanded foam	0.03

2.2. Steady-state heat flow simulation

The brick model was constructed and imported to ANSYS 19.2 (Student version). The thermal properties of brick were set according to Table 2. Then, the mesh (Figure 2) is generated by using hexahedral element (Hex 20 type). The minimum length of the hexahedral edge was set to 0.98175 mm to minimize errors. The skewness of the mesh was mainly between 0 and 0.13 which showed the closeness of the meshing to the ideal condition.

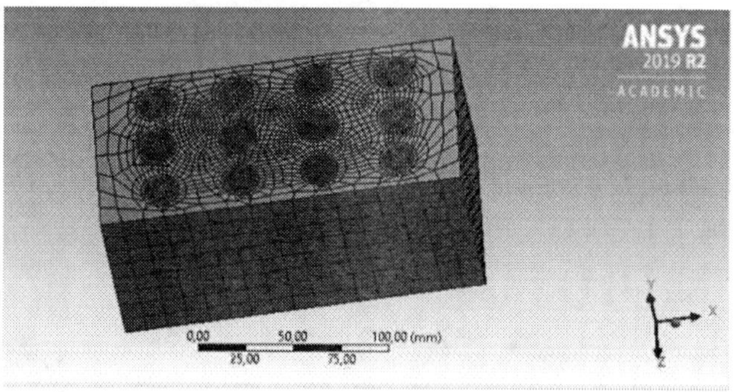

Figure 2. Mesh of 12 holes (3x4) hollow brick.

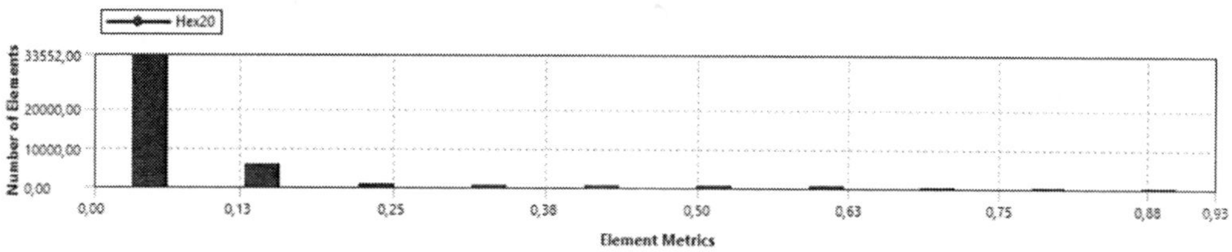

Figure 3. Distribution of the mesh skewness.

After the mesh was generated, two boundary conditions were defined for steady-state thermal analysis. One of the faces on the long edge of the brick was set to 32°C while another face was set to 24°C. This was to mimic internal and external temperature of building walls.

Besides the boundary conditions, commands "OUTRES,ERASE" and "OUTRES,ALL,ALL" have been inserted using the command object to obtain the gradient temperature and the thermal flux in the x-direction, which are necessary to obtain the thermal conductivity since by default, ANSYS did not display those results. Simulation was performed and 6 results are displayed: temperature distribution total thermal flux, directional thermal flux, gradient temperature in x-direction, thermal flux in x-direction, and the thermal conductivity after applying Fourier's Law of heat conduction.

3. Results

3.1. Temperature and heat flux distribution within brick

Figure 4 shows typical temperature distribution within the bricks. The temperature distribution is slightly distorted by the holes. This indicates the heat flux is not uniform for bricks with holes as in Figure 5. Heat flow is restricted by the hole, and to compensate for smaller cross-sectional area, the heat flux in this region is higher than other regions within the brick. For the solid brick, the heat flux is uniform.

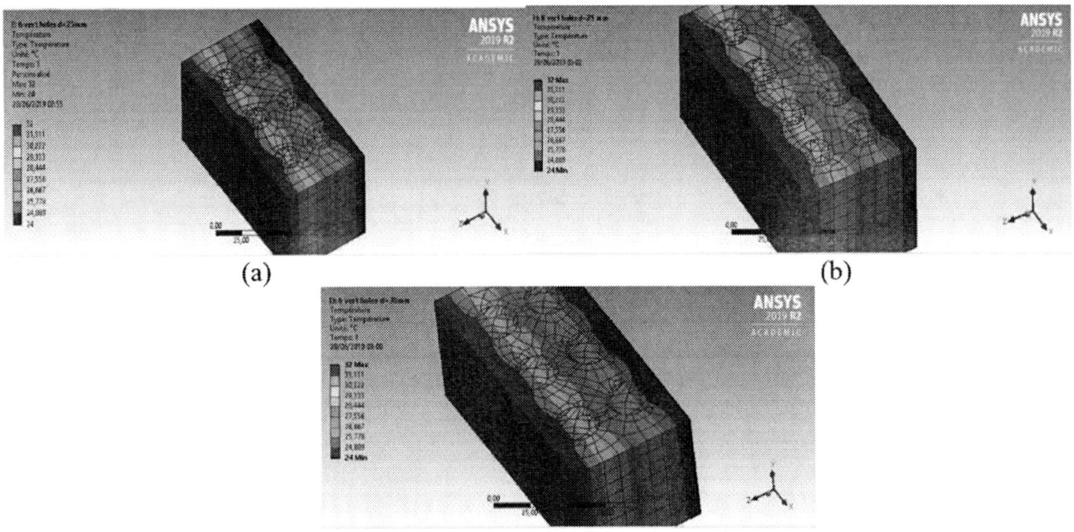

(a) (b) (c)

Figure 4. Temperature distribution in (a) 6 holes with 25mm diameter (b) 8 holes with 25mm diameter and (c) 6 holes with 30mm diameter.

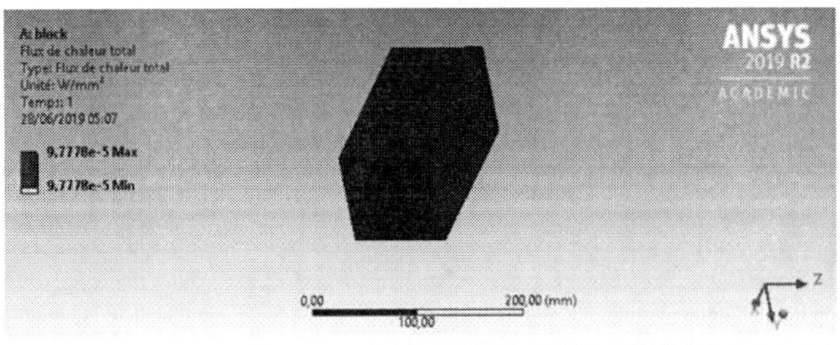

(a)

IOP Conf. Series: Materials Science and Engineering **834** (2020) 012021 doi:10.1088/1757-899X/834/1/012021

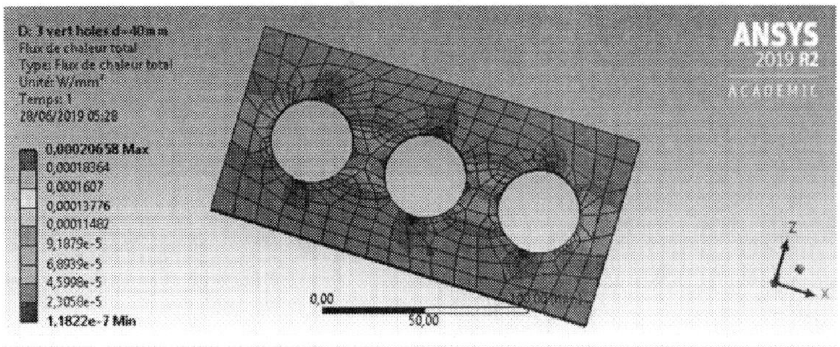

(b)

Figure 5. Variation of heat flux in (a) solid clay brick and (b) hollow brick with 3 holes.

3.2. *Effect of void volume on effective thermal conductivity*

The effective thermal conductivity (Figures 6 through 8) of bricks was reduced to almost half of the solid brick when the percentage of volume of holes reach 10%. At higher percentage, the effective thermal conductivity was reduced to its lowest at 0.4 W/mK, regardless of the filling materials. This may indicate that once the thermal conductivity of filling material is below a specific value, there are no significant differences can be observed in the effective thermal conductivity of brick.

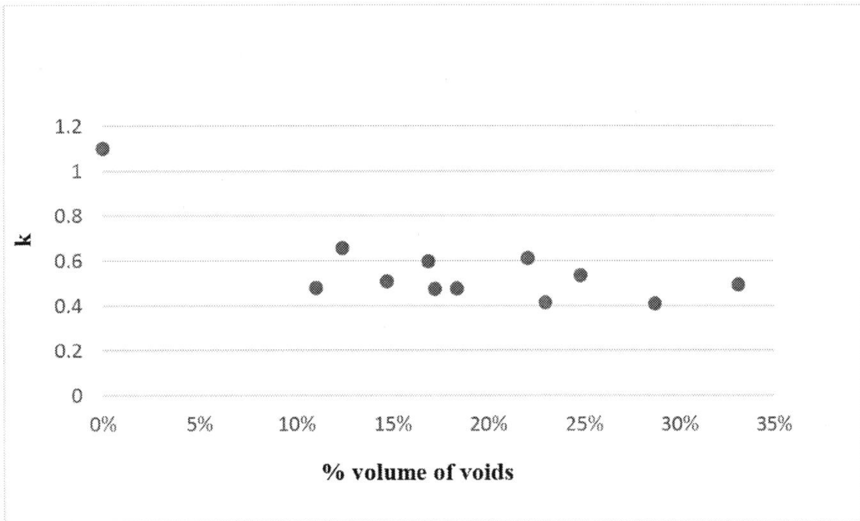

Figure 6. Thermal conductivity at various percentage of void volume for hollow bricks filled with air.

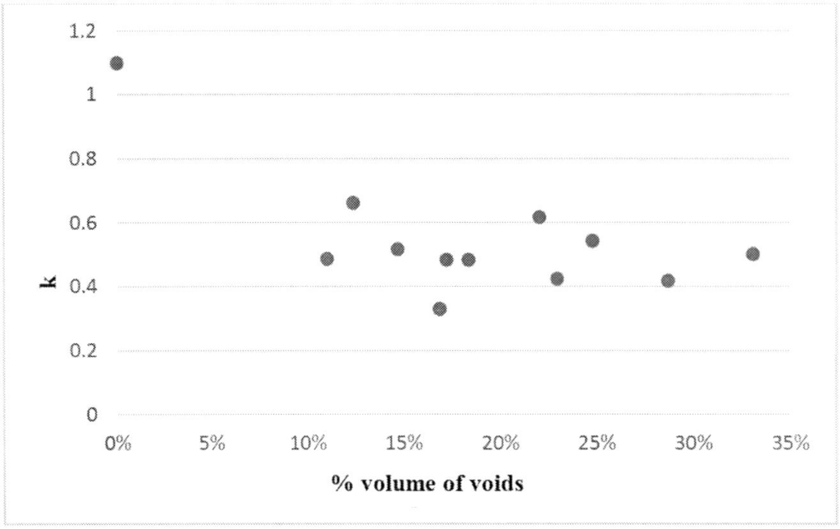

Figure 7. Thermal conductivity at various percentage of void volume for hollow bricks filled with wool.

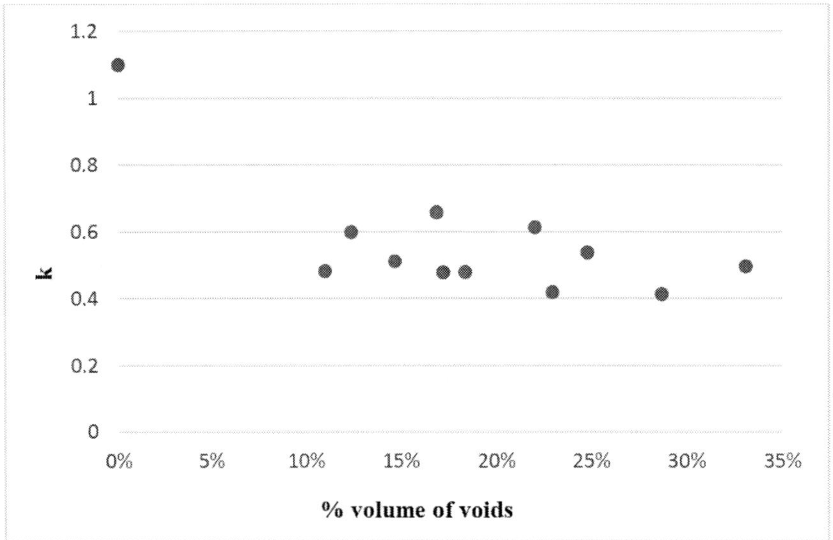

Figure 8. Thermal conductivity at various percentage of void volume for hollow bricks filled with expanded foam.

3.3. Effect of reduction in cross-sectional area on effective thermal conductivity

Figures 9 through 11 show the effective thermal conductivity of brick was reduced by almost 50% relative to a solid brick thermal conductivity when the cross-sectional area is reduced by 10%. This effect is similar to the trend shows by the effect of void volume on effective thermal conductivity of bricks. This finding shows that a higher reduction in cross-sectional area of bricks is not beneficial to its thermal properties. However, excessive reduction in cross-sectional area may compromise the mechanical properties of bricks.

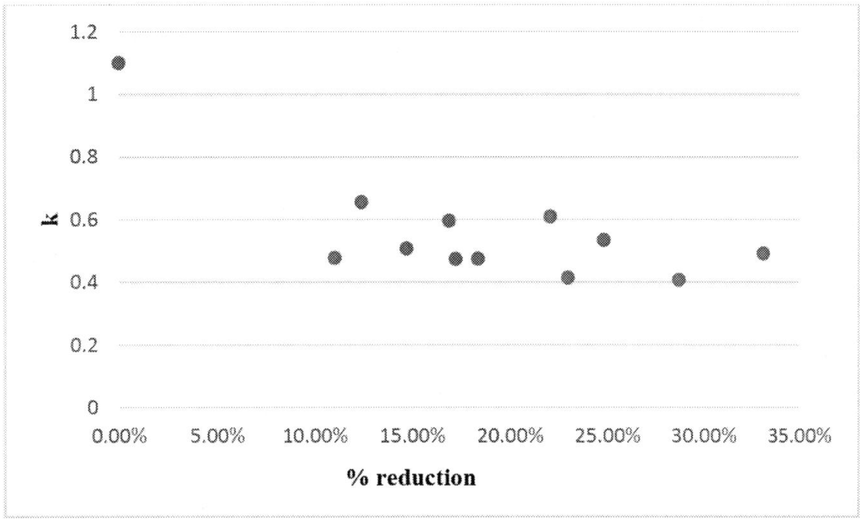

Figure 9. Thermal conductivity at various percentage of reduction in cross sectional area for heat conduction for hollow bricks filled with air.

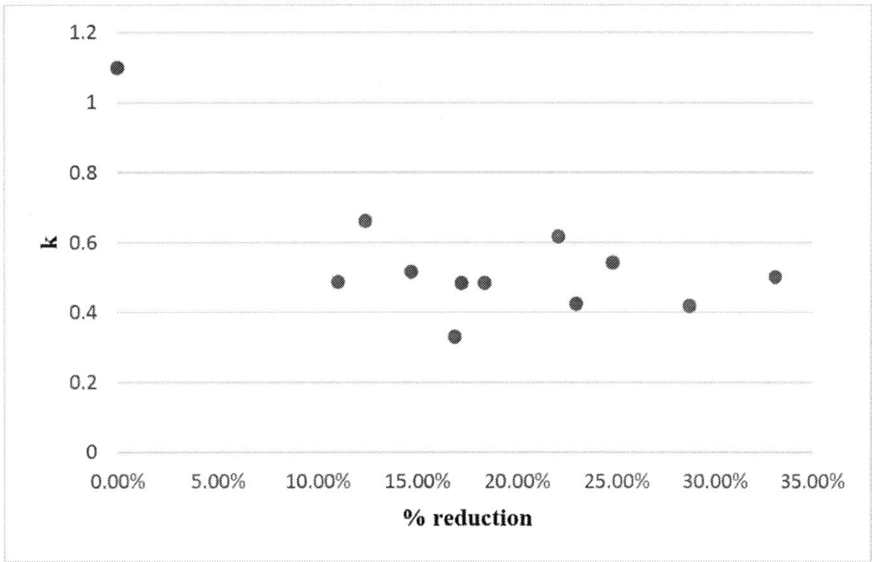

Figure 10. Thermal conductivity at various percentage of reduction in cross sectional area for heat conduction for hollow bricks filled with wool.

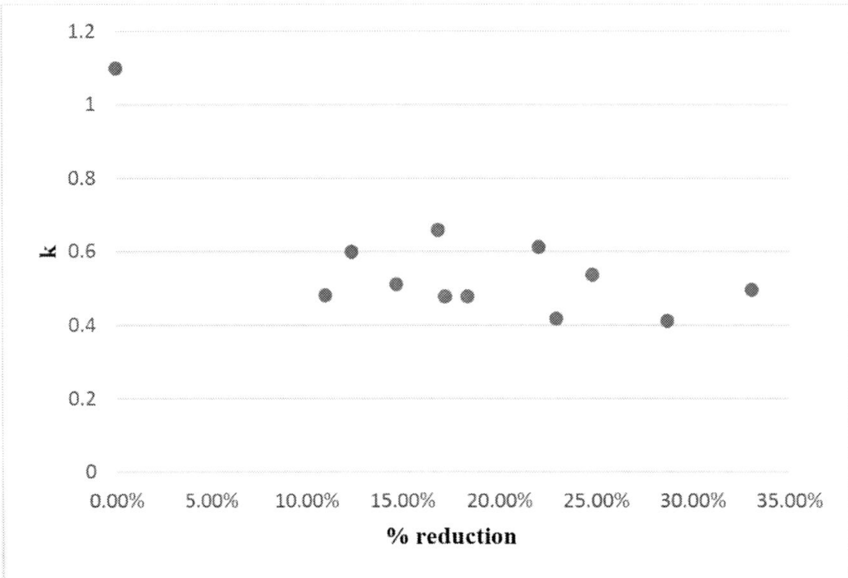

Figure 11. Thermal conductivity at various percentage of reduction in cross sectional area for heat conduction for hollow bricks filled with expanded foam.

4. Discussion

The results of this study show that the presence of holes in clay bricks can reduce its effective thermal conductivity by 50%. It is worth to note that this is based on steady state heat conduction analysis where other properties such as specific heat capacity and density is not considered. Thus, cautions should be taken in predicting the corresponding reduction in cooling load because the time lag for heat transmission through wall may altered the variation of cooling requirement. For instance, the highest cooling load may occur six hours after the external temperature of a wall reached its peak value [20]. The time lag between external peak temperature and highest cooling load is also influenced by the thermal mass and density of wall.

The pessimistic estimation of accuracy of effective thermal conductivity for clay bricks with its cavity filled with air obtained from the simulations is within 10% of its actual value. This is because there is no effect of heat convection included in the simulation. The assumption that there is no bulk movement of air is true for the cavity with a smaller diameter. For a cavity with a larger diameter, the effect of heat convection is more significant. For building's wall, heat transfer due to transport of moisture is expected to be more dominant and will alters the effective thermal conductivity of a wall. Thus, this may make the effect of heat convection is negligible. Further investigation on the effect of heat convection within the cavity filled with air is required in future work.

The effective thermal conductivity of hollow bricks was reduced by almost 50% when the void volume and reduction in cross-sectional area for heat transfer is more than 10%.This indicate that higher percentage of void volume and small cross-sectional area would not further reduce the effective thermal conductivity. This information is very useful to limit our intuitive desire in minimizing the percentage of solid clay since mechanical properties may be significantly affected if less solid is available to withstand compressive force.

5. Conclusion

The steady-state thermal performance of hollow bricks filled with air, wool and expanded foam was investigated. It can be concluded that the effective thermal conductivity is almost constant if the percentage of void volume and reduction in cross-sectional area was 10% or higher. It was found that the heat flux was non-uniform for hollow bricks, which lead to higher heat flux at several regions within the bricks. The findings presented in this paper is useful to construction industry where some reduction in energy consumption is feasible through usage of hollow brick as building wall. Further analysis is required to determine the effect of diameter of hole on the effective thermal conductivity of bricks. In future, it is worth to evaluate the transient thermal analysis to determine the effect of time lag on heat transmission, which is important to the evaluation of timewise cooling load of buildings. The mechanical properties of the bricks also need to be evaluated.

References

[1] Zhao H X and Magoulès F 2012 A review on the prediction of building energy consumption Renewable and Sustainable Energy Reviews vol. 16 pp 3586– 3592
[2] Mishra S, Usmani J A and Varshney S 2012 Energy Saving Analysis In Building Walls through thermal Insulation System International Journal of Engineering Research and Application, pp. 128-135
[3] Meng X, Luo T, Gao Y, Zhang L, Shen Q and Long E 2017 A new simple method to measure wall thermal transmittance in situ and its adaptability analysis Applied Thermal Engineering vol. 122 pp. 747-757
[4] Rhee-Duverne S and Baker P, "Research into the thermal performance of traditional brick walls," English Heritage research report, N. D..
[5] Muhaisen A, "Effect of wall thermal properties on the energy consumtion of buildings in the Gaza Strip," International Sustainable Building Symposium, N. D..
[6] Almujahid A and Kaneesamkandi A, "Construction of test room for evaluating thermal performance," International Engineering and Technology Journal of Innovative Research in Science, pp. 2000-2007, 2013.
[7] Zhang Y and Wang Q, "Influence of Hollow Block's Structural Configuration on the Thermal Characteristics of Hollow Block Wall," Procedia Engineering, p. 2341–2348, 2017.
[8] Lorente S, Petit M and Javelas R, "The effects of temperature conditions on the thermal resistance of walls made with different shapes vertical hollow bricks," Energy and Buildings, vol. 28, no. 3, pp. 237-240, 1998.
[9] Lorente S, Petit M and Javelas R, "Simplified analytical model for thermal transfer in vertical hollow brick," Energy and Buildings, vol. 24, pp. 95-103, 1996.
[10] Morales M P, Juárez M C, López-Ochoa L M and Doménech J, "Case Study: Study of the geometry of a voided clay brick using rectangular perforations to optimize its thermal properties," Applied Thermal Engineering, vol. 31, pp. 2063-2065, 2011.
[11] Morales M P, Juárez M C , Muñoz P and Gómez JA, "Study of the geometry of a voided clay brick using non-rectangular perforations to optimize its thermal properties," Energy and Buildings, vol. 43, no. 9, pp. 2494-2498, 2011.
[12] del Coz Díaza J J, García N P J, Díaz Péreza L M , and Fernández R P, "Nonlinear thermal analysis of multi-holed lightweight concrete blocks usedin external and non-habitable floors by FEM," International Journal of Heat and Mass Transfer, vol. 54, pp. 533-548, 2011.
[13] Al-Hadhrami L M and Ahmad A, "Assessment of thermal performance of different types of masonry bricks used in Saudi Arabia," Applied Thermal Engineering, vol. 9, p. 1123–1130, 2009.
[14] Hou S D, Liu F S, Wang S J , and Bian H B, "Coupled heat and moisture transfer in hollow concrete block wall filled with compressed straw bricks," Energy and Building, vol. 135, p. 74–84, 2017.
[15] Li J, Meng X, Gao Y, Mao W, Luo T and Zhang L, "Effect of the insulation materials filling on the thermal," Case Studies in Thermal Engineering, p. 62–70, 2018.
[16] Kikuchi K, Kang Y S, Kawasaki A, Nishida S and Ichida A, "Microstructural Modeling and Thermal Property Simulation of Unidirectional Composite," Materials Transaction, vol. 45, no. 2, pp. 542-549, 2004.
[17] Zbynek S and Marek K, "Numerical simulation of heat transfer through hollow bricks in the vertical direction," Journal of Building Physics, vol. 34, no. 4, pp. 325-350, 2010.
[18] Mahmouda A M, Ben-Nakhib A, Ben-Nakhi, and Rashed A A, "Conjugate conduction convection and radiation heat transfer through hollow autoclaved aerated concrete blocks," Journal of

Building Performance Simulation, vol. 5, no. 4, 2012.
[19] M. S. M. 1064, "Guide to Modular Coordination in Buildings: Part 8: Coordinating Sizes and Preferred Sizes for Masonry Bricks and Blocks," 2001.
[20] Ahamat M A , Abidin R, and Abdullah S M, "Effect of thermal mass and thickness of building envelope on cooling load," in ICGTEC, Malaysia, 2014.

Towards a smart simulation of surface modification by thermochemical treatments for stainless steel in the industrial revolution 4.0

Annissa Fanya[1], Esa Haruman[1*] and Mohd Shahriman Adenan[2]

[1]Bakrie University, Kuningan, Jakarta, 12920, Indonesia.
[2]Universiti Teknologi MARA, Shah Alam, Selangor, 45450, Malaysia.

Corresponding author's email: esa.haruman@bakrie.ac.id

Abstract. This paper provides a conceptual basis of the development of a virtual process control design using smart simulation in the thermochemical surface engineering design for stainless steel. A historical review as a foundation of further development in attaining a smart simulation on low temperature thermochemical treatments is highlighted. Both scientific platforms and commercial explorations are suggested to thoroughly study the possibility of attaining a smart simulation on this material manufacturing related field. It is concluded that three essential stages of future works need to be performed: the construction of base points, model and simulation development, and advancement of simulation related to additive manufacturing to create a smart simulation in the industry 4.0 era.

1. Historical perspective

Stainless steel is a group of metallic materials used in various kinds of industrial need. It was firstly invented by Harry Brearley, a metallurgist from Sheffield, UK, in 1913 [1]. A wide range of application of this material can be found everywhere, from domestic appliances, chemical industries, building structures, transport equipment, jewelries, medical equipment, and nowadays for surgical implanted parts. Stainless steel is one of the most commonly used biomaterials as internal fixation due to a great combination of its mechanical properties, biocompatibility, corrosion resistance and cost effectiveness [2].

The world's annual production of stainless steel has consistently increased over the last 50 years' period [Figure 1], thus demand for this material has continuously increased in various sectors of applications. The austenitic type of stainless steel constitutes around 70% of total stainless steel demand, with China being the highest producing country [Figure 2]. Regardless of its advantages pertaining to good manufacturability and a lower price, the austenitic type of stainless steel has the lowest values in hardness and wear resistance in comparison to other types of stainless steel.

Content from this work may be used under the terms of the Creative Commons Attribution 3.0 licence. Any further distribution of this work must maintain attribution to the author(s) and the title of the work, journal citation and DOI.
Published under licence by IOP Publishing Ltd

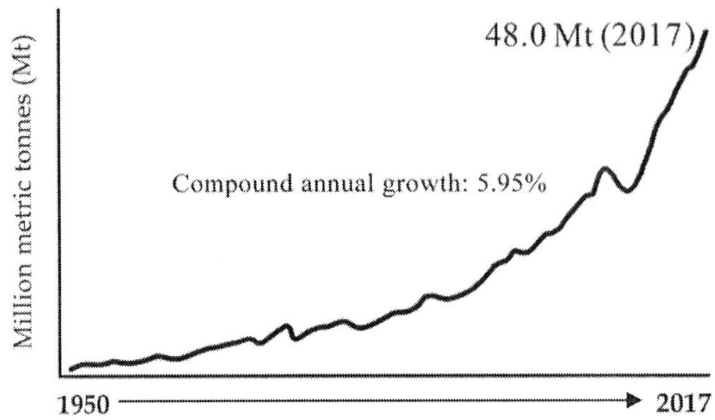

Figure 1. World's annual production of stainless steel [3]

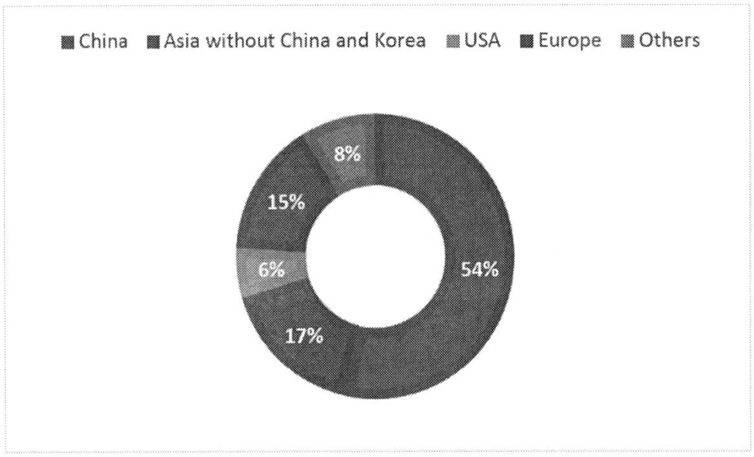

Figure 2. Stainless steel producing countries [3]

Efforts were made to improve the surface mechanical properties of the austenitic stainless steel in the early 1980's. The earliest publication on the surface modification of austenitic stainless steel was made by previous works [4-6]. The technique involved nitriding in a glow discharge assisted atmosphere or namely plasma nitriding to improve the surface hardness of the material. Improvement in the surface hardness was found to be five times significantly above the hardness of the untreated austenitic stainless steels substrate. Scientific evidence showed that surface hardness improvement was caused by the formation of expanded austenite, or identified as S-phase, formed in the nitrogen diffusion zone with thickness that depends on the treatment time [8-10].

Despite the marked improvement on the surface hardness of austenitic stainless steel AISI 316L, such investigations showed that the treatment causes a deterioration effect to the corrosion resistance of this material due to chromium depletion in the solid solution by forming CrN precipitates and consequently leading to a sensitivity to intergranular corrosion attacks [11]. Only after the nitriding treatment was performed below 500 °C the resultant nitride layer was free from nitrogen precipitation. These thermochemical treatments applied for stainless steel is technically termed as low temperature thermochemical treatment [12].

The development in low temperature thermochemical treatment in the early stages [1985-2000] was marked by various publications in exploring the characteristics of expanded austenite layers mostly produced by plasma nitriding and carburizing techniques [13]. The understanding of the physical metallurgy of S-phase or expanded austenite involving both nitrogen and carbon in low metastable solid solution was the main focus of research interests during that period.

In that period of time, various glow discharge techniques such as RF discharge, DC diode, plasma immersion ion implantation, low pressure plasma, and ion beam processing were the main attractions of investigations. Other surface modification techniques by solution nitriding [14], salt bath nitriding [15] and gaseous carburizing [16] were also studied with regard to their feasibilities and in agreement with precise control techniques using glow discharge assisted atmosphere.

A number of researches works from various parts of the world, shown in Table 1, were conducted to investigate the development of the thermochemical treatments. But only after the year 2005, low temperature nitriding became available for commercial applications. This excellent contribution by the international research communities in this specific area of surface heat treatments provided a sound scientific basis and technological solution to a required industrial need in the current progressive decade.

Table 1. World's research works

Year	Country
1985	United Kingdom
1990	Germany Denmark Korea Japan Australia
2005	Malaysia Singapore
2010	Brazil
2012	Iran
2014	Mexico
2018	Indonesia

Following the early years of research work, an introduction of hybrid thermochemical treatment was published [17] in which nitrogen and carbon are simultaneously diffused into the surface of stainless steel at temperatures below 500°C. As a result of this combined diffusion, a thick hard layer with smooth hardness gradient in the region adjacent to the layer-substrate interface were produced [Figure 3] thus providing a superior load bearing capacity than that resulted from individual nitriding and carburizing [18].

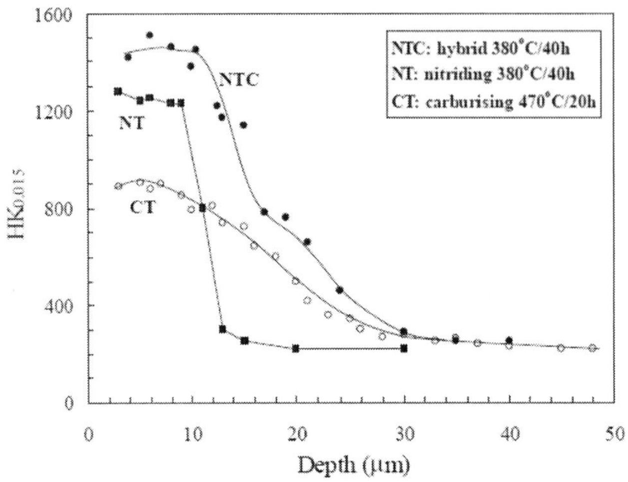

Figure 3. Typical hardness profiles resultant from various low temperature plasma surface alloying processes [17].

In the most recent years, thermochemical treatments to improve surface properties of stainless steel were also studied and was implemented to the duplex type of stainless steel [19]. The works using laboratory gaseous furnace showed that although the hardness of duplex stainless steels improved, difficulties were found in obtaining a thick layer of expanded austenite phase owing to the presence of ferrite phase in the structure. In the case of a hybrid process, the treatment did not produce a dual separate layer consisting of nitrogen expanded austenite (S_N-phase) on top and followed by carbon expanded austenite (S_C-phase) below it, as it was compared to layer formation on austenitic stainless steel surfaces. Figure 4 shows a comparison of resultant layers produced by thermochemical treatments for austenitic stainless steel and for duplex stainless steel.

The surface modification technologies of stainless steels in the explanation above are still the main attraction in the laboratory investigations done for further improvement associated with their commercial availability in industries. These technologies are finding a rapidly expanding niche market covering applications of human needs.

Today in the period of Industry 4.0, after three decades of research development in surface modification of stainless steel, the thermochemical treatments are feasible to be advanced by the usage of mathematical models, simulation, and optimization using computer assisted programming. This virtual technological approach may become a novelty for not only metallurgical related sciences but also for the innovation of additive manufacturing technologies to promote cost effectiveness particularly in manufacturing process design.

(a) (b)

Figure 4. Comparison of the resultant layers produced by thermochemical treatments for (a) austenitic stainless steels [17] and (b) duplex stainless steels [19].

2. Mathematical model and simulation

A mathematical model of thermochemical treatment was initially carried out for nitriding of low alloy steel by glow discharge technique [20]. The model performed mathematical formulations for various diffusional reactions to represent a complete model of nitriding and used a numerical solution to solve the model. The developed model was used to predict the distribution of nitrogen in solution and in the form of alloy nitride precipitates, the iron nitride layer thickness developed during plasma nitriding, the incubation time for the formation of iron nitride layer, and the nitride precipitation behavior of various alloying elements in the steel. It was concluded that the precision of the calculation results was strongly influenced by the accuracy of diffusion coefficient in various phases and the sputtering rate. The model however had a limitation that it only considered iron-alloy elements systems, whilst the influence of carbon in the substrate and reaction of excess nitrogen, as well as redistribution of carbon and carbides, were not included, hence the accuracy of the model was rather ambiguous.

Modifications of the above model were further proposed by other works [21,22]. In these two modified models, the diffusional coefficient of nitrogen across the diffusion layer was considered independent of the nitrogen content and can be applied to two-phase systems or more. But the models did not consider the impeding effect of dispersed nitrides on the overall diffusive flux of nitrogen. Corrections to the above models were then made by another work [23]. This adapted version of the nitriding model considered system was a continuous and single-phase system because the transition from the expanded austenite to the expanded austenite bulk was caused by the dissolution of nitrogen into the austenite lattice. The work also demonstrated that a strong concentration dependency exists for nitrogen diffusion with a case of expanded austenite stainless steel. The model therefore incorporates a compositional dependent diffusion in the austenitic phased stainless steels. This work however did not include the lattice distortion effect created by interstitial diffusion of nitrogen in the austenite phase during a diffusional process. Thus, the accuracy of the model was not considerable, and a more descriptive model needs to be developed. In the most recent work, the influence of lattice distortion was already considered by which the interaction between the composition-induced stresses and the diffusion was examined for both purely elastic and elastic–plastic stresses [24]. By taking the above considerations in nitriding [19-23], a hypothetical illustration of the mathematical model that can be utilized as a basal platform of further development in obtaining precise calculations for a simulation design on thermochemical treatments of stainless steels is given in Fig. 5. A more delicate mathematical computation might be expected to obtain a precise model for thermochemical treatments of stainless steels due to various parameter effects involved in the process.

Numerous computer software to simulate the manufacturing processes have been established [25, 26]. Simufact and Sorpas are two examples of commercial software that provide process design and simulation on forging and welding in manufacturing operations. In the view of thermochemical treatments for stainless steels the work on a computer simulation for this manufacturing related field has not been initiated elsewhere. The technological innovations driven by the advancement in artificial intelligence and cloud computing extends the opportunities to implement this information technology basis to the intelligent manufacturing. This will include surface properties enhancement and, in this particular interest, is the simulation of surface modification of stainless steels by thermochemical treatment methods. Such capabilities on applying the conceptual design into a valid simulation, for example the components with a choice of applications, and complementing the experimental research works with the virtual works of simulation, are the challenges to be resolved in order to attain a smart simulation.

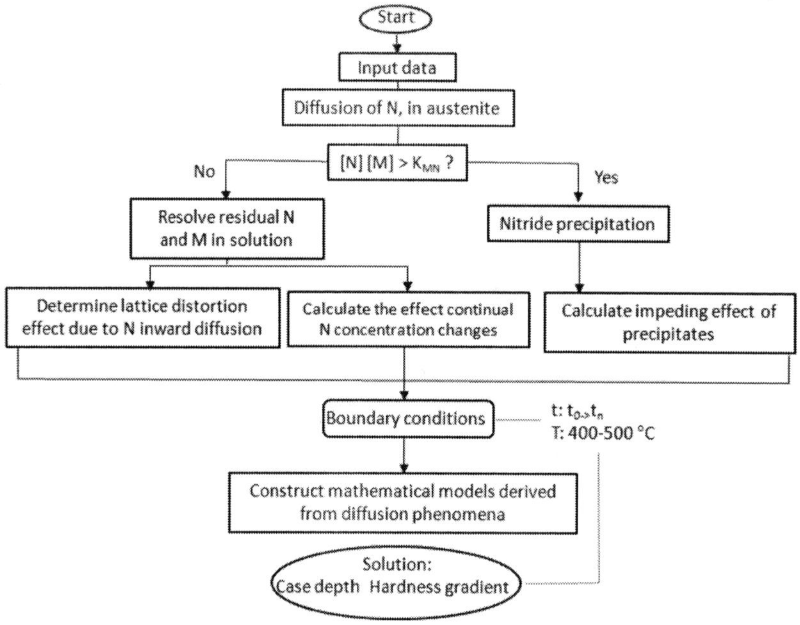

Figure 5. Mathematical Model Process.

3. Future works

Modeling and simulation of the process control in manufacturing operations have long been performed by use of numerical method analysis using computer assisted programming tools. Multi-disciplinary knowledge is needed to resolve complexities that arise in transformation of the physical and metallurgical phenomena to mathematical solutions and to process design simulations. This is the mutual practice that embraces the involvement of materials scientists, manufacturing engineers, as well as mathematicians. In developing the technological establishment of industry 4.0, the contribution of informatics engineers to translate mathematical domains into a virtual simulation becomes a key factor that results in a smart simulation with several user-friendly features. In view of present surface engineering for enhanced wear properties of stainless steels materials, such features like designing the process controls by a combination of working temperatures and treatment durations in a given treatment atmosphere to achieve a superior surface hardness with precision case depth are the main focus of simulation software development. Fig. 6 is a diagram that proposes the future works to be performed in connection to surface modifications of stainless steels by thermochemical treatments.

There are three essential stages that should be considered for the development of a smart simulation. These include construction of the base points, model and simulation development, and advancement of simulation. The four activities contained in the base points represent the need of balanced studies on both technological push factors and business pull factors. In the stage of model and simulation development, the controlling factors in achieving a smart simulation are the complete data of the base point activities. The third stage suggests the possibility of the simulation to be integrated with the additive manufacturing operations as a finishing step for the enhancement of surface integrity for the stainless steels' components. In the period of Industry 4.0 applying cloud computing to enable space less and flexible process controls will be the next stage in the development of smart simulation for additive manufacturing operations.

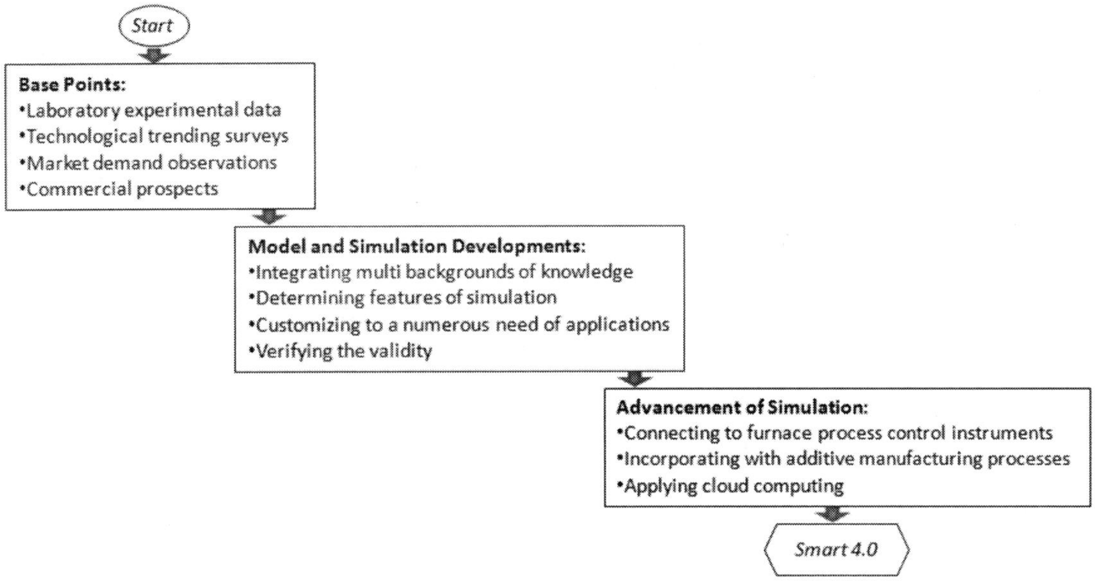

Figure 6. Future Works Diagram.

Acknowledgement
The authors would like to express their gratitude to the Ministry of Research, Technology and Higher Education, The Republic of Indonesia, for the financial support of the currently ingoing research work on mathematical modelling and simulation in thermochemical surface engineering.

References

[1] J. Capus, "100 Years of Stainless Steel", *Metal Powder Report,* vol. 68, no. 5, p. 12, 2013.
[2] A. J. Disegi and L., "Eschbach, Stainless steel in bone surgery", *Injury, Int. J. Care Injured,* vol.31, pp. 2-6, 2000.
[3] Anonymous, "International Stainless Steel Forum 2019", [Online]. Available: http://www.worldstainless.org. [Accessed: 08-05-2019]
[4] Z. L. Zhang and T. Bell T, " Structure and corrosion resistance of plasma nitrided stainless steel", *Surf. Eng.,* vol. 2, pp. 131-136, 1985.
[5] K. Ichii, K. Fujimura and T. Takase, "Structure of the ion-nitrided layer of 18-8 stainless steel", *Rep. Kansai Univ,* vol. 27, pp. 135 – 144, 1986.
[6] Rollinski, "Effect of plasma nitriding temperature on surface properties of austenitic stainless steels", *Surf. Eng.,* vol. 1, pp. 35 – 40, 1987.
[7] P. A. Deamley et al., "Some observations on plasma nitriding austenitic stainless steel", *Plasma Surf. Eng. (Conference Proceedings),* vol. 1, pp. 219 – 226, 1989.
[8] Y. Sun, X. Li and T. Bell, "Low temperature plasma carburising of austenitic stainless steels for improved wear and corrosion resistance", *Surf. Eng.,* vol. 15, pp. 49-54, 1999.
[9] Y. Sun et al., "The response of austenitic stainless steels to low-temperature plasma nitriding", *Heat Treatment of Metals,* vol. 1, pp. 9-16, 1999.
[10] Y. Sun, X. Li and T. Bell , "The formation and decomposition of nitrogen and carbot fct austenite in austenitic stainless steel", *Stainless Steel 2000,* vol. 263, 2000.
[11] P. A. Deamley, K. L. Dahm and G. Aldrich-Smith, "The nature of S phase coatings and their wear

and corrosion-wear behaviour when applied to 316L", *Stainless Steel 2000*, vol. 317, 2000.

[12] T. Bell and Y. Sun, "Low temperature plasma nitriding and carburising of austenitic stainless steels", *Stainless Steel 2000,* vol. 275, 2000.

[13] T. Bell and K. Akamatsu, "Stainless Steel", *Proceedings of an International Current Status Seminar on Thermochemical Surface Engineering of Stainless Steel*, 2000.

[14] H. Berns et al., "Solution nitriding of stainless steels – a new thermochemical heat treatment process", *Stainless Steel 2000,* vol. 117, 2000.

[15] M. Higashi, K. Shinkawa and K. Kurosawa,"Low temperature palsonite salt bath nitriding of austenitic stainless steel SUS304", *Stainless Steel 2000,* vol. 407, 2000.

[16] K. Aoki et al., "Low temperature gas carburising for austenitic stainless steels: the NV-pionite process", *Stainless Steel 2000, vol.* 389, 2000.

[17] Y. Sun and E. Haruman, "Low temperature plasma surface alloying of austenitic stainless steels", *Solid State Phenom.,* vol. 118, pp. 85-90, 2006.

[18] Y. Sun and E. Haruman, "Influence of processing conditions on structural characteristic of hybrid plasma surface alloyed austenitic stainless steel", *Surf. Coat. Techn.,* vol. 202, pp. 4069-4075, 2008.

[19] M. S. Adenan, M. N. Berhan and E. Haruman, "Formation of Expanded Austenite using Hybrid Low Temperature Thermochemical Treatment on 2205 Duplex Stainless Steel", *Advanced Materials Research,* vol. 970, 244-247, 2014.

[20] Y. Sun Y and T. Bell, "A numerical model of plasma nitriding of low alloy steels", *Mater. Sci. Eng.* vol. 224, pp. 33-47, 1997.

[21] R. E. Schacherl, P. C. J. Graat and E. J. Mittemeijer, "", *Metall. Mater. Trans. A.,* vol. 35A, p. 3387, 2004.

[22] J. D. Kamminga and G. C. A. M. Janssen, "", *Surf. Coat. Techn.,* vol. 200, p. 5896, 2006.

[23] T. Christiansen, K. V. Dahl and M. A. J. Somers, "Simulation of nitrogen concentration depth profiles in low temperature nitrided stainless steel", *Def. Diff. Forum,* vol. 258-260, pp. 378 383, 2006.

[24] F. N. Jespersen, H. J. Hattel and M. A. J. Somers, "Modelling the evolution of composition-and stress-depth profiles in austenitic stainless steels during low-temperature nitriding", *Modelling Simul. Mater. Sci. Eng,* vol. 24, p. 31, 2016.

[25] Z. Mikno and Z. Bartnik, "Heating of electrodes during spot resistance welding in FEM calculations", vol. 16, pp. 86-100, 2016.

[26] D. Qian et al., "Optimization for large spherical valve body based on FE simulation", *Procedia Engineering* Processing, vol. 81, pp. 2481 – 2487, 0000.

IOP Conf. Series: Materials Science and Engineering **834** (2020) 012023 doi:10.1088/1757-899X/834/1/012023

The effect of urban obstacles on the flow distribution at pedestrian area

M A Zainol[1], A A Razak[2]*, M F Mohamad[2]*, H F Pahroraji[1], A Sobri[2], H A Kasim[1] and A H Khalid[1]

[1] Faculty of Mechanical Engineering, Universiti Teknologi MARA (UiTM), Pasir Gudang Campus, Johor Darul Takzim,Malaysia.
[2] Faculty of Mechanical Engineering, Universiti Teknologi MARA (UiTM), Shah Alam, Selangor Darul Ehsan, Malaysia.

*azlirazak@salam.uitm.edu.my, faizal3744@salam.uitm.edu.my

Abstract. The pedestrian level is an important area where most of the human activities occur in this area such as walking, reading, and cycling. Along with the pedestrian activities, there are small street shops are builds and presences of the vehicle at the street canyon where people park their cars in this area. This study aims to investigate the effect of the presence of obstacles such as vehicles at the street canyon on wind flow structure and ventilation performance. A computational simulation study is carried out to analyze the contribution of obstacles to pedestrian ventilation performance. A steady Reynolds-averaged Navier-Stokes equation with the realizable standard k-ε model is used as a turbulence solver. The result showed presences of obstacle at a pedestrian level will effect on wind flow structure where the flow becomes more chaotic compared with no presences of obstacles at a pedestrian level. Furthermore, the air change rate (ACH) is reducing as the number of obstacles increase thus will affect the ventilation performance at the pedestrian area. The presences of an obstacle such as vehicle and street shops will contribute to the wind flow structure and ventilation performable at pedestrians' level.

1. Introduction

In modern city development, modern building design becoming a good choice for developer in order to attract customer or buyer. However, modern city development needs to consider the wind condition or air ventilation entering the city to provide good city breathability especially at pedestrian level. Due to increasing of population and market properties, high-rise building is designed and build inside the city to overcome the issue. The development of high building will effect on the wind condition inside the city, therefore wind condition and pedestrians safety become important requirements in urban planning and design[1]. At designing stage, the developer needs to get an information about wind condition inside the city such as wind speed and air flow around the buildings. This information is important to architects and urban planners to design the layout and to determine the suitable high of the buildings. Unplanning city layout or building height might be effect on the city breathability, air pollution and high wind speeds are regularly presented at a pedestrian level that can be experienced as uncomfortable, or infrequently even perilous [2]. The objective of this study is to investigate the contribution of urban obstacles such as car effects and small shops on wind flow structure at the pedestrian level. The condition of wind inside the pedestrian area are depending on the building shape, size and other obstacles that might blocking the movement of wind at pedestrian area [3].

Content from this work may be used under the terms of the Creative Commons Attribution 3.0 licence. Any further distribution of this work must maintain attribution to the author(s) and the title of the work, journal citation and DOI.
Published under licence by IOP Publishing Ltd

The high wind speed at a high level may be deflected down to ground level by tall buildings causing unpleasant and even dangerous conditions for pedestrians. The formation of vortex around buildings, between buildings or along avenues might accelerate wind speeds at the pedestrian level and giving rise to pedestrian discomfort. Besides that, the poor ventilation performance unable to disperse contaminated air from the urban area and lead to the accumulation of traffic fumes and other pollutants [4].

2. Methodology

2.1 Numerical method

Wind tunnel experiment is commonly used to investigate wind flow and ventilation performance [5], [6]. The result generated by the wind tunnel experiment is generally acknowledged in the scientific and engineering application. However, due to the time consuming and difficulties of constructing and modified the model, the Computational Fluid Dynamics (CFD) method becomes a choice for researchers instead of wind tunnel experiments. CFD method will solve fluid flow problems through numerical approximation and algorithm using the computer. For example, W.D. Jannsen et al. performed a CFD analysis to study the pedestrian wind comfort around the buildings [7]. The 3D Reynold-averaged Navier-Stokes (RANS) equation [8] with the renormalization group (RNG) k-ε turbulence closure model will solve the simulation alongside the continuity equation. The open source CFD software OpenFOAM 2.3 [9][10] is used to perform the CFD simulation [11]with the presumption of steady states, incompressible flow and isothermal condition during the simulation. The RNG k-ε turbulence model is a two-equation turbulence model where similar to the standard k-ε model but has been derived by using the renormalization Group method. This turbulence model varies from the standard k-εmodel just because of the alteration of ε to the equation. The time-averaged continuity of governing equation and momentum in RANS equation can be derived as follows [12][13]:

$$\frac{\partial u_i}{\partial x_i} = 0 \tag{1}$$

$$\frac{\partial}{\partial x_j}(u_i u_j) = \frac{\partial}{\partial x_j}\left[(v + v_t)\frac{\partial u_i}{\partial x_i}\right] - \frac{1}{\rho}\frac{\partial \rho}{\partial x_i} \tag{2}$$

The transport equation for turbulent kinetic energy (TKE) k and dissipation rate of turbulent kinetic energy ε can be written as below:

$$\frac{\partial}{\partial x_i}(k u_j) = \frac{\partial}{\partial x_j}\left[(v + \frac{v_t}{\partial_k})\frac{\partial k}{\partial x_j}\right] + v_t S_{ij}\frac{\partial u_i}{\partial x_j} \tag{3}$$

$$\frac{\partial}{\partial x_j}(\varepsilon u_j) = \frac{\partial}{\partial x_j}\left[(v + \frac{v_t}{\sigma\varepsilon})\frac{\partial \varepsilon}{\partial x_j}\right] + C_\varepsilon 1\frac{\varepsilon}{k}v_t S_{ij}\frac{\partial u_i}{\partial x_i} - C_\varepsilon 2\frac{\varepsilon^2}{k} - \frac{C\mu\eta^3\left(1 - \frac{\eta}{\eta_0}\right)\varepsilon^2}{1 - \beta\eta^2}\frac{\varepsilon^2}{k} \tag{4}$$

Where $u_i$andu_j are mean and fluctuating velocity components in the $x_i$andx_j direction respectively.pis the mean pressure and ρ is the fluid density. An isotropic eddy viscosity and mean strain tensor can be present as:

$$v_t = \frac{C\mu k^2}{\varepsilon} \tag{5}$$

$$S_{ij} = \frac{\left[\frac{\partial u_i}{\partial x_i} + \frac{\partial u_j}{\partial x_j}\right]}{2} \tag{6}$$

2.2. Simulation Setup

The validation study was performed to ensure the accuracy of the simulation setup and the reliability of the simulation result. The result obtained from the simulation was compared with the wind tunnel experiment done by Cheng et al.[14] and CFD simulation was done by Coceal et al. [5]. The dimensions of the domain for validation study is 0.1m x 0.1m x 0.1m (L x W x H) where the height of the building is 0.025 m represent as h. The cubical obstacles were arranged in a square layout with a 25% packing density (λp = 25%) as shown in Figure 1, packing density is defined as the ratio of building roof area over ground surface area. The streamwise wind velocity was set to be uniform at 2 m/s and cyclic condition was applied in a streamwise and spanwise direction to yield an infinity repeating model as shown in Figure 1. A slip condition was applied at top boundaries while wall function is imposed on the floor and building surface. Table 1 presents the boundary condition and Figure 2 illustrates the boundary condition set up in this study.

Table 1. Boundary Conditions

Domain	Boundary Conditions
Inlet, Outlet	Cyclic
Left, Right Sides	Cyclic
Floor	No-slip condition and wall function
Building Surfaces	No-slip condition and wall function
Top	Free slip

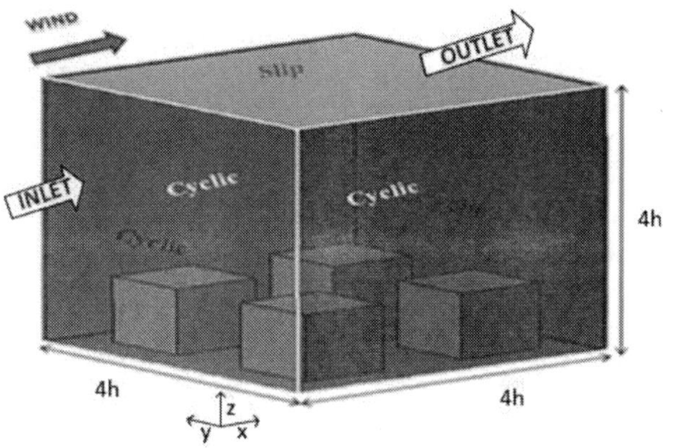

Figure 1. Boundary Conditions

3. Cases Study

In order to investigate the effect of building on flow structure at the pedestrian level, a cubical building with dimension h x h x h where h is 0.025 m is arranged in square arrangement and an obstacle in cubical shape are placed at pedestrian level in order to study the contribution on obstacle at pedestrian level on pedestrian flow structure. The detail position obstacles (case study) are illustrated in Figure 2.

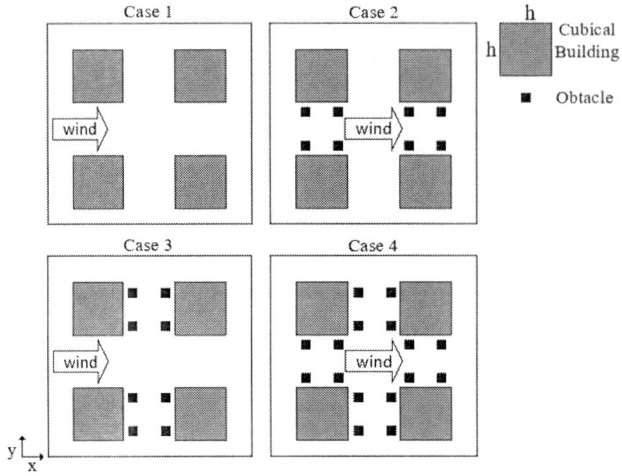

Figure 2. Case Study

4. Result and Discussion

4.1 Validation

The mean wind velocity is normalized with uref, where uref is taken at 4h from ground level and compared with simulation analysis performed by Coceal et al [5]. The point location of mean wind velocity is above the building surface as shown in Figure 3. Apart from the velocity profile, the flow structure has also been compared to ensure the accuracy of the simulation setup capturing the flow structure around the building as shown in Figure 4. The result shows the mean wind velocity has good agreement with mean wind velocity performed by Coceal et al [5]. Therefore, a similar simulation setup is used to investigate the contribution of obstacles at the pedestrian level.

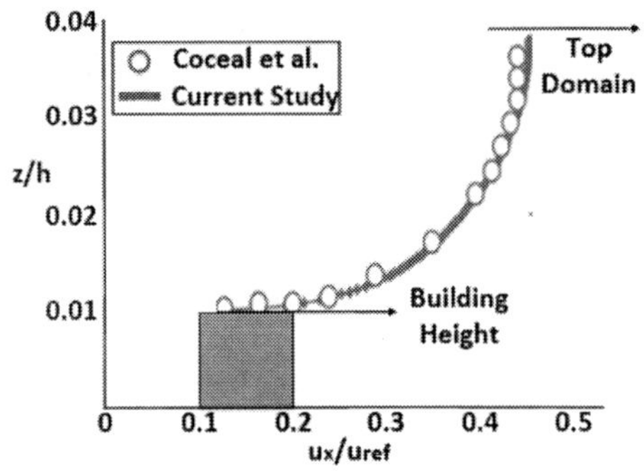

Figure 3. Mean wind velocity profile

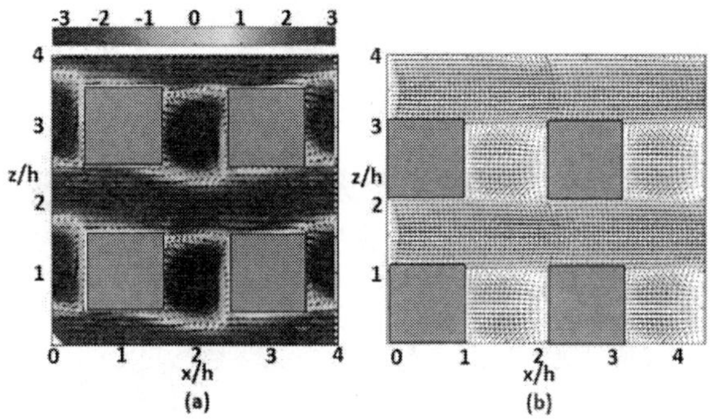

Figure 4. A comparison of vector plot between current study with Coceal et al. (a) Current study (Case 1) and (b) Coceal et al.

4.2 Simulation Result

4.2.1 Mean wind Velocity

The mean wind velocity profile was taken from the floor of the domain until the top of the domain for all three cases. Figure 5 shows the spatial average of normalized mean wind velocity, the mean wind velocity in the x-direction is normalized with u reference, where u reference is at 4h.There is no big difference between all three cases due to the addition of a car is slightly not affect the flow velocity in the domain.

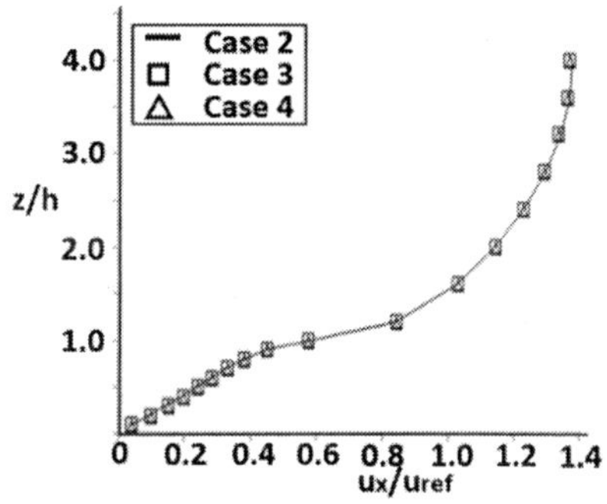

Figure 5. Spatially average profile of mean streamwise velocity ux normalized by reference velocity, uref.

4.2.2 Mean Flow Structure

The air flow structure is an important especially at the pedestrian level because it determines pressure drag and the dispersive stress. It also shows the accumulation of the pollutant that been trap at the vortex. Based on the result obtained in Figure 6, the airflow speed has low velocity inside the street canyon compared at the outside street canyon where blue region shows low velocity speed meanwhile red region shows high velocity speed. However, the airflow structure for all cases was different due to different obstacles orientations; the presence of an obstacle at the pedestrian level will affect the formation of airflow structure at the pedestrian level. From Figure 4 (case 1), the winds enter the urban area without any obstacles at the pedestrian level, the airflow through smoothly at the street canyon and the vortex only formed at the back of the buildings. However when an obstacles places behind the upfront building, at the street canyon and both places, the vortex formation occurred at the second row of the buildings, there are small and big vortex formed at this area (case 2), meanwhile as the obstacle position changed to street canyon, the vortex formation occurred at the first row of buildings (case 3). As the number of obstacles added to the pedestrian level as shown in case 4, there is no vortex big formed at this area as formed in case 2 and case 3, however, the wind flow movement for this case are not smooth compared to case1, case 2 and case 3. Based on the result, the number of obstacles and its positions will influence the flow structure formed at the pedestrian area, this might be contributing to the effects of pollutant disperse from the pedestrian area.

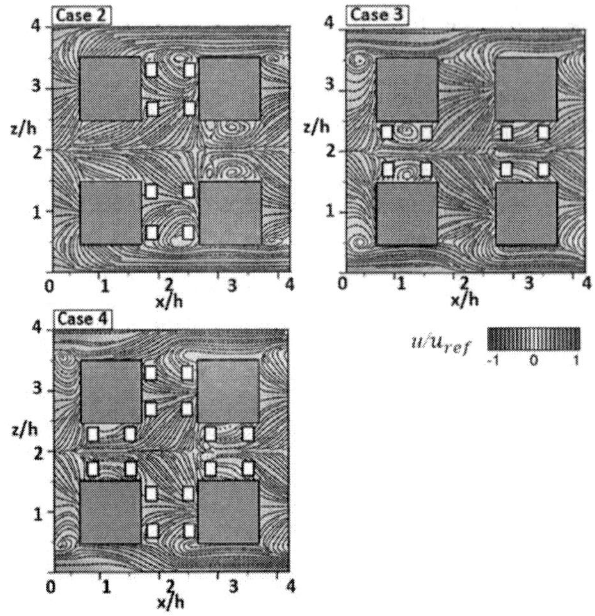

Figure 6. Flow structure of mean field (u,v) in a horizontal x-z plane close to the bottom surface at z = 0.05h for case 2, case 3 and case 4.

4.2.3 Air Change Rate (ACH)

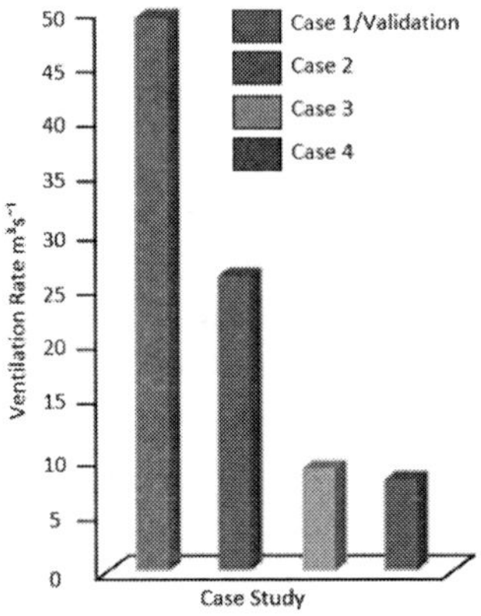

Figure 7. Average Ventilation Rate.

The pedestrian level is an important area in the urban area where all the people activities occurring in this area such as walking, reading, and resting. Therefore, in this area, the ventilation performance must be good to provide sufficient fresh air toward pedestrians. The ventilation at the pedestrian level will contribute to ventilation performance. Based on Figure 7 case 1 has a higher ventilation rate compared to other cases due to no obstacle at the pedestrian level. Meanwhile, as the number of obstacles increases at the pedestrian level the ventilation rate will be decreased. Based on the result, the presence of obstacles such as cars and street shops at the pedestrian's level will contribute to the ventilation performance.

5. Conclusion

The presence of an obstacle such as street shops and vehicles inside the street canyon will be contributed to the formation of wind flow structure formed and effect on the capability of wind to disperse pollutants from the pedestrian area. Besides that, as the number of obstacles inside the pedestrian area increase the flow structure formed becomes more chaotic and the air change rate (ACH) is reduced. These problems might affect breathability and ventilation performance inside the pedestrian level. The architect and town planner need to consider the contribution of obstacle at pedestrians' area at the design phase to avoid unpredictable ventilation problem after the development phase and at the same time will be an effect on the pedestrian's health.

Acknowledgement

The authors would like to express their sincere gratitude to Wind Engineering and Building Physics Center, Universiti Teknologi MARA (UiTM) and i-Kohza Wind Engineering for (Urban, Artificial, Man-made) Environment, Malaysia Japan International Institute of Technology (MJIIT) as the Research Institutes facilitating this research.

References

[1] I. Panagiotou, M. K. A. Neophytou, D. Hamlyn, and R. E. Britter, "City breathability as quantified by the exchange velocity and its spatial variation in real inhomogeneous urban geometries," *Science of the Total Environment*, 2013.

[2] N. Z. Saddok Houda, R. Belardi, "A CFD Comsol model for simulating complex urban flow," *Energy Procedia*, 2017.

[3] S. H. R. E. Britter, "Flow and Dispersion in Urban Area," *Annual. Review Fluid Mechanics*, 2003.

[4] Y. Huang, C. Long, J. Deng, and C.-N. Kim, "Impacts of Upstream Building Width and Upwind Building Arrangements on Airflow and Pollutant Dispersion in a Street Canyon," *Environment. Forensics*, 2014.

[5] O. Coceal, T. G. Thomas, I. P. Castro, and S. E. Belcher, "Mean flow and turbulence statistics over groups of urban-like cubical obstacles," *Boundary-Layer Meteorology*, 2006.

[6] O. Coceal and S. E. Belcher, "Mean winds through an inhomogeneous urban canopy," *Boundary-Layer Meteorology*, 2005.

[7] W. D. Janssen, B. Blocken, and T. van Hooff, "Pedestrian wind comfort around buildings: Comparison of wind comfort criteria based on whole-flow field data for a complex case study," *Building Environment*, 2013.

[8] Y. Jiang, C. Allocca, and Q. Chen, "Validation of CFD simulations for natural ventilation," *International Journal of Ventilation*, 2004.

[9] M. A. Zainol, A. A. Razak, N. M. Ali, and Q. J. Kwong, "Effect of Upstream Building Configurations on Mean Wind Speed Ratio at Urban Pedestrian Level Using LES," Science and Technology, 2017.

[10] H.K Versteeg and W.Malalasekera, *An Introduction to Computational Fluid Dynamics*. 2007.

[11] C. J. Greenshields, "OpenFOAM," , 2015.

[12] Y. Tominaga and T. Stathopoulos, "CFD simulations of near- field pollutant dispersion with different plume buoyancies," *Building and Environment*, 2018.

[13] B. Chen, L. Angeles, Y. Miao, and S. Wang, "Construction and Validation of an Urban Area Flow and Dispersion Model on Building Scales," *Journal of Meteorological Research*, 2013.

[14] H. Cheng and I. P. Castro, "Near-wall flow over urban-like roughness," *Boundary-Layer Meteorology*, 2002.

[15] A. Abd Razak, A. Hagishima, N. Ikegaya, and J. Tanimoto, "Analysis of airflow over building arrays for assessment of urban wind environment," *Building. Environment*, 2013.

Fatigue life investigation of non-load carrying fillet weld of structural offshore steel S460G2+M using experiment and FEM simulation

D Andud[1,3], S Saidin[2,3], Y H P Manurung*[3], M Faiz Mat[3], K Salim[1], A Mahmud[1], A A Dawam[1], M K M Said[1], S Bani[1], S Jalani[1], M Zaukah Ibel[1], M Y M Yasan[1], D Zainudin[1] and H Abas[1]

[1]University Kuala Lumpur Malaysia France Institute (UNiKLMFI), Bandar Baru Bangi, Selangor, Malaysia.
[2]Nusantara Technologies Sdn Bhd, Kota Kemuning, Shah Alam, Selangor, Malaysia.
[3]Universiti Teknologi MARA (UiTM), Shah Alam, Selangor, Malaysia.

*yupiter.manurung@uitm.edu.my

Abstract. This research deals with the fatigue life investigation using experiment and FEM simulation on longitudinal fillet weld of structural offshore steel S460G2+M with a thickness of 10mm. The experimental fatigue test is conducted based on the nominal stress (NS) approach while the simulation uses the effective notch stress (ENS) approach for fatigue life assessment. The investigation begins with the preparation of the longitudinal fillet weld fatigue specimen using the Milling Machine based on the IIW fatigue specimen design recommendation and joined using the semi-automatic GMAW process following the AWS D1.1 procedure. Fatigue testing on non-load carrying fillet weld is conducted using the Instron Fatigue Machine with a stress ratio of 0.1 with constant amplitude loading and stress loading from 50%-75% of the yield strength of the base material. For the simulation approach, the 3D longitudinal fillet weld geometry is created using CAD based on the ENS design procedures of IIW where the sizes and dimensions are similar to the experimental fatigue specimen. The static elastic stress analysis of the model is conducted using MSC Marc/Mentat FEM software. Based on the IIW fatigue data evaluation, it is found that the natural mean curve of the longitudinal fillet weld obtained 146 MPa of FAT class which exceeds approximately 106% from the IIW FAT class recommendation for a longitudinal fillet joint. However, the characteristic curve of 97.7% failure probability of the parts only attained 95 MPa of FAT class, but still exceeds approximately 33% form the IIW FAT class. In the ENS fatigue assessment of the 3D longitudinal fillet weld, it is found that the model obtained 195MPa of FAT class which is inferior approximately 15 % from the FAT class recommendation of IIW. Also, it is found that both S-N curves of NS after conversion to ENS system have a good agreement with the S-N curve of ENS of 3D longitudinal fillet weld model and the IIW recommendation due to only 16 % and 34% lower as compared with the IIW FAT class recommendation for ENS.

1. Introduction

Weld fatigue failure is the establishment and dissemination of cracks due to a repetitive load that might significantly below the yield strength of the base material. It starts with the formation of microscopic crack followed by macroscopic crack propagation that finally damages the welded joint component. Welded joints with poor design geometry are defenseless to fatigue damage when subjected to repetitive

Content from this work may be used under the terms of the Creative Commons Attribution 3.0 licence. Any further distribution of this work must maintain attribution to the author(s) and the title of the work, journal citation and DOI.
Published under licence by IOP Publishing Ltd

loadings. The welding heat, cooling process and the addition of filler material also give rise to fatigue failure. This phenomenon causes accidents and unbelievable economic harms since fatigue failure occurred without macroscopic plastic deformation [1]. Thus, it becomes the favorite issue among the scientists which motivates them to study this engineering problem using various fatigue assessment methods such as nominal stress (NS), hot spot stress (HSS), effective notch stress (ENS) and fracture mechanics (FM) for fatigue life assessment.

In this study, the non-load carrying longitudinal fillet weld of the structural offshore steel S460G2+M was assessed using NS and ENS approaches. The NS method is typically taken from the nominal stress amplitudes in the critical cross-section and matches them with the S–N curve of the acceptable nominal stress amplitudes [2]. It is practically an experimental test up to the final fracture. ENS approach is a fatigue life assessment that includes the stress raiser such as undercut and practically using the finite element method [3]. The ENS procedures were outlined in the IIW fatigue design recommendation [4].

Numerous research journals reporting the NS and ENS fatigue life assessment. Bertini et al. [5] studied the notch stress approach and the peak stress method for the fatigue life assessment of welded joints. They found that these local fatigue life assessment methods seemed to provide very similar outcomes in terms of scatter band and safety factors, with admiration for their corresponding reference design curve. Karakas et al. [6] investigated the fatigue performance of magnesium welded joints with the central objective to conclude the applicability of the critical distance method and compare the findings with Neuber's stress averaging method. They found that the scattered bands of the SN curve-lines of effective stresses resulting from the critical distance method are comparable with the scattered results from the Neuber's stress averaging method however expressively narrower than the scattered bands of SN-lines for notch stresses. Leitner et al. [7] studied the high-frequency mechanical impact (HFMI) treated joints using the master notch stress approach with over 230 experimental data tested using constant and variable amplitude. They found that the fatigue strength of the HFMI master notch stress approach, from constant amplitude loading, is fit for material strengths up to ultra-high-strength steels with a nominal yield limit of 1300 MPa. They also revealed that the notch stress-based service strength assessment of variable amplitude loaded HFMI-treated high-strength steel joints deliberating a specified damage sum of D=0.3 when using HFMI-treated joints. Nykänen et al. [8] studied the fatigue valuation of a welded joint with flexible amplitude loading using a novel notch stress approach. They found that the approach with modest residual stress guesses very similar linear damage sums with the damage sums from the nominal mean constant amplitude of S-N curves that derived experimentally from the specimens as used in variable amplitude tests. They also found that all data of constant amplitude notch stress range of failure were above the characteristic master S-N curve. Hyoung et al. [9] applied the effective notch stress approach on rib-to-deck joints and found a good correlation between the FEM results with the observed experimental results. Sonsino et al. [10] used the effective notch stress approach with a fictitious notch radius of 0.05 mm for thin-walled welded joints and 1 mm for thick-walled welded joints on the sandwich panels for ship decks, offshore K-nodes, an automobile door, and a trailing link. The FEM and experimental results were verified to check the consistency of the effective notch stress method to establish one single concept able to treat all possible weld geometry. Kaffenberger et al. [11] studied the effective notch radius of 0.2 mm on thin welded structures. They considered the radius from real scanned geometries and the effect of the radius and size effect. Malikoutsakis et al. [12] used the effective notch stress method in the 3D modeling and meshing of the weld toe and root on the motor truck with an effective notch radius of 0.2 mm for thin plates. They found that the ENS results have good agreement between the measured stresses of the strain gauge in the failure regions. In another investigation [13] they applied the effective notch stress approach on thick-walled rear axles and compared with numerical stress-strain results from experimental results under monotonic and cyclic loading to verify the accuracy of the modeling technique used. He obtained satisfactory results in complicated stress areas. Maljaars et al. [14] compared the nominal, structural and fictitious notch stresses for simple and complex welded joints with fillet welds. They found that three stresses difference is significant in complex cases and the fictitious notch stress approach showed good agreement with the experimental results. Park et al. [15] applied the effective notch stress approach on large-sized specimens of the diaphragm, cruciform, and out-of-plane gusset joints. They used a notch radius of 1 mm for the weld toes and weld roots to investigate the effects of the fixed effective notch radius on the weld size, plate thickness and width, and full penetration. They found that the effective notch stress approach could determine cracks initiate either at the weld toe or the weld root. Pang et al. [16] used a simplified 2D finite element model with the effective notch stress approach on tubular booms of draglines under different loading cases. They found that the critical location occurs at the weld toe and the root of the 2D model. Hongchao et al. [17] studied the fatigue life of the base material, butt weld, and cross fillet weld of a high-strength steel bar of the Q460D and Q690D using the standard design curve of AISC360, EC3, and BS7608. They found that the base material of high-strength steel owns high fatigue resistance. They also found that the fatigue life of butt welds Q460D have an adequate safety margin, but only suitable for low-fatigue life estimation of the Q690D butt weld. The AISC360 design curve, not only found to be suitable for the fatigue life analysis of Q460D and Q690D but also has enough safety margins, while EC3 and BS7608 codes own a relatively low fatigue limit.

Lately, many researchers have given special attention to the high strength steel material due to brilliant mechanical properties, fracture toughness, and fatigue performances as compared with the conventional steel [18]-[20]. Those benefits manifest the material open for fatigue life investigation. Based on the literature review above it is understood that fatigue exploration is enormously essential for the safety assessment of welded steel structures. Thus, driven by these facts, this research aims to determine and match the fatigue life of longitudinal fillet weld S460G2+M as well as its FAT class using NS and ENS approaches.

2. Experimental Methodology

2.1. Material Selection

In this investigation, the main material is structural offshore steel S460G2+M with a 10 mm thickness which is also classified as high strength steel. It was designed for the harsh environment for the oil and gas and marine applications [21]. Table 1 shows the mechanical properties and chemical composition of the material.

Table 1. Material properties of the S460G2+M [22].

HSLA S460G2+M								
Thickness, t				t<16				
Yield strength (Ys)				460 MPa(min)				
Tensile Strength (UTS)				540-700 MPa				
Ys/UTS				Max.0.93				
CVN (- 40°c)				>60J transverse				
Elongation. %				17%				
Chemical composition								
C%	**Cu%**	**Mn%**	**V%**	**Si%**	**Al%**	**Cr%**	**Nb%**	**Ni%**
0.19	0.11	1.62	0.10	0.6	0.032	0.10	0.012	0.09

2.2 Fatigue Specimen

The fatigue specimen for the investigation is a longitudinal fillet weld. It was prepared using a milling machine. The size and geometry of the specimen were in accordance to the IIW fatigue specimen design recommendation as depicted in Figure 1 [23].

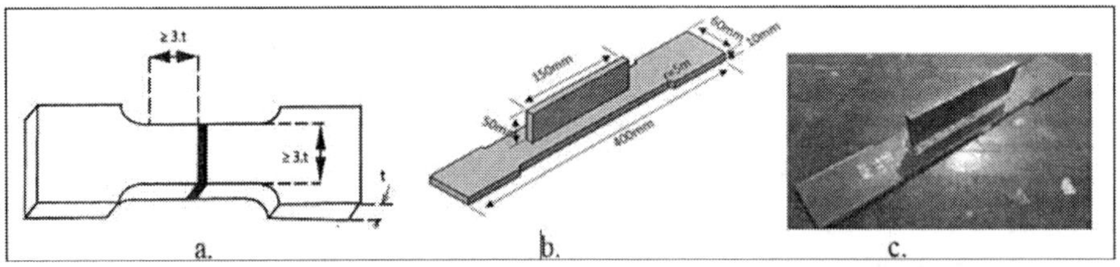

Figure 1. Fatigue specimen, a. Fatigue specimen recommendation by IIW, b. Size and dimension of longitudinal fillet joint, c. Longitudinal fillet weld fatigue specimen.

All the fatigue specimens were joined using gas metal arc welding (GMAW) by the non-certified welder but vast experiencing in joining metal using the suitable combination of welding parameter from [21]. However, its weld quality must comply with standard acceptance criteria of the AWS D1.1 [24]. Only one pass laid on the both sides of the fatigue specimen as shown in Figure 3c previously. The leg length of the welds is approximately 9-10mm. The current and voltage use is 200 Amp and 19 volts and joined in horizontal welding position using filler wire ER80S-N1 and mixed gases as shielding gas. The welding parameters for the GMAW process are summarized in Table 2.

Table 2. Welding parameters for GMAW

Variable	
Current	90-230 Amp
Voltage	18-20 V
Travel speed	20-25cm/min
Filler wire (diameter)	ER80S-N1(1.0mm)
Shielding gas	Mixed gases (80% Ar + 20% CO_2)
Welding position	Horizontal (2F)

2.3 Fatigue Loading

The stress loading for fatigue tests were calculated from variables of stress range (Δ_σ), mean stress (σ_{mean}), alternating stress (σ_a) and stress ratio (R) as presented in equation (1)-(4) [25]. The nominal stress ranges applied for fatigue test are 55%, 65% and 75% of the yield strength of the base material of S460G2+M. There were nine specimens tested for each nominal stress range.

$$\Delta_\sigma = \sigma_{max} - \sigma_{min} \tag{1}$$

$$\sigma_{mean} = \frac{\sigma_{\max} \mp \sigma_{min}}{2} \tag{2}$$

$$\sigma_a = \frac{\sigma_{\max} - \sigma_{min}}{2} \tag{3}$$

$$R = \frac{\sigma_{min}}{\sigma_{max}} \tag{4}$$

2.4 Effective Notch Stress (ENS) Modelling Procedure

The 3D longitudinal fillet weld geometry is modelled using the CATIA by assimilated the size and dimension of the experimental fatigue specimen based on the ENS procedures of IIW which includes the stress raisers of the fillet weld [4]. The stress raiser or notch is modelled with fictitious notch radius of 1mm since the thickness of the S460G2+M is 10mm. The fillet weld angle, θ is 45° respectively. The 3D model longitudinal fillet weld for ENS fatigue life assessment is shown in Figure 2.

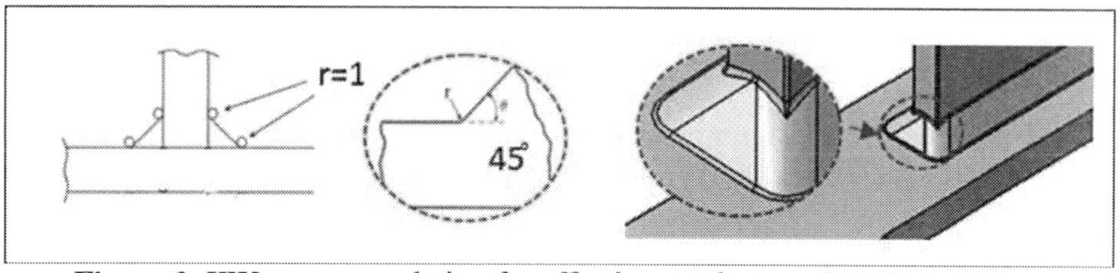

Figure 2. IIW recommendation for effective notch stress (ENS) modelling

The meshing procedure of the 3D longitudinal fillet weld is conducted using MSC. Marc/Mentat with 3D-solid quadratic tetrahedral element meshes. The mesh was applied to fictitious notch radius with minimum size of 0.15 mm. The completed 3D mesh modelling at the weld toe of the part is shown in Figure 3.

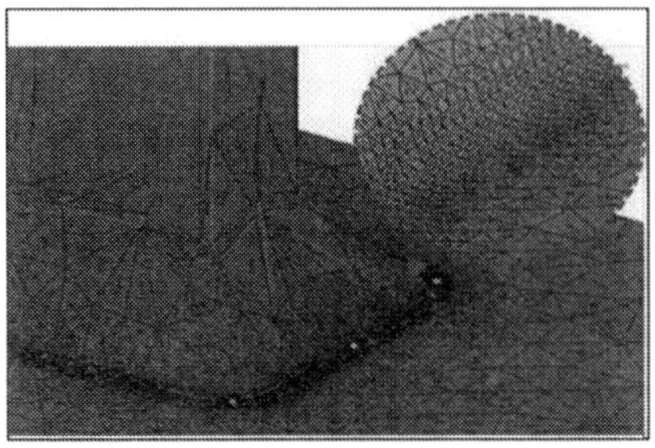

Figure 3. Mesh generation at the weld toe of longitudinal fillet weld.

2.5 Fatigue Life Evaluation of Experimental Data

The experimental fatigue data were evaluated in accordance with IIW commission XIII as presented in equation (5) - (8) [4]. This evaluation procedure is important in order to develop the S-N curves and estimate the FAT class. The evaluation procedures are as follows:

a) Calculate $\log_{10}$ of all data: Stress range **Δσ** and number of cycles **N**.
b) Calculate exponents' **m** and constant **logC** of the formulae using an equation (5).

$$\log N = \log C - m.\log \Delta\sigma \quad (5)$$

c) Calculate mean $\mathbf{x_m}$ and the standard deviation **Stdv** of **logC** using **m** obtained in **b** using equation (6) and (7).

$$x_m = \frac{\sum x_i}{n} \quad (6)$$

$$Stdv = \sqrt{\frac{\sum (x_m - x_i)^2}{n-1}} \quad (7)$$

d) Calculate the characteristic values $\mathbf{x_k}$ by the equation (8).

$$x_k = x_m - k.Stdv \quad (8)$$

2.6 Fatigue Life Assessment for Effective notch stress (ENS) Data.

The fatigue life estimation of the longitudinal fillet welds of S460G2+M using ENS approach requiring the static elastic analysis to determine the stress concentration factor ($\mathbf{K_f}$) in equation (9) as well as the maximum principal stress (σ_r) on the affected weld toe.

$$K_f = \frac{\sigma_{geom}}{\sigma_{Nom}} \quad (9)$$

The σ_{geom} or σ_r signifies as maximum principal stress and the σ_{Nom} signifies as nominal stress applied on the longitudinal fillet weld. The nominal stresses applied to static elastic stress analyses were similar to the nominal stress for experimental fatigue tests. Due to uni-axial stress of fatigue assessment, the maximum principal stress represents the effective notch stress (ENS). The fatigue life estimation based on ENS approach is using an equation (10) [24].

$$N = N_t \left(\frac{\Delta\sigma_{rd}}{\sigma r}\right)^m \tag{10}$$

The $\Delta\sigma_{rd}$ signifies the FAT class design recommendation which is FAT225 for 1mm radius by IIW. N_t is the required life, which is 2 million cycles, σ_r is the maximum principal stress form FEA and m is the gradient of the S/N curve, which is 3 for welded structure.

2.7 Experimental Data Conversion Procedure for Effective Notch system

To compare the NS fatigue data with the ENS approach, $\Delta\sigma_k$, conversion procedure using the equation (11) should be conducted by multiplying the nominal stresses, $\Delta\sigma_n$, of the experimental fatigue tests with the stress concentration factor obtained from the FEM simulation.

$$\Delta\sigma_K = K_f \cdot \Delta\sigma_n \tag{11}$$

3. Results and Discussion

3.1 Fatigue Life of Longitudinal Fillet Weld

The fatigue lives of the longitudinal fillet welds of S460G2+M are presented in Table 3. It is shown that the fatigue life of the part influenced by the applied fatigue loads on the parts. Based on the Table 3, it is found that as the fatigue loads decreases the fatigue life of the parts also increases. On an average, there are approximately 56 % of life increments when the fatigue loads decreased from 400 MPa to 374 MPa and 238% of life increments as the fatigue loads continue decrease to 294 MPa.

Table 3. Average number of cycles of longitudinal fillet weld at different fatigue loads

Fatigue specimen	No. specimen	Fatigue loads (MPa)	Fatigue life (N)
1. Longitudinal fillet weld	3	400	38895
2. Longitudinal fillet weld	3	347	61043
3. Longitudinal fillet weld	3	294	131634

The fatigue life (N) in Table 3 is evaluated based on the IIW evaluation procedures for experimental data in order to establish the S-N curve and to determine the FAT class of the longitudinal fillet weld. The S-N curve of the parts is depicted in Figure 4.

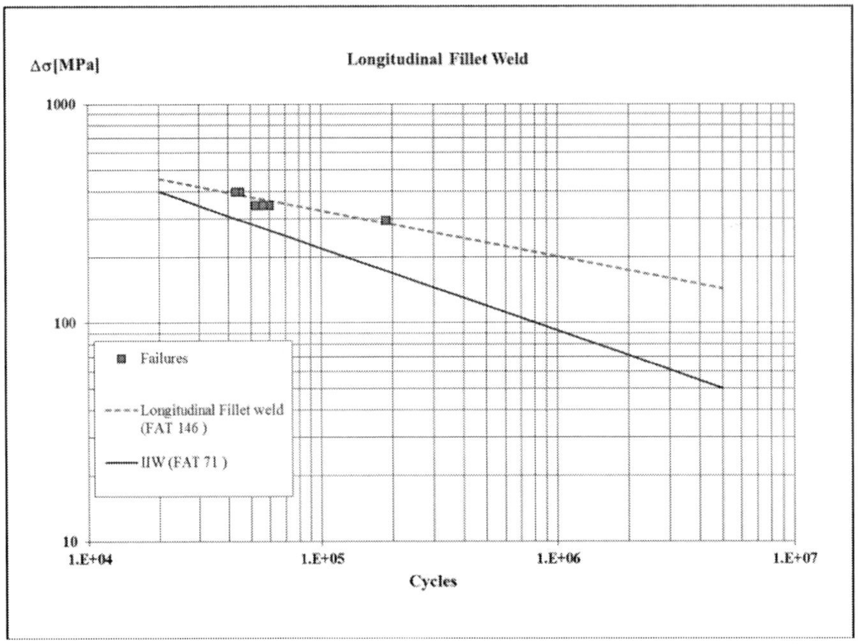

Figure 4. The S-N curve of the longitudinal Fillet welds versus S-N curve of IIW recommendation for longitudinal fillet weld.

From the diagram above, it is found that the FAT class of longitudinal fillet weld is reached 146 MPa which equivalents to 106 % increment from the IIW FAT class recommendation for longitudinal fillet weld and is equivalent to 2.0 of bonus factor. This result shows that the weld quality of the parts is exceptional even though was joined by the non-certified welder. In addition, the welding parameters of the GMAW process are suitable to produce the longitudinal fillet welds.

3.2 Static Elastic Stress Analysis Results

The maximum principal stresses of the different nominal stresses applied on the 3D longitudinal fillet weld under static elastic stress analysis using FEM software MSc. Marc/Mentat is shown in Figure 5.

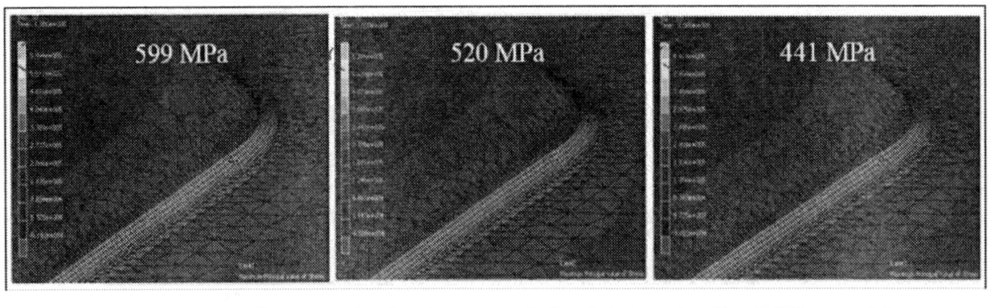

Figure 5. Maximum principal stress of the 3D longitudinal fillet weld.

Based on the Figure 5 above, it is found that the principal stresses in the model are perpendicular to the nominal stresses applied to the model due to its increase as the nominal stress increase. The weld toe region is found to have greater principal stresses as compared to the other regions of the model due to the location of the stress concentration. The principal stresses and intended stress concentration factors of the model are tabulated in Table 4.

Table 4. The maximum principal stresses and stress concentration factor of the 3D longitudinal fillet weld.

Nominal stress, (MPa)	Maximum Principal stress, σ_r (MPa)	Stress concentration factor, K_f
400	599	1.49
347	520	1.49
294	441	1.22

Based on the Table 4 above, it is found that the stress concentration factor of the nominal stress 400MPa and 347MPa is 1.49 while the nominal stress of 239MPa is 1.22. These results show that the stress concentration factor does not relent to the nominal stress applied to the model but rely on the quality of the welds. The stress concentration basically represents the stress raiser of the weld such as undercut which commonly the location of the crack initiation when the structure exposed to cyclic loading. However, fatigue life estimation is puzzling due to differential geometry and stress loading will produce varies stress concentration factor. Therefore, it is important to reduce the amount of stress concentration by means of eliminating the notches where possible.

3.3 Fatigue Life of the 3D Longitudinal Fillet using Effective Notch Stress Approach

The fatigue life of the 3D longitudinal fillet welds that is subjected with varying stress loading are shown in Table 5. All the variables such as ENS FAT, safety factor and slope were taken from IIW standard.

Table 5. The fatigue life of the 3D longitudinal fillet welds using ENS approach

IIW-ENS FAT (MPa)	Safety Factor, φQ	Slope, m	Principal stress, σ_r (MPa)	Fatigue life (N)
225	1.3	3	599	232877
225	1.3	3	520	355957
225	1.3	3	441	583568

Base on the Table 5 above, it is found that the fatigue life of the model influenced by the principal stresses on the component, where it increases as the principal stress decreased. There were 53% increment of fatigue life as the principal stress decreased to 520MPa and around 151% of increments when the principal stress decreases to 441MPa. The ENS approach includes the stress raisers from the local geometry, thus, according to IIW only one S-N-curve is required in respect to detail geometry. The S-N curve of the ENS of the longitudinal fillet weld is shown in Figure 6, which based on the characteristic mean curve of 95% survival probability. This curve is plotted with the FAT class of fatigue design strength of IIW for ENS approach.

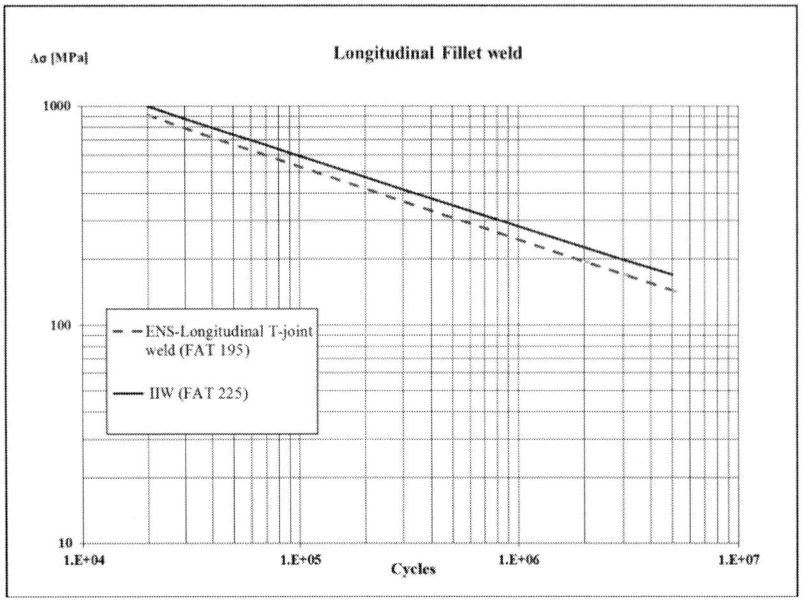

Figure 6. The S-N curve of the fatigue life of the longitudinal fillet welds using ENS assessment approach.

Based on the S-N curve in Figure 6 above, it is found that the S-N curve agree quite well with the FAT class 225 MPa S-N curve recommended for ENS assessment by the IIW. However, the FAT class of the 3D longitudinal Fillet welds model is found to be 15% lower as compared with the FAT class of IIW for ENS assessment.

3.4 Comparison of Fatigue Life of the Nominal Stress and ENS Assessment system

The conversion of the NS fatigue data to ENS system is essential to ensure the experimental fatigue data agreed with the FAT 225 by IIW for ENS assessment. The S-N curve comparison between NS data converted to ENS system and ENS approach are shown in Figure 7.

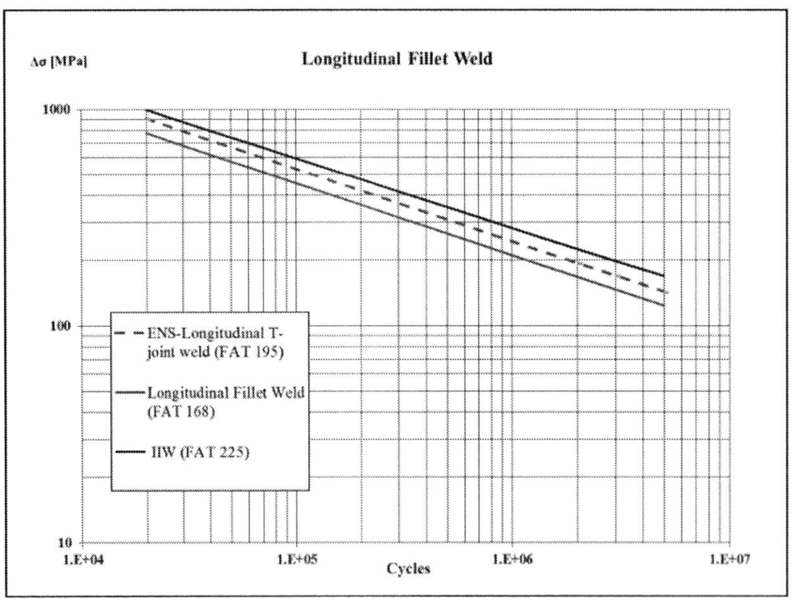

Figure 7. S-N curve comparison between NS data converted to ENS system and ENS approach.

Based on the Figure 7 above, it is found that both S-N curves of NS after conversion to ENS system have a good agreement with the S-N curve of ENS of 3D longitudinal fillet weld model and the IIW recommendation for ENS. It is also found that both FAT classes of the converted NS and 3D longitudinal fillet weld model are lower around 16 % and 34% as compared with the IIW FAT class recommendation for ENS. Even though, both FAT classes are below the FAT class of the IIW but it's higher than the FAT class of NS experimental due to ENS fatigue assessment provides superior fatigue life estimation as compared with the NS approach.

4. Conclusion

The fatigue life of the longitudinal fillet welds was successfully investigated using nominal stress and effective notch stress approach. Based on this investigation there are some conclusions which are as follows:

- The fatigue life of the longitudinal fillet weld depends on the fatigue loads, weld quality and suitable combination of welding parameters of GMAW.
- The stress concentration factor of longitudinal fillet weld does not depend to the stress applied to the structure but depend on the quality of the welds.
- The principal stress from FEM simulation for the 3D longitudinal fillet weld depends on the applied stress to the parts, where it increases as the stress increases.
- The single design S-N curve FAT 225 of IIW for ENS approach has good agreement with the S-N curves of NS converted to ENS system and ENS approaches and relevant for fatigue life comparison between the NS and ENS approaches due to ENS provides a superior fatigue life estimation

Acknowledgement

The authors would like to express their gratitude to the staff member of Research Interest Group: Advance Manufacturing Technology (RIG: AMT) and Advance Technology Excellent Centre (AMTEx) at Faculty of Mechanical Engineering, Universiti Teknologi MARA (UiTM) for encouraging this research. This research and conference participation are also financially supported by Technogerma engineering & Consulting Sdn. Bhd.

References

[1] Y. Q. Ni, X. W. Ye, J. M. Ko, "Monitoring-based fatigue reliability assessment of steel bridges: analytical model and application," Journal of Structural Engineering, vol. 136, no. 12, pp. 1563–1573, 2010.

[2] D. Radaj, C. M. Sonsino, W. Fricke, "Fatigue Assessment of Welded Joints by Local Approaches ", *Woodhead Publishing Limited*, 2006.

[3] D. Radaj, P. Lazzarin, F. Berto, "Generalised Neuber Concept of Fictitious Notch Rounding", *Int. J. Fatigue*, vol. 51, pp. 105–115, 2013.

[4] A. Hobbacher, "*Recommendations for Fatigue Design of Welded Joints and Components*", 2008 (IIW document XIII-2151-07 / XV-1254-07).

[5] L. Bertini, F. Frendo and G. Marulo, "Fatigue Life Assessment of Welded Joints by Two Local Stress Approaches: the Notch Stress Approach and the Peak Stress Method", *Int. J. Fatigue,* vol.110, pp. 246-253, 2018.

[6] O. Karakaş, G. Zhang, C.M. Sonsino, "Critical Distance Approach for the Fatigue Strength Assessment of Magnesium Welded Joints in Contrast to Neuber's Effective Stress Method", *Int. J.Fatigue*, vol. 18, pp. S0142-1123, 2018.

[7] M. Leitner, M Stoschka and M. Ottersböck, "Fatigue Assessment of Welded and High Frequency Mechanical Impact (HFMI) Treated Joints by Master Notch Stress Approach", *Int. J. Fatigue*, vol.17, pp. S0142-1123, 2017.

[8] T. Nykänen, H. Mettänen, T. Björk, A. Ahola, "Fatigue Assessment of Welded Joints under Variable Amplitude Loading using a Novel Notch Stress Approach", *Int. J. Fatigue,* vol. 16, pp. S0142-1123, 2016.

[9] H. B. Sim, "Effective Notch Stress Method for Fatigue Evaluation of Welded Joints in a Steel Bridge Deck", *Int. J. Fatigue*. vol. 5, no. 2, pp. 89-92, 2012.

[10] C.M Sonsino, W. Fricke, F.D. Bruyne, A. Hoppe, A, Ahmadi, and G Zhang, "Notch stress concepts for the fatigue assessment of welded joints - Background and applications", *Int. J. Fatigue*, vol. 34, pp. 2-16, 2012.

[11] M. Kaffenberger and M. Vormwald, "Fatigue Resistance of Weld ends - Analysis of the Notch Stress using Real Geometry", *Mat. Wiss. u. Werksttech.* vol.42, no. 10, pp. 874–880, 2011.

[12] M. Malikoutsakis and G. Savaidis, "Modeling and Fatigue Assessment of Weld Start-end Locations based on the Effective Notch Stress Approach", *Mat. Wiss. u. Werksttech.* vol. 42, no. 4, pp. 298–305, 2011.

[13] M. Malikoutsakis and G. Savaidis, "An Approach to the Effective Notch Stress Concept to Complex Geometry Welds focusing on the FE Modeling", *Mat. Wiss. u. Werksttech.* 2009.

[14] S. Maljaars, W.V. D. Bos, M. Bloem, and J. C. Rijsenbrij, "Application of the Fictitious Notch Stress Approach and Comparison with other Stress Based Approaches." 2012.

[15] W. Park, C. Miki, "Fatigue Assessment of Large-size Welded Joints based on the Effective Notch Stress approach," *Int. J. Fatigue*, vol. 30, no. 9, pp. 1556-1568, 2008.

[16] N.L. Pang, X. L. Zhao, F. R. Mashiri, P. Dayawansa, J.W.H Price, "Design of Welded Thick-Walled Tubular Dragline Joints using 2D Effective Notch Stress Method." Int. *J. of Steel Structures*, pp. 279–288, 2006.

[17] H. Guo, J. Wan, Y. Liu, J. Hao "Experimental Study on Fatigue Performance of High Strength Steel Welded Joints", *J. Thin-Walled Structures,* vol. 131, pp.45–54, 2018.

[18] S. Salina, A. Dahia, Y.P.H Manurung, M. F. Mat, N. Noridzwan, M. Leitner, "Effects of High Frequency Mechanical Impact on Fatigue Life of Semi-Automated Gas Metal Arc welding (GTAW) of HSLA butt weld", 2019, MATEC Web of Conferences. 269.06002.10.1051/matecconf/201926906002.

[19] U. Robert, N. Frantisek, "Fatigue Life of High Strength steel for cold forming" *Metallurgical sisak then zagreb*, vol.56, pp. 115-118, 2017.

[20] M. Benjamin, T. Alessio, W. Rainer, M.Tobias, "Cyclic Material Behavior of High-Strength Steels Used in the Fatigue Assessment of Welded Crane Structures with a Special Focus on Transient Material Effects", *AE Int. J. Engines*, vol. 10, pp. 331-339, 2017.

[21] D. Andud, M. F. Mat, N. Nordin, Y. H. P. Manurung, S. Saidin, "Parameters Identification for Weld Quality, Strength and Fatigue Life Enhancement on HSLA (S460G2+M) using Manual GMAW followed by HFMI/PIT" *J. Mech. Eng.*vol.5 pp. 205–222, 2018.

[22] A mittal 2013 *Stock Specifications - Plates High yield structural Plates for offshore S460G2 + M (*BS- EN 10225:2009).

[23] M H and S J M. H-P Lieurade 2012 *Recommendation on The Fatigue Testing of Welded Components* (Int. Inst. Weld.)

[24] American welding Society D1.1 2015 *Section 4 clause 4.9.3.5, the acceptance criteria (*American National Standard).

[25] A Lindqvist, H Nilsson, 2016 *Effective notch stress analysis of transverse attachments in steel bridges A parametric fatigue life assessment* (Sweden: Chalmers University of Technology Gothenburg).

Performance comparison of optimization techniques in EHS system using sliding mode controller

M S Chai[1], R Ghazali[1*], S H Chong[1], C C Soon[1], J Too[1] and Y M Sam[2]

[1]Centre for Robotics and Industrial Automation, Faculty of Electrical Engineering, Universiti Teknikal Malaysia Melaka, Hang Tuah Jaya, 76100 Durian Tunggal, Melaka, Malaysia.
[2]Control & Mechatronics Engineering, Faculty of Electrical Engineering, Universiti Teknologi Malaysia, 81310 Skudai, Johor Bahru, Malaysia

Corresponding author's email: rozaimi.ghazali@utem.edu.my

Abstract. Electro-Hydraulic Servo (EHS) system is a well-known nonlinear system in various types of industries due to its outstanding characteristics over other actuators. Despite its wide applications in industries, this system also suffering from nonlinearities and parameter uncertainties. Sliding Mode Controller (SMC) has been proven by many researchers to be effective in positioning control of the nonlinear system. However, SMC controller parameters must be tuned in appropriate ways to control the EHS system. In this research, an SMC strategy which combines with an enhanced PSO called Multi-Objective Particle Swarm Optimization (MOPSO) is proposed in positioning control of the EHS system. The conventional PSO is design as well for comparison purpose. The output performance is analysed and verified using different performance indexes such as rise time, settling time, overshoot percentage, and root mean square error. The results show that the MOPSO is better in tuning the controller parameters.

1. Introduction

Electro-Hydraulic Servo (EHS) systems are widely used in today's applications such as manufacturing industry, automotive, earthmoving vehicles, aerospace industry, and robotic manipulators [1]. EHS has always been chosen to be the actuator because of its advantages over other types of actuators such as electric motor and pneumatic actuator. The main advantages of the EHS system are high power to weight ratio and large force capabilities [2].

Despite the advantages and various applications of the EHS system, EHS system suffers from nonlinearities and parameter uncertainties which led to the degradation of the actuator performance. These are caused by fluid leakage in actuators, frictions, pressure-flow characteristics, oil temperature, and load stiffness [3]–[6].

In recent years, many researchers have proposed different types of control strategies in positioning control of the EHS system. However, Sliding Mode Controller (SMC) has been proven to be outstanding over other controllers in providing robust performance for the nonlinear dynamic system [7], [8]. Therefore, SMC will be chosen for EHS system positioning control in this research.

The aim of this research is to tune the best controller parameters in SMC in order to get the desired output performance of EHS system. According to [9], the tuning problem of controller gains can be considered as an optimization problem. In this research, two types of optimization techniques are used in tuning SMC parameters which are conventional PSO and MOPSO.

Content from this work may be used under the terms of the Creative Commons Attribution 3.0 licence. Any further distribution of this work must maintain attribution to the author(s) and the title of the work, journal citation and DOI.
Published under licence by IOP Publishing Ltd

The rest of the paper is structured as follows: Section 2 will discuss the EHS system model used in this research. Section 3 will explain the proposed control strategies. Section 4 will analyze and discuss the results and finally, a conclusion is to draw in section 5.

2. EHS System Modelling

The main objective of this research is to control the EHS system in terms of position control. The mathematical modeling is utilized in this research. Figure 1 shows the physical model of the EHS system used in this research.

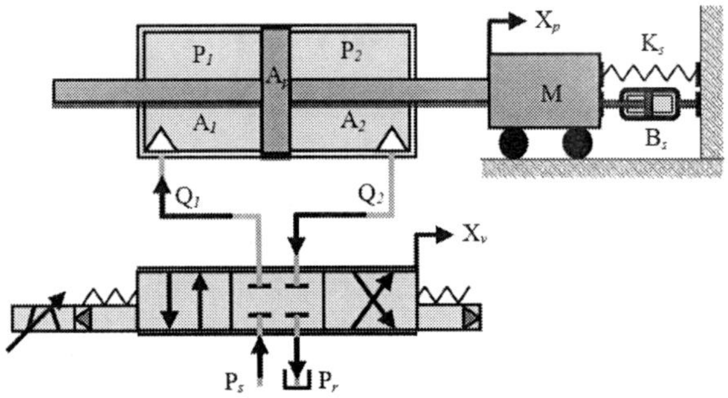

Figure 1. Physical model of the EHS system.

The details of the mathematical modeling can be found in [10]. The final total force produced from hydraulic actuator can be obtained in equation (1) which had included all the dynamic's equations of moving mass, damper, and spring.

$$F_p = A_p(P_1 - P_2) = M_p \frac{d^2x_p}{dt^2} + B_s \frac{dx_p}{dt} + K_s x_p + F_f \tag{1}$$

The EHS system parameters used in this research are tabulated in Table 1.

Table 1. EHS system parameters.

Symbol	Description	Value
I_{sat}	Torque motor saturation current	0.02 A
L_c	Servo-valve coil inductance	0.59 H
R_c	Servo-valve coil resistance	100 Ω
β	Hydraulic fluid bulk modulus	$1.4\text{x}10^9$ N/m^2
ω_n	Servo-valve natural frequency	543 rad/s
ξ	Servo-valve damping ratio	0.48
P_r	Return pressure	0 Pa
P_s	Pump pressure	$2.1\text{x}10^7$ Pa
K	Servo-valve gain	$2.38\text{x}10^{-5}$ m$^{5/2}$/kg$^{1/2}$
M_p	Total mass	9 kg
X_s	Total actuator displacement	0.1 m
B_s	Damping coefficient	2000 Ns/m

A_p	Piston area	$645\text{x}10^{-6}$ m^2
K_s	Spring stiffness	10 Nm

3. Proposed Control Strategies

In this research, a nonlinear SMC will be used in positioning control of the EHS system. Two optimization approaches which are PSO and MOPSO will be utilized and compared in this research for tuning the SMC parameters.

a. Sliding Mode Controller (SMC)

Sliding Mode Control is introduced in the early '60s in Russia. The most important step in designing an SMC is the construction of an appropriate sliding surface. The sliding surface, *s(t)* used in this research is expressed in equation (2).

$$s(t) = (\lambda + \frac{d}{dt})^{n-1} * e(t) \tag{2}$$

where λ = positive constant
n = model order of the system to be controlled
$e(t)$ = tracking error

The control signal of SMC is divided into two parts: equivalent control and switching control. The equivalent control and switching control in SMC are corresponding to the sliding and reaching phase. Both the equivalent control and switching control are expressed in the equations (3) and (4).

$$u_{eq}(t) = \frac{1}{C}(\dddot{x}_r + A_n * \ddot{x}_p + B_n * \dot{x}_p + 2 * \lambda * \ddot{e}(t) + \lambda^2 * \dot{e}(t)) \tag{3}$$

$$u_{sw}(t) = k_s * tanh(\frac{s}{\varphi}) \tag{4}$$

Finally, both control signals are added together to produce the total control signal of SMC.

$$u_{smc}(t) = \left(\frac{1}{C}(\dddot{x}_r + A_n * \ddot{x}_p + B_n * \dot{x}_p + 2 * \lambda * \ddot{e}(t) + \lambda^2 * \dot{e}(t))\right) + \left(k_s * tanh(\frac{s}{\varphi})\right) \tag{5}$$

From the equation (5), there are two parameters which are λ and φ that need to be tuned in SMC. The proposed optimization processes will be discussed in next subsections.

b. Conventional Particle Swarm Optimization (PSO)

PSO is proposed in this research to tune the SMC parameters, λ and φ. Conventional PSO is introduced by Kennedy and Eberhart in 1995. The idea of this optimization came from fish schooling and bird flocking while searching for foods [11].

In PSO, a random particles population called swarm is flying around a global searching space to search for the best particles using the position and velocity information of each particle. Equations (6) and (7) show the position and velocity used in this research.

$$x^{i+1} = x^i + v^{i+1} \tag{6}$$

$$v^{i+1} = \omega v^i + c_1 r_1 (P_{BEST}{}^i - x^i) + c_2 r_2 (G_{BEST} - x^i) \tag{7}$$

The position and velocity of each particle will change every iteration according to the given fitness equation. In conventional PSO, there is just one objective function in the fitness equation to be achieved. In this research, the error calculated from the EHS system performance will be the objective function in conventional PSO. The equation (8) shows the fitness equation used in this conventional PSO.

$$fitness\_equation = error \tag{8}$$

c. Multi-Objective Particle Swarm Optimization (MOPSO)

MOPSO is an enhanced version of conventional PSO. The only difference between the MOPSO and the conventional PSO is that MOPSO has two objective functions in its fitness equation. The objective functions to be achieved through MOPSO in this research will be the error and the overshoot calculated from the EHS system performance.

The intention to add in one more objective function is to make the conventional PSO become more effective when it selects the controller parameters. The fitness equation used in this MOPSO is expressed in equation (9).

$$fitness\_equation = error + overshoot \tag{9}$$

4. Results and Discussion

In order to test the performance of the EHS system using both PSO-SMC and MOPSO-SMC controllers, a step input is fed into the EHS system. The performance of the EHS system is analyzed using different types of the performance index.

Table 2. The output performance analysis of EHS system.

Analysis / Controllers	Rise Time (s)	Settling Time(s)	Overshoot Percentage (%)	Root Mean Square Error (mm)
PSO-SMC	0.0250	0.0460	0.9504	1.29
MOPSO-SMC	0.0380	0.0670	0.1518	1.45

Table 2 shows the output performance analysis of the EHS system. The analysis is done by using several performance indexes normally used in control system such as the rise time, settling time, overshoot percentage and finally, the root mean squared error. From the table, both the rise time and the settling time are smaller than 0.1 second which is in the acceptable range. For the root mean square error, PSO-SMC has a root mean square error value smaller than the MOPSO-SMC. This shows that the PSO-SMC is performed better and less error than the MOPSO-SMC.

However, it is obviously in table 2 that the overshoot percentage of the PSO-SMC is much higher than the MOPSO-SMC. PSO-SMC has a value of 0.9504% of overshoot percentage while the MOPSO-SMC only recorded 0.1518%. Higher overshoot means the EHS actuator will move beyond the desired value before settles down to the desired value. This proved that MOPSO-SMC will not have to move to a very high value as compared to PSO-SMC before settles down to the desired value. The biggest disadvantage of moving too much beyond the desired value will possibly cause damage to the EHS actuator.

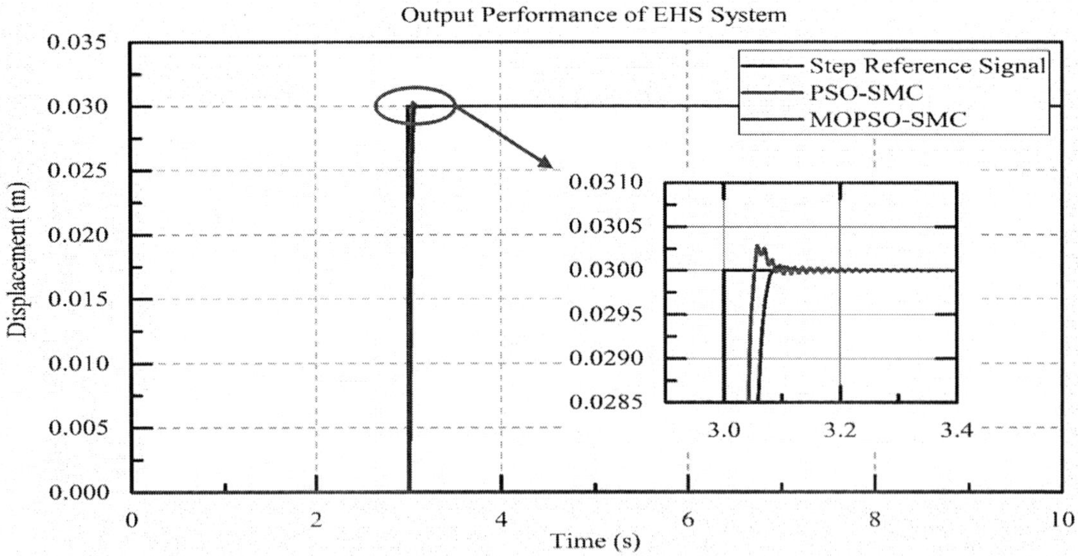

Figure 2. Output performance of EHS system.

Figure 2 shows the results of the output performance of the EHS system. The red line denoted the PSO-SMC while the blue line denoted the MOPSO-SMC. It is obvious from the figure that the overshoot of the PSO-SMC is higher than the MOPSO-SMC. There is a little chattering for the PSO-SMC as seen from figure 2 in which does not appear in the MOPSO-SMC.

5. Conclusion

Two types of optimization techniques are presented in this research to tune the SMC parameters in positioning control of the EHS system. One of it is conventional PSO combined with SMC and another one is an enhanced version of PSO called MOPSO paired with SMC. Both optimization techniques are applied in the simulation and output performance of the EHS system is analyzed. From the results, it shows that the MOPSO-SMC are performed better than the PSO-SMC. The additional objective function in MOPSO is shown to affect the final output performance of the EHS system in terms of overshoot percentage. The final output performance has an obvious degradation in overshoot percentage as compared to the conventional PSO which only has an objective function.

Acknowledgment

The support of Universiti Teknikal Malaysia Melaka (UTeM), Centre for Robotics and Industrial Automation (CeRIA), and Ministry of Education (MOE) are greatly acknowledged. The research was funded by the Fundamental Research Grant Scheme (FRGS) Grant No. (FRGS/1/2017/TK04/FKE-CeRIA/F00333).

References

[1] Y. Yoon, Z. Sun, and H. Du, "Inverse Modeling Approach for Parametric Frequency Domain Analysis of An Electrohydraulic System," *Mech. Syst. Signal Process.*, vol. 121, pp. 412–425,2019.

[2] W. Ma, W. Deng, and J. Yao, "Continuous Integral Robust Control of Electro-Hydraulic Systems with Modeling Uncertainties," *IEEE Access*, vol. 6, pp. 46146–46156, 2018.

[3] S. M. Othman, M. F. Rahmat, S. M. Rozali, and S. Salleh, "Review on Sliding Mode Control and Its Application in Electrohydraulic Actuator System," *J. Theor. Appl. Inf. Technol.*, vol. 77, no. 2, pp. 199–208, 2015.

[4] E. K. Gdoura, M. Feki, and N. Derbel, "Sliding Mode Control of A Hydraulic Servo System Position Using Adaptive Sliding Surface and Adaptive Gain," *Int. J. Model. Identif. Control*, vol. 23, no. 3, p. 248, 2015.

[5] Q. Guo, Y. Zhang, B. G. Celler, and S. W. Su, "Backstepping Control of Electro-Hydraulic System Based on Extended-State-Observer with Plant Dynamics Largely Unknown," *IEEE Trans. Ind. Electron.*, vol. 63, no. 11, pp. 6909–6920, 2016.

[6] S. Wang, Q. Xu, R. Lin, M. Yang, W. Zheng, and Z. Wang, "Feedback Linearization Control for Electro-Hydraulic Servo System Based on Nonlinear Disturbance Observer," in *Chinese Control Conference, CCC*, 2017, pp. 4940–4945.

[7] Y. Liu and H. Handroos, "Technical Note Sliding Mode Control for A Class of Hydraulic Position Servo," *Mechatronics*, vol. 9, no. 1, pp. 111–123, 1999.

[8] A. Bonchis, P. I. Corke, and D. C. Rye, "Experimental Evaluation of Position Control Methods for Hydraulic Systems," *IEEE Trans. Control Syst. Technol.*, vol. 10, no. 6, pp. 876–882, 2002.

[9] M. Y. Kim and C. O. Lee, "An Experimental Study on The Optimization Of Controller Gains for An Electro-Hydraulic Servo System Using Evolution Strategies," *Control Eng. Pract.*, vol. 14, no. 2, pp. 137–147, 2006.

[10] M. Kalyoncu and M. Haydim, "Mathematical Modelling and Fuzzy Logic Based Position Control of An Electrohydraulic Servosystem with Internal Leakage," *Mechatronics*, vol. 19, no. 6, pp. 847–858, 2009.

[11] J. Kennedy and R. Eberhart, "Particle Swarm Optimization," *Neural Networks, 1995. Proceedings., IEEE Int. Conf.*, vol. 4, pp. 1942–1948, 1995.

Improving Teaching and Learning Experience in Engineering Education using Sentiment Analysis Techniques.

Wandeep Kaur[1], Vimala Balakrishnan[1], Baljit Singh[2]

[1]Faculty of Computer Science & Information Technology, Universiti Malaya, Malaysia.
[2]Faculty of Mechanical Engineering, Universiti Teknologi Mara (UiTM), Shah Alam, Selangor, Malaysia.

Abstract. Students are the golden commodity when it comes to ringing in revenue for academic institution. Therefore, it is vital to ensure the opinions and feedback of students is taken seriously to ensure continuous improvement in the teaching and learning experience. In an era of digital information and rich opinionated text being easily available, it is crucial to look into different forms of data analysis from which vital information can be extracted. Sentiment and emotion analysis are one such area of research that looks to extract implicit information from written text and analyse data that would be able to provide a deeper insight compared to conventional measures. This paper is an extension of a previously conducted study of analysing emotion as well as sentiment of students' feedback taking Thermal Engineering (MEC551) from Universiti Teknologi Mara (UiTM). Supervised learning technique was adopted and data analysed revealed students were biased towards assignment and quizzes as these would help improve their carry forwards for the subject and the preference of chapters to the exam was more for conduction and convection compared to others which had more mathematical related calculation.

1. Introduction

Evaluating teaching and learning process at any education institution is not only crucial but also vital in determining the continuous growth of said institute. This is because such feedback from students help institutes to gain an insight as to how students are perceiving the subject and helps teachers to better improve the teaching process for future reference. In an age where opinions and feedbacks are deemed important, gathering information via a Likert scale would only cause important information to be lost in translation (Joseph & Varghese, 2019). Therefore, a new approach needs to be adopted when it comes to analysing student feedback and this can be done using sentiment and emotion analysis techniques.

Majority of higher learning education adopt the student response system (SRS) at the end of each semester in order to acquire student feedback. Online learning platforms such as Coursera (www.coursera.org) and Udemy (www.udemy.com) rely on such feedback as a measuring tool in order to improve their content. Students are constantly providing feedback on the free courses provided by sharing their opinion with millions of potential students who may be looking to sign up. Such is the era of rich text that even Malaysian students are referring to online forums such as EasyUni (www.easyuni.my) to comment on their educational experiences at local colleges and universities. Such feedback can directly impact the number of future student intakes who are potential customers for such higher learning institutes.

Content from this work may be used under the terms of the Creative Commons Attribution 3.0 licence. Any further distribution of this work must maintain attribution to the author(s) and the title of the work, journal citation and DOI.
Published under licence by IOP Publishing Ltd

Although sentiment analysis plays an essential role when it comes to effectively assessing student feedback (Esparza et al., 2017; Rani & Kumar, 2017), the analysis only classifies each feedback as either being positive, negative or neutral. With the availability of vast amounts of self-expressive and opinionated data, it is imperative to move forward from only classifying positive and negative sentiments to distinguishing discrete emotion that is layered within positive and negative text. This is because every positive or negative text cannot be treated as an equal (Sailunaz et al., 2018). For example, both anger and fear can be classified as negative sentiment; however, literature has shown anger is an emotion that provokes a positive outcome as people are motivated to make a change but fear on the other hand instils pessimistic responses (Sailunaz et al., 2018).

This paper is an extension of a preliminary study on student feedback (Kaur et al., 2017) that is looking to not only classify feedback extracted from students taking Thermal Engineering (MEC 551) course from the Faculty of Mechanical Engineering of University Teknologi Mara (UiTM) according to sentiment but also emotion with respect to Plutchik (2003) wheel of emotion. The extracted posts were subjugated to supervised machine learning algorithms such as Naïve Bayes (NB) and Support Vector Machine (SVM), implemented using Weka. The algorithms are compared for evaluation purpose using accuracy, precision and recall measures. The remainder of this paper is organized as follow; the next section discusses the literature followed by methodology, results, discussion and finally conclusion.

2. Literature Review

This section looks to provide a brief discussion on sentiment and emotion analysis followed by a discussion on various related work within the domain of student feedback analysis.

2.1. Sentiment Analysis

According to Rani et al. (2017), teaching and learning outcomes can be evaluated directly (assessment of students' work such as assignments, quizzes, exams etc.) and indirectly (students' learning experience and teaching quality). Sentiment and emotion analysis are a form of indirect assessment looking to gather information regarding student learning experience in order to improve teaching quality from the feedback provided.

A text can be classified as positive with respect to the number of positive related words extracted from it and negative similarly (Nithya & Maheswari, 2014). A neutral sentiment is assigned to a text when the algorithm is unable to detect positive or negative words from it. For example:

"The quiz on Chapter 3 had too many confusing questions"
"The open book exam today gave me hope to pass!"

As it can be observed from above, the first sentence clearly shows a negative sentiment towards the questions asked in the quiz (implied from the word *confusing*) while the second sentence shows a positive sentiment towards the given exam (implied from the word *pass*).

There are three approaches to conducting a sentiment analysis study on text; machine learning approach, lexicon-based approach and hybrid (Figure 1). The machine learning approach adopts machine learning algorithms in order to classify text while lexicon-based approach is more laborious with the creation of manual lexicon dictionaries. The hybrid approach is a combination of both machine learning and lexicon-based approaches. This paper looks into adopting the supervised learning

IOP Conf. Series: Materials Science and Engineering **834** (2020) 012026 doi:10.1088/1757-899X/834/1/012026

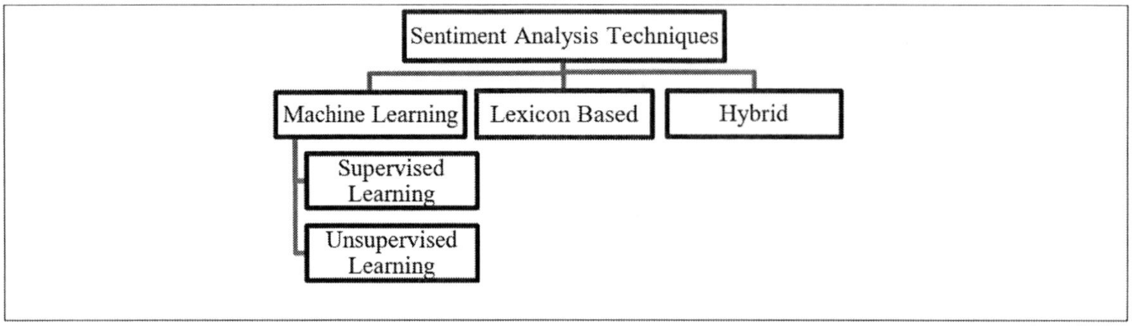

Figure 1. Sentiment Analysis Techniques (Kaur et al., 2017).

2.2. Emotion Analysis

Emotion analysis is an extension of sentiment analysis and defined as a linguistic process of identifying emotions expressed in written text (Sailunaz et al., 2018). Yadollahi et al. (2017) found a strong correlation between sentiment and emotion where an emotion may lead an individual to build an opinion on something and similarly an opinion may invoke an emotion in others. Furthermore, an extracted text may hold contradicting opinions and emotion as example:

"My supervisor thinks it's a good decision for me to take on the scholarship overseas although he is sad to see me go"

The example above shows a positive sentiment towards the possibility of pursuing studies overseas, however the emotion related to it shows *sadness*. Additionally, a text may have a sentiment attached to it yet not have an emotion. For example:

"I have not studied for the mid-terms"
"The results of the experimentation are disappointing"

As it can be observed, the first sentence carries a negative sentiment but there is no emotion attached to it. Similarly, the second sentence shows the same negative sentiment but the emotion detected is *disgust* as perceived by the usage of the word *disappointing*.

Literature has identified four main approaches to emotion detection (Figure 2). The keyword-based approach looks into the grammatical aspect of a text where Noun, Verb, Adjective and Adverb (words that have been identified as words that carry emotions) are matched to emotion dictionaries (Yadollahi et al., 2017). The machine learning, lexicon-based and hybrid approaches as similar to sentiment analysis approaches explained in section 2.1. This paper will look into adopting the lexicon-based approach adopting the NRC Emotion Lexicon for the purpose of emotion detection (Mohammad & Turney, 2013).

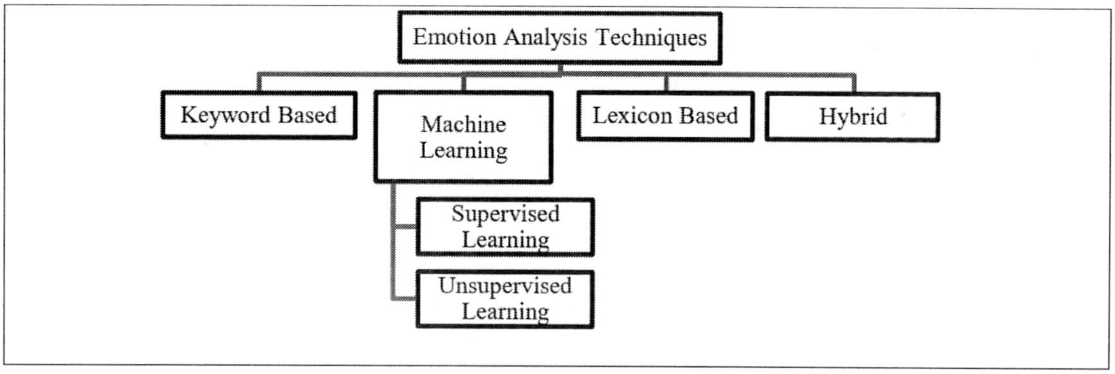

Figure 2. Emotion Analysis Techniques.

3. Methodology

This section will discuss the methodology adopted (Figure 3). Extracted feedback is classified according to sentiment polarity (positive, negative and neutral) and emotion detection with respect to Plutchik (2003) wheel of emotion (anger, anticipation, disgust, fear, joy, sadness, surprise and trust) using the NRC Emotion Lexicon (Mohammad et al., 2013). Each phase of the methodology will be discussed in detail in the following sub-sections.

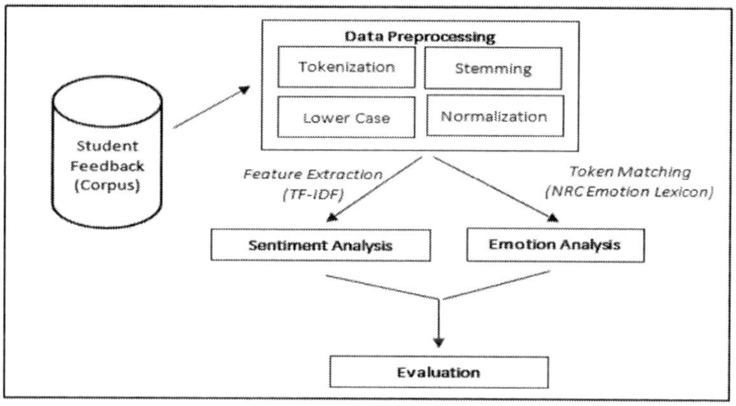

Figure 3. Framework overview.

3.1. Data Collection

A Google form was created with the intention of gathering feedback from students who registered for Thermal Engineering (MEC 551) course from the Faculty of Mechanical Engineering of University Teknologi Mara (UiTM). These students were a mixed bunch of Year 2 to Year 5 students who were attempting this subject between the first time to the third time. The questions on the feedback form were designed after a brainstorming session between the lecturers who were teaching said subject, verbal conversations with students attempting said subject as well as researchers working on sentiment analysis within the education domain. The google form link was distributed to the students during the last week of the lecture. After duration of 6 weeks, 121 students feedback forms were received. The cut-off point of the collection was decided upon as the number of feedbacks received started to dwindle and the number remained stagnant for a week. All feedbacks were treated with utmost confidentiality as no collection on names, student matric number or emails were collected during this phase in order to protect students' anonymity.

IOP Conf. Series: Materials Science and Engineering **834** (2020) 012026 doi:10.1088/1757-899X/834/1/012026

3.2. Data Preprocessing

The standard data pre-processing steps were adopted in this paper which include tokenization, stemming, normalization and changing all sentences to lower case. Apart from the aforementioned, the dataset also had to be cleaned from possible human typo errors, elimination of answers that were too short (less than 3 words in length) and removal of content that was irrelevant to analysing sentiment and emotions (punctuations, stop words such as the, a, an etc.). Although, this dataset came clean from errors, some of the feedbacks had to be removed as they were too short to perform analysis over. This brought the final count of the dataset to 103.

3.3. Sentiment & Emotion Identification

Two of the most widely used sentiment classification algorithm were used for this paper; Naïve Bayes (NB) and Support Vector Machine (SVM) (Sun et al., 2017). The algorithm was built using Weka 3.7. The cross-validation operator was set at 10-fold where the algorithm then proceeds to automatically split the fed dataset between testing and training. The corpus uploaded to the algorithm was analysed according to three major aspects namely feedback on the teaching and learning experience (includes classroom ambience, availability of materials online and feedback on lecturer), syllabus of the subject (breakdown on the chapters of the subject) and feedback on other student assessment measures (quizzes, mid-terms, exams and assignments). The training and testing phase of the algorithm was set manually as 80% for training and 20% for testing purposes.

For emotion identification on the other hand, Mohammad et al. (2013) built the NRC Emotion Lexicon by assigning each word in the lexicon dictionary to an emotion vector. Therefore, when the algorithm identifies a word in the student feedback, the corresponding emotion in the lexicon list is returned. For example;

Sample sentence 1: "*The quiz was so easy!*"

Would return the following emotion vectors:

Table 1. Emotion Vector for Sample Sentence 1

Anger	Anticipation	Disgust	Fear	Joy	Sadness	Surprise	Trust
0	0	0	0	1	0	0	0

The use of this lexicon was founded under the assumption that a text would be able to have both negative and positive emotion as well as showcase more than one emotion at a time. In case of such occurrence, the algorithm would then work to find the aspect upon which the emotion is attached to and assign the final emotion accordingly. For example:

Sample sentence 2:
"I was so scared going to the exam and I really thought I was going to fail the subject but thank God I managed to scrape through and pass"

The above sample would return the results as shown in Table 2 and based on the count, and the algorithm would identify the emotion *fear* (related to the keyword "*scared*" and "*fail*") is related to the aspect of *exam*.

Table 2. Emotion Vector for Sample Sentence 2

Anger	Anticipation	Disgust	Fear	Joy	Sadness	Surprise	Trust
0	0	0	2	1	0	0	0

4. Results and Discussion

This section will look into the results obtained during the phase of this study. Results will be displayed and discussed according to three features: chapter modules (breakdown on each chapter), feedback on other student assessment (exams, assignment, quizzes etc) and teaching evaluation (feedback on lecturer as well as classroom ambience). Results will be displayed for both sentiment and emotion respectively. The evaluation metrics used for this paper was pre-built within Weka itself, hence a direct comparison of f-measure scores of both Naïve Bayes (NB) and Support Vector Machine (SVM) are shown in Table 3. All results are displayed in percentage form.

Table 3. Result comparison NB and SVM.

	NB	SVM
Chapter Modules	94.5	92
Assessment Measures	87.1	91
Teaching Evaluation	75	70

From the table, it can be noted that SVM generally produced the best results for all features except teaching evaluation (70%). This could be possibly because the teaching evaluation feedback was wordier compared to the other features and NB has shown to perform better when classifying longer text compared to SVM (Sun et al., 2017).

4.1. Chapters Modules Results

Figure 4 shows sentiment analysis for each chapter thought in Thermal Engineering (MEC 551). As the chapters progressed within the subject, the number of negative feedbacks compared to positive feedbacks started to increase. Figure 5 shows the breakdown of emotion attached to said subjects.

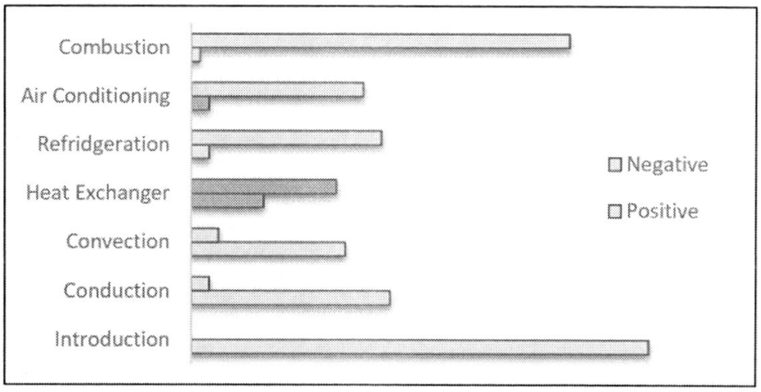

Figure 4. Sentiment classification based on chapter modules

Only negative emotions (anger, disgust, fear and sadness) results are displayed in Figure 5 as these emotions were more apparent when it came to classification process of this feature. Students were mostly fearful of the last few chapters as they claim it involved more calculation hence following the subject was difficult. Additionally, data also revealed students were more hesitant when it came to the last chapter because it involves chemistry knowledge which they do not have much exposure to.

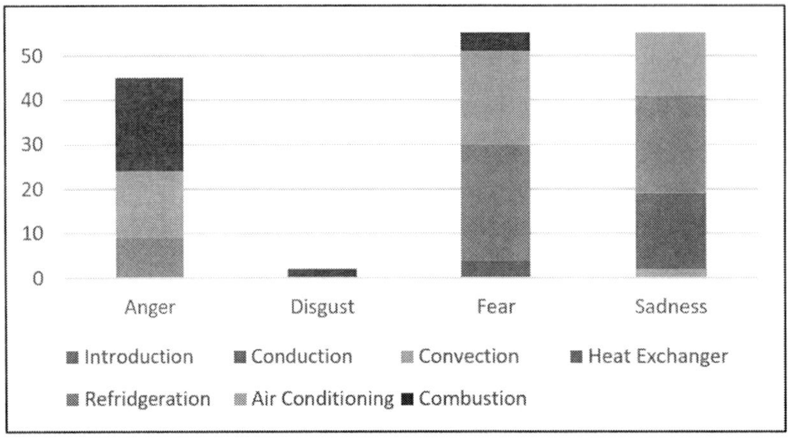

Figure 5. Emotion detection based on chapter modules

4.2. Assessment Measures

Figure 6 shows the emotion classification for assessment. Students favoured more assignments compared to tests and quizzes yet from the feedback they claimed the number of tests should be increased as this would help them obtain better carry forwards as they headed to finals.

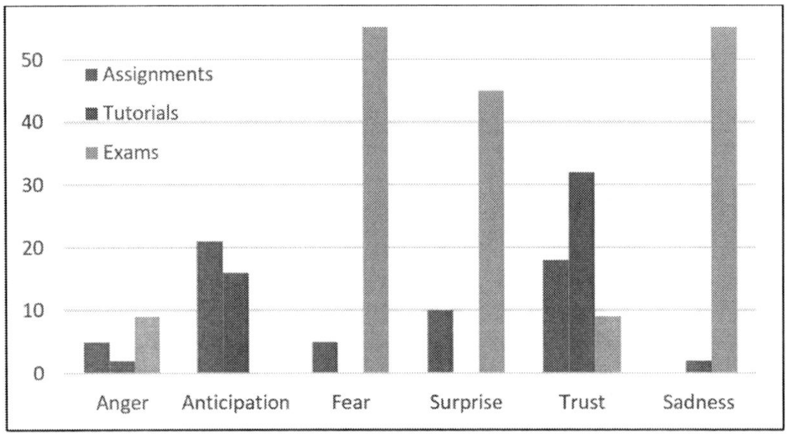

Figure 6. Emotion detection for assessment measures

4.3. Teaching Evaluation

When it came to evaluating lecturers, students generally had a very positive feedback. Although they have mentioned that they would prefer more time spent during lectures and tutorials so that they would be able to clear doubts that they would have within the class. Additionally, students did find the learning environment not to be conducive as the classrooms and lecture halls all were facing air conditioning issues. Figure 7 shows the sentiment classification of this feature.

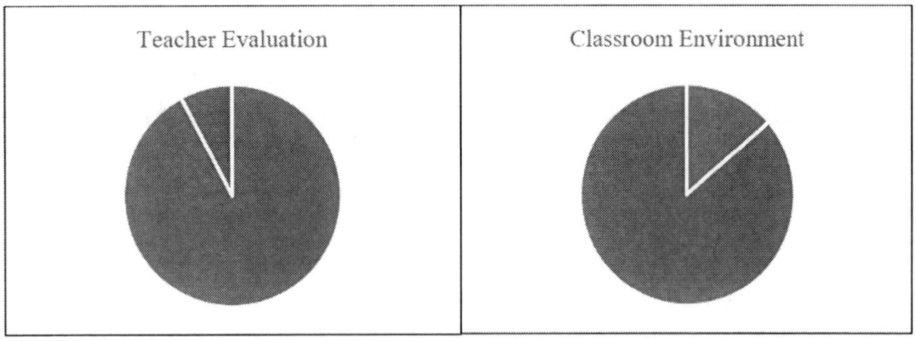

Figure 7. Sentiment classification for teaching evaluation

5. Conclusion

This study was an extension from a previously conducted study. More aspects from the feedback were extracted for analysis purpose and the format of collecting feedback was also changed to accommodate the objective of this study. Machine learning algorithms (Naïve Bayes and Support Vector Machine) were used for sentiment classification purpose and NRC Emotion lexicon was used for emotion analysis built using Weka 3.7 platform.

The results found in this phase of study met the objective of mining emotions as well as sentiment. This can be seen from the results discussed that students were more expressive when it came to providing feedback. Hence the results obtained can be used to provide an insight on how to better improve the teaching and learning experience in order to improve overall student quality.

For future research, this study would like to conduct a content analysis study to analyse the form of feedback that came through from the students. There are also intentions to look into how this study can be expanded to other subjects thought within the faculty.

References

[1] Esparza, G. G., de-Luna, A., Zezzatti, A. O., Hernandez, A., Ponce, J., Álvarez, M., . . . de Jesus Nava, J. (2017). *A sentiment analysis model to analyze students reviews of teacher performance using support vector machines.* Paper presented at the International Symposium on Distributed Computing and Artificial Intelligence.

[2] Joseph, G., & Varghese, V. (2019). Analyzing Airbnb Customer Experience Feedback Using Text Mining *Big Data and Innovation in Tourism, Travel, and Hospitality* (pp. 147-162): Springer.

[3] Kaur, W., Balakrishnan, V., & Singh, B. (2017). Social Media Sentiment Analysis of Thermal Engineering Students for Continuous Quality Improvement in Engineering Education. *Journal of Mechanical Engineering, Special Edition*(4(1)), 263-272.

[4] Mohammad, S. M., & Turney, P. D. (2013). Crowdsourcing a word–emotion association lexicon. *Computational Intelligence, 29*(3), 436-465.

[5] Nithya, R., & Maheswari, D. (2014). *Sentiment analysis on unstructured review.* Paper presented at the Proceedings - 2014 International Conference on Intelligent Computing Applications, ICICA 2014.

[6] Plutchik, R. (2003). *Emotions and life: Perspectives from psychology, biology, and evolution*: American Psychological Association.

[7] Rani, S., & Kumar, P. (2017). A sentiment analysis system to improve teaching and learning. *Computer, 50*(5), 36-43.

[8] Sailunaz, K., Dhaliwal, M., Rokne, J., & Alhajj, R. (2018). Emotion detection from text and speech: a survey. *Social Network Analysis and Mining, 8*(1), 28.

[9] Sun, S., Luo, C., & Chen, J. (2017). A review of natural language processing techniques for opinion mining systems. *Information Fusion, 36*, 10-25.

[10] Yadollahi, A., Shahraki, A. G., & Zaiane, O. R. (2017). Current state of text sentiment analysis from opinion to emotion mining. *ACM Computing Surveys (CSUR), 50*(2), 25.

A comparative study of various binarization schemes for flame-front detection in a S.I engine

S Shawal[1], J Saedon[1], M S Meon[1], N H Mohamad[1], H Husain[1], M Mahat[1] and M R M Nawi

[1]Faculty of Mechanical Engineering, Universiti Teknologi MARA, Malaysia.

syahar6595@salam.uitm.edu.my

Abstract. This paper formulates a new mathematical algorithm for automated flame detection in high-speed image. The proposed algorithm is simple, unsupervised and computationally inexpensive to segment a sequence of such images into two levels which correspond to flame and background. Flame images are used as input to calculate and predict the threshold value of the image based on Otsu's method. The threshold value of the current image is corrected to compute the flame boundary. The main idea is to exploit the images' correlation in time to predict a suitable binarization threshold from the previous image. The algorithm is expected to compute the projected burnt area, which will be used to obtain the flame characteristics in a fast, robust, and reliable way. Then the proposed scheme was compared with two famous different methods and explore their effects on detection characters in scene images. It is found that the proposed method outperforms the other two prior methods in detection flame edges with respect to the flame radius versus time.

1. Introduction

Edge detection is a critical step in image processing [1]. An image is considered to be effectively segmented if we can find the boundary of the object by locating all of its edges [2]. [3] in his paper, stated that simplification of image analysis can be achieved through edge detection process by a drastic reduction in the amount of data to be processed and preservation of structural information about object boundaries that are useful. In general, edges may be used to measure the size of the objects in the image; to isolate particular objects from their background; to recognize or classify objects [4]. Besides, edges are simply regions of intensity transition between one object and another, which makes the edge detection seems to be a relatively straightforward affair. However, despite its simplicity, gradient filters used in most edge detection such as (the Prewitt and Sobel operators) do not find edges per se but rather only give some indication of where they are mostly to occur [5]. In edge detection, feature extraction process and the criterion in the final edge map is the most important part. In the image extraction process, the first step in the process is based on a convolution of an image with a given operator, followed by the thresholding of the feature image in order to obtain an edge map [6]. Various image processing techniques exist for individual flame segmentation in single-shot images. The simplest type of approach is background subtraction, using a mean or median image computed from the overall set of images, and the global threshold. These techniques produce satisfying results of images with flames with intensity values that are significantly different than those of the background. However, they are very sensitive to noise and spatial changes and temporal illumination. Enhanced techniques provided by adaptive thresholds (local thresholds) or by Otsu's method increase segmentation accuracy, but this method suffers as far as the connection between neighboring pixels is not included [7]. Edge detection methods based on image gradients (first order derivatives), for example, Sobel, Prewitt, and so on, are used to

Content from this work may be used under the terms of the Creative Commons Attribution 3.0 licence. Any further distribution of this work must maintain attribution to the author(s) and the title of the work, journal citation and DOI.
Published under licence by IOP Publishing Ltd

detect cell boundaries. The methods involving the second derivative (LoG, DoH, DoG) used to detect the clumps and the advantages of linear filtration techniques are that it is quick and easy to calculate. However, the disadvantages of this method are that they usually need additional steps after processing. Edge thinning and connecting are examples of Canny's edge detectors. Non-linear filtering used in morphological operations relies heavily on morphology and flame form and requires knowledge of these features as the foregoing assumptions. Therefore the application potential for non-rigid objects is very limited [8].

In this paper, we explore the novel scheme which exploits the images' correlation in time to predict a suitable binarization threshold from the previous image. The threshold value of the current image is corrected to compute the flame boundary. The algorithm is expected to compute the projected burnt area, which will be used to obtain the flame characteristics in a fast, robust, and reliable way. The proposed method is able to find the flame edges accurately comparing some existing methods which fail to find these edges.

2. Related work

Comparative studies of various binarization methods have been carried out extensively. Most of these studies consist of two or three of the above methods (gradient and cluster-based method). This is to make use of the benefits of each method [9,10,11]. In our previous study [8], we applied this predictor-corrector scheme for 50 high-speed image sequences which be collected from different datasets. In this study, we improve the algorithm and compare the thresholding technique with other methods to show the effect of the benefits of the previous algorithm in this work.

[6] did a comparative study to evaluate the performance of the proposed fusion of Haar Wavelet and Prewitt operator compared to the gradient edge algorithms and Canny edge detector. The performance of these edge detection algorithms was tested in different cases which are in noisy and blurred conditions. The image results of applying each detection algorithm were then compared through direct visualization. Their study came up with a conclusion that the proposed Haar based Prewitt Edge detection does not perform better than the classical methods and Canny Edge Detector, but it performs better in blur condition. In contrast, Canny Edge Detector performs better in both noisy and blur conditions, but it depends highly on the adjustable parameters [6,13,12].

J. Saif et al. 2012 in their paper compare the two techniques of segmentation algorithms which are Canny edge detection and Otsu thresholding. Both methods were tested with a variety of color images. In their paper, the effectiveness of the two methods was evaluated and their suitability for natural as well as medical images was tested. As a result, they have found that Otsu's method is more suitable for those images with uniform background illumination and have a clear distinction between object and background. In contrast, Canny edge detector is more suitable for the medical images as the images have no clear distinction of objects from the backgrounds [8,14,16].

Neeta Nain et.al 2008, studied on adaptive thresholding based edge detection using morphological operators, which is by applying efficient adaptive efficient peak detection of the image histogram. Their study come up with a conclusion that their adaptive multi-level thresholding method extracts precise one pixel thick seamless, continuous (in a segment) image boundary which is very important to extract prominent and significant corners in images and also in computing image semantics compared to the existing edge detectors such as Canny and Robert's operator [9,15].

3. Methodology

In this paper, we propose a new binarization scheme to segment sequence of flame images into two classes; background and foreground (flame). Flame images are used as input to calculate and predict the threshold value of the image based on Otsu's method. The threshold value of the current image is corrected to compute the flame boundary. The main idea is to exploit the images' correlation in time to predict a suitable binarization threshold from the previous image. The algorithm is expected to compute the projected burnt area, which will be used to obtain the flame characteristics in a fast, robust, and reliable way. We summarize the entire process of the algorithm in Figure 1.

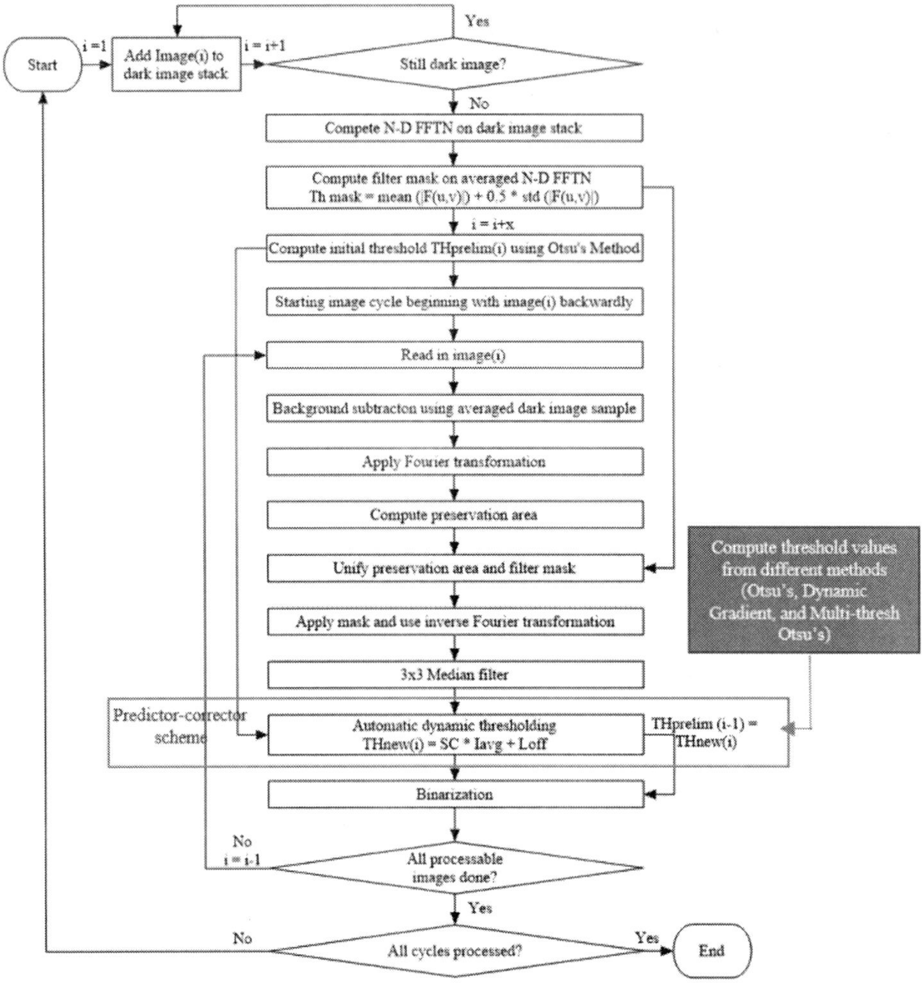

Figure 1. Predictor-corrector scheme algorithm flow chart

4. Results and discussion

In this section, two image series from endoscopic high-speed imaging was qualitatively discussed. Each image was binarized via predictor-corrector thresholding technique to obtain the instantaneous location of the projected burned area. The identification of the correct flame boundary allows calculating the projected flame area and thus flame growth rate. The accuracy of each detection scheme is presented in terms of the calculated flame radius versus propaging time.

4.1. Comparison between different binarization schemes

In this section, the proposed binarization algorithm is compared with three different algorithms in terms of detecting the boundary of the projected burned area based on the computed threshold value.

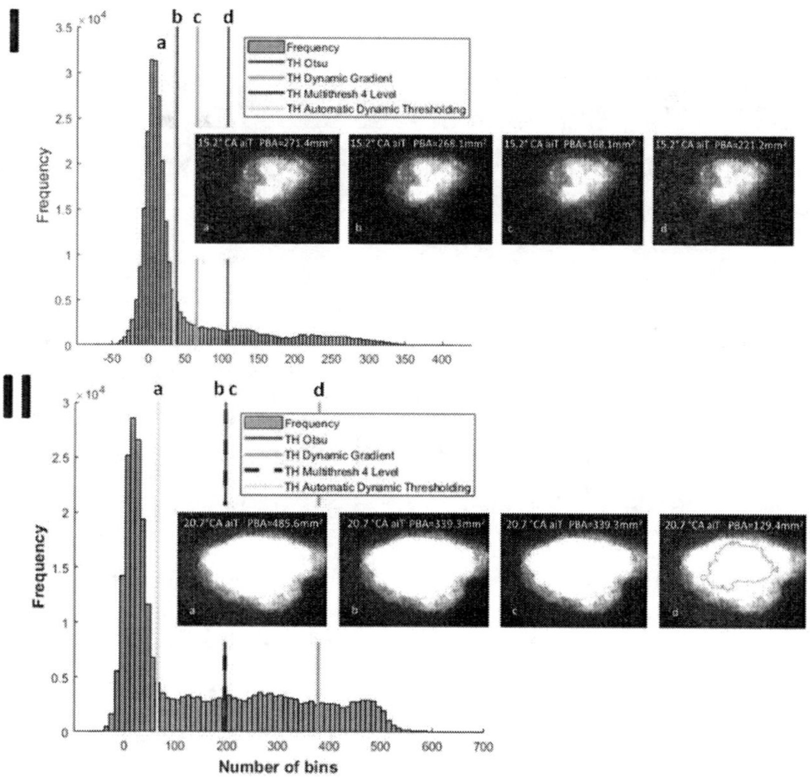

Figure 2. Computed threshold value from each binarization scheme for two sample images (I and II). The corresponding projected burnt area (PBA) is indicated in the upper right-hand corner of each image.

Figure 2 shows the histogram of the respective image (I and II). The computed threshold value from each thresholding algorithm is presented by a different color of solid lines on the histogram. In case I, it shows that the computed threshold values by the proposed algorithm and multi-thresh Otsu's give the most accurate values on detecting the projected burned area. The results of binarization are represented by the red line on the flame images as shown in Figure 2I(a-d). However, in case II, only the proposed algorithm gives the most accurate threshold value as compared to others. The detected flame boundary is shown in Figure 2II(a). The multi-threshold Otsu's and global Otsu's methods fail to detect the correct flame boundary in case II as the algorithm strongly depends on the shape of the image's histogram. [10]. Dynamic gradient method gives the worst result in terms of detecting the flame boundary as it is very sensitive to the image noise during the derivative of image gradient operation. Through direct visualization, it can be seen clearly that the automatic dynamic thresholding method is the best method of detecting the edge of the flame as compared to other schemes.

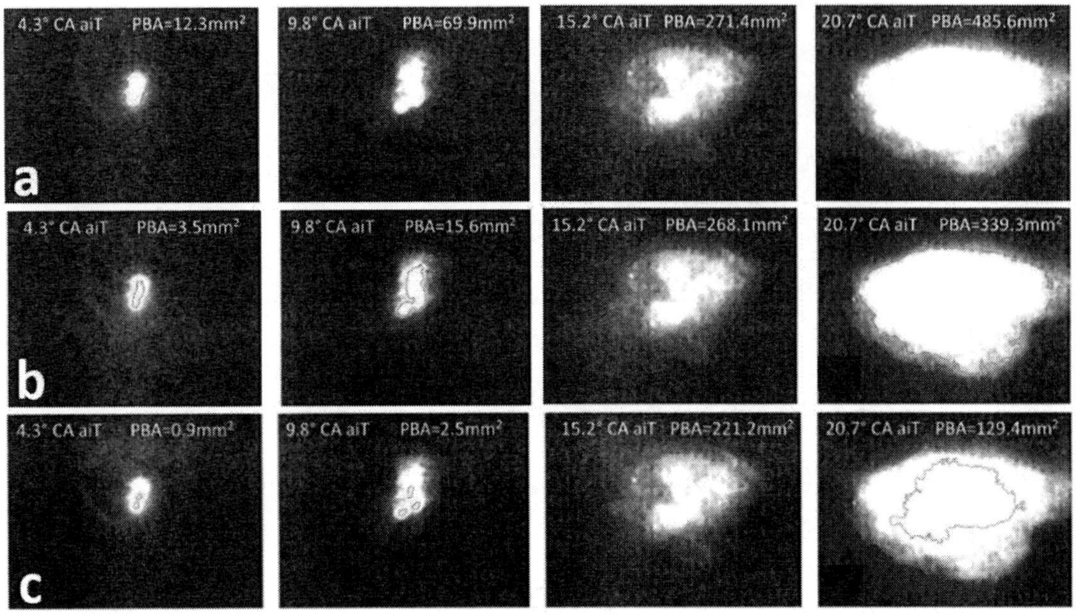

Figure 3. Sample of flame propagation image series. The red line in each image indicates the flame front as detected by the binarization schemes. (a) the proposed algorithm, (b) multi-level Otsu's, (c) dynamic gradient.

4.2. Equivalent flame radius

Figure 4 shows the equivalent flame radius as function of crank angle for 3 different binarization schemes. As it can be seen clearly in Figure 4c, the plot scatter is low, and lines mostly do not cross, which is physically plausible [11,8]. However, the other 2 methods show higher fluctuating, lines cross each other, and inconsistent over the propagating time. This is consistent with the visual appearance of the images in Figure 3(a-c).

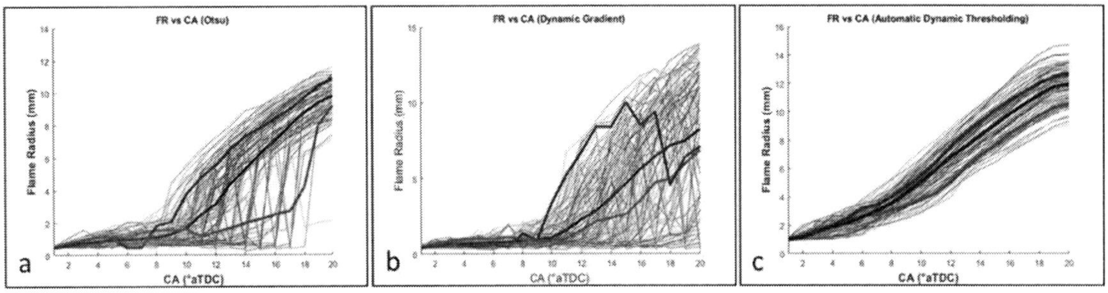

Figure 4. Flame radius for 100 consecutive cycles (coloured lines) and the multi-cycle mean of these 100 cycles (solid black line). Type of edge detection schemes is indicated in each graph. The blue and red lines in the graphs indicate the fast and slow cycles respectively: (a) global Otsu's method, (b) dynamic gradient method, (c) automatic dynamic thresholding (predictor-corrector).

5. Conclusion

This study presented a new edge detection scheme based on image's time correlation for detecting the early flame-front propagation in a spark-ignition engine. The projected burnt area was derived by morphological post-processing. The frequency filtering process was introduced to increase the robustness of the algorithm for detecting the flame boundary at low light conditions. The basic flame characteristics such as flame radius were computed based on the detected flame burned area and the results obtained from each edge detection schemes were compared with three different algorithms. By comparison, automatic dynamic thresholding method is the most accurate algorithm in detecting the flame boundary area in a spark-ignited engine, as compared to multi- level thresholding, global Otsu's and dynamic gradient methods. Thus, the proposed algorithm is seen to have a strong potential to accurately detect and measure flame boundary with a low chance of false edge detections.

Acknowledgments
This work was funded by the Rückkehrerprogramm of the NRW Ministry for Innovation, Science, and Research, the Deutsche Forschungsgemeinschaft, project Schu 1369/12-1, Universiti Teknologi MARA LESTARI grant, Fundamental Research Grant Scheme (FRGS) and the Ministry of Education of Malaysia.

References

[1] R. Medina-Carnicer, R. Muñoz-Salinas, E. Yeguas-Bolivar, and L. Diaz-Mas, "A novel method to look for the hysteresis thresholds for the Canny edge detector," Pattern Recognit., 2011.
[2] C. Solomon and T. Breckon, Fundamentals of Digital Image Processing: A Practical Approach with Examples in Matlab. 2011.
[3] J. Canny, "A Computational Approach to Edge Detection," IEEE Trans. Pattern Anal. Mach. Intell., 1986.
[4] A. McAndrew, "An introduction to digital image processing with matlab notes for scm2511 image processing," Sch. Comput. Sci. Math. Victoria Univ. Technol., 2004.
[5] R. Jain, R. Kasturi, and B. Schunck, "Edge detection," Mach. Vis., pp. 140–185, 1995.
[6] S. R. Joshi and R. Koju, "Study and comparison of edge detection algorithms," Asian Himalayas Int. Conf. Internet, 2012.
[7] P.K. Sahoo, S. Soltani, and A. K. C. Wong, "A survey of thresholding techniques," Comput. Vis. Graph Image Process, Vol. 41, pp. 233–260, 1988.
[8] Shawal, S., Goschutz, M., Schild, M., Kaiser, S. et al., "High-Speed Imaging of Early Flame Growth in Spark-Ignited Engines Using Different Imaging Systems via Endoscopic and Full Optical Access," SAE Int. J. Engines 9(2):704-718, 2016, https://doi.org/10.4271/2016-01 0644.
[9] Ali Abdo Mohammed Al-Kubati, J. A. M. Saif, and Murad A. A. Taher, "Evaluation of Canny and Otsu ImageSegmentation," Int. Conf. Emerg. Trends Comput. Electron. Eng., 2012.
[10] N. Nain, G. Jindal, A. Garg, and A. Jain, "Dynamic Thresholding Based Edge Detection," in Proceedings of the World Congress on Engineering, 2008, vol. 1, pp. 2–7.
[11] M. Krzanowski, "Improved noise reduction and edge detection in high-speed imaging of flame propagation in a spark ignited-engine Submitted by," no. April, 2017.
[12] A. Azeroual and K. Afdel, "Fast Image Edge Detection based on Faber Schauder Wavelet and Otsu Threshold," Heliyon, 2017.
[13] Liu, J. and C. E. Dumitrescu (2018). "Flame development analysis in a diesel optical engine converted to spark ignition natural gas operation." Applied Energy 230: 1205-1217.
[14] Mook Tzeng Lim, Ahmad Rashidi, Soong Der Chen, Noor Akma Watie Mohd Noor, Azmi Ahmad, Hamdan Hassan, et al., "Preliminary determination of flame velocities fro infrared images using image processing techniques in a 150 kWth coal-fired combustion test rig," Advances in Mechanical, Aeronautical and Production Techniques (MAPT),pp. 1-5, 2015.
[15] A. González-Cencerrado, B. Peña, and A. Gil, "Coal flame characterization by means of digital image processing in a semi-industrial scale PF swirl burner," Applied energy, vol. 94, pp. 375 384, 2012.
[16] N. Otsu, A Threshold Selection Method from Gray-Level Histograms, IEEE Transactions on systems, Man, and Cybernetics VOL. SC-9 (January 1979) NO. 1.

Capability analysis of Puck damage model in predicting the damage behavior of unidirectional composite laminates under different scenarios

M S Meon*[1,2], H Husain[1], J B Saedon[1], S Shawal[1], N H Mohamad Nor[1], M N Rao[2] and K-U Schröder[2]

[1]Faculty of Mechanical Engineering, Universiti Teknologi MARA, 40450 Shah Alam, Malaysia.
[2]Institute of Structural Mechanics and Lightweight Design, RWTH Aachen University, Aachen, Germany.

*msuhairil@uitm.edu.my

Abstract. Composite materials have been broadly used in many engineering fields because of their distinctive properties. They have advantages over metals as their properties offer significant advantages such as very good strength to weight ratio. However, failure in composite structures due to interlaminar and intralaminar raises a maintenance concern because it can lead to invisible damage. Therefore, this paper presents the capability of three-dimensional Puck failure criteria with gradual degradation rule to predict the structural responses, as well as the onset and propagation of failure due to variation of the situation in a composite laminate. The proposed damage model is achieved via the implementation of user-subroutines both in statics and explicit analysis using Abaqus software. Three different scenarios are investigated which were open-hole tension, low-velocity impact, and multi-bolt double-lap joint. The force-displacement and force-time curves are recorded and compared with experimental data taken from literature to measure the performance of such damage model. It is found out that the model adopted here responses well with test curves and demonstrates the high capability of predicting the damage in the direction of in-plane and out-of-plane in a composite laminate.

1. Introduction

A composite material is produced using at least two constituent materials with essentially different physical or chemical properties that, when combined, produce a material with characteristics different from the individual components. The common mixture of composite material consists of fiber reinforcement combined with matrix resin. The fiber reinforcement has a good character such as brittleness, high strength, and stiffness, while the matrix resin has high shear properties, low density and easy to fabricate where resulting in producing a composite which has very good strength to weight and stiffness to weight ratios. Furthermore, composite structures are fabricated with tailoring properties in a required or desired specification. They have been increasingly used in load-bearing structures such as aircraft and automobile industries as they offer significant advantages like strength to weight capacity over metals [1].

During the manufacture, maintenance, and service life, composites may suffer a variety of damages in tension, compression and out-of-plane direction. In composite laminates, damage phenomena can be divided into two main groups: (1) interlaminar failure (delamination), and (2) intralaminar failure (fiber

Content from this work may be used under the terms of the Creative Commons Attribution 3.0 licence. Any further distribution of this work must maintain attribution to the author(s) and the title of the work, journal citation and DOI.
Published under licence by IOP Publishing Ltd

and matrix breakage including fiber-matrix decohesion) [2]. The majority of research in composite damage is focusing on open-hole tension (OHT) [3,4], open-hole compression (OHC)[5–7], composite joints [8–10], and low-velocity impact [11–14]. From the previous researches published elsewhere, the most complicated works are emphasized on the development and validation of failure theory for the laminates. Hashin [15] started his ideas in developing the criteria than can identify different modes of failure which is a versatile formulation as compared to the polynomial formulation [16,17]. Puck and Schürmann [18] improvised the work of Hashin and introduced the Inter-fiber failure concept (IFF) where the prediction of failure is based on fracture plane resulted from stresses acting on the fracture plane. The enhancement of Puck damage criteria was done by Davila et al. [19] where they incorporated with a fracture mechanics approach to evaluate the in-situ strength. Further development of failure theories was concentrated on damage mechanics approach where the idea conceptually originates from research of Kachanov [20]. His idea later is adopted for in-plane failure damage criteria developed by Ladeveze and Le Dantec [21]. Despite having all those theories, the total failure of composite laminates cannot be achieved when the progression law is neglected. The most common technique to degrade the structural stiffness matrix is by using the ply-discount approach. The laminates become weaker whenever the reduction of elastic stiffness parameters or the stiffness matrix itself took place.

This paper presents a detailed Finite Element Model (FEM) of open-hole tension (OHT), low-velocity impact (LVI) and multi-bolt double-lap joint of composite laminates. All numerical analyses are carried out using Puck failure criteria together with gradual stiffness decay, whereas the efficiency of the proposed damage model is evaluated by comparing with experimental data. The FE-model is developed using Abaqus software and numerical calculation is realized by using user-subroutines UMAT (tension) and VUMAT (impact). The structural responses such as force-displacement and contact force-time histories are recorded and analyzed to measure the performance of the damage model.

2. Composite damage model

To simulate the onset and progression of failure in composite laminate, a versatile and efficient progressive damage model (PDM) consists of failure theory to predict the initiation of damage, and damage evolution law to monitor the damage progression since the damage is accumulated throughout the layers.

2.1. *Puck failure criteria and its constitutive law*

In this work, the onset of failure is detected using three-dimensional (3D) Puck failure criteria, while the degradation of stiffness is achieved via the ply-discount method using a gradual degradation scheme. Table 1 shows the formulation used in predicting fiber failure in both tension and compression, respectively.

Table 1. Fiber failure (FF) according to Puck failure criteria

Fiber failure (FF)			
FF in tension	$f_{FF}^{t} = \frac{1}{R_{\parallel}^{t}}\left[\sigma_{11} - \left(v_{\perp\parallel} - v_{\perp\parallel f} m_{\sigma f}\frac{E_{\parallel}}{E_{\parallel f}}\right)(\sigma_{22}+\sigma_{33})\right]$	for $[\ldots.] \geq 0$	(1)
FF in compression	$f_{FF}^{c} = \frac{1}{\lvert R_{\parallel}^{c}\rvert}\left[\sigma_{11} - \left(v_{\perp\parallel} - v_{\perp\parallel f} m_{\sigma f}\frac{E_{\parallel}}{E_{\parallel f}}\right)(\sigma_{22}+\sigma_{33})\right]$	for $[\ldots.] < 0$	(2)

Based on Puck and Schürmann [18], $R_{\parallel}^{t}$ and $R_{\parallel}^{c}$ are the tensile and compressive longitudinal strengths, while $v_{\perp\parallel}$ and $v_{\perp\parallel f}$ are the major in-plane Poisson's ratio, and fiber Poisson's ratio, respectively. The elastic properties of lamina in fiber directions and longitudinal directions are denoted as $E_{\parallel f}$ and $E_{\parallel}$, respectively, and $m_{\sigma f}$ is referred to magnification factor where the value of carbon fiber reinforced polymer (CFRP) and glass fiber reinforced polymer (GFRP) are taken as 1.1 and 1.3 [22], respectively.

Derivation for inter-fiber failure (IFF) has been made based on the assumption that stresses acting on the fracture plane (*fp*) causes the damage in the matrix compound. The action-plane fracture criteria have been formulated using stresses σ_n, τ_{nt} and τ_{n1} which act on a plane parallel to the fibers and at an angle θ. These stresses are calculated with the aid of the following transformation:

$$\begin{Bmatrix} \sigma_n(\theta) \\ \tau_{nt}(\theta) \\ \tau_{nl}(\theta) \end{Bmatrix} = \begin{bmatrix} c^2 & s^2 & 2sc & 0 & 0 \\ -sc & sc & (c^2-s^2) & 0 & 0 \\ 0 & 0 & 0 & s & c \end{bmatrix} \begin{Bmatrix} \sigma_{22} \\ \sigma_{33} \\ \tau_{23} \\ \tau_{31} \\ \tau_{21} \end{Bmatrix}; \quad c = \cos\theta, \quad s = \sin\theta \tag{3}$$

Table 2 summarises the functions used in forcasting the onset of matrix failure. Equation (3) was used to identify the matrix cracking in tension or compression.

Table 2. Inter-fiber failure (FF) according to Puck failure criteria

Inter-fiber failure (IFF)		
IFF in tension for $\sigma_n \geq 0$	$f_{IFF}^t(\theta) = \left[\left(\left(\frac{1}{R_\perp} - \frac{P_{\perp\psi}^t}{R_{\perp\psi}}\right)\sigma_n(\theta)\right)^2 + \left(\frac{\tau_{nt}(\theta)}{R_{\perp\perp}}\right)^2 + \left(\frac{\tau_{nl}(\theta)}{R_{\perp\parallel}}\right)^2\right]^{1/2} + \frac{P_{\perp\psi}^t}{R_{\perp\psi}}\sigma_n(\theta)$	(4)
IFF in compression for $\sigma_n < 0$	$f_{IFF}^c(\theta) = \left[\left(\frac{\tau_{nt}(\theta)}{R_{\perp\perp}}\right)^2 + \left(\frac{\tau_{nl}(\theta)}{R_{\perp\parallel}}\right)^2 + \left(\frac{P_{\perp\psi}^c}{R_{\perp\psi}}\sigma_n(\theta)\right)^2\right]^{1/2} + \frac{P_{\perp\psi}^c}{R_{\perp\psi}}\sigma_n(\theta)$	(5)

Other parameters such as Puck's inclination parameters ($p_{\perp\psi}^t$ and $p_{\perp\psi}^c$) used in IFF formulation have been extracted from Puck et al. [22] for a detailed explanation on additional parameters used here. The definition of the inclination parameters $p_{\perp\psi}^t$ and $p_{\perp\psi}^c$ at angle ψ can be written as:

$$\frac{p_{\perp\psi}^i}{R_{\perp\psi}} = \frac{p_{\perp\perp}^i}{R_{\perp\perp}}\cos^2\psi + \frac{p_{\perp\parallel}^i}{R_{\perp\parallel}}\sin^2\psi; \quad i = t, c \tag{6}$$

$$\cos^2\psi = \frac{\tau_{nt}^2(\theta)}{\tau_{nt}^2(\theta) + \tau_{nl}^2(\theta)}; \quad \sin^2\psi = \frac{\tau_{nl}^2(\theta)}{\tau_{nl}^2(\theta) + \tau_{nt}^2(\theta)} \tag{7}$$

$$R_\perp = R_\perp^t; \quad R_{\perp\parallel} = S_{21}; \quad R_{\perp\perp} = \frac{R_\perp^c}{2(1+p_{\perp\perp}^c)} \tag{8}$$

The parameter ψ denotes the shear angle in action plane, $R_\perp$ is failure resistance normal to fibers direction, while $R_{\perp\psi}, R_{\perp\perp}$ and $R_{\perp\parallel}$ are the fracture resistances of the action plane due to the shear stressing. In equation (8), $R_\perp^t$ and $R_\perp^c$ are the transverse tensile and compressive strength, respectively, while S_{21} is the in-plane shear strength of composite.

Whenever the fiber and inter-fiber failure functions reach unity (=1), then the structural stiffness matrix is updated using a reduced stiffness matrix, C^d:

$$C^d = \begin{bmatrix} \beta C_{11} & \kappa C_{12} & \kappa C_{13} & 0 & 0 & 0 \\ \kappa C_{21} & \kappa C_{22} & \kappa C_{23} & 0 & 0 & 0 \\ \kappa C_{31} & \kappa C_{32} & \kappa C_{33} & 0 & 0 & 0 \\ 0 & 0 & 0 & \beta\omega G_{12} & 0 & 0 \\ 0 & 0 & 0 & 0 & \beta\omega G_{13} & 0 \\ 0 & 0 & 0 & 0 & 0 & \beta\omega G_{23} \end{bmatrix} \tag{9}$$

$$\beta = 1 - d_{f,}\kappa = (1-d_f)(1-d_m), \omega = (1 - \mathrm{S_m^t d_m^t})(1 - \mathrm{S_m^c d_m^c})$$

In the constitutive law, d_f and d_m are used as the global damage variables corresponding to FF and IFF, respectively. Individual damage variables based on failure mode are represented by d_f^t, d_f^c, d_m^t and d_m^c for fibre failure in tension and compression and inter-fiber failure in tension and compression, respectively. The relationship between global and local variables is defined as $d_f = 1 - (1 - d_f^t)(1 - d_f^c)$ and $d_m = 1 - (1 - d_m^t)(1 - d_m^c)$. The value for control parameters ($S_m^t = 0.9$, and $S_m^c = 0.5$) were used for controlling the matrix cracking as recommended in Abaqus manual.

2.2. Cohesive law

One of the dominant types of failure in composite laminate is delamination or debonding. It occurs in the thickness direction and caused from out-of-plane type of loading. Since Puck formulation does not consider delamination effect, cohesive contact formulation was used in this numerical exercise, where the complete implementation was carried out using the embedded equations in Abaqus software itself. The prediction of delamination was achieved through the functions given in table 3 and equation (10).

Table 3. Failure criteria for delamination used in this article

Approach	Cohesive contact interface
Damage initiation	Quadratic nominal stress criterion $\left\{\frac{\langle t_n\rangle}{t_n^o}\right\}^2 + \left\{\frac{t_s}{t_s^o}\right\}^2 + \left\{\frac{t_t}{t_t^o}\right\}^2 = 1$
Damage evolution	Power law fracture criterion $\left\{\frac{G_n}{G_n^c}\right\}^\alpha + \left\{\frac{G_s}{G_s^c}\right\}^\alpha + \left\{\frac{G_t}{G_t^c}\right\}^\alpha = 1$

The traction vector can be defined as:

$$t = \begin{Bmatrix} t_n \\ t_s \\ t_t \end{Bmatrix} = \begin{bmatrix} K_{nn} & K_{ns} & K_{nt} \\ K_{sn} & K_{ss} & K_{st} \\ K_{tn} & K_{ts} & K_{tt} \end{bmatrix} \begin{Bmatrix} \varepsilon_n \\ \varepsilon_s \\ \varepsilon_t \end{Bmatrix} \tag{10}$$

From the formula, t is the nominal traction stress vector which consists of three components t_n, t_s, and t_t while $K_{ii}(i = n, s, t)$ is the stiffness coefficient corresponding to the three stresses in cohesive element and ε_i $(i = n, s, t)$ is the three strains at material orientation.

3. Model validation

3.1. Material properties

In this paper, three different case studies were performed to test the capability of proposed progressive damage model to predict the initiation and final failure of composite laminates: (a) Open-hole tension (b) low-velocity impact, and (c) bolted joint. These three cases used different material properties since the references are taken from different publications. The properties were listed in table 4 and 5 for material and cohesive properties, respectively.

Table 4. Elastic and strength properties adopted for this article [23–25].

Mechanical properties	T300/1034C	CCF300/Epoxy	T300/5228A
Longitudinal modulus, E_1 (GPa)	146.8	123.91	144.0
Transverse modulus, $E_2 = E_3$ (GPa)	11.4	9.72	9.31
The in-plane shear modulus, $G_{12} = G_{13}$ (GPa)	6.1	4.53	4.68

Transverse shear modulus, G_{23} (GPa)	3.8	2.56	3.0
The in-plane Poisson's ratio, $\upsilon_{12} = \upsilon_{13}$	0.3	0.288	0.31
Transverse Poisson's ratio, υ_{23}	0.5	0.347	0.31
Longitudinal tensile strength, $R_{\parallel}^{t}$ (MPa)	1730	1762.3	1633
Longitudinal compressive strength, $R_{\parallel}^{c}$ (MPa)	1379	1362.2	1021
Transverse tensile strength, $R_{\perp}^{t}$ (MPa)	66.5	71.1	53.8
Transverse compressive strength, $R_{\perp}^{c}$ (MPa)	268.2	218.3	212
The in-plane shear strength, $R_{\perp\parallel}$ (MPa)	58.2	83.5	80.4

Table 5. Cohesive properties used in LVI simulation [24].

K_{nn} (MPa)	K_{ss} (MPa)	K_{tt} (MPa)	t_n^0 (MPa)	t_s^0 (MPa)	t_t^0 (MPa)	G_n^c (J/m^2)	G_s^c (J/m^2)	G_t^c (J/m^2)
1 x 10^5	1 x 10^5	1 x 10^5	80	80	80	0.556	1.497	1.497

3.2. Finite element model

All finite element model (FEM) was developed using Abaqus 6-13 using solid reduced integration brick element (C3D8R) with enhanced hourglass control. This modeling part aims to mimic the test setup as close as possible.

3.2.1. Scenario I: Open-hole tension

The plies were modeled in stacked condition, which had one element in each ply in direction of thickness. Mesh size is quite fine at the vicinity of the hole, whilst coarser size further away from the notch as illustrated in figure 1. No end tabs required for this virtual test because of the existence of notch. The dominant failure was expected to occur in the area of the hole. The dimension of central notch composite laminates were length, L = 203.2 mm, width, W = 25.4 mm, thickness, t = 2.616 mm, and diameter of the notch, D = 6.35 mm. The laminate was tested in tension, and composed of plies made of T300/1034C with ply orientation of $[0/(\pm45)_3/90_3]_s$.

The simulation was performed using Abaqus/Standard solver and achieved via user-material subroutine (UMAT). One end of the specimen was clamped, and another end was displaced until the total failure.

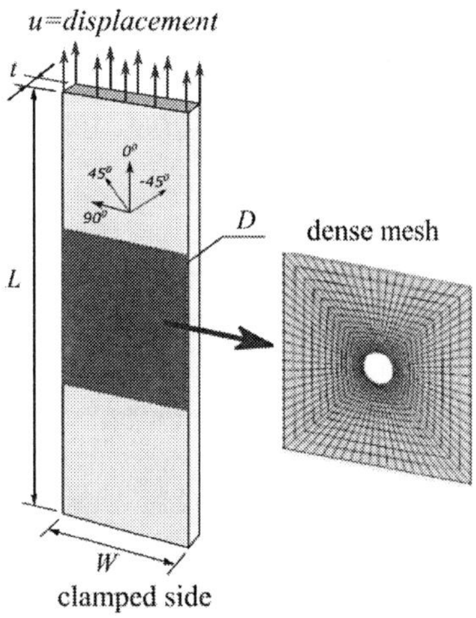

Figure 1. Dimension and boundary conditions of OHT coupon.

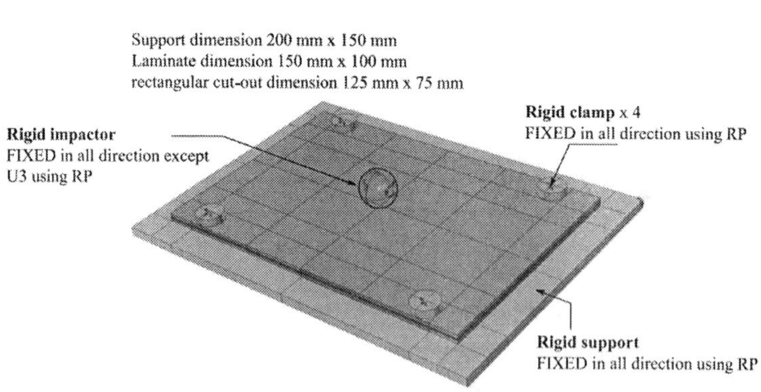

Figure 2. Complete boundary conditions and relevant dimensions in LVI specimen.

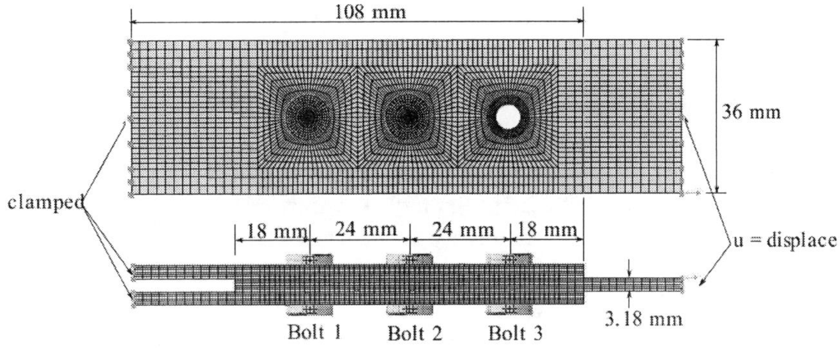

Figure 3. Dimension, mesh and boundary conditions of multi-bolt double lap joint set-up.

3.2.2. Scenario II: Low-velocity impact

A 3D-FE model of an LVI specimen was modeled based on Abaqus/Explicit requirement. The set-up used here mimicked the test set-up according to ASTM D7136/D7136M-15 standard. The laminate was meshed with enhanced solid reduced integration brick element (C3D8R), while the impactor, rubber-clamps and base support were meshed using rigid element (R3D4) since the stiffness of the items are very high when compared with the laminate. Boundary conditions were simulated by fixing all DOEs (degree of freedom) of the fixture base and clamps. Besides, the initial velocity of LVI was 2577 mm/s in the vertical direction was prescribed to the impactor in this 3D FE model. The impactor was locked in all directions except in the direction of loading as can be seen in figure 2.

The diameter of the spherical punch was 16 mm, and the lumped mass 5.36 kg was specified in reference point of the rigid, while the moment of inertia was set to be a large value to prevent rotation of the punch. The nominal thickness of CCF300/epoxy carbon fiber composite laminates was 4mm. The original stacking sequence of the actual composite layer $[45/0/-45/90]_{4S}$ was simplified to $[45_4/0_4/-45_4/90_8/-45_4/0_4/45_4]$ for the sake of computational cost. This approach is a so-called global-local approach and successfully implemented in Han et al. [26] works. The general contact was defined between the punch and composite laminate, base-support and bottom ply, rubber-clamps and top ply and between each ply, ignoring the effect of tangential friction force. The contact force between the punch and the laminate and the reaction force and displacement of compressive reference points were recorded and later was used for the failure analysis. Fiber breakage and matrix damage were predicted through the form of a user-defined variable (SDV).

3.2.3. Scenario III: Multi-bolt double-lap joint

The joint was modeled using the eight-node brick element (C3D8R) with hourglass controlled. The laminates were integrated into the model using a composite layup editor in Abaqus/CAE, where six-sub laminates were grouped with each layup contain four layers in the thickness direction. The composite specimens were made from T300/5228A carbon/epoxy composite material, having a stacking sequence of $[0/45/90/-45]_{3s}$. The material properties of this material are listed in table 3. Three bolts made of aerospace-grade Titanium alloy with a diameter of 6 mm each were arranged according to figure 3, where modulus and Poisson's ratio were 110 GPa and 0.3, respectively.

For the sake of simplification, washers between laminates and bolts were not modeled and the bolts were assembled with one body (bolt and nut are modelled as one part). No bolt torques were applied to the system due to the finger tight consideration. To simulate the contact interaction, the surface to surface contact was constructed, where "hard" contact was applied in the normal direction to transfer enough pressure between contact surfaces. The penalty formulation was utilized in the model with a friction coefficient of 0.2 between all contact surfaces.

The simulation was performed under static analysis and achieved by Abaqus/Standard platform using the user-material subroutine (UMAT) to predict damage behavior in such type of joint. The details of composite failure criteria and evolution law are shown in table 1.

4. Result and discussion

4.1. Analysis of OHT of T300/1034C composite laminate

The numerical model with progressive damage model is validated with experimental results taken from the literature [27]. The load-displacement history between simulation and experiment is shown in figure 4(a).

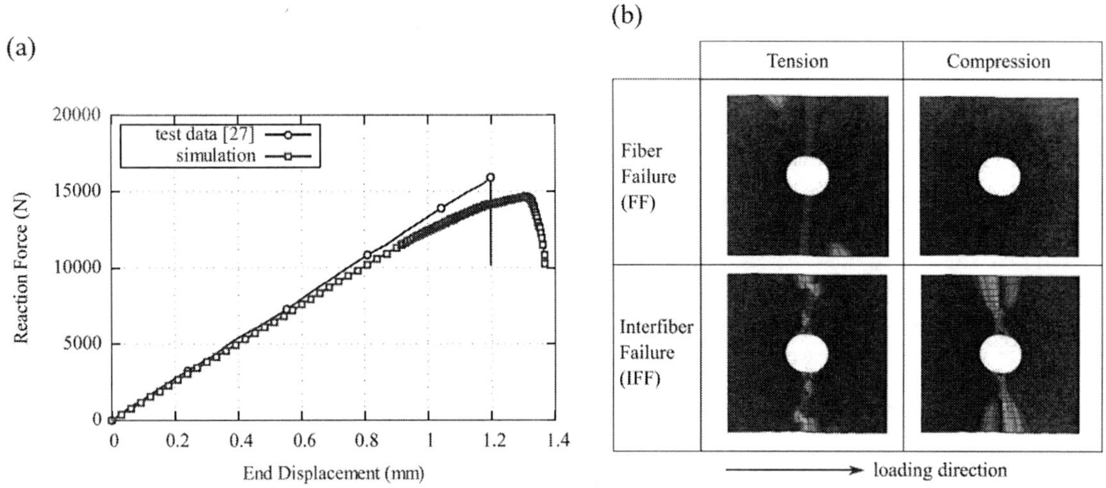

Figure 4. (a) Comparison of force-displacement curve of test data and simulation of OHT coupon, and (b) Damage pattern recorded at the outer layer for fiber breakage and matrix cracking in both direction

It can be said that the result of the simulation is highly acceptable since it followed the trend produced from the experiment. However, the proposed damage model is unable to precisely tracked the final pattern due to the non-linear functions in Puck formulation. Based on the experiment, the laminate lost its carrying capacity at 15.9 kN, while the FE model predicted the total failure at 14.64 kN which is 7.89% under-prediction from the benchmark.

Figure 4(b) demonstrates the failure patterns in fiber and inter-fiber(matrix) due to the tension load for the T300/1034C composite system. The red color represents total failure, whereby others represent partially fail and no failure conditions. The failure is concentrated in the direction of tension either in fiber or inter-fiber direction. It is observed clearly that both fiber and matrix cracking contributed to final failures to the structure as highlighted in red color. The onset of failure originated from the notch due to high concentration stress. The averaged crack paths are similar in both failure modes. The result also revealed that the net-tension failure obtained from the simulation was similar to the reference [27].

4.2. Structural response and damage analysis of LVI of CCF300/epoxy composite laminate

In this section, the CCF300/epoxy composite laminate is used as panel for low-velocity impact simulation under the velocity of 2.577 m/s. Based on figure 5(a), the contact force duration calculated from simulation is shorter than the test curve, which probably due to different stiffness and boundary condition applied to the model [26]. The result exhibited the performance of proposed damage model is sufficiently predict the peak force and the structural response where the difference was only 9.74%.

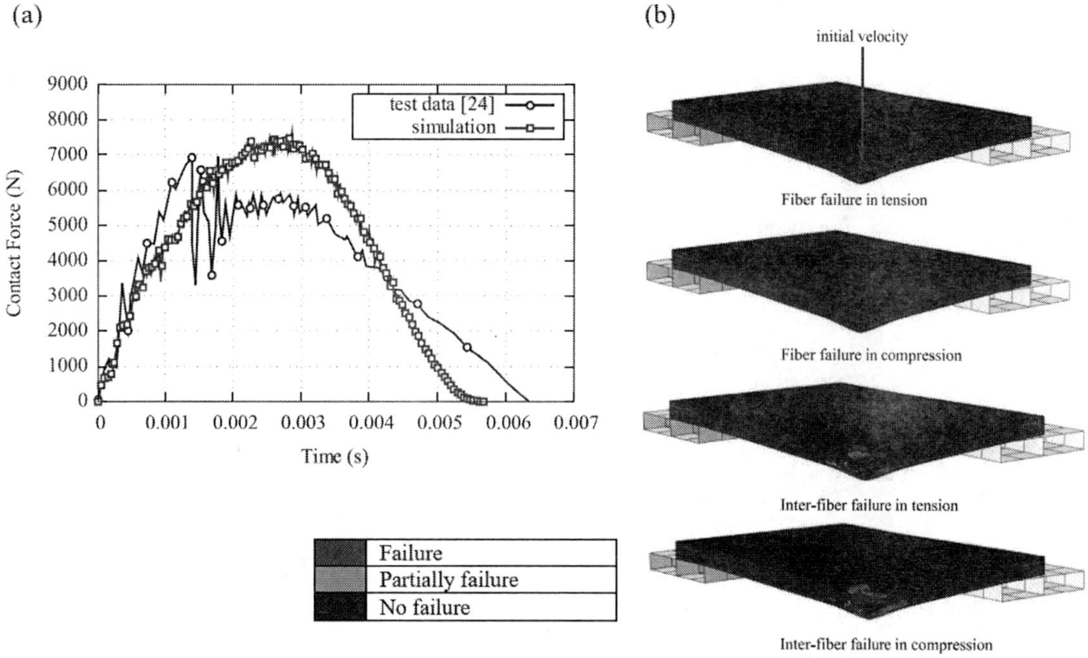

Figure 5. (a) Comparison of contact force-time curve of test data and simulation of LVI plate, and (b) Average damage patterns represent the fiber breakage and matrix cracking in both directions.

Under the impact energy of 4.45 J/mm (2.577m/s), No significant contribution of fiber failure as can be depicted in figure 5(b). This is because the energy applied to the structure is low to break the fiber either in tension or compression. The matrix failure dominates the monolayers having the orientation of 0°, 45°, and -45°, while almost no effect on 90° layers. Another type of failure occurs in this case is delamination due to high inter-laminar tension and shear stresses in the impact zone [13] as illustrated in figure 6. There were six interface layers developed in the model, which all layer shows the existence of delamination (red-colour). Delamination spreads away from the location under the impact point and propagates to the outward regions (top to bottom). The interface between 45° and 0° layers was the most critical zone where the damage is highly affected from the delamination. It is pointed out that, delamination area become greater when increasing the impact loading.

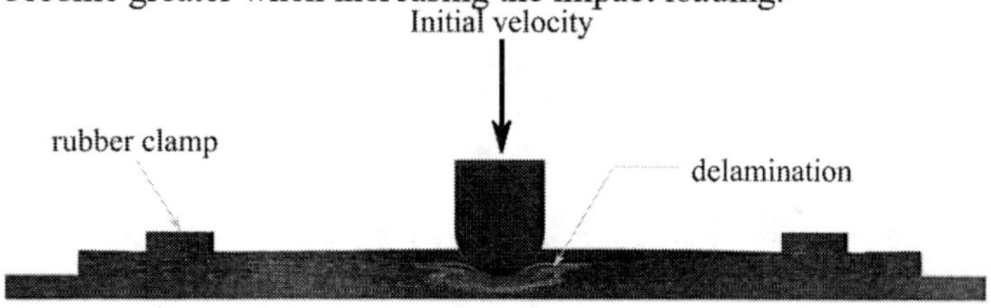

Figure 6. Delamination in-between the cohesive interfaces.

4.3. Structural response and damage analysis of T300/5228A carbon/epoxy composite under multi-bolt double-lap joint

A comparison of the force-displacement curve between experiment and simulation is shown in figure 7. The damage model agreed well in predicting the average maximum load of 25762 N. The structure exhibits linear deformation and non-linear pattern once it started to recognize the damage onset and continue with similar behavior until final failure. The FE model can predict well the linear stage and satisfactorily estimates the non-linear path.

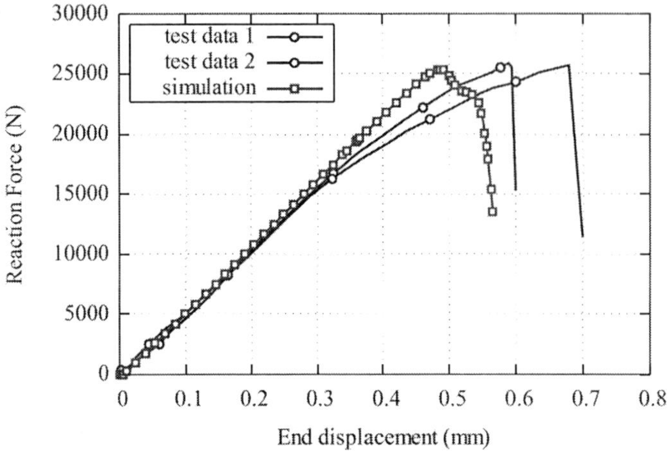

Figure 7. Comparison of reaction force-displacement curve of multi-bolt double lap joint between test data and simulation of T300/5228A composite system

The numerical result shows in figure 8 concluded that the initial damage, as well as the final failure type, is bearing failure. Bolt-3 (right) experiences larger notch size as compared with other two bolts, due to its location closer to load application area. Both fiber breakage and matrix cracking occur in compression direction, while failure in loading direction shows less significant in the double lap joint scenario.

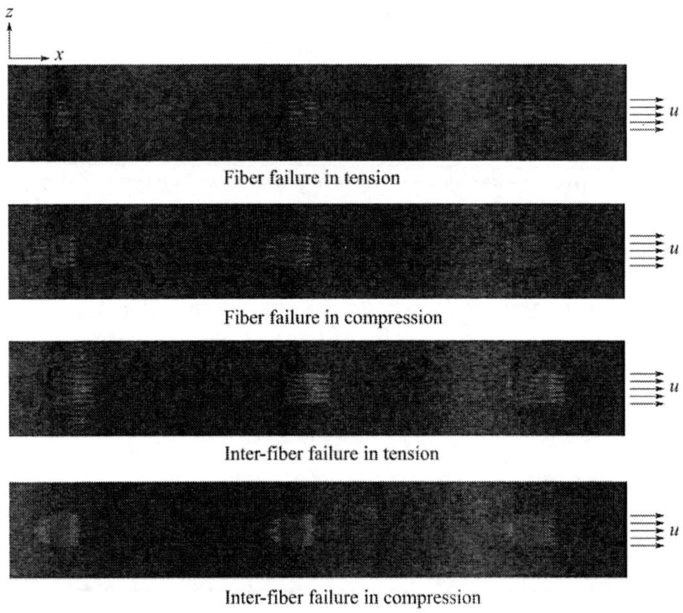

Figure 8. Damage patterns at the final stage (total failure).

5. Conclusion

In the present study, the numerical validations with three types of structural tests are implemented using 3D Puck failure theory incorporated with gradual degradation law. In the FE model, 3D solid elements are widely used to precisely predict the onset of failure as well as the damage accumulation in a composite laminate. The proposed failure model was able to predict the damage behavior of different loading scenarios and capture well the damage initiation and progression on composite laminates. It is hoped that the failure formulation can be a great tool for future analysis in another form of loadings.

References

[1] Hillermeier K 1984 Prospects of Aramid as a Substitute for Asbestos *Text. Res. J.* 54 575–80
[2] Reinoso J, Catalanotti G, Blázquez A, Areias P, Camanho P P and París F 2017 A consistent anisotropic damage model for laminated fiber-reinforced composites using the 3D-version of the Puck failure criterion *Int. J. Solids Struct.*
[3] Xiao M, Yongbo Z, Zhihua W and Huimin F 2017 Tensile failure analysis and residual strength prediction of CFRP laminates with open hole *Compos. Part B Eng.* 126
[4] Gu J and Chen P 2017 Some modifications of Hashin's failure criteria for unidirectional composite materials *Compos. Struct.* 182 143–52
[5] Wang J T, Pineda E J, Ranatunga V and Smeltzer S S 2015 3D Progressive Damage Modeling for Laminated Composite Based on Crack Band Theory and Continuum Damage Mechanics
[6] Zhou S, Sun Y, Chen B and Tay T E 2017 Progressive damage simulation of open-hole composite laminates under compression based on different failure criteria *J. Compos. Mater.* 51 1239–51
[7] Su Z C, Tay T E, Ridha M and Chen B Y 2015 Progressive damage modeling of open-hole composite laminates under compression *Compos. Struct.* 122 507–17
[8] Kapidžić Z, Ansell H, Schön J and Simonsson K 2015 Quasi-static bearing failure of CFRP composite in biaxially loaded bolted joints *Compos. Struct.* 125 60–71
[9] Egan B, McCarthy C T, McCarthy M A, Gray P J and Frizzell R M 2012 Modelling a single-bolt countersunk composite joint using implicit and explicit finite element analysis *Comput. Mater. Sci.* 64 203–8
[10] Ireman T, Ranvik T and Eriksson I 2000 On damage development in mechanically fastened composite laminates *Compos. Struct.* 49 151–71
[11] Riccio A, De Luca A, Di Felice G and Caputo F 2014 Modelling the simulation of impact induced damage onset and evolution in composites *Compos. Part B Eng.* 66 340–7
[12] Liu P F, Liao B B, Jia L Y and Peng X Q 2016 Finite element analysis of dynamic progressive failure of carbon fiber composite laminates under low velocity impact *Compos. Struct.* 149 408–22
[13] Zhang C, Duodu E A and Gu J 2017 Finite element modeling of damage development in cross-ply composite laminates subjected to low velocity impact *Compos. Struct.* 173 219–27
[14] Yu G C, Wu L Z, Ma L and Xiong J 2015 Low velocity impact of carbon fiber aluminum laminates *Compos. Struct.* 119 757–66
[15] Hashin Z 2009 Fatigue Failure Criteria for Unidirectional Fiber Composites *J. Appl. Mech.* 48 846
[16] Azzi V D and Tsai S W 2006 Anisotropic strength of composites *Exp. Mech.* 5 283–8
[17] Hoffman O 1967 Strength of Orthotropic Materials *J Comp Mat* 1 200–6
[18] Puck A and H 2004 Failure analysis of FRP laminates by means of physically based phenomenological models *Failure Criteria in Fibre-Reinforced-Polymer Composites* ed M J H S K D Soden (Oxford: Elsevier) pp 832–76
[19] Dávila C G, Camanho P P and Rose C A 2005 Failure criteria for FRP laminates *J. Compos. Mater.* 39 323–45
[20] Kachanov L M 1958 Time of the rupture process under creep conditions *Izv. Akad. Nauk SSSR. Otd. Tekhnicheskikh Nauk. Mekhanika i Mashinostroenie.* 8 26–31
[21] Ladeveze P and LeDantec E 1992 Damage modelling of the elementary ply for laminated composites *Compos. Sci. Technol.* 43 257–67
[22] Puck A, Kopp J and Knops M 2002 Errata to "Guidelines for the determination of the parameters in Puck's action plane strength criterion" [Composites Science and Technology 2001;62(3):371–8] *Compos. Sci. Technol.* 62 1275
[23] Lee C S, Kim J H, Kim S K, Ryu D M and Lee J M 2015 Initial and progressive failure analyses

for composite laminates using Puck failure criterion and damage-coupled finite element method *Compos. Struct.* 121 406–19
[24] Han G, Guan Z, Li X and Du S 2016 Failure analysis of carbon fiber reinforced composite subjected to low velocity impact and compression after impact *J. Reinf. Plast. Compos.* 35 727–46
[25] Cheng X, Wang S, Zhang J, Huang W, Cheng Y and Zhang J 2017 Effect of damage on failure mode of multi-bolt composite joints using failure envelope method *Compos. Struct.* 160 8–15
[26] Lou X, Cai H, Yu P, Jiao F and Han X 2017 Failure analysis of composite laminate under low-velocity impact based on micromechanics of failure *Compos. Struct.* 163
[27] El-Sisi A E D A, El-Emam H M, Salim H A and Sallam H E D M 2015 Efficient 3D modeling of damage in composite materials *J. Compos. Mater.* 49 817–28

IOP Conf. Series: Materials Science and Engineering **834** (2020) 012029 doi:10.1088/1757-899X/834/1/012029

Heat conductivity and permeability of aluminium foam

Ahmad Hilmi Khalid[1], Muhammad Hussain Ismail[2], Haszeme Bin Abu Kasim[1], Hazriel Faizal Pahroraji[1], Muhd Azhar Bin Zainol[1], Abdul Muzammil Bin Mohd Azaiauddin[2]

[1]Faculty of Mechanical Engineering Universiti Teknologi MARA (UiTM) Cawangan Johor, Kampus Pasir Gudang, 81750 Masai, Johor Darul Takzim, Malaysia
[2]Faculty of Mechanical Engineering Universiti Teknologi MARA (UiTM) Shah Alam, 40450 Shah Alam, Selangor, Malaysia

Corresponding author's email: hazriel@uitm.edu.my

Abstract. Aluminium foam is a material that structured by porous medium with certain characteristics such as good energy absorption, high heat conductivity and good heat transfer. These characteristics that aluminium metal foam provides, can be used in wide range of applications such as on heat exchanger, automotive and etc. The aim of this study is to evaluate the heat conductivity and permeability of the aluminium foam. The porosity of aluminium foam produced by NaCl replication are ranging from 60% to 75%. Besides, simulation by CFD software were carried out to compare with the effect of pore structure on heat conductivity and to determine the heat flux. At the end of the study, the heat conductivity value and permeability of experimental and analysis were identified. The value of heat conductivity k of aluminium foam is better compared to solid aluminium when the temperature increased. The permeability of the aluminium foam is dependent on the porosity of the aluminium foam. Higher porosity resulting to a higher rate of permeability.

1. Introduction

Metal aluminium foam is a metal that which provides new point of view in engineering application. Because of its characteristics, the demand for this aluminium foam is increasing from time to time. The outstanding features that aluminium foam provides is the weight. Metal aluminium foam weight is light because of the foam structure and the porosity. The definition of porous materials is as a holey structure, which are mainly characterized by porosity. Foamed materials, which are the porous materials with high porosity and low density are, resemble bubbling materials such as soda and soap.

The holey materials which have porosity less than70 %, are called porous materials [4]. Not only that, it strength also can be use in certain field of engineering which demands light but yet strong material. Because of the aluminium foam advantages, such as relatively high stiffness despite very low density, high corrosion resistance and etc., there is an increasing interest in using aluminium foams [11]. Metal foams are one of the most promising candidates for replacing traditional heat transfer augmentation techniques [3]. Recently, new research has found new method to manufacture aluminium foam with much less cost. Basically, the method is using NaCl as an agent holder to create the foam structure within the mold. The size of NaCl will determine the size of the porosity thus will also determine the permeability of the manufactured aluminium foam. This method is one of the recent method, and it still need to be explored thoroughly so a standard measurement can be determined.

These fabricated open cell-aluminium foam have different size of NaCl, ranging from 2 mm – 6 mm. Not only that, by using gas infiltration method are proven in creating a bigger size compared to

Content from this work may be used under the terms of the Creative Commons Attribution 3.0 licence. Any further distribution of this work must maintain attribution to the author(s) and the title of the work, journal citation and DOI.
Published under licence by IOP Publishing Ltd

other methods available. In this research, the focus of aluminium foam application will be on heat exchanger in industries. Aluminium foam have been identified as a good material that can distribute and absorb heat to replace the commercial fin and tubes design. But a standard measurement and data are still not have been decided yet as it still in research and development progress. Furthermore, the cellular structure material such as foam and other porous material are known with the special physical and mechanical properties [10].

There are a lot of researches done to set a standard measurement for aluminium foam in term of its characteristics. The characteristics that relate to the researches are the permeability inside a vessel, the heat conductivity and the mechanical properties of the aluminium foam. If all the data successfully analyzed, the application of the metal can be widening, thus, increasing the demand in the future.

2. Methodology

2.1. Sample Preparation

Because of the size of the fabricated aluminium foam was ranging from 2.6 cm to 10 cm, the size of the sample cut down to the size of 2.5 cm diameter and 3 cm of height. The process started with using band saw to cut it in smaller size before go through the EDM wire-cut machine for detailed cutting. Then after that, the sample were polished as finishing touch before the experiment started.

Figure 1. Sample of aluminium foam.

2.2. Heat Conductivity Test

As discussed in the literature review, the same method of axial flow method machine was used to determine the heat conductivity. The thermal conductivity machine located in the Thermofluid Laboratory in Faculty of Mechanical Engineering, UiTM Shah Alam. Heater is controlled by the amount of power [Watt] input. Ranging from 0 W - 80 W, the apparatus can produce heat at almost to 190°C. For this experiment, the maximum heat input was set around 120°C - 130°C to prevent the machine from being overheat.

$$Q = -kA\frac{\Delta T}{\Delta x} \tag{1}$$

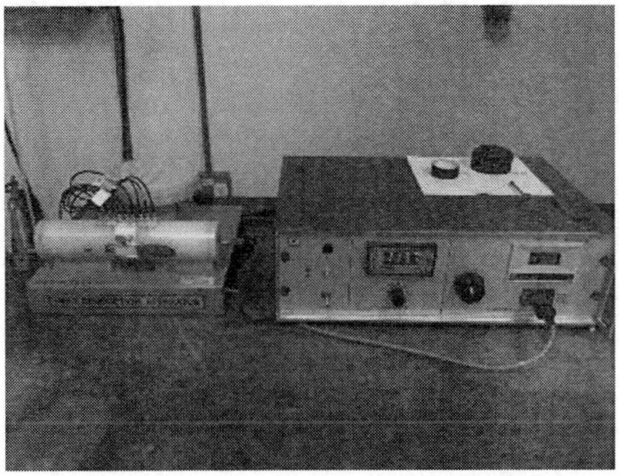

Figure 2. Thermal conductivity machine.

2.3. Permeability Test

By using the Constant Head Method, the experiment setup as close as to the actual experiment method. The experiment conducted in a simplified way. After the setup was done, the plain water was poured down into the funnel. The funnel was connected to the sample holder in order to channel the water to the water container underneath. With fixed amount of water poured, the time taken for plain water to drain gives different value between each sample. And also to determine the permeability for each sample. The equation 2 were used to determine the rate of permeability.

$$K_{con} = \frac{\mu QL}{\rho g h A \Delta t} \tag{2}$$

2.4. CFD Simulation Analysis

The image from 2D surface 3 was converted to simplified 3D model. And the geometry would go meshing process and analysis process. From here, the preliminary data could be obtained and the rough assumption could be determined. Software ANSYS v15.0 was used for the simulation purpose. Using FLUENT setup, the heat flux obtained from the simulation was calculated in the heat conductivity formula. By far, the software yielding a closest result compare to the actual experiment conducted by P. Ranout et al. (2014). Therefore, this method was the best option compared to other CFD analysis software. The model design was based on the 2D view of the aluminium foam. Then it was projected to a certain length to make sure the model geometry was identical to the experimental specimen.

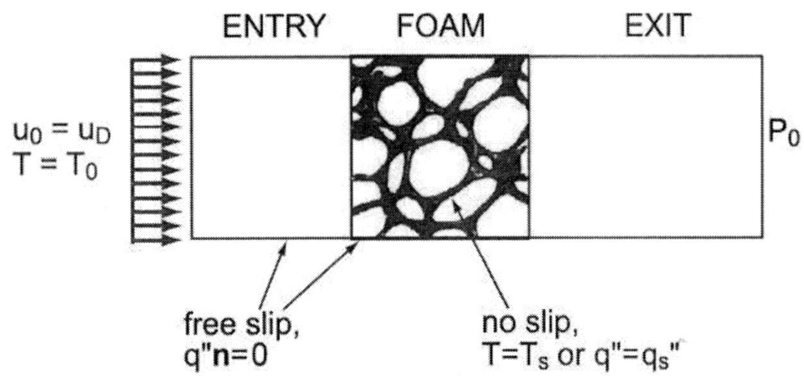

Figure 3. Schematic diagram of 2D simulation.

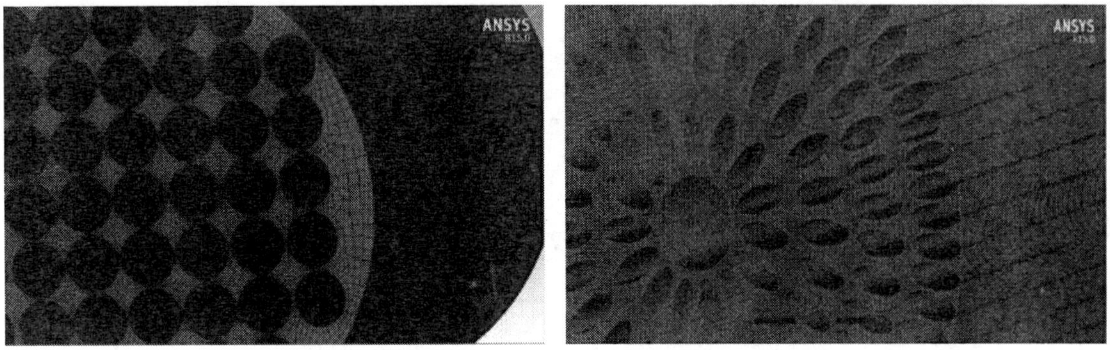

Figure 4. Cylindrical shape 3D geometry mesh (left) and ellipsoidal shape 3D geometry mesh (right).

2.5. Electron Observation

The observation of the pore structure of the aluminum foam is conducted in the Research Laboratory at the Faculty of Dentistry, UiTM Sungai Buloh. The machine used is as shown in Figure 2 above.

3. Result and Discussion

3.1. Density and Porosity of Aluminium Foam

Depending on its porosity and density, the quality of the foam is determined from equation 3.

$$\rho = \frac{m}{V} \tag{3}$$

The total open volume of interconnected and isolated shape can be defined as the porosity. Porosity percentage is calculated by rough measurement of the open volume equal to 100% minus the part density.

$$\%\,of\ porosity = \frac{\rho_{Al} - \rho_f}{\rho_{Al}} \times 100 \tag{4}$$

Where; ρ_{Al} = density of aluminium, ρ_f= density of foam.

Table 1. Density and porosity of aluminium foam tube

Sample	Mass [g]	Height [cm]	Volume [cm^3]	Density [g/cm^3]	Porosity [%]
1	41.5540	3	14.726	2.82	0
2	14.8113	3	14.726	1.69	62.75
3	13.6093	3	14.726	1.78	65.78
4	14.4507	3	14.726	1.72	63.66
5	13.1689	3	14.726	1.81	66.88
6	15.0171	3	14.726	1.68	62.24
7	11.5271	3	14.726	1.92	71.01
8	14.8123	3	14.726	1.69	62.75

From the graph, aluminium foam sample number 6 is the highest porosity recorded with 71.01%. And sample number 5 is the least porous with 62.24%. Based on the measurement recorded, the expected heat conductivity of aluminium foam will be affected by porosity ranging from 62.24% to 71.01%. Referring to research done by S. Dawood and S.S.M. Nazirudeen (2010), metal foam which consists of 60% - 70% porosity will have the best attributes in terms of mechanical and thermal properties.

3.2. Heat Conductivity

The purpose of heat conductivity experiments is to determine the value of heat conductivity of aluminium foam with different porosity. By doing the experiment, the most suitable value of porosity with the best heat conductivity value can be determined. The value of *k* can be obtained by using equation 1.

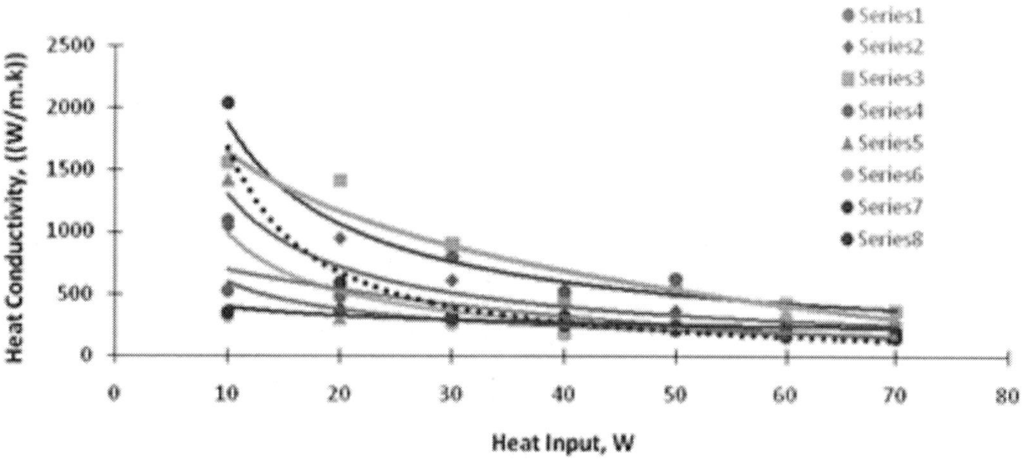

Figure 5. Heat conductivity-heat input experimental result

Based on the Figure 5, the value of heat conductivity, *k* for all samples are high, some are exceeding 500 W/m·K at the slightly above room temperature (32°C - 36°C). This is because of there is not enough heat or energy introduced in the sample, which cause the electron or molecule does not vibrate enough to transfer the energy. As the temperature increase, the heat conductivity decrease, this occurs because the molecular vibrations increase (in turn decreasing the mean free path of molecules). So, they obstruct the flow of free electrons, thus reducing the conductivity.

From research done by J.W. Paek et al. (2000), it is proven that the value of *k* will be decreased when the porosity decreased. The obtained data from this experiment shows that even though the heat conductivity is affected by the porosity of the sample, at the same time it is also affected by variations in the cell size at fixed. Both experiments give out the similar outcome. Usage of aluminium foam of 60% - 70% porosity as heat conductor is better compared to solid aluminium.

Based on the result shown in Figure 5 also, the heat conductivity value is decreasing as the heat input increased. At the maximum heat input; i.e. 70 W, most of the sample of aluminium foam including aluminium ingot have reached their constant state. For aluminium ingot, the value is 226 W/m·K . And as for the aluminium foam, the lowest heat conductivity value recorded at 70 W input is 150 W/m·K with the porosity of 63.66%.

E. Sadeghi et al. (2011) stated that the effective conductivity decreases with an increase in porosity. However, the effect of pore density seems to be insignificant. Lower porosity values are associated with a higher volume of solid ligaments which provide high conductive paths for the heat flow.

3.3. Simulation Result

In this simulation, there were a few range of temperature used, as shown in Figure 6. The simulation done by ANSYS v15.0 was to indicate that a different structure of porosity will affect the heat conductivity. There were three kind of shape that had been tested; Cylinders, Irregular and Ellipsoidal shape, as shown in Figure 7. All of the 3D model had the same porosity value at 60%. The heat input ranging from 94.4°C to 134°C with 20°C interval between. While the output was maintained throughout the simulation at 79.5°C.

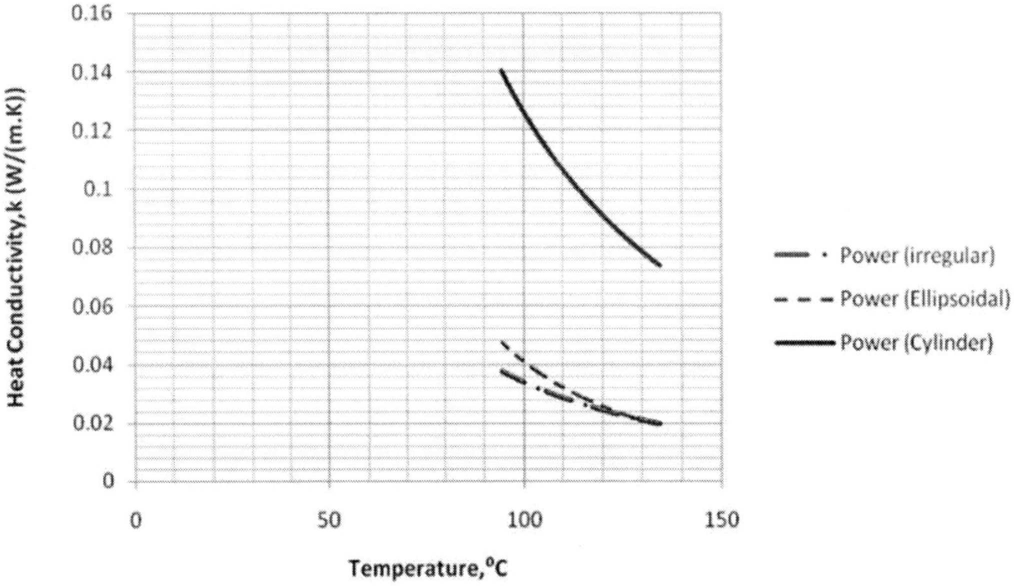

Figure 6. Heat conductivity-heat temperature simulation result.

The output from this simulation was total surface heat flux. With the value obtained, then the calculation using simple heat flux equation to get the value of heat conductivity was conducted. All of the simulation value gave out the same decreasing trends. The higher the temperature, the lower the heat conductivity value until it reached leveling off value.

From the simulation done in the software, the value was different according to its porosity shape. The highest was the uniform circle and the lowest is the irregular or random shape. From that, it could be confirmed that how the shape coefficient value effect the thermal conductivity on a substance with same porosity level. By referring the theory that was introduced by Zehner and Schlunder (1970), it was proven that the porous structures play an important part in determining the value of heat conductivity.

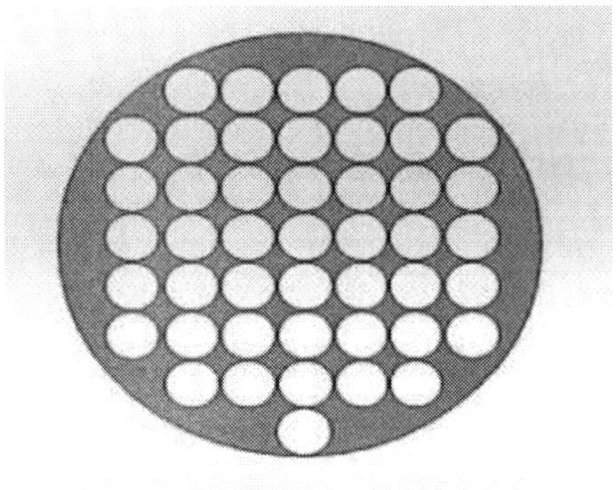

(i)

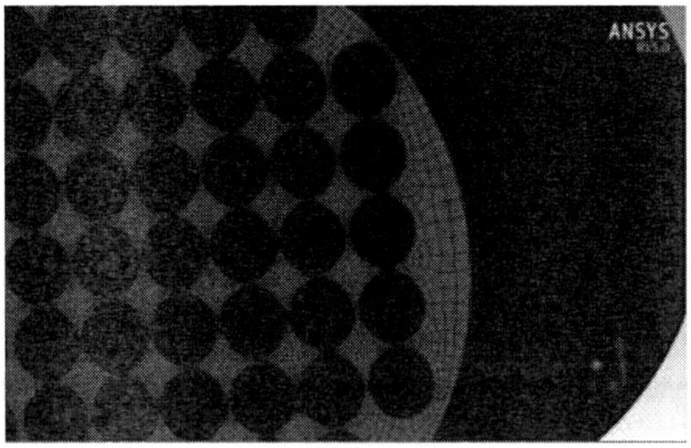

(ii)

IOP Conf. Series: Materials Science and Engineering **834** (2020) 012029 doi:10.1088/1757-899X/834/1/012029

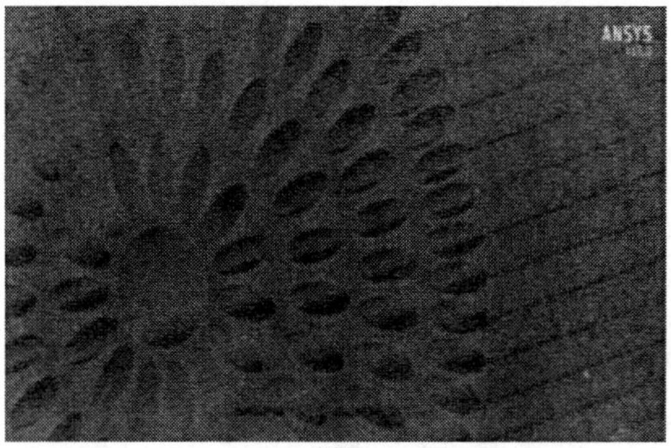

(iii)

(iv)

Figure 7. (i) 2D drawing of cylindrical shape pore structure, (ii) 3D geometry of cylindrical shape pore structure, (iii) 2D drawing of ellipsoidal shape pore structure, and (iv) 3D geometry of irregular shape pores structure

3.4. Permeability Test Result

The permeability test is a measure of the rate of the flow of water through aluminium foam. In this test, water was forced by a known constant pressure through a aluminium foam specimen of known dimensions and the rate of flow was determined and recorded.

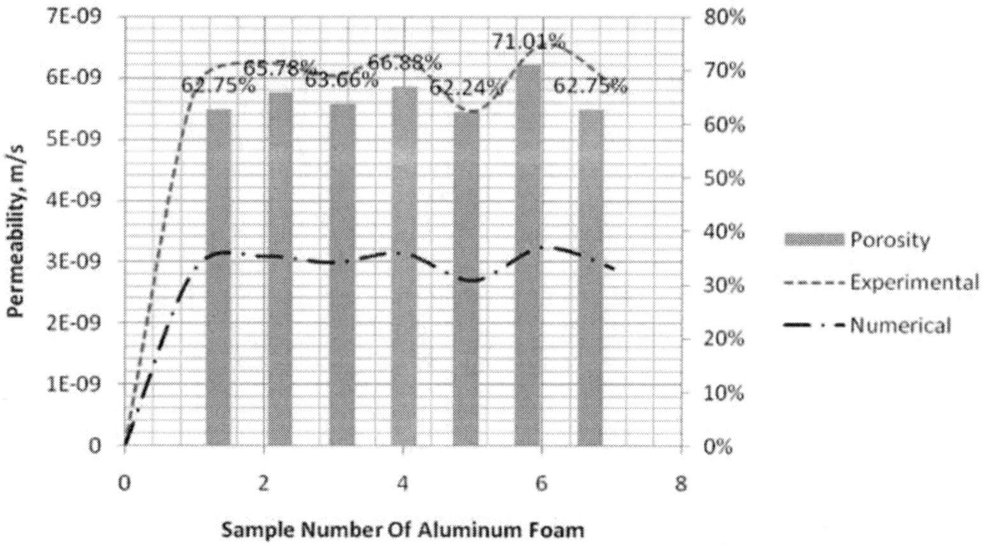

Figure 8. Experimental and numerical permeability result

Permeability are referred as the ease of water go through the porous medium. Based on Figure 8, it can be seen that the highest porosity drains the water the fastest. The expression for the permeability of replicated aluminium foam derived on the basis of the "bottleneck" model of a porous medium also agrees well with the experimental data. The expression can be applied successfully to design porous castings for filtering and noise reduction.

Based on the graph, the rate of permeability was the highest when the porosity was at 71.01%. The theory followed by the least permeable at the lowest porosity. Besides that, the result for experiment and numerical quite different in value. This is because of the different kind of characteristic that aluminium foam possesses compared to soil.

IOP Conf. Series: Materials Science and Engineering **834** (2020) 012029 doi:10.1088/1757-899X/834/1/012029

3.5. Pore Structure

For every sample of the aluminium foam tested, each has different pore structure level. These particular characteristics is very essential in determining the value of heat conductivity of aluminium foam. Clearer view of the pore structure of the aluminium foam is scanned by using the scanning electron microscope (SEM) TM3000 tabletop microscope machine. Each of the aluminium samples is scanned and magnified images of each samples are shown on the computer unit connected to the SEM.

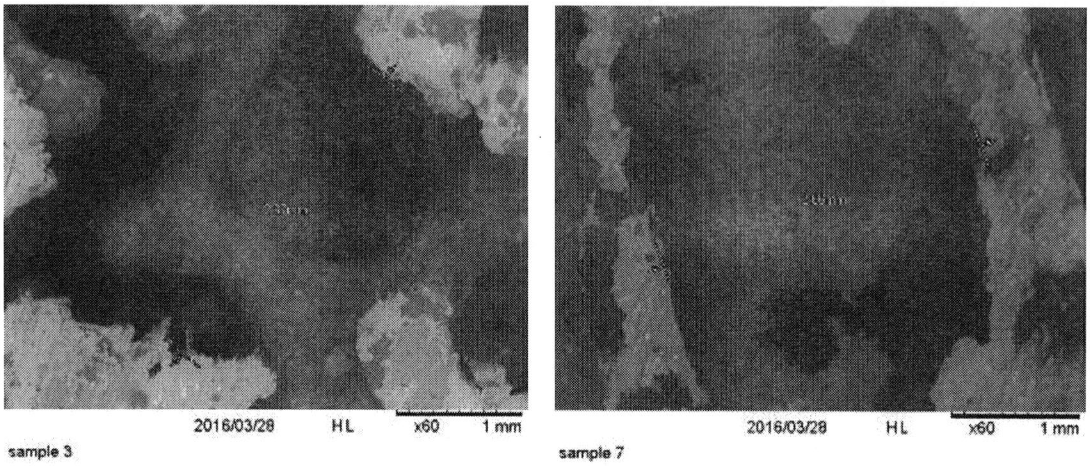

Figure 9. Poros structure of aluminium foam tube fabricated.

From the images obtained from SEM, it can be seen that there were some white fragments in the aluminium foam. This is because of the NaCl that are not fully removed. There is still certain method that need to be discovered to remove the NaCl completely from the fabricated aluminium foam. If the NaCl was removed completely, the structure of aluminium foam could be seen much clearly.

Most of all the sample were in range of 1.5 mm - 3 mm of their pore size. As seen in the table, the structures of the aluminium foam were different in term of pore size and shape. Image of each sample was slightly similar. From that, it is believed that the distribution of the porous structure for the entire sample was well distributed. In the image of porous structure, the pore size of sample (a) was slightly larger than the pore size of sample (b). It is shown that the percentage of porosity and density are depending on the pore size. The larger pore size will give a higher porosity percentage.

Besides that, all of the aluminium foam sample showed that the structures are interconnecting and have random shape. The sizes of the porosity depend on the NaCl used in the fabrication.

4. Conclusion

This study focused on heat conductivity and permeability of aluminium foam. From this study, it can be concluded that the objective is achieved. Aluminium foam does have better thermal conductivity compared to solid aluminium. The heat conductivity is a decreasing function of porosity, falling off very rapidly for the first few percent of porosity, and then leveling off. Other than that, permeability of aluminium foam depends on the porosity, the higher the porosity will give out the higher rate of permeability.

Acknowledgement
The authors acknowledge technical and financial support from UiTM Shah Alam to publish this paper. Besides that, special thanks go to authors group members of Paper to Publish (P2P) from UiTM Pasir Gudang because successfully make the journal paper become reality

References

[1] P. Ranout, E. Nobile, and L. Mancini, "High Resolution Microtomography-based CFD simulation of Flow and Heat Transfer in Aluminum Metal Foams", *Applied Thermal Engineering*, Volume 69, Issues 1-2, pp 230-240, 2014.

[2] Hideo Nakajima, *Porous Metals with Directional Pores*, Springer, (2013).

[3] Chin-Jye Yu, and John Banhart, "Mechanical Properties of Metallic Foams", *Proceeding of Fraunhofer USA Metal Foam*, 1998.

[4] Z. Hussain, and N.S.A. Suffin, "Microstructure and Mechanical Behaviour of Aluminium Foam Produced by Sintering Dissolution Process Using NaCI Space Holder", *Journal of Engineering Science*, Volume 7, pp 37-49, 2011.

[5] A.K. Shaik Dawood, and S.S. Mohamed Nazirudeen, "A Development of Technology for Making Porous Metal Foams Castings", *Jordan Journal of Mechanical and Industrial Engineering*, Volume 4, Number 2, pp 292 – 299, 2010.

[6] J.W. Paek, B.H. Kang, S.Y. Kim, and J.M. Hyun, "Effective Thermal Conductivity and Permeability of Aluminum Foam Material", *International Journal of Thermophysics*, Volume 21, Issue 2, pp 453 – 464, 2000.

[7] E. Sadeghi, S. Hsieh, and M. Bahrami, "Thermal Conductivity and Contact Resistance of Metal Foams", *Journal of Physics D: Applied Physics*, Volume 44, Number 12, 2011.

[8] P. Zehner, and Schlunder, "Thermal Conductivity of Granular Materials at Moderate Temperature (in German) ", *Chemie Ingr. Tech.*, Volume 42, pp 933-941, 1970.

[9] R. Dyga and S. Witczak, "Investigation of effective thermal conductivity aluminum foams", *20th International Congress of Chemical and Process Engineering CHISA 2010*, Procedia Engineering 42 (2012) 1193-1204, 2012.

[10] Guo-Yun, Bu-Yun Su, Zhi-Qiang Li, Zhi-Hua Wang, Wei-Dong Song and Hui-Ping Tang, "Thermal Properties of Closed-Cell Aluminum Foam with Circular Pores", *Journal of Thermal Science*, Volume 18, No. 5, pp. 1619-1624, 2014.

[11] Moran Wang and Ning Pan, "Modeling and prediction of the effective thermal conductivity of random open-cell porous foams", *International Journal of Heat and Mass Transfer*, Volume 51, pp. 1325-1331, 2008.

IOP Conf. Series: Materials Science and Engineering **834** (2020) 012030 doi:10.1088/1757-899X/834/1/012030

Computational fluid dynamics analysis on the effect of flow distribution on pedestrian in urban area

Haszeme Bin Abu Kasim[1], Mohd Faizal Bin Mohamad[2*], Siti Khadijah Alias[1], Ahmad Hilmi Khalid[1], Muhd Azhar Bin Zainol[1] and Nur Asyikin Binti Ahmad Tajudin[2]

[1]Faculty of Mechanical Engineering Universiti Teknologi MARA (UiTM) Cawangan Johor, Pasir Gudang Campus 81750 Masai, Johor Darul Takzim, Malaysia.
[2]Faculty of Mechanical Engineering Universiti Teknologi MARA (UiTM), Shah Alam Campus 40450 Shah Alam, Selangor, Malaysia.

*Corresponding author E-mail: mfaizal01@gmail.com

Abstract. Buildings in urban area consist of various heights and building arrangements. Meanwhile, the pedestrians at the area must be considered for their wind comfort and wind safety. This is because, near high-rise buildings, high wind velocities are often introduced at pedestrian level that can be experienced as uncomfortable or even dangerous. Nowadays feasibility Computational Fluid Dynamics (CFD) for the assessment in wind engineering such as wind comfort and safety of pedestrian are very demanding. Hence, this study is conducted using CFD to analyses the effect of different building heights on the wind flow distribution towards pedestrian. The study was done on two different cases based on two different buildings (uniform height and non-uniform height) with squared and staggered arrangements. The turbulent wind over the pedestrian was acquired by solving the 3D steady state Reynold's-Average Navier-Stokes (RANS) equations using RNG k-ε turbulence closure model. The results show that changes on the height of the buildings varies from the center of the urban canyon based on the vorticity of the flow distribution.

1. Introduction

In urban city centers, wind nuisance has become a more obvious problem as favored high-rise buildings are developed. Urban geometrical varieties do affect the wind flow whereby it alters the overall wind velocities [1]. This affects the strolling activity in the city. Hence, it is recommended to conduct a study on wind comfort. Responsible authorities like architects and urban planners should consider the wind flow pattern around high-rise buildings to create more pleasant and safer buildings. The environment nearby high-rise buildings can exert uncomfortable high wind velocities on pedestrian. Despite, due to difference in generally accepted comfort criteria in different regions, there is no commonly international standard for wind comfort criteria [2].

When a wind gust strikes a high-rise building, the wind is prone to be deflected towards the ground resulting to high wind velocities on the windward side and at the corners of the building. Occurrences on pedestrian area often associated with three types of flow which are vortex flow between buildings near to the ground level, descending airflows passing around leeward building corners and airflows passing through openings (passages) at ground level connecting the windward and leeward sides of buildings [3][4][5].

Generally, the friction and barrier produced by the urban canopy can lower the urban wind speed. The variations of building heights form an irregular surface area whereby these buildings apply a

Content from this work may be used under the terms of the Creative Commons Attribution 3.0 licence. Any further distribution of this work must maintain attribution to the author(s) and the title of the work, journal citation and DOI.
Published under licence by IOP Publishing Ltd

strong frictional drag of air moving over them leading to turbulence with quick sudden change in wind speed and direction [6][7][8].

It is inferable that the airflow within urban area is different from rural areas corresponding to the diversity of dispersion and concentrations of air pollutants for different wind directions and building shapes [9][10].

Based on past studies on urban wind, CFD is preferable to simulate the flow distribution due to its advantages and over-whelming demand to assess the wind comfort and safety of pedestrian [11][12]. To investigate the effect of flow distribution, it is examined throughout a standard condition whereby the flow is incompressible, turbulent, steady-state and isothermal where the temperature remains constant. Despite the Large-Eddy Simulation (LES) which based on a local filtering, the equations are unsteady to be solved and RANS is more preferred as less computational effort is required that gives time averaged mean value for velocity field [13][14].

In conclusion, CFD tool can be used to simulate the flow distribution in urban area. This study is intended to evaluate the distribution flow pattern on pedestrian within urban street canyon and to analyze the effect of high rise buildings in urban area using ANSYS Fluent software. The study is done on two different cases which are uniform height and non-uniform height based on two different building arrangements of squared and staggered. Table 1 shows for the case conditions and Figure 1 shows for the Schematics of the building geometry. Validation is done to ensure the reliability of the simulation.

2. Methodology

The numerical approach of CFD was used to simulate the flow distribution using ANSYS Fluent software. This simulation starts with pre-processing, solving and post-processing.

2.1 Pre-Processing Stage

In this stage, the geometries of the buildings were designed using CATIA V5R20 based on urban-like cubicles. The domain was then generated in ANSYS Design Modeler with domain size of 4H where H was the building height of 25mm. The boundary conditions were then assigned on the inlet, outlet, lateral sides, bottom, top and buildings.

In order to do meshing for the building geometries, tetrahedral mesh was selected due its appropriate for unstructured geometry. Refer to Figure 2 for the unstructured tetrahedral meshing. Meanwhile for the sizing of meshing, it was set up to fine mesh to get the conver-gence and accurate results such as Figure 3. The parameters of minimum orthogonal quality must be between 0.15 to 1.00 and maximum skewness between 0 to 0.94 had to be achieved. Refer to Figure 4 for the minimum orthogonal quality and Figure 5 for the max-imum skewness respectively.

Table 1. The case conditions.

Case	Condition
1	Uniform height with staggered arrangement (validation)
2	Uniform height with squared arrangement
3	Non-uniform height with staggered arrangement
4	Non-uniform height with squared arrangement

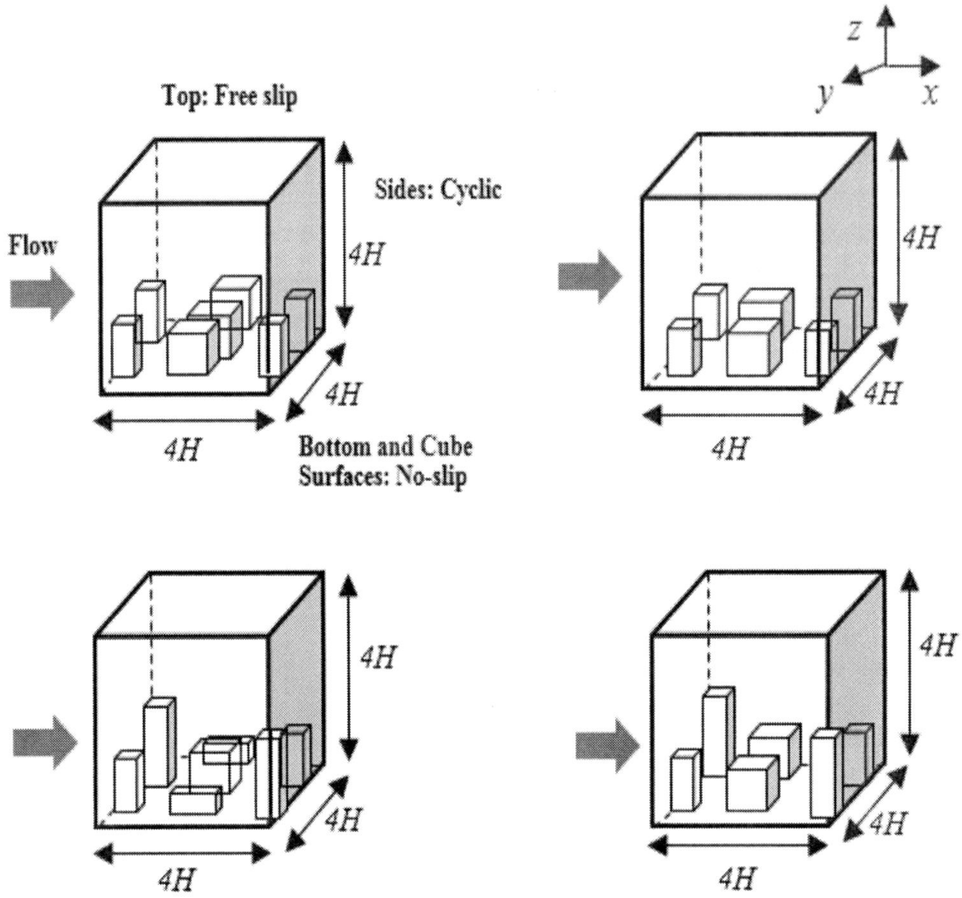

Figure 1. Schematics of the building geometry and their imposed boundary condition. (Upleft for case 1, upright for case 2, downleft for case 3 and downright for case 4)

2.2 Solver (Fluent)

In this study, the solver used was Fluent. This stage involves the inputs required to simulate the flow distribution. The viscous model and boundary conditions were applied here.

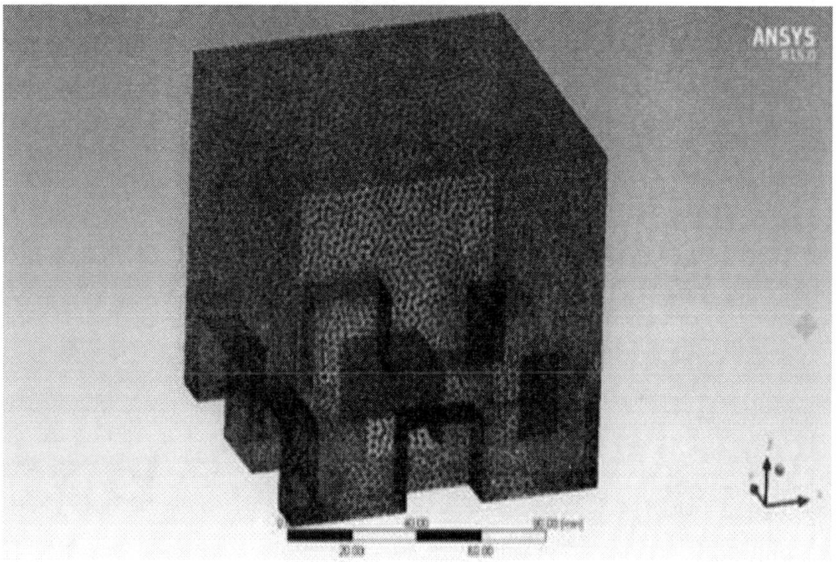

Figure 2. Unstructured tetrahedral meshing on case 1.

2.2.1 Viscous Model

The Reynold's Averaged Navier-Stokes (RANS) equation was used to solve the exact unknown solution because it was already sufficient for general engineering application (quick and reasonable) compared to Large Eddy Simulation (LES) which requires more time, memory and computing power. The turbulence closure model of RNG k-ε was then selected as it is suitable for complex flow of moderate swirl, vortices and locally transitional flow.

2.2.2 Boundary Conditions

For boundary conditions, the flow was maintained by streamwise pressure gradient of -0.874. At the top, free-slip boundary condition was applied. Meanwhile, at the wall of the buildings and floor, no-slip boundary conditions were applied. Interface meshing was then used to match the non-conformal walls and to impose the transla-tional periodic on streamwise and lateral faces.

Details of "Mesh"

Solver Preference	Fluent
Relevance	0
Sizing	
Use Advanced Size Fun...	On: Proximity and Curvature
Relevance Center	Fine
Initial Size Seed	Active Assembly
Smoothing	High
Transition	Slow
Span Angle Center	Fine
Curvature Normal A...	Default (18.0 °)
Num Cells Across Gap	10
Min Size	Default (2.5285e-002 mm)
Proximity Min Size	Default (2.5285e-002 mm)
Max Face Size	Default (2.52850 mm)
Max Size	Default (5.05710 mm)
Growth Rate	Default (1.20)
Minimum Edge Length	12.50 mm

Figure 3. Sizing of the meshing.

Statistics	
Nodes	77486
Elements	415789
Mesh Metric	Orthogonal Quality
Min	0.262798007573281
Max	0.997120586995978
Average	0.860174301578419
Standard Deviation	8.36466254913552E-02

Figure 4. Minimum orthogonal quality achieved is 0.26 (Good).

Statistics	
Nodes	77486
Elements	415789
Mesh Metric	Skewness
Min	1.36233643743044E-04
Max	0.799958294540251
Average	0.225555704090484
Standard Deviation	0.118402549396944

Figure 5. Maximum skewness achieved is 0.79 (Good).

2.2.3 Calculation Cycle
To acquire accurate result, the iterations done must be converged. For this simulation, the number of iterations was set to 20,000. This was to monitor and ensure the stability of the simulation.

2.3 Post-Processing
All the data were obtained after the iterations were done. The results of streamwise velocity and height (vertical direction) were obtained and exported to Excel Spreadsheet to get the velocity profile from the top of the building until top of the domain. The wind velocity streamline plot was also obtained to observe the wind flow pattern at the pedestrian height of 0.05H.

3. Result and Discussion

3.1 Results Case 1 (Validation)
To ensure the reliability and accuracy of the simulation, a validation was done on Case 1. This case was done based on the direct numer-ical simulation of turbulent flow over regular arrays of urban-like, cubicle obstacles [15]. The validation of velocity profile was com-pared as per Figure 6.

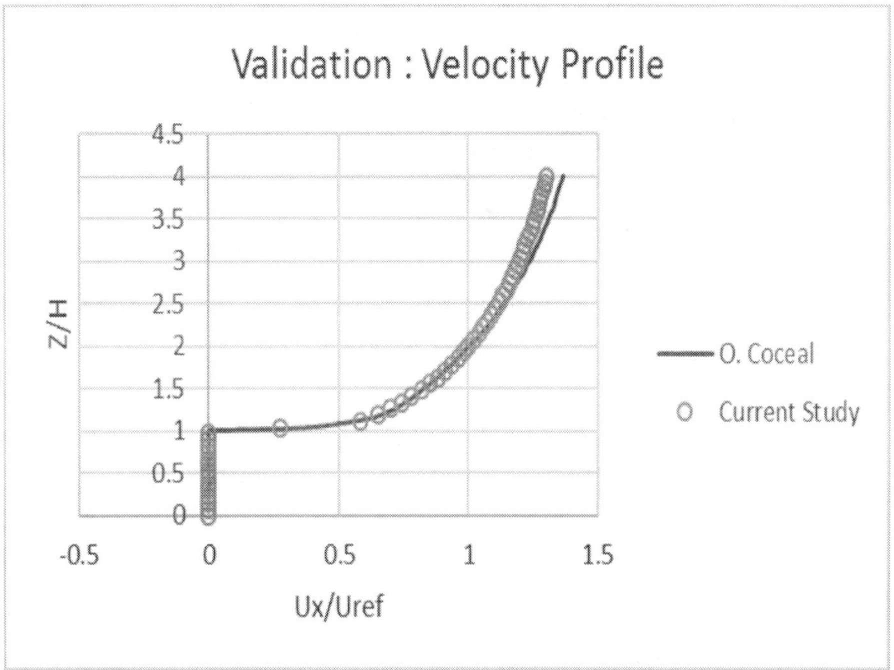

Figure 6. Validation of profile on mean streamwise velocity, Ux normalised by reference velocity, Uref.

The velocity profile is taken from the top of the building until to the top of the domain. It shows that the velocity profile is in good agreement with the current study. Hence, the simulation was con-sidered reliable.

3.2 Results Case Study (Case 2, Case 3 & Case 4)

Urban obstacles such as buildings can exert a relatively large drag force on the atmosphere. In the urban canopy sublayer, the flow at a specific point can be directly affected by local obstacles and in the level of roughness sublayer, the flow is still adjusting itself because of the previous obstacles.

In the case of non-uniform of heights, as the flow is closer to the buildings, individual velocity profiles deviate as they directly re-spond to the real surface due to the spatial irregularity.

For shear stress, at position of uniform height as shown in Figure 7, the maximum shear stress occurs approximately at the height of the obstacles. Lower than this height, the shear stress carried by the wind decreases as the buildings take up part of stress through the drag forces on them. Meanwhile for non-uniform height as shown in Figure 9 and Figure 11, the maximum shear stress occurs at the approximate-ly at the height of the highest building and de-creases to zero shear stress like uniform-height buildings.

It shows that in Figures 8, 10 and 12, at windward side of the building, when the ambient wind strikes the bluff body of the build-ing, a stagnation point is formed in front of the building with a sep-aration lines creating divergence alongside strong uplift and conver-gence at the leeward side.

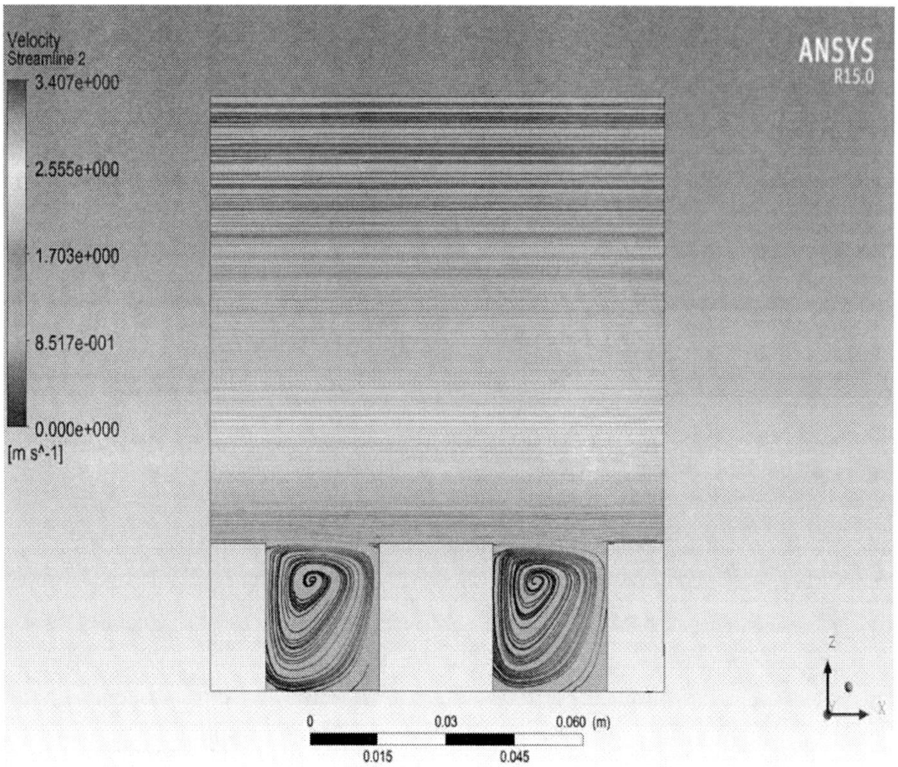

Figure 7. Velocity streamline of case 2 from side view (y-axis).

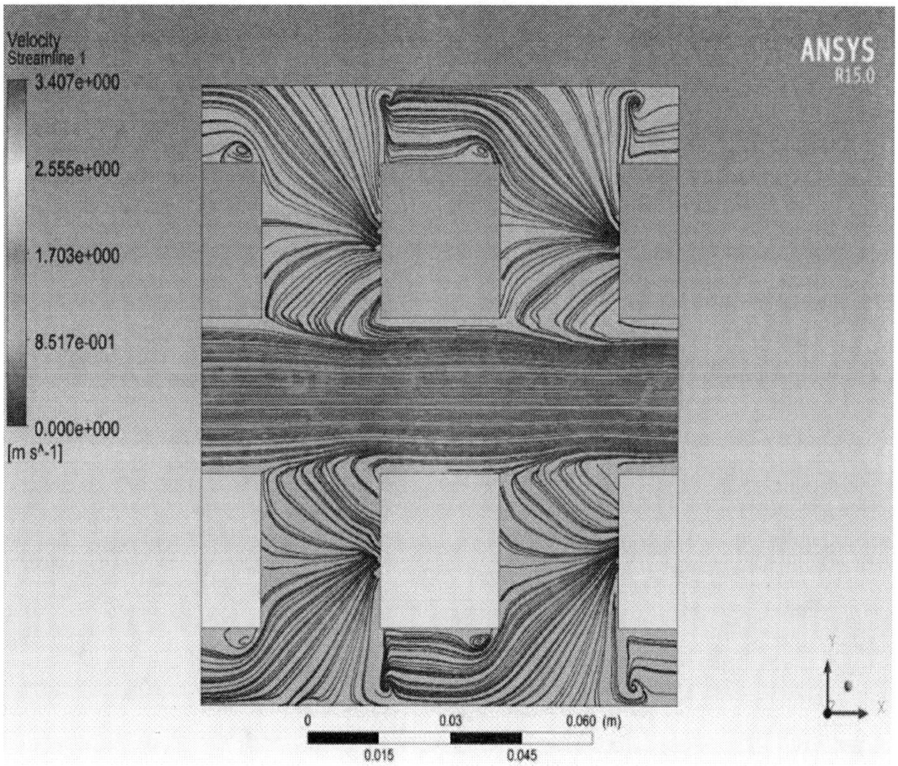

Figure 8. Velocity streamline of case 2 from top view (z-axis).

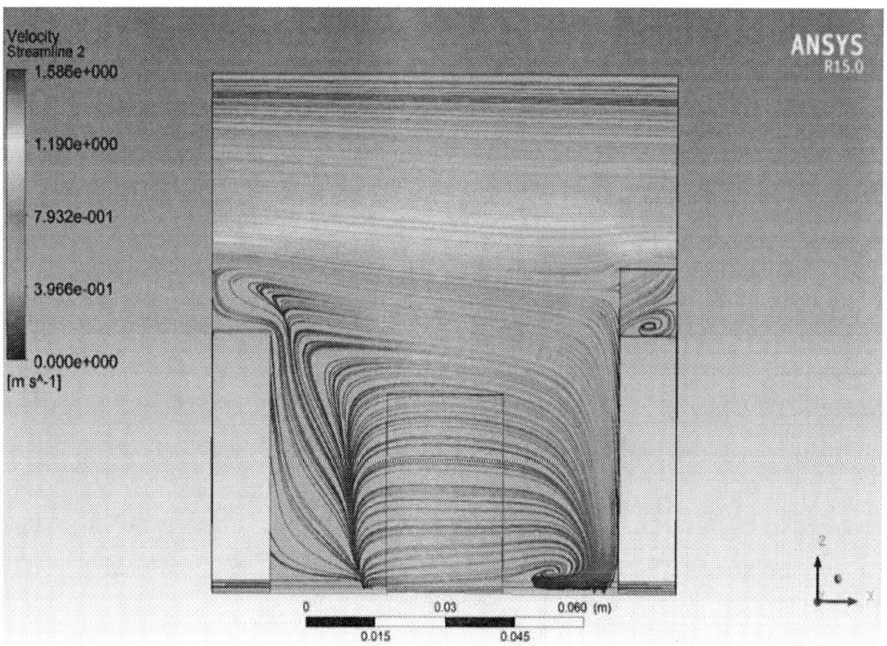

Figure 9. Velocity streamline of case 3 from side view (y-axis).

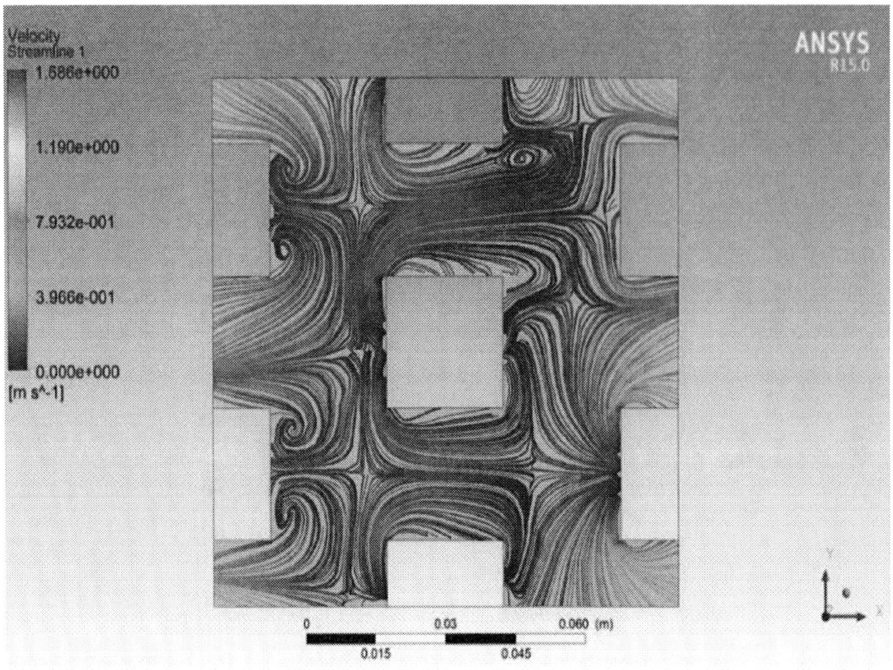

Figure 10. Velocity streamline of case 3 from top view (z-axis).

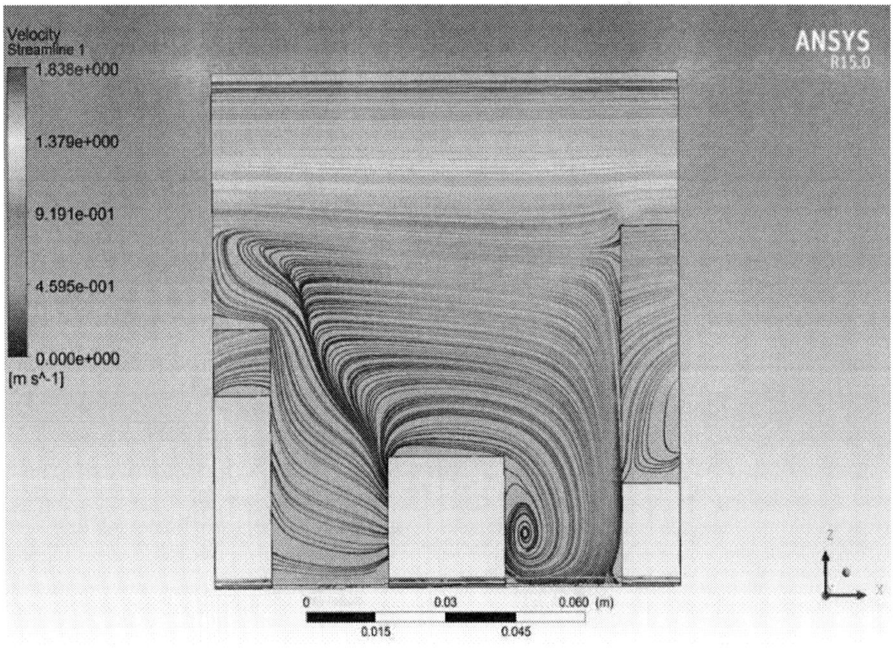

Figure 11. Velocity streamline of case 4 from side view (y-axis).

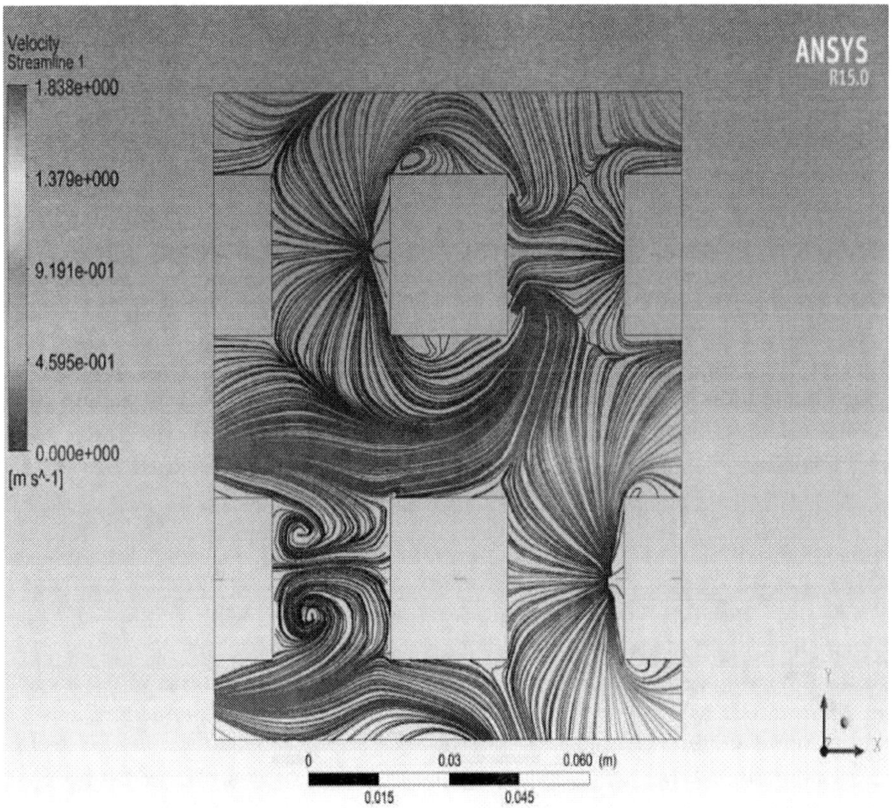

Figure 12. Velocity streamline of case 4 from top view (z-axis).

The horseshoe vortex is then developed down the bluff body for all cases. It is formed when producing a wake region which is charac-terized by periodic vortex formation with a circular cavity. In these areas, the pressure is lower than the stagnation point pressure. This cavity does not mix efficiently with other air and tends to circulate within the area. The effect of wake weakens as its distance from the building increases.

4. Conclusion & Recommendation

A compact city with geometrical inhomogeneity such as irregular building arrangements and heights means higher building density, higher population density, higher anthropogenic heat and reducing the wind flow. Despite this kind of buildings can be aesthetically pleasing, the reduced wind flow can lead to urban warming (2-3°C higher), poor air quality with pollutants accumulation, poor visibility due to dust, smokes and fogs trapped in urban canyon and thus reduced in productivity. In contrast to the rural area, wind cannot penetrate into a high-rise compact city due to large canopy drag [16]. The buoyancy driven flows along building walls (wall slope flows) will become dominant due to building heights and large wall areas.

To enhance the natural ventilation performance, a wind catcher or wind tower can be used to direct the atmospheric wind into urban street canyon [17]. Besides, through-building gap can be proposed to provide better ventilation thus making pedestrian to feel better and more comfortable.

Acknowledgement
The authors acknowledge technical and financial support from UiTM Shah Alam to publish this paper. Besides that, special thanks go to authors group members of Paper to Publish (P2P) from UiTM Pasir Gudang because successfully make the journal paper become reality

References
[1] K. M. Al-Obaidi, M. Ismail, and A. M. Abdul Rahman, "A study of the impact of environmental loads that penetrate a passive skylight roofing system in Malaysian buildings," Frontiers of Architectural Research, vol. 3, no. 2. pp. 178–191, 2014.
[2] H. Montazeri, B. Blocken, W. D. Janssen, and T. van Hooff, "CFD evaluation of new second skin facade concept for wind comfort on building balconies: Case study for the Park Tower in Antwerp," Build. Environ., vol. 68, pp. 179–192, 2013.
[3] J. Li and M. Donn, "The influence of building height variability on natural ventilation and neighbor buildings in dense urban areas," in Proc. Building Simulation 2017, 2017, pp. 2805–2813.
[4] M. Carpentieri, A. G. Robins, and S. Baldi, "Three-dimensional mapping of air flow at an urban canyon intersection," Boundary-Layer Meteorol., vol. 133, no. 2, pp. 277–296, 2009.
[5] M. S. Fadl and J. Karadelis, "CFD Simulation for Wind Comfort and Safety in Urban Area: A Case Study of Coventry University Central Campus," Int. J. Archit. Eng. Constr., vol. 2, no. 2, pp. 131–143, 2013.
[6] S. L. Ceccio, "Friction Drag Reduction of External Flows with Bubble and Gas Injection," Annu. Rev. Fluid Mech., vol. 42, no. 1, pp. 183–203, Jan. 2010.
[7] P. Sachs, "Wind Forces in Engineering," Wind Forces Eng., pp. 306–353, 1978.
[8] J. F. Manwell, J. McGowan, and A. Rogers, Wind Energy Explained: Theory, Design, and Application. 2009.
[9] M. A. Hart and D. J. Sailor, "Quantifying the influence of land-use and surface characteristics on spatial variability in the urban heat island," Theor. Appl. Climatol., vol. 95, no. 3–4, pp. 397–406, Mar. 2009.
[10] I. D. Stewart and T. R. Oke, "Local Climate Zones for Urban Temperature Studies," Bull. Am. Meteorol. Soc., vol. 93, no. 12, pp. 1879–1900, Dec. 2012.
[11] H. Liu, P. Li, and J. Van Lew, "CFD study on flow distribution uniformity in fuel distributors having multiple structural bifurcations of flow channels," Int. J. Hydrogen Energy, vol. 35, no. 17, pp. 9186–9198, 2010.
[12] M. Goutham, M. Shanmugam, M. Mihai, K. Maninder, K. Sid, and G. Ephraim, "Validation of computational fluid dynamics methodology used for human upper airway flow simulations," J. Biomech., vol. 42, no. 10, pp. 1553–1559, 2009.
[13] Y. Tominaga, T. S.-J. of W. E. and Industrial, and U. 2011, "Tominaga Y, Stathopoulos T. CFD modeling of pollution dispersion in a street canyon: comparison between LES and RANS. J Wind Eng Ind Aerodyn 2011; 99:340e8.," Elsevier, no. 99, pp. 340–348, 2011.
[14] M. L. Shur, P. R. Spalart, M. K. Strelets, and A. K. Travin, "A hybrid RANS-LES approach with delayed-DES and wall-modelled LES capabilities," Int. J. Heat Fluid Flow, vol. 29, no. 6, pp. 1638–1649, 2008.
[15] C. Yan, W.-X. Huang, S.-G. Miao, G.-X. Cui, and Z.-S. Zhang, "Large-Eddy Simulation of Flow Over a Vegetation-Like Canopy Modelled as Arrays of Bluff-Body Elements," Boundary-Layer Meteorol., vol. 165, no. 2, pp. 233–249, Nov. 2017.
[16] J. Barlow, "Urban climate," 2014.
[17] K. J. K. Ishugah T F, Li Y, Wang R Z, "Advances in Wind Energy Resource Exploitation In Urban Environment," Elsevier, vol. 47, pp. 1–5, 2017.

Simulation of thermal spray process of a three phase flow calculation with substrate temperature of 300K

Noh, M H M[1,2], Hamid, A H A[1], Sawada, N[2] and Mori, K[2]

[1]Faculty of Mechanical Engineering Universiti Teknologi MARA, 40450, Malaysia.
[2]Department of Aerospace Engineering, Nagoya University, Japan.

mhafiz_noh@yahoo.com

Abstract. We propose a particle method (MPS method) which can analyse the large deformation, splitting and coalescence of liquid as a new calculation method and a gas-liquid coupling method of finite volume method which can stably solve the compressible flow. By doing this, we aim to solve the interface resolution problem in the interface capture method and to establish a complicated gas-liquid multiphase flow analysis method that can deal with large deformation case. At the temperature of 300K, the validity of the flow field is consistent with the reference. The processes from the acceleration and deformation of a droplet to the solidification of the droplet on a substrate are successfully being observed. The shape of the splat is influence by substrate temperature.

1. Introduction

Thermal spray is one of the surface treatments that already established many years ago. It is a method that has been used for prevents corrosion, by a film that formed on the base material by using burning flame or electricity that spray the heated molten particles to the material. In the thermal coating process which the molten metal is sprayed, the formation rate of the surface film can be made extremely higher than other surface treatment methods such as plating and vapor deposition. Since the base material has low temperature, the thermal strain is relatively small, and any material can be used as base material such as ceramics, organic material, wood, clothes, and papers, not limited to metals only.

This have been influenced the substances that can be used for thermal spraying. As a result, the thermal spraying method has applied to high performance surface treatment technology such as corrosion prevention and rust prevention, rather than heat resistance, thermal insulation, wear resistance and corrosion resistance. Furthermore, it is now being used in devices that used in harsh environments such as thermal power boilers, aircraft jet engines, and marine diesel engines. Instead of having outstanding characteristics, there is a problem that related to low adhesion efficiency and required the roughening process for the substrate.

Droplets in thermal spray may experience three processes during surface treatment, which are acceleration, solidification and lamination process. Phenomena that can occur with these elementary processes are thought to be different. In the collision flattening solidification process, large deformation is notified mainly due to collision, splitting phenomenon and cooling and also solidification. In the lamination process, stacked droplets coalescence and mechanical interlocking can be observed. In the thermal spraying process, it is difficult to explain the phenomenon by combining these phenomena continuously. Once this droplets contact with the base material (substrate), we called

Content from this work may be used under the terms of the Creative Commons Attribution 3.0 licence. Any further distribution of this work must maintain attribution to the author(s) and the title of the work, journal citation and DOI.
Published under licence by IOP Publishing Ltd

it splat. Since this elementary process of this splat is a μmeter size and very rapid condition (like splash), so it is difficult to observe all this process via experiment, which has been proven by [1][2].

In multiphase flow simulation, an approach called Volume of Fluid (VOF) method is widely used. But this method has difficulties to capture in detail what happened in the interface. This interface is depends on the mesh, and if we want to analysis of droplet particles collide, flying and deformed at the substrate, it is necessary to use the entire domain between the nozzle and the substrate. Consequently, this will take longer computational time and demand very fine domain grid.

The higher speed droplets contact with the substrate to form a coating and flattened due to temperature difference. This flatness was relative to $Re^{0.2\text{-}0.3}$[3]. CFD analysis on sprayed particles and substrate has been successfully conducted by [1][2]. In this paper, we imposed the Moving Particle Semi Implicit (MPS) coupling together with Finite Volume Method (FVM) to answer the interface analysis problem.

2. Methodology

In this investigation, the Reynold-averaged three-dimensional unsteady Navier–Stokes equation is solved with 3rd order of MUSCL [4]. The air is assumed to be an ideal gas with SLAU is used for non-viscous flux and Sutherland relation is used for viscosity equation calculation. Van Albada limiter is amended and an explicit 3rd order Total Variation Diminishing Runge-Kutta method is used for time integration [5][6]. A structured grid with total grid points of 7,820,000 is being used in this calculation.

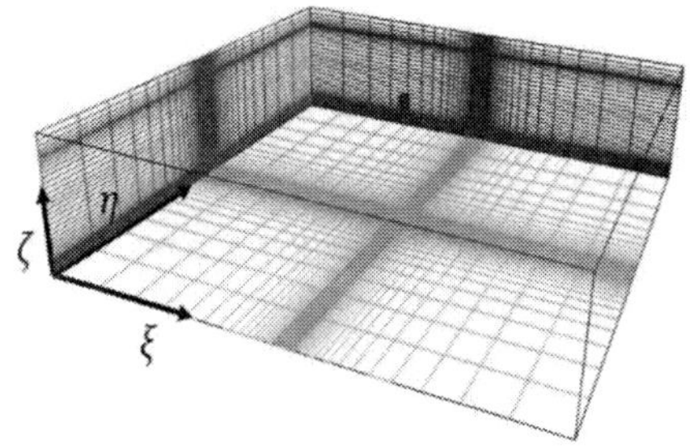

Figure 1. Structured grid ($7.82 \cdot 10^6$) for this study.

Figure 2 shows the droplet calculation conditions, which consists of 7800 particles. In this study, we simulate the conditions of flame spraying, which is the most common thermal spraying method among the actual thermal spraying methods.

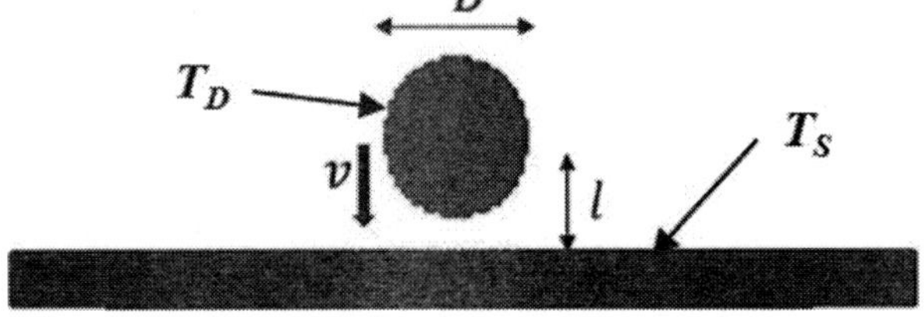

Figure 2. Initial placements of sprayed droplets.

In flame spraying, the velocity at which the spray droplets collide with the substrate changes depending on the distance between the nozzle and the substrate. Therefore, in the calculation of only the MPS method, the droplet velocity is given with reference to the literature [7]. Specifications of the thermal spray particles and the substrate are summarized in Table 1.

Table 1. Specifications of the thermal spray particles.

Droplet Substance	Copper
Substrate Material	SUS304
Distance between initial MPS particle	2.0 μm
Initial Droplet Diameter, *D*	50.0 μm
Initial Droplet Velocity, v	60 m/s
Initial Droplet Condition (height), *l*	1.0 mm
Initial Droplet Temperature, T_D	1700K
Substrate Temperature, T_S	300K
Droplet Solidification Temperature	1357K
Exit Nozzle Speed	100m/s
Exit Nozzle Temperature	273 K
Exit Nozzle Static Pressure	1.0 atm

For interface, the particle on the free domain must satisfy with equation (1)

$$n^* < \beta n^0 \tag{1}$$

Here, n^0 is the initial particle density, and n^* is the particle number density that is calculated after the explicit integration of the velocity field, while $\beta = 0.95$. In addition, Gotoh ASA conditions are imposed as an auxiliary judgment [8]. For solidification modelling, it has been control by solidification fraction, γ, which has been determined by enthalpy function:

$$\gamma = \begin{cases} 1 & (h < h_{s0}) \\ \dfrac{h_{s1} - h}{h_{s1} - h_{s0}} & (h_{s0} \le h \le h_{s1}) \\ 0 & (h > h_{s1}) \end{cases} \tag{2}$$

h_{s0} is initial enthalpy while h_{s1} is resultant enthalpy of the melting process. Furthermore, the droplet is in liquid form for $\gamma < 0.5$, and in solid form for $\gamma \geq 0.5$. After a particle is solidified on the substrate, then the particles are considered to be stick at substrate.

3. Results and Discussion

3.1 Validation on physical aspect

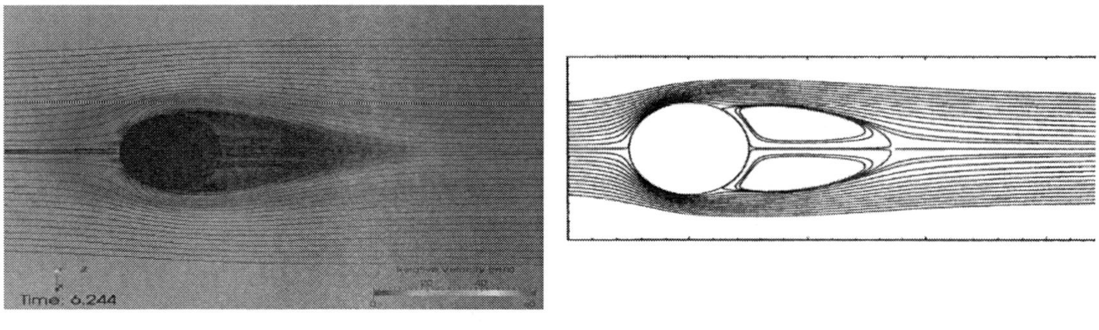

(a) (b)

Figure 3. (a) Streamline of the relative velocity of droplets: Re = 133. (b) Streamline around the sphere [13]: Re = 150

IOP Conf. Series: Materials Science and Engineering **834** (2020) 012031 doi:10.1088/1757-899X/834/1/012031

Figure 3 shows streamlines around a sphere at Re = 150 for which Johnson et al. [9] performed experiments and analysis. Comparing these, the separation position of the flow field obtained in this study is slightly forward, but the symmetrical vortices formed in the wake and the position where the separated streamline reappears are almost the same. . Therefore, the flow field formed by the droplets obtained in this study was able to analyze the real phenomenon and confirmed its validity.

3.2 Splat shape change by substrate temperature

We aim to deal with the higher interface domain requirement for bigger scale of deformation by coupling between gas-liquid multiphase flow analyses. Figure 4 shows the speed in term of time history from initial distance *l* until collide at the substrate.

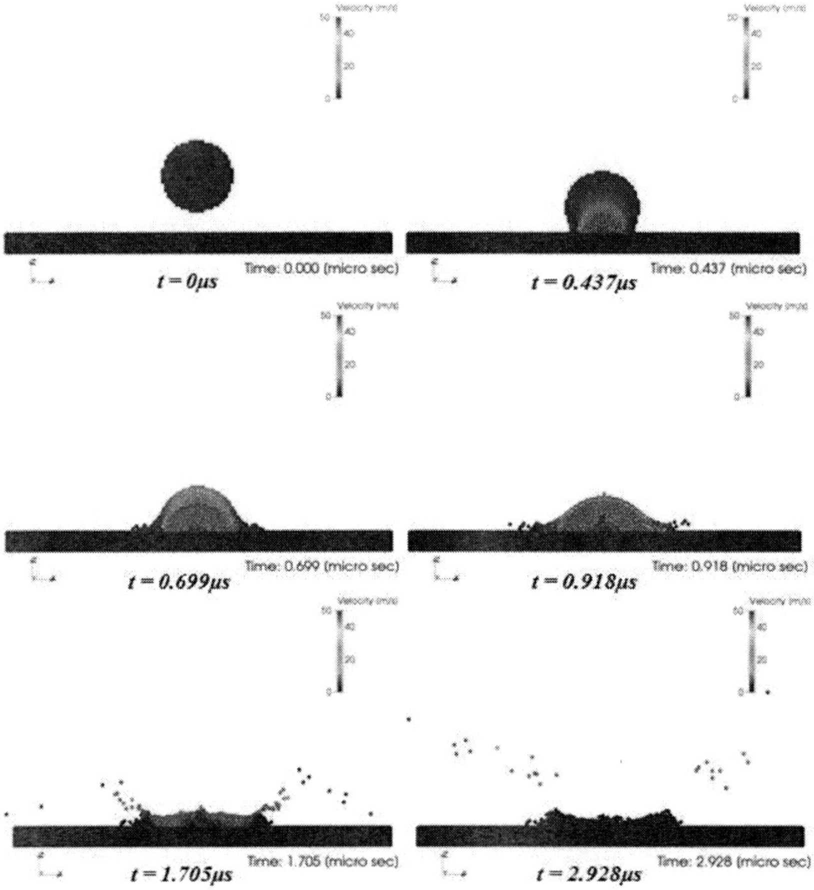

Figure 4. Speed history (300K).

As explained before, in order to observe the whole process from the initial droplet condition until contact with the substrate (acceleration, solidification and lamination process), it is necessary to calculate all surface between the nozzle and the substrate. As we can see in Figure 4, droplet started at distance *l* has initially first contact with the substrate approximately at $t = 0.4$μsecond, where the speed has been slightly change. The whole process can be simulating less than 3μsecond. Furthermore, the speed continuously to reduce and the droplet started to collide with the substrate. When moderate speed has been dominance for the entire droplets, the process of "flying" particles has been initially being occurred. Nearly at $t = 3$μsecond, most of the droplet has obtain the lamination process while the "flying" particle are maintained at the same speed. The two-dimensional cross-section is used since it will be easier to understand by visualization.

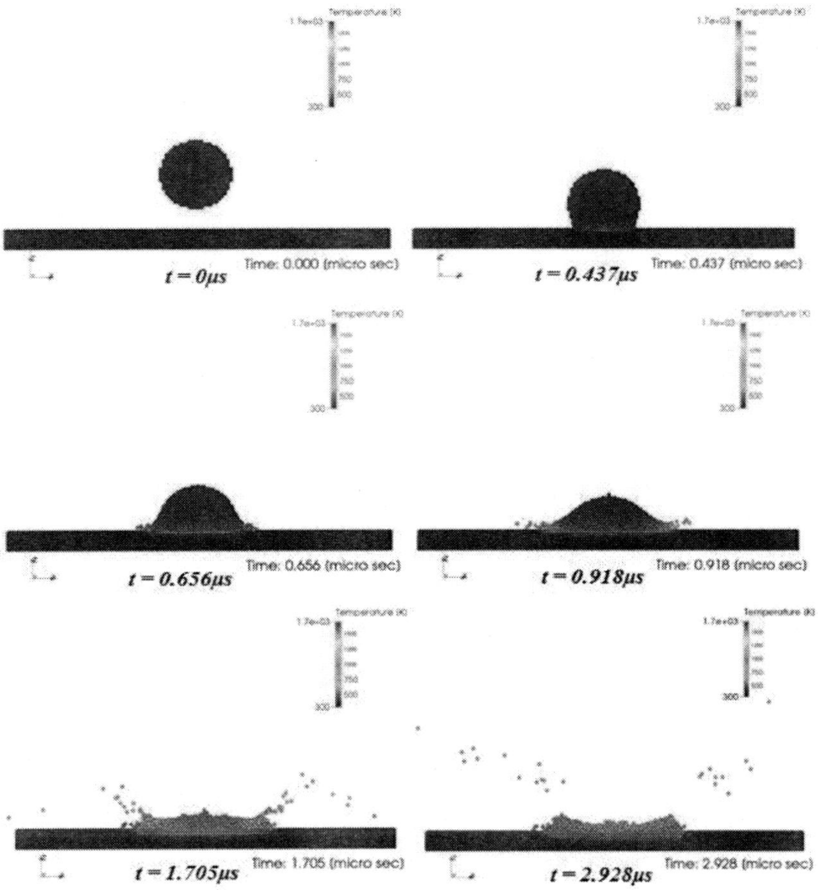

Figure 5. Temperature history (300K).

The movement of the droplets caused the solid droplets to be strongly cooled from the outer edge of the droplets in the process of collision and cooling of the high temperature droplets on the substrate. At the beginning, when the lower part of the droplet and the substrate come into contact, heat conduction occurs and the temperature is lowered by cooling, and conversely, the surface temperature of the substrate is increased (Figure 5). While the outer edge of the substrate is always cooled by the exchange of heat due to heat conduction with the low surface temperature substrate, the surface temperature of the substrate is high and the temperature gradient is large in the stationary droplet center. The temperature of the liquid is kept high because it cannot be obtained. Thus, a distribution was formed where the temperature increased in the direction from the outer edge to the center.

4. Conclusion

In this study, multi-phase flow analysis was performed by combining the finite volume method using a grid and the particle method using particles in the process of droplet flight and collision in the thermal spraying method. The conclusion can be are summarized as below.

1) Droplet particles are strongly cooled and solidified from the outer edge, causing deposition to become a factor in shape determination
2) It was confirmed by analysis that the substrate temperature is an important factor of the flat shape
3) Validity of the flow field by droplets obtained by the proposed coupled method was confirmed

References

[1] Toyonobu Yoshida, Tomohisa Okada, Hideki Hamatani, and Hisashi Kumaoka, Integrated fabrication process for solid oxide fuel cells using novel plasma spray- ing. Plasma Sources

Science and Technology, Vol. 1, pp. 195, 1992.

[2] K.Shinoda, T.Koseki and T.Yoshida, "Influence of Impact Parameters of Zirconia Droplets on Splat Formation and Morphology in Plasma Spraying", Journal of Applied Physics, Vol. 100-074903, 2006.

[3] Kentaro Shinoda, Toshihiko Koseki, and Toyonobu Yoshida. Influence of impact parameters of zirconia droplets on splat formation and morphology in plasma spraying. Journal of Applied Physics, Vol. 100, No. 2006, 2006.

[4] E.F.Toro, "Riemann Solvers and Numerical Methods for Fluid Dynamics", Springer-Verlag, Berlin, 2009.

[5] E.Shima and K.Kitamura, "Parameter-Free Simple Low-Dissipation AUSM-Family Scheme for All Speeds", AIAA Journal, Vol. 49, No. 8, pp. 1693-1709, 2011.

[6] C.W.Shu and S.Osher, "Efficient Implementation of Essentially Non-Oscillatory Shock Capturing Schemes", Journal of Computational Physics, Vol. 77, No.2, pp.439-471, 1988.

[7] Bram van Leer. Towards the ultimate conservative difference scheme. Journal of Computational Physics, Vol. 135, No. 2, pp. 229–248, 1997.

[8] Abbas Khayyer and Hitoshi Gotoh. A 3D higher order Laplacian model for enhancement and stabilization of pressure calculation in 3D MPS-based simu- lations. Applied Ocean Research, Vol. 37, No. 1, pp. 120–126, 2012.

[9] Johnson T. A., and Patel V. C., Flow Past a Sphere up to a Reynolds Number of 300, Journal of Fluid Mechanics, Vol. 378, pp. 19-70, 1999.

IOP Conf. Series: Materials Science and Engineering **834** (2020) 012032 doi:10.1088/1757-899X/834/1/012032

Evaluation of TiO_2 nanoparticles as viscosity modifier in palm oil bio-lubricant

N R Nik Roselina[1,2]*, Nor Syahirah Mohamad[2], Kasolang. S.[1,2],

[1]Centre of Tribology, Faculty of Mechanical Engineering, Universiti Teknologi MARA, 40450 Shah Alam, Selangor, Malaysia
[2]Faculty of Mechanical Engineering, Universiti Teknologi MARA, 40450 Shah Alam, Selangor, Malaysia

*roselina_roseley@salam.uitm.edu.my

Abstract. TiO_2 nanoparticles is one of the most studied nanomaterials as an additive in oil lubricant due to its outstanding chemical and physical properties as well as low toxicity. This study aims to investigate the effect of TiO_2 nanoparticles, as additive in renewable palm oil lubricant on the change of viscosity. Viscosity is one of the most vital physical parameters for assessing the performance of lubricant oil. The method for preparation of all samples involved adding of TiO_2 nanoparticles at concentrations of 0, 0.5 and 1.0 weight percentage in lubricants, followed by mixing using ultrasonication method for 30 minutes. Investigation on the viscosity showed that the viscosity of palm oil bio-lubricants is comparable to SAE 0W20 grade lubricant. The viscosity index tends to increase with the concentration of TiO_2 additive.

1. Introduction

In it commonly known that the main sources of wear and friction are owing to energy losses during high friction and failure in machinery engines, bearings and gears [1-3]. Friction is the main cause of energy loss in engine which covers approximately 15% from the total energy losses in engine [4]. Th value relies on the energy used to overcome the brake interaction, tyres rolling resistance, transmission systems and the rubbing of parts in the engine.

While many other countries used vegetable oil such as corn, canola and rapeseed as bio-lubricant oils, Malaysia also is seen to have started the work on using palm oil as bio-lubricant oil to promote our own homegrown resource. Palm oil is a sustainable option compared to the conventional petroleum-based lubricants. Studies carried out by Masjuki et al. revealed that chemically modified palm oil (CMPO) exhibits better performance compared to mineral oil in reducing friction with the addition of nanoadditives [5]. On the other hand, S.Syahrullail et al. reported that palm oil has a comparable tribological performance to paraffinic mineral oil to reduce the extrusion load [6].

Several types of nanoparticles have been used as additive to improve lubrication properties of bio-lubricant such as metal, metal oxide, carbon, carbonate, borate and organic materials [7-16]. The additives have been reported to act as an anti-friction and wear reducer. The performances of nanoparticles such as TiO2, ZnO, CuO and SiO2 as additives also have been investigated in numbers of works [8, 10, 17-23]. It was found that friction and wear reduction behaviour strongly relies on the characteristics of the additives such as size, morphology, hardness and concentration [8, 24, 25]. The size of particles used as additive is usually in the range of 2 to 120 nm [8, 12, 26-28]. Previous works also exposed that the smaller size of nanoparticles are more likely to interrelate with the surfaces forming thin film on the lubricating surface and behave as a protective film to reduce friction [28].

Content from this work may be used under the terms of the Creative Commons Attribution 3.0 licence. Any further distribution of this work must maintain attribution to the author(s) and the title of the work, journal citation and DOI.
Published under licence by IOP Publishing Ltd

However, one of the most challenging issues when using nanoparticles an additive, is the existence of attractive force between particles that promotes aggregation due to its high surface area [20]. This phenomenon limits the tribological performance of the nanoparticles as additives.

TiO2 nanoparticles are one of the well-known nanoadditives for synthetic lubricant. It was used as an additive in due to its low toxicity, anti-oxidant features and non-volatility behaviours [29-31]. Numberous of works have also reported the excellent performance of TiO_2 as an additive in lubricant oil. Nevertheless, most of the works were carried out with conventional synthetic oil. This work will provide initial information on the variation of viscosity with the addition of TiO_2 nanoparticles as additive in palm oil bio-lubricant. This paper will discuss how the viscosity of palm oil biolubricant changes with the addition of TiO2 nanoparticles as additive. It is well recognized that viscosity is one the the most critical physicochemical properties as a lubricant oil.

2. Experimental

2.1. Materials

In this study, palm oil and synthetic oil (SAE 10W-40) were selected as the base oil. Synthetic oil is used as comparison to investigate the performance of palm oil lubricant. The effect of adding nanoparticles additives at different concentration in palm oil and synthetic oil were investigated. The physicochemical properties of TiO_2 nanoparticles are shown in Table 1. Titanium oxide (TiO2) nanoparticles with average size of 21 nm with purity level of 99.5% were purchased from Sigma Aldrich.

Table 1. Physicochemical property of TiO2 nanoparticles.

Parameter	Value
Colour	White
Purity	99.5%
SSA	289 m2/s
Diameter	21 nm
Morphology	Nearly spherical

2.2. Preparation of lubricant samples

TiO_2 nanoparticles powder were added into the lubricant oils at different concentrations, range between zero to one weight percentage (wt%). The required quantity of nanoparticles powder is accurately weighted using a precision balance and mixed directly with the base oil. Surfactant ethylene glycol (EG) was then added into the lubricant. EG was used to disperse the nanoparticles in the base oil to avoid agglomeration and precipitation of the TiO_2 nanoparticles. The nanoparticles additive was mixed in the lubricating oil using ultrasonication method. The time of agitation was fixed at 30 minutes. Table 2 summarises the lubricant blends with its weight ratio percentage.

Table 2. Blend of lubricant samples prepared in this work.

Blend	**Type of base oil**	**TiO2 (wt%)**
SO	Synthetic oil	0.0
SO1	Synthetic oil	0.5
SO2	Synthetic oil	1.0
PO	Palm oil	0.0
PO1	Palm oil	0.5
PO2	Palm oil	1.0

2.3. Characterisation of the lubricants

The viscosity of all lubricant samples at 40°C and 100°C was measured using Kinematic Viscometer model Canon CT-500. Kinematic viscosity is expressed in centistokes (cSt) by multiplying the efflux time and viscometer constant. It is mathematically calculated using Eq. 1.

$$v = Ct \quad (1)$$

where
v = kinematic viscosity, mm^2/s
C = viscometer constant, mm^2/s
t = flow time, s

3. Results and Discussion

The viscosity of the samples was measured at temperature of 40°C and 100 °C for three concentrations of TiO_2 at 0, 0.5 and 1.0 wt%. The kinematic viscosity (VI) of the lubricant samples was then analysed according to the ASTM D445 and ASTM D2272 standard. Table 3 shown the recorded kinematic viscosity and their viscosity index (VI). The properties of some of the SAE grade lubricants are listed in Table 4 as a comparison [32]. The kinematic values are illustrated in Figure 1. Kinematic viscosity of all bio-lubricant samples (PO, PO1 and PO2) conforms to viscosity class of SAE 0W20 lubricant which indicate that the synthetic oil SAE 0W20 is a low-viscosity type of lubricant from group IV.

Table 3. Viscosity and Viscosity Index of lubricant samples.

Type of Lubricant	Viscosity at 100 °C (cSt)	Viscosity at 40 °C (cSt)	Viscosity Index
SO	19.11	112.11	192.31
SO1	19.33	112.93	193.52
SO2	20.44	113.20	206.04
PO	10.32	49.32	204.30
PO1	10.40	49.72	204.43
PO2	10.74	50.05	212.31

Table 4. Properties of typical SAE lubricant [32]

SAE grade	Viscosity (cSt)	
	40°C	100°C
0W20	45	9
5W30	65	13
10W40	100	18
40	160	18
50	200	19

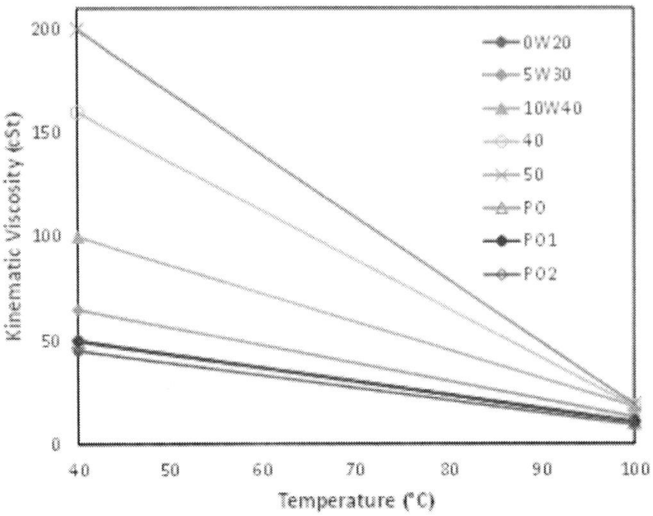

Figure 1. Variation of kinematic viscosity with temperature of SAE grade lubricants and bio-lubricant samples.

To further investigate the changes of kinematic viscosity of lubricant samples, the recorded values in Table 2 are presented in Figure 2(a and b). As can be seen in the figures, the value of kinematic viscosity increased for both synthetic and palm oil as the concentration of TiO2 nanoparticles rise. The viscosity of palm oil samples is approximately 50% lesser than the synthetic oil samples. Low viscosity of palm oil can minimize viscous friction [33]. Both synthetic and palm oil samples have the same trend of viscosity change with TiO_2 concentration. The results indicate that TiO_2 nanoparticles acted similarly in synthetic and palm oil lubricants. Especially at high temperature (Figure 2(b)), major increased of viscosity value can be observed at TiO_2 concentration of 1.0 wt%. It can therefore say that at high temperature TiO_2 effectively act as viscosity modifier in bio-lubricant samples.

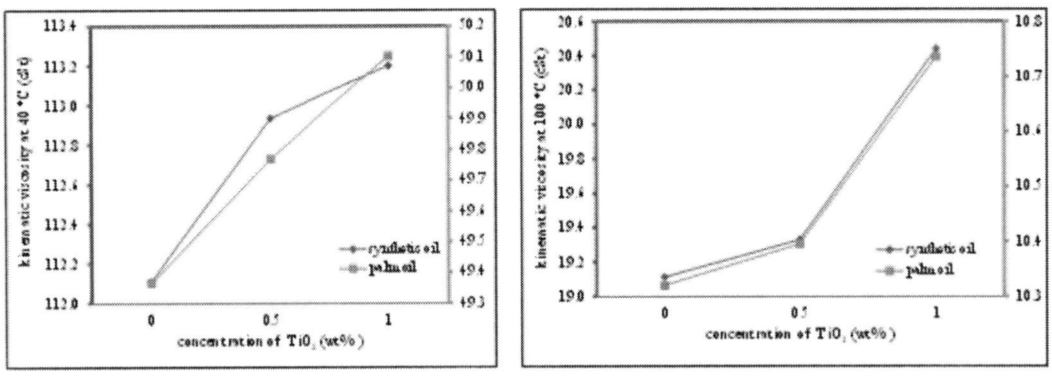

Figure 2. Kinematic viscosity of synthetic and palm oil at temperature of (a) 40°C and (b) 100°C respectively.

The similar finding has also been reported by many other works of TiO_2 in synthetic lubricant oil [34, 35]. However, the outcomes contradict with the earlier formulation developed by our group where the addition of TiO_2 reduces viscosity kinematic at 100°C [36]. Looking at the point of rheology, when TiO_2 nanoparticles were added to oil, it will act as obstacles that will disturb the oil flow thus increase the flow resistance like viscosity (Figure 3). In this work, ethylene glycol was added as a surfactant. It is widely known to enhance the dispersion of nanoparticles like TiO_2 by limiting aggregation and sedimentation of the particles. Addition of ethylene glycol may has produced ethylene glycol surface modified TiO_2 nanoparticles that are well dispersed in the lubricant oil samples as illustrates in Figure 4[37]. As the number of TiO_2 nanoparticles increased in the oil, there were more flow resistance in the oil and therefore increase the viscosity of the TiO_2 lubricant oil samples.

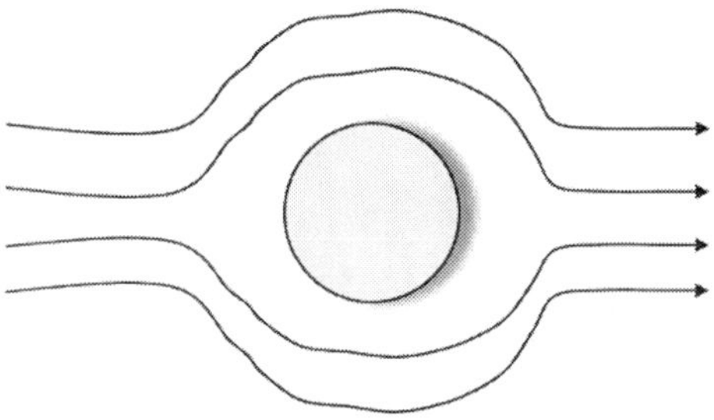

Figure 3. Schematic of lubricant oil flow with particle.

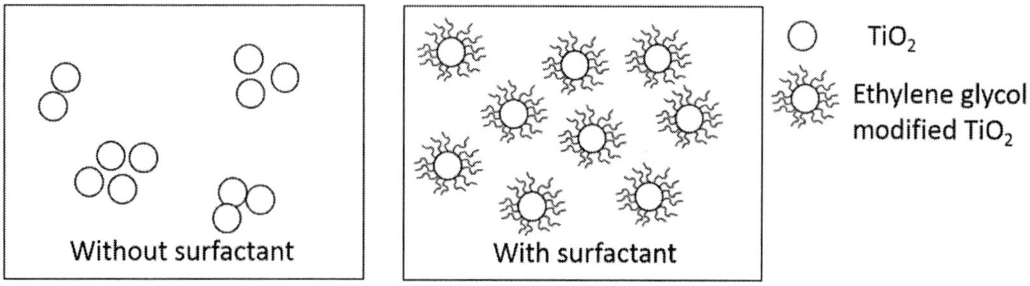

Figure 4. Schematic of TiO_2 dispersion in lubricant oil with and without surfactant ethylene glycol.

The viscosity index (VI) against the concentration of TiO2 is depicted in Figure 5. Higher viscosity index (VI) promotes a more stable viscosity across a range of temperatures (independent to temperature). An apparent change in the values of VI can be observed with the addition of 1.0% of TiO_2 by 4.1% and 7.3% for palm oil and synthetic lubricant samples, respectively. Lubricant samples with 0.5 wt% TiO_2 nanoparticles seem not to influence the VI for all lubricant samples. This confirms that the TiO_2 additive has successfully improved the stability of lubricant samples.

On the other hand, it can also be seen that the VI of palm oil is 5.9% higher as compared to the synthetic oil. It indicates that palm oil lubricant with 1.0% TiO2 will be able to provide good resistance to thinning of lubricant film for better fuel consumption in engine [33]. High VI of palm oil lubricants may attributed to the fact that the molecular weight of bio-lubricant is more stable than synthetic oil. This property makes the palm oil suitable for high-temperature applications, typically 250°C and above [38].

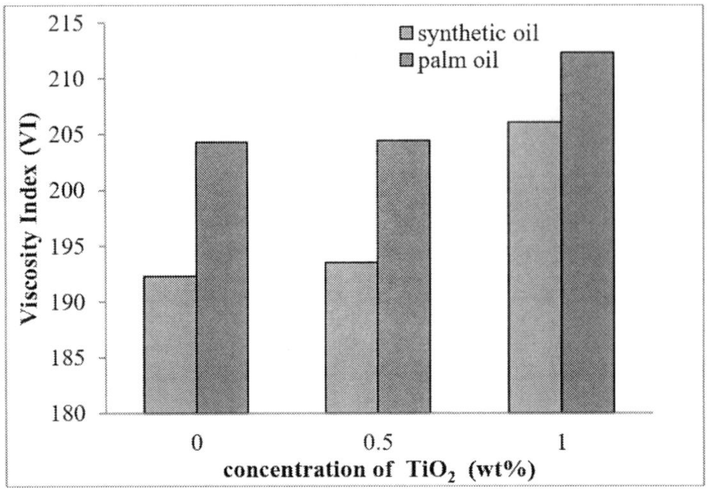

Figure 5. Viscosity index of synthetic oil and palm oil on difference TiO_2 nanoparticles concentration.

4. Conclusion

The current work investigates the viscosity change of palm oil bio-lubricant with the addition of TiO_2 nanoparticles as an additive. The experimental results show that palm oil bio-lubricant samples with and without TiO_2 additive exhibit comparable viscosity value to the SAE 0W20 grade at both temperature of 40°C and 60°C. Addition of TiO_2 as an additive in palm oil indicates improvement of the viscosity index of the bio-lubricant which may correlates to the well dispersion of TiO_2 with ethylene glycol. The viscosity index of palm oil bio-lubricant was increased by 4.1% using TiO_2 as an additive.

Acknowledgements

All experimental works were carried out in Centre of Tribology, Faculty of Mechanical Engineering, Universiti Teknologi MARA. The authors would like to thank the faculty for the financial support providing facilities for characterisation.

References

[1] Ferguson CR, *Internal Combustion Engine:applie thermosciences*:Wiley, 2001.
[2] S. C. Tung and M. L. McMillan, Automotive tribology overview of current advances and challenges for the future, *Tribol. Int.*, vol. 37, no. 7, pp. 517–536, 2004.
[3] Dmytro Demydov, Atanu Adhvaryu, Philip McCluskey, and Ajay P. Malshe, *Nanoscale Materials in Chemistry: Environmental Applications*, vol. 1045. Washington, DC: American Chemical Society, 2010.
[4] A. Vadiraj, G. Manivasagam, K. Kamani, and V. S. Sreenivasan, Effect of nano oil additive proportions on friction and wear performance of automotive materials, *Tribol. Ind.*, vol. 34, no. 1, pp. 3–10, 2012.
[5] N. W. M. Zulkifli, S. S. N. Azman, M. A. Kalam, H. H. Masjuki, R. Yunus, and M. Gulzar, Lubricity of bio-based lubricant derived from different chemically modified fatty acid methyl ester, *Tribol. Int.*, vol.93, pp.555–562, 2016.
[6] S. Syahrullail, B. M. Zubil, C. S. N. Azwadi, and M. J. M. Ridzuan, Experimental evaluation of palm oil as lubricant in cold forward extrusion process, *Int. J. Mech. Sci.*, *53*, 549–555, 2011.
[7] J. Zhou, Z. Wu, Z. Zhang, W. Liu, and Q. Xue, Tribological behavior and lubricating mechanism of Cu nanoparticles in oil, *Tribol. Lett.*, vol. 8, no. 4, pp. 213–218, 2000.
[8] Y. Y. Wu, W. C. Tsui, and T. C. Liu, Experimental analysis of tribological properties of lubricating oils with nanoparticle additives, *Wear*, vol. 262, no. 7–8, pp. 819–825,2007.
[9] S. Chen, W. Liu, and L. Yu, Preparation of DDP-coated PbS nanoparticles and investigation of

the antiwear ability of the prepared nanoparticles as additive in liquid paraffin, *Wear*, vol. 218, no. 2, pp. 153–158, 1998.
[10] G. Biresaw and G. Bantchev, Effect of chemical structure on film-forming properties of seed oils,‖ *J. Synth. Lubr.*, vol. 25, no. 4, pp. 159–183, 2008.
[11] Z. Q. Mamiyev and N. O. Balayeva, Preparation and optical studies of PbS nanoparticles, *Opt. Mater. (Amst).*, vol. 46, pp. 522–525, 2015.
[12] L. Rapoport, Y. Feldman, M. Homyonfer, H. Cohen, J. Sloan, J. . Hutchison, and R. Tenne, Inorganic fullerene-like material as additives to lubricants: structure–function relationship, *Wear*, vol. 225–229, pp. 975–982, 1999.
[13] Z. S. Hu and J. X. Dong, Study on antiwear and reducing friction additive of nanometer titanium borate, *Wear*, vol. 216, no. 1, pp. 87–91, 1998.
[14] H. D. Huang, J. P. Tu, L. P. Gan, and C. Z. Li, An investigation on tribological properties of graphite nanosheets as oil additive, *Wear*, vol. 261, no. 2, pp. 140–144, 2006.
[15] B. Bhushan, *Principles and applications of tribology*. John Wiley, 1999.
[16] E. F. Rico, I. Minondo, and D. G. Cuervo, The effectiveness of PTFE nanoparticle powder as an EP additive to mineral base oils, *Wear*, vol. 262, no. 11–12, pp. 1399–1406, 2007.
[17] L.R.Rudnick, *Lubricant additives :chemistry and applications*, 2009.
[18] P.K.Rohatgi, M.Tabandeh-Khorshid, E. Omrani, M. R. Lovell, and P. L. Menezes, *Tribology for Scientists and Engineers*, 2013.
[19] P.K.Rohatgi, M.Tabandeh-Khorshid, E. Omrani, M. R. Lovell, and P. L. Menezes, *Tribology for Scientists and Engineers*, 2013.
[20] H. Chang, C. Lan, C. Chen, M. Kao, and J. Guo, Anti-Wear and Friction Properties of Nanoparticles as Additives in the Lithium Grease, *Int. J. Precis. Eng. Manuf.*, vol. 15, no. 10, pp. 2059-2063, 2014.
[21] A. S. Pisal and D. S. Chavan, Experimental Investigation of Tribological Properties of Engine oil with CuO nanoparticles, no. 2, pp. 34–38, 2014.
[22] J. R. Greer, W. C. Oliver, and W. D. Nix, Size dependence of mechanical properties of gold at the micron scale in the absence of strain gradients, 2005.
[23] D. X. Peng, Y. Kang, R. M. Hwang, S. S. Shyr, and Y. P. Chang, Tribological properties of diamond and SiO2 nanoparticles added in paraffin, *Tribol. Int.*, vol. 42, no. 6, pp. 911–917, 2009.
[24] S. M. Alves, B. S. Barros, M. F. Trajano, K. S. B. Ribeiro, and E. Moura, Tribological behavior of vegetable oil-based lubricants with nanoparticles of 105 oxides in boundary lubrication conditions, *Tribol. Int.*, vol. 65, pp. 28–36, 2013.
[25] Z. W. Shan, G. Adesso, A. Cabot, M. P. Sherburne, S. A. Syed Asif, O. L. Warren, D. C. Chrzan, A. M. Minor, and A. P. Alivisatos, Ultrahigh stress and strain in hierarchically structured hollow nanoparticles, *Nat. Mater.*, vol. 7, no. 12, pp. 947–952, Dec. 2008.
[26] S. Chen, W. Liu, and L. Yu, Preparation of DDP-coated PbS nanoparticles and investigation of the antiwear ability of the prepared nanoparticles as additive in liquid paraffin, *Wear*, vol. 218, no. 2, pp. 153–158, 1998.
[27] F. Chi as-Castillo and H. A. Spikes, Mechanism of Action of Colloidal Solid Dispersions, *J. Tribol.*, vol. 125, no. 3, p. 552, 2003.
[28] J. Zhou, J. Yang, Z. Zhang, W. Liu, and Q. Xue, Study on the structure and tribological properties of surface-modified Cu nanoparticles, *Mater. Res. Bull.*, vol. 34, no. 9, pp. 1361 1367,1999.
[29] D. Johnson, M. Bachus, and J. Hils, Interaction between Lubricants Containing Phosphate Ester Additives and Stainless Steels., *Lubricants*, vol.*1*, pp.48–60,2013. 107
[30] S. Asadauskas, G. Bikulcˇius, R. Kreivaitis, and J. Padgurskas, Surfactant Influence on Stability and Lubrication Properties of Metal Nanoparticle Suspensions in Oil, *Surfactant in Triblogy*, vol.4, pp.151–182, 2014.
[31] W. Tuszynski, M. Szczerek, R. Michalczewski, E. Osuch-Slomka, E. Rogos, and A. Urbanski, The potential of the application of biodegradable and non-toxic base oils for the formulation of gear oils - model and component scuffing tests, *Lubr. Sci.*, *26*, 327–346, 2014.

[32] De Carvalho, M. J. S., Seidl, P. R., Belchior, C. R. P., & Sodré, J. R., Lubricant viscosity and viscosity improver additive effects on diesel fuel economy. *Tribology international*, *43*(12), 2298-2302, 2010.

[33] Ali, M. K. A., Xianjun, H., Mai, L., Qingping, C., Turkson, R. F., & Bicheng, C., Improving the tribological characteristics of piston ring assembly in automotive engines using Al_2O_3 and TiO_2 nanomaterials as nano-lubricant additives. *Tribology International*, *103*, 540-554, 2016.

[34] Suryawanshi, S. R., & Pattiwar, J. T., Effect of TiO_2 Nanoparticles Blended with Lubricating Oil on the Tribological Performance of the Journal Bearing. *Tribology in Industry*, *40*(3), 2018.

[35] Molea, A., Barabás, I., & Suciu, R., Influence of TiO_2 Nano-particles Content on Physicochemical and Tribological Properties of Lubricant Oil. In *International Congress of Automotive and Transport Engineering*, pp. 190-196, 2018.

[36] Shaari, M. Z., Roselina, N. R., Kasolang, S., Hyie, K. M., Murad, M. C., & Bakar, M. A. A., Investigation of tribological properties of palm oil biolubricant modified nanoparticles. *Jurnal Teknologi*, *76*(9), 69-73, 2015.

[37] Rahim, S., Ghamsari, M. S., & Radiman, S., Surface modification of titanium oxide nanocrystals with PEG. *Scientia Iranica*, *19*(3), 948-953, 2012.

[38] M. D. I. A, A. I. H, and Y. Y. A., Production of biolubricant from Jatropha curcas seed oil, *Journal of Mechanical Engineering and Material Science*, vol. 4, pp.72–79, 2013.

IOP Conf. Series: Materials Science and Engineering **834** (2020) 012033 doi:10.1088/1757-899X/834/1/012033

Assessment of Level-of-Service for On-Ramp Expressway using MHCM 2011 and Microsimulation Vissim

Muhammad Azim Muiz Bin Abu Bakar[1], Muhammad Akram Adnan[2], Norliana Sulaiman[2], Tuan Badrol Hisham Tuan Besar[3*]

[1] Endeavour Consult Sdn. Bhd, D-1-39, Block D, No. 8, Jalan PJU 1A/20B, Dataran Ara Damansara, 47301 Petaling Jaya, Selangor Darul Ehsan, Malaysia
[2] Faculty of Civil Engineering, Universiti Teknologi MARA, 40450 Shah Alam, Selangor Darul Ehsan, Malaysia
[3] Faculty of Business Management, Universiti Teknologi MARA Cawangan Selangor, 42300 Puncak Alam, Selangor Darul Ehsan, Malaysia

*tuan.badrol.hisham@gmail.com

Abstract The traffic congestion during peak hour in Malaysia is becoming worst from year to year due to the increase of traffic demand volume. At the year 2018, the total accumulated vehicles in Malaysia are 14,415,978 vehicles in 2018 based on 16-hours traffic volume [4]. Based on this figure, the normal growth of the vehicle has increased by 0.74% per year. It is crucial for traffic engineer in Malaysia to follow an appropriate standard and manual to determine the performance of highway facilities that will provide accurate result reflecting the existing operation of the highway. Thus, this paper provides the comparative study in evaluating the performance of the on-ramp segment between Malaysia Highway Capacity Manual 2011 (MHCM 2011) and microsimulation Vissim. The evaluation of performance measure for on-ramp expressway is based on the traffic density at merge influence area. Both performances measured using MHCM 2011 and Vissim was statistically compared with collected empirical data and the difference in the level of service (LOS) is also compared. The result from this study will determine which tools will be able to replicate the empirical data and the difference of grade of LOS between MHCM 2011 and Vissim.

1. Introduction

Ramp is a part of the expressway which served to connect between two highway facilities. A ramp can also be functioning as access to specific development. There are two important types of ramp known as on-ramp (merging ramp) and off-ramp (diverging ramp). On-ramp has caught attention from many traffic engineers as a faulty design of this type of ramp may lead to the severe congestion on the mainline expressway. Thus, it is essential for the traffic engineer to incorporate a proper procedure based on the Malaysian traffic condition [6,7]. It will ensure the result of the traffic analysis reflects the actual situation on the on-ramp.

There are various tools available for on-ramp performance analysis. Malaysia highway planning unit itself has released its standard and guideline known as Malaysia Highway Capacity Manual, 2011 to assist traffic engineer in obtaining the performance of on-ramp and other highway facilities [3]. Alternatively, the microscopic simulation program for modelling the traffic network has become a popular tool among the traffic engineer. This tool has become an alternative way to conduct highway performance analysis more efficiently. Therefore, the validation of vissim output and MHCM 2011 output with empirical data is required to ensure the reliability of both tools. This study will provide an

Content from this work may be used under the terms of the Creative Commons Attribution 3.0 licence. Any further distribution of this work must maintain attribution to the author(s) and the title of the work, journal citation and DOI.
Published under licence by IOP Publishing Ltd

in-depth view of the difference between vissim and MHCM 2011 output for on-ramp performance analysis.

2. Literature Review

Previous literature has provided the comparison study between HCM and various highway capacity analysis tools. Jolovic, Stevanovic, Sajjadi, & Martin, 2016 develop a comparison between observed data, FREEVAL and Vissim for on-ramp segment, basic segment and weaving segment [5]. FREEVAL is a macroscopic freeway analysis tool based on U.S Highway Capacity Manual 2010 [2]. The study reveals that the density at on-ramp segment, basic segment and weaving segment obtained from FREEVAL is generally higher than Vissim and empirical density while Vissim result can closely match with empirical density.

The previous study also provides the effect of calibration from the microsimulation model. Skabardonis and Kim, 2010 showed that lane changing parameter, car following parameter and percentage of through vehicle that yields to merging traffic had impacted the result of microsimulation [8]. Researchers found that the program mostly underpredicts the average speeds when adopting the default settings in the software. The calibrated model was successfully replicate the field conditions of the weaving area. However, the study is not considering lane changing behaviour.

Dowling et al., 2004 developed guidelines for calibration of microsimulation models [1]. The test bed consisted of a freeway section with two diamond interchanges and a parallel arterial with a signalised intersection. Authors divided calibration procedure into three steps, calibration for capacity at bottlenecks, route choice calibration and overall system performance calibration against field measurements such as travel time and delay, respectively. In case that one facility is calibrated, the second step in the procedure should be neglected. Parameters were classified into two groups: parameters chosen to be adjusted and ones left the default. Global and local parameters were classified based on how they affect the simulation process. Once global parameters were adequate, the adjustment of local parameters was deployed through the correct tuning process. Researchers gave an example of 13 mean headway as the main calibration parameter. Authors indicated that model satisfied calibration criteria.

3. Methodology

This study involves traffic survey on-site using automated traffic counting known as Trax Apollyon. A pneumatic road tube was installed at the mainline and on-ramp at KM 12, Grand Sepadu Expressway, and Selangor. Figure 1 shows the aerial photo of the location of installation of road tube at mainline and on-ramp segment. The site inventory study is conducted to obtain the geometric properties of on-ramp segment in order to be used as input value during traffic analysis using MHCM 2011 and mirosimulation Vissim. The geometric property of the on-ramp segment is shown in Table 1.

Figure 1. Aerial photo of the location of installation of road tube at mainline and on-ramp segment.

Table 1. The geometric properties of the on-ramp segment.

Parameter	Mainline	On-Ramp
Lane Width	3.7 meter	5.0 meter
Shoulder Width	-	1.0 meter
Speed Limit	90 km/h	50 km/h
Type of Terrain	Level	Level
Length of the Acceleration lane	122 meters	

The collected traffic data is sorted, filtered and arranged based on 15 minutes interval from 12.00 am to 12.00 pm. The traffic volume at mainline expressway and on-ramp is converted to the traffic flow rate in passenger cars per hour (pc/h) during the peak 15-minutes by using the rate from MHCM 2011. The empirical density is calculated simply by dividing the traffic flow with speed obtained from Trax Apollyon. Whilst the Eq. (1) is used to calculate the density of flow within the on-ramp influence area based on MHCM 2011:

$$D_R = 3.389 + 0.003369\, V_{12} + 0.005860\, V_R - 0.006396\, L_A \tag{1}$$

Where V_R is total flow rate at on-ramp expressway in pc/h, L_A is the length of acceleration lane in meter and V_{12} is the flow rate in lane 1 and lane 2 of the mainline expressway at the upstream of merge area. After determine the density of merge influence area, the level of service of on-ramp segment can be determined by referring to the criteria of level of service for on-ramp segment.

The density from Vissim is obtained by developing the microsimulation model using PTV Vissim software version 9.0. The link and connector are used to replicate the actual lane configuration with tapered acceleration lane. In this study, the right-side rule driving behavior with Wiedemann 99 type of car following model and slow lane rule is selected to suit with the existing site condition. With this option selected, vehicles with the higher speed will overtake of other vehicles in the right lane which replicates the driving behaviour at the expressway. The desired speed distribution is also calibrated based on the speed distribution pattern of each vehicle obtained from Trax Apollyon. The summary of the mean, standard deviation, minimum value, median, 85th percentile and maximum value of vehicle speed is shown in Table 2 for mainline and Table 3 for the ramp.

Table 2. Summary of the mean, standard deviation, minimum value, median, 85th percentile and maximum value of vehicle speed at mainline.

VARIABLE	MEAN (km/h)	MINIMUM (km/h)	MEDIAN (km/h)	MAXIMUM (km/h)	85TH PERCENTILE (km/h)
Car	86.186	67.920	86.582	123.919	94.66
Medium Lorry	82.422	54.718	82.076	106.216	90.67
Large Lorry	76.087	61.230	76.877	87.508	81.49
Buses	77.64	43.45	77.25	96.56	88.31
Motocycles	76.443	58.741	76.156	101.925	84.17

Table 3. Summary of the mean, standard deviation, minimum value, median, 85th percentile and maximum value of vehicle speed at mainline.

VARIABLE	MEAN (km/h)	MINIMUM (km/h)	MEDIAN (km/h)	MAXIMUM (km/h)	85TH PERCENTILE (km/h)
Car	64.87	43.45	65.98	87.71	74.52
Medium Lorry	59.63	40.23	57.94	90.12	77.89
Large Lorry	54.40	40.23	48.28	72.42	67.81
Buses	59.546	59.546	59.546	59.546	59.546
Motocycles	62.49	27.36	60.89	94.95	77.75

The simulation process is then run for three times with different random seed to get a different simulation result. The result of three simulation runs was averaged and compared to the empirical values to account for the stochastic nature of Vissim.

4. Result and Discussions

The summary of the mean, minimum value, median, and maximum value of density from Empirical, MHCM 2011 and Vissim is shown in Table 4. The result of traffic density from Vissim is presented in veh/km/ln as to follow the actual output of the software. Even though the unit is in veh/km/ln, the microsimulation models deal with heavy vehicle explicitly by assigning more sluggish characteristics to each of them. The frequency distribution of density for Empirical, MHCM 2011, and Vissim are shown in Figure 2. In short, the mean of density for MHCM 2011 is higher than empirical data while Vissim is closely match with empirical data. This indicates a potential different between the MHCM 2011 and empirical data.

Table 4. Summary of the mean, minimum value, median, and maximum value of density from Empirical, MHCM 2011 and Vissim

VARIABLE	MEAN	MINIMUM	MEDIAN	MAXIMUM
Empirical (pc/hr/ln)	4.294	0.228	4.973	10.041
MHCM 2011 (pc/hr/ln)	6.101	2.821	6.582	10.173
Vissim (veh/km/ln)	4.218	0.206	4.707	10.884

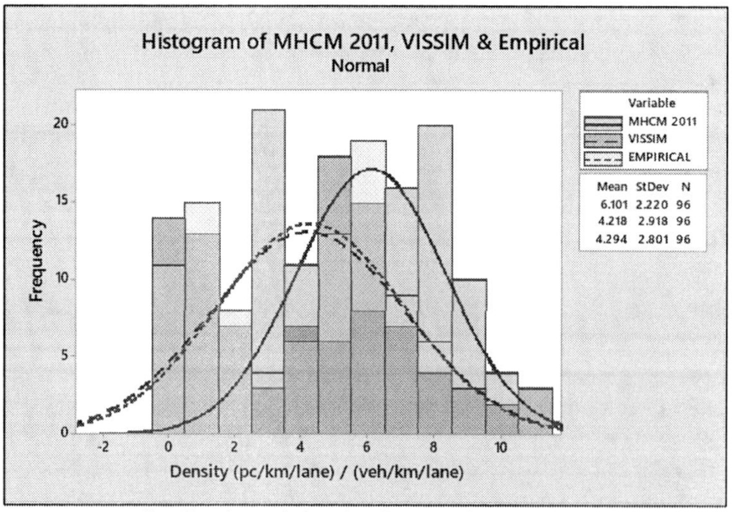

Figure 2. Frequency distribution of density for Empirical, MHCM 2011, and Vissim.

The Paired T-Test was conducted to determine if there is a significant difference between the density means of MHCM 2011, Vissim and empirical data. The test was conducted separately for MHCM 2011 and Vissim to compare the different mean with empirical data. The first test is conducted between MHCM 2011 and empirical data. The hypothesis for the first test was conducted as below:

Null Hypothesis, $H_{O(1)}$ = There is no significance different between the density mean of MHCM 2011 and empirical data.

Alternative Hypothesis, $H_{a(1)}$ = There is significance different between the density mean of MHCM 2011 and empirical data.

Table 5. Shows the result of paired T-test between MHCM 2011 and empirical data

SAMPLE	MEAN	STANDARD DEVIATION	STANDARD ERROR MEAN	P-VALUE
MHCM 2011	6.101	2.220	0.227	
EMPIRICAL	4.294	2.801	0.286	0.00
DIFFERENT	1.806	0.701	0.072	

Table 5 shows the result of paired T-test between MHCM 2011 and empirical data. The p-value is equal to 0.00 which is less than 0.05; it can conclude that the mean of density from MHCM 2011 differs from empirical data at the 95% confidence level. Thus, rejecting the null hypothesis $H_{O(1)}$.

The second test is conducted between Vissim and empirical data. The hypothesis for the second test was conducted as below:

Null Hypothesis, $H_{O(2)}$ = There is no significance different between the density mean of Vissim and empirical data.

Alternative Hypothesis, $H_{a(2)}$ = There is significance different between the density mean of Vissim and empirical data.

Table 6. The result of the paired T-test between Vissim and empirical data

SAMPLE	MEAN	STANDARD DEVIATION	STANDARD ERROR MEAN	P-VALUE
VISSIM	4.218	2.918	0.298	
EMPIRICAL	4.294	2.801	0.286	0.304
DIFFERENT	-0.076	0.721	0.074	

Table 6 shows the result of the paired T-test between Vissim and empirical data. The p-value is equal to 0.304 which is more than 0.05; it can conclude that there is no significant difference between the mean density of Vissim and empirical data at the 95% confidence level. Thus, accepting the null hypothesis $H_{O(2)}$.

Figure 3 shows the time series plot of density from MHCM 2011, Vissim and empirical data to graphically show the different pattern of density between MHCM 2011, Vissim and empirical data within 24-hours. The density from MHCM 2011 has the highest values at almost every 15-minutes interval evaluated. While, the Vissim density resulted in consistent estimates even though, Vissim estimates were slightly lower than empirical data. However, the Vissim density is observed to be highest during the maximum peak flow where the traffic volume is highest. Nonetheless, still, there is no significant difference between the mean density of Vissim and empirical data at the 95% confidence interval.

IOP Conf. Series: Materials Science and Engineering **834** (2020) 012033 doi:10.1088/1757-899X/834/1/012033

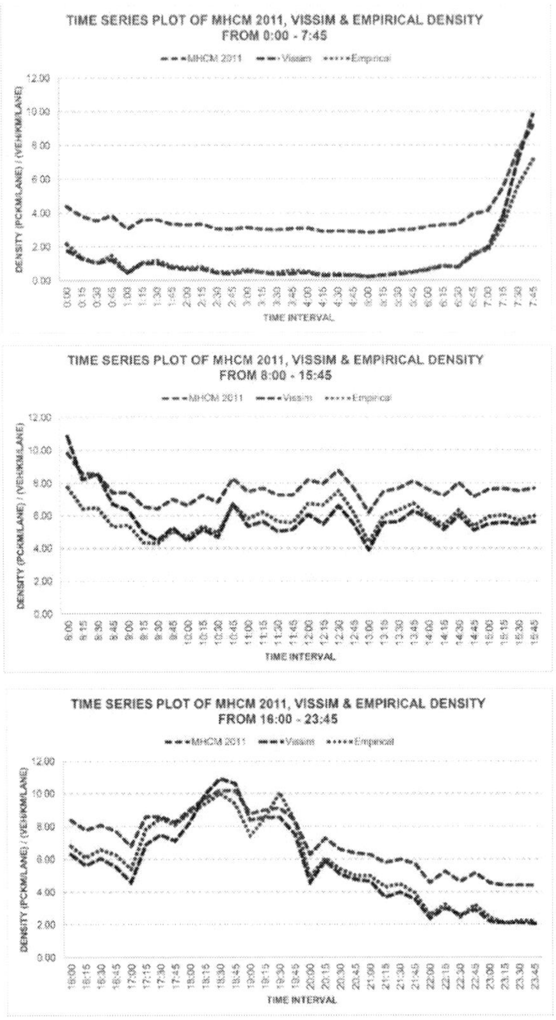

Figure 3. The time series plot of density from MHCM 2011, Vissim and empirical data.

Based on the trend shown by the density estimates from the two models, a more reasonable comparison could be drawn by using the scatter plot of density values with the 45-degree diagonal line for result validation. A 45-degree diagonal line plot is generated to see how the data points are scattered around the diagonal line, to examine and validation of the extent of the difference between the density estimates from the two models evaluated. Figure 4a shows the scatter plot graph of density MHCM 2011 vs Empirical and Figure 4b shows the scatter plot graph of density Vissim vs Empirical. In short, the MHCM 2011 density tend to overestimate the empirical data while Vissim is slightly underestimate the empirical data.

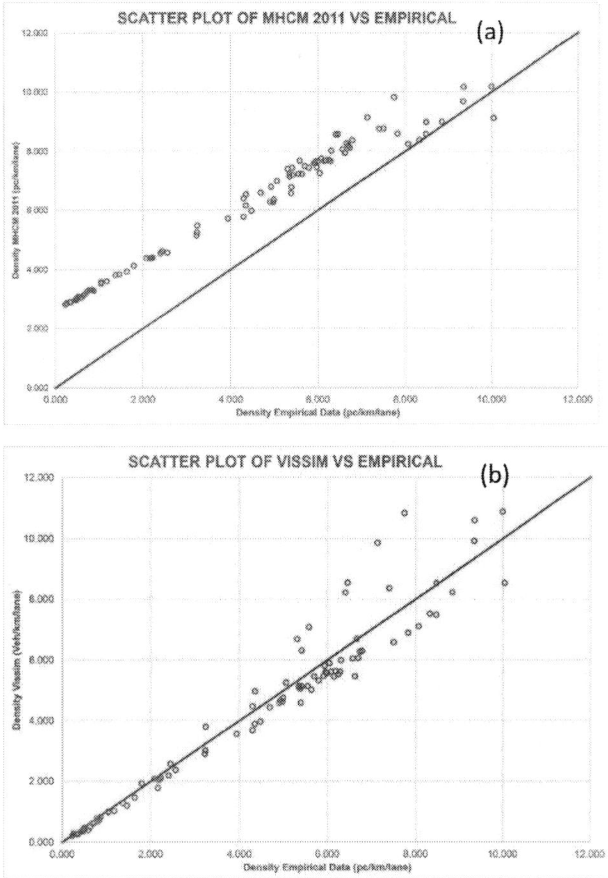

Figure 4. Scatter plot graph of density MHCM 2011 vs Empirical and Vissim vs Empirical.

The literature does not recommend the comparison of LOS obtained from the microsimulation model and HCM tools [2]. However, those recommendations do not state the magnitude of the discrepancy between the two methodologies. Understanding that government agencies and consultants use microsimulation and MHCM extensively, it would be beneficial to reveal the discrepancies between these tools.

Figure 5 shows the LOS comparisons between Vissim and MHCM 2011. One can observe that there is no different in LOS grade calculated based on MHCM versus Vissim analysis from 0.00 to 7.45 where the traffic flow during this time frame is considered lowest. The LOS between MHCM and Vissim shows inconsistent during the morning peak flow. At 7:30 to 8:00, the LOS from both tools is similar, and the LOS start to differ at 8:15 and similar again at 8:30. At evening peak, the LOS is consistently matched between both models and start to differ after evening off-peak from 20:00 to 21:00.

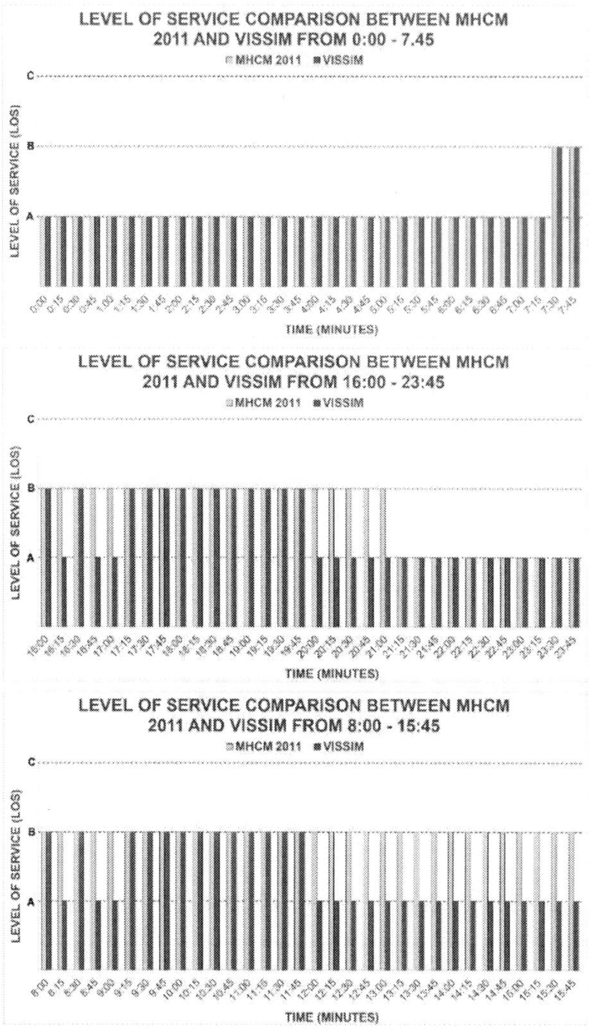

Figure 5*:* LOS comparisons between Vissim and MHCM 2011

5. Conclusion and Recommendation

Based on the results of paired T-test presented in Chapter 4, it was found that the null hypotheses for test 1 are rejected and conclude that the mean of density from MHCM 2011 differs from the empirical data at the 95% confidence level. The test 2 result shows that the null hypothesis is accepted and conclude that there is no significant difference between the mean density of Vissim and empirical data at the 95% confidence level.

The result is further validated with a scatter plot with a 45-degree diagonal line. It is concluded that MHCM 2011 tend to overestimate the density at most every 15-minutes interval within 24-hours of study. While the result of simulated density from Vissim is slightly underestimated and nearly matched with empirical data at most of the study interval although the result from Vissim can also be overestimated during peak flow where the traffic volume is highest.

It can be concluded that the level of service between MHCM 2011 and Vissim providing the same LOS during off-peak where the traffic flow is lowest. However, during the morning and evening peak, the LOS is inconsistently matched between MHCM 2011 and Vissim. After completed this study, several further studies are required in order to improve the comparison of density from both MHCM 2011 and Vissim with empirical data.

Another detail setting in Vissim needs to be further studied and adjust based on empirical data such as driving behaviour parameter and vehicle characteristic in order to consistently match with empirical data.

A traffic study needs to be conducted at the busy on-ramp expressway with a higher flow rate of vehicles. Thus, the result from Vissim and MHCM 2011 can be further evaluated and compared at such condition.

Vissim result will generally match the empirical density at most of the study interval. However, additional adjustments are necessary in order to 'translate' Vissim output to MHCM 2011 measurements as MHCM 2011 is using pc/km/lane as density unit which accounts for the passenger car unit factor. While, Vissim density unit is in veh/km/lane with the incorporation of sluggish, heavy vehicle behaviour and light vehicle such as motorcycle into the simulation.

Acknowledgements
This research was funded by the Universiti Teknologi 600-IRMI/MYRA 5/3/MITRA (007/2017)-1. The authors gratefully acknowledge the cooperation of staffs in Faculty of Civil Engineering and Faculty of Health Science for giving full support and services as well as handling to set up the tools and equipment for the driving and vision studies in the field. Thank you to Endeavour Consult Sdn. Bhd. for providing the licensed PTV Vissim software to accomplish this study.

References

[1] Dowling, R., Skabardonis, A., Halkias, J., McHale, G., & Zammit, G., "Guidelines for Calibration of Microsimulation Models: Framework and Applications", *Transportation Research Record*, Volume 1876 (1), 1-9, 2004 https://doi.org/10.3141/1876-01

[2] *Highway Capacity Manual 2010*. Fifth edition. Washington, D.C.: Transportation Research Board, 2010.

[3] Highway Planning Unit. *Malaysia Highway Capacity Manual 2011 (MHCM 2011)* (First Edition). Ministry of Works Malaysia, 2011.

[4] Highway Planning Unit. *Road Traffic Volume Malaysia 2018 (RTVM 2018)* (Vol. 2018). Ministry of Works Malaysia, 2018.

[5] Jolovic, D., Stevanovic, A., Sajjadi, S., & Martin, P. T., "Assessment of Level-Of-Service for Freeway Segments Using HCM and Microsimulation Methods", *Transportation Research Procedia*, *15*, 403–416. 2016, https://doi.org/10.1016/j.trpro.2016.06.034

[6] Malaysia Highway Authority (MHA). *Guidelines For Malaysia Toll Expressway System-Design Standards* (First Edit). 2011

[7] Ministry of Works Malaysia. *Arahan Teknik (Jalan) 8/86: A Guide on Geometric Design of Roads* (Pindaan 2015). 2015.

[8] Skabardonis, A., & Kim, A., "Weaving Analysis, Evaluation and Refinement". *California PATH Program*, UCB-ITS-PRR-2010-19 2010.

IOP Conf. Series: Materials Science and Engineering **834** (2020) 012034 doi:10.1088/1757-899X/834/1/012034

Comparative analysis on integrated digital forensic tools for digital forensic investigation

Jae-ung Lee, Woo-young Soh

Department of Computer Engineering, Hannam University Republic of Korea

leejaeung1990@gmail.com, wsoh@hnu.kr

Abstract. As a variety of digital devices have recently become widely commercialized, digital forensics, a process where electronic information stored in digital devices is accurately identified, collected, preserved and analysed and the information is submitted to the court as evidence, is gaining enormous popularity. This paper analyses characteristics, applications, limitations of digital forensic tools and compare the tool with others in terms of ease of use, thereby helping investigators to adopt integrated digital forensic tools in their investigation.

1. Introduction

Recently, the assassination of the Saudi dissident Jamal Khashoggi has drawn enormous attention from all over the world. To figure out what actually happened in his disappearance, investigators paid attention to the Apple Watch on his wrist. That is because the Apple Watch allegedly recorded the killing of Khashoggi and was connected to other devices. As a variety of digital devices have recently become widely commercialized, digital forensics, a process where electronic information stored in digital devices is accurately identified, collected, preserved and analyzed and the information is submitted to the court as evidence, is gaining enormous popularity. A comprehensive digital forensic tool integrates various functions such as analysis tools and reporting tools into one model. This paper will examine the characteristics, applications and limitations of these integrated digital forensic tools and further compares their ease of use with other tools, which will help investigators adopt the integrated digital forensic tools.

2. Relevant research

Digital forensics has traditionally been utilized in forensic science. In other words, it was used in fields such as fingerprinting, hair and DNA analysis and autopsies. However, with the advent of various digital devices, the concept of forensics has been expanded to include electronic evidence stored in digital devices, not to mention physical forms of evidence. Thus, digital forensics refers to the process of accurately identifying and collecting electronic information stored in digital devices, preserving and analyzing the information and submitting to the court as evidence. Depending on the subject of analysis, digital forensics includes disk forensics, system forensics, network forensics, Internet forensics, mobile forensics, database forensics, and cryptographic forensics.

2.1. Disk Forensics

Disk forensics is the science of extracting forensic information from physical digital storage media like hard disks, Floppy disks, CDs, USB devices etc.

Content from this work may be used under the terms of the Creative Commons Attribution 3.0 licence. Any further distribution of this work must maintain attribution to the author(s) and the title of the work, journal citation and DOI.
Published under licence by IOP Publishing Ltd

2.2. System Forensics
System forensics is a branch of digital forensics that extracts forensic information by analysing computer operating systems, application programs or computer processes. It is important to understand file system of various operating systems.

2.3. Network Forensics
Network forensics collects evidence by analysing data or password transmitted via network or network configurations. Most of the networks provide tracking tools for monitoring purposes.

2.4. Internet Forensics
Internet forensics collects information that uses internet protocols such as WWW or FTP.

2.5. Mobile Device Forensics
Mobile device forensics is a branch of digital forensics that extracts digital evidence or data from a mobile device. Given data from small handheld devices are easily concealable, mobile device forensics requires high level accuracy.

2.6. Database Forensics
Database forensics is a branch of digital forensics that collects and analyses information from databases. Considering companies store all their data in their databases, it can serve as a useful tool to secure evidence.

2.7. Password Forensics
It is a type of digital forensics that recovers password of files or systems. Files or systems are often encrypted in order to block access from unauthorized entities. Password forensics is used in such cases.

3. Integrated digital forensic tool comparative analysis
A digital forensic tool refers to a tool designed for digital investigation. Specifically, Integrated Digital Forensics combines functions such as analysis tools or reporting tools into one tool. Thus, it is classified into integrated analysis in dataset creation strategies that performs interpreting and search functions of the program interface. It can also be categorized as detailed analysis that performs analysis and carving for smart devices of which tool performance may vary.

3.1. EnCase
EnCase is an integrated digital forensic tool developed by Guidance Software and is available for about $ 2,995. Major functions include Triage, Collect, Process, Search, Analyse, and Report. The recommend systems settings for a computer's hardware and operating system is a CPU that is a quad core or higher, 16 GB RAM or higher, and Windows OS in 32-bit or 64-bit environment. It provides a wide range of features, including the operating system installation drive, installation images, portable data medium, physical memory and MacOSX analysis, as well as file restoration, extraction and reporting functions.

Thus, it can be widely used for digital forensics. Also, it is highly scalable, given that features that are not supported by the program can be dealt with by script programming. However, diverse functions of the tool make it complicated to handle. Therefore, it is necessary to fully familiarize oneself with EnCase before using it. It would be highly useful if an investigator who uses the tool to be familiar with it and capable of utilizing various functions of the tool.

3.2. FTK (Forensic Tool Kit)
Forensic Tool Kit (FTK) is an integrated digital forensic tool which is developed by AccessData and sold for about $ 5,144. Quad-core or higher CPU or Xeon-class CPUs, 32 GB of RAM, and a 64-bit Windows OS are recommended. Two available operating environments are server and desktop. FTK supports a wide range of analysis as well as operating installation drive and installation image,

portable storage medium, physical memory and encrypted disk volume. It uses collected artifacts for a database called Postgre to refine and select data. In addition, since it uses Tab, it is relatively easy to familiarize oneself with the tool. However, since dependency problems such as installation of database occur when the tool is installed, it is necessary to establish an investigation-friendly environment beforehand.

3.3. Forensic Explorer

Forensic Explorer is an integrated digital forensic tool developed by GetData and sold for about \$1,247.95. Quad-core or higher CPU, a minimum of 8 GB of RAM, a 32-bit and 64-bit Windows OS are recommended.

Forensic Explorer supports operating system installation drive and image, portable storage medium and a script for analysis. But the script should be written with Pascal, so learning Pascal is required. However, compared to Encase or FTK, it offers distinctive benefits such as operability in low-end PCs and an intuitive interface. Thus, even those who are not familiar with the tool can easily utilize Forensic Explorer. This makes it possible for low-skilled investigators to use the tool without impairing efficiency of investigation.

3.4. X-Ways Forensics

X-Ways Forensics is an integrated digital forensic tool developed by X-Ways and sold for about €1199. Quad-core or better-functioning CPU, 4GB of RAM or more, and Windows OS in 32-bit and 64-bit environment are recommended. X-Ways Forensics can be used in most forensic environments from media imaging to physical memory analysis. Collecting and aggregating artifacts and providing reporting functions, X-Ways Forensics has the lowest drive environment specification requirement among all the integrated digital forensic tools. It can also be operated without installation. Although it can run on such low specification devices, it still provides with analysis of image files as well. However, the functions it provides are relatively limited. In addition, since it mainly focuses on Hex View, it requires highly-skilled tool exploitation.

3.5. BlackLight

BlackLight is an integrated digital forensic tool developed by BlackBag and sold for about \$2,600. Quad-core or better-performing CPUs, 16GB of RAM or more, and a Window OS in 32-bit and 64-bit environment and MacOSX which has lower specification than Mavericks are recommended. BlackLight, analysing data with the use of user history in the MacOSX environment, provides multiple functions. It supports functions such as file system browsing and recovery functions for unallocated space as well as basic functions of Mac's operating system such as interpreting email, internet, message, schedule, maps and notes. In addition, it is capable of interpreting disk image files when interpreting HFS file systems. As for EFS, it interprets file systems with the use of keychain file. However, its operation is unstable in the Window OS environment and relatively stable in Mac OSX.

Thus, it can serve as an additional tool for the investigator. Except BlackLight, which is specifically designed for Mac OSX, the best performing tool is Forensic Explorer. It uses an intuitive interface that categorizes tabs by type of artifacts collected, Forensic Explorer is more convenient than other tools. Also, it provides with most of the functions of FTK or EnCase. Also, in terms of registry analysis, it uses automatic parsing to support outputting important information such as user information, installation and external storage medium. Other tools cannot be utilized unless one fully understands how to uses those tools. In Linux/Unix or Mac OSX, most of those tools mentioned above can interpret EXT, HFS, HFS+partitions. Thus, investigators can extract artifacts and analyse them.

BlackLight only analyses Mac OSX artifacts, and the supported partitions are limited to FAT partitions, HFS, and HFS + partitions. It automatically parses the key information of Mac OSX and outputs the result. The artifacts supported by the tool parse information such as user information, operating system information, Web history, e-mail, installation application, etc. and output the results. Thus, If one does not know the path of artifacts of Mac OSX or how to collect them, they can be expected to utilize BlackLight instead.

IOP Conf. Series: Materials Science and Engineering **834** (2020) 012034 doi:10.1088/1757-899X/834/1/012034

4. Conclusion

As a variety of digital devices have recently become widely commercialized, digital forensics, a process where electronic information stored in digital devices is accurately identified, collected, preserved and analysed and the information is submitted to the court as evidence, is gaining enormous popularity. The integrated digital forensic tool, a comprehensive tool that incorporates collection of digital evidence, reporting and analysis, offers various functions, which can help reduce the amount of time spent for analysis. In addition, you can get uniform results, given that it is capable of producing results of many artifacts in the form of a report.

However, it requires high-end PCs for investigative purposes and offers numerous functions which are hard to be familiar with. Also, since that most of them are commercial software, their license is quite costly. This paper will examine the characteristics, applications and limitations of these integrated digital forensic tools and further compares their ease of use with other tools, which will help investigators better adopt integrated digital forensic tools.

References

[1] Lovanshi M and Bansal P 2019 Comparative Study of Digital Forensic Tools *Engineering and Applications* pp 195-204

[2] Yagnik S.B and Panchal E.P 2019 AEACFE System—An Intelligent Digital Forensic System *Third International Congress on Information and Communication Technology* **797** pp 409 416

[3] Paul Joseph D and Norman J 2019 An Analysis of Digital Forensics in Cyber Security *First International Conference on Artificial Intelligence and Cognitive Computing* **815** pp 701 708

[4] Panchal E.P, Yagnik S.B and Sharma B.K 2019 Use of Machine Learning Algorithm on File Metadata for Digital Forensic Investigation Process *Third International Congress on Information and Communication Technology* **797** pp 401-408

[5] Mane D and Shibe K 2019 Big Data Forensic Analytics Data Management, Analytics and Innovation **839** pp 113-129

[6] Hemdan E.ED and Manjaiah D.H 2019 Digital Forensic Approach for Investigation of Cybercrimes in Private Cloud Environment *Recent Findings in Intelligent Computing Techniques* **707** pp 25-33

IOP Conf. Series: Materials Science and Engineering **834** (2020) 012035 doi:10.1088/1757-899X/834/1/012035

Electromyography Analysis of Lower Extremity Muscles during Squat and Stoop Movement

Noor Azlina Mohd Salleh[1], Muhammad Syukran Al-Baria Noor Sazali[1], Noor Ayuni Che Zakaria[*1], Nadia Mohd Mustafah[2], Aizreena Azaman[3],

[1]Faculty of Mechanical Engineering, Universiti Teknologi MARA Selangor, 40450 Shah Alam, Malaysia
[2]Faculty of Medicine, Universiti Teknologi MARA Sungai Buloh, Jalan Hospital, 47000 Selangor, Malaysia.
[3]School of Biomedical & Health Science, Faculty of Engineering, Universiti Teknologi Malaysia, 81310 Johor Bahru, Johor.

*ayuni8098@uitm.edu.my

Abstract. There are different styles of lifting objects of the ground, and each style gives out different percentage of contribution of each muscle. The purpose of this study is to investigate the muscle activity between stooping and squatting to avoid low back pain and how the muscle activities differ from a correct technique of squatting, and an incorrect one. The Rectus Abdominis, Erector Spinae, Quadriceps and Hamstrings muscle were chosen to be monitored for differences in this study. Selected subjects were two healthy males, with the same height and age, one subject has no experience in doing squats – define as amateur, and the other subject know the basic method of squat-ting – define as non-amateur. In order to capture and analyze the activity of the muscles, Myon EMG and ProEMG software were used. Participants performed, stooping and squatting movement for 12 repetitions with 5 minutes' rest in between. The muscle activation strength was obtained by calculating the Root Mean Square of the EMG data. Results shows that the non-amateur gave out greater RMS difference in muscle activation for the Erector Spinae and Rectus Abdominis among other muscles compared to amateur for both stooping and squatting. Another finding was that, the results of a non-amateur were more balanced between both left and right muscles activation compared to the amateur subject. These observations raised the need of mind muscle connection and muscle coordination in order to lift efficiently in order to avoid low back injury.

1. Introduction

According to M. Motamedzade, A. Dormohammadi, E. Zarei, and R. Dormohammadi (2016), under the Manual Material Handling, it is stated that load lifting is the most stressful activity and could cause work-related musculoskeletal disorders [1]. Both musculoskeletal disorders (MSD) and low-back pain (LBP) are common in industrial workplaces and has been acknowledge as a significant cost toward companies, society and the health care system [2]. Based on a similar study on lifting done by S. Hwang, Y. Kim, and Y. Kim (2009) [3] and J. H. Van Dieën, M. J. M. Hoozemans, and H. M. Toussaint (1999) [4], in avoiding low-back pain when load lifting, the squat technique is the commonly advised technique to be done. However, squat technique popularity is low among the general population due to its high energetic cost as stated in [5]. The technique that more preferred and commonly used is the stoop technique, where the lower back is bended more compared to the squat. Studies on squat and stoop

Content from this work may be used under the terms of the Creative Commons Attribution 3.0 licence. Any further distribution of this work must maintain attribution to the author(s) and the title of the work, journal citation and DOI.
Published under licence by IOP Publishing Ltd

movement have been conducted before, however very less in comparing muscle activities between the two movements, especially using Myon EMG (myon AG, Schwarzenberg, Switzerland). Therefore, there is a need to fill this gap in the literature.

Squatting is a fundamental human movement pattern that involves nearly every muscle in the body, mainly in the lower extremity [6]. It is considered fundamental because it is required in activities of daily living [6]. For examples, picking stuff up from the ground or even sitting. Sitting is considered to be similar to squatting due to both movements are involve in hip and knee flex-ion. Stoop on the other hand, involves in lumbar flexion or bend at the waist. S. Hwang, Y. Kim, and Y. Kim (2009) did a similar case study but focused on joint angles, moments, power, support moments and including dynamic EMG [3]. From the EMG results, the study found biceps femoris and rectus femoris showed large variances of activation during the squat lifting. While tibialis anterior, medial gastrocnemius and biceps femoris showed large variances of activation during stoop lifting. On top of that, the study also found rectus abdominis and lumbar erector spinae are not significant both during squat and stoop lifting. On the other hand, J. H. Van Dieën, M. J. M. Hoozemans, and H. M. Toussaint (1999) [4] also did a similar case study but with respect to the mechanical load on the back. The parameters used were muscle force, compression, shear and bending moments.

However, despite those studies on comparison between squat and stoop in which technique is superior from one another, there is no similar study found in comparing the correct squat technique and an incorrect one based on the muscle activities using Myon sEMG. Therefore, this study is to investigate the muscle activity between stooping and squatting to avoid low back pain and how the muscle activities differ from a correct technique of squatting and an incorrect one.

2. Methods

2.1 Participants

To reduce different strength capability when lifting weight, two young male university students with similar figures were selected for the study. Both participants were 24 years old, had similar height but different stature, and were non-athlete. The subjects were divided into a non-amateur and an amateur. A non-amateur is the person that has experienced in performing both stoop and a correct squat technique, while an amateur is someone that does not have any experience in performing both stoop and a proper squat technique. Both participants had no past injuries mainly in the lower extremity and were in good health.

2.2 Experiment Set-up

The experiment was set-up as shown in Figure 1. Participants changed to proper attire to prepare themselves for shaving and electrode placements. Before placing the electrodes, the skin was cleaned and wiped with alcohol swab. After preparation, conductive adhesive hydrogel disposable foam electrode (Kendall Product) with 1 and ½ inch diameter, and an inter-electrode distance of 2 cm were attached in parallel to the muscle fibres of the rectus abdominis (RA), erector spinae (ES), rectus femoris (RF) and bicep femoris (BF) according to the recommendation of a rehabilitation physician.

2.3 Experimental Procedure

Subjects were asked not to perform any exercise that may cause muscle leg fatigue or pain prior to the study because muscle fatigue may produce mislead outcome results. The experiment was started with stooping then squatting without warming up, to mimic real life event in the industry where no warming up prior to lifting. The experiment started out with amateur followed by non-amateur. For both stooping and squatting, the set up was similar, where the participant stood straight in front of the weight (dumbbell) with distance of 20 cm. Then the participants initiated the first movement (stooping), repetitively for 12 repetitions. A rest period of 5 minutes was given to the participant to allow full recovery, and then performed the second movement (squatting) directly after. The procedure was also conducted similarly with non-amateur subject.

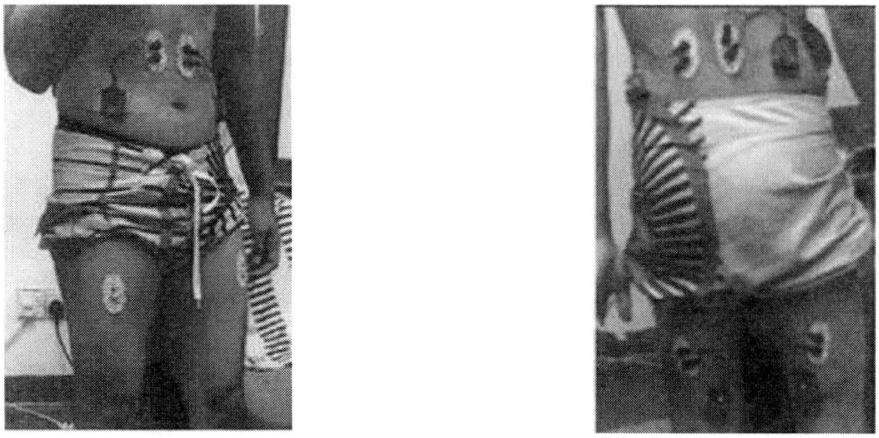

(a) (b)

Fig. 1. Picture of electrode positions (a) Rectus Abdominis and Rectus Femoris, (b) Erector Spinae and Bicep Femoris

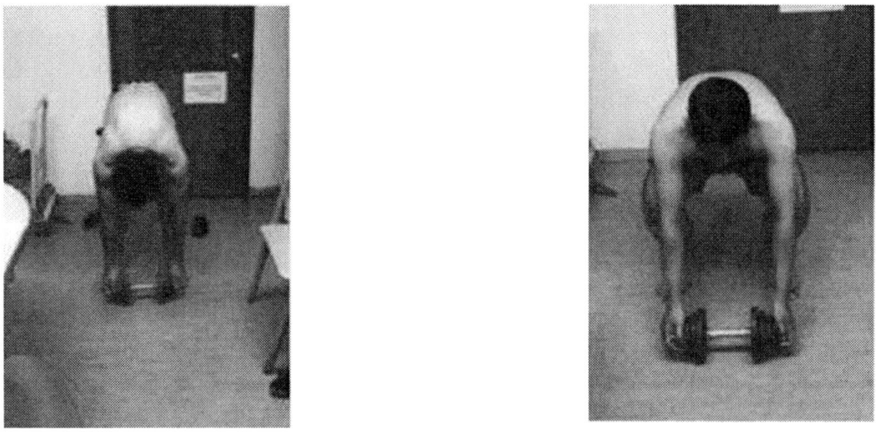

(a) (b)

Fig. 2. Pictures of non-amateur performing (a) squat, (b) stoop

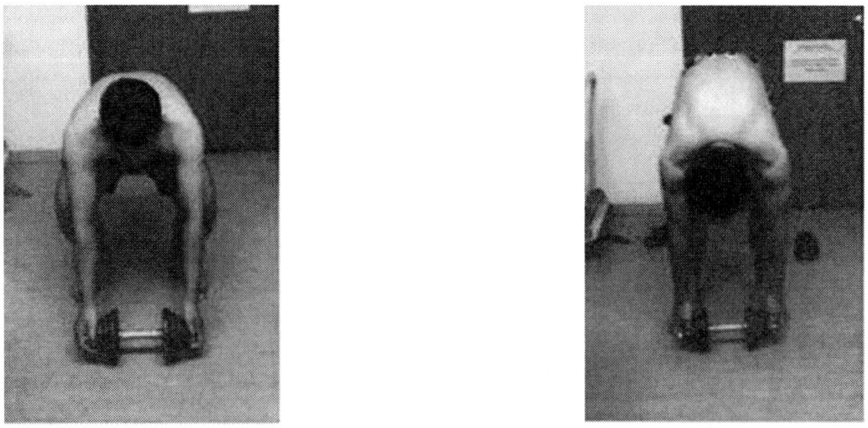

(a) (b)

Fig. 3. Picture of an amateur performing (a) squat, (b) stoop

IOP Conf. Series: Materials Science and Engineering **834** (2020) 012035 doi:10.1088/1757-899X/834/1/012035

2.4 Measurements
Surface electromyographic (sEMG) signals were recorded from erector spinae (ES), rectus abdominis (RA), rectus femoris (RF) and bicep femoris (BF) using Kendall™ 200 Series Foam Electrodes. Bipolar electrodes were used and placed on the muscle belly of each muscle, with distance of two centimeters between centers. Before placing the electrodes, alcohol swabs were used to cleaned of the skin surfaces. Any hair skin was shaved prior applying alcohol to increase accuracy of the data output. The transmitters that were attached to the electrodes are hold by using disposable adhesive tapes. The transmitter transmits the signal to the receiver then convert it from analog to digital through National Instruments data acquisition system then imported to processing computer. All signals were full-wave rectified. The EMG amplitude was normalised and smoothen before the measurement of the RMS and mean frequency values through ProEMG software. Smoothing operations are designed to make noisy signal smooth. The smoothing operations that were used were referred by L. R. Released et al. (2013) [7]:

- Moving Average Smoothing (Mean Frequency)

$$\boldsymbol{MAV_n} = \frac{\mathbf{1}}{\boldsymbol{N}} \sum_{n-N/2}^{n+N/2} \boldsymbol{S_i}$$

Where, N= number of samples and S= sample

- Root Mean Square

$$\boldsymbol{RMS_n} = \sqrt{\frac{\mathbf{1}}{\boldsymbol{N}}} \sqrt{\sum_{n-N/2}^{n+N/2} \boldsymbol{S_i^2}}$$

Where, N= number of samples and S= sample

Both RMS and MF were used in this study. Median Frequency and Integrated EMG were not used because the primary objective of this study is to investigate the muscle activity between stooping and squatting, and RMS and MF are sufficient for this study. As RMS will give info about muscle activation strength and MF to identify muscle fatigue.

3. Results and Discussion
The objective of this study is to identify the muscle pattern during squat and stoop between different subject ability. The muscle pat-tern can be described through the root mean square (RMS) of the muscle activity for both squat and stoop between non-amateur and amateur. The results are shown in figures below. The results were recorded and analysed for one selected session out of 12 repetitions with the same load of 7.5kg dumbbell being put 20cm in front of participants. The same selected session analysed for RMS is used for mean frequency analysis for better references.

Table 1. Pattern of EMG signals of a non-amateur for one cycle (7 seconds)

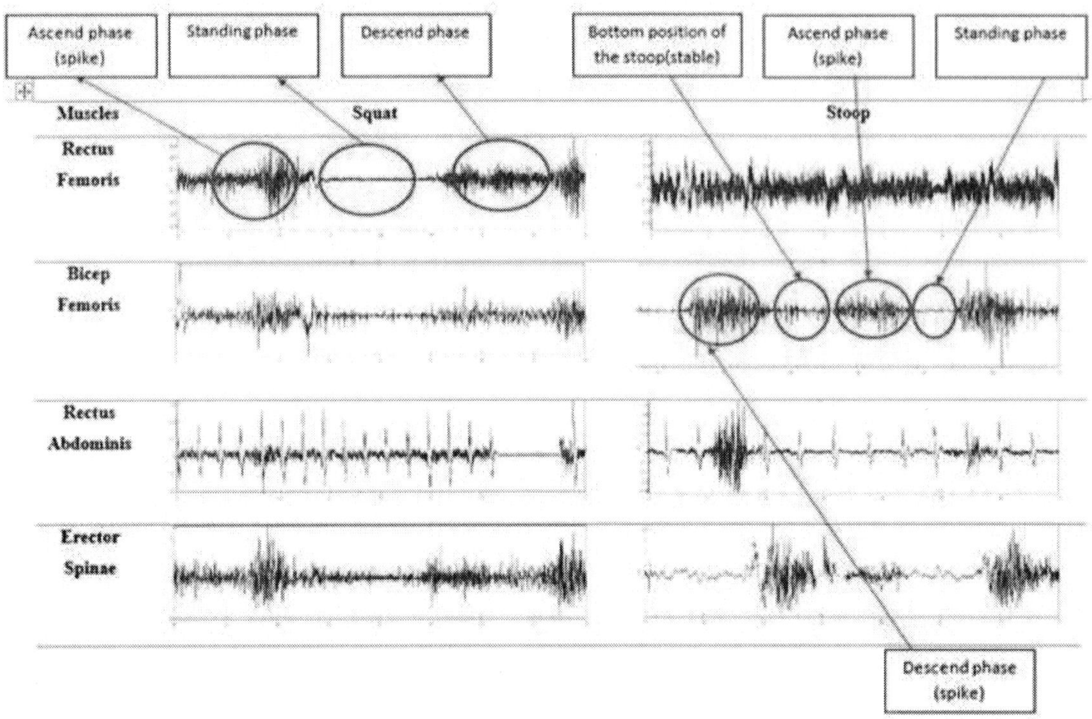

Table 2. Pattern of EMG signals of an amateur for one cycle (7 seconds)

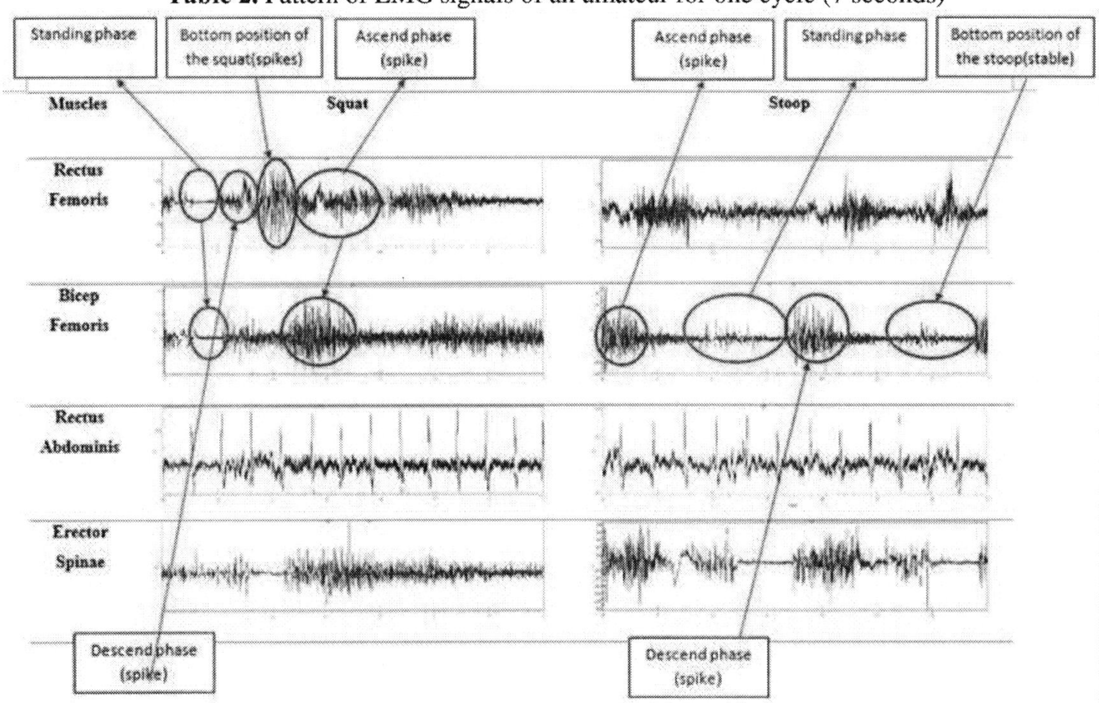

Table 1 and table 2 show the EMG muscle pattern in both subjects during squat and stoop movement. Different muscle signal can be seen during ascending, standing and descending phase. The signals are represented in volt versus time. For squat column, we analyse the movement by looking at the signal coming from rectus femoris. Reading the signal in the table from left to right, the subject began to descend into a squat position where the upper thighs are slightly below parallel from the ground. Once the subject starts to ascend, the EMG signal for rectus femoris spikes, then back to normal (stable) as the subject stands up straight. The signal for bicep femoris and erector spinae can be seen as similar to rectus femoris. It is similar for bicep femoris could be because of being as one of the biarticular muscle same as rectus femoris which is also one of the biarticular muscles. However, as for erector spinae, it is because it functions in keeping an upright torso or a neutral spine to avoid it from curving.

As for stoop column, we analyses the movement by looking at the signal coming from bicep femoris. Reading from left to right, the spikes mean that the subject starts to descend then the signal becomes stable for a while at the bottom position of the stoop. The signal spikes back as the subject starts to ascend back to standing position. Even though stooping means, there is no flexion, or bending occurs at the knees, the subject knows how to engage his muscles so that the leg muscles are being work during lifting in order to avoid straining the lower back.

As expected, overall for both RMS and MF results, the most significant finding in this study was the non-amateur gave out greater EMG activity compared to an amateur for each muscle and for both squat and stoop movements. Based on the data obtained, this may also mean that the non-amateur has better muscle coordi-nation and mind muscle connection due to producing more balanced EMG activity between left and right muscles compared to an amateur. However, this view may not be accurate, as EMG comparison between left and right extremities mostly caused by the dominance side of the individual. For example, left handed person with have higher RMS at left hand. On top of that, as stated in [8], EMG differences in left and right side is also due to gender differences.

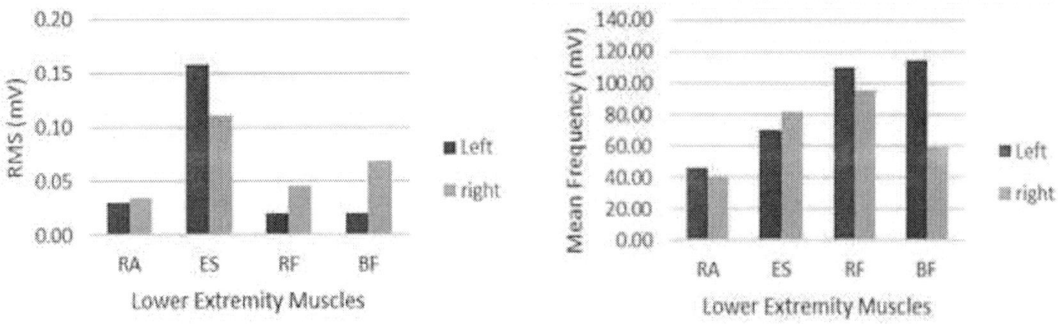

(a) (b)

Fig. 4. Squat analysis of a non-amateur (a) RMS data (b) MF data

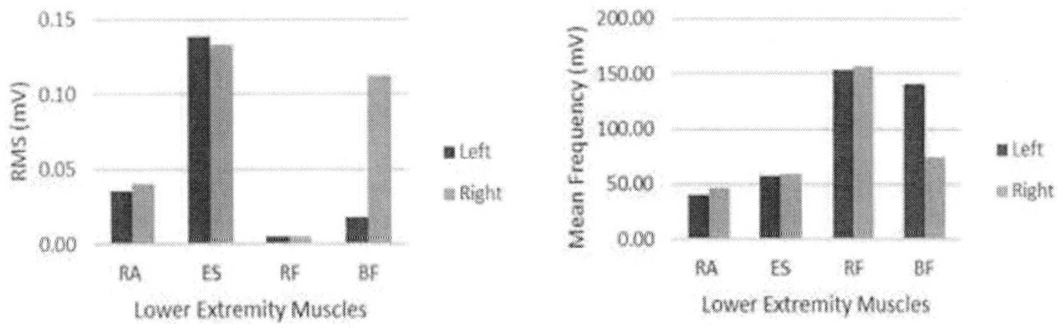

(a) (b)

Fig. 5. Stoop analysis of a non-amateur (a RMS data (b) MF data

Figure 4 and figure 5 show the squat and stoop analysis for both left and right muscles of a non-amateur respectively. For non-amateur, most of the muscles studied were shown to give balanced electrical (RMS and MF) signals for both right and left sides. However, the BF shows a higher RMS signal coming from the right side compared to the left when performing both squat and stoop. This could probably mean that the non-amateur has muscle imbalances in his rear thigh. However, as for the MF signal it's the opposite, the left side of BF muscle is higher than the right. Rea-son is because the RMS signal is actually used to look into muscle strength, whereas MF used to analyse muscle fatigue, therefore it explains why MF gave reverse data of the RMS. As for this study specifically, we are more focused on the RMS signal because the objective was to determined correct muscle usage i.e. strength, when performing the two movements, rather than to determine which muscles fatigued more. Furthermore, another significant finding would be that the non-amateur also gave out greater

EMG activity for ES during squat (RMS=0.159) than stoop (RMS=0.138) than the RA, hence these findings are in line with S. Hwang, Y. Kim, and Y. Kim, (2009) [3] and G. B. Shane Schwanbeck,Philip D.Chilibeck (2009) [9]. This could be due to the ES was working greater in keeping an upright torso rather than letting the torso bends, specifically at the lumbar region when performing the stoop.

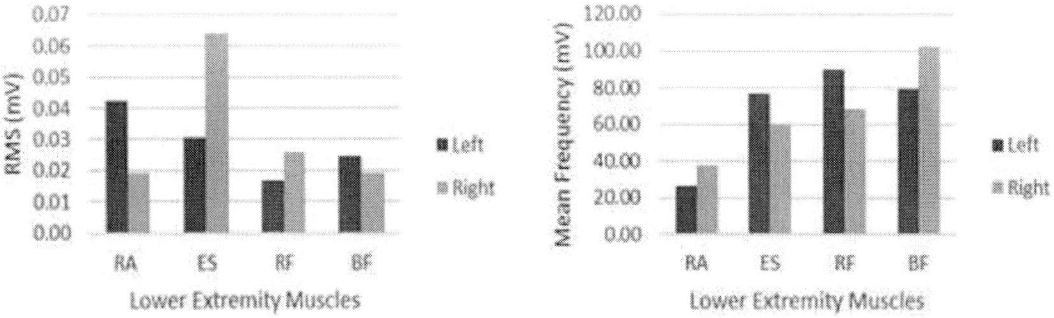

(a) (b)

Fig. 6. Squat analysis of an amateur (a) RMS data (b) MF data

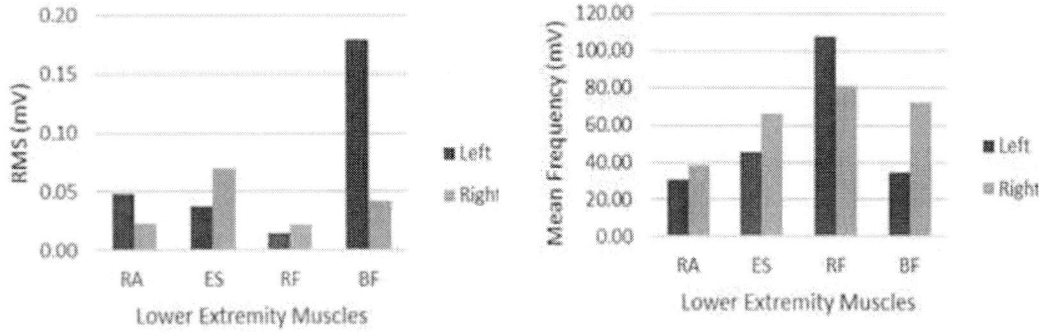

(a) (b)

Fig. 7. Stoop analysis of an amateur (a) RMS data (b) MF data

The data obtained from the amateur subject can be referred to figure 6 and 7, where muscle activities of the subject core muscles were significantly unbalanced specifically between left and right ES muscle for both squat and stoop. Another thing to notice is that it can be assumed that the left ES is weaker than the right, due to having lower electrical signal when performed both squat and stoop. This however has been compensated by left RA, which has worked greater than the RA. This is possibly due to the role between agonist and antagonist muscle, where for instance, when the right RA is relatively weak compared to the left, the right ES would work more to stabilize the upper body to avoid from twisting the lower back.

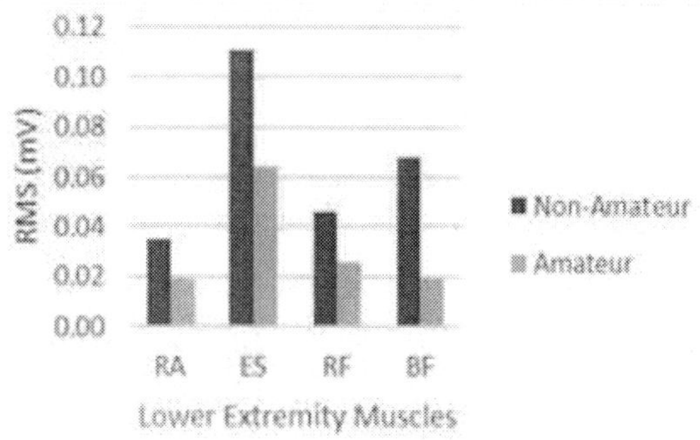

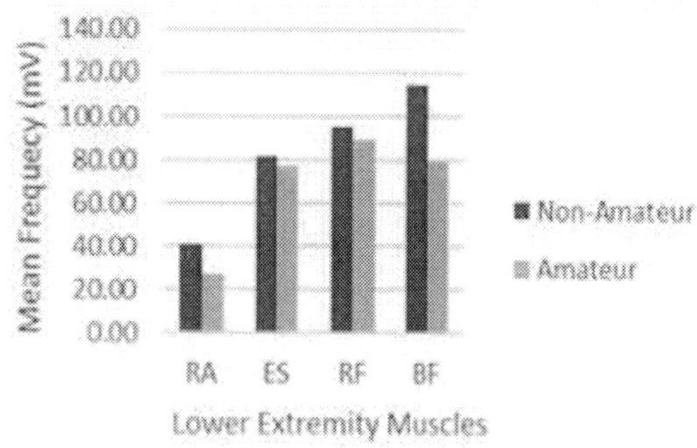

(a) (b)

Fig. 8. Squat analysis between amateur and non-amateur (a) RMS data (b) MF data

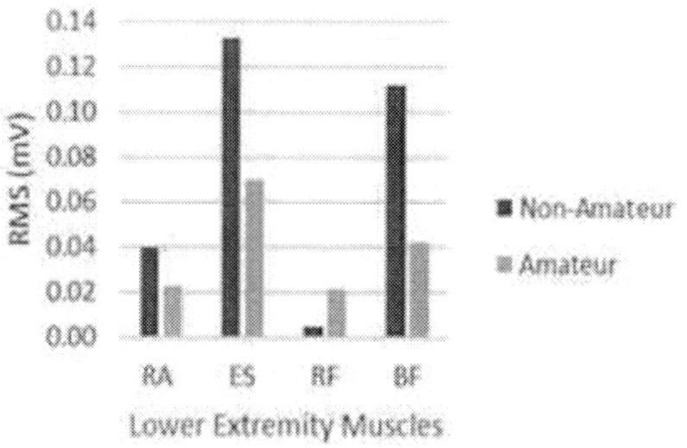

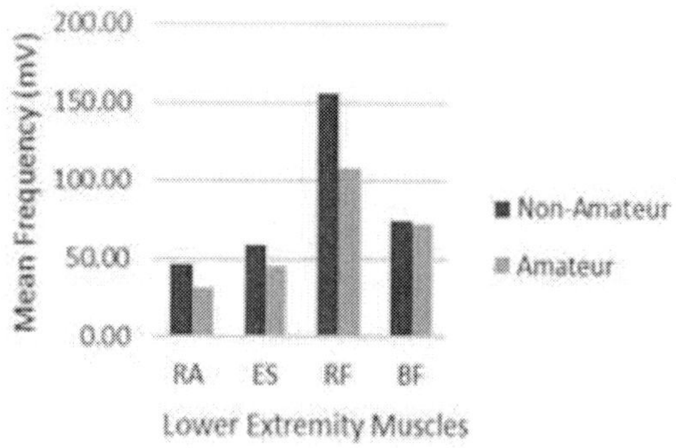

(a) (b)

Fig. 9. Stoop analysis between amateur and non-amateur (a) RMS data (b) MF data

Furthermore, from figure 8 and figure 9 we can see and distinguish the difference in muscles activities for squats and stoop between a non-amateur and an amateur. Non-amateur produced higher RMS and MF value for every muscle compared to an amateur. This makes sense because the non-amateur subject is more likely to be able to contracts each muscle proportionately and the whole muscles have to work together simultaneously so that proper technique and good posture can be attained. For example, the electrical signal of RA of a non-amateur would be higher com-pared to the amateur because non-amateur is able to utilized proper bracing technique when perform both squats and even during stooping to reduce the chance of getting low back pain.

Apart from that, there were also unexpected findings where the EMG activity shows a greater muscle electrical activity of BF than RF for squat. Contrary to previous findings, RF activity was usually higher than BF for squats, except for a study conducted by [10]. This may be due to several reasons, but mainly would be due to being inexperience in placing the electrodes accurately. However, it has been stated in a study conducted by Gregory. J. Hoyme. and J. J. J. James W. Youdas, John H. Hollman, James R. Hitchcock (2007) [11] that men are hamstring dominant and better at activating their hamstring more effectively.

4. Conclusion and Recommendation

In this study, the muscle activities of lower extremity were investigated through performing squat and stoop movements by using electromyography (EMG). The EMG data for erector spinae (ES), rectus abdominis (RA), rectus femoris (RF) and bicep femoris (BF) for both squat and stoop had been accumulated and normalized to root mean square (RMS) for ease comparison. The results show greater mean muscle activities for non-amateur than amateur. Apart from that, the non-amateur also gave out more balanced electrical signals in both sides for nearly every muscle except for bicep femoris (BF).

Even though the root mean square (RMS) value of the bicep femoris (BF) was slightly different from previous research, but the rectus femoris (RF) and core muscles data were in line with previous study.

Several past studies stated that there appears to be no single best lifting technique as stated in [12] and there is still no study able to prove and state that squat technique is superior than stoop for lifting. A correct squat technique is more preferable lifting technique as it requires to utilize the core muscles even more, for proper bracing. Hence, the chances of getting low back pain would be more likely to be reduced as the core muscles surrounding the lumbar area to protect and support it. As stated in L. Del Vecchio, (2017) [12], strengthening the muscles, ligaments and bones would increase the individual's ability to handle heavier loads and resist injuries, thus strengthening the core muscles are crucial in avoiding low back injury. Hence, recommendation to health practitioners and fitness enthusiast would be to create awareness and educate the general public in general lifting guidelines and also give exercise examples for corrective movement patterns from time to time through events such as "Hari Sukan Negara" con-ducted by the Ministry of Youth and Sports Malaysia. A well bal- ance posture and biomechanics are also important to avoid injury and preventing long term complication such as chronic low back pain.

References

[1] M. Motamedzade, A. Dormohammadi, E. Zarei, and R. Dormohammadi, "The role of ergonomic design and application of NIOSH method in improving the safety of load lifting tasks," ARAK Med. Univ. J., vol. 16, no. 6, p. 91, 2013.

[2] A. A. Godwin, J. M. Stevenson, M. J. Agnew, A. L. Twiddy, M. Abdoli-Eramaki, and C. A. Lotz, "Testing the efficacy of an ergonomic lifting aid at diminishing muscular fatigue in women over a prolonged period of lifting," Int. J. Ind. Ergon., vol. 39, no. 1, pp. 121–126, 2009.

[3] S. Hwang, Y. Kim, and Y. Kim, "Lower extremity joint kinetics and lumbar curvature during squat and stoop lifting," BMC Musculoskelet. Disord., vol. 10, no. 1, p. 15, 2009.

[4] J. H. Van Dieën, M. J. M. Hoozemans, and H. M. Toussaint, "Stoop or squat: A review of biomechanical studies on lifting technique," Clin. Biomech., vol. 14, no. 10, pp. 685–696, 1999.

[5] J. Wasilewski, "To stoop or squat, a thorough analysis. KNPE 450 - Ergonomics," 2014.

[6] and J. W. C. Jonathan C. Gullett, Mark D. Tillman, Gregory M. Gutierrez, "A Biomechanical Comparison of Back and Front Squats in Healthy Trained Individuals," pp. 284–292, 2008.

[7] L. R. Released et al., "proEMG 2.0 - How do you do?," no. March, pp. 1–51, 2013.

[8] M. Kaur, S. Nara, D. Shaw, and D. Bhatia, "EMG Asymetricity of Selected Knee Extensor Muscles in Sustained Squat Posture (A Yogic Posture) of Athletes in Relation to their Gender and Performance," vol. 7, pp. 1–6, 2016.

[9] G. B. Shane Schwanbeck,Philip D.Chilibeck, "a C Omparison of F Ree W Eight S Quat To S Mith," 2009.

[10] B. A. DeForest, G. S. Cantrell, and B. K. Schilling, "Muscle Activity in Single- vs. Double- Leg Squats.," Int. J. Exerc. Sci., vol. 7, no. 4, pp. 302–310, 2014.

[11] Gregory. J. Hoyme. and J. J. J. James W. Youdas, John H. Hollman, James R. Hitchcock, "comparison of hamstring and quadriceps femoris between men and women.pdf," J. Strength Cond. Res., 2007.

[12] L. Del Vecchio, "Choosing a Lifting Posture: Squat, Semi-Squat or Stoop," MOJ Yoga Phys. Ther., vol. 2, no. 2, pp. 1–7, 2017.

CFD analysis on velocity and temperature distribution of airflow inside model building with windcatcher

Nor Merlisa Ali*[1], Muyassarah Syahirah Mohd Yatim[1], Rosnadiah Bahsan[1]

[1]Faculty of Mechanical Engineering, Universiti Teknologi MARA, 40450 Shah Alam, Selangor, Malaysia.

**Corresponding email: merlisa@uitm.edu.my*

Abstract. Windcatchers are ventilation instruments used for obtaining the characteristic of cooling. They have been utilized in the buildings for a long time ago in countries with hot dry atmospheres especially in the Middle East. Natural ventilation is progressively utilized in the modern building to limit the use of non- sustainable supply. This study describes another alternative for adopting clean energy in buildings by applying windcatcher on buildings. This present study analyses between two types of windcatcher (one-sided and two-sided windcatcher) by varying the inlet velocity of airflow. This study aims to investigate wind velocity and temperature distribution in a model building with one outlet opening position using Computational Fluid Dynamics (CFD) simulation. Reynolds-Average Navier Stokes (RANS) approach with the turbulence model of standard k-ε was used to perform the simulation. It is shown that the two-sided windcatcher is more effective than the one-sided windcatcher. Two-sided windcatcher induces a high volume of airflow inside the building compared to the one-sided windcatcher.

1. Introduction

The exhibit of ventilation forte is available to the outside air causing unrestrained air flow in an enclosed space. Ventilation can be either completely regular or completely mechanical. Natural ventilation does not involve any mechanical appliances or devices to make the airflow inside the building. Meanwhile for mechanical ventilation, to make the air circulate inside the building, a mechanical device such as a blower or power drive is needed.

Recently windcatchers have undergone a considerable amount of research in terms of understanding better the effect of airflow through and over windcatchers as well as the ventilation rates by these systems [1]. Windcatcher is one of the methods used for natural ventilation. It can be characterized as an engineering component set on a building which gives natural air to the inside of living spaces and then discharges the stale air through windows or other depletes fragments [1]. It also could capture the wind at a high level and direct it to the building [3]. Windcatchers are generally classified into five groups which are one-sided, two-sided, four, six and eight-sided [4]. Numerous academic studies investigated the cooling performance that is executed by windcatchers. The result shows that windcatcher can play a significant role in creating natural ventilation and adjusting the thermal comfort of interior living spaces [5]-[6]. This present paper investigates the velocity and temperature distributions for both types of windcatcher by the means of CFD simulation.

The parameter that influences the performance of the windcatcher is the direction of the wind, the wind speed and the angle of incidence [2]-[3]. The distance between the windcatcher and opening outlet also is one of the factors for windcatcher performance. If the distance between the windcatcher and opening outlet is short, it will not give better-induced airflow for the windcatcher [1].

Content from this work may be used under the terms of the Creative Commons Attribution 3.0 licence. Any further distribution of this work must maintain attribution to the author(s) and the title of the work, journal citation and DOI.
Published under licence by IOP Publishing Ltd

2. Methodology

This study performed a simulation on a building with one-sided and two-sided windcatcher by comparing with two different inlet velocities which are 1.8 m/s and 3 m/s. This study aims to evaluate the velocity and temperature distribution inside the building with windcatcher. 3D model of the building was drawn using Computer-Aided Design (CAD), CATIA V5. The simulation of the airflow inside the model building was used Computational Fluid Dynamics (CFD), ANSYS 16.0 and Reynolds-Average Navier Stokes (RANS) approach with a turbulence model of standard k-ε was used. ANSYS FLUENT has been applied to various research works and found to be useful [9]–[12], which is adopted in this study which is to evaluate the airflow inside the building. Previous studies have also proved that these methods are reliable in terms of evaluating the ventilation performance of windcatchers [3].

2.1 Model Geometry and Building Configurations

Two different models were drawn using CATIA. The model building dimension is 6 m x 3 m x 8 m (width x height x length). Meanwhile, the dimension for the windcatcher is 1 m x 1.5 m x 1 m (width x height x length) and the size of the opening is 1 m x 1 m. Figure 1 shows the model building with one outlet opening at the opposite wall.

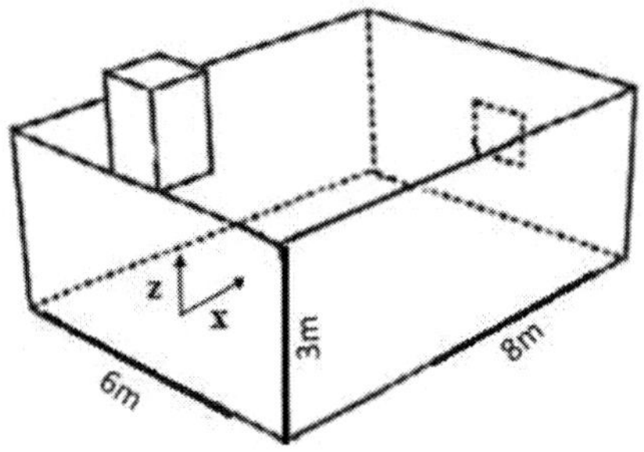

Figure 1. Model building with one outlet opening in CATIA drawing.

2.2 Numerical Setup and Model Validation

The turbulence model selected is the sstandard k-Epsilon; in short, k means the turbulent kinetic energy transported while Epsilon means turbulent dissipative energy. This turbulence model has been used widely for industrial purposes and performs better in low-speed air. Figure 2 shows the result of present study for normalized velocity in x-direction along with distance x, compared to validation from H. Montazeri et al [2].

In this study, the inlet velocity, U_{ref} used is 1.8 m/s and 3 m/s as it is the average wind speed in Malaysia and the temperature has been set to 300 K. The inlet velocity is in the x-direction. The velocity and temperature distribution were analysed in the next section of this paper.

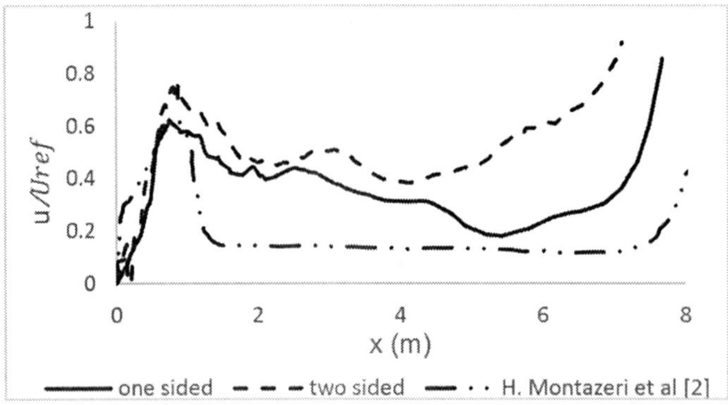

Figure 2. Model validation with result from H. Montazeri et al [2]

IOP Conf. Series: Materials Science and Engineering **834** (2020) 012036 doi:10.1088/1757-899X/834/1/012036

3. Results and Discussions

This section discusses the results obtained from CFD simulation with Reynolds-Average Navier Stokes (RANS) approach to perform the simulation. The result will be discussed in two sections: velocity and temperature distribution of the airflow inside the model building along the distance x, y, and z.

3.1 Velocity distribution in model building along x,y and z direction

Figure 3 (a) – (c) and Figure 4 (a) – (c) show comparisons between velocities inside the model building with one-sided and two-sided windcatchers. Both cases of the study investigated velocity and temperature distribution when the inlet airflow to windcatcher is 1.8 m/s and 3.0 m/s. It shows that the airflow velocity inside the model building fluctuated. It also shows that the two-sided windcatcher is more effective than a one-sided windcatcher. This result has been agreed based on a previous study [13] that shows that the two-sided windcatcher has higher efficiency than any other type, particularly in zero degrees wind incident angle which can induce a high volume of airflow into the room. Hence, this two-sided windcatcher is a typical conventional windcatcher in regions with predominant wind direction [14].

Figure 3(a) shows that the velocity of airflow inside the model building increases along distance x from the wall to the outlet of windcatcher. It is due to the source of the airflow leaves from the windcatcher to the model building. Then it shows that the velocity of airflow gradually reduces until it reaches near the wall with opening, and the velocity increases again due to the position of an opening which allows the airflow to speed up to enter the opening (which has a small area ratio to the wall). While Figure 3 (b) and (c) show quite similar profiles of airflow velocity inside the model building which increases at two streamwise positions: at the outlet of windcatcher and along with the position of opening at the opposite wall. But it also shows in Figure 3 (c) that there is evidence of reverse flow near the top of model building (about distance z is 3 m) and it shows also the airflow normalized velocity becomes almost unity inside the windcatcher. The airflow velocity in the model building can be seen as it increases very much at two significant locations: the outlet of windcatcher and near the outlet opening.

Figure 4 (a) – (c) show similar velocity profile to Figure 3 (a) – (c) except that the airflow velocity inside the model building is higher and will provide more flow rate since the inlet velocity is higher (3 m/s). The velocity reaches zero value from wall to wall (for y-direction) and also from bottom to top plane (for z-direction). But the airflow velocity reaches the maximum value at the outlet opening (for x-direction) for both inlet velocities: 1.8 m/s and 3.0 m/s.

IOP Conf. Series: Materials Science and Engineering **834** (2020) 012036 doi:10.1088/1757-899X/834/1/012036

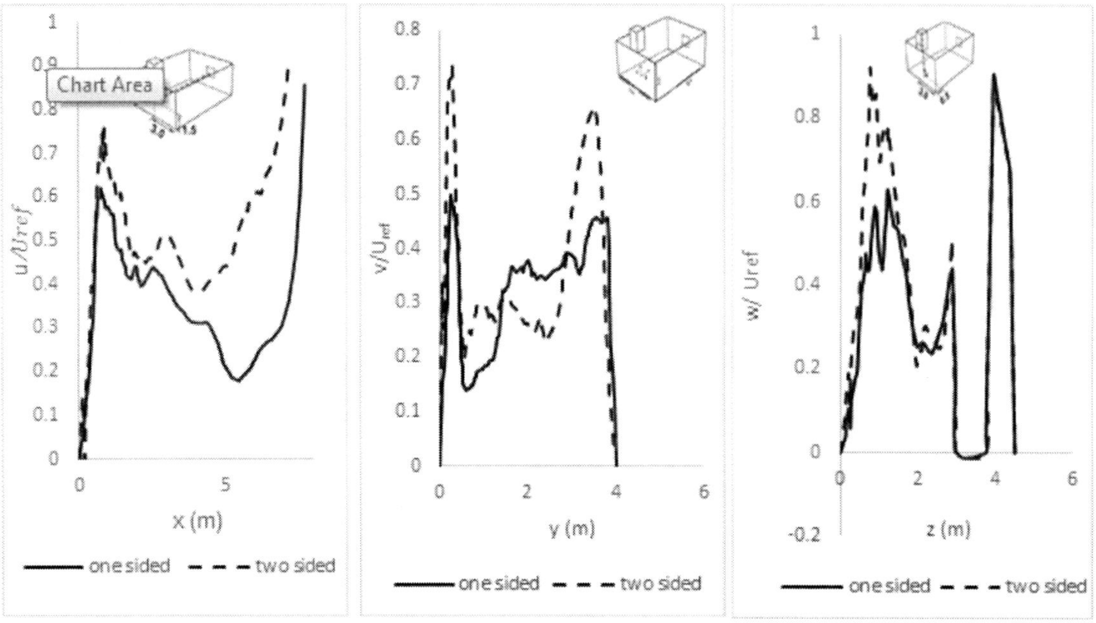

(a) (b) (c)

Figure 3. Velocity distribution in model building for inlet airflow 1.8 m/s along (a) distance x (b) distance y (c) distance z.

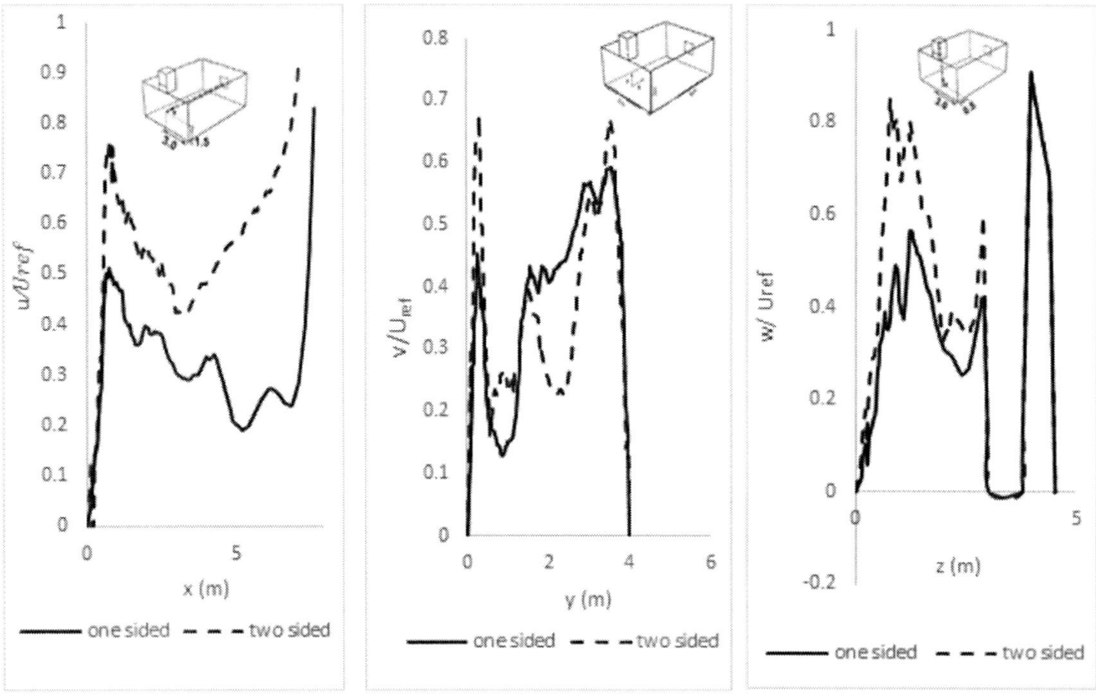

(a) (b) (c)

Figure 4. Velocity distribution in model building for inlet airflow 3.0 m/s along (a) distance x (b) distance y (c) distance z.

3.2 Temperature distribution in model building along x,y and z direction

Figure 5 (a) – (c) and Figure 6 (a) – (c) show comparison between temperature distribution inside the model building with one-sided and two-sided windcatchers. It shows that the temperature distribution inside the model building fluctuated and the profiles show similarity for both cases of velocity inlet to windcatcher (1.8 m/s and 3.0 m/s). It also shows that the temperature distribution inside the model building with two-sided windcatcher is lower than the one-sided windcatcher due to better airflow ventilation. The temperature also reduces in all direction inside the model building, except it, increases again when it reaches near boundaries (walls, top plane, and bottom plane) as the temperature is set to 300 K.

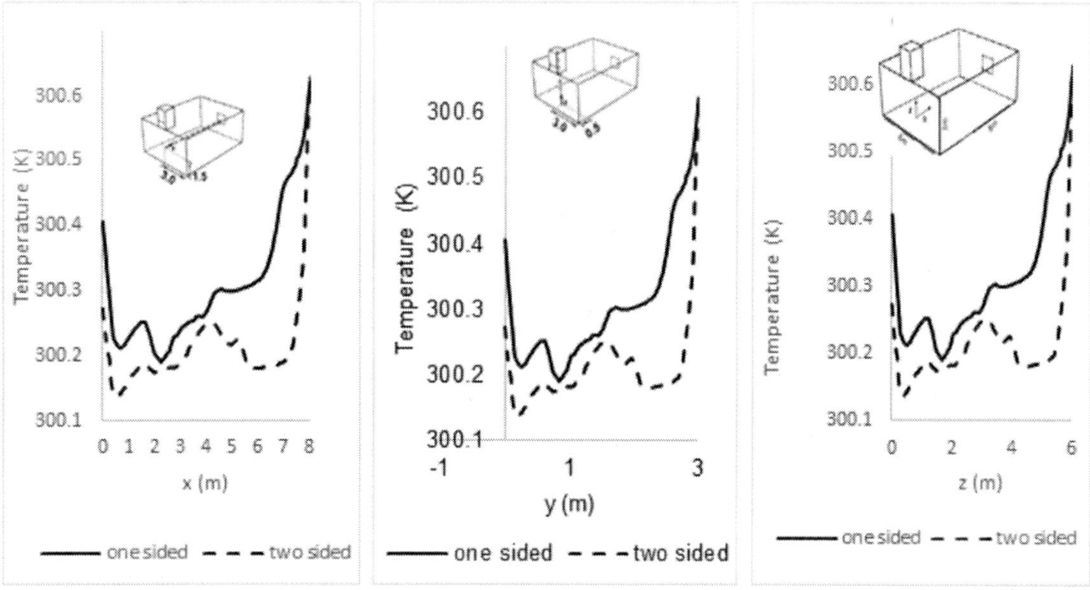

(a) (b) (c)

Figure 5. Temperature distribution along distance x, y and z in model building for inlet airflow 1.8 m/s.

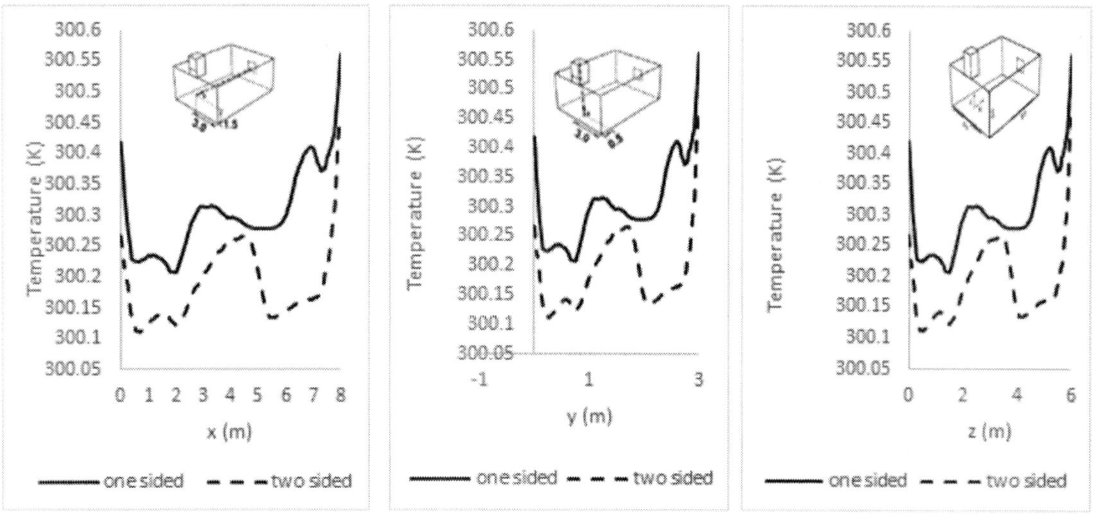

(a) (b) (c)

Figure 6. Temperature distribution along distance x, y and z in model building for inlet airflow 3.0 m/s.

4. Conclusions and Recommendation

In conclusion, the effects of one-sided and two-sided windcatcher to the model building with one opening at the opposite wall were examined in this present study by solving with the use of Reynolds-Average Navier Stokes (RANS) approach with standard k- ε turbulence model. The airflow velocity distributions inside the model building are affected by the velocity of inlet airflow, opening position, and also the type of windcatcher.

Based on the results of this present study, it can be concluded velocity and the temperature inside the simplified model building for both types of windcatchers are similar, except for small variations of normalized velocity at y-components which two-sided windcatcher has decreased at the middle plane of y-direction compared to a one-sided windcatcher. Meanwhile, the two-sided windcatcher gives a higher value of velocity or similar trend of the velocity profile at other location in x, y and z directions.

From the simulation result also can be concluded that a two-sided windcatcher induces a higher volume of airflow inside the building. Thus, it shows that both windcatchers are capable to provide better ventilation and thermal comfort inside the building but a two-sided windcatcher is preferable due to its more incoming airflow. Numerous studies have been done in order to evaluate the performance of windcatchers. For future research, the design of the windcatcher can be improved by adding humidifier or wet cloth or water spray so that it will be more effective. In terms of CFD modeling, researchers should investigate the impact of using other turbulence models to further improve the prediction of the models especially for complex flow inside the windcatcher.

References

[1] Calautit, J., O'Connor, D. and Hughes, B. (2014). Determining the optimum spacing and arrangement for commercial wind towers for ventilation performance. Building and Environment, 82, pp.274-287.

[2] Montazeri, H. and Montazeri, F. (2018). CFD simulation of cross-ventilation in buildings using rooftop wind-catchers: Impact of outlet openings. *Renewable Energy*, 118, pp.502-520.

[3] Dehghan, A., Esfeh, M. and Manshadi, M. (2013). Natural ventilation characteristics of one-sided wind catchers: experimental and analytical evaluation. *Energy and Buildings*, 61, pp.366-377.

[4] Nejat, P., Calautit, J., Majid, M., Hughes, B. and Jomehzadeh, F. (2016). Anti-short-circuit device: A new solution for short-circuiting in windcatcher and improvement of natural ventilation performance. *Building and Environment*, 105, pp.24-39.

[5] Kalantar, V. (2009). Numerical simulation of cooling performance of wind tower (Baud-Geer) in hot and arid region. *Renewable Energy*, 34(1), pp.246-254.

[6] Bahadori, M., Mazidi, M. and Dehghani, A. (2008). Experimental investigation of new designs of wind towers. *Renewable Energy*, 33(10), pp.2273-2281.

[7] Li, L. and Mak, C. (2007). The assessment of the performance of a windcatcher system using computational fluid dynamics. *Building and Environment*, 42(3), pp.1135-1141.

[8] Afshin, M., Sohankar, A., Manshadi, M. and Esfeh, M. (2016). An experimental study on the evaluation of natural ventilation performance of a two-sided wind-catcher for various wind angles. *Renewable Energy*, 85, pp.1068-1078.

[9] Toja-Silva, F., Kono, T., Peralta, C., Lopez-Garcia, O. and Chen, J. (2018). A review of computational fluid dynamics (CFD) simulations of the wind flow around buildings for urban wind energy exploitation. *Journal of Wind Engineering and Industrial Aerodynamics*, 180, pp.66-87.

[10] Toja-Silva, F., Kono, T., Peralta, C., Lopez-Garcia, O. and Chen, J. (2018). A review of computational fluid dynamics (CFD) simulations of the wind flow around buildings for urban wind energy exploitation. *Journal of Wind Engineering and Industrial Aerodynamics*, 180, pp.66-87.

[11] Mahdi, M. and Smaili, A. (2017). Numerical investigations of the thermal behavior of a HAWT nacelle using ANSYS FLUENT. *Energy Procedia*, 141, pp.394-398.

[12] Bojic, M., Lee, M. and Yik, F. (2002). Influence of a depth of a recessed space to flow due to air-conditioner heat rejection. *Energy and Buildings*, 34(1), pp.33-43.

[13] Montazeri, H. (2011). Experimental and numerical study on natural ventilation performance of various multi-opening wind catchers. *Building and Environment*, 46(2), pp.370-378.

[14] Montazeri, H., Montazeri, F., Azizian, R. and Mostafavi, S. (2010). Two-sided wind catcher performance evaluation using experimental, numerical and analytical modeling. *Renewable Energy*, 35(7), pp.1424-1435.

[15] Islam, M., Saidur, R. and Rahim, N. (2011). Assessment of wind energy potentiality at Kudat and Labuan, Malaysia using Weibull distribution function. *Energy*, 36(2), pp.985-992.

IOP Conf. Series: Materials Science and Engineering **834** (2020) 012037 doi:10.1088/1757-899X/834/1/012037

Numerical computation for prediction of grain growth on stainless steel 316L

Norasiah Muhammad[1], Yupiter HP Manurung[2], Muhd Faiz Mat[2], Mohamad Syakir Abdul Ghani[2], Marcel Graf[3], Tom-Eric Adams[4], Khairulnizam Kassim[1] and Shahriman Adenan[2]

[1]Mechanical Engineering Department, Politeknik Sultan Salahuddin Abdul Aziz Shah, Shah Alam, Selangor, Malaysia.
[2]Faculty Mechanical Engineering, Universiti Teknologi MARA (UiTM) Shah Alam, Selangor, Malaysia.
[3]Professorship of Virtual Production Engineering, Chemnitz University of Technology, Chemnitz, Germany.
[4]Chair of Welding Engineering, Chemnitz University of Technology, Chemnitz, Germany.

Corresponding author E-mail:yupiter.manurung@uitm.edu.my

Abstract. This research aims to predict the grain size growth using numerical computation. The material to be investigated is stainless steel 316L (SS316L). The mathematical modeling is derived into two cases, namely with the absence of precipitate (free growth) and with the presence of growing precipitates. The numerical computation involves ordinary differential equation using Runge-Kutta 4th order written with FORTRAN language. The experimental verification is carried out by using quenching and deformation dilatometers. It can be concluded that modified kinetic constant (Mo*) should be defined differently for certain temperature range.

1. Introduction

SS316L is an austenitic chromium-nickel stainless steel that contains between 2% and 3% molybdenum. The molybdenum content increases the corrosion resistance, improves resistance to pitting in chloride ion solutions, and increases strength at high temperatures [1]. Hence, this material is most familiar used in high resistance application such as heat exchanger, jet engine parts and also suitable in marine applications, in petrochemical reactor as well as gas industries [2]. This type of low carbon austenitic alloy is made to keep away from the formation of chromium carbides which result in an exhaustion of chromium from the austenite matrix and a loss in the resistance against corrosion. SS316L has a small thermal conductivity and a large thermal expansion coefficient and hence it is prone to produce welding residual stress and deformation [3].

The grain size has an influence on the material properties and a considerable effect on the mechanical properties. For instance at room temperature, the yield strength, hardness, fatigue strength, tensile strength and impact strength are increased with decreasing of grain size [4]. On the other hand, one of the most important factors controlling austenite grain structure at the end of cold-working and annealing operations is the austenite stability with respect to strain induced transformation [5]. Grain growth in stainless steel is enhanced by boron and to be increased with the temperature [3]. Consequently, to improve the strength of a material is to make the grains as small as possible by

Content from this work may be used under the terms of the Creative Commons Attribution 3.0 licence. Any further distribution of this work must maintain attribution to the author(s) and the title of the work, journal citation and DOI.
Published under licence by IOP Publishing Ltd

increasing the amount of grain boundary. Small grain sizes are accompanied by high volume of grain boundaries and as a result, the more grain boundaries that be present, the higher the strength becomes.

This research aim to predict the grain size using numerical computation of compiler Force 2.0 with Fortran programming languages by implementing the grain growth mathematical model. The grain growth for material SS316L was calculated in two cases with the absence of alloying elements or precipitates (free grain growth) and with the presence of growing precipitates. For comparison purpose, experimental investigation was carried out by using quenching and deformation dilatometers.

2. Numerical method for grain growth prediction

The flow process to calculate the grain size using numerical computation with Fortran programming languages is presented in figure 1. The input files contain input of temperature (T) and time (dt). The input data have heating and cooling temperatures while the condition of the temperature depends on the row that has been set by the user. The initial grain size used in this simulation is 20μm. In this programming, the grain size is considered as 20 μm during the heating temperature and the grain size will be calculated during the cooling temperature. The final grain size is calculated by using Ordinary Differential Equation (ODE)'s problem solving method.

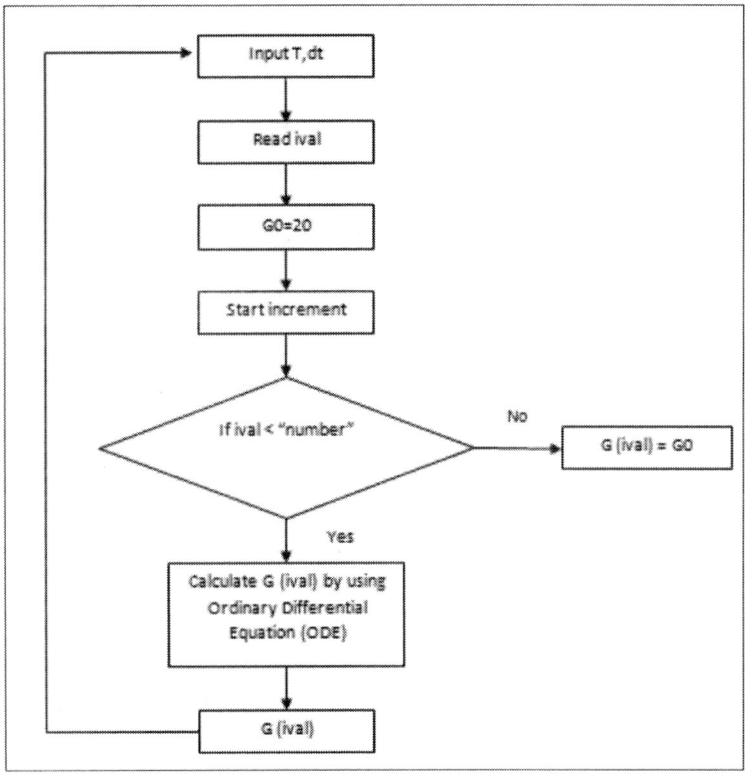

Figure 1. Flow chart for grain size calculation.

2.1. Basic equation for grain growth size calculation

In calculating the grain size for SS316L, there are two cases have been studied. Case 1 is a grain growth in the absence of alloying elements/precipitates (free grain growth) and case 2 is a grain growth in the presence of growing precipitates. A general formula for the grain growth as in equation (1) [6].

$$\frac{\partial \bar{g}}{\partial t} = M_0{}^{*}\exp\left[-\frac{Q_{app}}{RT(t)}\right]\left[\frac{1}{\bar{g}} - \frac{1}{\bar{g}_{lim}}\right]^{\left(\frac{1}{n}-1\right)} \quad (1)$$

Where;
$M_0{}^{*}$= Modified kinetic constant ($\mu m^2 s^{-1}$)
Q_{app}= Activation energy ($kJ\ mol^{-1}$)
R= Gas constant (8314 J $K^{-1} mol^{-1}$)
T= Absolute temperature (K)
$\bar{g}$= Average/initial grain size (μm) and $\bar{g}_{lim}$ depends on the cases.

The grain growth formula for the case 1 as in equation (2) below:

$$\frac{\partial \bar{g}}{\partial t} = M_0{}^{*}\exp\left[-\frac{Q_{app}}{RT(t)}\right]\left[\frac{1}{\bar{g}} - \frac{1}{\bar{g}_{lim}}\right]^{\left(\frac{1}{n}-1\right)} \quad (2)$$

Where;
$\bar{g}_{lim} = k\frac{r}{f}$

r = Radius of precipitate (μm)
f = Volume fraction of precipitate
k = Zener coefficient

The value of f = 0 due to absence of precipitate therefore $\bar{g}_{lim}$ will become ∞. As a result, the numerical equation for case 1 as in equation (3).

$$\frac{\partial \bar{g}}{\partial t} = M_0{}^{*}\exp\left[-\frac{Q_{app}}{RT(t)}\right]\left[\frac{1}{\bar{g}}\right]^{\left(\frac{1}{n}-1\right)} \quad (3)$$

The grain growth equation for the case 2 as in equation (4) below:

$$\frac{\partial \bar{g}}{\partial t} = M_0{}^{*}\exp\left[-\frac{Q_{app}}{RT(t)}\right]\left[\frac{1}{\bar{g}} - \frac{1}{\bar{g}_{lim}}\right]^{\left(\frac{1}{n}-1\right)} \quad (4)$$

Where;
$\bar{g}_{lim} = \left[\left(\overline{g_{lim}}^{0}\right) + \left(\frac{k}{f_0}\right)^3 I_2\right]^{\frac{1}{3}}$
$\overline{g_{lim}}^{0} = k\frac{r_0}{f_0}$
$I_2 = C_5\int_0^t \frac{1}{T(t)} exp\left[-\frac{Q_s}{RT(t)}\right]dt$
r_0 = Initial radius of precipitate (μm)
f_0 = Initial volume fraction of precipitate
I_2 = Kinetic strength of the thermal cycle particle coarsening (μm^3)
C_5 = Kinetic constant
Q_s = Activation energy for coarsening process ($kJ\ mol^{-1}$)

2.2. Ordinary differential equation (ODE)

In mathematics, ODE is a differential equation containing one or more functions of one independent variable and derivatives of those functions [7]. Since the equation for the grain size is a type of ODE equation, therefore the Runge-Kutta 4th order method was used to solve the grain size formula as in equation (5 – 9).

$$G_{i+1} = G_i + \frac{1}{6}(K_1 + 2K_2 + 2K_3 + K_4)h \tag{5}$$

$$K_1 = f(dt_i, T) \tag{6}$$

$$K_2 = f(dt_i + \frac{1}{2}h, G_i + \frac{1}{2}K_1 h) \tag{7}$$

$$K_3 = f(dt_i + \frac{1}{2}h, G_i + \frac{1}{2}K_2 h) \tag{8}$$

$$K_4 = f(dt_i + \frac{1}{2}h, G_i + K_3 h) \tag{9}$$

Where;
G_i = initial grain size
G_{i+1} = next grain size
h = time step
T = temperature
dt_i = time

3. Experimental validation of SS316L

In order to obtain the grain size of SS316L, experiment has been conducted using quenching and deformation dilatometers machine with model DIL 805A/D as in figure 2. The chemical composition of SS316L is tabulated in table 1. The specimen as presented in figure 3 had been through heat treatment process at temperature of 1000 °C, 1100 °C, and 1200 °C with the holding time of 30 s. Afterward, the epoxy has been used to mount the specimens. After 12 hours, the specimens were grinded by using forcipol grinding machine in order to get a better surface. For metallographic observation, the specimens were etched with V2A solution for 10 minutes and consequently the grain size of the SS316L was defined.

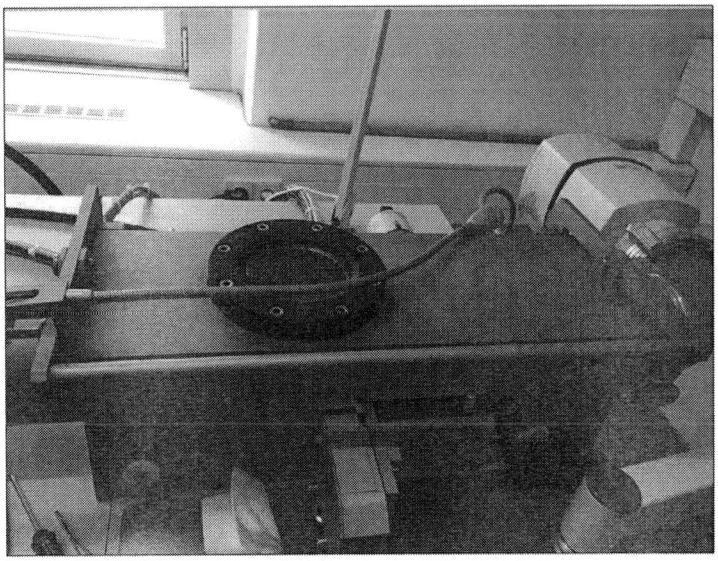

Figure 2. Quenching and deformation dilatometers.

Table 1. Chemical composition of SS316L.

%C	%Mn	%Si	%Cr	%Ni	%Mo	%P	%S
0.05	2.0	1.0	19	11	2.3	0.045	0.03

Figure 3. Specimen of SS316L.

4. Result and discussions

The grain growth of SS316L was predicted using numerical computation with Fortran programming languages by implementing the grain growth mathematical model. In this simulation, the temperature used to predict the grain size were 1000 °C, 1100 °C, and 1200 °C with the initial grain size was 20 μm. Other parameters for calculating the grain size such as activation energy, time exponent and gas constant were referred to Øystein G [6]. A simulation result in terms of grain size with the absence of alloying elements is presented in figure 4. From the graph, it can be seen that the grain grows from 20 μm to 27. 80 μm at temperature of 1000 °C, it grows from 20 μm to 40.54 μm at temperature of 1100 °C and grows bigger at temperature of 1200 °C which is from 20 μm to 64.28 μm.

Figure 5 presents the grain growth simulation results with the presence of growing precipitates. From the graph, it shows that the grain grows from 20 μm to 27.57 μm, 20 μm to 39.52 μm and 20 μm to 63.18 μm for the temperature of 1000 °C, 1100 °C, and 1200 °C respectively. Grain growth is a thermally activated process in which the grain size increases as temperature and time are increased [3], [8], [9]. This occurs when recovery and recrystallisation are complete and further reduction in the internal energy can only be achieved by reducing the total area of grain boundary. Thus, an easy way to improve the strength of a material is to make the grains as small as possible by increasing the amount of grain boundary. Smaller grains have better ratios of surface area to volume, which means a greater ratio of grain boundary to dislocations.

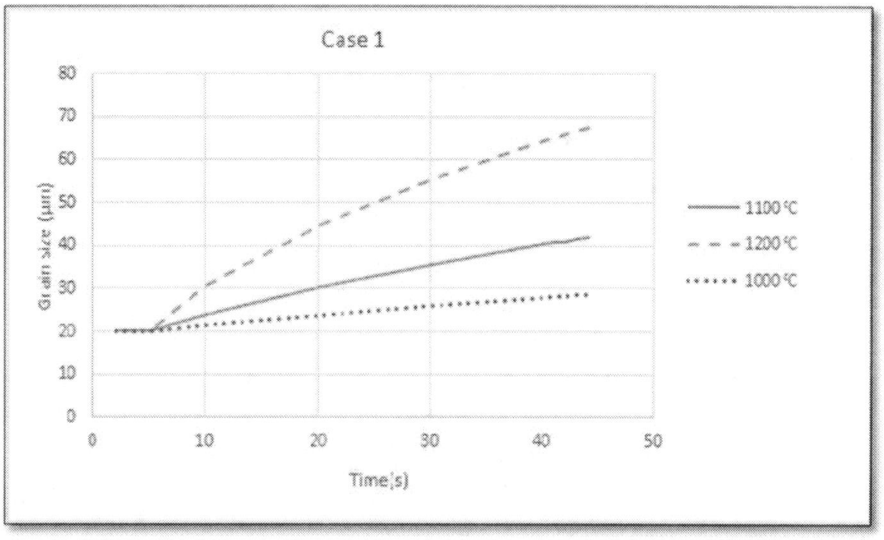

Figure 4. Grain growth with the absence of alloying elements.

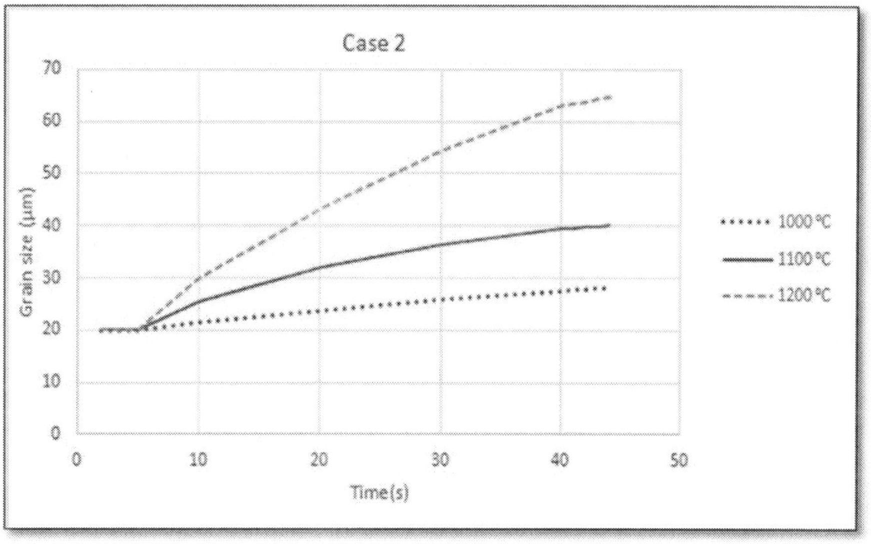

Figure 5. Grain growth with the presence of growing precipitates.

Grain growth occurs by boundary migration and the rate of boundary migration increases with driving force. The driving force for grain growth results from the decrease in the free energy of the system. The grain boundary area is the main source of energy in the austenitic microstructure. Consequently, during this process the system will evolved to reduce its grain boundary area. The larger grains will grow at the expense of the smaller ones. Microalloying elements such as vanadium, niobium and titanium have been employed in produce fine precipitation in the matrix. The austenite grain boundaries and dislocations are pinned by these precipitates, inhibiting their movement during the thermomechanical processing of steels [10]. The influence of these elements is shown by the way they delay the grain growth process and also the phase transformations, during subsequent cooling. As a result, their influence leads to fine-grained microstructures with improved the mechanical properties.

Table 2 displays the experimental result of the grain growth for SS316L. It was observed using optical microscope with 20X magnification. From the observation, it shows the grain size increases with increasing of the temperature. The grain growth is the increase of grains (crystallites) in a material at high temperature. The higher the temperature, the heating occur austenite grain size. When the heating temperature increased it will result in increasing of the grains size formed [11]. This is due to the increase of holding time and heating temperature that will increase the activation energy of growth which will go against previous dislocations that exists on the grain boundaries, so the growth of grain boundaries is not deferred by dislocation for growing.

Table 2. Experimental results.

Temperature	1000 °C	1100 °C	1200 °C
Grain size	28 μm	40 μm	63 μm
Micrograph	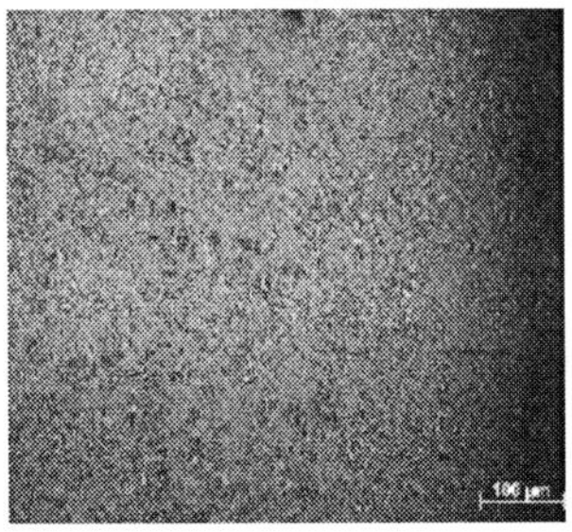		

Simulation and experimental results of the grain growth are depicted in Table 3. The results indicate the grain sizes increased with temperature, as predicted from the grain growth equations. An increase in temperature was found to have a larger effect on the grain size. Grain growth predicted with the presence of growing precipitates (case 2) is closed to experimental result. The approximate modified kinetic constant (Mo*) used in this simulation for the case 2 is tabulated in Table 4. The percentage error between simulation and experiment is presented in table 5. It was found out that the numerical computations of grain growth prediction are accurately correlated with the results of the experiment.

Table 3. Simulation and experimental results.

Temperature (°C)	Grain size (μm)		
	Case 1	Case 2	Experiment
1000	27.80	27.57	28
1100	40.54	39.52	40
1200	64.28	63.18	63

Table 4. Approximate modified kinetic constant.

Temperature (°C)	Mo* ($\mu m^2 s^{-1}$)
1000	5000×10^9
1100	3000×10^9
1200	2000×10^9

Table 5. Percentage error between simulation and experiment

Temperature (°C)	Percentage error (%)	
	Case 1	Case 2
1000	0.71	1.54
1100	1.35	1.20
1200	2.03	0.29

5. Conclusion

The fundamental study on prediction of grain size growth of SS316L was conducted by using Fortran based numerical computation. The comparison between results of numerical computation and experiment show good agreement, since different modified kinetic constant (Mo*) are considered for different temperature. As further recommendation, a strategy for fitting the Mo* should be discuss before implementing this model on manufacturing application.

Acknowledgments

The authors would like to express their gratitude to staff member of Research Interest Group: Advanced Manufacturing Technology (RIG:AMT) and Advanced Manufacturing Technology Excellence Centre (AMTEx) at Faculty of Mechanical Engineering, UniversitiTeknologi MARA (UiTM) for encouraging this research. This research and conference participation is also financially supported by DAAD Germany and Technogerma Engineering & Consulting Sdn. Bhd. A special thank is addressed to Jabatan Pengajian Politeknik for funding Applied Research Grantt Scheme.

References

[1] Song R B, Xiang J Y and Hou D P 2011 Characteristic of mechanical properties and microstructure for 316L austenitic stainless steel *Journal of Iron and Steel Research* **18** 53 - 59.
[2] Aryan S 2017 Hardness, wear and corrosion improvement techniques for austenitic stainless steel ss316L – a review *International Research Journal of Engineering and Technology* **04** 955 - 959
[3] Lichan L, Mengyu C, Yongquan L, Wenjie B and Quan D 2013 Effect of welding heat input on grain size and microstructure of 316L stainless steel welded joint *Applied Mechanics and Materials* **331** 578 – 582.
[4] Gladman T and Dulieu D 1973Grain size control in steel *Conference on Recrystallization in the Control of Microstructure* 167 – 175.
[5] Daniel H H 2005 Grain size and its influence on materials properties *The heat treat doctor* 20 –22.
[6] Øystein G 1997 Metallurgical modelling of welding *The Institute of Material* 338 – 386.
[7] Dennis G Z 2012 A first course in differential equations with modeling applications *Cengage Learning* ISBN 1-285-40110-7.
[8] Martin D S, Caballero F G, Capdevila C and Garcia de Andres C 2004 Kinetics of austenite grain growth during a continuous heating of a niobium microalloyed steel *2nd Joint International Conference on Recrystallization and Grain Growth* ISBN:978-0-87849-952-6.
[9] Wagner S, Kahraman D, Kungl H, Hoffmann M, Schuh C, Lubitz K. et al. 2005 Effect of temperature on grain size, phase composition, and electrical properties in the relaxor ferroelectric-system $Pb(Ni1/3Nb2/3)O_3$-$Pb(Zr,Ti)O_3$ *Journal of Applied Physics* **98**.
[10] Biyu Y, Yongchang L, Zejun W, Chenxi L, Yonghong S, Huijun L and Jianxing Y 2017 The effect of precipitate evolution on austenite grain growth in RAFM steel *Materials* **10** 1 – 11.
[11] Richard A M N 2017 Influence of heating rate and temperature on austenite grain size during reheating steel *Material Science and Engineering* **237** 1 – 6.

Experimental analysis on the effects of pigmentation in the defects formation of the polypropylene plastic injection moulding

Kausalyah V*[1], Nik Maisara E N M[2], Shasthri S[3]

[1,2]Faculty of Mechanical Engineering, Universiti Teknologi MARA, Shah Alam, Malaysia.
[3]School of Engineering and Physical Sciences, Heriott Watt University Malaysia, Precint5,62200, Putrajaya, Malaysia

*kausalyah@uitm.edu.my

Abstract. Pigment levels are known to have an effect on defect formation of the plastic injection moulding parts. Warpage, Shrinkage and Flash are common defects that were found during plastic injection moulding process. It was found that the pigment does play a role in the formation of defects on the plastic injection moulding parts. The main finding of this case study is to analyse and study the warpage, shrinkage and flash defects that were found in the Plastic Injection Moulding (PIM) parts which are moulded with different pigmentation percentages. It is also to identify the effects of the different pigments colours on the moulding quality of the parts when the pigment percentage is made constant. In this research, the dog bone design was used as the experimental specimen. The material used in this study is Polypropylene (PP). The other parameters such as injection pressure, melting temperature and cooling time were set to constant. The only parameter that varies is percentage of pigmentation. Warpage, Shrinkage and Flash defects are measured and calculated to obtain the value of defects. The pigmentation gives a huge effect on Warpage and Flash deformation. The warping and flash increase as the percentage of pigmentation increase. The highest value for warpage and flash are 3.0732 and 1.45304 respectively. Moreover, blue pigment gives the lowest value for Warpage and Flash defect. While yellow has the highest warping and flash defects. However, it was found that shrinkage is not affected by the pigmentation. The shrinkage result is less than 1%.

1. Introduction

The most common method of plastic processing technique is using the injection molding machine. It is the most practical and cost effective method to produce plastic product. There are a few common plastic injection molding defects that can be identified in the plastic injected parts such as warpage, sink marks, flash and shrinkage. Warpage is when the intended shape of the molded part is distorted during the cooling process. Mold warping can cause the part to fold, bend, twist or bow. Shrinkage is the contraction of the molded part as it cools after injection. Most of the part shrinkage occurs in the mold while cooling, while Flash is a molding defect that occurs when some molten plastic escapes from the mold cavity. Typical routes for escape are through the parting line or ejector pin locations.

Industry that has experienced rapid growth is the polymer manufacturing industry, in which polymer production has increased by 300 million tons from 1975 to 2014 [1]. The quality of the injection moldings depends on the material characteristics, the mold design and the process conditions [2].

Content from this work may be used under the terms of the Creative Commons Attribution 3.0 licence. Any further distribution of this work must maintain attribution to the author(s) and the title of the work, journal citation and DOI.
Published under licence by IOP Publishing Ltd

Polypropylene (PP) is a thermoplastic "addition polymer" formed from the combination of propylene monomers. Polypropylene is a thermoplastic polymer with chemical formula (C3H6)n and chemical name of poly(1-methylethylene) [3]. It is widely use due to it numerous properties of which comprise roughness, strong resistance to chemical solvents, as well as resistance to acids and bases. In addition, polypropylene is translucent and has a high fatigue and heat resistance. Polypropylene can be made translucent if uncolored but is not as readily made transparent as polystyrene, acrylic, or certain other plastics and it is reasonable economically [4].

The molding process parameters play an important role in the plastic injection molding [5]. Process parameters involved are melting temperature, holding time, cooling time and injection pressure. This process parameters will affect the quality of the product produce. It is important to optimize the process parameters to it best condition to reduce any defects in the plastic injection molding parts. It was found that holding pressure and melt temperature affect the tensile strength of material [6]. Since this research focusing on pigmentation, thus the process parameters are keep constant. The best parameters are chosen to avoid any defects formation caused by the parameters. Warpage and shrinkage are among the most significant defects. Rizwan and Gaurav said that process parameters can be modified at any stage of the process while other factors like mold design and material have certain limitations of cost and time [7]. Pujari and Naik found that optimized parameters combinations of different factors are considered as melt temp 235°c; holding pressure 35 bar, injection pressure 65 bar & cooling time 12 sec [8].

The most common problem for injection molded products is warpage [9]. Warpage is a form of distortion that can occur in some materials, such as wood or plastic. This usually results from uneven stresses that can be internal or external to the material being warped [10]. The main cause of warpage in injection molded parts is the uneven volumetric shrinkage from high to low temperatures [11].

Shrinkage is defined as the reduction in the size of part as compared to the size of the mold. Uniform shrinkage does not create part deformation and change in shape, but it becomes smaller [12]. High shrinkage occur when an inadequate amount of plastic is pushed into the mold and the pressure on the plastic is maintained for a short an interval of time [13].

Flash is excess plastic that has extended out past the edge of the part [14]. Flash happened due to clamp force is too weak, gap within the mold, high melt temperature, high injection pressure, very poor venting system [15].In this study the flash is measured as excess weight of the product. There are many factors contribute to the warpage, shrinkage and flash defects. The common parameters involved are cooling time, injection pressure, injection speed, holding time. However, in this study the factor that is being investigated is the pigmentation percentage in the specimen mixture. This research focuses on finding out how much pigmentation causes defect to the plastic injection molded parts such as warpage, shrinkage and flash. In the past, very few research work was done to analyse the effects of pigmentation on plastic injection molded part. James et.al. worked on similar concepts whereby it was identified that addition of colorants, fillers and process stabilizers affect the molded parts differently [16].

The main objective is to investigate if the concentration levels of pigmentation will produce warpage, shrinkage and flash defects. To ensure the defects produced are only cause by the pigmentation, other parameters such as injection pressure, melting temperature and cooling time were kept constant. The concentration level of the pigment varies from range 1% to 10%.

The other objective is to find out how the different pigmentation colors will effect plastic injection molding parts. There were 3 pigments tested blue, yellow and red. The concentration level of these 3 pigments is keep constant. The result will be obtained from the experiment and after the defects have been identified. However, different pigment will have different preparing process. So this research is to identify if different pigment will cause different effect on plastic injection molding parts. Thus, it is important to obtain the ingredient or material that were used and how does it affected the result.

Colorant is used in plastics injection molding to provide desired appearance properties which can help in selling the product. Pigment can be divided into two, inorganic and organic. Inorganic pigments have a good heat resistance and weathering properties compared organic pigments [16]. Moreover, inorganic pigments can be classified into standard metal oxides or complex inorganic colored pigments and effect pigments. Inorganic pigments provide durability and stability in extreme environments. [17]. In this research, inorganic pigment is used, yellow, blue and red. This pigment range from 1% until 10% are used to investigate the defects, warpage, shrinkage and flash.

The significance of this study is able to gain knowledge on the effect of small range percentage of pigmentation on the outer or physical appearance plastics injection molding part. Moreover, by completing this project it might be able to determine most suitable range for percentage of pigmentation in injection process in order to avoid the defects such as warpage and shrinkage from occurring. The outcome of this research will be able to assist the industry to differentiate the effect of different pigments on the physical appearance of the parts. Hence, this improves the quality of existing products and processes and simultaneously minimizes the costs significantly.

2. Methodology

2.1. Process Parameters

In this experiment, the process parameters that taken into account are melting temperature (210°C), injection pressure (100Bar) and cooling time (15s). All these parameters were set to constant. These parameters are chosen based on the default settings of the injection moulding machine used.

The first experiment investigates the effects of different percentages of pigments on the defects formation on the plastic injection moulding part. Hence, the only parameter that varies is percentage of pigmentation which ranges from 1% to 10%. The colour used in this experiment is yellow. Warpage, shrinkage and flash defects are then measured.

In the second experiment, different pigment colours are used to identify the colorant effect on the defect formation on the plastic injection moulding part. All process parameter such as injection speed, melting temperature, and cooling time are set to constant and percentage of pigmentation is set at 5%. Two other different pigment percentages were used to repeat the experiment such as 3% to represent small range percentage while 8% for high range percentage to gauge the results. The pigment colours used are blue, yellow and red.

2.2 Conducting Experiment

Selection of the injection moulding parameters and their levels was carried out on Plastic Injection Moulding machine VDC II series MMC II (Korea) as shown in Figure 1. The material used was Polypropylene resin and was injected into the dog bone mould as shown in Figure 3. First, the Polypropylene was mixed with colour additive, 10: 1 ratio in which 400 g and 4 g respectively. Then they were pre-heated in the hopper and ready for injection moulding shot.

Figure 1. Injection Moulding Machine

Figure 2. Polypropylene and master batch hand mixed.

The experiment was repeated from 1% until 10% of pigmentation by using yellow master batch. It was the same for the next experiment which used 3%, 5% and 8% for each pigment. The polypropylene and pigment were hand mixed before inserted into the machine.

2.3 Defects Investigation
The Figure 3 below is the specimen of the dog bone design that was produce by the machine. These specimens were then taken to be measured for their defects.

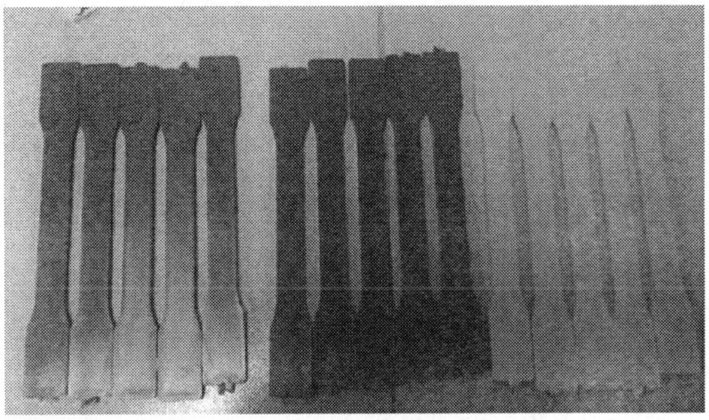

Figure 3. Dog bone specimens in yellow, blue and red.

2.3.1 Shrinkage. The first test was to measure the shrinkage defects. The shrinkage defect is measured by the thickness of the product. The actual thickness of the dog bone is 4.15 mm. Any value that is less than 4.15 mm is deemed as shrinkage. The thickness of the specimen is measured by using digital Vernier Callipers. 5 points were taken for each of the specimen to ensure the validity of the result.

2.3.2. Warpage. The following test is the warpage measurement. Warpage test is important in order to determine the allowable warping for the product in industry. Warpage test was carried out by using Quick Vision Ace 250 machine as shown in Figure 4. This machine is able to measure the surface of the specimen and give the exact value. The difference between the highest and lowest value is the value of warpage.

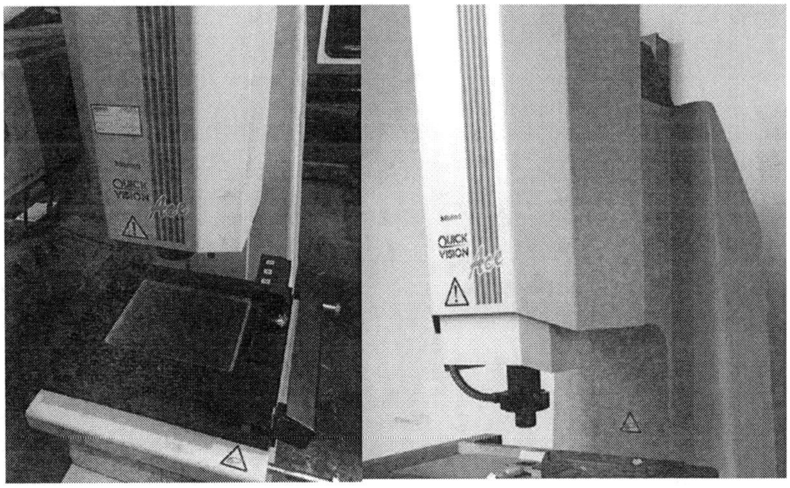

Figure 4. Non-contact Quick Vision Ace 250 machine.

*2.3.3. Flash.*Flash in this study is measured by the excess weight of the product. All the specimens are weighed using the electronic scale (Figure 5). The actual weight of the specimen is 8.20 gram. Any value more than 8.20 is counted as flash defect. The result is then tabulated.

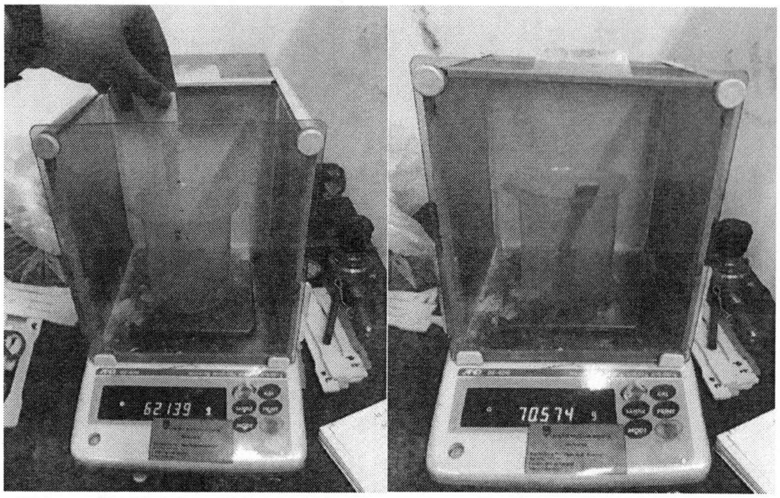

Figure 5. Weight of the specimen measured using an electronic scale.

3. Results and Discussion

3.1. Shrinkage

The thickness of the specimen is recorded and the average values of the 5 points are taken. The graph of Figure 6 shows the shrinkage rate at small range percentage of concentration level of pigmentation from 0% to 5%. The graph is constant and the shrinkage values are less than 1%. The highest shrinkage value is at 2% concentration level which is 0.530%. 1% and 3% was found to have negative shrinkage.

Next, the Figure 7 show the graph of shrinkage for 6% to 10% of pigmentation. The highest shrinkage value is at 7% which is 0.405%. However, the value still does not exceed 1%. Other concentration level such as 8% and 9% remain the same. This proves that from 0% of concentration level until 10%, it does not affect the plastic injection moulding parts in term of shrinkage defects.

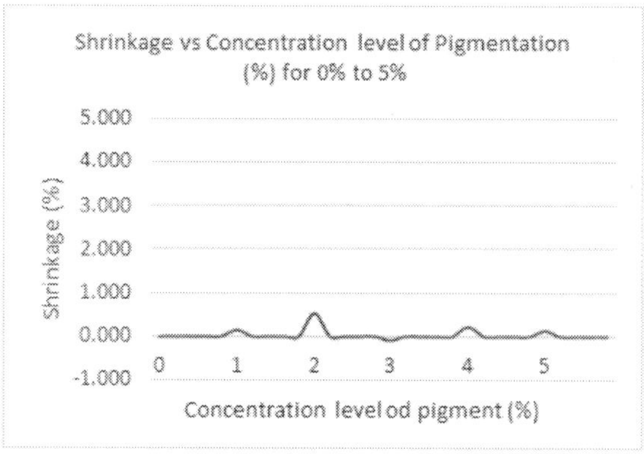

Figure 6. Graph of shrinkage against percentage of pigment for 0% to 5% pigment.

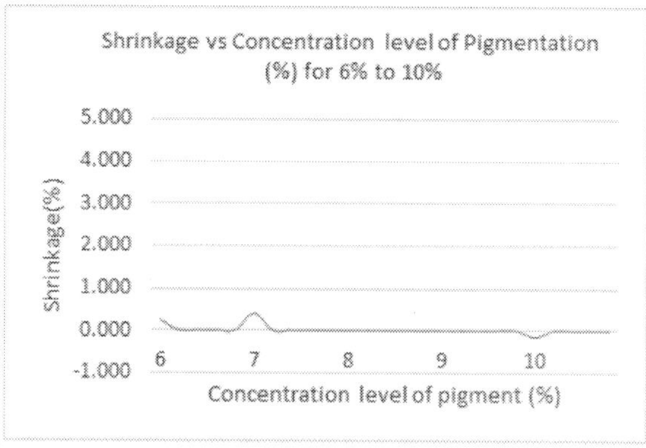

Figure 7. Graph of shrinkage against percentage of pigmentation for 6% to 10%.

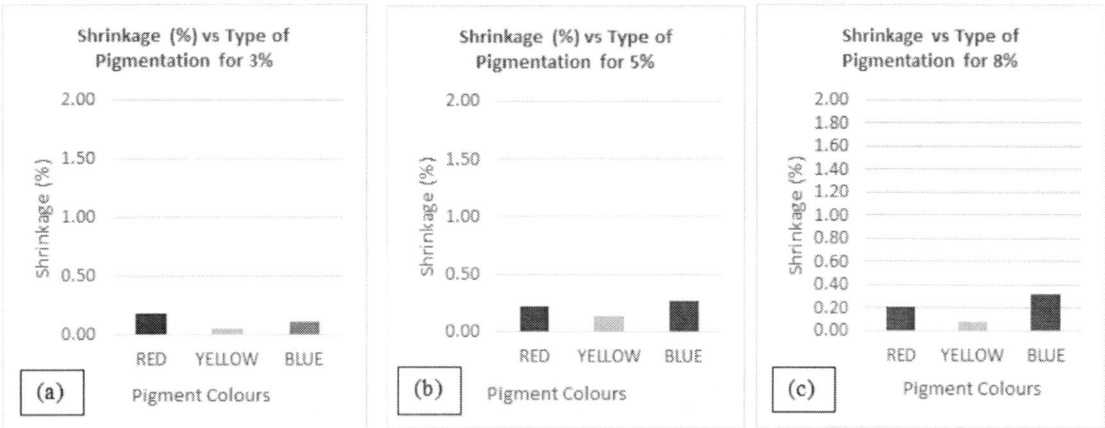

Figure 8. Graph of shrinkage against type of pigmentation for 3% (a), 5% (b) and 8% (c).

Figure 8,9 and 10 shows the graph of shrinkage for the three different type of inorganic pigmentation colours, red, blue and yellow. The first experiment used 3% of each pigment and then was repeated for 5% and 8%. As depicted in all the graphs, the value of the shrinkage defects for blue, yellow and red does not exceed 1% for three different percentages. Overall from these graphs, it can be concluded that inorganic pigment does not have a significant effect on shrinkage formation of plastic injection moulding part as they do not impact the crystallization rates like organic pigment.

3.2. Warpage

Figure 9 and 10 shows the graph of warpage against concentration level for 0% to 5%. It can be seen that as the concentration level increase, the warping also increases. The highest warpage value obtains at 5% of concentration level with 0.8783 mm. This is considered as low warping. However, at 4% the warpage suddenly dropped to 0.8278 mm from 0.84074 mm at 3%. At 4%, the data obtain was negative as it warped on the opposite side. Moreover, Figure 10 shows the warpage graph for the high range of concentration level, 6% to 10%. Based on the graph, it can be seen that the pattern is increasing. The warpage increase as the concentration level increase. At 6% of concentration, the warpage is 0.95036 mm. While at the 10% the warpage increase to 1.45304 mm. However, at 7% the warpage dropped down to 0.70854 mm.

Based on the result shown on Figure 11, the blue pigment has the lowest warping rate which is 0.30486 mm while yellow has the highest warping rate 0.84704 mm for 3% concentration level. This difference is caused by the content of the master batch that were used for both colours. Next, Figure 12 shows the graph of warpage for 5% concentration level of different pigment. The result shows in this experiment; the blue pigment has the lowest warpage formation which is 0.6690 mm while the yellow has the highest warpage rate which is 0.8783 mm.

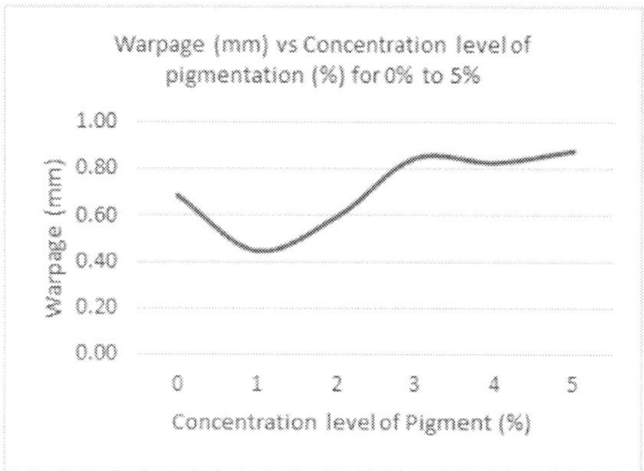

Figure 9. Graph of warpage against concentration level for 0% to 5%.

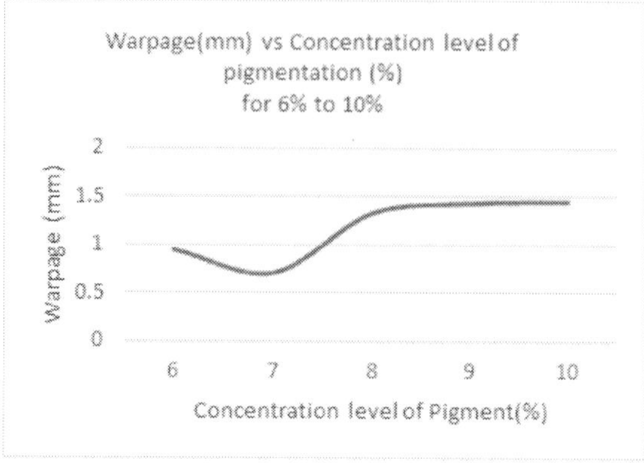

Figure 10. Graph of warpage against concentration level for 6% to 10%.

From both the graphs it can be clearly seen that yellow pigment fares poorer in comparison to the blue and red as it has very low warpage rate. However, Figure 13 shows that the Red specimen has the lowest warping rate which is 0.7332 mm. It is different compare to both result in 3% and 5% which stated blue as the lowest warping rate. But the yellow specimen still remained as the highest warping rate in this experiment with 1.32528 mm. From the warpage result it can be concluded that when the concentration level increase, the warping rate also increases. Moreover, the blue specimen has the lowest warping rate and yellow specimen has the highest warping rate. This is due to the content of pigmentation for both colours.

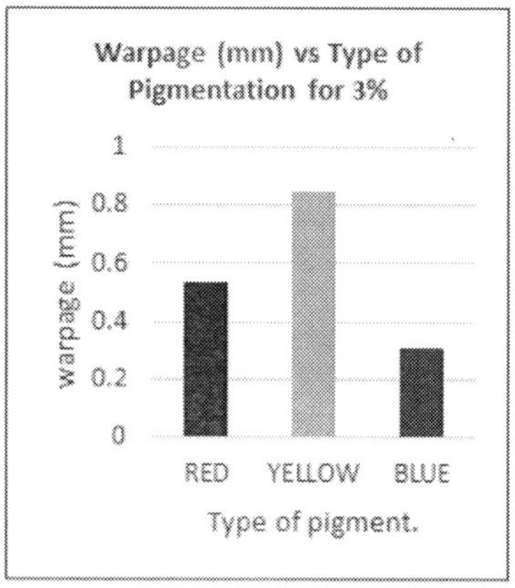

Figure 11. Graph of warpage vs type of pigmentation for 3% of concentration level.

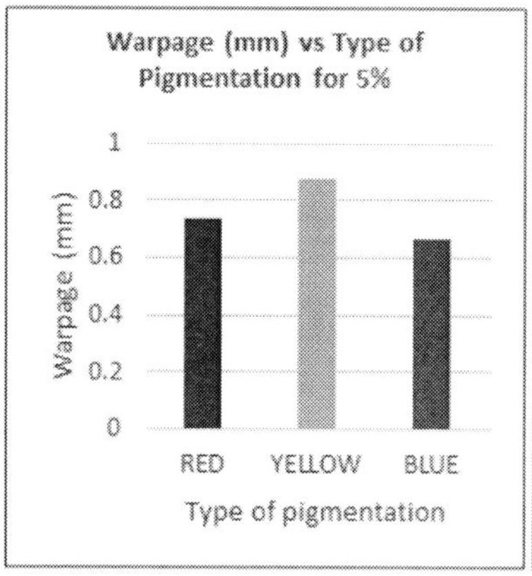

Figure 12. Graph of warpage vs type of pigmentation for 5% of concentration level.

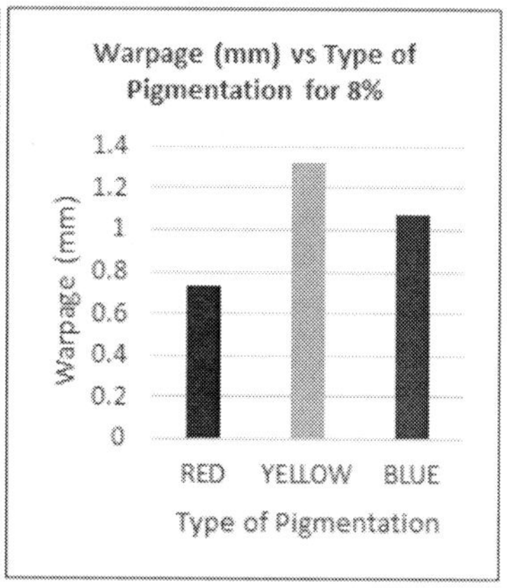

Figure 13. Graph of warpage vs type of pigmentation for 8% of concentration level.

3.3 Flash

The excess weight produce by the specimen is counted as flash. The actual weight for dog bone is 8.20 gram. Based on the Fig. 14, the graph show that as the concentration level increases, the percentage of flash also increases. The lowest flash rate is 1% concentration level which is 0.3098 %. The highest flash rate is at 5% concentration level, 1.80 (%). However, the result is differing for 2% and 4%. The weight is smaller which for 2% and 4% this might be due to error during injection process. Based on Fig. 15, the graph indicates the increase of flash defects as the concentration level increases. The highest flash rate is 3.0732 % for 10% concentration level. However, the graph indicates at 7% the flash rate decrease from 2.1268 to 0.7049. This was cause by the error during injection process.

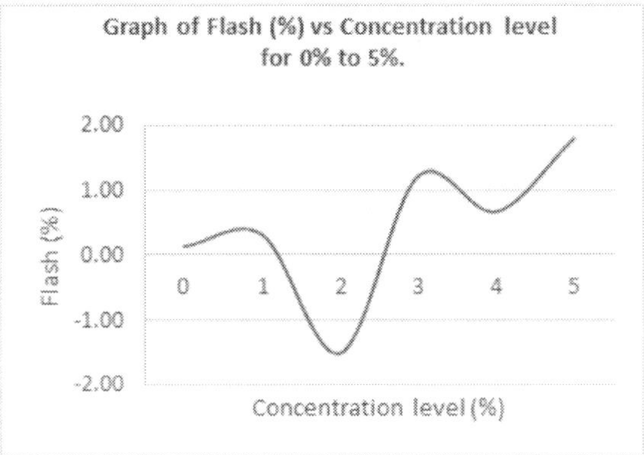

Figure 14. Graph of flash vs concentration level for 0% to 5%.

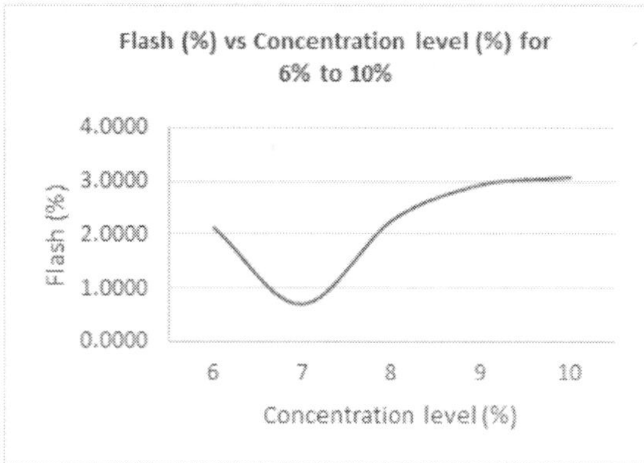

Figure 15. Graph of flash vs concentration level for 6% to 10%.

Figure 16 and Figure 17 shows the graph for flash defects against type of pigmentation for 3% and 5% respectively. The graph indicates that blue specimen has the lowest value of flash which is 0.5073% and 1.2902% respectively. While yellow has the highest value of flash for both experiment, 1.2195% and 1.80% respectively. However, Fig. 18 shows that red has the highest flash value compare to both previous experiment which indicates yellow as the highest flash defect. The value for red is 3.5098%. The blue specimen has the lowest flash value, 1.8756 %. This indicates that blue pigment can minimize the defects that cause by pigmentation.

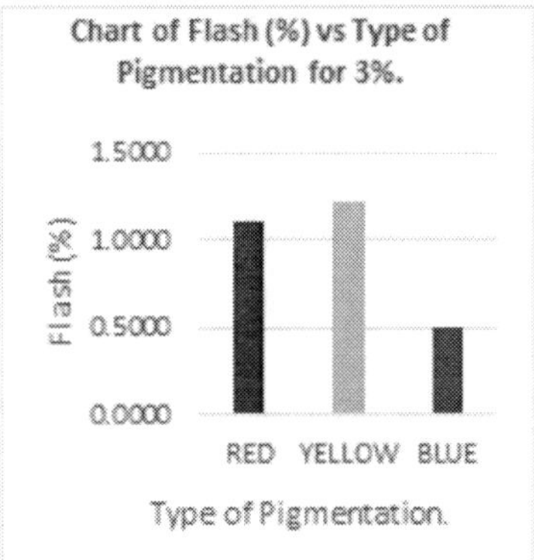

Figure 16. Graph of flash vs type of pigmentation for 3% concentration level.

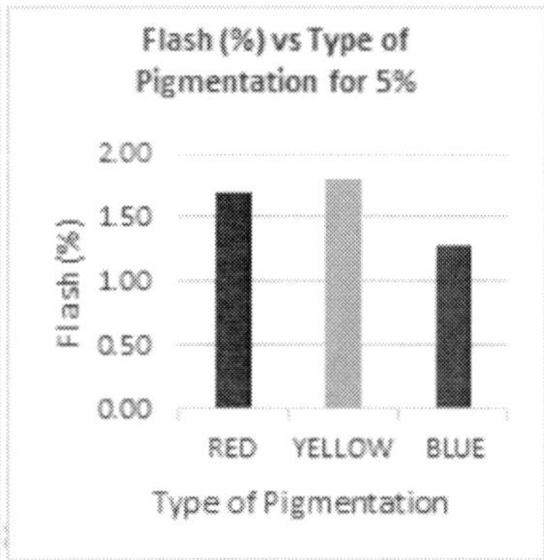

Figure 17. Graph of flash vs type of pigmentation for 5% concentration level.

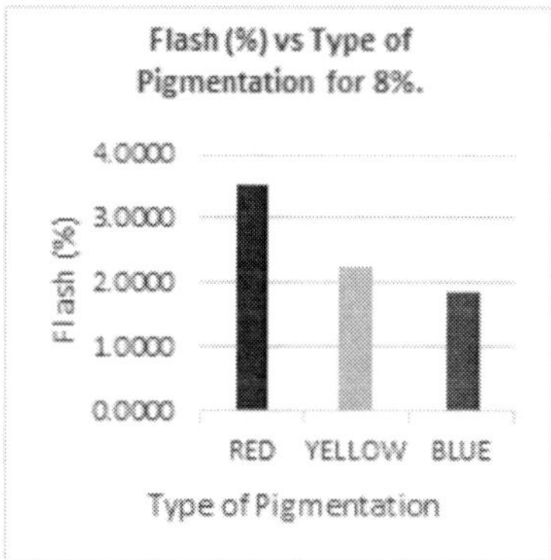

Figure 18. Graph of flash vs type of pigmentation for 10% concentration level.

3.4 Fourier Transform Infrared Spectroscopy (FTIR) Test

Infrared spectroscopy is the microscopy and spectroscopy of material which is the test used for analysis of functional groups. The FTIR technique were prepared and performed to study the surface chemistry of pigment colour that effect the properties of samples [14], leading to the various inconsistency as well variances amongst the pigments. Two samples from different pigment has been chosen to be tested in this FTIR test. Format of Infrared (IR) spectra is Transmittance (T) against Wavenumber (cm^{-i}). The FTIR spectrum in Figure 19, displays the blue spectrum for blue pigment pellet and the red for yellow pigment pellet. For yellow pellet, the spectrum showed strong broad of characteristics transmittance bands at 1409.63 cm^{-1} which corresponds to the Cobalt (II) Oxide (COO) but for blue pellet, the spectrum showed strong properties at 2915.29 cm^{-1}indicate methylene (CH_2) All FTIR spectrum of native hide is indicated in Table 1 and 2 with the functional groups respectively.

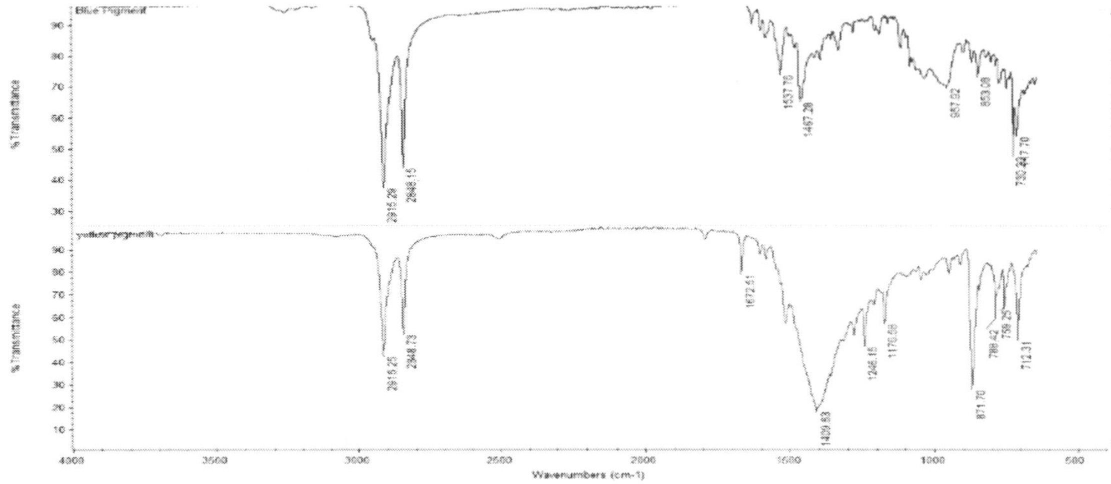

Figure 19. Graph of FTIR test

Table 1. Blue Pigment FTIR

Blue Pigment Pellet	
Wavenumber, cm^{-1}.	Functional Groups
2915.29	Methylene (CH_2)
2848.15	Aldehyde (CHO)
1537.76	Benzene
1467.28	Methylene (CH_2)
957.92	Carboxyl acid (-COOH)
853.08	Nitrogen dioxide (NO_2)
730.447	Benzene

Table 2. Yellow Pigment FTIR

Yellow Pigment Pellet	
Wavenumber (cm^{-1})	Functional Groups
2916.25	Methylene (CH_2)
2848.73	Aldehyde (CHO)
1672.51	Amide
1409.63	Cobalt (II) oxide (COO)
1246.15	Epoxide
1176.58	Ester
871.7	Epoxide
788.42	Alkyl halides
759.25	Meta-disubstituted. benzene
721.31	Monosubstituted Benzene

4. Conclusion and Recommendation

In this study, the effects pigmentation on the defects formation of plastic injection moulding has been conducted and investigated. The shrinkage, warpage and flash defects result has been obtained. The shrinkage rate is less than 1%. The warpage rate increases as the concentration level increases. The highest warpage occurs at 10% concentration level which is 1.45304 mm. It was found that the blue specimen has the lowest warpage value while yellow has the highest warpage value. The warping is considered as low warping as the defects does not exceed 10 mm. For flash defects, it was found that the highest flash occurs at 10% concentration with value 3.0732%. Blue specimen has the lowest flash rate and yellow has the highest flash rate. The result of current investigation is necessary to improve the quality of products from plastic injection moulding in industry where mass manufacturing takes place and a guideline is indeed necessary to minimize wastage, cost and increase production yield. The colorant analysis indicates here that the blue pigment is most stable and the factor is largely due to its chemical composition. To further analyse this factor, an in depth FTIR analysis can be done to understand the contribution of each chemical compound in the stability of the pigments.

IOP Conf. Series: Materials Science and Engineering **834** (2020) 012038 doi:10.1088/1757-899X/834/1/012038

References

[1] J. B. Tranter, P. Refalo, and A. Rochman, "Towards Sustainable Injection Molding of ABS Plastic Products," *J. Manuf. Process.*, vol. 29, pp. 399–406, 2017.

[2] K. Jain, S. R. Das, and D. Kumar, "IVth International Conference on Production and Industrial Engineering, CPIE-2016 Modeling and Optimization of Shrinkage in Plastic," no. 2010, 2016.

[3] H. Arora, D. P. Singh, and S. Kumar, "Parametric Effect during Plastic Injection Molding Process on Polypropylene Material," no. 04, pp. 118–121, 2015.

[4] S. S. Tomar, A. K. Sinha, and A. Shrivastava, "' Parametric Study of Injection Moulding Using Polypropylene H200mk Grade ,'" pp. 18–30.

[5] I. E. Manage, A. Hs, A. Mf, A. Mohib, and A. Ha, "Industrial Engineering & Management Minimization of Defects Percentage in Injection Molding Process using Design of Experiment and Taguchi Approach," vol. 4, no. 5, 2015.

[6] G. Singh and A. Verma, "A Brief Review on Injection Moulding Manufacturing Process," *Mater. Today Proc.*, vol. 4, no. 2, pp. 1423–1433, Jan. 2017.

[7] R. M. Khan and G. Acharya, "Plastic Injection Molding Process and Its Aspects for Quality: A Review," *Eur. J. Adv. Eng. Technol.*, vol. 3, no. 4, pp. 66–70, 2016.

[8] P. G. Student, M. Engineering, and M. Engineering, "Process Parameters Optimization for Development of Defect Free Injection," pp. 56–62.

[9] M. H. Othman, S. Hasan, S. Z. Khamis, M. H. I. Ibrahim, and S. Y. M. Amin, "Optimisation of Injection Moulding Parameter towards Shrinkage and Warpage for Polypropylene-Nanoclay-Gigantochloa Scortechinii Nanocomposites," *Procedia Eng.*, vol. 184, pp. 673–680, 2017.

[10] S. Selvaraj and P. Venkataramaiah, "Measurement of Warpage of Injection Moulded Plastic Components using Image Processing," vol. 2, no. 12, pp. 8071–8083, 2013.

[11] S. C. Nian, C. Y. Wu, and M. S. Huang, "Warpage Control of Thin-Walled Injection Molding using Local Mold Temperatures," *Int. Commun. Heat Mass Transf.*, vol. 61, no. 1, pp. 102–110, 2015.

[12] P. S. F. F. Alves, "Shrinkage and Warpage Behaviour on Injection Moulding Parts," 2008.

[13] I. Shrinkage, "2 Shrinkage and Warpage," *Library (Lond).*, p. 2.

[14] R. Kerkstra and S. Brammer, "30 - Flash BT - Injection Molding Advanced Troubleshooting Guide," Hanser, 2018, pp. 281–297.

[15] K. S.Kamble, B. Jadhav, and R. Kalmadi, "Investigation and Analysis of the Active Factors To Address the Defect in Plastic Injection Molding Process," *Int. J. Mech. Prod. Eng.*, vol. 2, no. 7, pp. 20–23, 2014.

[16] James. R. & Thomas C, "Impact of Pigments on the Dimensional Stability of Plastics Impact of Colorants,"BASF Corporation, pp. 325–334, 2016.

IOP Conf. Series: Materials Science and Engineering **834** (2020) 012039 doi:10.1088/1757-899X/834/1/012039

MXene-graphene hybrid nanoflakes as friction modifiers for outboard engine oil

Abdul Khaliq Rasheed[1*], Mohammad Khalid[1], Ariff Farhan Bin Mohd Nor[1], Wai Yin Wong[2], Tuerxun Duolikun[3,4], Varun Natu[5], Michel. W. Barsoum[5], Bey Fen Leo[3,6], Haizum Aimi Zaharin[7], Mariyam Jameelah Ghazali[7]

[1]Graphene and Advanced 2D Materials Research Group, School of Science and Technology, Sunway University, Sunway City, Petaling Jaya, 47500, Selangor, Malaysia
[2]Fuel Cell Institute, National University of Malaysia (UKM), Bangi, Selangor, Malaysia
[3]Nanotechnology & Catalysis Research Centre (NANOCAT), Institute for Advanced Studies, University of Malaya, 50603 Kuala Lumpur, Malaysia
[4]Department of Mechanical Engineering, Faculty of Engineering, University of Malaya, 50603, Kuala Lumpur, Malaysia
[5]Department of Materials Science and Engineering, Drexel University, Philadelphia, Pennsylvania 19104, USA
[6]Faculty of Medicine University of Malaya, 50603 Kuala Lumpur, Malaysia
[7]Centre for Materials Engineering and Smart Manufacturing, Faculty of Engineering and Built Environment, Universiti Kebangsaan Malaysia, 43600 UKM Bangi, Selangor, Malaysia

* rakhaliq47@yahoo.com

Abstract. MXenes are a relatively new type of two-dimensional materials which remain largely unexplored in terms of tribological applications. In this research, hybrid comprising Ti3C2-NG (MXene-Nitrogen-doped Graphene) is synthesized in an attempt to enhance the thermal and tribological properties of Outboard Marine Engine Oil. Oil sample were prepared using the two-step method which involved optimization of mixing technique and followed by tests according to ASTM standards. Results revealed that the thermal conductivity of the oil is enhanced by 6.62% for 15 minutes high shear blending whereas the viscosity is reduced by 4.71%. The decrease in viscosity could be as a result of pockets of debilitated intermolecular bonds in oil due to nanoparticles addition. Further, the Newtonian behaviour oil remains unchanged with the addition of nanoparticles. However, increasing shear rate revealed dilatant behaviour of the nanofluids corresponding to Taylor-Couette flow. The hybrid nanoflakes doesn't significantly alter the anti-friction and anti-wear behaviour of the oils although the coefficient of friction is decreased in the presence of 0.01 wt.% Ti3C2 additive nanoparticle by a marginal <1%.

1. Introduction

The According to the Food and Agriculture Organization (FAO), over 4.6 million fishing vessels operate in the world and out of which 64% are engine-powered in 2014. Fishing activity along coastal line of ASEAN countries contribute significantly to carbon emissions and affect aquatic life. At the same time, the rapid development of maritime transport, driven by international trade and seafood

Content from this work may be used under the terms of the Creative Commons Attribution 3.0 licence. Any further distribution of this work must maintain attribution to the author(s) and the title of the work, journal citation and DOI.
Published under licence by IOP Publishing Ltd

industry, has led to increase in the usage of marine engines. Furthermore, extreme engine conditions in rescue operations to reach the rescue area quickly through rough seas requires superior lubrication[1]. Hence, improvement of marine engine oil is vital to enhance engine performance and efficiency.

For better performance of engines, lubricants are required to have a strong film protection against metal-to-metal contact, avoid film rupture at high loads and high temperatures, reduced scuffing, pitting and rolling contact fatigue. Marine lubricants are also required to offer high residual fuel compatibility and total base number (TBN) reserve and retention properties. Lubricant additives which are chemical based improve lubricant properties, especially thermo-physical properties. Addition of nanometer sized (1-100 nm) particles to based oils is found to improve thermal and tribological properties of oils. Numerous works uing different kinds of carbon allotropes nanoparticles, for example, carbon nanotubes (single-walled or multi-walled [2, 3]), graphene [4, 5] and graphene oxide [6, 7] to produce nanolubricants or nanocoolants show excellent enhancement of thermal properties. Graphene has outstanding in-plane thermal conductivity that can reach as high as 5200 W/m.K. When graphene oxide was added at a concentration of 0.25wt% to water, thermal conductivity improved by 47.5% Since carbon family nanoparticles exhibit outstanding thermal conductivity and coefficient of heat transfer, their respective nanolubricants are expected to give better performance. Researchers have concluded that graphene that exists in two-dimensional (2D) nanostructures, possess better thermal conductivity, higher mechanical strength as well as improved electrical conductivity. Graphene, thus, indirectly contributes to more improvement in thermal conductivity in nanofluids, as compared with other analyzed nanofluids [8]. In previous study, graphene nanoparticles were synthesized and dispersed into API SN standard base engine oil to outperform conventional lubricants[9]. Results showed that graphene indicated enhancement in thermal and anti-wear properties by approximately 23% and 17%, respectively. At the same time, Graphene also plays an important role in improving the oxidative stability of the lubricants [10].

Besides graphene, development of other 2D nanoparticles is rapidly growing. Recently, a new family of transition metal nitrides, carbides or carbon-nitrides, known as MXene has shown properties better than existing 2D materials [11]. MXene originated from MAX phase, owing the formula $M_{n+1}AX_n$ ($n = 1, 2, 3$), where M stands for early *d*-block transition metals, A denotes main group *sp* elements (alumimium or silicon) and X can be either C or N atoms, has been synthesized by wet HF treatment at room temperature. During selective etching of Al from Ti_3AlC_2, Al atoms are substituted by O, OH and/or F atoms. The substitution of Al layer has weakened the M_{n+1} and X_n bonding, thus, preparing the layer to loss A element to become MXene. This young class of 2D materials has becoming research attention to explore further since MXenes is known as graphene analogue nanomaterials. Furthermore, this material offers great electrical properties, high mechanical strength and shows superior energy storage behavior [12-15]. Moreover, the recent discovery of MXene reveals that this material has the potential of becoming lubricant properties enhancer [16, 17].

Titanium carbide (Ti_3C_2) is the most common establishment for MXene type that has been studied, especially in structural composites applications [18, 19]. Ti_3C_2 belongs to ternary carbides group and possess appealing multi-functional behaviours. This novel material features stable thermal properties, as offered by ceramic-type materials, as well as electrical conductivity and machinability, as presented by metal-type materials. Barsoum *et al.* were the first researchers to report on thermal conductivity (~37-40 W/mK) of bulk Ti_3C_2 in 1999 [20]. Development of 2D-MAXene nanostructures and synthesized into nanofluid, with propylene glycol as the base fluid, exhibited around 45% improvement in thermal conductivity at the temperature of 323 K [21]. The outstanding properties of Ti_3C_2 suspended in the fluid has potential to outperform the general heat transfer fluids. Although many hybrid graphene nanoparticles [22, 23] have been explored as lubricant additives, hybrid graphene-MXene in lubricants has never been investigated.

Generally, wear mechanisms for spherical nano-inclusions include deposition in the ridges and valleys and nano-film formation between the contacting surfaces. Although adhesion of nanoparticles and agglomeration is common, the morphological transformation is rarely prominent for spherical particles. Existing reports on graphene-based lubricants have limited the investigation to evaluating the thermo-physical properties only. The morphological behaviour of graphene and other 2D materials after high-shear mixing still remains unexplored. Therefore, the present study aims to investigate the thermo-rheological performance of graphene-MXene based lubricants as well as their morphological transformation during high-shear mixing. The combined effect of both graphene and MXene could produce a far better lubricant with enhanced thermo-physical and wear properties. This research will focus on the synthesis of a stable nanolubricant using hybrid graphene and MXene, meeting certified FC-W grades of National Marine Manufacturers Association (NMMA). Thermo-physical properties of the nanolubricant will be examined in detail using ASTM standards as well as the impact of high-shear mixing on the morphology of graphene-MXene will be investigated.

2. Methods and Materials

2.1. *Hybrid Nanoparticles Synthesis and characterization*

Hummer's method was employed to produce graphene oxide (GO) and Nitrogen doping was carried out. Ti_3C_2 was exfoliated from MAX phase clay. As-prepared NG and MXene were mixed using an agate mortar with a mass ratio of 1:0.5. The powder was subjected to further pyrolysis at 900 °C for 1 hour under nitrogen flow. The pyrolysed sample was acid washed for 6 hour to remove unreacted species and the dried sample was labelled as NG-MXene. Transmission electron microscopy (HR-TEM) and X-ray Diffraction (XRD) pattern were used to characterize the prepared nanoparticles.

2.2. *Nanolubricant Preparation*

The samples were prepared by using the two-step method, as no surfactant was used. 0.01 wt.% of nanolubricants was prepared by dispersing each of the nanoparticles in marine engine oil meeting NMMA certified FC-W specifications.

2.3. *Thermo-physical and Tribological Characterization of Nanolubricants*

The KS-1 single-needle sensor (diameter: 1.3mm, length: 60mm), which connects to a microprocessor, was used to measure the thermal conductivity of the fluids [24]. The MCR 302 rheometer (Anton Paar, Austria), which is a stress-controlled rotational rheometer, was used to measure the dynamic viscosity of the nanolubricant. Four ball tribometer (Ducom) was used to measure the wear properties following the standard test ASTM D 4172. Weld load test was conducted for all lubricants using ASTM D 2783 at 1770 rpm speed, 25°C temperature, for 10 sec duration.

3. Results and discussion

3.1. *Material Characterization*

The synthesized material was characterized using XRD to validate the crystallinity and to identify the phase compositions of the G-MXene, Ti_3C_2 and Nitrogen doped graphene (N-G). Figure 1 shows the XRD pattern of the G-MXene, Ti_3C_2 and N-G nanocomposite. The XRD of G-MXene and Ti_3C_2 exhibit a reflection peak at $2\theta \approx 26°$ which corresponds to the weak graphite peak. This phenomenon might be due to the dense Ti_3C_2 coating on graphene, which could decrease the X-ray signal[25]. For both G-MXene nanocomposite and Ti_3C_2, their XRD characteristic reflection peaks exist at 9°, 17°, 26° and 62°, corresponding to the (002), (006), (008) and (110) planes of the Ti_3C_2[26]. These results further confirm the successful formation of hybrid nanocomposite. Furthermore, the XRD of N-G shows the presence of peak at $2\theta = 17°$ which confirms the attachment of MXene nanoparticles on the graphene sheet.

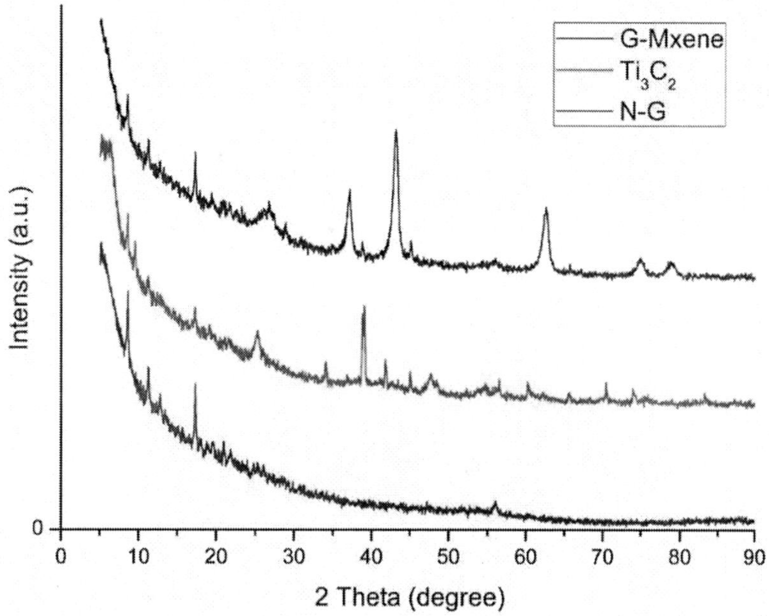

Figure 1. XRD pattern of G-Mxene, Ti_3C_2 and N-G samples

The microstructure and morphology of the G-MXene, Ti_3C_2 and NG powders were observed by high-resolution TEM (HRTEM), (Fig. 3). A sheet-like graphene structure was formed and consistent with the TEM analysis[27]. After nitrogen doping, as seen in Fig. 2, crumpled yet well-defined nanosheets are observed [28]. Fig.3 indicated a clear 2D layered structure of the Ti_3C_2 after undergoing treatment and depicting electron-transparent thin morphology and a few layer structures of Ti_3C_2 [25].The HRTEM image clearly shows that the G-MXene samples displayed a nanosheet morphology[29]. The diameter size distribution of samples unambiguously illustrates the formation of uniform ultra-small quantum dots. The average thickness of the multilayer nanosheet examined using TEM was approximately 1.5 nm and this implied that the composite was produced in nanosize [27, 30]. The size distribution was calculated by counting 100 particles, and the lateral sizes of the quantum dots were directly measured from the HRTEM images via Image J software.

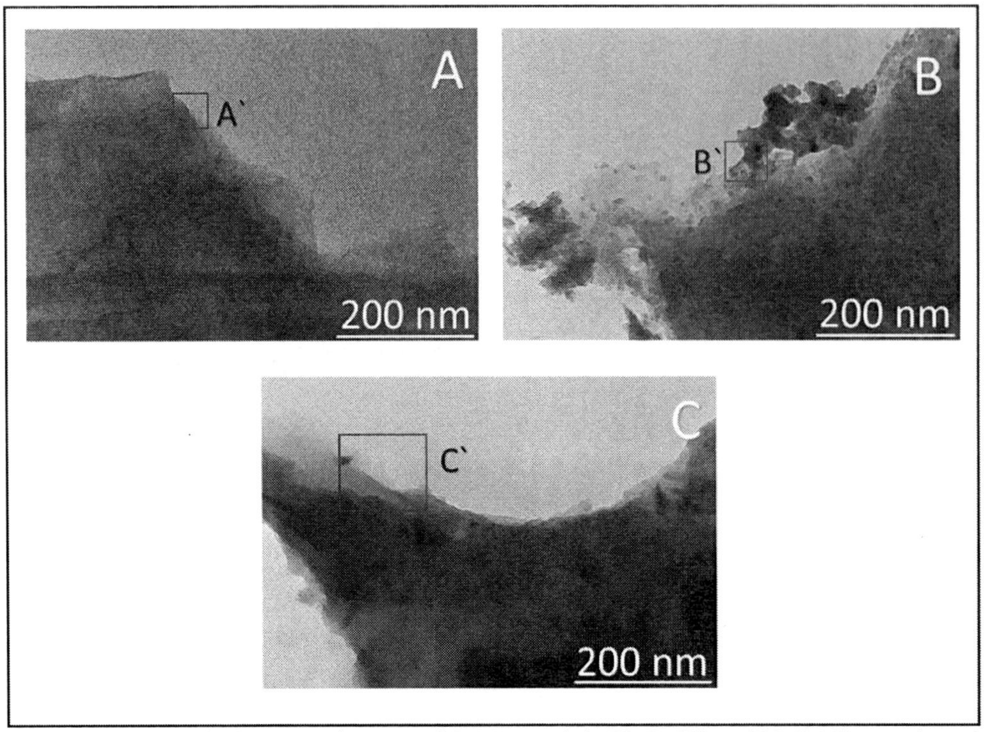

Figure 2. High resolution TEM images of G-Mxene (A), Ti_3C_2 (B) and N-G (C) samples with nanosheet thickness 1.16 nm, 1.14 nm and 1.53 nm, respectively.

3.2. Thermal conductivity

In this study, investigation on the effect of temperature on the nanofluid's thermal conductivity, containing the synthesized samples, was conducted as well as the impact of high shear blending time towards the thermal conductivity of nanofluids. It is known that the increasing temperature would increase nanolubricant thermal conductivity. As shown in Fig. 3, significant enhancements were observed to the k values of all nanofluid samples compared to the base oil at all temperature. This result indicates that nanoparticles have the potential to improve the thermal conductivity in engine oil as the temperature increase. This trend could be well explained by the Brownian motion [31], where the nanoparticles tend to collide with each other more aggressively at higher temperature due to their kinetic energy. For example, at 5 minutes HSB, k value enhancement at room temperature shows an improvement of 4.41%, 1.73% and 0.16% meanwhile at 60 °C, the thermal conductivity enhances up to 5.76%, 2.59% and 1.87% relative to the base oil for G-MXene, N-G and Ti_3C_2, respectively. It is evident that the k values exhibit higher sensitivity at a higher temperature. This phenomenon may be due to fast molecular colliding at high temperatures compared to lower temperature [32]. This improvement also illustrates that at higher temperature the thermal conductivity could increase further which indicates promising result for application in an outbound engine.

It can also be observed from Fig. 3, that G-MXene hybrid nanofluid exhibits the highest enhancement of thermal conductivity at 60°C among all HSB variations, with 6.62% increment at 15 minutes HSB (figure not included). N-G [33] and Ti_3C_2 [21] are well known as high thermal conducting nanoparticles for their high aspect ratio. Hence, the hybridisation of both nanoparticles enhances the overall thermal conductivity of the output nanoparticles. Also, these deviations could be associated to the dependence of thermal conductivity to the several factors e.g. nanoparticle type, nanoparticle size that complimented well with Paul et al [34]. Nevertheless, prior to introducing the G-MXene with these nanofluids, reliable stability data is required as it inherently influences the thermal conductivity. Further studies under G-MXene nanolubricant and functionalized methods would aid for better understanding.

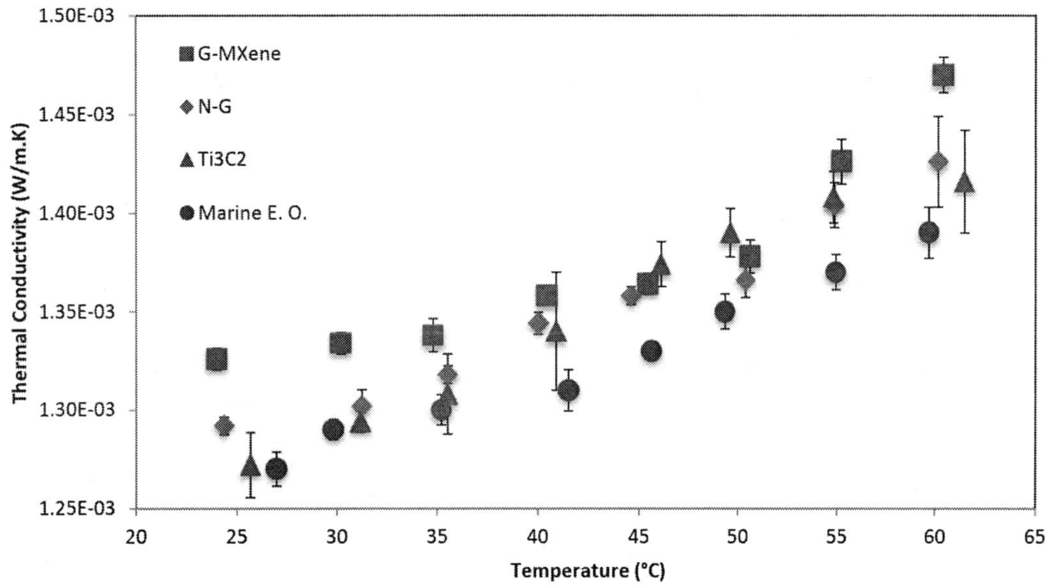

Figure 3. Thermal conductivity of nanolubricants for 5 minutes HSB

3.3. Rheology

The flow behaviors of hybrid nanofluids were explored at variable shear rates and temperatures as a function of shear stress, as shown in Fig. 4. A linear pattern of shear stress variation with shear rate can be seen for every hybrid samples at all temperatures. From Fig. 4, it can be deduced that nanolubricant showed Newtonian behavior at lower temperature. Comparable behaviors were seen in the study done by Motahari et al. [35] for MWCNT-SiO_2/20W50 motor oil and Afrand et al. [36] for Fe_3O_4-Ag/EG hybrid nanofluid. This is a vital criterion, especially for convective heat transfer. Although at a higher temperature (110 °C) all nanofluids show inclination towards non-Newtonian nature, however, it is insignificant in shear stress with respect to shear rates. The further increment of temperature anticipates a change from the Newtonian to non-Newtonian behavior of the nanofluids. Besides, the flow curves indicated that the difference in shear stress became larger between low and high temperature as the shear rates increase.

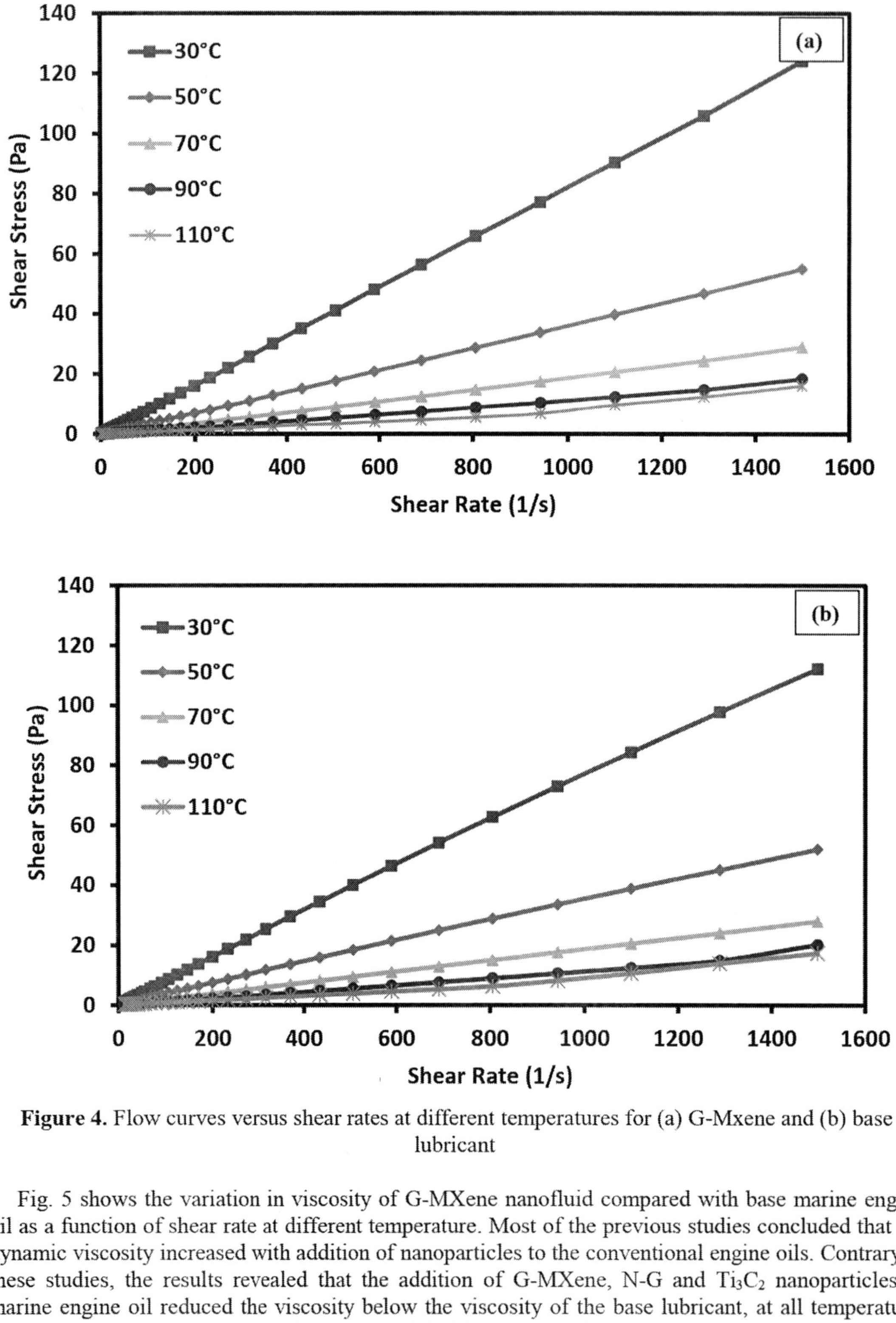

Figure 4. Flow curves versus shear rates at different temperatures for (a) G-Mxene and (b) base lubricant

Fig. 5 shows the variation in viscosity of G-MXene nanofluid compared with base marine engine oil as a function of shear rate at different temperature. Most of the previous studies concluded that the dynamic viscosity increased with addition of nanoparticles to the conventional engine oils. Contrary to these studies, the results revealed that the addition of G-MXene, N-G and Ti_3C_2 nanoparticles to marine engine oil reduced the viscosity below the viscosity of the base lubricant, at all temperature. Addition of G-MXene nanoparticles presented the highest reduction in term of viscosity up to 4.71%, 4.70%, 2.59%, 3.68% and 4.59% at 30 °C, 50 °C, 70 °C, 90 °C and 110 °C, respectively. The trends are consistant at all temperature tested over the entire shear rate of $1 s^{-1}$ to $1500 s^{-1}$.

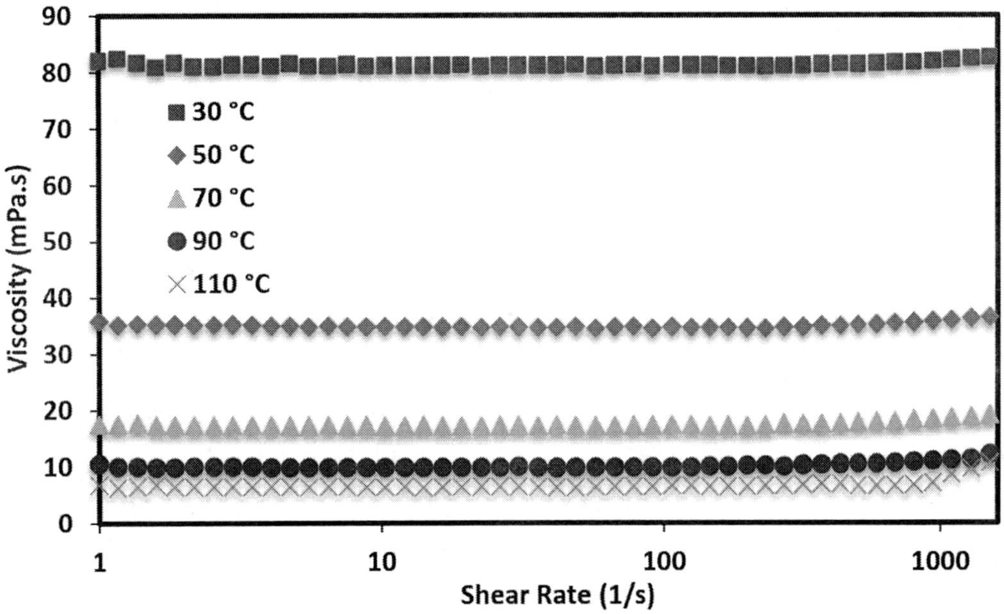

Figure 5. Viscosity vs shear rate for G-MXene hybrid nanofluid at different temperature

3.4. Anti-wear properties

Results indicate that addition of Graphene MXene hybrid additive does not have significant enhancement in anti-friction property of lubricant. The coefficient of friction for the lubricant and its variant also show similar behaviour. Friction coefficient is calculated according to IP-239. Increment in test load increase the coefficient of friction. The coefficient of friction for 0.01% G-MXene is found to be increased by 6.5%. Whereas the friction coefficient with N-G as lubricant additive is increased by 10.2%. On the other hand, the coefficient of friction for marine engine oil with 0.01% Ti_3C_2 additive has enhanced anti-friction property with reduction of μ by ~0.8%. The wear scar diameter (WSD) depends on the magnitude of the wear and the load bearing capacity of the lubricant. As observed, the presence of G-MXene nano particles resulted in small WSD compared to standalone N-G and Ti_3C_2 additive and base oil. Weld load test was conducted for all lubricants using ASTM D 2783 at 1770 rpm speed, 25°C temperature, for 10 sec duration. The result from ASTM D 2783 shows that there is no improvement in the load carrying capacity for marine engine oil with or without additive.

4. Conclusion

Marine engine oil with N-G, MXene and G-MXene additive with 0.01% wt are analysed for thermal conductivity, viscosity and frictional coefficient and compared with properties of conventional engine oil. It can be concluded that there is significant enhancement to the k values of all hybrid nanofluid samples compared to the base oil at all temperature and the conductivity improves as the temperature increase. G-MXene hybrid nanolubricant exhibits the highest enhancement of thermal conductivity with up to 6.62% increment in conductivity. All hybrid nanolubricants show shear thinning for nanofluids which is due to low shear rate, therefore reducing fluid viscosity. However, the changes are not significant to be considered as a non-Newtonian fluid. Results show that G-MXene nanoparticles presented the highest reduction in viscosity, up to 4.71%, compared to N-G and Ti_3C_2 nanoparticles to marine engine oil. Wear properties analysis show that oil with G-MXene hybrid additive depicts the frictional coefficient in between the oil with N-G and MXene additive, and higher than base engine oil. Result shows reduction in scar diameter with addition of NG-MXene hybrid, enhancing the anti-wear properties of lubricant. Nevertheless, application of NG-MXene with these nanofluids influences the thermal conductivity, viscosity and frictional properties of fluid. Further studies are required to investigate the discrepancy between the tribological findings of this study and the previous works.

Acknowledgement
This project was funded by Sunway University Internal Grant INT-2018-SST-RCNMET-03 and INT-2019-SST-RCNMET-01. Authors are grateful to Drexel University for supporting us with their MAX phase material for this work.

References
[1] Anand M, Hadfield M, Thomas B and Cantrill R 2017 The depletion of ZDDP additives within marine lubricants and associated cylinder liner wear in RNLI lifeboat engines *Proceedings of the Institution of Mechanical Engineers, Part L: Journal of Materials: Design and Applications* **231** 162-70
[2] Esfe M H, Firouzi M and Afrand M 2018 Experimental and theoretical investigation of thermal conductivity of ethylene glycol containing functionalized single walled carbon nanotubes *Physica E: Low-dimensional Systems and Nanostructures* **95** 71-7
[3] Khalil W, Mohamed A, Bayoumi M and Osman T 2017 Thermal and Rheological properties of industrial mineral gear oil and paraffinic oil/CNTs nanolubricants *Iranian Journal of Science and Technology, Transactions of Mechanical Engineering* 1-7
[4] Rasheed A, Khalid M, Javeed A, Rashmi W, Gupta T and Chan A 2016 Heat transfer and tribological performance of graphene nanolubricant in an internal combustion engine *Tribology International* **103** 504-15
[5] Kamel B M, Mohamed A, El Sherbiny M and Abed K 2016 Rheology and thermal conductivity of calcium grease containing multi-walled carbon nanotube *Fullerenes, Nanotubes and Carbon Nanostructures* **24** 260-5
[6] Ali M K A, Xianjun H, Abdelkareem M A, Gulzar M and Elsheikh A 2018 Novel approach of the graphene nanolubricant for energy saving via anti-friction/wear in automobile engines *Tribology International* **124** 209-29
[7] Mungse H P and Khatri O P 2014 Chemically functionalized reduced graphene oxide as a novel material for reduction of friction and wear *The Journal of Physical Chemistry C* **118** 14394 402
[8] Yarmand H, Gharehkhani S, Ahmadi G, Shirazi S F S, Baradaran S, Montazer E, Zubir M N M, Alehashem M S, Kazi S and Dahari M 2015 Graphene nanoplatelets–silver hybrid nanofluids for enhanced heat transfer *Energy conversion and management* **100** 419-28
[9] Rasheed A K, Khalid M, Javeed A, Rashmi W, Gupta T C S M and Chan A 2016 Heat transfer and tribological performance of graphene nanolubricant in an internal combustion engine *Tribology International* **103** 504-15
[10] Rasheed A K, Khalid M, Walvekar R, Gupta T C S M and Chan A 2015 Study of graphene nanolubricant using thermogravimetric analysis *Journal of Materials Research* **31** 1939-46
[11] Lei J-C, Zhang X and Zhou Z 2015 Recent advances in MXene: Preparation, properties, and applications *Frontiers of Physics* **10** 276-86
[12] Naguib M, Mochalin V, Barsoum M and Gogotsi Y G 2014 MXenes: A New Family of Two Dimensional Materials *Advanced Materials* **26**
[13] Naguib M, Mashtalir O, Carle J, Presser V, Lu J, Hultman L, Gogotsi Y and Barsoum M W 2012 Two-dimensional transition metal carbides *ACS nano* **6** 1322-31
[14] Tang Q, Zhou Z and Shen P 2012 Are MXenes promising anode materials for Li ion batteries? Computational studies on electronic properties and Li storage capability of Ti3C2 and Ti3C2X2 (X= F, OH) monolayer *Journal of the American Chemical Society* **134** 16909-16

[15] Lukatskaya M R, Mashtalir O, Ren C E, Dall'Agnese Y, Rozier P, Taberna P L, Naguib M, Simon P, Barsoum M W and Gogotsi Y 2013 Cation intercalation and high volumetric capacitance of two-dimensional titanium carbide *Science* **341** 1502-5

[16] Chakraborty P, Das T, Nafday D, Boeri L and Saha-Dasgupta T 2017 Manipulating the mechanical properties of Ti2C MXene: Effect of substitutional doping *Physical Review B* **95** 184106

[17] Hong Ng V M, Huang H, Zhou K, Lee P S, Que W, Xu J Z and Kong L B 2017 Recent progress in layered transition metal carbides and/or nitrides (MXenes) and their composites: synthesis and applications *Journal of Materials Chemistry A* **5** 3039-68

[18] Mahesh K, Balanand S, Raimond R, Mohamed A P and Ananthakumar S 2014 Polyaryletherketone polymer nanocomposite engineered with nanolaminated Ti3SiC2 ceramic fillers *Materials & Design* **63** 360-7

[19] Barsoum M W and El-Raghy T 2001 The MAX phases: Unique new carbide and nitride materials: Ternary ceramics turn out to be surprisingly soft and machinable, yet also heat tolerant, strong and lightweight *American Scientist* **89** 334-43

[20] Barsoum M, El-Raghy T, Rawn C, Porter W, Wang H, Payzant E and Hubbard C 1999 Thermal properties of Ti3SiC2 *Journal of Physics and Chemistry of Solids* **60** 429-39

[21] Mahesh K, Linsha V, Mohamed A P and Ananthakumar S 2016 Processing of 2D-MAXene nanostructures and design of high thermal conducting, rheo-controlled MAXene nanofluids as a potential nanocoolant *Chemical Engineering Journal* **297** 158-69

[22] Meng Y, Su F and Chen Y 2016 Supercritical Fluid Synthesis and Tribological Applications of Silver Nanoparticle-decorated Graphene in Engine Oil Nanofluid **6** 31246

[23] Soo Hui Q, Rashmi W, Khalid M, Gupta T C S M, Nabipoor M and Mohammad Taghi H 2017 Thermal conductivity and electrical properties of hybrid SiO 2 -graphene naphthenic mineral oil nanofluid as potential transformer oil *Materials Research Express* **4** 015504

[24] Hong K, Hong T-K and Yang H-S 2006 Thermal conductivity of Fe nanofluids depending on the cluster size of nanoparticles *Applied Physics Letters* **88** 031901

[25] Li M, Han M, Zhou J, Deng Q, Zhou X, Xue J, Du S, Yin X and Huang Q 2018 Novel Scale Like Structures of Graphite/TiC/Ti3C2 Hybrids for Electromagnetic Absorption *Advanced Electronic Materials* **4** 1700617

[26] Feng W, Luo H, Wang Y, Zeng S, Tan Y, Zhang H and Peng S 2018 Ultrasonic assisted etching and delaminating of Ti3C2 Mxene *Ceramics International* **44** 7084-7

[27] Muniandy S, Dinshaw I J, Teh S J, Lai C W, Ibrahim F, Thong K L and Leo B F 2017 Graphene-based label-free electrochemical aptasensor for rapid and sensitive detection of foodborne pathogen *Analytical and bioanalytical chemistry* **409** 6893-905

[28] Yang L, Zheng W, Zhang P, Chen J, Zhang W, Tian W and Sun Z 2019 Freestanding nitrogen doped d-Ti3C2/reduced graphene oxide hybrid films for high performance supercapacitors *Electrochimica Acta*

[29] Zhao D, Chen Z, Yang W, Liu S, Zhang X, Yu Y, Cheong W-C, Zheng L, Ren F and Ying G 2019 MXene (Ti3C2) Vacancy Confined Single-Atom Catalyst for Efficient Functionalization of CO2 *Journal of the American Chemical Society*

[30] Xu Q, Ding L, Wen Y, Yang W, Zhou H, Chen X, Street J, Zhou A, Ong W-J and Li N 2018 High Photoluminescence Quantum Yield of 18.7% by Nitrogen-Doped Ti3C2 MXene Quantum Dots *J Mater Chem C*

[31] Qing S H, Rashmi W, Khalid M, Gupta T, Nabipoor M and Hajibeigy M T 2017 Thermal conductivity and electrical properties of hybrid SiO2-graphene naphthenic mineral oil nanofluid as potential transformer oil *Materials Research Express* **4** 015504

[32] Akilu S, Baheta A T and Sharma K 2017 Experimental measurements of thermal conductivity and viscosity of ethylene glycol-based hybrid nanofluid with TiO2-CuO/C inclusions *Journal of Molecular Liquids* **246** 396-405

[33] Mehrali M, Sadeghinezhad E, Rosen M A, Akhiani A R, Latibari S T, Mehrali M and Metselaar H S C 2016 Experimental investigation of thermophysical properties, entropy generation and

IOP Conf. Series: Materials Science and Engineering **834** (2020) 012039 doi:10.1088/1757-899X/834/1/012039

convective heat transfer for a nitrogen-doped graphene nanofluid in a laminar flow regime *Advanced Powder Technology* **27** 717-27

[34] Paul G, Philip J, Raj B, Das P K and Manna I 2011 Synthesis, characterization, and thermal property measurement of nano-Al95Zn05 dispersed nanofluid prepared by a two-step process *International Journal of Heat and Mass Transfer* **54** 3783-8

[35] Motahari K, Moghaddam M A and Moradian M 2018 Experimental investigation and development of new correlation for influences of temperature and concentration on dynamic viscosity of MWCNT-SiO2 (20-80)/20W50 hybrid nano-lubricant *Chinese Journal of Chemical Engineering* **26** 152-8

[36] Afrand M, Toghraie D and Ruhani B 2016 Effects of temperature and nanoparticles concentration on rheological behavior of Fe3O4–Ag/EG hybrid nanofluid: an experimental study *Experimental Thermal and Fluid Science* **77** 38-44

IOP Conf. Series: Materials Science and Engineering **834** (2020) 012040 doi:10.1088/1757-899X/834/1/012040

Development of Kuala Lumpur driving cycle for the estimation of fuel consumption and vehicular emission

Muhammad Farhan Mohd Suhaimi*[1], Noor Azlina Mohd Salleh[1], Dato' Ts. Mohamad Madani Sahari[2], Mohd Sharulnizam Sarip[2]

[1]Faculty of Mechanical Engineering, Universiti Teknologi MARA, 40450 Shah Alam, Selangor, Malaysia

[2] Malaysia Automotive, Robotics and IoT Institute (MARii), 63000 Cyberjaya, Selangor, Malaysia

*mdfarhan.suhaimi@gmail.com

Abstract. Driving cycles is a series of data points representing the speed of vehicle verses time sequence profile developed for certain road, route, specific area or city. Generally, driving cycle is used to estimate the fuel consumption (FC) and vehicular exhaust emission. The common factors that affected the FC and exhaust emission are the engine operating conditions such as cold start, low engine loads and high engine loads which are found during actual driving. The research presents the measurement and analysis of engine conditions based on actual driving behaviours in an attempt to formulate representative driving cycle for Kuala Lumpur. The methodology used for this research is adapted from the Worldwide Harmonized Light-duty Test Cycle (WLTC) introduced by the working party on pollution and energy (GRPE) under the United Nations Economic Commission for Europe (UNECE). This research covered a wide area of paved road in Kuala Lumpur, about 62.34% by surveying the road condition on the eight selected routes. The used of On-board Diagnostic (OBD) and Global Positioning System (GPS) equipment are used to record the vehicle travel speed during data collection activity. The driving characteristics are then analysed from speed profile and its target statistic parameters are defined to develop representative driving cycle. The characteristics of the newly developed driving cycle are compared to the European driving cycle which is currently being used in Malaysia for new vehicle certification and registration. Therefore, the correlation of FC is derived to indicate the impacts of real-world driving conditions relative to domestic conditions.

1.0 Introduction

Due to the evolution of automotive industry in Malaysia, the rapid increase of number of passenger vehicles as well as the very limited control strategies of emission, motor vehicles are alleged to be one of the main source of air pollution. According to a report, Malaysia's Biennial Update Report (BUR), submitted to United Nations Framework Convention on Climate Change (UNFCCC) in 2015, road transport sector is accounted to be significant source of carbon emission, estimated to be the second largest sector following the industry sector which ranked at the first place [1]. Committed to reduce the emission level, Malaysia need to be very critical in developing the strategies in controlling the level of emission. Thus, Malaysia Automotive, Robotics and IoT Institute (MARii) is spearheading the initiative in addressing the vehicle emission level by conducting a study in estimating the fuel consumption based on driving behaviour. The study is to develop a representative Kuala Lumpur Driving Cycle (KLDC) for light-duty passenger vehicle. The idea is that, this driving cycle will directly reflect the actual fuel

Content from this work may be used under the terms of the Creative Commons Attribution 3.0 licence. Any further distribution of this work must maintain attribution to the author(s) and the title of the work, journal citation and DOI.
Published under licence by IOP Publishing Ltd

consumption and carbon emitted by the light-duty passenger vehicle based on actual local driving characteristics (road conditions, driver's behaviour, weather, etc).

Driving cycle is a representative plot of vehicle speed versus time obtained from real-world driving natures in a specific region or road network. Driving cycle also is defined to consists of sequence of several operating modes such as idling, cruising, acceleration and deceleration [2]. Generally, driving cycle derived from actual driving natures precisely represents a typical driving pattern for a particular population or region [3]. Thus, different region will result to different driving cycle due to dissimilarities of actual driving characteristics and regional traffic congestion level as well [4].

The fuel consumption is derived from carbon balance method which directly correlated with the emitted exhaust emission from a motor vehicle. The emission factors in turn depend on several other factors such as type of fuel, type of engine, driving attributes, utilization of equipment and weather condition. In short, it can be concluded that a driving cycle is directly related with the fuel consumption and exhaust emission. Thus, a representative of regional driving cycle is crucial in determining accurate estimations fuel consumption and emission model.

2.0 Methodology

By using the concept of Deming cycle; the Plan-Check-Do-Action (PDCA), the reasearch methodology is constrcuted accordingly in conducting the research. Basically, the Deming cycle is being used to ensure the project can run smoothly in order to achieve all the objectives. Figure 1 depicts systematic methodology process flow adapted from the PDCA cycle that had been used for this study.

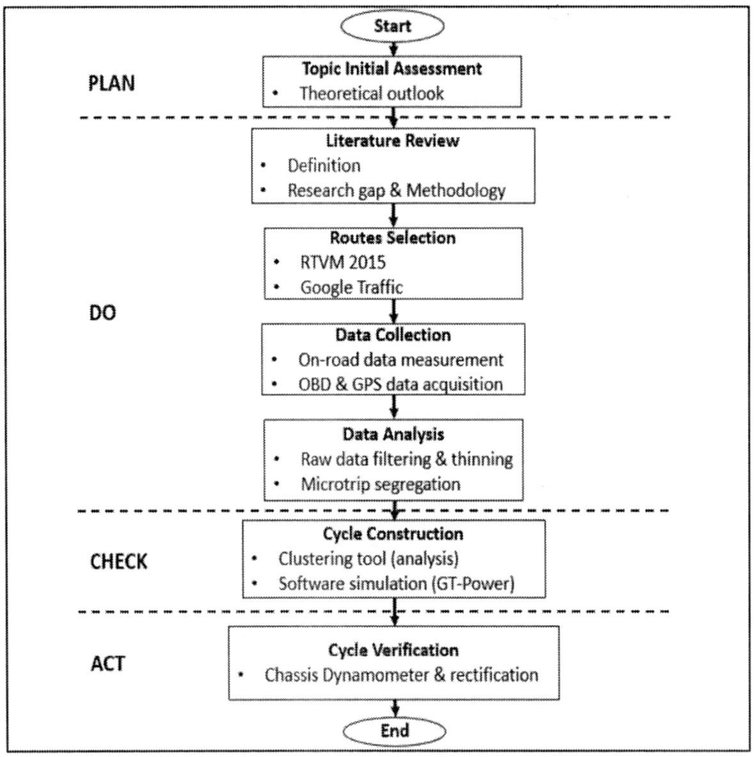

Figure 1. Research methodology.

3.0 Data Collection and Cycle Development

Prior to the data collection, considered as the key element in developing a driving cycle, route selection phase is very critical. It is because, the relationship between the routes selected and the driving behaviour is directly proportional. In that sense, it is important to ensure that selected routes of high utilization are included. In order to achieve that, there are few factors need to be accounted during the selection such as traffic density and peak periods. For this study, the traffic density and Level of Service (LOS) for a particular road network is assessed through two salient sources. A set of database provided by the Ministry of Work Malaysia namely RTVM 2015 is used as the prime source [5]. At the same time, live data on traffic density is extracted from the online platform, Google Traffic. Both data sources are used to obtain the most desired road conditions. Thus, it is important to ensure data are collected on the routes are representative of network-wide conditions instead of any few particular routes. Figure 2 illustrates the definition of LOS based on the RTVM 2015.

Figure 2. Definition of LOS in the RTVM 2015.

The choice of data collection methods (chase-car, instrumented vehicle with OBD and hybrid method combining with driving circulation) is based on suitability and appropriateness for the driving conditions. The salient considerations are data types and quality. Hence, considering to establish a driving cycle that will be used nationwide, the hybrid approach joined up with driving circulation will be used. It is because this method is superior in term of quality and accuracy compared to other methods. However, a few problems may arise from the GPS data which include loss of signal, stuck speed and false zero speed as well as correction during post-processed phase for filtering GPS data. With respect to eliminate this error, the data derived from OBD acts as the prime source for this study.

Driving conditions are analysed through the speed-time profile, generated from the road data measurement. For this purpose, the kinematic segments, known as microtrips are defined within the overall driving test. Typically, the microtrips are described by their idling duration and cross-contribution of the instantaneous speeds and accelerations. Correspondence analysis and clustering tools are used to classify the microtrips according to their speed-acceleration relation. The principle in assessing the relation is to group microtrips that are similar or close to each other and to differentiate groups that contrasted from each other. The unified speed profile obtained is set as the reference database. The homogeneity of the microtrips justify the class of each groups in order to describe the class through overall parameters. The parameters to be measured are including the average speed, average acceleration and deceleration, percentage of idling, percentage of cruising as well as number of stops. Therefore, the analysis makes it possible to develop a typology of local driving conditions into speed classes (i.e. low, medium, high) based on the speed profile characteristics.

For the cycle development, the selected microtrips are merged together into a set of trial driving cycle. The selection of the individual microtrips to be merged is done automatically by using clustering tool. Then, the Speed Acceleration Frequency Distribution (SAFD) is developed. The SAFD is used to illustrate the relationship of speed and acceleration in graphical manner. Generally, the bubble chart gives an idea on the driving pattern based on the real-world driving conditions. Once the trial driving cycle is constructed, it is then to be compared with the unified reference database. Then, statistical approach is used to compare the representativeness of the driving cycle with the reference database. An analysis named chi-square analysis is performed to analyse the similarity of the driving cycle to the reference database. In that sense, the final cycle shall replicate Kuala Lumpur actual driving behaviour constituting various driving operations like accelerating, cruising and idling. Figure 3 shows the example of SAFD to illustrates the relationship between speed and acceleration correspondence with the driving operations.

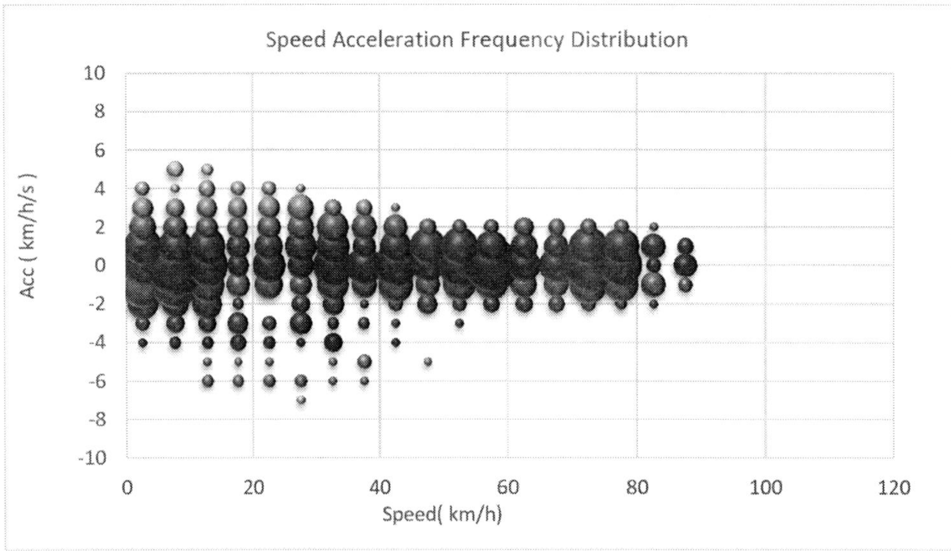

Figure 3. SAFD developed to determine the relationship of vehicle operations and driving behaviour.

4.0 Conclusions and Recommendation

The present research is attempted to establish a representative driving cycle for Kuala Lumpur territory. The methodology used is adopting the approach introduced by the GRPE under the UNECE for the development of WLTC. However, in order to ensure the novelty of this research, the method of route selection is a bit different in order to replicate the real-world driving behaviour based on domestic conditions. Google Traffic data is used to validate the current level of congestion by comparing the extracted data from RTVM 2015. By using the data from both traffic census as well as online platform, the assumption made for deciding the peak and off-peak periods could be justified by comparing both data.

To reflect the actual driving behaviour and fuel consumption for each vehicle category across nationwide, further study needs to be done in regions other than central, namely North, South, East Coast, Sabah and Sarawak. With the availability of the driving data nationwide, it is expected that the results will be more accurate and the assessment of emission as well as fuel consumption will be more stringent. Thus, it can be concluded that, better assessment for vehicular exhaust emission can be done with the driving behaviour data that are obtained from various regions across Malaysia.

Acknowledgement
The research is fully funded by the Government of Malaysia through Malaysia Automotive, Robotics and IoT Institute (MARii). The research is also conducted with the assistance from the Malaysian automotive players namely Perodua Manufacturing Sendirian Berhad and Proton Holdings Berhad in providing the expertise based on industrial perspectives and usages of driving cycle.

References

[1] Malaysia First Biennial Update Report to UNFCC 2015. Accessible through the website Ministry of Natural Resources and Environment Malaysia (www.nre.gov.my).
[2] André M., (2004), "The ARTEMIS European Driving Cycle for Measuring Car Pollutant Emissions", Sci. Total Environ. 334-335, 73-84
[3] Achour H. and Olabi A.G., (2016), "Driving Cycle Developments and Their Impacts on Energy Consumption of Transportation", Journal of Cleaner Production 112, 1778-1788.
[4] S.H. Ho, Y.D. Wong, V.W.C. Chang., (2014), "Developing Singapore Driving Cycle for Passenger Cars to Estimate Fuel Consumption and Vehicular Emissions", J. Atmos. Environ. 97, 353-362
[5] Ministry of Works Malaysia (2015), "Road Traffic Volume Malaysia (RTVM)", Purchased hardcopy through the Ministry of Works Malaysia, June 18, 2017.

Low Head Micro Hydro Systems for Rural Electrification

F M Tamiri*[1], M A Ismail[1] and W K Muzammil[2]

[1] Faculty of Engineering, University Malaysia Sabah, Malaysia
[2] Renewable Energy Research Unit, Faculty of Engineering, University Malaysia Sabah, Malaysia

*fadzlita@ums.edu.my

Abstract. Off-grid micro hydro is an economical and practical approach to generate electric power to fulfil energy demand in rural areas. Large numbers of successful high head micro hydro project have been reported however the application of low head micro hydro system is scarce. Low head micro hydro system commonly recognized using complex profile blades, larger civil structure and electromechanical components to extract energy. This leads to a more complex system and higher cost with respect to similar power rating. On the other hand, the potential to exploit low head micro hydro system is enormous because there is where population is concentrated. The aim of this paper is to present the main issues and technologies encircled the application of low head micro hydro power for rural electrifications. Reviews on relevant issues on micro hydro surrounding available turbines for low head sites, civil structures and electric motor are also presented. This is followed by accentuating the importance of information dissemination using the internet to potential users. The information can be used by researchers, plant managers, individual or rural communities to understand the potential of available technologies.

1. Introduction

Out of the total population of 630 million of ASEAN (The Association of Southeast Asian Nations) countries, 107 million are without access to electricity in 2015, and most of those affected live in remote areas and are far from electricity grids [1]. Myanmar and Cambodia were reported to be countries having the lowest percentage of electrification rate in 2017 with electrification percentage of 59.0% and 60.0% respectively [2]. Current statistics indicated an exponential increase in energy supply since 2005 with the implementation of new policies, strategies and experiences among ASEAN countries [2, 3]. One of the approaches is the use of appropriate technology solution to suit local condition, environment and resources. With that in mind, the main philosophy of rural electrification has been to provide cost effective and technical support, appropriate for remote isolated places.

The traditional way to generate electricity for rural areas is by extending existing national grid through distance and hilly terrain and this is prohibitively expensive. On the other hand, decentralized electric generation by building an off grid power generation system using diesel generators, solar power, wind power, biomass, hydropower or a hybrid system is increasingly being recognized and accepted. Considerable numbers of off-grid projects have been reported and highlighting the benefit and challenges faced by any types of power generation sources [4-7]. In many cases, it has become clear that micro hydro is a favourable source of energy if there is a potential site. With careful design and implementation, the capacity factor of a micro hydro system can reach up to 50%, beyond other types of renewable energy [8].

Content from this work may be used under the terms of the Creative Commons Attribution 3.0 licence. Any further distribution of this work must maintain attribution to the author(s) and the title of the work, journal citation and DOI.
Published under licence by IOP Publishing Ltd

IOP Conf. Series: Materials Science and Engineering **834** (2020) 012041 doi:10.1088/1757-899X/834/1/012041

Currently, common commercially available micro hydro in rural areas utilizes high head impulse type turbine such as Pelton Turbine and Crossflow Turbine. These impulse type turbines are widely used for high head micro hydro systems because they are cheaper and smaller than reaction turbines since pressurized water from head is used to generate power instead of flow rate. However, low head micro hydro is applicable to flat sites where the resident population is denser. Low head micro hydro systems typically require larger equipment and professional technical input to enable implementation and this is expensive. Consequently, applications of low head micro hydro in rural areas are rare and there is limited information available for potential users.

2. Hydropower formula

Power output from a hydropower is directly proportional to head and flow rate at the proposed site. For large hydropower, the hydraulic efficiencies are usually in the range of 80 -90 per cent however small hydropower systems are in the range of 60 - 80 per cent [9]. The potential installed power from the selected site can be calculated using equation

$$P = \eta \rho g Q h \tag{1}$$

where P (Watt) is the net power generated at the generator terminal, η (%) is the overall efficiency of the systems, ρ (kg/m^3) is the density of water, g (m/s^2) is the gravitational acceleration, Q (m^3/s) is the stream flow rate and h (meter) is the net head at the turbine inlet.

The selection of proper turbine to suit a potential site is dictated by the power generated by the turbine shaft, head, and flow rate at the turbine inlet. Specific speed as a dimensionless unit is used to characterize the type of flow and predict the turbine characteristic. Turbine specific speed, N_S is defined as

$$Ns = \frac{RPM\sqrt{P}}{H^{1.25}} \tag{2}$$

where RPM is the turbine rotation rate in RPM, P is power output (kW) and H is the water head (meter). Depending on the flow direction and path across the turbine wheel, each turbine type is specified as a radial, partial axial (mixed flow), or full axial turbine. Turbines with lower specific speed are often called impulse turbines and have a relatively higher head with lower flow rate.

For a low head system, the type of flow is categorized as mixed flow and axial flow and the specific speed is between 100 and 600 (mm/mm). In overall, high head site uses impulse type turbine and low head site uses reaction type turbines shown in table 1.

Table 1. Turbine Specific speed

Specific speed	Type of Flow	Types of turbine
15-65	Radial Flow	Turgo Turbine, Pelton Turbine
60-400	Mixed Flow	Cross flow, Francis
300-800	Axial Flow	Kaplan, Propeller

Impulse type turbine converts all the kinetic energy from high speed water jet exiting from the nozzle. The water jet makes contact with the turbine blade at atmospheric pressure, changing the momentum of the water and spins the turbine. Common impulse type turbines for micro hydro application are Turgo turbine, Pelton turbine and cross flow turbine. On the other hand, reaction type turbine uses the combination of kinetic and potential energy from the flowing water. The turbine is fully submerged in an enclosed casing, changing water pressure across the turbine. Draft tube is used to extend the tail water head and increase the pressure difference across the turbine. The range of typical commercial type turbines with respect to site head and flow rate is shown in Figure 1.

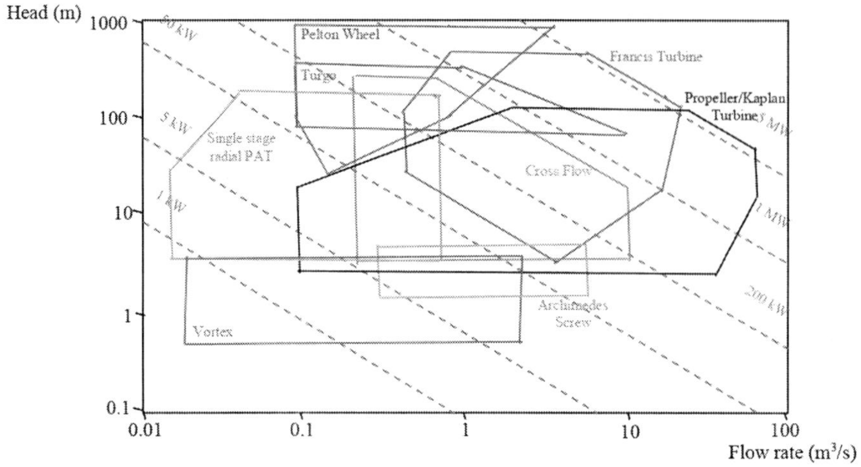

Figure 1. General micro hydro turbine with respect to operation range.
Adapted from [20, 23]

3. Micro hydro Turbines

Micro hydro refers to hydropower systems that produce power up to 100 kW; however this article will only be focusing on presenting a review for systems below 20 kW which is self-manageable by any individual or rural communities with adequate training. Very few commercial turbines are designed for the application of low head micro hydro in rural areas but low head micro hydro system is not a new technology. Several varieties of low head turbines may be found, including crossflow turbine, pumps as turbine, open flume turbine, bulb turbine, Archimedian screw, Kaplan turbine and the new turbine technology which is the Gravitational Water Vortex Hydropower Plant (GWVHP). The common commercial type turbines are shown in figure 2. The selection of turbines mainly depends on the pressure and flow available at the site.

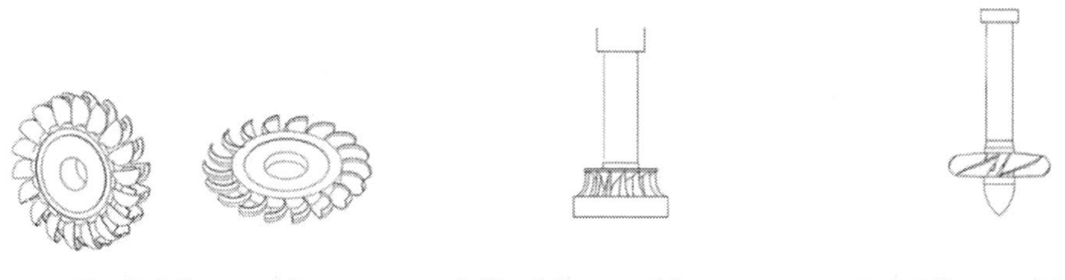

Radial flow turbine Mixed flow turbine Axial flow turbine

Figure 2. Commercial micro hydro turbines.

3.1. Francis Turbine

Francis turbine is used for medium and high head site. The turbine is positioned in a spiral volute. Water enters in a radial direction guided with adjustable guide vanes, tangential to the turbine. Pressurized water is converted into kinetic energy by spinning the turbine which is connected to the shaft. Nautilus LLC has manufactured Francis turbine that can operate between 1.2 meter and 12.2 meter generating power from 300 W to 6.4 kW. The main components of Francis turbine consists of spiral casing, guide vanes, runner blades and draft tube. Due to the complexity of the blade profile and application of guide vanes, this turbine is more expensive and less popular for micro hydro application in rural areas. Furthermore, with regard to the control mechanism for Francis turbine; guide vanes tend to trap debris from the flowing water and it mandates frequent maintenance.

3.2. *Open Flumes-Kaplan Turbine*

Pico-scale open flumes turbine, with a range between 200 W up to 3 kW is the most common turbine in the marketplace for application with heads between 1 and 3 meter [10-13]. This system uses a propeller type turbine and an AC or DC generator generating up to 1.5kW to charge a bank of batteries to supply electricity for a typical remote home [14]. Large numbers of these units are installed in over 60 countries [13]. These turbines are suitable for operation in rural areas due to their light weight, portability, simple installation and they also require minimal civil work. Due to their simple design, they have lower efficiency compared to spiral casing installation. Typical turbine efficiency is 50%.

3.3. *Archimedean Screw*

Archimedean screw offers fish friendly, with variable speed low head turbine generator. The system uses a special enclosed turbine casing to create tangential momentum to produce rotational movement. It uses non-regulated blade fixed with a permanent magnet variable speed generator and offers a good turbine performance over a large head range. The maximum single variable speed low head turbine generator can work for heads up to 8 meter with a flow rate ranging between 100 *l*/s up to 12000 *l*/s [15, 16]. This system requires considerable civil work to accommodate a large turbine which contributes to the high cost. Additionally, the Archimedean screw rotates at low speed, so a gearbox is used to increase the rotational speed to match the standard generators. For this reason, the Archimedes turbine is less popular for rural electrification.

3.4. *Pump as Turbine*

A pump as turbine (PAT) is a low cost option for generating electricity in terms of price and simplicity. An end suction centrifugal pump operating in turbine behaves similar to Francis turbine without guide vanes and the axial type pump behaves similar like axial type turbine. Mass production pump covers a wide range of medium-low head sites [17]. A standard induction pump with an induction motor can be used as an independent power generator in isolated micro hydro sites to avoid the use of belt or gear mechanism [18-20]. The application of PAT in Lao PDR for remote communities to substitute crossflow turbine to reduce electromechanical cost offers technical and economical alternative [18]. Turbine performance prediction can be presented by mathematical models; however it has been reported to give fairly inaccurate results. Typically, the performance of a standard PAT is best estimated from test results which are rarely available and not readily accessible [21]. This is a challenge to match the turbine characteristics to a particular micro hydro site.

3.5. *Gravitational Water Vortex Hydropower Plant (GWVHP)*

Gravitational water vortex power plant is one of the most recent turbine technology for low head micro hydro sites which was developed by Franz Zotlöterer in 2006. The system generates electricity by introducing a flow of water tangentially into a circular basin that creates a free vortex when the water drains from the bottom of the basin. A turbine thereof withdraws rotational energy from the gravitational vortex, which is converted into electric power in a generator [22]. The key difference of GWVHP arrangement compare to the classic Kaplan or Francis turbine arrangement is the fact that the chamber is 'open channel'. The fluid is subjected to atmospheric conditions with no large pressure differences built up in the system. Thus, the free surface develops a classis hyperbolic depression and air core, which results in a reduction of cross-sectional area at the orifice for water to pass, therefore resulting significant reduction of the relative flow rate and significant increase in both tangential and axial velocity of the vortex [23].

The turbine does not function based on pressure differential, but on the dynamic force of the water vortex. It uses both kinetic (run-of-river) and static potential energy (head) principles to generate electricity, and also aerates the water in a gentle way with minimum interference with the river and aquatic life. The plant can be installed with a water drop as little as 0.7 meters with relatively small and simple structure [22]. This technology is considered still in its infancy with limited literatures available for the technology to proceed beyond prototyping stage. The maximum efficiency obtained by researchers are approximately 30% while the commercial companies claimed about 50% of efficiency with 500 W to 20 kW of power generated. This is lower compare to conventional low-head turbines such as the Archimedean screw turbine. However, due to its ability to sustain relatively high efficiencies at low heads and small to medium flow rates, this technology addresses a gap in the current turbine application chart and demonstrates potential to be developed and optimized in both performance and hydraulic and mechanical efficiency [23, 24].

4. Electric Generator

Induction generator is desirable as electric generator generating continuous alternating current (AC) at rated frequency and voltage. A bank of capacitor supplies the reactive energy to build its own magnetism. With the advent of electric controller, induction generator controller (IGC) may be used to maintain good power quality at generator terminals. The most popular IGC for micro hydro application is Electric Load Controller (ELC). This type of IGC uses solid state electronic controller regulating the output power of a micro hydropower system, maintaining constant applied load at generator terminal by dummy load. The function of ELC is to eliminate hydraulic control mechanism and let the turbine and generator run at their optimum operating speed [20, 23]. Nowadays, three phase induction generators are widely used for most stand-alone micro hydro projects because of their availability and price. Direct current (DC) generated from dynamo or from rectifier should be avoided if the site has the potential to generate continuous reliable energy. Moreover, there are many losses related to energy storage and power conversion between AC to DC.

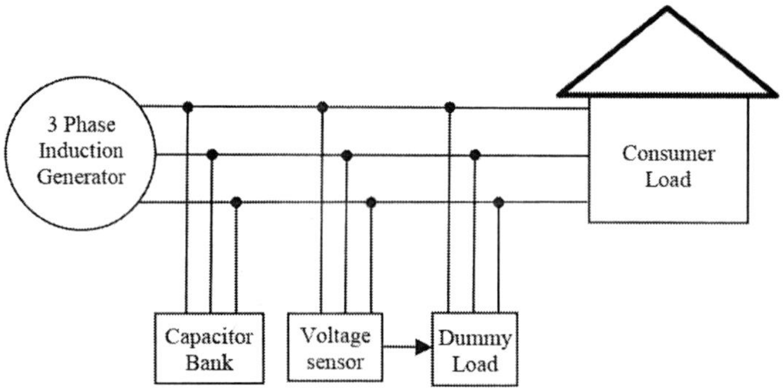

Figure 3. Schematic diagram of three phase induction motor [20, 23]

5. Civil Structure

Civil structure for any micro-hydro system is site specific and unique to the site itself. However, the main civil structure components are similar which comprise of diversion weir, channel, intake, forebay, regulating gates, powerhouse, penstock, spillways and penstock. The function of civil structure for a small scale hydropower is to regulate and control the water flowing from the intake to the turbine. Careful design and planning, exploiting locally available materials can reduce the civil structure components in micro hydro scheme.

For many cases, run-off river scheme is the most favourable system for micro hydro applications compared to gravity dam. A diversion weir is used to divert part of the flowing water of the river into a channel or penstock. The water is then transported to the silt basin to remove unwanted debris and suspended solids before being channelled to the forebay. The water flows downhill using the gravitational force to build up pressure. The pressurized water is used to rotate the turbine which is coupled with electric generator to generate electricity. The water is then returned to the river, so minimum water storage is required thus reducing the impact to the environment. With the absence of a dam, the ecosystem and natural habitats of the river is maintained and preserved. Furthermore, it is easier to control water flow to the turbine by using spillways. This will reduce the risk of damaging the powerhouse if flooding occurs. On top of that, the use of run-off river permits optimum power design for the electromechanical based on diverted flow from the river.

The micro hydro penstock is commonly the third in terms of the overall micro hydro cost. The selection criteria of a penstock should include pipe joints, valves, labour work, logistic and penstock anchoring. Therefore, the use of local available domestic pipe such as uPVC, MDPE and HDPE pipe is highly recommended to reduce the overall cost [25]. The selection of penstock and other building materials for isolated micro hydro sites is strongly influenced by the logistic.

6. Knowledge Dissemination

In order to make a low head micro hydro scheme relevant for rural electrifications, know how knowledge must be accessible and reachable by potential users. Most micro hydro information is available in books, magazines and web sites. With the advent of the Internet, online searches on micro hydro have become popular and widespread [26]. Generally, information with micro hydro in web portals is published by individuals, DIYers, renewable energy suppliers, non-governmental organization and research institutes. The first discussion web portal associated with micro hydro is microhydropower.net [27] which began when internet communities started sharing information related to micro hydro power technology in 1998. It was intended to be a dedicated information source for micro hydro enthusiast. The website was incorporated into Yahoo! Groups later on and provides information concerning micro hydro power updates such as events, news, database, downloadable files, literature overview, Internet links and case studies.

A non-government organization, Practical Action [28], a registered charity in the UK, provides information on innovative and practical solutions for poor communities to improve their quality of life and access technical options and knowledge by allowing anyone to access records and documents. Implementation guides on renewable energy are made available and free to download as an open source including micro hydro. Using best-practices cases for demonstration, technical standards for small isolated schemes are shared promoting technology and know-how transfer.

Pico-hydro is a web portal sharing hydropower for sizes below 5 kW [29]. It was pioneered by Arthur Williams, the director of a course taught in Nottingham University. This website shares information on ways to design, manufacture and install parts of pico hydro projects for low and high head sites. Case studies are presented, including technical and nontechnical information such as costing, manufacturing guidelines, electric distribution systems design, safety and financing support by academic research. Despite the fact that each hydropower site is unique, the author suggests the use of standardized parts and low cost approaches to make it relevant for the Third World countries. For low head turbines, this website focuses on the development of micro hydro turbine for site that have a head of 2.5m and a flow rate of 230 l/s which can produce 5kW.

Micro hydro suppliers [10-13, 30] have published a substantial amount of technical information, guides and manuals on their web site; covering installation of mechanical and electrical systems and include feasibility studies of the proposed sites. A web-based micro hydro power forum designed by a German authorized micro hydro power supplier combines most popular micro hydro power manufacturers' information concerning micro hydro topics including low and high head scheme. Currently, more than 3,000 active registered members discuss a broad range of issues on hydropower. Discussed topics are related to commercially available micro hydro power packs with strict discussion on listed hydropower products.

It is clear that by providing technical information through web portal, the potential users in flat areas who possess hydro sites may exploit low head micro hydro technology. Efforts to disseminate technology through the internet are valuable for rural electrification. The information should incorporate academic research and technical support to demonstrate reliable energy production. Based on the review on previously presented works, it is important to provide a sound web based information that uses internet as knowledge dissemination medium to help potential users exploit the potential of low head sites to generate electricity.

7. Conclusion

The potential of generating electricity from flowing water for rural remote communities should be considered as the focal power options where there is available suitable site. The study presented in this paper highlights the main issues relevant to low head micro hydro in rural areas. The main objective is to promote the application of available technologies and exploit the opportunity of generating electric power from low head sites. Current technologies allow micro hydro users to decide on applicable schemes, optimizing cost and efficiency. The dissemination know-how knowledge work should be prominence with the advent of internet.

Acknowledgement

This research was supported by the Ministry of Higher Education of Malaysia and Universiti Malaysia Sabah through their financial assistance (UMSGreat grant: GUG0172-2/2017) and facilities support.

References

[1] A. Gnanasagaran, “Electrifying Rural ASEAN”, [Online]. Available:https://theaseanpost.com/article/electrifying-rural-asean. [Accessed: 23-01-2018].

[2] Anonymous, “Southeast Asia Energy Outlook 2017: Special Report IEA”, [Onlne]. Available: https://www.iea.org/southeastasia/. [Accessed: 28-01-2019].

[3] "ASEAN Guideline on Off-Grid Electrification Approaches," 2013.

[4] M. Anyi, B. Kirke and S. Ali, “Remote community electrification in Sarawak, Malaysia”, *Renewable Energy,* vol. **35, pp.** 1609-13, 2010.

[5] S. Murni et al., “The Role of Micro Hydro Power Systems in Remote Rural Electrification: A Case Study in The Bawan Valley, Borneo”, *Procedia Engineering* vol. **49,** pp. 189-196, 2012.

[6] S. Murni et al., “Status of rural electrification in the ‘Heart of Borneo’: Role of micro hydro projects”, *Solar 2010 the 48th AuSES Annual Conference,* Canberra, Australia, 2010.

[7] T. McNish, D. M. Kammen and B. Gutierrez, “Clean Energy Options for Sabah: An Analysis of Resource Availability and Cost”, *RAEL,* University of California, Berkeley, 2010.

[8] R. Uhunmwangho and E. K. Okedu, “Small Hydropower for Sustainable Development”, *Pacific Journal of Science and Technology,* vol. **10, pp.** 535-543, 2009.

[9] S. Mishra, S. K. Singal and D. K. Khatod, “A review on electromechanical equipment applicable to small hydropower plants”, *International Journal of Energy Research, vol.* **36** pp. 553-571, 2012.

[10] Anonymous, “Crossflow Turbines”, [Online]. Available: http://www.renewablesfirst.co.uk/crossflow-turbines.html. [Accessed: 23-01-2018]

[11] Anonymous, “Energy System and Design”, [Online]. Available: http://www.microhydropower.com. [Accessed: 23-01-2018].

[12] Anonymous, “Exmork Homepage”, Exmork, [Online]. Available: http://www.exmork.com. [Accessed: 23-03-2018].

[13] Anonymous, “Powerpal -Electricity from Water with PowerPal”, 11-02-2013, [Online]. Available: http://www.powerpal.com. [Accessed: 13-03-2018].

[14] A. A. Lahimer et al., “Research and development aspects of pico-hydro power”, *Renewable and Sustainable Energy Reviews,* vol. **16**, pp. 5861-78, 2012.

[15] Anonymous, “Renewable First : Archimedian Screw”, [Online]. Available: http://www.renewablesfirst.co.uk/hydro-learning-centre/archimedean-screw. [Accessed: 23-01-2018].

[16] Anonymous, “Archimedes Hydro Screw”, [Online]. Available: http://www.archimedeshydroscrew.com/hydro-dynamic-screw-turbine/. [Accessed: 07-02-2018]

[17] H. Ramos and A. Borga, “Pumps as Turbines: An Unconventional Solution to Energy Production”, *Urban Water,* vol. **1, pp.** 261-263, 1999.

[18] M. Arriaga, “Pump as Turbine – A Pico-Hydro Alternative in Lao People's Democratic Republic”, *Renewable Energy,* vol. **35, pp.** 1109-15, 2010.

[19] A. A. Williams, “Pumps as Turbines for Low Cost Micro Hydro Power”, *Renewable Energy,* vol.**9, pp.** 1227-34, 1996.

[20] M. A. Ismail et al., “End Suction Centrifugal Pump Operating in Turbine Mode for Micro hydro Applications”, *Advances in Mechanical Engineering*, 139868, 2014.

IOP Conf. Series: Materials Science and Engineering **834** (2020) 012041 doi:10.1088/1757-899X/834/1/012041

[21] H. Nautiyal, Varun and A. Kumar, "Reverse Running Pumps Analytical, Experimental and Computational Study: A Review ", *Renewable and Sustainable Energy Reviews,* vol. **14, no.** 20, pp. 59-67, 2010.

[22] F. Zotlöterer, "2018 Gravitation Water Vortex Power Plant", [Online]. Available: http://www.zotloeterer.com/welcome/gravitation-water-vortex-power-plants/. [Accessed:19-02-2019].

[23] A. B. Timilsina, S. Mulligan and T. R. Bajracharya, "Water vortex hydropower technology: a state-of-the-art review of developmental trends", *Clean Technologies and Environmental Policy,* vol. **20, no.** 17, pp. 37-60, 2018.

[24] M. M. Rahman et al., "A Review on the Development of Gravitational Water Vortex Power Plant as Alternative Renewable Energy Resources" *IOP Conf. Series: Materials Science and Engineering,* vol. **217,** 012007, 2017.

[25] K. V. Alexander and E. P. Giddens, "Optimum penstocks for low head micro hydro schemes", *Renewable Energy,* vol. **33,** pp. 507-519, 2008.

[26] W. J. Klunne, "Using the internet to disseminate information on micro hydropower in ISES Solar World Congress", Kassel, Germany, 2011.

[27] Anonymous, "Microhydropower", [Online]. Available: http://microhydropower.net. [Accessed:23-01-2018].

[28] Anonymous, "Practical Action", [Online]. Available: http://practicalaction.org/micro_hydro_expertise. [Accessed: 13-03-2018].

[29] Anonymous, "Pico hydro", [Online]. Available: http://www.eee.nottingham.ac.uk/picohydro/index.html. [Accessed: 13-03-2018].

[30] Anonymous, "Nautilus Water Turbine", [Online]. Available: http://www.waterturbine.com. [Accessed: 07-02-2018].

IOP Conf. Series: Materials Science and Engineering **834** (2020) 012042 doi:10.1088/1757-899X/834/1/012042

Data Characterization of 4 Axis Shadow Robot Arm Integrated with Flex Sensor Mechanism

M. D. Amari*[1,2], Sukarnur Che Abdullah[1], Lee Yong Aik[2]

[1]Faculty of Mechanical Engineering,
University Technology MARA (UiTM) ,40150, Shah Alam, Selangor, Malaysia
[2]School of Engineering, KDU University College,40150 Shah Alam, Selangor, Malaysia

* dzulhelmy.a@kdu.edu.my

Abstract. Robotic arms have seen widespread use in various industries in the recent years. However, most robotic arms are controlled to move in a predetermined path. The goal of this study is to develop a wirelessly controlled robotic arm which is able to mimic the user's movements. The arm enables easier programming of robotic arms and increases ease of use of such arms. Methods of collecting the anthropometric measurements of the human body to be translated to digital data have been investigated. Integration of flex sensor for the fingertips move-ments were used to mimic by detecting the bending moment from the user. As for elbow detection, variable resistor was used to detect the changes by converting the changes of angle at the elbow structure to resistance value. Both of the data were mapped accordingly to the exo-skeleton arm receiver by using wireless Bluetooth to ensure all structure move accurately with even slight changes done by the user. The optimization and filtering of the recorded data was also conducted in the study to develop a robotic arm that is able to mimic the human arm motion to the highest accuracy possible.

1. Introduction

The use of robotics arms in various industries have given rise to a large number of methods and developments in robotics technology. Throughout the development of robotic technology, there have been an increased interest in developing humanoid robots. Humanoid robots are robots with anthropometric body plans and human like senses [1]. These efforts are motivated by the desire to create robots that are able to mimic human motion and work in close cooperation with humans. Compared to high-ly specialized industrial robots which can perform industrial tasks such as welding, pick and place etc. with high accuracy consistently, humanoid robots cater to a different set of needs. These robots are designed to be used for general purpose and is meant to be robust and easy to use. The applications for such robots can be in situations which are of high risk but require humans to complete a task such as fire rescue of in high heat/radiation environments.

To develop a humanoid robotic arm capable of mimicking the finger and elbow motions, the methods of collecting the anthropometric measurements and movements are of critical importance. Current techniques involve using MEMS sensors such as gyroscopes and accelerometer chips to track user movement. The aim of this study is to explore the possibility and feasibility of employing simpler and lower costs solutions in tracking movement. The main goal is to achieve readings with similar accuracies

Content from this work may be used under the terms of the Creative Commons Attribution 3.0 licence. Any further distribution of this work must maintain attribution to the author(s) and the title of the work, journal citation and DOI.
Published under licence by IOP Publishing Ltd

with these low-cost measuring techniques as the electronic techniques. Therefore, the circuitry set-up and coding required to achieve high accuracy was investigated.

The design of the arm is also crucial as it has to mimic all the motions of the human fingers and elbows and thus requires a total of 4-degrees of freedom. The ability for the actuators to move the arm to the desired position as detected by sensors were also investigated. Current technologies such as the robotic arm shown in [2] uses pneumatic muscles to achieve accurate movement. This study will employ a lower cost solution which is servo motors which are able to rotate with 1-degree accuracy and are highly consistent and repeatable as well. Several tests and simulations in software will have to be conducted to ensure the arm has the full range of motions a human arm has and also to verify that the sensors are able to generate the desired output.

2. Methodology

The conceptual design and programming development have been fulfilled by the integration of two main software which are Solid-Works and Arduino software (IDE). Both software had specific function where SolidWorks software is focused on achieving the optimized design and creating simulation for movement of the 4-axis robotic arm. For this project, Arduino Uno microcontroller has been selected to detect the voltage changes across flex sensor and then convert the input data into mechanical movement at the robotic arm. The methodology is specified into following stages:

Stage 1: Conceptual design of robotic arm and the controller
Stage 2: Prototype fabrication
Stage 3: Programming Input and Output

2.1. Conceptual Design

The design for the robotic arm was intended to mimic the mobili-ty of a human arm. The exoskeleton suit from 'Ironman' has the mobility of a human body and the robotic arm was mainly inspired by that [6]. Most of the parts of the robotic arm are curvature in shape which aided in increase in strength as curve shape has better load distribution compared to flat shape [7]. The strength of the design was further improved by designed the wrist in round-shaped to allowed the robot palm to support heavy load. The design for the fists was inspired by a robot called 'Atom' from the novel 'Real-Steel' which have two huge fingers and a thumb with approximate-ly similar size to the fingers [8]. The robotic arm was designed in SolidWorks software and the design consisted of 47 parts.

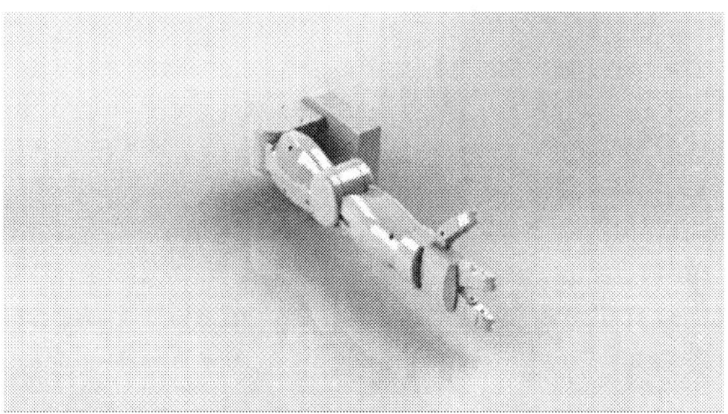

Figure 1. Isometric view for 3 fingered robotic arm

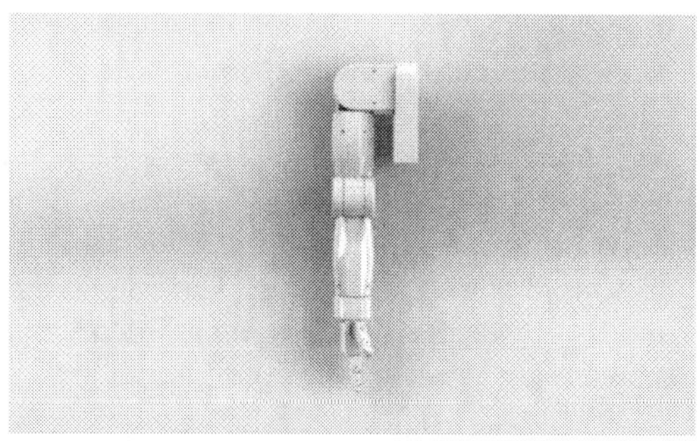

Figure 2. Top view for 3 fingered robotic arm

IOP Conf. Series: Materials Science and Engineering **834** (2020) 012042 doi:10.1088/1757-899X/834/1/012042

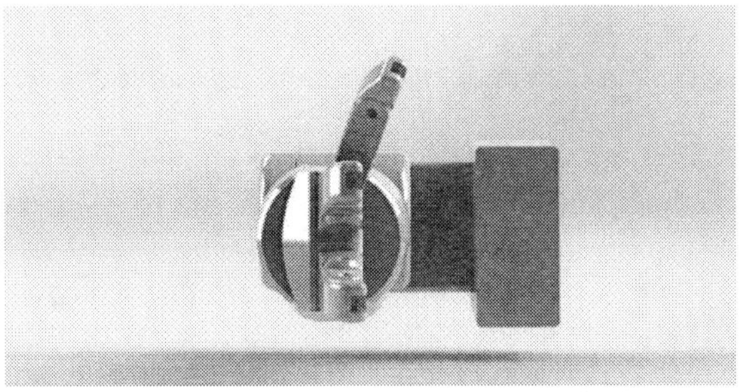

Figure 1. Front view for 3 fingered robotic arm

Figure 2. Side view for 3 fingered robotic arm

Figure 1 to 4 represent the assembly drawing for the robotic arm. The finger was designed to have only 3 units due to the pick and place testing that we will conduct in future works. The purpose of the design is to mimic the movement of actual hand.

2.2 Prototype Fabrication

The robotic arm is attached to the user's arm and is controlled by a microcontroller to replicate the mobility of the user's arm. However, there is a limitation found on the robotic arm as it only has 2 fingers with a thumb. There is no wrist rotation, but the elbow can be bent up to a maximum of 173°. The robotic arm was fabricated with 3D printer with the PLA plastic and 15% density. The material was chosen to be PLA due to the easier printing which able to produce finer details. As the robotic arm is still a prototype, the density of the material chosen to be 15% density due to lower cost to produce. In addition, 15% density of printed material won't affect the integrity of the structure. Since the future testing phase will be done by moving small object, 15% density won't affect the results too.

The robotic arm has 3-degrees of freedom (DOF) at the shoul-der part, 1 DOF at the elbow joint, 2 DOF at the thumb and 1 DOF for each finger. The fingers and thumb are both controlled by using SG90 servo which located at the forearm of the robotic arm. The fingertips and the tip of the thumb are tied with 0.1mm diameter fishing rope and 0.7mm diameter elastic cord while the other end for each fishing rope are tied with SG90 servo and the elastic cord tied with the interior part of the palm of the robotic arm.

SG90 servo was chosen to control the fingers and thumbs due to the small size of the motor which able to fit 3 of it in the forearm. Besides, the torque of each servo motor is 2.5kg.cm which is a decent torque for the grip of the robotic arm. The servo motor to control the movement of the elbow joint is directly placed in the elbow joint itself to maximize the force of the motor. The motor chosen for the elbow joint was MG996R with the torque of 10kg.cm which is optimal torque for the elbow joint to withstand heavy load. The movements of the shoulder joint required 3 servo motors, MG946R with torque of 13kg.cm, which is considered higher than MG996R as the shoulder joint has to carry higher weight from the robotic arm itself as compared to the elbow joint. The servo motors used for shoulder joint were placed in the bicep, shoulder and the stand or support of the robotic arm. The forearm, bicep and triceps of the robotic arm are in curvature shape mainly at the edge and has minor flat surface to increase the strength of the arm, thus allows it to increase the load limit carried by the robotic arm.

IOP Conf. Series: Materials Science and Engineering **834** (2020) 012042 doi:10.1088/1757-899X/834/1/012042

Figure 5. Isometric view for fabricated robotic arm

Figure 6. Side view for fabricated robotic arm

Figure 7. Front view for fabricated robotic arm

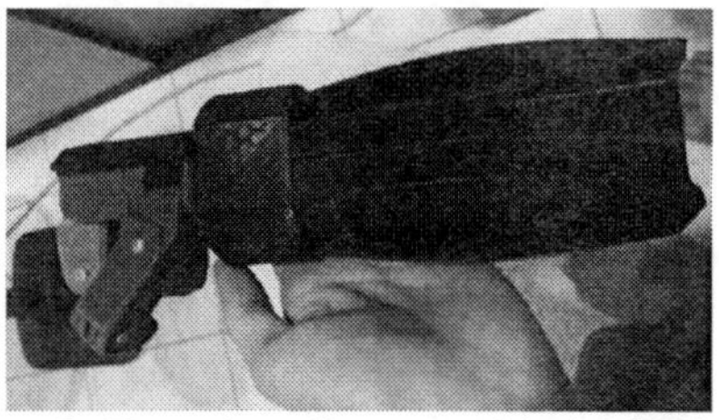

Figure 8. Top view for assembled robotic arm

Figure 5 to 8 represent the fabricated parts of the robotic arm. The robotic arm consists of 3 joints for two fingers as to mimic index finger and middle finger. The thumb finger will only have 2 joints as illustrated inside the figure.

2.3 Data Transmitter Communication Programming

Robotic arm and fingers (servomotors) were set to be controlled wirelessly by flex sensors through two Bluetooth modules. The first Bluetooth module was set as master device and second Blue-tooth module set as slave device [9]. The slave device (servomotors) must always executing the instruction given by the master device (flex sensors). Coding and circuit connection between two Blue-tooth modules on two separate Arduino Uno board must be well-configured in order to run the communication with minimum delay. Coding below shows configuration of data sender communication.

Master device

```
int finger1 = A0;
int finger2 = A1;
int finger3 = A2;
int elbow = A3;
int fpos1, fpos2, fpos3, epos;
int state =0;

void setup() {
Serial.begin(9600);
pinMode(A0, INPUT);
pinMode(A1, INPUT);
pinMode(A2, INPUT);
pinMode(A3, INPUT);
}
void loop()
```

```
{
if(Serial.available() > 0)
 {
  state = Serial.read();
 }
fpos1 = analogRead(finger1);
fpos2 = analogRead(finger2);
fpos3 = analogRead(finger3);
epos = analogRead(elbow);
fpos1 = map(fpos1, 0, 1023, 0, 180);
fpos2 = map(fpos2, 0, 1023, 0, 180);
fpos3 = map(fpos3, 0, 1023, 0, 180);
epos = map(epos, 0, 1023, 0, 180);
delay(100);
}
```

3. Results and Discussion

Analysis on the simulations and variations of voltage reading in mapping the value to the correct response mechanism. The micro-controller will send the data wirelessly to the shadow robotic arm to create similar movement done by the user.

3.1. Sensor Calibration Output

Servo motors only can be rotated at the desired angle based on the bending movement of the flex sensors. The range value of the flex sensor is from 0-1023 and the range value of servo motor is from 0-180. In master device, map () function is used to analyse and calculate the input data from flex sensor. The collected data will be converted into rotation angle as output and sent to slave device through Bluetooth communication. Once slave device is receiving the data, the servo motors will be started to rotate based on the cal-culated angle given by the flex sensor.

During calibrating, initial values for both flex sensors and servo motors were set at 0 respectively, assuming fingers, elbow and shoulder were initially at straight position. Any variations in value of flex sensors will be reflected to the value of servo motors. While both flex sensors and servo motors have reached maximum value of 1023 and 180 respectively. This shown fingers, elbow and shoulder were at maximum bending position. Based on the calibration, these predetermined values were matched correctly. Any errors found during calibration were solved accordingly.

3.2. Data Receiver Communication Programming

Slave device will be controlled accordingly based on the instruction given by the master device. Coding below shows configuration of data receiver communication.

3.2.1. Slave Device

```
#include <Servo.h>
Servo servo1, servo2, servo3, servo4;
int s1=3, s2=5, s3=6, s4=9;
int fpos1, fpos2, fpos3, epos;
int state= 0;

void setup()
{
servo1.attach(s1);
servo2.attach(s2);
```

```
servo3.attach(s3);
servo4.attach(s4);
//Set Servo write at 0 initially
servo1.write(0);
servo2.write(0);
servo3.write(0);
servo4.write(0);
//delay(100);
}
void loop()
{
 if(Serial.available() > 0)
 {
 state = Serial.read();
 }
 servo1.write(fpos1);
 servo2.write(fpos2);
 servo3.write(fpos3);
 servo4.write(epos);
}
```

3.3. Voltage Output Reading

The mapping between flex sensor and servo motor is very cru-cial to obtain the most accurate deflection. The servo motor was attached to the robotic arm to control the movement for each finger.

Table 1. Result for flex sensor and servo motor voltage reading when all fingers are fold

Transmitter				Receiver			
Finger 1	Finger 2	Finger 3	Elbow	Finger 1	Finger 2	Finger 3	Elbow
0	0	0	0	0	0	0	0
0	0	0	0	0	0	0	0
0	0	0	0	0	0	0	0
0	0	0	0	0	0	0	0
0	0	0	0	0	0	0	0
0	0	0	0	0	0	0	0
0	0	0	0	0	0	0	0
0	0	0	0	0	0	0	0
0	0	0	0	0	0	0	0
0	0	0	0	0	0	0	0
0	0	0	0	0	0	0	0
0	0	0	0	0	0	0	0
0	0	0	0	0	0	0	0
0	0	0	0	0	0	0	0
0	0	0	0	0	0	0	0
0	0	0	0	0	0	0	0
0	0	0	0	0	0	0	0
0	0	0	0	0	0	0	0
0	0	0	0	0	0	0	0
0	0	0	0	0	0	0	0

IOP Conf. Series: Materials Science and Engineering **834** (2020) 012042 doi:10.1088/1757-899X/834/1/012042

Table 2. Result for flex sensor and servo motor voltage reading when elbow is fold and the rests are open.

Transmitter			
Finger 1	Finger 2	Finger 3	Elbow
153	171	171	1
154	174	175	0
152	173	173	1
152	174	174	0
152	172	173	1
153	173	174	0
153	173	174	0
150	172	173	1
152	171	169	1
153	173	174	0
153	174	174	0
152	173	175	0
151	173	173	0
152	173	173	0
154	174	174	2
151	171	171	0
152	172	173	1
151	172	173	1
153	173	172	0
154	174	175	3

Receiver			
Finger 1	Finger 2	Finger 3	Elbow
152	173	174	0
153	173	174	0
153	173	174	0
151	171	173	1
152	172	173	1
153	174	174	0
153	171	171	1
154	174	175	0
152	173	173	1
152	174	174	0
152	172	173	1
153	173	174	0
153	173	174	0
150	172	173	1
152	171	169	1
153	173	174	0
153	174	174	0
152	173	175	0
151	173	173	0
152	173	173	0

Table 3. Result for flex sensor and servo motor voltage reading when elbow is open by 2 degree in angle

Transmitter			
Finger 1	Finger 2	Finger 3	Elbow
154	175	175	5
154	175	176	4
152	172	174	5
153	174	176	3
152	173	174	5
153	175	176	3
153	174	175	4
153	174	175	4
154	175	176	3
151	171	173	5
154	174	175	5
150	172	174	4
154	175	177	3
150	171	172	4
155	175	176	4
151	172	174	4
153	174	175	4
153	174	174	3
153	174	174	5
153	174	176	3

Receiver			
Finger 1	Finger 2	Finger 3	Elbow
153	174	174	3
154	175	176	4
154	175	176	3
154	175	175	5
153	174	176	3
154	175	175	5
154	175	176	4
152	172	174	5
153	174	176	3
152	173	174	5
153	175	176	3
153	174	175	4
153	174	175	4
154	175	176	3
151	171	173	5
154	174	175	5
150	172	174	4
154	175	177	3
150	171	172	4
155	175	176	4

Table 4. Result for flex sensor and servo motor voltage reading when all fingers and elbow are open

Transmitter			
Finger 1	Finger 2	Finger 3	Elbow
173	174	174	177
174	173	174	177
174	173	173	176
175	175	175	178
175	175	176	177
173	173	173	177
170	170	171	177
174	174	175	178
174	174	175	179
175	175	175	178
173	173	174	177
175	175	175	178
172	171	171	176
173	173	173	177
174	174	174	178
174	174	174	178
171	171	172	175
175	175	176	177
175	174	174	178
174	173	174	178

Receiver			
Finger 1	Finger 2	Finger 3	Elbow
172	171	171	176
175	174	175	178
170	171	172	173
174	174	175	178
174	174	175	178
172	173	173	176
175	174	174	178
173	174	174	177
174	173	174	177
174	173	173	176
175	175	175	178
175	175	176	177
173	173	173	177
170	174	171	177
174	174	175	178
174	175	175	179
175	175	175	178
173	173	174	177
175	170	175	178
172	174	171	176

After implementation of the coding, the voltage reading start from 0 and operate in live feed up to the voltage value of 179 when the bending mechanism is applied at maximum angle as shown in Table 1 to 4. The result shows reliable data obtained where the dif-ference value between transmitter and receiver is less than 5.

The mapping between transmitter and receiver are important to ensure that the bending movement created can mimic perfectly onto the robotic arm. Four mechanism including the flex and gyro mech-anism were integrated in one hand structure to detect the changes of the angle and the rotation of the 4 axis arm. The value is mapped to the movement on the servo motor to match the moving part of the sender mechanism. Hence, the receiving structure will move in similar accuracy as the sending structure

3.4. Characterization of Voltage changes in Different Angle

The next stage of the experimentation is to map the changes of voltage in transmitter and receiver of the robotic. The turnout of this stage is to verify the accuracy of the mechanism in capturing the slight changes of movement of the robotic arm as shown in Table 5 and 6.

The result also indicates that the accuracy is at 100% where the mapping between transmitter and receiver voltage are the same. The trend of the voltage changes needs to be refine where slight inconsistency voltage trend occurred from 120 to 180 degree angle of bending. This happened may due to the inconsistent placement of the flex sensor at the robotic structure which causes inaccurate trend at higher degree of bending.

IOP Conf. Series: Materials Science and Engineering **834** (2020) 012042 doi:10.1088/1757-899X/834/1/012042

Table 5. Result for transmitting voltage from 0 to 180-degree angle

Degree	Finger 1	Finger 2	Finger 3	Elbow
0	14	11	16	0
10	21	24	27	5
20	37	34	29	9
30	40	42	36	12
40	60	50	46	28
50	59	57	60	49
60	69	56	80	61
70	74	61	84	76
80	89	58	98	85
90	102	98	100	96
100	105	99	88	110
110	120	118	100	115
120	124	121	110	119
130	116	107	117	131
140	126	136	122	138
150	125	133	130	141
160	128	146	136	146
170	133	146	138	148
180	144	158	156	184

Table 6. Result for receiving voltage from 0 to 180-degree angle

Degree	Finger 1	Finger 2	Finger 3	Elbow
0	14	11	16	0
10	21	24	27	5
20	37	34	29	9
30	40	42	36	12
40	60	50	46	28
50	59	57	60	49
60	69	56	80	61
70	74	61	84	76
80	89	58	98	85
90	102	98	100	96
100	105	99	88	110
110	120	118	100	115
120	124	121	110	119
130	116	107	117	131
140	126	136	122	138
150	125	133	130	141
160	128	146	136	146
170	133	146	138	148
180	144	158	156	184

4. Conclusion

The mapping shows the flex sensor bending mechanism are mapped accordingly. Hence, the servo motor will capture the changes of the voltage reading accurately to exhibits even slight changes of movement. The work will be refined in future paper to comprehend the pick and place testing phase.

Acknowledgment

The authors wish to thank Universiti Teknologi MARA for supporting the research under the grant 600-IRMI 5/3/GIP (056/2018).

References

[1] S. Behnke, "Humanoid robots-from fiction to reality?", *KI*, vol. *22,* no.4, pp. 5-9, 2008.
[2] Wang et al., "A MEMS-based Air Flow Sensor with a Free-standing Micro-cantilever Structure", *Sensors*, vol. 7, no. 10, pp. 2389–2401, 2007.
[3] K. Toda, I. Sanemasa and K. Ishikawa, "Simple temperature compensation of thermal air-flow sensor", *Sensors and Actuators A: Physical,* vol. 57, no. 3, pp. 197–201, 1996.
[4] F. Magalhães and Á. Cunha, "Explaining operational modal analysis with data from an arch bridge", *Mechanical Systems and Signal Processing,* vol. 25, no. 5, pp. 1431–1450, 2011.
[5] Tuffield, Paul and H. Elias "The Shadow Robot Mimics Human Actions." *Industrial Robot: An International Journal*, vol. 30, no. 1, pp. 56–60, 2003.
[6] Humphries et al., "Reconstruction Part 1 of 6: Any Given Sunday", *Ultimate Comics: Ultimates* 19, Marvel Comics, February 2013.
[7] R. Matheson, "Steel: And Other Stories", Tom Doherty Associates, 2011.
[8] R. K. Megalingam et al. "IOP Conf. Ser." *Mater. Sci. Eng.,* vol. 225, 012204, 2017.
[9] Faravar, "Design, Implementation and Control of a Robotic Arm Using PIC 16F877A Microcontroller", 2014, [Online]. Available: https://pdfs.semanticscholar.org/10f8/ebe4038ba1363e813d4bb6a9004b1a99e88d.pdf.
[10] M. A. Doshi and S. J. Parekh, "Wireless Robotic Hand using Flex Sensors", *An International Journal,* vol. 6, no. 3, pp. 2229-5518, 2015.
[11] G. Saggio, "Shaping Resistive Bend Sensors to Enhance Readout Linearity", vol. 2012, 7 pages, 2012.

Electromyography (EMG) for Assessment in Low Back Pain; Erector Spinae Muscle

Nur Hajariah Che Hamid[1], Ayuni Che Zakaria*[1], Nadia Mohd Mustafah[2], Aizreena Azaman[3], Khairunnisa Johar[1]

[1]Faculty of Mechanical Engineering, Universiti Teknologi MARA Selangor, 40450 Shah Alam, Malaysia.
[2]Faculty of Medicine, Universiti Teknologi MARA Sungai Buloh, Jalan Hospital, 47000 Selangor, Malaysia.
[3]School of Biomedical & Health Science, Faculty of Engineering, Universiti Teknologi Malaysia, 81310 Johor Bahru, Johor.

*ayuni8098@uitm.edu.my

Abstract. The purpose of electromyography (EMG) study is to identify which position of surface EMG sensor attached to erector spinae muscle related with lower back pain by squat and stoop lifting technique. This is to avoid lower back pain (LBP) occur during Manual Materials Handling (MMH). There are only one types of upper extremity muscle were chosen to be monitored in this study which is erector spinae (ES) muscle with different electrodes placement on the surface electromyography (sEMG) sensor. However, each of the lifting styles come out with the different reading of root mean square (RMS) frequency for each muscle chosen. In this study, the two subjects consist of two females with normal body mass index (BMI) range from 18.5 to 24.9 with same physical measurement, was selected in order to perform both styles of lifting which are squatting and stooping. For every session the subject will undertake 15 repetitions with 15 minutes rest in between for each movement. In furtherance of to get the analysis muscle activity, proEMG software is used. The results of study for subject female 1 showed that the squat technique had higher levels of muscle activation compared to stoop technique on left erector spinae (LES) muscle. However, the LES muscle activation for subject 2 is greater at stoop activity. On the contrary, squat technique had lower fatigue analysis compared to stoop technique for both subjects on LES muscle. Conclusion, squat technique is better than stoop technique but stoop lift is more natural and spontaneously used for MMH.

1. Introduction

The widely recognized activity of manual material handling (MMH) in the workplace is the load lifting object [1]. Work associated musculoskeletal injuries are often associated with excessive work of the body at the workplace [2]. The musculoskeletal and LBP issue are usually related with overexertion of the body when the human operator attempts to fulfil the need of MMH activity [2]. The MMH activity is a main cause of musculoskeletal disorders (MSD). Low back pain (LBP) is the most common health disorder problem among all ages [3]. Most people who affected with LBP, they do not taking this disease as a serious problem and continue their life journey as a normal human being without taking any treatment or receive the advice from a doctor. Due to a common group of lack awareness about LBP, may lead to long-term back pain. Incorrect MMH method may result in LBP disease especially when lifting the object through squatting and stooping technique.

Content from this work may be used under the terms of the Creative Commons Attribution 3.0 licence. Any further distribution of this work must maintain attribution to the author(s) and the title of the work, journal citation and DOI.
Published under licence by IOP Publishing Ltd

Exploration for a safer lifting technique has attracted great attention due to the high risk of injury and LBP related to frequent lifting in the workplace [4]. Regardless of the well-recognized contribution of lifting in low back injuries [4], the literature on safer lifting techniques between squat versus stoop lifting remains controversial [5, 6]. However, the significance of the squat versus stoop lifting technique has been downplayed due to the lack of a clear biomechanical rationale for the preferment of either style may reduce the LBP [5, 6, 7]. According to [4], it is stated that ideal lifting methods, squat lifting technique is generally considered to be safer than the stoop lifting technique in bringing the load closer to the body and, also reducing the extra demand on back muscles. In spite of that, many workers prefer the stoop lift due to its easier operation and lower energy consumption in repetitive lifting tasks [4].

The purpose of this study is to identify the best technique of lifting an object as well as the best position of surface EMG sensor attached to erector spinae muscle by an invention of squatting rather than stooping technique. Therefore, In order to determine the ideal techniques, the sEMG will be used to support based on upper extremity muscle study which is erector spinae. From the data collection of sEMG, the conclusion or assumption can be made to respond the question of which exact position of sEMG sensor that should be working when doing proper squatting in order to avoid back pain as well as for stooping technique.

2. Methods

2.1 Participants

Two students volunteered for the present study has been selected, consist of two females with the same physical body measurement and normal body mass index(BMI) range between (18.5 – 24.9) were recruited in order to perform both stoop and squat movement techniques for this particular experiment and muscle. All subjects had no complaints of spine, or pathological since a decade ago for knee conditions which is they are in healthy condition throughout this experiment.

2.2 Experiment set-up

From Figure 1 show, the setup tools for placement of electrodes. Before starting the experiment, it is essential to charge the transmitter and the receiver first. The transmitter and receiver can be used up to two hours long after fully charged. Next, the erector spinae muscle palpation study based on SENIAM standard [15]. Then, the electrodes are attached along the erector spinae muscle for both subjects as illustrated in Figure 2.

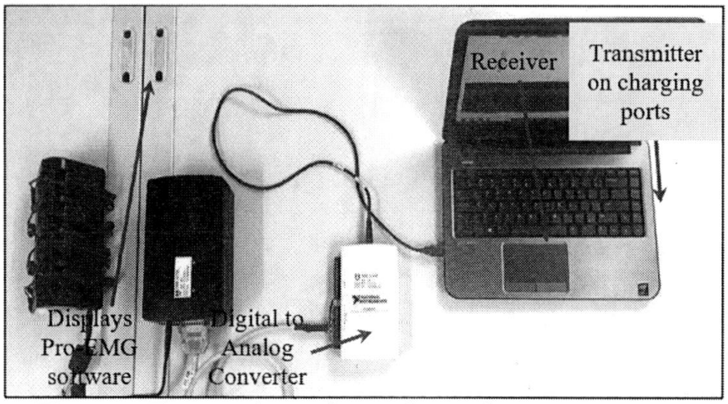

Figure 1. The experimental setup layout for EMG apparatus.

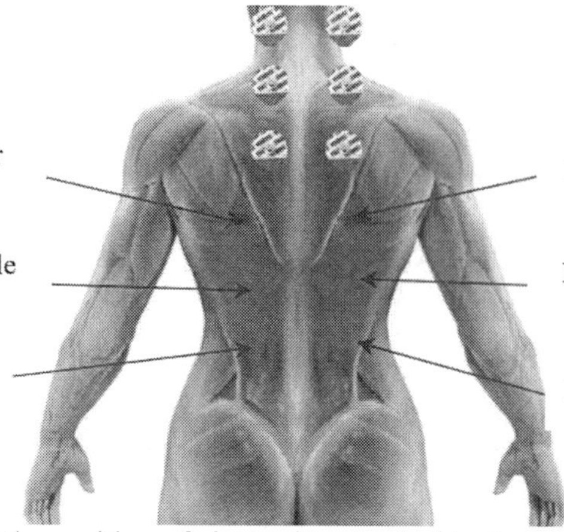

Figure 2. Illustration diagram for the position of electrodes attached to the Left Erector Spinae (LES) and Right Erector Spinae (RES) muscle.

2.3 Experimental procedure

The experiment session was conducted in a day. The subject will be performing the stooping and squatting techniques with 15 repetitions consecutively for every movement while carried the weight of 1.5kg which is acceptable weight from floor to knuckle height for female subjects. To avoid the possible biases resulting from the serious fatigue effect, a 15-min break was given among both movements [8]. From Figure 3 and Figure 4, that is how the stooping and squatting was performed during recording data. The stoop technique typically start with slanted trunk and almost broadened knees (knee flexion edge would be more than 135-degree with trunk flexion around 90-degree) [8]. The knee is kept straight, while the back and arms are forward to hold the object when lifting from floor level; this technique is known as lifting with back (back lift) as shown in Figure 3 [8]. A squat technique can be done with the initial position of deep knee flexion with the trunk near erect, as shown in Figure 4. Particularly the subjects when lifting from floor level this quantitative can be described as a knee flexion around 45-degree and trunk flexion not more than 30-degree [8].

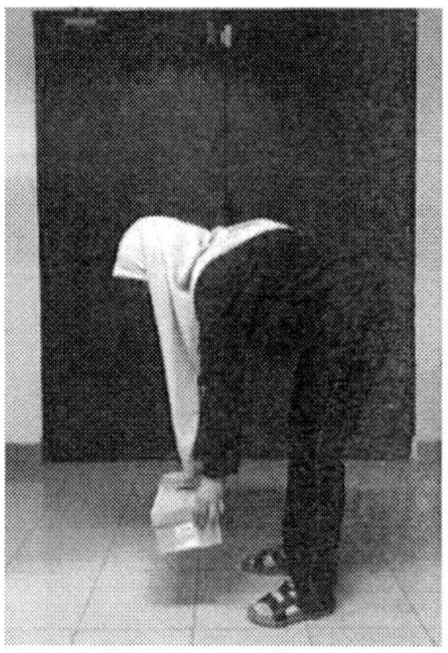

Figure 3. The stooping technique.

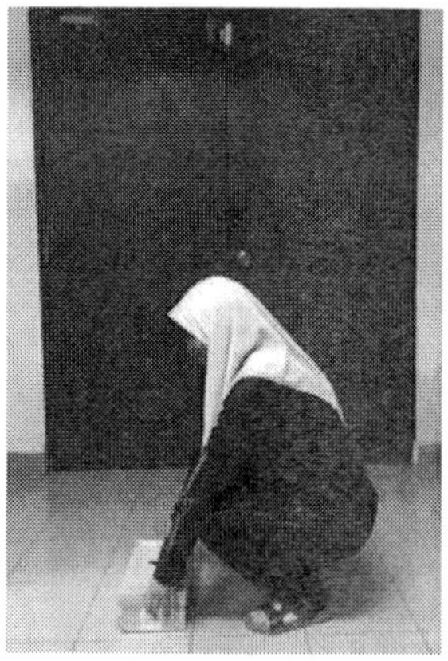

Figure 4. The squatting technique.

2.4 Data Collection

The data collected and will be compared from both sessions of stooping and squatting techniques for three positions of erector spinae muscle which is upper, middle and lower. The purpose of data analysis procedure is to compared muscle activities and fatigue analysis related for stooping and squatting technique. The root mean square (RMS) values consider as muscle activation and Mean Frequency (MF) values as fatigue analysis. Therefore, these can be accomplished by normalizing the raw EMG signal data analysis gathered and filtered by using peak and mean dynamic method.

The analysis data is done by using equation root mean square (RMS) smoothing (minimize artefacts left) recommended quantification method by Basmajian and DeLuca [9, 10] to filter the EMG signal data.

$$RMS_n = \sqrt{\frac{1}{N}}\sqrt{\sum_{n-N/2}^{n+N/2} S_{i^2}} \quad (1)$$

The two most important parameters for fatigue analysis are the median and mean frequency (MF). However, in our case study we only used MF to calculate fatigue analysis [6]. Analysis data by using equation Moving Average smoothing for fatigue analysis.

$$MAV_n = \frac{1}{N}\sum_{n-N/2}^{n+N/2} S_i \quad (2)$$

3. Result and Discussion

The measurement data for the activation stage of each electrodes placement on erector spinae muscle target for subject 1 and subject 2 females has been normalized against the root mean square (RMS) and Median frequency (MF). The measurement results for 1.5 kg lift load being put 20cm in front of the subject. Then, the result were analysed for 15 cycles consecutively for both technique stoop and squat are presented in table below. Based on the table, the EMG signal coming from the muscles of both subjects performing squat and stoop techniques. The signals represent the graph in volt versus cycle.

Table 1. Data on the rate of muscle activation for subject 1.

CYCLE	Subject 1 Stoop Technique, (RMS Average)						Subject 1 Squat Technique, (RMS Average)					
	LES Upper	RES Upper	LES Middle	RES Middle	LES Lower	RES Lower	LES Upper	RES Upper	LES Middle	RES Middle	LES Lower	RES Lower
1	0.0750	0.0893	0.0847	0.0853	0.1230	0.1160	0.0870	0.0684	0.0803	0.0786	0.120	0.114
2	0.0747	0.0749	0.0813	0.0822	0.1220	0.1160	0.0790	0.0827	0.0773	0.0760	0.138	0.143
3	0.0848	0.0944	0.0980	0.0978	0.1460	0.1280	0.0789	0.0776	0.0802	0.0764	0.124	0.125
4	0.1050	0.0966	0.0992	0.0994	0.1230	0.1140	0.069	0.0766	0.0775	0.0757	0.127	0.130
5	0.0795	0.0861	0.0884	0.0952	0.1310	0.1310	0.0933	0.0732	0.0866	0.0864	0.136	0.132
6	0.0977	0.0852	0.0933	0.0926	0.1350	0.1200	0.0781	0.0767	0.0788	0.0799	0.131	0.118
7	0.0894	0.0950	0.0851	0.0953	0.1290	0.1230	0.0931	0.0788	0.0858	0.0837	0.144	0.127
8	0.1100	0.1070	0.1030	0.1020	0.1310	0.1270	0.0818	0.0711	0.0736	0.0841	0.136	0.142
9	0.1120	0.1080	0.0955	0.1010	0.1220	0.1110	0.0770	0.0623	0.0776	0.0736	0.139	0.130
10	0.0878	0.0775	0.0806	0.0840	0.1070	0.0999	0.0837	0.0674	0.0692	0.0721	0.119	0.119
11	0.0937	0.1020	0.0844	0.0963	0.1150	0.0965	0.0743	0.0659	0.0752	0.0895	0.142	0.139
12	0.0932	0.1000	0.0864	0.0886	0.1120	0.1050	0.0757	0.0650	0.0799	0.0864	0.149	0.130
13	0.0819	0.0873	0.0671	0.0755	0.0865	0.0834	0.0666	0.0578	0.0774	0.0773	0.129	0.121
14	0.0761	0.0868	0.0704	0.0827	0.1080	0.0895	0.0843	0.0661	0.0808	0.0932	0.148	0.137
15	0.0896	0.0859	0.0799	0.0875	0.1060	0.1110	0.0577	0.0474	0.0671	0.0772	0.127	0.126

Table 2. Data on the fatigue analysis for subject 1.

CYCLE	Subject 1 Stoop Technique, (Median Average)						Subject 1 Squat Technique, (Median Average)					
	LES Upper	RES Upper	LES Middle	RES Middle	LES Lower	RES Lower	LES Upper	RES Upper	LES Middle	RES Middle	LES Lower	RES Lower
1	39.0	39.1	42.6	38.5	43.3	47.4	40.1	45.9	49.4	49.8	45.5	49.7
2	38.4	37.7	43.3	42.4	44.5	47.8	43.4	45.9	48.2	47.4	44.4	52.0
3	41.4	40.8	43.7	41.9	43.8	47.2	42.3	48.6	45.4	48.4	45.6	52.0
4	36.2	40.6	42.9	43.7	46.0	45.3	40.9	45.1	47.1	50.6	47.5	49.8
5	32.8	34.9	40.5	40.1	44.6	47.1	45.3	47.1	47.8	50.4	46.6	52.9
6	36.4	39.1	43.7	41.6	48.1	49.0	40.8	47.8	48.0	47.1	45.1	50.3
7	38.5	37.0	41.9	40.3	47.4	49.3	43.5	44.2	42.0	44.4	45.4	49.6
8	32.9	42.4	41.7	42.5	47.0	48.0	39.3	43.5	44.9	46.6	45.5	47.9
9	36.0	38.1	39.5	41.0	45.9	46.1	40.5	45.3	43.4	45.5	42.5	48.8
10	33.7	38.2	42.9	44.2	45.5	48.1	42.4	46.5	37.3	41.8	48.1	51.1

11	34.1	38.2	37.0	39.0	48.4	49.1	40.8	46.7	38.3	42.2	43.1	46.3
12	36.6	39.3	40.8	39.9	44.8	46.5	37.8	41.8	42.2	42.0	42.9	45.1
13	37.7	44.3	41.3	44.2	47.5	47.9	40.5	38.9	45.2	41.7	46.2	49.3
14	36.8	37.2	38.0	41.3	46.9	51.3	35.9	42.1	44.8	41.5	44.4	47.5
15	35.2	36.2	39.9	38.5	44.7	50.9	37.9	41.3	42.4	43.3	43.5	44.4

Table 3. Data on the rate of muscle activation for subject 2.

CYCLE	Subject 2 Stoop Technique, (RMS Average)						Subject 2 Squat Technique, (RMS Average)					
	LES Upper	RES Upper	LES Middle	RES Middle	LES Lower	RES Lower	LES Upper	RES Upper	LES Middle	RES Middle	LES Lower	RES Lower
1	0.0503	0.0885	0.0548	0.0521	0.557	0.0642	0.0419	0.0766	0.0563	0.067	0.912	0.0677
2	0.0503	0.1050	0.0538	0.0468	0.755	0.0537	0.0410	0.0832	0.0584	0.0598	0.818	0.0627
3	0.0427	0.0897	0.0496	0.0449	0.861	0.0527	0.0395	0.0772	0.0573	0.0672	0.911	0.0662
4	0.0410	0.0867	0.0470	0.0412	0.672	0.0525	0.0376	0.0606	0.0585	0.0596	0.835	0.0734
5	0.0517	0.1080	0.0500	0.0436	0.684	0.0476	0.0241	0.0476	0.0638	0.0482	1.010	0.0478
6	0.0392	0.0901	0.0436	0.0390	0.725	0.0456	0.0387	0.0694	0.0702	0.0626	0.603	0.0639
7	0.0504	0.0973	0.0346	0.0355	0.811	0.0395	0.0347	0.0648	0.0674	0.0594	0.623	0.0633
8	0.0474	0.1090	0.0415	0.0359	0.740	0.0399	0.0442	0.0854	0.0858	0.0670	0.608	0.0687
9	0.0484	0.1050	0.0435	0.0359	0.808	0.0432	0.0310	0.0587	0.0693	0.0575	0.730	0.0660
10	0.0504	0.1170	0.0408	0.0408	0.748	0.0465	0.0375	0.0585	0.0773	0.0620	0.878	0.0681
11	0.0383	0.0790	0.0466	0.0409	0.779	0.0504	0.0287	0.0570	0.0559	0.0567	1.040	0.0615
12	0.0528	0.1210	0.0381	0.0327	0.727	0.0420	0.0347	0.0686	0.0698	0.0514	0.940	0.0652
13	0.0439	0.0976	0.0443	0.0433	0.836	0.0497	0.0382	0.0799	0.0802	0.0509	0.857	0.0609
14	0.0470	0.1110	0.0415	0.0385	0.726	0.0430	0.0274	0.0510	0.0781	0.0518	1.070	0.0586
15	0.0472	0.0812	0.0472	0.0405	0.911	0.0451	0.0279	0.0505	0.0793	0.0475	0.855	0.0544

Table 4. Data on the rate fatigue analysis for subject 2 (female)

CYCLE	Subject 2 Stoop Technique, (Median Average)						Subject 2 Squat Technique, (Median Average)					
	LES Upper	RES Upper	LES Middle	RES Middle	LES Lower	RES Lower	LES Upper	RES Upper	LES Middle	RES Middle	LES Lower	RES Lower
1	44.5	44.6	40.8	44.7	39.7	40.7	35.9	42.2	43.8	50.4	37.0	41.0
2	49.8	46.8	39.9	44.2	41.2	40.7	47.4	46.0	38.2	41.0	50.9	40.8
3	40.7	40.4	42.2	41.2	39.5	40.6	44.3	46.4	41.8	41.9	32.6	42.2
4	45.7	47.3	40.0	43.6	47.4	43.4	26.5	42.7	39.5	37.9	30.9	38.4
5	39.6	39.4	39.4	39.2	47.5	39.9	44.8	46.0	38.3	43.6	37.6	36.8
6	50.8	51.9	39.2	37.8	49.9	37.6	40.0	35.0	40.5	37.7	41.0	36.6
7	50.7	43.2	34.1	35.4	42.0	36.0	42.6	41.2	39.1	38.1	36.7	35.9
8	41.1	42.7	36.9	38.9	49.0	38.7	38.6	39.3	40.0	41.5	43.2	34.0
9	47.3	48.7	32.7	37.4	54.2	38.0	39.5	43.4	42.9	45.9	35.3	37.1
10	45.3	47.3	39.7	37.5	41.6	35.4	40.4	37.4	41.0	38.1	42.5	41.3
11	44.3	42.7	39.9	39.9	45.1	40.3	47.9	46.3	41.9	45.7	38.5	43.1
12	47.6	51.1	34.3	35.3	54.7	38.8	43.9	40.4	39.2	41.8	39.8	38.5
13	41.9	40.8	38.0	40.3	39.7	39.3	51.3	49.4	43.4	39.9	41.2	37.4
14	42.0	45.2	34.7	34.2	48.7	33.2	49.1	47.8	42.2	47.6	44.2	42.2
15	48.2	48.5	39.0	37.1	60.2	38.6	44.2	37.6	40.1	41.0	47.6	42.7

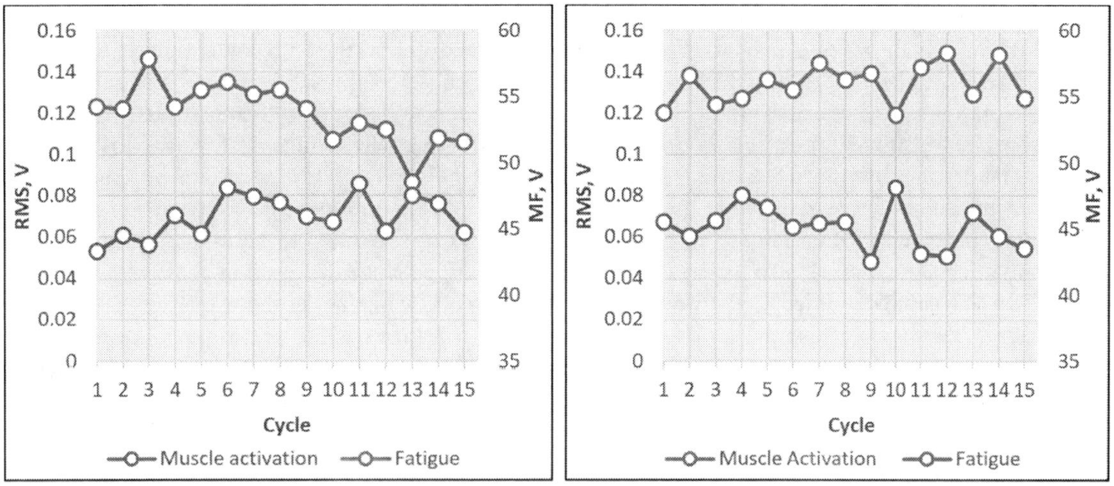

(a) (b)

Figure 5. The RMS average and MF analysis (a) stoop, (b) squat at LES lower muscle of subject 1.

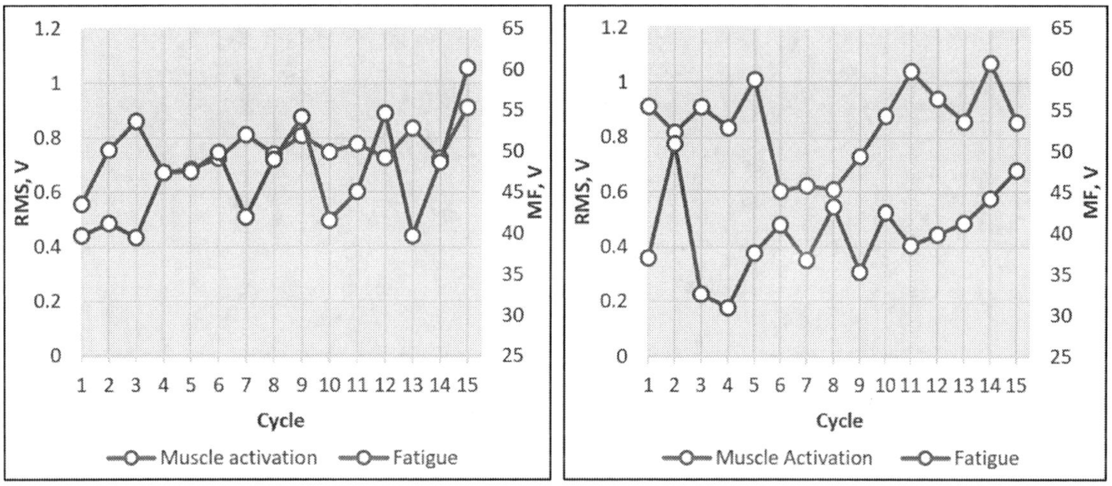

(a) (b)

Figure 6. The RMS average and MF analysis stoop (a) and squat (b) at LES lower muscle of subject 2.

The figure above shows the data of the average RMS and MF for both lifting techniques on electrodes placement at LES lower muscle for subject 1 and subject 2 female. RMS signal is used to look into muscle activation while MF is for fatigue analysis. From Figure 5(a) shows the muscle for subject 1 start to decrement from cycle 1 until 15 which indicates the decreases in the muscle activations. This shows that LES lower muscle which is representing the belly muscle of the ES requires more effort during stoop activity. However, from Figure 5(b) it shows different trends for squat activity. The muscle starts to spike from cycle 1 to 15 during squatting. This shows that LES lower muscle required less effort to lift weight during squat activity. Comparing both techniques the most significant finding in the study was the stoop technique understood to be wrong technique as the activation muscle at the belly muscle of ES can lead to low back pain. Nevertheless, the RMS value for subject 2 shows a different trend than subject 1 for both techniques. Based on Figure 6(a), from cycle 4 to 15 the LES lower muscle activation starts to ascend. However, the Figure 6(b) shows from cycle 1 to 6 the muscle activation start to show decrement before entering the stable phase along cycle 6 to 8 while squatting activity. Then, from cycle 8 to 15 the RMS value starts to show increment phase. Unfortunately, the RMS value for subject 2 is greater at stoop activity. Subject 1 showing stable data than subject 2. This is due to differ in lifting technique produce for each subject. Since, the significant studies have acknowledged that the stoop lift is widely and spontaneously used for "Bent-Over work" [11].

MF value is for fatigue analysis. From Figure 5(a) it shows that the LES lower muscle spike during stooping for subject 1. From cycle 1 to 15, the LES lower muscle shows the increment which represents muscle fatigue until the end of the cycle. When the muscle fatigue is increase, that means is not appropriate to lift weight especially during stooping technique. Nevertheless, it shows different trends for squat activity in Figure 5(b) the muscle starts to descend from cycle 1 to 15 during squatting. This shows muscle fatigue is reduced at the end of the cycle. Based on Figure 6(a) the subject 2 shows the LES lower muscle of data fatigue analysis starts to ascend from cycle 3 to 15 for stoop activity. Then, for squatting the subject 2 also have experience muscle fatigue since the trend increases from cycle 5 to 15. The MF value of subject 2 at 15 cycles for stooping (60.2V) is higher than squatting (47.6V). This indicates that subject 2 using a lot of muscle usage on LES lower while performing stooping activity and the effect by that can cause of muscle fatigue. Based on the previous study, squat is acknowledged as 'correct technique' for lifting low-lying objects. Thus, the decrement trend shows for squat activity proof that the squatting technique a little bit can reduce lower back pain [11].

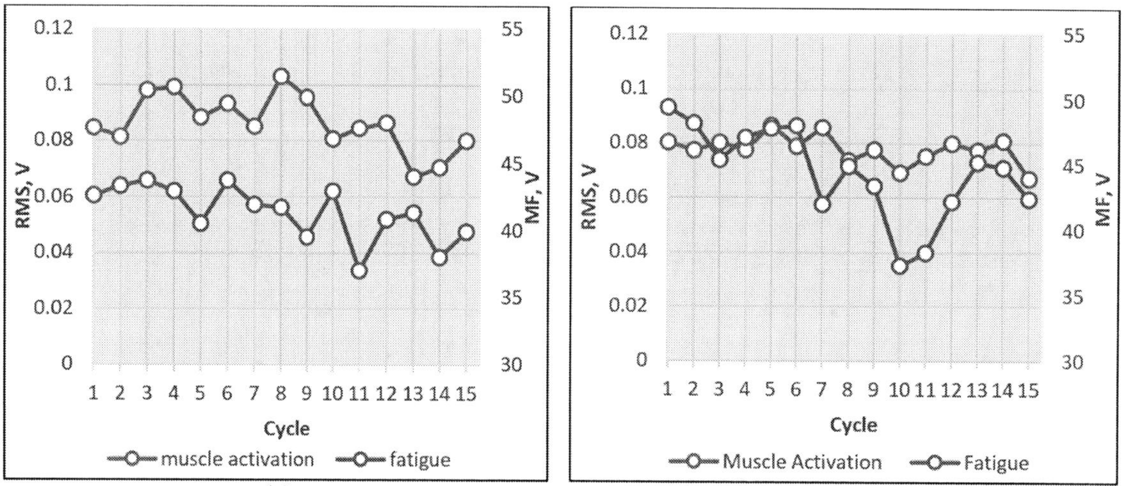

(a) (b)

Figure 7. The RMS average and MF analysis stoop (a) and squat (b) at LES middle muscle of subject 1.

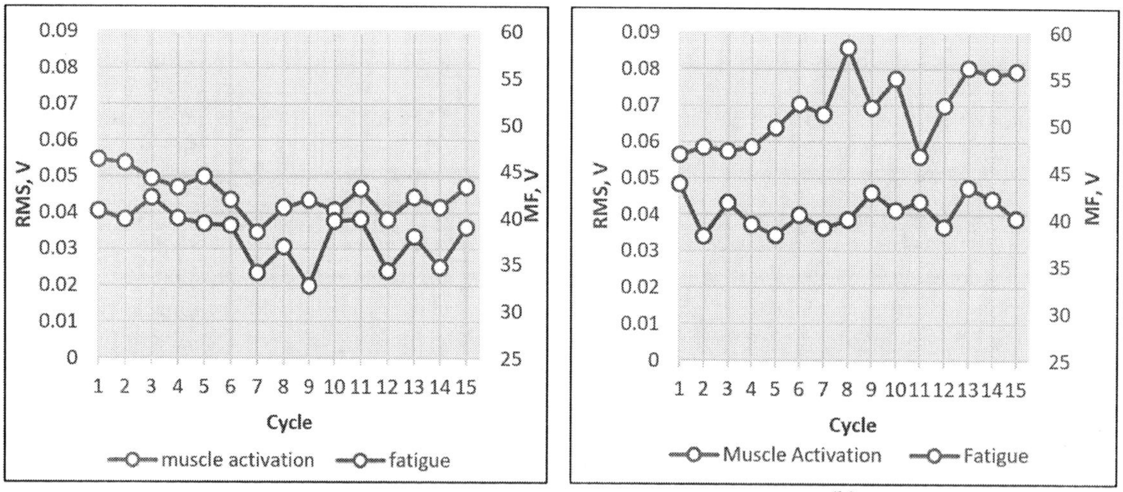

(a) (b)

Figure 8. The RMS average and MF analysis stoop (a) and squat (b) at LES middle muscle of subject 2.

From the Figure 7(a), 7(b) shows decrement MF value of electrode placement on LES middle muscle value of subject 1 start from cycle 1 to 15 for both techniques of stoop and squat activity. The significant value indicates subject 1 is using the same capacity of energy during both lifting techniques. As we can see that the muscle activation and fatigue analysis have the significant value between both techniques. Comparing to electrode placement on LES lower muscle the value is quite different. Since the value of MF for LES lower is much higher (0.120V) as compared to LES middle (0.0803V) for squat activity. Same as well as stoop activity where the MF value for LES lower (0.123V) and for LES middle (0.0847). So, the placement of electrodes at the LES lower is more effective than LES middle since it come out with more stable fatigue analysis.

Based on Figure 8(a), the subject 2 shows decrease in LES muscle activation through stoop technique from first to the end of cycle. However in Figure 8(b), RMS value increase through squat technique from cycle 1 to 15. Differ from the value of fatigue analysis, whereby for both technique shows decrement trends.

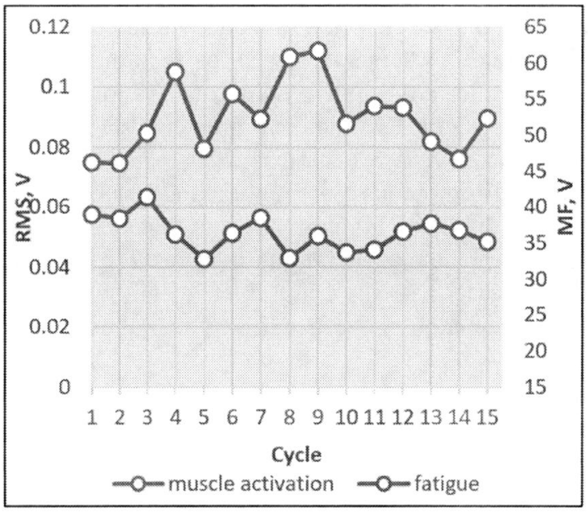

(a)

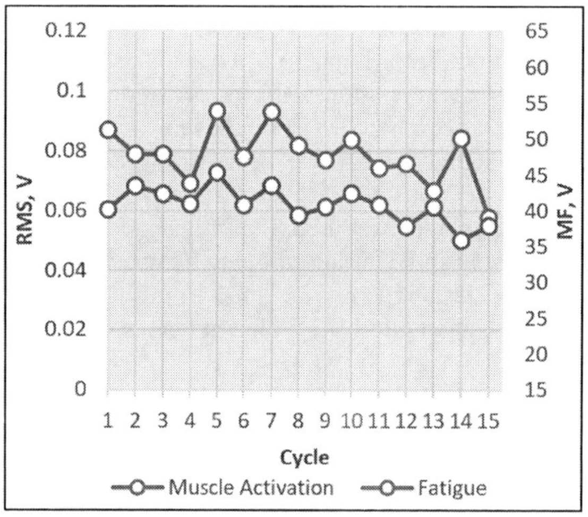

(b)

Figure 9. The RMS average and MF analysis stoop (a) and squat (b) at LES upper muscle of subject 1.

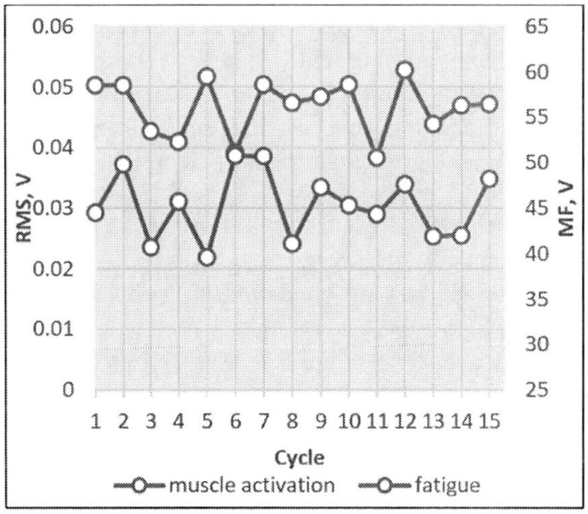

(a)

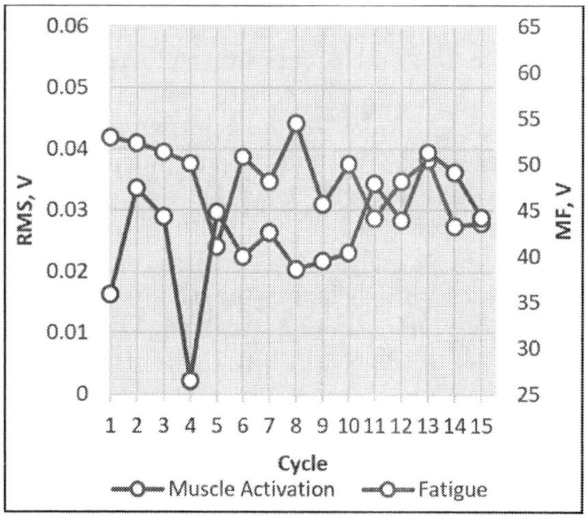

(b)

Figure 10. The RMS average and MF analysis stoop (a) and squat (b) at LES upper muscle of subject 2.

From Figure 9(a), 9(b) shows the MF for both lifting techniques on electrodes placement at LES upper muscle for Subject 1 and Subject 2 female. The Figure 9(a) shows the muscle activation for subject 1start to decrease from cycle 1 to 15 for stooping technique. However, the RMS value at cycle 1 is slightly differing than cycle 15 which is 0.075V to 0.0896V. Whereas for stooping technique Subject 2 as shown in Figure 10(a), the RMS value is stable but slightly increased from cycle 1 (0.0503V) to cycle 15 (0.0472V). This indicates that the placement electrode at the upper muscle of ES does not give any different in RMS value for stoop activity. But, for squat activity towards Subject 1, the muscle activation still can be seen. Since, the trend starts to show decrement from cycle 1 until 15. Similar with Subject 2 in Figure 10(b), the squat activity also shows decrement trends. The significant study acknowledge that, the electrodes placement at the LES upper muscle does not gives any different in RMS value for stooping technique. Although, for squatting technique still can be used.

4. Conclusion and Recommendation

From this study, we found the muscle activation activities of upper extremity were investigated through performing squat and stoop movements by using electromyography (EMG) to avoid LBP. The EMG data for different electrodes position of the ES muscle through stoop and squat technique had been accumulated for comparison. Both technique had been accumulated and normalized to root mean square (RMS) and Median Frequency (MF), where RMS represent muscle activation and MF indicate muscle fatigue. From Figure 5,6 the analysis of the muscle usage for subject 1, it can be seen that squatting movement is an ideal lifting technique as it promotes better and stable reading of LES lower muscle data as compared to stoop movement. However, for subject 2 the RMS value is greater at stoop activity. As stated in [12], the squat lifting technique appears to have lower lumbar shear pressure and less stressed places on the spinal passive tissue, while stoop lifting looks more natural and less fatigue as it is spontaneously used. For fatigue analysis for subject 1, based on Figure 5 it shows that the LES lower muscle spike during stooping. However, through squatting the muscle fatigue start to show decrement. Muscle fatigue during stoop lifting techniques can increase the risk of musculoskeletal injury during rapid disruption [13]. This indicates that, squat lifting technique is proper lifting technique to reduce LBP.

Although, some previous studies state that there is no single lifting technique that is best. There is still no significant study able to prove and state that squat technique is better than stoop for lifting weight to avoid LBP [11]. Besides that, from my point of view, I would suggest that squat technique is more ideal and proper lifting technique as it is more stable than stoop technique. For the recommendation, it is suggested that for the next experiment would involve at least 10 subjects as possible to have more data collection. The recommended subject consist of people who conduct stoop and squat movement as their daily basis working routine in order to prove the proper lifting technique. The other suggestion is from [14], it is recommended that to precisely designed and precise program aimed at the development of safe lifting techniques for employees involved in manual handling and lifting of work as a way to reduce the risk of injury in the back. Lastly, set up a variety of motion exercises that permit the lifters to use the correct biomechanics when lifting, preventing postural abnormalities and enhancing their movement network at the connection required performing regular manual handling tasks [14].

References

[1] N. Asadi et al., "A Comparative Assessment of Manual Load Lifting Using NIOSH Equation and WISHA Index Methods in Industrial Workers of Shiraz City", *Journal of Health Sciences & Surveillance System*, vol. 3, no. 1, pp. 8-12, 2015.

[2] C. J. Lin, S. J. Wang and H. J. Chen, "A Field Evaluation Method for Assessing Whole Body Biomechanical Joint Stress in Manual Lifting Tasks", *Journal of Industrial health*, vol. 44, no. 4, pp. 604-612, 2006.

[3] J. Ramdas and V. Jella, "Prevalence and Risk Factors of Low Back Pain", *International Journal of Advances in Medicine*, vol. 5, no. 5, pp. 1120-1123, 2018.

[4] B. Bazrgari, A. Shirazi-Adl and N. Arjmand, "Analysis of Squat and Stoop Dynamic Liftings: Muscle Forces and Internal Spinal Loads", *European Spine Journal,* vol. 16, no. 5, pp. 687-699, 2007.

[5] J. H. Van Dieën, M. J. Hoozemans and H. M. Toussaint, "Stoop or Squat: A Review of Biomechanical Studies on Lifting Technique", *Journal of Clinical Biomechanics*, vol 14, no. 10, pp. 685-696, 1999.

[6] S. M. Hsiang, G. E. Brogmus and T. K. Courtney, "Low Back Pain (LBP) And Lifting Technique—A Review", International Journal of Industrial Ergonomics, vol. 19, no. 1, pp. 59-74, 1997.

[7] S. M. McGill, "The Biomechanics of Low Back Injury: Implications on Current Practice in Industry and the Clinic", *Journal of biomechanics*, vol. 30, no. 5, pp. 465-475, 1997.

[8] Z. Wang et al., "Squat, stoop, or semi-squat: A comparative experiment on lifting technique", *Journal of Huazhong University of Science and Technology [Medical Sciences]*, vol. 32, no. 4, pp. 630-636, 2012.

[9] C. De Luca, "Electromyography", *Encyclopedia of Medical Devices and Instrumentation,* J. G. Webster, Ed, John Wiley Publisher, pp. 98-109, 2006.

[10] J. V. Basmajian and C. J. De Luca, "Muscles Alive: Their Functions Revealed By Electromyography", Baltimore, Williams & Wilkins, 1985.

[11] L. D. Vecchio, "Choosing a Lifting Posture: Squat, Semi-Squat or Stoop", *MOJ Yoga Physical Therapy. Department of Medical and Applied Sciences*, Central Queensland University, Australia, vol. 2, no. 2, pp. 56-62, 2017.

[12] R. Burgess-Limerick, "Squat, stoop, or something in between?", *International Journal of Industrial Ergonomics*, vol. 31, no. 3, pp.143-148, 2003.

[13] H. Hassanlouei et al., "Effect of Exercise Induced Fatigue on Postural Control of the Knee", *Journal of Electromyography and Kinesiology*, vol. 22, no. 3, pp. 342-347, 2012.

[14] GARG, A. and Saxena, U., 1979. Effects of Lifting Frequency and Technique on Physical Fatigue with Special Reference to Psychophysical Methodology and Metabolic Rate. American Industrial Hygiene Association Journal, 40(10), pp.894-903.

[15] F. Hermens, "SENIAM (surface EMG for a non-invasive assessment)", Roessingh Research and Development, 1996 [Online]. Available: http://www.seniam.org/. [Accessed 2018 November 22].

[16] D.Y. Patil and A. Shinde, "Pimpri, Pune and Maharashtra, "A Study of Soleus Muscle –Clinical and Embryological Consideration"", *International Journal of Biomedical and Advance Research*, vol. 7, no. 12, pp. 608-611, 2016.

IOP Conf. Series: Materials Science and Engineering **834** (2020) 012044 doi:10.1088/1757-899X/834/1/012044

Microstructural Characterization and Microhardness of Leucite Glass-Ceramics Synthesized from Sarawak Sand Reserves.

J. Qamaruzzaman[1], M. H. Ismail[2], A. Muchtar[3], M. Selamat[4] and R. Ahmad[1*].

[1]Centre of Restorative Dentistry Studies, Faculty of Dentistry, Universiti Teknologi MARA, Sg. Buloh Campus, Selangor, Malaysia.
[2]Faculty of Mechanical Engineering, Universiti Teknologi MARA, Shah Alam Campus, Selangor, Malaysia.
[3]Centre for Materials Engineering and Smart Manufacturing, Faculty of Engineering and Built Environment, Universiti Kebangsaan Malaysia, 43600 UKM Bangi, Selangor, Malaysia.
[4]Mineral Research Centre, Department of Minerals and Geoscience Malaysia, Ipoh, Perak, Malaysia.

*drrohana@uitm.edu.my

Abstract. The purpose of this study was to evaluate the effect of varying the sintering temperature and holding time on the crystallization and microhardness of the SiO_2 – Al_2O_3 –K_2O glass system using SiO_2 from Sarawak sand reserves. A starting glass composition of SiO_2 (from Sarawak sand), Al_2O_3, K_2O, Na_2O, CaO, LiO and TiO_2 was prepared by conventional melt quenching technique followed by ball milling. The glass powder was pressed under pressure into disk-shaped compacts with 2 mm thicknesses and 14 mm diameters. The samples were then subjected to differing heat treatment schedules of 750°C, 800°C and 850°C with 1, 3 and 6 hours holding times. They were characterized by X-ray Diffraction (XRD), Scanning electron microscopy (SEM) and Vickers microhardness. XRD showed the crystallization of tetragonal leucite with minor secondary sanidine phases. SEM showed the formation irregular shaped and non-uniformly dispersed crystals. The hardness values showed an increasing trend with longer holding time at temperatures of 750°C and 800°C. This pattern was inversed for sintering temperature 850°C. In conclusion, sintering temperatures and holding time combinations influenced the crystallization and microhardness values of leucite glass ceramics produced from Sarawak sand silica.

1. Introduction

The demand for aesthetics has driven the development of tooth colored restorative materials. Ceramic materials are the material of choice for aesthetic indirect restoration as it has the advantage of 'lifelike' match to the dentition due to its excellent optical properties [1]. In addition, ceramic restorations satisfy other pertinent clinical requirements for example: strength, wear resistance, accuracy of fit, color stability, biocompatibility and chemical inertness [2-5]. Leucite glass-ceramics is one of the earliest developed dental ceramic and is still widely used due to its outstanding aesthetics [6] and clinical

Content from this work may be used under the terms of the Creative Commons Attribution 3.0 licence. Any further distribution of this work must maintain attribution to the author(s) and the title of the work, journal citation and DOI.
Published under licence by IOP Publishing Ltd

durability [7]. The material is sufficiently strong to withstand occlusal forces when used as various types of indirect dental restorations such as veneers, anterior crowns, inlays and onlays. Furthermore, these restorations have excellent survival rates ranging from 96% to 91% at 4.5 years 7 years for inlays and onlays, 92% to 99% at 3 to 3.5 years for crowns, and 98 % after 5 years for veneers [7-9].

In dentistry, there are several possible methods of producing indirect dental restorations. Traditionally, porcelain is fused to metal substructure to produce the restorations. However, this compromises the aesthetics of restoration due to the greyish metal substructure showing-through the translucent overlying porcelain. High strength ceramics were developed to overcome the aforementioned problem. In addition, the conventional method of producing indirect dental restorations is labour extensive requiring the support of dental technicians and is time consuming. In recent years, computer-aided design/computer-aided manufacturing (CAD/CAM) system was introduced to shorten the production time of indirect ceramic restorations and the number of patient visits [10]. The procedure involves digitizing the patient's dentition and designing a virtual model of the dental restoration. This is immediately followed by direct milling of the ceramic block in the dental clinic and same-day cementation of the dental restoration. CAD/CAM ceramics restorations have been shown to have excellent aesthetics, good fit and clinical longevity [11-13].

At present, the dental industry in Malaysia relies heavily on imported machinable blocks for fabrication of all-ceramic restorations. These restorations are expensive and therefore beyond the means of a larger part of the population. A single leucite crown cost around RM2000-2500. Hence, a cheaper alternative is desired so that more patients can benefit. Malaysia has an estimated resource of 640 million tonnes of high-quality silica sand [14]. Approximately 148 million tonnes exist in natural forms and 492 million tonnes as tailing sand. The states with the highest production of silica by quantity percentage are: Selangor (36%), Sarawak (35%) and Perak (21%) [14]. These sources have the potential to be explored and processed for 'higher value' products such as advanced biomaterials. Fully utilised, this huge reserve of sand deposits could help to reduce our country's present import dependency and add value to our economy.

It is our aim to develop machinable leucite glass ceramics for dental restorations from the high-quality silica sand obtained from Sarawak. The purpose of this study was to explore the nucleation and crystallization kinetics of the glass system SiO_2 –Al_2O_3 –K_2O using raw sand as the source of SiO_2. In the present work, we investigated the microstructural characteristics and microhardness of the glass-ceramics which was heat treated at various temperatures and holding times. This knowledge will contribute to our final aim of designing and engineering machinable glass-ceramics for dental applications using local sand reserves.

2. Experimental procedure

Malaysian sand sample from Bintulu, Sarawak was used as the source of SiO_2 in this study. All the other chemical components were of analytical grade and purchased from Merck Germany; Unilab, Australia and Hamburg Chemical.

2.1. Glass powder preparation

A starting glass mixture composed of Al_2O_3, Na_2O, K_2O, CaO, TiO_2, Li_2O and sand from Bintulu, Sarawak (as the source of SiO_2) was prepared (Figure 1a). The mixture was melted in alumina crucible in an electric furnace at 10°C / min to 1450°C for three hours. The glass melt was quenched by pouring it into deionized water to produce glass frit (Figure 1b). The frit was dried and milled for 1 hour in the planetary ball mill (PM400, Retsch, Germany) using zirconia balls as the crushing media to obtain the glass powder (Figure 1c).

2.2. Glass-ceramic discs fabrication

The fine glass powder samples were pressed into disks (14x2mm) (Figure 1d) in a steel mould using hydraulic hand press (Carver, USA) followed by Cold Isostatic Press (American Isostatic Press, Inc., USA). The green glass compacts underwent heat treatment in an electric furnace (Multimat Touch & Press, Germany) to convert it to glass-ceramics by nucleation and growth of crystals. A heating rate of 10°C / min was used; with temperature of 750°C, 800°C and 850°C and soaking time of 1 hour, 3 hours and 6 hours. 750°C was selected as the starting point of the sintering temperature in accordance with other studies done on glass ceramics [15, 16].

2.3. Sample characterization

The crystalline phases were determined by a X-ray diffractometer (XRD; Rigaku Ultima IV, Japan) using Ni-filtered Cu Kα radiation at a scan speed of 2°/minute for 2θ from 10° to 80°. The diffraction patterns were compared with tetragonal leucite (ICDD: 01- 081-2221) and sanidine. Scanning electron microscope (su3500, Hitachi) was utilised to determine the morphologies of the leucite phase on the surfaces of the glass-ceramic samples. The indentation hardness of the glass-ceramics was measured by Vickers method at a load of 200 g for 15s. The disc shaped specimens were wet polished to a mirror finish. Five indentations approximately 0.5mm apart were made and the hardness calculated.

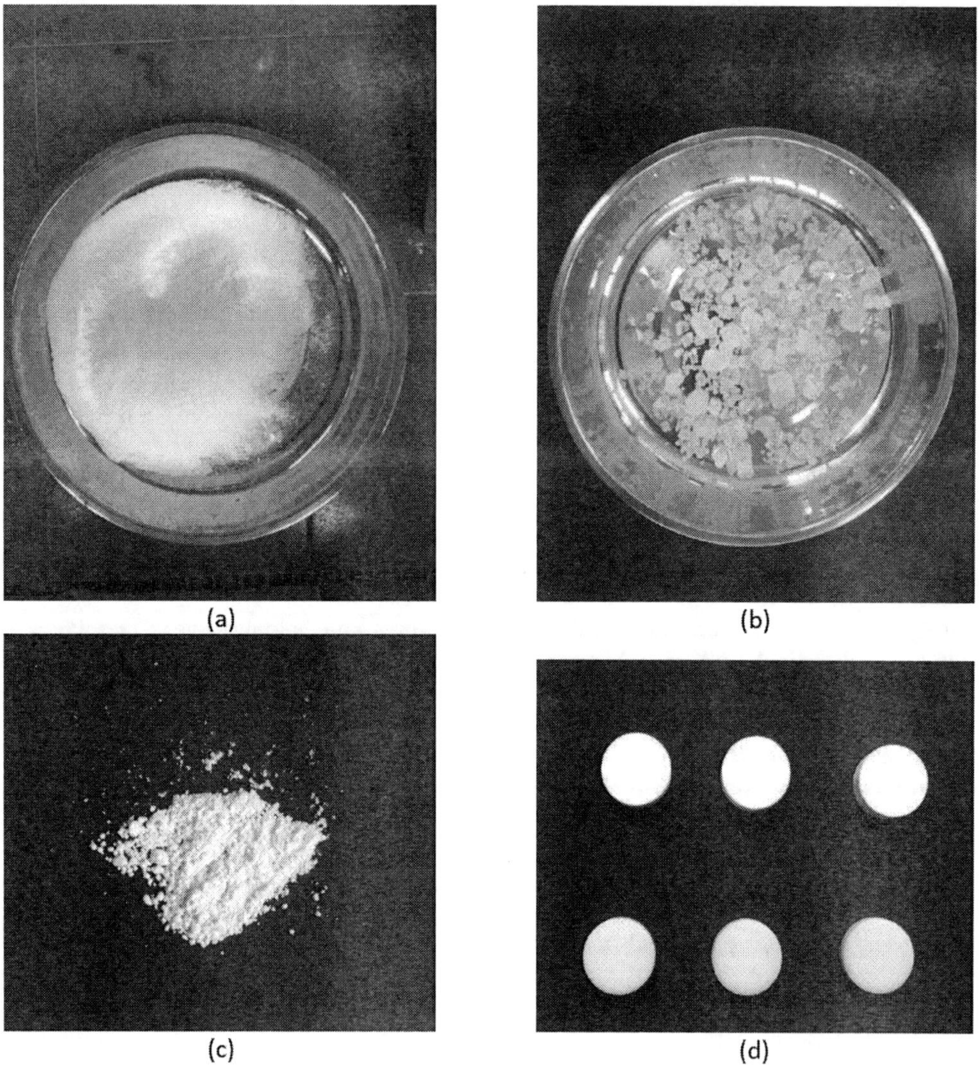

Figure 1. (a) Mixture composed of sand SiO_2, Al_2O_3, Na_2O, K_2O, CaO, TiO_2 and Li_2O (b) Glass frit produced by immediate quencing of the glass melt (c) Glass powder produced by ball milling and (d) Disk-shaped compacts before (upper row) and after (bottom row) sintering.

IOP Conf. Series: Materials Science and Engineering **834** (2020) 012044 doi:10.1088/1757-899X/834/1/012044

3. Results and Discussion

XRD revealed that the phases in the glass-ceramics obtained from the SiO_2 –Al_2O_3 –K_2O system using SiO_2 from Sarawak sand reserves are tetragonal leucite (major phase) and sanidine (minor phase) (Figure 2). Similar structures have been identified in the leucite glass-ceramic system [16]. In general, phase changes were observed with differing temperature and holding time combination. The minor sanidine phase developed at increased temperature and holding time, with higher temperature showing earlier development of the minor phase. This is in accordance with previous studies which showed that in leucite glass-ceramics, leucite and silica in the residual glass phase transformed to sanidine in the temperature range of 800°C - 950 °C at more than 60 min hold [17, 18].

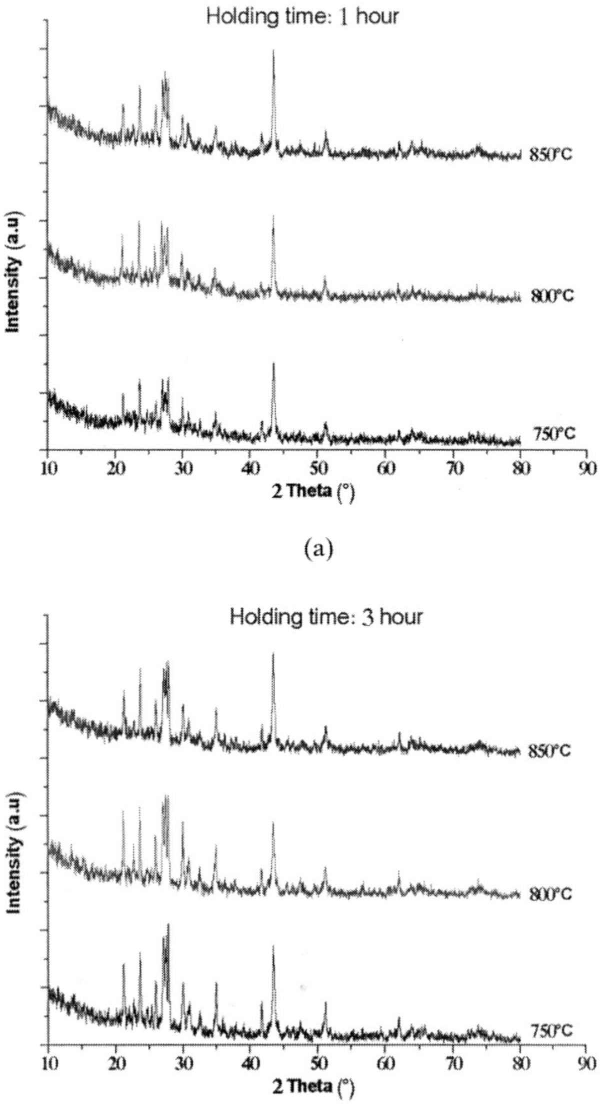

(a)

(b)

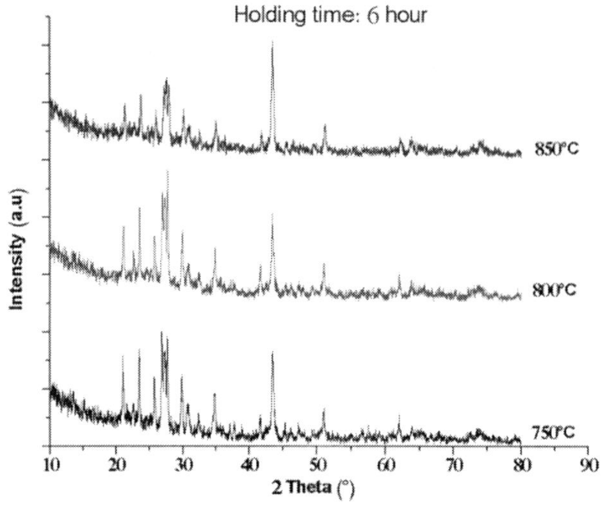

(c)

Figure 2. The XRD patterns of samples sintered at 750°C; 800°C and 850°C for (a)1.0 hour, (b) 3.0 hours and (c) 6.0 hours.

Scanning electron microscopy (SEM) was used to observe the surface microstructure of the samples. Irregular shaped crystals with non-uniform distribution (Figure. 3a) were seen in the glass-ceramics sintered at 750°C and 800°C held for 1, 3 and 6 hrs. Similar microstructure was observed with samples sintered at 850°C were held for 1 and 3 hrs. However, prolonged holding for 6 hours at 850°C led to a decreased crystal quantity and enlarged crystal size (Figure. 3b). This is in accordance with a recent study which showed that the average particle size of synthesized leucite increased with higher sintering temperatures [19]. This could be due to crystal growth via particle coalescence which is facilitated by reduction of glass viscosity at higher crystal growth temperatures [17].

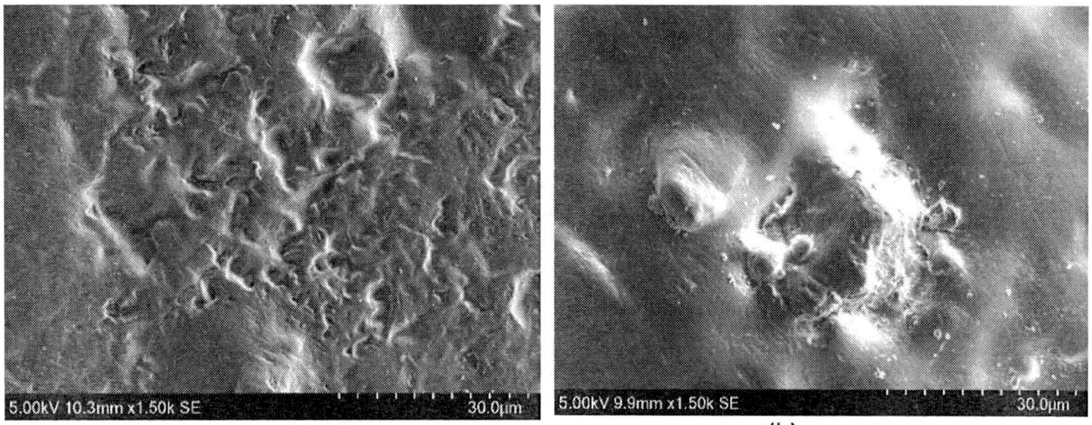

Figure 3. SEM micrographs of (a) sample sintered at 850°C; 1 hour holding time and (b) sample sintered at 850 °C; 6 hours holding time.

Figure 4 shows Vickers microhardness values of glass ceramics sintered at 750°C, 800°C and 850°C with holding times ranging from 1to 6 hours. The mean hardness values (VHN) of the glass-ceramics ranged from 335.16 ± 20.07HVN at 850°C held for 6 hours to 736.16 ± 77.95HVN at 850°C held for 1hour. The standard deviation of some of the experimental groups were relatively high showing a high variance of the data. This may be attributed to the low samples size numbers per group (n=5). The hardness values for leucite glass ceramics showed a clear increasing trend with increased holding time (from 1 hour to 6 hours) at sintering temperatures of 750°C and 800°C. However, the pattern is inversed for sintering temperature of 850°C. Longer holding times at high temperature led to the development of poorly dispersed leucite crystal agglomerations within the glassy matrix (Figure 3b) which may have contributed to the low microhardness values seen in these samples.

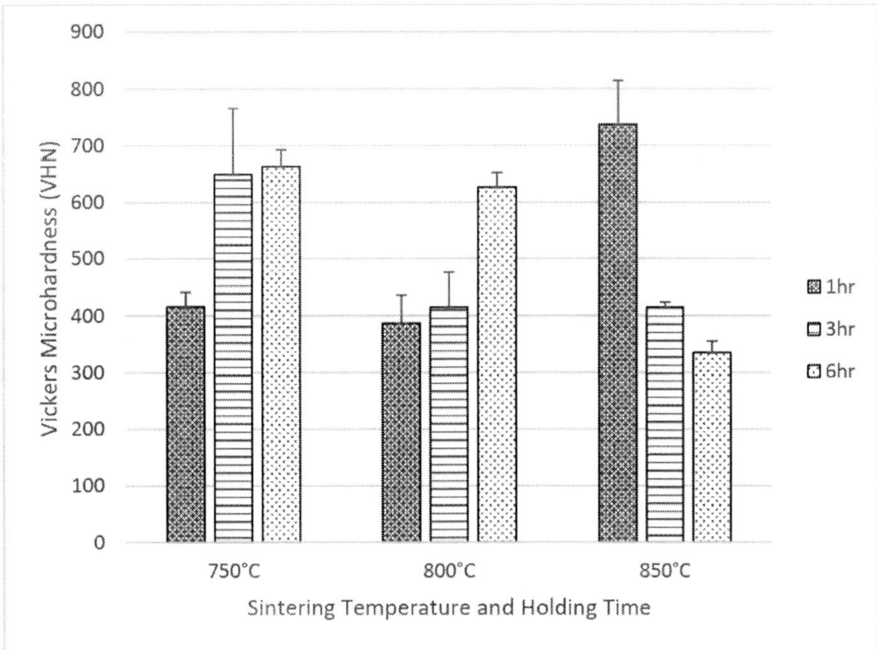

Figure 4. Mean Vickers microhardness of leucite glass-ceramics at 750°C, 800°C, 850°C with 1.0, 3.0 and 6.0 hr holding time.

4. Conclusion

It has been shown that the processing method adopted for producing ceramics greatly influences the composition, microstructure and microhardness of the products. The current study on glass mixture using Sarawak sand silica showed that leucite glass-ceramics could be produced. It was also found that 'sintering temperature and holding time' combination is a key factor to control leucite crystallization behaviour, morphology and dispersion with the corresponding hardness within the glass-ceramic material.

References

[1] Zahn T, Zahn B, Janko S, Weigl P, Gerhardt-Szep S and Lauer H 2016 Long-term behavior of double crown retained dentures with metal and metal-free secondary crowns and frameworks made of Vectris© on all-ceramic primary crowns: a prospective, randomized clinical trial up to 14 years *Clinical Oral Investigations* **20** 1087-100

[2] Stawarczyk B, Liebermann A, Eichberger M and Güth, JF 2016 Evaluation of mechanical and optical behavior of current esthetic dental restorative CAD/CAM composites *J Mech Behav Biomed Mater* **55** 1-11

[3] Contrepois M, Soenen A, Bartala M and Laviole O 2013 Marginal adaptation of ceramic crowns: a systematic review *J Prosthet Dent* **110** 447-54

[4] Montazerian M and Zanotto ED 2017 Bioactive and inert dental glass-ceramics. *J Biomed Mater Res A* **105** 619-39

[5] Atay A, Gürdal I, Bozok Çetıntas V, Üşümez A and Cal E 2019 Effects of new generation all ceramic and provisional materials on fibroblast cells *J Prosthodont* **28** e383-e94

[6] Della Bona A, Nogueira AD and Pecho OE 2014 Optical properties of CAD–CAM ceramic systems *J Dent* **42** 1202-9

[7] Belli R, Petschelt A, Hofner B, Hajtó J, Scherrer SS and Lohbauer U 2016 Fracture rates and lifetime estimations of CAD/CAM all-ceramic restorations *J Dent Res* **95** 67-73

[8] Nejatidanesh F, Savabi G, Amjadi M, Abbasi M and Savabi O 2018 Five year clinical outcomes and survival of chairside CAD/CAM ceramic laminate veneers—a retrospective study *Journal of prosthodontic research Journal of Prosthodontic Research* **62** 462-7

[9] Brochu JF and El-Mowaf O 2002 Longevity and clinical performance of IPS-Empress ceramic restorations-a literature review *Journal-Canadian Dental Association* **68** 233-8

[10] Sannino G, Germano F, Arcuri L, Bigelli E, Arcuri C and Barlattani A 2014 Cerec CAD/CAM chairside system *ORAL & Implantology* **7** 57

[11] Rödiger M, Schneider L and Rinke S 2018 Influence of material selection on the marginal accuracy of CAD/CAM-fabricated metal-and all-ceramic single crown copings. *BioMed Research International* 1-8

[12] Sareh H and Farshad D 2017 Achieving the desired esthetic with current CAD-CAM ceramics *Journal of Oral & Dental Health* **1**1-4

[13] Aldegheishem A, Loannidis G, Att W and Petridis H 2017 Success and survival of various types of all-ceramic single crowns: A critical review and analysis of studies with a mean follow-up of 5 years or longer *The International Journal of Prosthodontics* **30** 168-81

[14] Ali A 2003 The silica-based industry in Malaysia *Geological Society of Malaysia, Bulletin 46* 223-30

[15] Gualberto HR, Lopes RS, da Silva FA, Schnurr E, Poiate Junior E and Andrade MCD 2019 Influence of sintering temperature on mechanical properties of glass-ceramics produced with windshield waste *International Journal of Chemical Engineering* **2019** 1-8

[16] Nurddin SMASM and Selamat MB 2014 Crytallization and Properties of SiO2-Al2O3-K2O ZrO2 glass-ceramics for dental applications *Advanced Materials Research* **1024** 281-6

[17] Chen X, Chadwick TC, Wilson RM, Hill RG and Cattell MJ 2011 Crystallization and flexural strength optimization of fine-grained leucite glass-ceramics for dentistry *Dental materials* **27** 1153-61

[18] Barreiro MM and Vicente EE 1993 Kinetics of isothermal phase transformations in a dental porcelain *Journal of Materials Science: Materials in Medicine* **4** 431-6

[19] Tharajak J and Sanpo N 2019 The effect of sintering temperature on physical properties of leucite ceramics *Applied Mechanics and Materials* **891** 214-8

IOP Conf. Series: Materials Science and Engineering **834** (2020) 012045 doi:10.1088/1757-899X/834/1/012045

Effect of Welding Preheats on Metallurgical Analysis and Microstructural Development

Mazleenda Mazni*[1] , Muhammad Hussain Bin Ismail[2] , Hazriel Faizal Bin Pahroraji[2], Norjasween Abdul Malik[1], Mohd Ghazali bin Mohd Hamami[1], F. Sukarman[1], Nurulsaidatulsyida Sulong[1]

[1]Faculty of Mechanical Engineering, Universiti Teknologi MARA, UiTM Cawangan Johor, Pasir Gudang Campus, Jalan Purnama, 81750, Masai,Johor Darul Takzim, Malaysia.
[2]Faculty of Mechanical Engineering, Universiti Teknologi MARA, 40450 Shah Alam, Selangor, Malaysia

*mazleenda@johor.uitm.edu.my

Abstract. This project is to study the effect of welding preheat on metallurgical analysis and microstructural development. Variables such as current, speed of welding and size of specimen were fixed. In the present work, a mild steel plate with thickness of 100 mm and width size of 20 mm was used. SMAW (Shielded Metal Arc Welding) technique was chosen as it is the easiest way to perform and widely used in oil & gas and marine industries. Three different preheat temperatures were performed during the study; ambient temperature (no preheat), between 60 °C to 70°C and greater than 200 °C. The study emphasizes on the minimum preheat temperature that produce good quality welding by taking into account some of metallurgical aspects; microstructure and macrostructure development, hardness distribution at important areas in weld (Heat Affected Zone, parent metal and weld area) through thickness. From this study, code American Welding Society (AWS) D1.1 was used as a reference and it stated that for plate that has 100 mm thickness the preheat temperature should be in the range of 60 °C and 70°C. The result of microstructure and macrostructure showed that the depth of penetration was not vary too much. Hardness measurement, macro and microstructure observation were performed in order to obtain a good correlation exist between these parameters studied.

1. Introduction

Preheat is a heating procedure that applied to parent metal immediately before welding process is done. Preheat can be considered as one of the welding requirements that specified in many standards such as American Welding Society (AWS) D1.1structural welding code for steel [1]. This standard states that the requirement of preheat temperature depends on the type of steel and also the thickness of the material. Preheat involves heating the base metal, either in its entirety or just the region surrounding the joint, to a specific desired temperature, called the preheat temperature, prior to welding [2]. Many benefits can be obtained from preheat process such as; (1) Lower the cooling rate in weld and base metal, (2) Produce more ductile metallurgical structure with greater resistant to cracking, (3) Help to diffuse out hydrogen without causing crack, (4) To reduce shrinkage stresses in the weld and base metal and (5) Preheat can be used to help ensure specific mechanical properties such as notch toughness [2].

The choice of method of preheating depends on size of material, weldment size and heating equipment available before welding process. The methods can include furnace heating for small

Content from this work may be used under the terms of the Creative Commons Attribution 3.0 licence. Any further distribution of this work must maintain attribution to the author(s) and the title of the work, journal citation and DOI.
Published under licence by IOP Publishing Ltd

production of assemblies or gas torch, electric strip heater, induction heaters or radiation heaters for large structural components [3].

The crucial factor affecting from preheat is to slow down the cooling rate. Some alloys have properties that not allowed to cool quickly. This is due to fact that the development of hard or brittle phases in heat affected zone (HAZ) which are may lead to cracking and also reducing toughness in heat affected zone (HAZ). Preheating, is also one of the methods that widely employed in welding operation to prevent cracking [4]. Many welded steel has potential to hydrogen cracking due to fast cooling rate that leads to the formation of hard, susceptible microstructure and also lock hydrogen into the solidifying weld metal [5].

Even though preheat method is applicable in reducing cracking effects there are some material that should not be preheated such as austenitic manganese (13% Mn) steel, austenitic stainless steel, duplex stainless steel and also titanium alloys [5]. However, for many carbon steel and other alloy steels, preheat are required prior to welding process, particularly for thick surface. The thickness of each part is usually averaged over a distance of 75 mm from the weld line. However, the average thickness must depend on type of material, some material require to preheat when the length thickness is more or less 75 mm [6].

Effect of cooling rate on material after welding will determine the quality of welding; this is due to some defect of welding that may occur because of cooling rate such as cold cracking. Cooling rate is a rate of heat loss occurs from the surface of material. It gives an indicator to achieve some mechanical properties of material such as hardness, stress-relieve and also development of microstructure [7]. Cooling rate is significantly influenced from welding heat input and preheat temperature. As the heat input and preheats temperature increases, the rate of cooling decreases for a given base metal thickness. Other than input temperatures that changes the cooling rate, other variables such as material thickness, specific heat, density and thermal conductivity will influence the cooling rate. To simplify the effects of temperature with the cooling rate, Equation (1) shows the relationship between cooling rates, preheat temperature and heat input. Heat input is a relative measure of the energy transferred per unit length of weld. Heat input is typically calculated as the ratio of the power to the velocity of the heat source. To calculate heat input of welding Equation (2) can be used for governing information of welding current, voltage and travel speed [8].

$$\text{cooling rate, r} \propto \frac{1}{\text{preheat temperature,T} \times \text{heat input,H}} \tag{1}$$

$$\text{heat input, H} = \frac{\text{voltage,V} \times \text{current,I} \times 60}{\text{travel speed,S} \times 1000} \times \text{welding efficiency coefficient} \tag{2}$$

Metallurgy of carbon steel will give different of transformation according to the cooling time of material. Figure 1 shows the phase changes of Iron Carbon Diagram, which is significant for metallurgical analysis. At the higher temperature for 0.2 % carbon, the steel has austenite microstructure. When it is slowly cooled, the transformation will change into two-phase field of ferrite and austenite. If the temperature keeps decreasing the ferrite starts nucleating around the grain boundaries of austenite. On the slow cooling, the austenite get transform into pearlite. Pearlite is a lamellar mixture of ferrite and cementite. For each microstructure it has specific properties, for example ferrite has soft and ductile properties, while pearlite is hard and imparts mechanical strength to steel. Therefore, if the pearlite forms in material more than ferrite from result of rapid cooling the material will have hard and brittle properties. Therefore, slow cooling of material will help the material to maintain its behavior even after welding [9].

In this project, the effect of three different preheats temperature before welding was investigated. The material used for this project is mild steel because it is a common material used in fabrication of structure for oil and gas industry. Each specimen underwent metallurgical test such as hardness and the microstructural changes. This study was performed because to study the correlation between different preheat temperature and the quality of welding by referring metallurgical analysis, so that the change of material behavior can be seen especially in weld metal, heat affected zone (HAZ) and parent metal. The change of hardness of material also can be studied and compared. The purposes of this study were; (1)

to determine the effect of preheating on the time of cooling of welded structure, (2) to interpret welding quality in terms of depth of penetration and welding defect as a function of preheat temperature by performing macrostructural analysis, and (3) to analyze the metallurgical aspect of welded structure by performing microstructure analysis and hardness test.

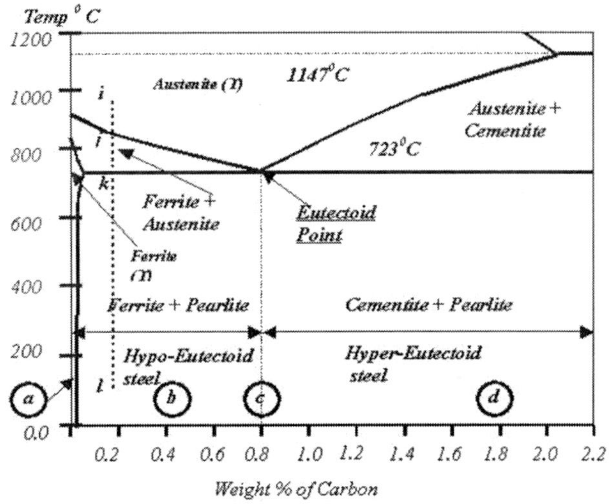

Figure 1. Iron carbon diagram of carbon steel [9]

2. Experimental Procedure

In this experiment, the effect of preheat before welding on mild steel plate was investigated in accordance with the standard AWS D1.1- Structural Welding Code for Steel that used at Labuan Shipyard and Engineering to fabricate steel structure for offshore applications. Material was prepared by the industry according to the standard welding practice that applied in fabrication of steel structure for oil and gas industry. Mild steel was used in this study because it is commonly used in fabrication of platform or other applications and due to its properties which are strength, ductility and ease of machining.

Based on preheat theory on welding, usually effect during welding of high carbon content or welding of thick material that more than 2 inch (50.8 mm). However, for other type of carbon steels, preheat is required if the thickness of weld part meet the requirement of the standard. The size for all specimens set as fixed value so that it will not affect the effect of preheats. The specimen has 75 mm X 20 mm X 100 mm of length, width and thickness as shown on Figure 2.

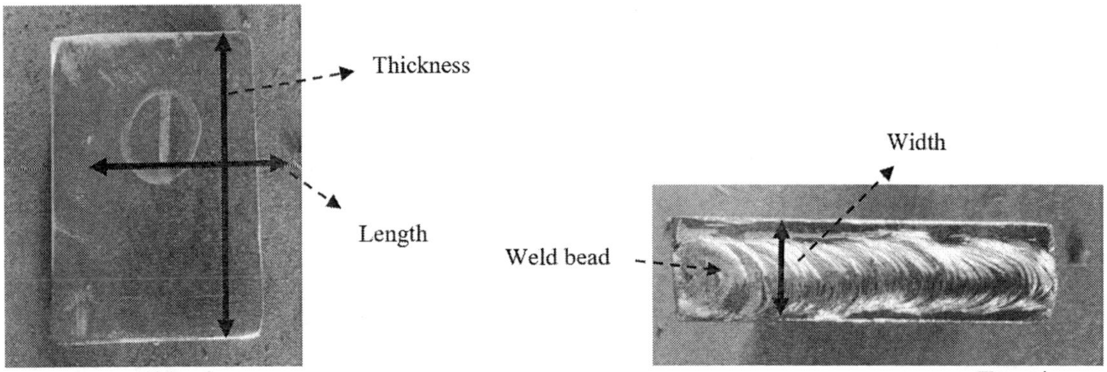

Side view Top view

Figure 2. Figure shows the size of specimen made from mild steel and the position of weld bead

Three conditions of preheat were studied; no preheat labelled as specimen 1, preheat temperature is 60°C to 70°C labelled as specimen 2 and lastly, specimen 3 when the preheat temperature is more than

200°C. The method of preheating used in this study is by using oxy-acetylene gas torch method. To measure the temperature, digital thermometer was used. To avoid heat loss to the surrounding which may affect the properties of welded structure, the welding process was started immediately after preheat process.

Welding process used in this study is shielded metal arc welding as shown schematically in Figure 3 using E7018 type of electrode that is low hydrogen electrode to avoid cracks as shows in Figure 3. Other parameters of welding process were shows in Table 1. The parameter of welding was same for all preheat temperature. After the welding process completed, the cooling time for weld bead was measured for 4 minutes with each 10 seconds the temperature drop recorded.

Table 1. Parameters of SMAW process

Properties	Parameter
Electrode material	Carbon Steel
Electrode size	4 mm
Polarity	DCEP
Current	145 A
Voltage	25 V

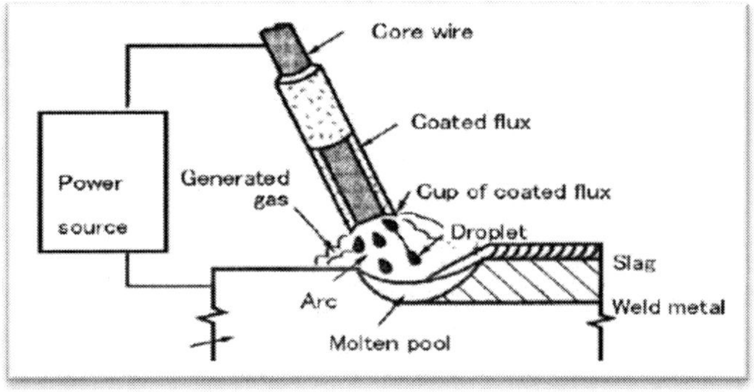

Figure 3. Shielded metal arc welding process [10]

To study the changes of macrostructure and microstructure, laboratory metallurgical analysis was carried out. In laboratory experiment, the tests conducted were:

1. Chemical composition test by using Arc Spark Spectrometer.
2. Macrostructure and microstructure analysis.
3. Hardness test through thickness material by using hardness Vickers's.
4. Image analyzer to determine quantity of grain structure for ferrite and pearlite.

For additional information, image analyzer was conducted to measure the quantity of microstructure developed for each preheat temperature. For etching process the solution used was 2% Nital. Nital contain of 90-99 mL of ethanol and1-10 mL of HNO_3. Etchants will reveal alpha grains boundaries and constituent. The specimen after polishing process was immersed for several seconds before macrostructural and microstructural analysis proceeded.

3. Results and Discussion

Cooling rate of weldment was investigated during preparation of specimen. The data of temperature drops after welding for 4 minutes were recorded. Figure 4 shows how the effects of preheat temperature on cooling rate. It is important to investigate the cooling rate of welding because it is one of the factors that contribute to crack. If the cooling rate too fast, it will help the formation of hardening structure such as cementite. Other than that, cooling rate also used to avoid hydrogen trapped in weld metal that will contribute the internal cracking. Lastly, cooling rate also helps to avoid cracks by rapid solidification of weld metal. Generally, to control the cooling rate either by using heat treatment method such as annealing or by controlling preheats and interpass temperature of welding. The relationship of cooling rate to the preheat temperature and heat input shown in Equation (1). From the relationship, the higher the temperature of preheat the lower the cooling rate of welding, the results of cooling of material shows in Figure 4. From Figure 4, the additional of preheat will give high temperature after welding and by referring the slope of the graph the temperature decrease slowly for preheat temperature between 60°C to 70°C and preheat more than 200°C, compare during no preheat temperature. If the graph continue until time is zero, the specimen with no preheat temperature will reach zero first followed by specimen 2 and 3. From Equation (2) the heat input of welding process with 0.8 welding efficiency coefficient is 1.16 kJ/mm.

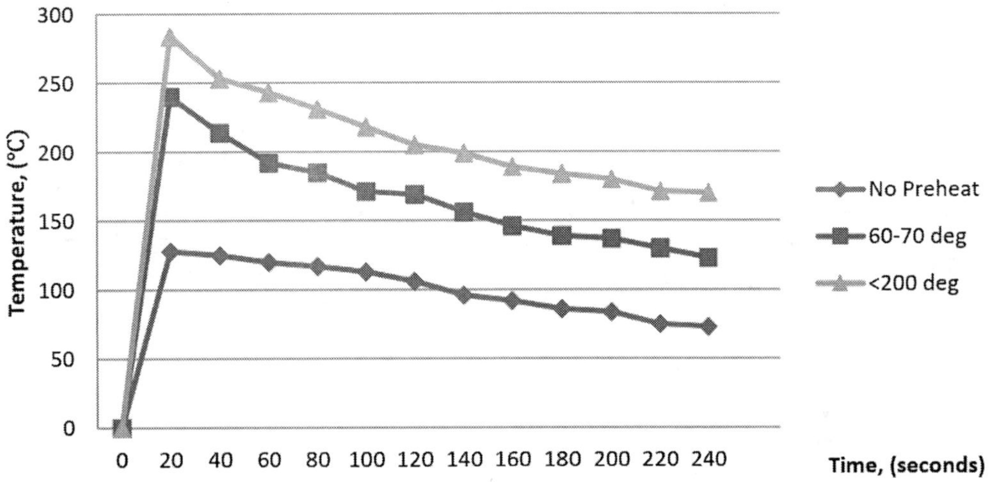

Figure 4. Relationship between cooling rate and temperature input of mild steel

Three different preheat temperature were applied to three different material having same size and material properties. Arc spark spectrometer were done to check the chemical compositions for each of material, hence calculation of carbon equivalent to check weldability each material can be made. Table 2 shows the chemical composition for each specimen. For specimen 1 and 3 it has nearly same of chemical composition, however for specimen 2 contain of carbon more than others specimen. Specimen 2 also can be considered as mild steel because carbon content less than 0.2 % with slightly different in material properties.

Table 2: Chemical composition for each specimen

Specimen 1							
C %	Mn %	Ni %	Cr %	Mo %	Cu %	V %	CE
0.152	1.84	0.012	0.119	0.013	0.027	0.0039	0.628
Specimen 2							
C %	Mn %	Ni %	Cr %	Mo %	Cu %	V %	CE
0.192	1.86	0.013	0.12	0.014	0.028	0.0041	0.674
Specimen 3							
C %	Mn %	Ni %	Cr %	Mo %	Cu %	V %	CE
0.157	1.86	0.012	0.121	0.015	0.028	0.0045	0.639

In this study, the results from arc spark spectrometer were used to define the classification and chemical contain in material. From Table 2 the carbon equivalent content of material is more than 0.5% therefore, according to AWS for each carbon steel that has more 0.5% of carbon equivalence content the material are crack sensitive, therefore to avoid cracks preheat of material before welding process can reduce the formation of crack in welding. Other than that, use of low hydrogen welding procedure also can help minimize the formation of cracks.

IOP Conf. Series: Materials Science and Engineering **834** (2020) 012045 doi:10.1088/1757-899X/834/1/012045

Other than weldability of specimen and carbon equivalent content, arc spark spectrometer can also determine the behavior of the base metal itself before welding. Composition of carbon in material shows the toughness properties of material, ductility and crack contributor. For other composition, manganese also will promote crack if the percentage of composition more than 1.4%, for material in this study the amount of manganese is more than 1.4% therefore it will allow cracks to occur if the welding procedure is not properly.

Examination of macrostructure for each different preheat temperature of weld across the thickness of material recorded so that the penetration of the weld area and quality of the weld can be determined. For all the different preheat temperature there is no defects or imperfections occur because of welding, therefore, the parameters used for welding are suitable for welding mild steel. However, the penetrations of weldment gives the different value for different preheat temperature. Figure 5 shows the macrograph of the specimen. From the Figure 5, the depth of penetration were varies with the different preheat temperature. For weldment with no preheat temperature the depth of penetration is smaller compare to during preheat at 60°C and preheat more 200°C. However, macrostructure between 60°C and 200°C preheat temperature there are only slightly different in depth of penetration.

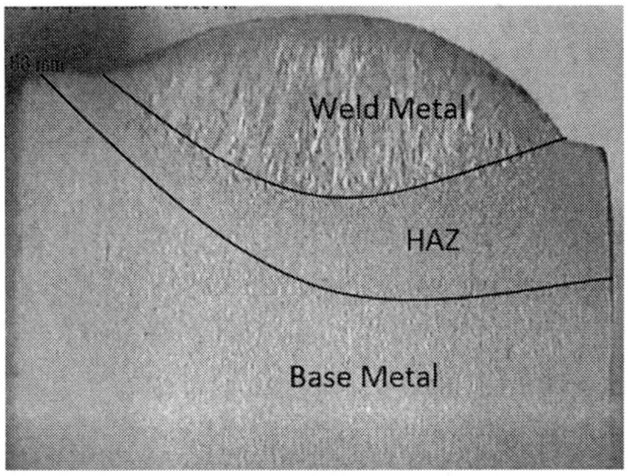

(a) Macrostructure when no preheat temperature

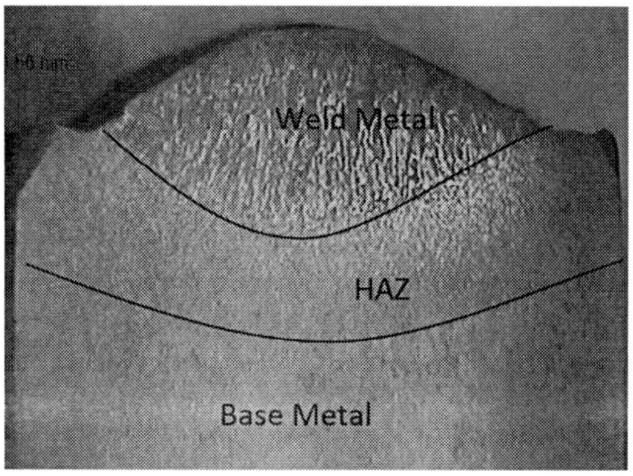

(b) Macrostructure when preheat between 60°C and 70°C

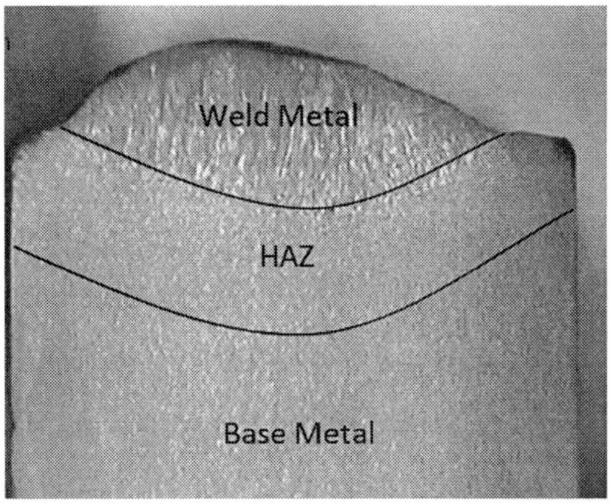

(c) Microstructure when preheat greater than 200°C

Figure 5. Macrostructure for; (a) no preheat temperature, (b) preheat between 60°C and 70°C, (c) preheat greater than 200°C

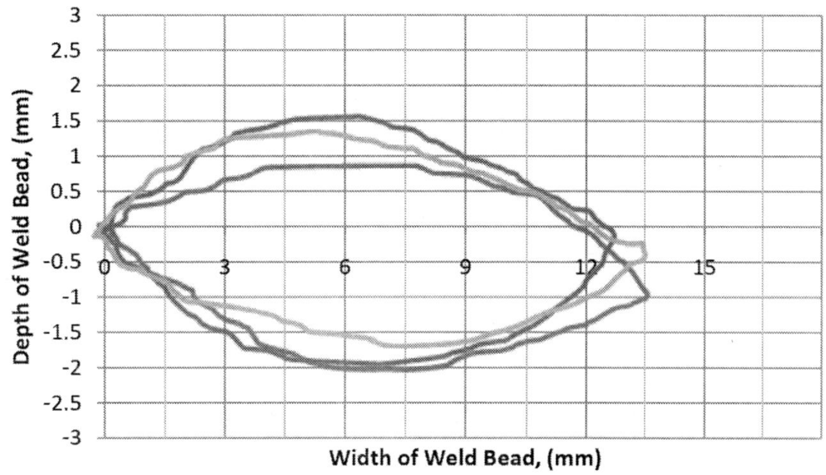

Figure 6. Comparison of weld bead size between different preheat temperature

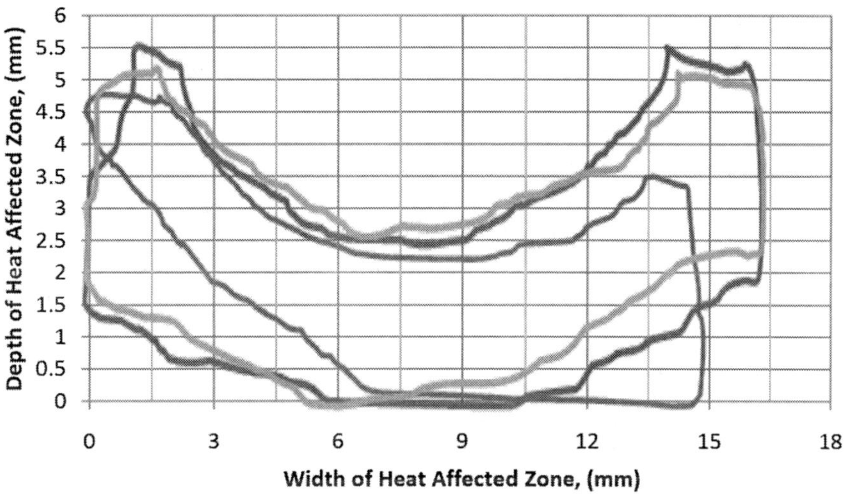

Figure 7. Comparison of heat affected zone area for different preheat temperature
(For Figure 6 and 7: blue color line for no preheat, red color for preheat between 60°C and 70°C and green color line for preheat more 200°C)

Microstructure examination of material with different preheat temperature were investigated to determine the changes of properties of material. The development of microstructure at base metal, heat affected zone and weld metal were identified to measure the quantity of ferrite and pearlite. For no preheat temperature the structure consist of pro-eutectoid ferrite in the grain boundary and bainitic structure of the columnar grains of deposited metal on the weld metal. At the base metal it consists of ferrite and pearlite where the pearlite is more looks like thin plate and close together. At heat affected zone the development of fine grain of ferrite produce the lower hardness compare to weld metal.

For preheat between 60°C to 70°C the microstructure of weld metal consist of pro-eutectoid ferrite in the grain boundaries. The array of bainitic laths in surround by ferrite grains can be seen. The structure at heat affected zone is shows the grains boundary more coarser structure possessing less toughness. Comparing microstructure from no preheat temperature with preheat temperature at 60°C to 70°C, preheat gives coarse grain structure and gives low toughness of material. For preheat temperature more than 200°C the size of grain is more coarse than grain structure during 60°C to 70°C. The development of microstructure is the same as during preheat temperature between 60°C to 70°C. In term of hardness during preheat more than 200°C the hardness is lower compare to other temperature of preheat. Table 3

IOP Conf. Series: Materials Science and Engineering **834** (2020) 012045 doi:10.1088/1757-899X/834/1/012045

shows the microstructure of base metal, heat affected zone and weld metal at different preheat temperature.

The study of preheat temperature also will lead to the changes of hardness of material; usually hardness of mild steel will be in the range of 140 HV to 230 HV. With the increase in temperature the more hardness of material will drop. The drop of hardness of material will gives the properties of ductile to the material. Figure 6 shows the different of hardness value through thickness on the different preheat temperature. Figure 8 shows that the hardness value varies with the temperature input of material, from the graph the higher temperature input will gives low hardness of material. However, for all specimen gives higher value of hardness at weld metal followed by heat affected zone then base metal. The value of hardness at the weld metal is high because of the grain size, amount ferrite and pearlite contain in material, thus it can conclude that the mechanical properties of material at weld metal differ from heat affected zone and base metal. In the base metal the grain size can be classified as fine-grain because it is less hardens, therefore it has fewer tendencies to crack.

Table 3. Microstructure at weld metal, heat affected zone and base metal for; (a) no preheat temperature, (b) preheat temperature between 60°C and 70°C, (c) preheat temperature greater than 200°C when using 2% Nital solution of etching solutions.

	Weld Metal	Heat Affected Zone	Base Metal
(a)		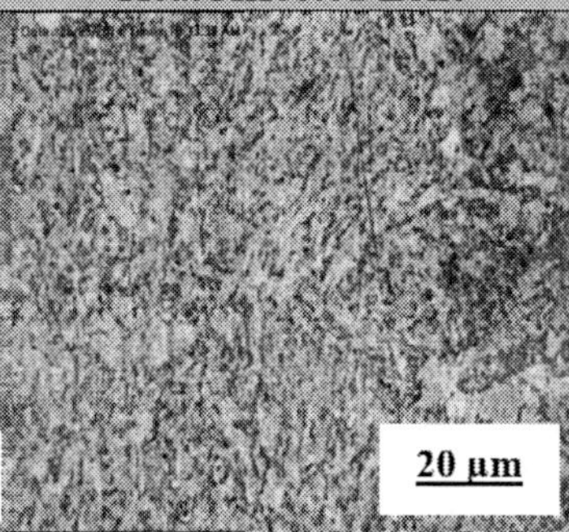	
(b)	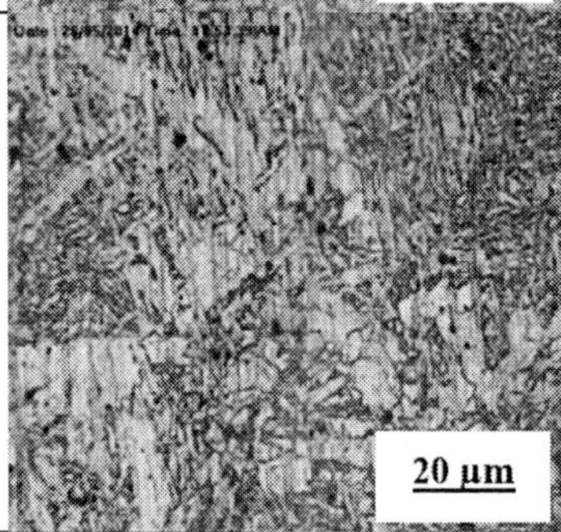		
(c)	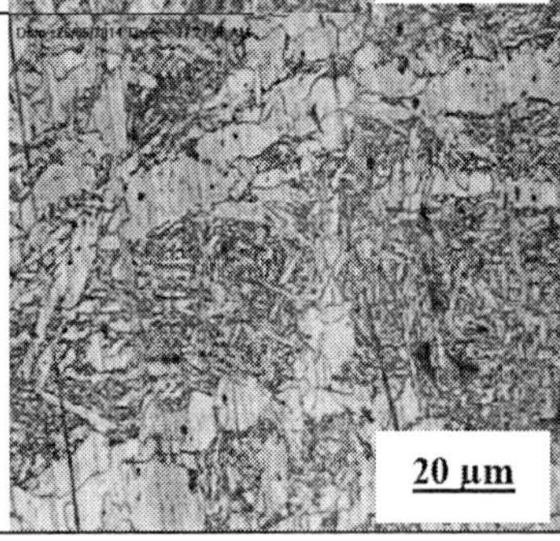	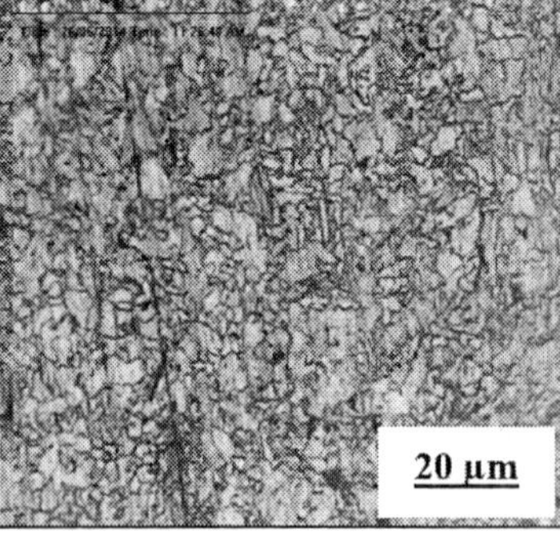	

IOP Conf. Series: Materials Science and Engineering **834** (2020) 012045 doi:10.1088/1757-899X/834/1/012045

Table 4. Microstructure at weld metal, heat affected zone and base metal for; (a) no preheat temperature, (b) preheat temperature between 60°C and 70°C, (c) preheat temperature greater than 200°C when Beraha's solution of etching used.

	Weld Metal	Heat Affected Zone	Base Metal
(a)		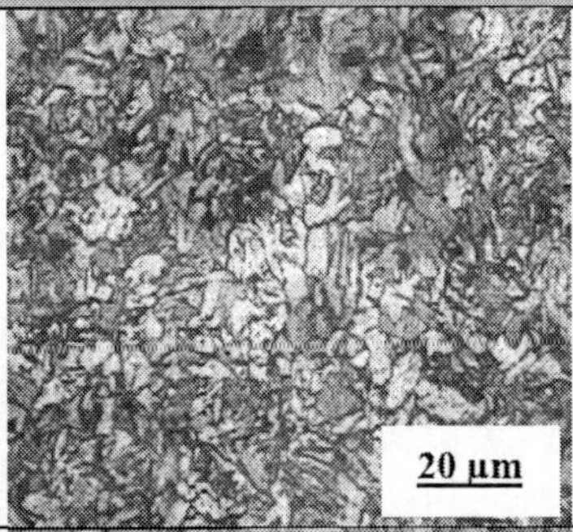	
(b)			
(c)			

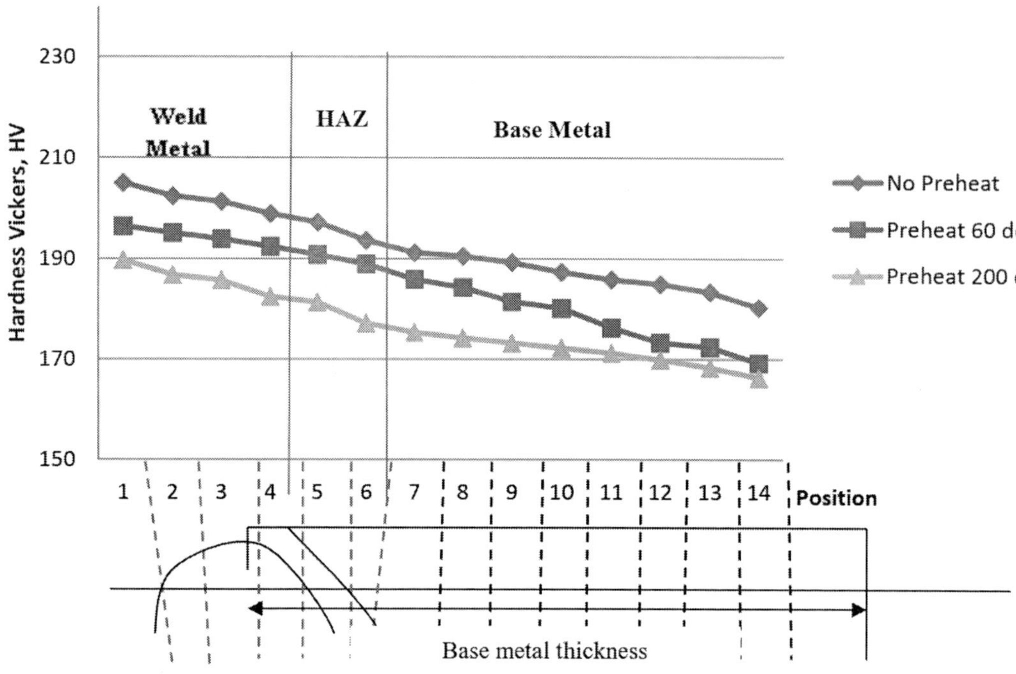

Figure 8: Comparison of hardness value for different preheat temperature through the thickness of mild steel

Conclusions

The effect of welding preheats temperature for mild steel material with 100 mm thickness plate on metallurgical analysis and microstructural development were investigated. The experiment results showed that different in welding preheat temperature effects the properties of material. Lowered the cooling rate will reduce the cold crack defects. The development of microstructure depended on the cooling rate of welding. Preheat controlled the cooling of material by slower the rate of cooling by increasing preheat temperature. Temperature input also controlled the hardness of material through thickness, where material has low hardness as preheats temperature increased.

Acknowledgement

Special dedication to the persons who involve in completing this research, especially to Mr. Agus Trang Winarno, Miss Mazlinda binti Sari Hassan, and others from the Labuan Shipyard and Engineering Sdn Bhd for giving fully support to finish this research. I would like also to thanks Mr. Rahimi, the UiTM Material Science lab technician for helping me completing this research.

References

[1] An American National Standard. AWS D1.1/D1.1M: 2008. Structural welding code – steel.
[2] Retrieved on (2013, Oct 21). What is preheat?. Retrieved from www .lincolnelectric.com/en-us/support/process-and-theory /pages/preheat-detail.aspx.
[3] Retrieved on (2014, Apr 7). Why is preheat used when arc welding steel, and how is it applied?. Retrieved from http://www.twi-global.com/technical-knowledge.
[4] Bipin Kumar Srivasrava, Dr. S.P. Teawri and Jyoti Prakash (2010). A review on effect of preheating and/or post weld heat treatment (PWHT) on mechanical behavior of ferrous metals. International Journal of Engineering Science and Technology, Vol. 2(4), 2010, 625-631.

[5] Retrieved on (2013, Oct 21). Preheating of material. Retrieved from http://www.boc.com.au/consumables/preheating of materials.
[6] Retrieved on (2013, Oct 21). Preheating of material. Retrieved from http://www.boc.com.au/consumables/preheating of materials
[7] B. A. Graville (1973). Welding Cooling Rates and Heat Affected Zone Hardness in Carbon Steel. Welding Journal, American Welding Society.
[8] N. Bailey, F.R. Coe, T.G. Gooch and others, (1973). Welding steels without hydrogen cracking, Woodhead Publishing Ltd.
[9] W. D. Callister. Jr, (2007). Material Science and Engineering, An Introduction. John Wiley & Son. Inc.
[10] Kalpakjian,S., Manufacturing engineering and technology, 10th Edition, Prentice-hall, 2010.
[11] George, S.B., Henry, R.C, & John, A.V., (2002). Material handbook 15th Edition. Mild Steel, pp 172: Mc-Graw Hill.
[12] R.C. Hibbeler. (2008). Mechanics of materials 7th Edition. Mechanical properties of materials, pp 85: Prentice Hall.
[13] Jianguo Lin (Ed.), Daniel, B. (Ed.), & Maciej, P (Ed.). (2012). Microstructure evolution in metal forming processes: Woodhead Publishing.
[14] David, B., & Wayne, D.K. (1999). Microstructural characterization of materials. John Wiley and Sons Ltd.
[15] S.M. Adedayo and S.O. Momoh (2010). Effect of initial elevated metal temperature on mechanical properties of an arc welded mild steel plate. Indian Journal of Science and Technology, Vol. 3, No. 12, 0974-6846.
[16] Enda Keehan (2004). Effect of microstructure on mechanical properties of high strength steel weld metals. Department of Experimental Physics, Chalmers University of Technology, 0346-718X.
[17] Pang, W., Ahmed, N. & Dunne, D. (2011). Hardness and microstructural gradients in the heat affected zone of welded low-carbon quenched and tempered steels. Australasian Welding Journal, 56(2), 36-48.
[18] G.E.Linnert (1985). Welding metallurgy of carbon and alloy steels, Volume 1, American Welding Society.
[19] R.Blondeau, (2001). Metallurgy and Mechanics of Welding, John Wiley and Son Ltd.
[20] E. G.Signes (1972). A simplified method for calculating cooling rates in mild and low alloy weld metals.
[21] T. Kasuya, N. Yurioka, and M. Okumura (1995). Methods for predicting maximum hardness of heat-affected zone and selecting necessary preheat temperature for steel welding. Nippon Steel Journal, UDC621.791.011/.02
[22] Retrieved on (2014, Jan 8). Arc welding. Retrieved from www.energymanagertraining.com/Journal/02112007/ArcWelding.pdf
[23] R.C. Hibbeler. (2008). Mechanics of materials 7th Edition. Mechanical properties of materials, pp 85: Prentice Hall.
[24] Zakaria Boumerzoug, Chemseddine Derfouf and Thierry Baudin (2010). Effect of welding on microstructure and mechanical properties of industrial low carbon steel. Journal of Engineering, 2010, 2, 502-506.

IOP Conf. Series: Materials Science and Engineering **834** (2020) 012046 doi:10.1088/1757-899X/834/1/012046

Welding Quality Assessment for Oil and Gas Industry by Comparison of Mechanical Testing Properties and Microstructure Analysis

F. Sukarman*[1], M. H. Ismail[2], M. A. M. Shah[1], N. H. Ramli[2], I. Tharazi[2]

[1]Faculty of Mechanical Engineering, Universiti Teknologi MARA, Kampus Pasir Gudang
[2]Faculty of Mechanical Engineering, Universiti Teknologi MARA, Shah Alam

*firdaus@uitm.edu.my

Abstract. WBS testing is the procedure of substantiation on material after a completed welding process to determine the material properties and behavior such as strength properties and forming properties. The conventional testing for determining strength properties of material involve parameters such as tensile strength, shear strength, toughness, brittleness and rigidity. On the other hand, the forming properties testing is performed to inspect elasticity, plasticity, ductility, malleability and hardness of the welded components. The importance of WPSs in the welding technology and how the WPSs is performed at the point around the welding procedure should be defined clearly. Thus, this project investigates the welding quality assessment in the industry by comparison of several testing methods (such as tensile and bend test) following determined acceptance criteria by American Welding Society (AWS D1.1). This research also aims to investigate the metallurgical aspects of welding joint by looking into different locations which as HAZ, weld metal and base metal. The investigated results are macro and microstructure observation, hardness distribution and also the density of the allocated area. This project will serve to develop a further understanding of welding and to find out the correlation between various testing methods that have been used. The welding quality assessment by comparison of mechanical testing properties and microstructure analysis are carried out to determine the property of the affected points. Then, the welding evaluation is determined by performing non-destructive and destructive testing methods. The comparison between the mechanical testing properties and microstructure analysis is made to determine welding quality of Shielded Metal Arc Welding. The acceptance criteria for both non-destructive and destructive testing were referred to American Welding Society (AWS) standard. Furthermore, this research is conducted to investigate the function of welding procedure specification in evaluation of welding quality assessment especially in oil and gas industry.

1. Introduction

Fabrication of products in manufacturing industry and construction companies requires some type of joining process for attach the components for structure completion. There are three main categories in material joining processes which is adhesive bonding, mechanical fastening and welding. Welding are different from others joining process because the joint process in welding is very strong and permanent [1]. The joint that made by welding are not easily separated and that is why welding has their procedure that must be followed. Welding also can be economical process than mechanical fastening and adhesive bonding because the capability to create joint permanently[2]. According to American Welding Society

Content from this work may be used under the terms of the Creative Commons Attribution 3.0 licence. Any further distribution of this work must maintain attribution to the author(s) and the title of the work, journal citation and DOI.
Published under licence by IOP Publishing Ltd

(AWS) there have about 94 different types of welding processes with different sources and method to joint metal. Then most commonly used electric arc welding today in fabrication service is Shielded Metal Arc Welding (SMAW), Gas Metal Arc Welding (GTAW), Gas Tungsten Arc Welding (GTAW), Flux-Cored Arc Welding (FCAW) and Submerged Arc Welding (SAW)[3]. All the welding has different purpose and application. Welding is one of the joining processes that commonly use in the industry, especially in oil and gas industry [4].

Welding can be defined as a joining process that is intended to make the material at the joint surfaces as homogeneous as possible with the base material. The homogeneity between base material and weldment area is according to the chemical composition, metallurgical, mechanical properties or physical characteristics. The source of the welding process is created by electric power supply. The welding machine will produce the electric arc that heat the material until its melting temperature. Established electric arc is created between the gap of an electrode and the metal to be welded. The electric arc will form a molten pool at the parent metal and dissolve with welding material. During this process, this molten pool must be protected from the atmosphere to prevent it reacts with the surrounding air that may cause the material at joining become brittle [5]. This can be avoided by using welding electrode with flux insertion or applying shielding gas around the molten arc while welding process is running. The use of shielding gas or flux coated on the electrode is depending on the application material and purposes the joints are used. However, in most cases, melting the material between the parts will not create a strong weld connection. Thus, the usage of filler compound is recommended to assure the mixing of electrode and welded parts. Filler compounds has been used commonly in industry to increase the overall effectiveness of the weld. However, it depends on the welding process whether the electrode is already purchased with filler metal in it or the filler metal is used separately during welding process. The other factor of welding effectiveness is the electric power supply itself. The type of current that usually use in the welding process may be either alternate current or direct current [6].

To achieve high welding quality is not only depending on the welder itself, but also depending on the parameter that have been used. The good parameter will give the parent metal matching with the filler metal. The good weld quality must achieve certain physical and chemical properties to meet the service requirement of the welded part.

2. Welding Evaluation

The quality of welding can be tested using several testing methods in welding quality assessment. Welding test is important to analyse the welding quality in term of the strength, mechanical properties, welding characteristics and the metallurgical aspects [4]. Weld testing is one of the procedure in welding evaluation and has its determined methods. There are three methods to evaluate the welding which is by visual inspection, non-destructive testing and destructive testing. Visual inspection must be done first; followed by non-destructive testing and destructive testing. The reason is non-destructive and destructive testing is a high cost evaluation method and required skilled person to perform the test[7]. Visual inspection does not require special equipment and tool but only need experience person to examine the quality of the welding. In visual inspection, only internal defect can be detected such as surface crack, surface irregularities, spatters, contour defect etc. The person that has a responsibility to perform visual inspection is called welding inspector.

There are two classes in welding test which is destructive and non-destructive test. Non-destructive testing is most popular and commonly used in welding evaluation because material in non-destructive testing does not undergoes destruction and can be reused. It's different from destructive testing which material use in destructive testing undergoes destructive and cannot be used again [5]. Non-destructive testing becomes necessary for welding evaluation, particularly in oil and gas fabrication because this testing method gives evaluation figure of the internal defect occurs in the welded joint. The common non-destructive test used today is Dye Penetrant Test (DPT), Magnetic Particle Inspection (MPI), Radiographic Test (RT), Ultrasonic Test (UT) and Eddy Current Test. The selection of test is depend on the material thickness, type of the welding used, and the possible defect occur [6]. Destructive test is carried out is to determine the strength of the material, material toughness and metallurgical aspect. Destructive test is usually made for the welding procedure and welder approval. The most popular test under destructive test categories is tensile test, bend test, hardness test and macroscopic test. Other test such as impact toughness test and nick break test also can be used to determine the fracture surface of the weld metal for flaws and to check the root penetration and fusion.

2.1. Tensile Test

Tensile test is the most widely used to determine the mechanical properties of materials. The tensile strength of an alloy is most commonly measured by placing a test specimen in the jaws of a tensile machine. According with American Welding Society (AWS B4.0:2007-Tension Test) the data obtained from tensile test may include the Ultimate Tensile Strength, Yield Strength, Yield Point if occurs, percentage of elongation, percentage of reduction area, stress strain diagram and mode of fracture. The tensile machine will apply stretching stress by gradually separating the jaws. The amount of stretching needed to break the specimen is measured and recorded. From this test, the yield strength also can be measured. Yield strength is the amount of stress a material can withstand without permanent deformation [8].

The relationship between average normal stress and average normal strain can be explained more on stress strain diagram. The stress-strain diagram consists of two regions which is elastic and plastic region. In elastic region, the material will return into the original shape when the load is removed while in plastic region; is the area where the material tends to fracture. Plastic region consists of the yielding, strain hardening, necking and fracture point. The tensile test guided is refer to American Welding Society (AWS B4.0:2007) standard[9]. It describes the methodology, specimen guidance and the calculation used for tensile test analysis. The specimen is called dog bone shape was prepared according to the thickness range and specimen shape whether tabular or plate shape. Different thickness must have a different dimension measurement.

3. Welding Quality Assessment

Several experiments were selected to get the answers regarding the objective of this project. Some experimental work has their standard and code that must be followed to ensure the result obtained is valid and standardized. In this project, the specimen and testing preparation were done in Labuan Shipyard and Engineering Sdn. Bhd. Shielded Metal Arc

Welding process is used for preparation of the welding sample and has been monitored by welding engineer. The specimen undergoes several testing techniques consist of non-destructive test and several destructive tests to evaluate welding quality of the specimen.

3.1. Sample Preparation

Selection of the material and the welding procedure specification (WPS) has been done according to the project given by Labuan Shipyard and Engineering (LSE). All the parameter and the procedure for welding have followed the WPS which were given and monitored by welding engineer. The material has been selected is A36 mild (low carbon) steel grade and have 15 mm thickness. The procedure including the parameter used in sample preparation is simplified from Welding Procedure Specification tabulated in Table 1.

The summary of welding procedure specification tabulated in Table 1 is taken from LSE Welding Procedure Specification. To control the influences of welding quality by human skills, welding processes were performed by a qualified welder in the company. The final welded specimen then undergoes visual inspection before proceed to other non-destructive testing and destructive testing [10]. The visual inspection only analyzes the weld surface to check any discontinuity and irregularities at the connections. Visual inspection is performed by qualified welding inspector in the company.

Table 1. Summary of Welding Procedure Specification.

1	**Material**	**A36 mild (low carbon) Steel**
2	Joints	Joints Design: Single Vee Multiple Pass: Yes 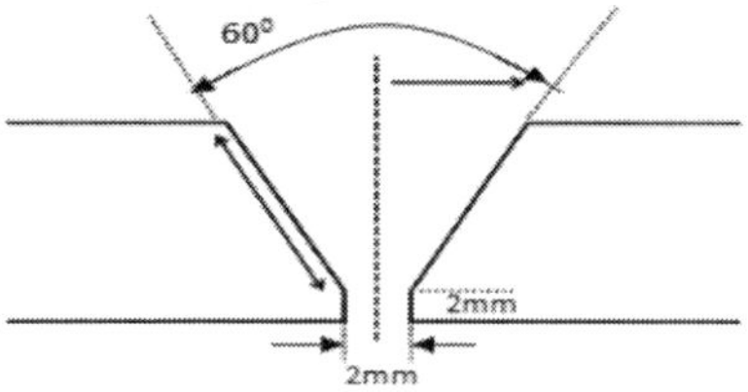
3	Filler Metal	Welding Process: SMAW Electrode: E7017 (root) : E7018 (2nd Pass)
4	Position	Position Groove: 1G Weld Progression: Dragging
5	Pre heat	Preheat temperature: 50oC Interpass Temperature Maximum: 250oC Preheat Method: Propane Gas
6	Electrical characteristics	Current polarity: DCEN/DCEP Welding technique: weaving Cleaning method: Grinding & Brushing Electrodes spacing 1.0mm – 3.0mm
7	Welding parameter	

Run no.	Filler metal		Current		Volt (V)	Heat input (kJ/mm)
	Size(mm)	class	Polarity	Amperage		
1	2.6	E 7016	DCEN	55 – 74	20 – 74	1.0 – 2.8
2 - n	3.2	E 7018	DCEP	110 - 150	24 – 26	1.0 – 2.8

3.2. Experimental works

In the second phase under methodology process, experimental work was performed to examine the welding quality assessment and to analyze the material characteristic before and after welding process. In this stage size and shape of the specimen must be prepared according to standard testing specification to obtain good and accurate results from the experiments. Non-destructive testing method must be done first followed by destructive testing.

3.2.1. Radiography Test

The radiographic test used x-ray as a source and automatic x-ray machine. The purpose of the x-ray is to identify defect occurs at the welding area. Parameter that used is 225KV of voltage and 3000 μA of current. The X-rays were emitted from the tube and pass through the workpiece to be inspected. Parts that have less obstruction to X-rays, such as cavities or inclusions, allow increased exposure of the film. The film is developed to form a radiograph with cavities or inclusions indicated by darker images. Section thickness increases (such as weld under-bead) appear as less dense images. Radiographic test must be conducted with qualified operator that have valid radiographic test certificate. The result has been recorded for reference.

The preparation of the specimen for radiographic test involve cleaning process and preparation of image quality indicator (IQI). Before performing the test, several criteria such as IQI sensitivity and defect discontinuity are set to determine the quality of the radiographic film. The sensitivity of film can be measured using image quality indicators (IQI's) tool.

IQI's or penetrameters are used to measure radiographic sensitivity and the quality of the techniques used. Several standards use for IQI's such as British Standard (BS EN 462), American Standard (ASME

SE 1025) and German Standard (DIN 62). For this experimental purpose the British Standard (BS EN 462-1) was used. Selection of IQI's must be suitable with the material of the specimen. Simple calculation is used to calculate the sensitivity of the film and how many wires should appear on the film to comply with the standard.

Thickness of thinner wire visible =sensitivity/(100 ×total weld thickness)

Thickness of thinner wire visible =2/(100 ×20mm)

Thickness of thinner wire visible =0.4,

Inspection results can be obtained by visually check the radiography film. From the calculation above, minimum of 4 elongated wires are permissible to pass the radiographic testing procedure.

3.2.2. Spectrometry Test

Spectrometry test is performed to determine the chemical composition of the material at the base and weld metal. The chemical composition of the material will determine the weldability of the material and influence the welding joint quality. There is no specific standard use for spectrometer test and the sample preparation is according the ability of the machine itself. The chemical composition of the mild steel has been compared at two different zone; weld metal and base metal using spectrometer.

3.2.3. Destructive Test (Tensile and Side Bend Test)

Tensile test is the common destructive testing method to determine the mechanical behaviour of the material. The welding quality also can be defined through the tensile test while observing the fracture location of the specimen. It also used to measure the transverse tensile strength under static loading of a butt joint employing butt welds.

Side bend test was carried out to determine the soundness of the weld metal and HAZ in a cross section. The selection of side bends instead of root and face bend test is according to the thickness range. This may be preferred to the transverse bend test on thick materials. It is also used on processes or procedures expecting lack of fusion. The acceptance criteria for bend test are in accordance to AWS D1.1 in which the length of flaw must not exceed 3mm at any direction by visual examination.

3.3. Microstructure Observation

Scanning Electron Microscopic (SEM) observation is performed in this project to observe the fracture surfaces of the material due to overloading failure. The specimen used for microscopic observation is taken from tensile test specimen. The fracture area of the specimen has been cut into small pieces to fit into Scanning Electron Microscope Machine (SEM). Figure 1 shows the specimen used for scanning electron microscopic analysis. The location of the specimen is set to produce clear image in the monitor. The resulted image resolution can be set until the final microstructure can be identified.

The purpose of macrostructure observation is to analyze the internal defects, surface defects and geometric defects with type, size based on previously determined acceptance criteria [11]. The sample of macrostructure and microstructure must undergo several processes such as grinding, polishing and etching. The purpose is to get a fine surface area so that the microstructure will appear under metallurgical microscopic. Procedure for microstructure and microstructure observation is almost similar except the magnification factor must be adjusted. Specimen is polished and etched to gain quality microstructure result. Next, microscopic optical viewer is acquired for microstructure analysis with suitable magnification level. The magnification started with 10x and 20x and lighting is adjusted to produce clear view of the sample microstructure. The microstructure picture of the test specimen is recorded. The same observation was taken for three different places of the surface area for comparison of the microstructure taken [12].

The samples shown in Figure 1 were placed under the lens and image of the microstructures was then captured by CCD camera.

Figure 1. Specimen for scanning electron microscopic observation

4. Result and Discussion

This chapter explains the detail of the experimental work that has been done in previous section. The detail of the result from non-destructive and destructive testing was discussed further. The information from the result have been used to make an analysis to evaluate the welding quality assessment. The discussion focused on the Shielded Metal Arc Welding Process and the effect on microstructure analysis in weld metal, heat affected zone (HAZ) and in base material. The discussion will be based on the results from all the testing techniques used for evaluation of welding quality assessment. In addition, the quality evaluation of the welding is based on welding procedure specifications provided. This chapter also explains the detail of the effect of applying the correct welding technique with the formation of microstructure analysis.

4.1. Welding Surveillance Check

The welding surveillance check was done during welding to control the important variable such as voltage, current travel speed and heat input. These variables have major effect in the formation of microstructure in weld metal and heat affected zone and the mechanical properties of material. For example, very high heat input control will increase the chances of distortion from occurs and will give heat affected zone become more brittle[13]. Table 2 shows the result of welding surveillance check for the welded sample. Is consists the information of the voltage, current, travel speed, and heat input.

Table 2. Welding Surveillance Check

Inter pass	Voltage (V)	Current (A)	Time /pass (min)	Travel speed (mm/min)	Heat input (kJ/mm)	Inter-pass Temperature (°C)
Root	20	65	9.11	31.6	2.47	243
2	24	123	3.33	86.5	2.05	310
3	25	125	2.29	125.8	1.49	329
4	28	124	3.42	84.2	2.47	330
5	26	123	2.19	131.5	1.46	350
6	27	126	1.54	187.0	1.07	381
7	25	123	2.23	129.1	1.43	390
8	24	122	2.06	139.8	1.26	400
9	24	122	2.18	132.1	1.33	421
10	23	124	2.19	131.5	1.30	431
11	25	123	2.16	133.3	1.38	381
12	26	126	2.22	129.7	1.52	390

4.2. Radiography Results

Typical defect can be detected and measured using x-ray radiographic testing (RT) technique such as porosity, slag inclusion, undercut, and surface crack etc[14]. Interpass slag inclusion about 2mm length as shown in Figure 2 was detected after performing a radiographic test. Interpass slag inclusion defect occurs due to lack of cleaning process after multiple pass welding. Darker area shows the lowest density while lighter areas indicate the high density of the material. Slag inclusion also due to the flux from filler metal that trapped into between passes of the welding.

Figure 2. Specimen for scanning electron microscopic observation

4.3. Chemical Composition Analysis

Chemical composition of the as-received sample was performed by using Arc Spark Optical Emission Spectrometer. It was taken after the specimen has been welded. The effect of the chemical composition of the specimen also has been compared with the mild certificate of the material supplied. The standard chemical compositions of S275JR mild steel grade is shown in Table 3. Three number of sparks that applied at the surface with different location which is base metal and weld metal are shown in Table 3.

Table 3. Comparison of the Chemical Composition

Chemical composition (S275JR mild steel grade)		**Percentage (%)**		
		Mild certificate* (base metal)	Experimental (spectrometer) Base metal	Experimental (spectrometer) Weld Metal
1	Iron (Fe)	98.4	97.9	97.5
2	Carbon (C)	0.18	0.22	0.165
3	Manganese (Mn)	1.05	1.4	1.56
4	Copper (Cu)	0.01	0.019	0.023
5	Phosphorus(P)	0.023	0.0034	Less than 0.001
6	Sulphur (S)	0.002	0.0029	0.0047
7	Silicon (Si)	0.29	0.256	0.53
8	Chromium (Cr)	0.01	0.022	0.030
9	Nickel (Ni)	0.01	0.012	0.012
10	Molybdenum (Mo)	0.004	0.013	0.013
11	Aluminium (Al)	0.042	0.095	0.137
12	Cobalt (Co)	0.00	0.0036	0.0023
13	Niobium (Nb)	0.001	0.0095	0.012
14	Vanadium *(V)*	0.001	0.0022	0.011
15	Stannum/Tin (Sn)	0.001	Less than 0.001	Less than 0.001

4.4. Macrostructure and Microstructure Analysis

Examination of the microstructure in various zone of the cross-sectional welding area often specified with taking a photo of micrograph for record purpose. The weld area must be examined to see the penetration of the specimen, but the information obtained is very limited due to the size of the specimen and the limitation of the microscope. The objective the macrostructure observation is to observe the penetration of the weld area and the quality of the weld. Figure 3 show the macrostructure development at different zone. It clearly shows the fusion line between heat affected zone, weld metal and the base metal of the specimen. From the macrostructure analysis, there are 12-layer welding interpass have been observed between the welding joint.

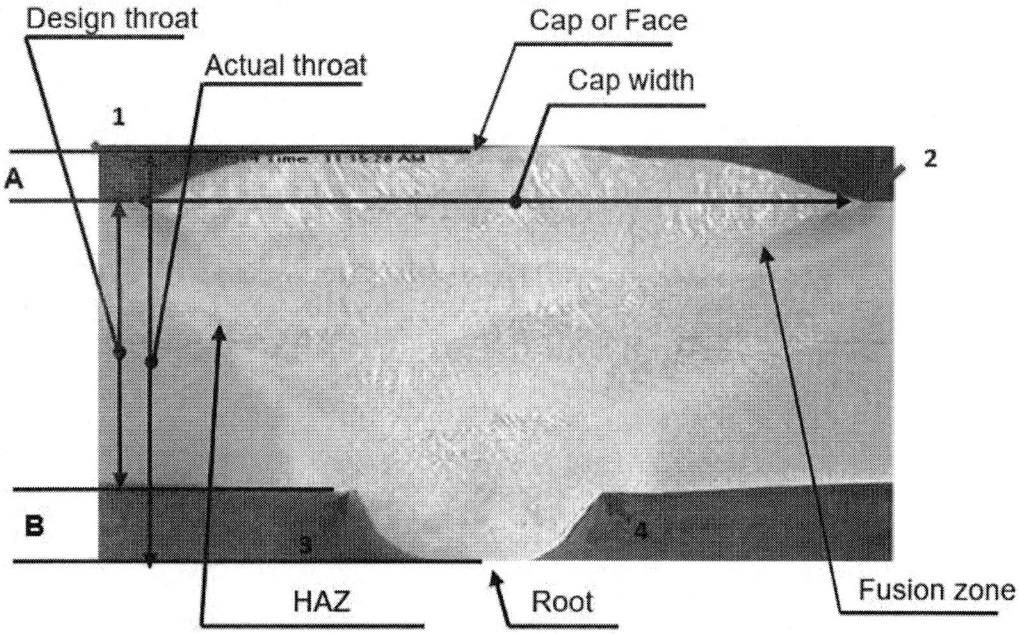

Figure 3. Specimen for scanning electron microscopic observation

Microstructure analysis is necessary to analyse the microstructural changes different region which is base metal, weld metal and heat affected zone. For the microstructure analysis purposes, the welding samples are observed to figure the changes in grain size, microstructure element and shape. It is observed that there are changes in microstructure at base metal and after welding process. The differences or changes in microstructure at base metal, heat affected zone and weld metal shown in Table 4. The images were magnified at 10x and 20x.

From the result of microstructure, there are no discontinuities or defect detected on the weldment area. The multiple-pass welding makes the weld metal change in formation of microstructure. This result will be used as reference in welding terminology and welding quality assessment evaluation. The microstructure observation is taking from 3 different points at the specimens. Table 4 shows the result of the microstructure formation at different location and different magnification of the specimen. From the observation, there are no changes in microstructure in parent metal. The properties of steel depend upon the microstructure. From the result, formation of polygonal ferrite and grain boundary ferrite (A) clearly shown in base metal as illustrated in Table 4. Formation of small region of pearlite (B) also can be seen in the base metal under microstructure observation at 10x magnification and 20x magnifications. Formation of pearlite and ferrite was clearly shown under 20x magnifications. Base metal shows course grain boundary compare to weld metal and heat affected zone. The size of the grain boundary will affect the strength, ductility and the toughness of the material.

IOP Conf. Series: Materials Science and Engineering **834** (2020) 012046 doi:10.1088/1757-899X/834/1/012046

Table 4. Microstructure analysis under different region in welded mild steel

Location	Microstructure of Mild Steel	
	10 x magnification	20 x magnification
Base Metal	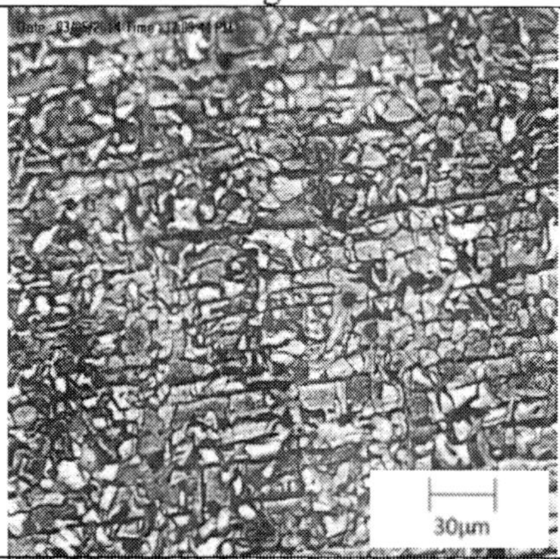	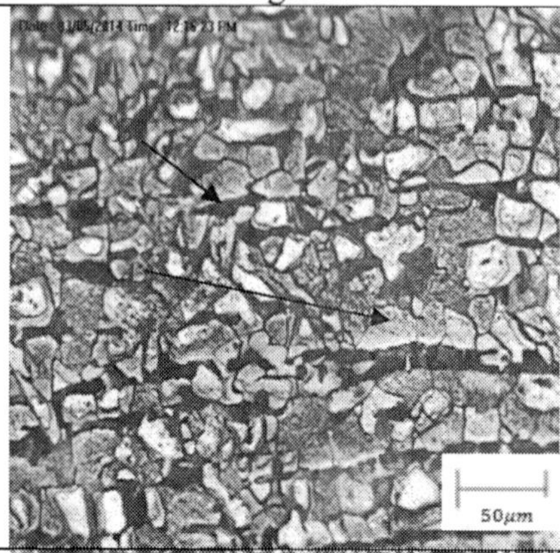
HAZ	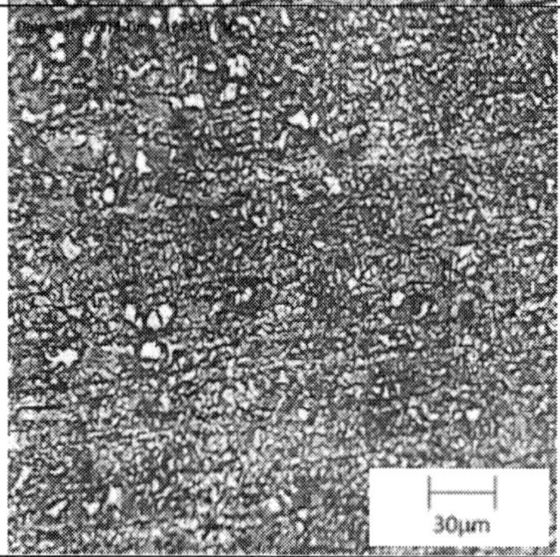	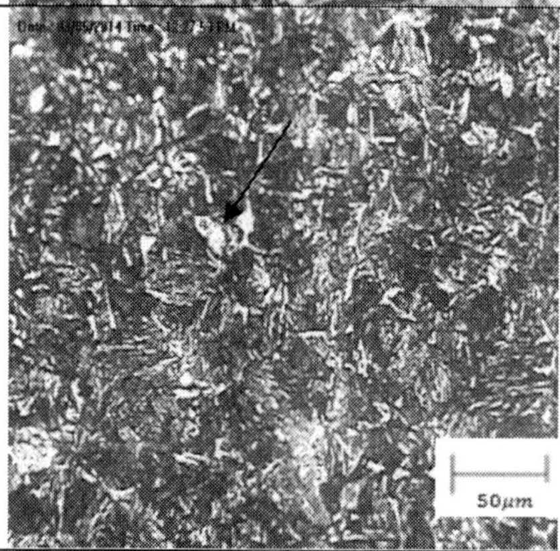
Weld Metal	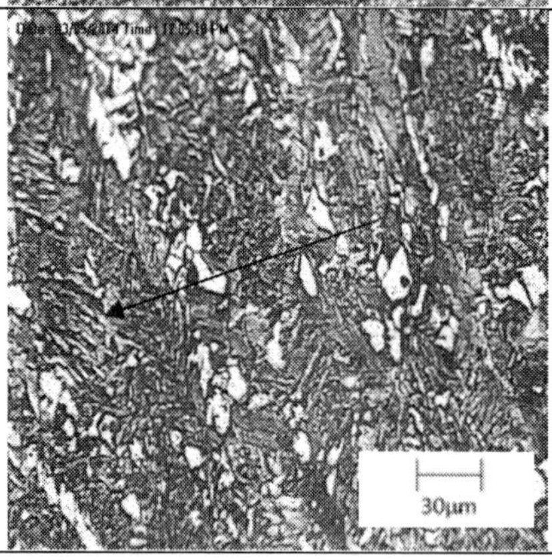	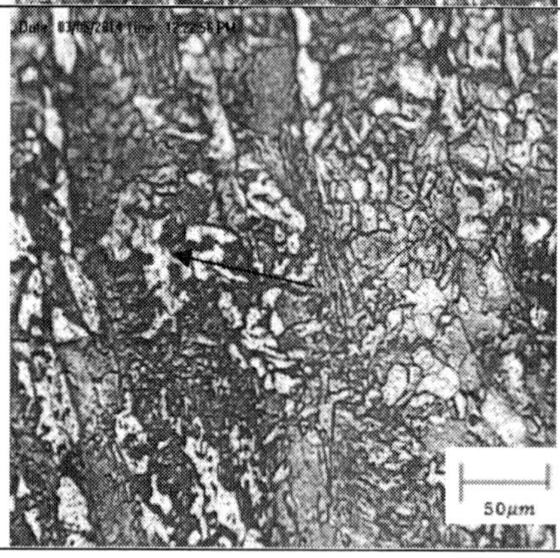

The microstructure in the weld metal is significantly changed due re-melting and solidification process. The metallurgy of the fusion zone of weld metal are totally different between the parent and base metal. Composition of the filler metal such as fine oxide particle was designed to create a good welding quality in term of the strength and toughness by producing a fine grain structure. Formation of acicular ferrite (D) in weld metal is due to solidification rate weld metal region. Based on the result, formation of martensite (D) at the weld metal and the degree of martensite formation is dependent on the temperature. Grain boundary ferrite (E) also appears in weld metal region but it appears in small region. It is due to carbon content between weld metal and base metal.

According to compositional analysis result, weld metal has less carbon content compare to base metal. Grain boundary ferrite and polygonal ferrite was decreased as increasing in carbon content. The manganese also gives big influence on the grain boundary size and polygonal ferrite formation. Increasing in manganese percentage in weld metal makes the grain boundary ferrite become smallest from base metal. Grain boundary ferrite (C) also have been appears on heat affected zone area. Other microstructure element has been identified in heat affected zone region which is grain boundary ferrite, small amount of martensite and bainite. There is no acicular ferrite in HAZ region. Acicular ferrite is important because it provide relative toughness and strong microstructure. This is the reason why the weld metal has high toughness compare to heat affected zone.

In addition, polygonal ferrite in weld metal, heat affected zone and base metal was consider not beneficial to toughness of mild steel. The major microstructure changes are on the grain size of the material. The changes in grain size due to temperature different from welding process.

4.5. Destructive Analysis

Figure 4 show the hardness value taken from different location of the specimen which is base metal, HAZ and weld metal. The hardness value is different at the different location according to the temperature different and chemical composition. Weld metal indicate the highest of the hardness value compare to HAZ and base metal. This was affected by the metallurgy of the weld metal and affected by the carbon content between weld metal and base metal. Theoretically martensite that exist in the weld metal will make the weld metal became harder and more brittle. This analysis was proven by tensile strength test where the fracture occurs at the base metal instead of weld metal.

5. Conclusion

From the observation and experimental performed, it can be concluded that the chemical composition in material have influence the mechanical properties of mild steel with the change in their microstructure during welding process. It effects the welding quality in terms of its strength and required lifespan to suit the joint purposes.

5.1. Chemical composition

From the spectrometry testing, the apparent changes are on the weight percentage of chemical composition of the mild steel before and after welding. The change is between base metal and weld metal. There are no additions of new or loss of the composition element in mild steel, but the change is on increasing and decreasing of the percentage of chemical composition.

5.2. Properties of material

The properties of mild steel are different between base materials and the weld metal. The major different is on the mechanical properties of the mild steel. The effect of mechanical properties is on the strength of the material. Tensile test result was proven that weld metal has high strength compare to base metal, but it depends on the use of correct welding procedure specification. The material undergoes welding process also increase in hardness value and it can be seen in different location of the material after performing welding process which is base metal, heat affected zone (HAZ) and the weld metal [15].

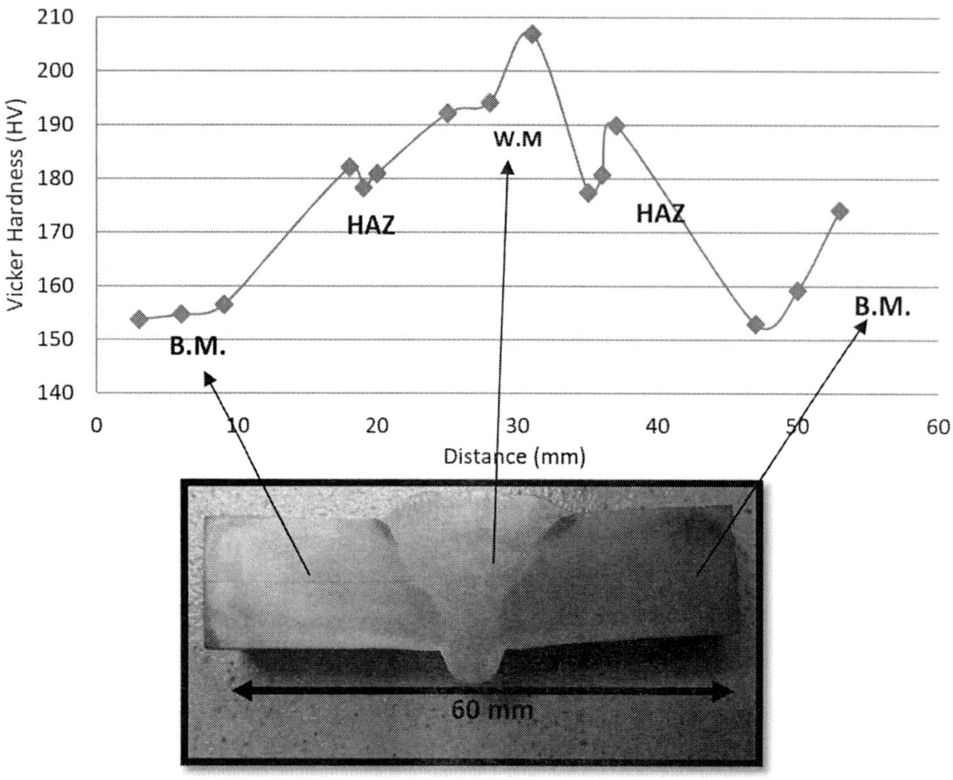

Figure 4. Micro hardness measurement on various regions

5.3. Macrostructure and microstructure analysis
The temperature increases during welding and solidification rate results in reducing of grain size of microstructure. The grain sizes of the structure become smaller and apparent under the microscope. Besides that, the elements of microstructure also change after welding process has been done. The changes are on the microstructure element from formation of ferrite and pearlite to another element such as bainite and martensite. It can be seen in heat affected zone and weld metal region. The microstructures depend on the temperature of welding and on the chemical composition of filler metal used. The high temperature will carry out the enlargement of grain size of the microstructure. The microstructure will influence the mechanical properties of the material.

5.4. Evaluation of welding quality
The quality of the welding can be evaluated by using non-destructive and destructive test methods and by comparison of mechanical properties of material and microstructure analysis. The changes in microstructure have influence in the quality of the welding by decreasing and increasing of the mechanical properties of the material. In addition, welding procedure specification play importance role in the quality of the welding because it gives the specific procedure of welding process from the parameter until the technique used.

Acknowledgments
The sample preparations of the welding are done successfully by following the Welding Procedure Specification (WPS) given. All the result analysis on microstructure and mechanical properties will be use as additional information to the Labuan Shipyard and Engineering company.

References
[1] R. S. Parmar, *Welding Engineering and Technology*. 2015.
[2] D. H. Phillips, *Welding Engineering: An Introduction*. 2015.

[3] A. N. S. Institute, "Structural Welding Code— Steel," *Am. Weld. Soc. D1 Comm. Struct. Weld.*, 2015.

[4] S. K. Sharma and S. Maheshwari, "A review on welding of high strength oil and gas pipeline steels," *Journal of Natural Gas Science and Engineering*. 2017.

[5] S.-H. M., C.-V. L., M. C.-V. J., M. F., J. R.-A. F., G.-V. B., A.-P. M., and D.-H. A., "Microstructure and mechanical properties correlation for the steel: A comparative methodology of educational research for physics and mechanical engineering trainings," *Int. J. Phys. Sci.*, 2017.

[6] R. Timings, *Fabrication and Welding Engineering*. 2014.

[7] S. Gholizadeh, "A review of non-destructive testing methods of composite materials," *Procedia Struct. Integr.*, 2016.

[8] C. Chen, S. P. Chiew, M. S. Zhao, C. K. Lee, and T. C. Fung, "Welding effect on tensile strength of grade S690Q steel butt joint," *J. Constr. Steel Res.*, vol. 153, pp. 153–168, 2019.

[9] I. Arrayago, E. Real, and L. Gardner, "Description of stress-strain curves for stainless steel alloys," *Mater. Des.*, vol. 87, pp. 540–552, 2015.

[10] J. M. Herrera, A. Laso, D. R. Salgado, I. Cambero, A. Marcelo, A. González, J. G. Sanz-Calcedo, and O. López, "Definition of a Procedure for the Comparative Analysis of Different Welding Technologies," *Procedia Eng.*, vol. 132, pp. 709–715, 2015.

[11] S. Ragu Nathan, V. Balasubramanian, S. Malarvizhi, and A. G. Rao, "Effect of welding processes on mechanical and microstructural characteristics of high strength low alloy naval grade steel joints," *Def. Technol.*, 2015.

[12] P. H. C. P. da Cunha, G. V. B. Lemos, L. Bergmann, A. Reguly, J. F. dos Santos, R. R. Marinho, and M. T. P. Paes, "Effect of welding speed on friction stir welds of GL E36 shipbuilding steel," *J. Mater. Res. Technol.*, no. x x, pp. 1–11, 2018.

[13] K. Sirin, S. Y. Sirin, and E. Kaluc, "Influence of the interpass temperature on t8/5and the mechanical properties of submerged arc welded pipe," *J. Mater. Process. Technol.*, vol. 238, pp. 152–159, 2016.

[14] N. Ghosh, P. K. Pal, and G. Nandi, "Parametric optimization of mig welding on 316L austenitic stainless steel by taguchi method," *Arch. Mater. Sci. Eng.*, vol. 79, no. 1, pp. 27–36, 2016.

[15] A. Moarrefzadeh, "Study of Heat Affected Zone (HAZ) in Resistance Welding Process," *J. Mech. Eng.*, 2012.

Thermal Conductivity and Permeability Studies of Recycle Copper Wire Mesh

Z. Michael[1], M. S. Martius[2], M. Khairulmaini[1], M. H. Ismail[2], M. N. Halmy[1], T. M. Jahi[2]

[1]Faculty of Mechanical Engineering, UiTM Cawangan Johor, Kampus Pasir Gudang, Jalan Purnama, 81750 Masai, Johor.
[2]Faculty of Mechanical Engineering, UiTM, 40450 Shah Alam, Selangor

zenomichael@uitm.edu.my

Abstract. Copper is well known as a valuable material, particularly for electrical industries due to its excellent properties. Besides, the price for copper scrap is also higher in comparison with other metal scrap materials found in many electrical appliances such as computer equipment, electric motor, transformer, and other components. This paper describes the potential of recycled copper wire shaped into a form suitable for heat sink application. The samples of wire mesh for this research were fabricated by compacting copper wire with different weight. Compaction was done to form a cylindrical shape with a diameter and height of approximately 25 mm and 30 mm, respectively. These samples were evaluated for porosity, thermal conductivity, and permeability. The experimental results showed that the sample with the lowest porosity exhibited great promising conductivity. In addition, lowering the amount of copper wire in compaction resulted in low density, thus improved its lightweight properties. The greater amount of porosity resulted in greater permeability. The optimum properties with a combination of the excellent thermal conductivity and permeability are necessary for heat sink application to ensure that the parts work effectively.

1. Introduction

Recently, electrical device has a huge usage in the world of industry. The production of electrical devices has been found mainly in electric and electronic devices, and one of the applications is in computer production. In this new era, computer or laptop is mainly used in daily life activity, as without computer most activities such as businesses and documentation cannot be done efficiently. Besides, a rapidly developing country like Malaysia needs computers which can sustain in processing more data with high speed [1].

Most of the computers can operate 24 hours on working day and they operate continuously which is mainly for industrial purpose. This situation may lead to greater thermal density and it increases heat dissipation, due to the increase of CPU temperature. As a result, the life cycle of electrical parts become shortened [2] and this will give adverse effects if the heat is not dissipated efficiently, and it will make the processor inefficient and fail to operate with optimum performance in processing data [1]. Due to that, it is important to maintain the operation of the processor without working above a certain level, otherwise, the reliability of the processor may be compromised [3].

Due to this problem, a component called heat sink basically is installed on a computer board which is used to dissipate heat from the processor itself. The main function of a heat sink is to decrease the temperature by dissipating heat [4]. Besides, the heat sink will work more efficiently with forced air

Content from this work may be used under the terms of the Creative Commons Attribution 3.0 licence. Any further distribution of this work must maintain attribution to the author(s) and the title of the work, journal citation and DOI.
Published under licence by IOP Publishing Ltd

convection, where a fan is used at the heat sink component to ensure that the process the dissipating of heat is more efficient for a processor to work at an ideal temperature condition. Hence, the existing heat sink could extend the life of circuit structures in Computer [5]. Therefore, this paper focused on the study of the recycle heat sink through experimental work on thermal conductivity and permeability of recycle copper wire mesh.

2. Literature Review

Nowadays, most electrical appliances were invented to facilitate human daily life. Generally, an electrical product will generate heat and affect the durability of the product. For many years, there are numerous improvements which have been made to increase the efficiency of a product and work at an ideal temperature condition. Excessive heat will form when there is no component used to dissipate heat or the designation of a heat sink is not working efficiently [6]. Due to that, when it does not have any hardware to dissipate the heat it will increase the temperature surface of a product and this will also affect their functionality.

One of the examples is a desktop computer set which contains a CPU heat sink. Basically, a desktop computer is used for personal or for business purposes. It has various functions and to operate all the system input, it needs a device called as central processing unit (CPU). Usually, these devices are the heart and brain for every computer as without them a computer cannot be switched on and will not function normally. The processors of a computer are used to interpret signals and codes from other devices such as Random Access Memory (RAM) and a keyboard.

2.1. Heat sink

The heat sink is the component attached to the processor that is placed on the motherboard [7]. This component will come together with a cooling fan. Most gaming computers use a gaming heat sink. Therefore, the heat sink was developed to control the temperature of the processor and to ensure that the components work normally to become an ideal temperature when the excessive workload is on it.

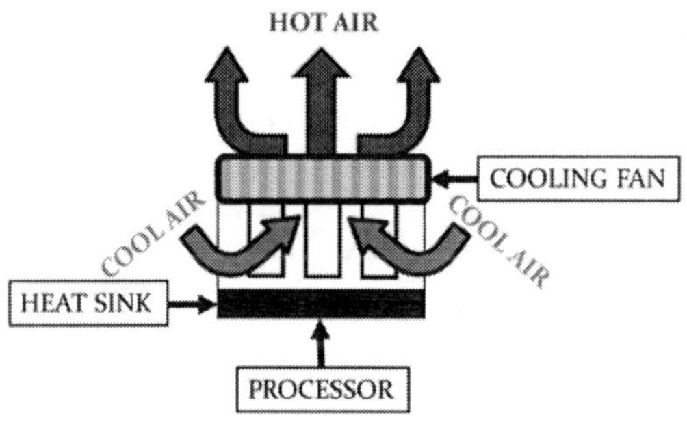

Figure 1. Illustration of heat sink application (Scrap Computer Trading Sdn Bhd)

Heat sink is known as a passive heat exchanger. The function of this component is to transfer the heat that is generated by an electrical, electronic or any mechanical appliances using coolant air in motion towards the ideal temperature. Therefore, it allows the regulation of the device temperature to remain at the physical feasible level. Figure 1 shows the illustration of a heat sink application, and this illustration explains how the heat sink removes the heat using forced convection or fan that is used to increase the rate of heat removal. The air from outside is being sucked into by using a fan, which is to carry out the hot air inside the processor. Due to this situation, a processor temperature will become the ideal temperature.

Most of the heat sink design is affected by the air velocity flow, the material used, design and surface treatment. The standard heat sinks are usually used in the CPU. This component comes together with a cooling fan which is used to increase the cooling rate with the air flow towards a heat sink. Usually, the standard cooling fan can achieve a speed between in range 8 000 RPM to 12 000 RPM. This speed can control the temperature of a processor that produces heat.

In addition, to improve the purpose of heat sink application, it needs some additional attachments, namely thermal interface materials. This interface material will make the temperature of integrated circuit decreases and remains in ideal temperature. The function of thermal adhesive or thermal grease is to ensure that there is no air gap between the heat sink and integrated circuits. Therefore, this can increase the performance of the heat sink's itself and the heat will be dissipating efficiently.

Thermal paste is usually placed on the top of the processor. This is important before a heat sink is attached on top of the processor. Processor reliability will decrease if a thermal paste is not placed before attaching a heat sink because heat cannot be transferred effectively towards a heat sink. This clearly shows the importance of thermal paste.

2.2. The significance of heat sink towards a Central Processing Unit (CPU)

Generally, a processor is installed together with a heat sink to dissipate heat away from the processor. This is crucial to increase and improve the reliability and prevent the component from forming premature breakdown [8]. Due to that, it can become the best heat sink that is able to control the temperature. A processor generally operates within a temperature between 70°C to 95°C [9], which is basically the range of temperature which is the normal result after applying a heat sink attached with a cooling fan.

Besides, a standard processor usually uses power around 80 W to 130 W. This range is used to ensure that a processor is able to operate without facing any problem [10], but in gaming specification, a manufacturer has developed the best specification of the processor to ensure that the processor operates in optimum condition.

The previous studies found that some parameters will affect the performance of a heat sink such as the shape of the fin, the base materials and the materials for the fin used [8]. The material used should have high thermal conductivity value and able to transfer the heat from one point to another point efficiently. As a result, it can reduce the failure of a mechanism and increase the reliability of a product.

3. Material Selection

Copper has been used in electrical component before the innovation of the electromagnet and telegraph in the 1820s [11]. Copper is commonly applied in electrical wiring and only react with atmospheric oxygen to form a layer of brown-black copper oxide, but it does not react with water and as a result, reaction with oxide had formed unlike rust occurs when the iron is exposed towards the air and protects the underlying copper from becoming extensive corrosion. Copper is the preferred and the predominant choice in the electrical industry because their advantage is higher in thermal conductivity. The new creation of the telephone in 1876 developed more copper wire used in an electrical conductor. Generally, copper mine is used to fabricate an electrical wire and a cable as conductors. Therefore, electrical wiring whether in the building or at home is the most vital use to increase the market in the copper industry.

3.1. Physical Properties of copper

Copper, silver, and gold consist of group 11 in the periodic table and they contribute certain attributes: on top of a filled d-electron shell they have one s-orbital electron and it is characterized by high ductility and electrical conductivity. It can simply be explained that the purity of copper has a good thermal conductivity which is 388 W/m.K [12]. Due to that, the resistivity of electrons to carry in metal at room temperature commonly begun from irregular of electrons on thermal vibrations of the lattice. This value should be made for the reference value for further experiment.

3.2. Electrical conductivity of copper

Kelvin meters per watt ($K{\cdot}m{\cdot}W^{-1}$) is the unit for thermal resistivity and the result of the reciprocal of the thermal conductivity. Electrical charge is formed in electrical conductivity flow process and this is important to determine the quality of the material used. Copper is categorized as non-precious metal and has the highest electrical conductivity rating. The electrical resistivity of copper is 1.72 Ω.m at room temperature [13]. There is a difference between the definition of the construction's thermal resistance and the reciprocal property and thermal conductance.

The electrical application mainly used electrolytic-tough pitch (ETP) copper (CW004A or ASTM designation C11040). This type of copper known as 99.90% pure and has an electrical conductivity of at least 101% IACS and contains a little percentage of oxygen (0.02 to 0.04%) [13].

These elements have a little resistance of conduction electrons flow under an electric field and exhibit high conductivity. In addition, copper has a large outermost electron and become an excellent conductor; about 100 atomic spacing between collisions.

3.3. Selection of Recycled Copper Wire

Nowadays, the largest growing waste stream is currently electrical and electronic waste which is normally known as e-waste. This situation occurs partly due to lack of legislation or enforcement. At present, there are proper ways for the e-waste to be discarded such as the general waste stream. Most of the developing countries like Bangladesh and Indonesia have an informal economy and lead them to recycle and reuse the materials by applying the rudimentary technique. These are the consequences to the production of wire rod and wire cable product. According to the world consumption of metals, copper metals is situated in the third rank after iron and aluminum.

In Malaysia, in August of 2015, scrap imported from Malaysia alone totalled to 14, 733 tones. According to the record, the largest supplier of the scrap material to Malaysia was the US in August of 2015 which was totalled at 5,609 tons. The second largest import of the scrap metal which is Singapore with hit total 4,326 tons. Lastly, Australia is in the third place with 1919 tons [14], [15]. Due to the increasing of scraps of copper wire, the heat sinks were developed by Computer Trading Sdn. Bhd.

4. Methodology

The flow chart in Figure 2 presents the preparation flow of the specimen until it is tested. It also illustrates from the scratch where the material is collected and fabricated.

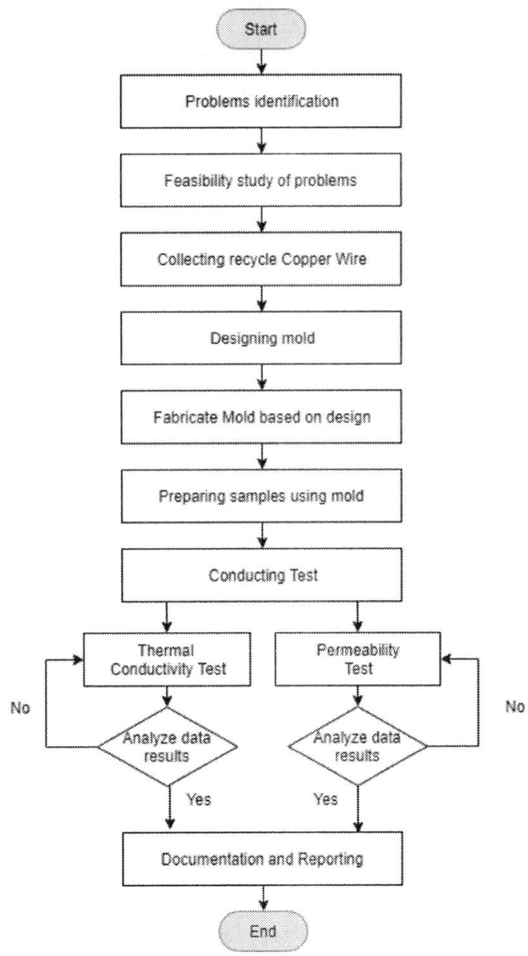

Figure 2. Flow chart of the specimen preparation

4.1. Collecting Recycled Copper Wire

The scrap of copper wire was collected from Scrap Computer Trading Sdn. Bhd. The company's main business is buying and selling of any kinds of recycled material such as copper, brass, aluminum, and steel.

The recycled copper wires were collected and selected based on their grade and quality. There are various types of recycled copper wire such as enamelled copper wire and standard copper wire.

4.2. Fabrication of mold

The mold is fabricated with the diameter of 25 mm and 30 mm height. There are several designs specification that need to be considered for making a mold.

Furthermore, the mold will be used to form a specimen through a compacting process. The complete shape of the mold is fabricated based on the proposed design shown in Figure 3. This mold were designed through several processes using lathe machine and milling machine.

Figure 3: Designated mold to produce sample

4.3. Fabrication of specimen

The specimen was fabricated by using the designated cylindrical mold shape with the assistance of hydraulic press machine. This machine will compact the recycled copper wire to a specified height required for testing. The compaction process is required to form the samples for the thermal conductivity test and permeability test. These specimens should be fit into the thermal conductivity apparatus. This can avoid error during testing. The recycled copper wire was loaded into a mold and a compression rod is placed on the top part of the middle rod. Furthermore, the load is applied to the mold to form the specimens.

4.4. Density and Porosity

Both density and porosity for each specimen in this study were considered. These factors are important to compare the specimens based on its weight. Equation (1) will be used to calculate the density for each specimen.

$$Density, \rho = \frac{Mass, m}{Volume,\ v} \quad (1)$$

Equation (2) is used to determine the percentage of porosity for the specimens. Each specimen has a different value of porosity due to the different value of mass.

$$Percentage\ of\ Porosity, \%_{porosity} = \left| \frac{\rho_{solid\ copper} - \rho_{specimen}}{\rho_{solid\ copper}} \right| \quad (2)$$

4.5. Thermal conductivity Test

At the early stage, an experiment was done for each specimen by measuring their efficiency as it transferred heat from one place to another. It is followed by the concept of Fourier's law of heat conduction. This concept is defined as the process of heat transfer from a hot region to a cold region. It

IOP Conf. Series: Materials Science and Engineering **834** (2020) 012047 doi:10.1088/1757-899X/834/1/012047

can be assumed that the cylindrical bar is the plane of wall thickness (Δx), temperature difference (ΔT) while uniform hot temperature (T1) is imposed on one end and a different temperature (T2) is imposed on the other. Thermal conductivity test needs to follow the amount of heat transfer during the test transfer which are 30 W, 40 W, 50 W, and 60 W.

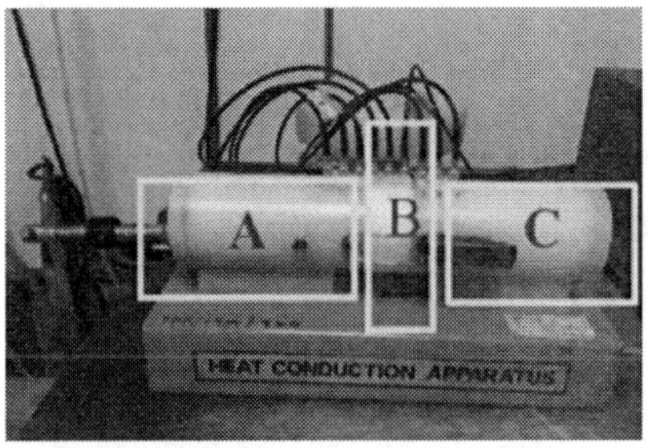

Figure 4. Heat Conduction Apparatus

Figure 4 shows the heat conduction apparatus connected to the power source. the wire connectors were connected into 3 sections which are the heater (A), specimen (B) and cooler (C). Heat flows from the heater section to the cooler section. Each section is connected with 3 thermocouple wires. Each wire is separated at 10 mm in between and the thermocouple provides the readings of temperature based on these distances. The apparatus consists of a power source, temperature display, heat transfer control, and heat transfer display. This section enables the researcher to control the amount of temperature when heat is applied to a specimen. The specimens were tested for 10 minutes for each test.

As a result, for the whole test a specimen needs to be tested 4 times but with different angles which are 0°, 90 °, 180 °, and 270°. The total time to complete a test for one specimen is 40 minutes. Due to that, all the data need to be recorded, and the average value is taken. The thermal conductivity was also tested.

The equation (3) used in this experiment is known as Fourier's Law of heat conduction, and the directional flow of heat in the axial direction.

$$\dot{Q} = -KA\frac{\partial T}{\partial n} \tag{3}$$

Where (Q̇) is the rate of heat conduction in the axial direction, K is the thermal conductivity to measure the ability of a material to conduct heat, $\partial T/\partial n$ is the temperature gradient and A is known as the cross-sectional area of the specimen. While a negative sign indicates the heat was transferred towards x direction and the temperature flow in decrease situation [16]. Generally, Fourier's Law comes out with a vector relationship which affects towards all directions of heat conduction and it is expressed in terms of its components as Equation (4):

$$\overline{\dot{Q}_{\vec{n}}} = \dot{Q}_x\vec{i} + \dot{Q}_y\vec{j} + \dot{Q}_z\vec{k} \tag{4}$$

Where $\vec{i}$, $\vec{j}$ and $\vec{k}$ known as a unit vector while $\dot{Q}_x$, $\dot{Q}_y$ and $\dot{Q}_z$ are the magnitude of heat conduction rates in the x-, y-, and z- direction which again to determine Fourier's Law as:

$$\dot{Q}_x = -KA_x\frac{\partial T}{\partial x} \tag{5}$$

$$\dot{Q}_y = -KA_y\frac{\partial T}{\partial y} \tag{6}$$

$$\dot{Q}_z = -KA_z\frac{\partial T}{\partial z} \tag{7}$$

For the experimental test, Equation (5), (6) and (7) are referred. To determine the value of thermal conductivity, K, these equations will be rearranged to form Equation (8). The thermal conductivity, K of the solid cylindrical copper value will be compared with the actual values from the other references.

$$K = -\frac{\dot{Q}}{A}\frac{\partial x}{\partial T} \tag{8}$$

The experimental errors value for the test can be calculated by using Equation (9).

$$Percentage\ error, \%_{error} = \left|\frac{k_{Theoritical\ value} - k_{Experimental\ value}}{k_{Theoritical\ value}}\right| \tag{9}$$

4.5.1. Figure captions.
Permeability was conducted to determine the efficiency of the specimen. To obtain the permeability of the specimens, water is used as medium and tested with different temperature which is 34°C, 40°C, 60°C, and 80°C. The schematic diagram of the permeability test setup is shown in Figure 5.

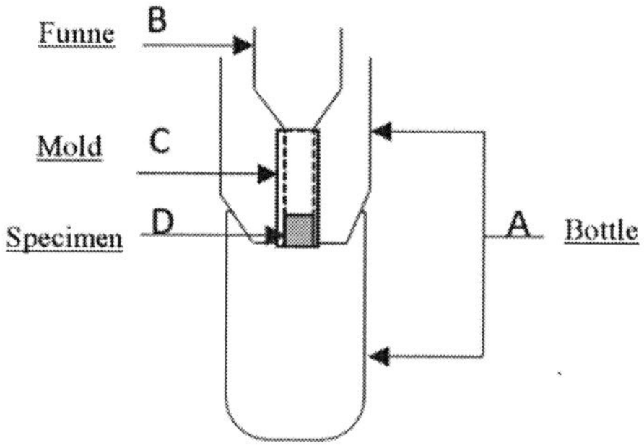

Figure 5. Permeability Test Setup

The effect of different temperature was used for identifying the correlation between the porosity and permeability of different temperature.

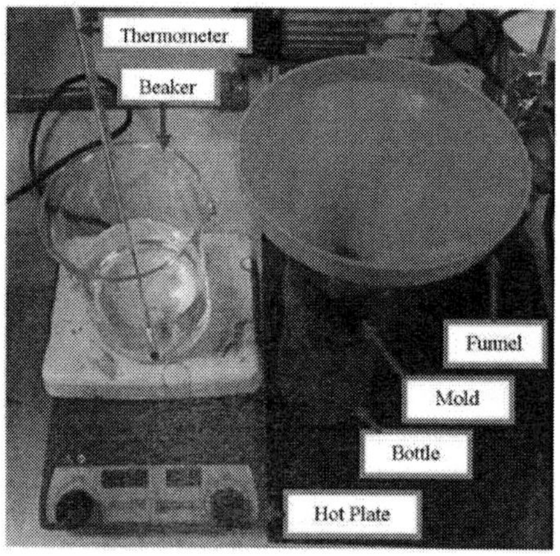

Figure 6. Permeability Test Apparatus

Figure 6 shows the apparatus used in the permeability test. A hot plate was used to undergo the heating process towards water from 34°C, 40°C, 60°C, and 80°C. This is crucial to get a varied reading for each specimen when in different temperature condition. This condition is needed to observe any effect occurs during the test. The value of the permeability test can be calculated using Equation (10). During the experiment, the amount of water and the time taken during the experiment must be considered and this is important as both criteria could give a good result of permeability.

$$Permeability = \left| \frac{Volume, V(liter)}{time, t_{final}(s) - time, t_{initial}(s)} \right| \quad (10)$$

5. Results and Discussions

In this section, the fabrication process, thermal conductivity test, and permeability test are discussed in detail.

5.1. Recycle copper wire mesh compaction

The specimens of the recycled copper wire mesh (as shown in Figure 7) after being compacted using a hydraulic press machine and different levels of porosity using different amount of copper wire can be observed. The specimens were evaluated to determine the value of thermal conductivity and permeability. This is crucial because the porosity adversely affects the result on thermal conductivity test and permeability test [17].

Figure 7. Compacted Recycle Copper Wire Mesh

The top view and side view of a specimen after the compaction process are shown in Figure 8. The cylindrical shape is selected because the equipment used in this experiment is cylindrical shape condition.

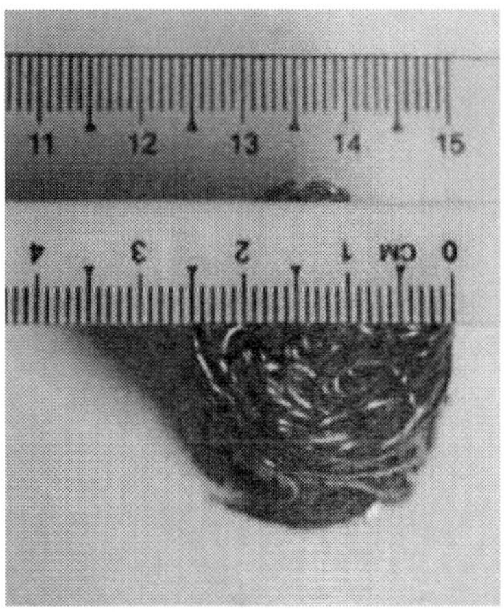

Figure 8. (a) Top view of specimen

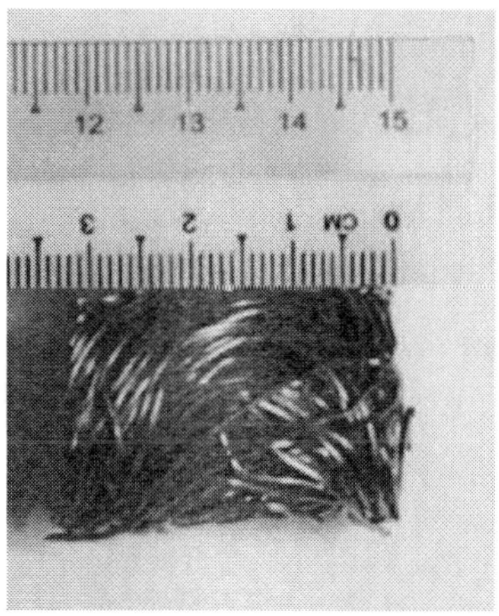

Figure 8. (b) Side view of specimen

5.2. Density and Porosity

Density is important in the heat sink application because it adversely affects the efficiency of the material in thermal conductivity activity [17]. Table 1 shows the density and percentage of porosity of compacted recycle copper wire based on the different weight. The density and percentage of porosity errors for the specimens could be obtained by using Equation (1) and Equation (2), respectively.

Table 1: The value of density (g/cm3) and porosity errors (%) with different weight.

Weight (g)	Density (g/cm^3)	Porosity (%)
30	2.04	76.37
40	2.72	68.49
50	3.40	60.61
60	4.07	52.74
127 (Solid Cylindrical Shape)	8.62	0.00

Basically, the highest weight of a solid cylindrical shape of copper wire obtained the highest density of 8.62 g/cm^3 with zero porosity and achieved the best value of thermal conductivity among the specimen.

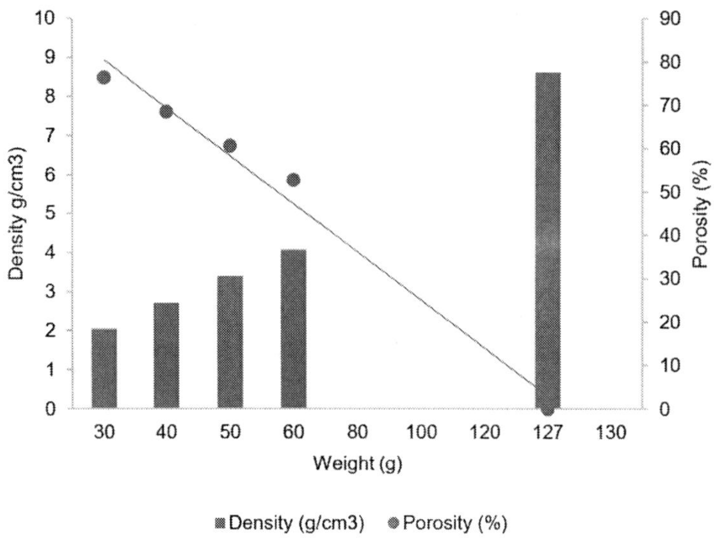

Figure 9. Weight (g) against Density (g/cm3) and Porosity (%)

The density and porosity were measured at various weights as shown in Figure 9. The porosity of specimens decreased as the weight increased, which resulted in the increase in density. This has been reported in the previous study, where the density of copper is affected with porosity [18]. Due to that, the value of thermal conductivity becomes more efficient when the value of density increases and this situation can be clearly seen where the value of thermal conductivity error decreases when the weight of specimen increases. Both criteria; density and porosity are important in heat sink application, where the lifespan of the processor maybe shortened and could not dissipate heat effectively.

5.3. Thermal Conductivity Test

The four sections of 0°, 90°, 180° and 270° angle as shown in Figure 10 were taken for the thermal conductivity test. The data is taken repeatedly with four-times for an average concept [19]. This is important due to the different shape of specimens, and different angle which resulted in different porosity. Figure 11 shows that the porosity of the same specimen has different porosity.

Each specimen requires 40-minutes to be completed in one section. Then, the apparatus needs to let the specimen reduce their temperature until it reaches room temperature.

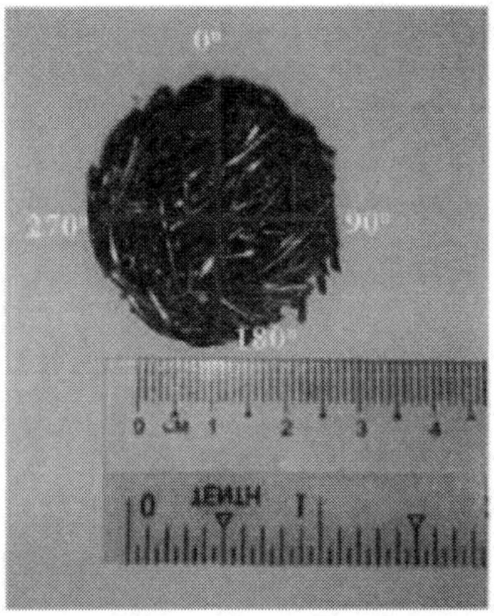

Figure 10. Four sections of angle

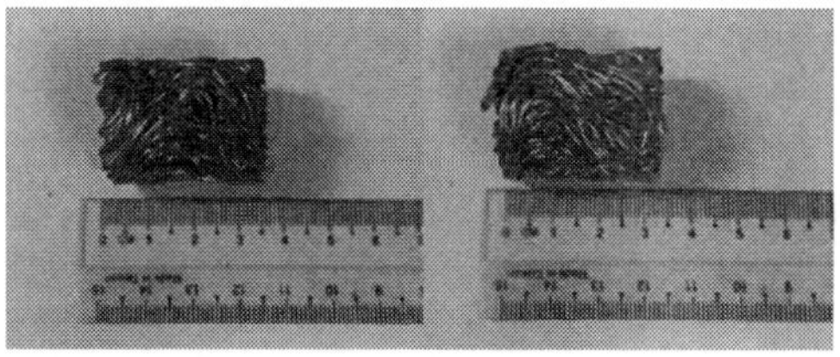

Figure 11. Similar specimen with different views

The collected data were presented in graphs as shown in Figure 12, Figure 13, Figure 14, Figure 15 and Figure 16. These graphs are used to indicate the trends of the temperature for each specimen.

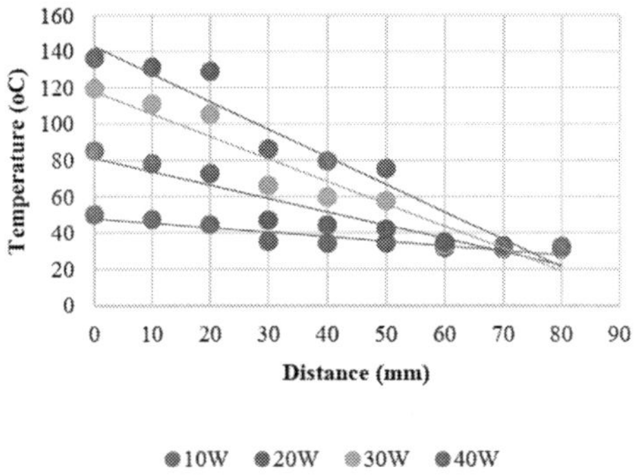

Figure 12. Distance (mm) against Temperature (°C) [Weight 30 g]

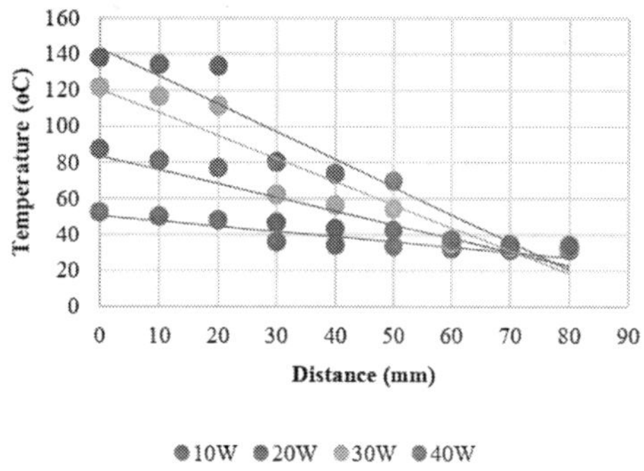

Figure 13. Distance (mm) against Temperature (°C) [Weight 40 g]

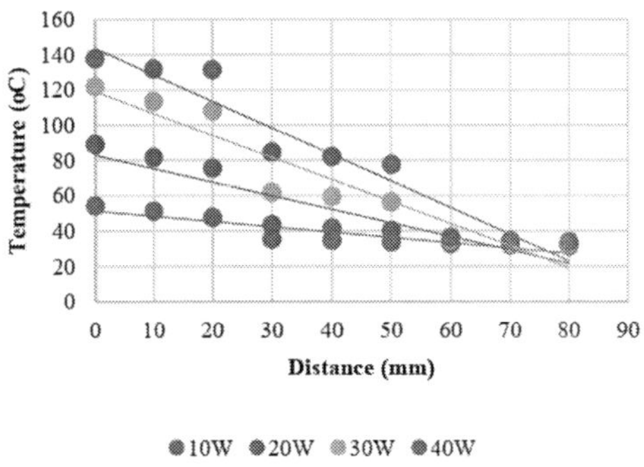

Figure 14. Distance (mm) against Temperature (°C) [Weight 50 g]

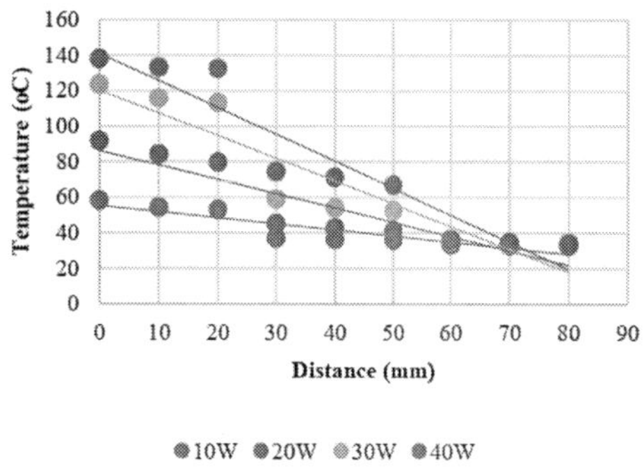

Figure 15. Distance (mm) against Temperature (°C) [Weight 60 g]

Based on these graphs, the evaluation of thermal conductivity is not counted in the heater section (from 0 to 20 mm) and cooler section (from 60 to 80 mm). Both sections do not require a calculation of thermal conductivity. It can be seen that the graphs patterns were quite the same for the whole specimen but different in temperature value. This situation occurred because of the different amount of porosity. When the amount of porosity is greater, the efficiency of heat transferred from one section to another section decreases [20].

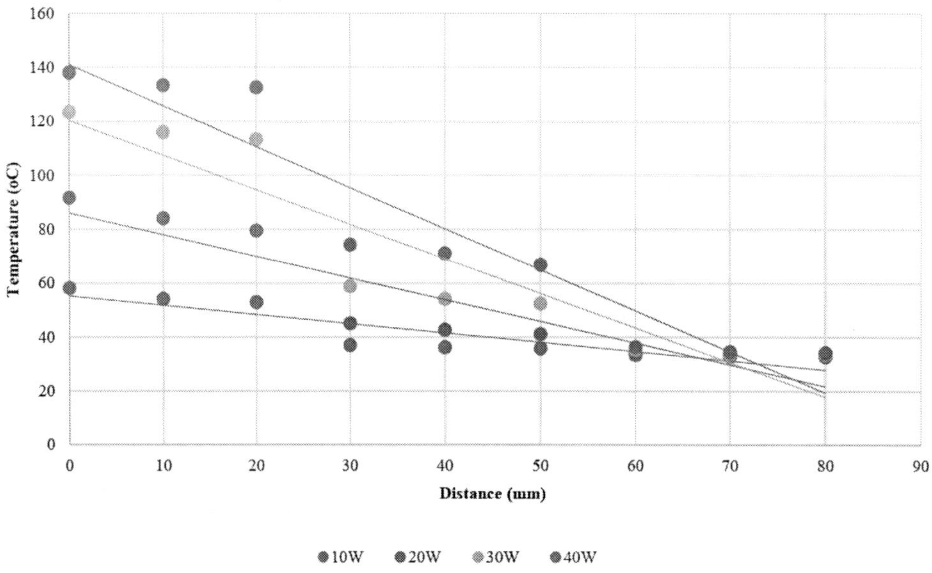

Figure 16. Distance (mm) against Temperature (°C) [Solid Copper Cylinder]

Table 2 below shows the thermal conductivity, k of the specimen.

Table 2: Thermal Conductivity Test Value (W/m.K)

Heat \ Weight	30 g	40 g	50 g	60 g	Solid Copper Cylinder
10 W	698.43	244.45	479.31	407.41	381.95
20 W	240.84	275.13	315.42	362.15	381.95
30W	218.91	233.55	291.01	349.21	392.16
40 W	255.30	235.05	324.85	350.46	384.96

The heat transfer efficiency of specimens was compared with a solid copper cylindrical shape and used as the benchmark for the other specimens.

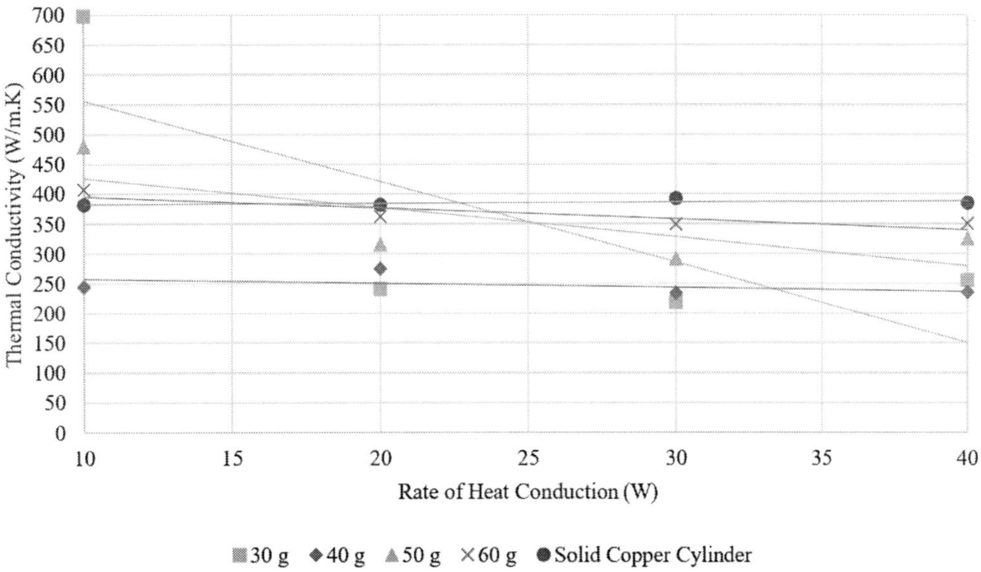

Figure 17. Rate of Heat Conduction (W) against Thermal Conductivity (W/m.K)

Figure 17 shows the linear graph of thermal conductivity and the linear graph of weight 60 g are around the value similar with the value of theoretical (388 W/m.K) which is between 350.46 W/m.K to 407.41 W/m.K. This situation occurred because the efficiency of the specimen to transfer the heat is high while the amount of porosity is smaller. Due to that, other specimens are not recommended for the thermal conductivity purposes especially for a specimen which weights 30 g. This is due to the highest error calculated. The error usually happened because of the different amount of porosity.

Table 3: Thermal Conductivity Test Error, (%)

Heat \ Weight	30 g	40 g	50 g	60 g	Solid Copper Cylinder
10 W	80.01	37.00	23.53	5.00	1.56
20 W	37.93	29.61	18.71	6.60	1.56
30 W	43.58	39.81	25.00	10.00	1.07
40 W	41.93	39.42	16.28	9.67	0.78

Table 3 shows the thermal conductivity test error. The theoretical value for thermal conductivity is 388 W/m.K, while the calculated experimental value for copper solid cylindrical shape is between 381.95 W/m.K to 384.96 W/m.K. This shows that the value of error of these specimens is about 0.78 % to 1.56 %, respectively. This error happened because of the lack of contact between the specimen with hotter section and cooler section. This situation can be clearly seen as according to the graph, it shows at the temperature distance between 20 mm to 30 mm, it fluctuated and dropping especially when a heat transfer achieves at 40 W. Due to the poor contact of specimen, thermal paste is required to attach them and to have a good value for heat transfer.

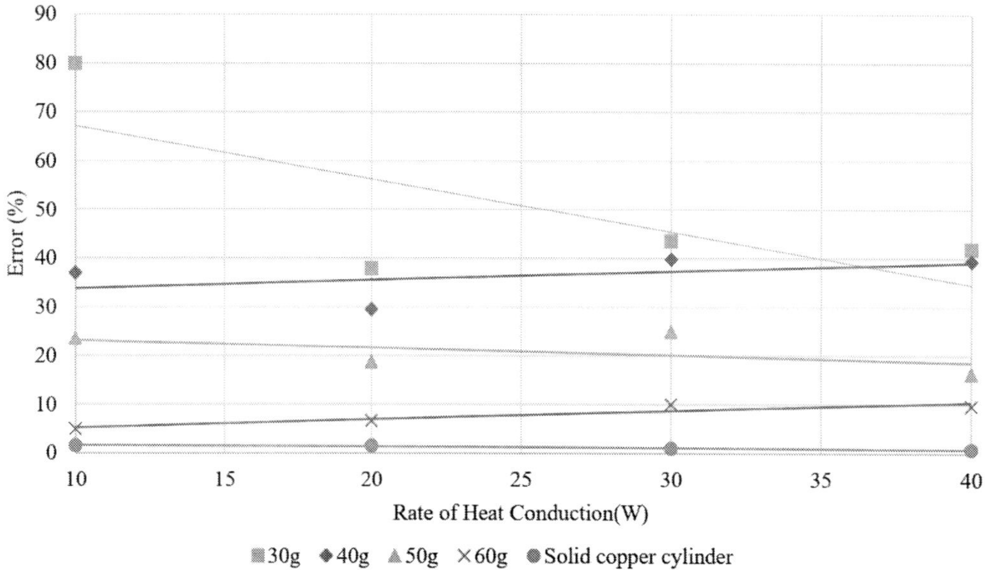

Figure 18. Rate of heat conduction(w) against error (%)

Figure 18 shows the linear graphs of error. A specimen of 60 g was the best among the others specimen, as the error values in between 5.00% to 10.00%. This indicated that 60 g is the lowest value of the error. This situation occurred due to the lowest amount of porosity itself. In addition, a porosity that in the specimen gives an effect to the thermal conductivity value [21] and when the percentage error of a specimen increases, the value of thermal conductivity increases.

Meanwhile, the highest value of error was found in 30 g of weight. This situation happened due to the formation of porosity in a specimen. A porosity gives bad effect toward heat transfer activity [17]. As a result, this type of weight is not suitable to be applied in thermal conductivity activity because the specimen is not able to transfer heat from one section to another section effectively due to the formation of pores.

5.4. Permeability Test

Permeability is the capability of substances to past through the obstacles which is in pore condition as an example such as soil, sponge, and fabric. This research had been made to determine the permeability of the specimen and the effect that occurred towards heat transfer [22]. As a result, it can determine the effectiveness of pore on the specimen and to prove the specimens which have a large amount of porosity and good in permeability. Due to the deformation porosity on the specimen, the value of permeability and the flow of the fluid is affected during the water past through the specimen. Table 4 shows the complete data of the permeability test for the whole specimen with different temperature condition.

Table 4: Permeability Test data, (l/min)

Weight \ Temperature	34 ^{0}C	40 ^{0}C	60 ^{0}C	80 ^{0}C
30 g	2.37	2.70	2.95	3.34
30 g	2.16	2.33	2.71	3.04
40 g	2.00	2.24	2.45	2.70
60 g	1.86	2.07	2.26	2.47

The medium used for the permeability test is water, and it was tested with different temperatures, which are 34°C, 40°C, 60°C, and 80°C. Figure 19 shows the permeability test results. When temperature increases the value of the permeability is increased. This situation happened because of the increase of water temperature, and the molecular kinetic energy of the water also will be increased. Due to that, it will affect the velocity of the molecule and the molecule easily moved between the pores that have deformed in the specimen.

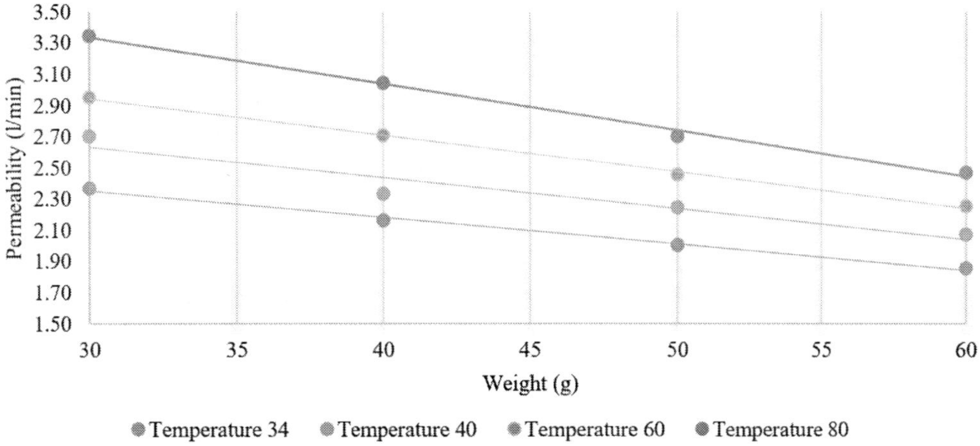

Figure 19. Weight (g) against Permeability (l/min)

Based on the results, the specimen with 30 g shows that the permeability is higher. This is caused by the amount of porosity that is higher when compared to the others. In addition, high porosity also leads to poor thermal conductivity.

5.5. Correlation Between Porosity, Thermal Conductivity, and Permeability

The efficiency in heat conduction, influence of porosity on the fluid flow rate and thermal conductivity of the specimen are illustrated in Figure 20. The graph shows that the increase of porosity could affect the fluid flow rate and thermal conductivity.

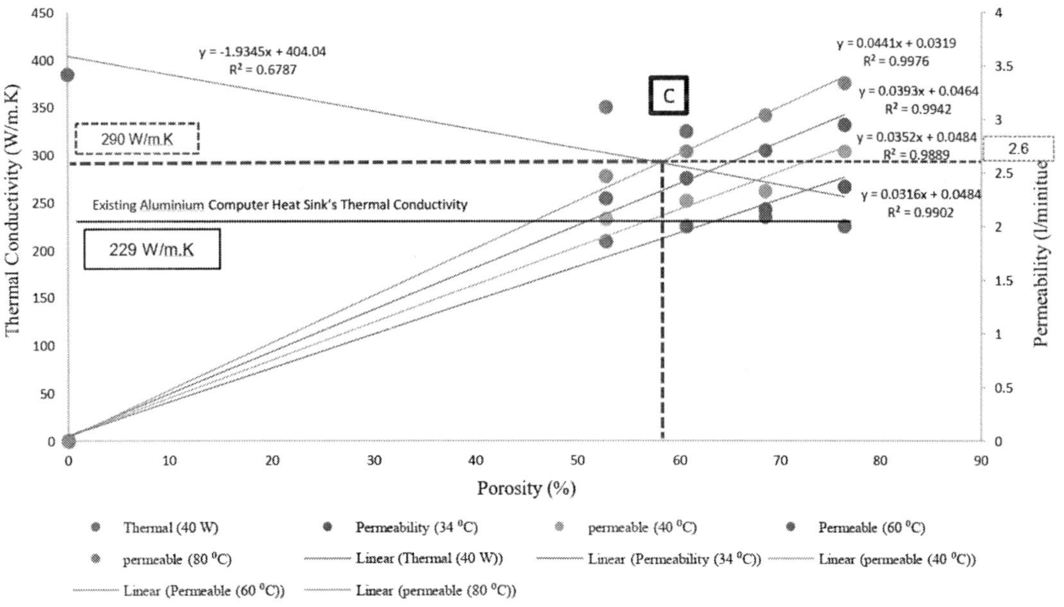

Figure 20. Porosity (%) against Thermal Conductivity (W/m.K) and Permeability (l/min)

The linear line drawn on this graph obtained several intersection points with the thermal conductivity and permeability. It is anticipated that the intersection at point C will exhibit porosity properties for heat sink application, thermal conductivity and its greater than the currently used aluminium heat sink. Higher permeability is necessary to ensure that the heat can be dissipated faster from the hot component.

6. Conclusion

A parametric study is experimentally conducted to evaluate the porosity and thermal conductivity characteristic of recycled copper wire. The effects of density with different weight of specimen are investigated based on the experimental results. The conclusion of this study can be summarized as follow:

- The best specimens have been identified and meet the standards and criteria.
- Copper wire mesh with different weight of porosities has been successfully fabricated by employing the compaction technique. The main goal is to study the thermal conductivity and permeability of recycled copper wire mesh. Besides, it indirectly gives some advantage to increase recycling activity.
- The porosity adversely affects the permeability and thermal conductivity.

Finally, the objective of this study has been achieved by determining the value of thermal conductivity and permeability of the copper wire mesh. The findings would be useful to assist researchers in recognizing which specimen has the least error in thermal conductivity.

Acknowledgement

The authors of this paper would like to express their sincere gratitude to the Scrap Computer Trading Sdn. Bhd and Faculty of Mechanical Engineering, UiTM Shah Alam for providing the raw materials and facilities for this research.

References

[1] T. Abbas, K. M. Abd-Elsalam, and K. H. Khodairy, "CPU thermal management of personal and notebook computer (Transient study)," in *2010 3rd International Conference on Thermal Issues in Emerging Technologies, Theory and Applications - Proceedings, ThETA3 2010*, 2010.

[2] S. Rittidech, A. Boonyaem, and P. Tipnet, "CPU Cooling of Desktop PC by Closed-end Oscillating Heat-pipe (CEOHP)," *Am. J. Appl. Sci.*, 2005.

[3] S. Belhardj, S. Mimouni, A. Saidane, and M. Benzohra, "Using microchannels to cool microprocessors: A transmission-line-matrix study," *Microelectronics J.*, 2003.

[4] M. Saini and R. L. Webb, "Heat rejection limits of air cooled plane fin heat sinks for computer cooling," *Intersoc. Conf. Therm. Thermomechanical Phenom. Electron. Syst. ITHERM*, 2002.

[5] V. G. Pastukhov and Y. F. Maydanik, "Low-noise cooling system for PC on the base of loop heat pipes," *Appl. Therm. Eng.*, 2007.

[6] F.-S. Jan and C.-T. Kwan, "An analysis of the extrusion process of the straight-fin heat sink in a computer," *Int. J. Adv. Manuf. Technol.*, vol. 26, no. 9–10, pp. 978–982, 2005.

[7] J. W. Klett and M. Trammell, "Addendum: Modular heat sinks for desktop computers and other electronics," *IEEE Trans. Device Mater. Reliab.*, vol. 4, no. 4, pp. 638–640, 2004.

[8] E. Ozturk and I. Tari, "Forced air cooling of CPUs with heat sinks: A numerical study," *IEEE Trans. Components Packag. Technol.*, 2008.

[9] B. P. Whelan, R. Kempers, and A. J. Robinson, "A liquid-based system for CPU cooling implementing a jet array impingement waterblock and a tube array remote heat exchanger," *Appl. Therm. Eng.*, 2012.

[10] C. M. Feroz and A. I. Uddin, "Cpu Cooling of Desktop Computer By Parallel," *J. Mech. Eng.*, vol. Vol. ME 40, no. No. 2, pp. 104–109, 2009.

[11] W. Applebaum, :"Groundbreaking Scientific Experiments, Inventions, and Discoveries of the Seventeenth Century," *Isis*. 2003.

[12] J. T. Al-Haidary, E. S. Al-Hassani, and S. R. Areef, "Corrosion behavior of copper-steel particulate composite," *Mod. Appl. Sci.*, vol. 5, no. 3, pp. 234–241, 2011.

[13] R. N. Wright, *Wire Technology: Process Engineering and Metallurgy: Second Edition*. 2016.
[14] Anil Mathews, "Malaysian steel scrap imports hit lowest level in 6 months," *Metal Recycling News*, 2017. [Online]. Available: https://www.scrapmonster.com/news/malaysian-steel-scrap-imports-hit-lowest-level-in-6-months/1/64432. [Accessed: 30-Jan-2017].
[15] J. L. Jolly, *The U.S. Copper-base Scrap Industry And Its By-products*. 2000.
[16] O. A. Ezekoye, "Conduction of heat in solids," in *SFPE Handbook of Fire Protection Engineering, Fifth Edition*, 2016.
[17] C. Vincent, J. F. Silvain, J. M. Heintz, and N. Chandra, "Effect of porosity on the thermal conductivity of copper processed by powder metallurgy," *J. Phys. Chem. Solids*, 2012.
[18] J. Yang, L. Yang, C. Xu, and X. Du, "Numerical analysis on thermal behavior of solid-liquid phase change within copper foam with varying porosity," *Int. J. Heat Mass Transf.*, 2015.
[19] S. W. Raudenbush, A. Martinez, and J. Spybrook, "Strategies for Improving Precision in Group-Randomized Experiments," *Educ. Eval. Policy Anal.*, 2007.
[20] A. A. Mehrizi, M. Farhadi, K. Sedighi, and M. A. Delavar, "Effect of fin position and porosity on heat transfer improvement in a plate porous media heat exchanger," *J. Taiwan Inst. Chem. Eng.*, 2013.
[21] J. Tian *et al.*, "The effects of topology upon fluid-flow and heat-transfer within cellular copper structures," *Int. J. Heat Mass Transf.*, 2004.
[22] G. Juncu, "The influence of the porous media permeability on the unsteady conjugate forced convection heat transfer from a porous sphere embedded in a porous medium," *Int. J. Heat Mass Transf.*, 2014.

IOP Conf. Series: Materials Science and Engineering **834** (2020) 012048 doi:10.1088/1757-899X/834/1/012048

A Case Study on Change Management Readiness for an Oil & Gas SME Company in Malaysia

I. S. Baharudin*[1], B. Abdullah[1], N. A. Mohd Salleh[1] and P. Shariffudin[2]

[1]Faculty of Mechanical Engineering, Universiti Teknologi MARA (UiTM), Shah Alam 40450 Selangor, Malaysia.
[2]Faculty of Business Management, Universiti Teknologi MARA (UiTM), Shah Alam 40450 Selangor, Malaysia.

*inderashaiful@gmail.com

Abstract. Change management is vital in sustaining and remain competitive in a volatile environment. A well-planned change management effort has been craved by organizations in need as being more profitable, reputable and sustainable is the ultimate achievement. These changes to the pillars of the organization normally are more transformative than incremental in nature. All arrows pointing to the single biggest step an organization is to embark if they choose to ride the path of a successful change management execution – the change management readiness. This paper discusses the aspects of change management readiness which entails the need for readiness, the activities within readiness, the factors of leadership, communication and awareness that embodies the readiness as a whole. A case study of an actual organization participated in the study using leadership profiling assessment, organizational communication survey, change management awareness survey and maturity assessment. The results from this study is realistic and impactful to its readers. This paper validates the underlying theory of the three factors in a real industry environment and how successful change management readiness can be obtained with diligent planning and execution.

1. Introduction

The sixth Prime Minister of Malaysia initiated the Economic Transformation Plan, which place high priority to the Oil and Gas industry, as it has been the most potential and lucrative industry at the moment. The concentration encourages new companies to join the industry, while the current businesses grow within rapidly. Many incentives are being given to support this effort but the biggest question remains whether these "Small Medium Enterprises" (SME) companies are ready for expansion, globalization or survival. Are they ready to compete on the international playing field in terms of deliverability, quality and price competitiveness? These are the biggest questions asked of a company in these situations. While they try to pursue change through generative learning, they see things and begin transforming its structures when waves of alarming needs corner them for change in process, people, structure and strategy [1] in order to survive.

This is the current problem and scenario that most companies in Malaysia are facing, as majority are still operating on local standards and deliverables. When Malaysia eventually opens up its market in complying to the Asian Free Trade Agreement (AFTA), many will feel the pinch and loss of business opportunity. Only sustainable model implementation and adaptation to the current market demand will enable them to survive this phenomenon.

Content from this work may be used under the terms of the Creative Commons Attribution 3.0 licence. Any further distribution of this work must maintain attribution to the author(s) and the title of the work, journal citation and DOI.
Published under licence by IOP Publishing Ltd

Since companies need to change in meeting the demands of the market, this study explores an element in change management for this concentrated industry in Malaysia – the readiness for change management in an organization. For the SME Construction companies within the Oil and Gas industry, they must be able to act fast and prevent the collapse of market competitiveness. Globalization will take its toll whether we like it or not, it's how these companies adapt and transform towards the external imposition that makes survival and growth possible.

This study has an objective of evaluating a company's readiness in executing change management through the factors of leadership, communication, and awareness. These three factors are essential in the completeness of change management readiness and have been proposed by previous research within the same field to study on the role of organizational communication processes, as well as the importance of information on levels of change readiness [2]. The discussion will be led on what change management is all about, its implementation, preparation for change management, leadership, communication, awareness, assessments, surveys and recommendation for the further impact the knowledge and industry stakeholders in the future.

2. Literature review

2.1 Change Management

Change Management was defined by Prosci as a process and set of tools to assist people/organization in achieving its desired outcome [3]. Change management in an organization relies heavily on human factor with leadership and communication prowess underlining its helm. Change Management is to be considered successful when the objective of the company is being met through people collaboration and successful execution of the program itself. It is glaring that change management is inevitable for companies in Malaysia as they need to respond to the volatile market. Many claim Malaysia [4] as one of the most open economies in the world [5], which makes change management and its readiness vital in any organization.

2.2 Organizational Readiness in Change Management

An organization's decision to embark in change management is as important as its decision to ensure the organization is ready for the change itself. Many top management fails to understand that the barriers to change are the employee themselves. There is a limit in imposing instruction on subordinates, therefore it is crucial to get the employee's willingness to improve, work better and embrace the changes within the organization.

While the majority of the change management effort failed due to the sponsor or the highest rank person with executive power in an organization didn't 100% on board with the change management program, the next contributing factor is the inability to remove or overcome the barrier of change itself. These barriers come in multiple forms from people resisting change, communication breakdown, insufficient time for training, and staff turnover during change [6]. These barriers boil down to three factors of leadership, communication, and awareness in change management respectively.

2.2.1 Leadership in Change Management.

Implementing change management is an easy feat with the right managerial leadership within the organization. Being exemplary and communicating strategies, information, updates, and instruction constantly towards people around you are the leaders of change management that have helped companies transformed successfully [7]. In a modern organization, the top-down approach is no longer valid as 360 relations, communications, interaction, evaluations come into the fray. This factor demands more from the manager to be more sociable, open and transparent with his/her subordinates. Studies have shown that two out of three transformation initiatives fail due to incompetent managers. We have seen many of these instances in companies struggling to compete and maintaining the high quality of services where lack of "movement" towards organizational needs.

This research emphasizes leadership role before, during and after the change management process. It is imperative to have the support, concern and committed managers to employee's welfare during the times of change [8]. Managers become leaders, acting as change drivers and change agents during these processes. They need to drive and gain the trust of their subordinates on why change is needed in the organization. Poor managers cum leaders will lead to a disastrous change climate within the organization and make it difficult to cope with the ever-increasing frequency of organizational change with it is already set in motion [9]. Developing managers with analytical, influential, emotional, political, interpersonal and technical attributes is essential, especially in contributing to the probability of change readiness success [10].

2.2.2 Communication in Change Management.
The purpose of employing communication as one of the key areas of change management is to disseminate information to the intended group of people. There are many forms of communication channels available within an organization including employee portals, newsletters, emails, static and non-static media as well as face to face communications. Many organization misunderstood the change communication as a process of mediating instead of intermediating [11]. To have effective communication, information needs to be interpreted and reinterpret them into ideas and processes that are easy to understand, thus applying the mediating technique. While using the communication channels, its best to incorporate latest technologies albeit social network and video conferencing to enhance outreach to employees. However, although it is the management's decision to establish any means of communication channels deemed effective, one must avoid a typical top-down and one way conduits approach while using these communication facilities to avoid a breakdown in communication [12].

2.2.3 Awareness in Change Management.
The first building block of one of the most popular change management model, Prosci's ADKAR, is awareness. [13]. An employee needs to know why change is needed, how it is affecting them and what are their roles in participating in this change. This is where awareness is crucial in getting their willingness and buy-in towards implementing change. Another factor of awareness is the believability of the information they receive. This relates much to the manager's role in relaying accurate information to build trust and enable the communicated messages being interpreted in the right manner. It is advisable to have constant top management engagement with employees, regular communication between managers and subordinates and continuous monitor of awareness level about the change management situation within the organization.

2.3 Situation of SME Oil & Gas Companies in Malaysia
Through the economic transformation plan launched by the government, oil and gas and renewable energy contribute a major portion of Malaysian economy with the increasing growth of 5% per year up to the year 2020 [14]. The concentration on oil and gas industry has grown over the years and more opportunities and potential are available in the market. This makes the growth of companies inevitable as when there's a pull factor strong enough to attract local SME companies, the same pull factor influences international companies in protruding the local construction scene. With the development and launching of Rapid Project, increasing construction and revamping works by big players like Petronas, Shell, ExxonMobil, and Petron [15] for their current and upcoming facilities, increase in competitiveness is the need for all companies battling to survive.

The largest contributor to our national income/business participation is through SME companies. These companies are companies within the range of 75-200 people [16]. The challenge of SME to react towards an open market and globalization is to meet the standards expected by all the current big players and new players in the Malaysian market through the influx of investment through the market. Companies like Tokuyama Corp., LION Group, Assar are all newly invested companies in Malaysia which can be considered as Foreign Direct Investment (FDI) [17]. The SMEs are facing an obstacle in the areas of manpower competency, infrastructure, business competition, technological and innovation infrastructure and government policies [18], [19]. It was identified that the key success factor for SMEs with more than 10 years in the business are including but not limited to the items mentioned below, which happens to be the biggest challenge for SME to compete and grow into an internationally competitive company [20]. The items are:

i. Sound management capability and integrity

ii. Sound business cultures and entrepreneurial spirit
iii. Prudent financial management
iv. High-quality products and services
v. An effective program for human resource development
vi. Strong support from financial institution
vii. Strong marketing strategies
viii. Continuously looking for opportunities to expand.

Most of the local SME are struggling with these key success factors and in order to overcome these challenges, especially in oil and gas, must undergo change management to be competitive, maintain quality, meeting on-time delivery while gaining significant profit at the same time. An SME company in Malaysia will need to transform by redirecting its strategies, internal competencies, structure, process, culture, and people in order to succeed [21].

3. Case study description, instruments, and method

3.1 Case organization

3.1.1 Background.
The case study for change management readiness is a medium sized oil & gas engineering construction company based in Kuala Lumpur. The company has been operating in Malaysia for more than 50 years with employee's headcount of 100 people at the point of the study (22 managerial level, 72 non-managerial level and 6 non-managerial/non-executive directors). Steadily growing since its establishment, the company's revenue for 2013 was RM153 million. Their previous clients are from international and local based company such as Petron, Petronas, ExxonMobil, and Tokuyama. The company specializes in "Engineering, Procurement, Construction and Commissioning" (EPCC) operations and provides its services with exceptional quality and standards. Company objective is to satisfy client through quality, safety and completion timeliness, maximizing profit for the shareholders as well as providing a rewarding career to all its employees.

3.1.2 Organizational issues and problems.
Change management was mooted by the managing director of the organization after a decline in client satisfaction in the "Customer Satisfaction Feedback (CSF)" for 4 of the last 7 projects, where they achieved below 2.5 of 5, completed by the company. The main reason for the decline in results was due to lack of manpower support, substandard engineering drawings quality and technical ability of the employees executing the projects.

3.1.3 The change management program.
In order to turnaround the company's performance, the company underwent organizational assessment and detailed analysis of every segment of the organization using 6-pillar model, introduced by Indera Shaiful and Bulan Abdullah [22]. The change management program was introduced based on these results and was carried out with the full support of board directors and company owners. The change management masterplan has 7 programs as follows:
i. Preparation for change programme
ii. Total organizational improvement programme
iii. Employees performance management system programme
iv. KPI tailoring programme
v. Organizational performance monitoring programme
vi. Compliance programme
vii. Work approach programme

The first change management program is to be conducted within the first three months and the next six programs to be completed within a year after the first program is completed. The preparation for change program was designed to minimize implementation risk and ensure success for the whole change management masterplan. This preparation is also called change readiness[23], where in order to overcome the barrier of change within the organization, the company has focused on leadership, communication, and awareness as the key towards achieving organizational change goals set by the management. Detail activities obtained from the company's masterplan are as follows:

i. Managerial training
ii. Readiness of communication channels
iii. Improvisation of communication channels and distribution of knowledge
iv. Leadership program
v. Communicating organizational strategies
vi. Change awareness initiatives

3.2 Instruments

There were 4 instruments/tools used to conduct the study by the researcher. These instruments/tools were carefully adopted and created in order to reflect the true nature of the organization through these assessments and surveys. The first instrument is the leadership profiling assessment, which identifies the managers in the organization on their sociability, dominance, and openness. This assessment was adopted from a book titled "Managers As Mentors: Building Partnerships for Learning" published in 2013 [24]. The second instrument is the organizational communication survey, which is a survey emphasizing to the communication condition within the organization through channels available internally. This survey was generated and distributed by Qualtrics, an online survey tool [25].

The third instrument was is the change management awareness survey, intended to gauge the awareness level of all employees on the objective, execution detail, roles and relationships and timeline of the change management plan the company intends to implement. This instrument was developed within the organization itself. The fourth instrument is the maturity assessment (overall change readiness assessment) which highlights the readiness of the organization and risk level in implementing change management as a whole. This instrument was adapted from November ag, a renowned consultant based in Switzerland [26].

4. Results and discussions

After the execution of the activities in preparation for change programme, the results of leadership, communication, and awareness was observed using the tools and measurement techniques in the previous section. It can be observed since total preparation and readiness of the organization is dependent on these 3 factors, these results determine whether there is more time required for change management preparation activities or it can be done immediately.

The results of the factors of leadership, communication awareness within the organization influence the outcome of the maturity assessment, which entails whether the organization is ready for change management implementation. All factors must comply with the readiness minimum requirement to reduce the risk of change management activities to be affected during execution.

4.1 Leadership

For the leadership factor, the assessment was done using the leadership profiling matrix as explained in section 3. The assessment was done successfully with 100% participation within the population of the managers in the organization. As explained in section 3 above, the 22 managers involved were subjected to 3 leadership factors of sociability, dominance, and openness. The Table 1 show the manager profile matrix.

Table 1. Manager profile matrix

Factors	Category	No. of Managers	Percentage
	Reserved	5	22.73%
Sociability	Outgoing	11	50.00%
	Moderate	6	27.27%
	Follower	4	18.18%
Dominance	Leader	8	36.36%
	Moderate	10	45.45%
	Guarded	8	36.36%
Openness	Expressive	12	54.55%
	Moderate	2	9.09%

In the sociability traits, 77.27% (17 of 22) of the managers are moderate and outgoing. At this stage, this is a positive indicator as the organization requires significantly "outgoing and sociable" managers to have higher chances of persuasive prowess over his/her subordinates in achieving the desired goal. A study on leadership attributes indicates that this influence the outcome of cohesiveness of a unit/department to have a leader who knows how to socialize within boundaries with its compatriots [27].

In the dominance traits, 8 person shows strong instructive influence while another 10 persons are acting based on situation and responses. This accumulated to 81.81% of desired traits as the communication although is more top-down but resulting in positive results in ensuring change happens within stipulated time [28].

However, in the openness traits, there is only a slight majority of 63.64% of managers who are moderate and expressive in nature. While this is not an alarming rate, there are risks of the fluency of change management activities to be embraced by the managers themselves where being guarded and not open to new ideas are one of the major barriers in the success of change management implementation [29].

4.2 Communication

Communication Survey was done with participation from all level of employees of 100 personnel within the organization. Figure 1 show the results for the communication survey held within the organization. This result shows that 88% of the respondents are happy with the existing communication channel and infrastructure available in the organization which allows smooth communication being done through all level of employment. This results also interpret that the use of these channels is well implemented and easy to use for employees to receive and transmit necessary information.

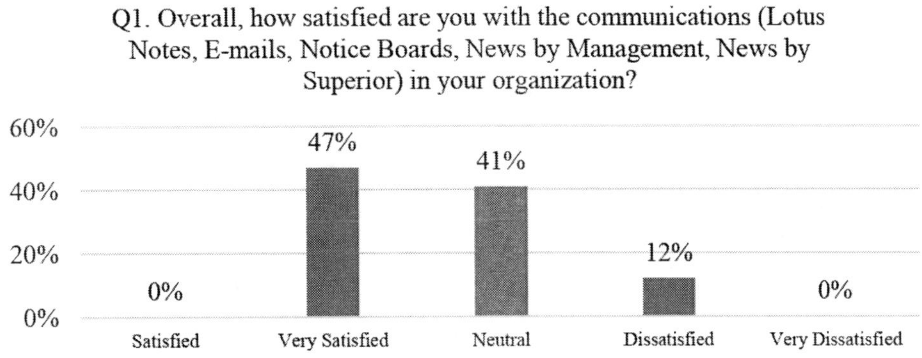

Figure 1. Results for question 1 in the communication survey

Figure 2 shows that 13% of the respondents are very satisfied and fully informed on all communications within the organization. It also highlights the regular communication that has been conveyed to the staff, underlining a further 60% satisfaction from the respondents. On another side of the spectrum, there are pockets of improvements needed, as 27% still feel isolated and uninvolved within the communication circle. This may due to the direct supervisor/manager handling them are having guarded behaviour and release information on a need to know basis.

Figure 2. Results for question 2 in communication survey

Figure 3 shows the trust of information which has been circulated to employees, with an overall percentage of only 12% have trust issues on the information they received. It is important to highlight that 66% of the respondents have high confidence in the information they receive and this may point to the positive transparency of information between staff, management, and interdepartmental relations.

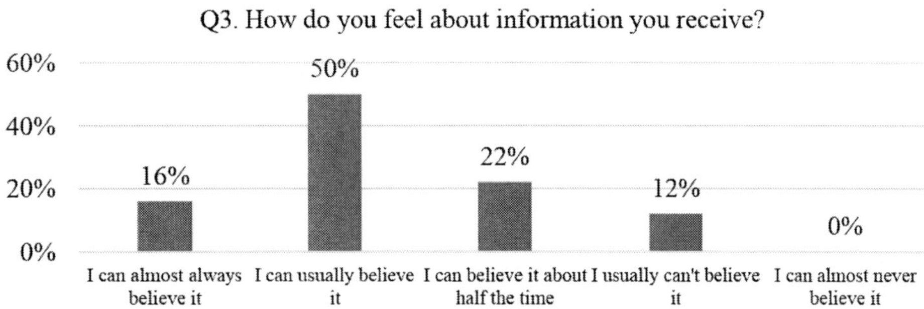

Figure 3. Results for question 3 in communication survey

The documentation, oral presentation, and direct supervisor's explanation have contributed to the 69% of employee knowing a fair amount and knowing very well about the organization as shown in Figure 4. Although this may be due to the longevity of employment in the organization, it doesn't dismiss the fact that information about the company has been flowing smoothly and reaching its intended targets.

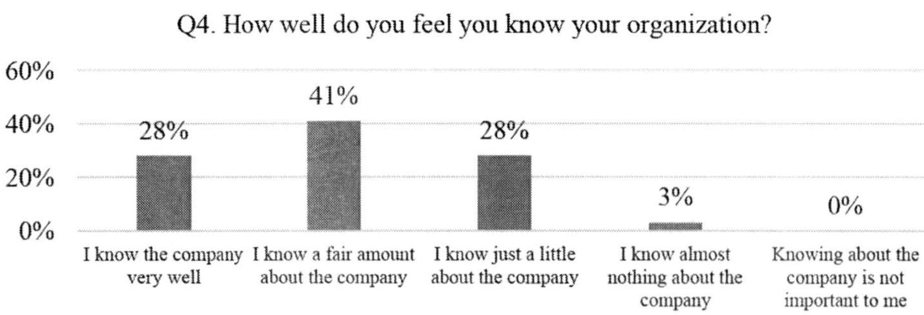

Figure 4. Results for question 4 in communication survey

While 81% of the employees expressing sufficient increase/retention of knowledge about the organization's strategies and accomplishments, 19% is still somewhat less knowledgeable as shown in Figure 5. This again points to the known issue but has to be addressed accordingly soonest to get total involvement from everyone in the organization. This results also in-line with the leadership profiling matrix results. The implication of this question is to increase the level of awareness of employees towards the company's objectives and strategies.

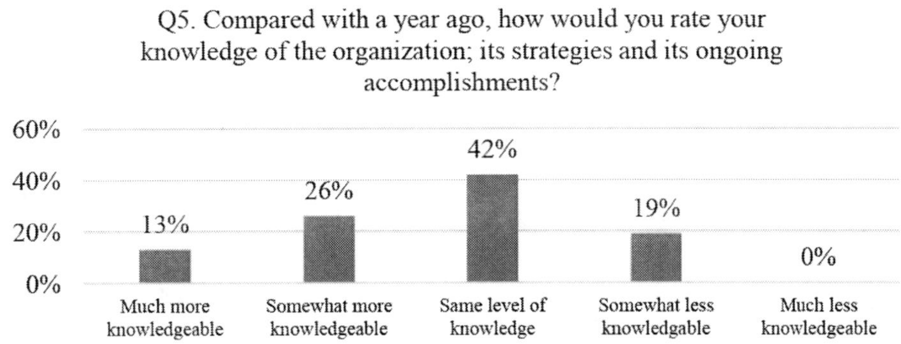

Figure 5. Results for question 5 in communication survey

These questions No 6 and 7 are referencing to how employee perceive their immediate superiors in day-to-day communication and while the results for their communication effectiveness is fairly balanced between excellent and poor, many find it is still easy to express their opinion to the manager. This trend indicates there are two-way communication between managers and subordinates on a positive level and a need for further training and improvement from management staffs in communicating effectively and in a timely manner as shown in Figure 6-7.

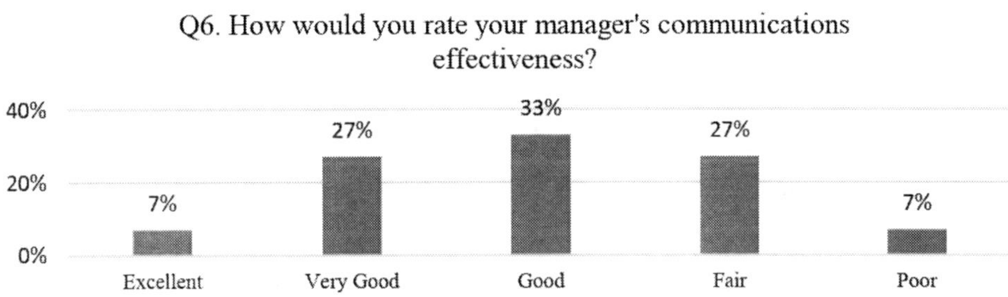

Figure 6. Results for question 6 in communication survey

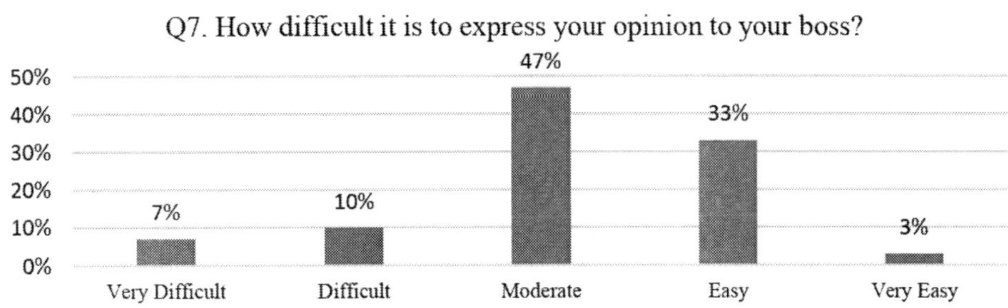

Figure 7. Results for question 7 in communication survey

From the data obtained from the seven (7) questions above, the general statistics is being calculated and shown in Table 2 below. The questionnaire scoring is based on 5 level of ratings, where rating 1 is the best and 5 is the worst. The statistic shows that all questions have mean rating of 2.00 and below. This concludes that the level of communication for change management in this organization is good and rated positively. The best response was answers to question 4, where the majority of the employees know and well informed of the organization. This means that the dissemination of information in a timely manner has been done properly and in an orderly fashion. However, there are mix responses for question 6 on the manager's communication skills. This is consistent with the results of leadership profiling matrix in section 5.1 above, concerning some managers not being sociable and lack of openness while interacting with their subordinates.

Table 2. Communication survey overall statistics

	Mean	Min	Max
Question 1	1.66	1	3
Question 2	1.97	0	4
Question 3	1.31	0	3
Question 4	1.06	0	3
Question 5	1.68	0	4
Question 6	2.00	0	4
Question 7	1.83	0	3

4.3 Awareness

Through all the activities done for the readiness of change management, employees below "managerial level" of 72 personnel took the survey on the awareness of change management activities to be done within the organization for the next year. Managerial level and above personnel were excluded because they are the change drivers and change agents respectively. Results are shown in Table 3 below:

Table 3. Change management awareness survey results

No	Questions	Response					Total
		1 (Poor)	2 (Below Average)	3 (Average)	4 (Good)	5 (Excellent)	
1	How do you feel about the change management awareness activities?	0	0	5	50	17	72
2	How do you rate the content of the activities informatively?	0	0	7	45	20	72
3	The content of the awareness activities was well organized.	0	0	7	49	16	72
4	I know more about organizational change in my organization after experiencing these activities.	0	0	17	41	14	72
5	The change management awareness activities answered all my questions.	1	2	18	42	9	72

From the awareness activities using promotion, presentations, forum, and other communication methods for the 3 months of change readiness activities, employees within the organization are rating high level of awareness and responding well to the change effort. Almost 99.99% commented average or better in all 5 questions combined and only three-person rates below average on question number 5. It can be said that the employees are well informed, satisfied and feel important being included in the change management effort from the early stage of implementation.

4.4 Maturity assessment (overall change readiness assessment)
Maturity assessment was conducted with the participation of 72% population of the organization and the results are shown in Table 4. With all the measurement tools, methods, activities, and surveys being done during the readiness towards change management in the organization, the ultimate test which reflects the overall outcome comes with the 31 questions of maturity assessment. It gives a clearer picture of change management implementation. Table 4 shows the results of the assessment based on the areas mentioned in the previous section. It can be seen that the mean for all areas was scoring a minimum of 7 out of 10. This indicates that the organization is ready for change and have a high success probability in ensuring change is being done with minimal risk.

Table 4. Maturity assessment survey results

No	Area	Topic	Mean
1	Develop Leadership	Leaders share a common vision of this change	9.24
2		Leaders are demonstrating their personal resolve for change through their actions	8.16
3		Leaders are creating a sense of urgency to change in the organization	8.35
4		Leaders understand trust and respect each other	8.27
5	Create Change Vision	There is a clear and compelling need for this change	9.13
6		It is clear what this change will mean in practice	8.28
7	Define Change Strategy	The approach taken to implement this change will work well in this organization	8.19
8		Change will happen fast enough to sustain interest	7.23
9		Authority and responsibility to achieve this change is clear	7.41
10		Project structures e.g. task forces, steering groups are keeping change on track	8.38
11		Decision making is not slowing this project down	8.11
12		Problems emerging in change are being solved quickly	7.05
13		Sufficient time to implement change is being allowed	8.13
14	Build Commitment	People expect this change to succeed	9.14
15		People will learn and apply the new skills and behaviour required for this change	7.98
16		People are being encouraged to be constructively critical of this change	7.35
17		Managers have the change skills necessary for this change	7.29
18		Change can be achieved without a significant shift in attitude and behaviours	8.42
19		There will be no losers (groups/individuals) through this change e.g. loss of job or status	8.21
20		There are rewards for participating in this change and punishments for failing to participate	7.27

21		Everyone involved is being well informed and kept up to date in this change	7.09
22		Those impacted by the change are clear about how change affects them and what to do differently	8.06
23		People's commitment is being built rather than change forced	7.37
24		The commitment of middle managers is being won before they are expected to lead change	8.06
25	Manage People Performance	Manages are being disciplined in this change. They are getting things done	8.33
26		This change will not be blocked by established organizational processes such as reward and performance management	7.25
27		Processes, policies, and rules are being changed where necessary to support this change	7.95
28		There is sufficient cross-functional co-operation for this change to succeed	7.14
29	Develop Culture	This change will be supported by the current management style and behaviour	8.00
30	Design Organization	This change will be supported by the current organization structure	8.26
31		This change will be supported by current the way jobs are structured	7.31

The risk chart in Figure 8 shows the level of risk the company will undertake in implementing the change management program in the organization. It can be seen that only 11 of the 31 points touches the border between medium and low risk, which can be mitigated and included within the revised change management program. All other points are well on the low-risk section and is a good indicator supporting the high probability of change management implementation success in this organization.

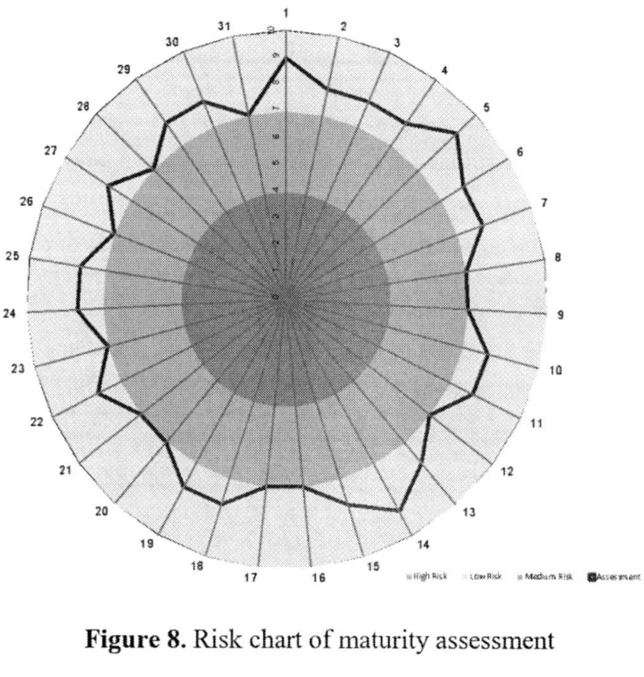

Figure 8. Risk chart of maturity assessment

5. Conclusion

The study on change management readiness efforts and assessment has demonstrated success in meeting its objective. The four objectives of determining leadership readiness, communication readiness, awareness, and overall change management readiness in the organization has been discovered. All 4 determinants are positive in nature as the leadership profile shows more than 60% have the desired leadership traits, organizational communication shows all feedback are good or better, awareness survey shows more than 99% are aware of the change management activities and maturity assessment shows minimum rating of 7/10. This shows that the organization is ready to embark in the change management journey with high success probability. Even with the limitation of willingness, cooperation, and time availability, the true and existing condition of the organization was assessed accurately and benefited the stakeholders involved.

From the finding, it can be clearly understood that it is important to have leaders in the organization who understands and execute the role in being a leader instead of just a mere manager who lack mentorship qualities. It can also be underlined that only with good communication flow, transparency and timeliness of information dissemination that you will get the involvement, satisfaction, and commitment for the population at large, in this case, the employees of the organization.

In future research, it is recommended to benchmark the outcome of an actual change management program implementation barrier results to the probability findings of this study, which will continuously improve the matrix of readiness within change management implemented organization in waiting.

Acknowledgement
First of all, I would like to thank the management of the organization involved in this case study for their willingness to disclose information and participate in this study. Next, highest gratitude to University Teknologi MARA (UiTM) and my supervisor for providing the platform for me to complete this study and contribute back to knowledge and society.

References

[1] G. Susman, K. Jansen, and J. Michael, "Innovation and Change Management in Small and Medium-Sized Manufacturing Companies," p. 55, 2006.
[2] T. Hallgrimsson, "Organizational change and change readiness: Employees' attitudes during times of proposed merger," pp. 1–98, 2008.
[3] Prosci, "The ADKAR Model for Change Management," *Chang. Manag. Tutor. Ser.*, 2002.
[4] M. Hashim, "Change Management," *Int. J. Acad. Res. Bus. Soc. Sci.*, vol. 3, no. 7, pp. 685–694, 2013.
[5] M. Zulkifli, A. K. Char, M. R. Yasoa', and Z. Hassan, "Small and Medium Enterprises (SMEs) Competing in the Global Business Environment : A Case of Malaysia," *Challenges*, vol. 3, no.1, pp. 66–75, 2010.
[6] J. Gigante, "Breaking Down the Barriers to Feedback," *Pediatr. Ther.*, vol. 02, no. 03, 2012.
[7] H. L. Sirkin, P. Keenan, and A. Jackson, "The Hard Side of Change Management," *Proj. Manag.J.*, vol. 45, no. 3, pp. 98–111, 2014.
[8] J. Kotter, "THE 8-STEP PROCESS FOR LEADING CHANGE Dr. Kotter's methodology of change leadership," 2012.
[9] R. R. Rehman, A. Khalid, and M. Khan, "Impact of employee decision making styles on organizational performance: In the moderating role of emotional intelligence," *World Appl. Sci. J.*, vol. 17, no. 10, pp. 1308–1315, 2012.
[10] C. Bond and M. Seneque, "Conceptualizing coaching as an approach to management and organizational development," *J. Manag. Dev.*, vol. 32, no. 1, pp. 57–72, 2013.
[11] A. Whittle, O. Suhomlinova, and F. Mueller, "Funnel of Interests: The Discursive Translation of Organizational Change," *J. Appl. Behav. Sci.*, vol. 46, no. 1, pp. 16–37, Mar. 2010.
[12] R. S. Bisel and J. K. Barge, "Discursive positioning and planned change in organizations," *Hum. Relations*, vol. 64, no. 2, pp. 257–283, 2011.
[13] A. M. Calder, "Organizational Change: Models for Successfully Implementing Change," *Undergrad. Honor. Thesis*, pp. 1–20, 2013.
[14] P. Chin and F. Kui, "Oil , Gas and Energy," 2011.
[15] P. Performance Management and Delivery Unit, "Oil , Gas and Energy NKEA Fact Sheet," 2011.
[16] S. C. Malaysia, "New Definition of SMEs - An additional 8,120 establishments set to be classified

as SMEs, bringing the share of SMEs to total establishments from 97," 2014. .
[17] M. B. I. Photovoltaic, "Tokuyama Corp to invest a further RM3 . 72bil in Sarawak," 2008. .
[18] F. Hashim, "JIBE," vol. 13, pp. 97–122, 2012.
[19] A. S. Saleh, P. Caputi, and C. Harvie, "Perceptions of business challenges facing Malaysian SMEs : some preliminary results Perceptions of business challenges facing Malaysian SMEs : some," vol. 2008, pp. 79–106, 2008.
[20] M. Zulkifli-Muhammad, "Small and medium enterprises (SMEs) competing in the global business environment: A case of Malaysia," *Int. Bus. ...*, vol. 3, no. 1, pp. 66–75, 2009.
[21] L. L. Tan, "The internationalisation of Malaysian manufacturing small and medium enterprises," 2008.
[22] I. S. Baharudin and B. Abdullah, "A critical analysis on organizational diagnostic models and new diagnostic model proposition for construction companies .," in *Mechanical Engineering and Sciences Postgraduate International Conference (MESPIC) 2018*, 2018, pp. 1–8.
[23] R. V. Riddell and M. Tofte, "Change Readiness," 2017.
[24] C. R. Bell and M. Goldsmith, *Managers As Mentors: Building Partnerships for Learning*, 3rd ed. Berrett-Koehler Publishers, 2013, 2013.
[25] "Qualtrics Experience Management Solutions." [Online]. Available: http://www.qualtrics.com.
[26] "Change Readiness Assessment." [Online]. Available: https://november.ch/resources/867/.
[27] D. T. Holt, a. a. Armenakis, H. S. Feild, and S. G. Harris, "Readiness for Organizational Change: The Systematic Development of a Scale," *J. Appl. Behav. Sci.*, vol. 43, no. 2, pp. 232–255, Jun. 2007.
[28] E. J. Nordin, "Communicating Organizational Change : Strategies For Communicating Change," pp. 135–145, 2014.
[29] J. Balogun, "From Intended Strategies to Unintended Outcomes: The Impact of Change Recipient Sensemaking," *Organ. Stud.*, vol. 26, no. 11, pp. 1573–1601, Nov. 2005.

Tool-path Generation for Groove Machining Feature

Ahmad Faiz Zubair, Mohd Salman Abu Mansor*

School of Mechanical Engineering, Engineering Campus, Universiti Sains Malaysia, 14300 Nibong Tebal, Pulau Pinang, Malaysia.

*mesalman@usm.my

Abstract. In this study, algorithm in generating machining sequences and tool-path for groove machining features are developed. The algorithm is based on ruled-based method and enhancing automatic feature recognition (AFR) via volume decompositions method. From the position of the vertex in z-axis and x-axis directions, machining sequences can be determined, and machining features can be sorted out. From the sub delta volume (SDV) generated from the recognized features of 3D part model, vertex positions of each SDV is determined thus positioning of the tool-path can be obtained. To validate the generated machining sequence and tool-path, one workpiece has been assigned to the algorithm. This workpiece is also tested with commercial software CAM function. The result shows that the developed algorithm is capable in generating the tool-path and can be visualized.

1. Introduction

Computer-aided Process Planning (CAPP) is an essential way to link computer-aided design (CAD) and computer-aided manufacturing (CAM). It works by determining manufacturing processes by identifying parameters that can lead to the justifications of machining purposes [1], [2]. Automation is the main keyword to address researches done by researchers to cultivate CAPP system since 1960's [3]. By automating the way toward supporting the machining parameters which done in CAM, numerous methodologies had been worked. The thoughts on connecting CAD and CAM need to begin by passing on the upheld configuration of record from CAD to CAM. Many commercial software either work with various sort of documents handling from CAD to CAM or these software's applications just have single CAD or CAM capacities. User need to import the files to process it. Therefore, information exchange for the software end up being progressively complicated. Moreover, groove recognition in processing the manufacturing information in CAM is always difficult as most software cannot perform it automatically.

Decomposed features from CAD model that were characterized as Sub-delta Volume (SDV) can be mapped towards machining features. Explicitly for cylindrical part models, Sub-delta Volume for Finishing (SDVF) were ordered dependent on its geometrical face definitions [4]. For instance, cylindrical SDVF which decomposed from cylindrical surfaces are correlated to straight turn and grooving process. Moreover, machining features for example taper and internal features can likewise be mapped. To make decomposed features beneficial, sequencing processes can be done in order to advance establish the SDVFs towards manufacturing purposes. Past studies shows by machining features' sequencing, energy consumption in machining system can be diminish [5], [6]. Furthermore, a study concentrating on interacting prismatic features by adopting knowledge-based and geometric reasoning [7] and graph-based tolerances [8] had been done. These investigations utilized features' information to produce more data towards machining highlights in various ways.

Content from this work may be used under the terms of the Creative Commons Attribution 3.0 licence. Any further distribution of this work must maintain attribution to the author(s) and the title of the work, journal citation and DOI.
Published under licence by IOP Publishing Ltd

This paper will utilize the SDVs to locate the particular vertices positions for the tool-path development task. By having SDVs, volume of materials that should be removed from the stock model can be determined [9]–[11], in this manner machining cost can be determined.

2. Related works

Machining with CNC can diminish production time in contrast to manual NC machine. Consequently, tool-path codes generation is required. The benefits of utilizing machining codes or bound to be called G-codes contrast with manual machining or manual parameter insert to the machine control unit (MCU) is perplexing geometry part model can be machine specifically and geometry related issue as a result of machine CAD lack of ability can be dismissed. This is because of the absence of sketching or drawing capacity at the MCU amid the data exchange from typically 2D attracting to the machine before machining is begun.

Related research lately demonstrates researchers were concentrating on obvious issues with obvious methodologies. Toolpath optimization for sharp corner pockets had been developed by Zhao et al [12], tool–path planning had been developed by Xiuzhi and Brent for circular hole based on STL file format [13] for milling machining. An important viewpoint in generating the tool-path is the vertex positions of the machining features. These positions will decide the situation of hardware to do the machining forms. Previous research shows tool-path can be produced by utilizing cloud data point [14]. These cloud data points are resolved from STL CAD file format to support reverse engineering.

Furthermore, groove machining feature was recognized in one of many machining features. One example is by using graph decomposition strategy that issues in multiple extreme faces [15]. Moreover in machining sequences, groove operation is considered after facing and straight turn operations [16] or secondary feature [17]. The recognition of groove operation is important as it will reflects the tooling selection and machining tool-path.

In this paper, the decomposed SDV are used to obtain the vertex point positions. Geometry-base data from the perceived features were inserted with manufacturing feature-based information to create the tool-path. The viability of the created algorithm is proved through tool-path simulation run through commercial CAD software SolidWorks 2018 CAM function.

3. Methodology

To produce tool-path for grooving, maximum diameter of each SDVF should be resolved. SDVF that have a similar most extreme point in x-axis will be the first segment to be removed from the stock model. This feature will be the first toolpath's profile distance. This toolpath will eliminate material from the edge of stock model in x-axis direction until the maximum point of the SDVF in x-axis direction with addition with the finishing thickness. Finishing thickness of the workpiece will be evacuated after roughing process is complete. The second SDV will be the feature with subsequent largest diameter. Minimum z-axis point of second feature is larger or the same with the first feature maximum point of z-axis direction. To create the grooving tool-path it is essential to measure each SDVF by determining its maximum and minimum vertex in x-axis and z-axis. This rule-based algorithm opt the IF THEN rule to acquire the information. The method of applying rules to get the desired information has the advantages of fast data acquittance compare to other method such as heuristic approach which used more complex formulations.

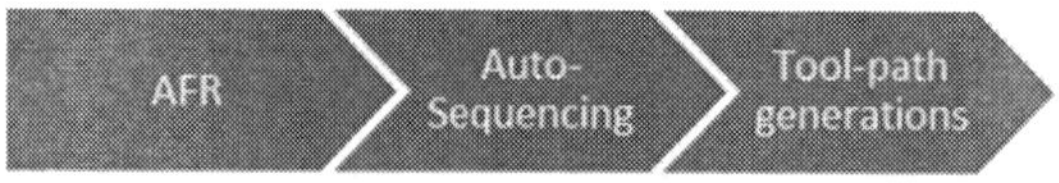

Figure 1. Process flow of the algorithm

3.1 Automatic feature recognitions

One important aspect before the recognition starts is the orientation of the 3D file. It has to be in correct positions before recognitions work can be done [18]. Cylindrical part model can be recognized by its features. This include internal features [19], chamfer and fillet [20]. Moreover, part model complexity can be measure by determining the topological data of the part model[21].

IOP Conf. Series: Materials Science and Engineering **834** (2020) 012049 doi:10.1088/1757-899X/834/1/012049

3.2 Auto-sequencing

For turning machining, sequence of machining operations will begin from facing and completed with parting. In the middle of these two activities, profiling, grooving, tapering, drilling and forming are done. These operations need to be established by automatic sequencing. By finding the data explicitly SDV's boundary positions, auto-sequencing should be possible. Before positioning determinations, SDVs which are decomposed in single body should be separated to ensure each single SDV's boundary positions can be resolved. Next to facing operations, SDV with the largest diameter is chosen and straight turn operation is allocated to it. Subsequently, it will rely on the second largest SDV positions.

If the next SDV is a cylindrical SDVF, profiling operation will be allocated. Algorithm will continue assigning profiling until the algorithm detect possible groove feature in the form of rule-based. Profiling will begin by roughing operation in bigger depth of cut and completed with finishing operation to all the SDV that been recognized.

3.3 Groove feature determination.

Whenever a depression detected on the surface of a workpiece, there is a possibility of grooving feature to be map for lathe machining part model [22]. However, these depression characteristics can be machined as a straight turn if it is located at front of the workpiece. This can save machining time as grooving required different tool from straight turn process. To save time, grooving tools can be selected as same as parting tool and grooving process can be planned to be the second last process before parting.

Therefore, this study is focusing on grooving tool-path generations. Figure 2 shows the process in designating groove tool-path generations. Volume removal of groove from the stock model can be expressed in Eqn. 1.

$$G_1 = SDVR_1 + SDVF_{C1} + SDVF_{P1} \quad (1)$$

Where $SDVR_1$ is the Sub-delta Volume for Roughing (SDVR) at the groove feature, $SDVF_{C1}$ is the cylindrical SDVF at the groove feature and $SDVF_{P1}$ is the planar SDVF at the groove feature. These SDVs are illustrated as Figure 3.

3.4 Sequencing of groove features

For the case of a workpiece that has more than one groove, the algorithm will detect groove feature which is less z-axis point value that the last groove feature to be selected. For example, in Figure 3, G2 will be evacuate prior to G1. This is to manage the tool change so that the last (G1) will be in sequences with parting process. This can be done by having rule that is designated as shown in the flow chart in Figure 2.

3.5 Tool-path generations of groove

Tool to do the groove process will start from home position. This home position is normally set at some point away from the (0,0). The tool will move in G00 mode towards the first z-axis point and an offset location of x-axis point that is detected from planar SDVF at the groove feature location. The tool will then be set towards the first cutting position that is normally depend on the tool width in lead angle that is set. In this study case the lead angle will be 45^0. Next the tool will move downward to the x-axis point that is set from the cylindrical SDVF at the groove feature. This movement will continue until the last point of the z-axis position of the cylindrical SDVF. The tool will return to home position and proceed to another groove feature.

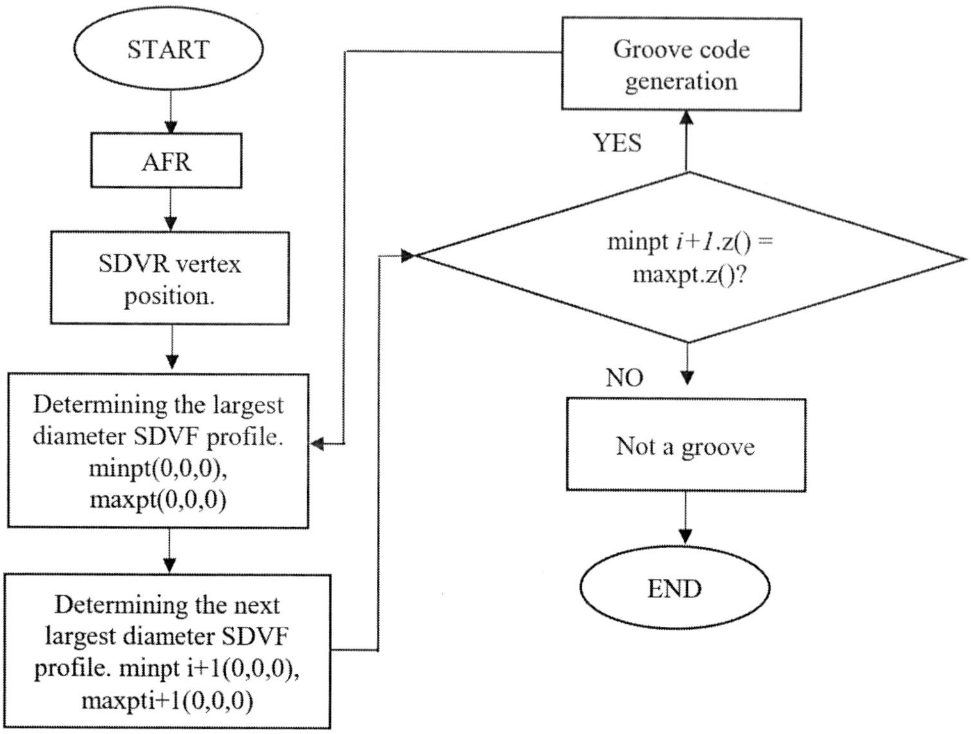

Figure 2. Groove determination process.

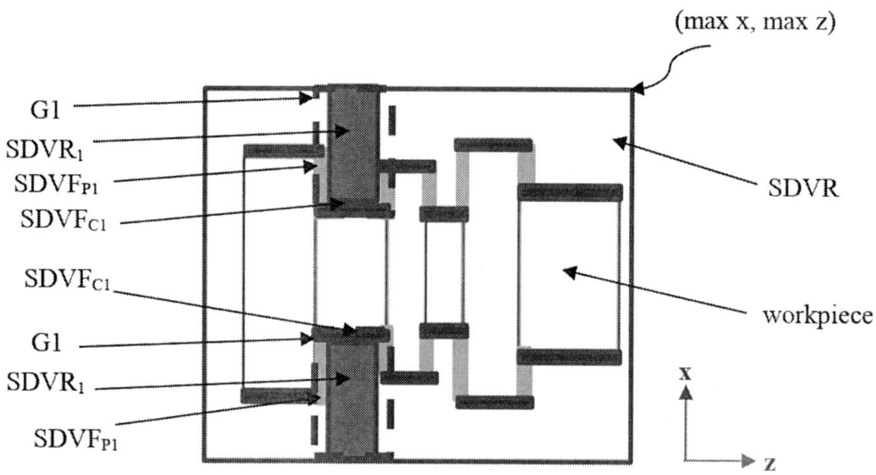

Figure 3. Example of part model with groove features.

4. Case Study

A 64-bit operating system with the performance of 8GB RAM Intel (R) Core i7-3770 CPU 3.40GHz is used to build the algorithm. It was built in Microsoft Visual Studio 2010 Professional Edition version 10.0.40219.1 SP1Rel with Microsoft.NET Framework version 4.6.01055 SP1Rel platform and ACIS solid modeller version R25 SP1 platform.

A workpiece consists of different regular faces type and position being verified by the developed algorithm. It is shown in Figure 4. This workpiece consists of five cylindrical features and a conical feature. These features are selected to verify the auto-sequences and tool-path generations that were developed. If depressions appear in the surfaces of the workpiece, there are possibilities of groove feature to be recognized.

Feature recognition process was successfully done, and Figure 5 shows the result of the generated SDV. The workpiece itself is in silver color, blue color indicates SDVR, green color indicates SDVF and yellow color indicate SDVF-FR.

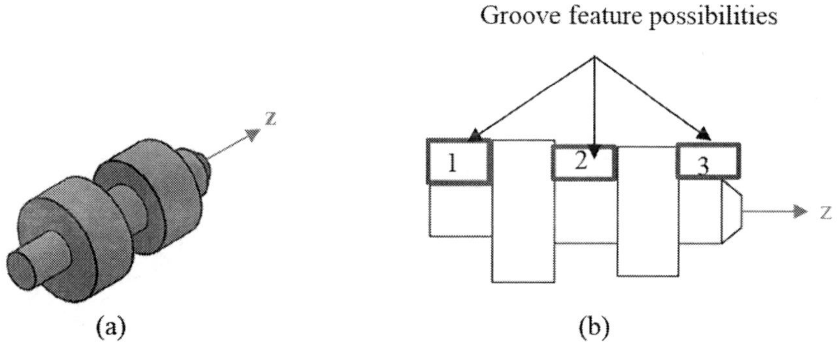

Figure 4. (a) Case Study workpiece and (b) the possibility of groove feature.

To compare the generated tool-path, the same workpiece was simulated via SolidWorks 2018 CAM function. Figure 4 shows the tool-path and sequences of the machining features determined by the developed algorithm. The toolpaths are expressed in different line colors, where red is for the facing, black for the first profiling, white line is the second profiling, purple line are for grooving and tapering and cyan color line is for parting. For visual purpose, Figure 6 is expressed in wireframe, blue color indicates the stock model or SDVR and green line indicates the SDVF. The algorithm successfully generates the groove tool-path and notice that the front groove is not generated. This happen as the algorithm considered it to be machined by straight turn and tapering that will use the same tool as the first profiling and facing. Figure 7 shows the toolpath simulations in SolidWorks 2018.

IOP Conf. Series: Materials Science and Engineering **834** (2020) 012049 doi:10.1088/1757-899X/834/1/012049

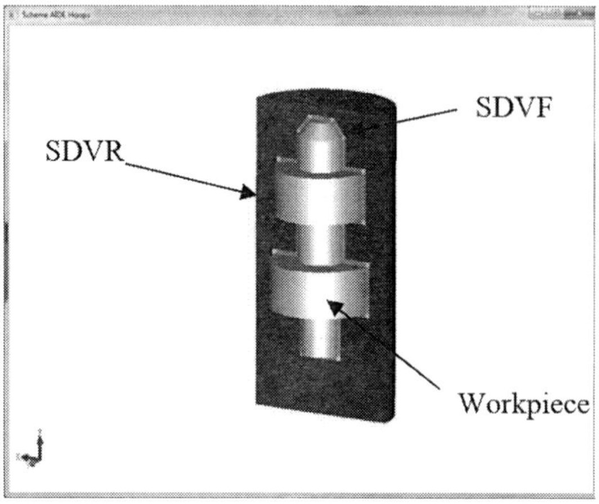

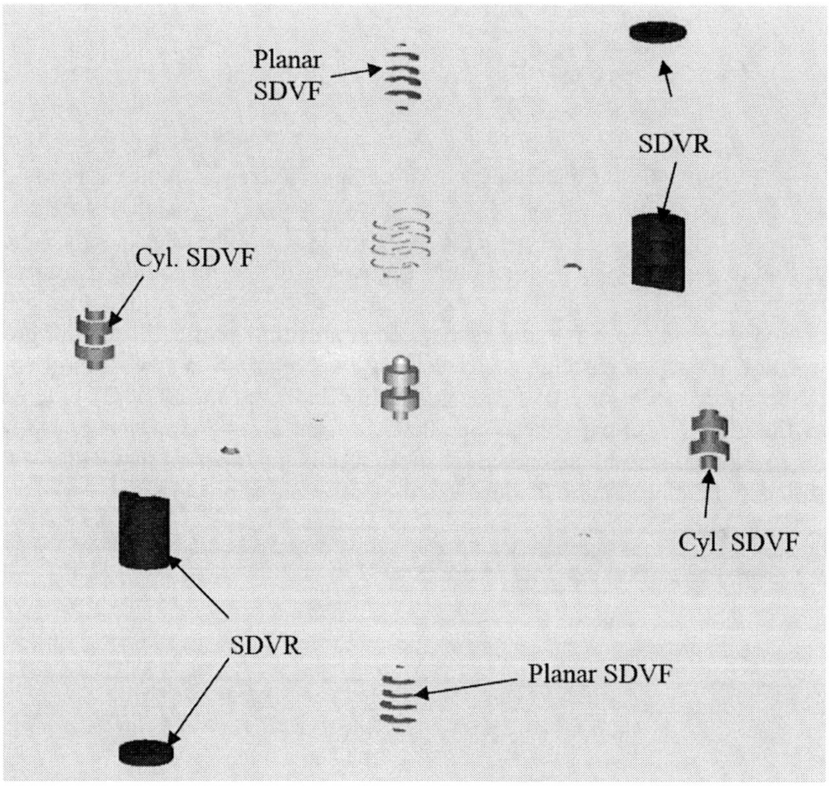

Figure 5. AFR of the workpiece. (a) Thee assembly of workpiece with half section of SDV and (b) the exploded view of all the SDV.

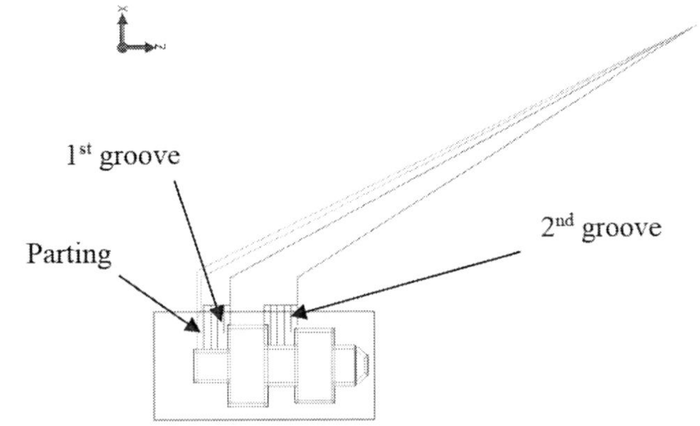

Figure 6. Tool-path for groove generated for the workpiece.

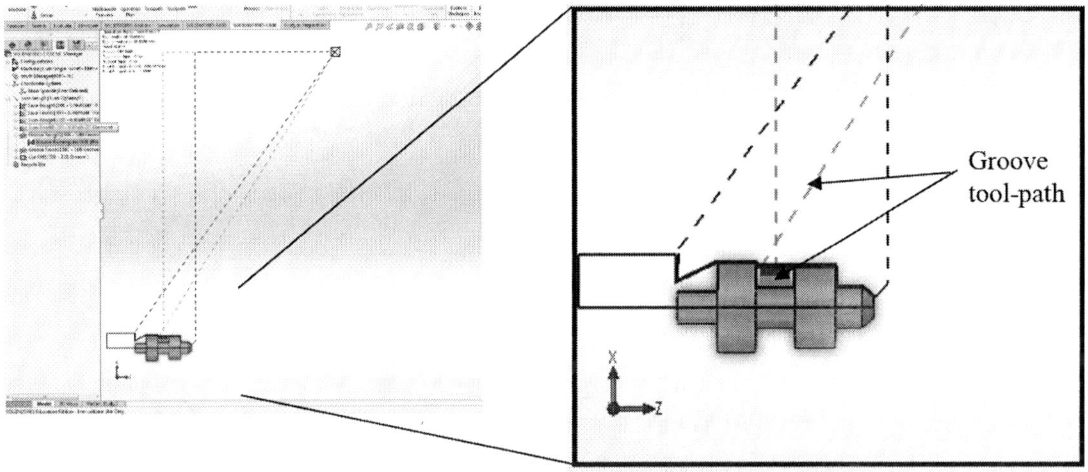

Figure 7. Tool-path simulations in SolidWorks 2018.

From the comparisons between the developed algorithm and SolidWorks 2018 CAM feature, it is founded that, the developed algorithm will generate the sequences by the SDV generated during the AFR process, while SolidWorks has its own machining feature recognitions. Groove features in SolidWorks 2018 were defined in the same sequence as generated by the developed algorithm rules. The sequences in SolidWorks can be seen from the NC manager product tree. Therefore the generated sequences is validated. The simulation in SolidWorks offer more options such as selections of display for the tools, stock model, chuck and more. This however can contribute to more options that can be equipped in the algorithm in future. Moreover, the develop algorithm suggest the toolpath in line based on different colours that depend on the machining features.

5. Conclusions

The developed algorithm is capable in recognizing groove feature, its operation sequences and generating the tool-path for the workpiece. The comparisons of generated tool-path are done with commercial software SolidWorks 2018 CAM function. The results shows that sequencing generated comply with the commercial technique used by SolidWorks. Further investigations can be done by implementing the G-codes generated to real machining.

Acknowledgement
This research is supported by the Ministry of Higher Education Malaysia and Universiti Sains Malaysia under the Fundamental Research Grant Scheme (FRGS) (Reference No. 6071369). The first author also would like to thank the support of the Universiti Teknologi MARA for staff's study sponsorship.

References
[1] S. Subrahmanyam and M. Wozny, "An overview of automatic feature recognition techniques for computer-aided process planning," *Comput. Ind.*, vol. 26, no. 1, pp. 1–21, 1995.
[2] M. Al-Wswasi, A. Ivanov, and H. Makatsoris, "A survey on smart automated computer-aided process planning (ACAPP) techniques," *Int. J. Adv. Manuf. Technol.*, 2018.
[3] X. Xu, L. H. Wang, and S. T. Newman, "Computer-aided process planning: a critical review of recent developments and future trends," *Int. J. Comput. Integr. Manuf.*, vol. 24, no. 1, pp. 1–31, 2011.
[4] A. F. Zubair and M. S. Abu Mansor, "Automatic feature recognition of regular features for symmetrical and non-symmetrical cylinder part using volume decomposition method," *Eng. Comput.*, vol. 34, no. 4, pp. 843–863, 2018.
[5] L. Hu, Y. Liu, C. Peng, W. Tang, R. Tang, and A. Tiwari, "Minimising the energy consumption of tool change and tool path of machining by sequencing the features," *Energy*, vol. 147, pp. 390–402, 2018.
[6] Z. Zhang, R. Tang, T. Peng, L. Tao, and S. Jia, "A method for minimizing the energy consumption of machining system: integration of process planning and scheduling," *J. Clean. Prod.*, vol. 137, pp. 1647–1662, 2016.
[7] Z. Liu and L. Wang, "Sequencing of interacting prismatic machining features for process planning," *Comput. Ind.*, vol. 58, no. 4, pp. 295–303, 2007.
[8] H. Samuel, "Automated setup planning for lathe machining," 1998.
[9] A. Y. Bok and M. S. Abu Mansor, "Generative regular-freeform surface recognition for generating material removal volume from stock model," *Comput. Ind. Eng.*, vol. 64, no. 1, pp. 162–178, 2012.
[10] P. S. Kataraki and M. S. Abu Mansor, "A novel classification of freeform volumetric features and generative CAPP approach for milling machine selection," *Int. J. Adv. Manuf. Technol.*, pp. 1–25, 2018.
[11] P. S. Kataraki and M. S. Abu Mansor, "Auto-recognition and generation of material removal volume for regular form surface and its volumetric features using volume decomposition method," *Int. J. Adv. Manuf. Technol.*, 2016.
[12] Z. Y. Zhao, C. Y. Wang, H. M. Zhou, and Z. Qin, "Pocketing toolpath optimization for sharp corners," *J. Mater. Process. Technol.*, vol. 192–193, pp. 175–180, 2007.
[13] X. Qu and B. Stucker, "Circular hole recognition for STL-based toolpath generation," *Rapid Prototyp. J.*, vol. 11, no. 3, pp. 132–139, 2005.
[14] A. Masood, R. Siddiqui, M. Pinto, H. Rehman, and M. A. Khan, "Tool path generation, for complex surface machining, using point cloud data," *Procedia CIRP*, vol. 26, pp. 397–402, 2015.
[15] L. Liu, Z. Huang, W. Liu, and W. Wu, "Extracting the turning volume and features for a mill / turn part with multiple extreme faces," *Int J Adv Manuf Technol*, 2017.
[16] M. C. Kayacan, I. H. Filiz, A. I. Sijnmez, A. Baykasoglu, and T. Dereli, "OPPS-ROT : An optimised process planning system for rotational parts," vol. 32, pp. 181–195, 1996.
[17] A. Oral and M. C. Cakir, "Automated cutting tool selection and cutting tool sequence optimisation for rotational parts," *Robot. Comput. Integr. Manuf.*, vol. 20, no. 2, pp. 127–141, 2004.
[18] A. F. Zubair and M. S. Abu Mansor, "Cylindrical axis detection and part model orientation for generating sub delta volume using feature based method," *ARPN J. Eng. Appl. Sci.*, vol. 11, no. 22, pp. 13415–13419, 2016.
[19] A. F. Zubair and M. S. Abu Mansor, "Hole Feature on Conical Face Recognition for Turning Part Model," *IOP Conf. Ser. Mater. Sci. Eng.*, vol. 328, p. 012026, 2018.

[20] A. F. Zubair and M. S. Abu Mansor, "Automatic Feature Recognition of Chamfer and Fillet for Turning Part Model by Volume Decomposition Method," *J. Mech. Eng.*, vol. 5, no. 2, pp. 204–216, 2018.

[21] A. F. Zubair and M. S. Abu Mansor, "Auto-recognition and part model complexity quantification of regular- freeform revolved surfaces through delta volume generations," *Eng. Comput.*, vol. 0, no. 0, p. 0, 2019.

[22] P. Gavankar and M. R. Henderson, "Graph-based extraction of protrusions and depressions from boundary representations," pp. 442–450, 1990.

IOP Conf. Series: Materials Science and Engineering **834** (2020) 012050 doi:10.1088/1757-899X/834/1/012050

Crash Performance of Double Hat Column Subjected to Compression Loading

Salwani MS*[1] and Mohamad Hisyam K[1]

[1]Faculty of Mechanical & Automotive Engineering Technology, Universiti Malaysia Pahang, 26600 Pekan, Pahang.

*salwani@ump.edu.my

Abstract. There are many types of structure designed to act like an energy absorber. Energy absorber is the one that responsible to absorb energy during crashes. An energy absorber is a system that converts, totally or partially, kinetic energy into another form of energy. Energy converted is either reversible or irreversible. One of the examples of the most important structure to absorb energy is the crash box. Crash box can be found in many shapes, and double hat column is the common one. The objective of this study is to study the effect of heat treatment and notch on the performance of the double hat column structure. The results of different specimens were compared in terms of tensile strength and also energy absorbed. It was found that the tensile strength and energy absorbed was slightly higher for specimens that were quenched in lower temperature and longer time and for notching distance of 20mm.

1. Introduction

Energy absorber is the one that responsible to absorb energy during crashes. Crash box is an example of energy absorber. Crash boxes can comes in various geometrical design and hat section among the common one [1]–[5]. Double-hat column design is an upgrade version of the single-hat column. The double-hat column looked like a single-hat column structure being attached together. On a single-hat column, the fin will be at the bottom of the structure. Meanwhile on a double-hat column, the fin points out at the center of the structure.

Nowadays boron steel, as the conventional lightweight materials that increasing use in automotive industry due to their high and ultra-high strength steel. The properties of the boron steel will increase when it go through the heat treatment process. The heat treatment process produced parts with extremely high strength and good ductility [6]. This fact inspires the use of boron steel to produce hat column in this study.

Combination of good crash box design and high strength material was expected to produce highly efficient energy absorber, thus reducing the risk of occupant injury during crash. The use of high strength steel such as boron steel however may increase the peak force. Thus, notch is introduced. In this study, boron steel sheet formed into double-hat column was analysed in terms of strength and energy absorption capability.

2. Methodology

2.1. Specimen preparation

From a large piece of sheet metal that we obtain from the manufacturers, we have to size it down to the planned size by using the shearing machine. The plate was undergone the bending process. The

Content from this work may be used under the terms of the Creative Commons Attribution 3.0 licence. Any further distribution of this work must maintain attribution to the author(s) and the title of the work, journal citation and DOI.
Published under licence by IOP Publishing Ltd

plate was bent to form the fin first. The size of the fin would be 20mm on both ends. When bending the sheet metal, be sure to bend it on line. Align the line well with the die of the bending machine. Make sure the angle is right if not, the shape will be crippled and we have to do it again starting from sheet metal cutting.

For specimen tagging, it consist of numbers. As example "2 10 3" means that the specimen thickness is 2mm and it is quenched in 10°C water and the quenching time is 3 seconds. Also note that "NN" stands for no notch and "NH" stands for no heat treatment.

2.2. Notching
For the next step, trigger was made for the specimens. This trigger is also known as notching. Different notching distance will give different effect. For this specimen, we have two distances for the trigger which is 10 mm and 20 mm. To make this notching, grinder was used. The width is about 1.3 mm and the depth is 0.5mm.

2.3. Welding
In this study, Metal Inert Gas (MIG) technique was used to join the plate together. The part that will be welded is at the fin. Note that we did not weld all the way through the fins. There are distances between each points of welding. Before welding, mark the distance that we need want to weld. There will be 13 points. Start with 6mm on the first point then 25mm until the last point. Note that last point distance is also 6mm. To finish, weld one spot at top of each fin with 10mm center.

2.4. Heat treatment
For the heat treatment process, it involved a furnace for heating up the specimens including the one for the tensile testing. For the tensile testing specimens, from a piece of boron sheet metal, we need to cut it into bone shape. The way the heat treatment is done is by heating the specimens at least 20 minutes and then quenched with certain temperature and time. We will need ten specimens for tensile testing which is five from 2mm thickness and five from 3mm thickness. For 2mm thickness, one specimen is not heat treated. The second one is quenched at 10C in a bucket of water with ice for 3 seconds. Third one is 10C for 8 seconds. The fourth one is quenched at 15C in a bucket of water with ice for 5 seconds. Fifth one was quenched at 15C for 8 seconds. For 3mm thickness, one specimen is not heat treated. The second one is quenched at 10C in a bucket of water with ice for 3 seconds. Third one is 10C for 8 seconds. The fourth one is quenched at 15C in a bucket of water with ice for 5 seconds. Fifth one was quenched at 15C for 8 seconds.

3. Result and Discussion
Figure 1 and Figure 2 shows stress-strain curve for 2mm tensile test and 3 mm sheet.

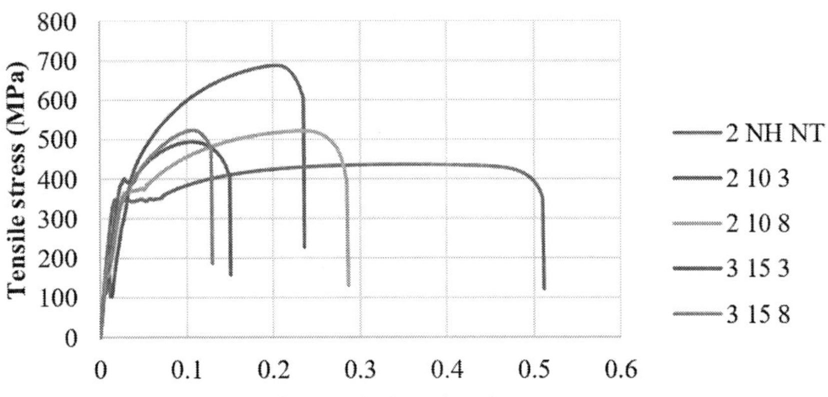

Figure 1. Graph of the 2mm stress strain

It can be seen from Figure 1, the highest tensile strength is specimen 2 10 3, which the tensile strength is 685.236MPa and the lower one is specimen 2 NN NT which just have 435.477MPa. The specimen 2 10 3 also have the higher break point compare to other. The specimen 2 10 3 have the higher tensile strength and break point may due to some error when running the test. If the tensile strength higher, the more force need to break the thing is higher.

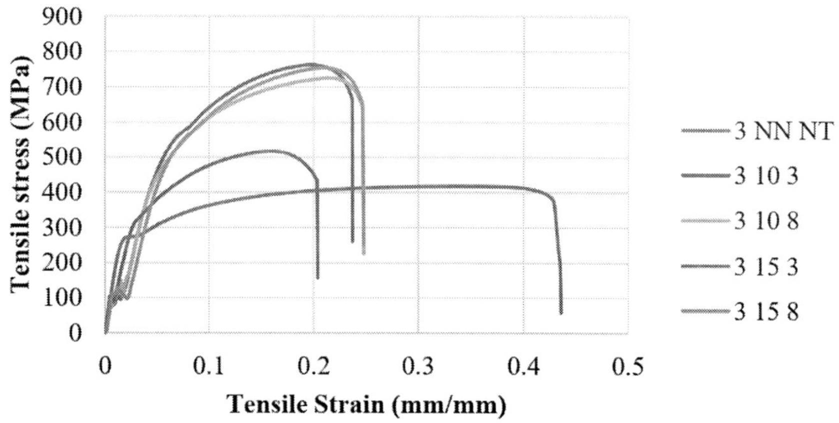

Figure 2. Graph of the 3mm stress strain

From Figure 2, the highest tensile strength is specimen 3 15 8, which the tensile strength is 752.782MPa and the lower one is specimen 2 NN NT which just have 417.709MPa. The specimen 3 15 8 also have the higher break point compare to other.

Based on Table 1 and Table 2, if we compare specimens without heat treatment and without notching, we could see that 3mm thick specimen have bigger energy absorbed value. So we can conclude that the thicker the specimen, the higher the value of energy absorbed. From the data obtained, supposedly the value of energy absorb would be bigger for the specimens that have longer quenching time.

Table 1. 10mm Notching Distance

Specimen	Energy absorbed (kJ)
2 NH 10	0.805
2 10 3	0.763
2 10 8	1.020
2 15 3	0.808
2 15 8	0.925
3 NH 10	0.945
3 10 3	1.403
3 10 8	0.846
3 15 3	1.426
3 15 8	0.982

Also from the data obtained, the energy absorbed should be slightly higher for specimens that were quenched in lower temperature. However, some specimens did not show such pattern also due to some errors during other process. If we compare between 10mm notching distance and 20mm notching distance, we could see that almost all specimens with 20mm notching distance has higher energy absorbed value. We can conclude that the distance of notching has effect on energy absorbed.

Table 2. 20mm Notching Distance

Specimen	Energy absorbed (kJ)
2 NH 20	0.792
2 10 3	0.956
2 10 8	1.144
2 15 3	1.210
2 15 8	0.959
3 NH 20	1.01
3 10 3	1.605
3 10 8	1.501
3 15 3	0.806
3 15 8	1.150

Table 3. No Notching, No Heat Treatment

Specimen	Energy absorbed (kJ)
3 NH NN	1.148
2 NH NN	1.007

Conclusion

Based on the works done, specimen heat treatment process and notching significantly affect the tensile strength and energy absorption of the material. It was found that the tensile strength and energy absorbed was slightly higher for specimens that were quenched in lower temperature and longer time and for notching distance of 20mm.

Acknowledgments

The author would like to acknowledge the support of the internal grant of Universiti Malaysia Pahang, RDU1803116 and support provided by the Ministry of Higher Education, Malaysia under Fundamental Research Grant Scheme, FRGS/1/2016/TK03/UMP/02/22.

Reference

[1] Yamashita M, Gotoh M. Control of crush mode of strengthening members with hat-shaped cross-section in axial impact. Mater Sci Forum 2004; 465: 21–26.
[2] White. A theoretical analysis for the quasi-static axial crushing of top-hat and double-hat thin walled sections. Int J Mech Sci 1998; 41: 209–233.
[3] Gümrük R, Karadeniz S, Gu R. A numerical study of the influence of bump type triggers on the axial crushing of top hat thin-walled sections. Thin-Walled Struct 2008; 46: 1094–1106.
[4] Yamashita M, Kenmotsu H, Hattori T. Dynamic axial compression of aluminum hollow tubes with hat cross-section and buckling initiator using inertia force during impact. Thin-Walled Struct 2012; 50: 37–44.
[5] Yamashita M, Kenmotsu H, Hattori T. Thin-Walled Structures Dynamic axial compression of aluminum hollow tubes with hat cross-section and buckling initiator using inertia force during impact. Thin Walled Struct 2012; 50: 37–44.
[6] Bingtao T, Fangxing W, Qingfeng W, Chenchen L, Jiyuan L, Hailong G. Numerical and experimental study on ductile fracture of quenchable boron steels with different microstructures. International Journal of Lightweight Materials and Manufacture 2019.

IOP Conf. Series: Materials Science and Engineering **834** (2020) 012051 doi:10.1088/1757-899X/834/1/012051

Effect of Additives on Performance and Emission Characteristic of Diesel Engine Fuelled with Biodiesel

Asiah Ab. Rahim[1], Hazim Sharudin[2], Mohd Aflah Quhairy Sapri[3], Helmisyah Ahmad Jalaludin[1], and Nik Rosli Abdullah*[3],

[1]Faculty of Mechanical Engineering, Universiti Teknologi MARA, 23200 Bukit Besi, Terengganu, Malaysia.
[2]Faculty of Mechanical Engineering, Universiti Teknologi MARA, 13500 Permatang Pauh, Pulau Pinang, Malaysia.
[3]Faculty of Mechanical Engineering, Universiti Teknologi MARA, 40450 Shah Alam, Selangor, Malaysia.

*nikrosli@salam.uitm.edu.my

Abstract. The depletion of fossil fuel resources and environmental pollution are two major crises that are currently faced by mankind. As the solution for the crises, an additive liquid like Tripmexx is introduced into automotive industries to increase performance of internal combustion engine, hence reducing bad emission to the environment. The objective of the study is to investigate the effect of various Tripmexx mixed into 80% diesel and 20% biodiesel (B20) with an amount of 0.1 ml (B20-0.1), 0.2ml (B20-0.2), and 0.3ml (B20-0.3) compared to conventional pure diesel. The experiment was carried out by using single cylinder, four-stroke diesel engine, and conducted with constant speed at range of 1000, 1500, and 2000 rpm with various load range from 2Nm to 6Nm. For constant speed 2000 rpm, the B20-0.3 is in lower brake specific fuel consumption value compared to the diesel fuel by 40% on average. The blend B20-0.3 produce higher brake thermal efficiency reading compared to diesel fuel by 5%. Besides, the carbon monoxide and hydrocarbon product of B20-0.3 is lower than diesel fuel by 92% and 0.002% respectively. However, the nitrogen oxide and carbon dioxide produced by B20-0.3 is higher than diesel fuel by 68% and 50% respectively. In overall, the B20-0.3 shows the best results for all measured parameters at all engine test due to consideration of green fuel.

1. Introduction

The continual use of fossil fuel resources by people in the world for transportation and industry which is involve in combustion process is a major contributor in air pollution. This may affect to unhealthy environment thus it may cause the global warming, ozone layer depletion, and jeopardize human respiratory system [1]. Moreover, air quality is a main concern to be improved to avoid air pollution to surrounding which may harmful to the human.

Alternative fuel such as biodiesel is receiving a lot of attention in the automotive industries due to the shortage in the hydrocarbon fuel source. The key issue in using vegetable oil-based fuels is oxidation stability, stoichiometric point, bio-fuel composition, antioxidants on the degradation, and much oxygen with comparing to diesel fuel [2]. So that the improvement of exhaust gas emission from biodiesel fuel engines is urgently required to meet the future stringent emission regulation. Biodiesel has higher cetane number than diesel fuel, no aromatic, almost no sulphur and contains 10-11% oxygen by weight. These

Content from this work may be used under the terms of the Creative Commons Attribution 3.0 licence. Any further distribution of this work must maintain attribution to the author(s) and the title of the work, journal citation and DOI.
Published under licence by IOP Publishing Ltd

characteristics of the fuel reduce the emission of carbon monoxide (CO) and hydrocarbon (HC) in exhaust gas. In addition, higher viscosity and specific gravity are found to increase the fuel quantity, injection timing, and spray pattern will strongly influence the degree of initial premixing and combustion process [3]. Numerous emissions using biodiesel and its blends show emission concentration varies depend on source of biodiesel and engine combustion system.

Fuel additive is chemical that are used to enhance the properties of the fuel. Currently, numerous chemical additives is used in transportation to improve the quality of biodiesel fuel and diesel fuel in order to convene up the most wanted performance level [3]. Additives will help petroleum to recover its engine combustion, performance, and emission environmental standards. Various studies have proven that adding additives in biodiesel enhancing engine performance and reduce engine emissions. According to [4], the result of feasibility of pure biodiesel and blended fuel at high blending ratio using different chemical additives by different researchers which evaluated the fuel properties trend, and engine performance and emissions with different chemical additives varied of chemical additives can be utilised with biodiesel fuel to improve the fuel properties. [5] investigated on the butanol/acetane (BA)-diesel and showed that BA can improved energy efficiency and reduce emissions. Moreover, [6] studied the effect of adding graphene oxide (GO) nanoparticles into Jatropha Methyl Ester (JME) and the result showed that there are increment in brake thermal efficiency by 17% while reduction in carbon dioxide (CO), unburned hydrocarbon (UHC), and nitrogen oxide (NOx) by 60%, 50%, and 15% respectively for JME-GO blends compared to JME fuel. Furthermore, [7] analysed the effect of silver oxide (Ag_2O) as an additive in various mass fractions, and the results are compared to conventional diesel. The investigation indicated that the addition of Ag_2O nanoparticles into palm oil mill effluent (POME) resulted in enhancement in ignition characteristics due to the enhanced surface area/volume ratio. In addition, it also resulted in enhancement in brake thermal efficiency (BTE) with a reduction in brake specific fuel consumption (BSFC), and improved level of HC, CO, smoke, and NOx emissions. [8] reported that additive blends B20 show an increased thermal efficiency and decrease in specific energy consumption compared to diesel and B20 fuel, also show significant reduction of CO and NOx. Meanwhile, recent research shows that Palm Oil Based 10% Biodiesel (PB10) with 0.8 ml additives produced the highest braking power and lowest fuel consumption as compared to the diesel. It resulted to not only improve the engine performance, but also led to the reduction of carbon emission [9].

The additives selection will be based upon the draw-backs of biodiesel fuel such as density, toxicity, viscosity, economic feasibility, additives solubility, auto ignition temperature, flash point, and cetane number for the fuel blending process. There are several types of additive which is mostly used in the ignition fuel which is metal based additive, oxygenated additive, cetane number improver, and antioxidant additive. Recently, [10] reported that the biodiesel stability can be obtained by using a reduced amount of additives, suggesting that biodiesel shelf life can be improved in association with a cost reduction when compared to the use of conventional antioxidants.

The objective of this study is to investigate the effect of fuel additive (Tripmexx) blended with B20 diesel-biodiesel fuel on engine performance and exhaust emission characteristics. Both were tested, evaluated, and compared with the diesel and bio-diesel fuel. In this study, Tripmexx is used as the fuel additive that acts as corrosion inhibitors or lubricant thus may create greater efficiency and power from the higher compression ratio. This additive is added into diesel-biodiesel blends in order to enhance engine performance, reduce emissions, and serve some other identified functional purpose beyond the basic fuel itself [11].

2. Methodology

In this study, the experiments were carried out using four-strokes, single-cylinder, and compression ignition (CI) diesel engine with specification as shown in Table 1. Single cylinder diesel engine test bench, Electric current dynamometer, Fuel estimation framework, voltage control and Gas Emission Analyzer mechanical assembly are utilized as shown in Figure 1. The engine type used is direct injection diesel engine with total displacement of 219cc. A 5 kW electric dynamometer is used to measure the revolution speed (rpm) and the torque from the engine.

Three phase power supply is plugged into the system. Next, the air ventilation for exhaust system, control panel and voltage control for dynamometer was set up. The engine was allowed to run about 10 minutes by using pure diesel fuel as a warm up. The exhaust port temperature was set to determine the exhaust temperature while gas analyser was set up on the exhaust tip. After exhaust temperature achieved 70°C, the gas analyser reading can be started and the data was taken after four minutes.

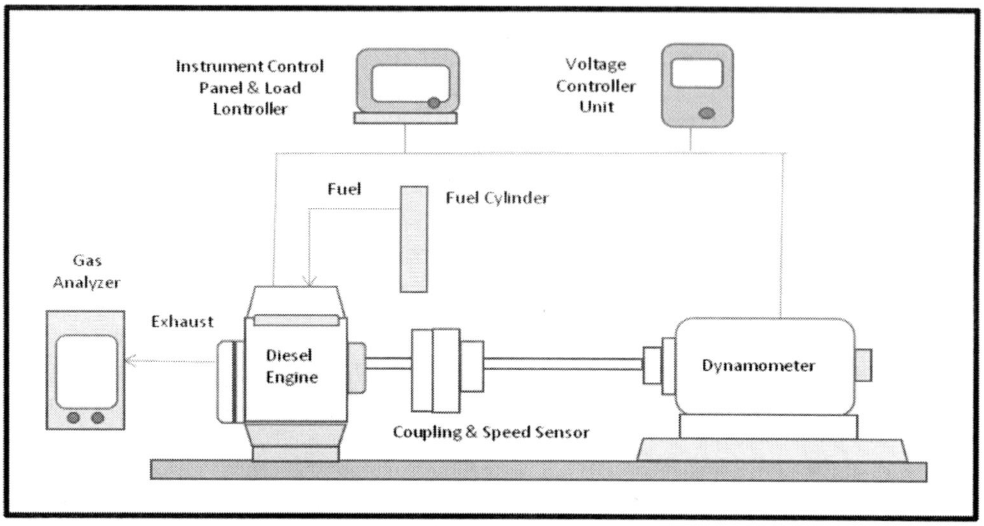

Figure 1. Layout of Experimental Test Rig

Table 1. Test engine specifications

Model	L48 Yanmar Italy
Engine type	Air cooled, 4-cycle, vertical cylinder
Fuel	Diesel
Combustion type	Direct injection
Engine cylinder	1
Total Displacement	219cc
BSFC at rated output	274gkW/hr
Max out put	6.8kW
Engine rated speed	3600rpm
Bore x Stroke	70 x 57 (mm)
Cooling system	Forced air cooling by flywheel fan

Table 2. Test Condition

Engine Speed	1000-2000 rpm (500 rpm increment)
Engine Load	2, 4, 6 N.m
Time Measurement for exhaust emissions	4 minute per testing
Time for fuel consumption	4 minute per testing

The engine will be tested at constant low engine speed range of from 1000rpm, 1500rpm, and 2000rpm with variation load 2 Nm, 4 Nm and 6 Nm by using dynamometer control as shown in Table 2. Hence, exhaust gas emission for each load and fuel consumption at varied speed will be recorded after 4 minutes. Two-cylinder fuel system with valve system is used to change sample used. This procedure will be repeated by using each fuel sample.

The mass flow rate for fuel consumption, engine power and torque generated are needed for the engine performance verification. The calorific value of each sample fuel is needed for the Brake Thermal Efficiency (BTE). Calorific value of the sample fuel was determined by using bomb calorie.

BSFC implies the ratio of fuel consumption rate to the brake power output. BSFC can be used as a convenient parameter in order to assess the engine performance with varied fuel blends. To calculate BSFC, the following formula used is;

$$BSFC = \frac{r}{P} \tag{1}$$

where;

BSFC is brake specific fuel consumption in (g/kW.h)
r Is the fuel consumption rate in kilograms per second (kg/s)
P Is the power produced in watts where $P = \tau\omega$
ω Is the engine speed in radians per second (rad/s)
τIs engine torque in Newton meters (N.m)

BTE can be additionally used to show on how well a motor ready to change over the thermal energy to deliver a valuable mechanical work. To ascertain BTE, the formula is;

$$\eta = \frac{BP \times 3600}{V \times cv} \times 100\% \tag{2}$$

where;
η is BTE in percentage (%)
BP is the brake power
V is the fuel's volume flow rate (kg/s)
cv is the calorific value (kJ/kg)

The exhaust gas analyser, MRU Gas Analyser VARIOplus is employed to measure the percentage of gas emission from exhaust such as O_2, CO_2, CO, NOx, and HC which give response time T90 at 30 seconds from analyser to inlet, 0.05% respectively 1ppm detection limit and both 1% FS for linearity error and repeatability. For data collection, the MRU gas analyser is allowed to run for 4 minutes for one data. The reading will be repeated for 3 times for the average value.

In this work, a fuel additive manufactured from German, Tripmexx is used. The additive is in a liquid form which is claimed to have oxygen rich molecule as fuel catalyst which works on the molecular level of fuel to enable better and more complete combustion to result in fuel saving, better engine performance with lesser emission of harmful gas to the environment.

3. Results and Discussion

BSFC implies the ratio of fuel consumption rate to the brake power output. BSFC can be used as a convenient parameter in order to assess the engine performance with varied fuel blends. Figure 2 shows the variations of the BSFC of different fuel blends against variation load condition at constant speed 2000 rpm which show the lowest BSFC. As shown in the graph, the BSFC of the modified blend decreased as the load condition increase from 2 Nm to 6Nm load condition. It can be clearly observed that all the diesel fuel and blend possessed lowest BSFC at the load of 6 Nm. At 6 Nm load, the diesel fuel possessed of BSFC of 396 g/kW.h, while the B20-0.3 blend possessed BSFC of 264 g/kWh which is 40% lower than the diesel fuel. It showed that B20-0.3 modified blend has better BSFC than the diesel fuel at this load. The B20, B20-0.1 and B20-0.2 blends exhibited 18%, 27% and 34% respectively lower BSFC as compared with diesel fuel. B20-0.3 blend possessed the lowest BSFC at this condition at value of 264 g/kWh. Addition of Tripmexx fuel additive enriches the oxygen content in the fuel mixture, higher oxygen content will increase the compression ratio of the engine. Thus, thermal efficiency also will be increased [2]. It will result lower BSFC and meet fuel economy regulations [1].

BTE can be used to indicate on how well an engine able to convert the thermal energy to produce a useful mechanical work. Some of researches have concluded that the BTE for the biodiesel fuel was lesser than the diesel fuel characteristics. The BTE decreased as the shorter ignition delay of biodiesel and heat loss of diesel engine. BTE also will decrease with biodiesel blend because of low calorific value, higher viscosity, high volatility and poor spray properties. In this study, BTE of the biodiesel increase as the fuel additive added into the fuel blend which promote to the increasing of engine heat due to complete combustion in the internal combustion chamber [12].

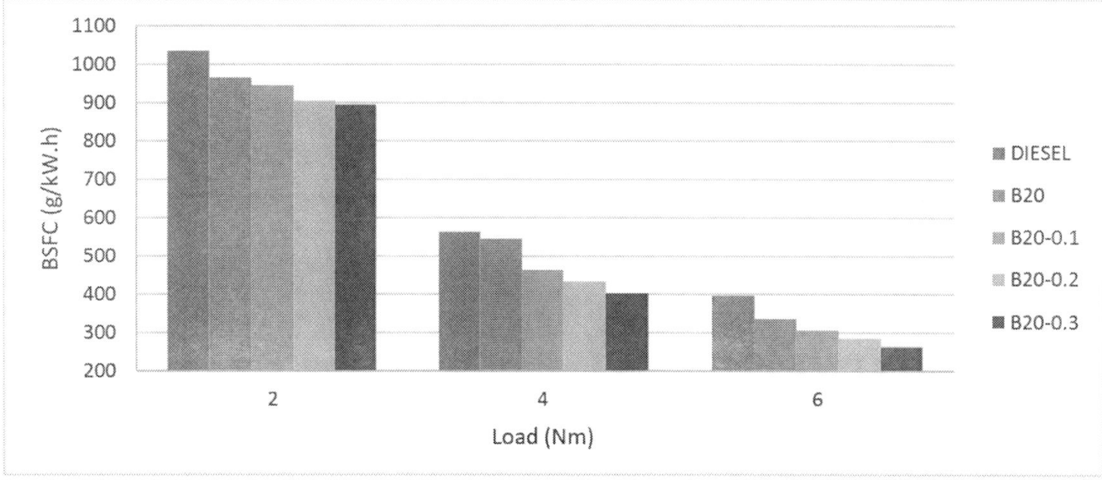

Figure 2. BSFCs vs Loads for base fuel and its modified blends at constant speed 2000 rpm

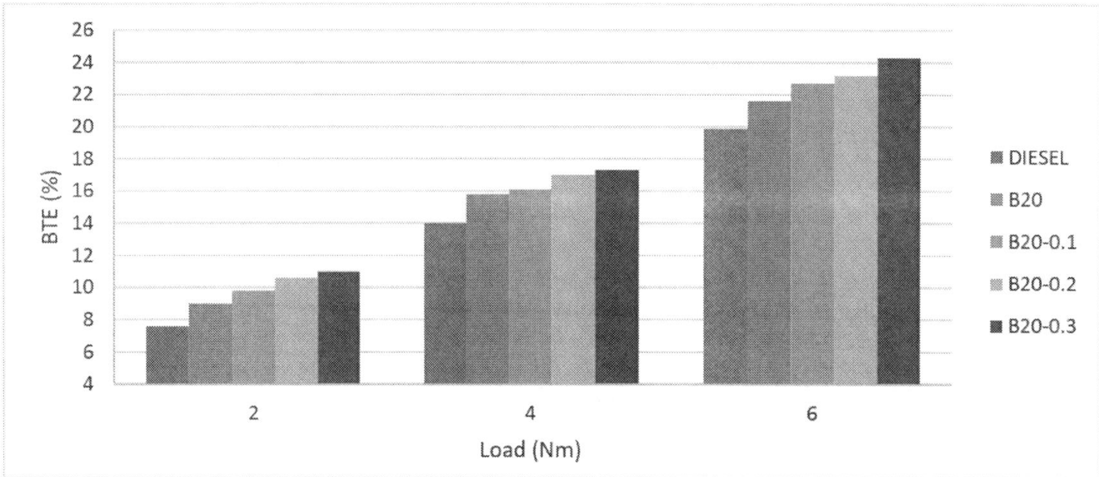

Figure 3. BTEs vs loads for base fuel and new formulated blends at constant speed 2000 rpm

Figure 3 shows the BTEs of the modified blends and diesel fuel against the load applied at constant speed of 2000 rpm. From the graph, the diesel fuel exhibited lowest BTE among the other blends on average 10% at 2 Nm load. This was 4% lower than the B20-0.3 blend at the same load. Moreover, the modified blends of B20, B20-0.1, and B20 0.2 improved BTE than diesel fuel on average 2%, 2.2%, and 3% at 2 Nm load. The reason behind this improvement is identical to the factors of reducing of the BSFCs. The higher compression ratio, the higher the thermal efficiency will be produced [8]. Combustion reactions will also be better due to the increases of BTE [13]. B20-0.3 blend at 6 Nm load condition gave the highest BTE in this condition which is 17%. It was 5% higher compared to the diesel fuel at the same load. Therefore, it can be concluded that the B20-0.3 give the higher BTE on average 13.2% compared with other fuel blend in all engine test.

CO is created because of incomplete combustion in the engine. CO is a colourless, odourless and poisonous gas that is promoted when the engine operates at fuel-rich equivalent ratio. It is produced when there is insufficient oxygen supply to convert all carbon to CO_2, resulting in some fuel does not get burned and some ends up as CO.

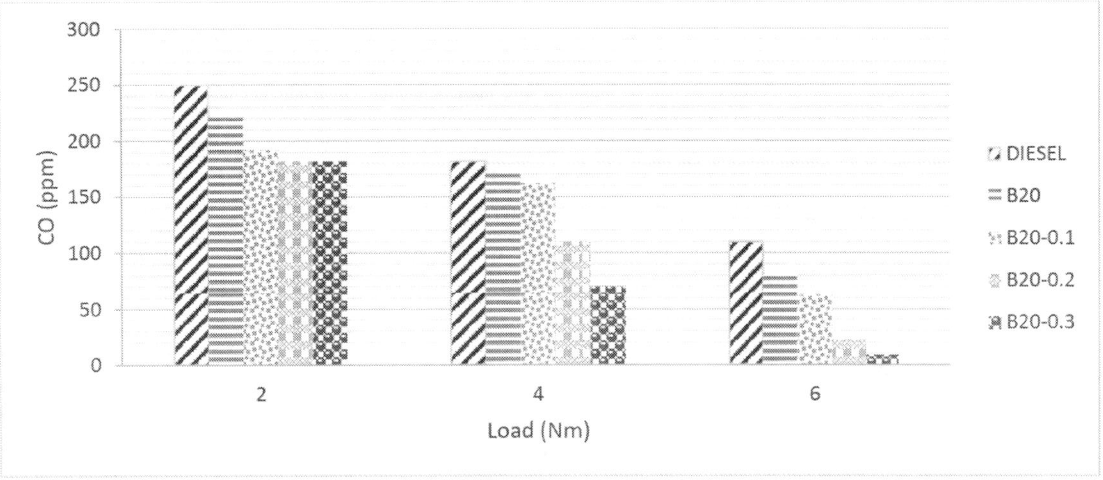

Figure 4. COs vs Loads for base fuel and its modified blends at constant speed 1500 rpm

Figure 4 exhibits the lowest CO of the tested fuel with respect to different loads at a constant speed of 1500 rpm. As shown in Figure 3.3, the CO of the tested fuels decreased as the load increased from 2Nm to 6Nm. The diesel fuels and formulated fuel possessed lowest CO at the load of 6Nm. From the graph, the diesel fuel exhibited highest CO among the other blends on average 110 ppm at 6Nm load. This was 92% higher than the B20-0.3 blends at the same load. Additionally, the modified blends of B20, B20-0.1, and B20-0.2 reduced CO than diesel fuel on average 26%, 43%, and 80% respectively at 6 Nm. The reason behind this enhancement is due to complete combustion is accomplished. The high content oxygen in the fuel helps the complete combustion reaction in the internal combustion chamber [3, 11, 13]. The CO formed is declined because CO is not product of complete combustion process. B20-0.3 produced lowest CO at constant speed 1500rpm for 6Nm of load with 9ppm. B20-0.3 possessed the lowest emission value of CO on also due to the good spray atomization and good air-fuel ratio in the internal combustion chamber [14]. In overall, B20-0.3 give the lowest value of CO on average 10% compared with other fuel blend in all engine tests.

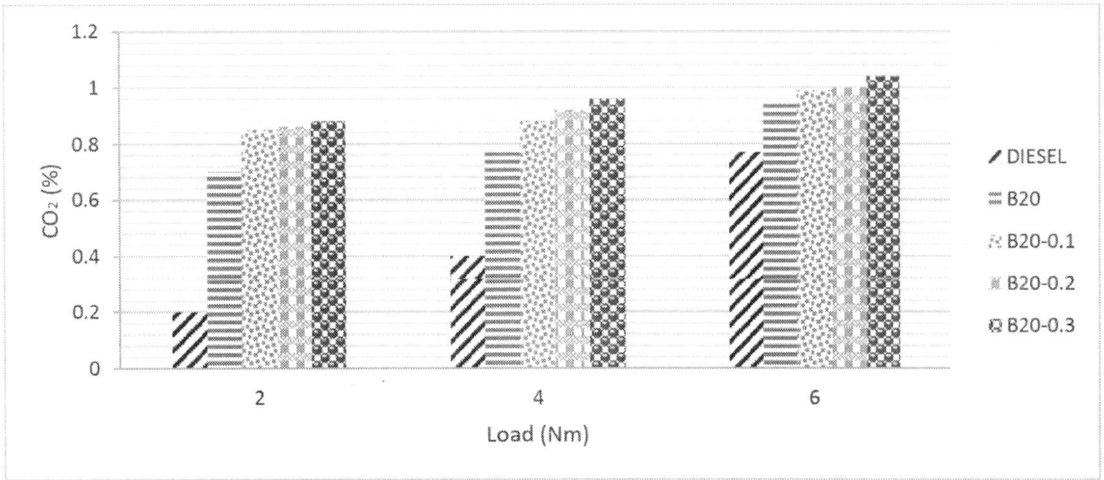

Figure 5. CO_2s vs Loads for base fuel and its modified blends at constant speed 1500 rpm

Carbon dioxide, CO_2 is colourless and odourless gas. The natural chemical compound is consisting of a carbon atom covalently double bonded to two oxygen atoms. Figure 5 exhibits the lowest CO_2 at a constant high-speed of 1500rpm compared other speed of the tested fuels with respect to different loads. It can be clearly seen that the diesel fuel exhibited lowest CO_2 among the other fuel blends on average

0.2% at the lowest load applied. This was 68% lower than the all fuel blends at 2 Nm. As the load increased, the CO_2 produced is also increased. This change was due to the complete combustion in internal combustion chamber happened. Complete combustion achieved in internal combustion engine because of oxygen enrichment and lesser quantity of carbon in the fuel blends in the present of biodiesel and fuel additive [13]. CO_2 at the 2 Nm for diesel fuel possessed lowest value, 0.2% compared to other blends at the same load. B20-0.3 at 6Nm produced 1.0%, which is the highest percentage of this condition. Therefore, it can be concluded that B20-0.3 fuel shows the highest value of CO_2 emission by 0.96% on average value CO_2 gas emission increased as the B20 fuel mixture with Tripmexx increase at every load applied.

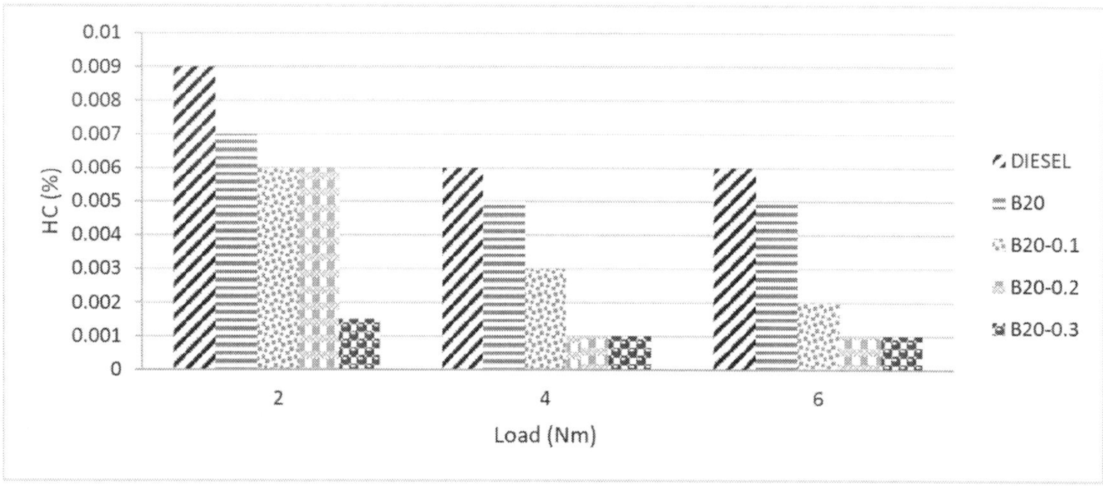

Figure 6. HCs vs Loads for base fuel and its modified blends at constant h speed 1500 rpm

Hydrocarbons can be found in the diesel and biodiesel fuel, which are its primary source of energy. Any hydrocarbon emitted from vehicle indicates unused fuel which results from incomplete fuel combustion. Figure 6 exhibits the lowest HC at a constant speed of 1500 rpm emission of the tested fuels with respect to different loads. From the graph, the diesel fuel exhibited highest HC among the other blends on average 0.009% at 2Nm. This was 0.002% higher than the B20-0.3 and 0.003% higher than B20-0.1 and B20-0.2 blends at the same load. The B20-0.3 and B20-0.2 blends possessed lowest HC emission with 0.001% emission at 4Nm and 6Nm load applied.

The formulated fuel B20, B20-0.1, and B20-0.2 reduce HC emission than diesel fuel on average 0.005%, 0.002%, and 0.001% respectively at 6Nm load. This improvement is due to complete combustion in the internal combustion chamber. High oxygen content in the fuels helps to achieve complete combustion in the engine [3]. HC is not a product of complete combustion; thus, the emission is able to be reduced. B20-0.3 produced the lowest percentage of HC at 2Nm, 4Nm and 6Nm. Hydrocarbon emission from internal combustion engine is resulted from unburned fuel. Higher cetane number and higher oxygen quantity and can reduced the Hydrocarbon emission. The quantity of the oxygen present in the biodiesel leads to complete combustion and the higher cetane number reduces the ignition delay, as shorter the ignition delay decreases the unburned hydrocarbon [15]. Therefore, it can be concluded that the B20-0.3 give the best result with the lowest emission of HC gas which is 0.0012% on average compared with other fuel blend.

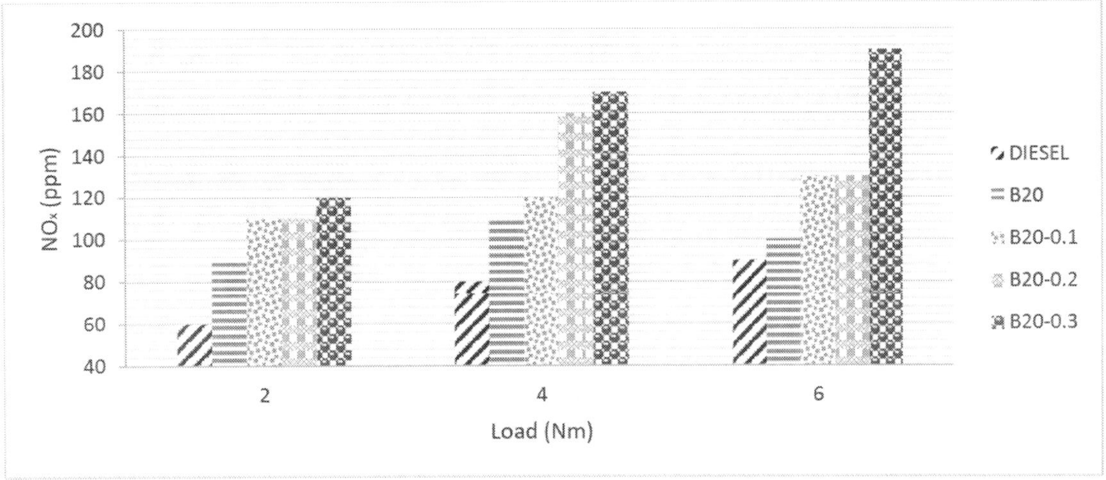

Figure 7. NOxs vs loads for base fuel and new formulated blends at constant speed 1000 rpm

NOx is formed due to the engine highly rely on the temperature. The higher the combustion reaction temperature, the more diatomic nitrogen, N_2 dissociates with monoatomic nitrogen, N and more NO_x will be formed. Pressure, reaction temperature, residence time of combustion products, premixed portion of combustion, availabilities of oxygen, ignition delay period, heat removal rate and operational parameters of the engine is part of the NO_x formation.

Figure 7 shows the lowest NOx of the tested fuels with respect to different engine loads at a constant speed of 1000rpm. The diesel fuel exhibited lowest NOx among other blends on average 6 ppm at 2Nm. This was lower than all blends at the same load. In addition, B20-0.3 blend at 6Nm produced 19ppm and increased NOx than diesel fuel on average 111%. The reason behind this improvement is due to complete combustion was achieved. The oxygen enrichment in the fuel aids the complete combustion reaction in the internal combustion chamber [3, 11, 13]. Complete combustion increasing the engine temperature and formed NOx. If the engine temperature is high, the emission of NOx will be increased. B20-0.3 at 6Nm for 1000rpm speed produced highest ppm value which is 19ppm compared to other blends for all speed. It can be concluded that fuel additive added significantly increase the formation of NO_x gas. Therefore, B20-0.3 fuel blend show the higher formation of NO_x which is 160ppm on average value compared with other fuel blend.

4. Conclusions and Recommendations

An inclusive study was performed to examine and understand the performance and exhaust emissions of diesel engine fuelled with diesel-biodiesel blends with different percentages of Tripmexx additives which were used on a single cylinder compression-ignition (CI) diesel engine. BSFC and BTE were measured and calculated to compare the performance characteristics of the engine. Exhaust emission such as NOx, CO, CO_2, and HC were measured for each selected fuel. Below are the summarised conclusions from the findings:

- BTE was enhanced by 5% and the BSFC was decreased by 40% with the addition of Tripmexx at B20-0.3 blends. The addition of Tripmexx additive into the fuel blends showed lower BSFC and BTE than the base fuel because oxygen content for the test fuels with Tripmexx is higher than the base fuel.
- The fuel additive has been improved some of the exhaust emission characteristics where it reduced the CO and HC while the CO2 and NOx increase significantly compared with base fuel. B20-0.3 give the lowest value for both CO (10%) and HC (0.0012%) on average compared with other fuel blend in all engine tests.
- In overall, the B20-0.3 blend shows the best result for all measured parameters at all engine test compared with other blended fuel.

Moreover, some recommendation and future study that can be made which related to this research which by using other fuel additive which can be compared with this study. The advantageous and disadvantageous of two different fuels will be compared to observe the best result produced. Hence, the use of other biodiesel Improved formulation of fuel blends using automatic stir system and using high calorific value, low viscosity and low volatility of biodiesel also will shows the different result to be compared with this study by using more advanced experimental apparatus which give more precise and accurate result. The experimental procedure also can be change by using high specification engine with more loads applied as a variable or at the constant load with various speed as a future study to investigate the effect of additive on its engine performance and emission characteristics.

5. References

[1] Geng P, Cao E Tan Q Wei L 2017 *Renew. Sust. Energy Rev.* **71** 523-34
[2] M.S.M Zaharin, N.R Abdullah, G Najafi, H.Sharudin, T.Yusaf 2017 *Renew. Sust. Energy Rev.* **79** 465-493
[3] Nadir Yilmaz, Alpaslan Atmanli, Miquela Trujillo 2017 *Fuel* **207** 461-469
[4] O.M.Ali, R.Mamat, N.R.Abdullah, A.A.Abdullah, F.Khoerunissa and R.E.Sardjono 2016 Effect of Different Chemical Additives on Biodiesel Fuel Properties and Engine Performance:A Comparison *Review MATEC Web of Conferences* 38, 03002 (2016)
[5] S.Jabbar, M.Algayyim, A.P.Wandel, T.Yusaf and S.Al-Iwayzy 2018 *Fuel* **227** 118-126
[6] I.El-seesy, H.Hassan and S.Ookawara 2018 *Energy Convers. Manag.* **166** 674-686
[7] Y.Devarajan, A.Mahalingam, D.B.Munuswamy and T.Arunkumar 2018 *Energy Fuels* **32** 8447 8452
[8] M.M.Musthafa, T.A.Kumar, T.Mohanraj and R.Chandramouli 2018 *Fuel* **225** 343-348
[9] C.B.How, N.M.Taib and M. R. A Mansor 2019 *International Journal of Automative and Mechanical Engineering (IJAME)* **16** 6173-6187
[10] S.S. Yahagi, A.C. Roveda, A.T.Sobral, I.P.Olivera, A.R.L.Caires, R.S.Gomes and M.A.G. Trinidade 2019 An Analytical Evaluation of the Synergistic Effect on Biodiesel Oxidation Stability Promoted by Binary and Terbary Blends Containing Multifunctional Additives *Int.Jour.of Analytical Chemistery* Vol.2019, Article ID 6467183.
[11] Arkadiusz, Jamrozik and Michal Sobieppanski 2017 *Thermal Science* **21** 591-604
[12] Karoon Fangsuwannarak, Ponrawee Wanriko, Thipwan Fangsuwannarak 2016 *Energy Procedia* **100** 227-236
[13] H. Kumar Patel, S. Kumar 2017 *Thermal Science and Engineering Progress* **4** 252-258
[14] Suthar Dinesh Kumar L, Dr Rathod Pravin P, Prof Patel Nikul K. 2013 Effect of Fuel Additive on Performance and Emission for Diesel Engine Fuelled with Blend of Naem Oil Biodiesel and Diesel *International Journal of Advanced Engineering Technology* Volume IV 66-70
[15] B.Ghobadian, H.Rahimi, A.M. Nikbakht, G.Najafi, T.F. Yusaf 2009 *Renew.Energy* **34** 976-982

Effect of TRIPMEXX Additives on Engine Performance of a 1.6L Multi-Cylinder Gasoline Engine

Hazim Sharudin[1], Nik Rosli Abdullah*[2*], A R Asiah[3], Helmisyah Ahmad Jalaludin[3], and Matthew EE[4]

[1]Faculty of Mechanical Engineering, Universiti Teknologi MARA, Cawangan Pulau Pinang, Kampus Permatang Pauh, 13500, Pulau Pinang.
[2]Faculty of Mechanical Engineering, Universiti Teknologi MARA,40450 Shah Alam, Selangor, Malaysia.
[3]Faculty of Mechanical Engineering, Universiti Teknologi MARA, 23200 Bukit Besi, Terengganu, Malaysia.
[3]Unit No.312, Block E, Phileo Damansara 1, No.9 Jalan 16/11, Off Jalan Damansara, 46350 Petaling Jaya, Selangor Malaysia.

* nikrosli@salam.uitm.edu.my

Abstract. This study discusses the findings of the project data obtained from the experimental work. It was based on the analysis of experimental output responses obtained which were Brake Specific Fuel Consumption (BSFC), power and engine torque. A 1.6L multi-cylinder gasoline engine was used to investigate the effect of TRIPMEXX on the fuel economy. The experiment was performed at variations of engine speed (1500, 2000, 2500 and 3000 rpm) with a constant engine load of 40 Nm. From the results obtained, it showed that brake specific fuel consumption (BSFC), power and engine torque for TRIPMEXX were better than gasoline fuel without TRIPMEXX fuel additive. Combustion efficiency of gasoline engine has improved with the use of TRIPMEXX fuel additives. Overall, Fuel Economy, Power and Engine Torque for 1 ml + 10L RON95 fuel sample showed the best results for all measured parameters at all engine test conditions.

1. Introduction

The fear from the exhaustion of petroleum oil and uncertainty in the oil prices for the transportation sector has raised concerns around the world. Approximately 25% of the total world energy consumption was dominated by the transportation of people and goods as per reported by the Energy Information Administration (EIA) [1]. In the current economic crisis, the requirement to develop high technologies in automotive industries especially to improve the combustion process is a key for future engine development. However, combustion efficiency is also possible to improve through reformulation of existing fuel. This prompt researchers to develop a fuel additive which can meet present requirements of energy supply and improve the combustion process in the transportation sector [2]. Alcohols such as methanol, ethanol and butanol have been actively used by the researchers as a fuel additive in the gasoline fuels for spark ignition engines [3-5].

S.B. Singh et al. [3] integrated butanol-gasoline blends with 5,10,20 50 and 75% volume percentage in a four cylinder spark ignition engine. The butanol-gasoline blend consumes slightly higher consumption of fuel compared with gasoline due to its lower calorific value than gasoline. In

Content from this work may be used under the terms of the Creative Commons Attribution 3.0 licence. Any further distribution of this work must maintain attribution to the author(s) and the title of the work, journal citation and DOI.
Published under licence by IOP Publishing Ltd

addition, combustion duration of butanol-gasoline blends become longer than gasoline due to an increase in turbulence and flame velocity [3]. L. Siwale et al. [5] prepares dual alcohol-gasoline blends with various proportions of methanol and n-butanol to be compared with single alcohol-gasoline blends (methanol) on engine performance, exhaust emission and combustion characteristics. The dual alcohol-gasoline blend displayed better brake thermal efficiency (BTE) and reduce exhaust gas temperature (EGT) compared with the single alcohol-gasoline blend. This is due to improved volumetric efficiency and also reduced the compression work [5]. Demonstration performed by Andersen et al. [6] shows that reduction of vapor pressure and evaporative emission on the fuel delivery system can be achieved by mixing lower alcohol (ethanol) with higher alcohol (butanol).

In terms of fuel additives fueled in the gasoline, a proper fuel formulation leads to a complete mixing process that results in complete combustion. One of the best methods to improve the combustion behaviour is by blending base fuel with additives. Fuel additives can contribute towards fuel economy either directly or indirectly. Some additives with extra lubricity properties are also possible to extend engine lifetime than usual. Fuel additives on the market come from different bases such as metal-based additives, octane booster, oxygenated additives, cetane number improvers, antioxidants and lubricity improvers. All of them are used to meet specifications and quality before entering the market. The selection of fuel additives is important because some additives resulted in higher octane number that leads to engine damage due to knocking and overpower beyond engine specification

This study involves advanced fuel additive called TRIPMEXX. Initially developed to enhance better fuel efficiency to boost engine power and later with additional developments resulted in fuel saving. TRIPMEXX now offers a complete fuel efficiency solution, providing improvement in power and fuel savings. There are several benefits offered by this additive such as enhanced power & torque, fuel economy, clean engine by reducing carbon & other deposits. Moreover, TRIPMEXX also able to reduce engine parts friction and prolong engine life-time, it is safe on the engine as it is non flamable and non corrosive. TRIPMEXX fuel additive is an oil-based compound with an effective combination of essential chemicals and has undergone a unique and special process to make it safe for engine as it is non-flammable and non-corrosive. Though many advantages have been made by using TRIPMEXX, there are still limited information to acknowledge the effect of TRIPMEXX on fuel-saving, power and torque improvement at different engine speeds and torques of a gasoline engine. Therefore, the aim of this study is to investigate the fuel economy, power and engine torque on 1.6L multi-cylinder gasoline engine fueled with gasoline and TRIPMEXX fuel additive.

2. Materials and Methods

2.1 Engine setup

Experimental work was carried out by using a 1.6L four stroke multi-cylinder gasoline engine attached with the dynamometer having a maximum power of 150 kW as illustrated in Figure 1. The detail of the engine specification is shown in Table 1. The engine was controlled and monitored by DaTAQ Pro V2 software where the input and output parameters were set and obtained automatically throughout the experiments.

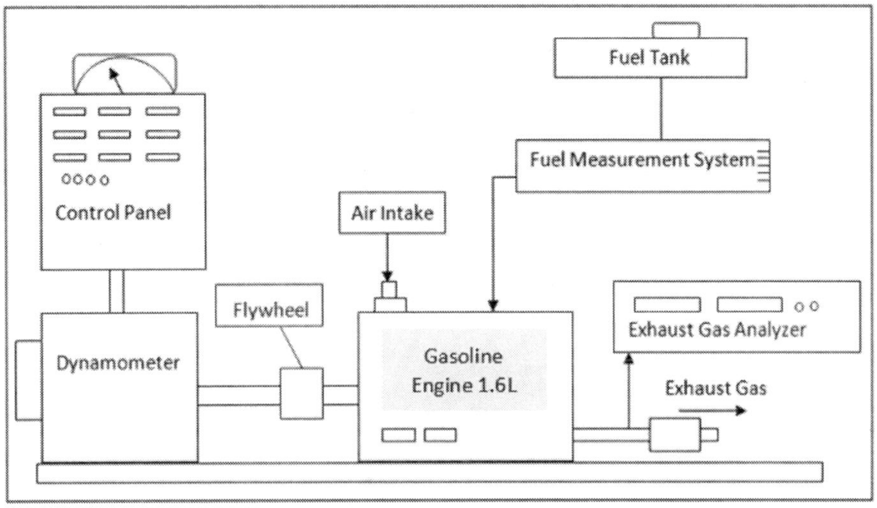

Figure 1. Schematic diagram of the engine setup.

Table 1. Engine specification.

Engine Parameter	Value
Engine Model	4 Cylinder, DOHC 16V
Injection Type	Fuel system multi-point injection (MPI)
Number of Cylinders	4
Configuration	In-line
Bore (mm)	76
Displacement (mm)	1597
Stroke (mm)	88

2.2 Engine test procedure
Engine testing was performed at a constant engine load of 40Nm with variations of engine speed from 1500 to 3000 rpm. During each test, the engine was warmed up with base fuel for 15 minutes to stabilize the engine temperature. The base fuel used in this study was Gasoline RON95 from Petronas Petrol Station. Then, every 10L of RON95 was blended with 1ml of TRIPMEXX additive as a tested fuel for this study. Two important parameters namely engine speeds and loads were controlled during the testing. Pro V2 software was used to communicate with engine and dynamometer. The engine performance was assessed based on the brake specific fuel concumption (BSFC), power and torque. The results presented in this study are the average of the measurements taken from three separate tests, with outlying points either removed or retested.

3. Results and Discussion

3.1 Brake Specific Fuel Consumption (BSFC)
Brake specific fuel consumption (BSFC) indicates how efficient the fuel consumption of the engine to perform work and the fuel is measured in terms of mass flow rate per unit of time for each engine speed. BSFC is one of the important parameters to evaluate engine performance and the quality of fuel combustion efficiency. The unit of BSFC is g/kWhr. The trends of BSFC on two types of different fuel at all engine test conditions are presented below.

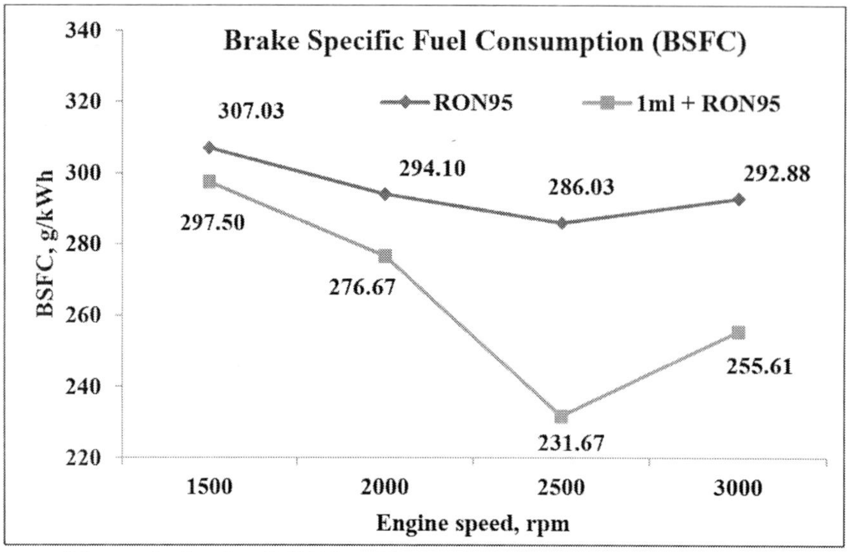

Figure 2. Graph of BSFC vs Engine Speed at constant 40 Nm load

Figure 2 shows the variation of BSFC of all test fuels (RON95 and 1ml + RON95) at different engine speed and constant engine load of 40Nm. The results showed a decreasing trend for both the base fuel and the TRIPMEXX additive blended fuels as the engine speed increased. Figure 2 shows that as the engine speed increases from 1500 rpm to 2500 rpm, the value of BSFC decreases. In average, base fuel (RON 95) has the highest value of BSFC as compared to 1ml + RON95. The average fuel saving when the engine operated with TRIPMEXX fuel additives was approximately 10.20% for the speed of 1500 - 3000 rpm. The average saving for the speed of 2000-3000 rpm was about 12.56%. However, this value increased up to 15.8 % when the engine operated at 2500-3000 rpm. Based on these data, the addition of 1ml TRIPMEXX into RON 95 has reduced the fuel consumption by (10.2% + 12.56% + 15.8%) at three different engine speed regions. The overall average fuel saving for all these regions was approximately 12.85% as tabulated in Table 2.

Table 2. Summary of fuel saving

Engine Speed (rpm)	**1500**	**2000**	**2500**	**3000**
RON95 (g/kWhr)	307.03	294.10	286.03	292.88
1ml + RON95 (g/kWhr)	297.50	276.67	231.67	255.61
Fuel Saving (%)	3.1	5.9	19.01	12.73

The reduction of BSFC occurs due to the ability of additive to promote complete mixing process resulted in homogenous combustion process [7]. At lower speed values, the presence of additives can act as anti-oxidant to accelerate the reaction and assist in producing complete combustions. It also showed that blends gave a better anti-knock performance during low speed acceleration than hydrocarbon fuels of the same octane range [7]. At higher speed of 3000 rpm, BSFC increased for both fuels due to the heat loss to the combustion chamber walls which was proportional greater while combustion efficiency was poorer. The volumetric efficiency was also reduced at higher engine speed that resulted in poor combustion process. Higher BSFC result also affect higher friction power at a rapid rate of engine speed. However, engine operating with TRIPMEXX fuel additive has lower BSFC as compared to RON95 without TRIPMEXX even at higher speed region.

3.2 Power Curve

The main performance envelope (power curve) for the engine is depicted in Figure 3. Power is the measure of how much work can be done in a specified time. The power of an internal combustion engine increases as engine speed increases. This is because power is dependent on torque and engine speed. However, the power of engine drops after reaching a certain engine speed. The maximum power for both fuels is almost identical. However, 1ml + RON95 showed a slightly higher power (approximately 3.0 % higher) at the same engine speed especially at higher speed region. It indicates that 1ml + RON95 burns less fuel in order to produce the same power at the same engine speed. This result also showed that the power produced by TRIPMEXX fuel additive is within engine capacity that protect the engine from damage due to higher pressure and severe knocking during combustion process [8].

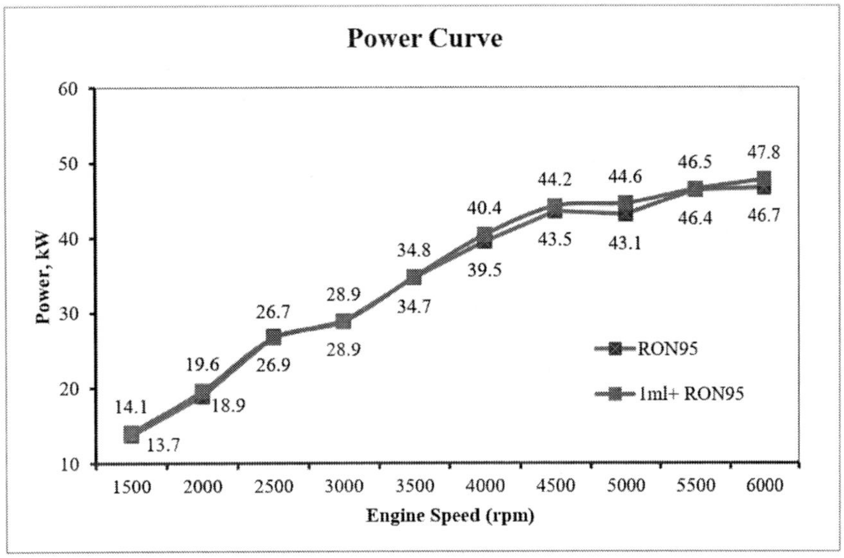

Figure 3. Power curve for engine speed of 1500-6000 rpm

3.3 Torque Curve

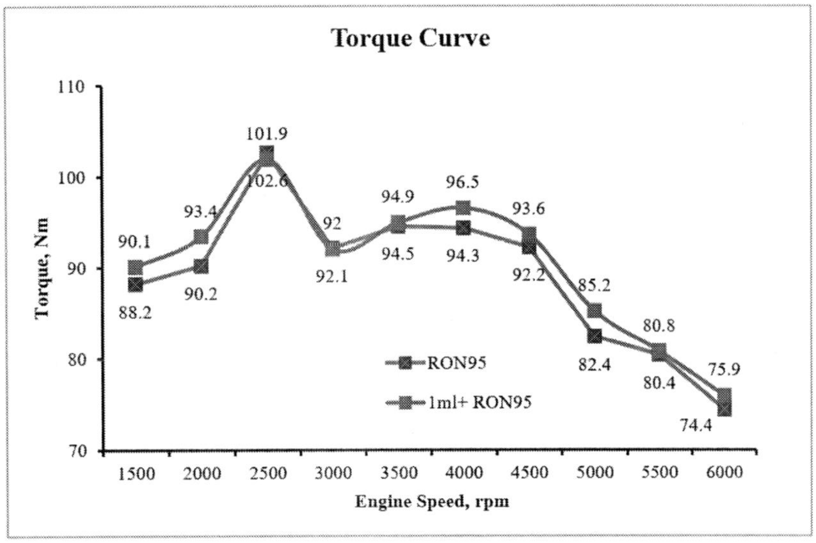

Figure 4: Torque curve for engine speed of 1500-6000 rpm

Torque and engine speed are the measured quantities of engine output. The amount of torque the engine can exert usually varies with engine speed. Figure 4 shows the variation of torque of all test fuels (RON95 and 1ml + RON95) at different engine speed. The results show a decreasing trend for both the base fuel and the additive blended fuel as the engine speed increases. This is because max torque occurs at the maximum volumetric efficiency (breathing ability of the engine) of the engine [9]. The volumetric efficiency of engine was higher at lower speed as compared to higher speed. The results showed that engine operated with 1ml TRIPMEXX fuel additive + RON95 produced higher torque (approximately is 3.0%) as compared to RON95 without fuel additive. Higher torque provides good acceleration and improves the power of an engine while driving. Torque is higher at lower engine speed region as compared to higher engine speed region. This is due to volumetric efficiency. This peak pressure is usually at the lower middle of the engine rpm range and hence resulted in peak torque in this range.

4. Conclusions

The results found that base fuel blended with TRIPMEXX fuel additive improved the BSFC, power and engine torque. The study proved that 1 ml TRIPMEXX fuel additive + RON 95 resulted in higher engine performance and good acceleration. The comparison between two types of fuel at constant engine load and various speed of the engine is listed below:

1. Base fuel blended with 1ml of TRIPMEXX fuel additive resulted in lower brake specific fuel consumption (BSFC) by approximately 3.1%, 5.9%, 19.01% and 12.73% at 1500 rpm, 2000 rpm, 2500 rpm and 3000 rpm respectively. Average fuel saving for engine operating with TRIPMEXX at engine speed of 1500 – 3000 rpm is approximately 10.2%.
2. Average fuel saving on engine when operating with 1ml TRIPMEXX fuel additive is approximately 12.56% at a higher speed of 2000– 3000 rpm. This speed region can be estimated as 80-110 km/hr on the road speed for a normal car.
3. Average fuel saving on higher engine speed (2500 - 3000 rpm) region that is possible to estimate as a highway driving cycle is approximately 15.8%.
4. Therefore, overall fuel saving for city driving and highway driving mode is approximately 12.85%.
5. The Power curve of 1ml TRIPMEXX fuel additive + RON95 resulted in increasing power by approximately 3% and torque by approximately 3% as compared to base fuel without TRIPMEXX fuel additive.

In conclusion, the base fuel blended with TRIPMEXX fuel additive showed better fuel economy, power and torque at all engine test conditions. During the test on power and torque, there is no evidence of knocking and jerking due to the fact that TRIPMEXX fuel additive does not alter the original fuel formulation (RON95), therefore we strongly believe that there is no damage done to the engine. This is due to the properties of TRIPMEXX fuel additive that improve fuel properties that resulted in complete and smooth combustion process.

Acknowledgement

The authors of this research would like to express their sincere gratitude to the Faculty of Mechanical Engineering, Universiti Teknologi MARA, Cawangan Pulau Pinang, Kampus Permatang Pauh, and Faculty of Mechanical Engineering, Universiti Teknologi MARA Shah Alam for the support and generous assistance throughout this research.

References

[1] J. Conti, P. Holtberg, J. Diefenderfer, A. LaRose, J. T. Turnure, and L. Westfall, "International energy outlook 2016 with projections to 2040," USDOE Energy Information Administration (EIA), Washington, DC, United States, 2016.

[2] Sangeeta, S. Moka, M. Pande, M. Rani, R. Gakhar, M. Sharma, *et al.*, "Alternative fuels: An overview of current trends and scope for future," *Renewable and Sustainable Energy Reviews,* vol. 32, pp. 697-712, 2014.

[3] S. B. Singh, A. Dhar, and A. K. Agarwal, "Technical feasibility study of butanol–gasoline blends for powering medium-duty transportation spark ignition engine," *Renewable Energy,* vol. 76, pp. 706-716, 2015.

[4] H. Sharudin, N. R. Abdullah, G. Najafi, R. Mamat, and H. Masjuki, "Investigation of the effects of iso-butanol additives on spark ignition engine fuelled with methanol-gasoline blends," *Applied Thermal Engineering,* vol. 114, pp. 593-600, 2017.

[5] L. Siwale, L. Kristóf, A. Bereczky, M. Mbarawa, and A. Kolesnikov, "Performance, combustion and emission characteristics of n-butanol additive in methanol–gasoline blend fired in a naturally-aspirated spark ignition engine," *Fuel Processing Technology,* vol. 118, pp. 318-326, 2014.

[6] V. F. Andersen, J. Anderson, T. Wallington, S. Mueller, and O. J. Nielsen, "Vapor pressures of alcohol− gasoline blends," *Energy & Fuels,* vol. 24, pp. 3647-3654, 2010.

[7] V. Ganesan, *Internal Combustion Engines*: McGraw-Hill Education, 2012.

[8] A. C. Alkidas, "Combustion advancements in gasoline engines," *Energy Conversion and Management,* vol. 48, pp. 2751-2761, 2007.

[9] H. Sharudin, N. R. Abdullah, M. F. A. A. Rashid, A. Pahmi, M. R. Arifin, and M. F. M. Salleh, "Performance and emissions characteristics of gasoline engine fueled with octane booster-gasoline blends," *Proceedings of Mechanical Engineering Research Day 2018,* vol. 2018, pp. 16-17, 2018.

IOP Conf. Series: Materials Science and Engineering **834** (2020) 012053 doi:10.1088/1757-899X/834/1/012053

X-Ray Diffractometer (XRD) and Scanning Electron Microscopy (SEM) of Silica Extracted from Banana Stems via Acid Leaching Treatment

N. J. Yusaidi*, S. A. Abdullah and N. A. Zarib

Faculty of Mechanical Engineering, Universiti Teknologi MARA (UiTM), 40450 Shah Alam, Selangor, Malaysia

*jannahyusaidi38@gmail.com

Abstract. Silica is known as the natural resources where it has the second most abundant mineral in Earth's crust. In order to help the expanding research more on agricultural waste for silica extraction, banana stems has been used for this experiment studies. The main aim of the study is to know the percentage of silica in banana stems via leaching process. Hydrochloric acid (HCI) and citric acid ($C_6H_8O_7$) was used as a leaching reagent for removing the inorganic impurities from the banana stems and the effect of leaching time, reagent concentration and the effect of acid on the banana stems are investigated. Both acids results used same concentration of 0.1, 0.5 and 1.0 mol while leaching time were 30, 60 and 90minute respectively. The results of hydrochloric acid (HCI) and citric acid ($C_6H_8O_7$) leaching treatment test using the Energy Disperse X-Ray Spectroscopy (EDX) spots shows silica can be produced at 85.5% and 99%, respectively. Lastly, the silica obtained from the experiment was crystalline silica using the analysis of X-Ray Diffraction (XRD) and the morphological surface of the banana stems were obtained using the Scanning Electron Microscopy (SEM).

1. Background of study

Silica is defines as a solid, unreactive, colourless compound which occurs as the mineral quartz same as the principal constituent of sandstone and other rocks. Amorphous silica has many uses in its application such as additive in painting and coatings, concrete and filter media [1], [2]. Silica can be found in agricultural waste such as rice husks and palm ash where both of these source has the highest silica production which is more than 90% [3], [4]. In 2018, Malaysia has planted bananas more than 110,000 metric tonnes where 47% of banana production were original from Johor followed by Pahang and Sabah [5]. This can be seen as one of the opportunities to use banana stems as the agricultural waste for silica demand. The large number of farmers disposed the banana stems using an open burning method where it can cause major environmental issue [6], [7]. For silica extraction, leaching treatment is the best method to do since it is less expensive [8]. Previous researcher has used hydrochloric acid (HCI), sulphuric acid (H_2SO_4), nitric acid (HNO_3) for strong acid and citric acid ($C_6H_8O_7C_6H_80_7$) for weak acid [8], [9]. According to [10], they discovered silica content in banana stem ash with the percentage of 16.5%. Therefore the focus of this project is to determine the potential of banana stems as a source of silica.

2. Methodology

In this study, the materials used was banana stems collected from Shah Alam, Selangor. Hydrochloric acid (analytical reagent, fuming 37) and citric acid (Anhydrous, chemically pure) where both product of

Content from this work may be used under the terms of the Creative Commons Attribution 3.0 licence. Any further distribution of this work must maintain attribution to the author(s) and the title of the work, journal citation and DOI.
Published under licence by IOP Publishing Ltd

R&M Chemicals with the total volume of 2.5litre and 500grams respectively. Distilled water are used for leaching and filtration process.

Banana stems were cut into small pieces and washed using the tap water to remove dirt and impurities. The sample was placed in the Drying Oven TYPE T3D-YT at 40°C for 24 hours to remove excessive moisture. Then the sample grind using Planetary Mono Mill PULVERISETTE 6 classic line for 300rpm and sieve using Vibratory Sieve Shaker Analysette 3 Spartan for 250μm (60 mesh) to become fine powder. There are 3 different acid concentration that was prepared for the leaching process. 15g of banana stems was mixed together into total volume of 500ml of hydrochloric acid and distilled water into the beaker. For 0.1mol of hydrochloric acid, the sample were stirred for 30minutes, 60minutes and 90minutes. The beakers was placed onto the hot plate magnetic stirrer (Protechno DS-30) using the magnetic stirrer bar. After the leaching was completed, the sample was rinsed using distilled water at room temperature to remove the hydrochloric acid content from the banana stems. The steps are then repeated using 0.1mol, 0.5mol, 1.0 mol for citric acid and hydrochloric acid for 0.5mol and 1.0mol. After the leaching process, the materials were dried at 85°C for 1 hour.

After the sample has been prepared, it was placed in the Scanning Electron Microscopy (SEM) with the model of Type Leo DSM 982 Gemini which was supplied from Hi-Tech Instruments (Malaysia). It is for examine the morphological surface at 15kV and the electron beam was directed at the sample with a range diameter of 5-10 nM. Before the sample were placed in the machine, it must be coating using the coating spatter where the material used for coating is platinum. The magnification range used for the banana stem samples was 20x to 1000x.

Lastly, the X-Ray Diffractometer (XRD) machines model used are Rigaku Ultima Model IV where the supplier are from Rigaku Corporation and its Global Subsidiaries. For sample preparation, the sample are placed onto the specific holder specimen (2cm x 2cm) by compacting it before placing it in the diffractometer. The initial angle was taken at 10° and end at 70° with reading rate of 0.5°/min.

3. Results and Discussion

Figure 1a) shows the SEM images of untreated banana stems powder at magnification of 1000x. It shows that the particles has smooth surface and irregular shape. This is because the sample was not treated with any acid leaching. Energy Dispersive X-ray spectroscopy (EDX) test are used to determine the measureable chemical analysis at certain spots on the sample image. The EDX results shows the SiO_2 has percentage of 85.8%. The SEM images of Figure 1b) shows there is presence of rough surface condition. This is because of the citric acid plays role on removing the metallic impurities during leaching treatment. The wt% of treated banana stems (1.0M of citric acid for 30 minute) using spot EDX is 99.9% which has increased from the untreated banana stems. Lastly, Figure 1c) shows the SEM images of treated banana stems (0.5M of hydrochloric acid for 90 minute). It can be seen the particle has rougher surface due to the acid leaching. The weight percentage of silica for 0.5M banana stems via HCL was 85.5% which are slightly lower than the untreated banana samples and 1.0M of citric acid. During the EDX test, there are might have higher silica content presence on other spot area. Noratiqah *et al.* reported in their study that they have obtained smooth surface and rough surface on the rice husk particle [11].

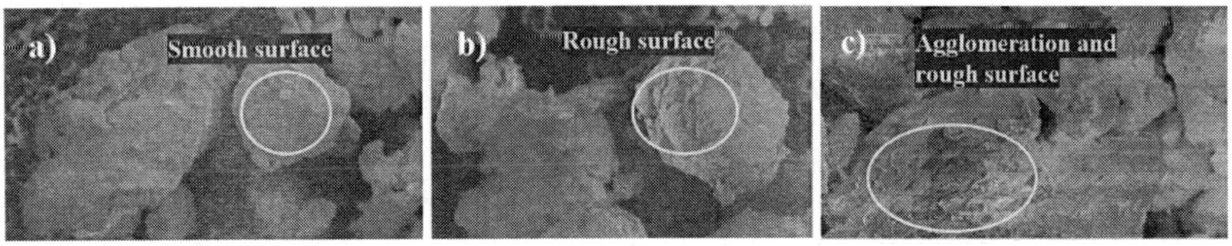

Figure 1. SEM images and EDX analysis of (a) untreated banana stems (b) treated banana stems of 1.0M citric acid (30 minute) (c) treated banana stems of 0.5M hydrochloric acid (90 minute) respectively at magnification of 1000x

IOP Conf. Series: Materials Science and Engineering **834** (2020) 012053 doi:10.1088/1757-899X/834/1/012053

Figure 2a) shows graph Intensity (a.u) versus Diffraction, 2θ/° for non-leaching sample. The banana stems sample was not added with any acids in order to know the element existed in the sample. The graph indicates the highest peak of silica was 22.04°, 35.135°, 37.81°, 43.33°, 52.49° and 57.44° of two theta. Figure 2b) shows the sample was added with 1.0 mol of citric acid for 30minutes of stirring time. The graph shows the silica are in crystalline form. It can be seen from the graph that the highest peak was range between 14.89° to 77.03° of two theta (degree). Lastly, the XRD pattern on figure 2c) shows the sample has undergo the leaching process. The sample was added with 0.5mol of hydrochloric acid for 90minutes of stirring time. Based on the graph below, the highest peak of silica in HCL was 22.34°, 25.56°, 35.14°, 43.29°, 52.61°, 57.45°, 66.44°, and 68.2° of two theta. It can be concluded that the silica existed in banana stems was crystalline silica. Y. Shinohara and N. Kohyama mentioned in their finding where they have also found the silica existed in crystalline form on the rice husk ash.

(a)
(b)
(c)

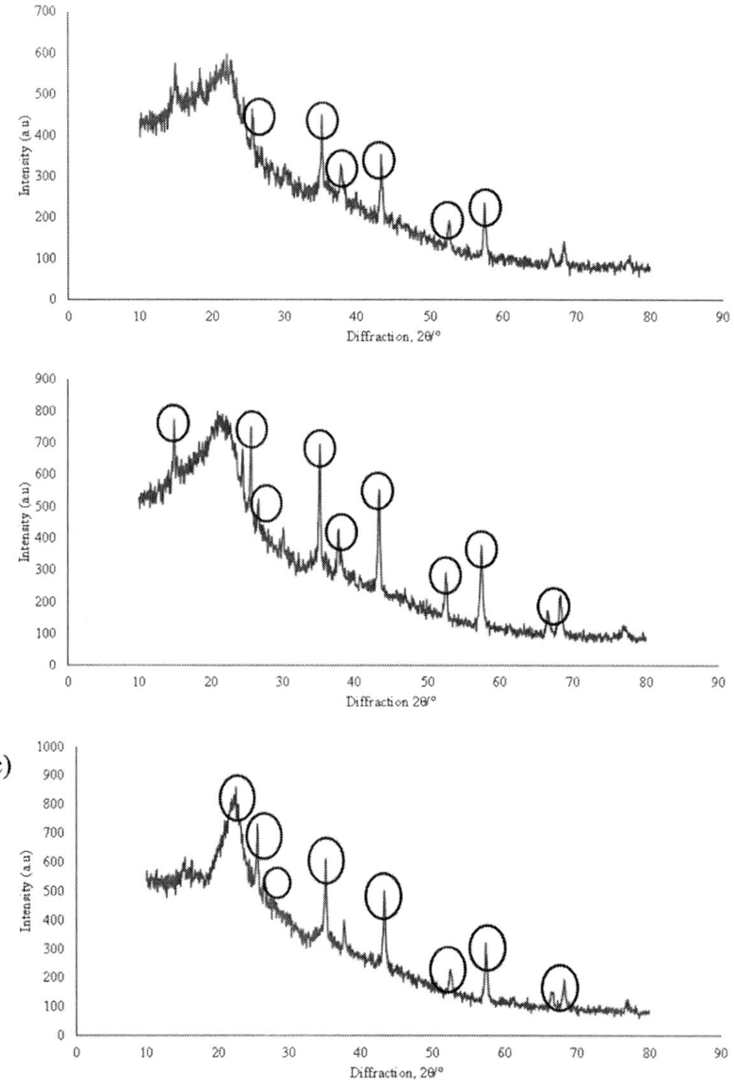

Figure 2: XRD Pattern of (a) untreated banana stems (b) treated banana stems of 1.0M citric acid (30 minute) (c) treated banana stems of 0.5M hydrochloric acid (90 minute)

4. Conclusion

In this study, the process for silica extraction has been conducted using acid concentration and leaching time for the parameters. Based on the SEM micrograph, as the acid concentration has increased, the surface of the sample material will become flakier and has rougher surface. Meanwhile for XRD graph, all of the sample shows the silicon dioxide was presented in the crystalline form and the higher peaks indicates it was silicon dioxide. The findings also shows the highest silica contained was in 0.5mol hydrochloric acid with 90 minute leaching time and 1.0mol citric acid with 30 minute leaching time which is 85.5% and 99%, respectively. The results of the current investigation is important to know the silica percentage in the natural fibre due to the need of expanding research in agricultural waste. Therefore, by expanding the research in agricultural waste via leaching treatment is important to reduce the unwanted waste to reach sustainable growth.

Acknowledgement

The authors would like to thank Universiti Teknologi MARA (UiTM) for financial support [Project Grant No: 600-IRMI/PERDANA 5/3 BESTARI (051/2018].

Reference

[1] L. Hao *et al.*, "Controllable fabrication and characterization of biocompatible core-shell particles and hollow capsules as drug carrier," *Appl. Surf. Sci.*, vol. 252, no. 24, pp. 8724–8733, 2006.
[2] B. Mistry, "Properties and Industrial Applications of Rice husk," *Int. J. Emerg. Technol. Adv. Eng.*, vol. 6, no. 10, pp. 2677–2679, 2016.
[3] K. G. Patel, R. R. Shettigar, and N. M. Misra, "Recent Advance in Silica Production Technologies from Agricultural Waste Stream – Review," *J. Adv. Agric. Technol.*, vol. 4, no. 3, pp. 274 279, 2017.
[4] C. P. Faizul, C. Abdullah, and B. Fazlul, "Extraction of silica from palm ashvia citric acid leaching treatment," *Adv. Environ. Biol.*, vol. 7, no. 12, pp. 3690–3695, 2013.
[5] M. of Agriculture, "Statistik Tanaman (Sub-Sektor Tanaman Makanan)," in *Buku Statistik Tanaman (Sub-Sektor Tanaman Makanan)*, pp. 44–58.
[6] A. Mohiuddin, M. K. Saha, M. S. Hossian, and A. Ferdoushi, "Usefulness of Banana (Musa paradisiaca) Wastes in Manufacturing of Bio-products: A Review," *Agric.*, vol. 12, no. 1, pp. 148–158, Aug. 2014.
[7] N. Abdullah, F. Sulaiman, M. Azman Miskam, and R. Mohd Taib, "Characterization of Banana (Musa spp.) Pseudo-Stem and Fruit-Bunch-Stem as a Potential Renewable Energy Resource," *Int. J. Energy Power Eng.*, vol. 8, no. 8, pp. 815–819, 2014.
[8] N. Permatasari, T. N. Sucahya, and A. B. D. Nandiyanto, "Indonesian Journal of Science & Technology Review : Agricultural Wastes as a Source of Silica Material," *Indones. J. Sci. Technol.*, vol. 1, no. 1, pp. 82–106, 2016.
[9] K. V. Selvakumar, A. Umesh, P. Ezhilkumar, S. Gayatri, P. Vinith, and V. Vignesh, "Extraction of silica from burnt paddy husk," *Int. J. ChemTech Res.*, vol. 6, no. 9, pp. 4455–4459, 2014.
[10] M. Marlinda, R. Ramli, and M. Irwan, "A Comparative study Of Catalityc Activity Of Heterogeneous Base Of Banana Stem Ash And Fly Ash On Production Of Biodiesel Byultrasonic Silica," *Int. J. Sci. Technol. Res.*, vol. 4, no. 08, pp. 169–172, 2015.
[11] N. Syahirah and S. Abdullah, "Effect of C 6 H 8 O7 Concentration on Silica Extraction of Rice Husk , Rice Husk Ash and Mixture of Rice Husk With Rice Husk Ash Via Acid Leaching Process," *Int. J. Eng. Technol.*, vol. 7, no. 4.18, pp. 190–195, 2018.

IOP Conf. Series: Materials Science and Engineering **834** (2020) 012054 doi:10.1088/1757-899X/834/1/012054

The Effects of Hydrostatic Pressure on Corrosion of Coated Stainless Steel

Nurulnatisya Ahmad[1], Bulan Abdullah*[2], Adib Ruzaini Mohd Najib[2], Siti Khadijah Alias[1], Syidatul Akma Sulaiman[1] , Freddawati Rashiddy Wong[2]

[1]Faculty of Mechanical Engineering, Universiti Teknologi Mara, Pasir Gudang
[2]Faculty of Mechanical Engineering, Universiti Teknologi Mara, Shah Alam

*bulan_abd@uitm.edu.my

Abstract. Underwater steel structure such as Remote Operated Vehicle (ROV) unit was exposed to corrosion element during its operated working hours and at different depth of sea level depending on its operational requirement. The different depth of each operational indicates the different pressure effects under water. The coated material was affected by the corrosion and may have changed in its micro substructure thus decreases the corrosion resistance. The aim of this study was to analyse the corrosion rate of the pressure effects on the coated materials through several testing. The type of coated material used is Commercial Anti-Corrosion spray. The tests done by pressurized the specimen at 2, 6 and 10 atm. The corrosion rates are determined by using immersion and weight loss. The specimens were soaked in the sea water with timeline of 24 hours. Scanning Electron Microscope (SEM) and microstructure observation are used to observed and determine the corrosion morphology of the specimen changes before and after immersion test. As a result, the pressure plays a major role on the corrosion effect and coating materials. The increments of pressure will deteriorate the corrosion resistance of the specimens. However, the coating materials can only slow down the deterioration process of the specimen.

1. Introduction

Autonomous submarine robots have been a rising interest in roaming the oceans freely and on an unprecedented scale. The robot is called Remotely Operated Vehicle (ROV) are safer, faster and efficient equipment for exploring especially in hazardous environment [1]. They are common in deep water industries such as oil and gas exploration. The deep sea is a complex environment, exposing any materials into high hydrostatic pressure and face huge challenges in the deep-sea environment [2]. High hydrostatic pressure has been becoming one of the biggest challenges in deep sea industries because it brought serious corrosion problem to the materials [3][4][5][6].

Traverso and Canepa (2014) studied corrosion performance in some of metallic materials that are commonly used in deep sea environments [2] and found out that all the alloys, including high strength steel, stainless steel, aluminium alloy etc., showed a decreased corrosion resistance in deep sea environments compared to ambient condition. Therefore, the reason for pressure promoted corrosion in deep water or sea environments are still arguing. Stainless steel has many types of grades in the industries, but not all of these are suitable for deep sea application. The most suitable type is authentic stainless steel and duplex stainless steel for deep sea structures. It is because of its good combination of corrosion resistance, high strength and wear resistance with good resistance to stress corrosion cracking. [7]

Content from this work may be used under the terms of the Creative Commons Attribution 3.0 licence. Any further distribution of this work must maintain attribution to the author(s) and the title of the work, journal citation and DOI.
Published under licence by IOP Publishing Ltd

There are two types of corrosion observed in deep sea piping, which are pitting and crevice. The most recognizable type of corrosion is pitting and can be caused by any number of factors. However, crevice corrosion is much harder to identify. Pitting is simply the breakdown of the passive layer, followed by localized corrosion that can cause cracking of steel vessel or pipes [8]. In 1985 DeBerry found that polyaniline electrochemically deposited on the ferritic stainless steel and provided a form of anodic protection that significantly reduce the corrosion rate [9]. The use of anti-corrosion coatings in one of the main strategies to prevent or minimize the corrosion effects of metallic structures [10].

Deep ocean exploration is harsh environments imply corrosive environments for structural materials because of the higher hydrostatic pressure. Thus, corrosion problems of materials in deep ocean condition must be considered. Some researchers have done some experimental laboratory and revealed that an increase in hydrostatic pressure improves that pitting susceptibility of metals. It is because of the increasing activity of chloride ions and their penetration into the passive film, the oxide and leads to the formation of pit initiation sites [11]. However, our knowledge is insufficient for the better understanding the effect of hydrostatic pressure on the corrosion of metals.

The Remote Operated Vehicle (ROV) unit was exposed to corrosion element most of the time of its operated working hours. The ROV also operated at different depth of sea level depending on its client requirement. Different depth of operational indicates the different pressure effects under water. Unfortunately, the corrosion was the main problem. The coated material was affected by the corrosion and may have changed in its micro substructure thus decreases the corrosion resistance. In this study, we focus on the type of coated material used which is Commercial Anti-Corrosion spray. The tests consist of by pressurized the specimen at 2, 6 and 10 atm. The test solution used is sea water.

Therefore, there are many factors that contribute to the formation of corrosion such as exposure to the salt, air and water. The aim of this study was to analyse the corrosion rate of the pressure effects on the coated materials through several testing.

2. Experimental Method

2.1. *Material*

The investigated material used was stainless steel 304L plate. The dimension of the specimen was 50mm (length) x 25mm (width) x 4.5mm (thickness). The stainless-steel specimen was sprayed or coated with anti-rusting primer on all the surfaces.

2.2. *Experimental Setup*

The pressure chamber was fabricated using air regulator, safety valve and on/off valve. The main cylinder was made from carbon steel materials with a 4 inches diameter x 6 inches long in dimension. The bottom of the pressure chamber was welded with a 120 mm x 4 mm pcd flange, while the top of the pressure chamber was removable 120 mm x 4 mm pcd flange. The air regulator and the safety bar were place at the top of the pressure chamber to set the pressure inside the pressure chamber. The O-ring was used to prevent air or water going out from the gap of the chamber and the top flange. The chamber was closed and tightens using 4 bolts and 8 lock nuts. The finished product was shown in Figure 1.

Figure 1. Finished Fabrication of Pressure chamber

2.3. Pressure Test

The specimen was put inside the pressure chamber with the sea water solution. The pressure was introduced inside the chamber using an air compressor. The tests were done by pressurized the specimen at 2 atm, 6 atm and 10 atm.

2.4. Immersion Test

The immersion test was used to obtain the corrosion rate by calculating the weight loss of the specimens before and after the immersion test. The soaking time of the specimen was in total of 24 hours. The solution that was used is sea water.

Table 1. Specification of Immersion test.

Apparatus	Pressure Chamber
Solution	Sea water
The time parameter	24 hours
Pressure rating	2 atm, 6 atm, 10 atm

The specimen has weighted before and after the test. The whole specimen is soaked with a solution to make sure to get accurate results. The corrosion rate has calculated using the formula given below:

$$Corrosion\ Rate = \frac{W \times t}{A \times 365}$$

Where;

W = Weight loss (g)
A = Total area exposure (mm^2)
t = Exposure time in hours
$g/mm^2/yr$ = gram per square mm per year (CR unit)

2.5. Microstructure Observation

The microscopic corrosion morphologies of the stainless-steel surface after 24 hours immersion in seawater solution at a different hydrostatic pressure were observed by Scanning Electron Microstructure

(SEM). The specimen was grinded, polished and etches before the testing was done. The specimen was also air dried clean before placing it in the SEM machine.

SEM process uses electron microscope that focused the beam of electron to produce images of the surface topography and composition of the specimen. The machine that was used for this testing is TM-3000. SEM can achieve resolution better than nanometers, and for this research it was used x500 of magnificent.

3. Result and Analysis

3.1 Corrosion rate from weight loss.

The specimens were expose to the liquid environment and pressure that can accelerate the process of corrosion. This is because of the chemical reaction occurs with oxygen and water that produce corrosive effects on metals by releasing metal ions to the water surrounding. The corrosion effects are boosted with the pressure supply inside the chamber that will create a hydrostatic pressure under the water. Table 2 shows that 2 atm pressurized specimen have loss 0.003 grams of weight after the immersion test. While, 6 atm and 10 atm pressurized specimen have loss 0.007 grams and 0.013 grams respectively. This reduction in weight is consistent with the increasing of pressure, which indicates the corrosion does occur.

Table 2. Corrosion rate calculated through weight loss.

Pressure, atm	Hours	Weight before, Wi (g)	Weight after, Wf (g)	Wi-Wf (g)	Corrosion rate mmpy
2	24	36.213	36.210	0.003	6.213e-8
6	24	37.264	37.257	0.007	1.45e-7
10	24	36.482	36.469	0.013	2.692e-7

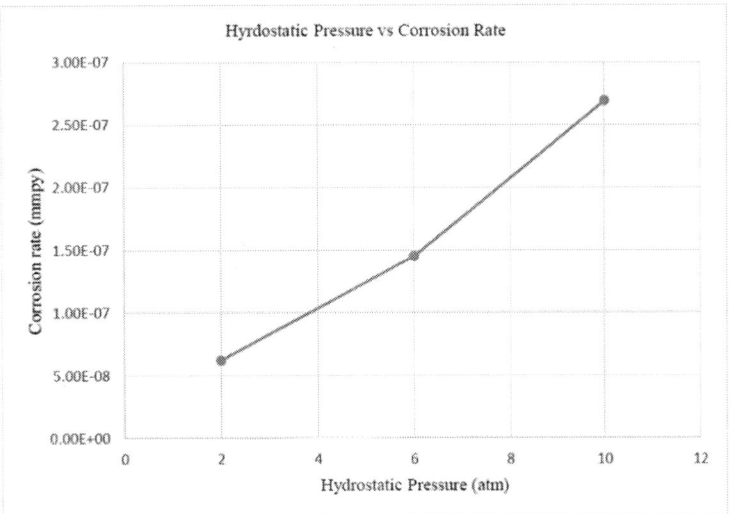

Figure 2. The corrosion rate of the specimen after the immersion test with different hydrostatic pressure which are 2 atm, 6 atm and 10 atm

The corrosion rate of coated material 304 stainless steel have increased from 6.213e-8 mmpy for 2 atm pressure rating to 1.45e-7 mmpy and 2.692e-7 mmpy for 6 atm and 10 atm pressure rating respectively. This demonstrates that the corrosion rate increases with the increase of pressure applied on the specimen as shown in Figure 2. This outcome is same as the polarization where corrosion resistance decreases in high pressure environment. Based on past research study, this result promises to

Yange Yang & et al during weight loss test which indicated that increasing hydrostatic pressure from 1 atm to 60 atm deteriorated the corrosion resistance of Ni-Cr-Mo-V steel [11].

3.2 Microstructure Observation

The surface coating of the stainless-steel corrosion morphology has shown in Figure 3. The most obvious type of corrosion present in the specimen is pitting corrosion. This can be observed in Figure 3(b) on the 2 atm specimen there are almost no presents of pitting with 20x magnifier. The coated surface does not change or develop any corrosion effects.

However, in Figure 3(c) on the 6 atm specimen, the present of pitting on the surface are minimal and hardly identified. Besides that, there are some spot on the surface where the colour of the stainless steel has seen through the small crack on the coating surface. This may cause by the rapture of the passive layer except for the small patch in contact with the electrolyte. This sighting is using 20x magnifier. Figure 3(d) shows that 10 atm specimen has a uniform pitting corrosion that clearly seen on the coating surface. The areas where the pitting occurs on the surface are larger and bigger than has seen under microscope. There are also in some spot where the sighting of the larger crack at the surface of the stainless steel has found.

The evidence found on the other researcher's studies, where the zinc coating was smooth, compact and completely cover the specimen surface. There is also no corrosion was found on the surface. Then after 14 days in a solution with different pressure shows that pitting corrosion occurred on the surface of specimen [12].

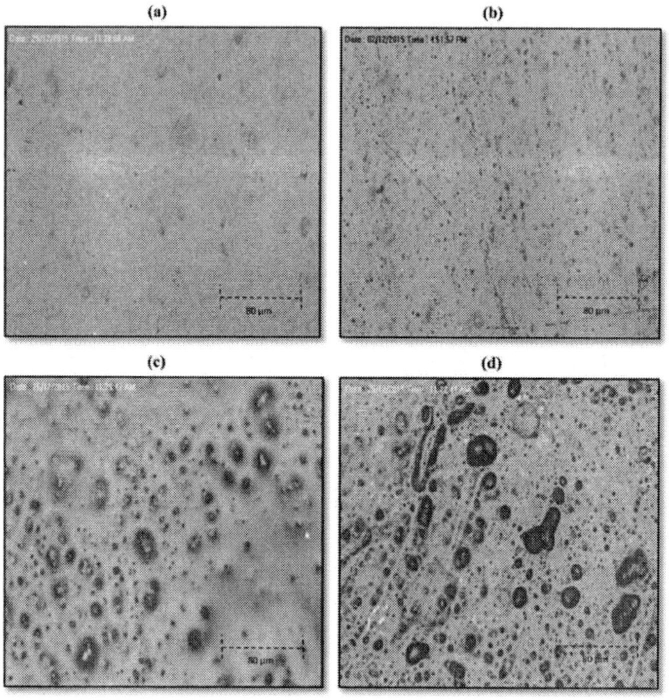

Figure 3. The microscopic corrosion morphology of coated stainless steel at (a) the surface before immerion, (b) with corrosion product at 2 atm, (c) with corrosion product at 6 atm and (d) with corrosion product at 10 atm.

3.3 Scanning Electron Microscope (SEM)

SEM has used to examine the corroded surfaces of specimens after 24 hours of immersion with corrosion products. The corrosion product can be seen in Figure 4(a) and Figure 4(b) where the nucleation stage of pitting is present at 2 atm. In this stage, the pitting corrosion is just initiated (nucleated). The nucleation of pits is influenced by surface defects that may be due to manufacturing or environment changes. The sight of pitting nucleation may be caused by the damaged protective layer, environmental factors that cause the layer to break down. These factors lead to anode and cathode site for corrosion.

Pitting growing sighted at the 6 atm specimen in Figure 4(c) and Figure 4(d). This stage of pitting shows the pits grow and have the potential to increase beyond the thickness of coating and lead to leaks. Pitting begins when the potential of the cell exceeds the pitting potential of the material in a given environment. This creates a self-propagating system where the increase acidity in the pit cavity increases the corrosion of the specimen.

This self-propagating clearly shows in Figure 4(e) and Figure 4(f) where the specimen has tested with hydrostatic of 10 atm. The coating fails to withstand the pitting corrosion occurred at 10 atm because the colour of the stainless steel has obviously seen through the pit. The area around the pit hole are surrounded by a uniform pitting and the pit area is getting bigger with increasing pressure. Furthermore, Figure 4(f) also shows the localized corrosion that can lead to accelerated failure of structure by acting as an initiation site of cracking. Based on Bin Liu & et al, the suggested that hydrostatic pressure did not change the mechanism of pitting formation. Besides that, the increasing in hydrostatic pressure accelerates the pitting generation rate [13].

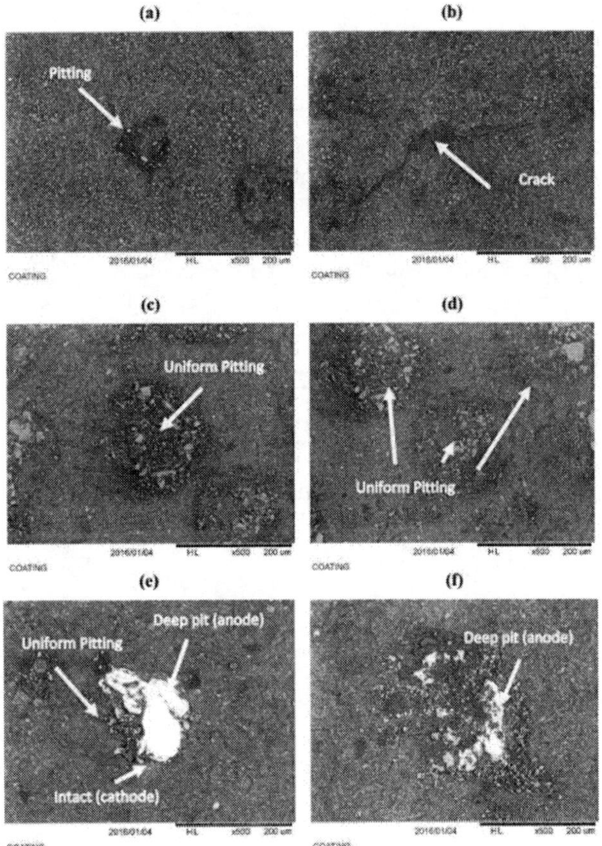

Figure 4. The SEM micrographic of coated stainless steel (a) and (b) at 2 atm, (c) and (d) at 6 atm and (e) and (f) at 10 atm with a corrosion product after the immersion in seawater for 24 hours

4. Conclusions

The aim of this study was to analyse the corrosion rate of the pressure effects on the coated materials through several testing. The type of coated material used is Commercial Anti-Corrosion spray. The tests done by pressurized the specimen at 2, 6 and 10 atm. The specimens have done immersions and weight loss test to determine the corrosion rates. As we know, there are two types of corrosion observed in deep-sea materials, which are pitting and crevice. These types of corrosion have observed in this research by using microstructure observations and Scanning Electron Microscope (SEM) test. The corrosion morphology of the specimen changes before and after immersion test has been determined. As a conclusion, the pressure plays a major role on the corrosion effect and coating materials. The increments of pressure will deteriorate the corrosion resistance of the specimens. However, the coating materials can only slow down the deterioration process of the specimen.

References

[1] Marzbanrad A., Eghtesad M., Sharafi J., & Kamali R. 2011 Design, Construction and Control of a Remotely Operated Vehicle (ROV). *ASME*, 7, 1295-1304
[2] Traverso P., Canepa E. 2014 A Review of Studies on Corrosion of Metals and Alloys in Deep Sea Environment. *Ocean. Eng,* 87, 10-15
[3] Sun H J., Liu L., Li Y., Wang F H. 2013 Effect of Hydrostatic Pressure on the Corrosion Behaviour of A low Alloy Steel. *J. Electrochem. Soc*, 160, 89-96
[4] Xiong X L., Tao X., Zhou Q J., Li J X., Volinsky A A., Su Y J. 2016 Hydrostatic Pressure Effects on Hydrogen Permeation in A514 Steel During Galvanostatic Hydrogen Charging. *Corros. Sci.* 112, 86-93
[5] Yang Z X., Kan B., Li J X., Su Y J., Qiao L J., Volinsky A A. 2017. A Statistical Study on the effect of Hydrostatic Pressure on Metastable Pitting Corrosion of X70 Pipeline Steel. *Materials.* 10, 1307
[6] Xiong X L., Ma H X., Tao X., Li J X., Su Y J., Zhou Q J., Volinsky A A. 2017 Hydrostatic Pressure Effects on The Kinetic Parameters of Hydrogen Evolution and Permeation in Armco Iron. *Electrochim Acta.* 225, 230
[7] Baddoo N., Barteri M., Burgan B., Knutsson H B., & etc 2006 *Design Manual for Structural Stainless Steel* (USA: Euro Inox and The Steel Construction Institute)
[8] Roberge P R. 2000 *Handbook of Corrosion in Engineering* 8th Edition (New York: McGraw-Hill)
[9] DeBerry D W. 1985 Modification of the Electrochemical and Corrosion Behavior of Stainless Steels with an Electroactive Coating. *J. Electrochem. Soc*, 132, 1022-1029
[10] Ambrosi A., & Pumera M. 2015 The Structural Stability of Graphene Anticorrosion Coating Materials is Compromised at Low Potentials. *Chem. Eur. J*, 21, 7896-7901
[11] Yang Y., Zhang T., Shao Y., Meng G., & Wang F. 2010 Effect of hydrostatoc pressure on the corrosion behaviour of Ni-Cr-Mo-V high strength steel. *Corrosion Science*, 52, 2697-2706
[12] Liu, S., Sun, H., Zhang, N., & Sun, L. 2013 The corrosion performance of galvanized steel in closed rusty seawater. *Int. J. Corros*, 1-10
[13] Liu, B., Zhang, T., Shao, Y., Meng, G., Liu, J., & Wang, F. 2012 Effect of hydrostatic pressure on the corrosion behavior of pure nickel. *Int. J. Electrochem. Sci, 7*, 1864-2883.

Investigation of Corrosion Rate for Different Type of Welding Joints using Shielded Metal Arc Welding (SMAW)

S. A. Sulaiman[1], B. Abdullah*[2], S. K. Alias[1], N. N. Ahmad[1], M. N. A Aziz[2]

[1]Faculty of Mechanical Engineering, University Teknologi MARA (UiTM) Kampus Pasir Gudang 81750 Masai Johor Malaysia

[2]Faculty of Mechanical Engineering, Universiti Teknologi MARA (UiTM), 40450 Shah Alam, Selangor Malaysia

*bulanabdullah@gmail.com

Abstract. Corrosion is the deterioration of a metal as a result of chemical reactions between metal and its surrounding. Different type of metal and environmental condition, particularly gasses that are in contact with the metal, determine the rate of corrosion. Different type of welding joints can be factors that contribute to the corrosion problems since different type of welding joint normally exposed to the corrosion and erosion agents like salt, air, water, material composition, dissolved oxygen content in the fluid, pH and fluid's velocity characteristics. Research was carried out in order to investigate the corrosion rate on mild steel at 3 different types of welding joint which are butt joint, lap joint and edge joint after immersed in Sodium Chloride (NaCl), Hydrochloric acid (HCl), distilled water and natural sea water for 30 to 50 days. The mild steel was joined using Shielded Metal Arc Welding. For each of the welding joint the parameters were fixed which are current, voltage and speed of travel. Immersion and hardness test were conducted to distinguish the corrosion rate. Results shown butt joint suffered the highest corrosion rate than lap joint and edge joint and morphology transformation differences occur during the corrosion process.

1. Introduction

Metal joining is a controlled process that is widely employed to fuse similar/dissimilar metals. Among the several metal joining techniques, welding is one of the most widely used for a variety of applications. Metallurgical, physical, and chemical changes caused by welding processes severely affect the corrosion resistance of welds [1]. SMAW occupies the most important position in the group of fusion welding processes, and due to its flexibility and cost effectiveness, it is an indispensable technology for the construction of steel-framed buildings, ship building, motor vehicle manufacture, power plants and other industries. Since the SMAW method is so versatile; it is the most suitable method to work in many situations [2]. Mild steel is the most commonly used steel for welding on account of relatively low and good material and mechanical properties that are suitable for many applications particularly in severe condition such as extreme weather, greenhouse effect, external massive loads and corrosive marine environment. [3] Unfortunately, the corrosion was crucial for the industries that used metal and using material such mild steel and add-ed to the welding process make the situation worst. Mild steel is among the metal that has lowest corrosion resistance and weld joint that was affected by corrosion changes the microstructures thus decrease the strength of the joining [4]. To make it worst, most of the structure and the welding joints were expose to the environment which contain the corrosion agent such as water, oxygen and other impurities thus accelerate the

Content from this work may be used under the terms of the Creative Commons Attribution 3.0 licence. Any further distribution of this work must maintain attribution to the author(s) and the title of the work, journal citation and DOI.
Published under licence by IOP Publishing Ltd

corrosion. When corrosion occurs, the weld joint will suffer critical damage as the behaviour of the corrosion. itself which deteriorates the affected metals. Furthermore, when the corrosion attacks, the hardness of the welding joints will be affected and weaken the microstructure due to corrosion. For critical applications like in construction field and automotive industries, knowledge related to welding corrosion must be considered. Localized and general corrosion reduces lifetime significantly and may cause failures [5]. The weld area just took a small area as compared to the base metal in welded joints. Thus, a selective weld attack will take place. Metal composition and microstructure alteration resulting from the heat cycle will affect the corrosion behaviour for the welded joint. Moreover, temperature gradient, microstructure variation, grain structure and chemical composition can create local electro-chemical potential variations thus leading to corrosion [6]. The corrosion behaviour at three adjacent regions in welded metal which were weld metal, heat affected zone and base metal is complex due to the heterogeneous microstructure over the welding zone [7]. Different welding joints may have different rate of corrosion. This study focused on 3 types of welding joints which were butt joint, lap joint and edge joint.

2. Experimental Procedure

2.1. Sample Preparation

The material used for this research was mild steel. The sample was cut into 200 mm length x 100mm width x 6mm thick. Before welding process, the surface of the mild steel was cleaned using metal grinder to clean the existing corrosion. The joining that performed in this study were butt joint, lap joint, and edge joint. All joint was done by using Shielded Metal Arc Welding (SMAW). Table 1 shows the parameter of welding process.

Table 1. Parameters of Shielded Metal Arc Welding (SMAW)

Parameters		Unit
Diameter of Rod		3.2 mm
Voltage		21-25 V
Current	Flat	90-100 (A)
	Vertical	80-95 (A)

2.2. Immersion Test

Corrosion rate was determined through immersion test when weight before and after was taken. The specimen was soaked on the sodium chloride 3.0M NaCl, hydrochloric acid (HCl) natural sea water and distilled water with a different timeline of 10 days, 30 days and 50 days. Corrosion rate was calculated with the formula:

Corrosion Rate (mm /y) = $\frac{w}{A}\frac{T}{365}$

w=Weight loss (gram)
A= Total area of exposure (cm^2)
T = Exposure time in hours g/mm^2/yr. = gram per square mm per year (corrosion rate units)

2.3. Hardness Test

Hardness test was conducted in order to see the hardness of the base metals, weld metals and heat affected zone (HAZ) area. The Vickers hardness test was used to measure the degree of hardness due to the material used as the sample.

Figure 1. Vickers hardness tester

2.4. Microstructure Test

This microstructure test was conducted to see the behaviour of the microstructure after the corrosion process. This is because, the corrosion also affects the microstructure between the base metals and weld metals. There are four preliminary steps need to be done to have the good image of the microstructure which are cutting, grinding, polishing and etching. Then the samples were soaked in the Nital solution which consists of 2% nitric acid and 98% alcohol. Lastly, the samples were observed under the Olympus BX 41 M microscope.

3. Result and Discussion

3.1. Immersion Test

Based on the result of immersion test, all the sample shows the deduction in weight. The weight of the samples was reduced with the time immersed. As the samples immersed in the solutions, there were a layer of corrosion developed on each sample, but this phenomenon not occurred in Hydrochloric acid (HCl) solution. The layer was known as insoluble graphitic layer of corrosion product left behind in the process. By figuring the graph of immersion, the corrosion rate directly proportional to the weight loss. Between all the solutions used, acidic solution recorded the highest loses.

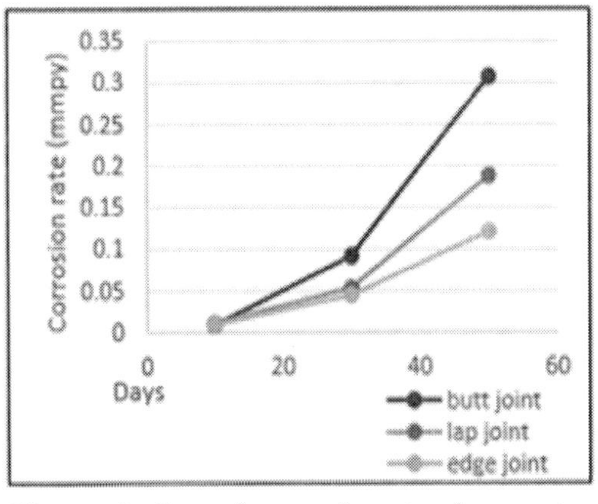

Figure. 2. Corrosion rate in natural sea water

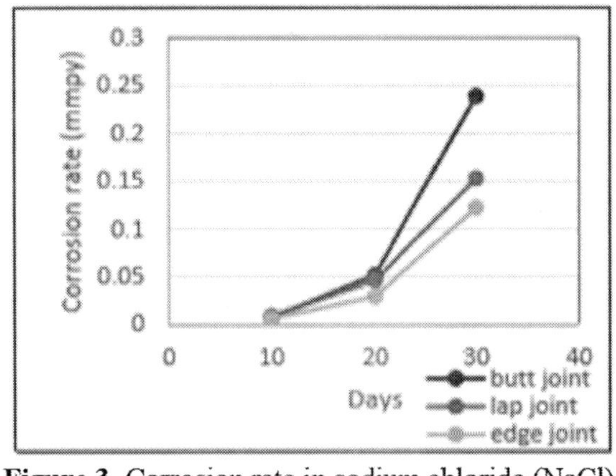

Figure 3. Corrosion rate in sodium chloride (NaCl)

Based on Figure 2 and Figure 3, natural sea water and Sodium Chloride (NaCl) has the corrosion agent which was salt, but the corrosion only occur and damaging the outer appearance of the sample only. The pH concentration was the one factor of decreased in corrosion rates of the specimens in the salt and alkali environment. The other factor may due to the increasing exposure time toward the electrochemical behaviour of SMAW process. Previous study shows that corrosion attack becomes slower due to alkaline environment. Higher the rate of alkalinity, the slower the chemical reaction of materials [8]. Corrosion was occurring due to reaction material and its surrounding. The chemical reaction between oxygen and water make the loss of electron. There are many factors that influencing

IOP Conf. Series: Materials Science and Engineering **834** (2020) 012055 doi:10.1088/1757-899X/834/1/012055

corrosion of weld section such as incomplete weld penetration, improper choice of filler metal etc. [9]. Longer time taken of immersion donated to the increasing of weight loss. Metals in seawater corrode by releasing metal ions into the water around them. Different environment, acid and salt environment with increasing exposure time effect due to electrochemical behaviour. The weight will loss due the corrosion.

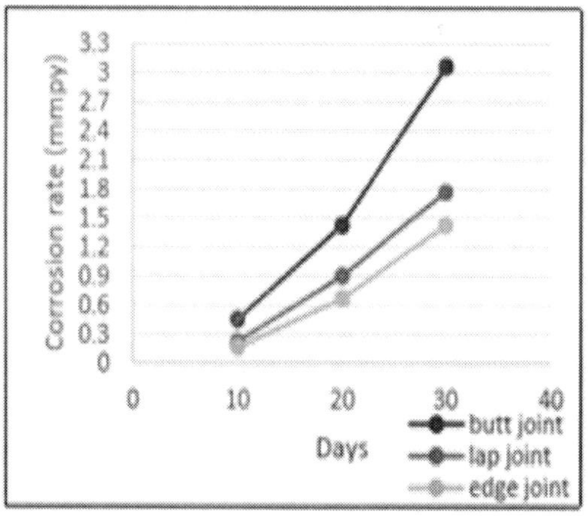

Figure 4. Corrosion rate in hydrochloric acid (HCl)

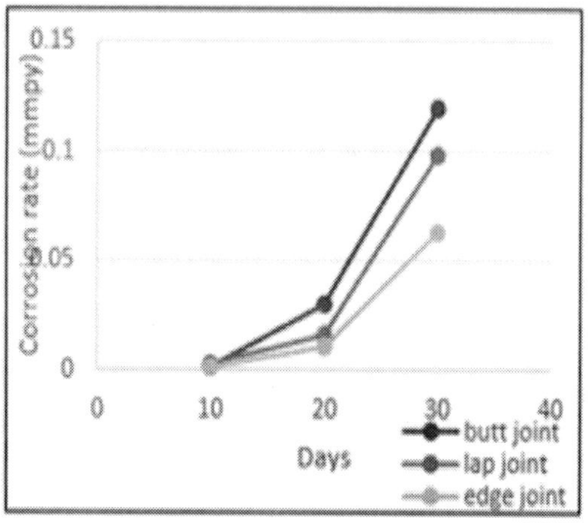

Figure 5. Corrosion rate in distilled water

The deduction of the weight loss caused by HCL was showed in Figure 4. Hydrochloric Acid (HCl) has the properties of deterioration of metals and non-metals. So once the samples were immersed in acidic solution, the acid will attack not only the physical appearance of the sample but also inside the sample itself. That the reason why the corrosion rate for mild steel immersed in acidic solution has the highest corrosion rate. While in Figure 5 the sample was immersed in distilled water. The rate of corrosion is the lowest among other solution since pH for distilled water is 7 which does not have ion or other calcium in the solution. Among those three types of joints that being immersed, butt joint recorded the highest percentage of weight loses. This due to the fabrication of the butt joint itself which the welding process was fast. Butt joint also severe the highest deterioration in Hydro-chloric acid (HCl) solution whereas lap and edge joint also deteriorate but the value is low.

3.2. Hardness Test

The hardness value was observed at three different spot which were base metals, weld metals and heat affected zone (HAZ). All the samples were immersed in natural seawater, Sodium Chloride (NaCl), Hydrochloric Acid (HCl) and distilled water for 50 days. Based on the graph the highest value of hardness recorded was lap joint. Whereas the lowest hardness value recorded was butt joint.

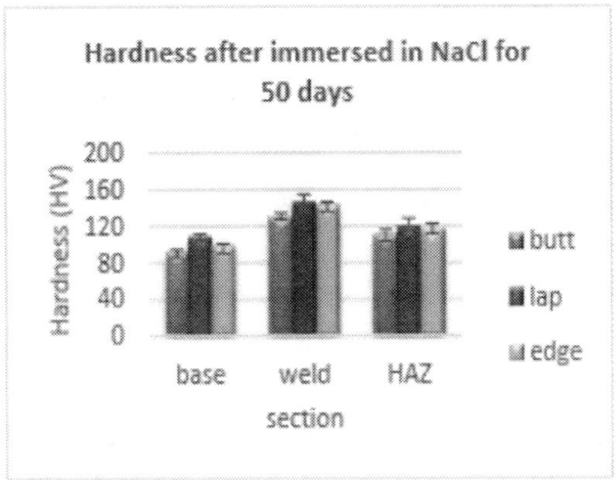

Figure 6. Sample immersed in NaCl

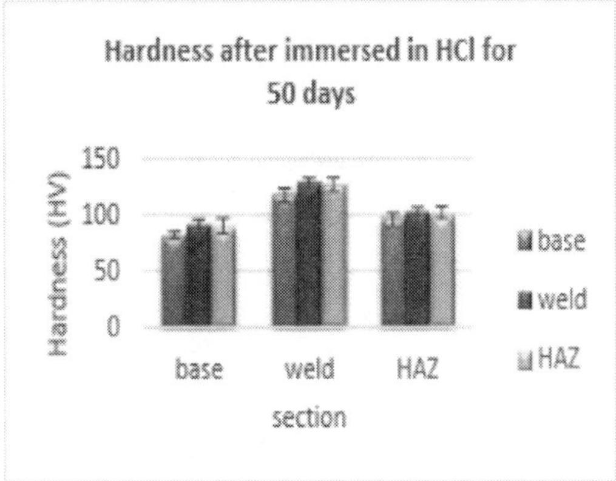

Figure 7. Sample immersed in HCl

Based on the data tabulated in Figure 6, Figure 7, Figure 8 and Figure 9 the sample which immersed in Hydrochloric acid (HCl) shows the highest deduction in the value of hardness, followed by natural sea water, Sodium Chloride (NaCl) and distilled water. Acidic solution such as Hydrochloric acid (HCl) has the properties of deteriorate metals thus the sample in acidic solution were suffered of metals deteriorate. This lead to the highest deduction of the hardness of the sample in Hydrochloric acid solution.

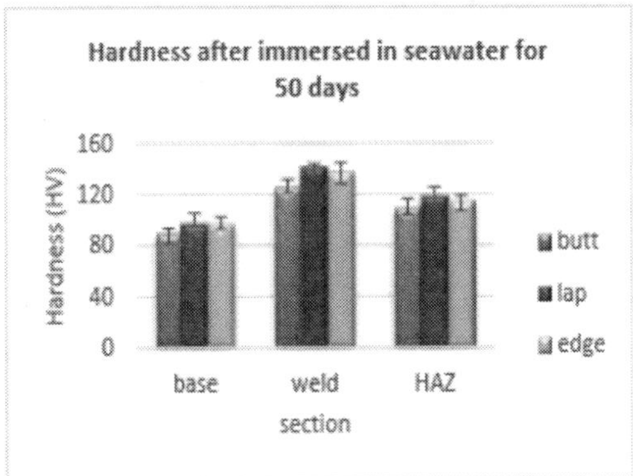

Figure 8. Sample immersed in natural sea water

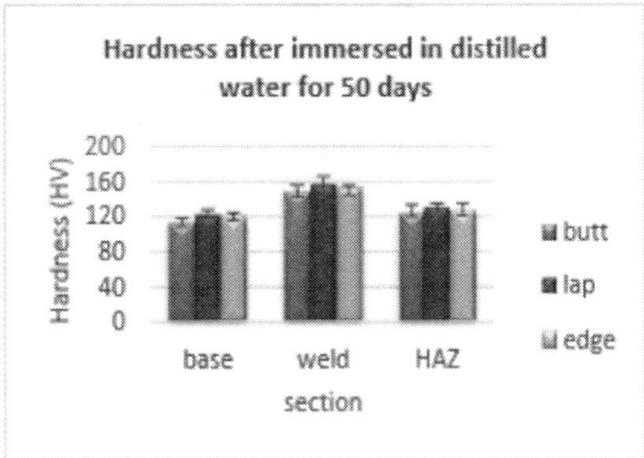

Figure 9. Sample immersed in distilled water

3.3. Microstructure Test

The microstructure of the welded sample was observed at three different sections which are at the base metals, weld metals and heat affected zone (HAZ). After exposed to the environment for 3 months, corrosion not only attacks on the surface of the sample but also in the inner part of the metal. That the reason why the strength of a metals decreasing when corrosion takes place. Corrosion start to attack on the perlite area which then continue to deteriorate metals from inside. Table 2 shows that typical microstructure of base metal is composed of ferrite and small regions of pearlite (α-Fe + Fe3C) at grain boundaries edges and corners. At Heat Affected Zone (HAZ) the Widmannstetter ferrite, and some colonies of pearlite were found. It is known that solid-state phase transformations, such as grain growth, recrystallization, phase transitions, annealing, and tempering, all occur in the HAZ of steel welds

Table 2. Microstructure for different type of joint

Type of Joint	Base Metal	HAZ	Weld Metal
Butt Joint	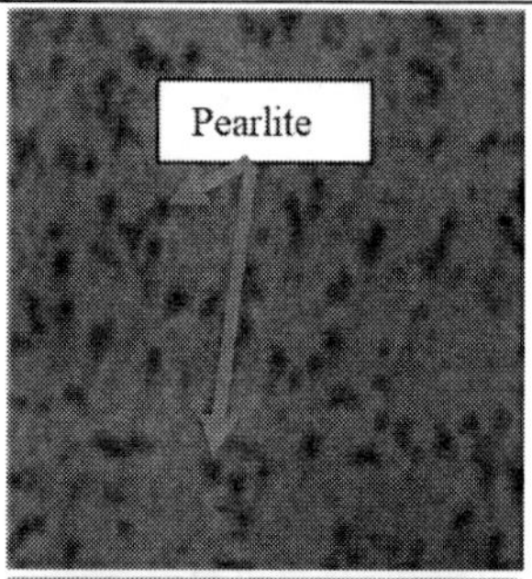		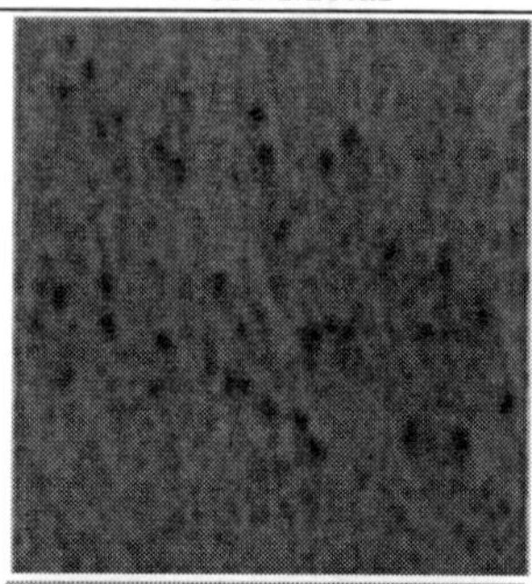
Lap Joint		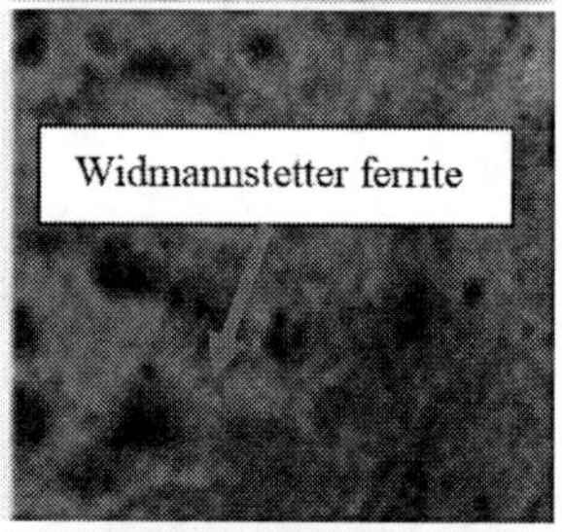	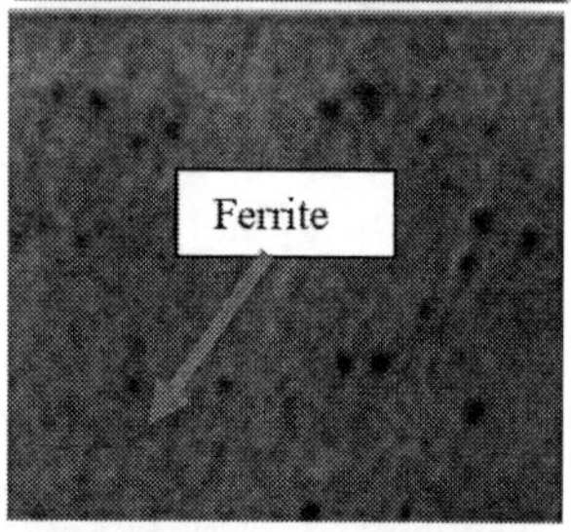
Edge Joint	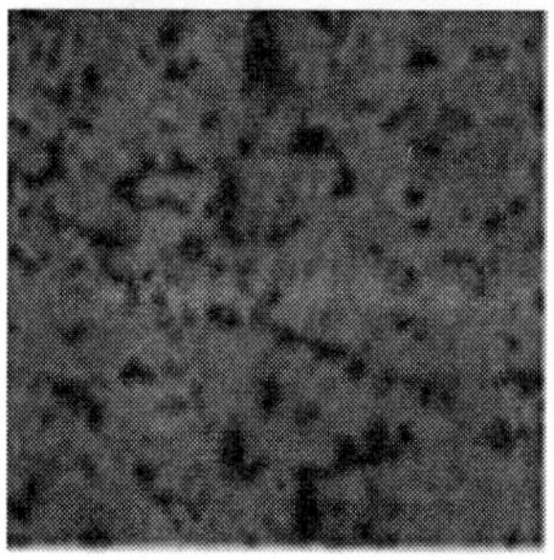		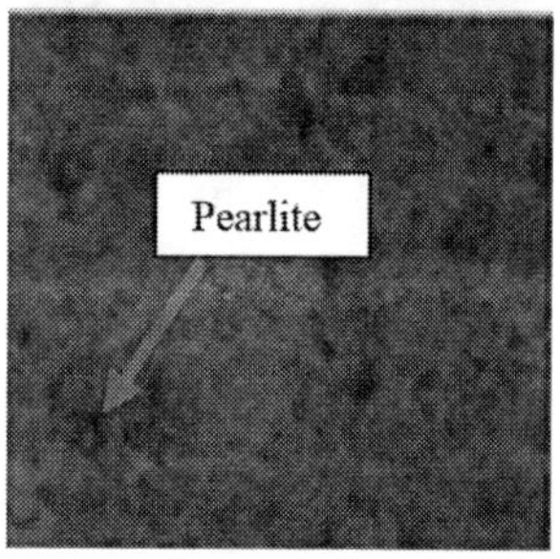

However, the microstructure of weld metal is totally different from the other zones. Due to rapid cooling it causes microstructural inhomogeneity. This zone contains mainly ferrite and some colonies of pearlite. The microstructure that evolved in the weld is heterogeneous due to the rapid cooling process and chemical reaction that evolve during the process [10].

Figure 6,7,8 and 9 shows that the hardness of weld metal is the highest mainly due to the strain induced hardening followed by HAZ and base metal. It can be clearly observed from Table 2 that the grain size for base metal was coarse as compared to weld metal and HAZ. That the reason why it being affected by corrosion in large scale. Most of carbon is dissolved in martensite and austenite phases, but there are still few carbides existing in the featherlike bainite [11]. Corrosion will start attack on the soft area (coarse grain) and as the time passed it will become more aggressive and attacked until the whole phase of the bainite [12].

4.0 Conclusion

Butt joint loss higher weight during the immersion test followed by lap joint. Meanwhile edge joint had the least loss weight and corrosion rate. Another contribution to the corrosion rate was the pH concentration between acidic, salt and alkali environment. The exposure time leads to the process of rusting metal from the chemical reaction oxygen and water make the loss of electrons. Moreover, the hardness of butt joint also the lowest. As observed under microscope, the grain structure for weld metal was the finest thus the maximum hardness values are situated in the area of weld metal and HAZ which indicates its specificity. It should be emphasized that obtained results could be applied for improving the possibility of constructional methods applied by engineers. Hence the next investigation area should be searching for the link between the material and structural design processes.

References

[1] Xizhang C, Arvind S, Serey K, Juergen RH, Kai W (2018), Corrosion of Materials after Advanced Surface Processing, Joining, and Welding. Internal Journal of Corrosion.
[2] Brajesh KS, Jha AK, Pravin KS (2015), Effects of Joint Geometries on Welding of Mild Steel By Shielded Metal Arc Welding (SMAW). International Research Journal of Engineering and Technology (IRJET) 02, 95-100.
[3] Busari YO, Ahmed II, Shuaib YL (2017), Effect of Heat In-put On the Mechanical and Corrosion Behaviour Of SMAW Mild Steel. Journal of Production engineering 20, 59-64.
[4] Abdullah B, Ismail S, Jumadin MH (2015), Investigation on Corrosion Rate Of Butt-Joint In Gas Metal Arc Welding, Flux Core Arc Welding And Submersible Arc Welding. Jurnal Teknologi 76, 61-65.
[5] Garcia MP, Mantovani GL, Kumar RV, Antunes RA (2017), corrosion Behavior of Metal Active Gas Welded Joints of a High-Strength Steel for Automotive Application. Journal of Materials Engineering and Performance
[6] Wang LW, Liu ZY, Cui ZY, Du CW, Wang XH, Li XG (2014), In Situ Corrosion Characterization of Simulated Weld Heat Affected Zone on API x80 Pipeline Steel, Corros. Sci. 85, 401–410
[7] Geng S, Sun J,Guo L, and Wang H (2015), Evolution of Microstructure and Corrosion Behavior in 2205 Duplex Stainless Steel GTA-Welding Joint. Journal of Manufacturing Pro-cessess. 19, 32–37
[8] Saliu Ojo Seidu, BJK. (2013). Effect of Heat Treatments on Corrosion of Welded Low-Carbon Steel in Acid and Salt Environments. Journal of Minerals and Materials Characterization and Engineering. 95-100
[9] Davis, J. 2006. Corrosion of Weldments. ASM International. 1-12.
[10] Gharibshahiyan E, Raouf A.H, Parvin N., Rahimian M. (2011). The effect of microstructure on hardness and toughness of low carbon welded steel using inert gas welding. Material and Design 32, 2042-2048
[11] Boumerzoug Z, Derfouf C, Baudin T. (2010) Effect of Welding on Microstructure and Mechanical Properties of an Industrial Low Carbon Steel. Scientific Research 2, 502-506.
[12] Vera R, Araya R, Garin C, Ossandon S, Rojas P. (2019) Study on the effect of atmospheric corrosion on mechanical properties with impact test: Carbon steel and Galvanized steel. Material & Corrosion.

IOP Conf. Series: Materials Science and Engineering **834** (2020) 012056 doi:10.1088/1757-899X/834/1/012056

Musculoskeletal Discomfort Evaluation using Quick Exposure Check (QEC) among Tower Crane Operators

N A Ibrahim[1], S A S A Rahman[1], S H Ismail[1] and H Abdullah[1]

[1] Razak Faculty of Technology and Informatics

Corresponding author's email: haslaile.kl@utm.my

Abstract. Tower crane operators are at risk of developing Musculoskeletal Disorders (MSD) due to work schedules that require prolonged sitting with minimal movements. The aim of this study was to assess the discomfort levels associated with the musculoskeletal disorder and its severity on affected human body areas of tower crane operators while driving and handling of materials. This study was conducted amongst 32 tower crane operators at the construction sites of a high-rise building in Klang Valley. A Quick Exposure Check (QEC) was used to assess the MSDs levels and the affected severity on human body areas. The results from the QEC showed the body parts that were associated with a very high risk of MSDs were the back and neck. Other factors contributed to the high risk of MSDs among tower crane operators were identified as driving and vibration. MSDs among tower crane operators were evident from the QEC analysis. Therefore, it is essential to include ergonomic intervention in the workplace in order to minimize the risk.

1. Introduction

MSDs are conditions that can affect muscles, bones, and joints [1]. MSDs normally affect any musculoskeletal system such as neck, shoulder, wrist, back, hips, legs, knees, and feet. Example of activities that could affect the musculoskeletal system are long sitting in the same position, repetitive movements and lifting of heavy materials.

A tower crane is an indispensable facility in construction sites of high-rise building [2]. It was used for transporting construction supplies and heavy materials to places over 20 to 400-metre-high from the ground. The crane operator was reported one of the sectors that have a higher risk of developing back and lower limb disorders [3]. It was followed by truck drivers, warehouse workers, airplane baggage handlers, construction trades, patient care workers and other large vehicle operators. The tower crane operators are at risk of developing musculoskeletal discomfort because of static or constrained body position, frequency of repetitive movement and concentrated force on small body parts such as hand or wrist and also the work pace [4, 5].

Generally, the tower crane operators were exposed to the neck, hip, upper and lower back of MSDs [6]. A study of tower crane operators in Iran found a high prevalence of musculoskeletal symptoms among the workers [7]. The study reported a majority of the operators disclosed some disorders in one or more parts of the body particularly on the lower back and neck area. Other studies on the crane operators have also recorded similar finding where 42% of operators were observed to suffer from neck pain [8] and 78% of the operators suffered pain at lower back, neck and shoulder [9]. In addition, approximately 70% of crane operators experience discomfort resulting from a bent forward sitting position, which was necessary to keep the payloads in view while executing lifts close to or directly below the operator's cabin [10]. Furthermore, the rubber-tired gantry crane operators were highly liable

Content from this work may be used under the terms of the Creative Commons Attribution 3.0 licence. Any further distribution of this work must maintain attribution to the author(s) and the title of the work, journal citation and DOI.
Published under licence by IOP Publishing Ltd

to MSDs due to awkward posture, long duration of the working hours, vibration and repeated or continuing of wrist force to handle joysticks [11, 12]. Psychosocial factors such as work shift and project deadline also found contributed to MSDs [13].

To date, there were various ergonomic tools such as QEC, ANOVA, NMQ, body discomfort map, REBA, and RULA that were used to identify the related MSDs factors among workers. in various industries such as sewing machine, rubber production, road building industry, electronic manufacturing, bumper tape-masking and oil palm harvesting. However, many studies suggested QEC was the most reliable tool to assess the workers' exposure to work-related musculoskeletal risk factors [14-17]. QEC developed by [18], is an observational tool developed for Occupational Safety and Health (OSH) practitioners to assess exposure to risks for Work-related Musculoskeletal Disorders (WMSD). It was used to provide a basis for ergonomic interventions. The checklist in QEC assesses the extent of exposure of the four main body areas to identify risk factors contributed to WMSDs. QEC covers many risk factors such as posture, load/force, movement/ frequency, duration, vibration, and others.

Therefore, the aim of this study was to assess the discomfort levels associated with the musculoskeletal disorder and its severity on affected human body areas of tower crane operators while driving and handling of materials using QEC measures.

2. Materials and Methods

The study participants comprised of 32 tower crane operators from eight construction sites in the Klang Valley. The participants were all competent and certified by the Department of Occupational Safety and Health. All participants were informed on the purpose of the study and completed the consent form before participating. All interviews were carried out either before beginning of their work, during lunch time or after working hours.

2.1. Participants

Table 1 summarizes the demographic information of 32 tower crane operators participated in this study.

Table 1. Tower crane operators demographic.

	Mean	Std. deviation
Age (years)	35.9	9.4
Working hour	11.8	4.6
Year of experience (years)	8.1	7.4

2.2. Data Collection

Data were collected through an interview which consisted of two parts. The first part of the questionnaire was on the demographic information and the second part was an assessment of musculoskeletal disorders risk among the tower crane operators by using a standardized tool known as QEC. This questionnaire involves both the practitioner (observer) and the workers (who have direct experience of performing the job) in conducting the assessment and identifying possibilities for change.

After the data score for section in the QEC was collected, the risk factor variables were tabulated to represent the level of MSD risk as outlined in Table 2. The exposure scores was categorized into four exposure levels stated as "low" , "moderate", "high" and "very high."

Table 2. Interpretation of QEC score [17]

	Exposure level			
Score	Low	Moderate	High	Very High
Back (static)	8-15	16-22	23-29	29-40
Back (moving)	10-20	21-30	31-40	41-56
Shoulder /Arm	10-20	21-30	31-40	41-56
Wrist / Hand	10-20	21-30	31-40	41-56
Neck	4-6	8-10	12-14	16-18
Driving	1	4	9	-
Vibration	1	4	9	-
Work Pace	1	4	9	-
Stress	1	4	9	16

3. Results and Discussion

Figure 1 shows the result of the QEC for MSD exposure risk among the tower crane operators for four main body areas namely back (static), shoulder, wrist band and neck. Shoulder was the only body parts recorded with a low-level risk of exposure scored of 28%. In the moderate exposure level category, back, shoulder and wrist were identified showing the score of 19%, 47% and 38% respectively. Interestingly, all four body parts were exposed to a high risk of MSD particularly on the wrist/hand with a score of 63% followed by the neck (38%), shoulder/arm (25%) and back (19%). Furthermore, both neck and back were categorized in the very high exposure level of MSD with value of 63%.

The results for other important MSD risk factors such as driving, vibration, work pace and stress is shown in Figure 2. Stress and work pace were found to be in the moderate risk category with a percentage of 66% and 44% respectively. However, majority of the tower crane operators were in a high-risk category due to driving and vibration with the score of 63% and 59% respectively. The overall results of this study demonstrated an increased risk of developing MSD among the tower crane operators due to working conditions and environment.

Our results in this study was comparable with other reported study where neck and lower back of crane operators were reported to have a MSD score of more than 60% [19]. In addition, a similar finding amongst the ready garment workers was also reported in [20]. The study found that 40–60% of operators within 12-month prevalence were having lower back pain (LBP) [20]. Furthermore, a cross-sectional study conducted in Italy shows that approximately 40% of crane operators working in the port of Venice had an overall prevalence of disorders in the lumbar spine [22]. Other studies [14, 23, 24] also recorded a high prevalence of LBF among drivers due to confined space, long working hour, limited movement and exposure of vibration. Therefore, intervention strategies to reduce the MSDs risk which include reducing the duration of exposure to the risk factors, such as vibration and static sitting is required.

The limitation of this study was small sample size due to limited project area which involves high rise building. It is highly suggested that the number of respondents should be greater in order to obtain a more significant and reliable result. Another limitation was time constraint to observe the work posture and interview amongst tower crane operators because the tower crane operators cannot be disrupted during the working hours as it has impact on the project progress. In addition, data and information of the medical leave related to MSDs was not able to be compiled because the tower crane operators belong to the outsourced companies. Adopting QEC has its limitation too since it is only be able to evaluate the upper body parts namely back, shoulder, wrist/ arm, and neck. Therefore, other ergonomics assessment tools that cover the entire body parts are essential to give a better result of MSDs of an individual crane operator.

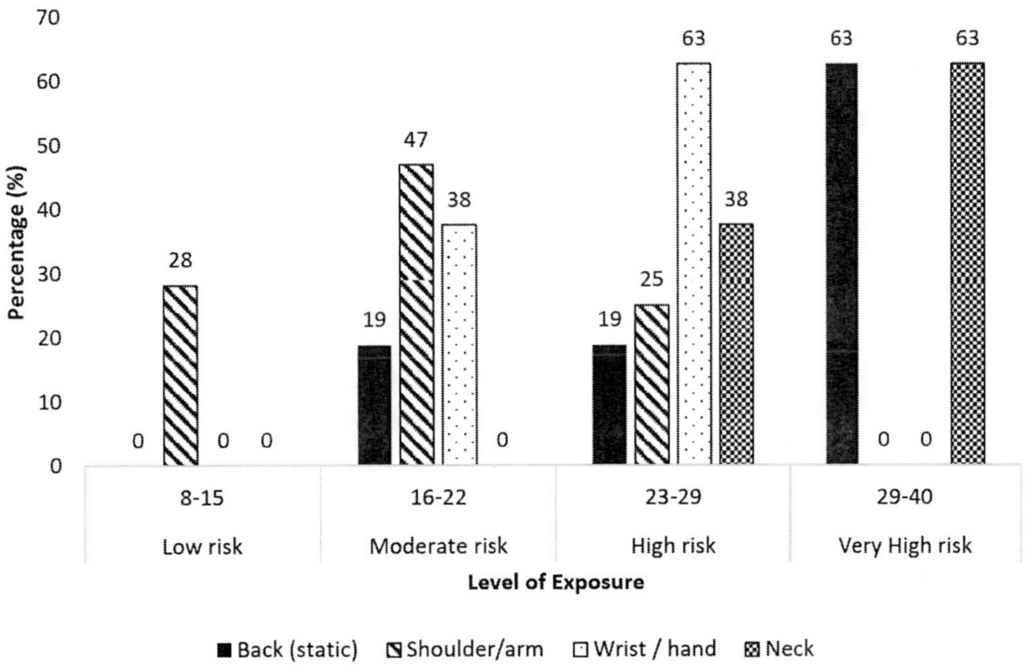

Figure 1. QEC Exposure level and percentage for back, shoulder, wrist, and neck.

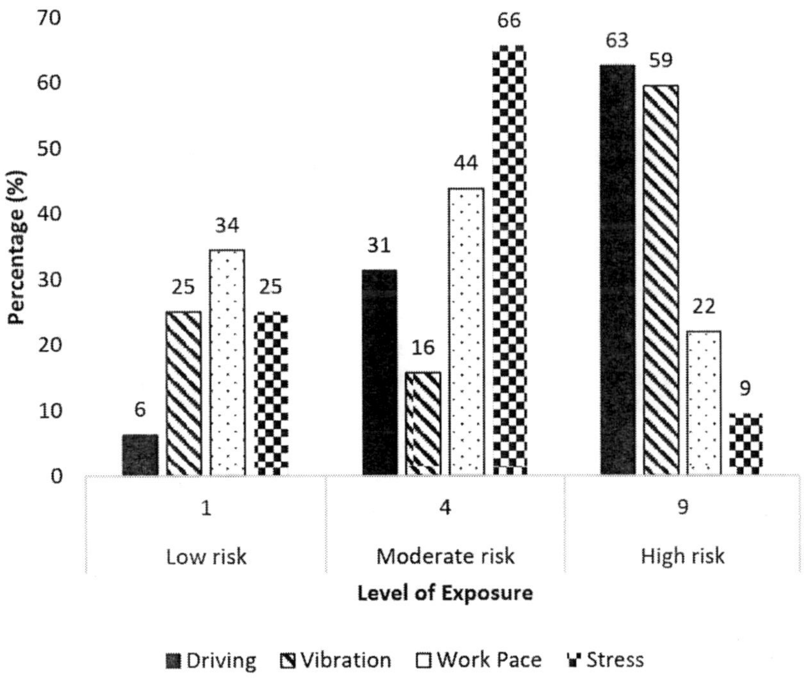

Figure 2. QEC Exposure level and percentage for driving, vibration, work pace and stress.

4. Conclusion

This study was conducted to assess the musculoskeletal discomfort amongst the tower crane operators during the driving and handling of materials of high-rise construction building. The body parts that were associated with a very high risk category of MSDs were the back and neck whereas the high risks for other factors were driving, vibration and stress. In order to minimize the occupational exposure to work-related musculoskeletal discomfort among the tower crane workers, an ergonomic intervention and control measures are essential.

Acknowledgments
This research was funded under Razak School Research Fund Grant (Vot No. 4J295) from Universiti Teknologi Malaysia. The authors would like to thank the Research Management Centre of UTM and the Malaysian Ministry of Education for their financial support and cooperation.

References

[1] Perruccio, A.V., Yip, C., Power, J.D., Canizares, M. and Badley, E.M., 2019. Discordance Between Population Impact of Musculoskeletal Disorders and Scientific Representation: A Bibliometric Study. *Arthritis care & research*, **71**(1), pp.56-60.
[2] Abdelmegid, M. A., Shawki, K. M., & Abdel-Khalek, H. 2015. GA optimization model for solving tower crane location problem in construction sites. *Alexandria Engineering Journal*, **54**(3), 519-526.
[3] Gustafson-Söderman, U. (1987). The effect of an adjustable sitting angle on the perceived discomfort from the back and neck-shoulder regions in building crane operators. Applied ergonomics, 18(4), 297-304.
[4] Health and Safety Security (HSE) (2006). Sector Information Minute: 05/2002/58.
[5] Health and Safety Executive (2017). Work-related Musculoskeletal Disorders (WRMSDs) Statistics in Great Britain.
[6] Punnett, L., & Wegman, D. H. (2004). Work-related musculoskeletal disorders: the epidemiologic evidence and the debate. Journal of electromyography and kinesiology, 14(1), 13-23.
[7] Darabad, M. N., Mazloumi, A., Saraji, G. N., Afshari, D., & Foroushani, A. R. (2017). Full shift assessment of back and head postures in overhead crane operators with and without symptoms. *Journal of occupational health*, **17**-0065.
[8] Burdorf, A., & Zondervan, H. (1990). An epidemiological study of low-back pain in crane operators. *Ergonomics*, **33**(8), 981-987.
[9] Krishna, O. B., Maiti, J., Ray, P. K., & Mandal, S. (2015). Assessment of risk of musculoskeletal disorders among crane operators in a steel plant: A data mining–based Analysis. Human Factors and Ergonomics in Manufacturing & Service Industries, 25(5), 559-572.
[10] Lis, A. M., Black, K. M., Korn, H., & Nordin, M. (2007). Association between sitting and occupational LBP. *European Spine Journal*, **16**(2), 283-298.
[11] Carlos, M. C., & Lucero, H. J. (2012). Risk assessment of the job tasks for heavy equipment operators. Risk Analysis VIII, 1(167), 115-26.
[12] Kadir, A., Mohammad R., Othman, N., (2015). Low Back Pain Problem amongst Port Crane Operator. Journal of Advanced Study in Applied Sciences and Engineering Technology. ISSN (online): 2462-1943 | Vol. 1, No. 1. Pages 13-26.
[13] Deeney, C., & O'Sullivan, L. (2009). Work related psychosocial risks and musculoskeletal disorders: potential risk factors, causation and evaluation methods. Work, 34(2), 239-248.
[14] Yasobant, S., Chandran, M., & Reddy, E. M. (2015). Are bus drivers at an increased risk for developing musculoskeletal disorders. An ergonomic risk assessment study. J Ergonom, 2015.
[15] Ericsson, P., Björklund, M., & Wahlström, J. (2012). Exposure assessment in different occupational groups at a hospital using Quick Exposure Check (QEC)–A pilot study. Work, 41(Supplement 1), 5718-5720.
[16] Farhadi, R., Omidi, L., Balabandi, S., Barzegar, S., Abbasi, A. M., Poornajaf, A. H., & Karchani, M. (2014). Investigation of musculoskeletal disorders and its relevant factors using quick exposure check (QEC) method among seymareh hydropower plant workers. Journal of Research & Health, 4(2), 714-20.
[17] Erdinc, O., & Vayvay, O. (2008). Ergonomics interventions improve quality in manufacturing: a case study. International Journal of Industrial and Systems Engineering, 3(6), 727-745.

[18] David, G., Woods, V., Li, G., & Buckle, P. (2008). The development of the Quick Exposure Check (QEC) for assessing exposure to risk factors for work-related musculoskeletal disorders. Applied ergonomics, 39(1), 57-69.

[19] Muthukumar, K et al. (2018). Study on Discomfort in Mobile Crane Operation. International Research Journal of Engineering and Technology (IRJET). Volume: 04 Issue: 12

[20] Hossain, M. D., Aftab, A., Al Imam, M. H., Mahmud, I., Chowdhury, I. A., Kabir, R. I., & Sarker, M. (2018). Prevalence of work-related musculoskeletal disorders (WMSDs) and ergonomic risk assessment among readymade garment workers of Bangladesh: A cross sectional study. PloS one, 13(7), e0200122.

[21] Bovenzi, M., Pinto, I., & Stacchini, N. (2002). Low back pain in port machinery operators. Journal of sound and vibration, 253(1), 3-20.

[22] Piccinni, S., Marchi, T., Lorusso, A., & Magarotto, G. (1992). The prevalence of spondylopathies among the crane operators in the port of Venice. La Medicina del lavoro, 83(2), 146-149.

[23] Hoy, J., Mubarak, N., Nelson, S., De Landas, M. S., Magnusson, M., Okunribido, O., & Pope, M. (2005). Whole body vibration and posture as risk factors for low back pain among forklift truck drivers. Journal of Sound and Vibration, 284(3-5), 933-946.

[24] Chen, J. C., Chang, W. R., Chang, W., & Christiani, D. (2005). Occupational factors associated with low back pain in urban taxi drivers. Occupational medicine, 55(7), 535-540.

IOP Conf. Series: Materials Science and Engineering **834** (2020) 012057 doi:10.1088/1757-899X/834/1/012057

Development of Heart Rate Monitoring System to Estimate Driver's Mental Workload Level

Ahmad Khushairy Makhtar*[1], Muhammad Izzat Sulaiman[1]
[1] Faculty of Mechanical Engineering, University Teknologi MARA (UiTM), Shah Alam Malaysia

*ahmadkhushairy@uitm.edu.my

Abstract. Driver's mental workload is affecting the driver's performance while driving. A high mental workload task proved that the driver's performance level decrease. The decreasing of driver's performance lead to driver carelessness that causes a road accident. Thus, the way to estimate and monitor the driver's mental workload has to be developed. The purpose of this project is to monitor the driver's mental workload and give a prompt warning to the drivers whether they are safe or not to drive, thus enhancing road safety. This project started by obtaining information about the driver's mental workload through a driver's heart rate. For early stage, this project conducted by using an electronic device for education, which is Arduino Kit and heart rate sensor. The heart rate sensor will detect the heart beat through the fingertips as an input. Then the input will be sent to Arduino, and it will monitor the reading of heart beat in beat per minutes (BPM) through a connected screen monitor. An experiment conducted in two conditions which are at rest condition and during Mathematical Arithmetic Task to estimate the driver's mental workload. By using the developed sensor system, the threshold value of mental workload was collected. The buzzer will produce a sound if the reading heart rate value is exceeding the maximum threshold heart rate reading for the safe driving condition.

1. Introduction

Road safety is one of the crucial things that must be considered by everyone. The carelessness of the driver may lead to an accident. According to the road safety annual report 2017, in 2015 there was a small increase of 0.5% from 2014 in the number of reported fatalities bringing the total to 6706. 20.3 % of the total are car occupants [1]. Then, based on provisional data published by the Malaysian Institute of Road Safety Research (MIROS), there were 7 152 reported road deaths in 2016, a 6.7% increase compared to 2015[2]. This shows that the problem of road safety still in critical condition as the year increases mean that the improvement of the safety system developed for vehicles also increases. However, the problem of road safety still cannot be solved as the number of road accident still in high number every year.

After looking at many previous studies about the causes of road accidents, approximately 23% of all crashes and near-crashes were caused by distraction due to the secondary task (task other than driving) [3]. It shows that distracted driving is a vital issue to be concerned. Driver distraction has a relationship to the driver's mental workload as distraction causing unstable driver's mental workload [4]. There are risky for the driver to drive in an unstable mental workload due to the driving performance begins to deteriorate and may lead to a road accident [3][4][7]. Realizing the importance of developing a method to estimate and detect the driver's high mental workload level, this project chooses the heart rate level as the indicator to estimate the mental workload.

There a lot of road accident happened due to human error, which leads by their mental workload. Not many cars have a safety system that focuses on the driver's driving condition. The main problem is

Content from this work may be used under the terms of the Creative Commons Attribution 3.0 licence. Any further distribution of this work must maintain attribution to the author(s) and the title of the work, journal citation and DOI.
Published under licence by IOP Publishing Ltd

it is difficult to estimate and monitor the driver's mental workload to prevent road accident. Currently, the detection of a driver's mental workload has been a challenging part of the research and become increasingly important in order to develop a precaution warning system for the driver. The studies on estimation of a driver's mental workload have been conducted using various ways and devices. However, the accuracy of the measurement of the devices is still vague.

There was various way to estimate the mental workload of human. Previously, there was a study about the mental workload that used pupillometry [8]. Pupillometry is the measurement of pupil size and reactivity toward task or work done by someone [9]. Gerhard Marquart and Joost de Winter used the smart Eye DR120 remote eye tracker to record the participant's pupil diameter, eyelid opening, and gaze direction while sitting behind a desktop computer [8].

In one of the studies conducted, the data were recorded by using human finger photoplethysmogram (PGT). Zeier (1979) measured heart rate in heavy city traffic while subjects drove a car with a manual transmission or a car with an automatic transmission or being a passenger of the car been driven by someone else. Both average heart rate and Heart Rate Variability (HRV) in time domain differed significantly between the manual-transmission condition and the other two conditions. Driving with automatic transmissions or riding as a passenger did not lead to a significant difference in heart rate measures [10]. Meister et al. have designed a model of mental workload and the relationship between the task demand and mental workload level as in Figure 1 [11]. This model divided into three regions, A, B and C. Region A indicated a low level of workload. Increase in workload is not accompanied by variations in performance since the operator has sufficient spare information-processing resources to compensate for a workload increase. Meanwhile, in region B, it consists of a higher level of workload that exceeds the capability of the operator to compensate. Then in region C, degradations in performance will be gradual during the initial stage of overload or total failure will occur at higher levels of load [12].

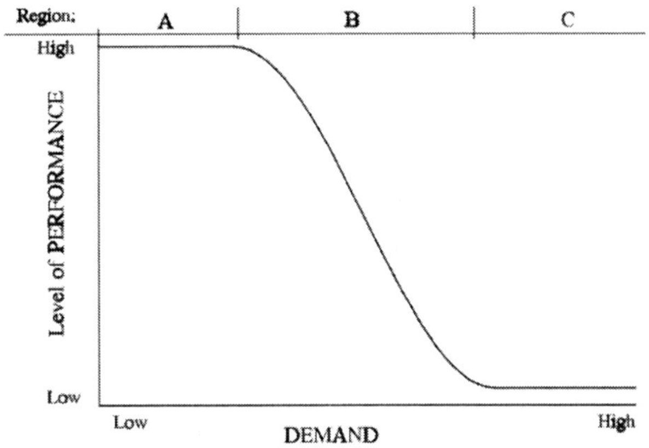

Figure 1. Hypothetical workload and performance relationship [11]

2. Material and Methods

2.1 Heart Rate Monitoring Device Component

Arduino software, nodeMCU and Pulse sensor are the main electric components to develop the heart rate monitoring device. After the device is set up, the heart rate data will be display through Blynk applications using a Wi-Fi connection. Table 1 shows the component to build the Heart Rate Monitoring Device and its functions.

IOP Conf. Series: Materials Science and Engineering **834** (2020) 012057 doi:10.1088/1757-899X/834/1/012057

Table 1: Heart Rate Monitoring Device Component

Component	Name and functions
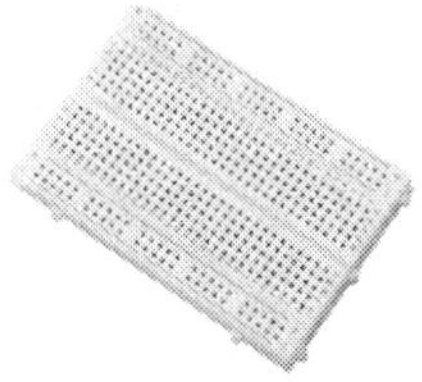	Mini Bread Board - As a platform to build a heart rate monitoring device
	Jumper Wire - As a connecting wire between the equipment
	Pulse Sensor - To sense the heart rate of the users
	NodeMCU - An open source of IoT platform - Act as a platform to build a device and as a Wi-Fi port.

2.2 Development of the heart rate monitoring system

Before starting the estimation of a driver's mental workload, the device to detect the heart rate of the driver has to be developed. In this project, the used of Arduino and heart rate sensor is compulsory to develop the heart rate sensor system to record the data, which is the driver's heart rate. The following Figure 2 shows how the heart rate sensor warning system was developed summarily.

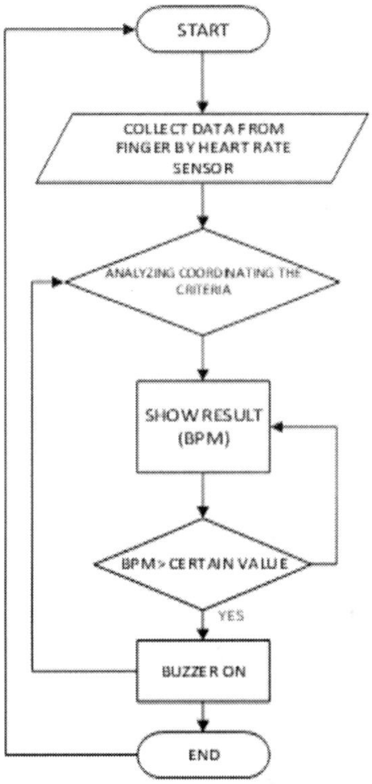

Figure 2 . Arduino Flow Chart

2.3 Experimental Setup

After developed the device that can detect the heart rate, the experiment on the change of the driver's heart rate have been conducted. Ten participants aged 20-35 who have a driving license and drive the car almost every day have been recruited for the experiment. The participants needed to sit on the driver seat in a selected car without driving that car. There were two conditions to be performed by the participants. The first condition was at rest, means relax and focus in front of the vehicle. For the second condition, the participants needed to perform a Mathematical Arithmetic Task (MAT) in 2 minutes. MAT task requires the participants to memorize a number presented before as well as solving of calculation. The participants were given a one-digit number. Three seconds will be given to the participants to solve that calculation. The reading of the participant's heart rates taken during performing the tasks. Figure 3 shows the setting of the devices during the experimental setup. While Figure 4 shows the example of MAT Task to be performed by the participants.

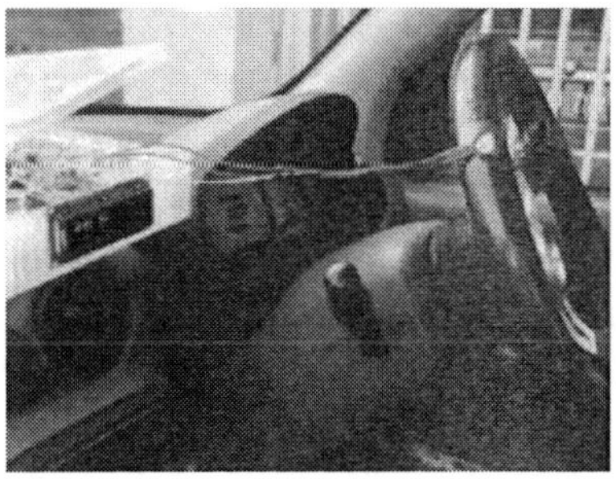

Figure 3. Setting of Devices

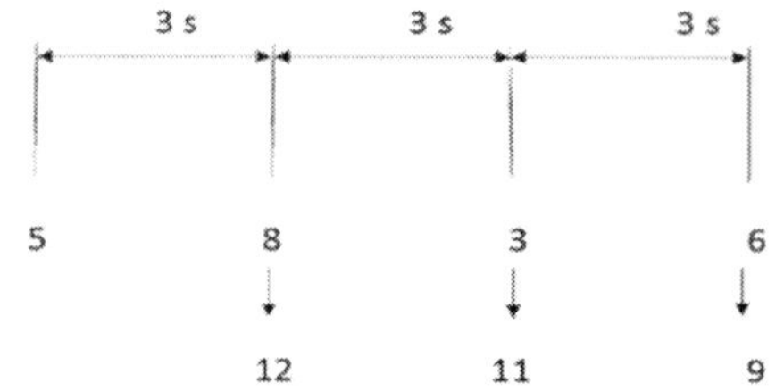

Figure 4. Example of MAT Task

The numbers of correct answer answered by the participants also recorded to evaluate their performance in the task. After both tasks completed, all participants needed to estimate the workload they experienced by NASA Task Load Index form [13]. After completed the experiment and the form filled by the participants, the data tabulated that consider psychological, performance, subjective of the participants.

3. Results and Discussion

3.1 Development of heart rate monitoring system

The development of the heart rate sensor system completed. After applying the correct circuit and the correct programming coding installed into the Arduino, this device can function well to detect human heart rate. This device can detect a human heart rate in beat per minute (BPM) and monitor the reading of BPM through LCD that connected with Arduino. If there is no pulse detected, the value on the LCD screen will be more than 200. By using this device, the heart rate data of each participant recorded. Figure 8 shows the setup of the heart rate monitoring system.

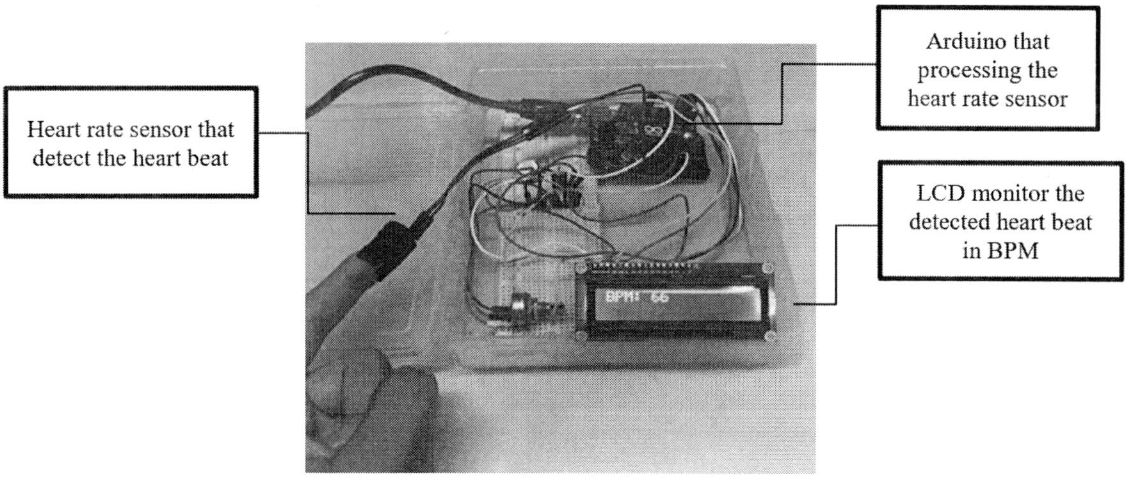

Figure 5. Heart Rate Monitoring System

The reading shown through the LCD is in the range of normal heart rates according to the United States National Institutes of Health (NIH) which are 60 PBM to 100 BPM for people age ten years old and above [14].

3.2 Result

3.2.1 Heart rate analysis of the driver.

The heart rate data were recorded through fingertips by using the developed system in order to determine the mental workload level. Table 2 and Figure 6 show the reading of heart rate for two tasks. The readings have been taken in 2 minutes for each person in order to get a more accurate value of participant' So basically, the data in these two figures are the average value of heart rate reading in 2 minutes.

Table 2: Heart rate reading

No	Heart Rate (BPM)	
	No task	MAT task
1	75	87
2	71	82
3	63	76
4	75	88
5	65	73
6	69	78
7	69	80
8	56	72
9	60	72
10	66	76
AVERAGE BPM	67	78

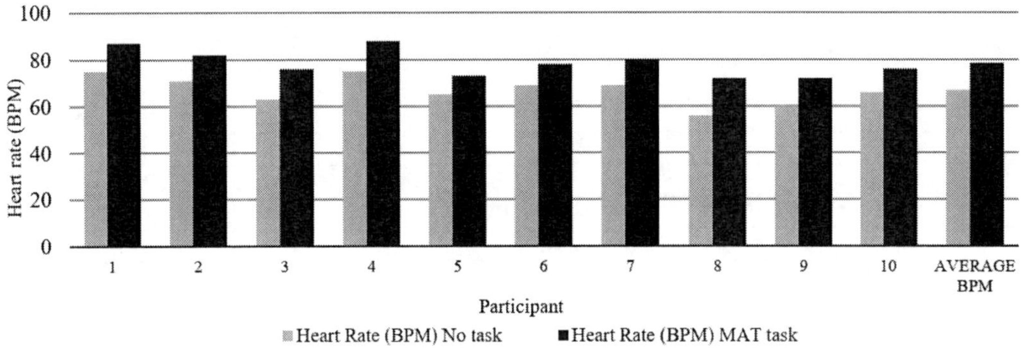

Figure 6. Heart rate reading

From Table 1 and Figure 6, the value of heart rate (BPM) for MAT Task is higher than during no task experiment. Averagely, the BPM reading for the mental workload was 67 BPM for no task and 78 for MAT task. So, the threshold value for the high mental workload was set as 78 BPM.

3.2.2 Performance analysis

The performance of the driver indicated by the correct number answer answered by the participants in the MAT task. Table 3 shows the performance analysis of the participants. All participant could answer correctly above 80% as the task in easy level. Based on the table, it can be seen that only 4 out of 10 participants did well, in the task which got correct answer 80% and above. Averagely, the performance of the participant is 78%, which did not meet the success value. This results showed that the participants were in high mental workload during the MAT task.

Table 3: Performance analysis

Participant	Corrected Answered	Performance (%)
1	35	90
2	22	56
3	33	85
4	29	74
5	29	74
6	33	85
7	30	77
8	29	74
9	31	79
10	32	82
AVERAGE	30	78

3.2.3 Subjective measurement (Questionnaire)
After completing all the task, all participants were requested to make a judgement about their feeling for the type of task they felt. The questionnaire was developed based on a standard measurement of mental workload which is NASA-TLX. [13]

Table 4 shows the result of the questionnaire. Most of them judged the task mentally demanded, and they were in high mental workload. This condition is not a good condition to drive a car because of the distraction caused by the task. In the real situation of driving, this task indicates the distraction for the driver that prevents the driver from focussing on the road. [3]

Table 4. Subjective Measurements Results (questionnaire)

Participant	Subjective (%)												Level	
	Mental demand		Physical demand		Temporal demand		Performance failure		Effort		Frustration			
	No task	Task	No task	Task	No task	Task	No task	Task	No task	Task	No task	Task	No task	Task
1	3	77	20	24	5	81	5	30	2	72	3	88	6	62
2	2	87	17	18	2	87	2	50	3	92	3	95	5	72
3	5	76	12	14	5	83	4	60	3	83	11	88	7	67
4	5	70	16	17	5	50	4	80	5	60	2	70	6	58
5	2	73	2	5	4	79	3	78	3	80	1	85	3	67
6	2	77	7	7	2	60	2	40	2	76	6	80	4	57
7	2	79	12	18	4	79	5	27	2	69	20	78	8	58
8	4	74	11	14	2	80	3	18	3	80	4	84	5	58
9	2	82	7	8	3	88	3	40	2	85	4	87	4	65
10	2	75	6	6	1	85	2	30	3	80	3	85	3	60
AVERAGE													5	62

4. Conclusions

The study carried out has been successfully met the goals of this project. The result of the installation of the heart rate device was successfully done. The device then used in the experiment. After the experiment, the result of the experiment met the expectation data which is the average value of heart rate, subjective measurement (questionnaire) of the participants for no task experiment is higher than the experiment with MAT task. The performance of the participants for the task also not in best as the average corrected answer answered by the participants is 78% only even though the level of the task is not tough.

For conclusion, from this experiment, the threshold value for the maximum mental workload is 78 BPM. 78 BPM is considered a high mental workload level for the driver and the precaution warning system have to be taken. In this stage, the warning system will be set 78 BPM as the condition that not suitable for driving activity.

For further study, it is recommended to recruit more participants to get a more accurate value, and the task should be levelled up to test the driver's mental workload level.

References

[1] Jabatan Keselamatan Jalan Raya Malaysia, "Statistik Keselamatan Jalan Raya," 2017. .
[2] V. Babulal, "Number of fatal road accidents up in 2016, more than 7,000 lives lost," *News Straits Times*, p. 1, 2017.
[3] T. Dingus and S. Klauer, "The relative risks of secondary task induced driver distraction," *SAE*, vol. 21, no. 0001, 2008.
[4] Ahmad Khushairy BIn Makhtar and Makoto Itoh, "An Analysis of Ear Plethysmogram for Evaluation of Driver's Mental Workload," in *10th AEARU Workshop on Computer Science and Web Technology*, 2015, pp. 2–3.
[5] K. L. Young and M. A. Regan, "Driver distraction : A review of the literature," in *Driver Distraction Conference*, 2005.

[6] S. Baker, "Driver's attitudes, awareness and knowledge about driver distractions : Research from two central Sydney communities," *Driv. Distraction Conf.*, pp. 255–268, 2007.

[7] Ahmad Khushairy Makhtar and Makoto Itoh, "DRIVER'S MENTAL WORKLOAD: TASK PERFORMANCE AND MENTAL WORKLOAD," *J. Teknol.*, vol. 7, pp. 31–39, 2015.

[8] G. Marquart and J. de Winter, "Workload assessment for mental arithmetic tasks using the task evoked pupillary response," *PeerJ Comput. Sci.*, vol. 1, p. e16, 2015.

[9] S. Sirois and J. Brisson, "Pupillometry," *Wiley Interdisciplinary Reviews: Cognitive Science*, vol. 5, no. 6. pp. 679–692, 2014.

[10] F. P. da Silva, "Mental Workload, Task Demand and Driving Performance: What Relation?" *Procedia - Soc. Behav. Sci.*, vol. 162, no. Panam, pp. 310–319, 2014.

[11] D. Meister, *Behavioral Analysis and Measurement Methods*. New York, New York, USA: John Wiley and Sons, 1985.

[12] D. de Waard and V. Studiecentrum, *The Measurement of Drivers' Mental Workload*, vol. 39, no. 4. 1996.

[13] S. G. Hart, M. F. California, and L. E. Staveland, "Development of NASA-TLX (Task Load Index): Results of Empirical and Theoretical Research," *Adv. Psychol.*, vol. 52, pp. 139–183, 1988.

[14] Y. Ostchega, D. Ph, K. S. Porter, J. Hughes, and C. F. Dillon, "Resting Pulse Rate Reference Data for Children, Adolescents, and Adults : United States, 1999 – 2008," no. 41, pp. 1999–2008, 2011.

IOP Conf. Series: Materials Science and Engineering **834** (2020) 012058 doi:10.1088/1757-899X/834/1/012058

Effect of surface attrition on hardness on the hardness and wear properties of 304 stainless steels

S K Alias[1*], M N Halmy[1], M A M Shah[1], N N Ahmad[1], S A Sulaiman[1], H.F. Pahroraji and B Abdullah[2]

[1]Faculty of Mechanical Engineering, Universiti Teknologi Mara Pasir Gudang
[2] Faculty of Mechanical Engineering, Universiti Teknologi Mara Shah Alam

*khadijah_alias@uitm.edu.my

Abstract. The occurrence of wear in austenitic stainless steel is inevitably unavoidable due to the presence of high chromium content and other alloying elements that hinder the implementation of surface treatment process. This process is usually use to improve the hardness and wear properties of steels and alloys by diffusing hard protected layer on the surface of the material. This paper investigates the effect of surface attrition on the hardness and wear properties of paste boronized 304 stainless steel. Surface attrition treatment was applied onto the surface of 304 stainless steel before paste boronizing was conducted at temperature of 850°C for 8 hours holding times. The microstructure of the boronized samples before and after surface attrition treatment was then observed and recorded in order to compare the phase constituent and boride layer thickness. The hardness of each phases was then evaluated using Vickers microhardness test and the wear resistance test of both paste boronized sample and surface attrited samples after boronizing was performed using pin on disk test. The microstructure results show that there are presences of both FeB and Fe2B phases on the surface of Pa-B850 and Pa-SB with boride layer thickness improvement of 3 times compared to untreated samples. This lead to enhancement of both hardness and wear resistance of the paste boronized samples due to better protection on the surface of the 304 stainless steel. The improvement of wear and hardness properties of 304 stainless steel could introduce new application that could be exposed to environment containing friction and wear.

1. Introduction

Generally, 304 stainless steel is the most common type of stainless steel that are widely used in automotive and oil & gas parts due their excellent corrosion resistance, high strength-to-weight ratio in conjunction to whole weight reduction and excellent formability [1-2]. Unfortunately, this material exhibited resistance to hardness and wear properties which restricted their usage in application that are exposed to constant loading and friction. Although there are many available surface treatments that enhanced the tribological behavior of steel, these treatments are ineffective to 304 stainless steel due to the presence of high alloying element contents that obstruct the formation of protective layer on their surface. Hence, a newly modified surface treatment should be implemented in order to develop dispersion of boride layer, thus promoting better wear resistance in 304 stainless steel.

Boronizing, which is a thermochemical method that formed protective boride layer that act as shielding layer on the steel's surface is used to enhance of hardness and wear behavior of steel [3]. However, the boride layer formation which exhibited combination of FeB and Fe_2B phase might not be

Content from this work may be used under the terms of the Creative Commons Attribution 3.0 licence. Any further distribution of this work must maintain attribution to the author(s) and the title of the work, journal citation and DOI.
Published under licence by IOP Publishing Ltd

attained in 304 steel as the existence of high alloying elements impede the creation of this phase, thus resulting in poor hardness and wear properties [4]. As there are presences of carbon and other alloying elements, for example chromium and vanadium in 304 stainless steels, the diffusion of boride layer phase may require longer time to form as the carbon and alloying elements will typically gathered at the boron diffusion zone [5-6]. In conjunction to that, the thickness of Fe_2B diffusion layer is significantly dependent on the selection of boronizing time and temperature which can be exceeded to more than 10 hours holding times. In conventional method, longer boronizing time are required in order to fully diffused Fe_2B and FeB layer with saw tooth morphology that have a minimum thickness of 50µm [7]. The alteration of surfaces layer by implementing shot blasting treatment are essential as it coul lead to reduction of boronizing parameters which are temperature and time. Decreasing the boronizing temperature and time are also important as it will effect in cost reduction without compromising the minimum thickness of FeB and Fe_2B layers.

Generally, there are many types of boronizing technique and each can be implemented with different medium and purposes [8]. In liquid boronizing, mixture of crystallized Boraks and boric acid was often used as the boronizing medium and improvement of wear resistance and mechanical properties could be achieved at lower and shorter boronizing temperature and time compare to pack boronizing [9-11]. Regardless to that, the limitation of this method is the need for additional process of removing the salt layers formed on the surfaces of boronized samples, which can be tedious and expensive. Past research on gas boronizing indicated that this method formed thicker boride layer than conventional boronizing methods at similar temperatures as the results of more chemical reactions were activated during the treatment [10]. Despite to the advantages given by this approach, high toxicity of the precursor and requirement of complex boronized unit set-up are the limitations of this method. Out of all the methods, paste boronizing is the most cost-effective method and offer better dispersion quality compared to powder form.

Paste boronizing will be implemented as boron in paste form offer better dispersion quality compared to powder form [6,12]. It is expected that with the successfulness of Fe2B formation on the surface of 304 stainless steel will lead to enhanced tribological behaviour, thus introducing new and innovative applications for this material. It would be beneficial if a new approach could be implemented, in order to increase both boron diffusion. Thus, this study focused on the effect of surface attrition on the improvement of hard-ness and wear properties of 304 stainless steel.

2. Research Work

The samples used in this study are Grade 304 stainless steel, which is in a form of rod with diameter of 10mm and plate with dimension of 75mm X 25mm X 8mm. Surface attrition treatment as conducted using Finimac shot blasting machine implementing ceramic Silison Carbide (SiC) ball with 2mm diameter and velocity of 70 m/s, in order to induce plastic deformation on the surface of stainless steel. In paste boronizing, the samples (Pa-850) was coated with the boron paste comprising of 5% B4C as donor, 90% SiC as the diluents and 5% KBF_4 which act as an activator with thickness of approximately 2mm and heated at temperature of 850°C for the 8 hours holding time.

The wear test was performed through Ducom Wear and Friction Pin on Disk TR-20LE tester using 125mm diameter rotating disk of hardened 316 stainless steel with speed of 200 r.p.m. load of 20N and sliding distance of 80 for 1 hour sliding time. The hardness value of all types of samples was evaluated and calculated through Mitutoyo MVX-H1 Vickers microhardness tester as per ASTM E-384 applying 10N load.

Table 1 shows the chemical composition of 304 stainless steel samples which was obtained from Spectro Maxx Spark emission machine. The microstructural observation and SEM observation analysis were performed via Olympus B X 41M microscope with IMAPs 4.0 edition and Scanning electron microscopy (SEM) analyzer using SMARTSEM software respectively. The energy dispersive X-Ray (EDX) spectrometry also conducted to determine the percentage of boron in each phase of the boride layer. The samples were compared before (Pa-B850) and after (Pa-SB850) surface attrition process.

Table 1. Chemical composition of 304 stainless steel samples.

Wt (%)	C	Mn	Si	P	S	Cr	Mo	Ni	Fe
304	0.065	1.8	0.75	0.04	0.03	18.5	0.0	8.5	Bal

3. Research Finding

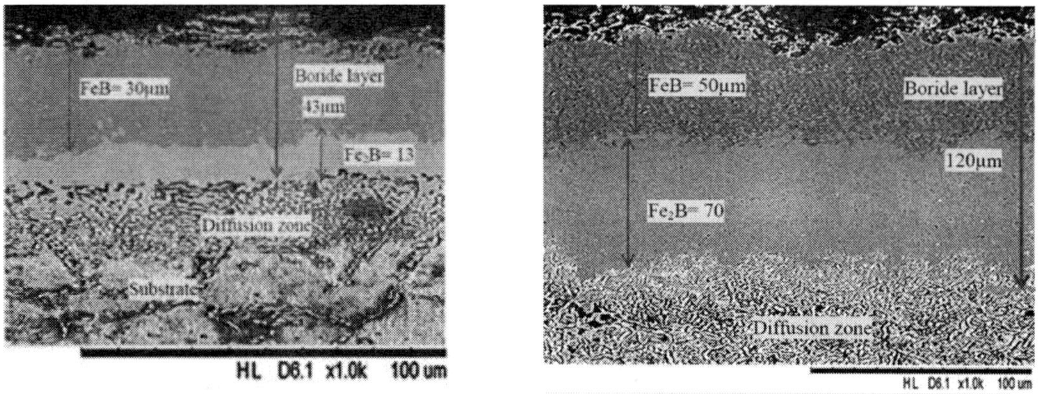

(a) (b)

Figure 1. SEM micrograph of powder boronized sample at 850°C (a) before ((Pa- B850) and (b) after (Pa-SB850) surface attrition at magnification of 1000X.

3.1. Microstructure Analysis

Figure 1 (a) shows the micrograph of 304 stainless steel sample (Pa-850B) after paste boronizing. It could be seen that there are presence of boride layer consisting of FeB and Fe_2B phases with thickness of approximately 43 μm was produced after the boronizing process. The microstructure of boronized stainless steel sample after surface attrition process was shown in Figure 1 (b). It could be observed that an improvement of boride layer thickness which is approximately 120 μm had been successfully achieved after implementation of surface attrition treatment before boronizing process. This value indicated that surface attrition treated samples attained boride layer thickness of more than 100 μm and three times the value of thickness layer of boronized stainless steel with the value of 43 μm without surface deformation at the same temperature. It was also noted there are vast enhancement on the thickness layer of the diffusion zone of boronized samples after surface attrition process.

Similar to paste boronized sample, implementation of surface attrition also produced FeB and Fe_2B phases. However, it could be seen that the formation of Fe_2B phase was thicker in surface attrition treated sample as compared to the as received sample. The thickness of FeB layer is 50 μm while Fe_2B layer attains a thickness of approximately 70 μm. The addition of surface attrition process created surface deformation such as atomic displacement and void which move the atom and allowing the boron atom to be diffuse onto the surface of boronized stainless steel. The formation of Fe_2B phase is essential in producing excellent strength and tough-ness to boride layer thickness, thus delivering prominent protection to the surface of boronized samples. A similar trend was obtained by past researchers [13-14] which indicated that alteration of the surface region by applying a permanent plastic deformation are beneficial in enhancing diffusivity of case hardening and thermochemical methods such as boronizing, carburizing and nitriding on both low alloy and high alloy steels. Other significant effects of surface deformation are occurrence of dislocation at both grain boundaries as well as within the grain, all in which are the justified reasons of major improvement in the case depth of boronized layer.

The presence of FeB and Fe2B phases was validated through XRD analysis at 2 Theta angles of 37°[1 1 9], 62°[4 3 4] and 80°[6 1 1] for FeB phase and 35°[2 0 8], 42°[0 1 1] and 82°[7 1 6] for Fe2B phase for Po-B850 samples . The validation of FeB phase in surface attrited boronized samples are performed by XRD analysis at 2 Theta angles of 37 [1 3 4], 42 [0 3 9] and 80 [5 2 3] while for Fe2B phase at angles of 35 [3 2 1], 58 [0 3 9] and 80 [2 5 2].

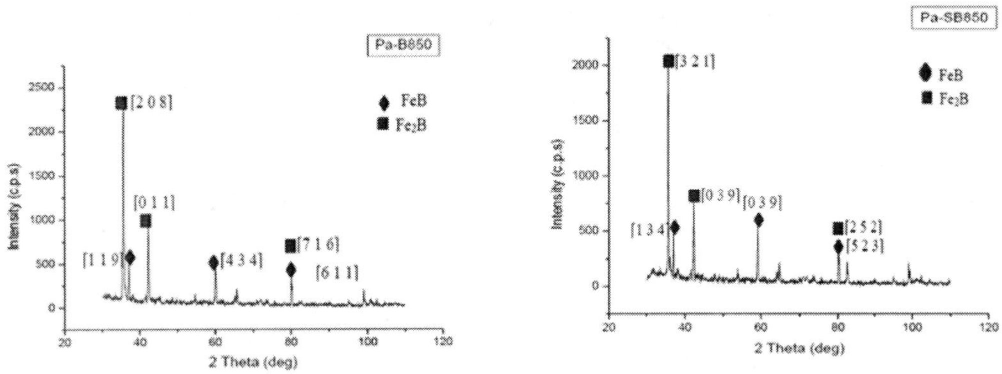

] (a) (b)

Figure 2. X-Ray Diffraction pattern of paste boronized stainless steel (a) before (Pa-B850) and (b) after (Pa-SB850) surface attrition process.

3.2. *Microhardness*

The microhardness values for paste boronized stainless steel before and after surface attrition is depicted in Figure 3. The microhardness of Pa-B850 sample was averaged at 1550 Hv. The diffusion layer started from the outer layer of FeB phase, which contained the highest microhardness value to Fe2B phases which contained adequate strength and toughness and diffusion layer which is the area in between the Fe2B phases and the substances. The substances generally portrayed the microhardness value of the material.

Pa-SB850 sample exhibited excellent microhardness with the maximum value of 1800 Hv at FeB phases with increment of 6 times or 600% the microhardness as compared to the substrate. Achievements of hard, but brittle nature of FeB phase is related with the boron composition of two times the value of Fe2B phase which is16.23 wt. %. This was also supported by past researchers indicating that FeB is much harder than more ductile Fe2B phase in boronized steels [15-16]. Reduction of microhardness values was observed on the Fe2B region of approximately 1550 Hv and at diffusion zone further decrement with average of 580Hv was perceived.

Pa-SB850 accomplished highest microhardness compared to other samples as thickest boride layer was obtained. The deformed surface acquired atom dislocations, thus creating small voids for the boron to diffuse at deeper thickness. Comparably, the microhardness values of boronized sample increased proportionately with the increment of boride layer thickness [17]. At substrate region, minimum microhardness values were obtained as compared to another region with an average of about 257 Hv for all types of samples. After surface attrition, there are noticeable improvements on microhardness as the values increased to an average of 330Hv. Although surface attrited samples attained the same austenitic structures, the size of grain plays important role in improvement of microhardness. After surface attrition, the grain sizes reduced tremendously through the impact given by blast media that created surface deformation.

Likewise, the improvement of residual stress at the applied surfaces through surface modification also leads to the microhardness enhancement apart from improvement of thicker boride layer [18]. Similarly, Arifvianto et al. [19] found that the steel slag ball blasting treatment increases the microhardness of samples surface and subsurface of 316L stainless steel. The slag balls are nearly spherical but with a bit irregular surface. Microhardness tester was used to measure the microhardness distribution on the samples cross-sectional area. The measurement was conducted from the closest point to the surface down into the bulk of the samples with an indenting load of 4.9 N for 0 to 600 sec.

Arifvianto et al. [19] demonstrated that the use of bigger slag balls for the treatment also produces a harder surface layer and thicker hard layer.

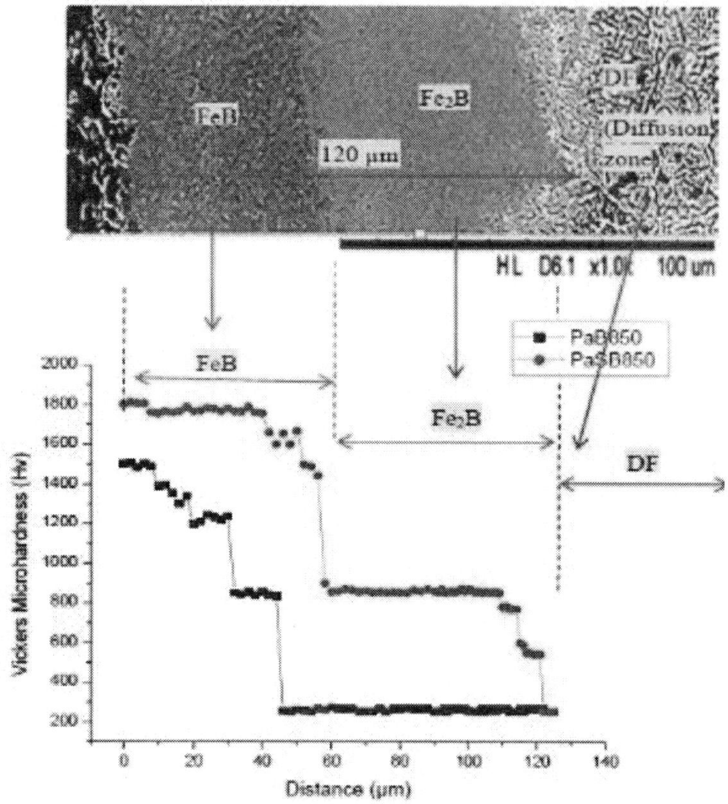

Figure 3. Microhardness of Pa-B850 and Pa-SB850 samples from the diffusion zone until the outer surface.

3.3. *Wear properties*

It was evaluated that boronizing using paste medium provides favourable benefits as the COF obtained was much lower with the average value 0.458 of than powder boronized samples that attained the value of 0.611 with improvement of 25%. This was associated with the formation of deeper boride layer of 43 µm on Pa-B (850°C) as compared to 15 µm samples. As paste medium produced higher activation energy, the diffusion occurs at a faster rate compared to powder boronized samples, thus provide better diffusion thickness of the boronized layer.

The formation of thicker FeB layer with thickness of 30 µm which is almost three times the thickness of Fe2B layer which is 13 µm. Although there are major improvement of the overall thickness of boride layer from 15 µm in Po-B850 sample to 43 µm in Pa-B850 sample, the enhancement of FeB layer thickness increased the brittleness of the layer, thus resulted in insignificant improvement of COF with the variation of only 25% as compared to Po-B850 sample. Paste medium exhibited smaller grain sizes than powder medium, the formation of boride layer of the outer surfaces are smoother, and thus it is harder for particles to be removed initially during the pin on disc test. Paste medium also produces a denser outer layer, which reduce the brittleness of the boride layer. This provides better protection to the surfaces, thus better wear quality was produced, indicated by the lower COF value.

Past researcher performed modifications on the properties of boronizing medium by adding chromium and rare earth elements and established that lower COF values was produced [20]. This was because the addition of these elements reduced the brittleness of the boride layer by enhancing the density of the layer. As the layers produced are denser, it restricted the action of wear during pin on disc test.

Figure 4 shows the results for coefficient of friction for Pa-SB850 sample as compared to Pa-B850 sample. The lowest COF values, indicating the highest wear resistances were achieved by combining surface attrition and paste boronizing at temperature of 850°C to the samples which is Pa-SB850 samples at the average value of 0.353 as compared to Pa-B850 sample with the value of 0.458 with improvement of 30 %. This was mainly because of enhancement of boride layer thickness of Pa-B850 sample from 43 µm to 120 µm in Pa-B850 after surface attrition process.

Additionally, the improvement of COF may also contribute by the formation of Fe2B phase is generally thicker with the thickness of 70 µm as compared to 50 µm in FeB phase. This indicates that a combination of paste boronizing and surface attrition successfully improved the wear resistance with the COF value of more than two times as compared to the COF as received grade 304 stainless steel which is 0.856, as more than 100 µm thicknesses of boride layer was produced. Theoretically, surface attrition enhanced the thickness of boride layer by enabling the boron atom to diffuse deeper, occupying by grain size refinement and density defect formation [21]. As a conclusion, it can be said that the values of COF increased in opposition to the boron diffusion layer thickness. It also relates that boronizing automotive parts with modification of the roughness parameters could enhance the wear resistances of the parts by 30 to 40 % [22].

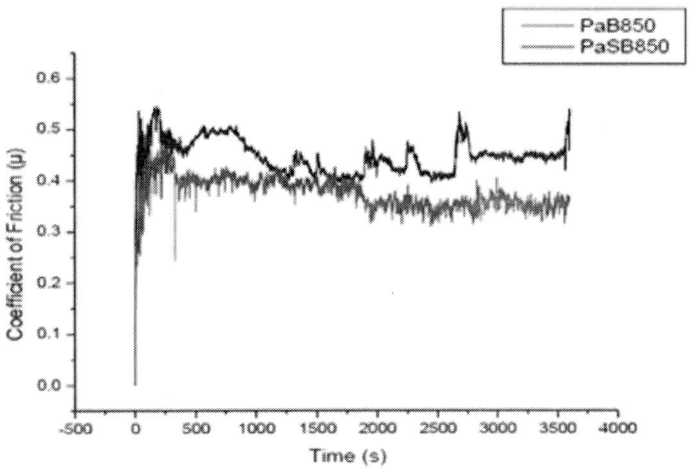

Figure 4. Coefficient of friction for Pa-SB850 sample as compared to Pa-B850 sample.

4. Conclusion

In this study, the influence of surface attrition process on the hardness and wear properties of boronized stainless steel was investigated and the following conclusion could be obtained:

Surface attrition process had increased the boride layer thickness of boronized stainless steel up to 3 times from 43 µm to 120 µm due to the surface deformation effect that allow the boron atom to be dispersed deeper into the deformed surface.

Pa-SB850 samples exhibited higher hardness and better wear resistance as the addition of surface attrition before boronizing process produced deeper and harder boride layer produce which allowed more surface protection to the material.

Acknowledgement

The authors would like to express their gratitude to the ministry of science and technology, Malaysia (MOSTI) and Ministry of Higher Education Malaysia (MOHE), and Universiti Teknologi MARA for the financial support given to this work through research grant of 600-IRMI/FGRS 5/3 (003/2019).

References

[1] Leffler , B., 2000. Stainless steel and their properties. Welding journal, 1 pp 5-9.
[2] Vaghani, .M, Vasanwala, S.A., Desai.A.K., 2014. Stainless Steel as a Structural Material: State Of Review. Journal of Engineering Research and Applications. 4(1), 657-662.
[3] Bartkowska, A., Swadźba, R., Popławski, M. and Bartkowski, D., 2016. Microstructure, microhardness, phase analysis and chemical composition laser remelted FeB-Fe2B surface layers produced on Vanadis-6 steel. Optics & Laser Technology. 86, 115-125.
[4] Mertgenc, E., Kesici, O.F. and Kayali, Y., 2019.Investigation of wear properties of borided austenitic stainless steel different temperatures and times . Materials Research Express. 6 (7) 076420
[5] Alias. S.K., Abdullah, B., Latip, S.A., Roseley, N.R., Jenal, N., Kasolang, S., 2013. Effect of temperature variations on wear properties of boronized stainless steel. Applied Mechanics and Material. 393, 217-221.
[6] Basir, M.H., Abdullah, B. and Alias, S.K., 2014. Wear properties of paste boronized 316 stainless steel before and after shot blasting process. Scientific Research Journal. 11, 2.
[7] Wang, D., Zhao, H., Wu, A., Wang, L., Zhang, X. and Wei-hua Huang, W., 2015. The effect of Na2S2O3 on RE-boronizing behavior of 45 carbon steel. Materials & Design. 67, 615-622.
[8] Naemchanthara, P. and Juijerm, P., 2018. Effects of Heat Treatment on Phase Transformation and Corrosion Resistance of Boride Layer on Austenitic Stainless Steel AISI 304. Metallurgical and Materials Transactions B: Process Metallurgy and Materials Processing Science. 49 (5) 2875-2880.
[9] Sert , H., Canb, A., Arıkan, H., Selcuk, B., Toprak. H., 2006. Wear behavior of different surface treated cam spindles. Wear. 260, 1013–1019.
[10] Gunes, I., 2013. Kinetics of borided gear steels, Indian Academy of Sciences. 38, 527–541.
[11] Taktak, S.,2007. Some mechanical properties of borided grade H13 and Grade 304 steel. Materials and Design. 28, 1836–1843.
[12] Yoon, J.H., Jee, Y.K. Lee, S.Y., 1999. Plasma paste boronizing treatment of the stainless steel grade 304. Surface and Coatings Technology, 112 71–75.
[13] Zhan, K., Jiang, C.H., Ji. V., 2012. Uniformity of residual stress distribution on the surface of S304 austenitic stainless steel by different shot peening processes, Materials Letters. 99, 61 64.
[14] Adachi. N and Ueda. N., 2013. Surface hardness improvement of plasma-sprayed AISI 316L stainless steel coating by low-temperature plasma carburizing Advanced Powder Technology. 24, 818–823.
[15] Ozbek, I., Bindal, C., 2011. Kinetics of borided grade M2 high speed steel. Vacuum. 86, 391-397.
[16] Xie , F., Sun, L., 2013. A novel approach to achieve thick single phase Fe2B coating by Alternating Current Field Enhanced pack Boronizing. Physic Procedia 50, 88-93.
[17] Zepon, G., Nascimento, A.R.C., Kasama, A.H., Nogueira, R.P., Kiminami, C.S. Botta, W.J., 2015. Design of wear resistant boron-modified supermartensitic stainless steel by spray forming process. Materials and Design. 83,214-223.
[18] Yilmaz, S.S., Unlu , B.S., Varol, R., 2010. Effect of boronizing and shot peening in ferrous based FeCu–Graphite powder metallurgy material on wear, microstructure and mechanical properties. Materials and Design .31 4496– 4501.
[19] Arifvianto, B., Mahardika, M., Dewo, P., Iswanto, PT., Salim, UA., 2011. Effect of surface mechanical attrition treatment (SMAT) on microhardness, surface roughness and wettability of Grade 316L. Materials Chemistry and Physics. 125 (3), 418-426.
[20] Yuan, X., Xu, B., Caib, Y., 2013. Study on Cr-Rare earth-Boronizing of the steel 45 at low temperature. Physics Procedia. 50 82 – 87.
[21] Yang, W.J., Zhang, M., Zhao, Y.H., Shen, M.L., Lei, H., Xu, L., Xiao, J.Q., Gong, J., Yu, B.H. Sun, C. 2016. Enhancement of mechanical property and corrosion resistance of 316 L stainless steels by low temperature arc plasma nitriding. Surface and Coatings Technology. 298, 64-72.

[22] Lubas, J., 2010. Practical application of boron-modified sliding pairs in IC.engine. Tribology International. 43, 2046–2050.

IOP Conf. Series: Materials Science and Engineering **834** (2020) 012059 doi:10.1088/1757-899X/834/1/012059

Effect of Palm Stearin binder to the Mechanical Properties of Alumina using Slip Casting Method

M N Halmy[1]*, S K Alias[1], M A Mat Shah[1], M K Maryam[2], M A Abu Bakar[2] and M H Ismail[2]

[1]Faculty of Mechanical Engineering, Universiti Teknologi Mara Pasir Gudang
[2] Faculty of Mechanical Engineering, Universiti Teknologi Mara Shah Alam

*halmy@johor.uitm.edu.my

Abstract. Alumina is one of bio-ceramics that have been used as bone replacement material in modern health care industries due to its good mechanical properties. Unfortunately, their potential as bone implant depends upon its strength and compatibility with the biological environment. Since palm stearin possesses low molecular weight, which can prevent residual stress and distortion with abundantly available in Malaysia (low cost). Furthermore, bone implant using palm stearin as a binder has never been studied before. Thus, a study has been carried out to investigate the effect of different contents of Palm Stearin (35 vol%, 36 vol%, 37vol % and 38 vol%) with different sintering temperatures (1300°C, 1350°C, 1400°C, and 1450°C) using a slip casting process to produce porous alumina like the human bone. The samples then were characterized in terms of physical and mechanical properties. The water impregnation test was used to determine the percentage of porosity of the sample whilst Scanning Electron Micrograph (SEM) to determine its pore structure. Three-point bend test and hardness test were used to determine its flexural strength and hardness respectively. It was found that by using 36 % volume of palm stearin and 64% volume alumina which sintered at 1400°C has porosity of 41.63% with a pore size measured 112μm in diameter. The flexural strength and hardness valued at 5.366 MPa.

1. Introduction

Since 1975, alumina ceramic has proven its bio inertness (it does not interact with the body's environment apart from causing an initial 'fibrous tissues'), high wear resistance and excellent biocompatibility (ability of surgical or dental implant to exist in the body with no adverse effect). Even though it has good properties physically and mechanically, but their potential as implant depends upon its strength and compatibility with the biological environment.

This research requires the sample produced from alumina and palm stearin binder using the slip casting method due to cheaper and wide variety of complex shapes can be produced. The samples will be sintered at a certain temperature. Before sintering, the sample will undergo thermal de-binding process to completely remove all binders. This temperature must be higher than the decomposition temperature of the binder. Hence, thermal de-binding and sintering temperature plays a major role to produce a strong and porous sample. According to [1], bone composed of two types i.e. cortical (compact) and cancellous (spongy) bones. Cortical bone has a density of ~2.0 g/cm3 whilst cancellous bone has much lower density; 0.2 g/cm3 [1]. Since alumina has proven its bio inertness, the usage of alumina in bone implant has widen and some examples are hip prostheses ball and sockets, knee prostheses, dental implants and bone screw [2]. The presence of the binder can affect the sintering and

Content from this work may be used under the terms of the Creative Commons Attribution 3.0 licence. Any further distribution of this work must maintain attribution to the author(s) and the title of the work, journal citation and DOI.
Published under licence by IOP Publishing Ltd

strength of alumina. Therefore, properties of palm stearin binder are studied and according to [3], palm stearin has a melting point of 61°C and its decomposition temperature is between 288°C - 463°C, environmentally acceptable and low cost [3]. One of the main properties that needs attention is the porosity of alumina that is like the human bone. As mentioned by [4], for bone regeneration, the ideal pore sizes would be in the range of 100 – 400 µm. Its mechanical properties are enough to withstand the in-vivo loads which is a basic requirement of a successful scaffold structure [4]. Porous materials have the advantage of biological anchorage for ingrowth of mineralized tissue into the pore spaces. The architecture of porous implant gives a great impact of bone in growth in the pore [4].

Moreover, the addition of binder will influence the shrinkage phenomenon. When this happened, more areas between the pores are in contact making the pore become smaller. Thus, making the bonding between particle stronger [5]. To test the mechanical properties of the Alumina, 3-point bend and hardness test is used. Brittle material strengths are statistical in nature and different specimen yield different strengths due to variations in the flaws and microstructure [6].

2. Research Work

Suitable methods or procedure is carried out according to the current standard. Flowchart below is the overview of the methodology.

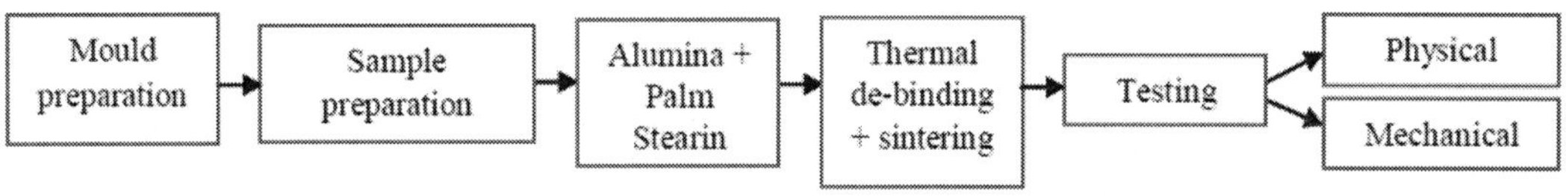

2.1 Mold preparation

Typical mold used by the slip casting method is made from plaster of Paris. However, plaster of Paris is not suitable and then it has been changed to silicon mold because the sample sticks to the mold.

2.2 Sample preparation

Important information for materials preparation is to know the pycnometer density of alumina and palm stearin, which are 3.625 g/cm3 and 0.9915 g/cm3 respectively.

2.3 Mixing and Slip casting

Alumina is mixed with Palm Stearin using the mechanical stirrer. Variables that are held constant during mixing pro-cess are:

Temperature: 60 - 64°C Speed: 10 rpm Time of mixing: 6 – 9 minutes.

Table 1. Different percentage of Alumina and Palm Stearin used.

Material	Sample 1	Sample 2	Sample 3	Sample 4
Alumina	62 vol. %	63 vol. %	64 vol. %	65 vol. %
Palm stearin	38 vol. %	37 vol. %	36 vol. %	35 vol. %

2.4 Thermal de-binding and sintering

Heating rate used for the sintering process is 1°C/min. Before the sample is sintered, it will undergo thermal de-binding process to completely remove all palm stearin in the sample.

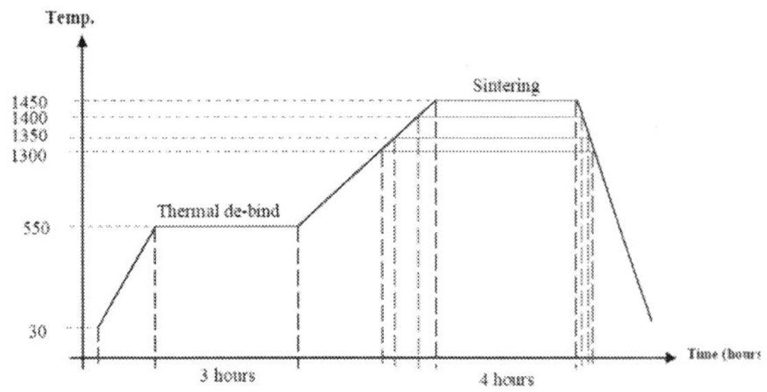

Figure 1. Thermal de-binding and sintering graph.

2.5 Physical testing

After all samples had undergone a sintering process, their porosity is calculated using equation (1):

$$P\,(\%) = \frac{\text{Palumina powder} - \text{Psample}}{\text{Palumina powder}} \times 100 \qquad (1)$$

The microstructure and pore size of alumina is observed using the HITACHI TM30000 Scanning Electron Micrograph (SEM) since different content of palm stearin and will result in different size and arrangement of pore structure.

2.6 Mechanical testing

The machine used for 3-point bend testing is INSTRON 54778 with a static rating ± 10kN and weight of 1.55 kg. The parameters that kept constant are the temperature (23°C), speed rate (20m/s) and humidity (50%).

3. Research Finding

3.1. Sintering

Figure 2 shows that the higher palm stearin content and sintering temperature gives greater shrinkage percentage. Greater shrinkage makes the sample become denser and the bonding between particle is stronger as the pore size becomes smaller. As mentioned by Nampi et. al. (2009), ceramic material shrinks linearly 20%. Hence, at 1450 °C with 38 vol. % of Palm stearin shows the highest percentage of shrinkage which is 25.62%, which is a little bit higher maybe due to the type of binder used.

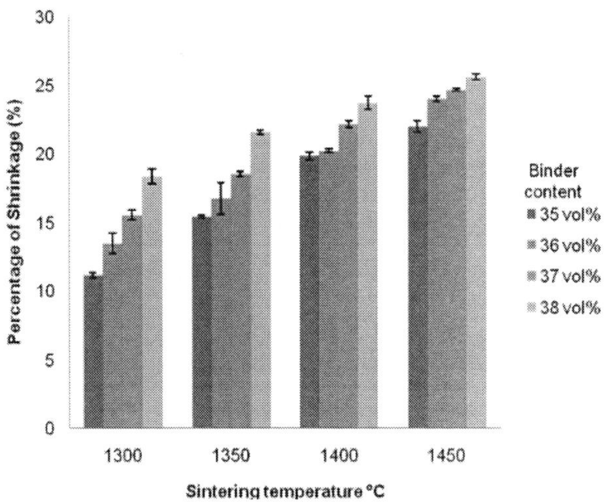

Figure 2. Effect of sintering temperatures to the percentage of shrinkage of alumina for different binder contents.

3.2. *Porosity*

One of the parameters that affect the porosity and pore size of Alumina is the sintering temperature. Porosity and pore size distribution are decreased when sintering temperatures increase [7]. This can be seen from Figure 3 where alumina that was sintered at 1300 °C with 38 vol. % of palm stearin is the most porous and as the sintering temperature increases, alumina became denser. Additionally, in this study, Additionally, the result shows that the porosity for all samples ranges from 35 – 51%. On the contrary, according to L. A. Stanciu et. al. (2001), fast heating rates favor densification instead of grain growth. Furthermore, sintering at higher temperatures can cause possible partial entrapment of the pores within grains [8].

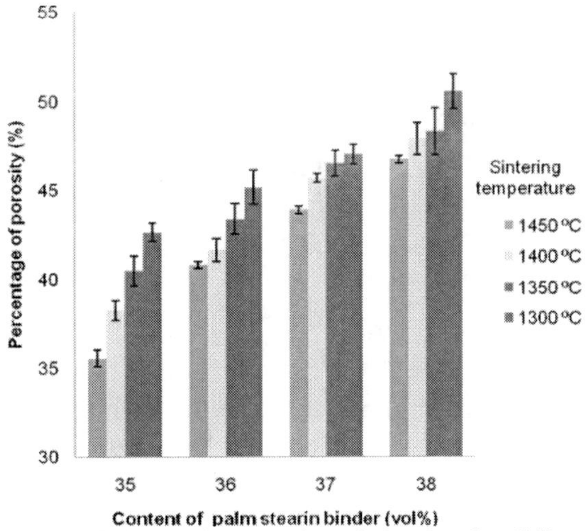

Figure 3. Effect of palm stearin binder to the porosity of alumina for different sintering temperatures.

3.3. *Flexural properties*

As shown in Figure 4, flexural strength increases with the sintering temperature and the highest flexural strength achieved was 14.51 MPa with 35 vol. % Palm stearin at 1450 °C. Furthermore, as more Palm stearin binder were added, the flexural strength decreases, and the lowest value recorded was 1.534 MPa.

The effect of sintering temperature really gives an impact towards the flexural strength of Alumina. Another factor that contributes to these changes is Palm stearin content. Basically, during sintering, all the binders will decompose leaving behind bars. More binders mean more pores created and hence, decreasing in flexural strength. Porosity has a significant role to influence the flexural strength of sintered Alumina and it is well known that the strength of porous Alumina increases with the decreasing of porosity [9].

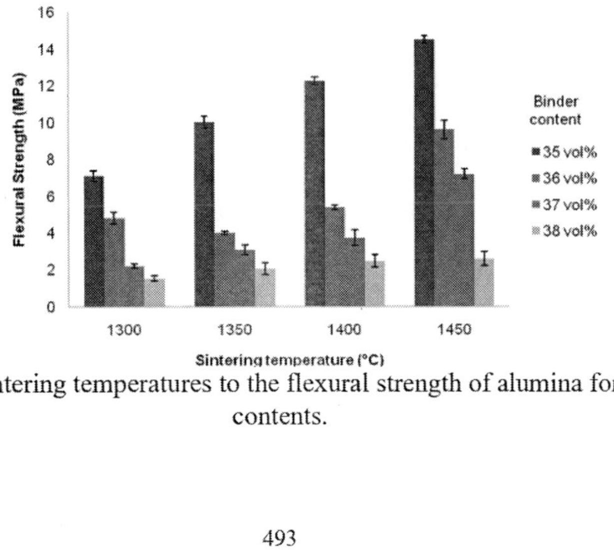

Figure 4. Effect of sintering temperatures to the flexural strength of alumina for different binder contents.

Flexural modulus is the ratio of stress to strain in flexural deformation. From the Figure 5, the modulus increases as sintering temperature increases, which means that, as the sintering temperature increases, Alumina become stronger and makes it harder to bend. The highest value obtained was 86.6269 MPa.

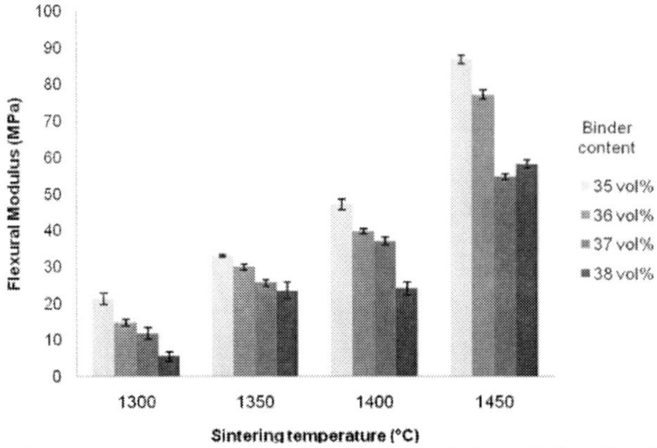

Figure 5. Effect of sintering temperatures to the flexural modulus of alumina for different binder contents.

3.4. *Hardness*

Hardness increases with the increase in temperature. As shown in the Figure 6 below, at 1450 °C with the least amount of Palm stearin binder (35 vol. %) is the highest recorded value for hardness which was 79.1 Hv and the lowest value was 17.6 Hv with the highest amount of Palm stearin. A similar trend has been observed by Rana (2009) and Basu et. al. (2005). Hardness value increases along with sintering temperature is because higher sintering temperature offers better densification [10].

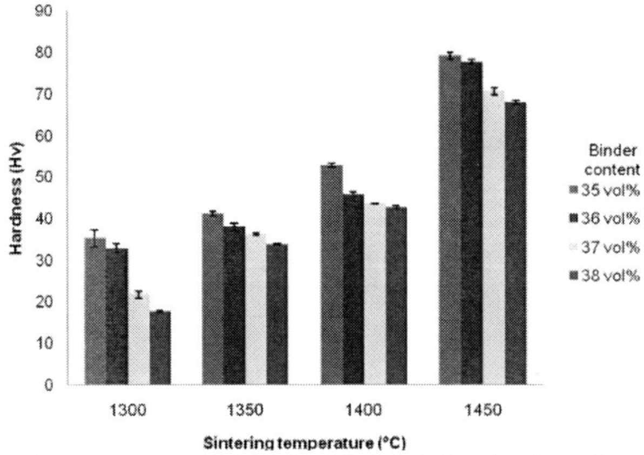

Figure 6. Effect of sintering temperature to the hardness of alumina for different binder content.

3.5. *Microstructural Analysis.*

The microstructure of Alumina mixed with vol. 38% of palm stearin sintered at 1400°C produce pore size of 348μm which is favorable for the bone to grow. Additionally, there's no presence of palm stearin in the sintered alumina which means that the 3 hours of soaking time is enough to decompose the binder.

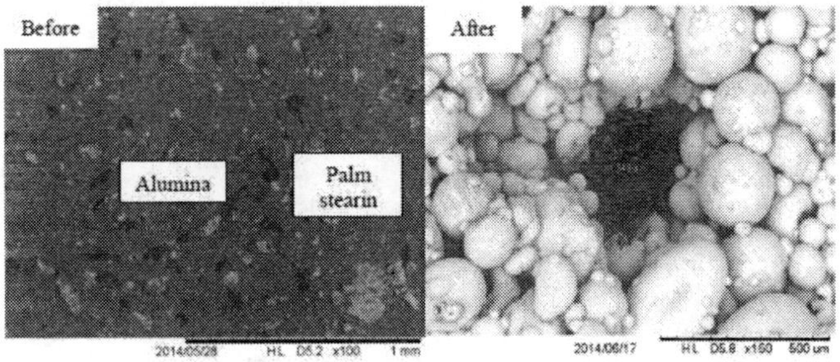

Figure 7. Microstructure of Alumina with 38 vol.% palm stearin before and after sintering with pore size 348 μm.

4. Conclusions

The first objective is to find out how the different binder contents can affect the porosity of Alumina; the preliminary result shows that the higher the binder, the porosity amount is increased. The ruptured samples after sintered at temperature 1400°C embedded by the Ytria was a good learning point to improve the heating method (thermal debinding and sintering).

Greater shrinkage makes the sample become denser and the bonding between particle is stronger as the pore size becomes smaller. Parameters that contribute to shrinkage are the Palm stearin content and sintering temperatures. Therefore, at 1450 °C with 38 vol. % of Palm stearin shows the highest percentage of shrinkage which is 25.62 %.

The porosity percentage increase as the amount of palm stearin increases. All samples produced porosity ranges from 35 – 51%, which lies in the range for bone porosity (40 – 60%). In terms of pore size, most samples have successfully produced pore sizes ranging from 100 – 350 μm which is ideal for bone regeneration.

Sintering temperature enhanced the strength and hardness of Alumina. It has been proved that the higher the sintering temperature, the stronger the Alumina. The flexural strength and hardness obtained by sample containing 35 vol. % of palm stearin are 14.51 MPa and 79.1 Hv respectively, is good for bone implantation.

Finally, it can be said that palm stearin has proved to be useful and more research can be done soon using palm stearin as a binder to know its full potential.

Acknowledgement

The authors would like to express their gratitude to the ministry of science and technology, Malaysia (MOSTI) and Ministry of Higher Education Malaysia (MOHE), and Universiti Teknologi MARA for the financial support given to this work through research grant of 600-IRMI/FGRS 5/3 (003/2019).

References

[1] S. Lee, M. Porter, S. Wasko, "Potential Bone Re-placement Material Prepared by Two Methods", National Science Foundation. 2012. p. 01 – 02.

[2] G. Carl, "The Advantages and Usage of Ceramics in Medicine", Journal of Orthosonic Ceramic Con-ference, 2002. p. 05 – 08.

[3] M. A. Omar, and I. Subuki, "Sintering Characteris-tics of Injection Moulded 316L Component using Palm-based Biopolymer Binder", ISBN: 978 – 953 – 51 – 0371 – 4, InTech, 2012. p.128.

[4] J. P. Li, S. H. Li, C. A. Van Blitterswijk, and K. de Groot, "A novel porous Ti6Al4V: Characteriza-tion of Cell Attachment", Wiley InterScience DOI: 10.1002/jbm.a.30278, 2004. p. 223.

[5] M. H. Ismail, A. Jumahat, A. R. M. Saed, and A. A. Wahab, "Effect of Compaction Pressure on Densification of Alumina Sintered Components", Journal of Mechanical Engineering Vol 4 No 1, 2007. p 33 – 43.

[6] K. Zeng, A. Oden, and D. Rowcliffe, "Flexure Tests on Dental Ceramics", The International Journal of Prosthodontics, 1996. p. 434 – 435.

[7] M. Khalili, S. Sabbaghi, H. Danehmand, M. M. Zerafat, "Synthesis and Characteization of α-Alumina Membrane Support and Effects of Bind-ers," International Journal of Chemical and Envi-ronmental Engineering, Vol 3 No 5, 2012. p. 329.
[8] L. A. Stanchu, "Effects of Heating Rate on Densification and Grain Growth of During Field-Assisted sintering of α-Alumina",Journal of Metal-lurgical and Materials, 2001. p. 01.
[9] H. A. Rahman and D. H. Yacob, "Effect of Double Sintering on the Properties of Porous Ce-ramic," Journal of Mechanical Engineering, 2006. p. 04.
[10] M. M. Hossen, F. U. Z. Chowdhury, M. A. Ga-fur, A. K. M. Abdul Hakim and S. Nasrin, "Inves-tigation of Mechanical Properties of Al2O3-20 wt % ZRO2 Composites as a Function of Sintering Temperature," European Scientific Journal, Vol 10 No 9, 2014. p. 407.

CAE Analysis for Disposable Mouth Mirror Based on Autodesk Moldflow Plastic Insight

J B Saedon[1], Mujahidah Zaira Azlan*[1], Mohd Shahriman Adenan[1], Azuddin M[2]
[1]Faculty of Mechanical Engineering, Universiti Teknologi MARA, Shah Alam
[2]Department of Mechanical Engineering, Faculty of Engineering, University of Malaya, Kuala Lumpur

*mujahidahzaira@gmail.com

Abstract. This study is carried out to focus on computer aided engineering analysis on dental toolkit for disposable mouth mirror by using Autodesk Moldflow Insight (AMI) software. Disposable mouth mirror designs contain with a very small and precise features for it handles and fix angle between shank and working area (mirror). A 3D cad model of disposable mouth mirror has been construct and imported to the simulation software. The part has been analysed with thermoplastic material of Polypropylene (PP) as a part material for its light weight and durability. The simulation and analysis by moulding software package is used in order to determine the best range of operating parameter for injection moulding process in furtherance to get the low manufacturing defect. The process parameters that has been selected includes mould surface temperature, melt temperature, flow rate and cooling time during the injection moulding process. At the end, the best range of operating parameters has been proposed and simulates to study the product behaviour.

1. Introduction

Computer aided engineering has been chosen as a one of the tools that is very common in plastic injection moulding process in performing the analysis by submitting the process input to obtain the expected result. Many researchers and manufacturers has been using this approach in order to analyse their product before the actual process take place. CAE analysis has been proven to be the best way in order to analyse finite element behaviour for the product with low time consuming and low cost. In 2011, W. Kuang et al. [1] implementing CAE technology to analysed out shell for digital camera. AS mention by Satoshi Kitayama et al. computer aided engineering in the plastic injection moulding is an alternative approach to determine the optimal process parameters. [2] Recently, computer aided design (CAE) is widely used in industries such as automotive, aeronautics, medical device and plastic manufacturing.

Autodesk Moldflow Insight (AMI) is one of the famous software in injection moulding process to determine product behaviour based on its thermoplastic properties. G. Singh et al. [3] find that most researchers believe that Moldflow Plastic Insight software generated simulation analysis report gave good results in comparison to any other software. AMI essentially can predict the part cycle time, fill time, pressure at end of fill, and give desired results to the researchers. Najiy et al.[4] used Moldflow software in order to analyse the flow of 20 g parison with different cavity numbers. R. Ramakrishnan et al. [5] simulates Acetal Polymer gear by using Moldflow to identify the main effect of volumetric shrinkage. The author obtain the results by controlling factors of melt temperature, mold temperature, packing time, packing pressure and cooling time in the process settings.

Content from this work may be used under the terms of the Creative Commons Attribution 3.0 licence. Any further distribution of this work must maintain attribution to the author(s) and the title of the work, journal citation and DOI.
Published under licence by IOP Publishing Ltd

Injection moulding process generally known as a forming process that involves in high pressure hydraulics injection system that melts down a plastic pallets and solidifies in mold under packing pressure. It is has been preferred in plastic manufacturing industry for its short cycle times and can produce a complex shape design in a large quantity. [6] In 2017, study shows that injection moulding is major net shape processes for thermoplastic polymers with over 30% of all plastic part has been using this method in processing plastic part. [3]

This paper discusses the application of computer aided engineering in simulating disposable mouth mirror by using Autodesk Moldflow Insight software package to obtain the sprue size, runner size, type of gating system, sequence analysis and material selection. Polypropylene (PP) has been selected as a thermoplastic material used for this study. This is followed by determining the best range of operating parameter without the existence of any error such as short shot. A full method in achieving this study will be discussed in the next section. Results are discussed and finally the paper concludes with the findings of the study

2. Experimental and Test Details

Based on objectives, the purpose of this study is to identify the best range of processing parameter for disposable mouth mirror by using simulation software package Autodesk Moldflow Insight. The process setting has been simulated by trial and error process. The best range of processing parameter has been obtained after several running simulation without the existence of any error. In order to get the best practise of parameters, there are several factor that has to be emphasized such as runner system, gating system, the position of sprue and many other criteria. The selection of thermoplastic material also is one of a major factor in developing good quality of the product. As said by previous researchers, the quality of the plastic part depend on its process parameter, material features and mould design [7].

2.1. Mould Design

In this study, Autodesk Moldflow Insight software is utilized to design a two plate mould with two samples of disposable mouth mirror. Single gating system has been used in this study as shown in Figure 1. 3D mesh is applied on the moulded part with 127288 of triangle elements, 63673 nodes and number of sprue, runner and gate elements are 52. The sprue diameter is 3 mm with the length of 60 mm while the diameter of runners are 6 mm.

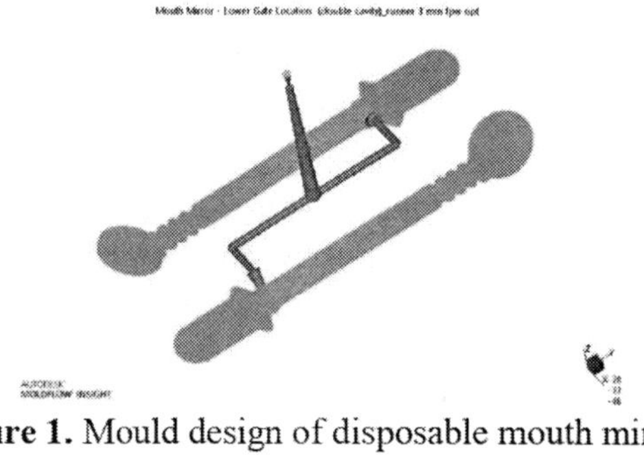

Figure 1. Mould design of disposable mouth mirror.

2.2. Mould and Thermoplastic material

The 3D solid part injection moulded thermoplastic model is made of Polypropylene (PP) and its material properties are given in Table 1. Tool steel P-20 was used as its moulded insert material. Previous researchers prove that this material is generally used for plastics that have lower viscosity [8]. Table 2 shows the properties of moulded insert material.

Table 1. Mechanical properties of polypropylene (PP).

Parameter	Value
Melt temperature (°C)	240
Injection mould temp. (°C)	20-60
Tensile strength (MPa)	55.2
Shrink rate (%)	0.1-0.3
Density (kg/m3)	728.28
Young modulus, E (MPa)	1340
Melt flow index (g/min)	0.2

All the parameters in Table 1 was based on technical data sheet of the material. P-20 steel has been selected as its mould material, it is chosen for its steel properties that is pre-hardened, machines well, high carbon and its general purpose steel. Table 2 shows the properties of the material.

Table 2. Properties of P20 mould steel [9].

Properties	Value
Mould density (g/cm3)	7.8
Mould specific heat (J/kgs)	460
Thermal conductivity, K (w/m^{0}C)	29
Elastic modulus, E (MPa)	2.0 x 10^5
Poisson's ratio	0.33
Mould coefficient of thermal expansion (1/C)	1.2 x 10^{-5}

2.3. *Injection Moulding Machine Setup*

The specification of injection moulding machine has been set as a default in the Moldflow software package to perform the simulation. Table 3 shows the specification of the machine during running the simulation.

Table 3. Injection moulding machine specification.

Description	Value
Maximum pressure limit (MPa)	180
Screw intensification ratio	10
Machine response time (s)	0.01
Maximum machine clamp force (tonne)	7000

2.4. *Process Parameter*

The selection of process parameter in this study was developed by defining the independent parameters and its range. A few test run has been carried in order to get the best range of operating parameter. The controlling process parameter in this study were mould surface temperature, melt temperature, flow rate and cooling time. Where, the hydraulics pressure was kept constant as 9 MPa. Table 4 shows the best range of operating parameter for disposable mouth mirror without the existence of any error. From this range, it can be proved that this range has the best operating parameter for this study.

Table 4. Range of process parameter.

Parameter	Range	
	Minimum	Maximum
Mold temperature (^{0}C)	30	70
Melt temperature (^{0}C)	180	220
Flow rate (cm^3/s)	162.4	243.6
Cooling time (s)	8	12

IOP Conf. Series: Materials Science and Engineering **834** (2020) 012060 doi:10.1088/1757-899X/834/1/012060

3. Simulation Results

After running simulation, the results can be obtained on result summary and analysis log. The results in term of bulk temperature, fill time, pressure, volumetric shrinkage and warpage has been tested on double cavity polypropylene disposable mouth mirror based on the best range of process parameter input as mention in Table 4. Test has been done for part that shown in Figure 1 with the analysis sequence of Fill+Pack+Warp.

3.1. Bulk Temperature

Bulk temperature is a weighted average temperature across the part. It also indicate an average temperature when the flow stops. Bulk temperature are an alternative way to examine the flow distribution [10] . As for this study, bulk temperature recorded for two samples are from 50^{0}C to 226^{0}C as shown in Figure 2. It starts with higher temperature and takes 10.09 s to cools down to 50^{0}C. Through this result, the bulk temperature has to be higher that degradation temperature to ensure that there is no non uniform shrinkage.

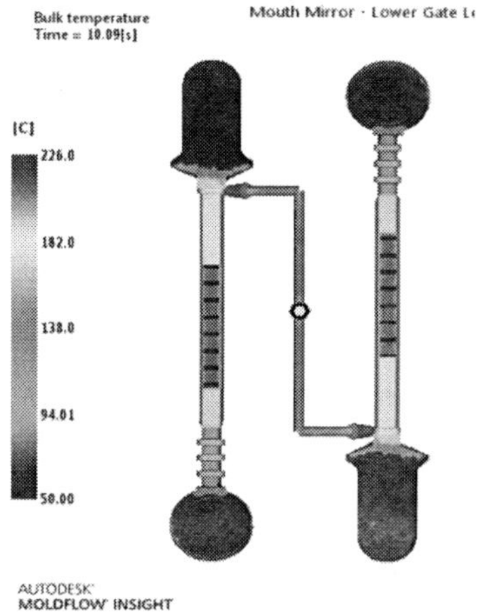

Figure 2. Bulk temperature.

3.2. Fill Time

Fill time results has been obtain from simulation study and gives results with 0.0950 s for both samples. It shows that the cavity fills in a regular intervals with the help of contour colours. From Figure 3 we can see that the last areas to fill are red, while if there is any occurrence of short shot, the cavity that did not fill has no colour. A good fill time result can be seen when the flow pattern is balanced for both part. This can be achieved by arranging the mould cavities in a balanced pattern [11].

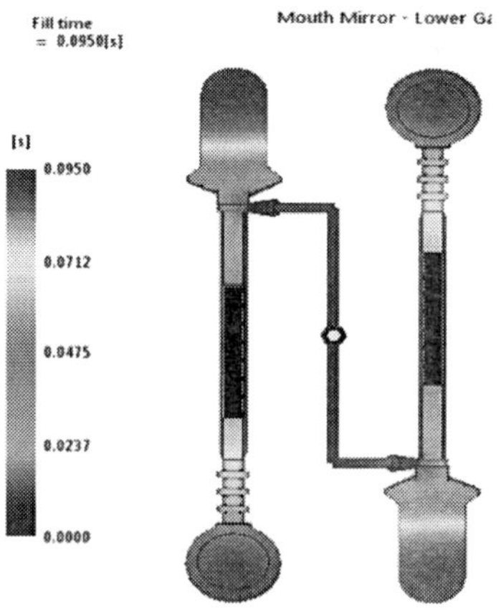

Figure 3. Fill time.

3.3. *Volumetric Shrinkage*

Volumetric shrinkage is the percentage increase in local density from the end of the packing phase to when the part has cooled to its reference temperature. Figure 4 shows the simulation results of volumetric shrinkage for two samples disposable mouth mirror with the values of 10.99%. Higher volumetric shrinkage indicates that the material of the plastic part has a greater transition of specific volume [12].

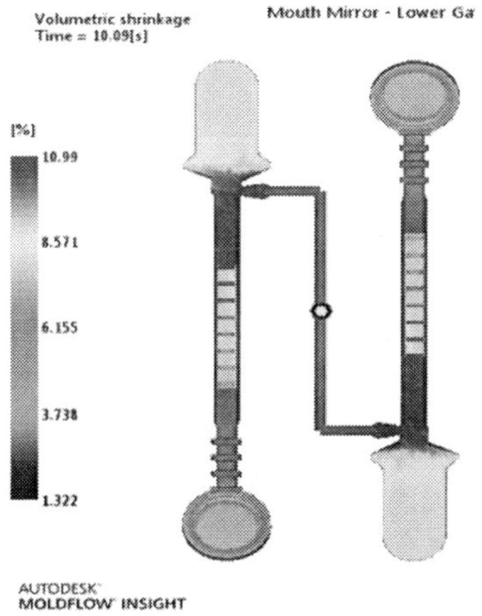

Figure 4. Volumetric shrinkage.

3.4. *Warpage*

As shown in Figure 5 below, the deflection for this plastic parts is from 0.0742 mm to 1.233mm. Higher deflection occurs at the end of the handle may cause by the lower thickness at the area compare to the other part. And it's depend on the phenomenon of the polymer shrinkage.

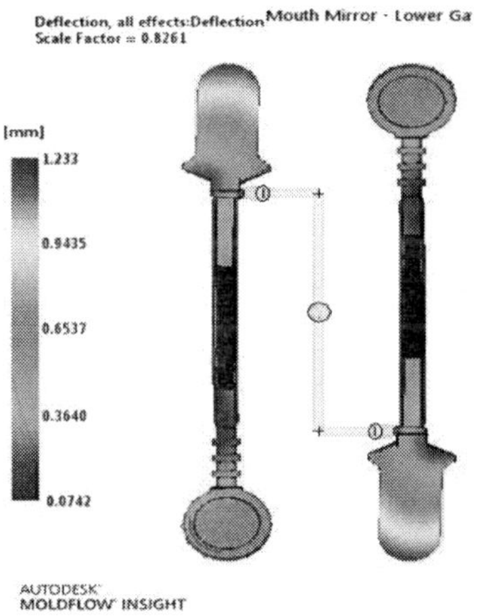

Figure 5. Deflection on all effects.

4. Conclusion

The use of computer aided engineering to obtain the best range of processing parameter has been reported in this paper. Using Moldflow simulation software package, four main parameters subsist of mold temperature, melt temperature, flow rate and cooling time has been set in order to achieve this study objectives. As the results, the operating parameter range of 30 - 70^{0}C for mold temperature, 180 - 220^{0}C for melt temperature, 162.4 - 243.6 cm3/s for flow rate and 8 - 12 s for cooling time shows the best operating parameter for this samples without the occurrence of any error.

IOP Conf. Series: Materials Science and Engineering **834** (2020) 012060 doi:10.1088/1757-899X/834/1/012060

References

[1] W. Kuang, L. Chen, Z. Xian, and Y. Chen, 'Implement of CAD/CAM/CAE integrated technology for out shell of digital camera', in *Procedia Engineering*, 2011.

[2] S. Kitayama and S. Natsume, 'Multi-objective optimization of volume shrinkage and clamping force for plastic injection molding via sequential approximate optimization', *Simul. Model. Pract. Theory*, vol. 48, 2014.

[3] G. Singh and A. Verma, 'A Brief Review on injection moulding manufaturing process', *Mater. Today Proc.*, vol. 4, no. 2, pp. 1423–1433, 2017.

[4] N. Rizal, S. Rizal, A. Mamat, and A. Jumahat, 'Analysis of Fill Time and Injection Pressure of Multiple 20 gram Parisons during Injection Moulding Process', vol. 25, pp. 233–240, 2017.

[5] R. Ramakrishnan and K. Mao, 'Minimization of Shrinkage in Injection Molding Process of Acetal Polymer Gear Using Taguchi DOE Optimization and ANOVA Method', vol. 4, pp. 72–79, 2017.

[6] M. S. Adenan, M. Z. Azlan, J. B. Saedon, and M. Azuddin, 'Design and Analysis on Disposable Mouth Mirror', *Int. J. Eng. Technol.*, vol. 7, pp. 1541–1544, 2018.

[7] M. H. N. Hidayah, Z. Shayfull, S. M. Nasir, M. Fathullah, and M. H. M. Hazwan, 'Warpage analysis in injection moulding process', *AIP Conf. Proc.*, vol. 1885, 2017.

[8] T. Erzurumlu and B. Ozcelik, 'Minimization of warpage and sink index in injection-molded thermoplastic parts using Taguchi optimization method', *Mater. Des.*, vol. 27, no. 10, pp. 853 861, 2006.

[9] J. M. Faiz, Z. Shayfull, S. M. Nasir, M. Fathullah, and M. M. Rashidi, 'Optimisation of process parameters on thin shell part using response surface methodology (RSM)', vol. 020071, p. 020071, 2017.

[10] Autodesk, 'Autodesk Simulation Moldflow Insight | HELP', *Autodesk Moldflow*. .

[11] J. J. Lau and M. Azuddin, 'Design, analysis and verification of 4 micro cavities for a standard runner system', in *Procedia Engineering*, 2013.

[12] E. Oliaei *et al.*, 'Warpage and Shrinkage Optimization of Injection-Molded Plastic Spoon Parts for Biodegradable Polymers Using Taguchi, ANOVA and Artificial Neural Network Methods', *J. Mater. Sci. Technol.*, vol. 32, no. 8, 2016.

IOP Conf. Series: Materials Science and Engineering **834** (2020) 012061 doi:10.1088/1757-899X/834/1/012061

AHP-QFD in assessing the outcome of gemba kaizen events

M H Mohd Rusli[1], **A Jaffar**[1], **S Kasolang**[1], **S M Kayat**[2]

[1] Faculty of Mechanical Engineering, Universiti Teknologi MARA, Shah Alam Selangor, Malaysia.
[2] Faculty of Electrical Engineering, Universiti Teknologi MARA, Shah Alam Selangor, Malaysia.

ahmedjaffar@uitm.edu.my

Abstract. The study shows the application of Analytical Hierarchy Process (AHP) and Quality Function Deployment (QFD) as a tool in assessing the outcome of the Gemba Kaizen Event. AHP-QFD method is used to analyze the process of prioritizing the Lean strategy, tools and technique for a Gemba Kaizen Event performed at a manufacturing company. The data obtained from the Gemba Kaizen Event were analyzed with AHP-QFD method in which the research evaluated the best combination and prioritization of lean elements to be deployed systematically. The analysis shows the linkage between the needs and lean strategies. The study found that the main strategy to implement Lean depends on the needs of the production situation of the company. The approach is mainly about improvement in process flow, standardization of the process, creation of Kanban Pull System and process visualization for control of the production.

1. Introduction

The aim of any manufacturer today is to bring down the costs and delivery time while maintain with the highest product quality possible [1]. More than 75% of organizations presently engage in some type of process improvement strategies to meet the expectations of the customers [3], and are forced to transform their pattern of manufacturing. The industries need the advantage required to skip ahead from the competition by having the ability to respond fast to fluctuating demands and immediate deliveries of the products required [4]. Manufacturing system transformation becomes a necessity to overcome increased competition [2]. One of the most popular approach to do this is through Lean Manufacturing. Lean Manufacturing is a derivation from the production system developed by Toyota Motor Corporation (TMC) in Japan which is called the Toyota Production System (TPS) [7]. It is a philosophy of continually improving the industrial production systems through continuously structuring, operating, controlling, and managing it [5]. Correct practices should be adapted to the best practices of Lean Manufacturing and the systematic deployment method to produce the desired results. The processes are re-engineer and improve on its operational characteristics towards a lean mode of production, ridding of waste and unnecessary expenses while enforcing quality at all stages [6].

Lean strategies have been developed to reduce or eliminate waste to improve operational efficiency in manufacturing processes. However, there are problems where manufacturers encounter difficulties in selecting appropriate Lean strategies based on their needs and within their limited amount of available resources [8]. Lean is not a one-size-fits-all concept, but it is a concept which needs to be adapted according to the industry situation and environment setting [9]. Implementation strategy differs between different areas of organizations, which makes it difficult to choose a proper strategy to address a company's problems [10]. Moreover, improper application of Lean strategies may lead to

Content from this work may be used under the terms of the Creative Commons Attribution 3.0 licence. Any further distribution of this work must maintain attribution to the author(s) and the title of the work, journal citation and DOI.
Published under licence by IOP Publishing Ltd

inefficient use of the resources of the organization and reduced employee confidence [11]. The existing Lean strategy selection methodologies unable to provide a comprehensive method for selecting appropriate Lean strategies within a manufacturing company. Moreover, the present methods lack a systematic approach and analytical model for selecting appropriate Lean strategies for the identified manufacturing wastes within manufacturers' resource constraints. This study dwells on how to select the appropriate Lean strategies and prioritize them to be applied to the manufacturing setting and scenario base on the needs. The study evaluates the Lean elements in the implementation model by analyzing the outcome from the Kaizen Events using AHP-QFD method.

2. Methodology

2.1. Gemba Kaizen Event

Gemba Kaizen Event is a structured improvement project in which Kaizen approach is used in a "short-term project focused on a specific process or set of activities, such as the work flow within a specific work center" [12]. In other definition of the term, Kaizen Event is a focused improvement event during which a cross-functional team of operators, engineers, line leaders spend several days analyzing and implementing improvements to a specific work area [13, 15]. The study adapted the TPS Standardized Work kaizen steps as the research steps to perform the Gemba Kaizen activities namely Kaizen Events. The data obtained from the Kaizen Events are the primary data for the research study.

STEP 1 Discussions & resources planning
STEP 2 Analysis of current situation
STEP 3 Define target
STEP 4 Develop strategy & tactics
STEP 5 Production & Instruction system kaizen
STEP 6 Confirm target achievement

For this research, a case study that has the Job shop configuration were taken into the study. In order to remain anonymous and maintaining the confidentiality of the company, the company is referred to as UAF. The Kaizen Event follows the Kaizen steps as explained above to perform the improvement activities. The results from the Kaizen Events are integrated and analyzed using AHP-QFD technique. The technique is applied to prioritize the lean strategies according to the needs of each Kaizen Event. It is also used to establish the linkage between needs and problem faced with the lean strategies.

2.2. AHP-QFD technique

Firstly, in the AHP-QFD technique, the AHP analysis was applied by following the items developed by Saaty [14]. In this research, the analysis was performed to select and prioritize the Lean strategies based on the findings from Gemba Kaizen Events. The overview of the process is shown in Figure 2.1.

IOP Conf. Series: Materials Science and Engineering **834** (2020) 012061 doi:10.1088/1757-899X/834/1/012061

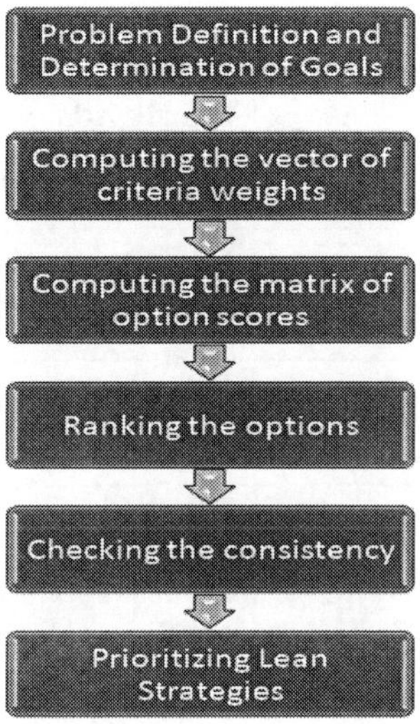

Figure 2.1. Steps in AHP Method to prioritize Lean Strategies.

2.2.1Problem definition and determination of goals

The hierarchy of the problem can be developed as shown in Figure 2.2. The goal is to prioritize the Lean strategies based on the findings from Gemba Kaizen Events.

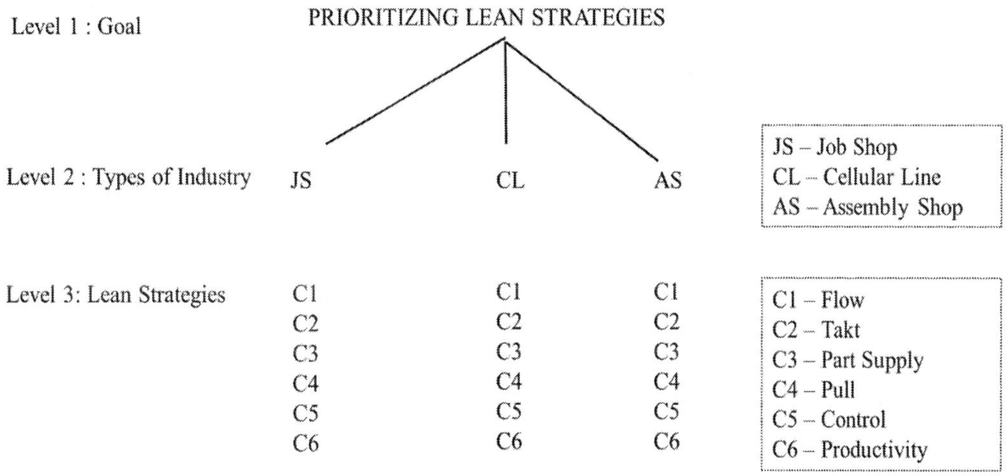

Figure 2.2. Hierarchy of the Problem.

2.2.2 Computing the vector of Lean Strategies weights

In order to compute the weights for the Lean strategies, a set of pair-wise comparison matrices was constructed. Each Lean strategy was evaluated by using the relative scale measurement shown in Table 2.1.

Table 2.1. Pair-Wise Comparison Scale for AHP Preferences

Definition	Index	Definition	Index
Equally important	1	Equally important	1/1
Equally or slightly more important	2	Equally or slightly less important	1/2
Slightly more important	3	Slightly less important	1/3
Slightly to much more important	4	Slightly to way less important	1/4
Much more important	5	Way important	1/5
Much to far more important	6	Way to far more important	1/6
Far more important	7	Far less important	1/7
Far more important to extremely more important	8	Far less important to extremely less important	1/8
Extremely more important	9	Extremely less important	1/9

Next in the AHP-QFD Methodology is to perform QFD analysis. The objective of QFD analysis in this research is to analyze the linkage between company needs and Lean strategies. The flow of the QFD analysis which consists of four main steps is depicted in Figure 2.3.

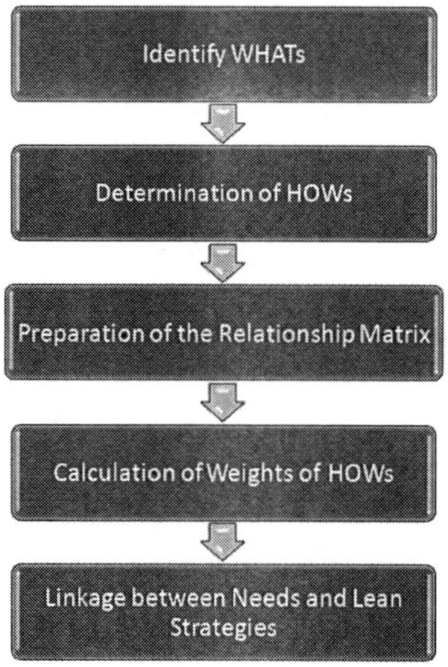

Figure 2.3. Steps in QFD method.

2.2.3 Preparation of the relationship matrix

The relationship matrix was performed to establish the relationship between "WHATs" and "HOWs". Figure 2.4 shows the relationship between Lean strategies (LS) and the company needs (N).

LEAN STRATEGIES, LS	IMPORTANCE RATING, IR	NEEDS, N		
		N_1	N_2	N_3
LS_1	IR_1	R_{11}	R_{12}	R_{13}
LS_2	IR_2	R_{21}	R_{22}	R_{23}
LS_3	IR_3	R_{31}	R_{32}	R_{33}
COLUMN WEIGHTS, CW		CW_1	CW_2	CW_3

Figure 2.4. Relationship Matrix.

2.2.4 Calculation of Weights of HOWs

Weights of HOWs is denoted as Column Weights (CW). CW represents the overall relationship between the needs (N) and Lean strategies (LS). The calculation for the value of CW_j is shown as follows:

$$CW_j = \sum [\, R_{ij} \ X \ IR_i \,] \tag{1}$$

IR_i denotes the value from Importance Ratings. To further explain on this formula, a few examples depicted from Figure 2.4 are given as follows:

$$CW_1 = [\, R_{11} \ X \ IR_1 \,] + [\, R_{21} \ X \ IR_2 \,] + [\, R_{31} \ X \ IR_3 \,]$$
$$CW_2 = [\, R_{12} \ X \ IR_1 \,] + [\, R_{22} \ X \ IR_2 \,] + [\, R_{23} \ X \ IR_3 \,]$$
$$CW_3 = [\, R_{13} \ X \ IR_1 \,] + [\, R_{23} \ X \ IR_2 \,] + [\, R_{33} \ X \ IR_3 \,] \tag{2}$$

3. UAF case study

UAF is a wholly owned subsidiary of a Malaysian conglomerate, located in Shah Alam, Malaysia. The company consists of three plants which are the Main plant, Stamping plant and Plastic plant. The manufacturing operations are divided into 2 main categories, the Plastic operations and Metal operations. For Plastic operations, the company produces air filter products which are the air filter modules, and Non-Air filter products such as charcoal canisters, cylinder head covers, fuel filters, brake reservoirs, and plate holders to be supplied to various auto manufacturer and vehicle assembler in Malaysia. The layout of the plant generally is of Job shop configuration. This study was done at the company's production line with a focus on air filter products ranging from Raw Material incoming until Finish Good delivery. The management wants to install the Kanban Pull System in the production area and selected the Plastic Operation value stream, which comprises of Delivery area, Warehouse, Plastic Injection lines and Plastic Assembly lines as the scope area or the targeted Value Stream to be kaizen. The major needs and goals for the Lean kaizen projects highlighted at this stage which are to improve material flow from plastic injection to warehouse, create a Pull system at air filter line, reduce the stock by 15%, increase productivity by 20% and to improve the quality at air filter line by achieving 0% customer claims.

The following are the kaizen points grasped from the MIFC before kaizen for UAF:

For the Plastic Injection lines, it is found out that production was run and scheduled according to monthly forecast and scheduled planning. The production was not tied up with sales and customer requirement thus over production of some models and parts shortage of some models frequently happen. Moreover, by planning, it leads to high raw material and finish goods, high work-in-process (WIP) stock resulting in stagnation of parts flow. The production lead time becomes long and creates a lot of non-value-added activities. The components were stored in wire mesh which leads to big production lot size of the models and creates too many WIP stocks. It also utilizes more space. The Plastic Assembly lines were arranged in 6 isolated cell layout configuration which produces a part number for each cell. The individual cell limits the sharing of workload between the operators. This is called the Fixed Manpower Configuration cell and it has no flexibility to adapt to demand changes. Hence, there was an unbalance in the workload distribution between the production cell which leads to low productivity and poor production visualization. The overall layout and material flow along the Value Stream was not smooth and the internal part supply method was not well established. The company practices Push production method which leads to long lead time and high WIP stagnation in the Value Stream.

The kaizen strategies done cover the following aspects:

Relayout for smooth flow
Smooth flow must first be established as a pre-requisite in creating the Kanban Pull System. The Production team members performed relayout of the Plastic Assembly area by grouping the cells between air filter lines and the Non-air filter lines. Rearrangement of production line according to the product group using Big Island concept was done. The machines are arranged according to process sequence. Relocation of FG storage near to plastic assembly line was done. A checking area for QC was also set up.

Standardized Work (SW)
SW for the operators was created. The Standard Work Chart (SWC) was developed for the production line and visualized at the shop floor. The quality key points are integrated into the SW. The Standard in Process Stock (SIPS) was set in the production line.

Productivity Kaizen
Improvement in productivity was done at the Plastic Assembly lines. The isolated production cells were combined thru the Big Island concept. By creating the Big Island cell, a multi model cells was created and thru Takt production, the manpower appointed to operate the cell was streamlined and given the correct workload. Waiting time and unbalance workload of the workers were eliminated thus improves productivity by improving the output per man hours.

Streamline the part supply system
The internal part supply system was improved by streamlining the part supply method between the Delivery area, Warehouse, Plastic Assembly lines and Plastic Injection lines. Flow racks and part boxes were used instead of wire mesh to store the Raw Material parts, WIP parts as well as the Finish Goods parts. The parts were stored in a fix quantity according to the determined lot size in the part boxes. Furthermore, the part supply route was streamlined by establishing Set Part Supply system and establishing the Standard Work for the material handler. Workload balancing for material handler was done. In the Set Part Supply system, the number of Transfer Person (TP), the parts that the TP need to transfer, the route the TP needs to follow in order to transfer the parts, the method and the timing of part transfer are synchronized according to the Takt production. By establishing the Set Part Supply system, material and information can be moved and transferred along the Value Stream according to JIT concept. The usage of forklift in the production area was eliminated. A Supply Part monitoring board was established to control and visualize the parts supply condition.

Kanban Pull System

The pulling mechanism was designed into the targeted Value Stream, starting from the Delivery area and Warehouse. The System team performed the Shipping management activities by improving the Shipping operations. The pulling effect start from the pulling from the customers, hence a clear shipping operations and control was vital in establishing the Kanban Pull System. In order to have a levelled pulling effect, *Heijunka* post was used. Line Store was designed into the system to place the FG and the Kanban chute was used to place the Kanban sequence. As for the internal Kanban, dual Kanban system was adopted, usage of PW (Parts withdrawal) and PI (Production Instruction) Kanban. The established Kanban Pull System is shown in Figure 3.1.

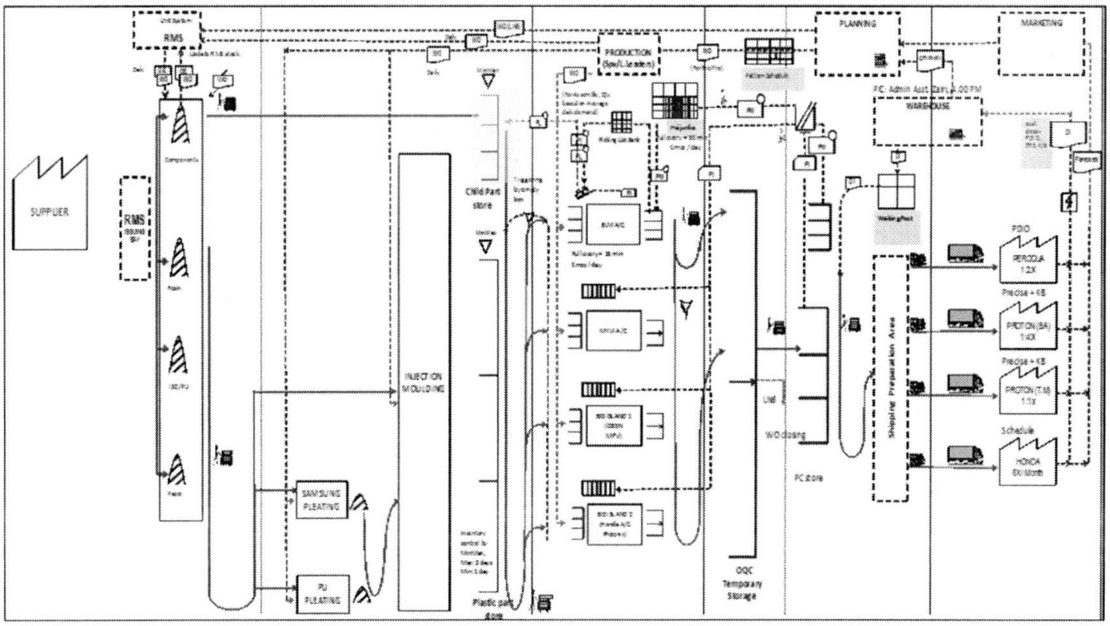

Figure 3.1. MIFC for the Kanban Pull System

4. Results and Discussion

4.1 AHP Analysis Results

The results for AHP analysis on Kaizen Event at UAF are divided into four sections: (i) Pair-Wise Comparison Matrix, (ii) Synthesized Matrix, (iii) Weighted Sum Matrix and (iv) Consistency Ratio as tabulated in Table 4.1, Table 4.2, Table 4.3 and Table 4.4.

4.1.1. AHP Pair-Wise Comparison Matrix. The results for pair-wise comparison matrix for UAF is shown in Table 4.1.

Table 4.1. Pair-Wise Comparison Matrix for UAF

JS-UAF	Flow	SW	Part Supply	Pull	V-SFE	Productivity
Flow	1.00	2.00	5.00	6.00	4.00	3.00
SW	0.50	1.00	4.00	5.00	3.00	2.00
Part Supply	0.20	0.25	1.00	2.00	0.50	0.33
Pull	0.17	0.20	0.50	1.00	0.33	0.25
V-SFE	0.25	0.33	2.00	3.00	1.00	0.50
Productivity	0.33	0.50	3.00	4.00	2.00	1.00

IOP Conf. Series: Materials Science and Engineering **834** (2020) 012061 doi:10.1088/1757-899X/834/1/012061

4.1.2. AHP Synthesized Matrix Results. The results for synthesized matrix for UAF are tabulated in Table 4.2. The results of priority vector show that Flow is the highest rank of Lean strategy for UAF. The lowest rank is Pull.

Table 4.2. Synthesized Matrix result for UAF

JS-UAF	Flow	SW	Part Supply	Pull	V-SFE	Productivity	Priority Vector
Flow	0.4082	0.4669	0.3226	0.2857	0.3692	0.4235	0.3794
SW	0.2041	0.2335	0.2581	0.2381	0.2769	0.2824	0.2488
Part Supply	0.0816	0.0584	0.0645	0.0952	0.0462	0.0471	0.0655
Pull	0.0680	0.0467	0.0323	0.0476	0.0308	0.0353	0.0434
V-SFE	0.1020	0.0778	0.1290	0.1429	0.0923	0.0706	0.1024
Productivity	0.1361	0.1167	0.1935	0.1905	0.1846	0.1412	0.1604

4.1.3.Weighted Sum Matrix Results. The result for weighted sum matrix for Kaizen Event for UAF is summarized in Table 4.3

Table 4.3. Weighted Sum Matrix results

	UAF
Flow	2.3562
SW	1.5459
Part Supply	0.3952
Pull	0.2634
V-SFE	0.6218
Productivity	0.9864

4.1.4. AHP Results on Consistency Ratio. Table 4.4 shows the results for Consistency Ratio (CR) of Kaizen Events for UAF. Based on the result, the CR for UAF is acceptable, as the ratio do not exceed 0.10. Thus, all judgments with regards to Lean strategies are consistent.

Table 4.4. Consistency Ratio

		Consistency Index, CI	Consistency Ratio, CR
UAF	6.1232	0.0246	0.0199

4.1.5. QFD Analysis Results

Table 4.5. QFD result for Company: UAF

S. No	LEAN STRATEGIES	Importance Rating	Ranking	Lead time reduction	Stock reduction	Lot size reduction	Productivity improvement	Man-power reduction	Change over time reduction	Quality improvement
				NEEDS						
1	Flow	0.3794	1	10	10	6	6	6	2	8
2	SW	0.2488	2	6	6	8	10	10	8	10
3	Part Supply	0.0655	5	8	10	10	8	8	6	2
4	Pull	0.0434	6	8	8	8	8	8	6	6
5	V-SFE	0.1024	4	2	2	2	2	2	6	6
6	Productivity	0.1604	3	6	8	6	10	10	10	10
Column Weights				7.33	7.78	6.44	7.45	7.45	5.62	8.13

Table 4.5 shows the QFD result for UAF. The highest point in the column weight is Quality improvement. UAF management has set the target to achieve 0% customer claim, which is very important for the company's credibility and image to the customers. Therefore, the main needs for UAF is on quality improvement. The second highest is on stock reduction. Productivity and manpower reduction both share the same points in the column weights. As a result, from the analysis, the linkage can be summarized and concluded as follows:

1) Reduction in production lead time - FLOW
2) Reduction in WIP stock - FLOW
3) Reduction in Lot Size – PART SUPPLY
4) Reduction in manpower usage - PRODUCTIVITY
5) Increment in productivity - PRODUCTIVITY
6) Improvement in changeover time - PRODUCTIVITY
7) Improvement in quality – STANDARDIZED WORK

SW and Productivity are strategies that have a close linkage to each other. Part Supply and Pull also links closely to each other. For strategies of Pull & V-SFE, both are ranked low in the priority vector because both strategies are usually deployed after all the other strategies are done. Pull binds all the lean elements, whereas V-SFE maintains the lean elements through abnormality control for sustain.

5. Conclusion

The AHP-QFD technique is used to analyze the data obtained from the Kaizen Event done at UAF. The technique is applied to prioritize the lean strategies according to the needs of the Kaizen Event. The result from the analysis is also used to establish the linkage between needs and problem faced with the lean strategies. UAF follows the guideline of the lean strategies prioritization in order to improve the production flow, part supply method and implementation of Kanban Pull System. The study outlines how an AHP-QFD approach, commonly used in engineering and manufacturing by virtue of selecting such as supplier selection and customer preferences prioritization, is well suited to developing a field-tested and theoretically grounded research instrument that can be useful and understandable for implementation in the industry. Other than using the AHP-QFD approach, the study can be expanded through the application of other techniques to analyze the model. Other techniques identified as suitable to be applied are Decision Tree (DT) and Discrete Event Simulation (DES).

References

[1] Tapping, D., Luyster, T., & Shuker, T. (2002), *Value stream management: Eight*
[2] Gadalla, M. (2010), A conceptual framework to excogitate agile from lean transformation, Int. J. Rapid Manufacturing, Vol. 1, No. 3, pp.308–322.
[3] Dhallin, A. J. (2012), The identification, validation, and modeling of critical parameters in Lean Six Sigma implementations. University of Southern California.
[4] Moser P., Isaksson O., Seifert R., 2017. How process industries can improve supply chain performance. CSCMP Supply Chain Quarterly: Strategy, Q3 2017.
[5] Detty, R., & Yingling, J. (2000), Quantifying benefits of conversion to lean manufacturing with discrete event simulation: A case study. *International Journal of Production Research, 38*(2), 429-445.
[6] Anthony, P., 2018. Lean in a Lean Economy. [online] Manufacturing Today. Available at: http://www.manufacturing-today.com/sections/columns/2001-lean-in-a-lean-economy
[7] Chen, J., Li, Y., & Shady, B. (2010), From value stream mapping toward a lean/sigma continuous improvement process: an industrial case study. *International Journal of Production Research, 48 (4), 1069-1086.*
[8] Nieto, M. and Perez, W. (2000), The development of theories from the analysis of the organisation: case studies by the patterns of behaviour", Management Decision, vol. 38, no. 10, pp. 723-733.
[9] Shah, R., & Ward, P. T. (2007), Defining and developing measures of lean production. Journal of Operations Management, 25(4), 785-805.
[10] Wan, H., & Chen, F. F. (2008), A leanness measure of manufacturing systems for quantifying impacts of lean initiatives. International Journal of Production Research, 46(23), 6567-6584.
[11] Marvel, J. H., & Standridge, C. R. (2009), Simulation-enhanced lean design process. Journal of Industrial Engineering and Management, 2(1), 90-113.
[12] Melnyk, S. A., Calantone, R. J., Montabon, F. L., & Smith, R. T. (1998), Short-term Action in Pursuit of Long-Term Improvements: Introducing Kaizen Events, Production and Inventory Management Journal, Vol. 39 No. 4, pp. 69-76.
[13] Kirby, K. & Greene B. (2003), How Value Stream Type Affects the Adoption of Lean Production Tools and Techniques, *Proceedings of the 2003 Industrial Engineering and Research Conference*, Portland, OR.
[14] Saaty, T.L. The analytic hierarchy process. New York: McGraw Hill, 1980.
[15] McNichols, T., Hassinger, R., & Bapst, G.W. (1999), Quick and Continuous Improvement Through Kaizen Blitz, *Hospital Materiel Management Quarterly*, Vol. 20 No. 4, pp. 1–7.

Production output real time data monitoring based on internet of things (IOT) at automotive part machining line

M S Shahar[1], N A Mohd Salleh[1], N S Shahrom[1] and S H Norsham[2]

[1]Faculty of Mechanical Engineering, Universiti Teknologi MARA, 40450 Shah Alam, Selangor, Malaysia
[2]DRB HICOM Berhad, 40000 Shah Alam, Selangor, Malaysia

syahmishaharkadir@gmail.com
noorazlinamohdsalleh@gmail.com

Abstract. DRB HICOM Berhad is currently facing challenge which is lack of accuracy and real time information on their machining output data. With a new era of the Internet of Things (IoT), it may reduce hassle for better productivity and quality. This paper proposes on Internet of Things (IoT) based real-time monitoring data production output at machining line in which the output parts will be monitored over a webserver using internet and will notify workers when a breakdown occurs through hand phone notification. The system is designed to collect data automatically such as plan vs actual goods, reject output and breakdown time and display them on OLED board as well as on webpage for the workers and management to see and monitor it easily without being in the working line area. The data from this should be transmitted to the management without human intervention. Moreover, it also offers a better solution to the workers and management to know the condition of the machining line and can take immediate preventive measures. The efficiency of the machining line can be improved gradually as a result of better system. Hence, the real-time monitoring data production output at machining line will help the company to generate higher revenue, obtain faster and accurate production data.

1. Introduction

The Internet of Things (IoT) is defined as an ecosystem of smart object, network and processing technologies working together in providing services to the end user.[1,2] This ecosystem allow industry to gain several benefits for their own businesses.[3] The manufacturing industry in Malaysia is undergoing a dramatic shift centered on new technologies that offer the promise of greater efficiency, speed and higher quality.[4] A central component of this shift toward Industry 4.0 is the Internet of Things (IoT), which serves as the frontline of data capture in the modern factory. The Internet of Things (IoT) intelligent systems enable rapid manufacturing of new products, dynamic response to product demands, and real-time optimization of manufacturing production and supply chain networks, by networking machinery, sensors and control systems together [5].

Company A is one of DRB HICOM subsidiaries that venture in the business of aluminum casting. Most of the part made in Company A are automotive part such as cylinder head, steering housing, pan oil and many more. It involves five (5) major processes which are melting, casting, fettling (secondary), machining and logistics. There has been some issue at this machining line whereby there is only one worker that need to do inspection and record the machining output data. Due to the high workload, most of the time the worker is unable to report the data on time, error in counting the finished products and sometimes error in classifying good and out of standard products. This causes delay in obtaining current output of the machining line especially if there is any quality issue that need immediate attention. A

Content from this work may be used under the terms of the Creative Commons Attribution 3.0 licence. Any further distribution of this work must maintain attribution to the author(s) and the title of the work, journal citation and DOI.
Published under licence by IOP Publishing Ltd

possible cause of this problem is due to the availability of only one operator involves in recording and monitoring goods and reject outputs and that person is also involves in inspection area. An Internet of Things (IoT) based real time data production output monitoring could be used as an alternative to resolve the situation. The objective of this study is; to design and develop IoT output data collection, and to validate the fabricated prototype.

2. Methodology

The methodology follows the Deming Cycle Plan-Do-Check-Act (PDCA) to breakdown the project into a more systematic flow. Table below shows the PDCA methodology for this project.

	Activities	Progress
PLAN	**STEP 1: TOPIC SELECTION** i. Meeting with UiTM supervisor and DRB-Hicom supervisor	i. Topic chosen: IoT Based Real Time Monitoring Data Production Output at Machining Line
	STEP 2: UNDERSTAND PROCESS AND SITUATION i. Study and understand about the topic chosen. ii. Define problem, literature review and information and data gathering.	i. Searched for previous study real time monitoring data ii. Site Visit iii. Do observation at die casting area
	STEP 3: PLAN ACTIVITIES i. Develop milestone of project	i. Draft milestone in completing the project.
	STEP 4: PROBLEM IDENTIFICATION AND IDEA GENERATION i. Identify problem and analyze data to generate idea in developing the system	i. Deep observation by taking pictures and videos. ii. Obtained historical downtime record from line supervisor.
DO	**STEP 5: CONSIDER AND IMPLEMENT COUNTERMEASURE CHECK RESULT** i. Proposed countermeasure action ii. Design and Develop iii. Propose Changes iv. Fabricate the prototype.	OBJECTIVE 1 Level 0: Sensors Level 1: Microcontroller Level 2: Communication Protocol Level 3: Cloud Level 4: Software Analytics Level 5: Software Interface
CHECK	**STEP 6: CHECK RESULT** i. Review feedback from management ii. Results Evaluation iii. Improve Prototype iv. Performance analysis	i. Install the Andon system in machining line ii. Ensure the system workability iii. Identify any problems occur to the system. iv. Compare the OEE analysis
ACT	**STEP 7: DOCUMENTATION AND REPORT** i. Documentation and report on this study.	i. Compile all information as a result of project.

3. Results and Discussion

The schematic diagram for the prototype is shown as in Figure 1. The system collects data and display it on the website as illustrated in Figure 2. The Overall Equipment Efficiency (OEE) have shown increase in its percentage after implementation. The comparison could be seen in the Figure 3. The application of this system have reduced the paper usage in daily data collection. It also helps in notifying the technicians in the case of breakdown.

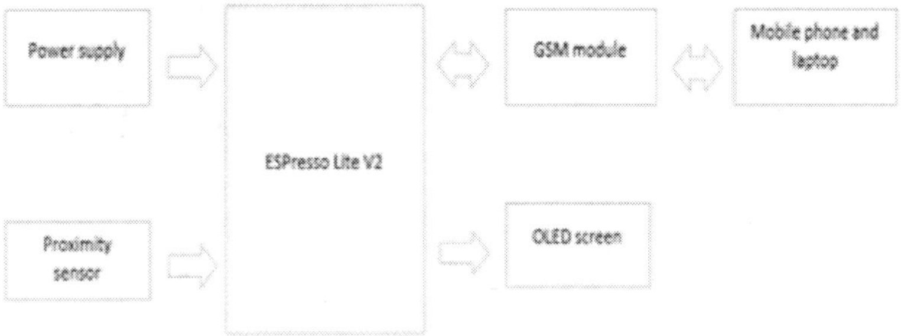

Figure 1. Internet of Things (IoT) system prototype

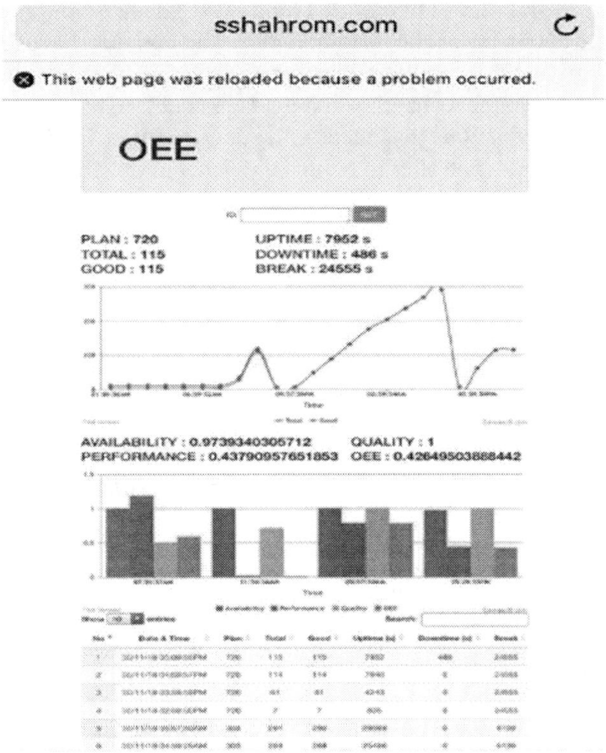

Figure 2. IoT System Website Data Display

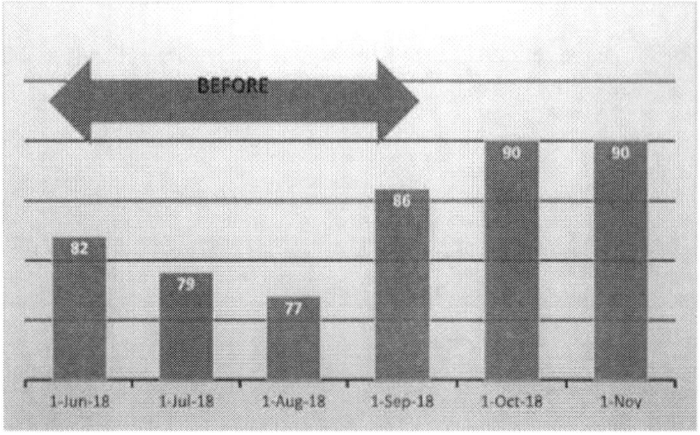

Figure 3. OEE Comparison

4. Conclusions and Recommendation

This project has successfully met the goals of developing a real time monitoring data production output system in line with the integration of Internet of Things (IoT) devices, improving the response time of the technician reacting to anomalies in the machining line and obtaining an operating status in real time.

The Internet of Things (IoT) system help in assessing the OEE performance of the machining line. The OEE performance can help the technician and management to identify which machining line needs better attention. Some recomendations that needs to be done to improve this projects are:-

- Use a faster processor microcontroller to reduce the delay in transferring data to the cloud.
- Use long - range proximity sensor suitable for industrial applications.
- A procedure standard (SOP) should be established to provide the worker with information on how to operate this system
- The system should be installed for a certain period of time to ensure that it can withstand the industrial environment in order to guarantee the efficiency of this system.
- The digital height should be readable from up to 50 m for simple counter version or tachometer version.

References

[1] P. Asghari, A. Rahmani and H. Javadi, "Internet of Things applications: A systematic review", Computer Networks, vol. 148, pp. 241-261, 2019.
[2] B. Dorsemaine, et al., Internet of things: a definition and taxonomy, Proc. – NGMAST 2015 9th Int. Conf. Next Gener. Mob. Appl. Serv. Technol. 2016 (2015) 72–77,
[3] A. A. Senthil Kumar et al., "An Industrial IOT in Engineering and Manufacturing Industries Benefits and Challenges", International Journal of Mechanical and Production Engineering Research and Development, vol. 9, no. 2, pp. 151-160, 2019.
[4] National Internet of Things (IoT) Strategic Roadmap. (2014). MIMOS. Retrieved June 13, 2018, from ://mimos.my/iot/National_IoT_Strategic_Roadmap_Book.pdf
[5] Ersue, M.; Romascanu, D.; Schoenwaelder, J.; Sehgal, A. (4 July 2014). "Management of Networks with Constrained Devices: Use Cases". IETF Internet Draft.

IOP Conf. Series: Materials Science and Engineering **834** (2020) 012063 doi:10.1088/1757-899X/834/1/012063

Analysis of distortion of coupled process forming to welding using fem-based virtual manufacturing with experimental verification

D P Ishak[1], K P Prajadhiana[2], Y H P Manurung*[2], M S Sulaiman[2] and M K A Saberi[2]

[1]Faculty of Industrial Engineering, University of Indonesia, Jakarta, Indonesia.
[2]Faculty of Mechanical Engineering, Universiti Teknologi MARA, Shah Alam, Malaysia.

*yupiter.manurung@uitm.edu.my

Abstract. In this paper, the numerical model of coupled process forming to welding by means of FEM is developed in order to analyse the displacement. Simufact.Forming and Simufact.Welding are utilized to execute a thermomechanical FEM analysis of forming to welding coupled process. Prior to welding process, a sheet metal bending process is performed in order to bend the 2mm plate into a U-shaped geometry. The material used for this process is S235 low carbon steel along with filler material ER70s. Goldak's double ellipsoid model is chosen as the heat source model for GMAW simulation. A series of sheet metal bending and GMAW experiment is executed in order to verify the distortion result produced by FEM analysis. ABB IRB 2400/16 is the assigned robotic welding equipment used in GMAW process while sheet metal bending machine SUNFLUID 800/2007 is utilized for sheet metal bending process. For the calibration model, a forming force between FE simulation will be compared with the forming force of experimental bending process. The final results have shown that the distortion result of FEM analysis of coupled process forming to welding shows a good agreement with the experimental coupled process.

1. Introduction

In today environment, the competition within engineering environment has reached a whole new level where a global integration between industries becomes the main concern, of which Virtual Manufacturing becomes one of the component that would able to tackle said problem by improving the state of the company by means of production ability and cost efficiency [1]. Since the last few decades, Virtual Manufacturing (VM) which covers information technology and computer simulation has been used to model real world manufacturing processes for analysing and understanding the physical behaviour of the product during the process. VM is also suitable for cost optimization of the integrated product, process and resource design, while reducing the time of development cycle, improving the production efficiency, as well as product quality [2-3].

VM can be defined by several hierarchies within its usage [4], one of the utilizations of VM is by implementing the Finite Element Method (FEM) on simulating the real manufacturing process. Finite Element Method has a purpose on separating a one big element into several similar smaller elements in order to simplify the total element calculation, which is a common method utilizes in VM cases [5].

Content from this work may be used under the terms of the Creative Commons Attribution 3.0 licence. Any further distribution of this work must maintain attribution to the author(s) and the title of the work, journal citation and DOI.
Published under licence by IOP Publishing Ltd

Two- or three-dimensional FEM analysis are the option that can be utilized in VM application of where the elements are divided into both planar and axial [6].

A forming process which aims to press the workpiece specimen into U or V shape geometry by utilizing punching die is called sheet metal bending process [7]. Bending Process, which is a part of forming process is divided by two methods of execution which are cold and hot forming process. In this study, cold forming process is selected which means the heat distribution within workpiece is static by means of maintaining the initial room condition while the workpiece is undergo a bending process. The usage of FEM on simulating the sheet metal bending process has been performed by engineers for the last several years. One of the examples is an engineering research performed by Panth et al [8] which utilizes FEM on spring-back prediction in a metal bending case.

Fusion welding is a process of joining metals by using adequate heat to produce coalescence in material to form a strong bonding. It is widely used in shipbuilding, construction, aviation and automotive industries. A stress phenomenon potentially leads to non-uniform expansion and contraction called distortion especially for materials that have low thermal conductivity. The most common distortion caused by heat is warping and buckling [9]. A type of fusion welding which utilizes a plasma arc that is enlisted the electrode and heat of a filler metal on top of workpiece while being covered by shielding gas is called Gas Metal Arc Welding (GMAW). The droplet that resulted by the arc heating would be transferred towards base material, thus embedding it with weld bead [10].

The welding process is often plagued with an unwanted deformation, of them is called Welding Distortion, which is created due to an unbalanced induced tress which often occurs within the region of the workpiece that unrestrained and often results in misalignment causing difficulties in assembly as well as poor appearance of the finished product or structure [11]. In welding process, distortion is caused by both local heating and cooling which affects the unstrained regions [12]. Several research over the past few years has investigated the distortion within welded components by means of Finite Element Method. An advanced engineering research performed by Andud et al [13] predicts distortion within the mild steel based GMAW process which shows that different material properties and pre-condition of welding affects the distortion value and its standard deviation in overall. The minimization of distortion of GMAW process by means of FE method utilizing Simufact.welding has performed by Vyas et al [14] with Voltage, current, travel speed, wire feed rate and gas flow rate as the welding parameters conditions which is similar to this study.

Multistage Manufacturing Process or also well known by the name Coupled Manufacturing Process is a combination between two or more manufacturing process which are utilized into creating an advanced product. Due to more complex product design arises in manufacturing environment today, a coupled manufacturing process has become a key on meeting the criteria of the demanded products. The CMP which concerns about the involvement of both forming and welding process, of which a FEM analysis also involved such as an analysis of CMP of forming and welding process by Zaeh et al [16], which talks about raw items being joined together by means of welding prior to be stamped during forming process. Bauer et al [17] has performed a study of forming to welding process of tube specimen by utilizing the two similar VM software. The material assigned for the mentioned research is mild steel DC04 which has less carbon compared to regular mild steel, which also offers better formability. The research explains the data management of transfer mechanism from Simufact.Forming to Simufact.Welding which covers the transition of result from the first manufacturing process towards the next one that includes displacements, temperature distribution as well as residual stresses. This study's main objective is to investigate the final distortion induced by coupled process forming to welding by means of VM simulation and experimental study.

2. FEM Simulation of Coupled Process Forming-Welding using VM software Simufact.

2.1 FEM for Forming process using Simufact.Forming

The VM software which utilizes to simulate the sheet metal forming process in this study is Simufact.Forming, of which a VM software issued by Simufact Engineering GmBh which specialized on developing a numerical model for forming processes, one of them is metal bending process. In this research, a sheet-metal bending process is chosen as the forming process operated within

Simufact.Forming, which falls under the cold forming module. The process flowchart of forming simulation within Simufact.Forming can be seen in Figure 1.

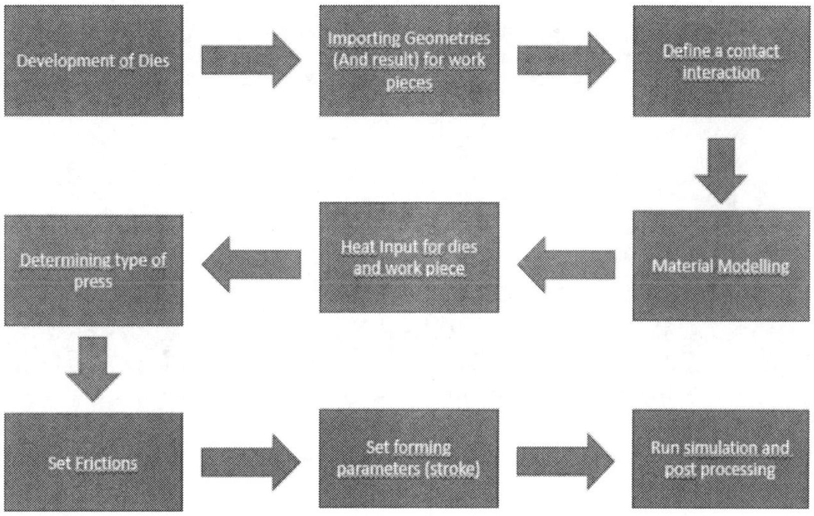

Figure 1. Forming simulation procedure using Simufact.Forming.

The dies geometry is designed within software CATIA while the weld plate which acts as a workpiece for sheet metal forming simulation is designed within FE software MSC Patran. The dimension of the dies, both lower and upper die are 150x92x40 while the weld plate possesses a 70x92x2 dimension. For the forming simulation, two workpieces are going to be formed by means of sheet metal bending process execution. The detailed engineering drawing for the forming dies is exhibited in Figure 2, while the forming simulation of the plain low carbon steel plate is exhibited in Figure 3.

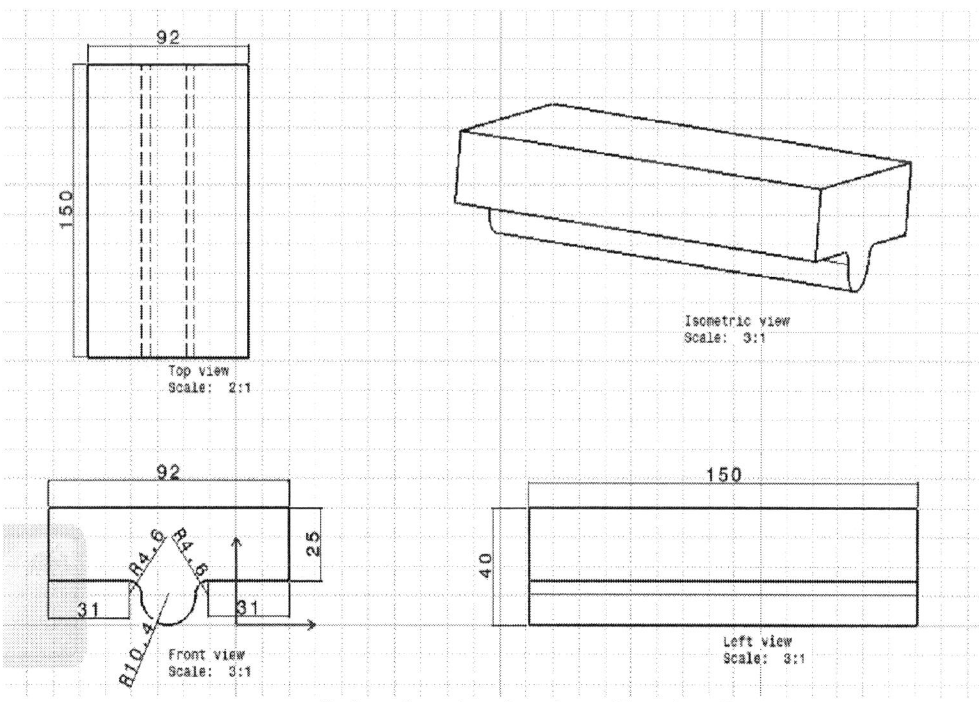

Figure 2. Detailed engineering drawing of forming die design.

Figure 3. Detailed setup on sheet metal bending simulation using Simufact.Forming.

For this research, a low carbon steel material S235 is chosen as the base plate material for the workpiece assigned in both Simufact.Forming and Simufact.Welding. This type of material has common usage in a manufacturing environment for decades especially in both welding and forming oriented researches due to its great formability and acceptable weldability. The temperature-dependent graph for S235 material is described within Figure 4 [18] which covers thermal conductivity, thermal expansion coefficient, young's modulus and specific heat capacity.

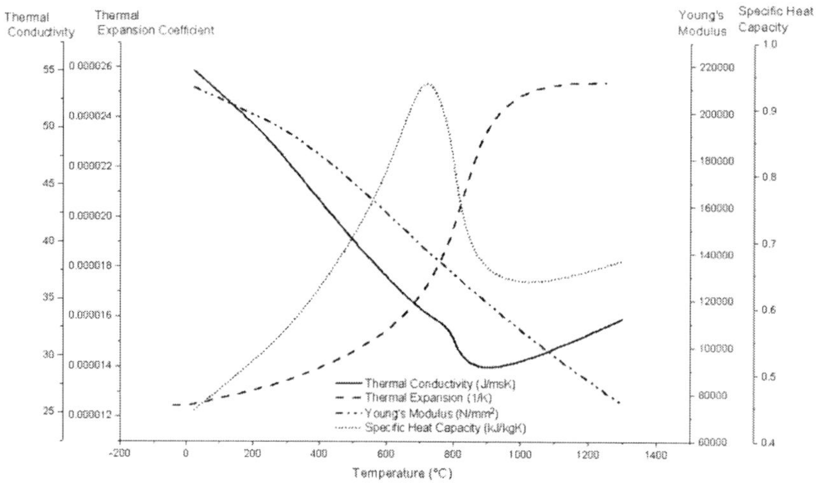

Figure 4. Temperature-dependent properties of S235 material.

The forming simulation parameters assigned in Simufact.Forming for this research is displayed by means of Table 1. These parameters consist of Forming Forces, press velocity and room temperature along with the type of press assigned within VM simulation of sheet metal bending process.

Table 1. Parameter of forming process in Simufact.Forming

Forming Parameters	Values
Stroke (mm)	15
Type of Press	Hydraulic press metal bending
Press Velocity (mm/s)	4.0
Room Temperature (´c)	25-35

IOP Conf. Series: Materials Science and Engineering **834** (2020) 012063 doi:10.1088/1757-899X/834/1/012063

2.2 FEM for Welding process using Simufact.Welding

Simufact.Welding is the VM software utilized to develop a FEM model of welding simulation in this study for distortion investigation, this VM software also incorporates a comprehensive clamping conditions which could mimic the real-life clamping condition placed within the weld component in order to ensure the precision of the result prediction. Simufact.Welding has a user - friendly pre-processing procedure which is exhibited by Figure 5 below.

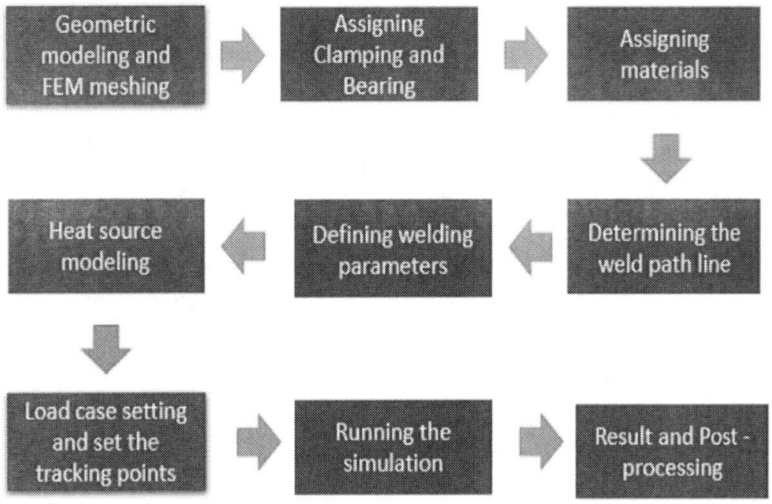

Figure 5. Welding simulation procedure using Simufact.Welding

The result of sheet metal bending simulation which was executed in Simufact.Forming would later undergo the proper transferring process to Simufact.Welding. The transferred forming results would bring the following results: Final displacement, heat input within the workpiece, final effective stress, material properties and the bended shape of the workpiece. The two bended work pieces would be aligned within the same coordinate in order to ensure the similar starting position for easier weld trajectory assignment. The material assigned within this welding simulation is still the low carbon steel S235 due to the data of S235 already transferred from the sheet metal bending result. The welding simulation setup which includes the clamping conditions implemented in Simufact.Welding is exhibited in Figure 6.

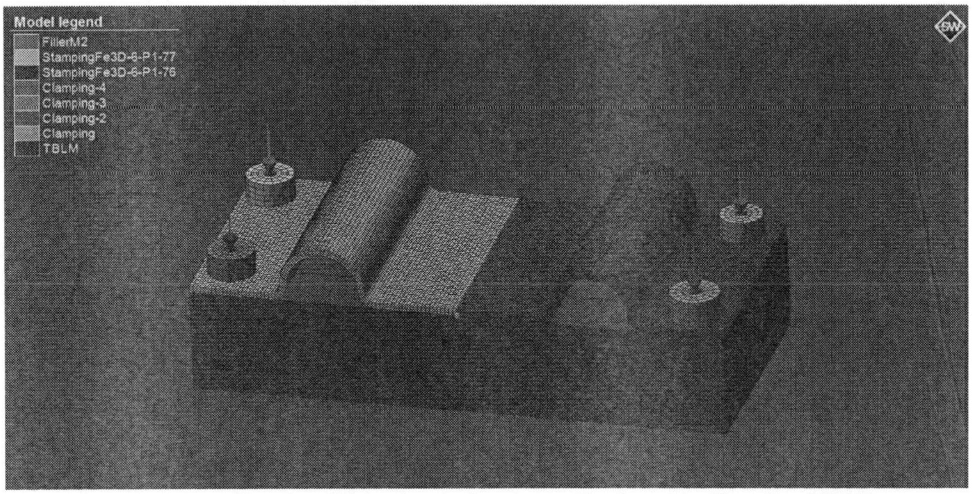

Figure 6. Details of welding simulation setup displayed in Simufact.Welding

The Goldak's double ellipsoid model is chosen as the heat source model involved within this welding simulation due to its capability to be implemented within GMAW process [18]. The Goldak's double ellipsoid parameters are set within the welding simulation in order to determine the expansion of the heat affected zone which occurs during the course of welding.

Heat input parameters which covers current, voltage, welding speed and heat efficiency are set within the thermal boundary conditions section within Simufact.Welding which would later use again for verification purpose by means of experimental study. Table 2 and 3 displays the welding parameters and heat source dimension parameters respectively.

Table 2. Parameter of welding process in Simufact.Welding

Parameters	Value
Current (A)	100
Voltage (V)	16.6
Efficiency (%)	80
Welding Speed (mm/s)	6.0

Table 3. Heat source parameters of T-Joint and Butt joint welding simulation using Simufact.Welding

Heat source dimension	Value
Width (mm)	3,0
Depth (mm)	3,0
Rear Length (mm)	5,0
Front Length (mm)	4,0

3. Experimental Setup and Procedures of Coupled Process Forming to Welding

A series of comprehensive forming and welding experimental processes are conducted in order to verify the result of the FEM simulation performed in both Simufact software. The first chain of manufacturing process executed is the sheet metal bending process which aims to bend the workpiece into a U-shaped geometry prior undergo the welding process. The equipment used for experimental process is the sheet metal bending machine SUNFLUID 800/2007 which is located in Advanced Manufacturing Laborotary in Faculty of Mechanical Engineering, Universiti Teknologi MARA, Malaysia. This forming machine incorporates the hydraulic fluid system on its mechanism. The sheet metal bending SUNFLUID 800/2007 is displayed in Figure 7.

Figure 7. Sheet metal bending machine SUNFLUID 800/2007

The forming parameters assigned for experimental metal bending process is within the similar range with the simulation of forming process in order to ensure the realistic comparison between experimental and simulation studies. For material assignment, a low carbon steel with code S235 is chosen as the workpiece material which is a similar material with the material assigned within forming simulation. The similarity of chemical composition between materials assigned in FE simulation and experiment has been verified by means of Ark Spark Spectrometer test. Table 4 displays the parameter of the experimental sheet metal bending process while Table 5 displays the chemical composition of the material used in experimental process for both forming and welding.

Table 4. Forming parameter used in sheet metal bending process

Forming Parameter	Value
Press velocity	3.5-4.5 mm/s
Bending stroke	14-15
Material used	S235 Mild steel

Table 5. Chemical compositions of material assigned in simulation and experiment

Alloying Elements (%)	C	Si	Mn	S	P	Cr	Ni
S235 in Simufact	0.18	1.6	0.55	0.035	0.035	0	0
Mild steel used in experiment	0.186	0.146	0.011	0.001	0.001	0.035	0.032

For welding experimental process, robotic welding equipment ABB IRB 2400/600 together alongside its power source KEMPI ProMig Evolution is utilized to make a joint between two bended plate. A shielding gas which consists of pure argon was activated during the course of welding process. The display of a robotic welding equipment is shown in Figure 8 while the configuration for the welding experimental which also displays the clamping condition is displayed by means of Figure 9. The parameters of experimental welding process are exhibited in Table 6.

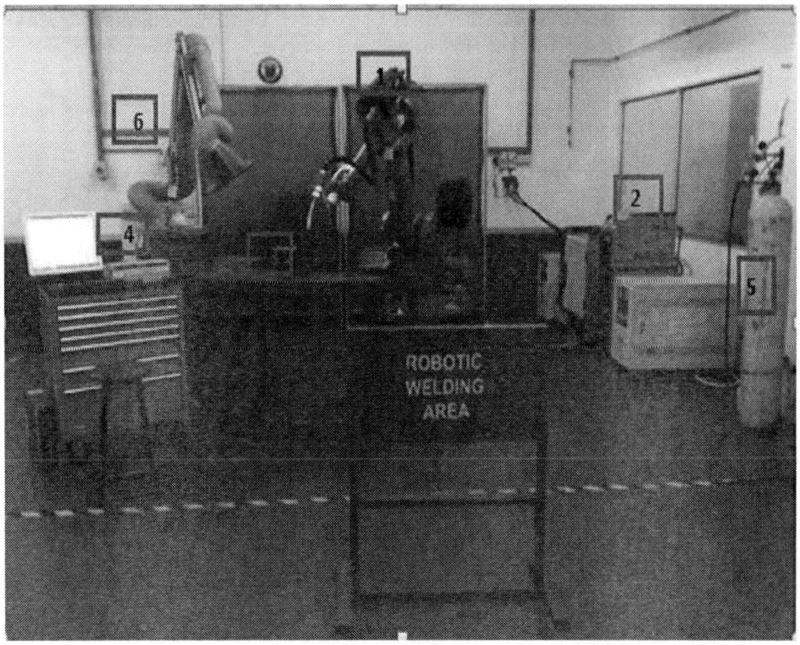

Figure 8. Robotic welding equipment ABB IRB 2400/60

Figure 9. Configuration for welding experimental process of a bended S235 plates

Table 6. Experimental parameter of welding process

Welding Parameter	Value
Current, I (A)	90 -105
Voltage, V (V)	15.8
Travel Speed, v (mm/s)	5.8
Wire Feed Speed (mm/min)	8.3

The distortion is measured prior and after welding process by using Coordinate Measurement Machine (CMM). By utilizing the CMM machine, the amount of displacement occurred during the course of welding process can be examined within certain nodes that become the focus of distortion measurement. CMM Machine Mitutoyo Beyond 707 is used to measure the distortion for experimental study and is displayed in Figure 10.

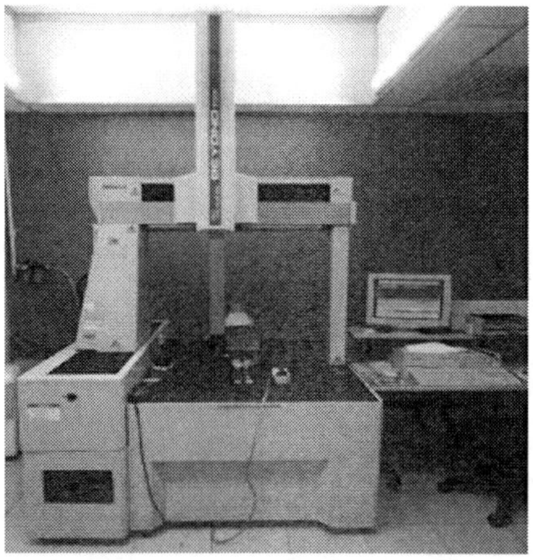

Figure 10. CMM Machine Mitutoyo Beyond 707

4. Results and Discussions

This section will cover the result presented by both simulation and experimental study of coupled process welding to forming, the first section will cover the displacement which causes the spring back effect within the specimen, thus serves as the initial condition for welding process later on. The spring back effect which was occurred within the work piece after sheet metal bending process is described in

Figure 11 below, which displays the value of workpiece radius. The spring back effect is calculated by involving both die radius and workpiece radius which is described in Table 7.

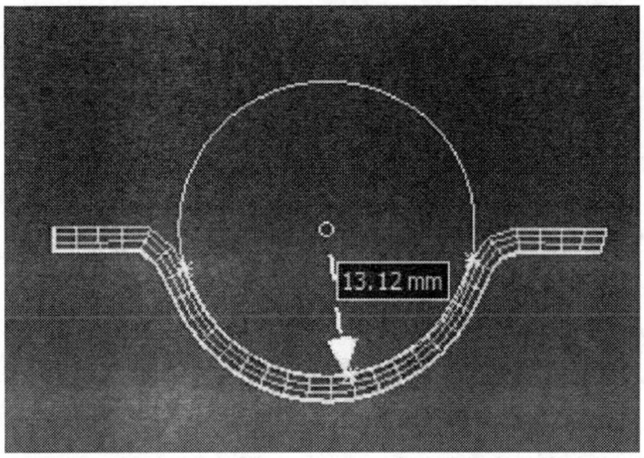

Figure 11. Spring back effect occurred within bended workpiece

Table 7. Spring back effect factor of metal bending process in Simufact.Forming

Die Radius	Workpiece Radius	Spring-back factor
13mm	13.12mm	1.008

The displacement caused by spring-back effect would later be transferred to Simufact.Welding as the initial condition prior welding process was executed. The angular distortion which symbolizes the displacement towards the vertical direction is the main aim for this research, which occurred after two bended specimens are joined together by means of welding process. The distortion result displayed in this research are represented by utilizing the contour band displays option, of which the colour legends are applied in order to describe the level of displacements of the welded component. Figure 12 and 13 displays the distortion result of the FE simulation executed in Simufact.Welding by means of front view and aerial view respectively.

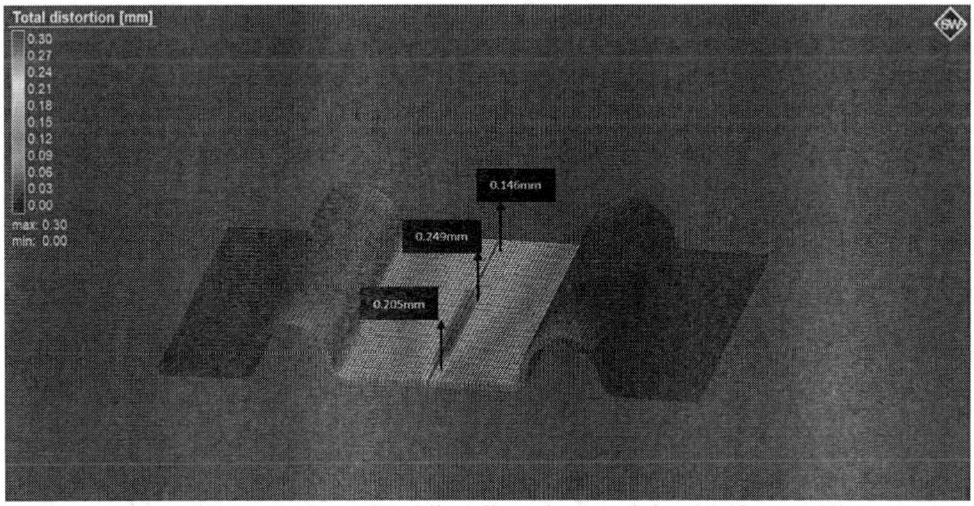

Figure 12. Front view of welding distortion result in Simufact.Welding

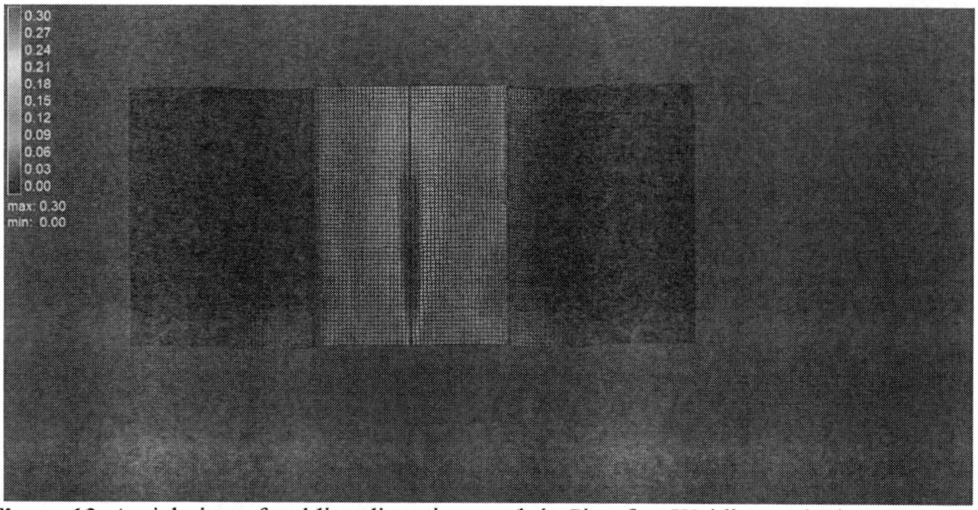

Figure 13. Aerial view of welding distortion result in Simufact.Welding and Distortion value

The result of the distortion caused by the coupled process forming to welding will be compared with the experimental distortion result. The final distortion value of experimental coupled process forming to welding is measured by utilizing the CMM machine. Table 8 exhibits the distortion result comparison between FE simulation of coupled process welding to forming and experimental coupled process. The measurement point for distortion analysis is taken in three measurement points within the middle of the component, which where the displacement occurs due to regions being unclamped which are marked in Figure 12 above.

Table 8. Distortion result comparison of coupled process based on simulation and experimental process

Coupled Process	Point 1 (mm)	Point 2 (mm)	Point 3 (mm)
Simulation	0.146	0.249	0.205
Experiment	0.198	0.197	0.201
Error percentage (%)	26.2	26.3	2.5

From the comparison distortion result which was shown in Table 7, it can be concluded that the distortion result which was resulted from the coupled process forming to welding by both Simufact.Forming and Simufact.Welding has been proven on producing a reliable result as the FE results displays almost similar result when it compared towards the experimental result within the range of error of 2.5% until 26.3%.

5. Conclusion

A study of investigation of distortion induced by coupled process forming to welding by utilizing FEM based VM simulation and experimental study has been successfully executed. The research covers both numerical and experimental analysis which was compared as final output of this research. The combination of FE software Simufact.Forming and Simufact.Welding are used for all numerical analysis. Some important points that need to be addressed in order to conclude this research are:

1. A good agreement of result in term of range of error has been achieved by looking at the result of coupled manufacturing process Forming and Welding using both Simfuact.Forming and Simufact.Welding
2. The final distortion result produced from coupled process simulation exhibits a similar result to the distortion of experimental process which possesses the error margin percentage for 26.2%, 26.3% and 2.5% for each measurement points,

3. The difference of the final result might be caused by pre-condition of manufacturing process as well as the non-homogeneous material assigned between simulation and experimental study.

As further recommendation, a more detailed metallurgical approach should be taken into account while assigning the material properties during VM simulation. Furthermore, the temperature-dependent properties of filler material should be created instead of using equivalent existing material.

References

[1] R. Radharamanan, "Virtual manufacturing: an emerging technology," Proceedings of the 2002 American Society for Engineering Education Annual Conference and Exposition, 2002.
[2] Q.S Liu, S.M Mahdavian, D. Aswin, S. Ding, "Experimental study of temperature and clamping force during Nd:YAG laser butt welding," Optics and laser Technology, Vol 41, pp. 794-799, 2009.
[3] M. Mochizuki, Y. Mikami, S. Okano, S. Itoh, "Computational simulation of weld microstructure distortion by considering process mechanocs," *International Conference on Advanced Structure and Functional Materials Design*, 2008.
[4] V.G Bharath, P. Rajashekar., "Virtual Manufacturing: A review," *International Journal of Engineering Research and Technology*, NCERAME-2015 Conference Proceedings, 2015.
[5] T.S Teng, C.C Fung, P.H Chang, W.C Yang, "Analysis of residual stresses and distortions in T-joint fillet welds," *Finite Element in Analysis and Design*, Volume 40, pp. 38-49, 2011.
[6] G.W Krutz, et al, "Finite element analysis of welded structures," *Welding Journal Research Supplement,* Vol. 57, pp. 211-216, 1998.
[7] S K. Gupta, "Sheet Metal Forming: Forming part families for generating shared press-brake setups, " *Journal of Manufacturing System*, Volume 21(5), pp 329-350, 2002.
[8] S.K Panthi, N. Ramakrishnan, M. Ahmed, S. S Singh, M.D. Goel, "Finite Element Analysis of sheet metal bending process to predict the springback," *Materials and Design*, Vol 31(2), pp. 657-662, 2010.
[9] J. A. Goldak, M. Akhlaghi, Computational Welding Mechanics, Springer Science+Bussiness Media, Inc., 2005
[10] S, Okano, M. Mochizuki, "Transient behaviour during TIG of thin steel plate," Journals of Material Processing Technology Volume 24, pp 103-110, 2017.
[11] W. Piekarskaa, M. Kubiaka, M Żmindak, "Issues in numerical modelling of phase transformation in welded joint," *Procedia Engineering*, Volume 117, pp. 141 – 148, 2017.
[12] S. Kalpakjian, S.R Schmid, "Manufacturing Engineering and Technology," 5th ed. Singapore: Prentice Hall, 2006.
[13] D. Andud, M.I Mihd Azhar, W.A Fahmy, et al, "Simulation of welding distortion on multi pass welding of fillet joint," *Journal of Science and technology*, Vol 03 (1), pp. 1-5, 2016.
[14] S.M Vyas, S.I Shah, G.D Archarya, "Minimization of distortion during gas metal arc welding process used for hydraulic shearing machine pressure plate square," *Journal of Experimental and Applied Mechanics*, Vol 8(3), pp. 48-53, 2017.
[15] D. Djurdjanovic, Y. Jiao, V. Majastorovic, "Multistage manufacturing process control robust to innacurate knowledge about process noise," *CIRP Annals*, Volume 66(1), pp. 434 – 440, 2017.
[16] M.F Zaeh, A.E Tekkaya, D. Bierman, A. Zabel, M. Lamghorst, A. Schober, "Integrated simulation of the process chain composite extrusion-milling-welding for lightweight frame structure," *Production Researh*, Volume 4, pp. 441-451, 2009.
[17] A. Bauer, Y.H.P Manurung, J. Sprungk, et al, "Investigation on forming-welding process chain for DC04 tube manufacturing using experiment and FEM simulation," *The International Journal of Advanced Manufacturing Technology*, Vol. 102, pp 1-10, 2019.
[18] Y.H.P Manurung, "Database of common material properties," Unpublished, 2015.

Anatomical modelling and simplified modelling in total hip replacement: difference in contact mechanics perspective

M F A Manap[**1], S Shuib**[*2], A F M Yamin**[3], A A Shokri**[4]

[1,3] Faculty of Mechanical Engineering, Universiti Teknologi MARA Pulau Pinang, Malaysia
[2] Faculty of Mechanical Engineering, Universiti Teknologi MARA Shah Alam, Malaysia
[4] Department of Orthopaedic, Universiti Sains Malaysia Kubang Kerian, Malaysia

*solehuddin2455@uitm.edu.my, **muhammadfaris@uitm.edu.my

Abstract. Total hip replacements (THR) is a surgical operation to replace defect bone at the hip joints. The rate of succession of THR post-operative still debatable as complication and failure rate of the prosthesis still exists. Edge-loading, dislocation, fracture and longevity are among the concerned issues with many studies were conducted via software analysis. This study aims to simulate the difference of anatomical and simplified modelling in finite element analysis (FEA) and investigate edge-loading effect at different inclination angle in both modelling conditions. A CT scan hemi-pelvic model was reshaped and converted into 3D model in SolidWorks and the next step, FEA was conducted in ANSYS Workbench V16 at different inclination angle. Anatomical and simplified model were run in ANSYS Workbench and the results were recorded. The anatomical modelling produced less contact pressure range 26% to 51% compared with simplified modelling at four inclination angle conditions. Von Mises stress and total deformation in anatomical also produced reduction of more than 65%. Both modelling conditions shows agreement that elevated inclination angle had induced higher contact pressure at superior region of acetabular cup. The inclusion of hemi-pelvic model gives lower value recorded in FEA as contact stress dispersed into the bone that already integrated with the implant given statistically significant ($p<0.05$). Noteworthy to include bone integration into implant during FEA study to produce unambiguous contact mechanics studies.

1. Introduction

A hip joint replacement known as THR was considered a successful intervention in orthopaedic field for hip joint related disease. Despite the successful outcomes, the clinical complications and unexpected failure of the prostheses remains concerned. Issues such as aseptic loosening, bone-implant interface, dislocations, edge-loading effect and superior region defect of the acetabular cup are among the top complication related to THR that required revision [1–4]. The contact of the femoral head on the edge of superior region acetabular cup was observed as the edge loading effect. In numerical aspect, the articulate contact patch between femoral head towards to the extended rim of the acetabular cup was known as true edge loading effect [5, 6]. The complications upon THR are interrelated among them as dislocation was induced by the edge-loading effect and excessive contact pressure inside the acetabular cup upon doing daily living activities (ADL). Studies shows that edge loading may reduce the efficiency of the tribological performance of the articulate contact [7–10].

Content from this work may be used under the terms of the Creative Commons Attribution 3.0 licence. Any further distribution of this work must maintain attribution to the author(s) and the title of the work, journal citation and DOI.
Published under licence by IOP Publishing Ltd

The occurrence of edge loading was being discussed with numerous previous studies [11, 12]. Influence of femoral head size [13], acetabular cup design[14], safe zone orientation[15, 16] and impingement[17] are among the issues that will leave adverse effect post-operational THR. There are three types of acetabular components; Metal-on-Metal (MoM), Ceramic-on-Ceramic (CoC) and Metalon-Polymer (MoP). Studies shows that edge loading in MoM and CoC could produce faster articulate wear, malpositioning and fracture of the materials [18]. Meanwhile, studies conducted on MoP indicated that edge loading does not particularly induced rapid wear, but stress and plastic strain substantial increasing will lead to fatigue and fracture [19].

The aims of the present study were, firstly to determine the difference of anatomical modelling and simplified modelling of THR contact analysis, and secondly, to inquire the edge loading effect of inclination angle in FE model analysis during normal walking (NW) gait cycle at both conditions.

2. Material and Methods

This study was conducted at hard on soft bearing combination of THR. A typical modular MoP total hip replacement system, consisting of metal shell, polyethylene liner and metallic femoral head, was analysed. The nominal diameter of femoral head and polyethylene liner were 28mm and 28.6mm respectively, giving a radial clearance of 0.3mm. The outer diameter of the acetabular component was at 50mm. A hemi-pelvic bone CT-scan format was obtained from one of the patients from USM Kubang Kerian with consent. The irregularity of the hemi-pelvic bone was removed by using MIMICS software.

The hemi-pelvic bone model consists of cancellous bone surrounded by cortical shell. The Hounsfield Unit used for the segmentation process was chosen to be 1035 to eliminate unnecessary element such as tissue or blood vessel. The thickness of the cortical shell was at 1.5mm. The acetabular subchondral bone was assumed to have been reamed completely prior to implantation.

All the materials in the FE model were modelled as homogenous, isotropic and linear elastic except the polyethylene liner which was modelled as non-linear elastic-plastic with the plastic stress-stain constitutive relationship [20]. The mechanical properties of the materials are shown in Table 1. The femoral head was considered as rigid material to reduce computational time. A mesh sensitivity studies were carried out at the contact condition of femoral head and acetabular cup in order to optimize the computational efficiency. The anatomical finite model was comprised of approximately 26133 nodes and 14436 elements with solid tetrahedral dominant element as it produces faster convergence.

Table 1. Material properties for the components used in the present study [21], [22]

Component	Materials	Young's modulus (GPa)	Poisson's ratio
UHMWPE cup	UHMWPE	0.95	0.3
Acetabular shell	Titanium	117	0.25
Cortical shell	Cortical bone	17	0.3
Cancellous bone	Cancellous bone	0.8	0.2

The contact interference of the anatomical model of hemi-pelvic and Titanium shell was assumed as fully bonded, while the contact of femoral head and UHMWPE liner was assumed frictionless. Full constrained at sacroiliac joint and pubic symphysis for the anatomical model where on contrary, the Titanium shell was assumed as fully constrained for the simplified model. The femoral head was constrained at rotational movement and only allowed at translational movement freedom for both models. The loading exerted was at 2450N in vertical component direction via gait studies conducted by [23] where the loading assumed at normal walking (NW) condition. The coordinate system of anatomical model was assumed to be aligned with in vivo study of pelvis coordinate system.

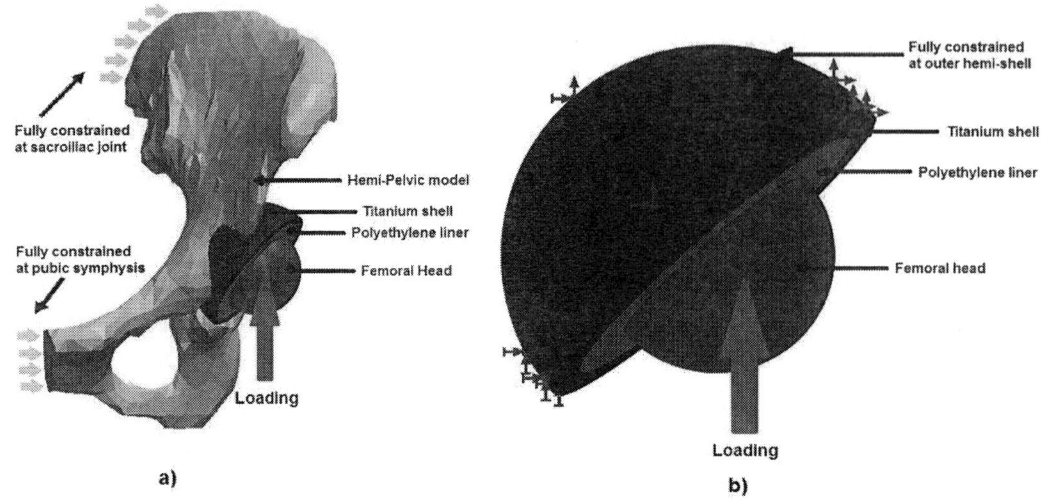

Figure 1. The FE modelling and the boundary condition. a) Anatomical model b) Simplified model

The FE simplified model was validated by Korduba et al.[21] and Hertzian Contact Theory at static dry contact. Eliminating the surface roughness, the notations of R_1 and R_2 are characterized by the head and cup radii, respectively, giving a radial clearance $c=R_2-R_1$ as shown in Fig. 2. Meanwhile, simpler configuration will yields the effective radius R' and elastic modulus E^* where E_1, v_1 and E_2, v_2 are the Young's Modulus and Poisson Ratio of the head and cup material, respectively. Elastic deflection of the surface under applied pressure is given by the following relation where is the elastic deflection.

$$u_z(x,y) = \frac{2\pi}{E'}\iint \frac{p\,(x',y')}{\sqrt{(x-x')^2+(y-y')^2}}dx'dy' \quad (1)$$

$$\frac{1}{R'} = \frac{1}{R_1} - \frac{1}{R_2} = \frac{c}{R_1(R_1+c)} \quad (2)$$

$$\frac{1}{E^*} = \left(\frac{1-v_1^2}{E_1} + \frac{1-v_2^2}{E_2}\right) \quad (3)$$

From the theory, theoretical Total Elastic Displacement d can be calculated based on Equation (4) given the amount of force exerted, F on that configuration.

$$F = \frac{4}{3}E^* R'^{1/2} d^{3/2} \quad (4)$$

Applying the Hertzian Theory for a static dry contact will yield the radius of the contact area, a as shown in Equation 5.

$$a = \sqrt[3]{\frac{3FR'}{2E^*}} \quad (5)$$

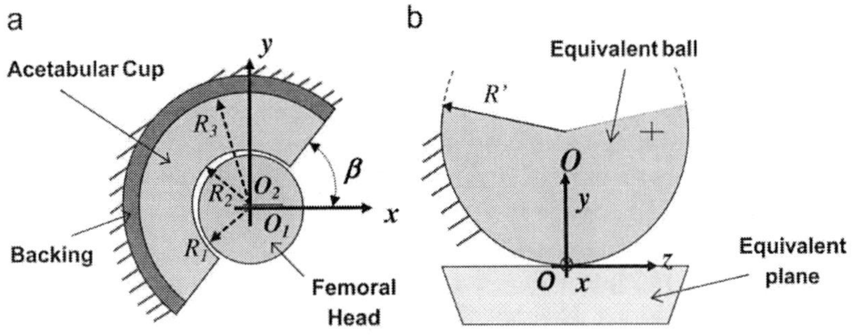

Figure 2. Ball-in-socket configuration a) ball-on-plane geometry b) setup for hip implants, where y is the vertical direction [24]

3. Results and Discussion

The contact pressure of Korduba Model and Simplified Model shows good agreement with less than 12% error which indicate the method was run as perfectly in well-mannered. From the analysis, anatomical model shows reduction of contact pressure at almost 30% for 40° and 50° inclination angle compared to simplified model as shown in Figure 3. On the other hand, the contact pressure reduction of more than 50% was recorded at 70° inclination angle for the anatomical model. Based on Hertzian Contact Theory, model also shows good agreement of less than 10% error. The values of simplified model at 0°, 40°, 50° and 70° inclination angle were 5.98MPa, 8.72MPa, 11.03MPa and 18.89MPa, respectively. Meanwhile, the values of anatomical model at 0°, 40°, 50° and 70° inclination angle were 3.41MPa, 6.43MPa, 7.82MPa and 5.90MPa, respectively.

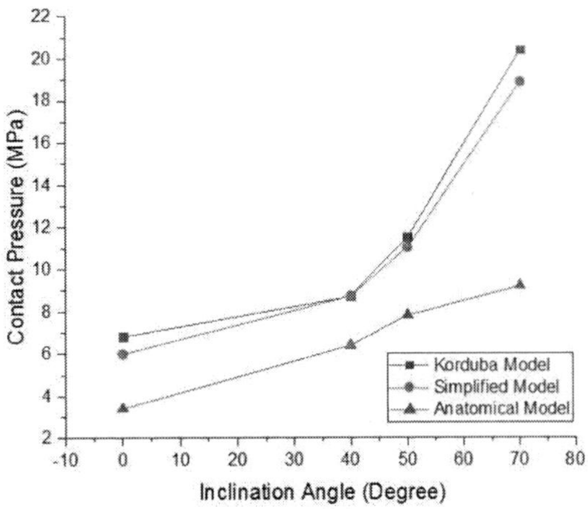

Figure 3. The contact pressure of acetabular components in three different model at different inclination angle

In term of Von Mises stress analysis, elevating the inclination angle of acetabular components showed Von Mises stress increased in both anatomical and simplified model. In spite, the reduction of Von Mises stress was higher in anatomical model compared to simplified model which recorded a reduction of 72.6%, 86.8%, 87.4% and 65.9% at 0°, 40°, 50°, and 70° inclination angle, respectively. The results show that including hemi-pelvis model apart from the analysis will reduce the stress data as the stress are not concentrated at the back side of acetabular shell only. FEA concluded that the stress was distributed towards the hemi-pelvis model as the shell and acetabulum was fully bonded. The stress was at maximum on the sacroiliac joint area in all inclination angle as shown in Figure 4.

Based on FEA analysis, it was indicated that the total deformation was almost insignificant difference recorded when inclination angle elevated in anatomical model. The results were recorded as contrary with simplified model, where a steep deformation was recorded from 50° to 70° inclination angle. In the Figure 5 shows that anatomical model gives reduction of total deformation at 85.2%, 91.2%, 94.2% and 97.0% for 0°, 40°, 50°, and 70° inclination angle, respectively. From these results, an assumption could be made; the small margin deformation analysis data recorded at the anatomical model proved the rational of human body could withstands almost 8 times of their own BW doing daily living activities. Although simplified model still applicable, it is suggested to include vital component of human body part to produce better accuracy in term of FE analysis. A small margin deformation produced from the FEA also suggested that the THR can be fit at the acetabulum without greater motion that will increase the THR stability.

An indication could be presumed that deformation of THR from FEA was considered as sub micrometre data which agreed with previous research that shown no significant difference in deformation but the value still affecting the implant lifespan [25]. The deformation occurring during intraoperative impaction and press-fit was not included in this study, thus the results valid at postoperative phase when the implant is fully integrated in the bone.

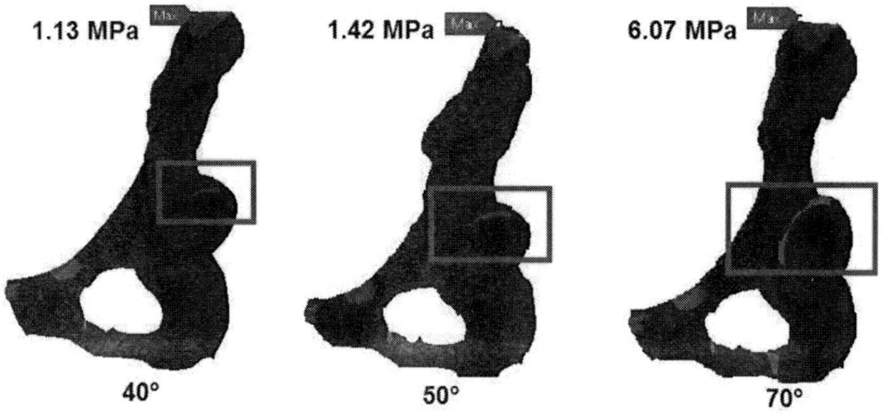

Figure 4. The equivalent Von Mises stress of anatomical model in three inclination angles. The red marked rectangular shape shows that the stress pattern at the superior region of the acetabular cup.

Note: The 0° inclination was not shown

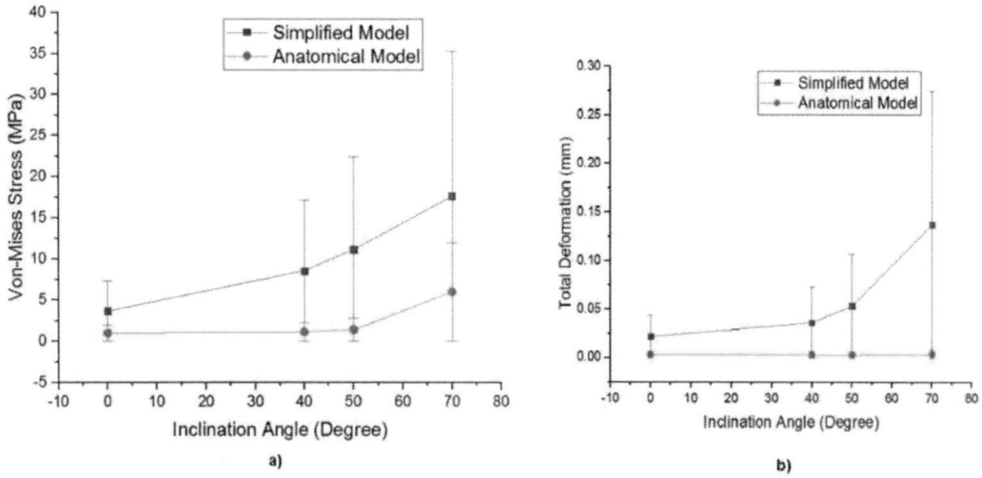

Figure 5. The difference of simplified model and anatomical model in term of a) Equivalent Von Mises Stress and b) Total deformation

4. Conclusions

There are hugely different data recorded in simplified model and anatomical model in term of contact pressure, equivalent Von Mises stress and Total deformation. Although the trend looks similar in both models which the value increasing with elevated inclination angle, the anatomical model produced lower results in all analysis that was conducted in this study. It is believed that the hemi-pelvic model affecting the FE analysis where the bone integration in FEA will disperse the pressure evenly compared to centralize concentrated at acetabular shell. Thus, analysis of THR using FEA should including anatomy model as the data will produce more accurate data. This study purposely for preliminary studies on improving the THR optimization in order to increase the lifespan from FEA aspect. Hence, further studies will be manifest in term of safe zone orientation, designs and material selection towards the longevity of the THR operation. Individual hemi-pelvic model results are varied and depending on many factors, yet the inclusion of the model in any analysis to improve the THR operations are vital.

Acknowledgments
A special gratitude for the University Technology of MARA (UiTM) and Ministry of Education for financial support via grants 600-IRMI/MYRA 5/3/MITRA (001/2017)-1 for this project.

References

[1] C. I. Esposito et al., 2012, Wear in alumina-on-alumina ceramic total hip replacements: a retrieval analysis of edge loading, *J. Bone Joint Surg. Br.* vol. 94, no. 7, pp. 901–907
[2] W. L. Walter et al., 2011, Retrieval analysis of squeaking alumina ceramic-on-ceramic bearings, *J. Bone Joint Surg. Br.* vol. 93-B, no. 12, pp. 1597–1601, 2011.
[3] R. De Haan, C. Pattyn, H. S. Gill, D. W. Murray, P. A. Campbell, and K. De Smet, 2008, Correlation between inclination of the acetabular component and metal ion levels in metal-onmetal hip resurfacing replacement, *J. Bone Joint Surg. Br.* vol. 90-B, no. 10, pp. 1291–1297
[4] N. Faiqa Ismail, S. Shuib, M. A. Yahaya, A. Z. Romli, and A. A. Shokri, 2018, Finite Element Analysis of Uncemented Total Hip Replacement: The Effect of Bone-Implant Interface, *Int. J. Eng. Technol*ogy pp. 230–234
[5] M. M. Mak, A. A. Besong, Z. M. Jin, and J. Fisher, 2002, Effect of microseparation on contact mechanics in ceramic-on-ceramic hip joint replacements, *Proc. Inst. Mech. Eng. Part H J. Eng. Med.* vol. 216, no. 6, pp. 403–408
[6] R. J. Underwood, A. Zografos, R. S. Sayles, A. Hart, and P. Cann, 2012, Edge loading in metalon-metal hips: Low clearance is a new risk factor, *Proc. Inst. Mech. Eng. Part H J. Eng. Med.* vol. 226, no. 3, pp. 217–226
[7] X. Hua, L. Wang, M. Al-Hajjar, Z. Jin, R. K. Wilcox, and J. Fisher, 2014, Experimental validation of finite element modelling of a modular metal-on-polyethylene total hip replacement, *Proc. Inst. Mech. Eng. Part H J. Eng. Med.* vol. 228, no. 7, pp. 682–692
[8] W. Li, M. Feng, X. Liu, M. Huang, and R. Ma, 2019, Ultra-High Molecular Weight Polyethylene Fibers / Epoxy Composites: Effect of Fiber Treatment on Properties, *Fibers and Polymers* vol. 20, no 2, pp. 421–427
[9] M. F. A. Manap, S. Shuib, A. Z. Romli, A.A. Shokri, 2017, Finite Element Study of Acetabular Cup Contact Region for Total Hip Replacement (THR), *J. of Mech Eng.* Vol. SI 2 (1), pp. 7182
[10] X. Hua, J. Li, Z. Jin, and J. Fisher, 2016, The contact mechanics and occurrence of edge loading in modular metal-on-polyethylene total hip replacement during daily activities, *Med. Eng. Phys.* vol. 38, no. 6, pp. 518–525
[11] S. J. Mellon, G. Grammatopoulos, M. S. Andersen, H. G. Pandit, H. S. Gill, and D. W. Murray, 2014, Optimal acetabular component orientation estimated using edge-loading and impingement risk in patients with metal-on-metal hip resurfacing arthroplasty, *J. Biomech.* vol. 48, no. 2, pp. 318–323

[12] I. Udofia, F. Liu, Z. Jin, P. Roberts, and P. Grigoris, 2007, The initial stability and contact mechanics of a press-fit resurfacing arthroplasty of the hip, *J. Bone Joint Surg. Br.* vol. 89-B, no. 4, pp. 549–556

[13] G. Tsikandylakis, M. Mohaddes, P. Cnudde, A. Eskelinen, J. Kärrholm, and O. Rolfson, 2018, Head size in primary total hip arthroplasty, *EFORT Open Rev.* vol. 3, no. 5, pp. 225–231

[14] J. B. Meding, S. R. Small, M. E. Jones, M. E. Berend, and M. A. Ritter, 2013, Acetabular cup design influences deformational response in total hip arthroplasty, *Clin. Orthop. Relat. Res.* vol. 471, no. 2, pp. 403–409

[15] G. Jung, Y. Kim, N. T. Khan, and Y. B. Kim, 2018, Risk evaluation method for range of motion after total hip arthroplasty applying monte-carlo simulation, *Procedia Comput. Sci.* vol. 141, pp. 351–357

[16] B.-H. Ko and Y.-S. Yoon, 2008, Optimal orientation of implanted components in total hip arthroplasty with polyethylene on metal articulation, *Clin. Biomech. (Bristol, Avon)* vol. 23, no. 8, pp. 996–1003

[17] M. Ghaffari, R. Nickmanesh, N. Tamannaee, and F. Farahmand, 2012, The impingementdislocation risk of total hip replacement: effects of cup orientation and patient maneuvers, *Conf. Proc. IEEE Eng. Med. Biol. Soc.* vol. 34, no. 1, pp. 6801–6804

[18] O. M. Posada, R. J. Tate, and M. H. Grant, 2014, Effects of CoCr metal wear debris generated from metal-on-metal hip implants and Co ions on human monocyte-like U937 cells, *Toxicol. In Vitro* vol. 29, no. 2, pp. 271–280

[19] X. Hua, B. M. Wroblewski, Z. Jin, and L. Wang, 2012, The effect of cup inclination and wear on the contact mechanics and cement fixation for ultra-high molecular weight polyethylene total hip replacements, *Med. Eng. Phys.* vol. 34, no. 3, pp. 318–325

[20] F. Liu, Z. M. Jin, P. Grigoris, F. Hirt, and C. Rieker, 2003, Contact mechanics of metal-on-metal hip implants employing a metallic cup with a UHMWPE backing, *Proc. Inst. Mech. Eng. H.* vol. 217, no. 3, pp. 207–213

[21] L. A. Korduba et al., 2014, Effect of Acetabular Cup Abduction Angle on Wear of UltrahighMolecular-Weight Polyethylene in Hip Simulator Testing, *Am. J. Orthop.* vol. 43, no. 10, pp. 466–471

[22] X. Hua, J. Li, L. Wang, Z. Jin, R. Wilcox, and J. Fisher, 2014, Contact mechanics of modular metal-on-polyethylene total hip replacement under adverse edge loading conditions, *J. Biomech.* vol. 47, no. 13, pp. 3303–3309

[23] G. Bergmann et al., 2001, Hip contact forces and gait patterns from routine activities, *J. Biomech.* vol. 34, no. 7, pp. 859–871

[24] L. Mattei, F. Di Puccio, B. Piccigallo, and E. Ciulli, 2011, Lubrication and wear modelling of artificial hip joints: A review, *Tribol. Int.* vol. 44, no. 5, pp. 532–549

[25] D. Kluess, I. Begerow, P. Goebel, W. Mittelmeier, and R. Bader, 2012, Analysis of Liner Deformation and Stress in Total Hip Replacements Considering Implant Materials, Position and Head Size, *Orthop. Proc.* vol. 94-B, no. SUPP_XXXVII, p. 209

IOP Conf. Series: Materials Science and Engineering **834** (2020) 012065 doi:10.1088/1757-899X/834/1/012065

Numerical-experimental approach to assess the deflection behaviour of carbon fibre fishing rod

Nur Aini Sabrin Manssor[1], A M M Ismail[1], Nurrul Amilin Zainal Abidin[1], Mohd Azran Aziz[2], Jamaluddin Mahmud[2,*]

[1]Faculty of Mechanical Engineering, Universiti Teknologi MARA, Cawangan Johor, Kampus Pasir Gudang, Johor, Malaysia.
[2]Faculty of Mechanical Engineering, Universiti Teknologi MARA, Shah Alam, Selangor, Malaysia.

[*]jm@.uitm.edu.my

Abstract. Deflection of a commercial hollow type fishing rod is investigated in this research. The fishing rod of interest is of Abu Garcia brand, made of carbon fibre. Analytical, experimental and numerical approach were performed to assess the bending of the rod under maximum 1.25 kg load located at the tip which represents the rod in pulling. The deflection is defined by longitudinal displacement of each point of guide rings located along the fishing rod. A mathematical model based on basic bending theory of cantilevered beam is formulated to obtain the deflection and then experiment was carried out by immobilizing one end horizontally on the wall and applying at tip force. The final position of the rings was marked, and the displacements were recorded. Finite element model was developed as an isotropic and elastic material via ANSYS software. Results from all approach were compared and it was found that analytical and numerical able to predict a similar trend of bending behaviour. However, large discrepancies were observed in the experimental curves due to several reasons.

1. Introduction

In this urban era, catching a fish is no longer an activity to earn a living, it also considered as a hobby and sport for some enthusiast. But similar to traditional way of fishing, an angler needs a fishing rod, line, and lure to fish whether in recreational or sport fishing. Although the literatures of fishing are abundance and dated back from the 80's, but there were mostly reported on the experiential studies while the scientific studies of fishing are still limited [1]. The improvements in fly fishing equipment have been rapid over the last 10-20 years. This improvement is due to increased knowledge about the physics of fly casting, development of more efficient casting techniques and technical development of the equipment.

The development of fishing rod manufacturing has led to light and stiff rods. These modern rods allow the caster to insert a high line speed, thus enabling a large amount of movement energy into the fly line[2]. The construction material of a rod affects the flexibility of the rod when a pressure is put at the tip of the rod. Rod can be created from different materials with different layers, all of which contributes to the responsiveness of the rod. Throughout the history, wood, bamboo, steel and fibreglass have been used as the fishing rod. Common material of modern fishing rods is high performance carbon fibers which can be classed in three categories, high resistance (HR), intermediate modulus (IM) and high modulus (HM) having modulus greater than 400 GPa [3].

Fishing rod design stage is a complex process that needs timely trial-and-error approach. In preliminary design work of a conventional rod fabrication, deflection board is used to gain sight of a

Content from this work may be used under the terms of the Creative Commons Attribution 3.0 licence. Any further distribution of this work must maintain attribution to the author(s) and the title of the work, journal citation and DOI.
Published under licence by IOP Publishing Ltd

finished rod that bends under load. One major disadvantage of this test is the rod must be finished before its flexural properties could be determined. Plus, it involves high pre-prototyping cost and enormous time effort. Such complications have led to the development of analytical analysis and numerical procedure to reduce unnecessary physical testing.

Theoretically, the bending or deflection of fishing rod is regarded as a tapered cantilever beam as employed by most studies [1], [4]–[6]. One of the earlier recognized works on modelling it mathematically is Leonard Euler in 1744 and succeeded later on by Jacob Bernoulli, Johann Bernoulli and L. Euler [4], [7]. Euler described the bending as large deflection of flexible rod where under loading, the free end will be subjected to a certain degree of deflection path. Ghuku and Saha had done a comprehensive literature of large deflections behavior of curved beams and reported that when an initially straight beam is subjected to pure bending moment, Euler-Bernoulli beam theory states that the bending moment is proportional to the change in curvature and was originally developed in geometrical nonlinear kinematic setting [7]. Thus, it is assumed that bending does not alter the length of the beam.

In the earliest reported research had done by Ohnishi and Matsuzaki. They analysed graphite type fishing rod to predict fracture location[8]. Both experimental and finite element approach were applied and they found a good agreement between those results. In a different study, Guan had numerically evaluated the bending of tubes with various cross-sectional shapes [9]. Finite element, FE models were developed to simulate the deformation of fishing rod by employing elastic constitutive model as well as nominal material properties. Guan introduced specific cross-sectional shape which is best for anglers called "Reuleaux Triangle".

One of important work specifically for a hollow, circular and tapered fishing rod was done by Otsuki in Ohtsuki [10] and [4]. Otsuki performed the experiment by fixing and tilting the rod by a certain degree and applying load at free tip. Later the experiment was compared to theoretical result based on large-deformation theory and both approach obtained good agreement. From the finding, the large-deformation theory is proven as best mathematical model to represent the deflection of fishing rod. However, complex calculation is required and in the above case, even the Runge-kutta-Gill method is adopted to carry out the solving of equation involved numerically. While Chang were focusing on the nonlinear mathematical and FE modelling of fishing rod under static and dynamic response.

Thus the objective of this study is to simulate the bending of a commercial fishing rod under loading by means of simple analytical and numerical approach via ANSYS software, and compare the performance with experimental setup. While previous researches utilized the complex-to-derived large-deformation theory, here the basic bending theory of tapered and cantilevered beam was applied. Comparison between all approaches will demonstrate the applicability of the bending theory for a simpler method, and the reliability of numerical analysis using a commercial FEA software.

2. Methodology

Investigation were done by replicating bending behavior of Abu Garcia™ fishing rod which is made of carbon fiber. The Young's modulus in 0° fiber direction is taken 175 GPa with Poisson's ratio of 0.3 based on the material's general properties [11]. As in Figure 1, the structure of the rod is a cylindrical hollow type and the diameter is decreasing from the butt towards the tip. The length of bending part is 1.525 m. Six points were marked from the end of bending part of the rod (at handle grip) to the tip which represents the string placers or guides. The position of guides were labelled as ∂n (where n = 1, 2, 3,... 6) their values are shown in Table 1. Load applied were ranged from 0.25 kg to 1.25 kg with the increment of 0.25 kg per step. Figure 2 shows the experimental setup of the fishing rod while loading and Table 1 lists the deflection points distance from the handle.

IOP Conf. Series: Materials Science and Engineering **834** (2020) 012065 doi:10.1088/1757-899X/834/1/012065

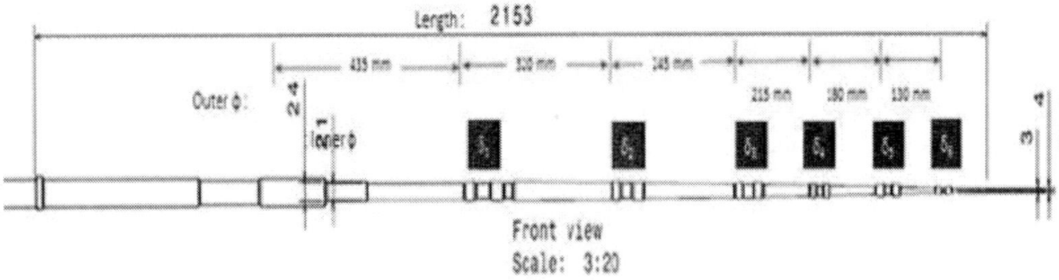

Figure 1. The detailed drawing of Abu Garcia fishing rod of total length 2153 mm. The length of the bending rod was taken as 1525 mm and six markers were placed along this length to represent guides as summarized in Table 1.

Table 1. Deflection points location from grip

Indicator	Deflection	Location (m)
Point 1	∂_1	0.435
Point 2	∂_2	0.755
Point 3	∂_3	1.000
Point 4	∂_4	1.215
Point 5	∂_5	1.395
Point 6	∂_6	1.525

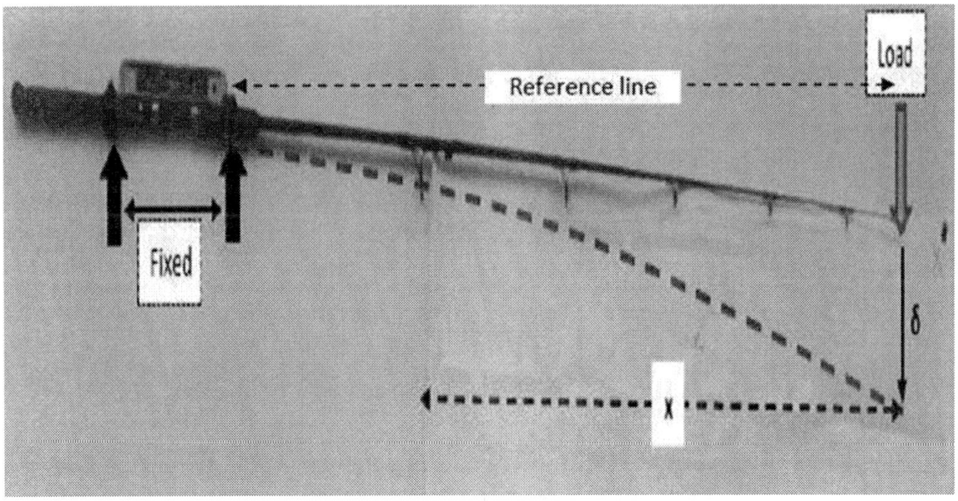

Figure 2. The experimental setup of fishing rod under loading. The reference line represents the origin line, δ is the longitudinal deflection of rod measured from reference, and x is lateral displacement measured from reference.

IOP Conf. Series: Materials Science and Engineering **834** (2020) 012065 doi:10.1088/1757-899X/834/1/012065

2.1. Mathematical Model of Fishing Rod

Fundamental bending moment, M of a beam is applied where the fishing rod is assumed as a rigid cantilever beam with one fixed end. Consider a long, thin cantilever leaf spring (Figure 3) where L *is* the length of the rod, x *is* the horizontal component of the displacement of the loaded end of the rod, ∂ is the corresponding vertical displacement, P *is* the concentrated vertical load at the free end, B the flexural stiffness of the rod's material mathematically expressed as $B = EI$ where E is the modulus of elasticity and I is the moment of inertia.

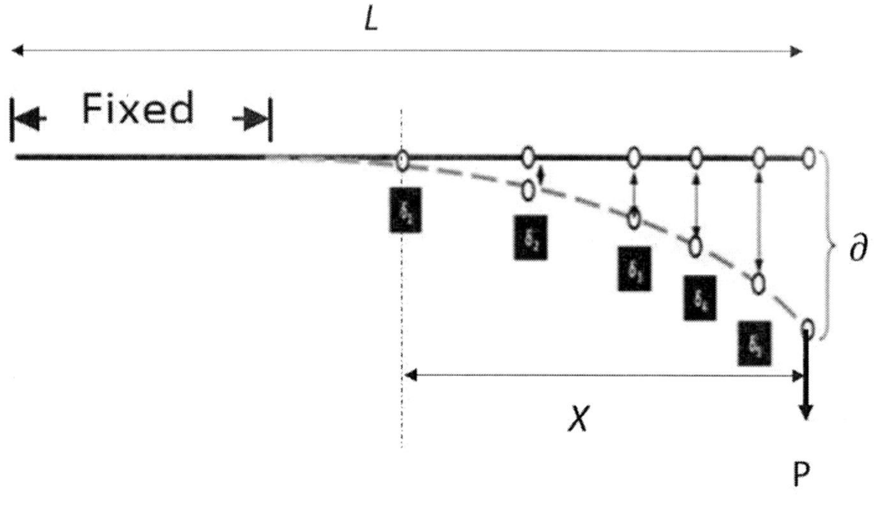

Figure 3. Free body diagram of fishing rod subjected to vertical load, P

The equation and derivation with elastic curve [12] is as in Equation 1 below:-

$$M = -Fx \tag{1}$$

By two order of integration and applying boundary conditions, Equation 1 becomes: -

$$\partial = \frac{-PL^3}{3B} \tag{2}$$

With this, Equation 2 serves as the theoretical model in predicting the deflection of all six points on the fishing rod.

2.2. Bending experiment of fishing rod

The experiment setup was employed from Ohtsuki [10], the fishing rod is treated as a cantilevered beam fixed at one end. A new Abu Garcia fishing rod was fixed on a wall, at the handle to represent the fixed end from the theory and resembles the angler's holding the rod (see Figure 4). Grid of 10 x 10 cm^2 made of A0 paper was pasted on the wall. The handle of the rod was fastened and immobilized so that only the free end of the rod could deflect under stress. The stress was transmitted to the rod by means of a thin nylon fishing line attached at the tip of the rod and pulled horizontally at a force measured by an attached digital scale calibrated accurately to 0.01 lb (0.0445 N). For every load from 0.25 kg to 1.25 kg (increment of 0.25 kg), the final position of each guide was recorded and marked on the paper.

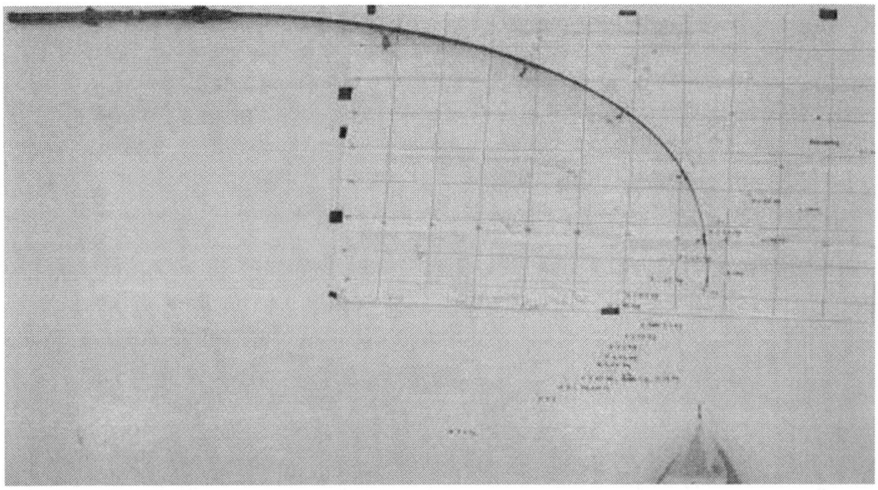

Figure 4. Experimental setup for bending of fishing rod consist of immobilized fishing rod, and A0 paper pasted on the wall to record the deflection of the rod free end.

2.3. Numerical Assessment of Fishing Rod

The deflection of fishing rod is simulated by ANSYS software as a linear, elastic and isotropic material. Material properties and actual dimensions including rod's handles as in Table 1 and Figure 1 respectively were utilized to generate the FE model. The analysis type is static (steady-state) with 1 x 1 meshing element size at all segments. The 3-noded element was used and the boundary condition is fixed end at left hand side as shown in Figure 5. The simulation was only considering the longitudinal deflection of the rod.

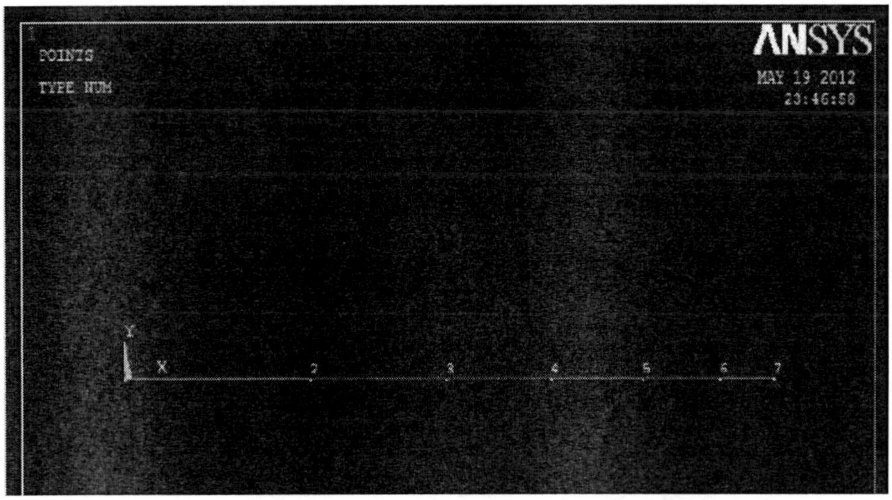

Figure 5. Simulation of the elastic curve calculated numerically using ANSYS software using E= 175 GPA for F=0.25 kg. The boundary condition is fixed end at left hand side.

IOP Conf. Series: Materials Science and Engineering **834** (2020) 012065 doi:10.1088/1757-899X/834/1/012065

3. Result and Discussion

A comparison of theoretical and experimental deflection curves for weight 0.25 kg (2.4524 N) through 1.25 kg (12.263N) is shown graphically in Figure 6 respectively. Generally, it is found that the bending deflection distributions of the fishing rod vary with the point location which shows the minimum bending strain travels from the tip to butt area with increasing load.

From the figures, it was shown that comparable profile for theoretical and numerical is achieved for load of 2.4524 to 12.263 N which presumably indicates that the rod satisfied the assumptions of basic large cantilevered beam under deflection. It was observed that at any specific location, the deflection differences for each loading are in a close range, except beyond point 4. This might suggest that the finite element model does not has suffice parameters to predict the deflection of such fishing rod. The current element model should be further refined for a proper meshing procedure, considering the tube type structure of the fishing rod. In addition, a gradient property is existing which is in this study, is the variation of flexural strength due to the decreasing of structural cross -sectional area towards the tips.

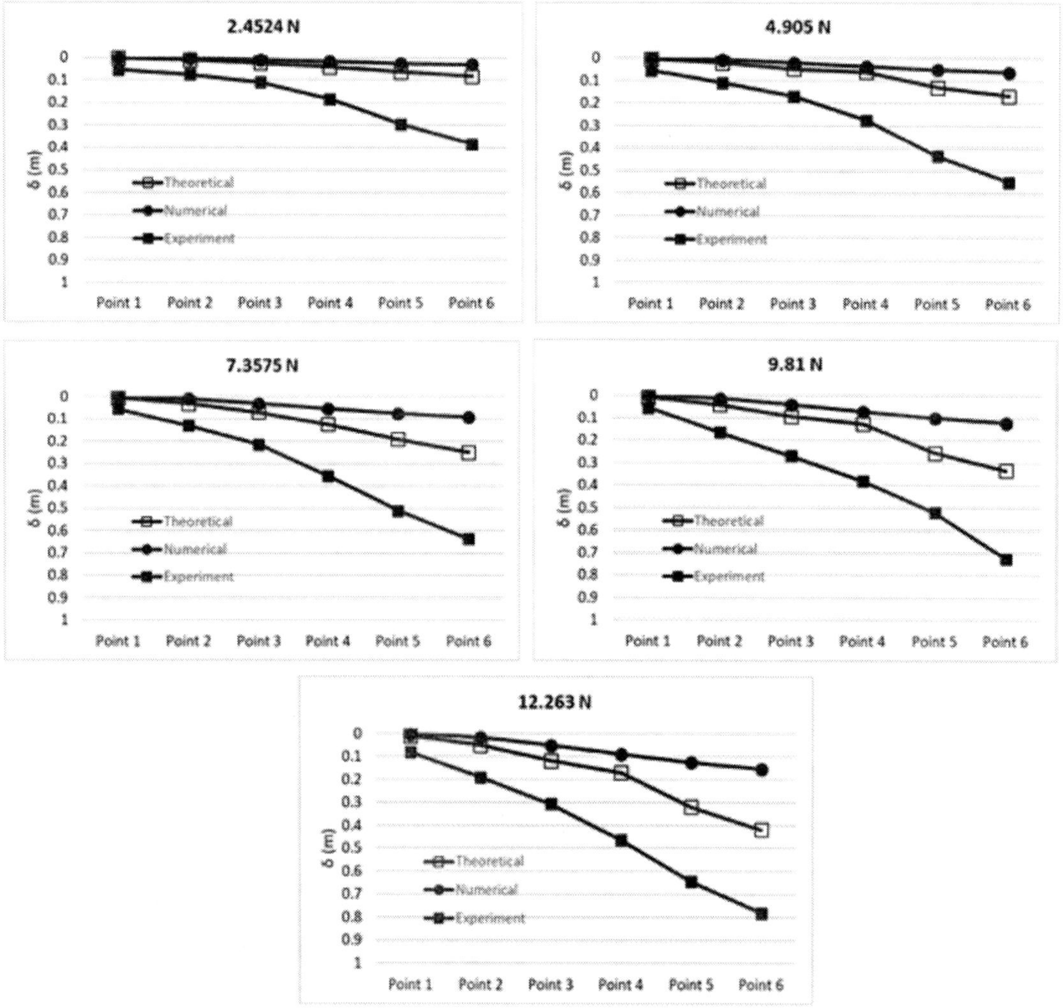

Figure 6. The deflection curves for theoretical, numerical (FEA) and experimental of bending fishing rod subjected to load 2.45, 4.905, 7.3575, 9.81 and 12.26 N respectively.

The comparable profiles of rods show an excellent fit of theory to numerical for approximately the initial 60% of the rod length and then a slight deviation for all five loads. The divergence seems like a phase change with discontinuous slope and presumably occurred neither from random or systematic error in the collection or processing of the data. The deviation pattern signifies a failure of the rod to fit to one or more of the assumed conditions for the validity of the theory for Equation (2) as in agreement with Silverman et. al [6].

The experimental graphs for all load levels show extreme deviations from both theoretical and numerical could imply that there was a significant systematic error involve in data processing such that the fishing rod was modelled as having a uniform cross-section throughout its length, but the rod used has a geometry variation at the tip end. The length of the rod is also assumed to be unchanged for each loading condition for the theoretical analysis, which is contradict with the experimental finding, as suggested by Ghuku and Saha. According to their research, it was suggested that the experimental load condition has an eccentricity that will change the effective length of the rod at each loading [13]. Besides, it is possible that the geometry variation along the rod could also come with variation of stiffness. Silverman et al suggested that the flexural rigidity or known as stiffness among the rod manufacturers may varies along the rod length [6] since stiffness is the product of a material parameter (E) and a geometric parameter (I), thus either or both can effect a change in flexural rigidity.

Plus, the variance occurred probably due to the inaccurate value of Young's modulus used in theoretical and numerical setting, as we have been unable to find any values of E reported by rod manufacturers or vendors for their rod. The Young's modulus value should be calculated using ANSYS by comparing the longitudinal deflections obtained from experiment with the numerically obtained values. It is possible that the value of modulus should be higher than 175 GPa.

In addition, it was observed that there exists slight deflection in experimental curves at point 1, whereas the deflection is missing in both models. This may due to the present of uniformly distributed load along the length of rod which is not considered in theoretical (Equation 2) and numerical analysis. Besides, the large deviation could be explained by the fact that the actual rod is using nylon string to hold the weight while the models eliminate this consideration in calculation. The actual bending of rod under loading is affected by the string not only at the tip, but other guide was pulled at the same time. Those guides (excluding the one at the tip), might had been introduced to extra load by the action of the string. Thus, larger deflections were observed experimentally.

4. Conclusion

Mathematical model and finite element model have been successfully developed to simulate the deflection of Abu Garcia fishing rod and compared to those obtained from the experimental setup. It is found that both approaches can predict a similar trend of bending behaviour. Large discrepancies were observed between experiment and theoretical prediction suggesting that the basic bending theory is less practical. Close range of errors are exhibited at their respective points which indicates that the FE model is lack of few parameters. Both methods could further produce more accurate deflection given that a material property of the fishing rod are reliably established by conducting physical experiments such as flexural test or numerically calculated. In addition, it is more reliable to take into consideration that the rod has a non-uniform geometry along its length, thus a better and accurate mathematical and numerical model can be approached. Plus, addition of horizontal displacement in x-axis is necessary for the best prediction of the deflection path.

Acknowledgement

The authors would like to thank to all personnel involved especially technicians in Computer Aided Design, Engineering and Manufacturing centre (CADEM) of Universiti Teknologi MARA (UiTM), Shah Alam, Malaysia.

References

[1] G. Wang and N. Wereley, "Analysis of fl y fi shing rod casting dynamics," vol. 18, pp. 839 855, 2011.

[2] S. Diridollou *et al.*, "In vivo model of the mechanical properties of the human skin under suction," *Ski. Res. Technol.*, 2000.

[3] N. Baral, H. Guezenoc, P. Davies, and C. Baley, "High modulus carbon fibre composites:

Correlation between transverse tensile and mode I interlaminar fracture properties," *Mater. Lett.*, vol. 62, no. 6–7, pp. 1096–1099, 2008.
[4] D. Chang, "Mathematical Analysis of Fly Fishing Rod Static and Dynamic Response," no. January, 2006.
[5] C. Gatti-bono and N. C. Perkins, "ARTICLE IN PRESS Numerical model for the dynamics of a coupled fly line / fly rod system and experimental validation," vol. 272, pp. 773–791, 2004.
[6] M. P. Silverman, J. Farrah, M. P. Silverman, and J. Farrah, "Bending of a Tapered Rod : Modern Application and Experimental Test of Elastica Theory Bending of a Tapered Rod : Modern Application and Experimental Test of Elastica Theory," 2018.
[7] S. Ghuku and K. N. Saha, "A Review on Stress and Deformation Analysis of Curved Beams under Large Deflection," no. July, 2017.
[8] A. Ohnishi, H Matsuzaki, "The Large Deformation Analysis of a Graphite Fishing Rod," *Comput. Struct.*, vol. 21, no. 2501, pp. 265–271, 1985.
[9] "D-Flex FEA Report," 2009.
[10] A. Ohtsuki, *Analysis of the characteristics of fishing rods based on the large-deformation theory*. 2001.
[11] P. K. Sharma, "Mechanics of materials," in *Studies in Health Technology and Informatics*, 2010.
[12] U. G. K. Wegst, "Bending efficiency through property gradients in bamboo, palm, and wood based composites," *J. Mech. Behav. Biomed. Mater.*, 2011.
[13] S. Ghuku and K. Nath Saha, "A theoretical and experimental study on geometric nonlinearity of initially curved cantilever beams," *Eng. Sci. Technol. an Int. J.*, 2016.

Emission characteristics of biodiesel ratios of 10%, 20%, and, 30% in a single-cylinder diesel engine

Helmisyah Ahmad Jalaludin[1], Nik Rosli Abdullah*[2], Hazim Sharudin[3], A R Asiah[1], Muhammad Firdaus Jumali[2]

[1]Faculty of Mechanical Engineering, Universiti Teknologi MARA, 23200 Bukit Besi, Terengganu, Malaysia.
[2]Faculty of Mechanical Engineering, Universiti Teknologi MARA,40450 Shah Alam, Selangor, Malaysia.
[3]Faculty of Mechanical Engineering, Universiti Teknologi MARA, 13500, Permatang Pauh, Pulau Pinang.

*nikrosli@salam.uitm.edu.my

Abstract. This research is devoted to investigate the effects of biodiesel ratios of 10%, 20%, and 30% to the emission of single cylinder diesel engine. Fuel blends comprise of 10% (B10), 20% (B20), and 30% (B30) with adding 0.1 ml of additives to each litre of fuel blend volume are put to the test on a diesel engine, comparing each aspect to differentiate the best fuel blend. A reference experiment with 100% diesel (D100) had been done to have a clear view of the working diesel engine emissions. These blends were used for testing in a four stroke, vertical single-cylinder, air-cooled diesel engine at a constant speed of 1500 rpm as with varying engine load of 2, 4, and 6 Nm. Exhaust emissions of nitrogen oxide (NOx), carbon monoxide (CO), carbon dioxide (CO_2), and unburned hydrocarbon (HC) are taken into account. The results show that the increase biodiesel blend in fuels does increase the NOx and CO_2 emission with steady increase of load. This is because these emissions relate to the complete combustion, high post-combustion and high oxygen content increase the amount of emission. Besides, the results prove that the higher percentage of biodiesel in the blends provides lower emission of CO and HC. Lower CO and HC emission due to better combustion efficiency as it reaches better overall combustion, producing higher energy output and better emissions. Thus, the findings provide helpful information for researchers to understand the technology of biodiesel and a possible substitute for the current diesel for a better fuel source.

1. Introduction

The global energy consumption has duplicated during the recent two decades. According to the Environment Information Assessment (EIA), fossil fuels has represent over 80% of the total energy supplies in the world and it is also known as the dominant in the global energy mix [1]. Due to rapid economic development, a further increase in the utilization of energy is being expected which will resulted in the fast depletion of fossil fuel reservoirs. Within a century, oil fossil fuel storage is eventually going to run out mainly because of an ever-growing industry with the constant need of fuel. Together with the increase in fuel prices per barrel has made people trying to find cheaper and possibly cheaper alternatives from the conventional fuel. Besides, the world today also faced with a serious environment pollution and global warming. The major gas that contributes to the greenhouse phenomenon are nitrogen oxide (NO_X) and carbon dioxide (CO_2) which is mainly emitted from the

Content from this work may be used under the terms of the Creative Commons Attribution 3.0 licence. Any further distribution of this work must maintain attribution to the author(s) and the title of the work, journal citation and DOI.
Published under licence by IOP Publishing Ltd

combustion of fossil fuel. This twin crisis on the depletion of fossil fuel and environment pollution that has been an issue in the last decades leads to the search of the alternative sources of liquid fuel energy. Therefore, there is an urgent to find alternative energy fuel that is clean, renewable, and reliable.

Biodiesel is one of the environmentally and renewable friendly alternative fuels that can be used in a diesel engine with little or no modifications to the engine. The term biodiesel refers to a non-petroleum-based diesel fuel which consists of the mono-alkyl esters of long chain fatty acids. Biodiesel commonly extracted from the reaction of vegetables oil or animal fat with methanol in the presence of a catalyst. Biodiesel are expected to lower the dependency on imported petroleum that correlates to economic vulnerability, reduce greenhouse gas emissions and other pollutants, and recuperate the economy by increasing demand and prices of agricultural products [2]. Thus, biodiesel is one of the best ways to replace current fuel source that has a good potential to perform up to par and fill in the demands in case of fuel depletion. Biodiesel can effectively reduce engine-out emissions, particularly unburned HC, CO, and NOx in a typical four-stroke compression-ignition engine. This is crucial in developing an innovative way to substitute or even better, upgrade the current fuel market to be greener without sacrificing performance too much. Higher oxygen content in biodiesel will increase the overall complete combustion in the engine hence higher value of Brake Thermal Efficiency (BTE). This will also help to ensure lower Brake Specific Fuel Consumption (BSFC) which is ideal in finding an alternative fuel. Diesel-biodiesel blends significantly reduce harmful emissions and increase performance of diesel engine, so that these studies should be researched even further to help in resolving current fuel problem.

A study by Nabi et al. [3] shows that biodiesel play an important role in overall emission of diesel engines. According to the study, CO emissions with biodiesel were found to be 11-51% lower relative to the reference diesel. The difference of particulate matter (PM) and particle number (PN) between biodiesels and conventional diesels are also being observed in the study. The result reported that there are reductions in PM and PN emissions with biodiesel blend relative to those of the reference diesel. Another study by Bueno et al. [4] states that both soybean and castor biodiesel increase the CO emissions caused by the biodiesel iodine value. Castor bean or castor oil plant is one of the promising non-edible oil crops since it has low implementation and production costs and relative resistance to hydric stress. However, the result of the study shows that castor oil biodiesel blending into diesel fuel did not advance fuel injection for the engine in use and continuously increased the NOx emissions. Other authors concluded that the effect of biodiesel on NOx emissions depends on the type of engine and its operating conditions. It will be easier to analyse the engine performance results by increasing the biodiesel ratio in the blended fuel because the measurement and evaluation of blended fuel property is an important indicator [5].

However, these studies differ from the origin of the biodiesel as well as the percentage of biodiesel-diesel blend. In this research, it will particularly focus on B10, B20, and B30 from palm oil-based biodiesel-diesel blend to standardize the results later. Some researchers [3, 6-8] suggested that B20 has a better density, viscosity, flash point, and calorific value similar to conventional diesel which makes it an ideal blend ratio for this research. Therefore, this research is carried out to compare the significance between the optimum biodiesel blend and the conventional diesel in current market. Indirectly, the study wants to help users to choose their fuels wisely since it will also affect environment and health despite of nowadays economic problems.

2. Methodologies

2.1. Sample Preparation

Sample of Palm Oil Methyl Ester (POME) biodiesel fuel blends were prepared through mechanical mixing and blending (based on volume). Briefly, each biodiesel blends are added with 0.1ml additives as a wet lubricant during combustion process. A 100% diesel (D100) will be a baseline reference for this research to compare later. The test fuels were 0.9 L diesel + 0.1 L palm oil + 0.1 mL additives (B10), 0.8 L diesel + 0.2 L palm oil + 0.1 mL additives (B20), and 0.7 L diesel + 0.3 L palm oil + 0.1 mL additives (B30). Each of the mixture was stirred continuously for 30 minutes by using an electric magnetic stirrer and left for 30 minutes to reach an equilibrium at room temperature before they were subjected to any test as recommended by previous research [9].

2.2. Measurement of Fuel Properties

The fuel properties play a big role in analyzing the results of its performance. For this research specifically, the density, calorific value, and viscosity are needed in the explanation of the results. It aim is to provide a clear understanding of the characteristics of blended biodiesel fuel with its additives. All the fuel properties of the samples as shown in Table 1 are tested out in Tribology Lab in the Faculty of Chemical Engineering, at Universiti Teknologi MARA Shah Alam.

Table 1. Measured properties of D100, B10, B20 and B30.

Fuel	Density @ 15°C g/ml	Calorific value (J/kg)	Kinematic viscosity @ 40 °C mm^2/s
D100	0.8270	45500	2.28
B10	0.8330	43825	2.49
B20	0.8350	43375	2.82
B30	0.8410	42925	2.85

2.3. Engine Setup and Specification

The engine used in this study is a natural aspirated, air cooled single cylinder diesel engine, namely the Yanmar TF-M by Yanmar Co. Ltd where Table 2 shows the specifications of the engine. An eddy dynamometer with a maximum power input of 20 kW at 2,450 rpm to 10,000 rpm will be attached to the engine. For further understands, a schematic experimental setup of the main components in detail is presented in Figure 1. The experimental procedures in this study were conducted in the Faculty of Mechanical Engineering, at Universiti Teknologi MARA Shah Alam.

Table 2: The Yanmar TF-M diesel engine specifications

Descriptions		**Specifications**	
No. of Cylinder		1	
Bore x Stroke (mm)		70x57	
Displacement (liter)		0.219	
Fuel Tank Capacity (liter)		2.4	
Continuous Rated Output	• Engine Speed rpm (min^{-1})	3600	3000
	• Output kW(hp)	3.1 (4.2)	2.8 (3.8)
Maximum Rated Output	• Engine Speed rpm (min^{-1})	3600	3000
	• Output kW(hp)	3.5 (4.7)	3.1 (4.2)

The analysis of exhaust emission is carried out by using a BOSH BEA-350 exhaust gas analyser to detect output of CO, CO2, HC, and NOx. The gas analyser is calibrated by the supplier company to assure the accuracy of the data collection. For the engine test operation conditions, the speed of the engine will be at a constant speed of 1500rpm with variation of engine load range of from 2, 4, and 6Nm. The engine is ran approximately for ten minutes with diesel fuel before the test, and the fuel is changed for the next test to ensure no traces of remaining previous fuel blend in the system, so that accuracy and consistency of measurement data could be achieved. Furthermore, the engine fuel tests were repeated three times for each fuel at the specified test conditions in order to have a reliable measurement.

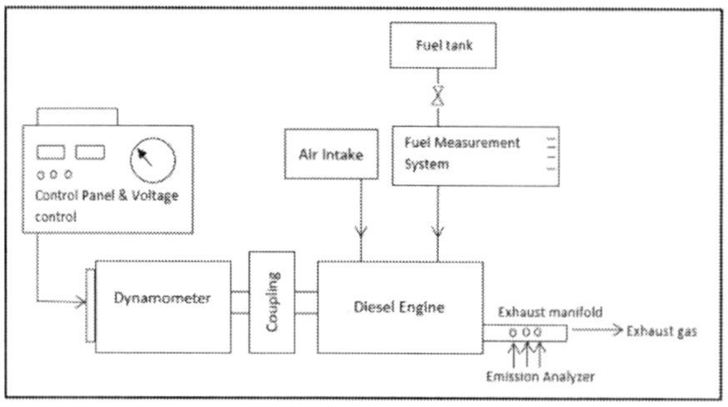

Figure 1. Schematic diagram of the experimental setup

3. Results and Discussion

3.1. Nitrogen Oxide Emission, NOx

The generation of nitrogen oxide (NOx) emissions is directly dependent on the temperature in the combustion chamber, lower enthalpy of vaporization and oxygen concentration [10]. Figure 2 shows the comparative results of NO_X emission between base diesel (D100), 10% biodiesel blend (B10), 20% biodiesel blend (B20), and 30% biodiesel blend (B30). All biodiesel fuel blends with additive generally produce higher emissions of NOx as the higher of the engine load is applied. The results obtained shows that B10 at 2 Nm results in the increase of 55% (62 ppm), B20 an increase of 115% (86 ppm) and B30 an increase of 140% (96 ppm) as compared to D100 (40 ppm). A maximum of NOx emission by D100 at 6Nm was 112 ppm, compared to the others blends, there is an increment of 11.6% (125 ppm) using B10, increase of 16.1% (125 ppm) on B20, and 25% increase on B30 (140 ppm). Many studies suggested the cause of higher NOx emission in biodiesel blends than normal diesel fuels are due to high burned gas temperature, concentration of oxygen, and equivalence ratio.

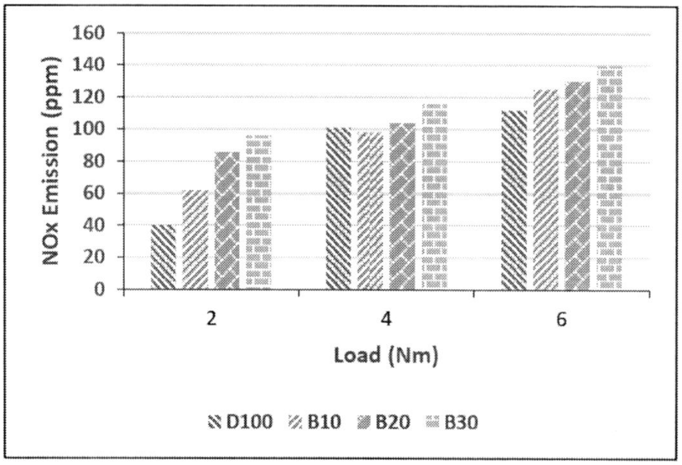

Figure 2. NOx emission vs engine load at constant engine speed of 1500 rpm

Based on a study by Agarwal et al. [11], the increase of NOx emission over base diesel and biodiesel blends is mainly due to the high concentration of oxygen in biodiesel and blends used. According to Bueno et al. [4], cetane number is a major parameter for nitrogen oxide formation. The study also said that in low temperature such as in 2 Nm load, carbon double bonds lowers in rate of reaction that reduces the cetane number in unsaturated biodiesel, while higher temperature condition as in 6 Nm, load increase the rate of reaction and provide higher cetane number, and overall NOx emission [4]. In this study, it proves that the increment of biodiesel blend percentage in fuel does increase the NOx emission with steady increase of load.

3.2. Carbon Monoxide Emission, CO

Carbon monoxide (CO) is colorless, odorless and considered as an intermediate product of hydrocarbon combustion. CO is a gas result from insufficient amount of oxygen reacting with all the fuel's carbon to produce CO_2, the unburned fuel is the cause of production of carbon monoxide [12]. Figure 3 shows the CO emissions for base diesel and blended biodiesel (10%, 20%, and 30%) at a constant engine speed of 1500 rpm and various engine loads. The results show that there is a slight decrement in CO emissions from using D100 to the other biodiesel blend fuels. At 2 Nm, a slight decrease has been recorded by using B10 of 11.6% (221 ppm) from using D100 (250 ppm). Furthermore, B20 gives a reduction of 12% (220 ppm) and B30 reduces for 20.8% (198 ppm) in comparison with the baseline diesel (250 ppm). There is a slight decrease of CO emission from 4 Nm from using diesel fuel (182 ppm) to 3.8% (175 ppm) using B10, 7.1% (169 ppm) using B20, and a drop of 14.3% (156 ppm) using B30. At 6 Nm, the results show that the CO emission of D100 is at 110 ppm. Upon further experimentation and observation, B20 shows a decrement of 8.2% (101 ppm), B20 decreases by 14.5% (94 ppm) and B30 decreases as much as 45.5% (60 ppm) from using diesel fuel.

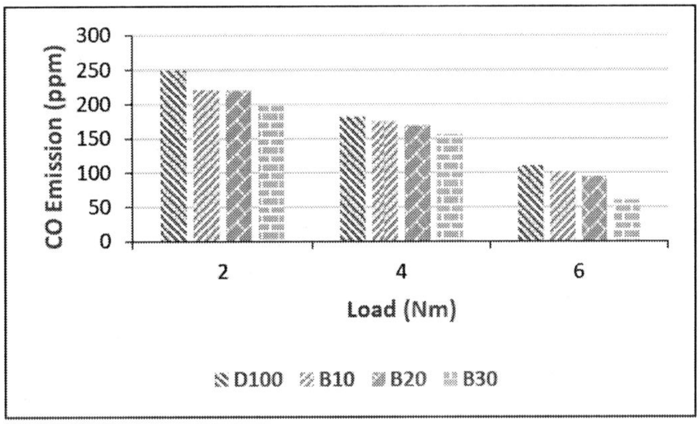

Figure 3. CO emission vs engine load at constant engine speed of 1500 rpm

All results obtained can be explained because of oxygen concentration in the biodiesel which leads to more complete combustion and results in conversion of CO to CO2. This result is in agreement with the previous study by Nabi et al. [3], which has found that the higher the engine speed, the lower the production of CO emission due to the increase in excess air as CO formed are factored by insufficient air supply. Other researches also reported that higher percentage of biofuel in the blends can decrease the amount of CO emission. This is because oxygen is inherently present in biofuels making them easier to be burnt at higher temperature in the chambers [13]. Thus, the results prove that the higher percentage of biofuel in the blends provide lower emission of carbon monoxide.

3.3. Carbon Dioxide, CO2

Generally, the ideal output of any combustion product of a hydrocarbon should be water and carbon dioxide (CO_2) emissions [14]. In that sense, between all the emissions, CO_2 emission should be the least concern as it contributes to less harm and can be counter-act with the increase of biodiesel crops. Figure 4 shows the variations of CO_2 emissions for base diesel and blended biodiesel (10%, 20%, and 30%) at a constant engine speed of 1500 rpm and various engine loads. The results show that biodiesel contents give higher CO_2 emissions compared to the base diesel at all engine loads. At 2 Nm, there is an increment to 0.54%, 0.82%, and 0.91% for B10, B20, and B30 respectively. The same trend of increment of CO_2 emission can be seen at all of the biodiesel fuel at every load. The maximum CO_2 emission is with B30 at all 2, 4, and 6Nm load with 0.91%, 0.97%, and 1.21% respectively.

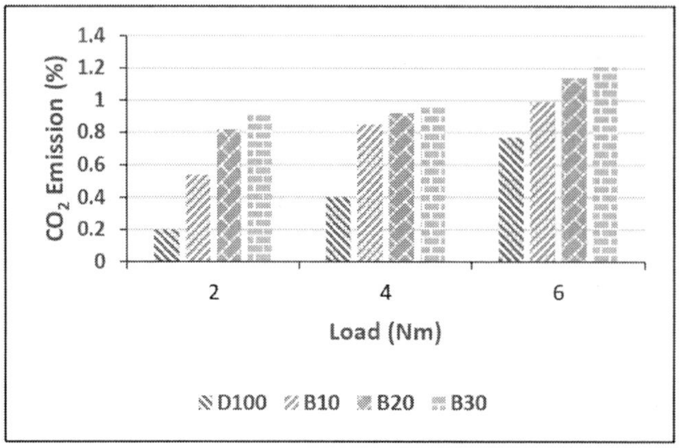

Figure 4. CO_2 emission vs engine load at constant engine speed of 1500rpm

This is mainly due to the present of oxygen content in the blended biodiesel fuels which interact with unburned carbon atoms during the combustion process, and produce greater amounts of CO_2. This result is in agreement with previous study which shows that the increase of biodiesel content gradually increases the percentage of CO_2 emission. It is due that CO_2 emission relates to the complete combustion, high post-combustion, and high oxygen content increases the amount of emission [15, 16].

3.4. Hydrocarbon Emission, HC

In general, unburned hydrocarbon emissions (HC) are hydrogen-carbon compound results from incomplete combustion. Combustion chamber geometry, deposits on walls, crevice volume, and engine operating parameters are also influence the HC emissions [17]. Figure 5 shows the variations of HC emissions with various engine loads at constant engine speed of 1500 rpm. The results show that D100 produces 0.009% at 2 Nm, 0.006% at 4 Nm, and 0.006% at 6 Nm of HC percentages from its emission. As seen in the Figure 5, there is a distinct decrement in comparison where B10 drops to 0.007% at 2 Nm, difference of 0.002%. Furthermore, B10 shows the same emission of 0.006%, same as the D100 results at 4 Nm but eventually drops to 0.005% at 6 Nm, a difference of 0.001% from the D100 results. B20 slightly decrease to 0.007% at 2 Nm, difference of 0.002%.

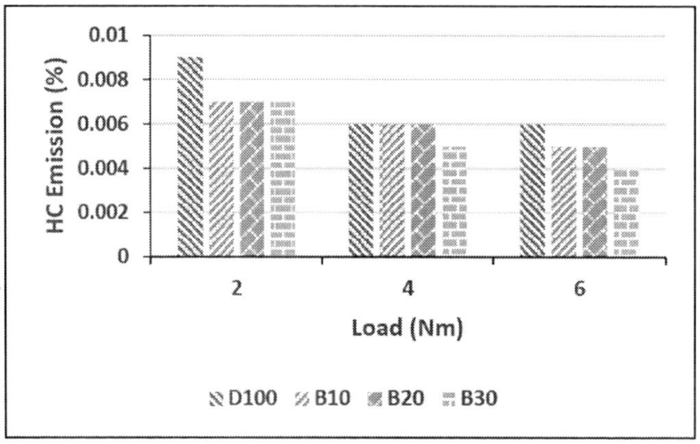

Figure 5. HC emission vs engine load at constant engine speed of 1500 rpm

However, B20 shows the same results with base diesel and B10 at 4Nm, and base diesel at 6 Nm at 0.006%. On the other hand, B30 shows HC emission of 0.007% at 2 Nm, a difference of 0.002% with base diesel, and it expelled 0.005% at 4 Nm, a difference of 0.001% from D100 at 4 Nm. At 6 Nm, B30 drops even lower to 0.004%, a difference of 0.002% compared to using D100. All the results obtained is in agreement with other researchers which proves that fuel blends that contain higher percentage of biodiesel would have lower HC emissions [18, 19]. Lower HC emission also suggests better combustion efficiency as it reaches better overall combustion, producing higher energy output and better emissions. Although the numbers are small but comparing using D100 fuel and other biodiesel blends proves that the HC emissions are lower using higher percentage of biodiesel.

From these experimental investigations, it demonstrates as well that methyl ester blends biodiesel can be used as a fuel in a diesel engine without any major change in the engine [20].

4. Conclusions

This work experimentally assessed the effects of biodiesel ratios of B10, B20, and B30 to the exhaust emission in a four-stroke, vertical single-cylinder, air-cooled diesel engine at a constant engine speed of 1500 rpm and various engine loads of 2 Nm, 4 Nm, and 6 Nm. Based on the results, the following are the conclusions that can be made:

1. Increase of palm oil methyl ester biodiesel content shows a higher nitrogen oxide (NO_x) and carbon dioxide (CO_2) emission in comparison with diesel fuel because of its higher cylinder temperature and high concentration of oxygen.
2. Carbon monoxide (CO) emission decreases with the increment in percentage of biodiesel blends. This is due to better combustion in the chambers cause by the higher presence of oxygen in biodiesel blends.
3. The emission of unburned hydrocarbon (HC) shows a slight decrement using biodiesel blends compared to diesel fuel caused by the higher cetane numbers and better overall combustion.

References

[1] A. Sieminski, "World energy consumption 1990-2040," 2013.

[2] M. Hasan and M. Rahman, "Performance and emission characteristics of biodiesel–diesel blend and environmental and economic impacts of biodiesel production: A review," *Renewable and Sustainable Energy Reviews,* vol. 74, pp. 938-948, 2017.

[3] M. N. Nabi, A. Zare, F. M. Hossain, Z. D. Ristovski, and R. J. Brown, "Reductions in diesel emissions including PM and PN emissions with diesel-biodiesel blends," *Journal of Cleaner Production,* vol. 166, pp. 860-868, 2017.

[4] A. V. Bueno, M. P. B. Pereira, J. V. de Oliveira Pontes, F. M. T. de Luna, and C. L. Cavalcante Jr, "Performance and emissions characteristics of castor oil biodiesel fuel blends," *Applied Thermal Engineering,* vol. 125, pp. 559-566, 2017.

[5] M. Zaharin, N. Abdullah, G. Najafi, H. Sharudin, and T. Yusaf, "Effects of physicochemical properties of biodiesel fuel blends with alcohol on diesel engine performance and exhaust emissions: A review," *Renewable and Sustainable energy reviews,* vol. 79, pp. 475-493, 2017.

[6] D. N. Thoai, S. Photaworn, A. Kumar, K. Prasertsit, and C. Tongurai, "A novel chemical method for determining ester content in biodiesel," *Energy Procedia,* vol. 138, pp. 536-543, 2017.

[7] M. Ventura, W. Deus, J. Silva, L. Andrade, T. Catunda, and S. Lima, "Determination of the biodiesel content in diesel/biodiesel blends by using the near-near-infrared thermal lens spectroscopy," *Fuel,* vol. 212, pp. 309-314, 2018.

[8] J. Xue, T. E. Grift, and A. C. Hansen, "Effect of biodiesel on engine performances and emissions," *Renewable and Sustainable Energy Reviews,* vol. 15, pp. 1098-1116, 2011.

[9] W. Lim, T. Ooi, and H. Hong, "Study On Low Temperature Properties Of Palm Oil Methyl Esters-Petrodiesel Blends," *Journal of Oil Palm Research,* vol. 21, pp. 683-692, 2009.

[10] H. Sharudin, N. R. Abdullah, G. Najafi, R. Mamat, and H. Masjuki, "Investigation of the effects of iso-butanol additives on spark ignition engine fuelled with methanol-gasoline blends," *Applied Thermal Engineering,* vol. 114, pp. 593-600, 2017.

[11] A. K. Agarwal, J. G. Gupta, and A. Dhar, "Potential and challenges for large-scale application of biodiesel in automotive sector," *Progress in Energy and Combustion Science,* vol. 61, pp. 113-149, 2017.

[12] E. Jiaqiang, M. Pham, D. Zhao, Y. Deng, D. Le, W. Zuo, *et al.*, "Effect of different technologies on combustion and emissions of the diesel engine fueled with biodiesel: A review," *Renewable and Sustainable Energy Reviews,* vol. 80, pp. 620-647, 2017.

IOP Conf. Series: Materials Science and Engineering **834** (2020) 012066 doi:10.1088/1757-899X/834/1/012066

[13] E. Buyukkaya, "Effects of biodiesel on a DI diesel engine performance, emission and combustion characteristics," *Fuel,* vol. 89, pp. 3099-3105, 2010.
[14] G. Knothe and L. F. Razon, "Biodiesel fuels," *Progress in Energy and Combustion Science,* vol. 58, pp. 36-59, 2017.
[15] H. Bayındır, M. Z. Işık, Z. Argunhan, H. L. Yücel, and H. Aydın, "Combustion, performance and emissions of a diesel power generator fueled with biodiesel-kerosene and biodiesel-kerosene-diesel blends," *Energy,* vol. 123, pp. 241-251, 2017.
[16] G. Najafi, "Diesel engine combustion characteristics using nano-particles in biodiesel-diesel blends," *Fuel,* vol. 212, pp. 668-678, 2018.
[17] V. Ganesan, *Internal Combustion Engines*: McGraw-Hill Education, 2012.
[18] H. Sharudin, N. R. Abdullah, A. Mamat, M. Zaharin, S. Alam, and M. A. Asiah, "Experimental Investigation on Physicochemical Properties of Iso-Butanol Additive in Methanol-Gasoline Blends," 2017.
[19] N.R. Abdullah, Z. Michael, AR Asiah, AJ Helmisyah, S. Buang, Effects of Palm Oil Methyl Ester (POME) on fuel consumption and exhaust emissions of diesel engine operating with blended fuel (fossil fuel+ Jatropha oil methyl ester (JOME)), Jurnal Teknologi, vol. 76(5), 2015.
[20] G. Goga, B.S. Chauhan, S.K. Mahla, H.M. Cho, Performance and Emission Characteristics of Diesel Engine Fueled with Rice Bran Biodiesel and n-butanol, Energy Reports 5 (2019), p78-83.

Effect of Acid Treatment on Extraction of Silica from Cogon Grass by Using $C_6H_8O_7$ and HCL Acid

N A Zarib*[1], S A Abdullah[1] and N A Jusri[2]

[1] Faculty of Mechanical Engineering, University Teknologi MARA (UiTM), 40450 Shah Alam, Selangor,Malaysia
[2] Faculty of Chemical Engineering, University Teknologi MARA (UiTM), 40450 Shah Alam, Selangor,Malaysia

*eiqasyahirah@gmail.com

Abstract. Cogon grass is one type of agricultural wastes that can easily be found around the world. A cogon grass contains with relatively low lignin (5.67%), ash (8.24%) and acceptable holo-cellulose and cellulose solubility of 64.9% and 37.1% respectively. The focus of this paper is to analyse the effect of acid treatment of silica extraction of cogon grass due to silica as a raw material that can be apply in industrial. Parameters such as concentration of acid and weight percentage used in cogon grass during organic and inorganic leaching method were also compared. In this paper, citric acid ($C_6H_8O_7$) and hydrochloric acid was used for production of silica from cogon grass. Samples were characterized using Scanning Electron microscopy (SEM), X-ray diffraction (XRD) and Thermal gravimetric Analysis (TGA). The result indicates that highest percentage silica of cogon grass after leaching was on 1.0M via HCL acid leaching and has an amorphous structure.

1. Background of study

Cogon grass known as Imperata cylindrica was popularized from Japan to the United States in 1912, then purposely introduced as a ravage crop in 1920 [1]. However, the plant was unsuccessful as a ravage crop due to high silica content on cogon grass it selves. Cogon grass is a C_4, rhizome producing perennial plant that can reach heights of three meters, but typically grows to heights of 1.2 meters. Cogon grass spreads lusty, through seeds, and not lusty, through rhizomes [2]. Many author mostly do not mentioned detailed in literature that silica content of cogon grass has not been recognized yet. However, other researcher stated that cogon grasses are existing a lot of silica impurities and can be used to produce high purity amorphous silica due to smooth blades were detected to be highly deposited at the boundaries of cogon grass [3]. In addition, high-purity silica may be obtained by removing minerals in the biomass through acid leaching treatment [4]. Cogon grass shows the lowest amount of holo-cellulose which is 64.9% than C.tataria (70.5%) and Palmyra palm fruit (68.5%). Commonly, holo-cellulose is the total content of hemicellulose and cellulose in dried materials and usually should accounts for 65 to 75% of the plant dry weight in order to be latent alternative fibre [5].

Chemical analyses were executed for each property under specific methods and normal values of cogon grass were summarized in Table 1. The measured data characterized by cogon grass with rather ash (8.24%), low lignin (5.67%) and acceptable holo-cellulose and cellulose solubility of 64.9% and 37.1% respectively. Chemically, holo-cellulose and cellulose are important parameters in measuring the suitability of material for paper production. Imperata cylindrical (64.9%) shows the lowest amount of holo-cellulose than Palmyra palm fruit (68.5%) and C.tataria (70.5%). Generally, holo-cellulose is the

Content from this work may be used under the terms of the Creative Commons Attribution 3.0 licence. Any further distribution of this work must maintain attribution to the author(s) and the title of the work, journal citation and DOI.
Published under licence by IOP Publishing Ltd

total content of cellulose and hemicellulose in dried materials and usually should accounts for 65 to 75% of the plant dry weight in order to be potential alternative fibre [6]. Therefore, the amount of holo-cellulose obtained in cogon grass is acceptable to be applied in paper production.

Table 1. Comparison of chemical composition of Cogon grass by published non-wood (Sridach, 2010)

Property (%w/w e.d)	Species			
	L.cylindrica (Kassim,2015)	C,tataria (Tutus,2010)	Switch grass (Madakazde,2010)	Palmyra grass (Sridach,2010)
Holo-cellulose	64.9	70.5	n.a	68.5
Cellulose	37.1	40.1	41.2	37.0
Lignin	5.67	24.5	23.9	18.5
Ash	8.24	7.83	4.83	0/64

2. Materials and method

Cogon grass was provided by the Universiti Teknologi MARA (UiTM) laboratory. Silica is obtained from the cogon grass in this research work. Firstly, 20 grams of cogon grass was prescribed become powder through the process of milling in 30 minutes to produce fine powder. Next, cogon grass was ground via sieving machine for 10 minutes to turn into fine powder before the acid leaching process. Then, cogon grass powder with a mean particle size of 45 μm was put into a 500 ml of HCL acid solution into the beaker. The concentrations of 1.0M were used for this research at different acid used. A constant temperature of 60°C was set up for hot plate magnetic stirrer and the beaker was placed on it. The water rinsing treatment of the cogon grass was carried out with the distilled water at room temperature after the acid leaching process to eliminate the both acid content from husk. The samples were dried for 60 minutes at 110°C in the furnace. Silica powder gained from the process of extraction underwent characterization technique. The acid leaching process then replace with HCL to citric acid ($C_6H_8O_7$)

2.1 Morphology Surface Analysis

The Hitachi SU3500 instrument was used. A diameter of 15mm each samples coated with aluminium in an aluminium sputter coater for 60 seconds at 15mA current output. The aluminium coating was necessary to ensure a conducting surface was obtained for electron bombardment and characterization. An elimination of electron charging effects the SEM was operated at 3kV and a working distance of 15mm. The selected areas of interest were focused and micrographs were taken. Software of Smart Insight EDAX was used for EDX analysis.

2.2 Thermal gravimetric analysis (TGA)

Thermal analyses were performed using thermogravimetric differential thermal analysis TG=DSC equipment (Mettler Toledo/ (TGA/DSC1) apparatus in alumina crucibles with ambient air supplement of 150 ml per minute and heating rate was 10°Cmin^{-1}. Nitrogen gas was used as ambient air supplement. The samples used had 20 mg under air flux and were heated from room temperature until 800°C.

3. Discussion

3.1 Morphology Surface Analysis

The morphology surface of the particle size and the agglomeration of nature at the ground samples were examined by Scanning Electron Microscope (SEM). In order to find out the effect of chemical treatment of silica and carbon. Energy dispersive x-ray spectroscopy (EDX) act as determination of the elements on images at different magnification from SEM. The silica and organic impurities distribution of untreated rice husk particle were analysed by using EDX. Figure 1 shows the image of (a) untreated cogon grass powder, (b) 1.0M of treated cogon grass via citric leaching treatment and (c) 1.0M of treated cogon grass via HCL leaching treatment respectively at magnification of 2000x.

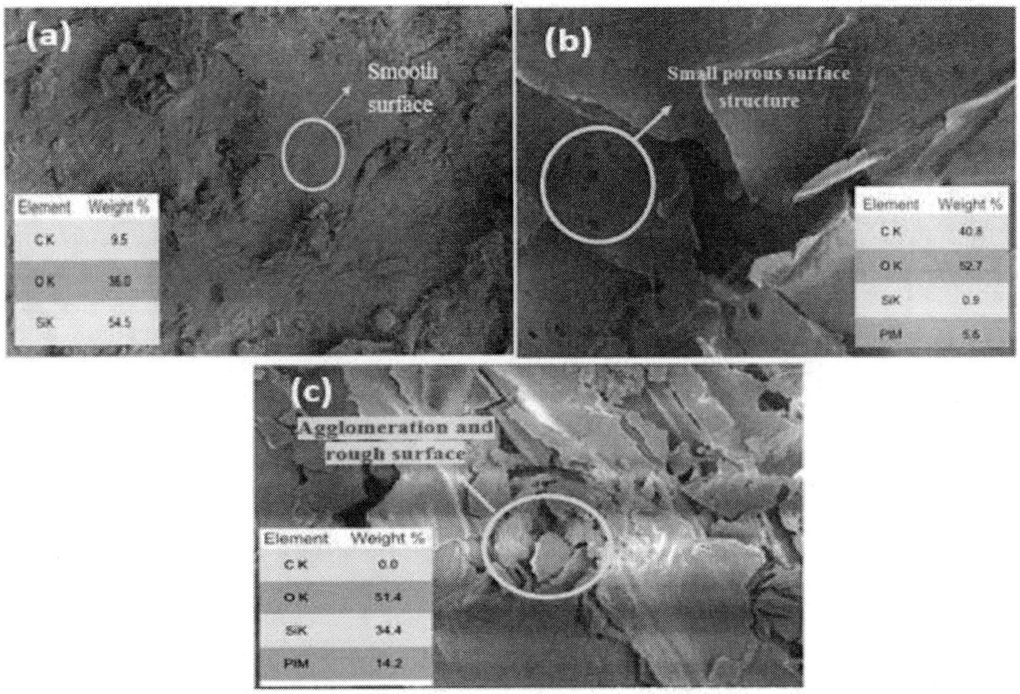

Figure 1: SEM images and EDX analysis of (a) untreated Cogon grass (b) 1.0M of treated cogon grass via citric leaching treatment and (c) 1.0M of treated cogon grass via HCL leaching treatment respectively at magnification of 2000x.

Figure 1(a) illustrates the SEM image of untreated Cogon grass at magnification of 2000x respectively. It shows that the particles have uneven porous surface (irregular shape), uniform and unblemished with a smooth surface. This is due to the sample is not treated with any chemical or any treatment. Energy dispersive X-ray spectroscopy (EDX) test also handled to make quantitative chemical analysis of Cogon at specific spot. It was detected that SiO_2 is the most concentrated in weight percent which is 49.6 wt % than other impurities. The inorganic impurities, mainly carbon (C) and platinum was also still present in cogon and concentrated in the surface. Platinum was present in SEM image due to coating process before. Figure 1(b) observed that cogon grass particles has an uneven small porous surface that giving the particles a disorganized cylindrical shape. The small porous appear when citric acid takes an action to remove metallic impurities and silica extract during leaching treatment. The weight percentage of treated cogon for citric acid increase compared with untreated rice husk. Citric acid is an organic and weak acid that insufficient or incapable to extract more silica or remove impurities metallic that resulting the metallic elements such as carbon and platinum have high weigh percent rather than silica. Figure 1(c) show SEM micrographs of the cellulose fibres exposed an agglomerated aspect of elongated fibres with thin extremities and persisted along the fibre surfaces, giving them a rough aspect HCL resulted in more degradation. At this concentration, the particle size of the sample was increased and rough surface consist of non-cellulosic materials was added. Some cracked fibre are also appear due to extraction process and it could also reduce the feature and strength properties of particles. Besides, weight percentage of silica for 1.0M cogon grass via HCL acid was 85.8% higher than untreated cogon grass and 1.0M of citric acid. It can be conclude that the morphology surface and weight percentages silica (SiO_2) of 1.0M cogon grass is higher by using HCL as acid leaching treatment. Jatinder, 2014, stated in his research that the surface of cogon grass is uneven with porous surface, similar to surface morphology of cogon grass in this research work [7].

3.2 Thermal gravimetric analysis

The thermogravimetric analysis of 1.0M cogon grass at 90 minutes samples of different acid used for leaching treatment in inert atmosphere at 10°C/ min is observed in Figure 2. The initial descending slope from the start of the curve to about 190°C correlated to loss of hygroscopic water at phase I (drying or evaporation. Hygroscopic is a substance is able to absorb or adsorb water from its surroundings. At phase II, mass loss from temperature of 190°C to 390°C can be isolate into two parts for 1.0M cogon grass via citric acid leaching treatment. First peak of the mass loss in the range of 190°C to 338°C was due to the volatilization and thermal decomposition of the organic part in cogon grass, while the mass loss of from 338°C to 390°C was due to oxidation process. Oxidation is the loss of electrons during a reaction by an ion, molecules or atom. Hence, thermal decomposition of untreated cogon grass starts at about 284°C which is quite late compared to acid-treated cogon grass which at range of 270°C.Besides,the HCL acid treated cogon grass sustained a greater mass loss than citric acid due to HCL acid is an inorganic acid that can remove metallic constituent. TGA curve also showed that decomposition of samples occur in stage of lignin decomposition (phase III) where the peak stable from 390°C to 800°C for all untreated and treated cogon grass respectively. Moreover, an increase in heating rate caused earlier consideration of thermal degradation which basically resulted in an earlier completion of mass loss phenomenon.

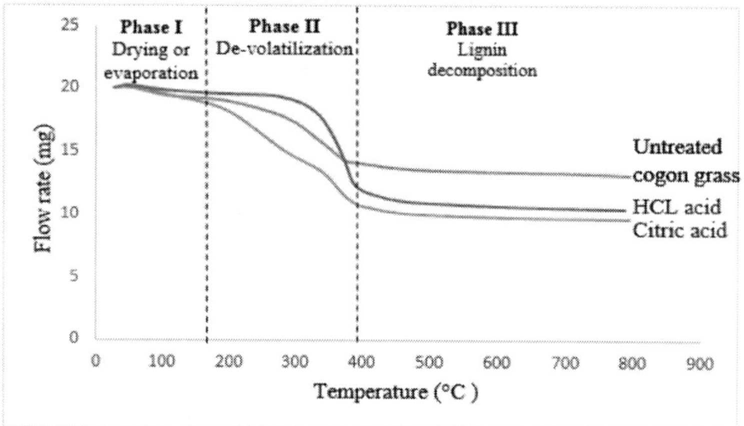

Figure 2: TGA curve of Untreated Cogon grass and 1.0M Treated cogon grass at 90 minutes of different acid leaching treatment

4. Conclusion

The properties of silica dioxide (SiO_2) formed via acid leaching treatment process has been deliberated. The samples were prepared and blend with 1.0M of $C_6H_8O_7$ for 60 minutes. 3 samples were evaluated using SEM and EDX through characterization techniques and it is recognized that highest silica content of 1.0M of cogon grass increased from 49.6% to 53.6% by citric acid solution while for HCL acid leaching treatment increase immensely from 49.6 % to 85.8%. The silica extraction and its morphological research are done by using SEM. Therefore, untreated cogon grass showed that SiO_2 enclose of smooth surface while for 1.0 M cogon grass of citric acid showed the small porous surface structure and for 1.0 M cogon grass of HCL acid show the particle is an agglomeration and rough surface. As a conclusion for TGA analysis can be summarize that between these two different acid used as a leaching treatments which is citric acid and HCL acid, the most effective acid used to this acid leaching treatment was HCL acid due to HCL is a strong acid compared with citric acid that can extract more silica content and remove others impurities. Hence, by using acid as a leaching treatment and increase the concentration, the degradation of thermal decomposition cogon grass is faster. Furthermore, through reuse techniques from waste materials; it is possible in improvise the environment quality and lower cost in the production of silica by using the waste materials.

Acknowledgement
Authors would like to show gratitude Universiti Teknologi MARA (UiTM) for financial support [Project Grant No: 600-IRMI/PERDANA 5/3 BESTARI (051/2018)].

Reference

[1] Kien W, Rozita Y and Abdul A. 2014. Characterisation of Bio-Silica Synthesised from Cogon Grass (Imperata cylindrica), *J. Powder Technology*. 254: 206-213.
[2] Yu C.C, Chi C.T and Chang H.C. 2012. Characterization of 21 Microsatellite Markers from Cogon Grass, Imperata Cylindrica (poaceae), and a Weed Species Distributed Worldwide, *J. American of Botany*. 428-430.
[3] Jhalique J.R.F and Ernesto J.D.R. 2013. Optimization of Pre-treatment and Enzymatic Saccharification of Cogon Grass Prior Ethanol Production, *J. Chemical and Molecular Engineering*.7, 5.
[4] Hagan, D.L., Jose, S., Lin, C. 2013. Allelopathic exudates of *I. cylindrical* (*Imperata cylindrica*): Implications for the Performance of Native Pine Savanna Plant Species in the South Eastern, *US. Journal of Chemical Ecology*. 39: 312-322.
[5] Holzmueller, E.J., Jose, S. 2011. Invasion Success of *I.* Cylindrical, an Alien C_4 Perennial Grass, in the South Eastern United States, *J. Exploration of the Ecological Basis Biological Invasions* 13: 435-42.
[6] Sridach, W. 2010. The Environmentally Benign Pulping Process of Non-Wood Fibers. *Journal of Science and Technology*. 17, 105-123.
[7] Jatinder, S. A., Stephen., F. E. & Nancy, J. L. 2014. Pushing towards Cogongrass (Imperata cylindrica) Patch Erication: The Influence of Herbicide Treatment and Application Timing on Cogongrass Rhizome Elimination. *Invasive Plant Science and Management*, 7, 398-407.

IOP Conf. Series: Materials Science and Engineering **834** (2020) 012068 doi:10.1088/1757-899X/834/1/012068

Thermal cutting analysis on grain size distribution using probabilistic FEM

Mohd Shahar Sulaiman[1], Yupiter HP Manurung[1], Marcel Graf[2], Mohamad Syakir AG[1] and Muhd Faiz M[1]

[1]Faculty of Mechanical Engineering, UiTM Shah Alam, Selangor, Malaysia
[2]Professorship of Virtual Production Engineering, Chemnitz University of Technology, Chemnitz, Germany

*Corresponding author E-mail: yupiter.manurung@salam.edu.my

Abstract. This paper investigates the capability of probabilistic FEM to predict grain size distributions due to thermal cutting process. The application of transient heat source causes non-uniform temperature distribution across the parent metal that can lead to non-uniform expansion and contraction during heating and cooling cycle. This phenomenon will induce thermal stresses to the workpiece that can subsequently lead to unwanted cutting deformation. Therefore, prediction of temperature distribution is important in order to control the amount of heat required in the cutting process. The simulation was computed through probabilistic FEM using Monte Carlo method based on non-linear thermo-elastic-plastic numerical analysis. The probabilistic FEM was carried out by varying the input power. In this study, the simulation method had been executed by using FEM software MSC MARC. The simulation analysis was also executed using customized material input module in order to take the grain growth into consideration during the cutting process. The additional subroutine of grain growth was introduced and implemented in order to predict the grain size distribution. The material used for the simulation was stainless steel 316L with the thickness of 2 mm. Based on the results obtained, it was found out that slight differences of the results were achieved between deterministic and probabilistic methods. The small observable differences occurred due to the probabilistic method was only executed to fluctuate the input power, while the other process parameters were still unchanged. Nevertheless, the Monte Carlo method was successfully integrated into the normal simulation which then transforming it into probabilistic analysis. Thus, through probabilistic method, reliability on prediction could be increased in which the prediction would be closer to reality.

1. Introduction

A heat treatment process, where a metal plate is subjected to a large amount of concentrated heat and the applied part of the metal plate melts or vaporizes away is known as thermal cutting. Thermal cutting is an exothermal process, which means that the burning process maintains itself with less heat required. There are several types of thermal cutting processes such as flame cutting, plasma arc cutting and laser cutting [1]. Plasma arc cutting (PAC) is a cutting method that uses extremely hot plasma gas that melts and partially vaporizes workpiece material. This plasma gas jet is formed by an arc and inert gas flow. Plasma arc is concentrated with a nozzle to produce more precise and even hotter arc, that operates in the region of 10 000 – 14 000 °C. Molten material is removed by the high velocity gas jet, like in oxyfuel gas cutting. Plasma arc is generated by heating gas with an arc. This gas then becomes partially ionized and thus, being able to conduct electricity. Heating the gas rapidly as it travels along

Content from this work may be used under the terms of the Creative Commons Attribution 3.0 licence. Any further distribution of this work must maintain attribution to the author(s) and the title of the work, journal citation and DOI.
Published under licence by IOP Publishing Ltd

the arc causes it to expand. Then the gas is accelerated through the nozzle towards the workpiece. Gases employed in plasma cutting comprise nitrogen, argon, oxygen, air and mixtures of nitrogen/hydrogen and argon/hydrogen [2-3].

Laser cutting is a non-contact thermal cutting process which is a state-of-the-art technology that eliminates such effects as machine vibration, mechanically induced thermal damage and tool wear [4]. It is the most extensively used for producing complex shapes and different geometries with narrow kerf in almost all categories of materials such as metals, non-metals, ceramics and composites. The cutting capability of laser primarily depends on the thermal and optical properties of the material instead of the mechanical properties [5]. The focused laser beam locally melts the material to be cut and produces a kerf when the molten material is blown away with the aid of assist gas, which flows coaxially with the laser beam.

Numerous numerical methods have been applied to perform thermal analysis. In [6], prediction of temperature distribution due to thermal cutting had been carried out on Inconel 718. The finite element analysis was computed based on thermo-mechanical analysis using temperature-dependent material properties. The results showed that the predicted temperature distribution well agreed to the experimental measurement. It was found out that the temperature trend in the specimen with 1 mm thickness was slightly lower than that of specimen with 2 mm thickness. Referring to the explanation in [7], arc-welding process is based on the same principles as arc-cutting process. A welding process normally adds up filler material to the base metal, whereas some materials are removed from the base metal in cutting process. RN Lidam et al. [8] conducted a simulation study on multipassed welding distortion of combined joint types induced by gas metal arc welding (GMAW) process. The simulation had been executed using SYSWELD based on thermo-mechanical-metallurgical analysis. The plasticity criterion was calculated based on isotropic strain hardening owing to non-cyclic loading in which the mechanical calculations were computed according to metallurgical history primarily following the constitutive equation proposed by Leblond. The outcomes uncovered that the angular distortions predicted using 2D and 3D analyses showed reasonable agreement. It was also found out that 2D analysis required shorter computational time as compared to 3D analysis. A research on residual stress field induced in pipe girth weld of 316 stainless steel had been carried out by J. G. Mullins et al. [9]. The residual stresses were simulated based on thermo-elastic-plastic mechanical analysis. Three different hardening models had been employed in the finite element analysis comprising isotropic, kinematic and mixed hardening models. The isotropic, kinematic and mixed hardening models were computed according to flow curves at different temperatures, bilinear-Ziegler rule and Lemaitre-Chaboche model respectively. By comparing the simulation results with the experimental verifications, the results revealed that the axial stresses were well predicted by means of isotropic hardening model. In addition, the hoop stresses calculated via isotropic hardening model were also in good agreement with the experimental measurements. The best predictions of hoop stresses were achieved through mixed hardening model. However, the predictions through kinematic hardening model had underestimated both magnitudes of stresses.

Even though all criteria have been taken into consideration in numerical simulation, but generally the FEA is deterministic by nature and hence is limited to describe typical characteristics of a system [10]. Different sources of uncertainty come up in the study of complex phenomena such as human error, dynamic loading, inherent randomness of the material and lack of data. The classic FEM has been integrated with other methodologies to develop a novel type of analysis in order to probe the systems with random variations and/or uncertainty in parameters. It has been known as a stochastic finite element method (SFEM) [11], a random finite element method (RFEM) [12] as well as a probabilistic finite element method (PFEM) [13]. Furthermore, to characterize the stochastic nature into a system, random fields are implemented in the classic FEM to describe and establish different stochastic scenarios. The effects of the random variations are evaluated by computing the statistical information of the response variables and evaluating the probability of an outcome of the system. A research on optimization of cutting parameters using probabilistic FEM had been performed by Miloš Madić et al. [14]. In this research, artificial neural network (ANN) based mathematical model using Monte Carlo method was established in order to connect the parameters in CO_2 laser cutting of 304 stainless steel consisting of laser power, cutting speed, assist gas pressure and focus position, as well as kerf taper angle. Monte Carlo method was an approach which manipulating random numbers in simulation algorithm. The Monte Carlo method could be regarded as a general mathematical tool for solving various problems. As compared to experiment, statistical results showed that the kerf taper angle could be calculated through ANN model within good accuracy. The optimum laser cutting parameters which minimized the kerf taper angle could be defined by implementing the Monte Carlo method. It was observed that concentrating the laser beam on around 2/3 of material thickness through low assist gas pressure (9 bar) at combination of high cutting speed (3 m/min) and low laser power (1.6 kW) resulted in acceptable kerf taper angle. By applying the Monte Carlo simulation, the optimization was resolved via the generation of random numbers ri,j which distributed uniformly in the range between 0 and 1. In order to meet the limitations of the laser cutting parameter values, the random numbers ri,j were manipulated to produce random numbers qi,j which uniformly distributed into the range of interest for each of the laser cutting parameter $[q_i^{min}, q_i^{max}]$. This was achieved using the following equation:

$$q_{i,j} = q_i^{min} + r_{i,j}.\left(q_i^{max} - q_i^{min}\right) \tag{1}$$

A lot of researches on grain growth predictions have been conducted beforehand due to welding processes [15,16,17]. In [18], the effects of cooling rates on grain growth had been observed in stainless steel material. The cooling rates were predicted through phenomenological model and finite element analysis, while the verification had been performed by measuring the secondary dendrite arm spacing (SDAS) which obtained from the microstructure of welded samples. The SDAS was measured using a scanning electron microscopy (SEM). Good agreement had been achieved between predicted results and experimental verifications. It could be seen that a reduction in cooling rate would lead to increasing grain size. Referring to research in [19], the major factors that influencing the grain growth in heat affected zone (HAZ) of stainless steel consisting of duration at elevated temperature, peak temperature and cooling rate. It was also found out that the grain size at the region experiencing higher peak temperature would have greater size as compared to the grain at the region experiencing lower peak temperature. The grain growth throughout the welding process could be occurring from the temperature of 900°C to the peak temperature during the heating cycle and continuing to 900°C during the cooling cycle. Besides, the grain growth in stainless steel material was influenced by thermal cycle, migration of grain boundary as well as particle precipitation.

2. Simulation Method and Procedure

2.1. Geometrical Modelling

A schematic illustration of solid FE model used in the simulation is displayed in Figure 1. The dimensions of the model are 100mm x 100mm x 2mm which are referring to the width, length and thickness of the model respectively.

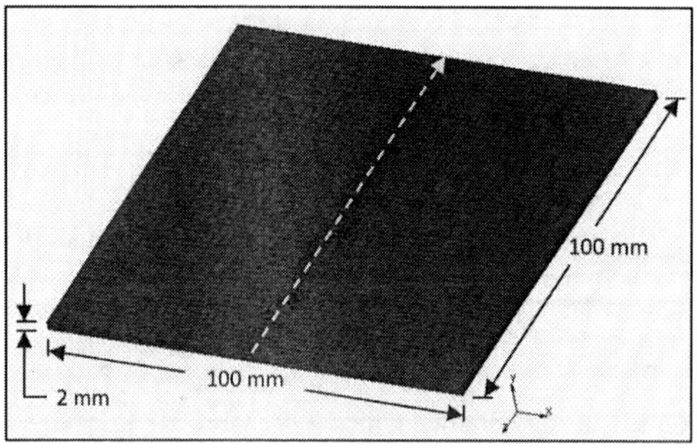

Figure 1. Geometrical model utilized for simulation

IOP Conf. Series: Materials Science and Engineering **834** (2020) 012068 doi:10.1088/1757-899X/834/1/012068

The cutting trajectory is represented by the yellow arrow in which the direction of cutting is determined by the arrow direction. The type of element used for this model is three-dimensional hexahedral element.

2.2. Material Modelling

In thermal cutting simulation, stainless steel 316 material had been used for analysing the temperature distribution. In simulation, all the computations were executed based on analysis using temperature-dependent material properties as shown in Figure 2. During cutting process, the workpiece was exposing to high working temperature which could be beyond its melting temperature especially at the focal cutting region and the heat would also be spreading over the heat affected zone (HAZ) area and dissipating into the base metal and the environment. Thus, due to this phenomenon, the magnitudes of values of material properties would change according to the temperature variations. Hence, the temperature-dependent material properties were very important since they would reflect the real situations that occurred to the workpiece during the cutting process.

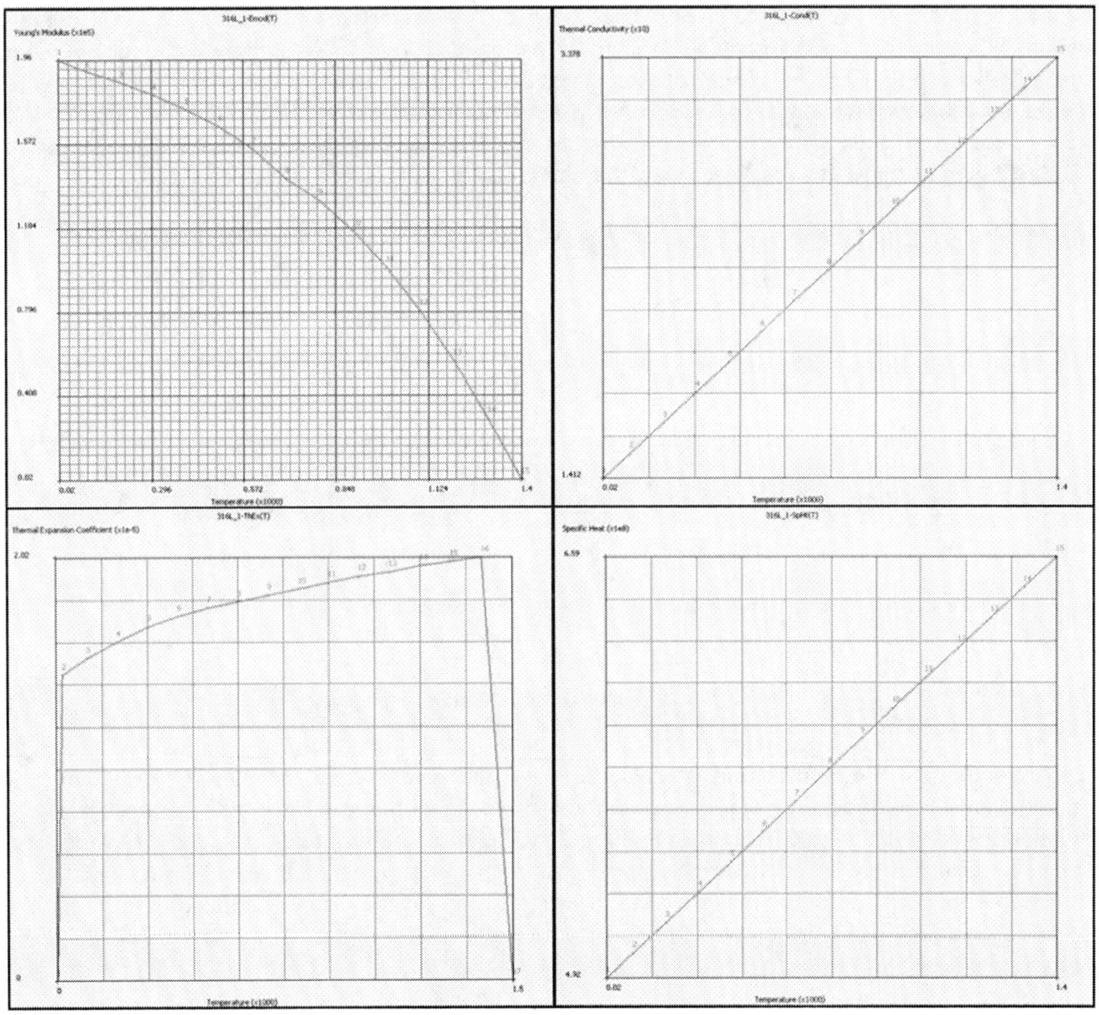

Figure 2. Temperature-dependent mechanical and physical properties

If the simulation executed using constant material properties, it will merely consider the material behaviour at room temperature which in turn can lead to incorrect prediction. However, the other two material properties which are mass density and Poisson's ratio were assumed to be 7850 kg/m3 and 0.3 respectively that were not dependent on temperatures.

In this study, the temperature analysis was carried out using non-linear thermo-elastic-plastic analysis in which mechanical analysis was also taken into consideration. Based on mechanical analysis, the total deformation was computed through the combination of elastic and plastic strains as expressed by Equation (2).

$$\varepsilon^{total} = \varepsilon^{e} + \varepsilon^{p} \quad (2)$$

When the stress in the specimen was below the yield stress of the material, the material behaves elastically and the stress in the specimen was proportional to the strain. However, when the stress in the specimen was greater than the yield stress, the material was no longer exhibiting elastic behaviour and thus, the stress-strain relationship became non-linear. The yield stress of a material is a measured stress level that separates the elastic and plastic behaviour of the material. The magnitudes of the yield stresses were generally obtained from the mechanical properties as defined through the Modulus of Elasticity and flow curve. However, the stresses in a structure are usually multiaxial. A measurement of yielding for the multiaxial state of stress is known as yield condition. In this analysis, for the application of isotropic material, von Mises criterion was selected to describe the yield condition which is commonly used for ductile materials particularly due to its suitability in characterizing metallic behaviour. The von Mises criterion states that yield occurs when the effective (or equivalent) stress, σ equals the yield stress, σ_y. The von Mises criterion can be described as expressed by Equation (3), where σ_1, σ_2, and σ_3 are the principal stresses that will be coupled with the strain hardening rule. Figure 3 demonstrates the von Mises yield surface in three-dimensional stress space.

$$\bar{\sigma} = [(\sigma_1 - \sigma_2)^2 + (\sigma_2 - \sigma_3)^2 + (\sigma_3 - \sigma_1)^2]^{1/2}/\sqrt{2} \quad (3)$$

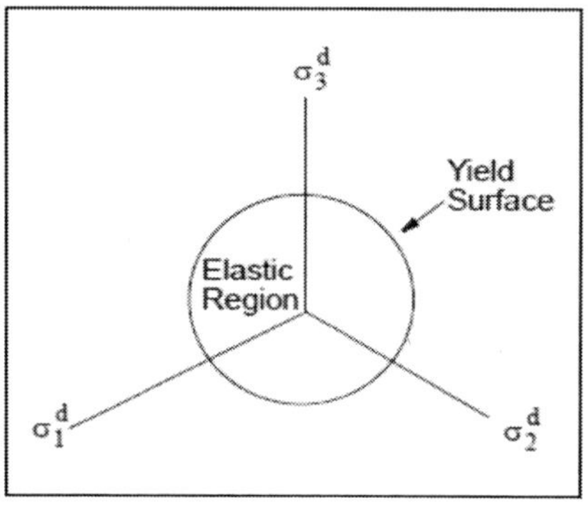

Figure 3. Three-dimensional stress space

In order to take hardening effects into account, the computational analysis was executed corresponding to the information on material data that characterize the plastic behaviour of the material in the form of flow curves. In addition, these quantities can vary with parameters such as temperature and strain rate. Referring to Figure 4, the slope of the total stress versus plastic strain curve is defined as the workhardening slope (H) of the material in which the workhardening slope is a function of plastic strain. The material data of flow stresses for the material utilized in this simulation were as exhibited in Figure 5. When it came to a state which was not defined by the available flow curves, the decision was being made through interpolation.

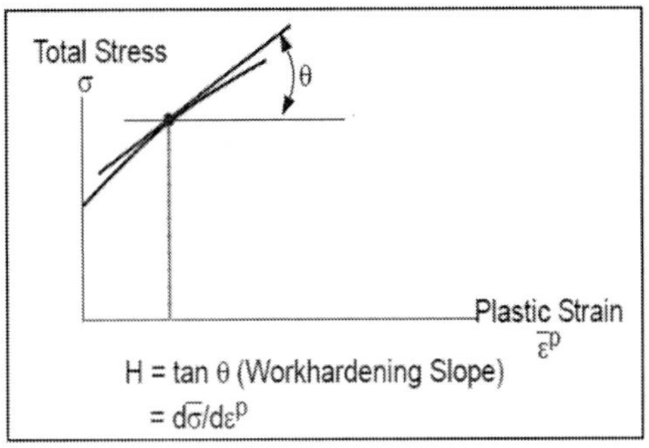

Figure 4. Definition of workhardening slope

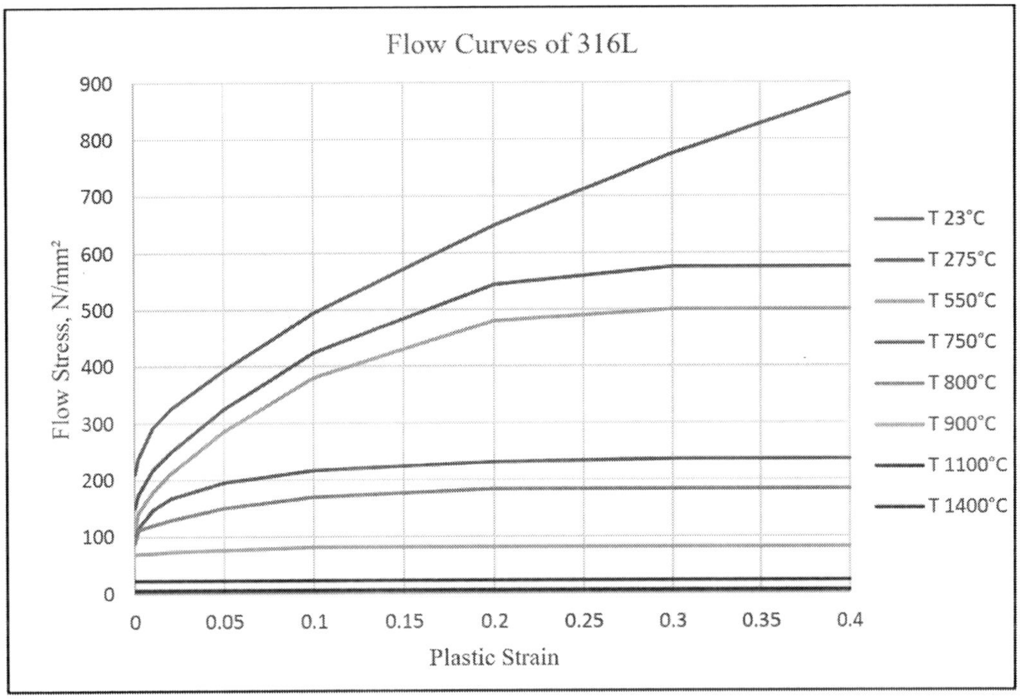

Figure 5. Flow curves of 316L at the same strain rate of 0.01 1/s

2.3. Heat Source Modelling

In cutting simulation, one of the most important steps is to model the heat source that can reflect the real process. Suitable heat source model must be selected in accordance with the thermal cutting process. Hence, for the thermal cutting process, a volumetric conical heat source model was employed to replicate the real heat source. The 3D conical heat source model used in the simulation was defined by C. S. Wu et al. [20] which was obtained through derivation from heat intensity distribution equation of thermal energy conservation. The Equation (4) was the model equation produced as a result of derivation which could characterize the through thickness decay of heat intensity distribution of the thermal arc. The subsequent Equation (5) was part of the preceding equation which determined the decrease of maximum heat source intensity throughout the thickness of working material. In these equations, Q_o represents the power input, r denotes the radial coordinate, z refers to the z-coordinate and r_o defines the distribution parameter of the heat intensity. Whereas, z_e and r_e represent the z-coordinate and radius of the top surface of the heat source model respectively. Similarly, the z-coordinate and radius of the heat source model at the bottom surface are defined by z_i and r_i correspondingly. The configuration of the conical heat source model is illustrated in Figure 6.

$$q(r,z) = \frac{9Q_o e^3}{\pi(e^3-1)} \cdot \frac{1}{(z_e-z_i)(r_e^2+r_e r_i+r_i^2)} \cdot \exp\left(-\frac{3r^2}{r_o^2}\right) \quad (4)$$

$$r_o(z) = r_i + (r_e - r_i) \cdot \frac{z-z_i}{z_e-z_i} \quad (5)$$

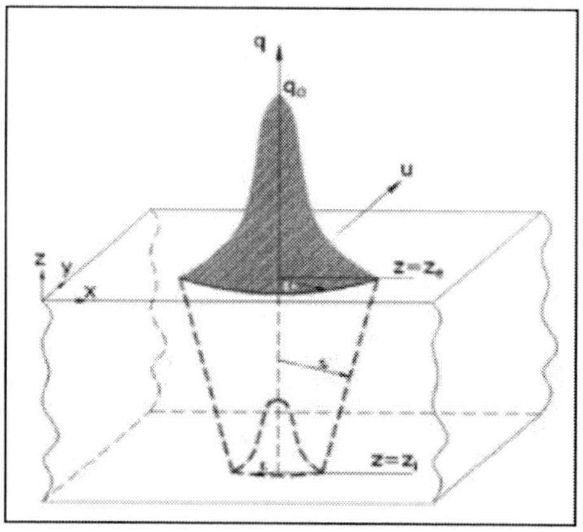

Figure 6. Conical heat source configuration

2.4. Grain Growth Modelling

When metals and alloys are subjected to thermal process, the grain growth can occur as a result of induced kinetic energy [21]. Normally, the grain growth is also involving the effect of impurity and precipitation. Precipitation is diffusion process that is dependent on time and temperature. In, thermal cutting as well as welding processes, since the heating cycle is normally too fast, the grain growth during heating could be ignored. Thus, the grain growth is considered and computed during the cooling cycle. The following general equation can be used to calculate the grain growth:

$$\frac{\partial \bar{g}}{\partial t} = M_o^* \exp\left[-\frac{Q_{app}}{RT(t)}\right]\left[\frac{1}{\bar{g}} - \frac{1}{\bar{g}_{lim}}\right]^{\left(\frac{1}{n}-1\right)} \quad (6)$$

Where,

M_o^* = Modified kinetic constant ($\mu m^{1/n} s^{-1}$)
Q_{app} = Activation energy ($kJ\ mol^{-1}$)
R = Gas constant (8.314 J $K^{-1} mol^{-1}$)
T = Absolute temperature (K)
$\bar{g}$ = Average/initial grain size (μm)
$\bar{g}_{lim}$ = Limiting grain size (μm)
n = Time exponent

In this study, during the cooling cycle, two mechanisms have been considered which are free grain growth and grain growth in the presence of growing precipitates. The free grain growth is considered due to there is no precipitates at high temperature near to solidus temperature, meanwhile below the temperature of approximately 1200°C to 900°C, the presence of precipitates especially Chromium Carbides in austenitic stainless steel are taken into account.

2.5. Boundary Conditions

There were two types of boundary conditions involved in the cutting simulation consisting of thermal and mechanical boundary conditions. The thermal boundary conditions were defined by heat input of volumetric heat flux as described previously, as well as heat losses due to convection and radiation as expressed by Equation (7) and Equation (8) respectively.

$$q_{conv} = h_{conv}(T_s - T_\infty) \tag{7}$$

$$q_{rad} = \sigma\varepsilon(T_s^4 - T_\infty^4) \tag{8}$$

In the above equations, h_{conv} is convective heat transfer coefficient, σ refers to Stefan Boltzmann's constant and ε represents thermal emissivity, while T_s and T_∞ are surface temperature and ambient temperature respectively. Whereas the mechanical boundary condition was defined by fixation on the model nodes to prevent body motion.

2.6. Monte Carlo Simulation

Monte Carlo simulation is a method that takes variability of the inputs into account. The simulation analysis could involve numerous of recalculations before it is complete. Through this method, it could be able to produce distributions of possible outcome values. In this study, to perform the probabilistic analysis, the heat input power of the heat source was fluctuated within a certain range while the other proses parameters had been maintained. To ensure it happened, the input power would be varied by using Gaussian random variables. Thus, in order to support this process, a source of random numbers should be produced. There are a lot of random number generators available, but the most extensively used is linear congruential generator (LCG) which is defined by the following formula [22,23]:

$$x_{i+1} = (a \cdot x_i + c) mod\, M, \qquad i \geq 0 \tag{9}$$

where M is the modulus, a and c are the multiplier and increment respectively while x_o is the seed or start value. Through this method, a sequence of pseudo-random integer numbers $\{x_i\}$ could be generated. Furthermore, in order to produce a random number X_i in the range between 0 and 1, the random number x_i generated previously should be divided by M as shown below:

$$X_i = \frac{x_i}{M}, \qquad i \geq 1 \tag{10}$$

However, the random number X_i produced previously was uniformly distributed, U (0,1). For this typical simulation, the Gaussian or normal random variable, N (0,1) is preferable. Hence, Box-Muller method [24,25,26] could be employed as a tool to transform the uniform random variable into the Gaussian random variable. Through this method, the transformation could be implemented by converting a pair of uniform random variables (X_1, X_2) into a pair of Gaussian random variables (Y_1, Y_2) via the subsequent equations:

$$Y_1 = \sqrt{-2ln(X_1)}cos(2\pi X_2) \tag{11}$$

$$Y_2 = \sqrt{-2ln(X_1)}sin(2\pi X_2) \tag{12}$$

For verification of the probabilistic method, the variation of the input power generated by the algorithm had been computed using general Fortran 77 program. This programming language was employed due to MSC MARC is built based on the same platform. In the cutting simulation, it was executed by using 200 values of different input powers for the variation instead of a single constant value. Therefore, the Fortran program was tested to generate a sequence of 200 random numbers. The

IOP Conf. Series: Materials Science and Engineering **834** (2020) 012068 doi:10.1088/1757-899X/834/1/012068

program outputs are exhibited in Figure 7. From the red bell curve as shown in the figure, it could be proven that the random input powers were generated according to normal distribution.

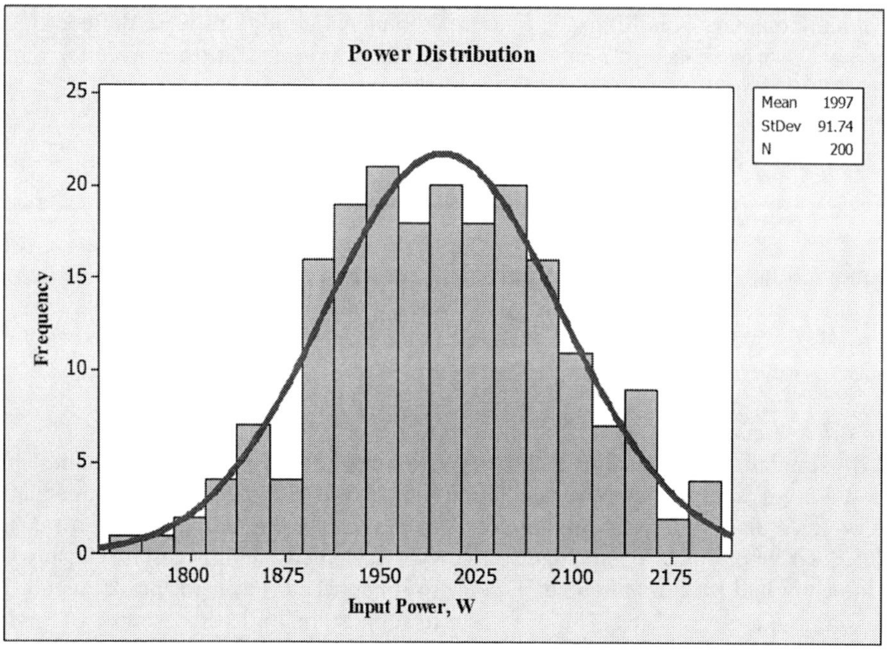

Figure 7. Distribution of random variables

In this study, both simulations of deterministic and probabilistic analyses were carried out using the process parameters as follows:

Table 1. Thermal cutting parameters

Parameters	Types of analyses	
	Normal (Deterministic)	Monte Carlo (Probabilistic)
Input power, P (W)	2000	$\mu = 2000$, $\sigma = 100$
Cutting speed, v (mm/s)	7.5	7.5
Heat source model	3D conical	3D conical

3. Results and Discussion

After the simulations were completed, the postprocessing analyses could be implemented in which the grain size distribution caused by the thermal cutting process could be observed. Figure 8 displays the predicted result of grain size distribution at the final time after the cutting process was finished (20 seconds). The grain size distributions induced by both simulation methods were almost similar. The maximum grain sizes produced by deterministic and probabilistic were 48.36 μm and 49.09 μm respectively.

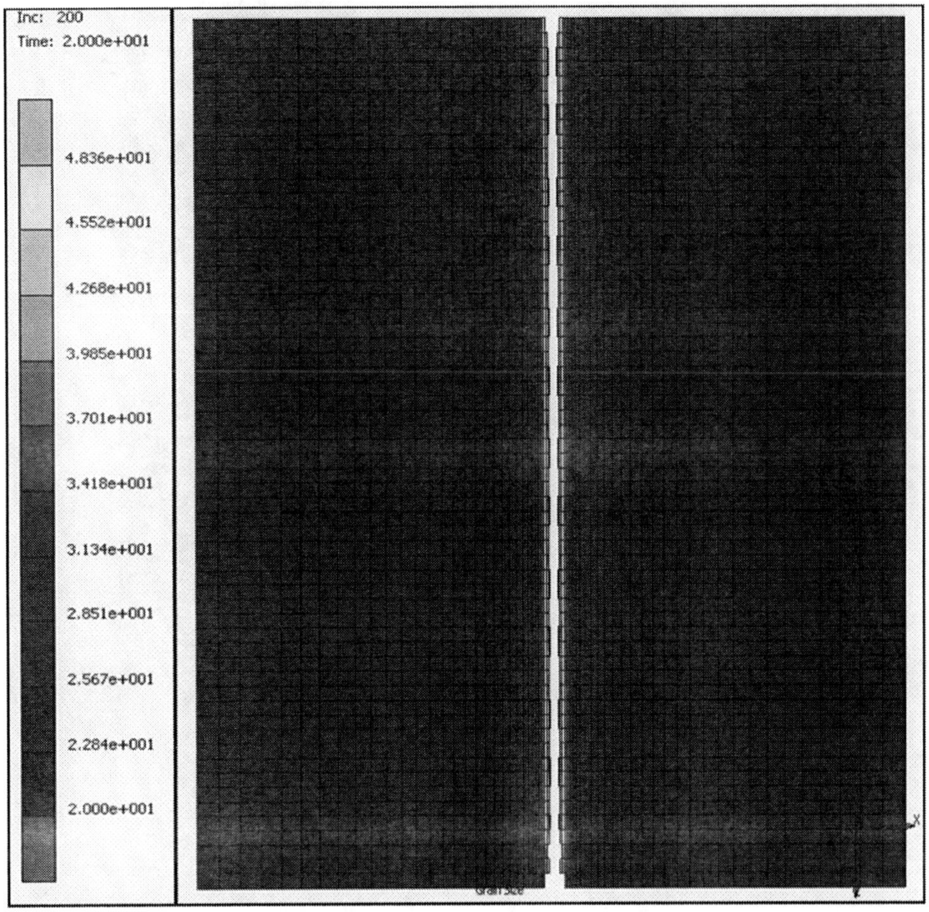

Figure 8. Grain size distribution after the thermal cutting process was finished

Figure 9 shows the comparison of grain size distributions through thickness at the middle length of the plate. From the cross sections, different grain size distributions could be obviously seen between both methods due to different temperature distributions. In order to see better comparison, the results produced at particular node which was Node 7851 has been chosen as shown in Figure 9 as well. The node was selected owing to its location which was in the region affected by the transient heat input.

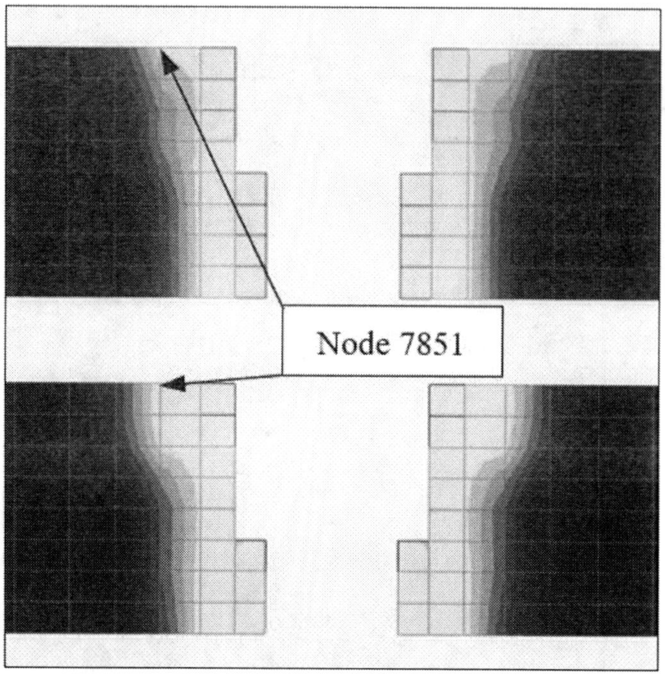

Figure 9. Cross sections of grain size distributions at 50 mm from the starting point (in the middle) obtained from probabilistic (upper) and deterministic (lower) methods

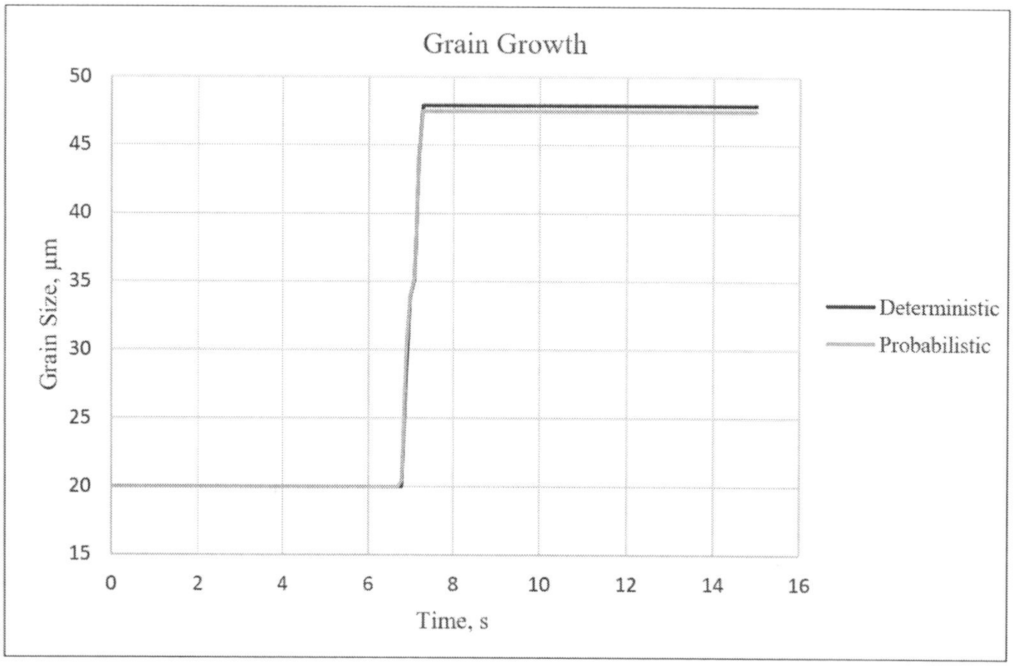

Figure 10. Grain growth during the thermal cutting process

Figure 10 displays the grain growth at the chosen node induced through both simulation methods. By comparison, the final grain size produced by probabilistic method was slightly lower. The grain growth was dependent on the thermal cycle.

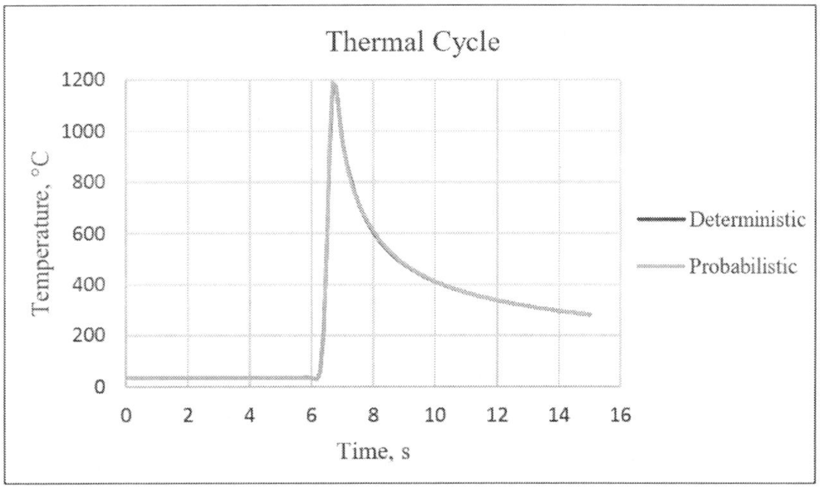

Figure 11. Temperature history at Node 7851

Table 2. Temperature distributions at Node 7851 between 6.0 and 8.0 seconds

Time, s	Temperature, °C	
	Probabilistic	Deterministic
6.0	34.1934	34.1379
6.1	30.3226	30.0667
6.2	29.7903	29.1721
6.3	64.8705	64.1524
6.4	222.808	217.599
6.5	538.16	534.246
6.6	989.793	950.039
6.7	1190.01	1178.3
6.8	1178.82	1176.29
6.9	1074.38	1077.44
7.0	963.501	967.173
7.1	892.93	898.364
7.2	835.921	845.478
7.3	789.109	800.912
7.4	753.46	759.156
7.5	724.039	724.636
7.6	697.035	694.947
7.7	671.368	668.092
7.8	647.949	644.124
7.9	626.745	622.68
8.0	607.337	603.232

The temperature history at the chosen node is exhibited in Figure 11. However, due to small difference, both thermal cycles seemed very similar. Thus, the numerical values of temperature distributions produced by both methods were displayed in Table 2 from 6 to 8 seconds. The range of time between 6 and 8 seconds was crucial in determining the grain growth. Based on the results listed in Table 2, the difference could be clearly seen between probabilistic and deterministic methods.

4. Conclusions

From the analysis of the results, it was found that slight differences in grain size distributions had been produced between probabilistic and deterministic methods. Even though the differences were not very significant, but more importantly the Monte Carlo or probabilistic method could be successfully implemented and produced different prediction from the deterministic analysis. In this research, the probabilistic method was just employed to variate the input power while the other parameters were still maintained and thus producing minor changes in the predicted results. More significant difference in results could be obtained through probabilistic method by varying more than one parameter. In addition, the capability of Monte Carlo method is dependent on the output random numbers. The random number generator used in this study was one of the simple generators that could be incorporated into the normal deterministic FEM analysis. Therefore, the effectiveness of probabilistic analysis is dependent on the type of random number generator. The probabilistic analysis had the capability to produce better prediction due to it could replicate nearly to the real process as compared to the deterministic analysis. In the real process, the supply of input power is normally fluctuated in certain range and quite impossible to maintain at certain value. Hence, probabilistic method is one of the simulation approaches that can be executed to obtain the prediction that is close to the reality.

Acknowledgement

The authors would like to express their gratitude to the staff members of the Advanced Manufacturing Technology Excellence Centre (AMTEx) at Faculty of Mechanical Engineering, Universiti Teknologi MARA (UiTM) Malaysia and Virtuelle Fertigungstechnik, Technische Universität Chemnitz Germany for supporting and encouraging this investigation. This research is also financially sponsored by Geran Inisiatif Penyeliaan (GIP) UiTM with reference number 600-IRMI 5/3/GIP (073/2019).

References

[1] Arttu Laitinen. Two-Dimensional Simulation of Thermal Cutting of Low Alloyed Steels, Master of Science Thesis, Tampere University of Technology, (2015)

[2] Teräslevyjen Terminen Leikkaus, *Metalliteollisuuden Kustannus-Oy*, ISBN 951-817-221-8, ISSN 0357-7368, Tekninen Tiedotus, Helsinki, (1985).

[3] Hendricks B.R. 1999 Simulation of Plasma Arc Cutting, M.Sc. Thesis, Faculty of Engineering, Peninsula Technikon, 1-42.

[4] Prasad G.V.S., Siores E., Wong W.C.K. 1998 Laser Cutting of Metallic Coated Sheet Steels *Journal of Materials Processing Technology* **74** 234–242

[5] Anil P. Varkey, Dr. Shajan Kuriakose, Prof. V Narayanan Unni 2014 Optimization of Edge Quality during CO_2 Laser Cutting of Titanium *Alloy International Journal of Innovative Research in Advanced Engineering* **1** 110-117

[6] Nyon K. Y., Nyeoh C. Y., Mohzani Mokhtar, Razi Abdul Rahman 2012 Finite Element Analysis of Laser Inert Gas Cutting on Inconel 718 *International Journal of Advanced Manufacturing Technology* **60** 995-1007

[7] Serope Kalpakjian, Steven R. Schmid. Manufacturing Engineering and Technology, Pearson Education South Asia Pte Ltd (2014)

[8] Lidam RN, Yupiter HPM, Redza MR, Rahim MR, Sulaiman MS, Zakaria MY, Tham G, Abas SK, Haruman E and Chau CY 2012 Simulation Study on Multipassed Welding Distortion of Combined Joint Types using Thermo-Elastic-Plastic FEM *The Journal of Engineering Research* **9** 1-16

[9] Jonathan G. Mullins and Jens Gunnars. Effect of Hardening Model on the Weld Residual Stress Field in Pipe Girth Welds, 20th International Conference on Structural Mechanics in Reactor Technology, Finland. (2009)

[10] José David Arregui-Mena, Lee Margetts, Paul M. Mummery 2014 Practical Application of the Stochastic Finite Element Method *Arch Computat Methods Eng* DOI 10.1007/s11831-014 9139-3.

[11] Der Kiureghian A, Ke J-B 1988 The Stochastic Finite Element Method in Structural Reliability *Probab Eng Mech* **3** 83-91

[12] Cassidy MJ, Uzielli M, Tian YH 2013 Probabilistic Combined Loading Failure Envelopes of a Strip Footing on Spatially Variable Soil *Comput Geotech* **49** 191-205

[13] Liu WK, Mani A, Belytschko T 1987 Finite Element Methods in Probabilistic Mechanics *Probab Eng Mech* **2** 201-213

[14] M. F. Ashby and K. E. Easterling 1982 A first report on diagrams for grain growth in welds *Acta Metallurgica* **30** 1969-1978

[15] B. P. Kashyap 1992 Grain growth behaviour of type 316L stainless steel *Materials Science and Engineering: A* **149** L13-L16

[16] Naoto Fujiyama and Akira Seki 2018 Austenite grain growth simulation in welding Heat Affected Zone *Materials Science Forum* **941** 620-626

[17] Debasish Das, Dilip Kumar Pratihar and Gour Gopal Roy 2018 Cooling rate predictions and its correlation with grain characteristics during electron beam welding of stainless steel *The International Journal of Advanced Manufacturing Technology* **97** 2241–2254

[18] Yichen Wang, Min Ding, Yong Zheng, Shisen Liu, Wenxian Wang and Zhaohan Zhang 2016 Finite-Element thermal analysis and grain growth behavior of HAZ on argon tungsten-arc welding of 443 stainless steel *Metals* **6** 77

[19] Miloš Madić, Miroslav Radovanović and Marin Gostimirović 2014 ANN Modeling of Kerf Taper Angle in CO2 Laser Cutting and Optimization of Cutting Parameters using Monte Carlo Method *International* Journal of Industrial Engineering Computations 6 1-10

[20] C. S. Wu, H. G. Wang and Y. M. ZHANG 2006 A New Heat Source Model for Keyhole Plasma Arc Welding in FEM Analysis of the Temperature Profile *Welding Journal* 284-291

[21] Øystein Grong. Metallurgical Modelling of Welding, The Institute of Materials, (1997).

[22] Timothy H. Click, Aibing Liu, George A. Kaminski 2010 Quality of Random Number Generators Significantly Affects Results of Monte Carlo Simulations for Organic and Biological Systems *Journal of Computational Chemistry* **32** 513-524

[23] Mark J. Durst, Using Linear Congruential Generators for Parallel Random Number Generation, Proceedings of the 1989 Winter Simulation Conference. (1989) 462-466

[24] William J. Thistleton, John A. Marsh, Kenric Nelson and Constantino Tsallis 2007 Generalized Box–Müller Method for Generating q-Gaussian Random Deviates *IEEE Transactions on Information Theory* **53** 4805-4810

[25] Adnane Addaim, Driss Gretete and Abdessalam Ait Madi 2018 Enhanced Box-Muller method for high quality Gaussian random number generation *Int. J. Computing Science and Mathematics* **9** 287-297

[26] J. Rodrigues Dias 2010 A Simple Generalization of the Box–Muller Method for Obtaining a pair of Correlated Standard Normal Variables *Journal of Statistical Computation and Simulation* **80** 953–958

IOP Conf. Series: Materials Science and Engineering **834** (2020) 012069 doi:10.1088/1757-899X/834/1/012069

Enhanced application of virtual manufacturing: influences of material parameters on numerical Simulation of forming process

M S Sulaiman[1], Y H P Manurung[1], M Graf[2], A Bauer[2], B Awiszus[2], A R Omar[1], W Kuntjoro[1] and M S Adenan[1]

[1] Faculty of Mechanical Engineering, UiTM Shah Alam, Selangor, Malaysia
[2] Professorship of Virtual Production Engineering, Chemnitz University of Technology, Chemnitz, Germany

*Corresponding author E-mail: yupiter.manurung@salam.edu.my

Abstract. This paper describes the sensitive influence of material parameters on the final numerical simulation result of forming process as one of virtual manufacturing applications. Although existing commercial software system provide predefined material properties, these can nevertheless differ with the actual material which might result in inaccuracies. Hence, material data needs to be enhanced to gain better simulation results. This investigation begins with development of flow curve of steel DC04 (1.0338) based on compressive test with various effective plastic strains under consideration of rolling direction. Further, Forming Limit Diagram (FLD) is to be defined by using Nakajima test to determine the strains up to the fracture for deep drawing process. In this research, the enhanced application on sheet forming process is modelled and simulated as well as verified under laboratory condition using an oil-hydraulic press. It can be demonstrated that the numerical simulation results are very sensitive due to the essential differences in material properties. However, this procedure will lead to time and cost intensive measurements of the material behaviour, which is mandatory for an accurate forecast of real forming process.

1. Introduction

Virtual Manufacturing (VM) is a system in which the abstract prototypes of manufacturing objects, processes, activities and principles evolve in a computer-based environment to enhance one or more attributes of the manufacturing process. The use of VM will provide a computer-based environment to simulate individual manufacturing processes or the total manufacturing enterprise. VM is also focusing on available methods and tools that allow continuous, experimental depictions of production processes and equipment using digital models. Areas that are concerned are (i) product and process design, (ii) process and production planning, and (iii) machine tools, robots and manufacturing system, and virtual reality applications in manufacturing [1]. Many aspects of products may be simulated, but for sheet metal products, mainly physical properties are of interest. A well-known method that is commonly applied in industry is Finite Element Method (FEM). The FEM is the general method to model and simulate the physical behaviour of bodies. In the process domain, the manufacturing processes are modelled and simulated. Examples of processes are laser cutting, punching, welding, and a variety of forming processes [2]. Nowadays, the increasingly growth in computer performance and rapid developments in numerical methods have allowed faster and more accurate predictions of computer modeling. VM can be also seen as an extension of CAD/CAM system in which computer is

Content from this work may be used under the terms of the Creative Commons Attribution 3.0 licence. Any further distribution of this work must maintain attribution to the author(s) and the title of the work, journal citation and DOI.
Published under licence by IOP Publishing Ltd

used in aiding design and product manufacturing. Through VM, manufacturing system is modeled using audiovisual or other sensory features to simulate or design alternatives for real manufacturing environment, and the proposed product can be manufactured using computers [3].

Deep drawing is an important sheet metal forming process which was developed in the 1700s. A punch presses a sheet metal blank into die cavity and formed into a cylindrical or box-shaped part. Parts such as beverage cans, pots, containers of all shapes and sizes, kitchen sinks and automobile panels are typical products that can be produced by this method [4]. In [5], the effects of blank holder forces were investigated in producing LPG bottles through deep drawing process. It was found out that high blank holder force restricted the material flow from the beginning of the process which led to maximum thinning. Low blank holder force in the initial stage managed to avoid necking failure between punch and die radius. The thickness distribution in drawn part was influenced by the magnitude of blank holder force at initial stage. Chandra Pal Singh et al. [6] performed a research on deep drawing process parameters. The important parameters were such as blank holder force, punch radius, die radius, material properties and coefficient of friction. Blank holder force could control the metal flow, thickness variation and wrinkling behaviour. The thickness of deformed material was dependent on the punch and die radii. Meanwhile, one of the vital parameters which was friction would affect relative thickness distribution and surface quality.

2. FEM Simulation Procedures using Simufact.Forming 13.3.1

Figure 1 shows the models used in the simulation that consist of punch, blank holder, matrix and specimen. In this research, all models are built in solid condition. The specimen model was meshed using sheetmesh that produces 3D geometry built of hexahedral elements of 1.5 mm in length which are suitable for geometry that has a sheet form.

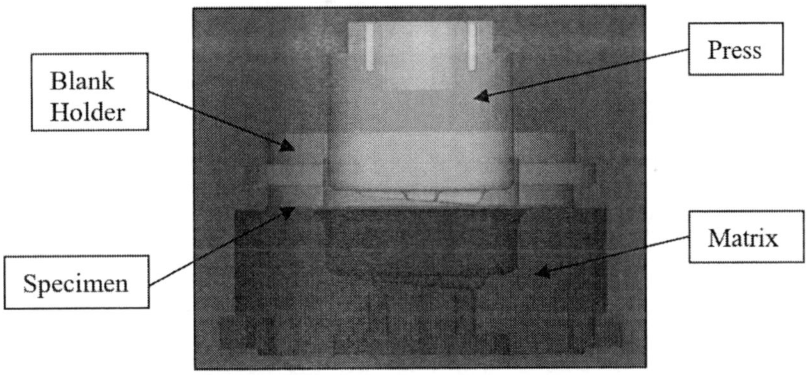

Figure 1. Complete models involved in simulation

In this simulation, DC04 or mild steel, has been employed to predict the strain of deformed part. The material was selected in accordance with experiment. This type of steel has been widely used in many applications due to good formability. There were many material properties involved in simulation process. However, for sheet metal forming, Forming Limit Diagram (FLD) and flow curve are two dominant material properties which are vital in characterizing the flow behavior of deformed material. In this study, both properties used for simulation purpose were obtained from experiments. The FLD was obtained from Nakajima Test using three specimens of different shapes but same in thickness. The FLD was drawn by plotting minor strain in horizontal axis and corresponding major strain in vertical axis. The subsequent Equation 1 and Equation 2 were applied to characterize the limit of deformation in which the plotted curve split up the safe region from the unsafe region. In these equations, C_0 is strain coefficient of y-intercept when the minor strain is zero, while D_1 and C_1 are strain coefficients of slopes of graph for negative and positive minor strains respectively. The graph of FLD is illustrated in Figure 2.

$$FLD(\varepsilon_2) = C_o + D_1\varepsilon_2 \qquad \varepsilon_2 < 0 \tag{1}$$

$$FLD(\varepsilon_2) = C_o + C_1\varepsilon_2 \qquad \varepsilon_2 > 0 \tag{2}$$

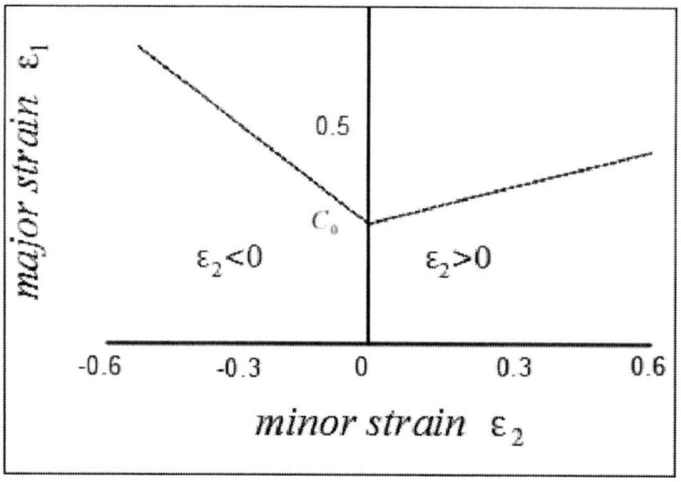

Figure 2. Forming Limit Diagram (FLD) of DC04

There are a few methods that can be conducted in order to extract the flow curve. In this case, compressive test has been selected and carried out. The flow curve was constructed by plotting plastic strain in horizontal axis and corresponding flow stress in vertical axis. In constructing the flow curve, only plastic strain region was interested instead of elastic strain region. Figure 3 exhibits the flow curve used for computational analysis. The forming parameters that have been employed during the simulation and experimental verification were tabulated in Table 1.

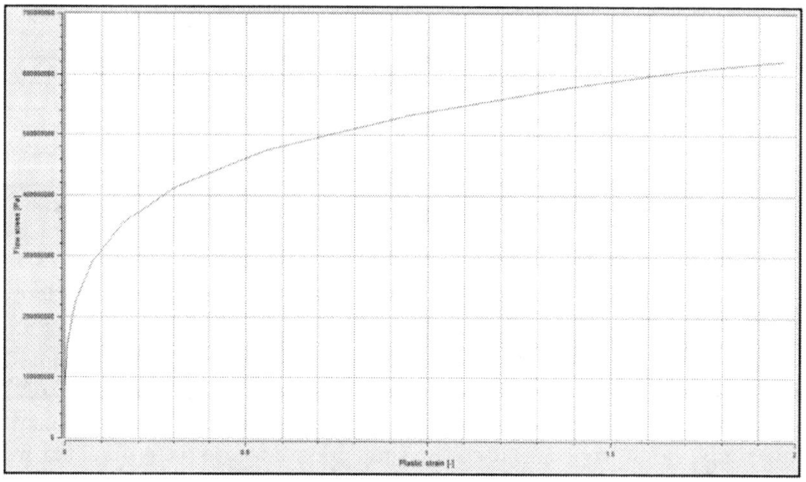

Figure 3. Flow curve of DC04

Table 1. Forming parameters

Parameters	Values
Punch velocity, v (mm/s)	10
Coefficient of friction, μ	0.05
Blank holder force, F (kN)	25

3. Influences of Material Parameters on Forming Simulation

The most important material-specific properties, which influence the quality of the simulation results in the present case, are the flow curve and the forming limit diagram to describe the deformation behaviour of the sheet material.

For the calculation of the flow curve, a series of plane strain compressive tests were applied (Figure 4a). The sample dimensions were 20 mm, 50 mm and 2 mm in width, length and thickness respectively, which was mechanically removed from the batch of the component sheet. These samples were deformed in universal testing machine Galdabini Quasar 50 kN and the force-displacement-curves were recorded. The plain strain compressive test is very well suited for the characterization of sheet material as the stress condition is analogous to the real test by taking into account the real process parameters (temperature, strain rate). The punch width is an important factor in the plain strain compressive test, because this should be smaller than the sample width (punch:sheet > 1:5) so that the spreading can be ignored.

By means of the force-displacement-curves, the flow curve could be calculated by using the equivalent stress hypothesis of von Mises under consideration of an additional friction and heat development during the deformation. The corrections are necessary since increasing friction requires higher forces for the deformation of the material. The temperature correction results from the generation of dissipation heat, which means that this material test does not operate under isothermal conditions. As a reason for the fact, that the plain strain compressive test causes a plain stress state, the equivalent stress hypothesis of von Mises is necessary as the flow curve is defined for uniaxial stress states [7]. The following flow curve (see Figure 4b) was obtained for the sheet material DC04.

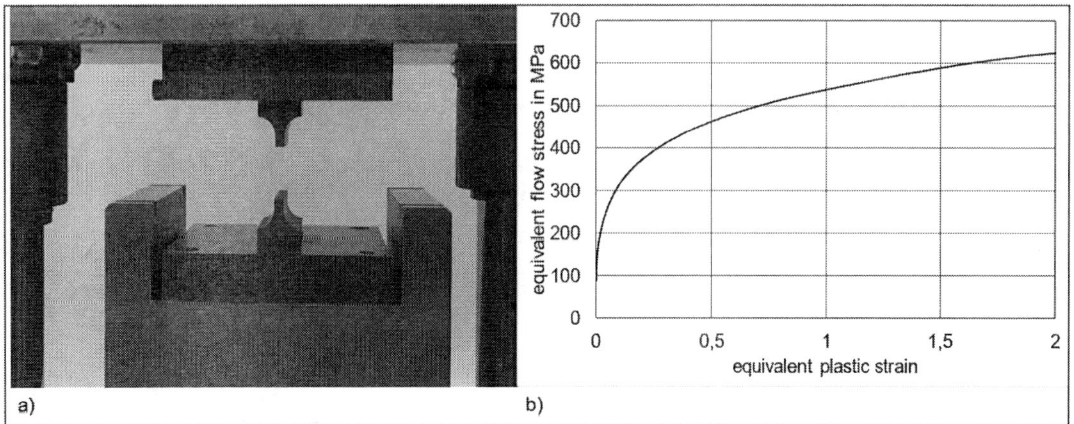

Figure 4. Experimental setup of the plain strain compressive test (a) and flow curve of DC04 for room temperature and strain rate of 0.1 s^{-1} (b)

Because of a deep-drawing process should be designed, the determination of forming limit diagram according to DIN EN ISO 12004-2 has a great importance. Nakajima-Test was carried out using a universal sheet testing machine BUP 600 (see Figure 5) at the Institute of Metal Forming in Freiberg, Germany. The shape changes were analysed by means of the AutoGrid system.

IOP Conf. Series: Materials Science and Engineering **834** (2020) 012069 doi:10.1088/1757-899X/834/1/012069

Technical data:	
drive:	hydraulic
maximum test load:	600 kN
maximum clamping force:	600 kN
maximum punching force:	600 kN
test temperature:	RT ... 350°C
heating:	RT ... 350°C
test atmosphere:	air
maximum testing speed:	20 mm/s
maximum ram stroke:	120 mm
maximum sheet thickness:	10 mm

Figure 5. Sheet Forming Machine BUP 600

The test was also carried out on samples from the batch of 2 mm sheet thickness. By evaluating the occurrence of cracks, it was possible to determine the maximum forming behaviour of this material (see Figure 6).

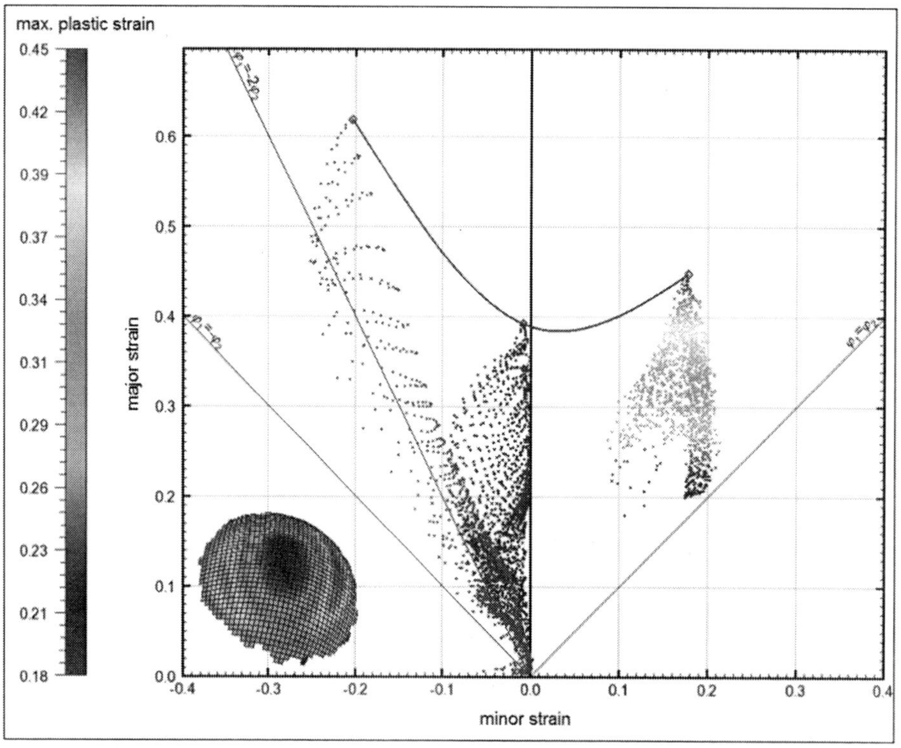

Figure 6. Forming limit diagram with three measurements for steel DC04

With the help of these sensitive material data and their implementation in the FEA software Simufact.Forming 13.3.1, the deep drawing process could be specifically modelled, analysed and validated with the real tests for the investigated material. Further material data were used from the material database of the FEA program.

4. Results and Discussion

Figure 4 to Figure 7 present the actual results obtained from experiment. In order to measure the deformation, the results of plastic strains were calculated using geometric detecting system as shown in Figure 8 (left). However, the degrees of deformation were measured in logarithmic plastic strains (φ). After the simulation was finished, the post-processing analysis could be implemented in which the plastic strains (ε) of the deformed part occurred due to the deep drawing process could be observed. In order to compare the simulated results with the experimental measurements, the relation between both units was stated by Equation 3 as below:

$$\varphi = \ln(1 + \varepsilon) \tag{3}$$

When comparing the results as displayed in Figure 8, the plastic strains obtained from simulation analysis were in good agreement with the experimental values. The computational analysis was successfully executed to deform the part until full stroke without thinning effect and even the cross shape has been successfully formed which similar to that of part produced by means of experiment.

Figure 7. Deep drawn part produced through experiment

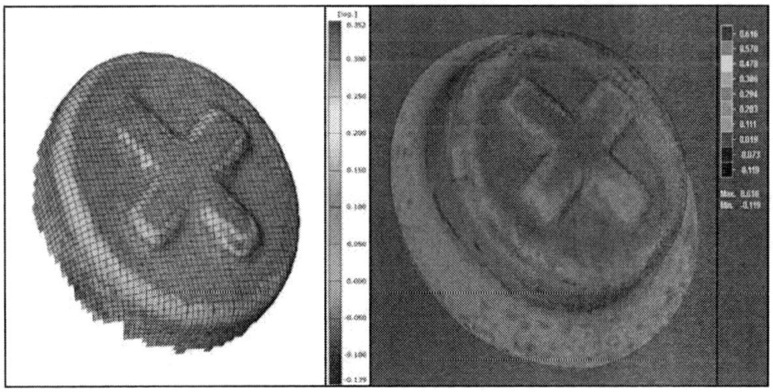

Figure 8. Results from experiment (left) and simulation (right)

5. Conclusions

From the analysis of the results, it was found that a good agreement has been achieved between finite element prediction and experimental verification. This outcome revealed that the capability and reliability of virtual manufacturing has been obviously proven in forecasting the degree of deformation sensibly. In order to obtain a good prediction, the flow behavior of material must be well characterized to reflect the real deep drawing process. In addition, the specimen model should be properly meshed since it would affect the computational time as well as the accuracy of prediction. Therefore, significant impact from this study shows that the material parameters will have great influence towards the precision of numerical simulation.

Acknowledgements
The authors would like to express their gratitude to German Academic Exchange Service (DAAD) for the financial support. The authors would also like to thank to the staff members of Virtuelle Fertigungstechnik, Technische Universität Chemnitz Germany and Advanced Manufacturing Technology Excellence Centre (AMTEx) at the Faculty of Mechanical Engineering, Universiti Teknologi MARA Malaysia for encouraging this project. Furthermore, special thanks go to the Institute of Metal Forming at the Technische Universität Bergakademie Freiberg for the determination of the forming limit diagram of DC04. This research is also financially sponsored by Geran Inisiatif Penyeliaan (GIP) UiTM with reference number 600-IRMI 5/3/GIP (073/2019).

References

[1] P. Depince, D. Chablat, E. Noel and P.O. Woelk. The virtual manufacturing concept, scope, socio-economic aspects and future trends, Proceedings of 2004 ASME Design Engineering Technical Conferences, Utah, USA. (2004) 1-10

[2] Amos H.C., Ng Josef Adolfsson, Martin Sundberg and Leo J. De Vin 2008 Virtual manufacturing for press line monitoring and diagnostics *International Journal of Machine Tools & Manufacture* **48** 565-575

[3] Oluleke Bamodu and Xuming Ye 2013 Virtual manufacturing and components of virtual reality Applied Mechanics and Material 318 83-86

[4] Serope Kalpakjian and Stevan R. Schmid. Manufacturing Processes for Engineering Materials, Prentice Hall, Singapore, (2008)

[5] R. Padmanabhan, M.C. Oliveira, J.L. Alves and L.F. Menezes 2008 Numerical simulation and analysis on the deep drawing of LPG bottles *Journal of Materials Processing Technology* **200** 416-423

[6] Chandra Pal Singh and Geeta Agnihotri 2015 Study of deep drawing process parameters *International Journal of Scientific and Research Publications* **5** 1-15

[7] Pöhlandt K., Becker H.-J. And Becker N. Werkstoffe und Werkstoffprüfung für die Kaltmassivumformung, Kontakt & Studium, Band 427, Expert Verlag, Herausgeber: W. J. Bartz (1994) ISBN: 978-3816909187

IOP Conf. Series: Materials Science and Engineering **834** (2020) 012070 doi:10.1088/1757-899X/834/1/012070

Effect of Different Preheating's Temperature towards the Integrity of Weldment AISI 1045

M. A. Mat Shah[1*], F. H. Rosli[2], B. Abdullah[2], S. K. Alias[1], G. Hamami[1] and M. N. Halmy[1]

[1]Faculty of Mechanical Engineering, Universiti Teknologi Mara Pasir Gudang, Jalan Purnama, Bandar Seri Alam, 81750 Masai, Johor Darul Takzim, Malaysia.
[2] Faculty of Mechanical Engineering, Universiti Teknologi Mara Shah Alam, 40450 Shah Alam, Malaysia.

*amir8776@uitm.edu.my

Abstract. The purpose of this paper is to show the effect of different preheating's temperature for different thickness of AISI 1045 before it was welded by using Shield Metal Arc Welding (SMAW). The temperature of preheating approach that will be use are 200oC, 250oC and 300oC and were applied to pair of base plate metal with different thickness of 16mm, 22mm and 28mm. Then the pair samples had been welded by using Shield Metal Arc Welding (SMAW). To monitoring the temperature, non-contact type instrument which is thermomapping was used in this study. Time was recorded after preheating was applied and then was stopped when the desired temperature has been obtained. As to evaluate the weld quality, the macrostructure observation, dye penetration, hardness, and tensile test were conducted. As the result, it found that the electrode rod with size16mm and 22mm required preheating's temperature around 200° C while for 28mm required temperature up to 250°C. This paper will give a deeper understanding about the effect of preheating in term of strength of joining. The result obtained from this study can be used to assist welder while performing welding on medium carbon steel.

1. Introduction

Welding has been widely used in industry due to its strength to last longer in period. However, when it comes to weld material with higher carbon content, the possibilities of defects to occur is high. One of the defects that always associated with medium carbon steel is crack. The crack occurs when the hydrogen trapped inside the material, do not manage to fully escape. The crack occur at the weld bead obviously will affect the weld joint thus will decrease the strength and quality. As a response to that problem, preheating is applied to diminish the occurrence of crack. However, after doing some research, the suggested preheat provided just mentioned 200°C-300°C. The different thickness of medium carbon steel, for sure require different heat input. There were no duration time and temperature have been stated when preheating this material. To those welders who were not using temperature indicator like thermocouple or non-contact thermomapping, they were just rely on their experiences when performing preheat which sometimes not reliable. For this study, the scope was narrowed just to find the sufficient time taken and temperature for preheating towards different thickness (16mm, 22mm and 28mm) of AISI before it was welded by using Shield Metal Arc Welding (SMAW). Then, all the tests have been conducted to determine if there were differences in term of integrity of weldment for different thickness.

Content from this work may be used under the terms of the Creative Commons Attribution 3.0 licence. Any further distribution of this work must maintain attribution to the author(s) and the title of the work, journal citation and DOI.
Published under licence by IOP Publishing Ltd

2. Research Work

The sample composition material need to first by undergoing Arc Spectrometer Test. Then sample will be preheating at several temperatures before weld. There are three stages need to be go through to get the good result in the research. The first stage is ensuring only the good welded sample will go through mechanical properties test. To determine the good welded sample, die penetration test will be perform to the sample. Two mechanical properties will be performed which are tensile test and hardness test. The last stage is to verify the sample that tested in mechanical test is good sample in term of material structure. Where in this test, the sample need to be cut and prepare for macrostructure observation. The material that had been used was medium carbon steel, AISI 1045. This research was done by applying preheat with temperature 200°C, 250°C and 300°C to the different thickness of 16mm,22mm and 28mm before it was welded by Shield Metal Arc Welding.

2.1. Sample Preparation

All the specimen material with different thickness of 16mm, 22mm and 28mm was cut by using cutter into desired dimension for the welding requirement which is 10mm x 80 mm and also suitable use for testing requirement.

2.1.1. Arc Spectrometer Test. Some of the specimen material were cut, cleaning and do a facing process by using lathe machine. The material composition of this specimen will be checking to ensure it composition is following the composition of element that provided by AISI 1045. The Spectrometer test was conducted to get the chemical composition in medium carbon steel. The testing was carried out at foundry lab by using Arc Spark Optical Emission Spectrometer Tester. Measurement done by the specimen was placed on the support and then the Arc Spark Optical Emission Spectrometer Tester will spark on a few areas of the surfaces. As the specimen standard is confirm, then it can be proceed to preheating process.

2.1.2. Preheating and Welding. Preheating of the specimen conducted by using oxy- acetylene flame combine with a rotating chamber machine were the specimen will be place inside it. The temperature of preheating that applied to the specimen will be monitor by using thermomapping device. This to ensure the entire specimen will absorb the same heat level that already set in this research. Shield Metal Arc Welding (SMAW) was using to joining the specimen pair. The welding technique and position for welding used was stringer 5G. The constant parameter for SMAW are shown at the below:

Table 1. Constant Parameter.

Type of Parameter	Parameter Setting
Current	100A
Voltage	25V
Type of Weld Joint	Double Butt Joint
Type of Electrode	Low hydrogen E7016

Preheating process begin by set the consumption of combination oxygen and acetylene to get the desired flame as in table 2. Then the torch was pointed to the material with a fixed distance 15mm. As to get the uniform temperature, the specimen was clamped to the machine which will assist the specimen to rotate. The torch was maintained at fixed point for 2 minutes. Then, non-contact thermomapping will measure the temperature as it dropped time by time. This same process was performed for 200°C, 250°C and 300°C towards different thickness of 16mm, 22mm and 28mm.

After that, the welding process can be proceed to join the specimen pair. The power supply of the machine was turned on and the equipment was set up as mention in table 1. For leading direct current straight polarity (DCSP), the electrode was clamped at positive (+ve) terminal while the base metal was clamped at negative (-ve) terminal. The current and voltage was set to 110 A and 25 V .For the materials, the root distance was set to 1mm. For a good quality weld bead, the rod was positioned at angle of 45 from the horizon. The rod then was tapped to produce a welding arc before welding took place. The arc gap was constantly maintained 1.5mm as possible and the rotating machine's speed was controlled. The stringer technique was used along the welding process. Once finished, the slag was removed out by using chipping hammer and wire brush.

Table 2. Preheating Parameter Setting.

Preheating Parameter	Thermal Setting
Type of gas	Oxygen and Acetylene
Pressure	2 bar (Oxygen), 0.5 bar (acetylene)

2.2. Checking sample condition
Once all the samples were completely welded, second stage of testing then was performed to check and select the good for mechanical properties test. This test is known as NDT (Non Destructive Testing) because it does not destruct the physical properties of the specimen. Basically, it will show up any defects where a very fine dye can penetrate. The steps taken involving pre-cleaning, apply the penetrant and post cleaning.

Dye penetrant examination is one of the most popular nondestructive testing (NDT) methods in the industry. It is economical, and just require minimal training when compared to other NDE methods. The purpose is to check for material flaws open to the surface. Welds are the most common item inspected. Once cleaned, apply the penetrant (a bright red liquid dye). Spray plenty on the welds. Leave the dye to penetrate, normally for about 20 minutes. The time scale is usually printed on the can or on the product data sheet. Cleaning again. It need to remove excess dye from the area. Use a dry tissue to remove excess dye, then spray cleaner onto a clean cloth for final cleaning. Once it's all clean again, it need to spray on the developer, a white substance that dries slightly powdery after a few minutes. The developer draws the remaining dye from any flaws, showing them as very red marks in the brilliant white developer. This particular test had no issues to resolve. The edges of the cleaned area showing a slight redness. If there had been a hole in the weld it would have shown up very clearly as a pink-red spot.

2.3. Mechanical Properties Test

2.3.1. Tensile Test and Hardness Test. The test was performed using Instron 8802 with ±500 kN (112,500 lbf) axial force capacity. According to ASTM E8 Standard Test Methods, the specimen must be shaped is Tension Testing of Metallic Materials. Setting up the specimen by gripping both ends parallel with software set up which include the speed, dimensions and type of material. The test begins by continually increasing uni-axial 34 load until failure occurs. The stress-strain graph was obtained once completed this test. The tensile test is probably the simplest and most widely used test to characterize the mechanical properties of a material.

2.4. Macrostructure observation
This macrostructure observation test is a test to ensure the welded sample that perform in mechanical properties test is a good sample where there is no porous or any defect inside the material structure after welded. The purpose of using a Microscope after the mechanical properties test is to let the specimen test first. It not possible to run the mechanical properties test for the cut specimen. In this test, it required to cut and prepare the specimen surface with mirror finish and place under microscope. The observation test was performed by using Olympus BX 60 M microscope with NK Vision NZM7045-T1 Microscope and supported by software IMAPS version 4.0 Professional. The mechanical properties test data that produce in tensile and hardness test is depending on this observation test. Only specimen without any microstructure defect will be used to analyse.

3. Research Finding
The data and result obtain in this experiment were recorded and tabulated in table and visualized in the graph from which all the result arranged consecutively section by section as followed:

3.1. Arc Spectrometer

The first step of this research, was to ensure the material used was as required. By demonstrating arc spectrometer test, the carbon composition obtained can be compared with standard composition. The number 45 of AISI 1045 indicated 0.45%of carbon. From the result obtained, the carbon composition was still in range as shows in table below:

Table 3. Result of composition of AISI 1045 specimen.

Element	C	Si	Mn	P	S	Cr	Ni	Mo	Fe
%	0.448	0.255	0.65	0.0040	0.0056	0.0032	0.0.064	0.0037	98.5

3.2. Hardness Test

The hardness test was conducted to determine the hardness of the joining. The value was collected at three spots which are base metal, heat affected zone and weld metal. The test is to show the quality of the welding zone and quality of effected area near to welding zone. The result of the Hardness test is shows in table 4, 5 and 6.

Table 4. Tensile Test for 16mm AISI 1045.

Temperature	Base Metal		Heat Affected Zone		Weld Metal	
	Value	Average	Value	Average	Value	Average
200	232.5	224.5	219.1	217.5	187.3	176.3
	221.7		223.0		176.8	
	219.4		210.5		164.7	
250	222.1	233.7	216.6	209.6	178.1	183.3
	245.9		203.1		183.4	
	233.0		209.3		188.4	
300	241.8	257.4	207.4	203.6	210.1	202.1
	277.5		203.8		193.4	
	252.9		199.5		202.7	

Table 5. Tensile Test for 16mm AISI 1045.

Temperature	Base Metal		Heat Affected Zone		Weld Metal	
	Value	Average	Value	Average	Value	Average
200	239.6	227.5	209.5	214.6	168.9	169.4
	231.8		212.6		160.9	
	211.0		221.7		178.4	
250	222.1	232.2	210.6	203.2	168.4	176.7
	245.9		197.3		183.7	
	228.6		201.7		178.1	
300	233.8	250.5	207.4	202.1	210.6	197.1
	273.5		195.8		192.8	
	244.3		203.1		187.7	

Table 6. Tensile Test for 28mm AISI 1045.

Temperature	Base Metal		Heat Affected Zone		Weld Metal	
	Value	Average	Value	Average	Value	Average
200	228.6	226.8	207.1	209.7	175.2	175.8
	233.9		223.8		163.5	
	218.0		198.2		188.7	
250	219.7	229.2	218.3	206.9	175.3	162.6
	240.1		205.5		158.4	
	227.7		197.0		154.2	
300	224.1	233.7	197.3	201.5	208.8	193.7
	230.8		206.5		196.6	
	246.2		200.7		175.7	

3.3. Tensile Test Result

Table 7, 8 and 9 show the result for tensile test for welded specimens with different preheat. For 16mm, the tables show that material of 200°C preheating temperature was the highest tensile stress, followed up by material with 250°C and 300°C preheating temperature. The highest point for tensile stress is 428.49 MPa while the other two appeared just 10.1%, lower which the strength owned by 300°C is 273.318 MPa.

The 200°C preheat was also be the highest tensile strength with 22mm diameter and the value is 399.662MPa. According to the table, it can be stated that the material with preheat 250°C and 300°C was less about 18.8% and 28.2 %. Both of the material merely recorded 324.47 MPa and 286.99 MPa respectively.

The tensile test for 28mm, the preheat 200°C was no longer recorded as the maximum tensile strength which only 349.882 MPa, 7.9% less than the highest tensile strength which recorded with preheat 250°C. It obtained 379.945 MPa, distinguished the minimum strength with 7.9%. There were not much different in term of strength of material with 200°C and 300°C when there differed by 0.06%.

Preheat temperature is highly effected the changes of tensile strength value for each thickness. It can be a benchmark if the data recorded at macrostructure and it obviously signified that the material with a maximum tensile strength was not having any or less defects compared to other material with different temperature. It clearly implies that the defect occurred has correlation with the tensile strength of material.

Table 7. Tensile Test for 16mm AISI 1045.

Temperature	Tensile Stress at Maximum load (MPa)	Tensile Strain at Maximum load (mm/mm)	Modulus (MPa)	Tensile Extension at Maximum Load (mm)
200	428.788	0.026	49782.551	1.586
250	304.176	0.020	51498.619	1.208
300	273.318	0.009	65559.948	0.555

Table 8. Tensile Test for 22mm AISI 1045.

Temperature	Tensile Stress at Maximum load (MPa)	Tensile Strain at Maximum load (mm/mm)	Modulus (MPa)	Tensile Extension at Maximum Load (mm)
200	399.662	0.014	68418.881	0.899
250	324.466	0.001	74764.340	0.615
300	286.99	0.001	67705.461	0.609

Table 9. Tensile Test for 28mm AISI 1045.

Temperature	Tensile Stress at Maximum load (MPa)	Tensile Strain at Maximum load (mm/mm)	Modulus (MPa)	Tensile Extension at Maximum Load (mm)
200	349.882	0.010	77531.209	0.625
250	379.945	0.015	69521.831	0.930
300	349662	0.013	77515.805	0.796

4. Conclusion

After preheated for 1 minute, the time taken the material to accomplish a desired temperature will give a different value according to the thickness. The experiments show that the thickness of material tend to possess less time to get the desired temperature. Also, the least flaw occurred during dye penetration test was at thickness 28mm when the flaw merely appeared on the specimen at 250°C. The effectiveness of this method relies on many factors that involve cleanliness of the part, skill of the welder, and the procedure that was used to perform welding. This test is appropriate to be performed at the earlier stage, so that the welder can examine the occurrence of imperfection prior.

In this study, the influence of niobium addition on the microstructure of austempered ductile iron was investigated and the following conclusion could be obtained:

The microstructure of austempered ductile iron consist of nodular graphite surrounded by retained austenite and acicular ferrite while the addition of niobium to austempered ductile iron promoted the formation of Niobium carbide (NbC) to the microstructure.

The addition of niobium resulting in much finer retained austenite and acicular ferrite structure but reduced the nodularity, nodule count and graphite area due to the presence of niobium carbide.

Acknowledgement

The authors would like to express their gratitude to the ministry of science and technology, Malaysia (MOSTI) and Ministry of Higher Education Malaysia (MOHE), and Universiti Teknologi MARA for the financial support given to this work through research grant of 600-IRMI/FGRS 5/3 (003/2019).

References

[1] M. Bala Chennaiah, P. N. Kumar and K. P. Rao, “Influence of Heat Input and PWHT on the Microstructure and Mechanical Properties in Dissimilar (IS2062-EN8) Welded joints”, *Procedia Computer Science,* vol. 85, pp. 54-61, 2016.

[2] B.L. Luk and Alan H.S. Chan, "Human Factors and Ergonomics in Dye Penetrant and Magnetic Particles Nondestructive Inspection Methods", Engineering Letters, (ISSN: 1816-093X) Volume 15, Issue 1, pp 163-169, 2007.

[3] C. Chuenarrom, P. B., “Effect of Indentation Load and Time on Knoop and Vickers Microhardness Tests for Enamel and Dentin”, *Materials Research*, vol. 473, 2009.

[4] The Lincoln Electric Company, “DC vs. AC Polarity for SMAW”, (n.d.). [Online]. Available: https://www.lincolnelectric.com/en-us/support/welding-solutions/Pages /polarity-for-smaw.aspx [Accessed: 20- June-2018].

[5] G. S. Sidhu and S. S., “Role of Shielded Metal Arc Welding”, *International Journal of Emerging Technology and Advanced Engineering*, vol. 746, 2012.

[6] S. Zhang and Q. X., “Mechanical characterization of flat specimens in tensile test and numerical simulation”, *Journal of Mechanical Science and Technology* 26(2): 401 409, 2012.

[7] T. P. Hung, E. S. Hao and H. K. Jao, “Temperature Modeling of AISI 1045 Steel during Surface Hardening Processes”, *Materials*, vol. 11, p. 1815, 2018.

[8] X. C. Hea, F. S. Gu and A. Ball, “A review of numerical analysis of friction stir welding”, *Progress in Materials Science*, vol. 65, pp. 1-66, 2014.

[9] K. Masubuchi, “Analysis of Welded Structures: Residual Stresses, Distortion, and Their Consequences” *International Series on Materials Science and Technology*, vol. 33, p. 628, 2013.

IOP Conf. Series: Materials Science and Engineering **834** (2020) 012071 doi:10.1088/1757-899X/834/1/012071

Development of Child Safety Car Alert System Using Arduino and GSM Module

A. Hazizan[1*], N. A. Md Lazam[1] and N. I. Hassan[2]

[1] University Malaysia of Computer Science & Engineering (UNIMY)
[2] Heriot-Watt University Malaysia

*azrulhazizan@rocketmail.com

Abstract. The Child Safety Car Alert System by Arduino is an integrated device that sends alerts to the driver if a child is left unintendedly in the car. The system is developed using the Arduino board which incorporates the integration between sensors and GSM module. This system uses pressure and motion sensors to detect the presence of a child located at the back seat of the vehicles. Meanwhile, the GSM or Global System for Mobile Communication allows the system to send an alert to the driver within a short period. The GSM is chosen due to its ability to lower the energy consumption per bit while providing higher data rates. Two sensors are used; Force Sensitive Resistor (FSR) Sensor and Pressure Infrared (PIR) Sensor. The alert system is triggered when both sensors detect the presence of a child at the back of the car seat which will then notify the parents or a driver immediately through a message sent via a mobile phone.

1. Introduction

Every few days in the world, there are reports of young children dying from heatstroke after being left in parked cars [1-5]. With cases of clear neglect, the justified intervention and local solution must be addressed accordingly.

Figure 1 shows the percentages of the child being left in vehicles. 54% is due to the "forgotten in-vehicle" element. Even though an emphasis on education and awareness focused at parents or drivers and to the public, in general, have been given, this could not stop the number of a child end up in this kind of tragedy. This shows that it's very important to develop a device or system that can notify the driver or alert them on the issue. To summarize this section, the death could have been prevented if the parent being notified in case, they leave a child behind. To prevent this tragedy, Arduino-based-system is developed to send out a warning message once the system detects the presence of the children based on pressure and motion sensors placed at the back seat of the car after the driver left.

This project aims to develop a lab prototype of a low-cost and simple implementation of a Child Safety Car Alert System (SCCAS) using Arduino kits and Arduino IDE 1.8.2 with intended for the system that sends alerts to the drivers

The technology has been in the market for a while. Studies and reviews on three existing products similar to this project development have been made to choose the most suitable detection mechanism for building SCCAS [4,7,8,10,11]. Based on the comparison, a Passive Infrared (PIR) sensor is chosen to detect the motion of the child through IR radiation emission. Adding to this is a pressure sensor. Both motion and pressure sensor are used to determine whether the heuristic detected is a human being.

Content from this work may be used under the terms of the Creative Commons Attribution 3.0 licence. Any further distribution of this work must maintain attribution to the author(s) and the title of the work, journal citation and DOI.
Published under licence by IOP Publishing Ltd

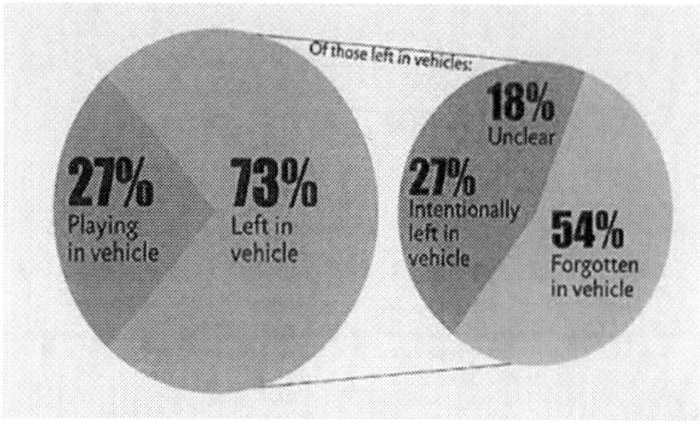

Figure 1. Percentages of child left by circumstances

For the notification alert, the dedicated fob is mainly used to send a text alert to the parents or the driver when an incident happens. This fob will beep when it is separated from the UCRS that place on the CRS (Child Restraint System) at a certain distance [12].

2. Research and Method

2.1 System Overview

Arduino is an open-source physical computing platform which implements the Processing Language on a simple input/output (I/O) board. This technology was initially introduced in Interaction Design Institute, Italy on 2005 [3,7]. The Safety Child Car Alert System is an Arduino-based system. It is a standalone system where the system does not interact with the vehicle's internal system. This system is controlled by Arduino Uno R3, a small microcontroller that turns out to be a control unit. This proposed system has two major parts which are; the detection mechanism and the prevention mechanism connected to the control unit.

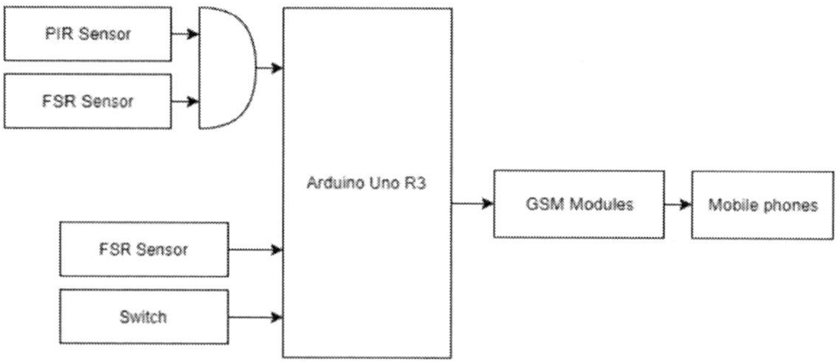

Figure 2. Block Diagram of Safety Child Car Alert System

The detection mechanism is used to detect the presence of a child at the back seat. Passive Infrared (PIR) is used to detect the motion based on IR radiation emission produce from the child [11]. For Force Sensitive Resistor (FSR), it is used to detect the force applied on the infant seat based on the average children's weight. Since both sensors are important to detect and confirm the presence of a child, the AND gate is used for this detection mechanism. Force Sensitive Resistor (FSR) is also used to detect the presence of the driver while the switch is used to simulate the car ignition ON/OFF.

Whenever the detection mechanism detects and confirms the presence of a child while the driver left, a signal will be sent to the control unit to trigger the GSM modules [9]. As the GSM modules get triggered it will send text alert to parents.

IOP Conf. Series: Materials Science and Engineering **834** (2020) 012071 doi:10.1088/1757-899X/834/1/012071

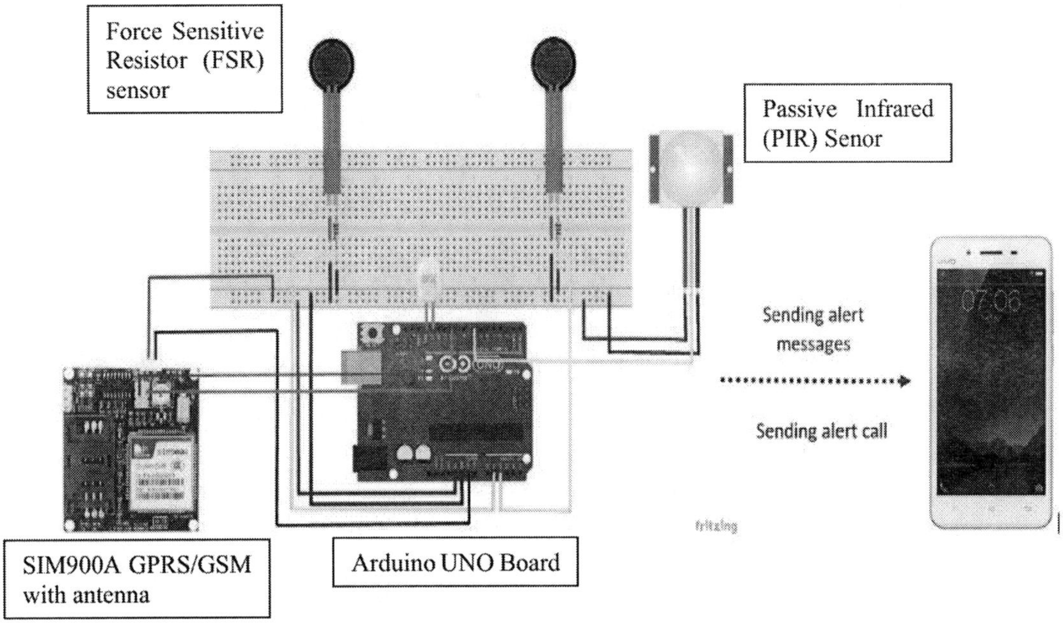

Figure 3. Final Module Integration for SCCAS

The final module of this development is shown in Figure 4. Force Sensitive Resistor (FSR) sensor placed at the driver seat and on the infant seat to detect the presence of the driver and to sense a weight produced by a child. FSR sensor will then be integrated with the GSM module. Concurrently, the Passive Infrared (PIR) sensor will be placed on the top chassis of the car between the driver and infant seat. It will be positioned at a 45° degree facing the infant seat. This sensor is required for the confirmation process to distinguish whether the weight detected by the pressure sensor is a human being or not by revealing the amount of Infrared emits from the child's motion [6,10].

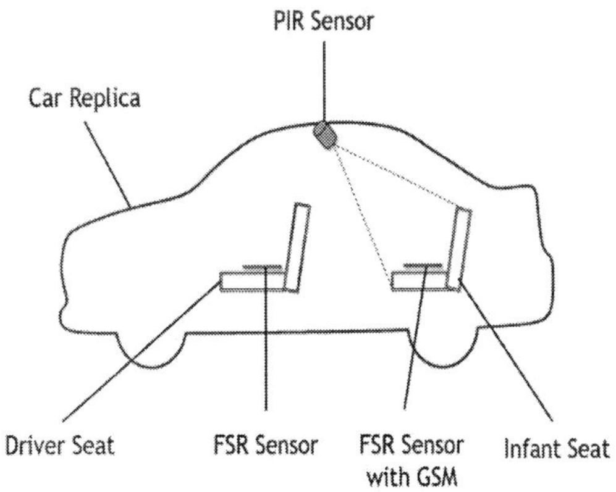

Figure 4. Implementation of SCCAS

3. Result and Discussion

In the scenario where parents left a child in a parked car after a minute passes, initialization of the process would start, and this will activate the Safety Child Car Alert System. Several seconds later, if the pressure sensor detects a weight, the detection mechanism would be triggered and as in Figure 3. The presence of a child would need a confirmation mechanism where motion produced by a child due to prolonged heat exposure in the car would be detected. The system then sends out alert messages to parents to notify them as in figures below. The system idle once the child being removed from the seat.

Figure 5. FSR sensor detecting the presence of driver in the car

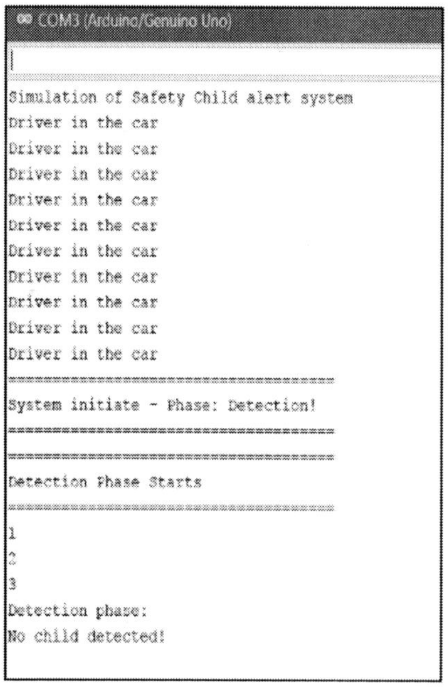

Figure 6: No child detected left in the car after the driver left

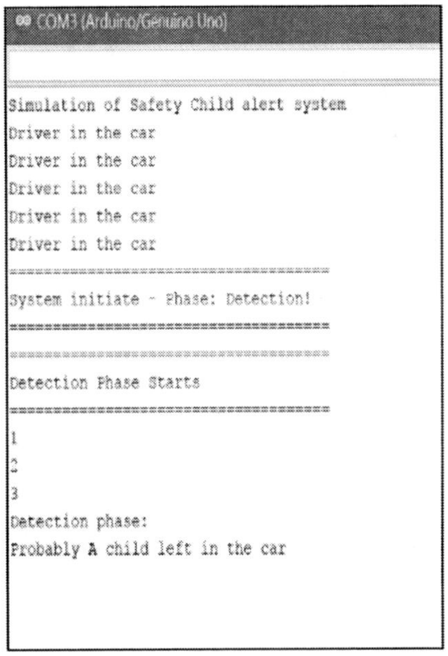

Figure 7. A child left in the car after the driver left

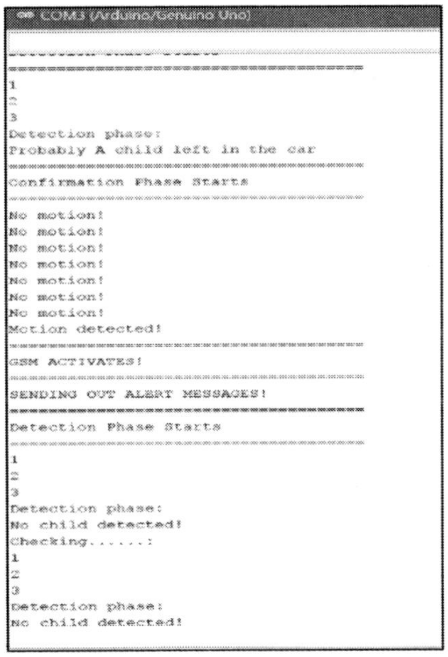

Figure 8: Alert messages send out and child being removed from seat

The results are summarized in Table 1 according to certain aspects. These assessments are being accessed for its basic system operations such as how the system activates, how detection mechanism and confirmation mechanism work and how the alert text is sent out.

Table 1. The functional assessment of SCCAS

Operational Features	Response or Units	Result	Notes
CRS Model		Safety Child Car Alert System	Lab working prototype using Arduino Development Kits.
Orientation	FF = Front Facing		Use motion sensor so, not applicable for Rear Facing (RF) child seat.
Audible Presence Detection Confirmation	Yes or No	No	The green LED flashes when the parents left the driver seat.
User Action Required for Activation	Yes or No	Yes	The parents left the driver seat when no pressure detected by the FSR sensor
User Action Confirmation	Yes, No or n/a	Yes	The green LED flashes when no pressure on driver seat and turn off when no child on the seat.
End-of-Trip Reminder Notification	Yes or No	No	The system is deactivated when the child is removed from the CRS
Left-Behind Notification	Yes or No	Yes	Text alert is triggered "15 seconds after parent left the car"
Left-Behind Notification Recipients	User, Vehicle Surroundings, Telecommunication	User	Text alert sent out if detect motion.
Left-Behind In-Vehicle Cancellation	Yes, No or n/a	No	When the child is removed from the seat, the green LED will be turn off
Snooze Function	Yes, No or n/a	No	

4. Conclusion

The Safety Child Car Alert System may not accommodate all possible technologies; however, it shows great potential for parents/driver to leverage on its benefit to prevent any fatal incident from happening. The system intends to provide reminders to parents who left their child in car unintentionally. This effort is more applicable and relevant for children in a dominant age range between newborn to toddler. The add-on confirmation detection mechanism allows a rigid and precise detection of the entity mechanism to be working perfectly. Issuance of notifications/alerts to parents on add-on integrated GSM module fulfilled the objective of the project. Nevertheless, this system does not address the effectiveness of notification once issued. Despite low-cost, the system's easy set up make SCCAS a user-friendly device. In all, the cost of this project sum up to RM 350. Yet there are always rooms for improvement. For future work, the system should ideally be equipped with robust operating capabilities such as battery life, feedback to indicate proper function, provide end-of-trip convenience reminder and minimal additional user action.

References

[1] Perrucci, G. P., Fitzek, F. H. P., Sasso, G., Kellerer, W., Widmer, J., "On the Impact of 2G and 3G Network Usage for Mobile Phones' Battery Life", *2009 European Wireless Conference*, 255-259, 2009.

[2] KidsandCars.org, "Child Vehicular Heat Stroke Summary", *Summary of heat stroke issue,* 3-12-15, 2012.

[3] Guard, A., Gallagher, S.S., "Heat related deaths to young children in parked cars: an analysis of 171 fatalities in the United States, 1995–2002", *Injury Prevention,* 11:33-37. 2005.

[4] Chilton A., "The Working Principle and Key Applications of Infrared Sensors", *Infrared Radiation Theory.* Azo Sensors ArticleID=339, 2014

[5] Arbogast, K. B., Belwadi, A., Allison, M., "Reducing the Potential for Heat Stroke to Children in Parked Motor Vehicles: Evaluation of Reminder Technology". *National Highway Traffic Safety Administration, Washington, DC,* 2012.

[6] Hashim N. M. Z., Basri, H. H., Jaafar, A., Aziz, M. Z. A. A., Salleh, A., Ja'afar, A. S., "Child in Car Alarm System Using Various Sensor". *ARPN Journal of Engineering and Applied Sciences,* 9 (9), 1653-1658, 2014.

[7] Evans, M., Hochenbaum, J., Noble, J., Arduino In Action. Shelter Island, NY: Manning., 2013.

[8] S.I. Yaniger, "Force Sensing Resistors: A Review Of The Technology", *Electro International,* 10.1109/ELECTR.1991.718294, 2002

[9] Prinima, Pruthi, J., "Evolution of Mobile Communication Network: from 1G to 5G", *International Journal of Innovative Research in Computer and Communication Engineering, 4*(4), 224-227, 2016.

[10] Tejaswini, Koteswar, P., Srinivasulu, M., "A Child-Left-Behind Alerting System Based On Capacitive Sensing Principle", *International Journal of Scientific Engineering and Technology Research,* 4(32), 6329-6332, 2015.

[11] Rao, K. V., Rao, P. K., B Ravi, B., "PIR Sensor Based Robot with HMH Applications", *International Journal of Engineering Research and Development,* 12(5), pp. 23-28, 2016.

[12] Rudd, R., "Functional Assessment of Unattended Child Reminder Systems", *National Highway Traffic Safety Administration, Washington, DC,* 2015.

IOP Conf. Series: Materials Science and Engineering **834** (2020) 012072 doi:10.1088/1757-899X/834/1/012072

The Efficacy of Visual and Auditory Bird Scaring Techniques using Drone at Paddy Fields

Wan Mazlina Wan Mohamed[1,2]*, Mohamad Nizar Mohd Naim[1], and Afiq Abdullah[2]

[1]Faculty of Mechanical Engineering, Universiti Teknologi MARA, 40450 Shah Alam, Malaysia
[2]Malaysia Institute of Transport, Universiti Teknologi MARA, 40450 Shah Alam, Malaysia

*wmazlina@uitm.edu.my

Abstract. Birds can cause damages to crop and financial losses to farmers and able to spread exotic diseases such as Avian influenza and Newcastle disease to human. Numerous methods have been used in controlling birds, e.g. chemical, auditory, visual and lethal. In this research, visual and auditory methods were applied, where reflective boards and speaker were installed on a drone and were flown at different altitudes and intervals. The scaring approach were divided into five (5) categories; reflective board only, ultrasonic sound, predator sound, combination of reflective board and ultrasonic sound and combination of reflective board and predator sound. The test was conducted at a paddy field plot, Botanical Park, Shah Alam. The findings indicate that combination of visual and auditory techniques scared the highest number of birds. However, the effectiveness gradually decreases as altitude increases. The interval of flight does not show any correlation with the number of birds that flew away. It is recommended to conduct future research far from residential area due the sound from the drone and auditory devices which cause disturbances to residents. Future experiments should be conducted in a control environment, free from human and ground vehicles interruption in order to have better observations on the efficacy of the chosen methods.

1. Introduction

Damages to crops cause by birds can cause substantial losses to agriculture industry. Farmers perceived the losses are more than the estimates made by surveys of crop damage bird food requirements [1], [2]. In the USA, it was estimated a loss of $50-$200 per acre for apple crops and from $1000-$5000 per acre for seasonal crops like blueberries [3]. The estimated losses were based on consumption, product losses which include partially eaten, crops felt on ground, or contaminated by fecal droppings.

Besides that, birds also consume fish in aquaculture ponds, bees from hives, livestock food from feed lots, and pose risk to aircraft strike. They also can spread exotic diseases to human such as Avian influenza and Newcastle disease [4]. Birds are able to spread disease to other animals who consumed the infected fruits. Starlings and other pest birds are implicated of transferring disease to cattle [5]. These issues cause concern not only to farmers but also consumers, because the disease could be transmitted through the consumption of contaminated food products.

There are several methods that have been used in bird management control either to reduce the population, deter or scare them from eating and damage the crops. These methods can be categorised

Content from this work may be used under the terms of the Creative Commons Attribution 3.0 licence. Any further distribution of this work must maintain attribution to the author(s) and the title of the work, journal citation and DOI.
Published under licence by IOP Publishing Ltd

into; lethal control, habitat modification, physical exclusion, chemical deterrents, auditory deterrent, and visual deterrents [6].

Lethal control methods permitted are by shooting, a cage trap, and/or net (hand-propelled). Shooting are divided into two different approaches; shooting to kill and shooting to scare. Shooting efficacy depends on a number of factors; the target species, site characteristics and shooting regime. It is more effective at smaller site, and the number of consecutive days and shooting parties have an effect on the reduction of bird numbers [6]. However, the reduction in numbers is limited, because birds able to evade shooters, and this technique is expensive, and time-consuming [7]. Live trapping is not cost effective either. Trapping birds to help in reducing the population but not significantly able to reduce the damages to crop [4].

Habitat modification involves removal and/or alteration of habitat features. Typical actions include trimming or cutting trees, removal of puddles or ponds, revegetation of barren areas and allowing grass to grow taller especially near airports. Brough and Brigman [8], reported that longer grass was found to be successful at reducing the number of birds [8]. This technique is expensive, especially in grass maintenance. It is effective in reducing the number of birds rather than elimination of birds [6].

As for physical exclusion methods, netting is commonly use in vineyards. It is also used to prevent birds eating the fishes from aquaculture farm. However, it is not suitable if the covered area is large because the cost of netting will be expensive. Another disadvantage of nets is that they are easily damaged by wind or bad weather and if there are any rips or tears, it will allow birds or other small animals to exploit the crops. Netting also requires frequent maintenance. Overall, netting is an effective method in preventing birds damaging the crops but it is dependent on proper installation and maintenance, thus it is only practical for small scale agricultural area [6].

Chemical deterrent can be categorized as tactile, behavioral and taste repellents. Tactile repellents are sticky substance which repel birds from sitting on perches, ledges, antennas or signboards. To apply the sticky substances on the identified ledges and perches are laborious, and the stickiness is not long lasting [9]. Behavioural repellents such as Avitrol can be classified as toxicants which cause disorientation and erratic behaviour to birds. They are usually added to bait and upon consumption, the birds will emit distress call and exhibit erratic behavior which frighten other birds. It can be lethal if too high dose was ingested. Permits are required and agent must be administered by the authority [9]. As for taste repellents, they are divided into primary and secondary repellents. Primary repellents smell or taste offensive, and secondary repellents can cause illness after ingestion. Primary repellants are derived from natural products while secondary repellents are from synthetic agrochemical pesticides. Secondary repellents are considered more effective form of deterrent [6].

Auditory deterrent methods are scaring stimulus such as loud noises detonated from guns, cannons, modified rockets, firecrackers, or high intensity sound transmitted from sonic devices such as recorded bird alarm, distress calls, ultrasonic and predator sounds. The auditory methods found to be effective due to loud noises which provoke fear response from birds [10] and birds' instinct to avoid danger. However, these methods are labour intensive and noise transmitted from explosions and audible bird scare devices are considered as nuisance to residents and also birds can become habituated to them unless they are varied regularly [11].

The common visual deterrent methods used are scarecrows, reflectors and reflecting tapes, predator model, gull model, kites and balloons. Based on the literature compilation of stationary visual deterrents by Transport Canada [9], their effectiveness are short term. Introduction of falconry, where trained falcons and hawks are used to chase birds from specific areas was found successful and effective [12] but must be conducted by well-trained dedicated falconers, and not suitable during high winds or heavy rains and at night [9].

Based on the research findings by Bishop et.al. [9], in terms of effectiveness of different categories mentioned above, habitat modification, lethal techniques and exclusion appeared to be more consistently effective than auditory, visual and chemical methods. However, since only one investigation was conducted on habitat modification and five experiments on exclusion, therefore it required further investigation to conclude the efficacy of these methods. As for cost effectiveness, exclusion method was rated to be most expensive followed by chemical techniques. Visual and auditory methods are less costly. It was also highlighted that combinations of techniques, used in integrated control strategy are considered more effective than techniques apply using single method [6], but lack of references to back up the scientific evaluation.

The latest technology used in deterrent and scaring strategies is the utilisation of unmanned aerial vehicles (UAVs) or drones. The experiments using UAVs or drones in bird management are becoming popular among researchers, industries and farmers. UAVs have been used to replicate natural predation. Falconry as mentioned above was found effective but expensive control method since it requires professional falconers and multiple predatory birds, making it becomes uneconomical to utilize this method to manage large area. Whereas, UAVs now are becoming more affordable to own and the cost of operators can be reduced with autonomous technologies. UAVs can easily be programmed and has the advantage of covering large vineyards or agriculture land in within a short duration compared to ground vehicles. Biomimetic design was incorporated on UAVs such as appearance mimicking, replicating wing flaming motion, broadcasting predatory cries and imitating predatory behavior [13].

Wandrie et. al [14] experimented with fixed-wing and rotary-wing UAVs at different altitude flown over captive and free-ranging blackbirds. Fixed-wing results show that the birds were not afraid of fixed-wing but responded to approaches by rotary-wing at low altitudes [14]. A hexacopter multirotor drone equipped with horn tweeter and crow taxidermy was used to scare the birds psychologically as though a crow was caught by the UAV and emits distress call. The drone was flown manually, and the effectiveness was evaluated based on the number of birds before and after the flights in relation to radius covered and duration of influence. The findings indicate that there was significant reduction of silvereye birds after the UAV was flown but it could not completely chase away all the birds after each flight [15].

The aim of this research is to conduct an experiment on birds' activity at the paddy fields in Malaysia, and evaluate the efficacy of drone equipped with speaker to broadcast the predator and ultrasonic sound, and reflective tape in chasing the birds away after flight. The important parameters to be determined are the effectiveness of the scaring methods, the influence of flight intervals and altitude.

2. Methods

The bird scaring experiments were performed in 2018 at paddy field plot, Botanical Park, Shah Alam, Malaysia. The size of each plot is about 65 x 45 meters (refer Figure 1). Paddy field is a place of interest of sparrows since rice is the main crop eaten by sparrows. Sparrows are categorised as major pest birds due to their ability to cause localised damage to the grain fields as they feed in large numbers over a small area. Sparrows are host of some parasites and diseases, like Chlamydiosis, Salmonellosis, Mycoplasma diseases, protozoal diseases and internal parasites like roundworms and tapeworms [16].

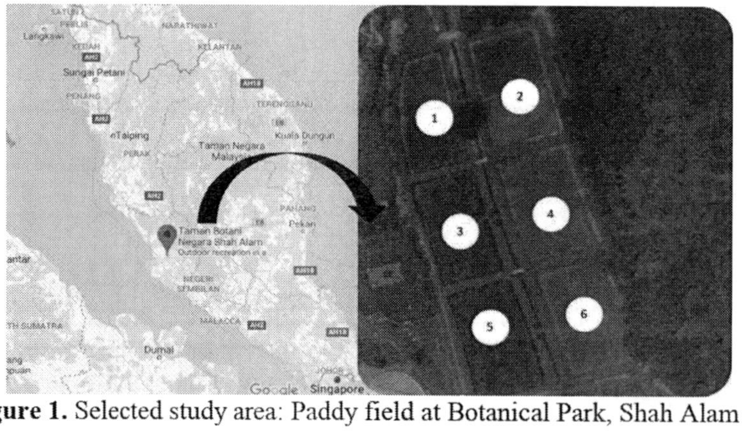

Figure 1. Selected study area: Paddy field at Botanical Park, Shah Alam

The drone used in this study was a multirotor DJI Phantom 3. It was selected due to its capabilities of built in applications and camera which can integrate with smartphones to display the altitude and the captured images. The auditory and visual deterrents devices were selected for this study. Ultrasonic and predator sound were used as the auditory deterrents and for the visual deterrent, a reflective board of a size 0m x 0.1m with a thickness of 2mm was used. The JBL Go speaker which is light (130 grams) and small was found suitable to be used as the medium to broadcast the ultrasonic and predator sound and attached on the drone.

The experiments were divided into five (5) categories, and Figure 2 shows the attachment of the reflective boards and speaker on the drone:

i. Reflective boards only - The drone was attached with reflective boards at each end of the drone landing gear. They were suspended freely using paper rope.
ii. Ultrasonic sound only - The speaker was connected with MP3 player using 3.5mm audio jack cable and playing ultrasonic sound for a period of time. It was suspended using a double-sided tape glued on the drone
iii. Predator sound only - The cried predator sound was broadcasted from the speaker.
iv. Reflective boards and ultrasonic sound - Both reflective board and speaker were attached to the drone. The speaker broadcasted the ultrasonic sound and the board reflected the sunlight
v. Reflective boards and predator sound - Same as in iv, except the speaker was broadcasting predator sound.

Figure 2. The application of drone with, (a) reflective board only, (b) reflective board and speaker, (c) speaker only

2.1. Experiment design

The initial experiment was conducted for six (6) consecutive days in late April 2018 between 9.00 am -12.00 pm. These hours were chosen because from earlier observations and advice from the wildlife authority, the birds were most active during those hours. No baits or feed were used to attract the birds to the testing site. The preparation of the drone equipped with the scaring devices was done early in the morning before the birds became active at the paddy field. Once the drone was ready to fly, the investigator measured and recorded the surrounding ambient sound level using the digital meter and the initial number of birds around the paddy fields were also recorded.

There are three (3) flight time intervals of flying the drone; every four (4) minutes, 10 minutes and random. The drone was flown at the height of five meters (m) and 10 m. For the case of four (4) minutes interval, the first time the drone flown in the air, it is recorded as 0 time, and the number of birds flown away were recorded. The investigators observed and recorded the number of birds at the paddy field for the next four (4) minutes. The drone was deployed again after four (4) minutes and the number of birds was flown recorded again. As the for 10 minutes interval, the first time the drone flown was set as 0 minute and the number of birds flew away are were counted. The investigator recorded the number of birds observed at the paddy field. The drone was flown again after the next 10 minutes and the number of bird flown were recorded again. As for random intervals, the drone was flown randomly and the number of birds flew away were recorded every time the drone was flown. Figure 3 shows the process flow involved in the experiment. The same process flow was repeated again at 10 m height. The difference between the number of birds before and after flights provided an indication of the effectiveness of these methods.

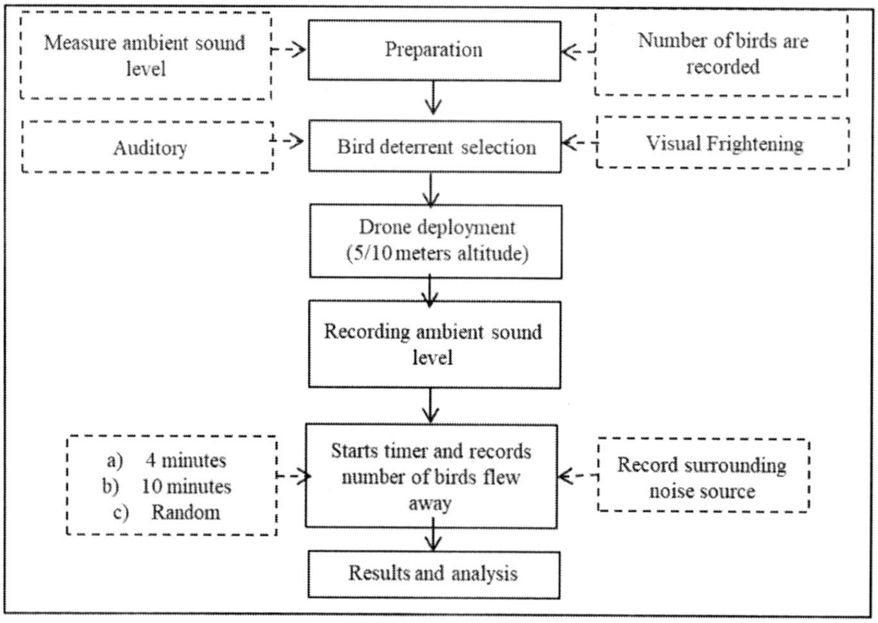

Figure 3. The process flow of the experiment

The number of birds flew away caused by other disturbances were also recorded during the 10 minutes observation. As the paddy field is situated in a public park, it was not possible to stop public from passing through the test area. However, it was interesting to see whether ground disruptions have greater impact in chasing the birds away.

3. Results

It was difficult to have the same number of birds to begin with as the reference since it was free-ranging study. As the initial number of birds observed were not the same for each type of experiments, therefore observed number of birds flown away were tabulated in percentage. Figure 4(a) shows the percentage of bird that flew away at the height of 5m, whereas Figure 4(b) is the result for the altitude of 10m. It is quite apparent that at higher altitude the effectiveness of scaring the bird for all the methods has decreased.

The results shown indicated that random flights for both altitudes seem to record lower percentage of birds flew away. This could be the factor that the birds are becoming immune to the drone flying around the paddy fields, and learnt that it would not harm them. Among all the five (5) devices used in this experiment, combination of reflective boards and ultrasonic/predator sound showed a slightly consistent result on the percentage of birds flown away except for the random flight at 10m height.

As the paddy fields are accessible to public, therefore other interferences on the ground such as movement of human, wind, and vehicles e.g. lorry, car, motorcycle, van or bicycle affect the number of initial birds during observation period.

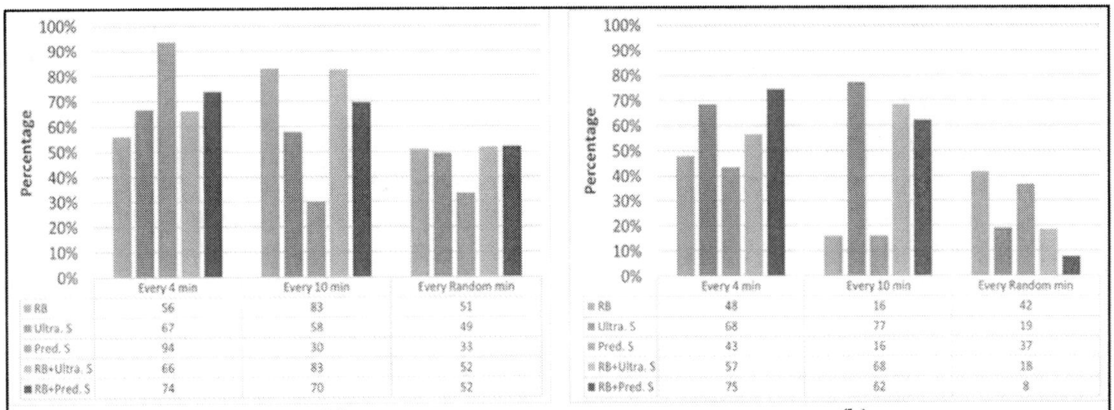

(a) (b)

*RB = *Reflective Boards*
**Ultra. S* = *Ultrasonic Sound*
**Pred. S* = *Predator Sound*
**RB+Ultra S* = *Reflective Boards and Ultrasonic Sound*
**RB+Pred. S* = *Reflective Boards and Predator Sound*

Figure 4. The percentage of bird flew away at different flight interval at the height of (a) 5m (b) 10m

The percentage of birds flew away for 5m and 10m heights were combined and the average was calculated to reflect which methods is most effective in scaring the birds away. Figure 5 shows the average percentage of the number of birds flew away compared to the percentage of birds flew away caused by ground disturbances.

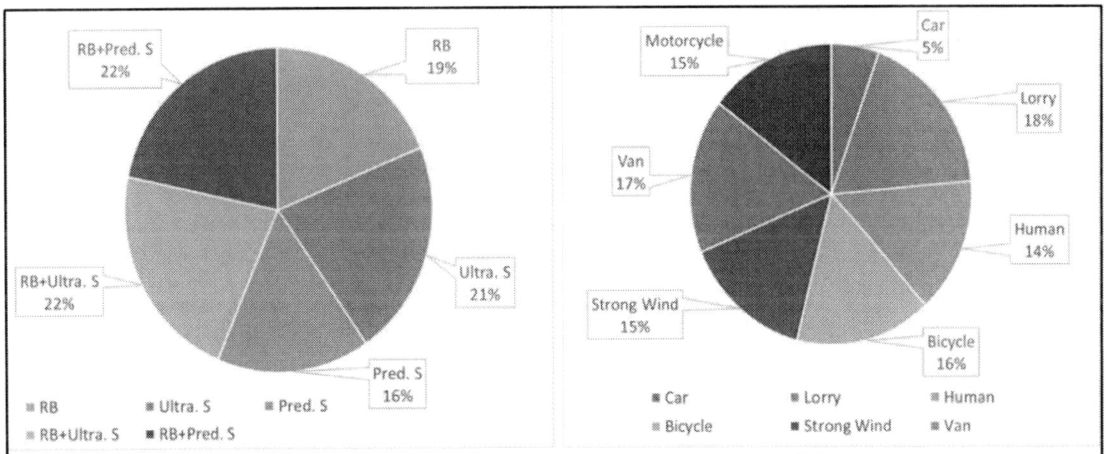

(a) (b)

*RB = *Reflective Boards*
**Ultra. S* = *Ultrasonic Sound*
**Pred. S* = *Predator Sound*
**RB+Ultra S* = *Reflective Boards and Ultrasonic Sound*
**RB+Pred. S* = *Reflective Boards and Predator Sound*

Figure 5. The mean percentage of birds flew away (a) for different methods b) different types of ground disturbances

Figure 5(a) shows the highest percentage of bird flown away are from the combination of reflection boards and auditory sound (ultrasonic and predator) with 22%, and followed by ultrasonic sound (21%). The predator sound seems to score the lowest (16%), but the result cannot be conclusive to state that this method is not effective because predator sound recorded the highest percentage during the testing of four (4) minutes interval (refer Figure 4a). Figure 5(b) shows that ground disturbances seem to have almost equal percentage among the sources. This probably mean that scaring the birds from the ground is more effective due to the sound emitted by large ground vehicle such as lorry is louder compared from the air, thus it is able to scare more birds.

3.1. Sound Level Measurement Analysis

The sound level measurement is usually used to limit the noise level from harming human. However, in this research, the sound level is measured to classify its subjective effect and community respond toward the sound produced from bird scaring methods. In order to measure the sound level difference between ambient sound and noise source, the following formula (1) is used [17].

$$L_{diff} = 10\log\left[10^{\frac{L_{p1}}{10}} - 10^{\frac{L_{p2}}{10}}\right] \tag{1}$$

L_{diff} is calculated by subtracting the sound level recorded with noise source and ambient sound. L_{p1} is the noise source produced by each method, while the L_{p2} is the ambient sound level. Once the value of sound level difference has been obtained, the subjective effects are referred to Table 1 [18].

Table 1. Subjective effect of changes in noise levels

Change in level of dBA	*Subjective Effect*
3	Just perceptible
5	Clearly perceptible
10	Twice as loud
20	Much louder

Referring to Table 2, at the height of 5 m, the sound produces to scare the bird can be considered as very loud, but at the height of 10m, the sound level has been slightly reduced. Thus, it shows that the effectiveness of scaring techniques have been reduced as the drone's flying height increased.

Table 2. The difference of sound level with height of drone at 5m and 10m

Noise Source with Drone Application	L_{diff1} (dBA)	L_{diff2} (dBA)	L_{diff3} (dBA)	Subjective Effect
h = 5m				
Reflective Board	12.9*	12.8*	9.7	*twice as loud (≥10 dBA)
Ultrasonic Sound	11.5*	13.6*	19.9*	
Predator Sound	16.5*	12.1*	10.3*	
Reflective Board and Ultrasonic Sound	16.8*	9.6	10.4*	
Reflective Board and Predator Sound	13.8*	15.4*	11.3*	
h = 10m				
Reflective Board	11.7	8.2*	6.9*	*clearly perceptible (≥ 5dBA but <10dBA)
Ultrasonic Sound	9.2*	8.2*	3.2*	
Predator Sound	6.3*	3.7*	7.1*	
Reflective Board and Ultrasonic Sound	9.1*	10.8	4.8*	
Reflective Board and Predator Sound	8.7*	13.4	5.5*	

*L_{diff1} = Change in level of dB(A) between ambient sound and noise source when drone flying every 4 minutes
*L_{diff2} = Change in level of dB(A) between ambient sound and noise source when drone flying every 10 minutes
*L_{diff3} = Change in level of dB(A) between ambient sound and noise source when drone flying every random minutes

4. Conclusion and Future Research

This study proposes a low-cost and non-lethal bird scaring methods to scare the birds away at the paddy fields. In conclusion, incorporating visual and auditory devices on a drone create a new dynamic approach for bird control method. The short-term responses from the birds indicate that the drone can potentially be used to scare birds away, but it must be carried out continuously. It was found that ground vehicles interferences have immediate effect in scaring the birds at the paddy fields, thus combination of air and ground methods should be looked further in the future.

For the percentage analysis, results indicated that combination technique of reflective board and ultrasonic/predator sound are more effective compared to single device. However, with increased of flight altitude to 10m, the effectiveness showed reduction. For the drone's flight time interval, it was found that the flight times have no significant difference to the number of bird that had been chased away. This is perhaps due to uncontrolled environment, where the number of initial birds during the experiment for each technique was different. The weather such as wind and rain also have an effect on the number of birds, where certain days during the investigation have less number of birds and some days have more birds.

The ground vehicles or human that passed through the park, chased away birds during the observation periods, thus affect the number of birds counted. Finally, sound level measurement showed that the noise emitted by the drone and the auditory devices can cause noise disturbance and annoyance to humans.

Therefore, the bird scaring method using drone are recommended to be implemented far from residential areas. For future studies, it is recommended that the bird scaring method to be carried out in a controlled environment where the initial number of birds can be assumed constant, and no interferences from ground vehicles or human. It is recommended that future research employed ground and air methods simultaneously equipped with several other visual and auditory scaring devices.

Acknowledgement

The authors would like to thank Universiti Teknologi MARA (UiTM) and Malaysia Institute of Transport (MITRANS), UiTM for supporting and sponsoring the author to present this paper at ICAME 2019.

References

[1] S. Noske, "Aspects of the behaviour and ecology of the white cackatoo (cacatua galerita) and galah (c.roseicapilla) in crops land in north-east New South Wales" *M.Sc thesis,* University of New England, Armidale, 1990.

[2] P. J. Jarman, "Bird pest research: The gap between research and application", *Proc. Natural Bird Pest Workshop,* NSW Agriculture and Fisheries, Armidale, pp. 7-12, 1990.

[3] A. Anderson et al., "Bird damage to selected fruit crops: the cost of damage and benefits of control in five states", *Crop Protection,* vol. 52, pp. 103-109, 2013.

[4] M. Bramord, "Review of research on control of bird pests in Australia", *Proc. of 15th Vertebrate Pest Conf.*, University of Nebraska-Lincoln, pp. 93-96, 1992.

[5] J. LeJeune et al., "Role of the European starling in the transmission of E.coli O157 on dairy farms", *Proc. 23rd Vertebrate Pest Conf.*, University of California, Davis, pp. 31-34, 2008.

[6] J. D. Bishop et al., "Review of international research literature regarding the effectiveness of auditory bird scaring techniques and potential alternatives", *Dep. Food Rural Aff.*, pp. 1-52, 2003.

[7] P. Fleming, "Some other bird control techniques", *National Bird Pest Workshop,* NSW Agriculture and Fisheries, pp. 17-22, 1990.

[8] T. Brough and G. J. Bridgman, "An evaluation of long grass as a bird deterrent on British airfields", *J. Applied Ecology,* vol. 17 pp. 234-253, 1980.

[9] Transport Canada, "Evaluation of the efficacy of the products and and techniques for airport bird control", 03-1998, TP 13029, *Government of Canada website,* [Online]. Available: https: /. [Accessed: 00-00-20XX].

[10] Z. Wang et al., "Physcologicl warfare in vineyard: Using drones and bird psychology to control

bird damage to wine grapes", *Crop Protection,* vol. 12, pp. 163-170, 2019.

[11] Narural Resources Wales (NRW), "Detering birds from farms, fisheries, airfields, landfill sites and urband environments" *NRW website,* [Online]. Available: http:// /. [Accessed: 00-00-20XX].

[12] O. Daugovish and M. Yamamoto, "Control of bird pests in the strawberry field with falconry", *Proc. 23rd Vertebrate Pest Conf.*, University of california, Davis, pp. 80-83, 2008.

[13] Z. Wang and K. C. Wong, "Biomimetic design for pest bird control UAVs: A survey", *17th Australian Aerospace Congress,* 26-28 February, 2017.

[14] L. J. Wandrie, P. E. Klug and M. E. Clark, "Evaluation of two unmanned aircraft systems as tools for protecting crops from blackbird damage", *Crop Protection,* vol. 117, pp. 15-19, 2019.

[15] Z. Wang et al., "Psychological warfare in vineyard: Using drones and bird psychology to control bird damage to wine grapes", *Crop Protection,* vol. 120, pp. 163-170, 2019.

[16] J. D. Brown et al., "Infectious and lethal doses of H5N1 highly pathogenic Avian influenza virus for house sparrows (Passer domesticus) and rock pigeons (Columbia livia)", *J. Vet. Diagn. Invest,* vol. 21, pp. 437-445, 2019.

[17] P. J. Guyer, "Fundamentals of acoustics, Course no. M02-022, Continuing Education and Development", *Inc.*, p. 8, 2009.

[18] E. Murphy and E. A. King, "Principles of Environmental Noise", *Environmental Noise Pollution,* pp. 9- 49, 2014.

IOP Conf. Series: Materials Science and Engineering **834** (2020) 012073 doi:10.1088/1757-899X/834/1/012073

Effects of Material Properties in Spinal Fusion Cage for Lumbar Vertebrae

Rusnani Yahya[1,2], Muhammad Lukman Shudin[2], Muhammad Hazli Mazlan[3], Solehuddin Shuib[2], Abdul Halim Abdullah*[2]

[1]Department of Electrical Engineering, Politeknik Sultan Salahuddin Abdul Aziz Shah, Seksyen U1, 40150 Shah Alam, Malaysia
[2]Faculty of Mechanical Engineering, Universiti Teknologi MARA, 40450 Shah Alam, Malaysia
[3]Microelectronics and Nanotechnology-Shamsudin Research Center (MiNT-SRC), Faculty of Electrical and Electronic Engineering, Universiti Tun Hussein Onn Malaysia, 86400 Batu Pahat, Malaysia

*halim471@uitm.edu.my

Abstract. Vertebral endplate subsidence failure will increase the possibility of mechanical instability in the lumbar spinal fusion. The spinal implants of cages are many type of design that can be use in spinal fusion and the type of material also been considerate. Posterior lumbar interbody fusion (PLIF) cage is believed to maintain the stability and to promote fusion between vertebrae. Different type biomaterial of cage was made by metal, ceramic and polymer, also composite. The Posterior Instrument (PI) such pedicle screw and rod was made by Titanium Alloy. The aim of this study was to examine the effects of different materials of cage in lumbar fusion. 3D vertebrae model (L3-L4) with interbody fusion using different material properties such as PEEK, PLA, Cobalt Chromium, Titanium Alloy and Stainless Steel. A fusion model with pedicle screw systems L3–L4 levels were reconstructed based on the respective surgical protocols. The stress distribution and total deformation of the cage were measured under different compressive loading conditions and motion. Results show that the titanium alloy is the best material for metal categories while PLA (Poly lactic acid) for composite category.

1. Introduction

Spinal fusion surgery is one of the most common procedure in treating spinal disorders. Although it has been implemented for many years, there are still many important challenges and issues need to be highlighted especially the level of unity. The finite element analysis is part of the current approaches in predicting the influence of different cage designs on the fusion process [1, 2]. Spinal fusion is a surgical technique performed to combine two or more vertebrates in the spine. It is usually performed to treat a number of conditions such as trauma, spinal defect (eg, scoliosis or kyphosis), degenerative diseases, and infectious or neoplastic diseases [3]. The procedure of this surgery is by using the interbody fusion technique which the cages implant will be inserted between the vertebra bones to fill the disk space with supported by rods and screws to hold the vertebra bones. There are many type of interbody cages implant design available in market and some model had been approved by the United States Food and Drug Administration (FDA) [4].

Interbody implants are developed for a rigid axial mechanical support which will improve the initial stability. The grafting material housed within and around the spacer could achieve a better biological

Content from this work may be used under the terms of the Creative Commons Attribution 3.0 licence. Any further distribution of this work must maintain attribution to the author(s) and the title of the work, journal citation and DOI.
Published under licence by IOP Publishing Ltd

fusion [5-7]. It will correct deformity and achieve nerve root decompression through restoration of lost disk space with implantation of cages [8]. The cages are designed to restore normal disc height, improve construct stiffness and thus reduce posterior instrumentation failure. Cages also have several type of material which is PEEK, PLA, Titanium Alloy, Cobalt Chromium and Stainless Steel. The most commonly used materials used for interbody cages are titanium metal and polymer polyetheretherketone (PEEK). Poly lactic acid (PLA) is assumed to be the alternative material which could provide cheaper material and lower production cost.

The aim of this study is to investigate the effect of material properties of intervertebral fusion cage and the posterior instrumentation on compressive load, and four loading condition of a treated spinal lumbar motion segment.

2. Material & Methods

2.1. Development of 3D model

Finite Element Analysis was applied to evaluate the cage subsidence phenomenon in the lumbar vertebra. The 3D finite element model of third to fourth lumbar vertebrae, known as L3 and L4, respectively as shown in Figure 1.

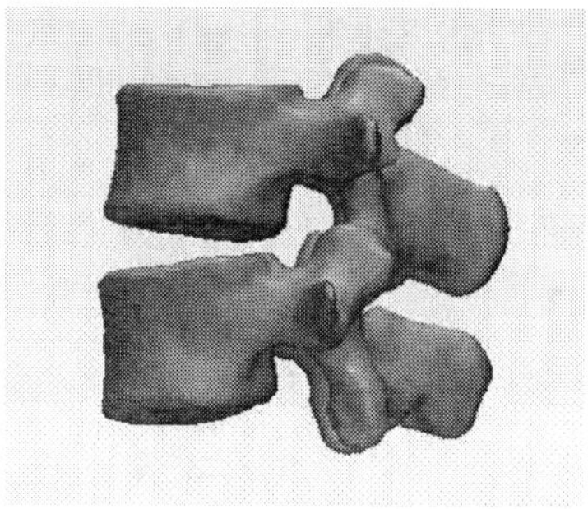

Figure 1. 3D model of L3-L4 bones

The CT scan data of vertebra bone of patient is transfer to 3D model. The CT scan data of lumbar vertebra cutting process is performing by biomedical software. The part of lumbar vertebra that needs to develop is L3 and L4 lumbar vertebra bone only. So, the vertebra need to be slice by each CT scan data to ensure it is not connected to lower and the upper part of lumbar vertebra bones. The other part of vertebra bones and pelvis will be deleted by slices of CT scan data, shown in Figure 2 whole region of the vertebra bone.

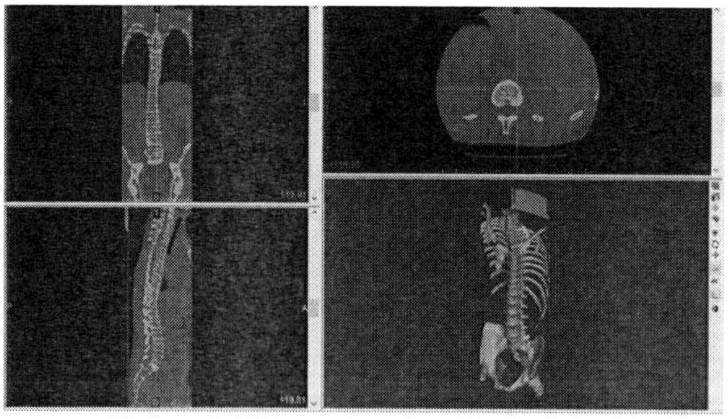

Figure 2. The green region of whole part

After all part that connected to L3 and L4 lumbar vertebra bone have been deleted by edit 3D mask of the model, the part left will be import to dedicated design software to edit the surfaces and shells of the model. Finally, it will be imported back to biomedical software to gets the model of L3 and L4 vertebra bones completely edited and export it to igs file.

Five different of material of fusion cages were simulated. A conventional monobloc cage namely convex-shaped (Figure 3) was modelled to better fit the disc space anatomy. The cage was modelled as a material composite of PEEK and PLA. The all implant design that written in .stl file will be converted into igs by using the CAD Converter software. Then, import the .igs file in computational software to get the detail of the implant design. Then, design software (CATIA) was used to convert the written file as .sat for the model and further to perform the FE analysis.

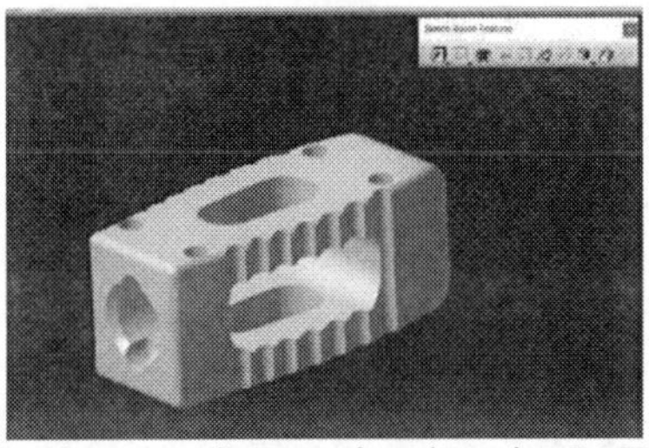

Figure 3. The imported cages implant in CATIA software

The spinal implant will be placed at the L3 and L4 lumbar vertebra to complete the spinal fusion model which it will be import to CATIA software to perform analysis step. The screws will be implanted bilateral at each lumbar vertebra bone and the rods will be placed at each two screw to connect it together. The cages implant will be inserted between lumbar vertebrae with same spot (Figure 4).

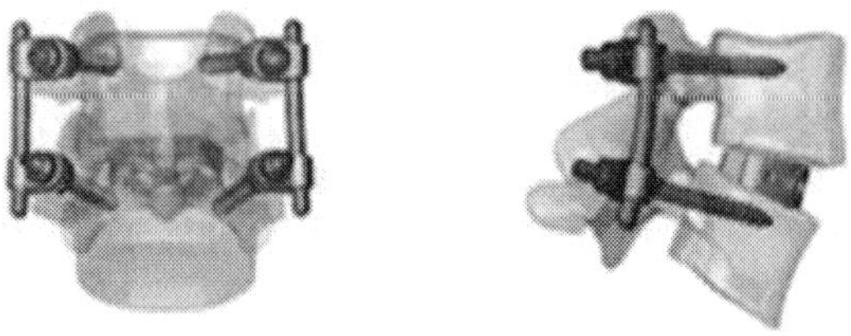

Figure 4. Example of reconstructed model of spinal fusion on lumbar vertebra at different views

2.2 Finite Element Modelling

The mechanical properties of the lumbar model and implant was assigned as linear, isotopic and homogeneous material as shown in Table 1.

Table 1. Material properties of the FE model

Material	Young's Modulus (MPa)	Poisson ratio	Yield's Strength (MPa)
Cancellous Bone	350	0.25	-
Cortical Bone	12000	0.3	-
Intervertebral disc	8.4	0.45	-
Facet Joint	11	0.2	-
PLA	3.620×10^3	0.39	130
PEEK	1.459×10^3	0.4	231.2
Titanium Alloy	105×10^5	0.34	795
Stainless Steel	210×10^5	0.29	190
CoCr	200×10^5	0.3	450

The spinal fusion on lumbar vertebra model of L3 and L4 will be import to computational software to perform the analysis on it with the applying distributed loading of compressive force at the top of the lumbar vertebra bones. Boundary condition applied in this study represent the normal physiological activities of the spine. The FE models were loaded with compressive load of 1000N. The results of FE analysis for each material properties of cages implant on spinal fusion will be compared each other to get the different effect on using different type of material properties cages implant.

3. Results and Discussion

The effects of different material properties of the cages were discussed in the resulting von-mises stress and displacement. Simulation result demonstrated good slip behaviour under axial rotation, flexion, extension and lateral bending, as compared with the different material properties interbody cages, even with the additional posterior fixation. However, the flexion of Titanium Alloy is highest, showed that material suitable used for cages and also for posterior instrument but lowest extension for this material. But had not much difference under axial rotation and lateral bending for all type of cages material properties.

Findings on total displacement at the human spine FE model for the load case of a 1000N in Figure 5 and Figure 6. Showed that relation between spinal fusion with posterior instrument and L3-L4 vertebrae are not much affected on displacement, cage is stable in the position and also the vertebrae.

Figure 5. Displacement for L3-L4 (unit: mm) at different materials of cages

PLA are most affected on displacement when apply the material properties on L3-L4 vertebrae model in simulation. However, these materials are cheaper material but give high quality of implant biomaterial. Cage from these categories must in complete set with posterior instrument for more stability. Usually, posterior material is from metal categories such as Titanium Alloy are used for pedicle screw and rod.

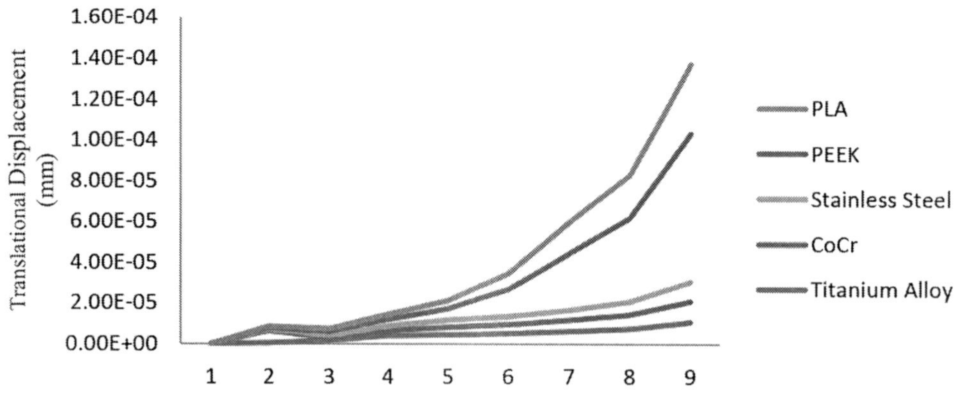

Figure 6. Translational Displacement of L3-L4 on 1000N compressive loading

The simulation study suggested that the different material properties of spinal fusion and posterior instrument important to get better stability and also less stress at the bone when doing physiology activity such as walking or climbing/jumping.

4. Conclusion

A computational modelling study was conducted to evaluate the effect of material properties for different cage material with pedicle screw and rod. However, the results of our study demonstrated that the posterior lumbar interbody fusion (PLIF) with different categories of implant material was important stabilization system with low stiffness not only produced desirable stabilization but also allowed near-intact motion. FE study has predicted changes in stress due to flexibility of stabilization systems. The FE result clearly showed that the vertebrae with hinged screw permit a better load sharing with lower stresses at the bone-implant interface. however, further analysis of the construct in the presence of follower compression pre-load and long-term assessment are essential for overall assessment of the stress and displacement of vertebrae in all condition of physiological motion.

Acknowledgments

This research was supported by Universiti Teknologi MARA, UiTM under Grant No. 600-IRMI/PERDANA 5/3 BESTARI (103/2018). We thank and acknowledge our colleagues from Kyushu University, Japan who provided insight and expertise that greatly assisted the research.

References

[1] Jalil, M. H., Mazlan, M. H., & Todo, M. (2017). Biomechanical Comparison of Polymeric Spinal Cages Using Ct Based Finite Element Method. *International Journal of Bioscience, Biochemistry and Bioinformatics*, *7*(2), 110–117.

[2] Bashkuev, M., Checa, S., Postigo, S., Duda, G., & Schmidt, H. (2015). Computational analyses of different intervertebral cages for lumbar spinal fusion. *Journal of Biomechanics*, *48*(12), 3274–3282.

[3] Manbachi, A., Cobbold, R. S. C., & Ginsberg, H. J. (2014). Guided pedicle screw insertion: Techniques and training. *Spine Journal*, *14*(1), 165–179. https://doi.org/10.1016/j.spinee.

[4] Cho, W., Wu, C., Mehbod, A. A., & Transfeldt, E. E. (2008). Comparison of cage designs for transforaminal lumbar interbody fusion: A biomechanical study. *Clinical Biomechanics*, *23*(8), 979–985.

[5] Davis W. *et al.*, "Modern spinal instrumentation. Part 1: Normal spinal implants," *Clin. Radiol.*, vol. 68, no. 1, pp. 64–74, 2013

[6] Mazlan, M. H., Todo, M., Takano, H., & Yonezawa, I. (2016). Effect of Cage Insertion Orientation on StressProfiles and Subsidence Phenomenon in Posterior Lumbar Interbody Fusion. *Journal of Medical and Bioengineering*, 5(2), 93–97.85–94.

[7] Provatidis, C., Vossou, C., Petropoulou, E., Balanika, A., & Lyritis, G. (2009). A finite element analysis of a T12 vertebra in two consecutive examinations to evaluate the progress of osteoporosis. *Medical Engineering and Physics*, *31*(6), 632–641.

[8] Tsuang, Y., Chiang,Y., Hung,C., Wei,H., Huang, C., & Cheng, C.,(2009). Comparison of cage application modality in posterior lumbar interbody fusion with posterior instrumentation-A finite element study, *Med. Eng. Phys.*, 31(5), 565–570.

IOP Conf. Series: Materials Science and Engineering **834** (2020) 012074 doi:10.1088/1757-899X/834/1/012074

The Effects of Sintering Schedule on Alloy Powder ASC 100.29 Compacts Formed at Elevated Temperature

Z Hasan*, M A Ismail and M M Rahman

Department of Mechanical Engineering, Universiti Tenaga Nasional,
Putrajaya Campus, Jalan IKRAM-UNITEN, 43000 Kajang, Selangor, Malaysia

*zaimah@uniten.edu.my

Abstract. This paper presents the outcomes of an experimental investigations on the effect of sintering schedule i.e. holding time and temperature to the final properties of alloy powder ASC 100.29 formed through uniaxial die compaction process at elevated temperature. A lab-scale powder compaction rig with uniaxial compression and installation of heater was designed and fabricated in order to produce green compact from powder mass at elevated temperature. Alloy powder ASC 100.29 (99.7 wt%) and lubricant, zinc stearate (0.3 wt%) were mixed mechanically for 30 min at a rotation of 30 rpm. The powder mass was compacted at 150°C via simultaneous upward and downward uniaxial compression load of 425 MPa. Defect-free green compacts were generated from the compaction and were subsequently sintered at 1000°C using a custom argon gas fired furnace at three different sintering rates, i.e., 5, 10 and 15°C/min for three holding times, i.e., 30, 60 and 90 min, respectively. The final products were characterized through their physical, electrical and mechanical properties and their microstructures evaluated. The results revealed that the sample sintered for a longer period of time, i.e. 90 minutes at the slowest sintering rate, i.e. 5°C/min obtained the better final characteristics.

1. Introduction

The rapid developments in powder metallurgy have led material processing route into a new phase especially in past few decades since these techniques offer alternative way to mitigate the deficiencies of the traditional methods of material processing such as forging, casting, stamping, forming and machining. Powder metallurgy process involved three main components, i.e., powder mixing, powder compaction into desired shape and sintering of the products generated from compaction [1]. Nowadays, the demand of the alloy in industries has increased due to its ability to improve the properties of the material in many aspects, i.e., physical, mechanical, chemical, electrical and thermal properties, thus capable of resisting damage at critical environment especially under high mechanical stress at higher temperature and in corrosive atmosphere, as well as becoming more versatile [2-4].

There are several methods available to produce alloy components but these methods are not economical as these methods require series of linked lengthy processes, hence the long production time, inefficient use of energy and more extra scrap losses produced. This is the especially evident when most of the alloys are transformed in billet form in the first place before undergoing various processing technologies to convert it into required alloy components [5]. Alternatively, utilization of pre-alloyed powder through warm powder compaction process has been widely chosen to produce alloy components where the process begins with feedstock preparation involving formulated pre-alloyed powder and mixing before compacting the powder mass at elevated temperature into desired shape and to the final process which is sintering of green compacts in controlled atmosphere [6]. Alloy components equipped with high strength and near-net-shape can be achieved by applying this technique since it is relatively new although has been initiated in 1998 [6].

Content from this work may be used under the terms of the Creative Commons Attribution 3.0 licence. Any further distribution of this work must maintain attribution to the author(s) and the title of the work, journal citation and DOI.
Published under licence by IOP Publishing Ltd

Several research activities are found on the warm compaction process which practically used iron ASC 100.29 powder as main powder constituent to form alloy components with other different elemental powders in order to fulfil the demands [7-9]. The results revealed that high quality green compact alloy components can be achieved through compaction at elevated temperature and pressing simultaneously from both upward and downward direction compared to cold compaction method. Green compacts generated with higher density and green strength are expected to have higher flexural, tensile and fatigue strength yet better dimensional tolerances [10]. The frictional force is unavoidable during compaction due to friction between the die wall and the powders as well as among the powder particles itself. Zinc stearate is used as lubricant to reduce the frictional force and to extend tool life. The optimum value of lubricant is important to generate high quality alloy components since it was able to influence the density of the product [11-12].

A lot of research activities are reported on green compact generation at elevated temperature but no thorough study on characterization of sintering schedule especially on near-net-shape alloy components production has been investigated. Sintering is a process of heat treatment on the green compact products conducted in controlled environment in order to complete a full cycle of powder compaction process. Sintering process tends to reduce the surface contact among particles powder after material transport has been initiated through change of pore geometry, particles growth and shrinkage of pore volume [13]. The attraction of atomic and molecular bonding in powder compacts during heat treatment allows the development of a solid coherent body and subsequently, increases the strength of particles bonding as well as the rate of the desired degree of bonding among the particles [14].

Sintering is controlled by three main parameters, i.e., sintering temperature, sintering rate (heating and cooling) and holding time. These three parameters can be manipulated in order to improve the microstructure and reduce the number of pores inside the samples. There are three phases in the sintering process which started with heating phase, then holding phase (high temperature phase) and lastly, cooling phase which is assisted by transport mechanisms, i.e., plastic or viscous flow, diffusion (surface, volume, and grain-boundary) and atoms evaporated on surfaces, thus affecting the recrystallization and densification [14]. Densification process can be interpreted through the relationship of the sintered density to the mass changes and the volume (dimensional) since they are directly related to each other [15].

Proper sintering schedule is important to the powder metallurgy process as sintering generally can improve the properties of green compact product in many aspects. Correct sintering schedules need to be identified in order to produce near net shape of final alloy powder ASC 100.29 product since the current practice only focusses on foundry process and mechanical alloying. Thus, the objective of this paper is to investigate experimentally the effect of sintering schedule to the alloy powder ASC 100.29 compacts formed at elevated temperature.

2. Materials and Methods

The experimental process was conducted by applying four consecutive steps, i.e., a) feedstock preparation, (b) compaction and shaping, (c) sintering, and (d) sample characterization. In this experiment, 99.7 wt% of alloy powder ASC 100.29 manufactured by Höganäs AB was utilized and the balance was zinc stearate powder as lubricant. Alloy powder ASC 100.29 and zinc stearate were mixed mechanically for 30 min at a rotation of 30 rpm. The mixed powder mass was filled into cylinder shaped die cavity and compacted by using T-15 compaction rig at 150°C through axial loadings from both upward and downward direction at 425 MPa simultaneously to generate defect-free green compact with sufficient strength. Sintering process was executed by using a custom argon gas fired furnace (Model: HT3-1400-SIC, S/N: LT007) and all defect-free green compacts were sintered at a constant sintering temperature of 1000 °C at different sintering rates (5, 10 and 15°C/min) and different holding times (30, 60 and 90 min), respectively.

Both the green samples and sintered products were characterized for their physical and electrical properties. Afterwards, sintered products were subsequently characterized for their mechanical properties and their microstructures were analysed. The density of each sample was calculated by using the mass and volume (cylinder) of the sample. The density of the product was divided by the pore free density in order to find the relative density of the product. The electrical resistivity was calculated by measuring the resistance (ohm) using a digital multimeter (Brand: Fluke, Model: Fluke 115, S/N: 28341190ws) and the bending strength was measured by using three point bending test machine (Brand:

Instron, Model: Instron 3365, S/N: SAA61569) based on the standard (ASTM E290-09). The microstructures of the sample were evaluated by analysing the image of the fractured surface via scanning electron microscopy (Brand: JEOL, Model: JSM- 6010PLUS/LA).

3. Results and Discussion

The results of relative densities for final products are recorded in figure 1 after sintering via 5°C/min, 10°C/min and 15°C/min of sintering rate for 30 minutes, 60 minutes and 90 minutes of holding time, respectively. Relative densities of whole samples escalated after all green compacts were sintered using these parameters. The highest relative density was acquired by the samples sintered at 15°C/min of sintering rate for 90 minutes holding time. The samples sintered at prolonged holding time are exposed with longer high temperature phase which is the most important phase in sintering process. The maximum sintering temperature was continuously applied here based on the holding time and thus, the powder particles have fully reacted to initiate inter-particle bonding where the grains and necks started to grow and this has resulted to a decrement in size of pores inside the samples [16]. As for the sintering rate, 10°C/min and 15°C/min seemed to have the capability to provide higher relative density when the holding time is prolonged. Slower sintering rate might have affected the duration of sintering process where the time will be increased especially during heating and cooling phase.

After sintering, the powder compacts might experience either shrinkage or swelling due to various circumstances for example green compact density, the size of particle, heating rate, processing temperature, material compatibility and other factors [17-18]. In figure 2, the result indicated that the final products have gone through shrinkage after sintering schedule was applied on green compacts. Sample sintered at 15°C/min of sintering rate for 60 minutes of holding time is recorded as having highest percentage of volumetric shrinkage. Sample sintered using 15°C/min of sintering rate seemed to have both highest and lowest percentage of volumetric shrinkage at different holding times which are 60 minutes and 30 minutes, respectively. Higher percentage of volumetric shrinkage is not recommended although it is able to give high relative density near-net-shape products. The effects of sintering parameters to the volumetric shrinkage are assisted by the number of pores left from the zinc stearate that disappeared upon full burning during sintering.

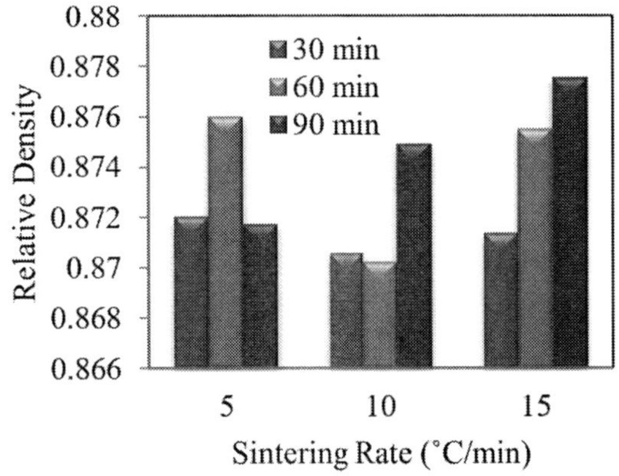

Figure 1. Relative density of sintered product at different holding time and sintering rate

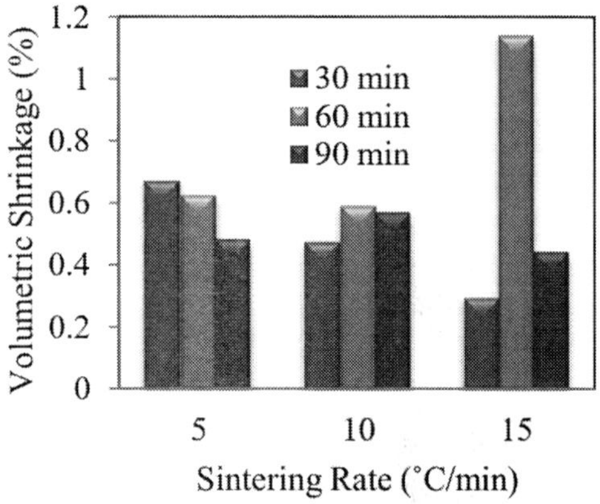

Figure 2. The percentage of volumetric shrinkage after sintering at different schedule

The electrical resistivity of the sample has been influenced by sintering parameters resulting to the final products to have obtained lower electrical resistivity compared to green compacts. The electrical resistivity of the final products is shown in figure 3 and the result indicated that all samples have almost similar value except for 3 samples which have higher resistivity value. Highest electrical resistivity is obtained by the sample sintered at 5°C/min of sintering rate for 90 minutes of holding time. Theoretically, lower electrical resistivity caused electrical conductivity to be higher due to the current flowing inside the sample without having difficulty [19].

Bending strength is defined as the ability of the material to resist deformation under load and acted as a key factor for mechanical stability. Bending strength of the sample sintered under different sintering parameters is shown in figure 4. Highest bending strength is obtained by sample sintered at 10°C/min of sintering rate for 90 minutes of holding time and the lowest is acquired by sample sintered at 10°C/min for 30 minutes. Based on the results, longer holding time is capable to produce higher bending strength of final products. Bending strength of final products might be affected by the number of pores left by zinc stearate since it was able to initiate crack propagation among interconnected pores when load is applied.

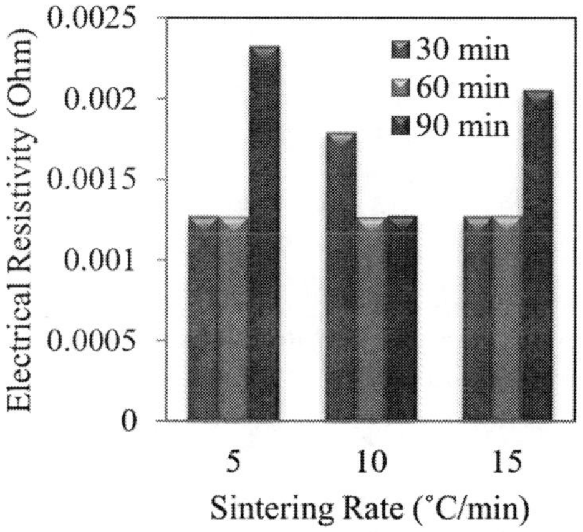

Figure 3. Electrical resistivity for sintered products under different rates and holding times

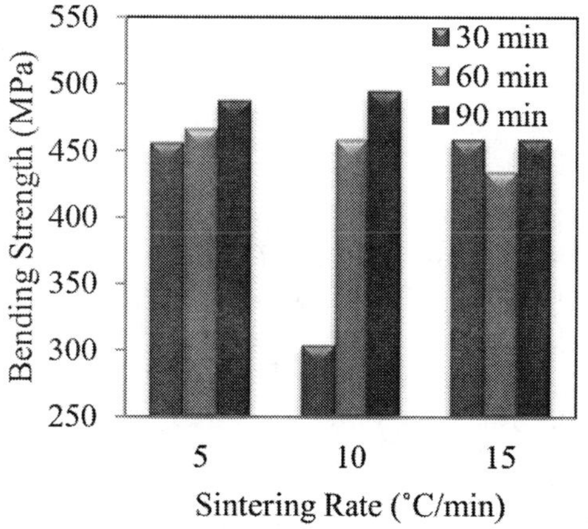

Figure 4. Bending strength of sintered products

The microscopic images of final products at 1500 of scale magnification are shown in Figure 5. The result indicated that 5°C/min of sintering rate acquired a better quality of microstructure compared to 10°C/min and 15°C/min of samples since the grain boundary area have been reduced due to grain growth and simultaneously reduced surface area. Thus, the number of pores have been reduced and only consisted of small sized pores but at certain area, the small pores are connected to each other to develop a link of interconnected pores. The grain growth in alloy powder mass under slow sintering rate may avoid development of micro-cracks due to the thermal shock and subsequently, reduced the load bearing capability of the products [20].

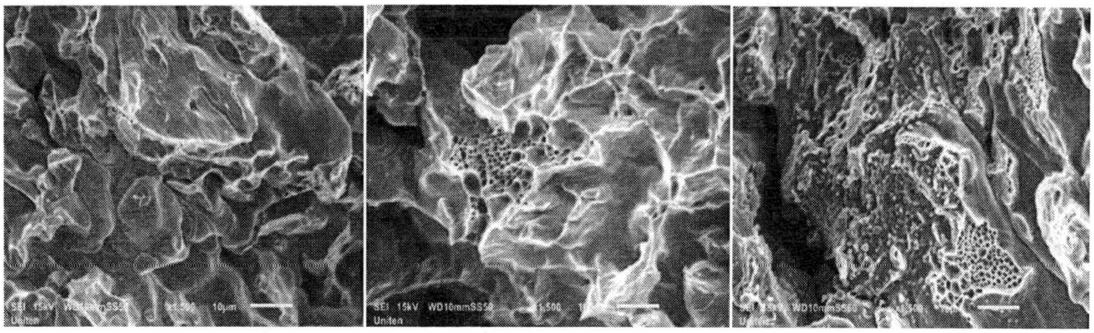

(a) 5°C/min, 1000°C (b) 10°C/min, 1000°C (c) 15°C/min, 1000°C

Figure 5. Microstructures of fractured surface for products sintered at 1000°C at different sintering rates

4. Conclusion

The effects of sintering schedule in producing cylindrical shaped products from alloy powder ASC 100.29 compacts formed at 150°C was investigated. The results revealed that higher relative density, bending strength and electrical resistivity with moderate percentage of volumetric shrinkage can be achieved by using 90 minutes of holding time. As for sintering rate, 5°C/min can be considered to be suitable for ASC 100.29 alloy in order to generate microstructure with good inter particle bonding with reduced number of pores and higher electrical resistivity in final products.

Acknowledgement
This research is financially supported by Universiti Tenaga Nasional through the research project UNITEN Internal Research Grant (UNIIG 2018) J510050871.

References
[1] Yamaguchi K, Takakura N, and Imatani S 1997 *Journal of Materials Processing Technology* **63** 364-369
[2] Rahman M M and Talib A A A 2015 *Materials Today Communications* **4** 42-49
[3] Rahman M M and Yogaswerarow S R 2016 In *Advanced Materials Research* **1133** 264-268
[4] Akca E and Gürsel A 2015 Periodicals of Engineering and Natural Sciences (PEN) **3**
[5] Pusavec F, Deshpande A, Yang S, M'Saoubi R, Kopac J, Dillon Jr O W, and Jawahir I S 2014 *Journal of Cleaner Production* **81** 255-269
[6] Sonsino C M and Ratzi R 2004 *Journal of Powder metallurgy* **47** 352-357
[7] Rahman M M, Ariffin A K, Nor S S M and Rahman H Y 2011 *Materials & Design* **32** 264-271
[8] Rahman M M and Nor A M 2014 *Advanced Materials Research* **879** 245-250
[9] Rahman M M, Nor S S M and Rahman H Y 2012 International Journal on Advanced Science, Engineering and Information Technology **2** 241-245
[10] Rahman M M, Rahman H Y and Zabri N H 2015 *Scientific Journal of PPI-UKM* **2** 108-111
[11] Rahman M M, Nor S S M and Ariffin A K 2013 *Procedia Engineering* **68** 425-430
[12] Nor S S M, Rahman M M, Tarlochan F, Shahida B and Ariffin A K 2008 *Journal of Materials Processing Technology* **207** 118-124
[13] Thummler F and Oberacker R 1993 *Introduction to Powder Metallurgy* (Oxford: Oxford Science Publications)
[14] Upadhyaya G S *2000 Sintered metallic and ceramic materials: preparation, properties, and applications* (Chichester NY: Wiley)
[15] Wang Y U 2006 *Acta materialia* **54** 953-961
[16] Narula G K, Narula K S and Gupta V K 1989 *Materials science* (New York: Tata McGraw-Hill Education)
[17] Kang H Z and Hu C T 2004 *Materials chemistry and physics* **88** 264-272
[18] Rahman M M, Rosli M H and Namasivayam D 2016 *International Journal of Automotive and Mechanical Engineering* **13** 3168
[19] Rahman M M, Ismail M A and Rahman H Y 2016 *Scientific journal of PPI-UKM 3* 156-161
[20] Rahman M M, Ismail M A, Sopyan I and Rahman H Y 2018 *Materials Science and Engineering* **290**

IOP Conf. Series: Materials Science and Engineering **834** (2020) 012075 doi:10.1088/1757-899X/834/1/012075

Study on Hardness and Shear Strength with Microstructure Properties of Sn52Bi/Cu + 1% Al_2O_3 Nanoparticles

Amares Singh*[1,2] and Rajkumar Durairaj[2]

[1]Centre for Advanced Materials Research and Manufacturing Processes, SEGi University, No 9, Jalan Teknologi, Taman Sains Selangor, Kota Damansara PJU 5, 47810 Petaling Jaya, Selangor Darul Ehsan, Malaysia.
[2]Lee Kong Chian Faculty of Engineering and Science, Universiti Tunku Abdul Rahman, Jalan Sungai Long, Bandar Sungai Long, 43000 Kajang, Selangor, Malaysia

*amaressinghgill@segi.edu.my

Abstract. The Sn-58Bi (SB) lead-free solder was added with 1% nanoparticles (NP) of aluminium oxide (Al_2O_3) to investigate the shear strength based on microstructure and joint properties. The microstructures of the SB added with 1% Al_2O_3 NP containing Bi precipitation and β-Sn grains. The spacing between the lamellar structures of a SB solder alloy was narrower. Proper arrangement of the microstructure of β-Sn, Bi and with the presence of Al_2O_3 as discrete particles (dispersion strengthening) contributes to the increase on the shear strength and hardness value. The shear strength of the SB added with 1% Al_2O_3 NP solder joint was enhanced by 50% compared to the bare SB solder joint while there was a 33% increase in the microhardness as well. The shear test was conducted based on ASTM D1002 specification of single lap joint method. The presence of Al_2O_3 NP along the interface (Hall-Petch effect) contributes to the improvement in the shear strength by reducing the thickness of the IMC layer and retarding the growth of the Cu_6Sn_5 IMC layer. This study provides evident results for the enhancement of the mechanical properties of the 1% Al_2O_3 NP added SB solder.

1. Introduction/Literature

The electronic packaging industry vastly developed in current era creates more importance to the solder as it plays a crucial role in providing integrity electronic assemblies. Unfortunately, the common Sn-Pb solder is harmful to the environment and human due to lead (Pb) [1,9]. To replace this solder, the alternative solder alloys should be competing with the credible mechanical property of the lead solder alloy, especially in the shear strength. The SnPb solder alloy provides a thin IMC layer between the solder and the copper substrate in the printed circuit board, which contributes to the shear strength increment. Therefore, replacing this solder requires the alternative solder alloys to be practically compatible with the traditional solder. Numerous lead free solder alloys such as SnCu [1], SnAg [2], SnZn [3], SnBi [4], SnAgZn [5], SnAgCu [6] and the research materials are vastly available. On the other hand, some major drawbacks are found upon testing these solder alloys, for example poor wettability (SnZn) [7], thick intermetallic (IMC) layers (SnCu, SnAg) [8], low shear strength (SnAgZn, SnAg) [9], disposed to corrosion (SnZnAg, SnZn) [10] and many more. Another key aspect is to provide a low-temperature solder alloy and in that context, the SnBi solder is a potential candidate. Based on the phase diagram, a eutectic composition of SnBi solder provides melting temperature of, T_M=139°C which is lower than the most used SnPb, T_M=183°C [11] and SnAgCu, T_M=227°C [12]. At the same time, the electronic products are currently minimized, making the mass and density to be a vital aspect to select

Content from this work may be used under the terms of the Creative Commons Attribution 3.0 licence. Any further distribution of this work must maintain attribution to the author(s) and the title of the work, journal citation and DOI.
Published under licence by IOP Publishing Ltd

a solder alloy. As commonly practised, adding a new element to the mother solder alloy usually increases the shear strength, but then the mass increases as well making the electric components to be heavy. Such a problem should be resolved and the current researches lean to enhance the properties of a solder alloy with additions of nanoparticles. These nanoparticles can influence by affecting the solubility of the solder alloy and being discrete particles to the mechanical strength. It is crucial to provide and understand the influences of these nanoparticles to a solder alloy [13].

2. Methodology

The preparation of bare Sn58Bi (SB) solder alloys was based on the weight percentage of 42% Tin (Sn) and 58% Bismuth (Bi). An overall weight of 20g of SB chosen to replicate the usage of small consumption of solder paste used in the electronic industries. The Al_2O_3 NP (size < 50nm,) of 0.6 gram (1%) was weighed and melted together in the SB solder alloy using an induction furnace for an hour and thirty minutes hour at a constant temperature of 600°C to ensure a homogeneous mixture. To ensure all nanoparticles are properly mixed, the SB added with the Al_2O_3 was re-melted using a hot plate at 250°C with stirring for 30 minutes. The SB + 1% Al_2O_3 solder alloy was formed into billets of 5mm × 1mm for experiment purposes. The SB + 1% Al_2O_3 billet undertakes the standard metallographic sample preparation to produce clean and mirror-like surface finish. The Vickers micro hardness test with 1kgf load were induced to the solder billets to discover the hardness value. The shear strength were evaluated based on the ASTM D1002 (single shear lap) as the billets were soldered onto the Copper (Cu) substrate with a dimension of 30×10×1.5mm at a soldering temperature of 250°C for 30 seconds. The Zinc Chloride (ZnCl) flux applied during soldering to avoid oxidation.

3. Results and Properties Discussion.

3.1 Microstructure Properties

The microstructure of the SB + 1% Al_2O_3 NP was investigated using the high-resolution optical microscope as shown in Figure 1. This investigation provides an initial sight on the morphology of the SB + 1% Al_2O_3 NP that can be related to the mechanical properties especially the hardness of the solder. Figure 2 shows the microstructure of SB + 1% Al_2O_3 NP with a clear presence of lamellar type structure observed relating to the common structure of a Sn-Bi solder alloy system [14], which was also investigated as a purpose if comparison. Adding 1%, Al_2O_3 NP shows light phases (Bi) as well dark phase (Sn) and nearer lamellar gap structure compared to the bare SB solder alloy in Figure 2 (a). This confirms that the presence of Al_2O_3 NP is the main factor that influences the different configuration of the microstructure. The Al_2O_3 NP will be avoiding a self-diffusion process with the SB solder matrix due to its high melting temperature, T_M=2072°C. This consequently enables the nanoparticles to accumulate at the grain boundaries of the solder alloy [15]. Literally, this is a typical characteristic of any nanoparticles as they will accumulate at the high-energy site of in a molten solder (at grain boundary), as reported by [16]. The presence of the nanoparticles along the grain boundary will stimulate nucleation site for primary Sn matrices and produces smaller and finer Sn matrices by limiting the growth velocity of the plane. This phenomenon can be explained using the surface absorption theory as researched by [9] which defines that the plane with the maximum surface tension grows fastest (acting as the nucleation site) with an increasing adsorption element (Al_2O_3.) An EDX analysis proved the presence of the Al_2O_3 in the SB solder alloy as in Figure 2 (e).

3.2 Mechanical Properties (Hardness and Shear Strength)

The addition of 1% Al_2O_3 NP produces average micro hardness value of 16.3HV taken from five readings. Comparing to the other common solders (Figure 2), the increment in hardness was evident with the presence of Al_2O_3 NP. These increments are purely due to the presence of the nanoparticles, which acts as an obstacle for the motion of dislocations. The physical property of these Al_2O_3 NP, which, which does not take part in any diffusion process, acts as discrete particles, which segregate at the grain does not take part in any diffusion process, acts as discrete particles, which segregate at the grain boundary and impedes the dislocation. These characteristics are the key to the increment in hardness [17]. Furthermore, the uniform microstructure with fine Sn-matrices with the wider eutectic area (Figure

1) is another factor that increases the hardness in the Al_2O_3 NP reinforced SB solder alloy. Theoretically, a larger surface area of the nanoparticles per unit volume in the solder alloy can increase the hardness of a material [18].

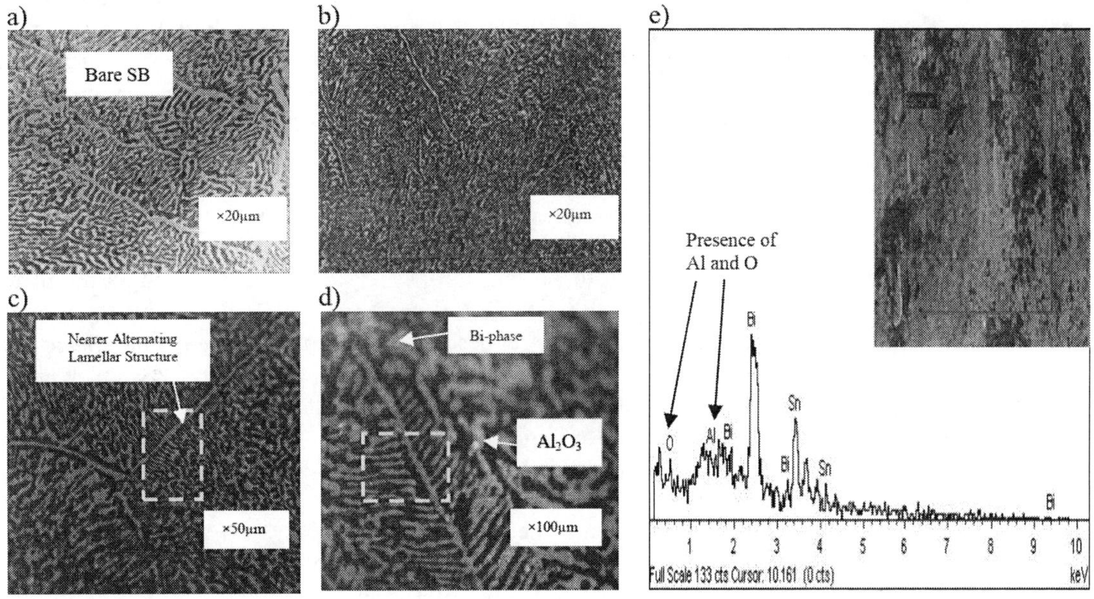

Figure 1. Microstructure of a) Bare SB solder alloy, b), c) d) SB + 1% Al_2O_3 and e) EDX analysis of SB+ 1% Al_2O_3

The shear strength result of the 1% Al_2O_3 NP added SB solder alloy was 63.45MPa which was 35% higher compared to the bare SB solder alloy. The shear strength results of the bare SB, SB + 1% Al_2O_3 from this work and few other compiled results are shown in Table 1. The presence of these Al_2O_3 NP in the solder alloy contributes to increase in the strength of the solder joint as a dispersion strengthening mechanism [15, 19]. Obeying the theory of dispersion, these Al_2O_3 NP will not take part in any activity and exist as a discrete particle that increases the strength of the solder alloy by pinning along the grain boundary [11, 16]. As more dislocations are pinned at the solder joint, the stress has to bend over these dislocations or penetrate through these high dense dislocations [8]. Such was the findings and explanation in this research too.

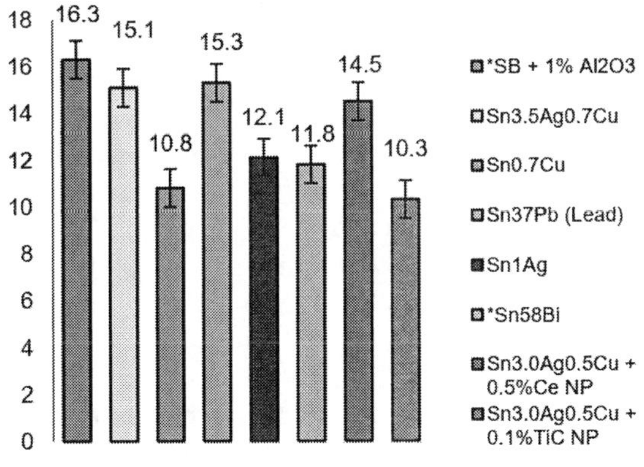

Figure 2: Vickers microhardness comparison.

Table 1: Shear strength comparison.

Solder	Shear Strength (MPa)	References
SB+1% Al_2O_3	63.45	This work
SB	41.08	
Sn3.8Ag0.7Cu	44.68	[12]
Sn3.5Ag0.5Cu + 1Ti NP	57.6	[20]
Sn98.5Ag1Cu0.5/Fe + 30Ni	53.98	[5]
Sn9Zn+1% Al_2O_3	39.00	[19]

4. Conclusion

The aims of this research is to provide the hardness and shear strength properties of the SB + 1% Al_2O_3 NP. High hardness value (16.3Hv) and shear strength value (63.45MPa) in the SB + 1% Al_2O_3 NP were produced compared to the unreinforced SB. This increment obviously proven by the microstructure morphology of the SB solder alloy, which was altered by the Al_2O_3 NP. The SB + 1% Al_2O_3 NP can be operated by the desired soldering temperature in the electronics industry (250°C) as this was the temperature used for soldering purpose in this project. In general, the SB + 1% Al_2O_3 NP solder in this study shows positive outcome in terms of the mechanical properties and further investigation could be potential to produce an alternative solder alloy to accommodate current miniaturization electronics.

References

[1] M.A.A. Mohd Salleh et al (2012). Mechanical properties of Sn-0. 7Cu/Si3N4 lead-free composite solder. Materials Science & Engineering A 556
[2] L. Zhang et al (2012). Microstructures and properties of SnZn-xEr lead-free solders. Microelectronics Reliability 52
[3] Chen Wenxue et al (2010). Effects of Ag on Properties of Sn-9Zn Lead-Free Solder. Rare Metal Materials and Engineering 39
[4] L. Shen et al (2012). Elastic modulus, hardness and creep performance of SnBi alloys using nanoindentation. Materials Science & Engineering A 558
[5] L. Zhang and K.N. Tu (2014). Structure and properties of lead-free solders bearing micro and nano particles. Materials Science and Engineering R 82
[6] S.Y. Chang et al (2011). Strengthening mechanism of nano-Al 2 O 3 particles reinforced Sn3.5Ag0.5Cu lead-free solder. Materials and Design 32
[7] Ervina Efzan M.N and Tan S.Y(2013). Wettability of Molten Sn-Zn-Bi Solder on Cu Substrate. Applied Mechanics and Materials 315.
[8] Y. Yao et al (2016). Interfacial structure and growth kinetics of intermetallic compounds between Sn-3.5Ag solder and Al substrate during solder process. Journal of Alloys and Compounds 682
[9] Ervina Efzan Mhd Noor and Amares Singh (2014). Review on the effect of alloying element and nanoparticle additions on the properties of Sn-Ag-Cu solder alloys. Soldering & Surface Mount Technology 26
[10] G. Ren et al (2016). Alloying influences on low melt temperature SnZn and SnBi solder alloys for electronic interconnections. Journal of Alloys and Compounds 665
[11] Amares Singh and Rajkumar Durairaj (2018). Effect on the wettability, hardness and shear strength properties of 3%-nano Titanium Oxide (TiO2) added Sn-3.8 Ag-0.7 Cu (SAC)/Copper (Cu) solder joint MATEC Web of Conferences 237, 02013.
[12] Amares Singh et al (2018). Reliability Study of Lead Free Sn-3.8Ag-0.7Cu and Copper (Cu) Substrate based on the Microstructure, Physical and Mechanical Properties. Journal of Mechanical Engineering 5(2).
[13] Anattiya Wattanakornphaiboon et al (2018). Effect of SnO2 reinforcement on creep property of Sn–Ag–Cu solders Materials Today: Proceedings 5
[14] H.F.Zou, et al (2012). Interfacial microstructure and mechanical properties of SnBi/Cu joints by alloying Cu substrate. Materials Science and Engineering A 532
[15] R. Tian et al (2017). Mechanical properties and fracture mechanisms of Sn-3.0 Ag-0.5 Cu solder alloys and joints at cryogenic temperatures Materials Science & Engineering A 709
[16] A.E Hammad et al (2014). Mechanical deformation behavior of Sn-Ag-Cu solders with minor addition of 0.05 wt.% Ni. Materials and Design (43)
[17] Tang et al (2014). Effects of TiO2 nanoparticles addition on microstructure, microhardness and tensile properties of Sn–3.0 Ag–0.5 Cu–xTiO2 composite solder. Materials and Design 55
[18] Law et al (2006). SnAgCu Microstructure, solderability, and growth of intermetallic compounds of Sn-Ag-Cu-RE lead-free solder alloys. Journal of Electronic Materials 35

IOP Conf. Series: Materials Science and Engineering **834** (2020) 012075 doi:10.1088/1757-899X/834/1/012075

[19] M. Ding et al (2018). Effect of micro alumina particles additions on the interfacial behavior and mechanical properties of Sn-9Zn-1Al2O3 nanoparticles on low temperature wetting and soldering of 6061 aluminum alloys. Journal of Alloys and Compounds 739
[20] Chuang et al (2012). Effects of small amount of active Ti element additions on microstructure and property of Sn3. 5Ag0. 5Cu solder. Materials Science & Engineering A558

IOP Conf. Series: Materials Science and Engineering **834** (2020) 012076 doi:10.1088/1757-899X/834/1/012076

Ergonomic Interventions in Lighting Products Manufacturing Plant

Amir Mazli Masahuling, Alias Mohd Saman*
Faculty of Mechanical Engineering, Universiti Teknologi MARA (UiTM), Shah Alam

*aliasms@uitm.edu.my

Abstract. This study emphasizes on ergonomic interventions at an assembly line in lighting products manufacturing plant located at Shah Alam, Selangor. Subjective assessment and direct observation methods were used to the operators and the workplace at manufacturing plant to uncover the problems. Subjective assessment used included interview and questionnaire via modified Nordic Musculoskeletal Questionnaires (NMQs). Direct observation involved site observations throughout the manufacturing plant completed with Rapid Upper Limb Assessment (RULA). The problems found that workers handled heavy lamps between the work station and poor design of lamp burning test station. Two ergonomics interventions were implemented. Wheel tray was applied for transporting the lamps in the assembly line, and new burning test station was introduced on the table. The implemented designs were simple and cost-effective. The ergonomics intervention was analysed and shown reduced in industrial risks and improved the working condition and occupational safety and health.

1.0 Introduction

Ergonomics is a science concerned with the interaction between humans and the work environment. It helps to improve the working surroundings for the use of the workers by fitting jobs to the employees to bring about industrial improvements [1]. Companies seek on various approaches such as ergonomics intervention to increase performance and achieve competitive benefits in terms of quality, productivity, cost savings and workers' well-being [2].

Ergonomics interventions can be defined as a real life experiment was conducted. It is also one of the efforts to reduce the risk of ergonomics such as the musculoskeletal disorders (MSDs) in the industries. It can be done by improving the existing workstation design to promote the comfort of the workers. Ergonomic interventions involve adjusting a workers' environment, behaviour, and other long-term educational approaches to treat and prevent further damage due to work-related musculoskeletal disorders [3].

In Malaysia, Social Security Organization (SOCSO) reported that issues of MSDs have increased sharply in recent years. The number of cases has shown 107 times increments from 2006 to 2016 [4]. The rise in MSDs also resulted a negative impact on the economic growth. Employees' compensation for occupational diseases was recorded increased from RM2.65 million in 2009 to RM14.05 million in 2014 [5].

In this research, an ergonomics study was done at a lighting products manufacturer located at Shah Alam, Malaysia. The works at assembly section demands the workers physical strength and endurance as they deal with heavy lamp that is required to be handled carefully. Figure 1 shows assembly line at the manufacturing plant, while Figure 2 shows a worker assembling a big size of lamp. This study aims to implement an ergonomics intervention at the manufacturing plant. The problems at the

Content from this work may be used under the terms of the Creative Commons Attribution 3.0 licence. Any further distribution of this work must maintain attribution to the author(s) and the title of the work, journal citation and DOI.
Published under licence by IOP Publishing Ltd

workplace have been detected by subjective assessment and direct observation. Two main problems were identified. Workers were handling big and heavy lamps between the workstations and poor design of lamp burning test station.

Figure 1. Assembly line at the manufacturing plant

Figure 2. Assembly works of lamp

2.0 Methodology

The purpose of this project is to design for human factors in the manufacturing plant of lighting products. The major focus is to minimize the ergonomics risk in the manufacturing plant. Series of steps were executed in order to complete the study.

2.1 Subjective Assessment

The assessment use to study the people involved in the system as measuring instruments. In essence, we rely on workers to come to some kind of conclusion about the system, and then access that conclusion as a measurement of the system. This assessment contains any method that gets its data from the psychological contents of people's heads [6]. Typical methods used in subjective assessment are ranking methods, rating methods, questionnaire methods, interviews and checklists. A subjective assessment was analyzed using modified Nordic Musculoskeletal Questionnaires (NMQs). The questionnaire consists of 19 questions and distributed to 12 respondents at the workplace. All respondents were guided during answering the questionnaires to ensure understanding during the study.

2.2 Direct Observation

Direct observation is the method where the person who conducts the study merely walking around an operating human or machine system to see what goes on. Observation methods are useful in collecting not only quantitative data but also in collecting qualitative data on product routings, task sequences, causes of delays and error taxonomies [7]. The direct observation was conducted on the working process at the manufacturing plant to understand overall process and to investigate the problems that occur at the workstations. Observation data was analysed using Rapid Upper Limb Assessment (RULA) ergonomics tool. RULA is a survey method use in ergonomics investigations where work-related upper limb disorders are reported. This method requires no particular equipment in providing a fast appraisal of the postures of the neck, body and upper limbs along with muscle function and the external loads experienced by the body.

2.3 Implementing intervention and review

The proposed designs for intervention were introduced at the final assembly line where the problems are identified. After three months, the intervention results were reviewed again using NMQs and RULA methods. The outcomes were analysed and compared to reveal the effectiveness of the intervention.

3.0 Result and Discussion

The case study was done in assembly line at the manufacturing plant. The analysis was done based on the modified NMQs. Table 1 specifies the 9 working stations involved in the line. Based on the interviewed, the workers mentioned about the requirement of physical strength while performing the works. They deal with heavy lamp and have to withstand stress over prolonged periods of time. Two tasks have been highlighted as the most challenging, handling of heavy lamps from one workstation to another and lifting the lamp during burning test.

Figure 3 shows the most tasks respondents hate doing. The burning test is the most hated tasks the respondents hate doing as agreed with 33% of the respondents. All four tasks required manual handling of heavy lamp especially when it moves toward the end of assembly line.

Table 1. Working stations in assembly line

No.	Working station
1	PCB LED assembly
2	Wiring and PCB LED assembly
3	Body and Wing assembly
4	Glass assembly
5	Glass plate assembly
6	Top and rubber cover screw assembly
7	Inspection and top cover sheet assembly
8	Roof assembly and packaging
9	Burning test

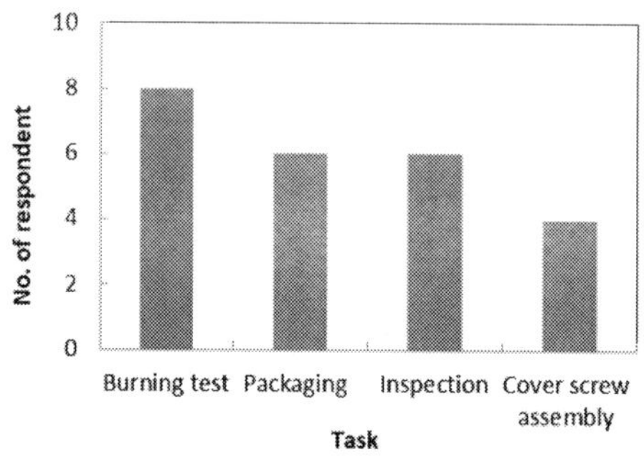

Figure 3. Task Respondents Hate Doing

Figure 4 shows the respondents' opinion on the weight of the product, while Figure 5 shows the needs of respondents to bend their body part while performing their task. Majority of the workers complained about the lamp weight and the needs to bends their body while handling the product in assembly line.

Figure 6 shows respondents' body part discomfort. From the total of 12 respondents, data demonstrates 10 complaints of wrist/hand discomfort and 9 complaints of hips/thighs/buttocks, elbows and upper back. 8 complaints were found for neck, low back and knees, while 7 discomforts complaints were recorded for shoulder and ankle/feet.

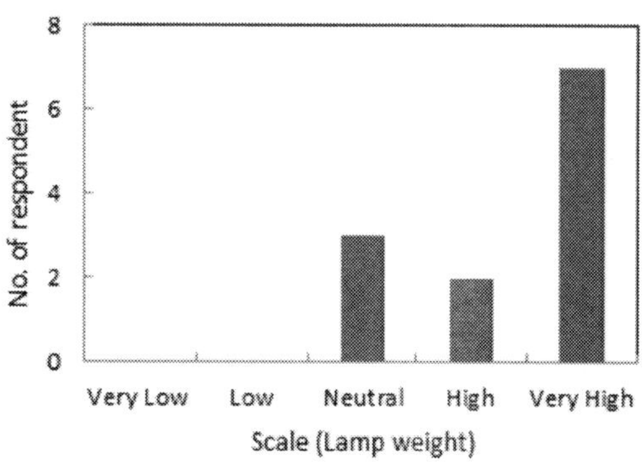

Figure 4. Lamp Weight

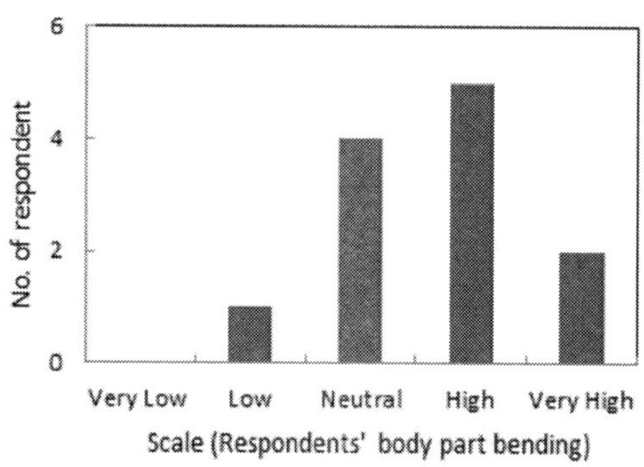

Figure 5. Respondents' Body Part Bending

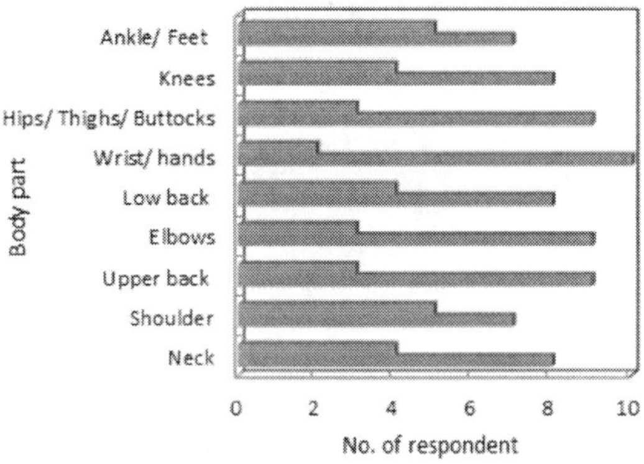

Figure 6. Respondents' Body Part Discomfort

Figure 7. Intervention 1: Wheel Tray

3.1 Ergonomic Intervention 1: Wheel Tray

Transporting of the lamp between the workstations is handled manually by the workers at the assembly line. The task is being identified for improvement by ergonomic intervention due to its criticality. The weight of the lamp that is being handled by the workers is up to 20kg. Muscle effort increase in response to high load required which increases fatigue and risk of MSDs. Furthermore, the workers were working in the same position for a long time involving bending or twisting the back and working in awkward position.

With respect to the ergonomic analytical study, the wheeled tray has been decided to be used for the workers in the assembly line to move the product from one station to another. This wheel tray eases the workers to transport the lamp and reduce the unnecessary load that being carried by them daily. It also will improve operators' working conditions, occupational health and safety, productivity and quality of the products. The wheeled tray consists of aluminium platform and four wheels as shown in Figure 7. Figure 8 shows a worker handling the heavy lamp manually, while Figure 9 displays after the intervention .The worker just needs to push the lamp on the wheel tray from station A to B to be received by the next worker.

3.2 Ergonomic Intervention 2: Burning test table

The lamp was being tested by switching on for about 30 minutes to one hour to ensure the condition of the lamp before it can be packed for delivery. Before the intervention, burning test was done by positioned the lamp on the test station as shown in Figure 10. The task is physically stressful because the worker needs to lift the heavy lamps (up to 20 kg) to be slotted to the testing pole. Workers are doing the task repeatedly for a long time, involving bending and twisting the back, and working in awkward position. The task can cause the pain to the worker's neck, shoulder, wrist and low back.

Figure 8. Before intervention : Handling the heavy lamp manually

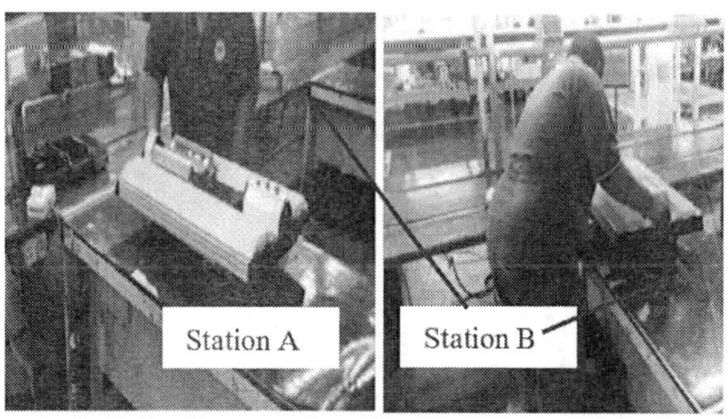

Figure 9. After intervention; The lamp is positioned on the wheel tray and push from station A and B

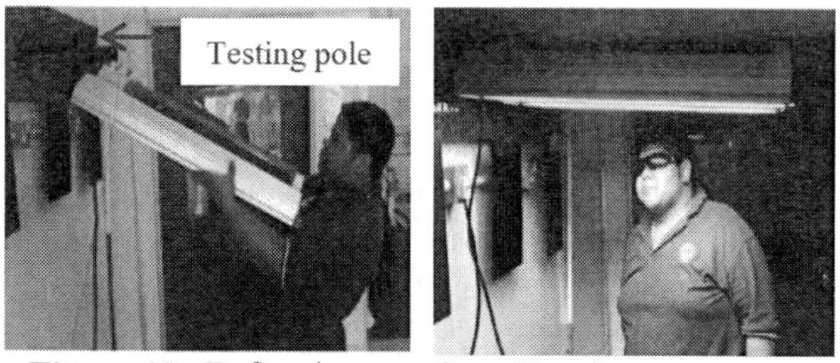

Figure 10. Before intervention : Burning test process before intervention

Figure 11. After intervention : The burning test table

The new intervention burning test table was made of a wooden platform, steel framework and four wheels. This table will ease the burning test process by eliminating the needs to lift the lamp. The lamp is pushed on the wheel tray from the assembly station to the test table and can instantly undergo the burning test. A maximum of five lamps can be tested concurrently. The new user friendly test table improve operators' working conditions, occupational health and safety, productivity and quality of the products. Figure 11 shows the burning test process on the table.

3.3 RULA Score Analysis

The postural analysis was done using RULA method. A musculoskeletal loading associated with the worker's posture were calculated for the posture of workers body part by giving a grand score. The analysis involving two main groups. Group A includes arm and wrist analysis. While group B represents neck, trunk and legs analysis. The posture scores of group A and B were then considered with muscle used and force load to obtain each group final score. The combination of these scores generating a grand score that represents the level of MSD risk. The RULA scores in Table 2 estimate the different levels of MSD risk and the resulting requirements for action.

Table 2. RULA score, level of MSD risk, and requirements for action. [8]

Score	Level of risk and Requirements for Action
1 or 2	Negligible risk, no action required
3 or 4	Low risk, changes may be required
5 or 6	Medium risk, investigation and changes are required
7	Very high risk, changes are required immediately.

Table 3 shows the comparison of RULA score before and after intervention 1. The grand score of RULA before the intervention is 7 indicates very high risk in handling heavy lamp to transports manually to the next station. The task required for engineering and/or work method changes to reduce or eliminate MSD risk immediately as outlined in the Table 2. The intervention was successfully reduced the risk of MSD with score 3.

The introduction of wheel tray on the assembly line minimizes industrial risk such as the repetition of awkward posture and reduces excessive force on the workers while doing their tasks. The workers feel less tired compared to task before intervention. The workers are now more motivated as they feel comfortable.

Table 4 shows the comparison of RULA score before and after intervention 2. The grand score before the intervention is 7 indicates that task to lift the lamp for burning test is very high risk and required immediate action. Huge reductions for final score A and B indicates the task before intervention demand highly for workers strength and energy. Only big size of workers fit for the task.

The introduction of burning test table has improved the workers working condition and occupational safety and health. The issues such as lower back and neck pain have been eliminated because burning test process after intervention removed the unnecessary lifting process. The problem of workers fails to do/ skip the burning test due of tiredness was eliminated and the workers can now focus on their daily task.

Table 3. Comparison of RULA score before and after intervention 1

Part of RULA assessment	Before Intevention	After Intervention	Reduction
Score A	5	3	2
Final Score A	8	4	4
Score B	4	2	2
Final Score B	7	3	4
Grand Score	7	3	4

Table 4. Comparison of RULA score before and after intervention 2

Part of RULA assessment	Before Intevention	After Intervention	Reduction
Score A	6	2	4
Final Score A	9	2	7
Score B	4	2	2
Final Score B	7	2	5
Grand Score	7	2	5

3.4 Respondents' Body Part Discomfort (After Intervention)

After three months of running the assembly line with the new intervention, workers body part discomforts were evaluated again. Figure 12 shows respondents' body part discomfort after intervention. The finding was compared to data before the intervention. The neck discomfort experienced by the workers have decreased from 8 to 2. Shoulder discomfort complaints were reduced from 7 to 3. Only 2 upper back problem recorded compared to 9 previously. Only 3 workers complaints about low back and knees pain. 4 workers mentioned about elbows and hips/thighs/buttocks discomfort, while 5 complaints regarding wrist/hands and ankle/feet discomfort. The result in overall shows ergonomics improvement for workers' wellbeing after the intervention.

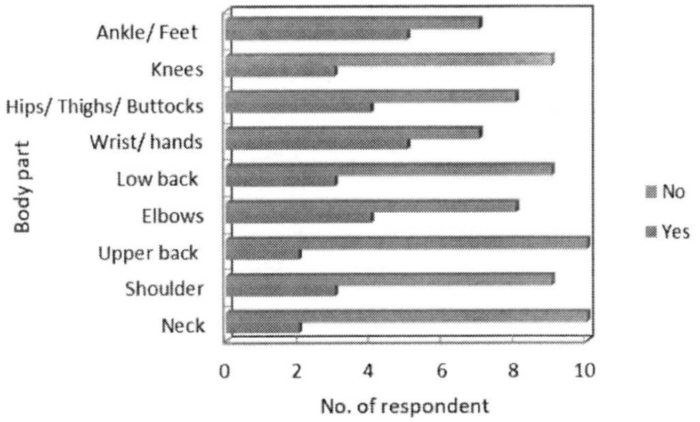

Figure 12. Respondents' Body Part Discomfort (After Intervention)

4.0 Conclusion

This project executed an ergonomics study to improve operators' working conditions, occupational health and safety and quality in the lighting products manufacturing plant. Two interventions have been applied to reduce the industrial risk and improve ergonomics condition on the assembly line. Besides direct observation and interview, modified questionnaires have been used to study the ergonomics in the manufacturing plant. Analyses conducted give the results that the workers in the assembly line were exposed to awkward working postures due to poor design of work station and the manual handling of heavy lamp. Wrist/hand, elbow and lower back were indicated as having the most work-related musculoskeletal disorder problem among the manufacturing plant workers. The results of RULA analysis after ergonomic interventions in the assembly line shows that industrial risks were minimized and working condition and occupational safety and health were improving.

References

[1] H.S. Loo, N. H. Saad, M. R.M. Redza, 2014. *Ergonomics Intervention in Steel Panels Handling for Improving Workers' Well-Being Outcomes*, Journal of Mechanical Engineering Vol. 11, No. 2, 93-10

[2] J. Dul, W.P. Neumann, 2009. *Ergonomics contributions to company strategies*, Applied Ergonomics. 40, 745-752

[3] E.K. Sylvia, J. Chun, J. Hong, 2013. *Ergonomic Interventions as a Treatment and Preventative Tool for Work-Related Musculoskeletal Disorders*. International Journal of Caring Sciences. Vol. 6, Issue 3.

[4] Noorhashirin H., Huda B.Z., Anisah B., Hayati K.S., Mohd Rafee B.B. 2018. *Prevalence of Musculoskeletal Disorder among Office Workers in Lembaga Tabung Haji Kuala Lumpur*, International Journal of Public Health and Clinical Sciences. Vol. 5:No.5

[5] M.S.S. Mohamed, I. Halim, A.H. Azani, L.Y. Ling, 2018. *Ergonomics Intervention to Improve Plastic Roll Handling Process at Production Area in Plastic Manufacturing Industry*, Human Factors and Ergonomics Journal. Vol. 3 (1)

[6] M.A. Sinclair, 1995. *"Subjective assessment" In: Wilson, J.R., Corlett, E.N. (Eds.)*, Evaluation of Human Work- A practical Ergonomics Methodology. Taylor & Francis, London, pp. 69-100

[7] C.C. Drury, 1995. *"Methods for direct observation of performance" In: Wilson, J.R., Corlett, E.N. (Eds.)*, Evaluation of Human Work- A practical Ergonomics Methodology. Taylor & Francis, London, pp. 45-6

[8] S. Digiesi, F. Facchini, G. Mossa, G. Mummolo, 2018. *Minimizing and Balancing Ergonomic Risk of Workers of an Assembly Line by Job Rotation: a MINLP Model*. International Journal of Industrial Engineering and Management 9(3):129-138

IOP Conf. Series: Materials Science and Engineering **834** (2020) 012077 doi:10.1088/1757-899X/834/1/012077

Parameter Optimization of WC-TaC-6Co Green Part in Injection Moulding using Taguchi Method

Hazriel Faizal Pahroraji*[1], Mohd Halim Irwan Ibrahim[2], Prathabrao Muniandy[3] Sri Yulis M Amin[3], Siti Khadijah Alias[1]
[1]Faculty of Mechanical Engineering, Universiti Teknologi MARA Cawangan Johor, Kampus Pasir Gudang, 81750, Masai, Johor, Malaysia
[2]Advanced Manufacturing & Material Center (AMMC) Universiti Tun Hussein Onn Malaysia, 86400, Parit Raja, Batu Pahat, Johor, Malaysia
[3]Universiti Tun Hussein Onn Malaysia, Department of Engineering Mechanics, Parit Raja, Johor, Malaysia

*hazriel@uitm.edu.my

Abstract. This study represents the injection moulding parameters optimization via Taguchi method on density and porosity of the green part of cemented carbide WC-TaC-6Co. The experiment commences with the preparation of WC-TaC-6Co feedstock mixed with 60 % palm stearin (PS), and 40 % of low-density polyethylene (LDPE). The important parameters for this study are the percentage of grain growth inhibitor (GGI), temperature of injection, injection pressure and injection speed. Utilizing orthogonal array L_9 (3^4), signal to noise ratio (S/N) was used to determine the significant levels and its contribution to the responses of density and porosity. The study signified that for density and porosity response, injection pressure and GGI is the most influential parameters. Based on that, it was found that the best parameter combinations for density and porosity is, GGI at 1.2 wt. %, temperature of injection at 155°C, injection pressure at 55% and 50%, and injection speed at 40% and 30%. Thus, it is determined that by controlling the setting of the best parameter, the optimal quality of the desired product can be accomplished and sustained without much complications during the process.

1. Introduction

Metal injection moulding (MIM) is a process of producing a near net-shape moulding products in which combining the procedure of both injection and powder metallurgy [1]. It is usually considered as a new processing technology used in powder metallurgy processing industries. This process is considerably beneficial and cost effective specifically for manufacturing large quantities of small parts and complex components. This process has a promising potential as in present time, most researchers implement this method in producing small and intricate part with better production cost [2-5]. During the process of injection moulding, several green parts were identified in terms of densities, tensile strength, defects such as shrinkage, and warpage among others. Adjustment of moulding process parameters in an intelligent way will lead to the acceptable way of maximizing or minimizing the characteristics required. Impartial optimization methods or the integration of it with other methods proposed a highly operative practices in finding the best development parameter values in which leads to least defects such as warpage, shrinkage, distortion and other associated defects [5-8].

Tungsten carbide (WC) is known as alloy with excellent combination of hardness and wear resistance. The average of its grain size usually is in between 0.2 μm and 20 μm and the normal

Content from this work may be used under the terms of the Creative Commons Attribution 3.0 licence. Any further distribution of this work must maintain attribution to the author(s) and the title of the work, journal citation and DOI.
Published under licence by IOP Publishing Ltd

proportion of carbide phase is in between 70% to 97% of the total weight of the composite. Meanwhile, Tungsten carbide and Cobalt (Co) alloy are the most common hard phase and binder metal phase respectively. The underlying reasons for the main role of Cobalt are some of the exceptional properties of Co and the Co-W-C ternary system. It was known that the solubility of WC in Co is considerably high, and it varies strongly depending on the applied temperature. Moreover, the use of Co alloy in hard metals properties is controlled by crucial properties including the high hardness, yield stress, toughness, and strength of Cobalt [9]. The basic cemented carbide structure and grade are formed by the two stated materials, which is often based on this concept which referred to as the straight grades in a simplified term. Titanium Carbide (TiC), Tantalum Carbide (TaC) or Niobium Carbide (NbC) and others might successfully act as a grain growth inhibitor in cemented carbide.

Taguchi parameter design is a tool of importance for this robust design. The simple and systematic method of the design is vital in optimizing design for cost, performance and quality. In robust design, signal to noise ratio and orthogonal array are used as dominant tools. Signal to noise ratio (S/N) concentrate on determine quality with the stress on variation and orthogonal arrays, which simultaneously hold various design factors [10]. Design of experiment (DOE) technique contribution the researchers to identify the quality parameter which demands to be controlled. Based on the studies by Ji et al. [11], the investigation of the consequence of sintering factors on the properties of sintered part was found to be the main factors for their studies. Khairur et al. [12] on the other hand discovered the effects of injection parameters on the green part quality characteristics such as green density, strength and defects and utilized classical Design of Experiment (DOE) approach. The other researchers who have also used Taguchi as a channel to optimize their parameter are Ghani et al. [10], Ahmad et al. [13], Chen et al. [14], Tuncay et al. [15] as well as Oktem et al. [16].

2. Methodology

2.1. Starting Materials

WC-6%Co were used as the main metal powder combined with 0.4 wt. %, 0.8 wt. % and 1.2 wt. % of TaC as Grain Growth Inhibitors. The characteristics of the metal powder is as illustrated in Table 1.

Table 1. Characteristics of the metal powder

Powder	WC-Co	TaC
Manufacturer	US Research Nanomaterials, Inc.	US Research Nanomaterials, Inc.
Grain Size	40-80 nm	950 nm
Morphology	Nearly spherical	Cubic
True Density	14.7 g/cm^3	13.9 g/cm^3

A multicomponent binder system was utilized in preparing the feedstock. The characteristics of the binder components are as stated in Table 2. In this study, major fraction (60%) of the binder system consists of palm stearin (PS) and minor fraction (40%) consists of polyethylene (LDPE).

Table 2. Some properties of binder

Type	Density (g/cm^3)	Melt Temperature (°C)	Decompose Temperature (°C)
Palm Stearin	0.891	54.3	398.5 – 598.8
LDPE	0.95	127	389.6 – 501.6

Figure 1 and Figure 2 shows the powder microstructure for WC-6Co and TaC that was evaluated via Scanning Electron Microscopy (SEM). From the SEM, the powder particles for WC-6Co and TaC are exhibit irregular shapes and some agglomerations.

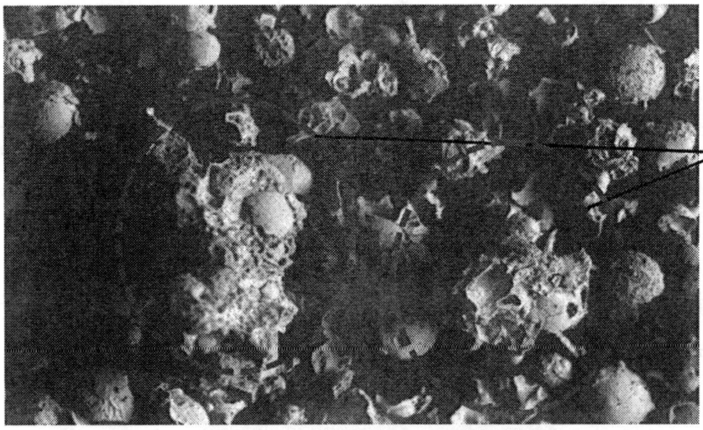

Figure 1. SEM image of WC-6Co

Figure 2. SEM image of TaC

2.2. Feedstock Preparation

In this study, Brabender Plastograph mixer is used to mix the composition of WC-6Co-TaC with the powder loadings of 43 %. The duration taken for this process is 1 hour at 140 °C and 40 rpm of speed until a homogenous mixture was obtained. Shimadzu is used to perform the rheology analysis of the feedstock to identify which flow characteristics of the feedstock are suitable for injection processes. Nissei NP7-1F type are chosen as the injection moulding machine.

2.3. Design of Experiment (DOE)

In this research, L9 (3^4) orthogonal array which consists of 4 columns and 9 experiment trials are utilized as DOE with four parameters involved are GGI (A), injection temperature (B), injection pressure (C), and injection speed (D). Table 3 and 4 present Taguchi's orthogonal array which demonstrates the parameter of this study. There are 27 samples for 9 trials were injected to be tested.

Green strength optimization in MIM is a significant process as it focuses on investigating the performance and superiority of the green part. The data of signal to noise ratios were obtained by using Statistical software-Minitab version 16. In line to achieve the optimal factors, the signal to noise response graph were constructed to determine the highest level based on density and porosity. The optimum value of the SN ratio stipulates the best setting for the selected factors. The ranks of the most prominent factors have also been obtained from the software.

For an engineering analysis, S/N ratio characteristics normally can be divided into three; the small-the better, the nominal-the better, and the larger-the better. In this study, the smaller-the better and larger-the better are used to find the optimal parameters.

Table 3. Injection Parameters of Taguchi Design

Level	GGI (%)	Injection Temp. (°C)	Injection Pressure (%)	Injection Speed (%)
	A	B	C	D
0	0.4	145	45	30
1	0.8	150	50	35
2	1.2	155	55	40

** Unit conversion: 1 % of injection pressure equal to 1.61 MPa, 1 % of injection speed equal to 3.50 rpm

Table 4. Orthogonal Array (OA)

Level / Trials	Injection Moulding Parameter Level			
	GGI (%)	Injection Temperature (°C)	Injection Pressure (%)	Injection Speed (%)
	A	B	C	D
1	0.4	145	45	30
2	0.4	150	50	35
3	0.4	155	55	40
4	0.8	145	50	40
5	0.8	150	55	30
6	0.8	155	45	35
7	1.2	145	55	35
8	1.2	150	45	40
9	1.2	155	50	30

The deficiencies of the composite including the porosity, shrinkage and warpage, unwanted particulates in the deposition process were determined by the Taguchi's method implementing S/N ratio for smaller-the-better quality. The logarithmic function which is mainly constructed through the mean square deviation [12] derived from the equation which is used to calculate the signal-to-noise ratio for the smaller-the-better quality characteristics of the composite. The S/N ratio equation:

$$S/N = -\ 10\ Log\ MSD \tag{1}$$

The equation mean square deviation for a smaller-the-better characteristic can be written as:

$$MSD = \frac{{y_1}^2 + {y_2}^2 + \cdots + {y_n}^2}{n} \tag{2}$$

The S/N equation can be written as:

$$\frac{S}{N} = -10\ log \left| \frac{{y_1}^2 + {y_2}^2 + \cdots + {y_n}^2}{n} \right| \tag{3}$$

Meanwhile, Taguchi's S/N-Ratio for larger-the-better quality characteristics is applied for a desired output such as strength and critical current. The equation for the signal-to-noise ratio for the larger the better quality characteristic is written and applied similar to the smaller-the-better S/N formula [12]. The equation applied for the larger-the-better mean square deviation is written as follows:

$$MSD = \frac{1/{y_1}^2 + 1/{y_2}^2 + \cdots + 1/{y_n}^2}{n} \tag{4}$$

The larger-the-better S/N equation can be written as:

$$\frac{S}{N} = -10\ log \left| \frac{1/{y_1}^2 + 1/{y_2}^2 + \cdots + 1/{y_n}^2}{n} \right| \quad (5)$$

3. Result and Discussion

3.1. Density and Porosity Results

The performance and capabilities to obtain superior green part was investigated through green strength optimization in MIM process. The data of signal to noise ratios were attained to achieve the optimized factors by via Statistical software-Minitab version 16. The data obtained from the density and porosity which can be used to interpret the optimal level from the four factors and three levels by constructing the signal to noise response graph. The highest value of the SN ratio indicates the best setting for the selected factors. From the software, the ranks of the most influential factors successfully achieved.

The density and porosity of the samples were measured by using Archimedes method based on Metal Powder Industries Federation (MPIF) Standard 42 [17]. Figure 3 shows the experimental results for density where the optimum parameter will be based on the highest peak at each parameter. For the result, the optimal configuration for obtaining optimal density for green part would be A3 (1.2 wt. % of GGI), B3 (155 °C of injection temperature), C3 (55% of injection pressure) and D3 (40% of injection speed), respectively. The conditions were explained as below:

i. The injection temperature (B) selected must be sufficient to encourage smooth flow of the molten feedstock. This is due to the fact that the viscosity behaviour will be depended on the selected temperature, where high temperature of the feedstock will defer initial freezing at the early stage of the process before it fills up the cavity [1].

ii. Injection pressure (C) will largely affected the final properties as it has an imperative rule of increasing the green density of the samples. However, inappropriate selection of injection pressure will eventually leads to powder binder separation, thus leading to defected end product [1].

iii. Injection speed (D) is used as one of the parameter in producing good quality green parts. The ram will push the feedstock into cavity to fill up the mould with variable speed.

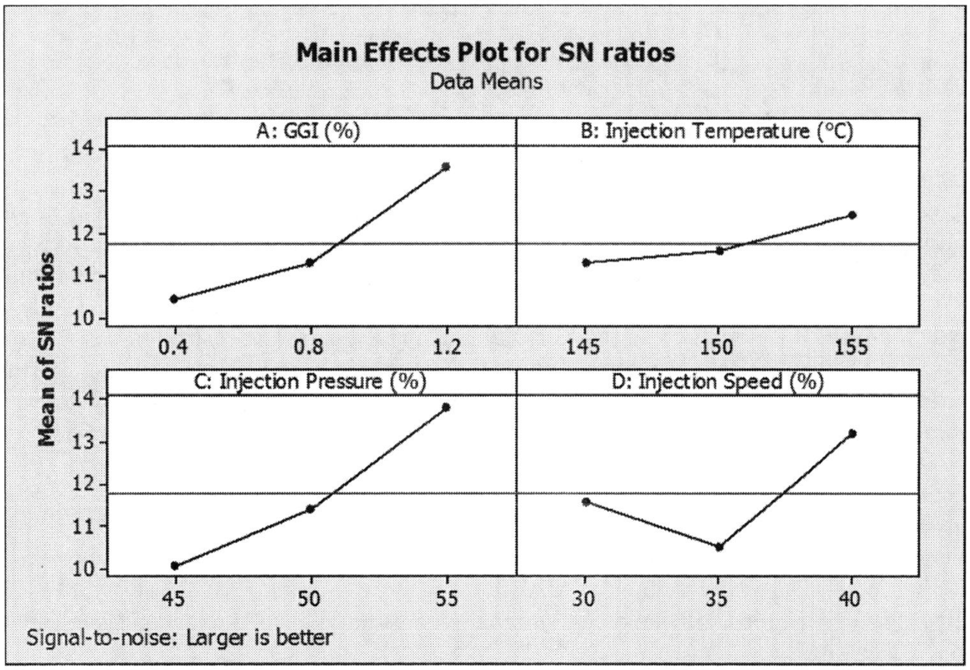

Figure 3. S/N ratio main effects plot (data means) for density of green part

Figure 4 shows the main effects plot (data means) for S/N ratio for porosity where at each parameters, the optimal parameter is based on the highest peak. It was found that the optimum configurations for optimal density selection for green part was found in samples A_3, B_3, C_2 and D_1. It was inferred as 1.2 wt. % of GGI percentage (A), 155 °C of injection temperature (B), 50% of injection pressure (C) and 30% of injection speed (D). This is because the presences of pores will create voids between the particles that is considered as in the green body and will significantly reduce sintered strength. It was found that the GGI percentage and injection temperature resulted in the highest mean ratio .The selection of GGI at high percentage together with high injection temperature will act to cover smaller pores in the green part. The selection of high heat will facilitate in producing a defect-free green part by covering the pores. As the smaller pores filled the void, the size of pores will increase which lead to reduction of the porosity in the samples. This led to increment of the bulk density which improved the hardness value as the porosity volume fraction decreased [18]. Generally, the injection pressure and injection speed provide a moderate effect on the porosity response.

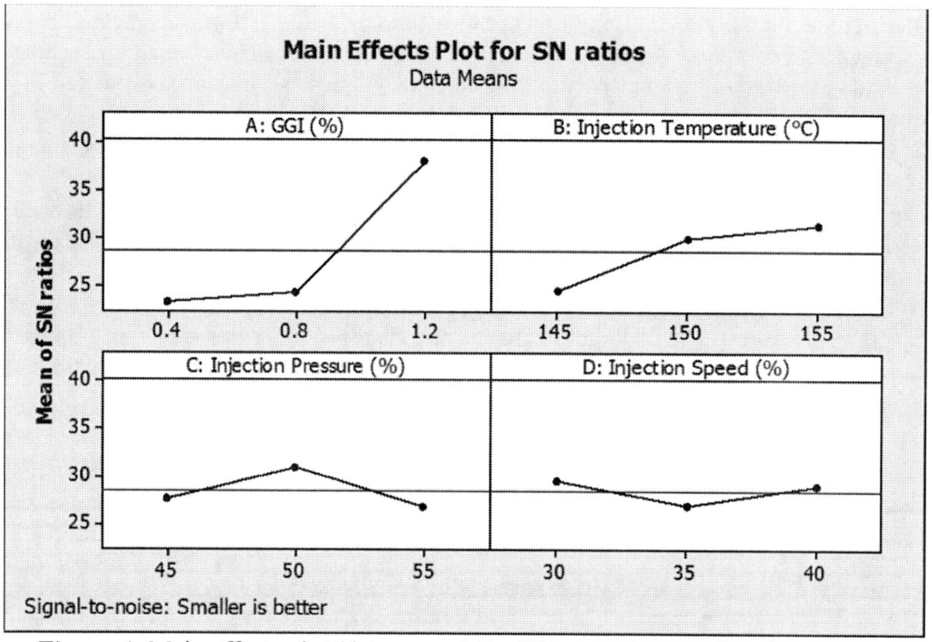

Figure 4. Main effects plot (data means) for S/N ratio for porosity of green part

4. Conclusion

The objectives of this study were successfully achieved. Many factors might be affected the performance of quality of products and processes in Taguchi's orthogonal array. Based on the results, it is concluded that for nominal density, the optimum parameters to be used are, GGI (1.2 wt. %), injection temperature (155°C), injection pressure (55%) and injection speed (40%). For ideal situation of this study, smaller-the-better was used to determine the least porosity in the green part. The best parameter that have been found is GGI (1.2 wt. %), injection temperature (155°C), injection pressure (50%) and injection speed (30%). Based on the findings, it is concluded that by controlling the optimum parameter setting, the optimum quality product desired can be continuously produced. The input from this research is hoped to be useful for further testing to be done in the future.

Acknowledgements

The authors would like to thank the Ministry of Higher Education Malaysia, Universiti Tun Hussein Onn Malaysia and University Teknologi MARA for a research grant (FRGS-1590, GPPS-U828 and 600-IRMI/FRGS 5/3 (166/2019)) and laboratory apparatus provided at Advanced Manufacturing and Material Center (AMMC) lab.

References

[1] R. M. German and A. Bose 1997 "Powder Injection Molding of Metal and Ceramics: Metal Powder Industries Federation," *Princeton, NJ,*

[2] H. F. Pahroraji and M. H. I. Ibrahim 2016 "Characterization of sustainable binder with hydroxyapatite via powder metallurgy route," *ARPN J. Eng. Appl. Sci.*, vol. 11, no. 12.

[3] M. Prathabrao, S. Y. M. Amin, and M. H. I. Ibrahim 2017 "Review on Sintering Process of WC-Co Cemented Carbide in Metal Injection Molding Technology," in *IOP Conference Series: Materials Science and Engineering,*

[4] S. Y. M. Amin, N. Muhamad, K. R. Jamaludin, and others 2013 "Optimization of injection molding parameters for WC-Co feedstocks," *J. Teknol. (Sciences & Eng.)*, vol. 63, no. 1, pp. 51–54

[5] M. H. I. Ibrahim, N. Muhamad, A. B. Sulong, K. R. Jamaludin, N. H. M. Nor, S. Ahmad, M. R. Harun, and H. Zakaria 2010 "Parameter Optimization towards Highest Micro MIM Density by Using Taguchi Method," *Key Eng. Mater.*, vol. 443, no. October 2015, pp. 705–710

[6] N. Mustafa, M. H. I. Ibrahim, A. M. Amin, and R. Asmawi 2017 "Parameter Optimization Of Natural Hydroxyapatite / SS316l Via Metal Injection Molding (MIM),"

[7] K. R. Jamaludin, N. Muhamad, M. N. Mohd, S. Ahmad, M. H. I. Ibrahim, and N. H. M. Nor 2009 "Optimizing the injection parameter of water atomised SS316l powder with design of experiment method for best sintered density," *Chiang Mai J. Sci.*, vol. 36, no. 3, pp. 349–358

[8] N. Ünal, F. Kern, M. L. Öveolu, and R. Gadow 2011 "Optimization of processing conditions and characterization investigations of ceramic injection molded and sintered ZrO2/WC composites," *J. Alloys Compd.*, vol. 509, no. 14, pp. 4797–4804

[9] G. Gille, J. Bredthauer, B. Gries, B. Mende, and W. Heinrich 2000 "Advanced and new grades of WC and binder powder - their properties and application," *Int. J. Refract. Met. Hard Mater.*, vol. 18, no. 2, pp. 87–102

[10] J. A. Ghani, I. A. Choudhury, and H. H. Hassan 2004"Application of Taguchi method in the optimization of end milling parameters," *J. Mater. Process. Technol.*, vol. 145, no. 1, pp. 84–92

[11] C. H. Ji, N. H. Loh, K. A. Khor, and S. B. Tor 2001 "Sintering study of 316L stainless steel metal injection molding parts using Taguchi method: Final density," *Mater. Sci. Eng. A*, vol. 311, no. 1–2, pp. 74–82

[12] K. R. Jamaludin, N. Muhamad, M. N. A. Rahman, and S. Yulis, "Analysis of Variance on the Metal Injection molding parameters using a bimodal particle size distribution feedstock."

[13] S. Ahmad, N. Muhamad, A. Muchtar, J. Sahari, K. R. Jamaludin, M. H. I. Ibrahim, and N. H. M. Nor 2010 "Optimisation of processing parameters of titanium foams using Taguchi method for compressive strength," in *Key Engineering Materials*, vol. 447, pp. 671–675.

[14] R. S. Chen, H. H. Lee, and C. Y. Yu 1997 "Application of Taguchi's method on the optimal process design of an injection molded PC/PBT automobile bumper," *Compos. Struct.*, vol. 39, no. 3–4, pp. 209–214

[15] T. Erzurumlu and B. Ozcelik 2006"Minimization of warpage and sink index in injection-molded thermoplastic parts using Taguchi optimization method," *Mater. Des.*, vol. 27, no. 10, pp. 853–861

[16] H. Oktem, T. Erzurumlu, and I. Uzman 2007 "Application of Taguchi optimization technique in determining plastic injection molding process parameters for a thin-shell part," *Mater. Des.*, vol. 28, no. 4, pp. 1271–1278

[17] W. Y. Ning, N. Muhamad, A. B. Sulong, A. Fayyaz, and M. R. Raza 2015 "Effects of vanadium carbide on sintered WC-10%Co produced by micro-powder injection molding," *Sains Malaysiana*, vol. 44, no. 8, pp. 1175–1181

[18] Q. Yang, J. Yang, H. Yang, and J. Ruan 2016 "The effects of fine WC contents and temperature on the microstructure and mechanical properties of inhomogeneous WC-(fine WC-Co) cemented carbides," *Ceram. Int.*, vol. 42, no. 16, pp. 18100–18107

IOP Conf. Series: Materials Science and Engineering **834** (2020) 012078 doi:10.1088/1757-899X/834/1/012078

Effect of Sintered Alumina Blended Polyethylene Glycol Binder on the Pin on Disc Test Parameter

Siti Norazlini Abd Aziz, Mimi Azlina Abu Bakar*, Salmiah Kasolang, Muhammad Hussain Ismail

Centre of Advance Materials Research (CAMAR)
Faculty of Mechanical Engineering, Universiti Teknologi MARA,
UiTM 40450 Shah Alam, Selangor, Malaysia

mimi_azlina@salam.uitm.edu.my

Abstract: This study presents the effects of sintering temperature and binder content of polyethylene glycol (PEG) on wear properties of compacted alumina using pin on disc wear test under dry sliding condition. Wear rate was measured at 20N load and sliding speed of 200rpm at 0.28 m/s sliding velocity. The wear test sample was tested at 3 different sintering temperatures (1000^{0}C, 1100^{0}C, and 1200^{0}C) to produce different microstructures. The results show that a dry pressed alumina ceramic with 5 wt. % of PEG content and sintering temperature at 1200^{0}C recorded the lowest wear rate and volume lost compared to other samples. SEM was carried out to investigate the microstructure how it affects the wear property. It was found that at high sintering temperature of 1200° C the particle formation was strengthen to each other hence promote highest wear resistance.

1. Introduction

These days because of a unique properties of ceramic phases, for example, high wear resistance of the ceramic has developed the enthusiasm among researches about. Regardless of high hardness and chemical inertness, ceramic get high compressive quality and temperature opposition which permit great execution under outrageous conditions. This especially under sliding circumstance with present of high speeds and temperatures, accordingly unlubricated contact and forceful condition [1-3]. Ceramic material such as Al_2O_3,(alumina), ZrO_2, Si_3N_4 and SiC have been utilized broadly in metal-use segments like metal roller, mechanical water seals, cutting devices, aides and dies in manufacturing industries have been traded for materials formed by ceramic phases.

Alumina is the most commonly used material in applications requiring wear resistance because of novel properties in a wide scope of uses and great cost-viability [3]. The wide range of Alumina is also in the ceramic composites and nanocomposites as Al_2O_3-TiC, Al_2O_3-ZrO_2 and Al_2O_3-SiC. Subsequently, numerous researchers about have been directed so as to upgrade the preparing of those ceramic production and especially on microstructure for development of ceramic wear properties. Wear is expulsion of material from a strong surface by the sliding activity of another strong and is brought about by friction, fatigue or vibration [3]. Subsequently, it causes dynamic harm including material loss and happens on the surface of the segment because of sliding.

Content from this work may be used under the terms of the Creative Commons Attribution 3.0 licence. Any further distribution of this work must maintain attribution to the author(s) and the title of the work, journal citation and DOI.
Published under licence by IOP Publishing Ltd

In addition, the tribological properties of alumina and ceramic composites have present huge contrast in spite of the fact as the wide researches develop on friction and wear. This might be influenced by their properties, for example, hardness, fragility, thermal conductivity and chemical inertness. Many publications presented results wear behavior of alumina, and much of the focus has been on the role of wear parameter for example, contact geometry, sliding speed and distance, surface conditions, connected load, relative humidity or kind of condition, surface temperature, substance similarity, test device and specimen technique preparation. [4]. Other than that, there are some issue while analyse the alumina wear information is that habitually over the referenced test conditions factors are not depicted or even monitored. This paper will be discussed the particular effect of alumina pin-on-disc test to the effect of the wear rate test parameter including characteristics of the samples worn surface subsequently.

In addition, the effect test condition of alumina wear behaviour and wear properties of alumina has been studied comprehensively. Alumina or also known as Al_2O_3 is an engineering ceramic have relatively high density and purity, exhibit very high hardness about in the range of (Hv > 16 GPa) and the value of toughness is prompted considerable for tribological applications in the range of ($KIC \geq 4$ $MPa.m^{1/2}$). Furthermore, alumina properties such as its brittleness results low friction coefficient (μ) of the unlubricated sliding contact pair of similar alumina. Bajwa [5] reported that the friction coefficient values in the range of 0.2 to 0.8 [1, 2, 4] considering the characteristics of alumina itself, environment conditions as well as the surface roughness and geometry.

For tribological applications, wear of alumina may be categorized into mild' and 'severe' [1, 4-8]. There are several research associated from testing parameter where the wear mode result from normal load, temperature sliding velocity and distance and etc. Ravikiran [7] reported on the wear coefficient is reduced as the apparent area of contact increases. In addition, for such ordinary load and sliding rate, a mild wear mode which is lower than 10^{-6} $mm^3/N.m$ produces a generally smooth worn surface thus the severe wear mode demonstrating a separate higher value of roughness which creates a harsh worn surface. Furthermore, since the particular wear rate is $<10^{-6}$ $mm^3/N.m$ it is demonstrated that a ceramic will be helpful as a business tribological part. Therefore, surface examination is required to determine the wear mechanism. For pin on disc wear test the fundamental assessment is the friction procedure that repeated section of the pin on a similar wear track. Concerning the fragility of ceramic, the presence of the wear debris on the worn surface and powder formation is vital factor to get wear mechanism. [1, 4].

The contact stresses and the contact pressure involve in the formation of mechanical wear debris. The stress became related by normal load and surface roughness and frictional heating from the contact thermal stress. For the most part, the smooth wear surface related to a mild wear mode coming about because of flattening or polishing of surface grains and delicate crack. Thus, decreasing the wear rate and sustain in mild wear mode creates a preventive tribofilm with the repetition cycles of the pin subsequently.

2. Experimental procedure

The alumina powder (NM9620) with sub micrometer average particles size of 0.6μm and high purity (99,999% purity) used in this study was supplied by Vistec Technology. The surface area were $3m^2/g$ and density 3.98 g/cm^3 respectively. The pin and the disc sample was produce using mixing process of alumina powder blended with polyethylene glycol (PEG) and the formulation were varies in the percentage of PEG binder apply with 3 different sintering temperatures respectively. The cylindrical pin produced with a semispherical tip (radius of 6.5 mm) and was compacted by uniaxial followed by isostatic pressing at room temperature under the pressure of 24MPa according to ASTM G99-05. The sintering was performed for 20 h at 3 different temperature of 1000°C, 1100° C and 1200°C with a lower heating rate of 1°C/min. The disc was produced using the same method using compaction with PEG binder using isostatic pressed and form a disc with 40 in diameter and 8 mm of thickness and subsequently it was sintered at 1000°C,1100°C and 1200°C for 20 h process as in Table 1.

IOP Conf. Series: Materials Science and Engineering **834** (2020) 012078 doi:10.1088/1757-899X/834/1/012078

Table 1. Properties of alumina powder and percentage used.

Formulation	% of PEG	Sintering Temperature (°C)
F1	1	1000
F2	5	
F3	1	1100
F4	5	
F5	1	1200
F6	1	

The wear test was carried out using a pin-on-disc test machine (Model: Wear & Friction Monitor TR-20; Brand; MAGNUM). As in Figure 1, the wear test for all specimens was carried out under the normal loads of 20N and a test rpm of 200 for about 10 minutes. In this experiment all specimen were prepared for the wear tests for an absolute sliding distance of around 251 m and about 0.28 m/s sliding speed. The resulting wear rate was determined from the weight loss by deciding its volume loss per unit sliding distance. The wear volume loss was estimated by optical microscopy where it was resolved from the pin worn scar distance diameter. Sample tests with worn surface delivered in the mild wear mode and in the area of progress to severe wear, recognized by the wear rate in the scope of 10^{-6} to 10^{-5} mm^3/N.m were chosen for evaluation and microstructure characterization were examined by scanning electron microscopy (SEM).

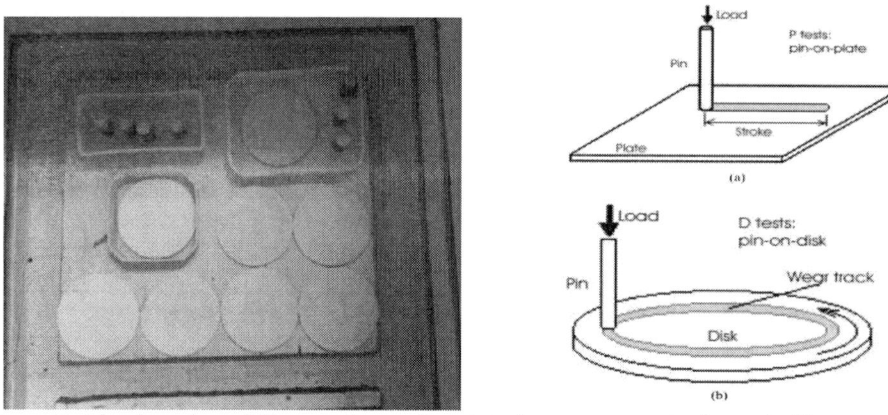

Figure 1. Alumina pin and disc specimen after sintering process and schematic diagram of pin on disc test

3. Result and discussion

3.1. Wear Properties of Dry Pressed Ceramic

Figure 2 demonstrate the wear rate of progress transition from gentle to extreme or severe wear mode as capacity of consistent sliding speed and steady ordinary load as a primer endeavor to distinguish the change conditions. The test was implemented at constant load of 20N showed wear rate values for different formulation of alumina blended PEG. The wear rate at sintering temperature of 1000 °C for F1 about 2.89x10^{-6} mm^3/N.m and decrease trend for highest sintering temperature at 1200 °C about 0.084x10^{-6} mm^3/N.m. This result is supported by Todesco et al. [6], where the pin-on-disc test of alumina was carried out at 3 different load with 20N, 30N and 40N showed wear rate values between in the range of 2 to 25.$10^{-6}$$mm^3$/N.m. The increasing wear at sliding speed of about 0.4 m/s for load of 30N result an ordinary transition from mild to severe wear was appeared.

Therefore, as can be observed for load of 20N in Figure 2, the typical transition was shifted from mild to severe wear where the wear rate showed an abrupt increase at a sliding speed of about 0.28 m/s. The chose zone for microstructure characterization in the middle of the mellow wear mode and

IOP Conf. Series: Materials Science and Engineering **834** (2020) 012078 doi:10.1088/1757-899X/834/1/012078

in the area of progress to severe wear, found by the wear rate between in the range of 10^{-6} to 10^{-5} mm^3/N.m. However, higher the wear rates will cause error increases too. This phenomena results an uncertain behaviour of the pin-on-disc apparatus that effect the wear volume measurement and also variation caused by load slip and controlled conditions [6]. On the other hand, the wear rates below 10^{-6} mm^3/N.m will exhibit smooth surfaces are corresponded with mild wear mode, while if there are grooves appear, micro-fracture and rate above 10^{-6} associated to the severe wear mode [2].

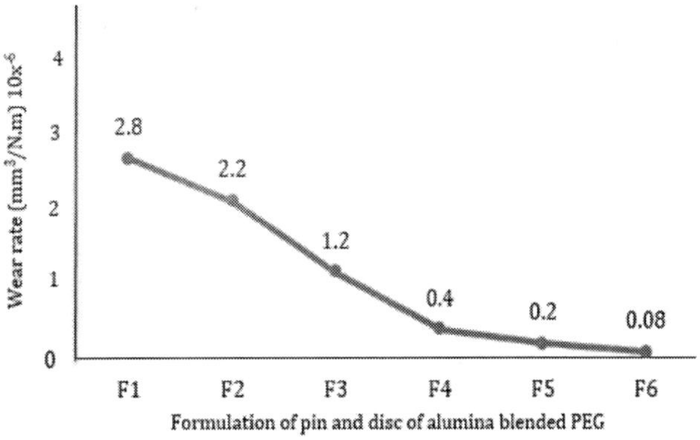

Figure 2. W[illegible]r rate as a function of constant sliding speed for different formulation of alumina blended P[illegible] (F1,F2,F3,F4,F5,F6) of alumina/alumina in a pin-on-disc configuration.

As illustrate in Figure 3 show the volume loses of alumina samples loaded with 1 wt. % PEG binder (F1, F3, and F5) & 5wt. % PEG (F2, F4, F5) show an decreasing trend associated with higher sintering temperature. The sample weighed before and after the test. The sample unable to withstand the 20 N load for 10 minutes, thus permit little evidence possessed a lot of weight lost with sliding speed of 0.28 m/s. This phenomena have a contradiction as reported by Ahmad et al. [8] in their finding where the weight loss of AMC (Alumina matrix composite) is increased when the load applied 100 N to 150 N accordingly. Hence, higher sintering temperature results the lowest volume loss of samples. The decrease in weight losses was due to highly formation of alpha alumina via transition alumina (gamma alumina) which occurs by nucleation and growth process during heating.

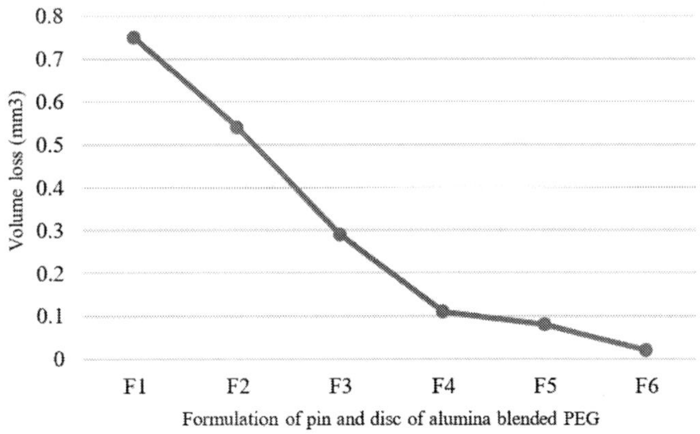

Figure 3. Volume loss as function of constant load of 20N for different formulation of alumina blended PEG (F1,F2,F3,F4,F5,F6).

From the graph in Figure 4 shows bar chart of different wear resistance calculated from wear rate measured by pin-on disc testing. Based on these values, the profound wear resistance was come from F6 where 5 wt. % of binder was used and sintered at 1200ºC. The wear resistance can be defined as it is inversely proportional to the wear rate results. At F6 formulation increase wear resistance trend can be observed, this is due to the uniformity of alumina microstructure and results high mechanical properties. The use of fine alumina powder with bi-modal particle size distribution and higher specific surface, promotes microstructure uniformity with lower closed porosity. This is supported by Milak, [10] where the wear resistance is consequently of the fracture mode and mechanical properties. The greater wear resistance of the alumina occurs because the fraction of material removed during severe wear through the alteration of the intergranular fracture mode in Al_2O_3 to the transgranular mode undergoes less wear where it results a smoother surface. Therefore, the F6 formulation results in higher mechanical properties and wear resistance of the range studied ceramics.

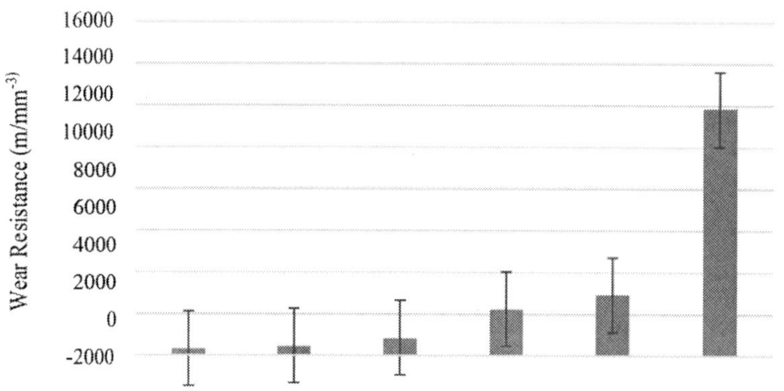

Formulation	F1	F2	F3	F4	F5	F6
wear resistance	334.7	467.4	863.53	2238.4	2925.3	11819

% of PEG

Figure 4. Wear Resistance with Variation of Formulation

Figure 5 shows the worn surfaces analyses under SEM characterization of the samples was carried out in order to observe and interpreted the transition phenomena. For F1 formulation at 1% PEG binder at 1000° C sintering temperature where the wear rates of about 2.9×10^{-6} shown a rough worn surface and increase the wear rate. In analysis at Figure 5 (F2) at 5% PEG binder, the grooves started to be found and produce wear scars. Both specimens show existence of grooves at disk and pin, and the peak and valley roughness merge well with the matting surfaces since the sliding pair is nearly the same hardness.

It can be seen that the scar produced at Figure 5 (F3) at 1% PEG binder at sintering temperature of 1100° C. However, the explanation should take into account that the scars depths have been found to be much greater. Figure 5 (F4) SEM micrograph show a lump defect was observed due to a not well recognize behaviour with the increasing of sintering temperatures, the worn surface. Increasing the sintering temperature Figure 5 (F5), at 1% PEG binder at 1200° C, it is clearly seen that the surface material is pushed to the sides of the scratch, as well as the formation of wear debris expelled to the edge of the scar. While Figure 5 (F6) show the amount of wear particles is much smaller, and the particles remaining adherent in scratching scars are small.

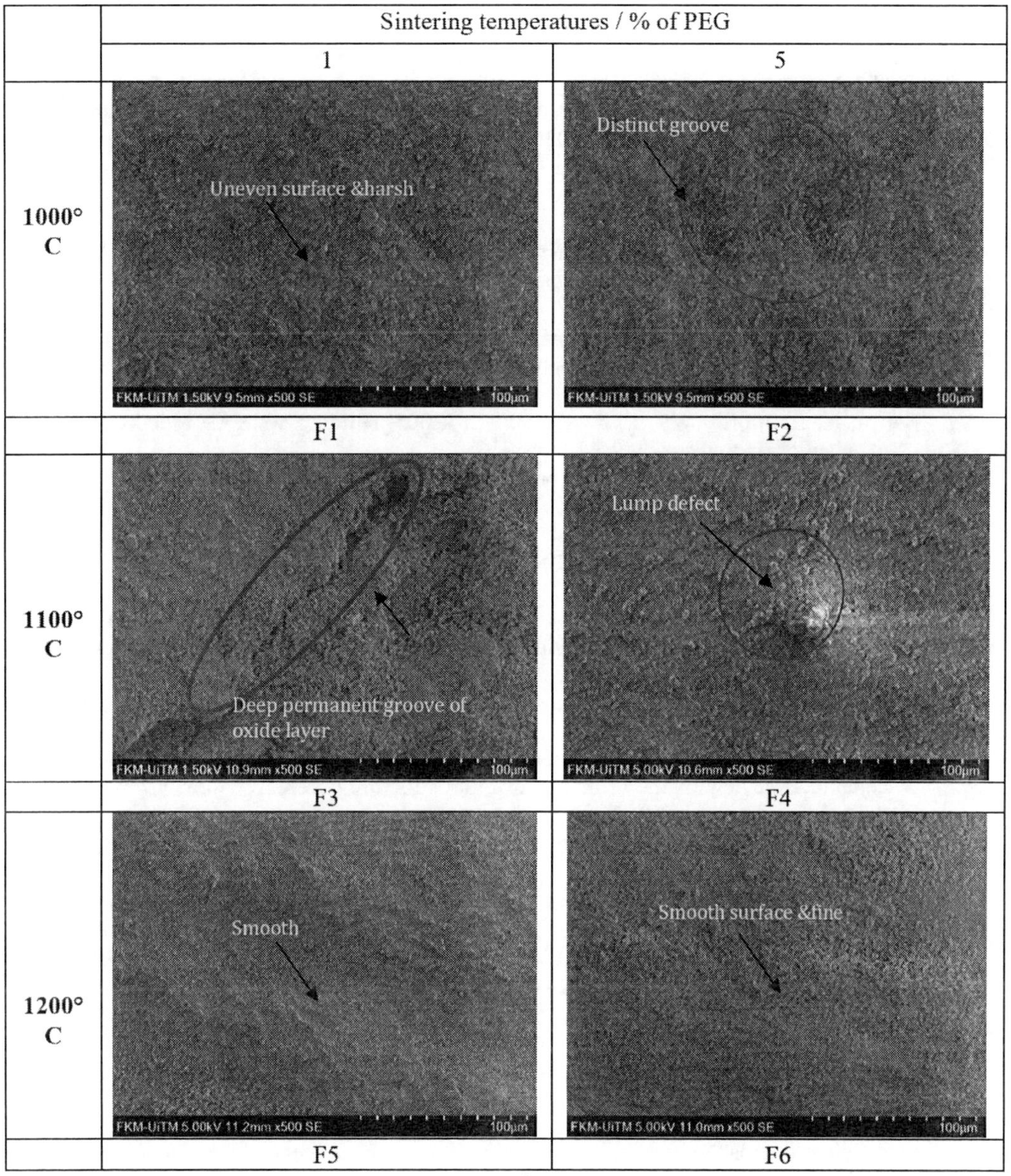

Figure 5. SEM Micrograph of worn surface for all formulation from F1, F2, F3, F4, F5 and F6.

4. Conclusion

Based on the wear rate and volume loss of alumina tested at 20N load and sliding speed of 0.28 m/s, it was concluded that the increment of sintering temperature of alumina blended PEG results to lower wear rate between 2.9 to 0.08 $x10^{-6}$ $mm^3/N.m$ in terms of the specific wear rate and worn surface analysed. However, the ratio of PEG binder give less significant effect to the pin on disc wear test due to the less amount of binder have been introduced in the test. It was observed the lower sintering temperature permit higher wear rate and the presence the grooves at the surrounding area. This phenomena probably happen that due to the surface roughening and existent of the wear debris. The microstructural studies show at high sintering temperate of 1200°C, the smooth contact surface was noted and contain strengthening of alumina particle that resist the wear. Applied sintering temperature may affect the wear mechanism and need further improvement in projection with high content of PEG binder for comparable studies onwards.

Acknowledgement
The authors would like to thank and express appreciation to Geran Insentif Penyeliaan (GIP) Universiti Teknologi MARA no file IRMI 5/3/GIP (030/2018), Malaysia for the financial support in completing this research.

References
[1] Standard, A. S. T. M. (2010). G99-05, 2010,". *Standard Test Method for Wear Testing with a Pin-on-Disk Apparatus," ASTM International, West Conshohocken, PA.*
[2] Kato, K., & Adachi, K. (2002). Wear of advanced ceramics. *Wear*, *253*(11-12), 1097-1104.
[3] Medvedovski, E. (2001). Wear-resistant engineering ceramics. *Wear*, *249*(9), 821-828.
[4] Rainforth, W. M. (2004). The wear behaviour of oxide ceramics-A Review. *Journal of materials science*, *39*(22), 6705-6721.
[5] Bajwa, S., Rainforth, W. M., & Lee, W. E. (2005). Sliding wear behaviour of SiC–Al2O3 nanocomposites. *Wear*, *259*(1-6), 553-561.
[6] Tedesco, N. R., Pallone, E. M., & Tomasi, R. (2010). Effects of the pin-on-disc parameters on the wear of alumina. In *Advances in Science and Technology* (Vol. 65, pp. 39-44). Trans Tech Publications.
[7] Ravikiran, A., & Jahanmir, S. (2001). Effect of contact pressure and load on wear of alumina. *Wear*, *251*(1-12), 980-984.
[8] Ahmad, F., Rafi Raza, M., Ab Rani, A. M., & Jason Lo, S. H. (2011). Wear properties of alumina particles reinforced aluminium alloy matrix composite. *Journal of Applied Sciences*, *11*(9), 1673-1677.
[9] Davis, K. (2010). Material Review: Alumina (Al 2 O 3). *School of Doctoral Studies European Union Journal*, (2).
[10] Milak, P. C., Minatto, F. D., & Montedo, O. R. K. (2015). Wear performance of alumina-based ceramics-a review of the influence of microstructure on erosive wear. *Cerâmica*, *61*(357), 88-103.
[11] Souriou, D., Goeuriot, P., Bonnefoy, O., Thomas, G., & Doré, F. (2009). Influence of the formulation of an alumina powder on compaction. *Powder Technology*, *190*(1-2), 152-159.
[12] Ekka, P. (2012). *Effect of binders and plasticisers on alumina processing* (Doctoral dissertation).
[13] Daraei, A., Nabipoor Hassankiadeh, M., & Golmohammadi, M. (2016). Effect of Polyethylene Glycol (PEG) Powder on Compressibility and Microstructural Properties of Sintered α-Alumina. *Chemical Engineering Communications*, *203*(1), 47-52.
[14] Eskandari, A., Aminzare, M., Aboutalebi, S. H., & Sadrnezhaad, S. K. (2012). Effect of high energy ball milling on compressibility and sintering behavior of alumina nanoparticles. *Ceramics International*, *38*(4), 2627-2632.
[15] Ghaderi, M. J., Afarani, M. S., & Roudini, G. (2013). Synthesis of alumina porous supports via different compaction routes: vibration and pressing. *Journal of Chemical Technology and Metallurgy*, *48*(3), 289-295.
[16] Olofsson, J. (2011). *Friction and Wear Mechanisms of Ceramic Surfaces: With Applications to Micro Motors and Hip Joint Replacements* (Doctoral dissertation, Acta Universitatis Upsaliensis).
[17] Li, X., Olofsson, U., & Bergseth, E. (2017). Pin-on-Disc Study of Tribological Performance of Standard and Sintered Gear Materials Treated with Triboconditioning Process: Pretreatment by Pressure-Induced Tribofilm Formation. *Tribology Transactions*, *60*(1), 47-57.
[18] Burris, D. L., & Sawyer, W. G. (2006). Improved wear resistance in alumina-PTFE nanocomposites with irregular shaped nanoparticles. *Wear*, *260*(7-8), 915-918
[19] Kim, H. J., Lee, S. M., Oh, Y. S., Yang, Y. H., Lim, Y. S., Yoon, D. H., ... & Ruoff, R. S. (2014). Unoxidized graphene/alumina nanocomposite: fracture-and wear-resistance effects of graphene on alumina matrix. *Scientific Reports*, *4*, 5176.

[20] Wang, Y., Jiang, S., Wang, M., Wang, S., Xiao, T. D., & Strutt, P. R. (2000). Abrasive wear characteristics of plasma sprayed nanostructured alumina/titania coatings. *Wear*, *237*(2), 176-185.

[21] McElwain, S. E., Blanchet, T. A., Schadler, L. S., & Sawyer, W. G. (2008). Effect of particle size on the wear resistance of alumina-filled PTFE micro-and nanocomposites. *Tribology Transactions*, *51*(3), 247-253.

[22] Das, S., Das, S., & Das, K. (2007). Abrasive wear of zircon sand and alumina reinforced Al–4.5 wt% Cu alloy matrix composites–A comparative study. *Composites Science and Technology*, *67*(3-4), 746-751.

[23] Sawyer, W. G., Freudenberg, K. D., Bhimaraj, P., & Schadler, L. S. (2003). A study on the friction and wear behavior of PTFE filled with alumina nanoparticles. *Wear*, *254*(5-6), 573-

IOP Conf. Series: Materials Science and Engineering **834** (2020) 012079 doi:10.1088/1757-899X/834/1/012079

Exploring Colour Change Properties in 3D Print Photochromic PLA: towards Designing Central Cyanosis Actuator in a Baby Manikin

Azmi Nur Fatihah*[1, 2], **Frank Delbressine**[1], **Loe Feijs**[1] and **Peter Andriessen M.D.**[3]

[1] Eindhoven University of Technology, 5600 MB, Eindhoven, The Netherlands.
[2] Universiti Teknikal Malaysia Melaka, 76100, Melaka, Malaysia.
[3] Department of Neonatology, Máxima Medical Center, Veldhoven, The Netherlands.

*n.f.b.azmi@tue.nl

Abstract. A high fidelity baby manikin with a colour changing mechanism is required for medical training, especially for cyanosis skin colouration assessment in a newborn baby. However, commercially available baby manikin simulators do not have the real colour-change of cyanosis skin. The efforts in increasing the fidelity of the manikin might improve the accuracy of cyanosis evaluation among the novices, thus enhances the quality of the medical training. This study contributes towards designing a colour change actuator for central cyanosis simulation in a baby manikin for medical training. The actuation method of the 3D print photochromic PLA using a low-voltage UV LED and the evaluation of its colour properties using a colour measurement device are described. The colorimetric properties of the photochromic PLA were evaluated in CIE $u'v'$ colour space. Results are presented, showing a strong relationship between the colour intensity of the UV LED and the colour of the photochromic PLA. We reflect on the potential of the photochromic PLA as an actuator in imitating the cyanosis in a baby manikin. The specific PLA we used still cannot simulate the cyanosis well. However, the idea of using PLA photochromic is still considered promising.

1. Introduction

Photochromism is a reversible and repeatable colour changing mechanism, typically from colourless to a coloured state. It reacts to the stimuli of light, for instance to the UV light or the sunlight. When we remove the UV source, the colour state will revert to the original state. In medical simulation area, 3D printing (3DP) holds great promise because of its commercially accessible 3DP material and ease of manufacturing [1]. Thielen demonstrated a 3DP ribs cage for the realistic infant's manikin for cardiopulmonary resuscitation (CPR) procedure training [2]. However, they focus on exploiting mechanical properties of the new material and their production methods (not colour). Another 3D printed models in medical training are the lung which realistically mimicking such organs that allows students and trainees to practice the lung related surgery [3]. Nevertheless, this promise has not been yet realized for a colour change 3DP thermoplastic for realistic colour changes in skin colour, especially for imitating the central cyanosis in a newborn baby. It reacts to the stimuli of light, for instance to the UV light or the sunlight. When we remove the UV source, the colour state will revert to the original state. The recent development of 3DP PLA chromic material enables digital printing models with the ability to change the colour reversibly with the UV light stimuli. The UV LEDs responsive 3DP photochromic actuator makes it an attractive platform for the application of designing a colour- changing artificial

Content from this work may be used under the terms of the Creative Commons Attribution 3.0 licence. Any further distribution of this work must maintain attribution to the author(s) and the title of the work, journal citation and DOI.
Published under licence by IOP Publishing Ltd

skin. To the best of our knowledge, no work so far has presented a light-responsive baby manikin for medical training in imitating the colour change of central cyanosis.

The colour change of the baby's skin is one of the immediate evaluation made after the baby was born. The assessment gives significant indication of the newborn's condition [4] [5]. For example, when the lips and mouth area of the baby remained blue after 10 minutes, it is recognized as central cyanosis [6], and urgent treatment from a medical team is needed. In various lighting condition in the baby's delivery room as well as the colored objects around the babies makes central cyanosis assessment in a newborn quite challenging [7]. Hence, the cyanosis simulating baby manikin is needed for the training of cyanosis visual evaluation in a newborn among new medical trainees. The design parameters for a cyanosis actuator in baby manikin are; the colour change of an actuator should be controllable and reversible in a specific timing. As central cyanosis normally visualized in the thinnest epidermis area such as around the mouth and lips, hence, the sizing and dimensions of the actuator must be in the correct size and mimics the mouth and lips of the full term baby. Safety and power consumption are important e.g. 7.4W when the 3.7V lithium-ion battery pack is used [8].

The idea of using photochromic material is attractive because we could avoid many weaknesses and side effects of traditional light sources. Comparing to conventional light sources such as visible-spectrum LEDs the strong point of the photochromic material is that the visual effect is independent of the external light source (the room's illumination). If there would be purple LEDs inside the manikin, the purple LED light would have to go up when the light in the training room gets brighter. In addition, the purple LED must dim when the outside light goes down. Even the shadow of a doctor's hand would reveal the LED nature of the colour. As a solution, the purple LED could be made adaptive, but that involves designing a sensor-based feedback loop. Of course, the photochromic PLA needs an LED too, but that is only for changing the PLA's subtractive colour behavior. The latter is like the real baby skin behavior. When increasing the room's illumination, pink lips stay pink, and purple lips remain purple. We can also compare to thermochromic materials, which are environment-independent too. However, working with temperature is difficult: we need to compensate for the environment's temperature. Next, the thermochromic matter has a specific heat capacity: heat accumulates in the manikin's head, and we need yet another feedback control loop to manage these effects. Even worse, the head gets hot, so the trainee could feel the heat, which is weird (it breaks the spell of suspended disbelief). The UV light, by contrast, does not accumulate by the heat and it is invisible. Of course, we have to see how the photochromic PLA can make these promises come true. That is why we undertake the design exploration of this chapter. Figure 1 illustrates the cyanosis colour change upon the UV LEDs stimuli in the baby manikin. This 3D printed photochromic baby's head exhibits a purple colouration upon the exposure to UV LED (activation state) with an applied potential of a low voltage [9] and returns to the original colour when the UV LED is off (deactivation state).

Figure 1. The colour change of the photochromic baby's head during deactivation and activation

Quantifying colour values in a photochromic material is a challenging task as the colour transition time of from colourless to coloured state very fast within seconds [10]. Moreover, the data of colour changing time upon the stimuli UV light is lacking from the manufacturer making such evaluation more difficult. Viková et al. investigated the colour measurement in photochromic for textiles application, using the solution spectrophotometer ACS ChromaSensor CS-5 with spherical measuring geometry [11]. In this study, we use the Spectroradiometer Specbos 1201 [12] to quantify the colouration of photochromic PLA. Here, we first examine the colouration of a 3DP PLA by modulating the light intensity (in %) of the UV LEDs by the adjustment of the potentiometer. From this experiment, the colour change of during activation and deactivation state of photochromic PLA was obtained in the luminous, L $[cd/m^2]$ unit. The chromaticity data are also measured in the International Commission on Illumination (CIE) $u'v'$ 1976 colour space [13].

The main contributions of this work are 1) a controllable 3D printed colour change actuator using Pulse Frequency Modulation (PFM) for central cyanosis application, and 2) a quantifiable photochromic PLA colouration in CIE colour space. Although it does not resulted in the correct colour values of cyanosis in a real baby as reported in [8], we believe that the effort required to provide the colorimetry data of the photochromic PLA in CIE system is necessary and will provide a more direct path to the results we wished to present in the future. We adopt the working hypothesis that these four elements can be achieved by implementing the photochromic PLA in a cyanosis baby manikin. Thus, the findings provide new insight into the structure of research and offer a new platform for future research.

2. Creating a 3D printed baby's head

We found two manufacturers Go-3D Print, based in San Francisco, California and RepRapWorld BV, Nootdorp, The Netherlands. The photochromic PLA we got from Go-3D Print, changes from white to blue and the photochromic PLA we got from RepRapWorld BV changes from white to purple. We noticed that the materials were rather pale for the baby's face, so we considered spray- painted. Our initial experiment based on spray- painting from the outside was a failure because the UV induced colour change was hardly noticeable. Spray painting from the inside still considered to be options. Our design space includes at least four combinations because we can choose from two different materials and two different spray- paint options: spray- painted from the inside and not spray- painted. Rather than trying all combinations, we took two combinations, which are shown, in Table 1 and in Figure 2.

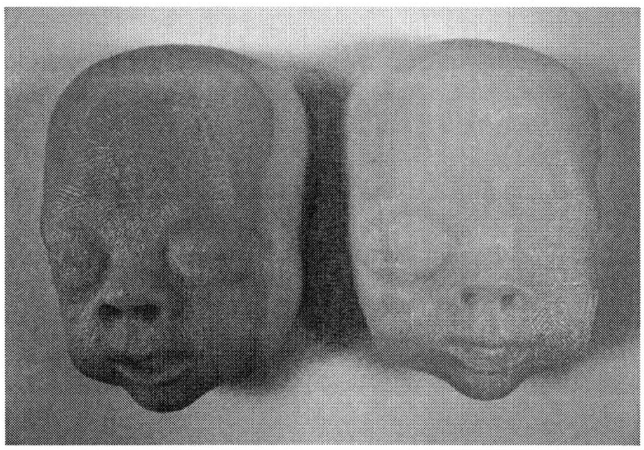

Figure 2. To the left: Spray-painted from the inside. To the right: Original colour, not spray-painted.

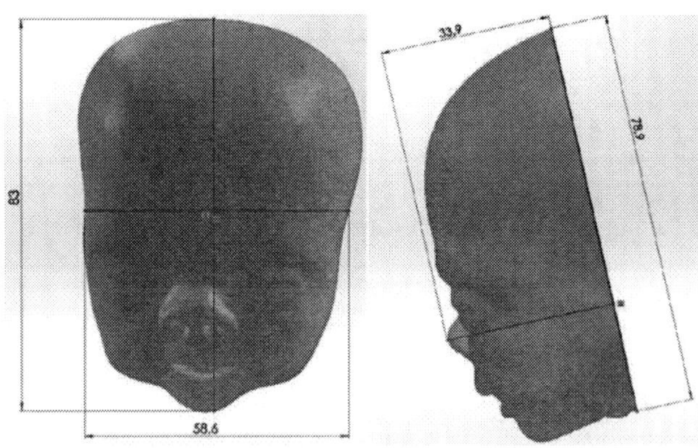

Figure 3. The dimension of the 3D printed baby's head.

Table 1. Commercial Photochromic PLAs.

Head	Diameter	Spray- painted	Colour Change	Trades Name
1	1.75mm	Not spray	Original to purple	RepRapWorld BV
2	1.75mm	Spray from inside	Original to blue	3D PLA

One of the two models was spray-painted with MOTIP carat lacquer spray in 'BABY' colour to imitate the colour of a real human's skin [14]. The 'BABY' colour was chosen as in Figure 3, based on the reference colour of the Pantone 108-8 C from the Humanae project which a collection of different human skin colours in a chromatic range [15]. The photochromic filaments were printed into the identical shapes of preterm babies in the dimension of 58.6 x 83 x 33.9 mm (See Figure 3). Both filaments were printed using a Craftbot-PLUS 3D printer, from CraftUnique in MakerPoint Eindhoven with the thickness: 1.0mm.

3. Material activation by UV LEDs

The plan is to illuminate the photochromic baby head from the inside using a particular type of LEDs, emitting UV light, which is by itself invisible to the human eye. First, we present how classical LEDs behave, then we move on to deploy UV LEDs. The concept of visible light's brightness from LEDs is relatively straightforward. For instance, the brightness of LED perceived by the eyes usually measured in units of luminous intensity or candelas, I (cd). The total power output of an LED is a measurement of the number of Lumens, lm (Φ) [16]. These two measurements are the photometric quantities which relate to how human eyes perceived colours. All LEDs have the property that increased current leads to increase luminosity (brightness). An application notes of NICHIA white LED gives an example where relative luminosity goes from 1 to 1.7 if the current goes from 20mA to 40mA [17]. In the same application note, for a white LED NSCW215, we find the luminosity of 600mcd at 20mA and 1020mcd at 40mA, respectively. We assume UV LEDs work similarly. We acquired four UV LEDs of type PLCC-2 manufactured by Vishay Semiconductors, Malvern, Pennsylvania [18]. According the data sheet it should produce radiant power of 6.8mW at 20mA and the radiant intensity of 2.5mW/sr. The usage of mW/sr is common for UV and IR. For normal LEDs the unit mcd is used, a human-eye corrected measurement.

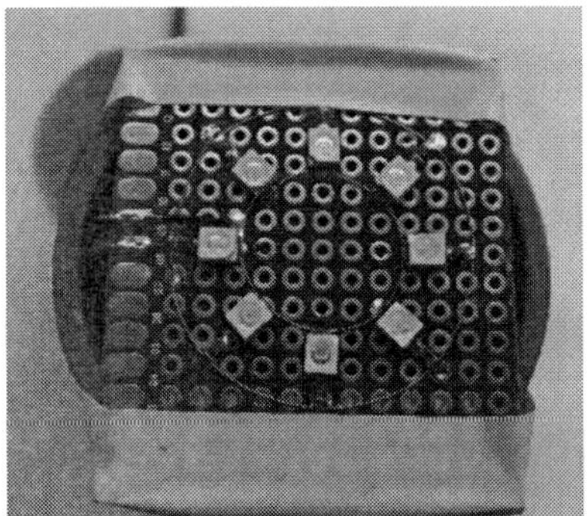

Figure 4. Eight parallel-connected ring-shaped UV LEDs

One single UV LED maybe not enough to generate the proper illumination, so we decided to put multiple LEDs in parallel. We tested several configurations, and we show one of the test results in Figure 5. The figure contains the measurement results where we used eight LEDs. We connected each LED in series with a resistor of 56 Ohm. Then eight of these LED-resistor pairs were put in parallel. In Figure 5, we see a knee Voltage of about 2.9V whereas at higher Voltages the behavior is mostly linear, which is the effect of the resistors. We thus also found that we could send high current through the LEDs, up to 230mA.

It is a requirement for the design of the baby manikin that the colour can be adjusted continuously. There are two ways of doing this; the first is to have a variable series-resistor connected to the LEDs (potentiometer or well-regulated transistor). The second way of doing this is by pulse modulation, which means that we switch the LEDs on and off regularly. The first solution has a severe disadvantage viz. the dissipation in the series resistor.

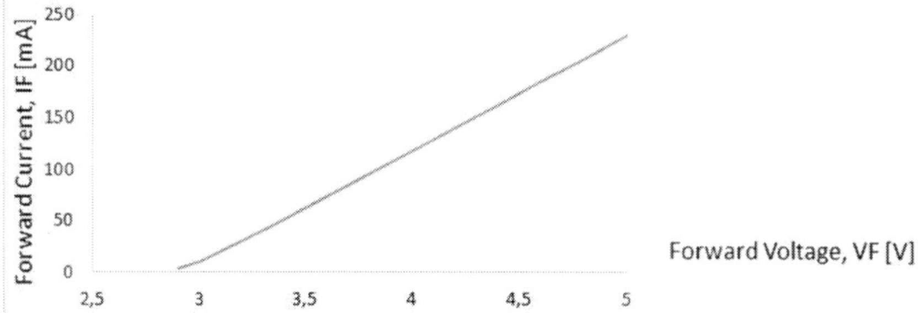

Figure 5. Forward Current vs. Forward Voltage

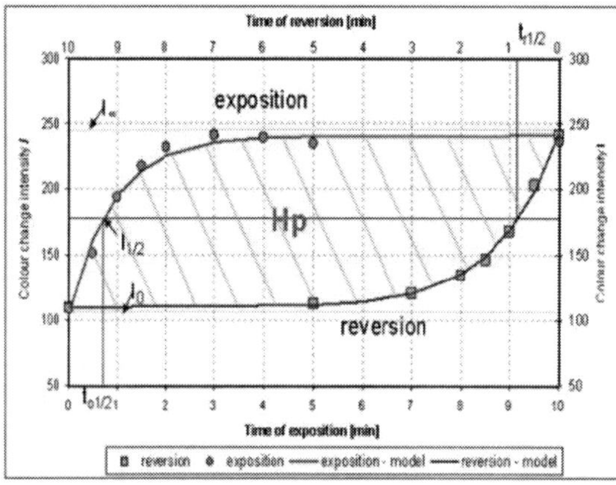

Figure 6. Slow colour changes observed in photochromic materials (Retrieved from [11]).

It would cause unnecessary heat development in the manikin and exhaust batteries. Now, we turn our attention to the second solution. We ask ourselves whether the pulses would appear in the colour of the PLA (the baby's face). Fortunately, we found the result of Viková , which we learn that photochromic materials behave rather slowly, see Figure 6.

We decided to apply very short pulses of maximum intensity. We design a mouth-shaped configuration of four LEDs with no series-resistors. This allows us sending more than 250mA through the combined LEDs (put in parallel). The UV LEDs were arranged in the mouth- shape to illuminate the photochromic PLA, resulting in evenly spread colouration of central cyanosis area. The distance of LEDs to photochromic's surface is approximately 31.6mm.

From the datasheet of the LED, we learn that we can send high currents through the LED provided the pulses are short (less than 10μs). Therefore, we fixed our pulse width and varied the time between the pulses. This is called pulse frequency modulation (PFM). Looking back, we believe that PWM would also be a valid solution as well. In Figure 7, an Arduino compatible microcontroller, Teensy 3.2 board was used to generate the frequency signal of PFM. We mapped the Arduino linearly to read a potentiometer whose position to the between-pulses waiting time, ranging from 25μs (high intensity) to 1ms (low intensity). In our experiment, we refer to this highest intensity as 100%. The Arduino reads the percentage to the serial port, and from there it goes to a PC for inspection during the experiment.

In this experiment, we applied a DC voltage of 5V to the UV LED to activate the photochromic's head. For each measurement, we allow the UV LEDs to be activated at least 20s to let the PLA change colour. We noticed that the colour change stabilized after few seconds (1 or 2s at high intensity and less than 10s at low intensity). At low intensity, we waited 20s.

4. Instrumental Methods

4.1. Instruments and Colour Measurement Setup

We considered using a handheld spectrophotometer. We have the Ci6x X-Rite handheld portable spectrophotometer in our lab. The initial measurement revealed that it is not possible because of the built-in lamp in the measuring device. The device's integrated illuminant source is a gas-filled tungsten lamp and UV LEDs [9]. Because of that, the colour measurement using this spectrophotometer will activate the photochromic properties and itself already induced a colour change from normal colour to the purple colouration of the printed model. Even if we would add a UV filter or remove its UV LEDs, we cannot focus it on the cyanosis-mouth area, and it is not meant to measure curved surfaces. Therefore, we did not use the X-Rite spectrophotometer.

Another option would be to measure CIELAB values using an ordinary camera and our colour correction algorithm (we did those measurements, but we switched to our third option). Our third option has the advantage that we need no correction at all. The third option is to use the miniaturized and fast broadband Specbos 1201 spectroradiometer (with a wavelength of 380 to 780nm). This third option has its difficulties too, namely that the UV light will be measured by the spectroradiometer as well. Please note that the output of the UV LEDs is between 400 and 410nm with the typical value of 405nm as is specified by Vishay Semiconductors, which is in the range of the spectroradiometer. One solution would be to remove this part of this spectrum after the measurement. We show the problem in Figure 8.

Another solution is to take advantage of the fact that the PLA material does not react instantaneously. We can activate it for a number of seconds, then deactivate and measure immediately (< 1s). This is the solution we adopted. Figure 9 illustrates the position of the photochromic head in the experimental setup.

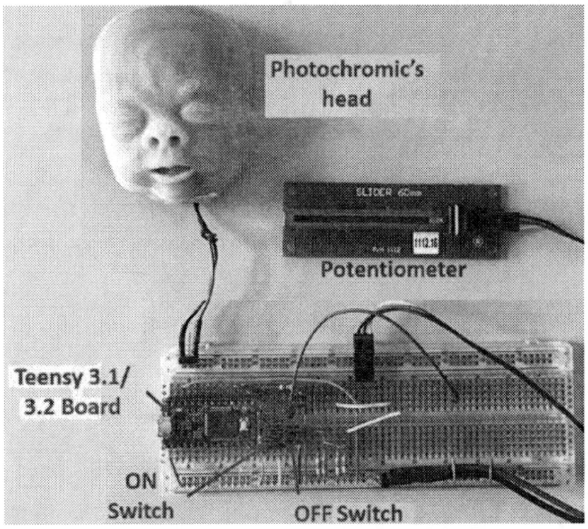

Figure 7. Set up for controlling the colour change of head with the UV LEDs backlight actuation.

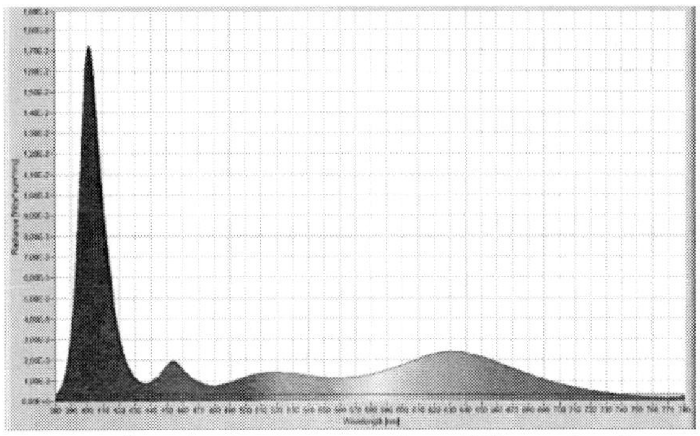

Figure 8. Measured spectrum showing a strong peak at 402nm clearly, caused by the UV LEDs.

The cyanosis observation lamp, Philips Master LEDspot LED light (940 type lamp; Philips) [19] was set up as the source of the illuminant. We performed the colour measurement in a white box with the cyanosis observation light thus avoiding the unknown lighting conditions. The measurements took place in a dark room. In this way, we eliminated any influence from other types of light sources, which could affect the colour properties in the photochromic's head (such as sunlight or other UV sources). We measured the light box with a white background which functioned as diffusers and helps to spread the light evenly and reduce unnecessary shadows [20]. A colour characterization of the PLA photochromic measurements was carried out with a distance of 30cm from the measured object as in Figure 9. At this distance, a target diameter of a spot laser is similar to cyanosis region's size (approximately 10mm).

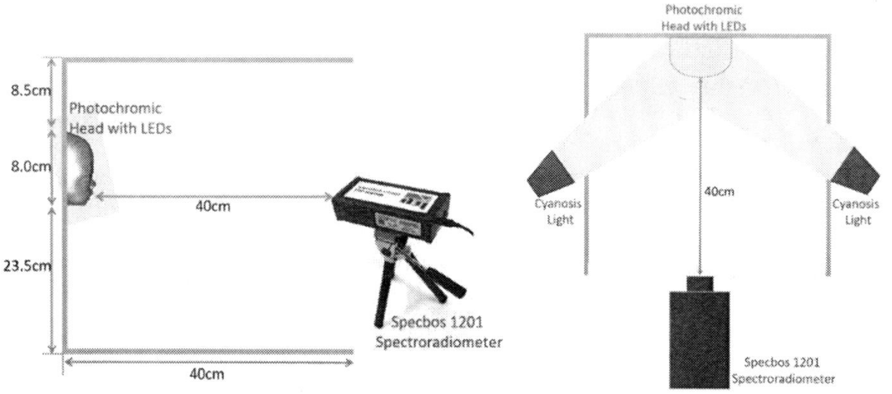

Figure 9. Colour measurement setup for measuring the colour value of the baby manikin head.

4.2. Colour Measurement of 3D Print PLA Photochromic

There is a notion of luminance, L $[cd/m^2]$ in the radiometric quantities which would be used when testing light source, but this is not what we need because our baby head is not a light source (unlike baby heads whose colour change would be implemented by visible-light LEDs). We need to apply the external light sources to see and measure the manikin heads like a real baby. We need to work in one of the CIE colour spaces. We found that the CIELAB implementation of our spectroradiometer was not functioning properly, and therefore we use the option of the spectroradiometer to measure the u' and v'according to CIE $u'v'$1976 (2-degree observer) [13] [21]. There are conversion formulas between CIE $u'v'$and CIELAB, which later can be used to compare the PLA heads to the colour of the real babies (with and without cyanosis). Figure 10 and 12 shows the cyanosis observation light illuminating the photochromic's head, which is then, observed by the human eye. The photochromic head transmitted the illumination spectrum and weighted with its spectral reflection behaviour. If the human eye receives this signal, and the pectral convolution with its own three sensitivity curves, $\bar{x}, \bar{y}, and\ \bar{z}$ will took place here. The result shows how the colour is perceived in the photochromic head, expressed by the tristimulus values X, Y and Z (See Equation 2) [22]. The CIE XYZ (1931) colour system is the root of all colorimetry. Here, the Y is refers to luminance, and all visible colours can be defined using only positive values, and, the Y value is luminance. The chromaticity diagram is highly non- linear and X, Y and Z tristimulus values do not contribute any information about lightness, hue and saturation. Therefore, a transformation into other colour systems which is quantifiable is needed.

The chromaticity values of CIE L*u*v*are calculated according to the equations below [23]:

$$L^* = 116(Y/Y_n)^{\frac{1}{3}} - 16, \text{when } Y/Y_n > 0008856$$
$$L^* = 903.3(Y/Y_n)^{\frac{1}{3}} - 16, \text{when } Y/Y_n \leq 0.008856$$
$$u^* = 13L^*(u'/u'_n) \text{ and } v^* = 13L^*(v'/v'_n)$$

where Y is the tristimulus value, u' and v' are chromaticity coordinates from the CIE $u'v'$1976 diagram. Y_n, u'_n and v'_n are tristimulus values of Y_n and the chromaticity coordinates of u'_n and v'_n of the perfect reflecting diffuser, respectively. The transformation from CIE $u'$and $v'$from CIE L*u*v*is:

$$u' = \frac{u^*}{13L^*} + u'_n v', \quad v' = \frac{u^*}{13L^*} + v'$$

5. Results and Discussion

5.1. Photographs of the effect

Figure 10 and Figure 11 show the colour change of Head 1 and 2, respectively when the UV LEDs intensity output was changed by adjusting the duty cycle to 2%, 20%, 40%, 60%, 80%, and 100%. Upon increasing the intensity from 2% to 100% brightness range, we measure the colour values of the head during activation and deactivation state. The photographs were taken immediately after the activation of UV LEDs. The purple colouration can be seen for both heads very clearly. Head 1 is not very realistic, mostly because of the white colour when the UV LEDs are off. Head 2 appears more promising. The reversible colour change from non-cyanosis to cyanosis colouration based on the UV LED actuation allows users to control the colour based on the intensity. 2 to 100% is corresponding to the relative position of the potentiometer and it's corresponding to the specific frequency in PFM.

5.2. Measuring Luminance

The basic idea of this measurement is to illuminate the baby head by the Cyanosis Light described before (two lamps), and measure how much light comes back and meets the observer's eye. Here, we focus on luminance, L, not yet a colour. As before, during activation, we allowed the material to become stabilized before taking the measurement. The deactivation values are measured immediately after switching off.

The luminance output, L during both states was recorded in Table 2 and Table 3 for Head 1 and Head 2, respectively. Referring to Table 2, during the activation state of Head 1, the luminance change was decreasing when the intensity was increased from 2% to 100%, indicating that the activation state already started at the lowest frequency of the PFM. The maximum L was visible during 2% with 148[cd/m2] and the minimum L was visible at 100% with 123$[cd/m^2]$. The luminance changes started from 2% with 148[cd/m2] to the maximum intensity of 100% with 141[$[cd/m^2]$.

To get an impression of the time constant involved, we measured the luminance immediately after switching off the LEDs (after full activation, which is 100%). The results can be seen in Table 2. Photochromic PLA's colour stabilization happened in about 20s.

Referring to Table 3, during the activation state of Head 2, the luminance of a colour change also was decreasing when the intensity was increased from 2% to 100%. The maximum L was also visible during 2% with 110$[cd/m^2]$, and the minimum L was visible at 100% with 80.2$[cd/m^2]$. Moreover, in the deactivation state of Head 2, when the UV LEDs is switched off, the colour change also decreased from 119$[cd/m^2]$ to 74.6$[cd/m^2]$. The colour of the photochromic heads change in parallel with the change of the LEDs intensity, starting from 2% to 100%.

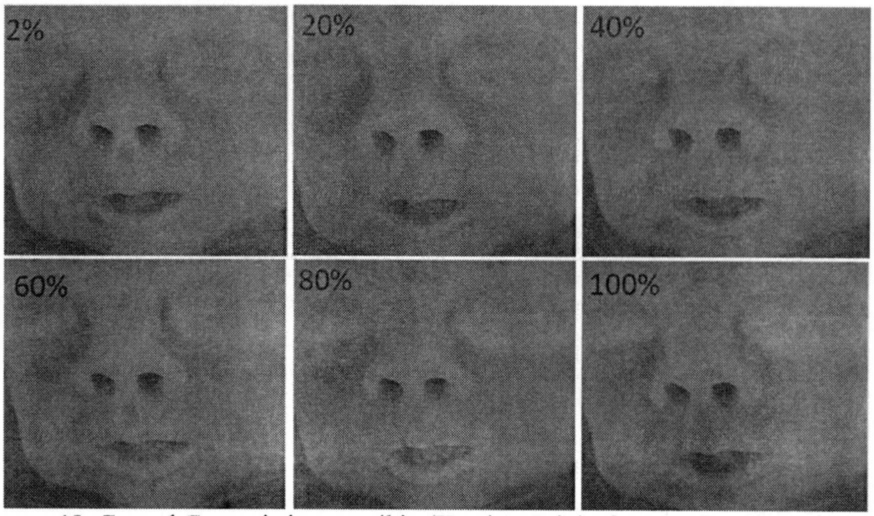

Figure 10. Central Cyanosis in a manikin (Head 1: Original to purple colour change).

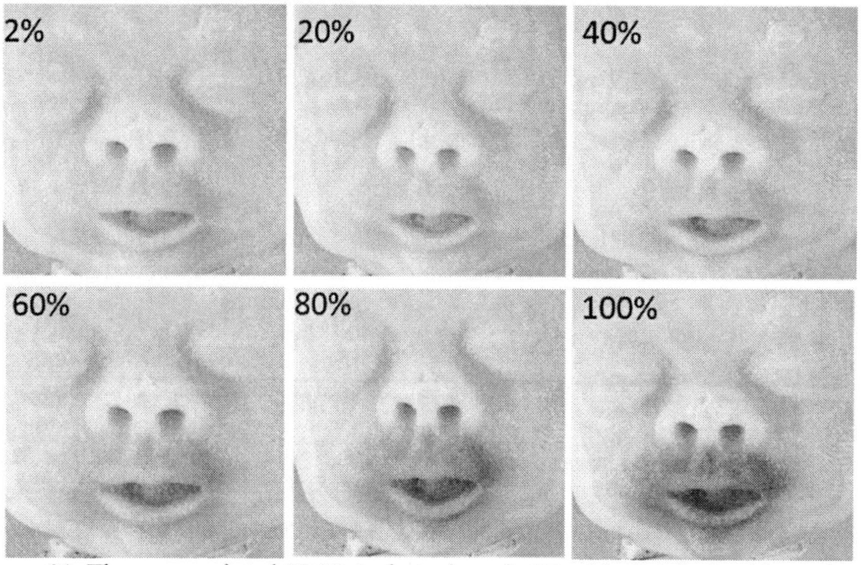

Figure 11. The spray-painted 3D PLA photochromic (Head 2: Original to blue colour change)

Table 2. Head 1: Not spray-painted. Original to purple colour change.

	Luminance, L[cd/m^2]	
	Activation	Deactivation
2%	148	148
20%	145	148
40%	144	147
60%	141	146
80%	137	146
100%	123	141
LED OFF	156	

To get an impression of the time constant involved, we measured the luminance immediately after switching off the LEDs (after full activation, which is 100%). The results can be seen in Figure 12. Stabilization happened in about 50s.

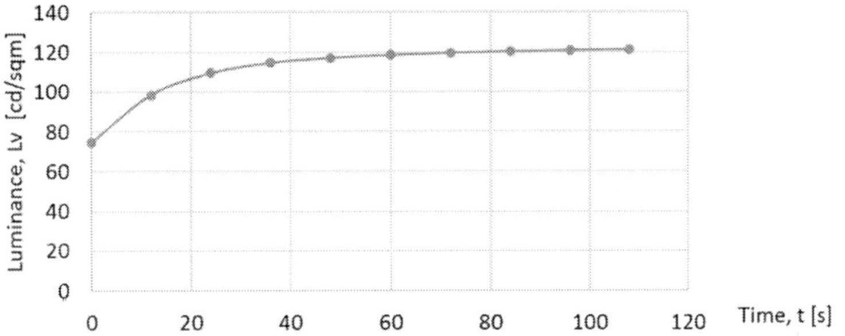

Figure 12. Luminance change of the Head 1 (White) after switching off the LEDs.

Table 3. Head 2: Original to blue colour change.

	Luminance, L[cd/m^2]	
	Activation	Deactivation
2%	110	119
20%	105	119
40%	105	111
60%	103	106
80%	102	96.4
100%	80.2	74.6
LED OFF	121	

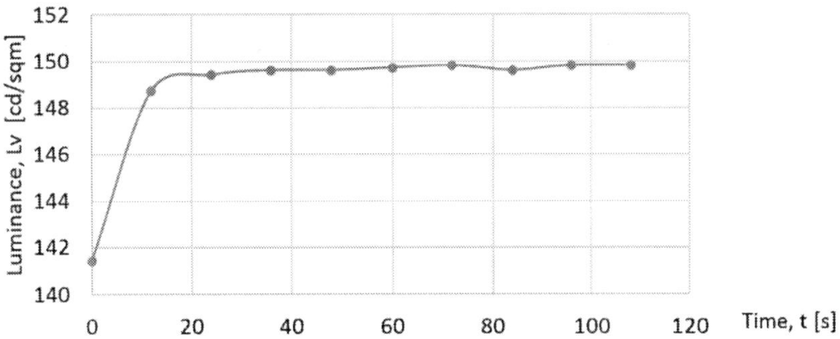

Figure 13. Luminance change of the Head 2 (Spray-printed) after switching off the LEDs.

5.2.1. Discussion If we try to interpret this result, we see that the luminance goes down when the UV LED is more active. This is what we expected, the PLA goes from white to purple, subtracting certain part of the visible spectrum. The PLA photochromic Head 2 reflects more light than Head 1. We also see that there is a difference between the activated and deactivated state, despite our protocol of fast measuring immediately after going from activation to deactivation. There appears to be a fast reaction by the material. We note that at high LED activation percentages, there is an opposite effect (illumination goes down from 80.2 to 74.0 when the LED is switched off). We conjecture that this is UV light shining through and picked by the spectroradiometer (which is more sensitive to UV than our human eyes). However, we want to focus on the objective colour change, more than on the luminance. In the next stage, we will focus on the CIE $u'v'$ colour space.

5.3. Colorimetric Measurement

To illustrate the colouration of photochromic PLA as perceived by human eyes, the colour change was plotted in CIE $u'v'$space. We decided to work in the CIE $u'v'$space rather than in CIELAB because we found that the L^* values given by the Spectroradiometer are sometimes above 100, so we believed that possibly the a^* and b^*also could be unreliable. Both CIE $u'v'$and CIELAB are designed to respect the perceptibility of colour differences by the human eye [24].

Before zooming in, we give the big picture of the entire CIE $u'v'$ colour space. First, we present the colour change of Head 1 in Figure 14 (left). We see all the measurement points are not too far from the "white point". Please note the lowest point, measured during 100% of activation. It is clear that the UV light itself is dominant here. We consider the left-most peak, which we observed in Figure 8 as evidence that indeed the UV light itself shines through the PLA material. In the same way, we present Head 2 in Figure 14 (right). We notice that there is no deep blue outlier anymore. It can also be observed that the changes are smaller. We conjuncture that the spray-painted partially inhibits the penetration of the UV lights.

Figure 15 shows the colour value shift in CIE $u'v'$during the activation and deactivation states of Head 1. During activation and deactivation states, there is a slight increase in CIE u' from 0.2440 to 0.2478 and from 0.2428 to 0.2584, respectively. Meanwhile, as for CIE v' during activation state, there is a decrement in from 0.4759 to 0.2399 when going from 2 to 100%. A slight decrement in the deactivation state was observed in CIE v' from 0.5045 to 0.4878.

Figure 17 shows the colour change of Head 2. During activation and deactivation states, there is a decrement in CIE u' from 0.2536 to 0.2363 and from 0.2550 to 0.2511, respectively. Meanwhile, as for CIE v' during activation state, there is a decrement from 0.4972 to 0.4218 from 2 to 100% of dimming brightness. A slight decrement in the deactivation state was observed in CIE v' from 0.5015 to 0.5000.

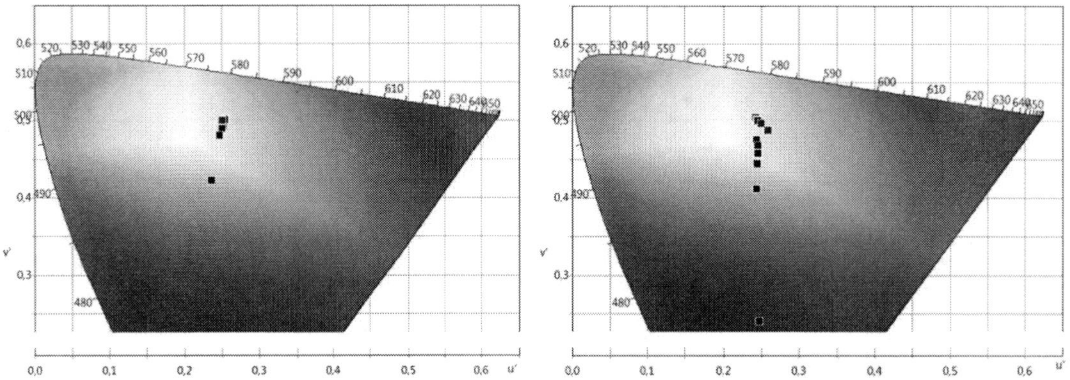

Figure 14. Colour change of the Head 1 (Left) and Head 2 (Right) : White photochromic PLA, change colour from original to purple in CIE $u'v'$. The black series show the measurement points of the activation state. The blue series show the measurement points of the deactivation state

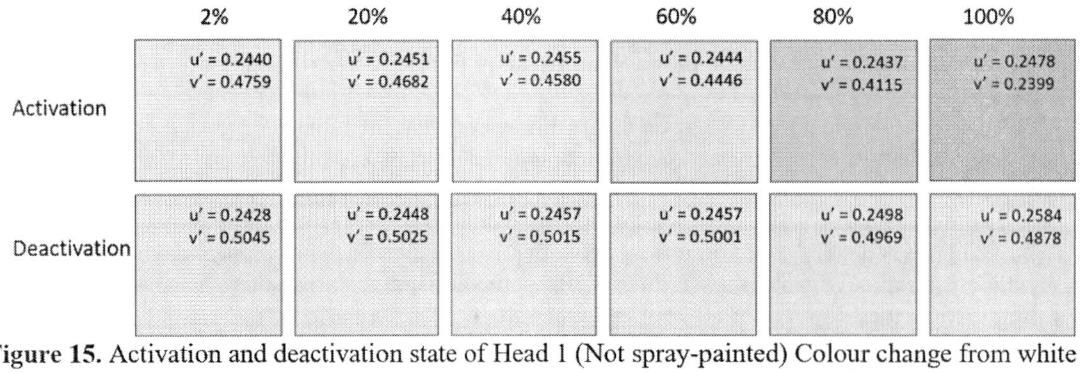

Figure 15. Activation and deactivation state of Head 1 (Not spray-painted) Colour change from white to purple in CIE $u'v'$colour space.

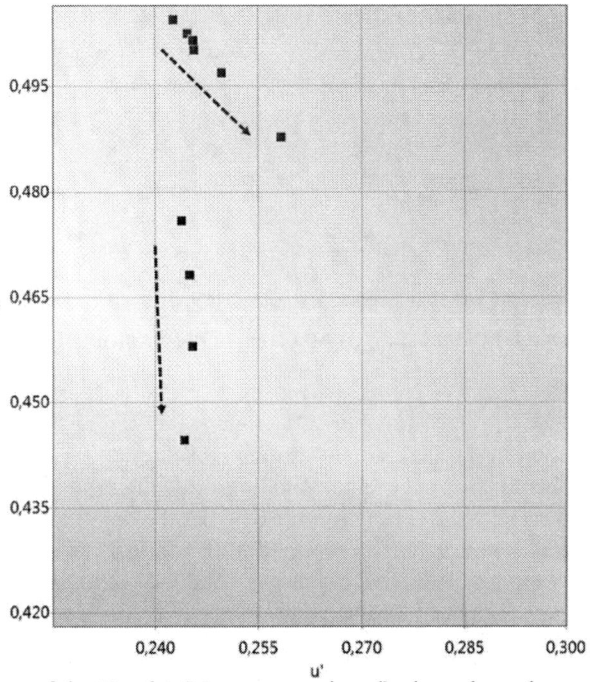

Figure 16. Colour change of the Head 1 (Not spray-painted): the colour change from original to purple photochromic PLA in CIE $u'v'$.

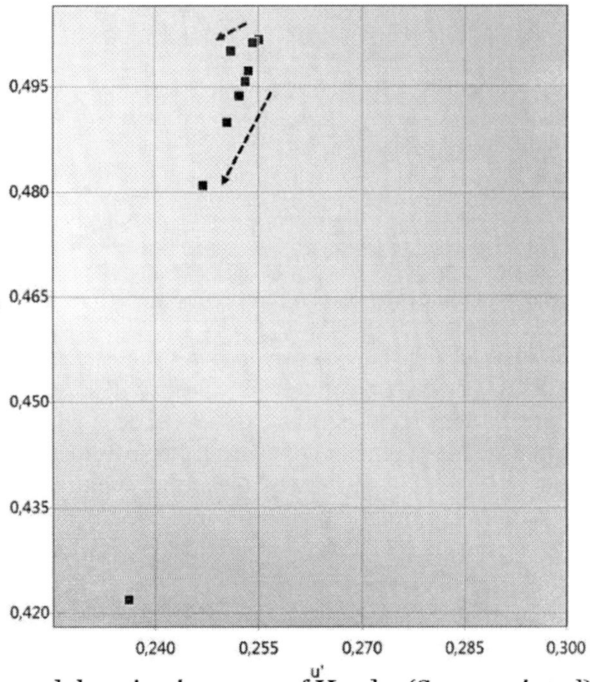

Figure 17. Activation and deactivation state of Head 2 (Spray-painted): Colour change from white to blue in CIE $u'v'$colour space.

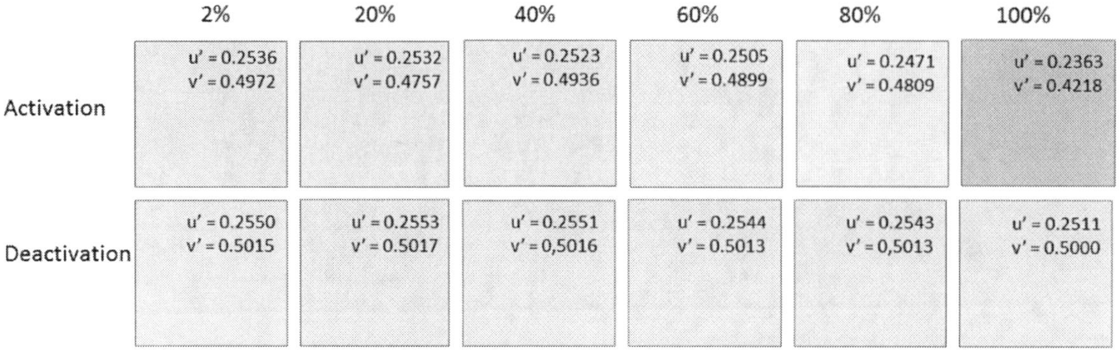

Figure 18. Colour change of the Head 2: Spray-painted photochromic PLA, change colour from original to blue in CIE $u'v'$.

5.4. Comparing Baby Data

The colour change recorded in the experiment of the previous section shows a clear colour difference between activation and deactivation state of the photochromic material and we can shift by changing the UV LED's intensity. Nevertheless, it is not obvious whether the colouration achieved in this experiment approximates real cyanosis colour values as reported in [8]. In this section we compare them.

In Figure 19, we show the data from real babies, as obtained in the earlier study. We then convert the CIE a^*b^*values to the CIE $u'v'$space. The calculation of the colour space conversion from CIELAB to CIE $u'v'$ as discussed in Section 4.2.

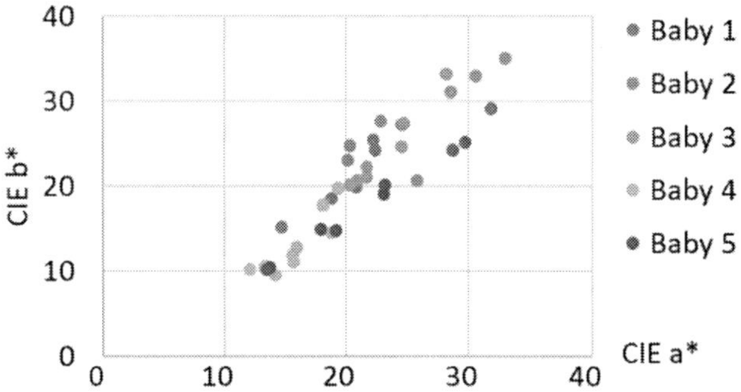

Figure 19. Colour change of cyanosis to non cyanosis colouration from the data of five real baby images

Figure 20 depicts the colour change of cyanosis to non-cyanosis colouration from the data of five real baby images in CIE $u'v'$. The colour values of cyanosis will be minimum at approximately CIE $u'v'$of (0.2380, 0.4856) and the non-cyanosis colour value will be maximum at 0.2896 and 0.5130. The arrows in the figure illustrate the measurement points from cyanosis to non-cyanosis. If we compare the colour change show in Figure 17 and Figure 19, we can see that colour changes in the deactivation state of Head 2 (by changing the intensity from 2 to 100%) could mimic the changes from cyanosis to non-cyanosis lips colour as in the database of Baby 4 in Figure 19. As for Baby 4, the cyanosis starts at (0.2639, 0.4953) in comparison with the (0.2536, 0.4972) at the maximum activation of the LEDs.

6. Conclusion

In this study, we examine the effects of the colour change and the UV LED's intensity on the 3D print photochromic PLA filaments. The evaluation of the PLA photochromic was done by calculating the colour value for each of the colouration from 2% to 100 % intensity of the colour change upon applying UV LEDs. Large colour differences were found in chromaticity values during the activation mode but smaller in the deactivation mode. The specific PLA we used still cannot simulate the cyanosis well enough for training the precise colour observation skill, but the combined effect of the reduced lightens and the colour shift gives a remarkable impression of cyanosis. Hence, the idea of using PLA photochromic is consider promising. The main contributions of this work are 1) a 3D printed controllable colour change actuator for central cyanosis application and 2) a quantifiable colouration of the PLA photochromic in CIE $u'v'$colour space.

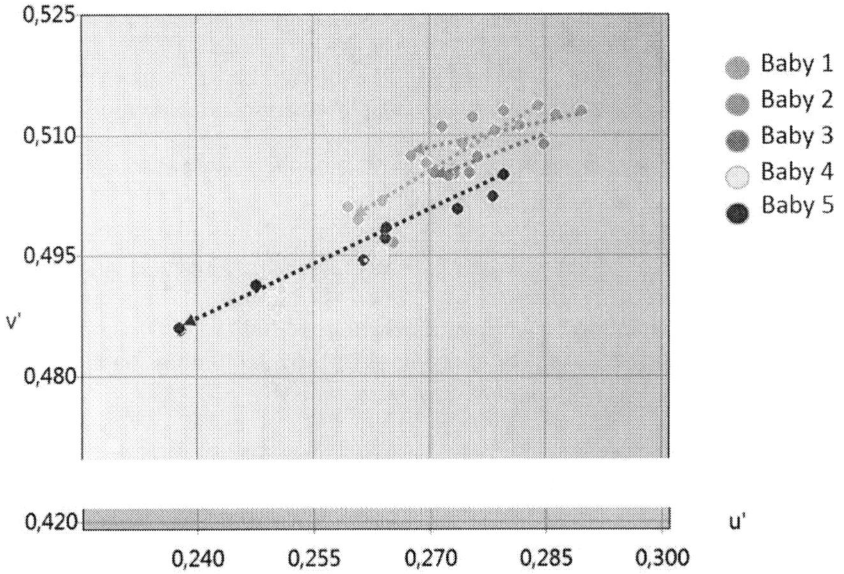

Figure 20. Colour change of cyanosis to non-cyanosis colouration from the data of five real baby images in CIE $u'v'$.

Acknowledgments

The author of Nur Fatihah Azmi (N.F.B. Azmi) is funded by the funds from Ministry Education of Malaysia and Universiti Teknikal Malaysia Melaka. The author would like to acknowledge the Department of Industrial Design, Eindhoven University of Technology for their excellent support throughout this research activity. Also the acknowledgment for Peter Andriessen M.D and Tessa Pols from Neonatal Intensive Care Unit (NICU) of Máxima Medical Center, Veldhoven for their support and contribution towards this project.

References

[1] J. R. Fredieu, J. Kerbo, M. Herron, R. Klatte, and M. Cooke, "Anatomical Models: a Digital Revolution," *Medical Science Educator*. 2015.

[2] M. Thielen, R. Joshi, F. Delbressine, S. Bambang Oetomo, and L. Feijs, "An innovative design for cardiopulmonary resuscitation manikins based on a human-like thorax and embedded flow sensors," *Proc. Inst. Mech. Eng. Part H J. Eng. Med.*, vol. 231, no. 3, pp. 243–249, 2017.

[3] "Realistic 3D Printed 'Biotexture Wet Models' for Surgical Training." [Online]. Available: https://3dprint.com/49992/fasotec-3d-print-wet-model/. [Accessed: 24-Oct-2018].

[4] Robin H. Steinhorn, "Evaluation and management of the cyanotic neonate," *J. Clin Pediatr Emerg Med.*, vol. 9, no. 3, p. (169-175), 2008.

[5] J. S. Everett, M. Budescu, and M. S. Sommers, "Making sense of skin color in clinical care.," *Clin. Nurs. Res.*, vol. 21, no. 4, pp. 495–516, 2012.

[6] P. C. Frommelt and M. A. Frommelt, "Cyanosis," in *Practical Strategies In Pediatric Diagnosis and Therapy*, First., W.B. Saunders, 2004, p. (166--180).

[7] S. J. Dain, "The visual recognition of cyanosis and the influence of lighting and color vision," in *Advances in Dermatology Research*, Nova Science Publishers, 2015, p. (131-140).

[8] N. F. B. Azmi, F. Delbressine, L. Feijs, P. Andriessen, and T. Pols, "Color Changing Actuators: A Review Towards Designing Cyanosis in a Baby Manikin Simulator," p. V002T06A016, 2018.

[9] S. Stitzel, R. Byrne, and D. Diamond, "LED switching of spiropyran-doped polymer films," *J. Mater. Sci.*, vol. 41, no. 18, pp. 5841–5844, 2006.
[10] "Sunlight / UV Sunlight Color Change PLA 3D Print Filament." [Online]. Available: http://www.go-3dprint.com/products/sunlight-uv-sunlight-color-change-pla-3d-print-filament-value-pack-2x-225g.html. [Accessed: 24-Oct-2018].
[11] M. Viková, "Methodology of measurement of chromic materials," in *Applied Science Innovations Pvt. Ltd.*, 2009, pp. 509–534.
[12] "Spectroradiometer Specbos 1201," *JETI Technische Instrumente GmbH*, 2018. [Online]. Available: https://www.jeti.com/cms/index.php/instruments-55/radiometer/specbos/specbos-1211.
[13] D. Pascale, D. Malacara, Munsell, and H. S. & Y. R. Jenny Preece, *Color Vision and Colorimetry: Theory and Applications*, 4th Editio. 2002.
[14] M. D. B.V., "Carat Lacquer (BABY color)."
[15] A. Dass, "Humanae Project." [Online]. Available: https://www.angelicadass.com/humanae-project/.
[16] T. I. Rich Rosen, "Dimming Techniques for Switched-Mode LED Drivers Literature," 2009.
[17] Excelsys Technology Ltd, "Application Note – AN1101 : Dimming control of LEDs."
[18] V. Semiconductors, "Vishay Semiconductors Bicolor SMD LED PLCC-3 SYMBOL," no. 83042, pp. 1–7.
[19] N. A. Midolo and L. Sergeyeva, "Lighting for Clinical Observation of Cyanosis," *Aust. Hosp. Eng.*, vol. 30, no. 2, p. (38-46), 2007.
[20] H. Davis, *Creative Close-Ups: Digital Photography Tips and Techniques*. John Wiley & Sons, 2011.
[21] M. Wang, "The influence of the skin colour on the perceived attributes," The University of Leeds, 2017.
[22] M. A. P. P. Vik, *CHROMIC MATERIALS Fundamentals, Measurements, and Applications*, vol. 1st. 2018.
[23] M. R. Luo, "Development of colour-difference formulae," *Rev. Prog. Color. Relat. Top.*, vol. 32, pp. 28–39, 2002.
[24] M. Tkalcic and J. F. Tasic, "Colour spaces - Perceptual, historical and applicational background," in *IEEE Region 8 EUROCON 2003: Computer as a Tool - Proceedings*, 2003.

IOP Conf. Series: Materials Science and Engineering **834** (2020) 012080 doi:10.1088/1757-899X/834/1/012080

Experimental research and modelling of the response of magnetorheological elastomers to cyclic load

Iva Petríková [1], Bohdana Marvalová [1]

[1]Faculty of Mechanical Engineering, Technical University of Liberec, Studentská 2, 461 17 Liberec, Czech Republic

iva.petrikova@tul.cz, bohda.marvalova@tul.cz

Abstract. Magnetorheological elastomers (MREs) belong to the new group of functional materials called "smart". The MR effect is increased by choosing the material of the particles with high permeability. The dependence of dynamic moduli and damping of MREs on the external magnetic field and the applied frequency of cyclic shear deformation was studied. Samples of MRE were made of the silicon rubber matrix filled with micron-sized particles. The dynamic stiffness of MRE depends on magnetic flux density and rises with increasing frequency. The loss factor of MRE samples is controlled by the magnetic flux density and it depends also on the testing frequency. The behaviour of viscoelastic materials under uniaxial loading may be represented using rheological models composed of elastic and viscous elements. The dynamic variables storage and loss moduli and loss tangent were determined on the basis of the fitted parameters of the fractional model.

1. Introduction

Magnetorheological elastomers (MREs) are a group of materials with rheological properties very quickly modified by the application of a magnetic field. Magnetorheological elastomers be classed to the group of functional materials called "smart". Smart materials are intelligent materials that have the ability to change their material properties under the influence of an external impetus such as stress, temperature, moisture, electric or magnetic fields. The rheological properties of MREs (the deformation and flow behaviour under stress) are modified by the application of an external magnetic field. Magnetorheological materials have their viscoelastic and rheological properties altered in the presence of a magnetic field. They demonstrate a quick response to external stimuli [1, 2].

Many factors influence on characteristic response, e.g. size, composition, distribution and volume of the iron particles, the isotropic or anisotropic arrangement of microparticles, elastomeric matrix etc. The MREs consist of magnetically permeable micro particles, dispersed in a non-magnetic matrix.

The unique characteristic of MREs is the shear modulus controlled by the external magnetic field continuously [3]. The MR effect is raised by choosing the material of the particles with high permeability.

The matrix materials of the MREs are silicon rubber, natural rubber, carbon black filled rubber. MRE composites adopt basic properties of the elastomeric matrix such as large deformations, Moulinse effect, frequency and amplitude dependencies, a decrease of stiffness at cyclic loading and viscoelastic time dependent characteristic [4].

Content from this work may be used under the terms of the Creative Commons Attribution 3.0 licence. Any further distribution of this work must maintain attribution to the author(s) and the title of the work, journal citation and DOI.
Published under licence by IOP Publishing Ltd

1.1 Response of MREs to dynamic loading

The MR effect implies not only an instantaneous and reciprocated increase of the modulus and stiffness of a MR elastomers but also MREs exhibit a field-dependent damping [5]. The loss factor at low frequencies was found to increase by about 30% in the magnetic field [6]. The damping and stiffness of aligned MREs depend on the magnetic field, directions of load and the particle aligning in the elastomer [5, 7]

The response of MREs under dynamic loading has been studied experimentally by many researchers. Jolly et al. [8] presented a model of the magneto-viscoelastic effect of aligned anisotropic MREs in shear taking into account shear stress caused by the magnetic forces of particle dipoles. The results of their experiments showed the material moduli increasing monotonically with the applied field until magnetic saturation of the composite arises between 0.6 T and 0.8 T. The dynamic modulus increases slowly with frequency but the change of the modulus induced by the magnetic field is relatively insensitive to frequency.

All magnetorheological elastomeric composites show Payne effect, i.e. the storage modulus decreases with increasing strain amplitude because of the filler network failure. The Payne effect increases with raising concentration of carbonyl iron microparticels in the composite [5]. The loss factor of the shear modulus also strongly depends on the strain amplitude even for very small amplitudes [9]. The loss factor exhibits the increasing tendency with increasing of the strain amplitude of dynamic loading.

1.2. Modelling of dynamic behaviour of MREs

Main time-dependent features of MREs are derived from the viscoelastic properties of the elastomeric matrix. The dynamic properties depend obviously on the external magnetic field and the content of particles. Several phenomenological models developed as the connections of classical rheological models. The parameters of the model depending on the magnetic flux density are reported Li Y et al and Cantera et al [10, 11].

The four-parameter viscoelastic model based on the classical standard solid model combined with one additional parallel spring which represents the dependence of modulus on the magnetic field was developed by Li W H et al. [9]. Their experimental study contains the harmonic strain-controlled shear loading with different strain amplitudes and frequencies at various magnetic fields. The presented hysteresis loop have elliptical shapes, Authors showed a good agreement between experimental data and the results calculated by the model.

A complex linear viscoelastic model for isotropic MREs was presented by Xin et al. [12]. The viscoelasticity of MREs was divided into a mechanical part and magnetic part. The shear storage and loss moduli are developed using the Kraus model. The magnetic shear storage and loss moduli are derived based on the magneto-elastic theory with consideration of the magnetic saturation. The proposed model was evaluated by the experimental data.

Fractional rheological models are frequently applied recently to describe basic characteristics of different materials, rheological behavior of linear viscoelastic substances. There have been numerous tries to simulate damping properties of MREs using a fractional rheological element. This leads to a significant reduction in the number of parameters that is necessary to determine from experiments.co

The model of magneto-sensitive isotropic rubber related to small strain was presented by Blom and Kari [13]. Their constitutive model characterises the amplitude dependence of magnetic sensitivity and viscoelasticity of isotropic MRE. The viscoelastic dependence is described by Abel type integral equation solved via fractional derivative calculus. The model includes the influence of changes in the magnetic field on the parameters.

The constitutive model of isotropic MREs formed on fractional derivatives, which are matrix-particle interaction and the magneto-induced modulus model was developed by Agirre-Olabide et al. [14]. The viscoelastic behavior was modelled using a four-parameter fractional derivative model. The limits of application of the linear viscoelasticity for the magneto-dynamic characterisation of MREs were defined by author Agirre-Olabide et al. [15]. The linear viscoelasticity region limits were examined depending on the particle content, frequency, external magnetic field, the inner structure of the samples and the temperature.

2. Experimental

The magneto-rheological elastomers were produced from room temperature vulcanization silicone rubber as the matrix matter and carbonyl iron microparticles as the filler. Globular carbonyl iron particles of diameters from 2 to 5 μm were supplied by Sigma-Aldrich company. The matrix material of silicone rubber is mixed of ZA 22 base and RZA 22 curative at a ratio of 1:1. Carbonyl iron microparticles of 27 vol% were mixed into the silicone rubber mixture. The polymerization process was performed at room temperature in a vacuum in order to remove bubbles. MRE samples are isotropic.

Dynamic viscoelastic measurements were carried out on an Instron Electropuls 3kN, [1]. The magnetic field induced by an electromagnet was utilized during the cyclic loading on double-shear samples of MRE under controlled shear strain. The detailed information about samples is described in Table 1.

Table 1. Parameters of MRE samples.

Sample	
Material of matrix	RTV silicone rubber
Material of filler	Globular carbonyl iron microparticles
Sample shape	Rectangular
Sample dimensions	20x20x5 mm^3
Glue	Cyanoacrylate glue with activator
Volume of carbonyl iron powder	27%

The amplitude of the cyclic shear strain was 0.1 and the frequency was changed from 1 to 10 Hz with the step 10 Hz. The samples were cyclically loaded under the magnetic field from 0.1 to 0.8 T. The samples were tested without and with the magnetic field to receive the basic dynamic MRE properties, specifically storage and loss moduli and the loss angle. Dynamic moduli and the loss factor were found out as the function of the magnetic flux density (magnetic field) and the frequency and amplitude of the cyclic deformation in shear. Two methods were used for evaluation of experiments: Phase shift angle determination and evaluation of loss modulus from the area of the hysteresis loop.

2.1 Phase shift angle determination

We suppose that the raw signals, i.e. displacement u(t) and force response F(t), are harmonic functions approximately and we use the Fast Fourier Transform (FFT) in order to determine the phase shift δ at the main excitation frequency [16]. Amplitudes of shear strain and shear stress were determined from raw recorded signals, eq. (1).

$$\gamma(t) = \gamma_0 + \gamma_a \sin(\omega t) \rightarrow \tau(t) = \tau_0 + \tau_a \sin(\omega t + \delta) \quad (1)$$

Dynamic modulus G* was determined as the ratio between the amplitudes of stress and strain. The storage and loss moduli were calculated as follows

$$G^* = \frac{\tau_a}{\gamma_a}, \quad G' = G^*\cos(\delta), \quad G'' = G^*\sin(\delta) \quad (2)$$

2.2. Loss modulus evaluation from the area of a hysteresis loop

We calculated the loss modulus from the dissipated energy density which corresponds to the area of the hysteresis loop. The dissipation of mechanical energy in the course of one cycle can be expressed as

$$D = \int_0^T \tau(t)\dot{\gamma}(t)dt = \pi\gamma_a^2 G^* \sin\delta = \pi\gamma_a^2 G'' \quad (3)$$

The density of the dissipated energy was calculated numerically as the mean value of the areas of twenty consecutive steady hysteresis loops in each experiment. The raw signal evaluation can be crucial for successful modelling of the MRE's response.

3. Modelling

The behavior of viscoelastic properties of MRE samples under uniaxial loading may be described by means of the rheological models containing elastic and viscous elements. The rheological Zener model is a specific case of the generalized Maxwell model assembled of a Maxwell branch in parallel with a spring. Such a model of three parameters is not sufficient for the quantitative representation of the behavior of the MRE. In order to improve the similarity, the number of parameters needs to be increased. This is done by connecting a number of springs and dashpots - to create a so-called ladder model.

The dependency of dynamic moduli can be characterized by a minimum of material constants using the fractional calculus. The fractional Zener model is an adequate uniaxial model for small strains [14, 17]. It corresponds to a linear spring with an elastic modulus in parallel with a fractional Maxwell element. The fractional differential equation is adequately similar to the 1st order differential equation of the classic Zener element.

$$\sigma(t) + \tau^\alpha \frac{d^\alpha \sigma}{dt^\alpha} = (G_m + G_e)\tau^\alpha \frac{d^\alpha \varepsilon}{dt^\alpha} + G_e \varepsilon(t) \quad (4)$$

The model has 4 parameters (elastic part of the viscoelastic member of the model G_e, the stiffness of the spring in elastic branch G_m, the relaxation time τ and the fractional parameter α) which should be fitted to the experimental data.

The shift from the time domain to the frequency domain is realized by the Fourier Transform. Storage and loss moduli are obtained by separating the real and imaginary part of the dynamic modulus.

$$G' = G_e + G_m(\tau\omega)^\alpha \frac{\cos(\alpha\pi/2) + (\tau\omega)^\alpha}{1 + 2(\tau\omega)^\alpha \cos(\alpha\pi/2) + (\tau\omega)^{2\alpha}} \quad (5)$$

$$G'' = G_m(\tau\omega)^\alpha \frac{\sin(\alpha\pi/2)}{1 + 2(\tau\omega)^\alpha \cos(\alpha\pi/2) + (\tau\omega)^{2\alpha}}, \quad \tan\delta = \frac{G''}{G'} \quad (6)$$

The loss tangent values evaluated from experiments were fitted to the fractional Zener model.

4. Results and Discussion

4.1. Results of experimental measurements

The dynamic stiffness of MRE depends on magnetic field and rises with increasing testing frequency, see Figure 1 & 2. The loss angle of MRE is controlled by the magnetic field and it also depends on the testing frequency. The dependency of all variables on frequency is slightly increasing in the frequency range of 1 to 10 Hz. The increments of storage and loss moduli between the zero-field and 0.78 Tesla are 32% and 38% resp., the loss angle increase was around 6%.

4.2. Fitted parameters of fractional model

The loss tangent values evaluated from experiments were fitted to the fractional Zener model. We obtained an exceptionally good agreement even for different intensity of the magnetic field (ranging from 0 to 0.6 Tesla) as can be seen in Figure 3. The fitted parameter alpha of the fractional order derivative decreases slightly with the rising magnetic field intensity (Figure 4). This indicates that with the increasing intensity of the magnetic field, the stiffness of the MRE sample increases.

The hysteresis loop record carries all the information we can harvest from a dynamic experiment. It is, therefore, appropriate to fit the model parameters directly to the recorded raw data of strain and stress. The fitting was carried out in Matlab by nonlinear least squares. The dynamic moduli G', G'' and loss tangent were determined on the basis of the fitted parameters of the fractional model.

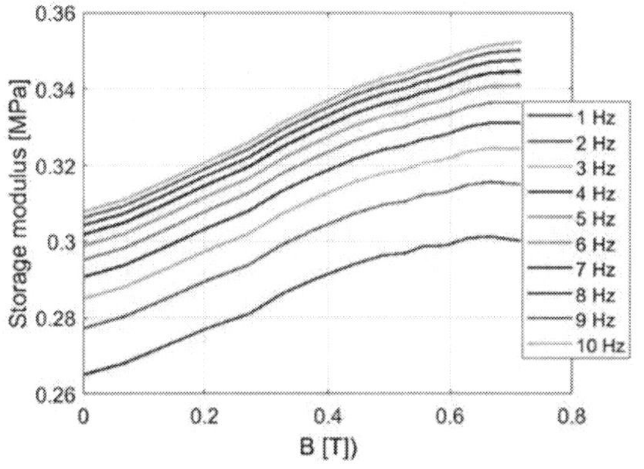

Figure 1. The storage modulus dependency on the magnetic field

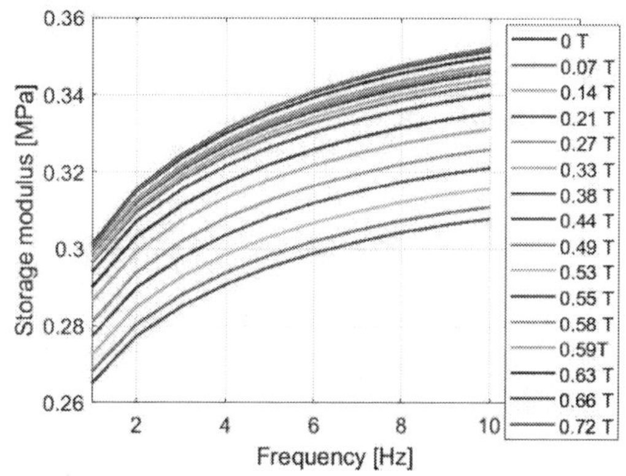

Figure 2. The storage modulus dependency on the frequency

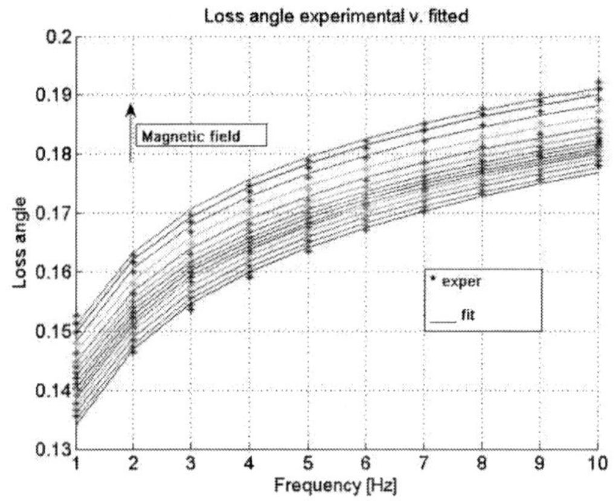

Figure 3. The loss angle evaluated from experiments and fitted for different intensity of the magnetic field.

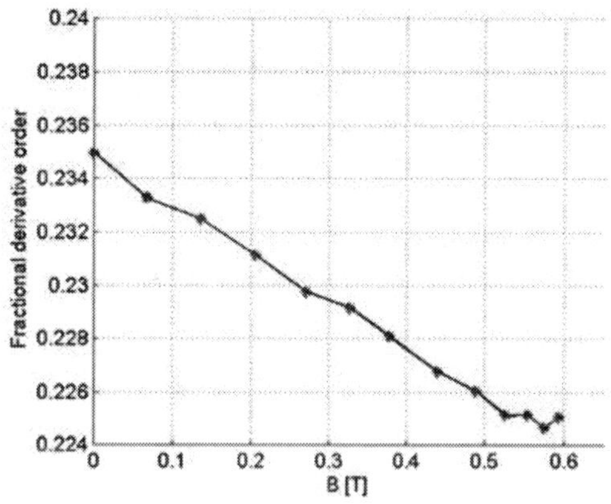

Figure 4. The fitted parameter alpha of the fractional order derivative.

5. Conclusion

In this contribution, we presented a study of the mechanical behavior of magneto-sensitive elastomers with the isotropic distribution of the magnetic particles in an external magnetic field and we presented the phenomenological model of .the dynamic response based on fractional Zener derivative rheological element. The hysteresis loop record carries all the information we can harvest from a dynamic experiment. Therefore, it is appropriate to fit the model parameters directly to the recorded raw data of strain and stress. The fitted parameter alpha of the fractional order derivative decreases slightly with the rising magnetic field intensity. This indicates that with the increasing intensity of the magnetic field, the stiffness of the MRE sample increases.

Acknowledgements

This work was supported by the Ministry of Education, Youth and Sports of the Czech Republic and the European Union - European Structural and Investment Funds in the frames of Operational Programme

Research, Development and Education - project Hybrid Materials for Hierarchical Structures (HyHi,Reg.No.CZ.02.1.01/0.0/0.0/16_019/0000843).

References

[1] Petrikova I and Marvalova B 2017 Experimental research and numerical simulation of the damping properties of magnetorheological elastomers, *Proc. of the 10th Europ. Conf. on Constitutive Models for Rubber*, ed A Lion A and M Johlitz, (London:Taylor & Francis Group) pp 11-18

[2] Ruddy C, Ahearne E and Byrne G 2012 A review of magnetorheological elastomers: properties and applications *Advanced Manufacturing Science (AMS) Research.* http://www. ucd. ie/mecheng/ams/news_items/Cillian% 20Ruddy. pdf Accessed, 20

[3] Jolly M R, Carlson J D and Munoz B C 1996 A model of the behaviour of magnetorheological materials *Smart Materials and Structures* **5**(5) 607

[4] Lion A 1998 Thixotropic behaviour of rubber under dynamic loading histories: experiments and theory *J. of the Mechanics and Physics of Solids* 46(5) pp 895-930

[5] Kallio M 2005 The elastic and damping properties of magnetorheological elastomers, *PhD thesis* p 149 VTT

[6] Kallio M, Lindroos T, Aalto S, Järvinen E. Kärnä T and Meinander T 2007 Dynamic compression testing of a tunable spring element consisting of a magnetorheological elastomer *Smart Material and. Structures* **16** 506

[7] Ivaneyko D, Toshchevikov V and Saphiannikova M 2015 Dynamic moduli of magneto-sensitive elastomers: a coarse-grained network model. *Soft Matter* **11** 7627

[8] Jolly M R, Carlson J D, Muñoz B C and Bullions T A 1996 The magnetoviscoelastic response of elastomer composites consisting of ferrous particles embedded in a polymer matrix. *J. of Intelligent Material Systems and Structures*, **7** 613

[9] Li W H, Zhou Y and Tian T F 2010. Viscoelastic properties of MR elastomers under harmonic loading. *Rheologica acta* **49** p 733

[10] Li Y, Li J, Li W and Du H 2014 A state-of-the-art review on magnetorheological elastomer devices *Smart materials and structures* **23** 123001

[11] Cantera M A, Behroo M, Gibson R F and Gordaninejad F 2017 Modeling of magneto-mechanical response of magnetorheological elastomers (MRE) and MRE-based systems. *Smart Materials and Structures* **26** 23001

[12] Xin F L, Bai X X and Qian L J 2016. Modeling and experimental verification of frequency-, amplitude-, and magneto-dependent viscoelasticity of magnetorheological elastomers. *Smart Materials and Structures* **25** 105002

[13] Blom P and Kari L 2012 The frequency, amplitude and magnetic field dependent torsional stiffness of a magneto-sensitive rubber bushing. *Inter. J. of Mechanical Sciences* **60** 54

[14] Agirre-Olabide I, Lion A and Elejabarrieta M J 2017 A new three-dimensional magneto-viscoelastic model for isotropic magnetorheological elastomers *Smart Materials and Structures* **26** 035021

[15] Agirre-Olabide I, Berasategui J, Elejabarrieta M J and Bou-Ali M M 2014 Characterization of the linear viscoelastic region of magnetorheological elastomers *J. of Intelligent Material Systems and Structures* **25** 2074

[16] Petrikova I, Marvalova B and Nhan P T 2011 Influence of thermal ageing on mechanical properties of styrene-butadiene rubber. In *Proc. of the 7th Europ. Conf. Constitutive Models for Rubber VII* ed S Jerrams and N Murphy (London:Taylor and Francis Group) pp 77-83

[17] Lion A 2001 Thermomechanically consistent formulations of the standard linear solid using fractional derivatives. *Archives of Mechanics* **53** 253

IOP Conf. Series: Materials Science and Engineering **834** (2020) 012081 doi:10.1088/1757-899X/834/1/012081

Synthesis and Tribological Characterization of Copper Based Composites Containing Carbide

Rajnesh Tyagi, Rohit Kumar Singh Gautam*

Indian Institute of Technology (BHU), Varanasi, Mechanical Department, India, 221005

*rohitsingh200873@gmail.com

Abstract: Keeping in view the importance of the Cu based composites in sliding electric contact applications, the aim of the study is synthesise copper-chromium alloy based composites containing different wt. % of TiC by using a cost effective stir casting technique and to compare their mechanical and tribological behaviour. The study also proposes to analyse the friction and wear behaviour of the formed composites under the sliding speed of 1m/s and load varied 5 N, 10 N, 15 N and 20 N. The hardness of the composite Cu-1wt%Cr-1wt%TiC and Cu-1wt%Cr-2wt%TiC has been found to increase slightly when compared with Cu-Cr alloy and as received Cu. The wear rate was found to show linear trend with increasing normal load following the Archard law. However, composites were found to show low wear as compared to that observed in pure copper. This might be due to high hardness of the composite, the compact transfer layer of wear debris and due to presence of hard TiC particles whicg results in a lower real area of contact which further causes lower wear. There was found no cracks in the composite. The addition of chromium was found to increase the wettability and mechanical properties. The composite hardness and strength was found to increase with addition of TiC and chromium element in the matrix phase up to certain content.

1. Introduction

Copper is used in many industrial applications because it possess very interesting features like excellent electrical and thermal conductivities, low price, cheaply available, easy to fabricate, and resistance towards corrosion. However, it possess low hardness, low tensile strength and low resistance towards wear. These features restrict its applications. To enhance the mechanical properties and resistance towards wear in the copper there are two ways either by an age hardening mechanism or to incorporate the hard phase in the matrix. In age hardening method, the low wt. % of chromium or zirconium in the copper matrix lead to the precipitation of a hard second phase which does not dissolve properly in the copper matrix mainly in low temperature regimes. In other alternative method to reinforce hard phase particles such as carbides, nitrides, oxides, borides into the copper base matrix to develop copper base metal matrix composites.

Due to low harness, and high ductility copper has been found to show low resistance towards wear. Copper has been found to show severe adhesive wear and a large extent of wear debris comes out of the surface during sliding on a hard steel counterpart. It is desirable to enhance the wear resistance in the copper and to understand wear behaviour of Cu-base composites. Tjong and Lau [1] investigates the tribological behavior of Cu–SiCp (5–20%) composites under velocity of 1 m/s and loads varied from 15–55 N. From the study they showed that as the SiC particles added in the copper matrix they reduced the extent of strain localization in the sub-surface region leading to reduction in volume loss. Sharma

Content from this work may be used under the terms of the Creative Commons Attribution 3.0 licence. Any further distribution of this work must maintain attribution to the author(s) and the title of the work, journal citation and DOI.
Published under licence by IOP Publishing Ltd

and co-workers [2] found that for Cu–SiCp composites as the load increases with speed the wear transform from mild to severe wear and then to seizure. The tribological behavior of copper–graphite (8–20%) composites by Moustafa and co-workers [3] shown that copper coated graphite composites shows wear resistance behavior in comparision to pure copper. Kennedy et al [4] examined the tribological behavior of Cu–Cu coated SiCp composites using pin on disk configuration under speed of 1 m/s and at a load of 27 N. They show the importance of interface bonding between the matrix and the reinforcing phase. They were use Co-B coating on the SiC particles which increses the interface bonding and thus helps in reduced pullout of the SiC particles from the copper matrix and thus makes the composite more wear resistance. Wan et al. [4] shows that copper coated and nickel coated alumina particle reinforced in Cu–alumina composites exhibited more wear resistance behavior as compared to uncoated ones. This was attributed to strong interfacial bonding between reinforce phase and the matrix.

Zhan and Zhang [2] found that reinforcement of graphite particles to Cu–SiCp composites, fabricated via powder metallurgy process, diminished the wear rates and lifted the transition from mild to severe to 723 K. The COF of copper hybrid composite was more stable and lower than that of SiCp/Cu. However, the wear test was done at 45 N load under 0.15 m/s speed. Gonzalez et al. [5] reported that modeling effort was required for SiCp reinforced composites to validate the results also. It was shown that testing temperature has a greater influence on wear rate of 8090 Al alloy reinforced with SiCp. Rice and Moslehy [6] showed an interface model to show the probable dynamic effects in dry sliding wear. Muller and Van [7] talked about the role of the fatigue and ductility in labeling the wear phenomenon. The Taguchi method was assumed by Sahin [8] to examine the influence of abrasive particle properties and testing conditionings on wear rate of 2014 aluminium alloy in which as a reinforcing phase was 10% and 20% SiCp. Axen et al. [9], formulate a friction model which was able to explain the discrepancies in the prediction of the COF multiphase materials by linear rule of mixture and COF was dominated by the COF of the most wear resistant phase.

In this present investigation, an attempt made to develop Cu-base composite using 2 wt. % TiC as reinforce into the Cu-Cr alloy. The titanium carbide is an exciting engineering material for the reinforce into copper matrix because it possess some exciting features like high melting point, high hardness and great extent of resistance towards wear. Optimal control of the reinforcement particles will give the best properties for sliding contact applications.

2. Experimental Procedure

2.1 Materials Selection

In this investigation pure copper (99.9%) used as matrix and reinforcing materials was pure chromium (99.8% and < 50 μm) which improve the wettability and finally titanium carbide added to prepare cast composites.

Stir Casting is composite material fabrication method in a molten state, in which a reinforcing phase (ceramic particles, short fibers) is mixed with the matrix material in molten state by means of mechanical stirring. It is very simple and straight forward process and also cheap method for the production of particulate matrix composites. To achieve better mechanical properties uniform dispersion of the reinforcement is very essential. Porosity has to be control with casting technique. The matrix material was heated by the furnace in a crucible. At the time of melting the reinforcing phase added with in the matrix and simultaneously the mechanical stirring done to disperse the material in all the direction. After stirring directional solidification takes place to obtain the desired product.

Furnace Setup and Stirrer: An electric muffle furnace of Kanthal wire as shown in Figure 1 was used to develop the cast composite. It has a PID controller which controls the temperature in the range of 0°C to 1200°C. The required temperature is set by changing temperature on digital display. Once the required temperature is reached it is maintained until furnace is turned off. Stirrer is made of long circular rod of stainless steel and blades welded at the end. A variable rpm motor is used for motion of stirrer.

Figure 1. Stir casting Furnace

2.2 Melting and Casting

The present work is aimed to investigate the mechanical properties along with tribological behaviour of cast Cu–1wt.%Cr–1wt.%TiC and Cu–1wt.%Cr–2wt.%TiC composite synthesized by alloying molten copper with Cr and dispersion of TiC particles. By stir casting route metal matrix composite was prepared. Pure copper of 2 kg was melted as a base matrix material in a muffle furnace. The melting process was commences in a graphite crucible. The temperature was raised up to 1200°C and after that it was allowed to stabilize the temperature then after 1wt.% of chromium was added to increase the wettability and then after 1wt% (and 2wt%) of titanium carbide was added in the metal matrix. When the temperature raises around 1215°C, pure TiC was mixed mechanically into melted Cu matrix alloyed with Cr. Chromium carbides—Cr–C, $Cr_{23}C_6$, Cr_3C_7 etc., were formed, through an exothermic reaction between TiC and Cr in the melt. Then the melt was stirred at a uniform rate with the help of a stainless steel turbine stirrer. The stirring was done at an impeller speed of 200 rpm. At the time of mixing the temperature was maintained 1200°C during addition of chromium and titanium carbide. The dispersion of all reinforced particles were achieved by the vortex method. The melted matrix with reinforcing phase were poured into the permanent metallic mold. The melt was allowed to solidify in the mold. After some period of time when temperature cools down to room temperature desired casting is obtained of the mold shape.

(a)

(b)

Figure 2. Casting of (a) Cu-1wt%Cr-1%TiC composite and (b) Cu-1wt%Cr-2wt%TiC composite

2.3 Friction and Wear Testing

Typical cylindrical wear testing specimens of 10 mm were produced from all the composites. There are number of testing machine available to evaluate wear but pin-on-disc type is most popular for tribological studies. Tribological test were conducted by sliding pins which are made of the composite material rubbing against a counter face which are made of steel disk made of SAE 4615 steel. The hardness of the steel was 174 BHN. The tests were conducted under a normal load of 5, 10, 15 and 20 N. The speed was 1 ms^{-1}. The flat surface of testing pin and steel disc were grounded of the order of 0.40 μm and these were thoroughly degreased, cleaned and dried before commencement of each test run. The cylindrical testing pin specimen was of 10 mm in diameter and 15 mm in height and the counter face disk was of 120 mm diameter and 10 mm thickness. Before conducting the test all the specimen of the composite material were ground with 4/0 grade emery paper, cleaned in solution of acetone to remove any particle which may stick on the surface. After testing, the weight loss of the specimens were measured, after covering the total sliding distance of 1800 m. The difference in weights of the pin before and after the experiment gave the weight loss from which volume of wear rate was calculated. Tests were conducted for various compositions for different combinations of loads and a fixed sliding speed. The total sliding distance covered in a test was 9000 m. Average value has been reported after conducting the wear test at least three times.

3. Results and discussions:

3.1 Metallographic Analysis:

The microstructure of the composites was examined under an optical microscope, Olympus, PME3, Tokyo, Japan. Figure 3 shows the optical microstructure of the as received copper, cast Cu-Cr alloy, Cu-1%Cr-1% TiC and Cu-1%Cr-2% TiC respectively. It is observed in Fig3 (a) that a fine tuned copper structure is occurs in the pure commercial copper, which is basic structure of the cooper. The Cu-Cr alloy as shown in Figure 3 (b) shows the presence of fine particles of chromium which imparts the characteristics of corrosion as well as wettability of the alloy. Figure 3 (c) and (d) shows the optical microstructure of Cu-1%Cr-1% TiC and Cu-1%Cr-2% TiC respectively, it is clearly observed the presence of reinforcing TiC and Cr particles. It is also observed that some formation of carbides resulting the reaction between Cu-Cr alloy and TiC and it increases as wt.% of TiC increases as depicted in the Figure 3 (c) and (d).

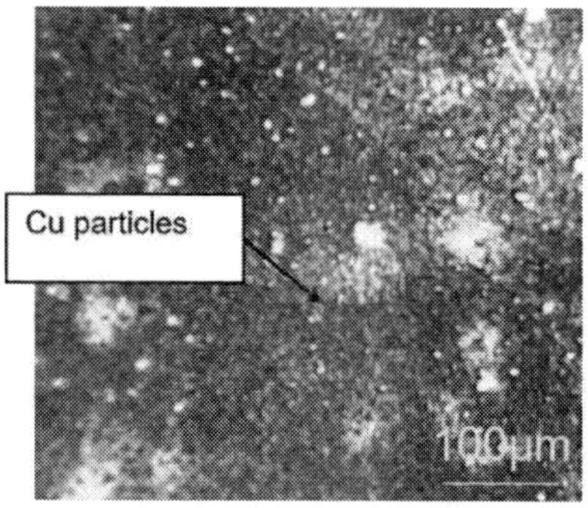

(a) As received copper

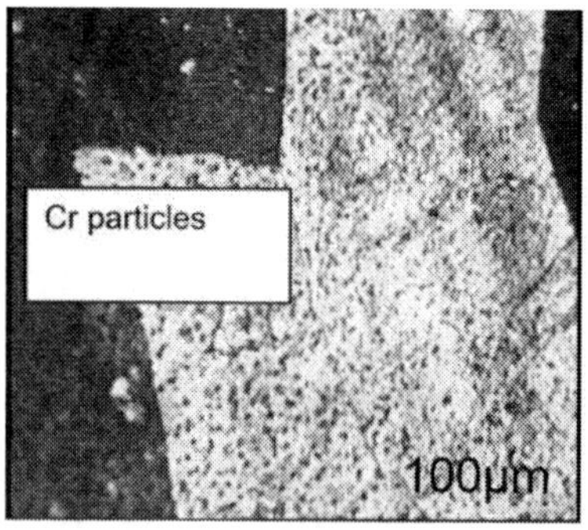

(b) Cu-Cr alloy

IOP Conf. Series: Materials Science and Engineering **834** (2020) 012081 doi:10.1088/1757-899X/834/1/012081

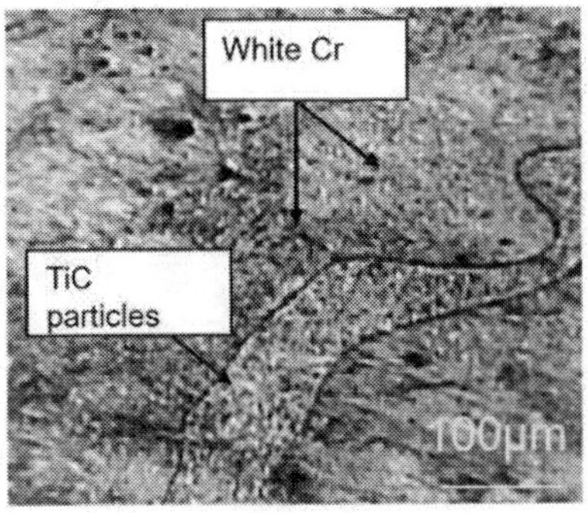

(c) Cu-1% Cr-1% TiC

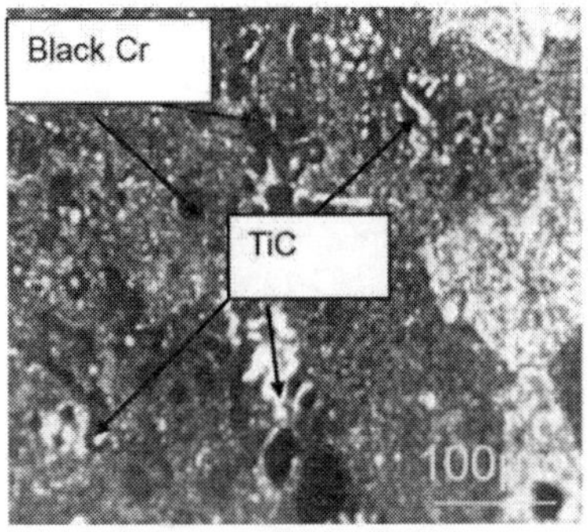

(d) Cu-1% Cr-2% TiC

Figure 3. Optical Micrographs of the composites (a) pure Cu (b) Cu-Cr alloy (c) Cu-1wt%Cr-1%TiC Composite (d) Cu-1wt%Cr-2%TiC Composite

3.2 Brinell hardness test

Hardness test was performed on the surface of the composites to check the hardness of the composites. Harness test was performed by applying 500 Kg load on the composites. Indentation size on the surface of the composites are given in the Table 1. As the TiC wt. % increases in the pure copper matrix the indentation size decreases as can be seen in Figure 4(a) and Figure 4(b). Indentation size directly reflect the hardness of the composites. Lower size of indentation is meaning of high resistance and high hardness.

Brinell hardness test was performed with the help of a Hardness Tester.

Brinell hardness number, BHN= $\frac{2P}{\pi D\sqrt{(D^2-d^2)}}$

P=load applied (in kg)
D=diameter of indenter (in mm) = 10 mm
d=diameter of indentation (in mm)

The hardness of the different materials used in the present investigation is listed in Table 1.

Table 1. Variation of Hardness with TiC content

Materials	Load applied (in kg)	Indentation (in mm)	Average Hardness (BHN)
Pure copper	500	5.2	35
Copper Chromium alloy	500	4.8	36.12
Cu-1wt%Cr-1wt%TiC Composite	500	4.0	38.12
Cu-1wt%Cr-2wt%TiC Composite	500	3.7	44.85

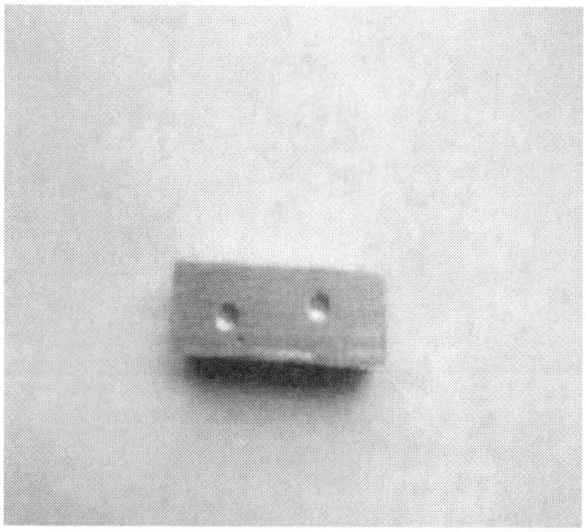

Figure 4(a). Hardness indentation of Cu-1wt%Cr-1%TiC Composite

Figure 4(b). Hardness indentation of Cu-1wt%Cr-2%TiC Composite

Table 2. Variation of Tensile Strength with TiC content

Material	% Elongation in gage length	% Reduction in area	Tensile Strength (MPa)
Pure Copper	55	59	220.15
Copper Chromium alloy	50	57	230.3
Cu-1wt%Cr-1wt %TiC Composite	47	55	249.82
Cu-1wt%Cr-2wt %TiC Composite	33	35	257.62

It could be observed from the table 2 that as the alloying addition increases the strength of Cu increases but that occurs at the expense of ductility, which is not surprising as the hard TiC particles induce some hardness reducing thereby the ductility.

The elongation in gauge length of Cu was observed to be 55% and its tensile strength was calculated as 220.15 MPa. For Cu-Cr alloy, elongation in gauge length was observed to be 50% and its tensile strength was calculated as 230.3 MPa. For Cu-1wt%Cr-1wt%TiC Composite, elongation in gauge length was observed to be 47% and also its tensile strength increases in comparison to copper and copper chromium alloy. In this specimen necking was observed before failure and elongation in gauge length was significant. For Cu-1wt%Cr-2wt%TiC composite, elongation was less than former specimen, showing increase in concentration of TiC decreases its ductility and improved tensile strength. In this specimen, fracture was expected to occur with necking. But it was observed to fail without necking. This may be because of some voids or surface cracks present or due to some other impurities

4. Wear and Friction Behaviour

4.1. Variation of cumulative volume loss with Normal Load

Figure 5 shows the effect of normal load on volume loss shows that volume loss increases continuously with load in a linear manner irrespective of sliding velocity. But in case of composites volume decreases with increase in TiC content in the present investigation.

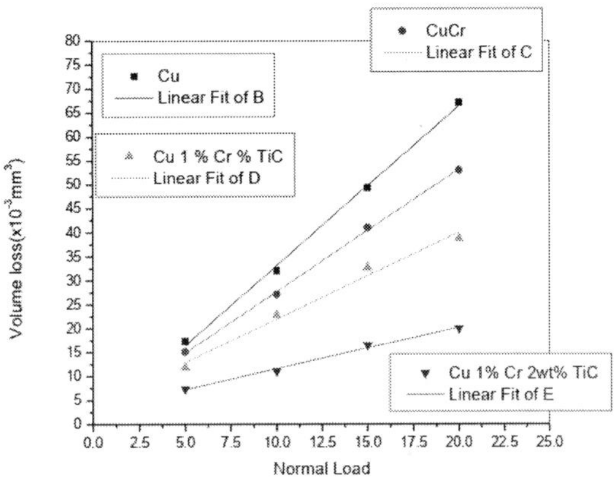

Figure 5. Variation of volume loss with normal load for cast copper, copper chromium alloy, Cu–1 %(by wt.)Cr–1 %(by wt.) TiC and Cu- 1 %(by wt.)Cr-2 %(by wt.) TiC composite at sliding speed of 1 m/s.

4.2. Variation of Cumulative Volume Loss with sliding distance at 5 N Load

The variation of cumulative volume loss with sliding distance was investigated at different loads of 5, 10, 15 and 20 N and at a constant sliding speed of 1 m/s and the same has been depicted in Figure 6-9 for different materials used in the study. All the data points have been fitted with linear least square fit line. It could be seen that the variation has a linearly varying relationship at all the loads following Archard's law and for all the specimens used in the study. It could also be observed that the at each load the composite containing 2 wt.% TiC shows the lowest volume loss whereas the pure Cu shows the highest indicating thus the beneficial effect of addition of TiC which raises the hardness of the Cu and improves its wear resistance.

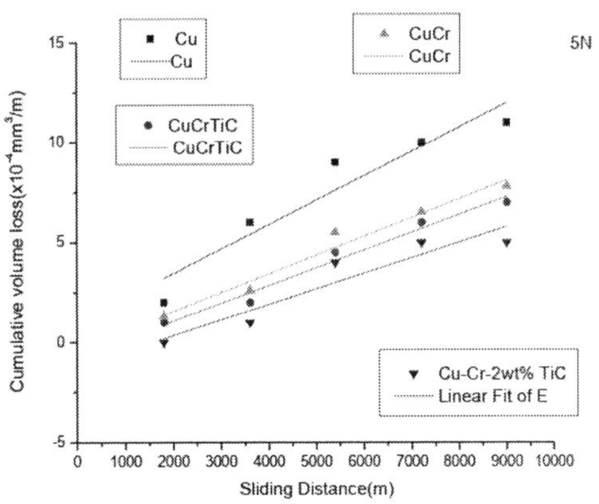

Figure 6. Cumulative volume loss with sliding distance at normal load of 5 N for cast copper, copper chromium alloy, Cu–1 %(by wt.)Cr–1 %(by wt.)TiC and Cu- 1 %(by wt.)Cr-2 %(by wt.) TiC composite at sliding speed of 1 m/s.

4.3. Variation of Cumulative Volume Loss at 10 N Load

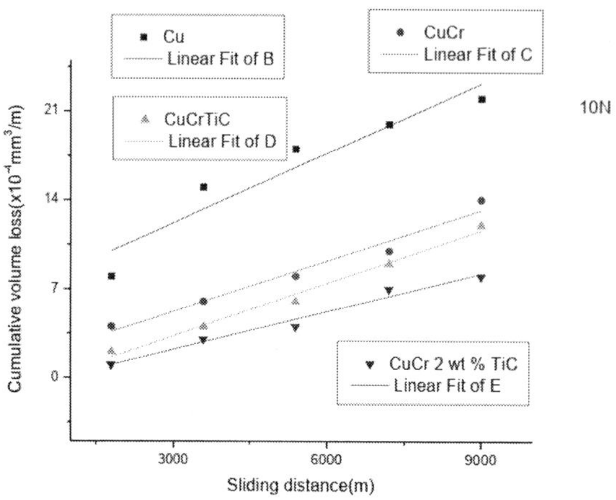

Figure 7. Cumulative volume loss with sliding distance at normal load of 10 N for cast copper, copper chromium alloy, Cu–1 %(by wt.) Cr–1 %(by wt.) TiC and Cu- 1 %(by wt.) Cr-2 %(by wt.) TiC composite at sliding speed of 1 m/s.

4.4. Variation of Cumulative Volume Loss at 15 N Load

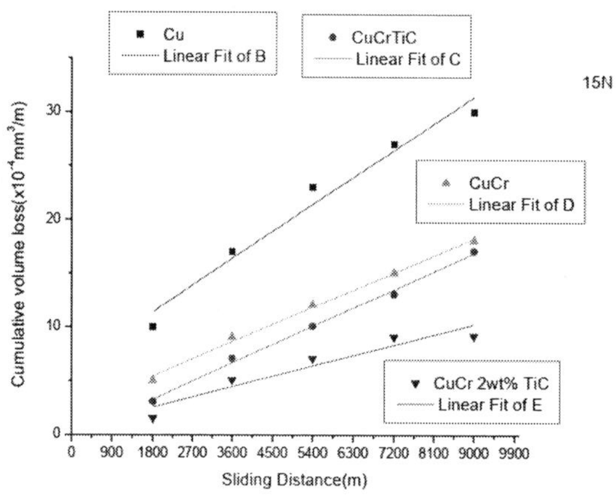

Figure 8. Cumulative volume loss with sliding distance at normal load of 15 N for cast copper, copper chromium alloy, Cu–1 %(by wt.)Cr–1 %(by wt.)TiC and Cu- 1 %(by wt.)Cr-2 %(by wt.) TiC composite at sliding speed of 1 m/s.

4.5. Variation of Cumulative Volume Loss at 20 N Load

It could also be observed that the at each load the composite containing 2 wt.% TiC shows the lowest volume loss whereas the pure Cu shows the highest indicating thus the beneficial effect of addition of TiC which raises the hardness of the Cu and improves its wear resistance.

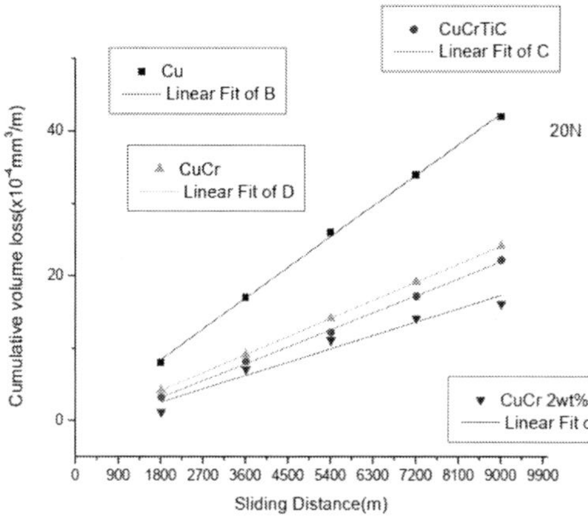

Figure 9. Cumulative volume loss with sliding distance at normal load of 20 N for cast copper, copper chromium alloy, Cu–1 %(by wt.)Cr–1 %(by wt.)TiC and Cu- 1 %(by wt.)Cr-2 %(by wt.) TiC composite at sliding speed of 1 m/s.

4.6. Variation of wear rate at Normal Load

The wear rate, i.e., volume loss per unit sliding distance has been estimated from the slope of the linear fit curves of variation of volume loss with sliding shown in Figure 6-9. Figure 10 shows the variation of wear rate with applied normal load for all the materials investigated in the present work. It can be observed that the wear rate increases linearly with normal load for all the materials. However, the composite containing 2 wt. % TiC shows the lowest wear rate and the Cu shows the highest wear rate. The wear rate decreases by a factor of 6 approximately at lowest load and 2 at the highest load by incorporation of 2 wt. % TiC.

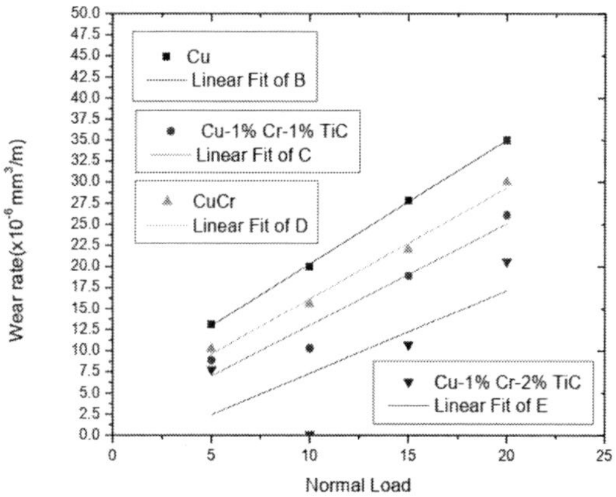

Figure 10. The variation of wear rate with normal load for as-received copper, cast copper, copper chromium alloy, Cu–1 %(by wt.)Cr–1 %(by wt.)TiC and Cu- 1 %(by wt.)Cr-2 %(by wt.) TiC composite at sliding speed of 1 m/s.

5. Friction Behavior

The friction force was constantly recorded on regular intervals which are displayed on the controller attached with the machine. Friction coefficient was obtained by dividing the friction force by normal load.

5.1. Variation of coefficient of friction with sliding distance at 5 N load

The variation of friction coefficient with sliding distance has been shown in Figure 11-14 at different loads of 5, 10, 15 and 20 N for all the specimens tested under the present study. All the variations show the fluctuating trend with respect to sliding distance. The value of COF for all the composites shows a decreasing trend as the applied load increases from 5 to 20N. The fluctuations in the COF may be not better contact in the early stage of testing then after sample and the counter face try to develop better surface conformity with each other.

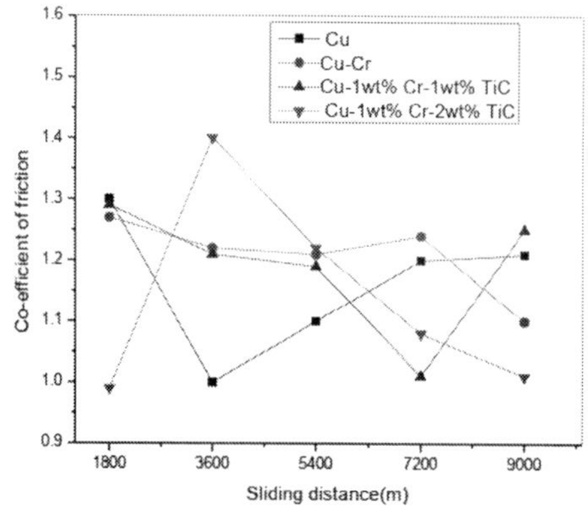

Figure 11. COF with sliding distance at normal load of 5 N at sliding speed of 1 m/s for cast copper, copper chromium alloy, Cu–1 %(by wt.)Cr–1 %(by wt.)TiC and Cu- 1 %(by wt.)Cr-2 %(by wt.) TiC composite at sliding speed of 1 m/s.

5.2. Variation of COF with sliding distance at 10 N load

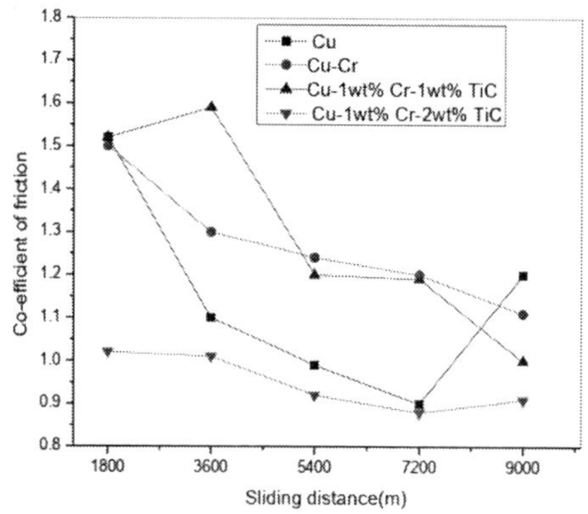

Figure 12. COF with sliding distance at normal load of 10 N at sliding speed for cast copper, copper chromium alloy, Cu–1 %(by wt.)Cr–1 %(by wt.)TiC and Cu- 1 %(by wt.)Cr-2 %(by wt.) TiC composite at sliding speed of 1 m/s.

5.3. Variation of COF with sliding distance at 15 N load

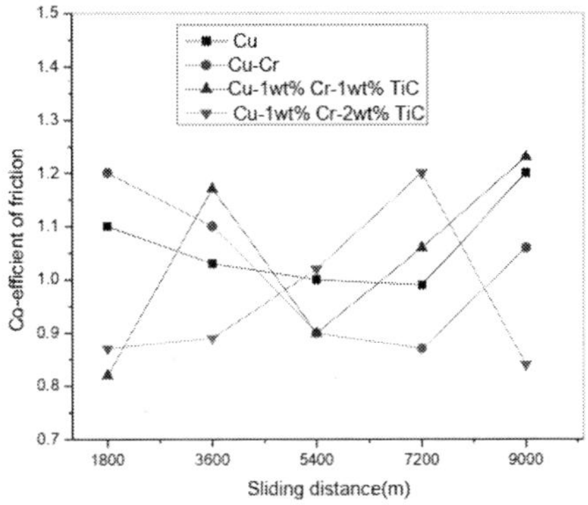

Figure 13. COF with sliding distance at normal load of 15 N at sliding speed for cast copper, copper chromium alloy, Cu–1 %(by wt.)Cr–1 %(by wt.)TiC and Cu- 1 %(by wt.)Cr-2 %(by wt.) TiC composite at sliding speed of 1 m/s.

5.4. Variation of COF with sliding distance at 20 N load

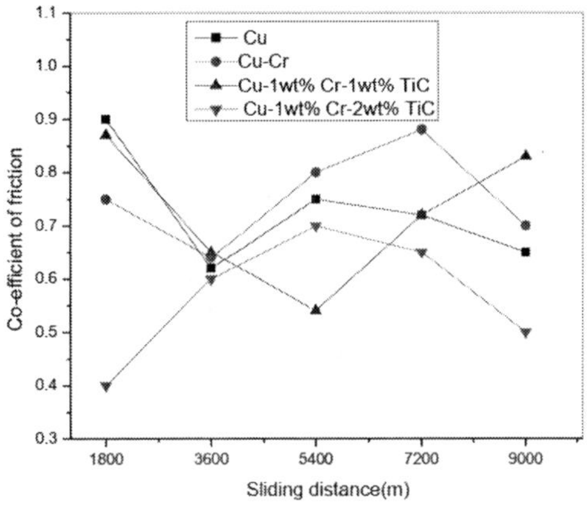

Figure 14. COF with sliding distance at normal load of 20 N at sliding speed for cast copper, copper chromium alloy, Cu–1 %(by wt.)Cr–1 %(by wt.)TiC and Cu- 1 %(by wt.)Cr-2 %(by wt.) TiC composite at sliding speed of 1 m/s.

5.5. Variation of average COF with normal load

The variation of average COF with normal load is shown in Figure 15. It can be seen that the COF decreases with increasing normal load and increasing content of TiC. The Cu-Cr-2 wt. % TiC composite shows the lowest COF whereas the copper shows the highest at all the loads used in the present study.

5.6. Variation of average COF with compositions

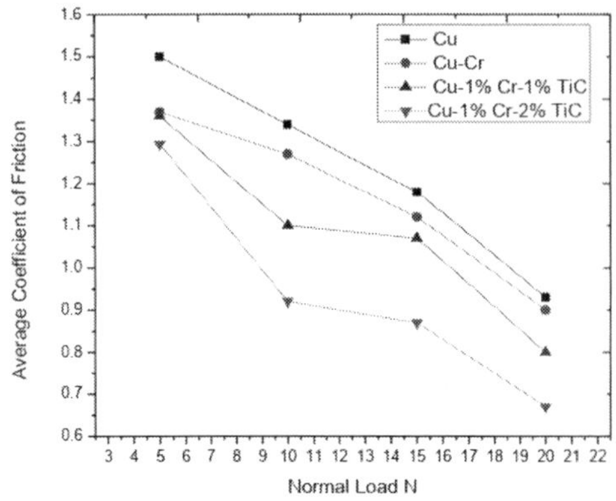

Figure 15. The variation of average COF with normal load at sliding speed of 1 m/s and sliding distance of 9000 m for cast copper, copper chromium alloy, Cu–1 %(by wt.)Cr–1 %(by wt.)TiC and Cu- 1 %(by wt.)Cr-2 %(by wt.) TiC composite.

Figure 16 shows the bar chart for the variation of average COF with the materials having different compositions. One could observe that the COF of Cu is more in comparison to other three and composite containing 2 wt. % TiC shows the lowest COF.

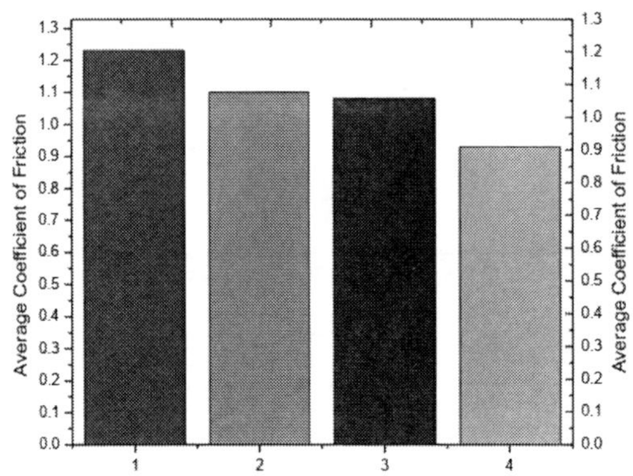

Figure 16. The variation of average COF with compositions at sliding speed of 1 m/s and sliding distance of 9000 m. (1) cast copper, (2) copper chromium alloy, (3) Cu–1 %(by wt.)Cr–1 %(by wt.)TiC composite (4) Cu- 1 %(by wt.)Cr-2 %(by wt.) TiC composite.

5.7. Examination of Worn Surfaces

The worn surfaces of the composites have been examined by SEM to explore the mechanism of wear and to correlate it with the observed friction and wear behavior in the materials. Figures 17 (a-d) shows the SEM micrographs of the materials tested at a load of 20 N and at a constant sliding speed of 1 m/s. All the specimens show the wear tracks and smooth surface covered with a layer of compacted wear debris which is called as transfer layer. However, the transfer layer appears to be fractured in case of Cu as shown in Figure 17 (a) whereas it well compacted in case of composites containing TiC as could be seen from Figure 17 (c ,d). Deeper groves could be observed in the micrograph shown in Figure 17 (b) which indicates a relatively larger loss of material from Cu-Cr alloy in comparison to composites containing TiC.

A relatively lower rate of wear observed for the composite containing 2 wt. % TiC as shown in Figure 17 could be attributed to the relatively higher hardness of the composite which lowers the real area of

IOP Conf. Series: Materials Science and Engineering **834** (2020) 012081 doi:10.1088/1757-899X/834/1/012081

contact and in turn lowers the wear volume loss and wear rate. The other contributing factor may be the cover of transfer layer over the sliding surface which reduces metal-metal contact resulting in reduced mass loss and wear rate. Figure 17 (a-d) shows the presence of the transfer layer over the worn surface but the extent of cover provided appears to be more in case of 2 wt. % composite in comparison to Pure Cu, Cu-Cr and 1 wt. % TiC composite. Hence it shows the lowest wear rate under the prevailing conditions of tribo-testing in the present work. The increase in wear arte with normal load is in consonance with the prediction by Archard's law so it not a surprising features. As the load increases the real area of contact also increase which results in more loss of material.

The reason behind relatively lower COF in the composites might be acknowledged to the lower real area of contact and the larger extent of cover of transfer layer. The other reason may the thickness of the transferred layer rather it has not been quantified in the present study. A harder substrate is able to hold a thicker layer of compacted wear debris which prevents the digging from the harder asperities of counter face inhibiting thus, metal-metal contact resulting in a lower coefficient of friction. Decreasing trend of COF was observed with increasing load might be due to better compaction of the wear debris of the transfer layer because some extra frictional heating also provided at the higher load.

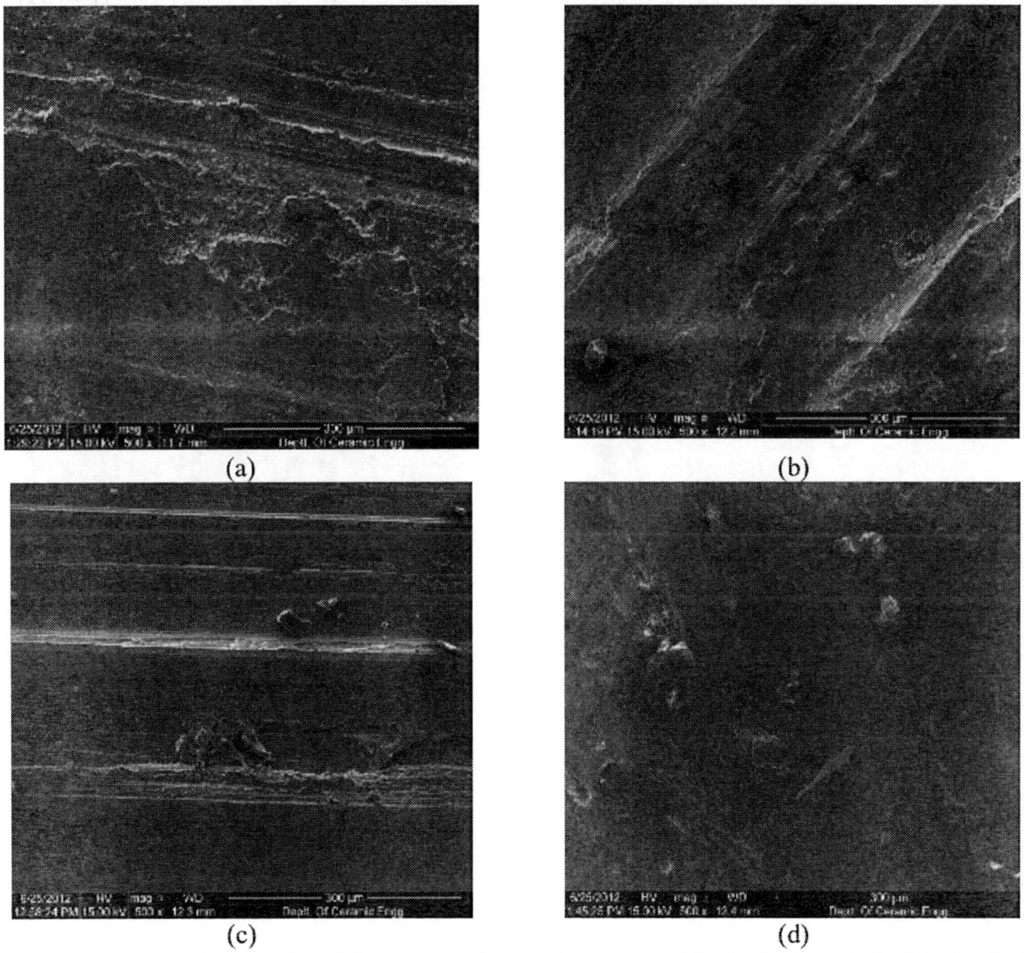

Figure 17. SEM micrographs of the worn suface under a normal load of 20 N (a) Cu (b) Cu-Cr (c) Cu-Cr-1 wt.% TiC and (d) Cu-Cr.2 wt.% TiC

6. Conclusions

- A cast Cu-Cr-TiC composite for the application of sliding contact and wear resistant was developed by using by cost effective stir casting method.
- Different type of intermetallic formation takes place which is seen by optical microscope.

- Addition of TiC in Cu matrix increased its hardness considerably due to which mechanical properties improves.
- For the given load, the cumulative wear volumes of composites, copper and copper chromium alloy increase linearly with sliding distance under dry sliding conditions.
- The wear rate follow the linear trend as the load increases. Composites were found better wear resistant materials.
- The average COF decreases with load in both pure copper and composites and copper chromium alloy. However composites show a lower COF than that observed in pure copper.

It may be finally concluded that Cu-Cr-TiC composite is a promising engineering materials for wear resistance applications.

References

[1] Rice S L and Moslehy F A, Modeling friction and wear phenomenon 1997 *Wear* 206, 136–46.
[2] Zhang Z, Zhang L and Y Mai, Modeling steady wear of steel/Al_2O_3–Al particle reinforced composite system 1997 *Wear* 211, 147–50.
[3] Kennedy F E, Balbahadur A C and Lashmore D S, The friction and wear of Cu based silicon carbide particulate metal matrix composites for brake applications 1997 *Wear* 203, 715–21.
[4] Wan Y Z, Wang Y L, Chemg G X, Tao H M, and Cao Y, Effect of processing parameters, particle characteristics and metallic coatings on properties of Al_2O_3 copper alloy matrix composites 1998 *Powder Metall.* 41, 59–63.
[5] Saka N, Pamies-Teixeira J J, and Suh N P, Wear of Two Phase Metals 1977 *Wear* 44, 77-86.
[6] Sahin Y, The prediction of wear resistance model for the metal matrix composites 2005 *Wear* \ 258, 1717–1722.
[7] Muller N, and Van K, Numerical simulation of the sliding wear test in relation to material properties 1997 *Wear* 203, 180–86.
[8] Sahin Y, Wear behaviour of aluminium alloy and its composites reinforced by SiC particles using statistical analysis 2003 *Mater. Des.* 24, 95–103.
[9] Axen N , Hutchings I M, and Jacobson S, A model for the friction of multiphase materials in abrasion 1996 *Tribology International* 29, 467–75.

IOP Conf. Series: Materials Science and Engineering **834** (2020) 012082 doi:10.1088/1757-899X/834/1/012082

Simulating tribological characteristics of Palm Methyl Ester (PME) lubricated contact

S H Hamdan*[1] and W W F Chong[2,3]

[1]Section of Bioengineering Technology, University Kuala Lumpur Malaysia Institute of Chemical & Bioengineering Technology, Taboh Naning, 78000 Alor Gajah, Melaka, Malaysia
[2]School of Mechanical Engineering, Faculty of Engineering, Universiti Teknologi Malaysia, 81310 Johor Bahru, Johor, Malaysia
[3]UTM Centre for Low Carbon Transport in Cooperation with Imperial College London, Universiti Teknologi Malaysia (UTM), Johor Bahru, Johor, Malaysia

*sitihartini@unikl.edu.my

Abstract. The study simulates the lubricant Stribeck curve for Palm Methyl Ester (PME) by coupling modified Reynolds solution with Greenwood and Tripp's rough surface contact model. The predicted lubricant Stribeck curves for PME is validated with measured data from a pin-on-disc tribometer. The Reynolds equation is modified to accommodate for the lubricant properties of PME (e.g. viscosity and density), which is mathematically described using Gibbs energy additivity approach. Solving the modified Reynolds equation would then provide the fluid film formation behaviour of PME, such as the contact pressure and film thickness. These fluid film parameters are then used as the input to determine the boundary and viscous friction of the investigated lubricated contact using Greenwood and Tripp's rough surface contact model. The proposed mathematical solution correlates well with experimental data and is shown to be capable of predicting lubricant Stribeck, capturing the frictional behavior for the whole range of lubrication regimes. The findings of the present study prepare for a mathematical foundation to further explore the use of biodiesel as an alternative biodegradable lubricant to mineral-based ones.

1. Introduction

The global energy demand is predicted to increase significantly from 557 quadrillion BTU in the year 2014 to 703 quadrillion BTU in the year 2040. This could lead to environmental concerns related to greenhouse gas (GHG) emissions [1, 2]. It is now required that vehicle manufacturers develop more efficient components and system to counter such concern. One of the ways to improve efficiency of vehicles and at the same time rectify the concerns related to emissions of GHG is to have a better lubrication system, which is less dependent on mineral oil-based lubricant. A type of biodegradable lubricant that could be considered is biodiesel, namely the mono-alkyl esters of vegetable oils or animal fats or triacylglycerol-containing feedstock [3]. Biodiesels can be derived from many sources either from vegetable oil or animal fat such as palm oil, soybean oil, coconut oil and canola oil [4, 5]. Biodiesel is known to have low or no sulfur content with no aromatics content, high flash point, good lubricity

Content from this work may be used under the terms of the Creative Commons Attribution 3.0 licence. Any further distribution of this work must maintain attribution to the author(s) and the title of the work, journal citation and DOI.
Published under licence by IOP Publishing Ltd

and biodegradability [2, 6]. Usage of biodiesel in compression ignition engine is also known to be able to significantly reduce most regulated exhaust emissions [7].

Friction behavior of a lubrication system can be typically assessed using the classical lubricant Stribeck curve [8]. There are four main lubrication regimes in a Stribeck curve, namely hydrodynamic lubrication (HL), elastohydrodynamic lubrication (EHL), mixed lubrication (ML) and boundary lubrication (BL) regimes. Transitions between these lubrication regimes are affected by the relative sliding speed between the rubbing opposing surfaces, applied normal load, contact geometry and also lubricant properties. Stribeck curves are usually measured using tribometers [8], which proved to be costly, with limited predictive tools available in generating such curve, especially for biodiesel [9]. Therefore, the current study intends to simulate the lubricant Stribeck curve and also to predict frictional behaviour of palm oil derived biodiesel or palm methyl ester (PME). The simulation is conducted by solving for the Reynolds equation and Greenwood and Tripp's rough surface contact model. To validate the derived mathematical model, a pin-on-disk type tribometer is used to measure the friction generated by PME.

2. Methodology

2.1 Mathematical approach

2.1.1 Modified Reynolds equation

Reynolds equation is used to predict fluid film formation of PME. The equation relates sliding velocity of two surfaces, contact geometry, lubricant properties (density and viscosity) and applied normal load with lubricant fluid film formation properties, such as contact pressure distribution and lubricant film thickness. The contact pressure distribution is governed by the partial differential equation along the lubricated conjunction. The modified Reynolds equation is given as follow [10]:

$$\frac{\partial}{\partial x}\left(\frac{h^3}{\eta}\frac{\partial P}{\partial x}\right)+\frac{\partial}{\partial y}\left(\frac{h^3}{\eta}\frac{\partial P}{\partial y}\right)=12\,\chi\left\{\frac{\partial}{\partial x}[(U)\rho h]+\frac{\partial}{\partial y}[(V)\rho h]+\frac{\partial(\rho h)}{\partial t}\right\} \tag{1}$$

Where U represents the entrainment motion along the sliding direction and V is the speed of lubricant side leakage. The term χ is introduced to the modified Reynolds equation as a correction factor in considering methyl ester type lubricants. This term will be determined by correlating the simulation results with experimentally measured data. The film profile, h is determined using the following equation:

$$h = h_o + h_s + \delta \tag{2}$$

Where h_o refers to the initial undeformed central separation gap, h_s refers to the contact geometry and δ refers to the surface deflection as a result of generated fluid film contact pressure. The deflection is computed as given in reference [20]. For biodiesel, Chum-in et al. [11] and Phankosol et al. [12] showed that the lubricant rheological properties of methyl ester (viscosity and density) can be described using Gibbs energy additivity approach. This viscosity and density relation with pressure and temperature can be calculated using the average carbon number (z_{ave}) and the average number of double bonds (nd_{ave}). The relevant equations are given below for the dynamic viscosity and the density:

$$\ln\eta = \ln\eta_o + \frac{P}{P_{atm}}\left[0.0006 - 0.000011.z_{ave} - 0.00022.nd_{ave} + \frac{0.0415 + 0.0103.z_{ave} + 0.054.nd_{ave}}{T}\right] \tag{3}$$

$$\ln \rho = \ln \rho_o + \frac{P}{P_{atm}}\left[0.000228 - 0.0000026.z_{ave} - 0.000006.nd_{ave} + \frac{0.003171.z_{ave} + 0.00223.nd_{ave} - 0.0416}{T}\right] \quad (4)$$

The above-mentioned Reynolds equation is then solved numerically based on the method described in reference [13].

2.1.2 *Friction modelling*

The solution from the modified Reynolds equation provides the information with relation to the contact pressure distribution and the lubricant film thickness. These are then used as the input to predict the friction force along the tribological conjunction lubricated with PME. The total friction is assumed to consist of boundary friction (F_b) and viscous friction (F_v) components. Boundary friction is due to a direct surface asperity contact interaction while lubricant shearing leads to viscous friction along a lubricated conjunction [14]. Hence, the total friction can be expressed as follow:

$$F_f = F_b + F_v \quad (5)$$

The viscous friction force can be computed as:

$$F_V = \int \tau_v \, dA = \sum \tau_v \, (A - A_a) \quad (6)$$

Where τ_v is the viscous shear stress ($\tau_v = \frac{\eta_o U}{h}$). The boundary shear can then be calculated using:

$$F_b = \sum A_a \, (\tau_b) \; ; \quad \tau_b = \tau_o + m\left(\frac{W_a}{A_a}\right) \quad (7)$$

Where τ_o is the Eyring shear stress of the lubricant and m is the pressure coefficient of the boundary shear strength. The term W_a is the load carried by the asperities, A is the apparent contact area and A_a is the actual contact area. These parameters are determined using Greenwood and Tripp's rough surface contact model [14].

2.2 *Friction measurement*

In order to characterize the frictional properties of PME, an experimental friction test has been conducted under pure sliding motion that covers the range of three main lubrication regimes such as, EHL, ML and BL. This test is carried out using a tribometer pin-on-disc machine. A stainless-steel wear disc is rotated against a stationary cast iron pin with spherical cap of 10 mm diameter. The friction tests are conducted under room temperature condition. Throughout the test, the wear disc and pin are loaded with a constant normal load of 20 N, where the wear disc is rotated at a speed ranging from 2000 rpm to 20 rpm. The wear track is set at 20mm, giving sliding velocities of wear disc between 4.2ms^{-1} to 0.042 ms^{-1}. The friction test is conducted for three minutes for each of the set speed, which is in accordance to the method proposed in reference [5].

2.3 *Biodiesel production*

Transesterification process is used to derive PME from commercially available palm oil. This process requires the chemical reaction between triglyceride reactants and short-chain alcohol at sufficiently high temperature in the presence of acid, alkaline or lipase catalysts. By the end of this process, methyl esters are obtained with crude glycerol as the by-product. Among the types of catalysts, alkaline catalyst is a better option when high yield of the biodiesel is required [15]. The production of PME is based on the method proposed by Hamdan et al. [5]. Table 1 shows the measured rheological properties of the produced PME at the simulated room temperature of 25°C.

Table 1. Experimentally measured rheological properties of Palm Methyl Ester (PME).

Type of Methyl Ester	Bulk viscosity, η_o (mPa.s)	Bulk density, η_o (g/m^3)
PME	5.1459	873.0

3. Result and discussion

Using the derived numerical scheme for solving Modified Reynolds equation mention earlier, the frictional behaviour for the pin-on-disc tribometer tribological conjunction lubricated with PME is simulated. Figure 1 shows the simulated Stribeck curve for PME lubricated conjunction. By using χ value of 53, the simulated curve is depicted to correlate well with experimental data. It is to note that the simulated curve is capable of capturing the transitions of lubrication regimes with R^2 equal 88.8%, following the ones observed experimentally with R^2 equal 86.6%.

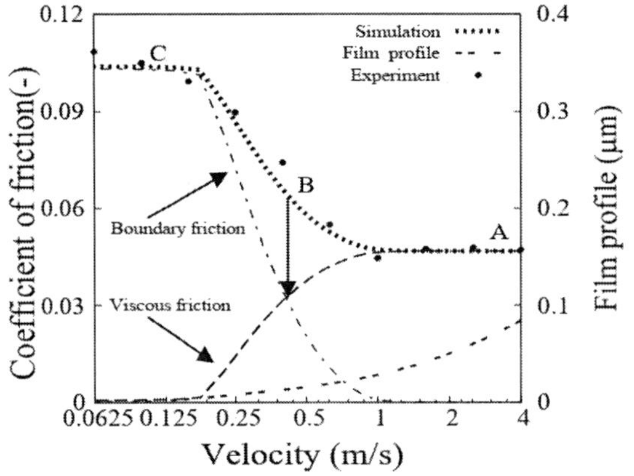

Figure 1. Comparison of simulated and experimentally measured Stribeck curve for PME

To investigate further the simulated properties of the PME lubricated conjunction, Figure 2 shows the contact pressure distribution and contact film profile for the investigated lubrication system. Three different sliding velocities was chosen at locations A, B and C (refer Figure 1), representing the sliding velocity at 4 ms^{-1}, 0.4 ms^{-1} and 0.1 ms^{-1}, respectively. The contact pressure resembles the Hertzian contact pressure at location A with the contact pressure being in the range of GPa but with a secondary peak. As the magnitude of film profile becomes smaller at location B and C, the probability of having surface asperity contact increases between the rubbing surfaces. Thus, this could lead to boundary friction.

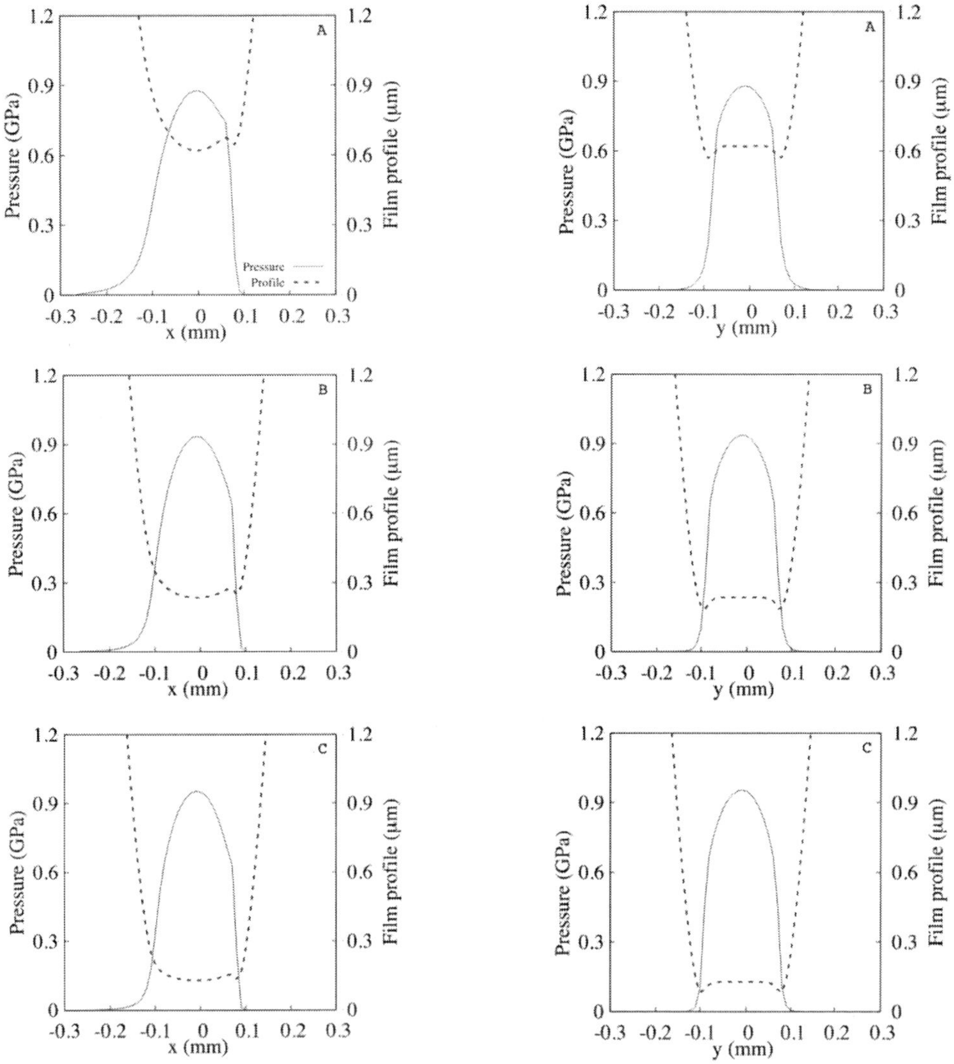

(a) X-axis (b) Y-axis

Figure 2. Distributed contact pressure and fluid film profile simulated for PME at three different sliding velocities at locations A, B and C (refer Figure 1)

Figure 3 shows that the frictional property at location A is dominated by viscous shearing of PME fluid film with no boundary component observed. This is synonymous to the lubrication regime indicated in Figure 1 at location A, where fluid film lubrication of elastohydrodynamic nature is expected. It is observed that shear properties at location B indicate a fair mix of boundary shear and viscous shear components. This corroborates with the transition of lubrication regime, which is expected to occur as being highlighted in Figure 1, giving rise to the ML regime. Finally, at location C, the simulated contact shows that the boundary shear component begins to significantly dominate over the viscous shear component, indicating BL regime. By reviewing the shear properties of these selected locations, it is demonstrated that the proposed numerical scheme is capable of capturing its tribological behaviour at various lubrication regimes for the investigated PME lubricated contact, similar to the measured regimes given in Figure 1.

IOP Conf. Series: Materials Science and Engineering **834** (2020) 012082 doi:10.1088/1757-899X/834/1/012082

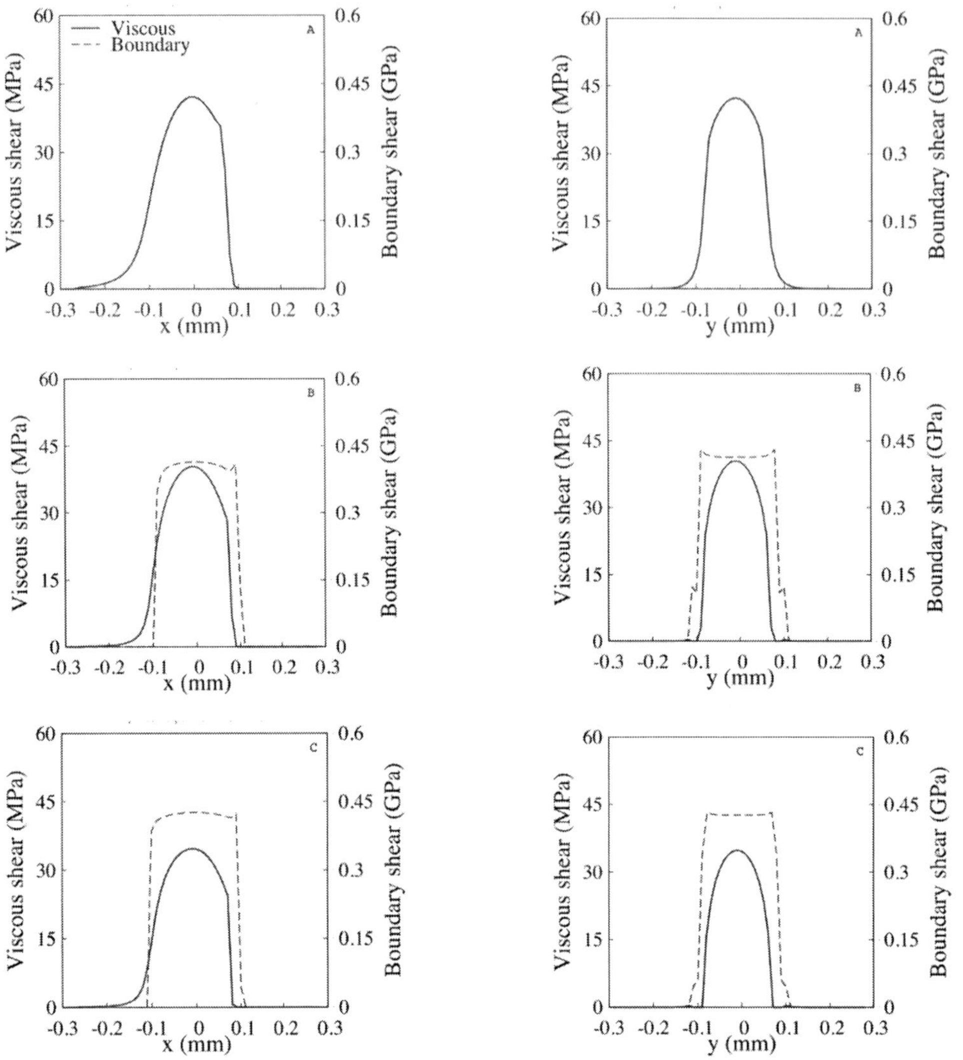

(a) X-axis (b) Y-axis

Figure 3. Simulated viscous shear and boundary shear of PME at three different sliding velocities at locations A, B and C (refer to Figure 1).

4. Conclusion

A lubrication Stribeck curve has been simulated and validated with measured friction using a pin-on-disc tribometer. The proposed mathematical tool is demonstrated to be capable of capturing the lubrication behavior of PME using a correction factor, χ. The present study sets in place a mathematical foundation to further investigate the relevance of the introduced correction factor, χ in the Reynolds equation for biodiesel derived from other feedstocks with varying fatty acid compositions. This is in hope that the correction factor could then be correlated to the average carbon number (z_{ave}) and average number of double bonds (nd_{ave}), providing an effective predictive tool to better explore the use of biodiesel as an alternative lubricant.

Acknowledgment

The authors would like to acknowledge the financial support provided by the Majlis Amanah Rakyat (MARA), Universiti Kuala Lumpur (UniKL) and Universiti Teknologi Malaysia through the Transdisciplinary Research Program (TDR) of "Innovative Materials for Resilient Contact Systems" (project vot. no. 06G19 and 05G56).

IOP Conf. Series: Materials Science and Engineering **834** (2020) 012082 doi:10.1088/1757-899X/834/1/012082

References

[1] Chong WWF, Ng J-H, Rajoo S, Chong CT. Passenger transportation sector gasoline consumption due to friction in Southeast Asian countries. Energy Conversion and Management. 2018;158:346-58.

[2] Hamdan SH, Chong WWF, Ng J-H, Chong CT, Rajoo S. A study of the tribological impact of biodiesel dilution on engine lubricant properties. Process Safety and Environmental Protection. 2017;112:288-97.

[3] Knothe G, Razon LF. Biodiesel fuels. Progress in Energy and Combustion Science. 2017;58:36-59.

[4] Tang H, De Guzman RC, Salley SO, Ng KS. Formation of insolubles in palm oil-, yellow grease-, and soybean oil-based biodiesel blends after cold soaking at 4° C. Journal of the American Oil Chemists' Society. 2008;85(12):1173-82.

[5] Hamdan SH, Chong WWF, Ng J-H, Ghazali M, Wood RJ. Influence of fatty acid methyl ester composition on tribological properties of vegetable oils and duck fat derived biodiesel. Tribology International. 2017;113:76-82.

[6] Andreae M, Fang H, Bhandary K. Biodiesel and fuel dilution of engine oil. SAE Technical Paper; 2007.

[7] Knothe G, Steidley KR. Lubricity of components of biodiesel and petrodiesel. The origin of biodiesel lubricity. Energy & fuels. 2005;19(3):1192-200.

[8] Maru MM, Trommer RM, Cavalcanti KF, Figueiredo ES, Silva RF, Achete CA. The Stribeck curve as a suitable characterization method of the lubricity of biodiesel and diesel blends. Energy. 2014;69:673-81.

[9] Zhu D, Wang J, Wang QJ. On the Stribeck curves for lubricated counterformal contacts of rough surfaces. Journal of Tribology. 2015;137(2):021501.

[10] Hamdan SH. Multi-scale Friction Modelling of Lubricant-Surface Interaction in Tribological Contacts. University of Southampton, 2018

[11] Chum-in T, Sudaprasert K, Phankosol S, Lilitchan S, Aryusuk K, Krisnangkura K. Gibbs energy additivity approaches to QSPR in modeling of high pressure dynamic viscosity of FAME and biodiesel. Journal of Molecular Liquids. 2016;223:1006-12.

[12] Phankosol S, Sudaprasert K, Lilitchan S, Aryusuk K, Krisnangkura K. Estimation of density of biodiesel. Energy & Fuels. 2014;28(7):4633-41.

[13] Chong WWF, De la Cruz M. Elastoplastic contact of rough surfaces: a line contact model for boundary regime of lubrication. Meccanica. 2014;49(5):1177-91.

[14] Chong WWF, Teodorescu M, Rahnejat H. Mixed thermo-elastohydrodynamic cam–tappet power loss in low-speed emission cycles. International Journal of Engine Research. 2014;15(2):153-64.

[15] Chhetri AB, Watts KC, Islam MR. Waste cooking oil as an alternate feedstock for biodiesel production. Energies. 2008;1(1):3-18.

IOP Conf. Series: Materials Science and Engineering **834** (2020) 012083 doi:10.1088/1757-899X/834/1/012083

FEM Simulation Procedure for Distortion and Residual Stress Analysis of Wire Arc Additive Manufacturing

Siti Nursyahirah Ahmad[1], Yupiter HP Manurung*[1], Muhd Faiz Mat[1], Zaidi Minggu[1], Ahmed Jaffar[1], Simon Pruller[2], Martin Leitner[2]

[1]Faculty of Mechanical Engineering, Universiti Teknologi MARA (UiTM), Malaysia
[2]Chair of Mechanical Engineering, Montanuniversität Leoben, Austria

*yupiter.manurung@uitm.edu.my

Abstract. This fundamental research presents an investigation of Wire Arc Additive Manufacturing (WAAM) process on distortion and residual stress using numerical simulation. Further, WAAM geometry was modeled using simplified rectangular bead shape with five layers and three strings. The thermo-mechanical numerical simulation is conducted under consideration of non-linear isotropic hardening modeled by using general purposed FEM software MSC Marc/Mentat. In this simulation, different heat source models (Goldak's double-ellipsoid and rectangular) are implemented and compared, whereby the later model was to be developed by using subroutine provided in the software. For analyzing the residual stress, separation technique between wall and substrate was introduced after the cooling down period. The outcome of this research is to develop an effective procedure to analyze the distortion and residual stress of wire arc additive manufacturing of stainless steel.

1. Introduction

In recent years, numerous studies have predicted that Additive Manufacturing (AM) plays a profound role in the manufacturing industry of the future [1]. Metal AM has become a crucial advanced-manufacturing process of custom-built metal workpieces. There are various advantages in comparison to conventional processes. For instance, AM is one method to replace conventional moulding technologies like casting and forging that often require a higher rate of material usage in machining and cannot meet the continuously demands of a sustainable, lower cost and friendly modern industry. A basic AM discipline comprises a combination of a heat source, motion strategy, and feedstock [2]. In addition, AM is a promising way-out for manufacturing parts or components fabricated of high-end materials such as titanium or aluminium alloys due to the high value of the buy-to-fly ratio [3]. Hence, metallic additive manufacturing processes are subdivided into three different categories which are powder-feed, powder-bed, and wire-feed systems [3, 4].

One of a promising breakthrough for AM techniques is Wire Arc Additive Manufacturing (WAAM) due to its high deposition rate, which enables fast and highly efficient production processes on a large-scale structural parts of medium complexity geometry with lower volume metal workpieces [6]. WAAM is made up of electrical arc heating sources with a metallic wire feeding scheme [2]. In WAAM, the building strategy allows the production of metallic structural components in a layer by layer fashion by utilizing Gas Metal Arc Welding (GMAW) welding process. A component manufactured by a wire-based additive manufacturing method can achieve the tensile strength of a casting or forging complex component [7].

Content from this work may be used under the terms of the Creative Commons Attribution 3.0 licence. Any further distribution of this work must maintain attribution to the author(s) and the title of the work, journal citation and DOI.
Published under licence by IOP Publishing Ltd

However, WAAM gives poor surface finish and low accuracy in comparison with powder-bed or beam-melting among the AM technologies so it is crucial to predict the interaction of each assigned welding process parameter with the material properties such as microstructure and residual stresses in order to produce a high quality product. An initial forecast can be made using numerical computation taking thermal-mechanical properties and different boundary conditions into consideration. The finite element modelling (FEM) simulation of WAAM process carries out the similar techniques applied in the multi-passes welding simulation by simulating the heat transfer from the arc to the molten pool of metal workpiece by means of heat source model. Thus, the heat propagation in GMAW-based additive manufacturing is profoundly non-linear transient because of the presence of solidification, multiple fusion and phase transformation [8].

In this paper, an efficient FEM procedure is to be developed in order to analyse the thermal-mechanical behaviour of WAAM process on a five layers and three strings wall structure. A numerical computation is carried out with two different type of heat source models, which are Goldak's double ellipsoidal and rectangular heat source model by using algorithm configuration of subroutine. A comparison between these two models will be observed and discussed in terms of distortions and residual stresses.

2. Nonlinear FEM Simulation Procedure

In order to make a prediction of distortion of the WAAM structure and possible residual stresses, the general purposed FEM software MSC Marc/Mentat was utilized to conduct the coupled thermo-mechanical numerical simulations.

2.1 Geometrical and Material Modelling

A full 3D-solid FE model of WAAM structure was developed which consists of three sets of geometries as shown in Figure 1(a): a table, a base plate, and multi-layers weld beads. The base plate was modelled with dimension of 200 mm (L) x 97.5 mm (W) x 8 mm (T). Meanwhile for the multi-layered deposited walls with a rectangular-shaped weld bead of 150 mm length was modelled and consistently sketched from the first layer until a complete set of five layers with three strings.

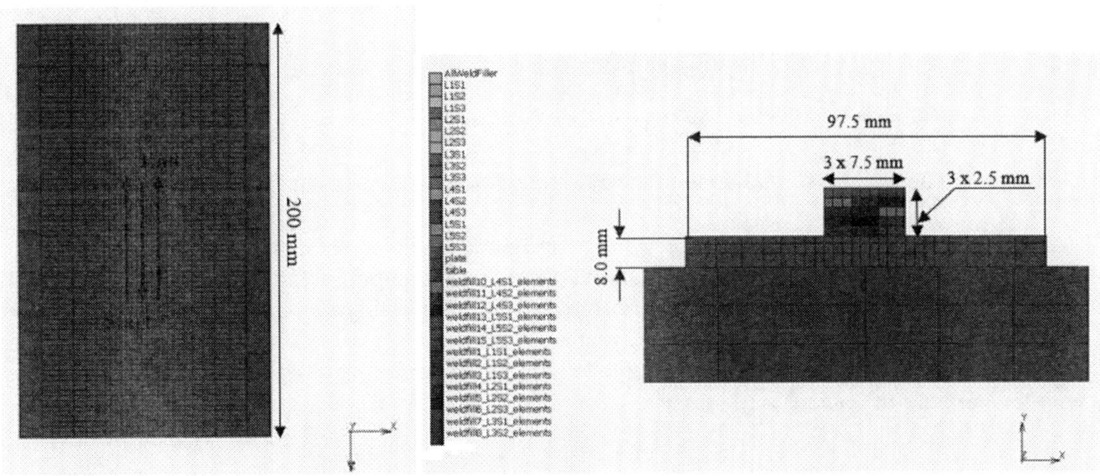

Figure 1. Weld direction and geometrical model of WAAM

A bead geometry modelling of WAAM structure plays an important role to yield a metallic structure with a high degree of geometrical accuracy and smoothness of numerical computation process. In this research, the bead modelling strategy utilizes a much simpler rectangular shape which gives a smaller margin of relative percentage errors and shorter CPU-Time [9]. Hence, this WAAM model applied 'element birth technique' to simulate the weld filler material [10]. This technique works as at the initial status of analysis, all the elements of the weld bead are deactivated and then the elements are activated consecutively following the heat source movements [10, 11].

For mechanical boundary conditions, Figure 2 shows the 400N clamping force exerted on the base plate in negatively Y-directions. Furthermore, there are three basic equations that shall be considered in mechanical analysis which are the equilibrium equations, constitutive stress–strain relations and geometric compatibility equations [12]. The variations in the temperature distribution contributes to the deformation of the body via thermal strains and influences the material properties. A large strain was activated as the nonlinear procedure for analysis option which large plastic strains and large deformations are accounted for meanwhile the additive decomposition of elastic, plastic and thermal strain contributions is employed for the stress recovery process [13]. In addition, phase transformation effects in the numerical analysis are not taken into account. Thus, the modelling of the fluid flow is also not comprised because the effect of the fluid flow on the body deformation and stress field can be considered as insignificant [14].

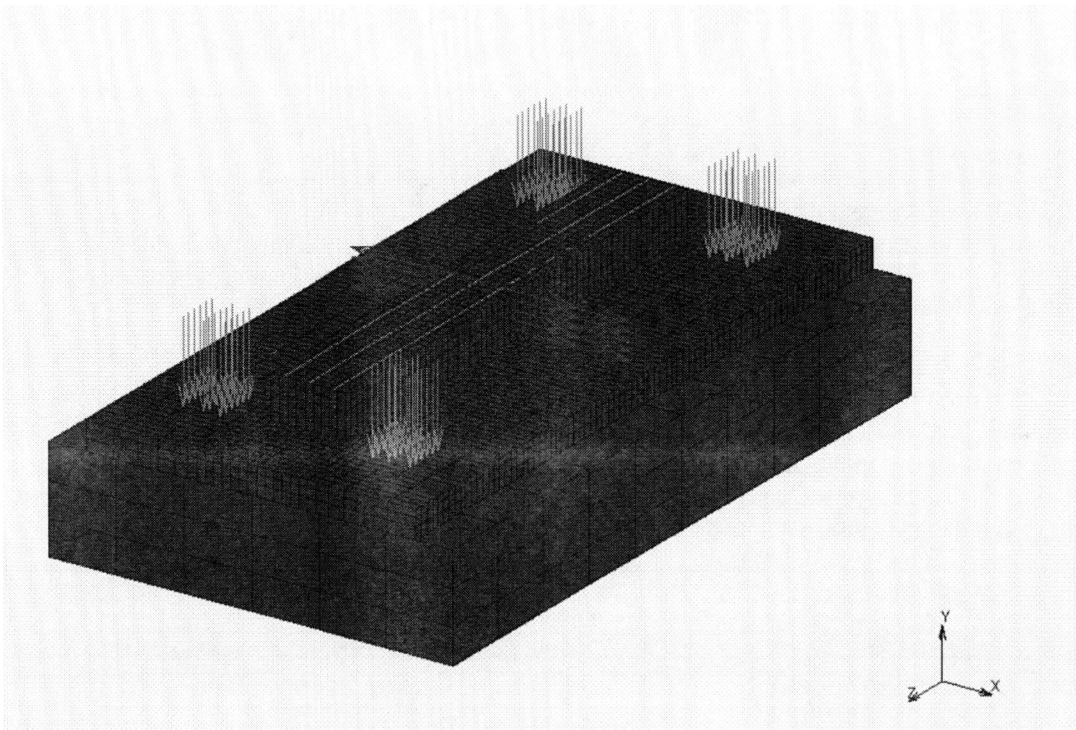

Figure 2. The boundary conditions of the clamping force exerted on the base plate in Y-directions

In this simulation, a low carbon steel S235 is designated as substrate material with S316L is applied as filler material. Both thermo-physical material properties are imported into FEM simulation using the existing databases from the MSC Marc/Mentat material properties library. The thermo-physical material properties are predicted equivalent with the real application, whereby C15 is assigned as the base metal with X5Crnimo18_10_1 as the filler material. Figure 3 shows the material assigned for the FE model with temperature-dependent variables.

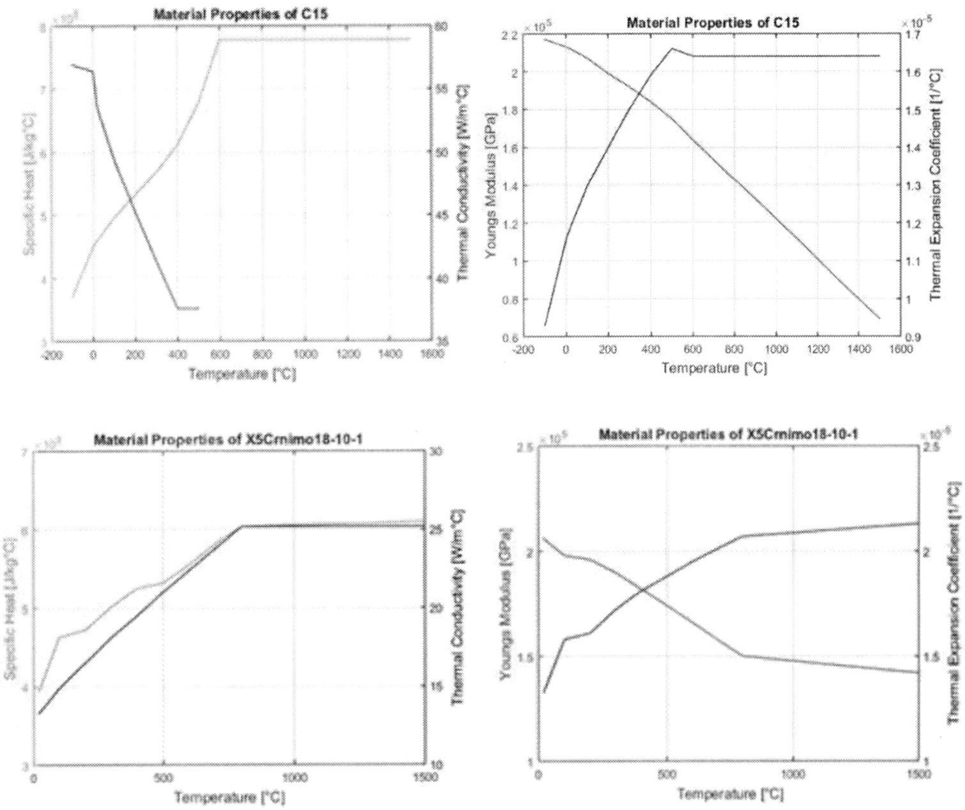

Figure 3. Mechanical and thermal properties of C15 and X5Crnimo18_10_1.

In addition, in a way to have an accurate estimation of residual stresses after substrate removal from the weld filler material, a separation technique was introduced and applied on the WAAM model. After the cooling down period at the fifth layer, another loadcase was added to deactivate the substrate material and the table. The mechanical behaviour due to residual stresses will be discussed in the Results and Discussions section.

2.2 Heat Source Model: Rectangular and Goldak's Double Ellipsoid

Heat source modelling is the most interesting phase in the numerical simulation procedure of generating welding process. Suitable heat source model must be chosen as the welding process accordingly. Hence, double ellipsoidal model was executed to imitate the real heat source of GMAW process. Goldak et al. defined the double ellipsoid model [15]. Nowadays, the Goldak's double ellipsoidal model is extensively utilized as the heat source model for GMAW welding process in manufacturing background [16]. The double ellipsoidal heat source has been revealed to accurately represent the heat power density from an electric arc penetrating the surface of a flat workpiece of plate. Figure 4 presents Goldak's double ellipsoid heat source model with the geometrical conditions but the geometrical parameters a, b and c can be modified in values for the front or rear quadrants.

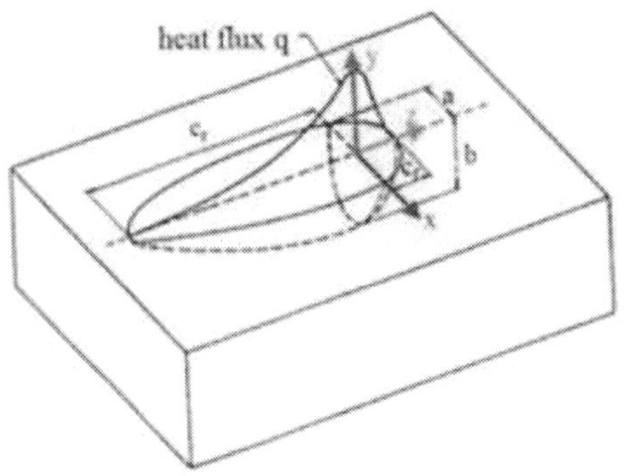

Figure 4. An illustration of Goldak's double ellipsoidal heat source model [5]

The subsequent equations were applied to characterize the power density distribution inside the front and rear quadrants of the heat source along the welding path (z-axis). The power density of the heat flux in front section (q_{vf}) of heat source can be modelled with equation (1) and the rear section (q_{vr}) with equation (2) [17].

$$q_{vf}(x,y,z) = \frac{6\sqrt{3}f_f Q}{abc_f\pi\sqrt{\pi}}e^{-3\frac{x^2}{a^2}}.e^{-3\frac{y^2}{b^2}}.e^{-3\frac{z^2}{c_f^2}} \quad (1)$$

$$q_{vr}(x,y,z) = \frac{6\sqrt{3}f_f Q}{abc_r\pi\sqrt{\pi}}e^{-3\frac{x^2}{a^2}}.e^{-3\frac{y^2}{b^2}}.e^{-3\frac{z^2}{c_r^2}} \quad (2)$$

The distribution of heat fluxes in the double ellipsoid model is determined by 4 directions: Width (a), Depth (b), Rear Length (c_r) and Front Length (c_f), whereas c_r and c_f are the heat deposited fractional factors in the rear and front quadrant respectively and its sum is equal to 2. Detailed values for each direction are demonstrated in Table 1.

Table 1. Goldak's Double Ellipsoidal Heat Source Dimension in FEM simulation

Heat Source Dimensions (mm)	Value
Width	7.5
Depth	5
Rear Length, c_r	3
Front Length, c_f	12

Furthermore, in this research the primary objective is to compare the thermo-mechanical behaviour between two different heat source models which are Goldak's double ellipse and rectangular. Hence, the relevant thermal analysis for WAAM application begins with the solution of a moving plane heat source of appropriate shape and heat intensity distribution [18]. The moving rectangular heat source model is generated with the modified user 'uweldflux' subroutine in the MSC/Marc Mentat code. The heat intensity for various heat sources with uniform distribution is given by

$$Q = \int_0^\infty \int_0^\infty \int_0^\infty q\,(x,y,z)\,dxdydz \quad (3)$$

$$Q = 4qacb \quad (4)$$

Figure 5 presents the rectangular heat source model with the geometrical conditions with the distribution of heat fluxes is determined by three directions: a, b, c which are classified as width, depth and length respectively. The power density of the heat flux of rectangular heat source can be modelled with equation (4). The values for each parameters are shown in Table 2.

IOP Conf. Series: Materials Science and Engineering **834** (2020) 012083 doi:10.1088/1757-899X/834/1/012083

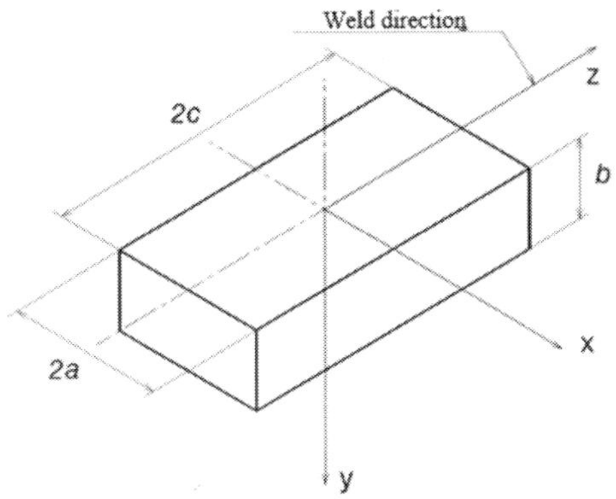

Figure 5. An illustration of rectangular heat source model

Table 2. Rectangular Heat Source Parameters in FEM simulation

Parameters	Value
Width, a (mm)	4.0
Depth, b (mm)	3.0
Length, c (mm)	6.0
Power, q (kW)	3.4

3.0 Results and Discussions

3.1 FE Mechanical Results

3.1.1 Distortion of WAAM Model

The distortion result produced by both models are illustrated on Figure 6 below. The predicted angular distortion along the side edge wall from the two mechanical models are compared.

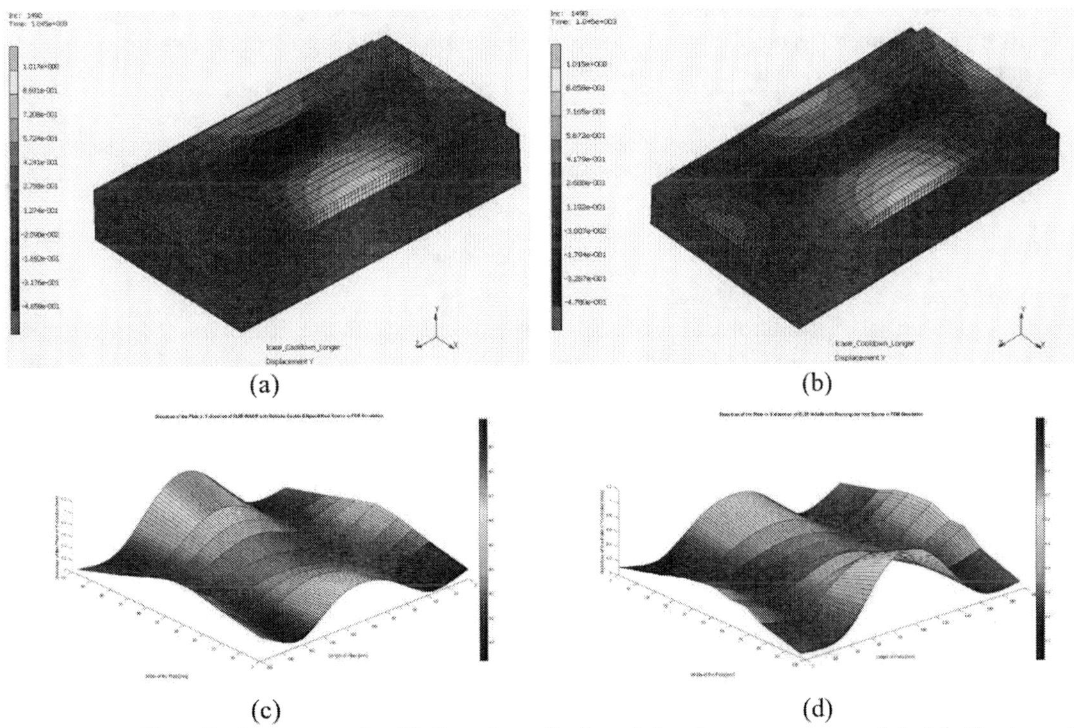

Figure 6. The displacements in Y-direction displayed by heat source model (a) Rectangular (b) Goldak's double ellipse. An angular distortion characteristic of 5Layers 3Strings WAAM model with different heat source model (c) Rectangular (d) Goldak's double ellipse

Figure 7 shows how the distorted shape from both models is nearly identical. However, WAAM model with Goldak's double ellipsoidal heat source shows higher distortion compared to rectangular model.

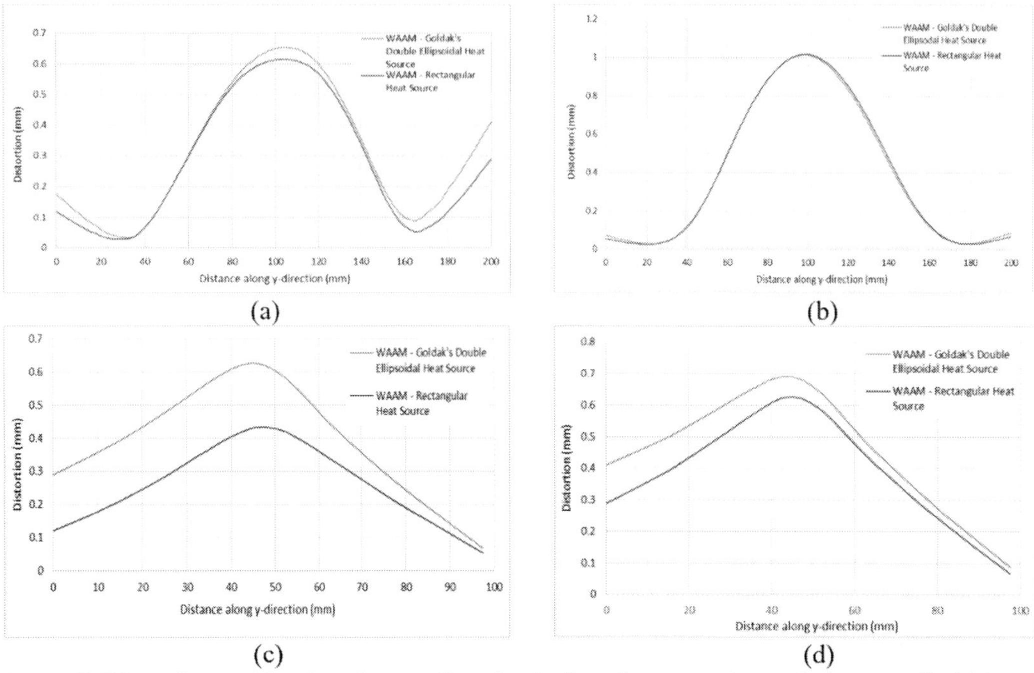

Figure 7. Distortion verification along y-direction before the separation technique applied (a) Left side edge (b) Right side edge (c) Front side edge (d) Rear side edge

Furthermore, a computational time for both models is also compared. Computational time is an overall time consumed to complete the numerical simulation. Table 3 presents the comparison of total computational time between these two models.

Table 3. Total computational time of two FEM model

Heat Source Model	**CPU-Time (Hrs)**
Rectangular	34.9
Goldak's Double Ellipsoidal	35.8

3.1.2 Residual Stress of WAAM Model

Figure 8 presents the distribution of longitudinal residual stresses on the fifth layer, the final deposited layer of WAAM model. Based on these results between the rectangular heat source and Goldak's double ellipsoidal heat source of WAAM model, the following comprehensive observations can be made and analysed. The heat input caused by multi-passes welding process gives an effect on a tensile residual stress along the deposited weld filler as a result of the material contraction during solidification, which indirectly a balancing compressive residual stress occurs on the base plate. Based on these results, it show that the high degree of the compressive residual stress where the clamping force has been exerted and this is due to the restraint in the region locally.

In addition, around the top layer of deposited weld bead has the maximum tensile residual stress. It is shown in Figure 8 (a) and (c) that the higher tensile residual stress accumulates at the top layer after a completion of the deposition of the fifth layer. However, these stresses are relieved partially caused by the repetitive of reheating and cooling process while depositing the following upper layers [19]. After a long cooling down period, the residual stress changes from tensile to compressive at the interfaces of the successive layers, meanwhile on the base plate the longitudinal stresses go into tension as indicated in the Figure 8 (b) and (d). Thus, in WAAM, residual stresses can be as high as the yield strength of the material, the mechanical properties can be negatively affected with leading to distortions and decreased tolerances [20]. For a better understanding of the deviations of these stresses, the residual stress was plotted along x-direction along the line shown in Figure 9 (a).

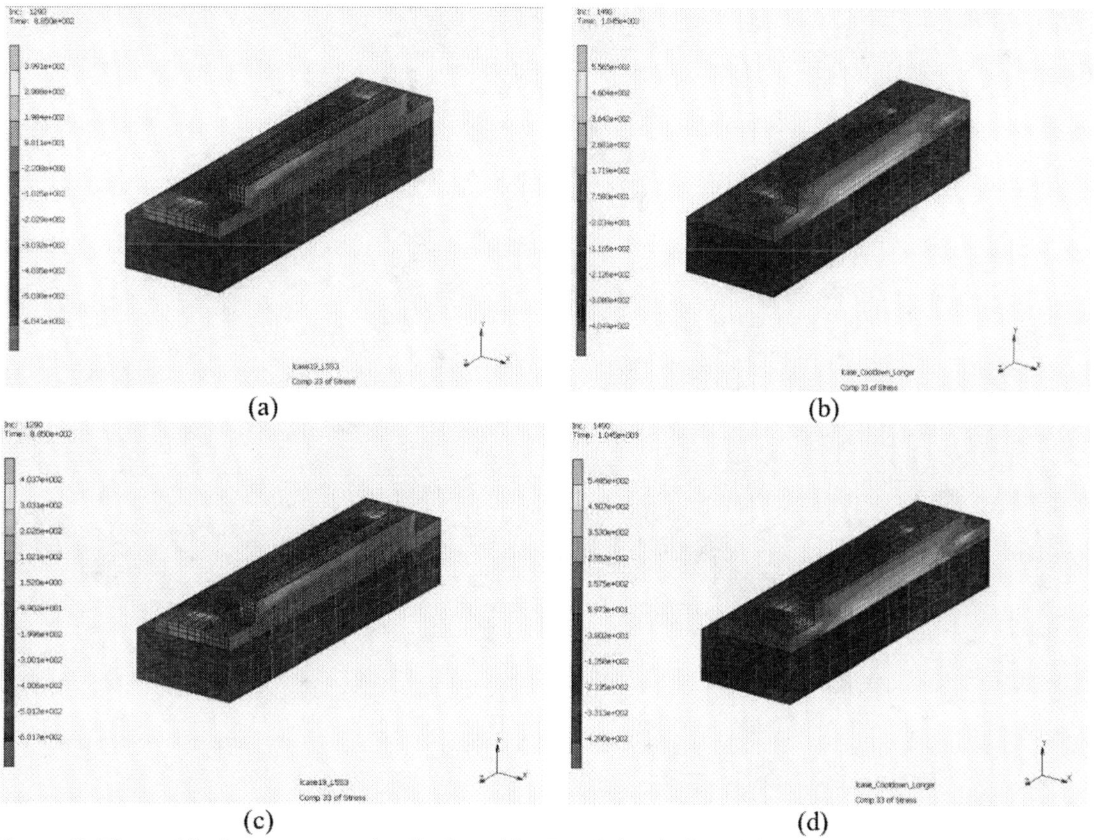

Figure 8. The residual stresses results displayed by FEM simulations (a) A complete weld deposited of 5layers 3Strings of rectangular heat source model (b) After a cooling down period of 5layers 3Strings of rectangular heat source model (c) A complete weld deposited of 5layers 3Strings of Goldak's double ellipsoid heat source model (d) After a cooling down period of 5layers 3Strings of Goldak's double ellipsoid heat source model

As expected, Figure 10 (a) shows the highly non-uniform distributed of all three residual stresses components. These stresses induce repetitive expansion and contraction of the material simultaneously. As shown in Figure 10 (a), mostly compressive residual stresses occur on the longitudinal direction of the substrate due the contractions of the materials as getting farther from the heat input. The significant drop in the residual stress which is lower than the yield strength after the substrate removal which the separation technique applied, is also shown in Figure 10 (b). This is due to the residual stresses are relieved as well when the substrate is detached from the deposited weld fillers. Hence, the reducing these residual stresses give a massive impact to its mechanical behaviour in terms of distortions particularly. However, figure 9 (c) demonstrate the WAAM model applied Goldak's double ellipsoidal heat source has higher deformations compared to the rectangular heat source model, where the residual stress at the edge deposited walls changes from compressive to tensile as getting farther to the centre of the deposited weld fillers.

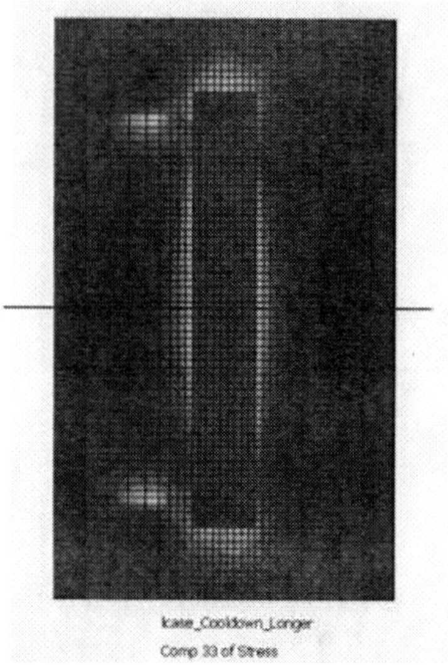

(a)

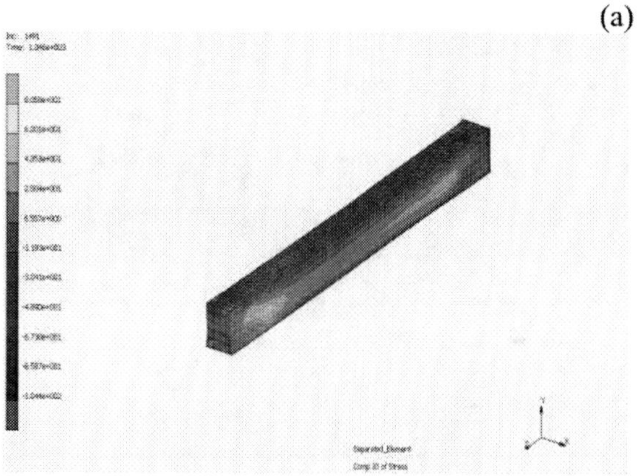

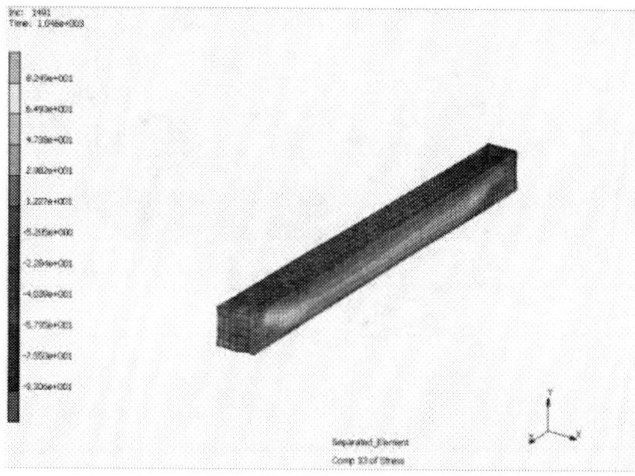

(b) (c)

Figure 9. (a) The A-A cross section on the top view of WAAM model. The residual stress distribution along longitudinal direction with contour band setting view after separation technique applied (b) rectangular heat source (c) Goldak's double ellipsoidal heat source

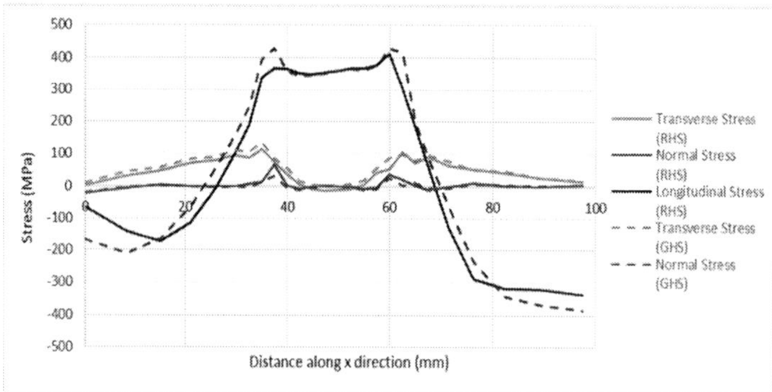

(a)

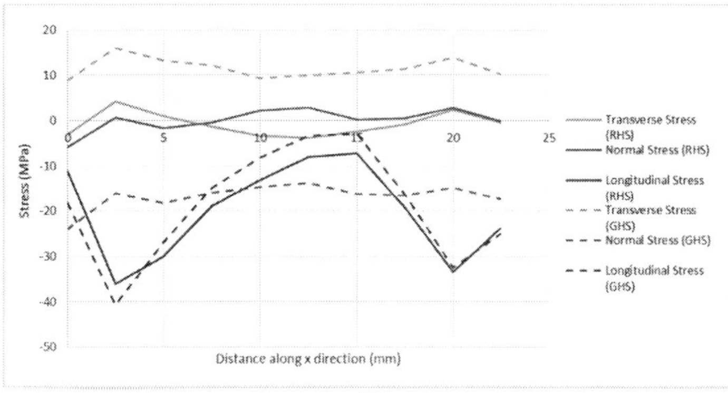

(b)

Figure 10. The residual stresses along x-direction on the cross section line A-A (a) Before separation technique applied (b) After separation technique applied

In general, a highly non-uniform distributed stresses generated during WAAM process can cause big distortions on the structures after the removal of table and substrate or separation technique to be exact. The FEM models introduced in this research gives an initial estimate of the residual stresses and angular substrate distortions of WAAM process that later will be beneficial for future work in order to make comparison with experiments.

4.0 Conclusions

In this research, the numerical computation model is employed under consideration of non-linear isotropic hardening modelled by using general purposed FEM software MSC Marc/Mentat to simulate the thermo-mechanical properties of five layers and three strings WAAM structures. The numerical simulation are carried out with two different heat source models which are rectangular and Goldak's double ellipsoidal. The thermo-mechanical analysis in terms of distortions and residual stresses is compared and discussed. As a conclusion, it can be drawn for this research as follows:

1. A comprehensive numerical study for five layers and three strings of WAAM components had been successfully executed and the simulation procedure is clear as well as structured.
2. Both heat source models namely the rectangular heat source and Goldak's double ellipsoid can be implemented for repetitive heating and cooling cycles during WAAM process.
3. In terms of FE mechanical results, it shows that the distorted shape from both models are nearly similar. Thus, the deviations of the residual stresses depends on the transient temperature distribution during cooling phase and the application exerted on the WAAM structure after the cooling process.
4. A significant drop of the residual stresses distribution can be observed after the removal of the substrate and the table. As a result, the residual stress at the edge deposited walls changes from compressive to tensile as getting farther to the centre of the deposited weld fillers.
5. The rectangular heat source model displays a short CPU-Time which caused by a simpler rectangular shape according to the weld bead modelling geometry to be calculated. This will be a significant advantage where the WAAM process will be optimised to calculate the predicted residual stresses and distortions.
6. From the knowledge perspective, crucial information through simulation is acquired which can be utilized as a planning tool in the design phase or prior to actual experimental process.

As further research, a five layers and three strings WAAM model will be carried out by experimental work and the FE model will implement the evoluted filler material instead of utilizing equivalent default material.

Acknowledgement

The authors would like to express their gratitude to staff member of Smart Manufacturing Research Institute (SMRI), Research Interest Group: Advanced Manufacturing Technology (RIG:AMT) and

Advanced Manufacturing Technology Excellence Centre (AMTEx) at Faculty of Mechanical Engineering, UniversitiTeknologi MARA (UiTM) for encouraging this research. This research and conference participation are also financially supported by ASEA-UNINET grant with the project number ASEA 2019/Montan/1, ERASMUS+ (Montan University in Leoben) and Technogerma Engineering & Consulting Sdn. Bhd.

References

[1] S. Ford and M. Despeisse, "Additive manufacturing and sustainability: an exploratory study of the advantages and challenges," *J. Clean. Prod.*, 2016.
[2] S. W. Williams, F. Martina, A. C. Addison, J. Ding, G. Pardal, and P. Colegrove, "Wire + Arc Additive Manufacturing," *Mater. Sci. Technol.*, vol. 32, no. 7, pp. 641–647, 2016.
[3] M. Liberini *et al.*, "Selection of Optimal Process Parameters for Wire Arc Additive Manufacturing," *Procedia CIRP*, vol. 62, pp. 470–474, 2017.
[4] Y. Ogino, S. Asai, and Y. Hirata, "Numerical simulation of WAAM process by a GMAW weld pool model," *Weld. World*, 2018.
[5] W. E. Frazier, "Metal additive manufacturing: A review," *Journal of Materials Engineering and Performance*. 2014.
[6] D.Wood, "Additive Layer Manufacturing at Airbus – Reality check or view into the future?," *TCT 17*, pp. 23–27, 2000.
[7] E. Brandl, B. Baufeld, C. Leyens, and R. Gault, "Additive manufactured Ti-6A1-4V using welding wire: Comparison of laser and arc beam deposition and evaluation with respect to aerospace material specifications," in *Physics Procedia*, 2010.
[8] J. Xiong, R. Li, Y. Lei, and H. Chen, "Heat propagation of circular thin-walled parts fabricated in additive manufacturing using gas metal arc welding," *J. Mater. Process. Technol.*, 2018.
[9] K. P. Prajadhiama *et al.*, "Development of Bead Modelling for Distortion Analysis Induced by Wire Arc Additive Manufacturing using FEM and Experiment," *MATEC Web Conf.*, vol. 269, p. 05003, 2019.
[10] X. Y. Shan, M. J. Tan, and N. P. O'Dowd, "Developing a realistic FE analysis method for the welding of a NET single-bead-on-plate test specimen," *J. Mater. Process. Technol.*, 2007.
[11] J. Ding *et al.*, "Thermo-mechanical analysis of Wire and Arc Additive Layer Manufacturing process on large multi-layer parts," *Comput. Mater. Sci.*, vol. 50, no. 12, pp. 3315–3322, 2011.
[12] L. E. Lindgren, *Computational Welding Mechanics: Thermomechanical and Microstructural Simulations*. 2007.
[13] "Help Reference," *Computer (Long. Beach. Calif).*, 2005.
[14] J. A. Goldak and M. Akhlaghi, *Computational welding mechanics*. 2005.
[15] J. Goldak, A. Chakravarti, and M. Bibby, "A New Finite Element Model for Welding Heat Source_Goldak_1984.pdf," vol. 15, no. June, 1984.
[16] P. Son Minh and T. Viet Phu, "Study on the Structure Deformation in the Process of Gas Metal Arc Welding (GMAW)," *Am. J. Mech. Eng.*, vol. 2, no. 4, pp. 120–124, 2014.
[17] J. Goldak, A. Chakravarti, and M. Bibby, "A new finite element model for welding heat sources," *Metall. Trans. B*, 1984.
[18] Z. B. Hou and R. Komanduri, "General solutions for stationary/moving plane heat source problems in manufacturing and tribology," *Int. J. Heat Mass Transf.*, vol. 43, no. 10, pp. 1679–1698, 2000.
[19] T. Mukherjee, W. Zhang, and T. Debroy, "An improved prediction of residual stresses and distortion in additive manufacturing," vol. 126, pp. 360–372, 2017.
[20] P. A. Colegrove *et al.*, "Microstructure and residual stress improvement in wire and arc additively manufactured parts through high-pressure rolling," *J. Mater. Process. Technol.*, 2013.

AUTHOR INDEX